OPERATION OF WASTEWATER TREATMENT PLANTS

Volume II

Seventh Edition

A Field Study Training Program

prepared by

Office of Water Programs
College of Engineering and Computer Science
California State University, Sacramento

in cooperation with

California Water Environment Association

Kenneth D. Kerri, Project Director
Bill B. Dendy, Co-Director
John Brady, Consultant and Co-Director
William Crooks, Consultant

2007

Cover: Santa Cruz Wastewater Treatment Plant, Santa Cruz, California.
Photo courtesy of Paul Cockrell Photography,
paulcockrellphoto@yahoo.com

In recognition of the need to preserve natural resources, this manual is printed using recycled paper. The text paper is composed of 10% post-consumer waste and the cover is composed of 10% post-consumer waste. The Office of Water Programs will strive to increase its commitment to sustainable printing practices.

Funding for this operator training manual was provided by the Office of Water Programs, California State University, Sacramento. Mention of trade names or commercial products does not constitute endorsement or recommendation for use by the Office of Water Programs or California State University, Sacramento.

First edition published 1971.

Printed in the United States of America

22 21 20 19 18 5 6 7 8 9

ISBN
978-1-59371-038-5

www.owp.csus.edu

OFFICE OF WATER PROGRAMS

The Office of Water Programs is a nonprofit organization operating under the California State University, Sacramento, to provide distance learning courses for persons interested in the operation and maintenance of drinking water and wastewater facilities. These training programs were developed by people who explain, through the use of our manuals, how they operate and maintain their facilities. The university, fully accredited by the Western Association of Schools and Colleges, administers and monitors these training programs, under the direction of Dr. Ramzi J. Mahmood.

Our training group develops and implements programs and publishes manuals for operators of water treatment plants, water distribution systems, wastewater collection systems, and municipal and industrial wastewater treatment and reclamation facilities. We also offer programs and materials for pretreatment facility inspectors, environmental compliance inspectors, and utility managers. All training is offered as distance learning, using correspondence, video, or computer-based formats with opportunities for continuing education and contact hours for operators, supervisors, managers, and administrators.

Materials and opportunities available from our office include manuals in print, CD, or video formats, and enrollments for courses providing CEU (Continuing Education Unit) contact hours. Here is a sample:

- Industrial Waste Treatment, 2 volumes (print, course enrollment)
- Operation of Wastewater Treatment Plants, 2 volumes (print, CD, course enrollment)
- Advanced Waste Treatment (print, course enrollment)
- Treatment of Metal Wastestreams (print, course enrollment)
- Pretreatment Facility Inspection (print, video, course enrollment)
- Small Wastewater System Operation and Maintenance, 2 volumes (print, course enrollment)
- Operation and Maintenance of Wastewater Collection Systems, 2 volumes (print, course enrollment)
- Collection System Operation and Maintenance Training Videos (video, course enrollment)
- Utility Management (print, course enrollment)
- Manage for Success (print, course enrollment)
- and more

These and other materials may be ordered from:

Office of Water Programs
California State University, Sacramento
6000 J Street
Sacramento, CA 95819-6025
(916) 278-6142 – phone
(916) 278-5959 – FAX

or

visit us on the web at www.owp.csus.edu

ADDITIONAL VOLUMES OF INTEREST

Operation of Wastewater Treatment Plants, Volume I

The Treatment Plant Operator
Why Treat Wastes?
Wastewater Treatment Facilities
Racks, Screens, Comminutors, and Grit Removal
Sedimentation and Flotation
Trickling Filters
Rotating Biological Contactors
Activated Sludge (Package Plants and Oxidation Ditches)
Wastewater Stabilization Ponds
Disinfection and Chlorination

Advanced Waste Treatment

Odor Control
Activated Sludge (Pure Oxygen Plants and Operational Control Options)
Residual Solids Management
Solids Removal from Secondary Effluents
Phosphorus Removal
Nitrogen Removal
Enhanced Biological (Nutrient) Control
Wastewater Reclamation and Reuse
Instrumentation

Industrial Waste Treatment, Volume I

The Industrial Plant Operator
Industrial Wastewaters
Regulatory Requirements
Preventing and Minimizing Wastes at the Source
Industrial Waste Monitoring
Flow Measurement
Preliminary Treatment (Equalization, Screening, and pH Adjustment)
Physical–Chemical Treatment Processes (Coagulation, Flocculation, and Sedimentation)
Filtration
Physical Treatment Processes (Air Stripping and Carbon Adsorption)
Treatment of Metal Wastestreams
Instrumentation
Safety
Maintenance

Industrial Waste Treatment, Volume II

The Industrial Plant Operator
Fixed Growth Processes (Trickling Filters and Rotating Biological Contactors)
Activated Sludge Process Control
Sequencing Batch Reactors
Enhanced Biological Control
Anaerobic Treatment
Residual Solids Management
Maintenance

Treatment of Metal Wastestreams

Need for Treatment
Sources of Wastewater
Material Safety Data Sheets (MSDSs)
Employee Right-To-Know Laws
Methods of Treatment
Advanced Technologies
Sludge Treatment and Disposal
Operation, Maintenance, and Troubleshooting
Polymers
Oxidation-Reduction Potential (ORP)

Pretreatment Facility Inspection

The Pretreatment Facility Inspector
Pretreatment Program Administration
Development and Application of Regulations
Inspection of a Typical Industry
Safety in Pretreatment Inspection and Sampling Work
Sampling Procedures for Wastewater
Wastewater Flow Monitoring
Industrial Wastewaters
Pretreatment Technology (Source Control)
Industrial Inspection Procedures
Emergency Response

PREFACE TO THE SEVENTH EDITION

The technology for treating wastewater and the knowledge and skills required for plant operators and managers to efficiently operate and maintain their plants continues to advance. This volume, a continuation of Volume I, shifts emphasis from smaller treatment facilities toward larger conventional treatment plants. It will also be helpful to operators in supervisory and management positions.

In Chapter 13, we recognize that plant effluent is too valuable for disposal. Plant effluent today is becoming an important resource and is being reclaimed and reused. Also, effluent may be discharged to receiving waters and used directly or indirectly to meet the many demands for beneficial uses of water. Computers are being used more than ever by operators and managers to operate and maintain plants as well as store, retrieve, analyze, and present data. In Chapter 18, procedures are provided on how to develop prediction or forecasting equations to help operators and managers make appropriate decisions. Chapter 20 contains new information on plant security and a new section on ethics. The ethics section helps operators and managers identify ethical situations, evaluate alternative solutions, and select an ethical response.

Kenneth D. Kerri
Office of Water Programs
California State University, Sacramento
6000 J Street
Sacramento, CA 95819-6025
(916) 278-6142 – phone
wateroffice@csus.edu – e-mail

2007

USES OF THIS MANUAL

Originally, this manual was developed to serve as a home-study course for operators in remote areas or persons unable to attend formal classes either due to shift work, personal reasons, or the unavailability of suitable classes. This home-study training program used the concepts of self-paced instruction where you are your own instructor and work at your own speed. In order to certify that a person has successfully completed this program, objective tests and special answer sheets for each chapter are provided when a person enrolls in this course.

Once operators started using this manual for home study, they realized that it could serve effectively as a textbook in the classroom. Many colleges and universities have used the manual as a text in formal classes often taught by operators. In areas where colleges are not available or are unable to offer classes in the operation of wastewater treatment plants, operators and utility agencies can join together to offer their own courses using the manual.

Occasionally, a utility agency has enrolled from three to over 300 of its operators in this training program. A manual is purchased for each operator. A senior operator or a group of operators are designated as instructors. These operators help answer questions when the persons in the training program have questions or need assistance. The instructors grade the objective tests, record scores, and notify California State University, Sacramento, of the scores when a person successfully completes this program. This approach eliminates any waiting while papers are being graded and returned by the university.

This manual was prepared to help operators run their treatment plants. Please feel free to use it in the manner that best fits your training needs and the needs of other operators. We will be happy to work with you to assist you in developing your training program. Please feel free to contact:

Project Director
Office of Water Programs
California State University, Sacramento
6000 J Street
Sacramento, CA 95819-6025
(916) 278-6142 – phone
(916) 278-5959 – FAX
wateroffice@csus.edu – e-mail

TECHNICAL CONSULTANTS
OPERATION OF WASTEWATER TREATMENT PLANTS

(previous editions and other volumes)

Russ Armstrong
William Garber
George Gardner
Larry Hannah
Mike Mulbarger
Joe Nagano
Carl Nagel
Al Petrasek
Frank Phillips
Warren Prentice
Ralph Stowell
Larry Trumbull

INSTRUCTIONS TO PARTICIPANTS IN HOME-STUDY COURSE

Procedures for reading the lessons and answering the questions are contained in this section.

To progress steadily through this program, you should establish a regular study schedule. For example, many operators in the past have set aside two hours during two evenings a week for study.

The study material in Volume II is contained in 10 chapters. Some chapters are longer and more difficult than others. For this reason, many of the chapters are divided into two or more lessons. The time required to complete a lesson will depend on your background and experience. Some people might require an hour to complete a lesson and some might require three hours; but that is perfectly all right. The important thing is that you understand the material in the lesson.

Each lesson is arranged for you to read a short section, write the answers to the questions at the end of the section, check your answers against suggested answers; and then *YOU* decide if you understand the material sufficiently to continue or whether you should read the section again. You will find that this procedure is slower than reading a typical textbook, but you will remember much more when you have finished the lesson.

Some discussion and review questions are provided following each lesson in the chapters. These questions review the important points you have covered in the lesson. Write the answers to the discussion and review questions in your notebook.

In the appendix at the end of this manual, you will find some comprehensive review questions and suggested answers. These questions and answers are provided as a way for you to review how well you remember the material. You may wish to review the entire manual before you attempt to answer the questions. Some of the questions are essay-type questions, which are used by some states for higher-level certification examinations. After you have answered all the questions, check your answers with those provided and determine the areas in which you might need additional review before your next certification or civil service examination. Please do not send your answers to California State University, Sacramento.

You are your own teacher in this program. You could merely look up the suggested answers at the end of the chapters or comprehensive review questions or copy them from someone else, but you would not understand the material. Consequently, you would not be able to apply the material to the operation of your plant or recall it during an examination for certification or a civil service position.

You will get out of this program what you put into it.

SUMMARY OF PROCEDURE

OPERATOR (YOU)

1. Read what you are expected to learn in each chapter; the major topics are listed at the beginning of the chapter.
2. Read sections in the lesson.
3. Write your answers to questions at the end of each section in your notebook. You should write the answers to the questions just as you would if these were questions on a test.
4. Check your answers with the suggested answers.
5. Decide whether to reread the section or to continue with the next section.
6. Write your answers to the discussion and review questions at the end of each lesson in your notebook.

ORDER OF WORKING LESSONS

To complete this program, you will have to work all of the lessons. You may proceed in numerical order, or you may wish to work some lessons sooner.

OPERATION OF WASTEWATER TREATMENT PLANTS
COURSE OUTLINE

VOLUME II, SEVENTH EDITION

CHAPTER 11

ACTIVATED SLUDGE

Operation of Conventional Activated Sludge Plants

by

John Brady

Revised by

Ross Gudgel

Further information related to this topic may be found in:

OPERATION OF WASTEWATER TREATMENT PLANTS

Volume I, Chapter 8
Package Plants and Oxidation Ditches

ADVANCED WASTE TREATMENT

Chapter 2
Pure Oxygen Plants and Operational Control Options

TABLE OF CONTENTS

Chapter 11. ACTIVATED SLUDGE

OPERATION OF CONVENTIONAL ACTIVATED SLUDGE PLANTS

Page

LESSON 3

LESSON 4

LESSON 7

OBJECTIVES

Chapter 11. ACTIVATED SLUDGE

OPERATION OF CONVENTIONAL ACTIVATED SLUDGE PLANTS

The activated sludge process is a very important wastewater treatment process. For this reason, the chapters on activated sludge have been divided into three parts and will be presented in three separate manuals.

I. Package Plants and Oxidation Ditches (Volume I)

II. Operation of Conventional Activated Sludge Plants (Volume II)

III. Pure Oxygen Plants and Operational Control Options *(ADVANCED WASTE TREATMENT)*

If you are the operator of a package plant or oxidation ditch, Volume I will provide you with the information you need to know to operate your plant. Volume II and the *ADVANCED WASTE TREATMENT* manual will help you better understand your plant and do a better job. If you operate a conventional activated sludge plant or a modification, Volume I will help you understand the activated sludge process and Volume II will tell you how to operate your plant. The *ADVANCED WASTE TREATMENT* manual will explain to you alternative means of operational control that may work very well for your plant. If you operate a pure oxygen plant, the *ADVANCED WASTE TREATMENT* manual will tell you what you need to know to operate the pure oxygen system. All three parts contain information important to the proper operation of your plant. The *ADVANCED WASTE TREATMENT* manual also contains information helpful to operators using the activated sludge process to treat special wastes such as industrial wastes.

The following objectives apply to the treatment plants covered in each of the three parts. After completion of the appropriate part on activated sludge, you should be able to:

1. Explain the principles of the activated sludge process and the factors that influence and control the process.
2. Inspect a new activated sludge facility for proper installation.
3. Place a new activated sludge process into service.
4. Schedule and conduct operation and maintenance duties.
5. Collect samples, interpret lab results, and make appropriate adjustments in treatment processes.
6. Recognize factors that indicate an activated sludge process is not performing properly, identify the source of the problem, and take corrective action.
7. Conduct your duties in a safe fashion.
8. Determine aerator loadings and understand the application of different loading guidelines.
9. Keep records for an activated sludge plant.
10. Identify the common modifications of the activated sludge process.
11. Review plans and specifications for an activated sludge plant.
12. Describe each of the process stages used to treat wastewater in a sequencing batch reactor (SBR).
13. Place a new sequencing batch reactor in service.
14. Collect and analyze samples and make appropriate process adjustments during start-up and normal operation.
15. Safely operate and maintain a sequencing batch reactor.
16. Review plans and specifications for a sequencing batch reactor.

WORDS

Chapter 11. ACTIVATED SLUDGE

OPERATION OF CONVENTIONAL ACTIVATED SLUDGE PLANTS

ABSORPTION (ab-SORP-shun) ABSORPTION

The taking in or soaking up of one substance into the body of another by molecular or chemical action (as tree roots absorb dissolved nutrients in the soil).

ACTIVATED SLUDGE ACTIVATED SLUDGE

Sludge particles produced in raw or settled wastewater (primary effluent) by the growth of organisms (including zoogleal bacteria) in aeration tanks in the presence of dissolved oxygen. The term "activated" comes from the fact that the particles are teeming with bacteria, fungi, and protozoa. Activated sludge is different from primary sludge in that the sludge particles contain many living organisms that can feed on the incoming wastewater.

ACTIVATED SLUDGE PROCESS ACTIVATED SLUDGE PROCESS

A biological wastewater treatment process that speeds up the decomposition of wastes in the wastewater being treated. Activated sludge is added to wastewater and the mixture (mixed liquor) is aerated and agitated. After some time in the aeration tank, the activated sludge is allowed to settle out by sedimentation and is disposed of (wasted) or reused (returned to the aeration tank) as needed. The remaining wastewater then undergoes more treatment.

ADSORPTION (add-SORP-shun) ADSORPTION

The gathering of a gas, liquid, or dissolved substance on the surface or interface zone of another material.

AERATION (air-A-shun) LIQUOR AERATION LIQUOR

Mixed liquor. The contents of the aeration tank, including living organisms and material carried into the tank by either untreated wastewater or primary effluent.

AERATION (air-A-shun) TANK AERATION TANK

The tank where raw or settled wastewater is mixed with return sludge and aerated. The same as aeration bay, aerator, or reactor.

AEROBES AEROBES

Bacteria that must have dissolved oxygen (DO) to survive. Aerobes are aerobic bacteria.

AEROBIC (air-O-bick) DIGESTION AEROBIC DIGESTION

The breakdown of wastes by microorganisms in the presence of dissolved oxygen. This digestion process may be used to treat only waste activated sludge, or trickling filter sludge and primary (raw) sludge, or waste sludge from activated sludge treatment plants designed without primary settling. The sludge to be treated is placed in a large aerated tank where aerobic microorganisms decompose the organic matter in the sludge. This is an extension of the activated sludge process.

AGGLOMERATION (uh-glom-er-A-shun) AGGLOMERATION

The growing or coming together of small scattered particles into larger flocs or particles, which settle rapidly. Also see FLOC.

AIR LIFT PUMP AIR LIFT PUMP

A special type of pump consisting of a vertical riser pipe submerged in the wastewater or sludge to be pumped. Compressed air is injected into a tail piece at the bottom of the pipe. Fine air bubbles mix with the wastewater or sludge to form a mixture lighter than the surrounding water, which causes the mixture to rise in the discharge pipe to the outlet.

ALIQUOT (AL-uh-kwot) ALIQUOT

Representative portion of a sample. Often, an equally divided portion of a sample.

ANAEROBES | ANAEROBES

Bacteria that do not need dissolved oxygen (DO) to survive.

ANOXIC (an-OX-ick) | ANOXIC

A condition in which the aquatic (water) environment does not contain dissolved oxygen (DO), which is called an oxygen deficient condition. Generally refers to an environment in which chemically bound oxygen, such as in nitrate, is present. The term is similar to ANAEROBIC.

BOD (pronounce as separate letters) | BOD

Biochemical Oxygen Demand. The rate at which organisms use the oxygen in water or wastewater while stabilizing decomposable organic matter under aerobic conditions. In decomposition, organic matter serves as food for the bacteria and energy results from its oxidation. BOD measurements are used as a surrogate measure of the organic strength of wastes in water.

BACTERIAL (back-TEER-e-ul) CULTURE | BACTERIAL CULTURE

In the case of activated sludge, the bacterial culture refers to the group of bacteria classified as AEROBES and FACULTATIVE BACTERIA, which covers a wide range of organisms. Most treatment processes in the United States grow facultative bacteria that use the carbonaceous (carbon compounds) BOD. Facultative bacteria can live when oxygen resources are low. When nitrification is required, the nitrifying organisms are obligate aerobes (require oxygen) and must have at least 0.5 mg/*L* of dissolved oxygen throughout the whole system to function properly.

BATCH PROCESS | BATCH PROCESS

A treatment process in which a tank or reactor is filled, the water (or wastewater or other solution) is treated or a chemical solution is prepared, and the tank is emptied. The tank may then be filled and the process repeated. Batch processes are also used to cleanse, stabilize, or condition chemical solutions for use in industrial manufacturing and treatment processes.

BIOCHEMICAL OXYGEN DEMAND (BOD) | BIOCHEMICAL OXYGEN DEMAND (BOD)

See BOD.

BIOMASS (BUY-o-mass) | BIOMASS

A mass or clump of organic material consisting of living organisms feeding on wastes, dead organisms, and other debris. Also see ZOOGLEAL MASS and ZOOGLEAL MAT (FILM).

BOUND WATER | BOUND WATER

Water contained within the cell mass of sludges or strongly held on the surface of colloidal particles. One of the causes of bulking sludge in the activated sludge process.

BULKING | BULKING

Clouds of billowing sludge that occur throughout secondary clarifiers and sludge thickeners when the sludge does not settle properly. In the activated sludge process, bulking is usually caused by filamentous bacteria or bound water.

CARBONACEOUS (car-bun-NAY-shus) STAGE | CARBONACEOUS STAGE

A stage of decomposition that occurs in biological treatment processes when aerobic bacteria, using dissolved oxygen, change carbon compounds to carbon dioxide. Sometimes referred to as first-stage BOD because the microorganisms attack organic or carbon compounds first and nitrogen compounds later. Also see NITRIFICATION STAGE.

CATHODIC (kath-ODD-ick) PROTECTION | CATHODIC PROTECTION

An electrical system for prevention of rust, corrosion, and pitting of metal surfaces that are in contact with water, wastewater, or soil. A low-voltage current is made to flow through a liquid (water) or a soil in contact with the metal in such a manner that the external electromotive force renders the metal structure cathodic. This concentrates corrosion on auxiliary anodic parts, which are deliberately allowed to corrode instead of letting the structure corrode.

CHEMICAL OXYGEN DEMAND (COD) | CHEMICAL OXYGEN DEMAND (COD)

A measure of the oxygen-consuming capacity of organic matter present in wastewater. COD is expressed as the amount of oxygen consumed from a chemical oxidant in mg/*L* during a specific test. Results are not necessarily related to the biochemical oxygen demand (BOD) because the chemical oxidant may react with substances that bacteria do not stabilize.

COAGULATION (ko-agg-yoo-LAY-shun) | COAGULATION

The clumping together of very fine particles into larger particles (floc) caused by the use of chemicals (coagulants). The chemicals neutralize the electrical charges of the fine particles, allowing them to come closer and form larger clumps.

COMPOSITE (PROPORTIONAL) SAMPLE — COMPOSITE (PROPORTIONAL) SAMPLE

A composite sample is a collection of individual samples obtained at regular intervals, usually every one or two hours during a 24-hour time span. Each individual sample is combined with the others in proportion to the rate of flow when the sample was collected. Equal volume individual samples also may be collected at intervals after a specific volume of flow passes the sampling point or after equal time intervals and still be referred to as a composite sample. The resulting mixture (composite sample) forms a representative sample and is analyzed to determine the average conditions during the sampling period.

CONING — CONING

Development of a cone-shaped flow of liquid, like a whirlpool, through sludge. This can occur in a sludge hopper during sludge withdrawal when the sludge becomes too thick. Part of the sludge remains in place while liquid rather than sludge flows out of the hopper. Also called coring.

CONTACT STABILIZATION — CONTACT STABILIZATION

Contact stabilization is a modification of the conventional activated sludge process. In contact stabilization, two aeration tanks are used. One tank is for separate reaeration of the return sludge for at least four hours before it is permitted to flow into the other aeration tank to be mixed with the primary effluent requiring treatment. The process may also occur in one long tank.

DECIBEL (DES-uh-bull) — DECIBEL

A unit for expressing the relative intensity of sounds on a scale from zero for the average least perceptible sound to about 130 for the average level at which sound causes pain to humans. Abbreviated dB.

DENITRIFICATION (dee-NYE-truh-fuh-KAY-shun) — DENITRIFICATION

(1) The anoxic biological reduction of nitrate nitrogen to nitrogen gas.

(2) The removal of some nitrogen from a system.

(3) An anoxic process that occurs when nitrite or nitrate ions are reduced to nitrogen gas and nitrogen bubbles are formed as a result of this process. The bubbles attach to the biological floc and float the floc to the surface of the secondary clarifiers. This condition is often the cause of rising sludge observed in secondary clarifiers or gravity thickeners. Also see NITRIFICATION.

DIFFUSED-AIR AERATION — DIFFUSED-AIR AERATION

A diffused-air activated sludge plant takes air, compresses it, and then discharges the air below the water surface of the aerator through some type of air diffusion device.

DIFFUSER — DIFFUSER

A device (porous plate, tube, bag) used to break the air stream from the blower system into fine bubbles in an aeration tank or reactor.

DISSOLVED OXYGEN — DISSOLVED OXYGEN

Molecular oxygen dissolved in water or wastewater, usually abbreviated DO.

DOGS — DOGS

Wedges attached to a slide gate and frame that force the gate to seal tightly.

ELECTROLYSIS (ee-leck-TRAWL-uh-sis) — ELECTROLYSIS

The decomposition of material by an outside electric current.

ENDOGENOUS (en-DODGE-en-us) RESPIRATION — ENDOGENOUS RESPIRATION

A situation in which living organisms oxidize some of their own cellular mass instead of new organic matter they adsorb or absorb from their environment.

F/M RATIO — F/M RATIO

See FOOD/MICROORGANISM RATIO.

FACULTATIVE (FACK-ul-tay-tive) BACTERIA — FACULTATIVE BACTERIA

Facultative bacteria can use either dissolved oxygen or oxygen obtained from food materials such as sulfate or nitrate ions. In other words, facultative bacteria can live under aerobic, anoxic, or anaerobic conditions.

FILAMENTOUS (fill-uh-MEN-tuss) ORGANISMS — FILAMENTOUS ORGANISMS

Organisms that grow in a thread or filamentous form. Common types are *Thiothrix* and *Actinomycetes*. A common cause of sludge bulking in the activated sludge process.

FLIGHTS FLIGHTS

Scraper boards, made from redwood or other rot-resistant woods or plastic, used to collect and move settled sludge or floating scum.

FLOC FLOC

Clumps of bacteria and particles, or coagulants and impurities, that have come together and formed a cluster. Found in flocculation tanks, sedimentation basins, aeration tanks, secondary clarifiers, and chemical precipitation processes.

FLOCCULATION (flock-yoo-LAY-shun) FLOCCULATION

The gathering together of fine particles after coagulation to form larger particles by a process of gentle mixing. This clumping together makes it easier to separate the solids from the water by settling, skimming, draining, or filtering.

FOOD/MICROORGANISM (F/M) RATIO FOOD/MICROORGANISM (F/M) RATIO

Food to microorganism ratio. A measure of food provided to bacteria in an aeration tank.

$$\frac{\text{Food}}{\text{Microorganisms}} = \frac{\text{BOD, lbs/day}}{\text{MLVSS, lbs}}$$

$$= \frac{\text{Flow, MGD} \times \text{BOD, mg/}L \times 8.34\text{ lbs/gal}}{\text{Volume, MG} \times \text{MLVSS, mg/}L \times 8.34\text{ lbs/gal}}$$

or by calculator math system

$$= \text{Flow, MGD} \times \text{BOD, mg/}L \div \text{Volume, MG} \div \text{MLVSS, mg/}L$$

or metric

$$= \frac{\text{BOD, kg/day}}{\text{MLVSS, kg}}$$

$$= \frac{\text{Flow, M}L\text{/day} \times \text{BOD, mg/}L \times 1\text{ kg/M mg}}{\text{Volume, M}L \times \text{MLVSS, mg/}L \times 1\text{ kg/M mg}}$$

FREEBOARD FREEBOARD

(1) The vertical distance from the normal water surface to the top of the confining wall.

(2) The vertical distance from the sand surface to the underside of a trough in a sand filter. This distance is also called AVAILABLE EXPANSION.

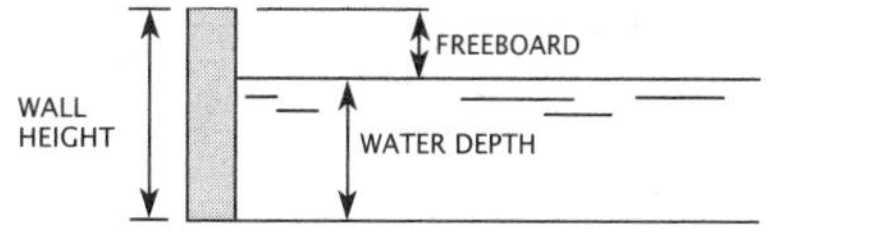

HEADER HEADER

A large pipe to which the ends of a series of smaller pipes are connected. Also called a MANIFOLD.

JAR TEST JAR TEST

A laboratory procedure that simulates coagulation/flocculation with differing chemical doses. The purpose of the procedure is to estimate the minimum coagulant dose required to achieve certain water quality goals. Samples of water to be treated are placed in six jars. Various amounts of chemicals are added to each jar, stirred, and the settling of solids is observed. The lowest dose of chemicals that provides satisfactory settling is the dose used to treat the water.

MCRT MCRT

Mean Cell Residence Time. An expression of the average time (days) that a microorganism will spend in the activated sludge process.

$$\text{MCRT, days} = \frac{\text{Total Suspended Solids in Activated Sludge Process, lbs}}{\text{Total Suspended Solids Removed From Process, lbs/day}}$$

or

$$\text{MCRT, days} = \frac{\text{Total Suspended Solids in Activated Sludge Process, kg}}{\text{Total Suspended Solids Removed From Process, kg/day}}$$

NOTE: Operators at different plants calculate the Total Suspended Solids (TSS) in the Activated Sludge Process, lbs (kg), by three different methods:

1. TSS in the Aeration Basin or Reactor Zone, lbs (kg)
2. TSS in the Aeration Basin and Secondary Clarifier, lbs (kg)
3. TSS in the Aeration Basin and Secondary Clarifier Sludge Blanket, lbs (kg)

These three different methods make it difficult to compare MCRTs in days among different plants unless everyone uses the same method.

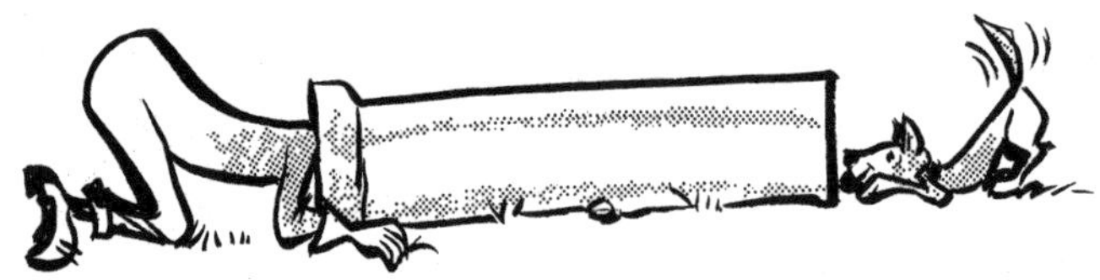

MANIFOLD MANIFOLD

A large pipe to which the ends of a series of smaller pipes are connected. Also called a HEADER.

MEAN CELL RESIDENCE TIME (MCRT) MEAN CELL RESIDENCE TIME (MCRT)

See MCRT.

MECHANICAL AERATION MECHANICAL AERATION

The use of machinery to mix air and water so that oxygen can be absorbed into the water. Some examples are: paddle wheels, mixers, or rotating brushes to agitate the surface of an aeration tank; pumps to create fountains; and pumps to discharge water down a series of steps forming falls or cascades.

MEG MEG

(1) Abbreviation of MEGOHM.

(2) A procedure used for checking the insulation resistance on motors, feeders, bus bar systems, grounds, and branch circuit wiring. Also see MEGGER.

MEGGER (from megohm) MEGGER

An instrument used for checking the insulation resistance on motors, feeders, bus bar systems, grounds, and branch circuit wiring. A megger reads in millions of ohms. Also see MEG.

METABOLISM METABOLISM

All of the processes or chemical changes in an organism or a single cell by which food is built up (anabolism) into living protoplasm and by which protoplasm is broken down (catabolism) into simpler compounds with the exchange of energy.

MICROORGANISMS (MY-crow-OR-gan-is-ums) MICROORGANISMS

Living organisms that can be seen individually only with the aid of a microscope.

MIXED LIQUOR MIXED LIQUOR

When the activated sludge in an aeration tank is mixed with primary effluent or the raw wastewater and return sludge, this mixture is then referred to as mixed liquor as long as it is in the aeration tank. Mixed liquor also may refer to the contents of mixed aerobic or anaerobic digesters.

MIXED LIQUOR SUSPENDED SOLIDS (MLSS) MIXED LIQUOR SUSPENDED SOLIDS (MLSS)

The amount (mg/*L*) of suspended solids in the mixed liquor of an aeration tank.

MIXED LIQUOR VOLATILE SUSPENDED SOLIDS (MLVSS) MIXED LIQUOR VOLATILE SUSPENDED SOLIDS (MLVSS)

The amount (mg/*L*) of organic or volatile suspended solids in the mixed liquor of an aeration tank. This volatile portion is used as a measure or indication of the microorganisms present.

NITRIFICATION (NYE-truh-fuh-KAY-shun) NITRIFICATION

An aerobic process in which bacteria change the ammonia and organic nitrogen in water or wastewater into oxidized nitrogen (usually nitrate).

NITRIFICATION (NYE-truh-fuh-KAY-shun) STAGE NITRIFICATION STAGE

A stage of decomposition that occurs in biological treatment processes when aerobic bacteria, using dissolved oxygen, change nitrogen compounds (ammonia and organic nitrogen) into oxidized nitrogen (usually nitrate). The second-stage BOD is sometimes referred to as the nitrification stage (first-stage BOD is called the carbonaceous stage).

OXIDATION OXIDATION

Oxidation is the addition of oxygen, removal of hydrogen, or the removal of electrons from an element or compound; in the environment and in wastewater treatment processes, organic matter is oxidized to more stable substances. The opposite of REDUCTION.

PLUG FLOW PLUG FLOW

A type of flow that occurs in tanks, basins, or reactors when a slug of water or wastewater moves through a tank without ever dispersing or mixing with the rest of the water or wastewater flowing through the tank.

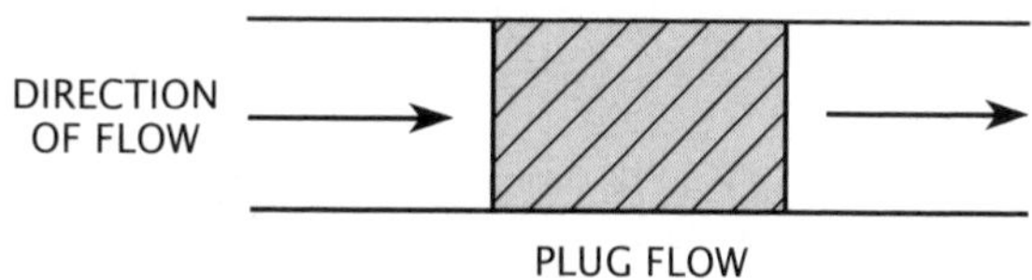

POLYELECTROLYTE (POLY-ee-LECK-tro-lite) POLYELECTROLYTE

A high-molecular-weight (relatively heavy) substance, having points of positive or negative electrical charges, that is formed by either natural or synthetic (manmade) processes. Natural polyelectrolytes may be of biological origin or obtained from starch products or cellulose derivatives. Synthetic polyelectrolytes consist of simple substances that have been made into complex, high-molecular-weight substances. Used with other chemical coagulants to aid in binding small suspended particles to larger chemical flocs for their removal from water. Often called a POLYMER.

POLYMER (POLY-mer) POLYMER

A long-chain molecule formed by the union of many monomers (molecules of lower molecular weight). Polymers are used with other chemical coagulants to aid in binding small suspended particles to larger chemical flocs for their removal from water. Also see POLYELECTROLYTE.

PROGRAMMABLE LOGIC CONTROLLER (PLC) PROGRAMMABLE LOGIC CONTROLLER (PLC)

A microcomputer-based control device containing programmable software; used to control process variables.

PROTOZOA (pro-toe-ZOE-ah) PROTOZOA

A group of motile, microscopic organisms (usually single-celled and aerobic) that sometimes cluster into colonies and generally consume bacteria as an energy source.

REDUCTION (re-DUCK-shun) REDUCTION

Reduction is the addition of hydrogen, removal of oxygen, or the addition of electrons to an element or compound. Under anaerobic conditions (no dissolved oxygen present), sulfur compounds are reduced to odor-producing hydrogen sulfide (H_2S) and other compounds. In the treatment of metal finishing wastewaters, hexavalent chromium (Cr^{6+}) is reduced to the trivalent form (Cr^{3+}). The opposite of OXIDATION.

RISING SLUDGE RISING SLUDGE

Rising sludge occurs in the secondary clarifiers of activated sludge plants when the sludge settles to the bottom of the clarifier, is compacted, and then starts to rise to the surface, usually as a result of denitrification, or anaerobic biological activity that produces carbon dioxide and/or methane.

SECCHI (SECK-key) DISK SECCHI DISK

A flat, white disk lowered into the water by a rope until it is just barely visible. At this point, the depth of the disk from the water surface is the recorded Secchi disk transparency.

SEIZING or SEIZE UP SEIZING or SEIZE UP

Seizing occurs when an engine overheats and a part expands to the point where the engine will not run. Also called freezing.

SEPTIC (SEP-tick) or SEPTICITY SEPTIC or SEPTICITY

A condition produced by bacteria when all oxygen supplies are depleted. If severe, the bottom deposits produce hydrogen sulfide, the deposits and water turn black, give off foul odors, and the water has a greatly increased oxygen and chlorine demand.

SHOCK LOAD SHOCK LOAD

The arrival at a treatment process of water or wastewater containing unusually high concentrations of contaminants in sufficient quantity or strength to cause operating problems. Organic or hydraulic overloads also can cause a shock load.

(1) For activated sludge, possible problems include odors and bulking sludge, which will result in a high loss of solids from the secondary clarifiers into the plant effluent and a biological process upset that may require several days to a week to recover.

(2) For trickling filters, possible problems include odors and sloughing off of the growth or slime on the trickling filter media.

(3) For drinking water treatment, possible problems include filter blinding and product water with taste and odor, color, or turbidity problems.

SLUDGE AGE SLUDGE AGE

A measure of the length of time a particle of suspended solids has been retained in the activated sludge process.

$$\text{Sludge Age, days} = \frac{\text{Suspended Solids Under Aeration, lbs or kg}}{\text{Suspended Solids Added, lbs/day or kg/day}}$$

SLUDGE DENSITY INDEX (SDI) SLUDGE DENSITY INDEX (SDI)

This calculation is used in a way similar to the Sludge Volume Index (SVI) to indicate the settleability of a sludge in a secondary clarifier or effluent. The weight in grams of one milliliter of sludge after settling for 30 minutes. SDI = 100/SVI. Also see SLUDGE VOLUME INDEX.

SLUDGE VOLUME INDEX (SVI) SLUDGE VOLUME INDEX (SVI)

A calculation that indicates the tendency of activated sludge solids (aerated solids) to thicken or to become concentrated during the sedimentation/thickening process. SVI is calculated in the following manner: (1) allow a mixed liquor sample from the aeration basin to settle for 30 minutes; (2) determine the suspended solids concentration for a sample of the same mixed liquor; (3) calculate SVI by dividing the measured (or observed) wet volume (m*L*/*L*) of the settled sludge by the dry weight concentration of MLSS in grams/*L*.

$$\text{SVI, m}L\text{/gm} = \frac{\text{Settled Sludge Volume/Sample Volume, m}L/L}{\text{Suspended Solids Concentration, mg/}L} \times \frac{1{,}000\text{ mg}}{\text{gram}}$$

SOFTWARE PROGRAM SOFTWARE PROGRAM

Computer program; the list of instructions that tell a computer how to perform a given task or tasks. Some software programs are designed and written to monitor and control treatment processes.

SOLIDS CONCENTRATION SOLIDS CONCENTRATION

The solids in the aeration tank that carry microorganisms that feed on wastewater. Expressed as milligrams per liter of mixed liquor volatile suspended solids (MLVSS, mg/*L*).

STABILIZED WASTE STABILIZED WASTE

A waste that has been treated or decomposed to the extent that, if discharged or released, its rate and state of decomposition would be such that the waste would not cause a nuisance or odors in the receiving water.

STEP-FEED AERATION STEP-FEED AERATION

Step-feed aeration is a modification of the conventional activated sludge process. In step-feed aeration, primary effluent enters the aeration tank at several points along the length of the tank, rather than at the beginning or head of the tank and flowing through the entire tank in a plug flow mode.

SUBSTRATE SUBSTRATE

(1) The base on which an organism lives. The soil is the substrate of most seed plants; rocks, soil, water, or other plants or animals are substrates for other organisms.

(2) Chemical used by an organism to support growth. The organic matter in wastewater is a substrate for the organisms in activated sludge.

SUPERNATANT (soo-per-NAY-tent) SUPERNATANT

Liquid removed from settled sludge. Supernatant commonly refers to the liquid between the sludge on the bottom and the scum on the surface.

TOC (pronounce as separate letters) TOC

Total Organic Carbon. TOC measures the amount of organic carbon in water.

THRUST BLOCK THRUST BLOCK

A mass of concrete or similar material appropriately placed around a pipe to prevent movement when the pipe is carrying water. Usually placed at bends and valve structures.

TURBIDITY (ter-BID-it-tee) METER TURBIDITY METER

An instrument for measuring and comparing the turbidity of liquids by passing light through them and determining how much light is reflected by the particles in the liquid. The normal measuring range is 0 to 100 and is expressed as nephelometric turbidity units (NTUs). Also called a turbidimeter.

TURBIDITY (ter-BID-it-tee) UNITS (TU) TURBIDITY UNITS (TU)

Turbidity units are a measure of the cloudiness of water. If measured by a nephelometric (deflected light) instrumental procedure, turbidity units are expressed in nephelometric turbidity units (NTU) or simply TU. Those turbidity units obtained by visual methods are expressed in Jackson turbidity units (JTU), which are a measure of the cloudiness of water; they are used to indicate the clarity of water. There is no real connection between NTUs and JTUs. The Jackson turbidimeter is a visual method and the nephelometer is an instrumental method based on deflected light.

VOLUTE (vol-LOOT) VOLUTE

The spiral-shaped casing that surrounds a pump, blower, or turbine impeller and collects the liquid or gas discharged by the impeller.

ZOOGLEAL (ZOE-uh-glee-ul) FILM ZOOGLEAL FILM

See ZOOGLEAL MAT (FILM).

ZOOGLEAL (ZOE-uh-glee-ul) MASS ZOOGLEAL MASS

Jelly-like masses of bacteria found in both the trickling filter and activated sludge processes. These masses may be formed for or function as the protection against predators and for storage of food supplies. Also see BIOMASS.

ZOOGLEAL (ZOE-uh-glee-al) MAT (FILM) ZOOGLEAL MAT (FILM)

A complex population of organisms that form a slime growth on the sand filter media and break down the organic matter in wastewater. These slimes consist of living organisms feeding on wastes, dead organisms, silt, and other debris. On a properly loaded and operating sand filter, these mats are so thin as to be invisible to the naked eye. Slime growth is a more common term.

CHAPTER 11. ACTIVATED SLUDGE

OPERATION OF CONVENTIONAL ACTIVATED SLUDGE PLANTS

(Lesson 1 of 7 Lessons)

11.0 THE ACTIVATED SLUDGE PROCESS

11.00 Wastewater Treatment by Activated Sludge

When wastewater enters an activated sludge plant, the preliminary treatment processes (Figure 11.1) remove the coarse or heavy solids (grit) and other debris, such as roots, rags, and boards. Primary clarifiers remove much of the floatable and settleable material. Normally the activated sludge process treats settled wastewater, but in some plants the raw wastewater flows from the preliminary treatment processes directly to the activated sludge process.

11.01 Definitions

ACTIVATED SLUDGE. Activated sludge consists of sludge particles produced in raw or settled wastewater (primary effluent) by the growth of organisms (including zoogleal bacteria) in aeration tanks in the presence of dissolved oxygen. The term "activated" comes from the fact that the particles are teeming with bacteria, fungi, and protozoa.

ACTIVATED SLUDGE PROCESS (Figure 11.2). The activated sludge process is a biological wastewater treatment process that uses *MICROORGANISMS*[1] to speed up decomposition of wastes. When activated sludge is added to wastewater, the microorganisms feed and grow on waste particles in the wastewater. As the organisms grow and reproduce, more and more waste is removed, leaving the wastewater partially cleaned. To function efficiently, the mass of organisms *(SOLIDS CONCENTRATION*[2]*)* needs a steady balance of food *(FOOD/MICROORGANISM RATIO*[3]*)* and oxygen.

11.02 Process Description

Secondary treatment in the form of the activated sludge process is aimed at *OXIDATION*[4] and removal of soluble or finely divided suspended materials that were not removed by previous treatment. Aerobic organisms do this in a few hours as wastewater flows through an aeration tank. The organisms *STABILIZE*[5] soluble or finely divided suspended solids by partial oxidation forming carbon dioxide, water, and sulfate and nitrate compounds. The remaining solids are changed to a form that can be settled and removed as sludge during sedimentation.

After the aeration period, the wastewater is routed to a secondary settling tank for a liquid-organism (water-solids) separation. Settled organisms in the final clarifier are in a deteriorating condition due to lack of oxygen and food and should be returned to the aeration tank as quickly as possible. The remaining clarifier effluent is usually chlorinated and discharged from the plant.

Conversion of dissolved and suspended material to settleable solids is the main objective of high-rate activated sludge

---

[1] *Microorganisms* (MY-crow-OR-gan-is-ums). Living organisms that can be seen individually only with the aid of a microscope.

[2] *Solids Concentration.* The solids in the aeration tank that carry microorganisms that feed on wastewater. Expressed as milligrams per liter of mixed liquor volatile suspended solids (MLVSS, mg/*L*).

[3] *Food/Microorganism (F/M) Ratio.* Food to microorganism ratio. A measure of food provided to bacteria in an aeration tank.

$$\frac{\text{Food}}{\text{Microorganisms}} = \frac{\text{BOD, lbs/day}}{\text{MLVSS, lbs}}$$

$$= \frac{\text{Flow, MGD} \times \text{BOD, mg}/L \times 8.34\text{ lbs/gal}}{\text{Volume, MG} \times \text{MLVSS, mg}/L \times 8.34\text{ lbs/gal}}$$

or by calculator math system

$$= \text{Flow, MGD} \times \text{BOD, mg}/L \div \text{Volume, MG} \div \text{MLVSS, mg}/L$$

or metric

$$= \frac{\text{BOD, kg/day}}{\text{MLVSS, kg}}$$

$$= \frac{\text{Flow, M}L\text{/day} \times \text{BOD, mg}/L \times 1\text{ kg/M mg}}{\text{Volume, M}L \times \text{MLVSS, mg}/L \times 1\text{ kg/M mg}}$$

[4] *Oxidation.* Oxidation is the addition of oxygen, removal of hydrogen, or the removal of electrons from an element or compound; in the environment and in wastewater treatment processes, organic matter is oxidized to more stable substances. The opposite of REDUCTION.

[5] *Stabilized Waste.* A waste that has been treated or decomposed to the extent that, if discharged or released, its rate and state of decomposition would be such that the waste would not cause a nuisance or odors in the receiving water.

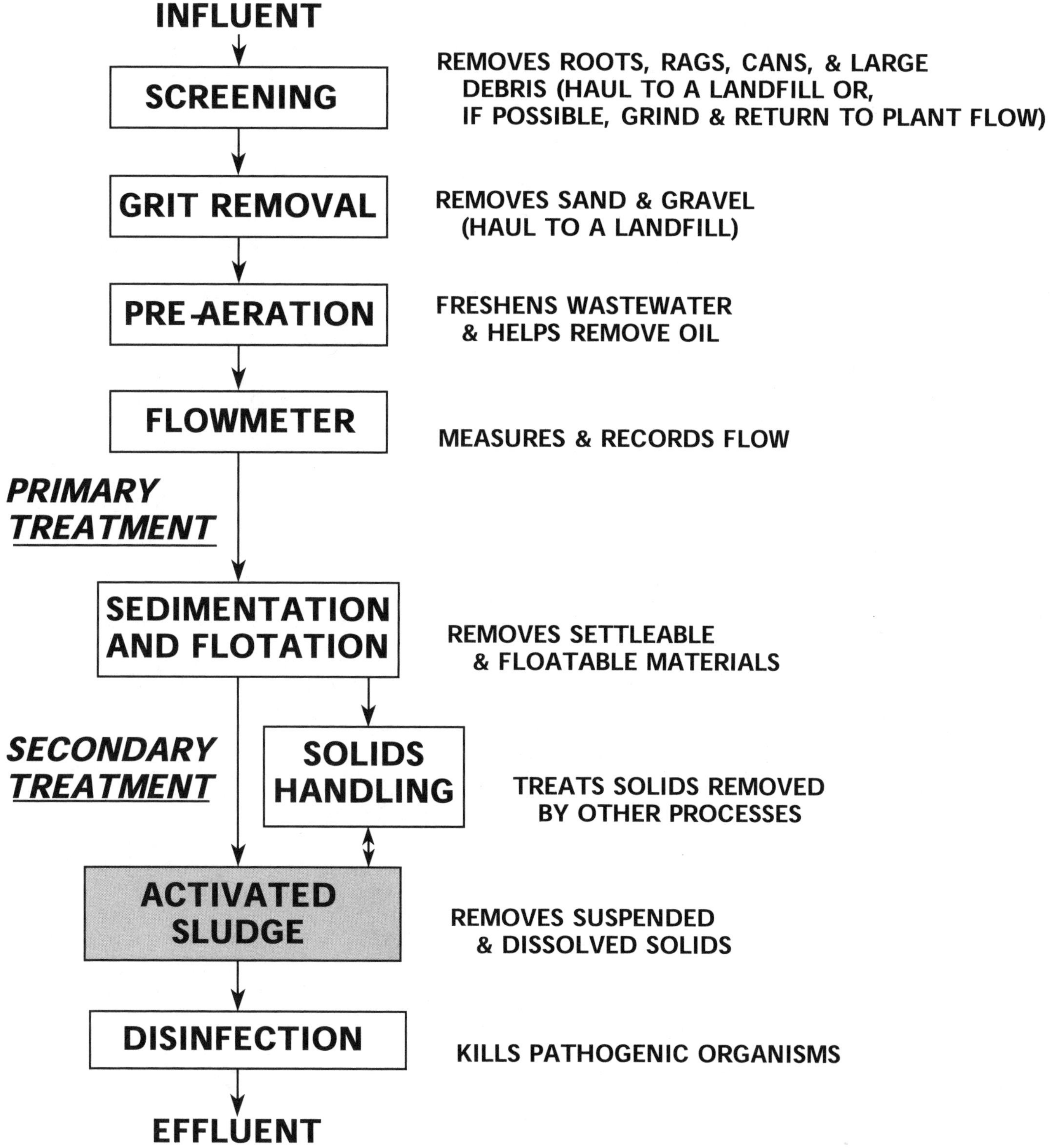
TREATMENT PROCESS
FUNCTION
PRELIMINARY TREATMENT
INFLUENT
SCREENING
REMOVES ROOTS, RAGS, CANS, & LARGE DEBRIS (HAUL TO A LANDFILL OR, IF POSSIBLE, GRIND & RETURN TO PLANT FLOW)
GRIT REMOVAL
REMOVES SAND & GRAVEL (HAUL TO A LANDFILL)
PRE-AERATION
FRESHENS WASTEWATER & HELPS REMOVE OIL
FLOWMETER
MEASURES & RECORDS FLOW
PRIMARY TREATMENT
SEDIMENTATION AND FLOTATION
REMOVES SETTLEABLE & FLOATABLE MATERIALS
SECONDARY TREATMENT
SOLIDS HANDLING
TREATS SOLIDS REMOVED BY OTHER PROCESSES
ACTIVATED SLUDGE
REMOVES SUSPENDED & DISSOLVED SOLIDS
DISINFECTION
KILLS PATHOGENIC ORGANISMS
EFFLUENT

Fig. 11.1 Flow diagram of a typical plant

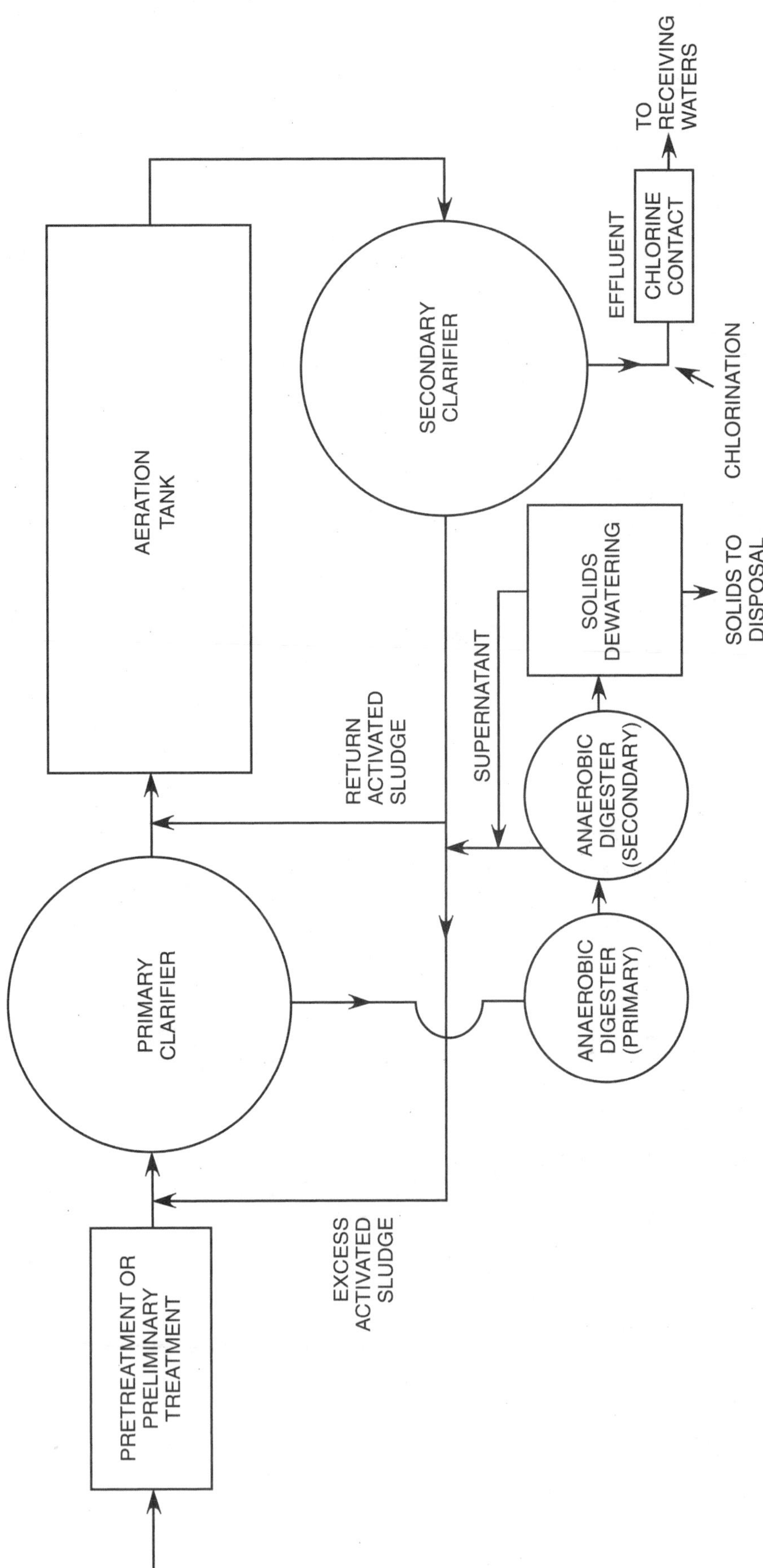

Fig. 11.2 Plan layout of a typical activated sludge plant

processes, while low-rate processes stress oxidation. The oxidation may be by chemical or biological processes. In the activated sludge process, the biochemical oxidation carried out by living organisms is stressed. The same organisms also are effective in conversion of substances to settleable solids if the plant is operated properly.

When wastewater enters the aeration tanks, it is mixed with the activated sludge to form a mixture of sludge, carrier water, and influent solids. These solids come mainly from the discharges from homes, factories, and businesses. The activated sludge that is added contains many different types of helpful living organisms, which were grown during previous contact with wastewater. (Lesson 7, Section 11.10, "Microbiology for Activated Sludge," describes the most common types of activated sludge organisms.) These organisms are the workers in the treatment process. They use the incoming wastes for food and as a source of energy for their life processes and for the reproduction of more organisms. These organisms will use more food contained in the wastewater in treating the wastes. The activated sludge also forms a lacy network or floc mass that entraps many materials not used as food.

Some organisms will require a long time to use the available food in the wastewater at a given waste concentration. Many organisms will compete with each other in the use of available food (waste) to shorten the time factor and increase the portion of waste stabilized. The ratio of food to organisms is a primary control in the activated sludge process. Organisms tend to increase with waste (food) load and time spent in the aeration tank. Under favorable conditions the operator will remove the excess organisms to maintain the required number of workers for effective waste treatment. Therefore, removal of organisms from the treatment process (sludge wasting) is a very important control technique.

Oxygen, usually supplied from air, is needed by the living organisms as they oxidize wastes to obtain energy for growth. Insufficient oxygen will slow down aerobic organisms, make *FACULTATIVE*[6] organisms work less efficiently, and favor production of foul-smelling intermediate products of decomposition and incomplete reactions. (Aeration systems are described more fully in Section 11.1, "Aeration Systems.")

An increase in organisms in an aeration tank will require greater amounts of oxygen. More food in the influent encourages more organism activity and more oxidation; consequently, more oxygen is required in the aeration tank. An excess of oxygen is required for complete waste stabilization. Therefore, the dissolved oxygen (DO) content in the aeration tank is an essential control factor. Some minimum level of oxygen must be maintained to favor the desired type of organism activity to achieve the necessary treatment efficiency. If the DO in the aeration tank is too low, *FILAMENTOUS ORGANISMS*[7] will thrive and the sludge *FLOC*[8] will not settle in the secondary clarifier. Also, if the DO is too high, pinpoint floc will develop and not be removed in the secondary clarifier. Therefore, the proper DO level must be maintained so solids will settle properly and the plant effluent will be clear.

Flows must be distributed evenly among two or more similar treatment units. If your plant is equipped with a splitter box or a series of boxes, it will be necessary to periodically check and estimate whether the flow is being split as intended.

Activated sludge solids concentrations in the aerator and the secondary clarifier should be determined by the operator for process control purposes. Solids are in a deteriorating condition as long as they remain in the secondary clarifier. Depth of sludge blanket in the secondary clarifier and concentrations of solids in the aerator are very important for successful wastewater treatment. Centrifuge tests will give a quick estimate of solids concentrations and locations in the units. Precise solids tests should be made periodically for comparison with centrifuge solids tests. Before any changes are made in the mode of operation, precise solids measurements should be obtained. Settleability tests show the degree and volume of solids settling that may be obtained in a secondary clarifier; however, visual plant checks show what is actually happening.

Primary clarifiers are designed to remove material that settles to the bottom or floats to the top. Activated sludge helps this process along by collecting and *AGGLOMERATING*[9] the tiny particles in the primary effluent or raw wastewater so that they will settle better. If for some reason the organisms fail to make this change in the soluble solids, then the secondary clarifier effluent quality will not be satisfactory. For the activated sludge process to work properly, the operator must control the number of organisms, the dissolved oxygen in the aeration tanks, and the treatment time. When these factors are under proper control, the organisms will convert soluble solids and agglomerate the fine particles into a floc mass.

A floc mass is made up of millions of organisms (10^{12} to 10^{18}•100 m*L* in a good activated sludge), including bacteria, fungi, yeast, protozoa, and worms. When a floc mass is returned to the aerator from the final clarifier, the organisms grow as a result of taking food from the inflowing wastewater. The surface of the floc mass is irregular and promotes the transfer of wastewater pollutants into the solids by means of mechanical entrapment, absorption, adsorption, or adhesion.

---

[6] *Facultative* (FACK-ul-tay-tive) *Bacteria.* Facultative bacteria can use either dissolved oxygen or oxygen obtained from food materials such as sulfate or nitrate ions. In other words, facultative bacteria can live under aerobic, anoxic, or anaerobic conditions.

[7] *Filamentous* (fill-uh-MEN-tuss) *Organisms.* Organisms that grow in a thread or filamentous form. Common types are *Thiothrix* and *Actinomycetes*. A common cause of sludge bulking in the activated sludge process.

[8] *Floc.* Clumps of bacteria and particles, or coagulants and impurities, that have come together and formed a cluster. Found in flocculation tanks, sedimentation basins, aeration tanks, secondary clarifiers, and chemical precipitation processes.

[9] *Agglomeration* (uh-glom-er-A-shun). The growing or coming together of small scattered particles into larger flocs or particles, which settle rapidly. Also see FLOC.

Many substances not used as food also are transferred to the floc mass, thus improving the quality of the plant effluent.

Material taken into the floc mass is partially oxidized to form cell mass and oxidation products. Ash or inorganic material (silt and sand) taken in by the floc mass increases the density of the mass. Mixing the contents of the aerator causes the floc masses to bump into each other and form larger clumps. Eventually these masses become heavy enough to settle to the bottom of the secondary clarifier where they can be removed easily. This sludge now contains most of the organisms and waste material that had been mixed in the wastewater.

The next step in the activated sludge process is removal of sludge from the secondary clarifier. Some of the material is converted and released to the atmosphere in the form of stripped gases (carbon dioxide or volatile gases not converted and released in the aeration tank). That leaves water and sludge solids. A certain amount of the solids (return activated sludge) will be returned to the aerator to treat incoming wastewater. The operator must pump these solids to the aerator. The rest of the solids (waste activated sludge) must be removed and disposed of so that it does not continue in the plant flow. After the sludge solids have been removed from the final clarifier, the treated wastewater moves to advanced waste treatment processes or the disinfection process.

The successful operation of an activated sludge plant requires the operator to be aware of the many factors influencing the process and to check them repeatedly. To keep the organisms working in the activated sludge, you *MUST* provide a suitable environment. High concentrations of acids, bases, and other toxic substances are undesirable and may kill the working organisms. Uneven flows of wastewater may cause overfeeding, starvation, and other problems that upset the activated sludge process. Failure to supply enough oxygen can cause an unfavorable environment, which results in decreased organism activity.

While successful operation of an activated sludge plant involves an understanding of many factors, actual control of the process as outlined in this section is relatively simple. Control consists of maintaining the proper solids (floc mass) concentration in the aerator for the waste (food) inflow by adjusting the waste sludge pumping rate and regulating the oxygen supply to maintain a satisfactory level of dissolved oxygen in the process.

11.03 Requirements for Control

Effective control of the activated sludge process depends on the operator's understanding of and ability to adjust several interrelated factors. Some of these factors are:

1. Wastewater flow, concentration, and characteristics of the wastewater received
2. Amount of activated sludge (containing the working organisms) to be maintained in the process relative to inflow
3. Amount of oxygen required to stabilize wastewater oxygen demands and to maintain a satisfactory level of dissolved oxygen to meet organism requirements
4. Equal division of plant flow and waste load between duplicate treatment units (two or more clarifiers or aeration tanks)
5. Transfer of the pollutional material (food) from the wastewater to the floc mass (solids or workers) and separation of the solids from the treated wastewater
6. Effective control and disposal of in-plant residues (solids, scums, and supernatants) to accomplish ultimate disposal in a nonpolluting manner
7. Provisions for maintaining a suitable environment for the organisms treating the wastes to keep them healthy and happy

Effluent quality requirements may be stated by your regulatory agency in terms of percentage removal of wastes. Current regulations frequently specify allowable quantities of wastes that may be discharged. These quantities are based on flow and concentrations of significant items such as solids, oxygen demand, coliform bacteria, nitrogen, and oil as specified by your regulatory agencies in your NPDES permit.

The effluent quality requirements in your NPDES permit usually determine what kind of activated sludge operation you can use and how tightly you must control the process. For example, if an effluent containing 50 mg/*L* of suspended solids and BOD (refers to five-day BOD) is satisfactory, a high-rate activated sludge process will probably meet your needs. If the limit is 10 mg/*L,* the high-rate process would not be suitable. If a high degree of treatment is required, very close process control and additional treatment after the activated sludge process may be needed. Today secondary treatment plants are expected to remove 85 percent of the BOD and provide an effluent with a 30-day average BOD of less than 30 mg/*L.* (Section 11.8, "Modifications of the Activated Sludge Process," describes several modifications of the activated sludge process. One additional variation of the process, use of sequencing batch reactors (SBRs), is described in Section 11.9.)

QUESTIONS

Write your answers in a notebook and then compare your answers with those on page 130.

11.0A What is the purpose of the activated sludge process in treating wastewater?

11.0B What is a stabilized waste?

11.0C Why is air added to the aeration tank in the activated sludge process?

11.0D What happens to the air requirement in the aeration tank when the strength (BOD) of the incoming wastewater increases?

11.0E What factors could cause an unsuitable environment for the activated sludge process in an aeration tank?

11.0F What two different ways may effluent quality requirements be stated by regulatory agencies?

11.04 Basic Variables

Wastewater flows and contents change daily. The activated sludge plant operator attempts to maintain the process at some balanced state that will be capable of handling the minor variations in flows or wastewater characteristics and produce the desired quality of effluent. To accomplish this goal you must operate the activated sludge process on known data and knowledge obtained at other plants and relate them to your plant. After your plant becomes operational, you then must relate the control procedures to actual plant experience. The variations that affect the operation are derived from two sources: (1) changes in wastewater characteristics and collection system flows, and (2) in-plant operational variables.

11.040 Variables in Collection System

During storms, treatment plants served by combined sewer systems will receive an increase in flow, which may cause the following problems:

1. Reduced amount of time wastewater spends in treatment units (hydraulic overload)
2. Increased amounts of grit and silt, which lower the volatile (food) content of the solids
3. Increased organic load during initial washout of accumulated sewer deposits
4. Rapid changes in wastewater temperature and solids content

Collection system maintenance activities can also affect treatment plant operation. If a lift station has been out of service for a period of time, large volumes of septic wastewater could cause a shock load on your treatment processes. Similar problems could be created when a blockage in a line is cleared or a new line is connected to the system. Analysis of inflow quantities and characteristics when these flows reach a treatment plant can indicate whether or not they will cause a serious problem. Advance notice of collection system maintenance crew activities can be very helpful.

Various industries, businesses, and cesspool or septic tank service firms can cause uneven flows and changes in the types of wastewater entering a plant. You should become acquainted with the managers of businesses and facilities whose discharges could upset your treatment processes. Convince these people in a friendly manner how vital it is to your plant processes and the receiving waters for you to be notified of any potentially harmful discharges. Try to obtain their cooperation and request them to notify you whenever an accidental spill, a process change, or a cleaning operation occurs that could cause undesirable waste discharges. This requires diplomacy to obtain cooperation from dischargers to regulate their own discharges and to reduce the number of midnight dumps. If your plant has an industrial pretreatment facility inspection program, the pretreatment inspector may help you make arrangements with industrial dischargers for waste disposal, pretreatment, or controlled discharges to be sure that substances harmful to your treatment plant are diluted before they enter the plant.

11.041 Operational Variables

An activated sludge plant may be operated in any one of three ranges or operational zones on the basis of *SLUDGE AGE,*[10] which is an expression of pounds of solids added per day per pound of solids maintained in the particular process. Sludge age is a control guide that is widely used and is an indicator of the length of time a pound of solids is kept under aeration in the system. If the amount of solids under aeration remains fairly constant, then an increase in the influent solids load will decrease the sludge age. Use of this measure of sludge age is recommended for the new activated sludge plant operator because of the ease in understanding this approach. The experienced operator may not accept this method of control because it ignores the soluble BOD that is related to the solids production but not measured by suspended solids tests on the influent.

The following values are typical sludge ages for different types of municipal activated sludge plants with very little industrial waste. Actual loadings must be related to the type of waste and local situation.

1. *HIGH-RATE.* A high-rate activated sludge plant operates at the highest loading of food to microorganisms; the sludge age ranges from 0.5 to 2.0 days. Due to this higher loading, the system produces a lower quality of effluent than the other types of activated sludge plants. This system is more easily upset than others and requires tighter control and frequent testing.
2. *CONVENTIONAL.* Conventional activated sludge plants are the most common type in use today. The loading of food to microorganisms is approximately 50 percent lower than in a high-rate plant, and the sludge age ranges from 3.5 to 7.0 days. This method of operation produces a high quality of effluent and is able to absorb some shock loads without lowering effluent quality.
3. *EXTENDED AERATION.* Extended aeration is often used in smaller package-type plants (Volume I, Chapter 8) or so-called complete oxidation systems. These are the most stable of the three processes due to the light loading of food to microorganisms, and the sludge age is commonly greater than 10 days. Effluent suspended solids commonly are higher than found under conventional loadings.

Several different types of activated sludge plants have been built using various flow arrangements, tank configurations, or oxygen application equipment. However, all of these variations are essentially modifications of the basic concept of conventional activated sludge. Section 11.8 describes the contact stabilization, step-feed, Kraus, and complete mix variations, and sequencing batch reactors (SBRs) are discussed in Section 11.9.

Continual review of laboratory test results is essential in determining whether a treatment plant is meeting discharge requirements in terms of such water quality indicators as BOD, COD, suspended solids, coliforms, nitrogen, and oil. If the desired and required quality of the plant effluent is not achieved,

---

[10]

$$\text{Sludge Age, days} = \frac{(\text{Suspended Sol in Mixed Liq, mg}/L)(\text{Aerator Vol, MG})(8.34\text{ lbs/gal})}{(\text{Suspended Sol in Primary Effl, mg}/L)(\text{Flow, MGD})(8.34\text{ lbs/gal})}$$

$$= \frac{\text{Suspended Solids Under Aeration, lbs}}{\text{Suspended Solids Added, lbs/day}}$$

$$\text{or metric} = \frac{(\text{Suspended Sol in Mixed Liq, mg}/L)(\text{Aerator Vol, M}L)(1\text{ kg/M mg})}{(\text{Suspended Sol in Primary Effl, mg}/L)(\text{Flow, M}L\text{/day})(1\text{ kg/M mg})}$$

$$= \frac{\text{Suspended Solids Under Aeration, kg}}{\text{Suspended Solids Added, kg/day}}$$

the operator must determine what factor or factors have changed to upset plant performance and thus reduce efficiency.

Important factors that could have changed include:

1. Higher COD or BOD load applied to the aerator (influent load)
2. More difficult-to-treat wastes have caused a change in influent characteristics
3. Unsuitable mixed liquor suspended solids concentration in the aerator
4. Lower or higher rate of wasting activated sludge
5. Unsuitable rate of returning sludge to the aerator could adversely influence mixed liquor suspended solids
6. Higher solids concentrations in digester *SUPERNATANT*[11] returned to the plant flow, or return too rapid
7. Dropping of oxygen concentration in the aerator below desirable levels
8. Increase or decrease in wastewater temperature

Examination of plant records should reveal the items that have changed that could have upset the treatment process.

QUESTIONS

Write your answers in a notebook and then compare your answers with those on page 130.

11.0G What two major variables affect the way an activated sludge plant is operated?

11.0H What problems can be caused in an activated sludge plant when excessive stormwater flows through the process?

11.0I Besides excessive stormwater flows, what other variables in the collection system can affect the operation of an activated sludge plant?

11.0J Write the formula for calculating sludge age.

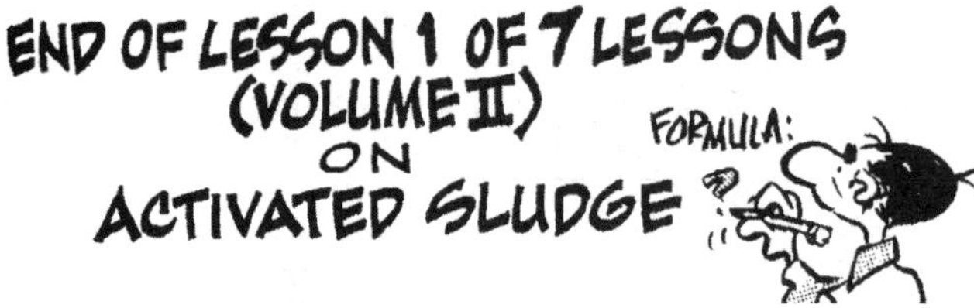

Please answer the discussion and review questions next.

DISCUSSION AND REVIEW QUESTIONS

Chapter 11. ACTIVATED SLUDGE

(Lesson 1 of 7 Lessons)

At the end of each lesson in this chapter, you will find some discussion and review questions. The purpose of these questions is to indicate to you how well you understand the material in the lesson. Write the answers to these questions in your notebook.

1. Define activated sludge.
2. Briefly describe the activated sludge process.
3. Why should activated sludge in the final clarifier be returned to the aeration tank as quickly as possible?
4. How can an operator control the activated sludge process?
5. During storms, an activated sludge plant will receive an increased inflow that may cause which of the following problems?
   1. Dilution of wastes, which makes them easy to treat
   2. Reduced amount of time wastewater spends in treatment units
   3. Increased amounts of grit and silt
   4. Increased organic loading
   5. Fluctuating wastewater temperatures and solids content
6. How can maintenance activities in a collection system cause operational problems in an activated sludge treatment plant?
7. How can the operator attempt to reduce problems caused by waste discharges into the collection system?
8. What can the operator determine from laboratory test results on the plant effluent?

[11] *Supernatant* (soo-per-NAY-tent). Liquid removed from settled sludge. Supernatant commonly refers to the liquid between the sludge on the bottom and the scum on the surface.

CHAPTER 11. ACTIVATED SLUDGE

OPERATION OF CONVENTIONAL ACTIVATED SLUDGE PLANTS

(Lesson 2 of 7 Lessons)

11.1 AERATION SYSTEMS

11.10 Purpose of Aeration

Aeration serves the dual purposes of providing dissolved oxygen and mixing of the mixed liquor and wastewater in the aeration tank. Two methods are commonly used to disperse oxygen from the air to the microorganisms—*MECHANICAL AERATION* (Figure 11.3) and *DIFFUSED AERATION* (Figure 11.4). Both methods are mechanical processes with the major difference being whether the mechanisms are located at or in the aerator or at a remote location. Oxygen also may be provided to the microorganisms by *PURE OXYGEN* systems (see Chapter 2, *ADVANCED WASTE TREATMENT* manual).

11.11 Mechanical Aeration Systems

Mechanical aeration devices agitate the water surface in the aerator to cause spray and waves by paddle wheels, mixers, rotating brushes, or some other method of splashing water into the air or air into the water where the oxygen can be absorbed.

Surface aerators use a motor-driven rotating impeller (Figure 11.5) or a brush rotor (Volume I, Chapter 8, Figure 8.11). Both devices splash the mixed liquor into the atmosphere above the aeration tank. Oxygen transfer to the mixed liquor is achieved by this method of aeration as the mixed liquor passes through the atmosphere. Surface aerators either float (for use in aerated ponds) or are mounted on supports in or above an aeration basin. In ponds they are equipped with draft tubes to improve their mixing characteristics.

A surface aerator's oxygen transfer efficiency is stated in terms of oxygen transferred per motor horsepower per hour. Typical oxygen transfer efficiencies are about two to three pounds of oxygen per hour per motor horsepower (1.2 to 1.8 kg/hr/kW). The oxygen transfer efficiency increases as the submergence of the aerator is increased. However, power costs also increase because more power is required to move the aerator impeller or agitator through the mixed liquor due to greater submergence and increased load on the drive motor. Mechanical aerators in the tank tend to be lower in installation and maintenance costs. Usually, they are more versatile in terms of mixing, production of surface area of bubbles, and oxygen transfer per unit of applied power.

The turbine aerator is another type of motor-driven mechanical aerator. Turbine aerators are frequently used in the complete mix or pure oxygen activated sludge process. An outside source of air is supplied to the aerator, usually from a blower or oxygen-generating system. Turbine aerators are more efficient and use less horsepower than standard surface aerators because the extra air supply creates turbulence in the immediate area of the rising air bubbles.

11.12 Diffused Aeration Systems

Diffused aeration systems are the most common type of aeration system used in the activated sludge process. A diffuser (Figure 11.4) breaks up the air stream from the blowers into fine bubbles in the mixed liquor. The smaller the bubbles, the greater the oxygen transfer due to the greater surface area of rising air bubbles surrounded by water. Unfortunately, fine bubbles will tend to regroup into larger bubbles while rising unless broken up by suitable mixing energy and turbulence.

The aeration tank distribution system consists of numerous diffusers attached to the bottom of air *HEADERS*[12] and located near the bottom of the aeration tank. Diffusers located in this position maximize the contact time of the air bubbles with the mixed liquor. Also, this location encourages mixing and discourages deposits on the tank bottom.

Although diffused aeration is used in the aeration tanks of activated sludge plants, it may also be found in aerated grit chambers, pre-aeration systems, aerated flow channels, and return sludge wet well aeration.

11.120 Air Filters

Filters (Figure 11.6) remove dust and dirt from air before it is compressed and sent to the various plant processes. Clean air is essential for the protection of:

1. Blowers
   a. Large objects entering the impellers or lobes may cause severe damage.
   b. Deposits on the impellers or lobes reduce clearances and cause excessive wear and vibration problems.
2. Process systems
   a. Clean air is required to protect downstream equipment.
   b. Clean air prevents fouling of air conduits, pipes, tubing, or dispersing devices on diffusers.

The filters may be constructed of a fiber mesh or metal mesh material that is sandwiched between a screen material and encased in a frame. The filter frames are then installed in a filter chamber. Other types of filters include bags, oil-coated traveling screens, and electrostatic precipitators.

---

[12] *Header.* A large pipe to which the ends of a series of smaller pipes are connected. Also called a MANIFOLD.

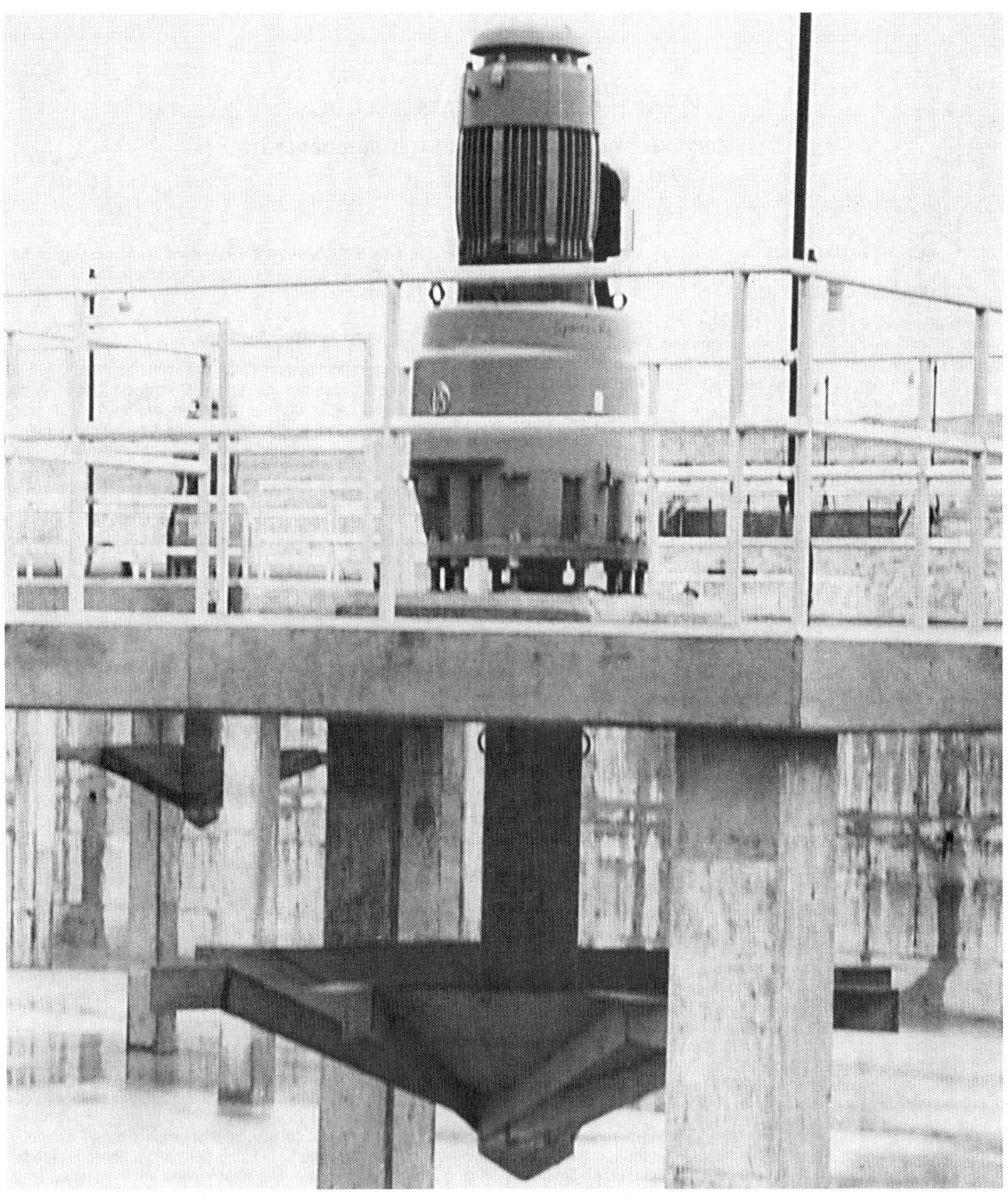

Fig. 11.3 Mechanical aeration device

(Courtesy INFILCO INC.)

NOTE: Diffuser is not level and most of the air is leaving on the left side of the diffuser.

Fig. 11.4 Air diffuser (fine bubbles)

(Courtesy Paul Halbach, National Training Center, Water Quality Office/EPA)

Fig. 11.5 Bridge-mounted surface aerators. Motor, gear reducer, and impeller shown
(Courtesy of Chino Basin Municipal Water District)

Dust and dirt filters in a filter chamber

1. Filter elements
2. Filter chamber
3. Filter chamber inlet pressure tube to manometer or water column gauge
4. Filter chamber outlet pressure tube to manometer or water column gauge

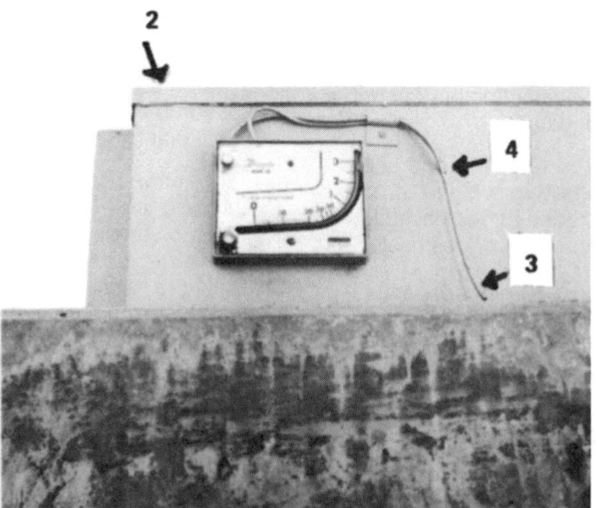

Manometer on a filter chamber (not a U-tube manometer) reading inches of pressure difference

Fig. 11.6 Air filters

The process air is usually drawn directly from the atmosphere. Some treatment plants have pretreatment and primary treatment process tanks covered for odor control. In plants of this type, odorous air is also drawn from under the roofs of these covered tanks for use as process air.

11.121 Blowers (Process Air Compressors)

Blowers are of the positive displacement type or the centrifugal type that provide air to the various plant processes through a pipe or conduit header system. Usually positive displacement blowers operate at low RPMs (revolutions per minute) and produce less than 20,000 CFM (cubic feet per minute) (570 cu m/min) while centrifugal blowers operate at high RPMs and range from 20,000 to 150,000 CFM (570 to 4,300 cu m/min).

POSITIVE DISPLACEMENT BLOWERS (Figures 11.7, 11.8, and 11.9)

The positive displacement blower provides a constant volume (cubic feet or cubic meters) output of air per revolution for a specific set of rotors or lobes. Blower output is varied by changing rotor or lobe speed (RPMs or revolutions per minute). The higher the RPM the greater the air output.

Small positive displacement blowers ranging from 100 to 1,000 cubic feet per minute (CFM) (3 to 28 cu m/min) are usually installed to be operated at a fixed volume output. These smaller units are directly driven by electric motors through a direct coupling or through sheaves and belts.

If a change in air volume output is required, it is accomplished by changing the motor to one with a higher or lower RPM or by changing sheaves to increase or decrease blower rotor or lobe rotation (RPM), thus increasing or decreasing air output.

NOTE: These small units are commonly used with package plants, pond aeration systems, small aerobic digesters, gas mixing in digesters, and gas storage compressors.

Large positive displacement blowers (2,000 to 20,000 CFM or 57 to 570 cu m/min) also may be driven by internal combustion engines or variable-speed electric motors in order to change blower volume outputs as required in activated sludge plants. By increasing or decreasing engine or motor RPM, the positive displacement blower output can be increased or decreased.

The air lines are connected to the blower through a flexible coupling to keep vibration to a minimum and to allow for heat expansion. When air is compressed, heat is generated, thus increasing the discharge temperature as much as 100°F (56°C), or more.

A check valve follows next, which prevents the blower from operating in reverse should other blowers in the same system be operating while this blower is off.

The discharge line from the blower is equipped with an air relief valve, which protects the blower from excessive back pressure and overload. Air relief valves are adjusted by weights or springs to open when air pressure exceeds a point above normal operating range, around 6.0 to 10.0 psi (0.4 to 0.7 kg/sq cm) in most wastewater treatment plants.

An air discharge silencer is also installed to provide *DECIBEL*[13] noise reduction. Ear protective devices should be worn when working near noisy blowers.

The impellers are machined on all exterior surfaces for operating at close tolerances; they are statically and dynamically balanced. Impeller shafts are made of machined steel and are securely fastened to the impellers. Timing gears accurately position the impellers.

Lubrication to the gears and bearings is maintained by a lube oil pump driven from one of the impeller shafts. An oil pressure gauge monitors the system oil pressure. An oil filter is located in the oil sump to ensure that the oil is free from foreign materials. An oil level is maintained in the gear housing so that gears and bearings will receive splash lubrication in case of lube oil pump failure. Air vents are located between the seals and the impeller chamber to relieve excessive pressure on the seals.

CENTRIFUGAL BLOWERS

The centrifugal blower (Figure 11.10) is a motor connected to a speed-increasing, gear-driven blower that provides a variable air output. Minimum through maximum air output is controlled by guide vanes, which are located on the intake side of the blower. These vanes may be positioned manually by operating personnel or may be controlled by plant instrumentation based on either dissolved oxygen levels in the aeration tanks or the plant influent flows.

The blower consists of an impeller, *VOLUTE*[14] casing, shaft and bearings, speed-increasing gear box, and an electric motor or internal combustion engine to drive the unit. Air enters the volute casing through an inlet nozzle and is picked up by the whirling vanes of the impeller where it is hurled by centrifugal force into the volute casing. Air enters the volute in its smallest section and moves in a circular motion to the largest section of the volute where it is discharged through the discharge nozzle.

Air lines are connected to the blower through flexible couplings in order to keep vibration to a minimum and to allow for heat expansion. The air suction line is usually equipped with a manually operated butterfly valve. Air bypass and discharge valves are usually electrically or pneumatically operated.

---

[13] *Decibel* (DES-uh-bull). A unit for expressing the relative intensity of sounds on a scale from zero for the average least perceptible sound to about 130 for the average level at which sound causes pain to humans. Abbreviated dB.

[14] *Volute* (vol-LOOT). The spiral-shaped casing that surrounds a pump, blower, or turbine impeller and collects the liquid or gas discharged by the impeller.

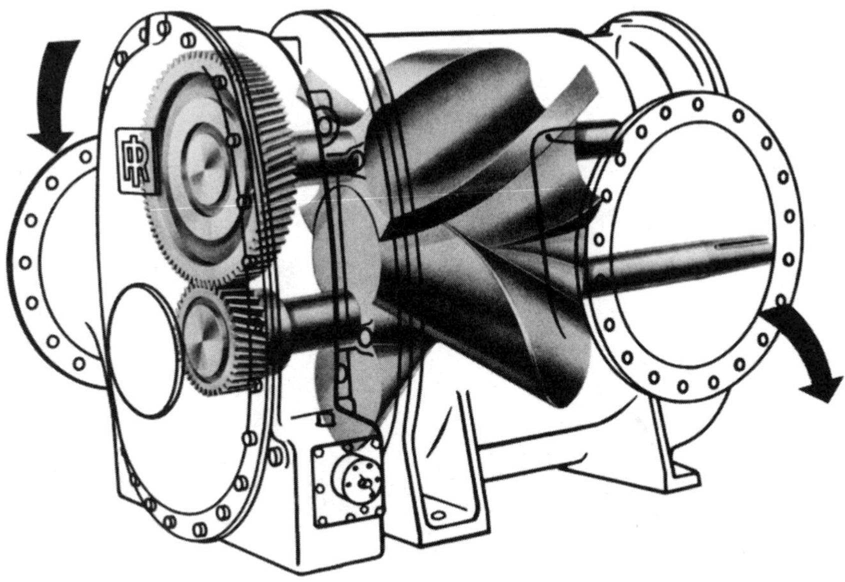

1. Air enters the compressor at the left and leaves at the right.

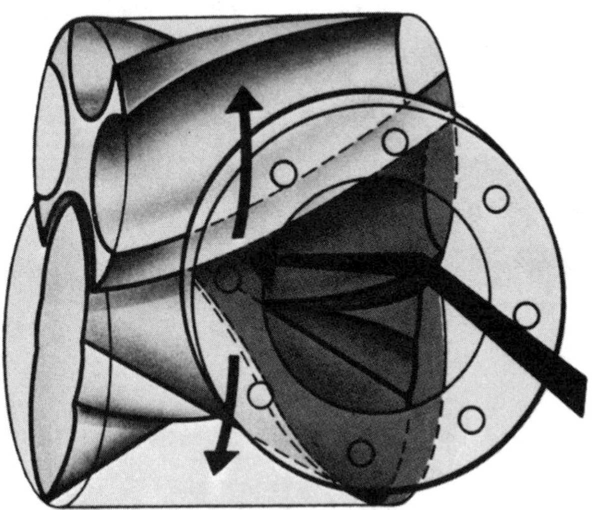

2. As the rotors turn away from each other on the inlet side, air is drawn into pockets formed between the rotors and the casing wall.

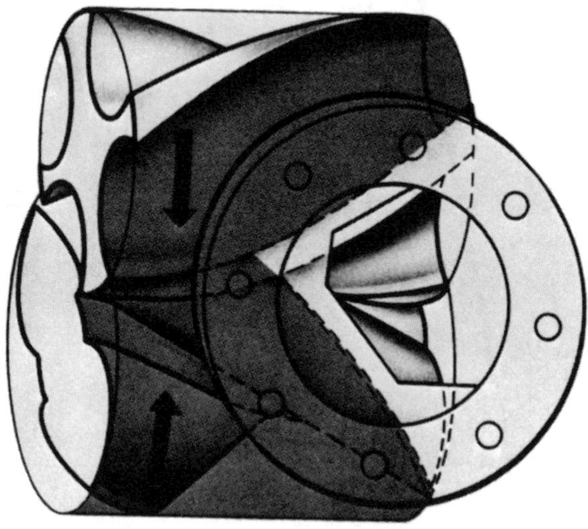

3. The pockets of air move along and around the rotor axes, then join and diminish in volume, thus compressing the air. The air flow is smooth and continuous, without pumping or surging.

Fig. 11.7 How a rotary positive displacement compressor works

(Permission of Chicago Pump)

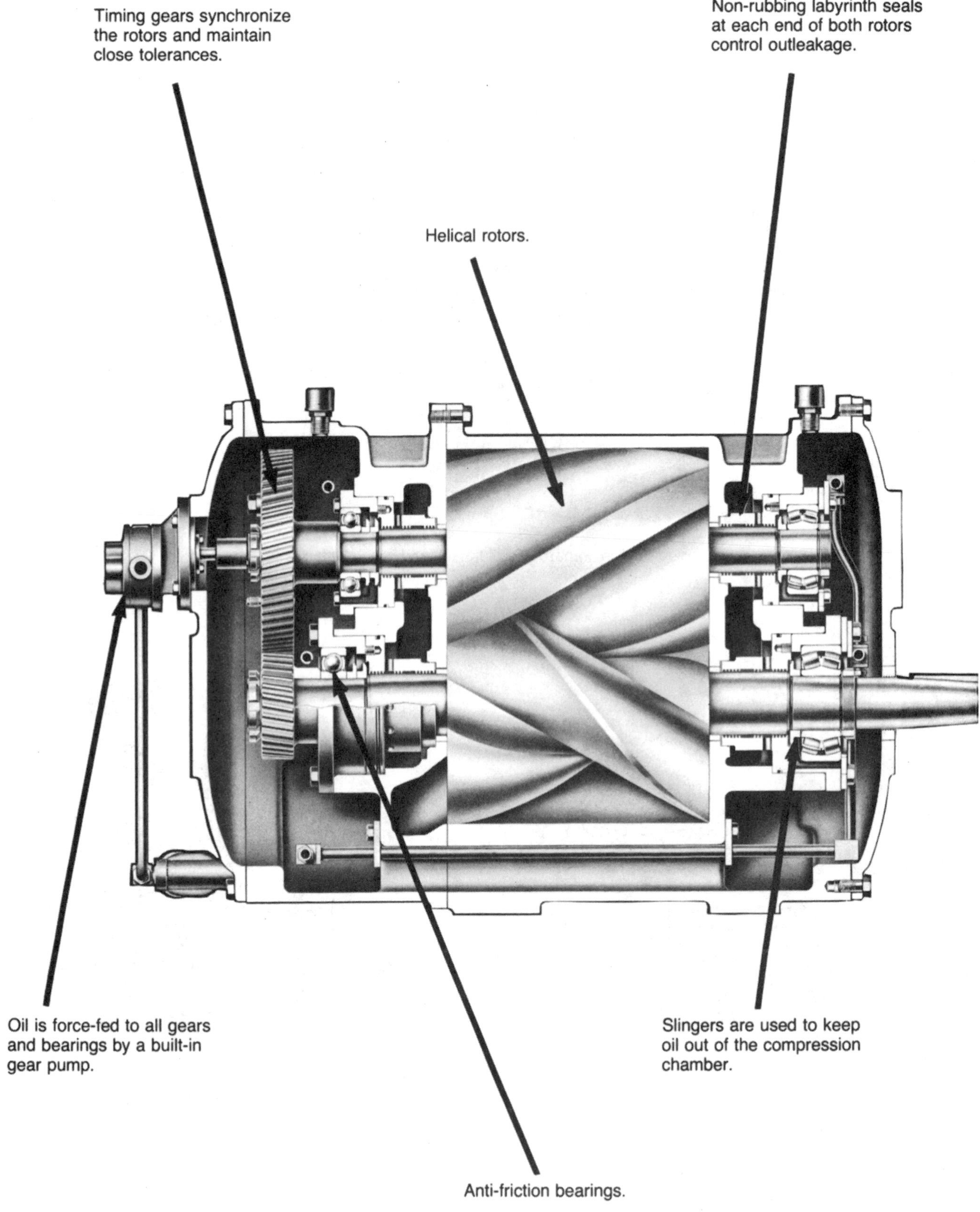
Timing gears synchronize the rotors and maintain close tolerances.
Non-rubbing labyrinth seals at each end of both rotors control outleakage.
Helical rotors.
Oil is force-fed to all gears and bearings by a built-in gear pump.
Slingers are used to keep oil out of the compression chamber.
Anti-friction bearings.

Fig. 11.8 Cut-away view of rotary positive displacement compressor

(Permission of Chicago Pump)

1. motor
2.* belts
3.* sheaves
4.* air line (suction)
5. air line (discharge)
6. flexible coupling
7. check valve
8. air relief valve
9. air discharge silencer
10.* impellers
11. lube oil pump
12. oil pressure gauge
13. oil filter
14. oil level indicator and reservoir
15. oil seal air vents
16. inlet valve
17.* discharge valve
18. oil temperature gauge
19. belt guard

* Not marked on photos

Fig. 11.9 Positive displacement blower

1. impeller casing
2. motor
3. speed-increasing gears
4. guide vanes
5. intake nozzle
6. discharge nozzle
7. flexible coupling
8. manual suction line valve
9. bypass valve
10. discharge valve
11. bearing stand
12. main oil pump
13. auxiliary oil pump
14. oil reservoir
15. couplings

Fig. 11.10 Centrifugal blower

The impeller is machined on all surfaces for operating at close tolerances and is statically (not moving) and dynamically (moving) balanced. The impeller shaft is supported in a shaft bearing stand, which contains a thrust bearing and journal bearings.

Lubrication to the bearings and gears is maintained by a positive displacement main oil pump that is driven by the speed-increasing gear unit. An auxiliary, electrically operated, centrifugal oil pump also is used to provide oil pressure in the event of failure of the main oil pump and to lubricate the blower shaft bearings before start-up and after shutdown. The oil reservoir is located in the blower baseplate. Cartridge type or disc-and-spacer type oil filters are provided to ensure the oil is free of foreign materials. The selection of either type filter is based on the degree of filtration required.

Due to the very high speeds at which these blower units operate and resultant high oil temperatures, an oil cooler unit is installed. This unit, in most cases, is a shell and tube, oil-to-water heat exchanger.

11.122 Air Distribution System

The air distribution system consists of pipes, valves, and metering devices that deliver air from the blowers to air headers in the aeration tanks and other plant processes (Figure 11.11).

An air metering device should be located in a straight section of the air main on the discharge side of the blower. This device consists of an orifice plate inserted between two specially made pipe flanges. The orifice plate is made of stainless steel with a precision hole cut through the center. The diameter of the hole will vary according to the flow rates to be measured. The plate is made of 1/8-inch (3-mm) thick material and is slightly larger than the inside diameter of the pipe. A rectangular handle is attached to the plate. The plate is installed between the flanges, blocking the pipeline except for the hole in the center of the plate. One side is bevelled, leaving a sharp edge on the opposite side. The handle of the orifice plate will have numbers stamped into it giving the orifice size. These numbers on the handle are stamped on the same side as the sharp edge of the orifice opening. When viewing the plate to read the numbers, the blower should be behind you. The sharp edge of the plate and the numbers must be on the side toward the blower for the meter to operate properly.

On top of each pipe flange holding the orifice plate will be a tapped hole. Tubing connected to the hole leads to the instrument that indicates the rate of air flow. There may be more than one orifice plate and metering device in the distribution system to monitor the air to the various plant processes.

Condensate traps are located at each meter and at the lowest point(s) of the distribution system.

11.123 Air Headers

Air headers are located in or along the aeration tank and are connected to the air distribution system from which they supply air to the diffusers. The two most common types of air headers are the swing header and the fixed header.

The swing header is a pipe with a distribution system connector fitting, a valve, a double pivot upper swing joint, upper and lower riser pipes, pivot elbow, leveling tee, and horizontal air headers. An air blowoff leg, as an extension of the lower tee connection, is fabricated with multiple alignment flanges, gaskets, and jack screws for leveling of the header.

The swing joint and pivot elbow allow the header to be raised from the aerator with an electric or manual hydraulic hoist so the header or diffusers may be serviced (Figure 11.12).

The fixed header is a pipe with a distribution system connector fitting, a valve, a union, a riser pipe, horizontal air headers, and header support "feet." These headers are generally not provided with adjustable leveling devices, but rely on the fixed leveling afforded by the "feet" attached to the bottom of the horizontal air headers. The fixed header is commonly found in package plants, channel aeration, and grit chamber aeration (Figure 11.13).

Header valves are used to adjust the air flow to the header assembly and to block the air flow to the assembly when servicing the header or diffusers. Headers are designed for a maximum air flow in cubic feet per minute (or cubic meters per minute) at a total maximum head loss measured in inches (or millimeters) of water.

11.124 Diffusers

Three types of diffusers are commonly in use today, fine bubble diffusers, medium bubble diffusers, and coarse bubble diffusers. Plate and tube diffusers and dome diffusers are classified as fine bubble diffusers. Medium bubble diffusers are commonly porous nylon or Dacron socks, and fiberglass- or saran-wrapped tubes (Figure 11.14). Fine bubble diffusers are easily clogged because these diffusers are self-sealing from the inside if dirty air is pumped into them. They may clog from the outside due to biological growths. These diffusers typically have an oxygen transfer efficiency of around 6 to 15 percent.

Coarse bubble diffusers are generally made of plastic and are of various shapes and sizes (Figure 11.15). Although these types of diffusers have lower oxygen transfer efficiencies (about four to eight percent), they are lower in cost and relatively maintenance free.

In most applications many of the same type diffusers are mounted to the horizontal air header. The number of diffusers mounted to the air header is determined by the desired mixing (diffusion pattern) and oxygen transfer requirements desired for the aeration tank (Figure 11.16).

Air distribution piping entering Y-wall of a large activated sludge plant.

1. orifice plate between pipe flanges
2. meter tubing
3. meter
4. air headers
5. air regulating/isolation valve
6. condensate traps
7. meter manifold valves
8. meter air supply tubing take-off valves

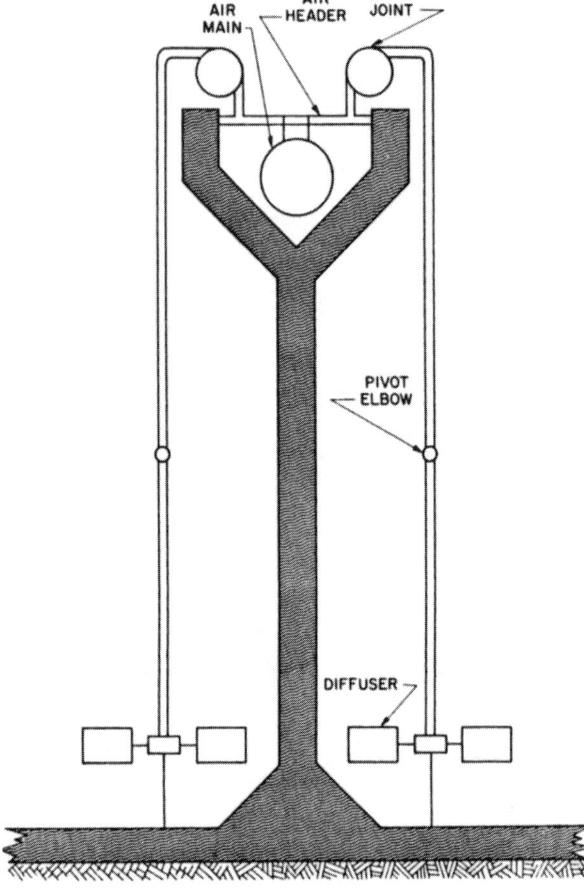

Air distribution piping in the Y-wall of a large activated sludge plant connected to air headers.

Fig. 11.11 Air distribution piping

Air headers

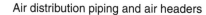

Air distribution piping and air headers

Swing header

1. distribution system connector fittings
2. header valve—regulating and isolation
3. double pivot upper swing joint
4. upper riser pipe
5. lower riser pipe
6. pivot elbow
7. leveling tee
8. horizontal air header
9. air blowoff leg
10. hoist

Fig. 11.12 Swing headers

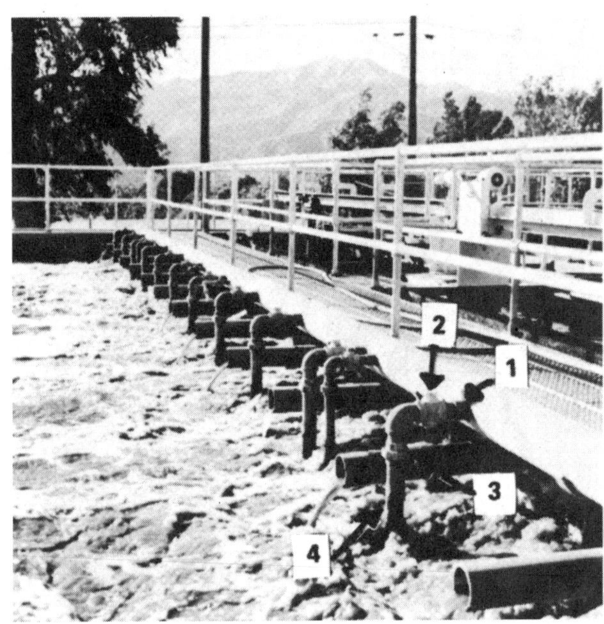

Air headers (package plant)

Air header (grit chamber)

1. distribution system connector fitting
2. header valve—regulating and isolation
3. union
4. riser pipe
5. horizontal air header
6. diffuser

Air header (channel aeration)

Fig. 11.13 Fixed headers

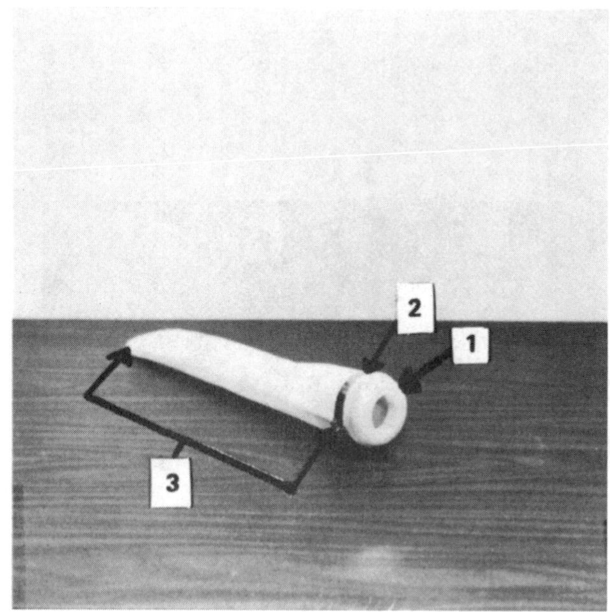

1. sock to support frame and threaded header mount
2. sock to support frame clamp
3. porous sock (air comes out between ends indicated by arrows)

Nylon sock

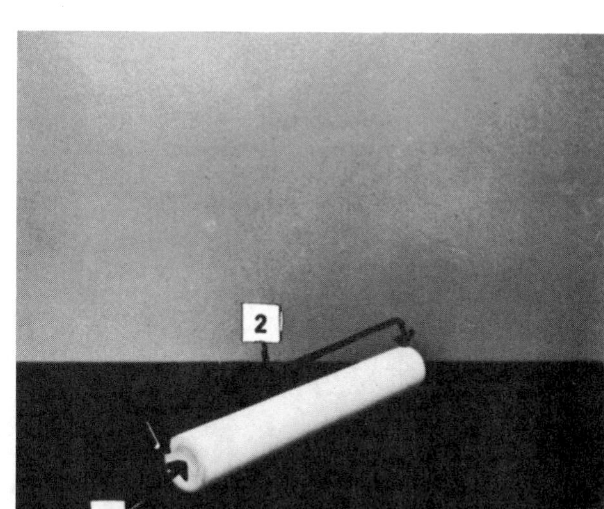

1. header mounting hole (fastening bolt not shown)
2. porous fiberglass (air comes out between ends indicated by arrows)

Fiberglass-wrapped tube

Fig. 11.14 Medium bubble diffusers

1. threaded stub for header mounting
2. air outlet orifices (four)

1. threaded stub for header mounting
2. air outlet orifices (four)
3. air deflector crown

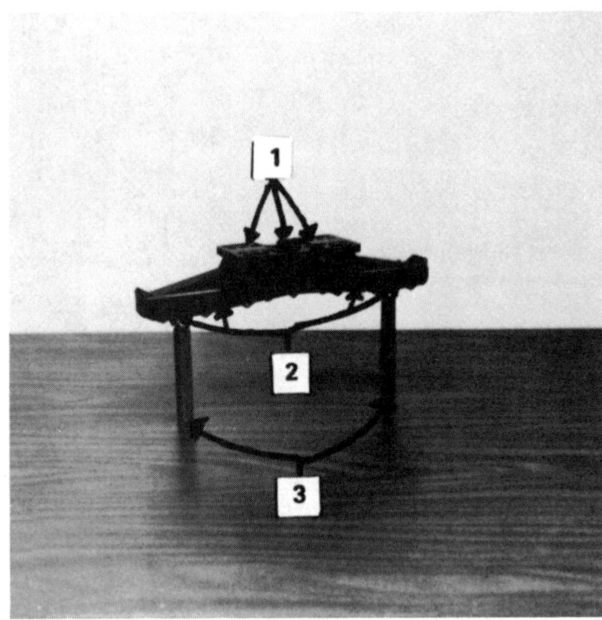

1. air inlet and header mounting surfaces
2. air outlet orifices (eight)
3. plunge tubes

Fig. 11.15 Coarse bubble diffusers

Narrow band pattern
(Diffusers installed on only one side of air header)

Wide band pattern
(Diffusers on both sides of air headers)

Fig. 11.16 Diffuser patterns

QUESTIONS

Write your answers in a notebook and then compare your answers with those on page 130.

11.1A What are the purposes of aeration?

11.1B List two methods of supplying oxygen from air to bacteria in the activated sludge process.

11.1C Why are diffusers located near the bottom of the aeration tank?

11.1D How are the guide vanes positioned that control air output from centrifugal blowers?

11.1E Why are fine bubble diffusers easily clogged?

11.2 SAFETY

For your own safety and to avoid damaging equipment, always read the manufacturers' manuals and thoroughly understand them before operating or maintaining any piece of equipment.

11.20 Aeration Tanks and Clarifiers

Whenever you must work around aeration tanks and clarifiers, use safe procedures and exercise extreme caution at all times.

1. Wear safety shoes with steel toes, shanks, and soles that retard slipping. Cork-inserted composition soles provide the best traction for all-around use.
2. Wear a Coast Guard-approved life jacket when working around an aeration tank where there are no guardrails to protect you. If you fall into the aerator, air will tend to collect in your clothing and may help to keep you afloat. Drownings in aerators have occurred apparently because a person was overcome by the initial shock or there was nothing to grab hold of to keep them afloat. A flotation device can save your life.
3. Slippery algal growths should be scrubbed and washed away whenever they appear on walkways.
4. Keep the area clear of spilled oil or grease.
5. Do not leave tools, equipment, and materials where they could create a safety hazard.
6. Adequate lighting should be permanently installed for night work, especially for use during emergencies.
7. Ice conditions in winter may require spiked shoes and icy areas should be sanded if ice cannot be thawed away with wash water.
8. Remove only sections of guardrails necessary for the immediate job. Removed sections should be properly stored out of the way and secured against falling. The area should be properly roped off or barricaded to prevent the entry and possible injury of personnel.

11.21 Surface Aerators

If maintenance or repair is required on the aerator, the aerator must be shut down and the *main power breaker must be opened (shut off), locked out, and properly tagged.* The lockout should be accomplished with a padlock and you should keep the key in your pocket. Tag the breaker with a lockout tag and note the date the aerator was locked out, the reason for the lockout, and the name of the person who locked out the aerator.

If an electrical problem exists with the aerator drive, *only qualified electricians* should be allowed to troubleshoot and repair the problem. Serious damage has occurred to equipment and to unqualified people who were "just trying to fix it."

Surface aerators are located directly over the aeration basin and caution is required when working in that area. If the basin is empty, a 15- to 40-foot (4.5- to 12-m) fall could be fatal. The worker should be protected by a fall arrest system that will safely suspend the worker in case of a fall. Requirements for fall arrest systems can be found in Chapter 14, "Plant Safety," and Section 14.10 of this manual. When the basin is full of water you could drown if you fell into the water. Whenever any work must be done on a surface aerator over a basin, the work should be done by two persons wearing approved flotation devices or fall arrest systems depending on the status (full or empty) of the basin.

11.22 Air Filters

When cleaning the air filters, shut down and secure the blower system you will be working on, even if it means shutting down the entire blower system. A 30- to 60-minute shutdown will not adversely affect the activated sludge process. Do not take chances by trying to operate the blower system while cleaning the filters if you cannot bypass the filters being cleaned. If the blowers are operating while trying to remove or install filters, foreign material can be drawn into the filter chamber and ultimately into the blower unit where extensive damage to the blower will result.

Wear gloves when removing and installing filters to protect your hands from cuts. Safety goggles should be worn when cleaning the filters to keep foreign matter out of your eyes. An approved dust and mist respirator should be worn to prevent ingestion or inhalation of filter dust. Persons should not be assigned to tasks requiring use of respirators unless it has been determined that they are physically able to perform the work and use the equipment. Check with your local safety regulatory agency for specific physical and training requirements.

11.23 Blowers

Before starting any blower, be sure all inlet and discharge valves are open throughout the system. Remove all foreign matter that might enter the blower. All personnel must be clear of the blower before starting. Always wear appropriate hearing protection when working near an operating blower. Hearing protectors must attenuate (reduce) your noise exposure at least to an eight-hour time weighted average of 90 decibels (see Chapter 14, Section 14.191, "Noise," page 277). Whenever a blower must be shut down for maintenance or repair, be sure the main power breaker is *opened (shut off), locked out, and properly tagged.* If an electrical problem exists with the blower drive motor, *only qualified and authorized electricians* should be allowed to troubleshoot and repair the problem.

11.24 Air Distribution System

The aeration tank areas where the distribution piping is located are hazardous and caution is required when working on distribution systems. If the aeration tank or channel is empty, a 10- to 20-foot (3- to 6-m) fall could be fatal. The worker should be protected by a fall arrest system that will safely suspend the worker in case of a fall. When the aeration tank or channel is full of water you could drown if you fell into the water. When working on air distribution system piping near an aeration tank or channel, have at least two operators present and have everyone wear approved flotation devices or fall arrest systems depending on the status (full or empty) of the basin.

11.25 Air Headers and Diffusers

Air headers are located in areas with hazards similar to those encountered when working on the air distribution system. Exercise care to avoid falling into empty tanks or tanks full of wastewater.

Before using an electric or manual hydraulic hoist, become fully familiar with all of the electric and hydraulic controls. The hydraulic fluid flow-control valve must be adjusted to allow the header lifting arm to descend at a safe rate of speed. Descending too fast could be hazardous.

Other cautions that should be observed when using the hoist include:

1. Never lift or lower a header pipe until you are sure the hoist is firmly and properly anchored and that its capacity is sufficient. If it is not, the hoist may jackknife into the tank when lifting or lowering starts and you could be knocked into the tank.
2. Never lift or lower a header pipe until the double pivot upper swing joint locking pin is removed. Lifting or lowering the header with the locking pin in place will cause the pivot to crack.
3. Make sure that the hoist support foot transmits the load to the concrete tank structure and not to the removable decking. This decking is designed to support only a few hundred pounds of weight.
4. Use the double pivot upper swing joint locking pin to secure the header assembly above the walkway. Failure to do this will result in the header assembly lowering itself into the tank if the hoist hydraulic system fails.

QUESTIONS

Write your answers in a notebook and then compare your answers with those on page 130.

11.2A What precautions must be taken before attempting to maintain or repair a surface aerator?

11.2B What information should be on a lockout tag?

11.2C What precautions should be taken when working with blowers?

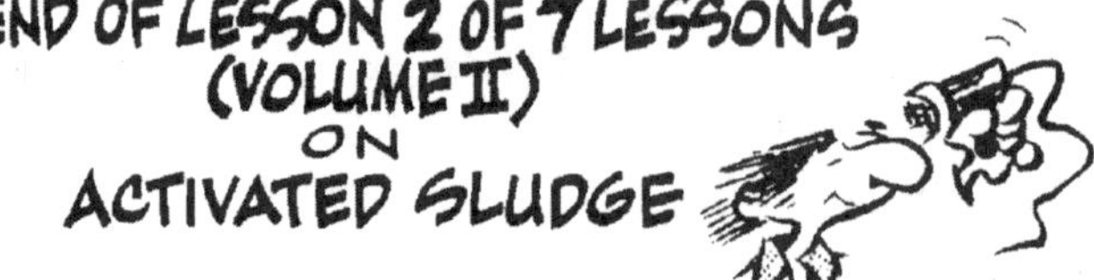

Please answer the discussion and review questions next.

DISCUSSION AND REVIEW QUESTIONS

Chapter 11. ACTIVATED SLUDGE

(Lesson 2 of 7 Lessons)

Write the answers to these questions in your notebook. The question numbering continues from Lesson 1.

9. What is the difference between mechanical aeration and diffused aeration?
10. What happens to power costs when the submergence of a surface aerator is increased and why?
11. Why is clean air essential for the protection of blowers and process systems?
12. Why should you always read the manufacturer's manuals and thoroughly understand them before operating or maintaining any piece of equipment?
13. What can happen to blowers if someone attempts to remove or install the air filters while the blowers are running?
14. What safety precautions must be taken when removing, cleaning, and installing air filters?
15. What hazards could be encountered while working on air headers?

CHAPTER 11. ACTIVATED SLUDGE

OPERATION OF CONVENTIONAL ACTIVATED SLUDGE PLANTS

(Lesson 3 of 7 Lessons)

11.3 CHECKING OUT A NEW PLANT

There is nothing more disappointing than starting a new plant and having to shut down after two or three days because of some small item that would have required only an hour or so to correct on a preliminary check-out. The main reasons an operator should completely check the equipment and structures before start-up are to: (1) familiarize the operator with plant equipment and locations of piping, and (2) to be reasonably sure that everything will function properly when the plant is put into service. When each item is checked, be sure that you know what it is supposed to do, how it is done, and how to service it correctly and safely. If possible, have a contractor or manufacturer's representative present during check-out and start-up. When each item of equipment is examined, check your manual of instructions to be sure your equipment is discussed. Start now with an orderly file on each piece of equipment if you have not already started.

Read the manufacturer's manual and thoroughly understand it before any piece of equipment is ever started. This manual should be provided by the manufacturer and can be obtained by writing to the company indicated on the equipment nameplate. Usually, in the construction specifications, the equipment suppliers are requested to supply three or four copies of instructions for each piece of equipment. If they are lost, a new one may be obtained by writing to the manufacturer and giving the name, size, serial number, and the treatment plant location and contract number, if possible. Each manufacturer normally supplies a one-year warranty on their equipment. If the equipment is damaged due to improper operation, the warranty becomes void and your community or district must absorb the cost of repairs or replacement. Proper operation prolongs the life of equipment, reduces the cost of repairs and replacement, and helps prevent operators from being injured.

This section outlines the steps you should follow when checking out a new plant. The descriptions are based on a particular type of plant layout (Figure 11.2). Your plant may vary from this layout, but by following this example and obtaining assistance during check-out from the design engineer and representatives of the equipment manufacturers, many initial operating problems can be eliminated.

11.30 A New Activated Sludge Plant: Description

Imagine that your old trickling filter plant has just been replaced with a diffused air activated sludge plant. You will continue to use your old screens, grit channel, and primary clarifiers. The new aeration tank is 100 feet (30 m) long, 45 feet (13.5 m) wide, and 16.5 feet (5 m) deep, and has a Y-wall dividing it down the center. Air headers are located along the full length of the tank on each side of the "Y" and spaced approximately 10 feet (3 m) apart. Air bubbles come out at the bottom of the tank through the headers equipped with diffusers (Figures 11.4, 11.12, and 11.13). Drawings of the aeration tank and secondary clarifier are provided in Figure 11.17.

The new secondary clarifier is circular with an 80-foot (24-m) inside diameter and a 12-foot (3.7-m) side wall depth sloping from the wall to the center of the tank. The tank is equipped with a sweep mechanism fitted with suction devices to collect the sludge after it has settled.

The resident engineer informs you that the contractor has completed the work and the new system may be put into service. First item on the agenda is to completely recheck the equipment and structures. Normally the contractor and the construction inspector are charged with this responsibility, but many times important items are overlooked due to time schedules, negligence, or oversight.

11.31 Aerator

11.310 Control Gates

In Figure 11.17 (plan) there are four storm flow gates (S) and four pedestal lift gates (P). The storm flow gates are marked by circles and the pedestal gates by rectangles. Open and close the gates even if it does permit the entry of some wastewater and check them for ease of operation and access. They must travel smoothly with no binding or jumping during opening and closing. Now is the time to check your means of noting the gate or valve location in the open, closed, or partially opened position. If you do not have a rising stem or other visible indicator of valve position, count the valve turns and record.

Lines and channels should be cleaned before start-up to prevent debris from occupying space or damaging equipment. When you have the gates open, check the aerator influent line or channel for debris such as rocks, sand, timber, waste concrete, or other foreign material. Short pieces of 2 × 4 (5 cm × 10 cm) boards and other form lumber can suddenly appear during initial flow and this material can jam a pump or stall a clarifier mechanism. Cleaning also will familiarize the operator with the flow routes. After the line or channel is clean, close the gate, making sure there is no foreign material along the side guides and *DOGS*.[15] When the gate closes, be sure it is properly seated.

[15] *Dogs.* Wedges attached to a slide gate and frame that force the gate to seal tightly.

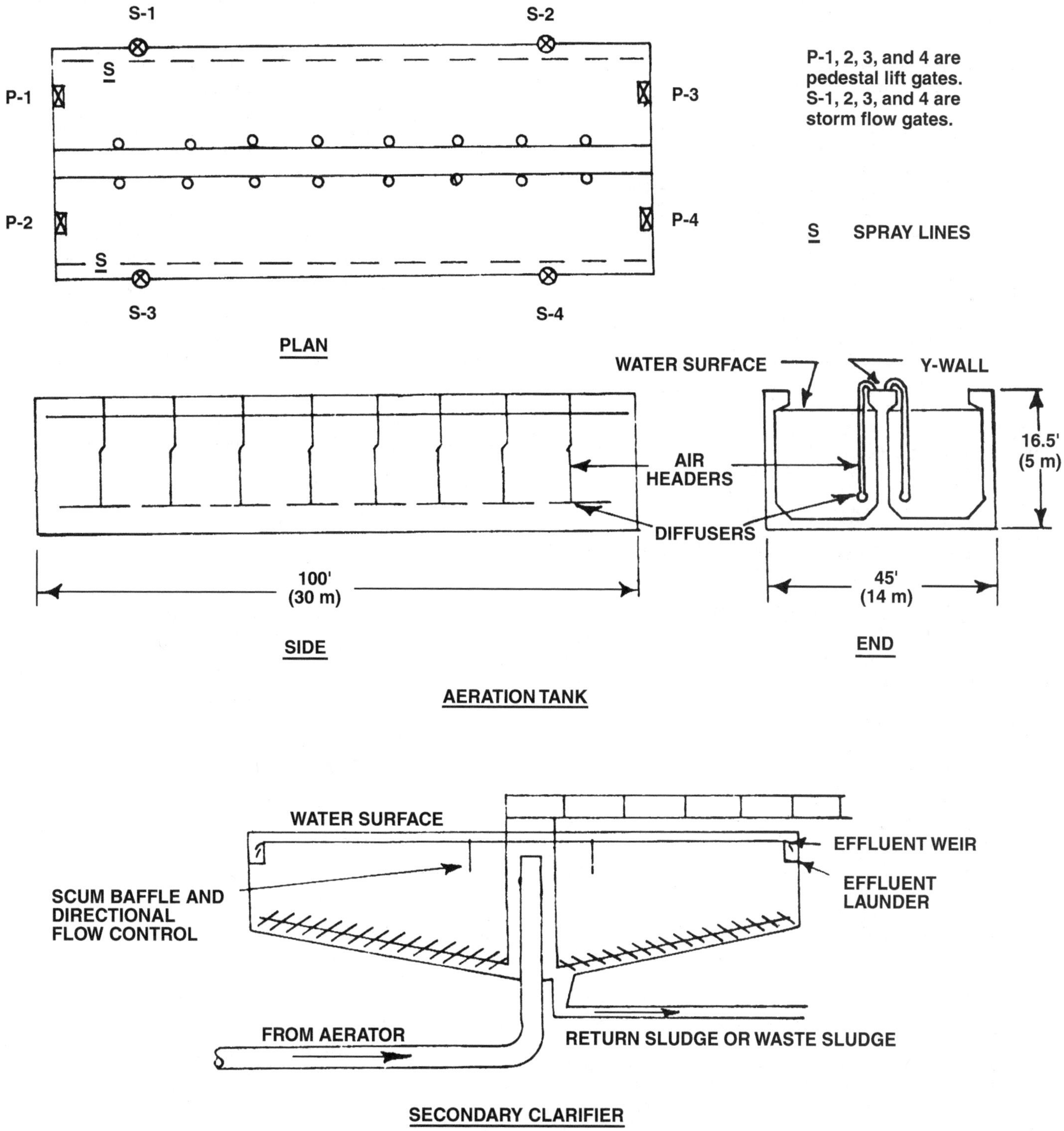

S-1
S-2
P-1
P-3
P-2
P-4
S
S-3
S-4
P-1, 2, 3, and 4 are pedestal lift gates.
S-1, 2, 3, and 4 are storm flow gates.
S SPRAY LINES
PLAN
WATER SURFACE
Y-WALL
AIR HEADERS
DIFFUSERS
16.5' (5 m)
100' (30 m)
45' (14 m)
SIDE
END
AERATION TANK
WATER SURFACE
EFFLUENT WEIR
EFFLUENT LAUNDER
SCUM BAFFLE AND DIRECTIONAL FLOW CONTROL
FROM AERATOR
RETURN SLUDGE OR WASTE SLUDGE
SECONDARY CLARIFIER

Fig. 11.17 Activated sludge process units

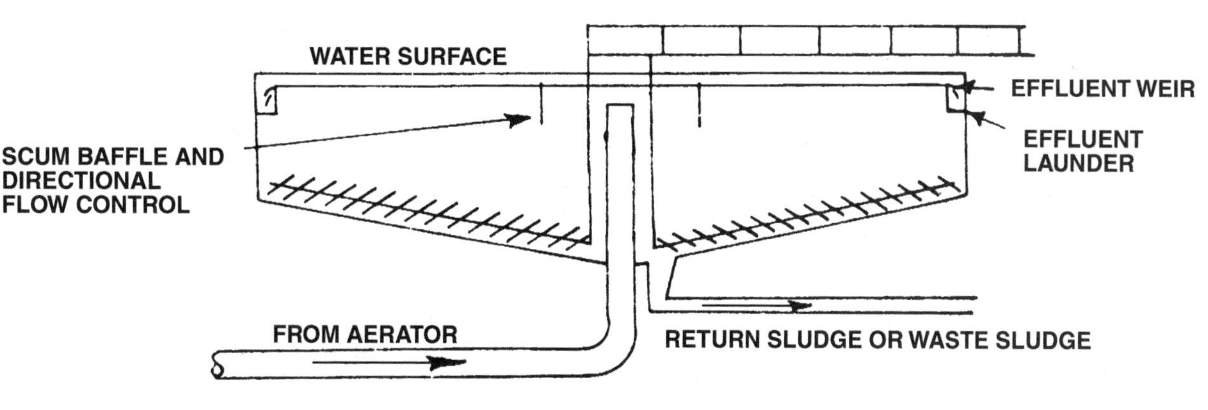

The gate should have been painted for rust and corrosion protection, and now is an excellent time to touch up any chips or scrapes. This will ensure as long a life as possible for plant equipment. Remove the housing protecting the threaded stem and check to be sure there is lubricant on the stem threads, and see if there is a stop-nut at the top of the stem. The main purpose of the stop-nut is to limit the length of travel. Without a stop-nut, the gate could fall into the aeration tank if it is opened too far or the stem could be bent if an operator attempts to close the gate farther than necessary.

11.311 Storm Flow Gates and Pedestal Lift Gates

Storm flow gates may be used for normal conventional type aeration tank feeding or abnormally high flows. Pedestal lift gates are commonly used for *STEP-FEED AERATION.*[16] Mud valves and step-feed valves are types of valves also found in aeration tanks. Open and close each valve, checking for ease of operation and proper seating. Give each stem a generous coat of heavy-duty, waterproof grease because the stems are normally exposed directly to the wastewater.

QUESTIONS

Write your answers in a notebook and then compare your answers with those on page 131.

11.3A Why should the operator completely check the equipment and structures before start-up?

11.3B If your plant does not have a manufacturer's manual, how would you obtain one?

11.3C Why is it important to read the manufacturer's manual before starting any piece of equipment?

11.3D Why should proper procedures be followed when starting, operating, and stopping equipment?

11.3E Why should lines and channels be cleaned before start-up?

11.3F Why should chips and scrapes on aerator control gates be painted?

11.3G What is the purpose of stop-nuts on aerator control gate stems?

11.312 Weirs

Weirs may be used instead of pedestal lift gates to control flow at the outlet or effluent end of the aeration tank. Check these with a surveyor's level to be sure one end is not higher than the other. Recheck after the tank is filled. If they are not level, the effluent will not be evenly distributed over the weir. This could cause short-circuiting and an uneven distribution of solids in the effluent. The weirs should have an adequate protective paint coating unless they are made of a corrosion-resistant material.

11.313 Movable Gates

Check the guide slots of bulkheads for nicks or rocks and make sure the gate seats and operates properly.

11.314 Water Sprays for Froth Control

Check each water spray head to see if it has a nozzle that will form a fan of water. The fan of water should have an angle to the tank surface of approximately 45 degrees. Turn the water on for a few minutes and check to be sure the spray properly covers the desired area.

> The nozzles must be properly installed to cover the intended area of the aerator or they will be ineffective, spraying the rails and walks. This creates a slipping hazard from algal growths in the summer or ice formation in the winter, and also does not dissipate much foam.

There should be no leakage in the piping system, and the water fans should overlap one another for effective control. At the dead end of each pipeline, check the operation of the valve that allows flushing the whole line when opened. A "Y" strainer should be installed at the inlet end of the system to filter out large material, thus saving future maintenance time.

QUESTIONS

Write your answers in a notebook and then compare your answers with those on page 131.

11.3H Why should an effluent weir be level?

11.3I What dangers are created if a foam-control spray nozzle sprays water on a rail or walk?

11.315 Air System

Check the air system by following it through in the direction of air flow to the aeration tank.

[16] *Step-Feed Aeration.* Step-feed aeration is a modification of the conventional activated sludge process. In step-feed aeration, primary effluent enters the aeration tank at several points along the length of the tank, rather than at the beginning or head of the tank and flowing through the entire tank in a plug flow mode.

11.3150 AIR FILTERS

Filters remove dust and dirt from the air before it is compressed and sent to the aeration tank.

First check the access doors or hatches to the filter chamber. These should be lined around the edges with rubber or some material to form a tight seal. If there is a gap or torn place, repair it. Check the inside of the filter chamber floors for cleanliness and debris—remove any sand, dirt, paper, or other debris.

If bag filters are installed, check the filter bags for proper coating of filter media. Be sure that they are securely installed and that one of the workers did not leave a tool or some item that could be drawn directly into one of the blowers.

A manometer is installed to read the difference in pressure between the inlet side of the filters and the outlet side. The U-tube manometer is mounted on the outside of the filter housing with two small copper lines running to the intake and outlet side of the filters.

To check the manometer, remove the glass U-tube from the manometer housing, taking care not to break it. Blow air through each line back to the manometer, checking to see that the lines are clear and not plugged or accidentally crimped during installation. If the lines are equipped with fittings, check for tightness and leakage. If the lines are OK, fill the manometer U-tube with the required fluid. This could be a specified oil, or water may be acceptable. Fill the U-tube approximately half the distance from the top of the tubes to the bottom of the "U"; both columns should be the same height. If water is permissible to use, a drop or two of red or blue chart ink added to the tube before the water is added gives a good color and an easy indicator for reading the manometer. Before adding the ink, experiment with laboratory glassware to be sure the ink will not permanently stain the manometer glass and make future readings impossible. (See Chapter 18, Section 18.2, "Manometer and Gauge Reading," for instructions on how to read manometers.)

The difference in pressure recorded by the manometers will be small when the filters are clean, but as they become dirtier, the manometer reading will increase and each column will move farther away from the manometer zero mark. When this difference reaches approximately 2 to 3 inches (5 to 8 cm) of water column more than the initial difference, it is probably time to clean or change the filters. The manufacturer's operating manual should be reviewed for the recommended maximum allowable pressure differences as well as the procedure for cleaning the screens or bags.

Check the air duct valving from the filters to the inlet side of the air blowers (compressor) for proper installation. Also, check these ducts or pipes for dirt and debris because they should be very clean.

QUESTIONS

Write your answers in a notebook and then compare your answers with those on page 131.

11.3J How is air cleaned before being compressed and sent to the aeration tank?

11.3K How can you determine when the air filters need cleaning?

11.3151 BLOWERS (COMPRESSORS)

Read the manufacturer's manual and thoroughly understand it before the blowers are ever started.

The blowers are of the positive displacement type. Special attention must be directed toward starting and stopping procedures. Prepare a list of items to be checked before the start button is pushed. Check switches, indicators, and pump connectors with drawings. Check the type of oil and oil level. Start the oil circulation pump and check the circulation system and oil level before the blowers are started.

Check inlet and discharge valves by opening and closing them and noting ease of operation. Many of these units are to be started under no load (the discharge air is vented to atmosphere until the unit is running properly) and then valved in slowly to put the air into the system and load the unit. When the unit is shut down, the reverse of the starting procedure is followed. Be absolutely positive of the correct procedure for starting and stopping the unit, including procedures for starting and stopping one unit while the others are operating. Improper procedures can shorten the life of the equipment considerably. Your supervisor will not be happy when equipment breaks down or is worn out because of improper procedures.

Next, check the driver (electric motor or engine), the blower base plates, and the coupling for alignment. Usually the contractor's millwright sets the equipment and aligns the couplings with a dial indicator. Request the dial readings from the contractor and file them with the rest of the equipment data. These readings are invaluable in case of equipment failure, or for future checks on proper alignment. These should be no more than 0.005 of an inch (0.125 mm) off, and larger equipment calls for even closer tolerances. Check for base plate bolts or nuts because if one corner is loose, the whole alignment will be thrown out on start-up. Do not attempt to tighten base plate nuts or bolts because this also will cause misalignment. If one nut is loose or there is a gap between the equipment mount and the base plate, the coupling must be realigned and the equipment mounts shimmed.

If the blower and driver are securely anchored, all lubrication points are in good condition and coupling alignments are satisfactory, then motor and compressor will turn with a reasonable pull on the outside pulleys. When all safety guards are installed and well clear of the moving parts, then you are ready to check the main air lines or "air mains."

On both the inlet and discharge sides of the blower, air mains (air lines) are connected through a flexible coupling to keep vibration at a minimum and allow for heat expansion because, when air is compressed, heat is generated. This can increase the discharge temperature as much as 100°F (56°C) or more. Check to be certain there is sufficient room for movement in the flexible couplings.

The discharge air line from the blower is equipped with an air relief valve that protects the blower from excessive back pressure and overload. It is adjusted by weights or springs to open when air pressure exceeds a point above normal operating range, around 6.0 to 10.0 psi (0.4 to 0.7 kg/sq cm). Check this valve to ensure its free operation by manually lifting the valve off the seat.

Between the air relief valve and the discharge side of the blower is a pressure gauge to read discharge pressure. Check the gauge for proper installation by looking for air leaks and for easy accessibility to read.

An air metering device should be located in a straight section of the air main on the discharge side of the blower. This device consists of an orifice plate inserted between two specially made pipe flanges. The orifice plate is made of stainless steel with a precision hole cut through the center. The diameter of the hole will vary according to the flow rates to be measured. The plate is made of 1/8-inch (3-mm) thick material and is slightly larger than the inside diameter of the pipe. A rectangular handle is attached to the plate. The plate is installed between the flanges, blocking the pipeline except the hole in the center of the plate. One side is bevelled, leaving a sharp edge on the opposite side. The handle of the orifice plate will have numbers stamped into it giving the orifice size. These numbers on the handle are stamped on the same side as the sharp edge of the orifice opening. When viewing the plate to read the numbers, the blower should be behind you. The sharp edge of the plate and the numbers must be on the side toward the blower for the meter to read properly.

On top of each pipe flange holding the orifice plate will be a tapped hole. Tubing connected to the hole leads to the instruments that indicate the rate of air flow. Check these connections to see that there are no air leaks. The instruments themselves should have been installed and calibrated correctly, but occasionally an orifice plate is installed backward or an instrument line is left disconnected. When the meter is properly connected to a dial or totalizer, a zero reading should be recorded when the blower is off.

Check the condensate trap or drain located near the aeration tank at the lowest point of the air main. Open and close the drain line and remove any dirt or sand in the air main at this location.

Inspect the air main from the blowers to the air headers for leaks, tightness, and expansion allowances.

11.3152 AIR HEADERS

The air main runs down the center Y-wall of the aeration tank, distributing air to the headers along the tank. This plant is equipped with "swing headers." A swing header is merely a pipe with movable couplings so the header may be raised from the tank with a hoist to service the diffusers.

Take the hoist that is to be used to raise and lower the air headers out to the center Y-wall and make sure it properly fits each air header. The hoist is anchored by the air header opposite the one to be lifted. Sometimes the correct spacing is missed and the hoist is either too short or too long. In either case, you would not be able to lift those two headers, for there is no other safe way to anchor the hoist. If the hoist is not anchored properly and a header is being raised, the hoist could catapult into the tank, possibly taking the operator with it and surely damaging the hoist (and probably the diffusers and headers too).

Each header is equipped with a valve to shut the air off when removing it from the tank to replace or clean the diffusers. Consequently, the other headers may be kept in service, thus not disrupting the air supply to the tank contents. Inspect each air header valve for proper operation, and determine which position of the stem indicates open and which position indicates closed.

> **The center Y-wall is a hazardous area and caution is required when working here. If the aeration tank is empty, the 16-foot (5-m) fall could be fatal. When the tank is in service, you could drown if you fell into it. A good policy is that *any work on the center Y-wall calls for two operators and approved flotation devices or fall arrest systems.***

Other safety precautions that should be observed at all times when working on center Y-walls include:

1. Wear safety shoes with steel toes, shanks, and soles that retard slipping. Cork-inserted composition soles provide the best traction for all-around use.
2. Slippery algal growths should be scrubbed and washed down whenever they appear.
3. Keep the area clear of spilled oil or grease.

4. Do not leave tools, equipment, and materials where they could create a safety hazard.
5. Adequate lighting should be permanently installed for night work.
6. Ice conditions in winter may require spiked shoes and sanding icy areas if ice cannot be thawed away with wash water.
7. Remove only sections of removable guardrails necessary for the immediate job. Removed sections should be properly stored out of the way and secured against falling. The area should be properly roped off or barricaded to prevent the entry and possible injury of unauthorized personnel.

When the aeration tank is empty or nearly empty, place a ladder in the tank and continue the check-out. Remember, working with a ladder can be very hazardous. Make sure the ladder is positioned properly, secured, and that it is appropriate for the job. Portable ladders are rated for the weight they can hold; make sure your ladder is rated for your weight and the weight of any tools or equipment it may have to support. Also, remember that the aeration tank is a confined space, and confined space procedures are required for entry and occupancy. (See Chapter 14, "Plant Safety," Section 14.12, "Confined Spaces.")

First, remove any debris, including sand or dirt, from the bottom of the tank. Next, check the air headers. This is the continuation of the air system to the bottom of the tank with the pipes creating an inverted "T." The horizontal run of pipe forming the cross-bar on the top of the "T" should be perfectly level. If one end is 1/2 inch (1.2 cm) higher than the other end, then more air will escape through the high end, causing poor air distribution. The contractor should have set them with a surveyor's level to ensure that they are all at the same elevation, but it is advisable to check them again. Be sure diffusers are secure and header or diffuser plugs are in place; replace damaged or defective items. The pipe should have been flushed with water to remove dirt, dust, and scale.

QUESTIONS

Write your answers in a notebook and then compare your answers with those on page 131.

11.3L How should the sharp edge of a metering orifice be placed in a pipe?

11.3M Why must the hoist used to lift the air headers be properly anchored?

11.3N What precautions would you take when working on the center Y-wall?

11.3O Why must the horizontal pipes containing the air diffusers all be at the same elevation (level)?

11.3153 DIFFUSERS

Thread-type diffusers are installed in our example plant. Before the diffusers were installed, the pipe should have been flushed clean with water. Inspect the diffusers. A light application of grease on the threads should have been applied as a protective coating and for ease in removing the diffusers when they need cleaning. Check to be sure that none of the lubricant touches the opening into the diffuser. Do not use a wrench when installing threaded diffusers; normal "hand tight" is sufficient, and even this can create a problem when the time arrives to remove them.

11.3154 BLOWER TESTING

After you have inspected the air headers and diffusers in the aeration tank, check the blower operation. Start with the blowers discharging air directly to the atmosphere. Review the stop and start procedures for the blowers and turn them on. If at all possible, let them run for three or four hours to check on any heating problems or vibrations. Check temperatures, amperage readings from the electric motors, air flow rates, and differential pressures across the filter system, and record the readings.

Repeat these checks after the process tanks are filled. Now is the time to check the air relief valves for correct settings by closing several valves out on the headers until the air relief valves open. Also take another set of amperage readings on the blower motor to see if it is overloaded under operating conditions. If everything functions properly for four hours, you can feel fairly certain that you will not have any immediate problems from the blowers.

QUESTIONS

Write your answers in a notebook and then compare your answers with those on page 131.

11.3P Why should a light application of grease be applied to the threads of a diffuser?

11.3Q While the blowers are running, what items should be checked, and why?

11.32 Secondary Clarifiers

Circular clarifiers—check items:

1. Control gates for operation
2. Clarifier tank for sand and debris
3. Collector drive mechanism for lubrication, drive alignment, overload alarm, and complete assembly
4. Squeegee blades on the collector plows for proper distance from the floor of the tank
5. Connecting lines or channels between aerator and clarifier for debris
6. Pump suction line assembly and controls
7. Inlet baffles and discharge weirs for level
8. Scum control mechanism

Once the collectors are started in a new tank, each scraper should be checked for a clearance of one to two inches (2.5 to 5 cm) between the wall and the end of the scraper. If a scraper is too long, it may rub the tank wall and break the scraper, jamming other scrapers and breaking them. Once a broken scraper is detected, it should be replaced or removed from the support structure.

11.33 Return Sludge and Waste Sludge Pumps

Since these two items are identical except for size, the same checks may be used.

Clean out trash in all sludge hoppers, lines, valves, gates, sweeps, and drive mechanisms before checking pumps.

> To check out pumps, first lock out and tag the pump motor at the power panel so it cannot be started. On the tag, indicate who locked out the motor and when.

Remove the handhole cover, clean out the pump casing or volute, and check the impeller for debris and secure fit to the pump shaft. Wear gloves when cleaning pump casings to protect your hands from dangerous, sharp objects. Replace the handhole cover. Install or check pressure gauges on the suction and discharge lines of the pump. Check inlet and outlet valve operation and connecting lines for clearance.

Check pump and motor bearings for lubrication. Inspect coupling alignment and base plate anchor bolts. Turn on the seal water to the pump. Since the pump has not operated, back off the packing gland nuts and rotate the pump shaft by hand through several revolutions. Leave the packing nuts loose and adjust them properly after start-up. If everything checks out satisfactorily, then run some water into the final clarifier hopper or return sludge well where the pump suction is located so it may be operated for a few minutes. While the water is filling the hopper or wet well, check the suction and discharge valves and check (flap) valves on the system. Check the other valves on the return sludge line for proper operation. Open both inlet and discharge valves before starting the pump so the water may flow during the pump test.

There should be several hundred gallons of water available in the final clarifier for the pump test. Prime the pump if necessary and, if everything has checked out properly, unlock the controls at the power panel and turn the pump on. First, check pump rotation and the packing gland for water, but let it flow freely for now. Read pressure gauges on both suction and discharge and the level of the water in the clarifier or wet well; record all three measurements so that the operational characteristics and efficiency of the pumps may be checked. Take an amperage reading on the motor and record. If possible, make sure that the flowmeter is functioning and record meter reading too. The water available will not afford a very long run, but it will indicate any major problems. When the pump is shut off, make sure the check valve closes and seats.

Run the waste sludge pump now to be sure it will operate properly when you are ready to start wasting activated sludge.

The following list contains a summary of the items that should be recorded when testing pumps.

1. Name of pump: ____________________
2. Location of pump: ____________________
3. Pump suction pressure: ____________________
4. Pump discharge pressure: ____________________
5. Water level in wet well: ____________________
6. Motor amperage: ____________________
7. Pump discharge (flow): ____________________
8. Date of test: ____________________
9. Name of operator: ____________________

QUESTION

Write your answer in a notebook and then compare your answer with the one on page 132.

11.3R List all the information that should be recorded for future reference when you test the return sludge and waste sludge pumps.

11.4 PROCESS START-UP PROCEDURES

11.40 General

Procedures for starting the activated sludge process are outlined in this section. Procedures and example calculations will be for the plant checked out in Section 11.3. An initial average daily flow of 4.0 MGD (15,140 cu m/day) will be assumed, and the plant will be operated as a conventional activated sludge plant.

Start-up help should be available from the design engineer, vendors, nearby operators, or other specialists. The equipment manufacturers or contractor should be under contract for start-up instruction and assistance. During start-up, they should be present to be sure that any equipment breakdowns are not caused by improper start-up procedures.

The operator may have several options in the choice of start-up procedures with regard to number of tanks used and procedures to establish a suitable working culture in the aeration tanks. The method described in this section is recommended because it provides the longest possible aeration time, reduces chances of solids washout, and provides the opportunity to use most of the equipment for a good test of its acceptability and workability before the end of the warranty.

11.41 First Day

First, start the air blowers and have air passing through the diffusers before primary effluent is admitted to the aeration tanks. This prevents diffuser clogging from material in the primary effluent and is particularly important if fine bubble diffusers are used.

Fill both aeration tanks to the normal operating water depth, thus allowing the aeration equipment to operate at maximum efficiency. Using all of the aeration tanks will provide the longest possible aeration time. You are trying to build up a microorganism population with a minimum amount of seed organisms, and you will need all the aeration capacity available to give the organisms a chance to reach the settling stage.

When both aeration tanks have been filled, begin filling the two secondary clarifiers. Use of all the secondary clarifiers will provide the longest possible detention time to reduce washout of light solids containing rapidly growing organisms and will encourage solids buildup.

When the secondary clarifiers are approximately three-fourths full, start the clarifier collector mechanism and return sludge pumps. During start-up, return sludge pumping rates must be adjusted to rapidly return the solids (organisms) to the aeration tanks and to keep the sludge blanket in the secondary clarifier as low as possible. The solids should never remain in the secondary clarifiers longer than 1.5 hours. Trouble also may develop if the return sludge pumping rate is too high (greater than 50 percent of the raw wastewater flow), because the high flows through the clarifier may not allow sufficient time for solids to settle to the bottom of the clarifier. A conventional activated sludge plant usually operates satisfactorily at return sludge rates of 20 to 30 percent of raw wastewater flow; however, current designs often provide for return capacities of 50 to 100 percent of the wastewater flow at larger plants. The return rate selected should be based upon returning organisms to the aerator where they can treat the incoming wastes. A thin sludge will require a higher return percentage than a thick one. However, increasing the return percentage with a thin sludge can produce a thinner sludge. Addition of a coagulant or flocculant aid at the end of the aeration tank will hasten solids buildup and improve effluent during start-up.

When the secondary clarifiers are full and begin to overflow, start effluent chlorination to disinfect the plant effluent and to protect the health of the receiving water users.

Filling the aeration tanks and aerating the wastewater starts the activated sludge process. The *AEROBES*[17] in the aeration tank have food and are now being supplied with oxygen; consequently this worker population will begin to increase.

After two or three hours of aeration, you should check the dissolved oxygen (DO) of the aeration tanks to see if enough air is being supplied. (See Chapter 16, "Laboratory Procedures and Chemistry," for procedure to run the DO test.)

Check the DO throughout the aerator. If possible, use a DO probe and determine the DO along both sides of the aerator at both the water surface and the bottom of the tank. Oxygen must be available for the aerobes throughout the tank. If the DO is less than 1.0 mg/*L*, increase the air supply. If the DO is greater than 3.0 mg/*L* the air supply may be decreased, but not to the point where the tank would stop mixing. There will probably be an excess amount of DO at first due to the limited number of organisms initially present to use it.

After a biological culture of aerobes is established in the aeration tanks, sufficient oxygen must be supplied to the aeration tank to overcome the following demands:

1. DO usually is low in both influent wastewater and return sludge to the aerator.
2. Influent wastewater may be septic, thus creating an immediate oxygen demand.
3. Organisms in the presence of sufficient food create a high demand for oxygen.

The effluent end of the aerator should have a dissolved oxygen level of at least 1.0 mg/*L*. DO in the aerator should be checked every two hours until a pattern is established. Thereafter, DO should be checked as frequently as needed to maintain the desired DO level and to maintain aerobic conditions in the aerator. Daily flow variations will create different oxygen demands. Until these patterns are established, you will not know whether just enough or too much air is being delivered to the aeration tanks. Frequently excess air is provided during early mornings when the inflow waste load is low. Air supply may be too low during the afternoon and evening hours because the waste load tends to increase during the day.

Turn on the water spray system as soon as possible because foaming will be severe until there is a sufficient buildup of activated sludge solids. If spray water is not available at start-up or the water spray pump is not working, you may need to apply commercial defoamers to the surface of the aerator to control foam.

11.42 Second Day

Collect a sample from the aeration tank and run a 30-minute settleability test using a 1,000-m*L* graduated cylinder. If possible, use a 2,000-m*L* cylinder with a five-inch (125-mm) diameter[18] to obtain better results. Results of the 30-minute settleability test indicate the flocculating, settling, and compacting characteristics of the sludge. Observe the sludge settling in the sample for approximately one hour. It will probably have the same color as the primary effluent during the first few days. After a few minutes in the cylinder, very fine particles will start forming with a light buff color. The particles remain suspended, but settling, similar to fine particles of dust in a light beam. After one hour, a small amount of these particles may have settled to the bottom of the cylinder to a depth of 10 or 20 m*L*, but most are still in suspension. This indicates that you are making a start toward establishing a good condition in the aeration tank, but many more particles are needed for effective wastewater treatment.

11.43 Third Through Fifth Days

During this period of operation the only controls applied to the system usually consist of maintaining DO concentrations in the system and maintaining proper sludge return rates. A sampling program should be started in accordance with Section 11.58, "Recordkeeping," to develop and record the necessary data required for future plant control.

Aeration of wastewater to maintain DO will require some time before settling will produce a clear liquid over the settled solids. Time is required for organisms to grow to the point where there are sufficient numbers to perform the work needed—to produce an activated sludge organism culture. Usually within 24 to 72 hours of aeration you will note that the settleable solids do not fall through the liquid quite so rapidly, but the liquid remaining above the solids is clearer.

The active solids (organisms) are light and may wash out of the clarifier to some extent. Try to retain most of them because a rapid solids buildup will not occur unless they are retained. A

---

[17] *Aerobes.* Bacteria that must have dissolved oxygen (DO) to survive. Aerobes are aerobic bacteria.

[18] Mallory Direct Reading Settlometer (a 2-liter graduated cylinder approximately 5 inches (125 mm) in diameter and 7 inches (175 mm) high). Obtain from Wilmad-LabGlass, 1002 Harding Highway, PO Box 688, Buena, NJ 08310. Catalog No. LG-5601-100. Price: $167.84 each.

good garden soil will add organisms and solids particles for start-up. Mix the soil with water and hose in the lighter slurry, but try to avoid a lot of grit. A truckload of activated sludge from a neighboring treatment plant also will help to start the process. Hopefully, you will not have to treat design flows during plant start-up. More time is needed both for aeration and clarification until you have collected enough organisms in your return sludge to enable you to produce a clear effluent after a short period of mixing with the influent followed by settling.

QUESTIONS

Write your answers in a notebook and then compare your answers with those on page 132.

11.4A Why should the blowers be started before primary effluent is admitted to the aeration tanks?

11.4B How is the return sludge rate selected during initial start-up?

11.4C Why should chlorination equipment be put in service when effluent starts leaving the plant?

11.4D At what locations in the aeration tank would you check the DO, and why?

11.44 Sixth Day

A reasonably clear effluent should be produced by the sixth day. Solids buildup in the aeration tank should be closely checked using the 30-minute settleable solids test during the first week. Results of this test indicate the flocculating, settling, and compacting characteristics of the sludge. Suspended solids buildup is very slow at first but increases as the waste removal efficiency improves. This buildup should be carefully measured and evaluated each day.

Microorganisms in the system are so varied and small that it is impossible to count them. To obtain an indication of the size of the organism population in the aeration tank, the solids are measured either in mg/*L* or in pounds of dry solids. Suspended solids determinations for aerator mixed liquor will give the desired information in mg/*L,* and the total pounds of solids may be calculated on the basis of the size of the aerator.

$$\text{Total Susp Solids, lbs} = \text{Susp Solids, mg}/L \times \text{Aer Vol, MG} \times 8.34\ \text{lbs/gal}$$

or

$$\text{Total Susp Solids, kg} = \text{SS, mg}/L \times \text{Aer V, cu m} \times \frac{1\ \text{kg}}{1{,}000{,}000\ \text{mg}} \times \frac{1{,}000\ L}{1\ \text{cu m}}$$

The suspended solids test (see Chapter 16, "Laboratory Procedures and Chemistry") conducted on activated sludge plant mixed liquor normally requires a grab sample obtained at the effluent end of the aerator. The sample should be collected at the same time every day, preferably during peak flows, in order to make day-to-day comparisons of the results. Collect the mixed liquor sample approximately 5 feet (1.5 m) from the effluent end of the aeration tank and 1.5 to 2 feet (0.4 to 0.6 m) below the water surface to ensure a good sample. A return sludge sample also should be collected at this time every day to determine its concentration.

With information from the lab tests, estimates of the mass (weight) of organisms in the aerator can be calculated.

Information needed:

1. Aeration Tank Dimensions

   100 feet long, 45 feet wide, and 16.5 feet deep

2. Results of Laboratory Tests

   Mixed Liquor Suspended Solids, 780 mg/*L*

Steps to calculate pounds of solids in aeration tank[a]:

1. *DETERMINE AERATION TANK VOLUME.*

$$\text{Volume, cu ft} = \text{Length, ft} \times \text{Width, ft} \times \text{Depth, ft}$$

$$= 100\ \text{ft} \times 45\ \text{ft} \times 16.5\ \text{ft}$$

$$= 74{,}250\ \text{cu ft}$$

2. *CONVERT CU FT TO GALLONS.*

$$\text{Aerator Volume, gal} = 74{,}250\ \text{cu ft} \times 7.48\ \text{gal/cu ft}$$

$$= 555{,}390\ \text{gal}$$

or $= 555{,}000$ gal (approximately)

or $= 0.55$ MG

3. *CALCULATE POUNDS OF SOLIDS UNDER AERATION.*

$$\text{Solids, lbs} = \text{MLSS, mg}/L \times \text{Aer Vol, M Gal} \times 8.34\ \text{lbs/gal}$$

$$= 780\ \text{mg}/L \times 0.55\ \text{M Gal} \times 8.34\ \text{lbs/gal}$$

$$= \frac{780\ \text{mg}}{\text{M mg}} \times 0.55\ \text{M Gal} \times 8.34\ \text{lbs/gal}$$

$$= 780 \times 4.6^{b}\ \text{lbs}$$

$$= 3{,}588\ \text{lbs}$$

[a] All calculations and examples in this manual are also provided on a similar basis using metric units in the Arithmetic Appendix, Section A.5, *TYPICAL WASTEWATER TREATMENT PLANT PROBLEMS (METRIC SYSTEM),* page 797.

[b] The factor 4.6 lbs is equivalent to 0.55 × 8.34, a constant for your plant. You will use this value every day as long as you use the same aeration tank capacity. Only a change in the suspended solids concentration will cause a change in the pounds of solids in the aeration tank.

Close observation of the suspended solids buildup and results from the 30-minute settleability test will indicate the solids growth rate, condition of solids in the aerator, and how much sludge should be returned to ensure proper return of the organisms to the aerator. It will be necessary to return all of the sludge for 10 to 15 days or longer if the wastewater is weak.

Results from the 30-minute settleability test can be used to calculate the return sludge rate. If the volume of settled sludge in the cylinder is indicative of the amount of sludge settling in the secondary clarifier, the volume of return sludge should be approximately equal to the ratio of the volume occupied (in milliliters) by the settleable solids from the aeration tank effluent to the volume of the clarified liquid (in milliliters) after settling for 30 minutes in a 1,000-milliliter graduated cylinder.

Estimate the return sludge pumping rate.

Information needed:

1. Flow to Aerator from Primary Clarifier = 4.0 MGD
2. Return Sludge Flow, 1.0 MGD, or 700 GPM
3. Volume of Mixed Liquor Solids Settled in 30 Minutes, 180 m*L* in 1.0 liter

EXAMPLE:

Flow to Aerator from Primary Clarifier	= 4.0 MGD
Return Sludge Flow to Aerator	= 1.0 MGD
Total Flow Through Aerator	= 5.0 MGD

$$\text{Return Sludge Flow Ratio} = \frac{\text{30-Min Settleable Solids, m}L}{\text{Clear Liquid, m}L}$$

$$= \frac{180 \text{ m}L}{1{,}000 \text{ m}L - 180 \text{ m}L}$$

$$= 0.22$$

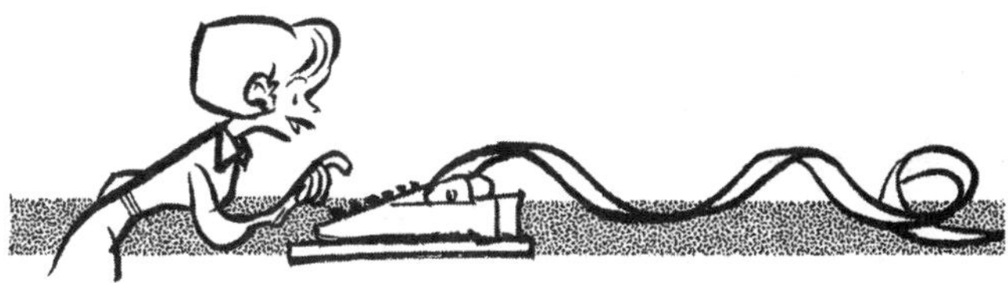

$$\text{Return Sludge Rate, MGD} = \text{Aerator Flow, MGD} \times \text{Return Sludge Flow Ratio}$$

$$= 5.0 \text{ MGD} \times 0.22$$

$$= 1.1 \text{ MGD or } 1{,}100{,}000 \text{ GPD}$$

$$\text{Return Sludge Rate, GPM} = \frac{1{,}100{,}000 \text{ gal/day}}{1{,}440 \text{ min/day}}$$

$$= 764 \text{ GPM } (48.2\ L\text{/sec})$$

Therefore, the initially selected 700 GPM return sludge rate is acceptable at this time. It ensures that most solids are being returned to the aeration tank. A return sludge pumping rate slightly higher than calculated is recommended to return the organisms as fast as possible to the aerator. However, a return sludge rate that is too high must be avoided because the resulting high flows reduce the detention time in the aerator and secondary clarifier.

If the return sludge rate is too low, the following undesirable conditions may develop:

1. Insufficient organisms will be in the aerator to treat the influent waste (food) load. This normally occurs during the first week or two of start-up.
2. Septic sludge could develop if the detention time in the secondary clarifier is too long.
3. Accumulation of sludge in the clarifier creates a deep sludge blanket, which will allow solids to escape in the effluent.
4. If aeration is sufficient to produce nitrate in the aeration tanks, denitrification may occur resulting in rising sludge and subsequent effluent solids.

QUESTIONS

Write your answers in a notebook and then compare your answers with those on page 132.

11.4E When and where should solids samples be collected to provide the operator with a record of solids buildup in the aeration tank?

11.4F Determine the pounds of solids in an aeration tank with a volume of 0.25 MG and a Mixed Liquor Suspended Solids (MLSS) concentration of 640 mg/*L*.

11.4G Estimate the return sludge pumping rate (GPM) if the plant inflow is 2.0 MGD and the return sludge flow is 0.5 MGD. The results of the 30-minute settleability test indicate the volume of solids settled to be 170 m*L* in 1 liter, or 17%.

Please answer the discussion and review questions next.

DISCUSSION AND REVIEW QUESTIONS

Chapter 11. ACTIVATED SLUDGE

(Lesson 3 of 7 Lessons)

Write the answers to these questions in your notebook. The question numbering continues from Lesson 2.

16. Why should proper procedures be followed when checking equipment?
17. Why should the equipment manufacturer's manual be thoroughly read and understood before any equipment is started?
18. What should the operator look for when checking out control gates?
19. What happens when spray nozzles are not installed properly and spray falls on rails and walks?
20. What safety precaution should be exercised before checking the impeller of a return sludge pump?
21. When starting a new activated sludge plant, who might the operator contact for assistance and advice?
22. When starting the activated sludge process, why should you use all of the aerators and all of the secondary clarifiers?
23. What level of dissolved oxygen should be maintained in the aeration tank during start-up?
24. What essential laboratory tests and on-site tests are recommended when starting the activated sludge process, and what are they used for?
25. What is the purpose of the 30-minute settleability test?

CHAPTER 11. ACTIVATED SLUDGE

OPERATION OF CONVENTIONAL ACTIVATED SLUDGE PLANTS

(Lesson 4 of 7 Lessons)

11.5 ROUTINE OPERATIONAL CONTROL

11.50 Operational Strategy

An excellent effluent can be produced by the activated sludge process if the operator has a plan of operation or operational strategy and understands how the microorganisms remove wastes from the water being treated. This lesson provides a brief plan for operating an activated sludge plant on a day-to-day basis. If problems develop, refer to Section 11.6, "Abnormal Operation (Operational Problems)," for possible solutions. There are three areas of major concern for the operator of an activated sludge plant:

1. What enters the plant
2. The environment for treating the wastes
3. What leaves the plant

All three of these areas are closely related or are influenced by the other two areas.

11.500 Influent Characteristics

As the influent flows and waste concentrations change, the environment changes in the aeration tank and secondary clarifiers where the wastes are treated. If the activated sludge process is in balance (a good secondary effluent with BOD and suspended solids levels less than 20 or 25 mg/*L*) for routinely experienced high flows, then the plant should perform as expected. If flows increase significantly, a switch to step-feed aeration (Section 11.83) may be necessary to avoid loss of the activated sludge bacteria. This solution will be effective if it causes a reduction in the mixed liquor suspended solids fed to the secondary clarifiers. If the influent solids loading increases, the mixed liquor suspended solids may have to be increased to treat these wastes by reducing the sludge wasting rate.

11.501 Aeration Tank Environment

If a good effluent is being produced, maintain a DO of 2 to 4 mg/*L* and thorough mixing throughout the entire aeration tank. If significant changes occur in the influent solids, adjust the mixed liquor suspended solids by regulating the waste sludge rate as described in Section 11.53. If there is a white foam on the aeration tank surface, reduce sludge wasting rates. If there is a thick, dark foam on the tank surface, increase the sludge wasting rates. The higher the F/M, the more food and therefore a higher DO is needed.

11.502 Secondary Clarifier

The settleability test is a good indication of how solids will settle in a clarifier. Adjust the return sludge rate so the sludge blanket will stay as low as possible in the clarifier. Remember that the sludge blanket can increase or rise with either an increase or a decrease in the return sludge rate as a result of either an increase or decrease in detention time. This increase or decrease in detention time also can be caused by changes in the influent flow. For these reasons you must keep good records and experiment to find the best return rate for the conditions in your plant. See Section 11.6, "Abnormal Operation (Operational Problems)," for different types of solids that may be observed on the surface of the clarifier or the plant effluent and possible solutions to the problems.

11.503 Plant Effluent

The turbidity test using a *TURBIDITY METER*[19] is a quick way to determine the quality of your plant effluent. When your plant is operating properly, try to determine why. Plot trend charts describing influent characteristics, aeration tank and secondary clarifier conditions, and effluent characteristics. When problems start to develop, try to determine why and correct the situation. Remember that influent and process environmental conditions (such as temperature) are continuously changing and you must adjust for these changes.

11.504 Typical Lab Results for an Activated Sludge Plant

Typical results of lab tests (Table 11.1) for an activated sludge plant are provided to assist in the evaluation of lab results and plant performance. Remember that every plant is different and is influenced by different conditions.

[19] *Turbidity* (ter-BID-it-tee) *Meter.* An instrument for measuring and comparing the turbidity of liquids by passing light through them and determining how much light is reflected by the particles in the liquid. The normal measuring range is 0 to 100 and is expressed as nephelometric turbidity units (NTUs). Also called a turbidimeter.

TABLE 11.1 TYPICAL LAB RESULTS

Test	Location	Common Range	
COD	Influent	250–1,000	mg/*L*
	Primary Effluent	200–400	mg/*L*
	Final Effluent (Conv. Act. Sl.)	30–70	mg/*L*
BOD	Influent	150–400	mg/*L*
	Primary Effluent	100–280	mg/*L*
	Final Effluent (Conv. Act. Sl.)	10–20	mg/*L*
SUSPENDED SOLIDS	Influent	150–400	mg/*L*
	Primary Effluent	60–160	mg/*L*
	Mixed Liquor	1,000–4,500	mg/*L*
	Return Sludge	2,000–10,000	mg/*L*
	Final Effluent (Conv. Act. Sl.)	1–20	mg/*L*
DISSOLVED OXYGEN	Mixed Liquor	0.5–4	mg/*L*
	Final Effl. (Outfall)	2–6	mg/*L*
CHLORINE RESIDUAL (30 min.)	Final Effluent	<0.1–2.0[a]	mg/*L*[b]
COLIFORM GROUP BACTERIA, MPN	Final Effluent (Chlorinated)	2–700/100 m*L*	
CLARITY (Secchi Disc)	Final Effluent	3–8 ft	
TURBIDITY (Turbidimeter)	Final Effluent	1–3 NTU	
pH	Influent	6.8–8.0	
	Effluent	6.9–8.5	

[a] < means less than. For example, less than 0.1 mg/*L* chlorine residual.
[b] Regulatory agencies normally specify a chlorine residual remaining after a certain time period.

11.51 How to Control the Process

The effectiveness of the activated sludge treatment process in reducing the waste load depends on the amount of activated sludge solids in the system and the health of the organisms that are a part of the solids. Successfully maintaining control of the solids and health of the organisms requires continuous (seven days a week) observation and checking by the plant operators. *SLUDGE AGE*[20] is one of the methods used by operators to determine and maintain the desired amount of activated sludge solids in the aeration tank. Sludge age is recommended for operational control because suspended solids are relatively easy to measure. In addition, sludge age considers two factors vital to successful operation: (1) solids (food) entering the treatment process, and (2) solids (organisms) available to treat the incoming waste (food). A critical point to recognize is that the solids test is capable of indicating both the amount of food carried by the inflow to the process and the number of organisms available to treat the waste.

NOTE: The activated sludge process we are describing in this example plant is controlled on the basis of sludge age, but other process control options may be used. Refer to Section 11.55, "Aerator Loading Guidelines for Process Control," for information about how to control an activated sludge process based on food/microorganism (F/M) ratio or mean cell residence time (MCRT).

Our example conventional activated sludge plant has an allowable effluent BOD of 20 mg/*L*. A sludge age of five days will serve as a satisfactory loading target during start-up for this plant. After the plant is in operation, various sludge ages may be tried in an effort to improve the quality of the plant effluent.

How activated sludge is wasted can have an important impact on the best sludge age for your plant. If activated sludge is wasted to the primary clarifiers, the wasted organisms could be mistaken for food entering the treatment process. If activated sludge is wasted to a gravity, belt, or flotation thickener, not as many organisms would be in the effluent from the primary clarifier. Therefore, plants wasting to the primary clarifier may have a lower sludge age than plants wasting to a gravity, belt, or flotation thickener. Temperature also influences sludge age. The warmer the weather the more active the organisms, so the sludge age can be lowered by reducing the MLSS (mixed liquor suspended solids).

Always remember that you must maintain the DO in the aerator and more air will be required when aeration tank solids increase in concentration and activity.

11.52 Determination of Sludge Age

Whether a new plant is being started or the operation of an existing plant is being checked, the sludge age is used to indicate when activated sludge should be wasted and, if necessary, to calculate the waste sludge pumping rate.

Information needed to determine sludge age includes:

1. Mixed Liquor Suspended Solids = 2,380 mg/*L*
2. Primary Effluent Composite Suspended Solids = 72 mg/*L* (average of daily values for past week)
3. Average Daily Influent Flow = 4.0 MGD
4. Aerator Factor = 4.6 (factor from Section 11.44, page 50)

[20]
$$\text{Sludge Age, days} = \frac{(\text{Suspended Sol in Mixed Liq, mg}/L)(\text{Aerator Vol, MG})(8.34\text{ lbs/gal})}{(\text{Suspended Sol in Primary Effl, mg}/L)(\text{Flow, MGD})(8.34\text{ lbs/gal})}$$

$$= \frac{\text{Suspended Solids Under Aeration, lbs}}{\text{Suspended Solids Added, lbs/day}}$$

or metric
$$= \frac{(\text{Suspended Sol in Mixed Liq, mg}/L)(\text{Aerator Vol, M}L)(1\text{ kg/M mg})}{(\text{Suspended Sol in Primary Effl, mg}/L)(\text{Flow, M}L\text{/day})(1\text{ kg/M mg})}$$

$$= \frac{\text{Suspended Solids Under Aeration, kg}}{\text{Suspended Solids Added, kg/day}}$$

1. *DETERMINE POUNDS OF MIXED LIQUOR SUSPENDED SOLIDS IN AERATOR.*

$$\text{Solids in Aerator, lbs} = \text{MLSS, mg}/L \times \text{Aerator Vol, MG} \times 8.34 \text{ lbs/gal}$$

$$= 2{,}380 \text{ mg}/L \times 4.6 \text{ (factor from Section 11.44, page 50)}$$

$$= 10{,}948 \text{ lbs, or approximately } 11{,}000 \text{ lbs}$$

2. *DETERMINE POUNDS OF SOLIDS ADDED PER DAY TO SYSTEM BY PRIMARY EFFLUENT.*

$$\text{Solids Added by Primary Effluent, lbs/day} = \text{Prim Effl SS, mg}/L \times \text{Flow, MGD} \times 8.34 \text{ lbs/gal}$$

$$= 72 \text{ mg}/L \times 4.0 \text{ MGD} \times 8.34 \text{ lbs/gal}$$

$$= 72 \times 33.4 \text{ lbs/day}$$

$$= 2{,}405 \text{ lbs/day, or approximately } 2{,}400 \text{ lbs/day}$$

3. *CALCULATE SLUDGE AGE IN DAYS.*

$$\text{Sludge Age, days} = \frac{\text{Suspended Solids in Aerator, lbs}}{\text{Suspended Solids in Primary Effluent, lbs/day}}$$

$$= \frac{11{,}000 \text{ lbs}}{2{,}400 \text{ lbs/day}}$$

$$= 4.6 \text{ days}$$

If the results of lab tests and calculations indicate a sludge age of 4.6 days when the target sludge age is 5 days, no sludge should be wasted. If a sludge age of 4.6 days was obtained during the start-up of the example plant, the operator should continue to allow the solids to increase in the aerator. In an existing plant, if the sludge age is below the desired level, any sludge wasting should be reduced or stopped.

A simple way to find the desired pounds of aerator solids to be maintained is to multiply the average daily pounds of primary effluent solids added per day by the desired sludge age.

EXAMPLE:

Find the desired pounds of solids to be maintained in the aerator.

Information needed:

Desired Sludge Age, days = 5 days

Solids in Primary Effluent, lbs/day = 2,400 lbs/day

Calculate pounds of solids desired in aerator using the following equation:

$$\text{Sludge Age, days} = \frac{\text{Suspended Solids in Aerator, lbs}}{\text{Suspended Solids in Primary Effluent, lbs/day}}$$

Rearrange the equation to obtain:

$$\text{Suspended Solids in Aerator, lbs} = \text{Sludge Age, days} \times \text{SS in Prim Effl, lbs/day}$$

$$= 5 \text{ days} \times 2{,}400 \text{ lbs/day}$$

$$= 12{,}000 \text{ lbs}$$

Suspended solids in the mixed liquor in the aerator should be allowed to build up until 12,000 pounds (5,455 kg) of solids are in the aerator. Sludge wasting may be started when the desired level is exceeded.

Some activated sludge must be wasted to prevent an excessive solids buildup in the aerator. To determine when an excess of activated sludge is in the aerator and some should be wasted, many operators calculate the desired mixed liquor suspended solids concentration in the aerator. When this concentration is exceeded, some of the excess activated sludge is wasted. The desired mixed liquor suspended solids concentration in mg/*L* can be determined using either of the two formulas listed below (they are the same).

$$\text{Desired Mixed Liquor Suspended Solids, mg}/L = \frac{\text{Desired Susp Sol in Aerator, lbs}}{\text{Weight of Water in Aerator, million lbs}}$$

or

$$= \frac{\text{Desired Susp Sol in Aerator, lbs}}{\text{Vol of Aerator, M gal} \times 8.34 \text{ lbs/gal}}$$

$$= \frac{12{,}000 \text{ lbs}}{4.6 \text{ M lbs}} \text{ (factor from Section 11.44, page 50)}$$

$$= 2{,}608 \text{ mg}/L$$

$$= 2{,}600 \text{ mg}/L \text{ (target concentration)}$$

Wasting of activated sludge from the example plant should not start until the mixed liquor suspended solids concentration exceeds 2,600 mg/*L*.

11.53 Wasting Activated Sludge

The amount of activated sludge wasted may vary from 1 to 20 percent of the total incoming flow. Normally, waste activated sludge is expressed in gallons per day or pounds of solids removed from the aeration system. Continuous wasting is preferred. Try not to change your sludge wasting rate by more than 10 or 15 percent from one day to the next. The main purpose is to maintain a sludge age that produces the best effluent.

Wasting (Figure 11.18) is normally accomplished by diverting a portion of the return sludge to a primary clarifier, gravity thickener, gravity belt thickener, dissolved air flotation thickener (DAFT), aerobic digester, or anaerobic digester. Normal operations in a conventional activated sludge plant will concentrate return sludge and waste sludge solids three to four times as much as the solids concentration of the mixed liquor. This may provide return sludge with a concentration of 2,000 to 10,000 mg/*L*, or 0.2 to 1.0 percent in terms of total solids. If the waste sludge line discharges directly to the anaerobic digestion system, it would contain 10 to 20 times as much water as

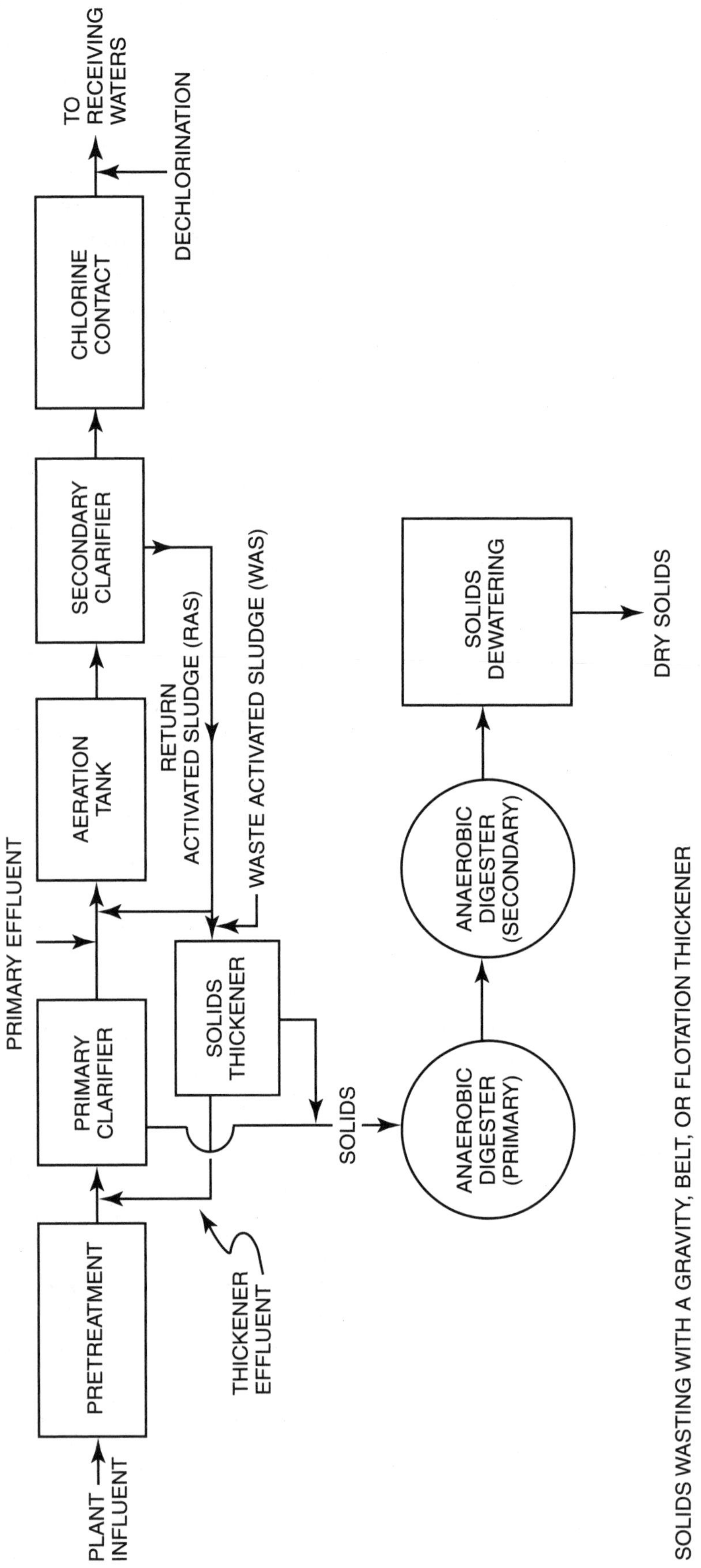

PLANT INFLUENT
PRETREATMENT
PRIMARY CLARIFIER
PRIMARY EFFLUENT
AERATION TANK
SECONDARY CLARIFIER
CHLORINE CONTACT
DECHLORINATION
TO RECEIVING WATERS
RETURN ACTIVATED SLUDGE (RAS)
WASTE ACTIVATED SLUDGE (WAS)
SOLIDS THICKENER
THICKENER EFFLUENT
SOLIDS
ANAEROBIC DIGESTER (PRIMARY)
ANAEROBIC DIGESTER (SECONDARY)
SOLIDS DEWATERING
DRY SOLIDS

SOLIDS WASTING WITH A GRAVITY, BELT, OR FLOTATION THICKENER

Fig. 11.18 Waste sludge flow diagram

should be entering the anaerobic system with that amount of solids. Operating an anaerobic digester would be difficult under this condition. It would be wiser to waste to the primary clarifiers where combining with primary sludge minimizes the addition of excess water to the digester.

Wasting activated sludge will occur in the effluent whether or not it is controlled. In all activated sludge plants, wasting must be controlled by the operator. Mixed liquor suspended solids that need to be wasted accumulate from two sources. The first is the suspended solids in the plant flow from the primary clarifiers or raw wastewater. The second and main source is the new cell production by the microorganisms.

For every pound of BOD or solids removed by treatment in the activated sludge system, a part of that pound will remain in the system as microorganisms. The rate of production of excess sludge will depend on the type of process being operated and the nature of the waste load. The high-rate activated sludge plant is capable of producing 0.75 pound of sludge volatile matter for every pound of BOD removed. The conventional plant runs around 0.55 pound of sludge volatile matter per pound of BOD removed in the activated sludge system. The extended aeration plant drops down to about 0.15 pound of sludge volatile matter per pound of BOD removed. Excessive silt or inert material may increase sludge production beyond that indicated by the BOD test.

11.54 Determination of Waste Sludge Pumping Rate

To illustrate the determination of a waste sludge pumping rate, assume the following plant data:

1. Mixed Liquor Suspended Solids = 2,985 mg/*L*
2. Return Sludge, Suspended Solids = 6,200 mg/*L*
3. Primary Effluent, Suspended Solids = 72 mg/*L*
4. Average Daily Flow = 4.0 MGD
5. Current Waste Sludge Pumping Rate = 0.050 MGD

Using the procedures outlined earlier in this lesson, the following information can be calculated.

1. Solids in Aeration Tank = 13,731 pounds
2. Solids Added by Primary Effluent = 2,400 lbs/day
3. Sludge Age = 5.7 days

The results of the calculations indicate that the sludge age is too high (5.7 days instead of 5 days) and the solids in the aeration tank also are too high (13,731 pounds instead of 12,000 pounds). To reduce the sludge age and solids in the aerator, some of the activated sludge removed by the secondary clarifier should be pumped to the inlet of the primary clarifier or to the solids thickener, depending on facilities in the plant for wasting activated sludge.

The formula to calculate the additional waste return sludge pumping rate is:

Additional Waste Return Sludge Pumping Rate, MGD

$$= \frac{\text{Additional Solids to be Wasted, lbs/day}}{\text{Return Sludge Conc, mg/}L \times 8.34\text{ lbs/gal}}$$

$$= \frac{(13{,}700\text{ lbs} - 12{,}000\text{ lbs})/\text{day}^{a}}{6{,}200\text{ mg/}L \times 8.34\text{ lbs/gal}}$$

$$= \frac{1{,}700\text{ lbs/day}}{6{,}200\text{ mg/}L \times 8.34\text{ lbs/gal}}$$

$$= \frac{1{,}700}{51{,}700}$$

$$= 0.032\text{ MGD}$$

[a] Biological cultures should be subject to slow changes rather than rapid ones; therefore, the pounds wasted will be removed during a 24-hour period.

Calculate the total waste return sludge pumping rate.

$$\text{Total Return Sludge Pumping Rate, MGD} = \text{Current Rate, MGD} + \text{Additional Rate, MGD}$$

$$= 0.050\text{ MGD} + 0.032\text{ MGD}$$

$$= 0.082\text{ MGD}$$

Usually the waste return sludge pumping rate is expressed in gallons per minute instead of MGD.

$$\text{Pump Waste Rate, GPM} = (\text{Pumping Rate, MGD})(694\text{ GPM/MGD})$$

$$= (0.082\text{ MGD})(694\text{ GPM/MGD})$$

$$= 57\text{ GPM } (3.6\ L/\text{sec})$$

Set the waste pumping rate at 50 or 55 GPM (3.2 or 3.5 *L*/sec) for the next 24-hour period. It is better to waste a little less activated sludge than the theoretical calculation. Also try not to make large changes in the waste pumping rate from one day to the next.

QUESTIONS

Write your answers in a notebook and then compare your answers with those on page 132.

11.5A Calculate the sludge age for an activated sludge process if the aerator volume is 0.5 MG and the mixed liquor suspended solids concentration is 2,100 mg/*L*. The influent flow is 4.0 MGD and the primary effluent suspended solids concentration is 70 mg/*L*.

11.5B Why must some activated sludge be wasted?

11.5C How is the wasting of excess activated sludge normally accomplished?

11.55 Aerator Loading Guidelines for Process Control

The activated sludge process can be controlled by changes in the aerator loadings based on sludge age, food/microorganism ratio, or mean cell residence time (MCRT). All three methods of control are mathematically similar. In each case, the operator begins operations with aerator loadings based on data and the experience of other similar plants. This value (loading rate) is then adjusted by the operator until the operating range is found that produces the best quality effluent for the plant.

The critical factor in any method of aeration tank control is the food/microorganism relationship and this cannot be precisely estimated for any specific plant. The operator tries to keep enough solids (microorganisms) in the aeration tank to use up the incoming waste (food). Neither too many organisms nor too few organisms should be in the aeration tank in relation to the incoming food. Operation of the activated sludge process requires removing the organisms (settled activated sludge) from the secondary clarifier as quickly as possible. The organisms are either returned to the aerator to use the incoming food or they are wasted. This procedure has been discussed and an example provided in Sections 11.53 and 11.54 using sludge age to determine the aerator loading value. Select a method to operate your plant and stick with it. Do not continually try to switch from one method to another.

11.550 Food/Microorganism Ratio

The food-to-microorganism loading ratio is based on the food provided each day to the microorganism mass in the aerator. Food (waste) provided is preferably measured by the COD of the influent to the aerator. COD is recommended because test results are available within four hours and process changes can be made before the process becomes upset. Many operators load aerators on the basis of the BOD test, but results five days later are too late for operational control. A comparison of influent BOD and COD values will give the operator, over a period of time, a fairly accurate picture of the organic matter that is available for use by the microorganisms as long as industrial waste is either constant (highly unlikely) or an insignificant contributor.

If you know the amount of waste (food) entering the aerator, you can provide sufficient organisms in the aerator to treat (eat) the incoming waste. You estimate the amount of organisms in the aerator by measuring the suspended solids in the mixed liquor. The mixed liquor suspended solids (MLSS) consist of both organic (volatile) and inorganic material. The organic portion represents the organisms available to treat the incoming waste. Thus, by measuring the mixed liquor volatile suspended solids (MLVSS), you have a more accurate measure of the organisms present to treat the incoming waste. MLSS is a measure of organisms and inorganic or inert materials present, but MLVSS is a more accurate measure for control purposes because only organisms and organic materials are measured.

Typical loading guidelines (ranges) have been established for the three operational modes of activated sludge and are summarized as follows:

1. *HIGH-RATE*

   COD: >1[a] lb COD per day/1 lb of MLVSS under aeration

   BOD[b]: >0.5 lb BOD per day/1 lb of MLVSS under aeration

2. *CONVENTIONAL*

   COD: 0.5 to 1.0 lb COD per day/1 lb of MLVSS under aeration

   BOD[b]: 0.25 to 0.5 lb BOD per day/1 lb of MLVSS under aeration

3. *EXTENDED AERATION*

   COD: <0.2[c] lb COD per day/1 lb MLVSS under aeration

   BOD[b]: <0.10 lb BOD per day/1 lb MLVSS under aeration

---

[a] > means greater than. Greater than 1 lb COD.
[b] For untreated domestic wastewater, BOD = 0.4 to 0.8 times COD.
[c] < means less than. Less than 0.2 lb COD.

11.551 Calculation of Food/Microorganism Aerator Loading

Determine the amount of mixed liquor volatile suspended solids to be maintained in the aerator of the conventional plant studied in this chapter. Assume a food/microorganism ratio of 0.5 lb COD per day/1 lb of mixed liquor volatile suspended solids under aeration. Frequently, this loading is expressed as 50 lbs COD per day/100 lbs of MLVSS.

Information needed:

1. Average COD of Primary Effluent, mg/*L* = 150 mg/*L*
2. Average Daily Flow, MGD = 4.0 MGD

Find pounds of COD provided to aerator per day.

$$\text{Aerator Loading, } \frac{\text{lbs COD}}{\text{day}} = \text{Prim Effl COD, mg/}L \times \text{Daily Flow, MGD} \times 8.34 \text{ lbs/gal}$$

$$= 150 \text{ mg/}L \times 4.0 \text{ MGD} \times 8.34 \text{ lbs/gal}$$

$$= 5{,}004 \text{ or } 5{,}000 \text{ lbs COD/day}$$

Find desired pounds of mixed liquor volatile suspended solids (MLVSS) under aeration, based on 0.5 lb COD per day/1 lb of MLVSS.

$$\text{MLVSS, lbs} = \frac{\text{Primary Effluent COD, lbs/day}}{\text{Loading Factor in lbs COD/day/1 lb MLVSS}}$$

$$= \frac{5{,}000 \text{ lbs COD/day}}{0.5 \text{ lb COD/day/lb MLVSS}}$$

$$= \frac{5{,}000}{0.5 \text{ lb MLVSS}}$$

$$= 10{,}000 \text{ lbs MLVSS Under Aeration}$$

The MLVSS is a measure of the organisms in the aerator available to work on the incoming waste. When operating your plant on the basis of MLVSS, you should note any fluctuations that may occur during the week and make appropriate adjustments.

If the COD load applied to the aerator increases or drops to a significantly different level for two consecutive days, a new mixed liquor solids value should be calculated and activated sludge wasting adjusted to achieve the new value of solids desired under aeration. Calculation of waste sludge rates is outlined in Sections 11.53 and 11.54.

11.552 Mean Cell Residence Time (MCRT)

Another approach for solids control used by operators is the Mean Cell Residence Time (MCRT) or Solids Retention Time (SRT). This is a refinement of the sludge age. Both terms are almost the same. The equation for MCRT is:

$$\text{MCRT, days} = \frac{\text{Pounds of Suspended Solids in Aeration System*}}{\text{lbs of Susp Sol Wasted/day} + \text{lbs of Susp Sol Lost in Effl/day}}$$

The most desirable MCRT for a given plant is determined experimentally just as with the use of sludge age or the mixed liquor volatile suspended solids concentration. The desired MCRT for conventional plant operation usually will fall between 4 and 15 days. (Do not confuse this time with the recommended range for sludge age of 3.5 to 10 days.)

**NOTE:* There are three different ways to calculate the pounds of suspended solids in the aeration system:

1. MLSS, mg/*L* × Aerator Vol, MG × 8.34 lbs/gal
2. MLSS, mg/*L* × (Aerator Vol, MG + Final Clar Vol, MG) × 8.34 lbs/gal
3. (MLSS, mg/*L* × Aerator Vol, MG + Return Sludge SS, mg/*L* × Sludge Blanket Vol, MG) × 8.34 lbs/gal

The first method is illustrated in this section, but you may use whichever method produces satisfactory results for your plant; just be consistent in the method you use.

A way of determining MCRT for the example plant in this chapter would be as follows:

Required data:

1.	Aerator Volume, gal	= 1,000,000 gal
2.	Final Clarifier Volume, gal	= 500,000 gal
3.	Wastewater Flow to Aerator, MGD	= 4.0 MGD
4.	Waste Sludge Flow for Past 24 hr, MGD	= 0.075 MGD
5.	Mixed Liquor Suspended Solids Concentration, mg/*L*	= 2,400 mg/*L*
6.	Waste Sludge (or Return Sludge) Suspended Solids Concentration, mg/*L*	= 6,200 mg/*L*
7.	Final Effluent Suspended Solids Concentration, mg/*L*	= 12 mg/*L*

11.553 Calculation of Mean Cell Residence Time

$$\text{MCRT, days} = \frac{\text{Suspended Solids in Aeration Systems, lbs}}{\text{Susp Sol Wasted, lbs/day} + \text{Susp Sol Lost in Effl, lbs/day*}}$$

$$\text{MCRT, days} = \frac{\text{Susp Sol in Mixed Liq, mg/}L \times \text{Aerator, MG} \times 8.34\text{ lbs/gal}}{(\text{Susp Sol in Waste, mg/}L \times \text{Waste Rate, MGD} \times 8.34\text{ lbs/gal}) + (\text{Susp Sol in Effl, mg/}L \times \text{Plant Flow, MGD} \times 8.34\text{ lbs/gal})}$$

$$= \frac{2{,}400\text{ mg/}L \times 1.0\text{ MG} \times 8.34\text{ lbs/gal}}{(6{,}200\text{ mg/}L \times 0.075\text{ MGD} \times 8.34\text{ lbs/gal}) + (12\text{ mg/}L \times 4.0\text{ MGD} \times 8.34\text{ lbs/gal})}$$

$$= \frac{20{,}016\text{ lbs}}{3{,}878\text{ lbs/day} + 400\text{ lbs/day}}$$

$$= \frac{20{,}016\text{ lbs}}{4{,}278\text{ lbs/day}}$$

$$= 4.7\text{ days}$$

**NOTE:* Some operators use volatile suspended solids instead of suspended solids.

If you are operating the plant on the basis of MCRT and the plant operates satisfactorily at the MCRT of 8, 9, 10, 11, or even 15 days, the main method of control is to adjust the waste sludge rate to maintain the MCRT at the desired number of days.

Rearranging the equation above, calculation of the sludge waste rate from the system merely means plugging in the chosen MCRT (use 4.7 days) and solids figures.

EXAMPLE:

$$\text{Waste Sludge, lbs/day} = \frac{\text{Susp Sol in System, lbs}}{\text{MCRT, days}} - \text{Susp Sol in Effl, lbs/day}$$

$$= \frac{(2{,}400\text{ mg/}L \times 1.0\text{ MG} \times 8.34\text{ lbs/gal})}{4.7\text{ days}} - (12\text{ mg/}L \times 4.0\text{ MGD} \times 8.34\text{ lbs/gal})$$

$$= \frac{20{,}016\text{ lbs}}{4.7\text{ days}} - 400\text{ lbs/day}$$

$$= 4{,}259\text{ lbs/day} - 400\text{ lbs/day}$$

$$= 3{,}859\text{ lbs/day}$$

The waste sludge pumping rate of 3,859 lbs/day appears to be correct to maintain a Mean Cell Residence Time of 4.7 days.

The best target MCRT for your plant is determined the same way you select the best sludge age or food/microorganism (F/M) ratio, on the basis of circumstances, process objectives, and experience gained operating your plant. If you are aiming for a high degree of nitrification, use higher MCRT values. Lower MCRT values are used during periods of warmer weather. Operate at lower MCRTs if you have limited aeration (oxygen) capacity or are trying to save energy. If you have good secondary clarification and waste activated sludge thickening capacity, then you can operate at lower MCRT levels.

QUESTIONS

Write your answers in a notebook and then compare your answers with those on page 132.

11.5D If you calculate that your plant has 12,000 pounds of mixed liquor volatile suspended solids under aeration and you need 9,000 pounds under aeration, how many pounds should be wasted?

11.5E What should be the waste sludge pumping rate (GPM) if a plant should be wasting 3,000 pounds per day and the concentration of return sludge is 6,000 mg/*L*?

11.5F Estimate the waste sludge rate (lbs/day) from an activated sludge plant operating at an MCRT of 10 days. The system contains 40,000 pounds of suspended solids and the effluent has a suspended solids concentration of 10 mg/*L* at a flow of 5 MGD.

11.56 Actual Operation Under Normal Conditions

This section describes the operational strategy of an efficiently operating activated sludge plant. Remember that every plant is different and that you must develop a strategy that works for your plant.

The example plant is an activated sludge plant that operates in the contact stabilization mode with diffused aeration providing complete mixing in the aeration tanks. It was designed to treat a flow of 20 MGD (76 M*L*/day). During the month studied the peak flow was 29.2 MGD (111 M*L*/day) due to a storm and the peak influent BOD was 580 mg/*L* as a result of an industrial waste spill of methanol. In spite of the abnormal events, the operator produced an average effluent BOD of 9 mg/*L* with a maximum BOD of 16 mg/*L*. The effluent suspended solids averaged 13 mg/*L* and the maximum was 27 mg/*L*. This excellent effluent was produced because every operator on every shift followed the same operational strategy.

11.560 Initial Inspection

At the start of every shift, the shift operator goes through the following steps.

1. Review the log book.
   a. Has anything unusual happened? If so, were any corrections required and has the problem been corrected?
   b. Check the status of the return sludge pumps and waste sludge pumps.
   c. What are the levels of DO in the aerators? Maintain a DO of 1.0 to 3.0 mg/*L* in all aerators.

      NOTE: Higher DO levels may be required to maintain DO in the secondary clarifiers.

   d. If chlorinating the return sludge flow for control of filamentous growths, be sure to check the pounds of chlorine applied per day.
2. Visually inspect the activated sludge process.
   a. Is there any foam on the surface of the aerators? If so, is it white or a thick, dark foam?
   b. Do the secondary clarifiers appear normal? Are they turbid or are there any solids on the surfaces? Check the depth of the sludge blanket in each clarifier. If the sludge blanket is high, be sure the return sludge pumps are working properly.

      NOTE: In this plant the sludge blanket is kept as low as possible in order to maintain aerobic conditions in the secondary clarifiers.

   c. Inspect the effluent from the secondary clarifiers. The effluent should be clear and free of suspended solids and floatables.
3. Review the lab results.
   a. *SVI*[21] and microorganisms.

      If the SVI is high, something is wrong. Collect a sample of aerator mixed liquor and examine under a microscope for the typical numbers, types, and condition of the mixed liquor microorganisms. If there is an excess or lack of any type of microorganism, carefully inspect the laboratory data. Look for unusual values for: (1) pounds of volatile suspended solids under aeration, (2) sludge wasting rates, (3) air usage, and (4) dissolved oxygen levels in the reaerator and aerators. If the microscopic examination of the mixed liquor indicates an excess of normal filamentous microorganisms, then chlorination of the return activated sludge should be started.

   b. Pounds of volatile matter under aeration.

      Check the suspended solids levels in mg/*L* and percent volatile matter in the aerators.

   c. Sludge wasting rate.

      Adjust the sludge wasting rate to maintain the desired pounds of volatile matter under aeration.

   d. Air usage.

      Examine air usage in millions of cubic feet per day needed to maintain desired DO levels in the aerators. Air required serves as an excellent indication of influent BOD.

11.561 Process Adjustment

Table 11.2 contains a portion of the data available to the operator for controlling the activated sludge process. Figures 11.19 and 11.20 show the results of plotting the lab data. You must decide what information to plot that will help you carry out your operational strategy. The remainder of this section describes how the operator of this plant identified a potential problem and corrected it before there was any major deterioration of the effluent quality.

---

[21] *Sludge Volume Index (SVI).* A calculation that indicates the tendency of activated sludge solids (aerated solids) to thicken or to become concentrated during the sedimentation/thickening process. SVI is calculated in the following manner: (1) allow a mixed liquor sample from the aeration basin to settle for 30 minutes; (2) determine the suspended solids concentration for a sample of the same mixed liquor; (3) calculate SVI by dividing the measured (or observed) wet volume (m*L*/*L*) of the settled sludge by the dry weight concentration of MLSS in grams/*L*.

$$\text{SVI, m}L\text{/gm} = \frac{\text{Settled Sludge Volume/Sample Volume, m}L/L}{\text{Suspended Solids Concentration, mg}/L} \times \frac{1{,}000 \text{ mg}}{\text{gram}}$$

TABLE 11.2 MONTHLY LAB DATA FOR EXAMPLE ACTIVATED SLUDGE PLANT

Date	Day	Flow, MGD	BOD		COD	Suspended Solids		Suspended Solids		Volatile Solids		Total Vol, lbs[b]	F/M Ratio	Waste, lbs[c]/day	Air, MCF[d]/day	SVI
			Infl, mg/L	Effl, mg/L	Infl, mg/L	Infl, mg/L	Effl, mg/L	Reaer, mg/L	Aer,[a] mg/L	Reaer, %	Aer,[a] %					
1	W	19.9	440	8	992	446	8	8060	2150	83	82	95786	.43	44298	80.2	270[e]
2	T	22.3	430	11	957	540	18	8860	2160	80	80	98453	.66	50544	80.2	384
3	F	29.2	580	9	1420	682	11	9830	2510	81	80[f]	109930	.38	59929	64.8	295
4	S	24.0	270	12	838	542	27	7930	2170	77	74	87353	.60	53041[g]	51.2	248
5	S	18.8	280	10	770	586	16	8880	2380	73	72	98932	.17	39474	39.0	109
6	M	20.0	360	12	894	456	12	9470	2140	72	72	91818	.38	59156	56.6	98
7	T	20.0	380	8	992	502	26	6090	3050	71	72	85877	.56	36112	61.8	125
8	W	21.7	350	6	881	476	11	8690	2180	74	77	91710	.75	50152	72.5	92
9	T	20.0	340	8	770	380	14	8370	1750	75	79	83716	.44	42931	66.3	–
10	F	25.0	360	9	846	430	16	6070	2040	76	78	74570	.56	24350	68.2	98
11	S	19.9	400	6	710	476	16	8580	2390	76	79	97214	.38	33130	43.7	106
12	S	17.1	210	6	650	380	6	7820	1790	77	75	80672	.25	21131	32.0	156
13	M	21.3	380	9	930	424	6	8220	1990	76	78	87313	.45	23720	65.8	106
14	T	21.2	400	9	920	430	8	8960	2160	78	82	94866	.47	38335	72.4	113
15	W	20.9	440	16	1000	472	21	8720	2680	80	67[h]	98913	.49	44798	77.5	101
16	T	19.1	380	8	927	450	12	8410	2080	81	84	96448	.43	46362	75.3	135
17	F	16.3	350	6	907	428	11	7070	1950	82	87	86835	.38	34965	66.0	133
18	S	16.5	320	5	720	418	14	6120	1830	84	89	79820	.31	23785	42.0	131
19	S	15.4	200	6	620	360	11	5460	1800	84	85	72937	.25	15665	36.4	138
20	M	16.3	400	9	1040	470	8	5430	1590	82	84	67433	.54	13812	51.5	151
21	T	15.4	460	14	1080	416	8	6300	1940	84	86	81982	.50	20176	71.4	155
22	W	15.8	420	6	986	432	9	6620	2120	85	86	88288	.37	28213	69.4	189[e]
23	T	15.1	400	6	888	418	12	6410	1990	85	85	83811	.33	29510	62.2	347
24	F	14.8	360	8	848	414	13	5940	1900	86	90	81395	.21	22689	57.5	289
25	S	16.2	220	8	797	400	14	5490	1800	85	84	73218	.30	15567	42.0	183
26	S	14.2	260	8	700	396	14	5850	1710	84	85	74451	.22	14442	38.4	292
27	M	20.3	420	16	1010	489	14	6330	1510	83	93	75674	.54	18530	58.9	152
28	T	20.4	440	7	961	428	12	6960	1770	85	88	84673	.52	25889	82.1[i]	158
29	W	20.3	480	7	1120	438	8	7180	1810	86	85	86398	.63	30360	94.2	138
30	T	20.8	500	8	1200	616	7	7870	1880	86	88	93977	.54	40300	97.1	144
31	F	21.9	410	7	952	320	10	7710	1870	87	91	74545	.60	38709	92.0	128
Avg		19.4	375	9	914	455	13	7410	2035	81	82	86536	.44	33551	63.5	172

[a] Middle of three aeration tanks.
[b] Multiply lbs × 0.454 to obtain kg.
[c] Multiply lbs/day × 0.454 to obtain kg/day.
[d] Multiply MCF/day × 0.0283 to obtain Million cu m/day.
[e] Filamentous growths. Increase chlorine dose to return sludge.
[f] Drop in percent volatile matter due to storm.
[g] Carefully decrease solids wasting rate due to storm.
[h] Lab error.
[i] Increased air requirements and increased solids wasting due to industrial dump of methanol.

Lab data in Table 11.2 indicated that the SVIs were high during the first four days of the month. An inspection of the mixed liquor under a microscope on the first day revealed excessive filamentous growths. Chlorination of the return sludge was started and continued until the SVIs dropped down to normal and the excessive filamentous growths disappeared on the fifth day.

On the second day of the month a storm caused the flow to peak on the third day at 29.2 MGD (111 ML/day). The percent volatile matter in the aerator suspended solids dropped below 80 percent. The waste sludge rate had to be carefully decreased to maintain the desired pounds of volatile matter under aeration and to build the percent volatile matter under aeration back up to 80 percent.

An industrial waste spill containing methanol hit the plant on the 27th day of the month. The air requirement immediately increased drastically due to the COD of the methanol and the increased activity of the microorganisms. The waste sludge rate had to be increased the next day to maintain the desired pounds of volatile matter under aeration and to prevent solids from escaping in the effluent.

In spite of three different problems resulting from filamentous growths, a storm, and an industrial spill, Figure 11.19 reveals that by sticking to this plant's operational strategy, the operators produced a consistently high-quality effluent. The key to success was the fact that the process was adjusted before the effluent deteriorated by identifying the causes of changes in the aeration tank environment and taking appropriate corrective action immediately.

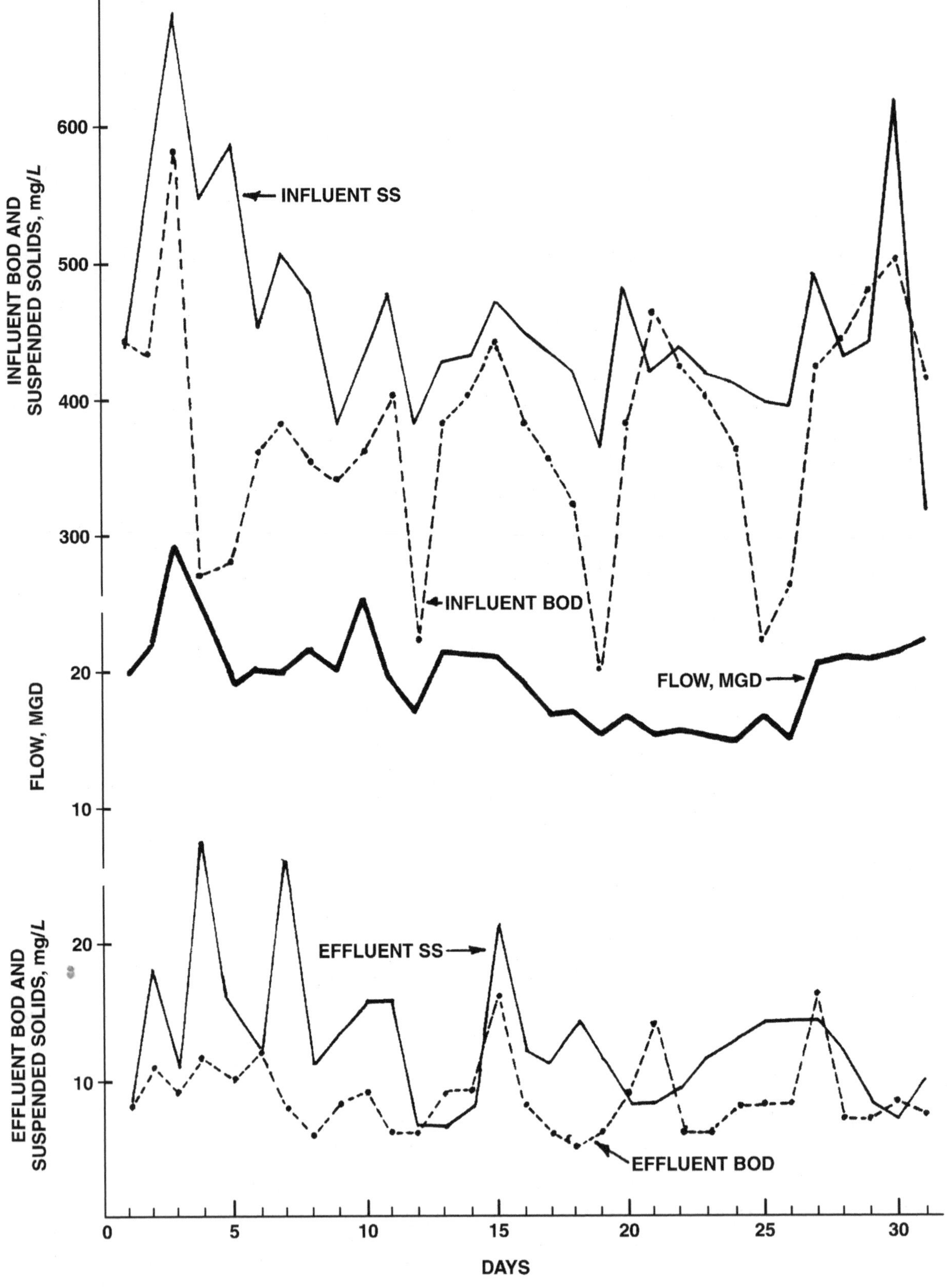

Fig. 11.19 Flow and influent and effluent BOD and suspended solids (Plots include actual daily flows and influent and effluent SS and BOD values.)

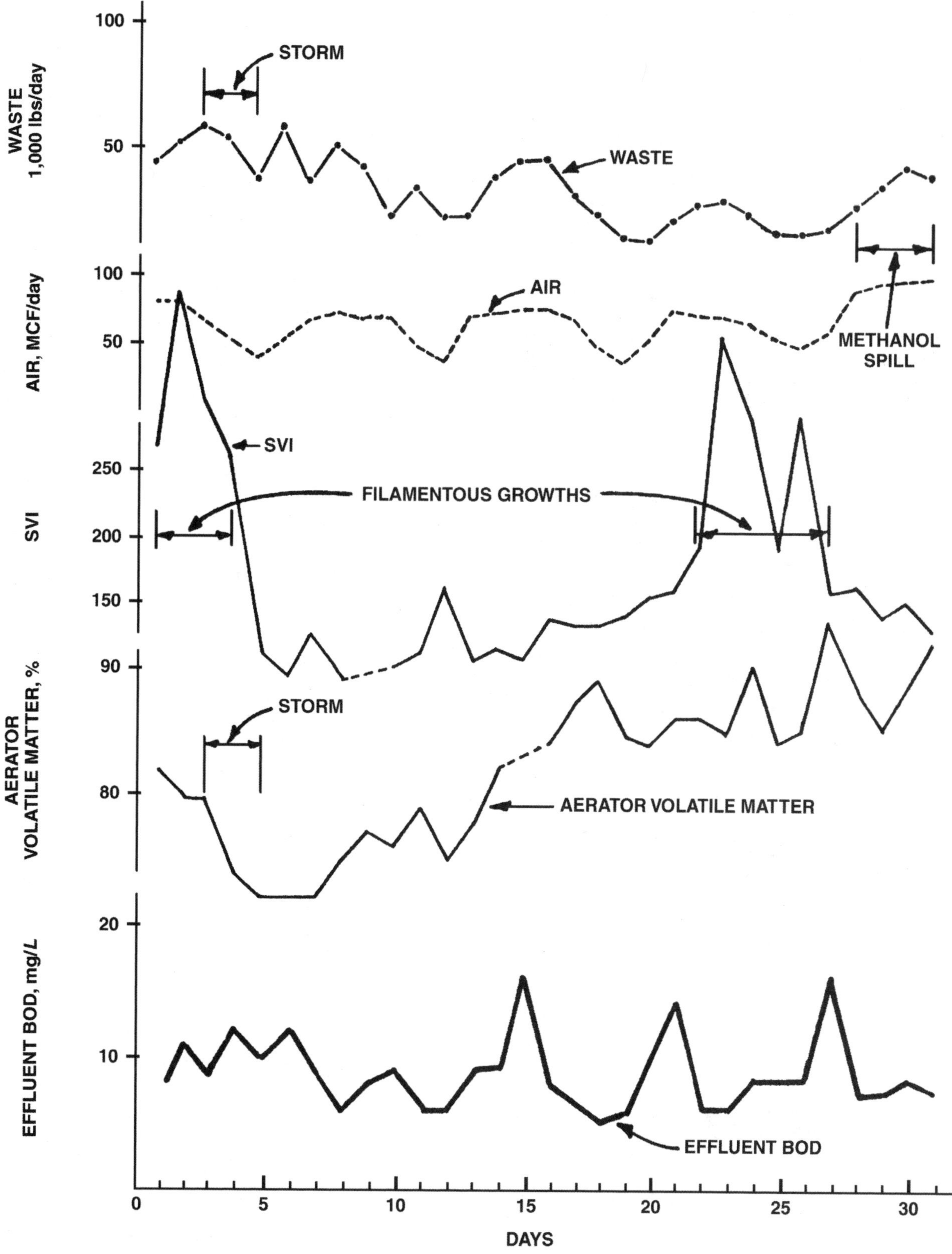

Fig. 11.20 Process adjustment information

11.57 Actual Operation Under Abnormal Conditions

Some wastewater treatment plants become seriously overloaded during certain times of the year, such as during the canning season. Figure 11.21 shows the extent of overload our example activated sludge plant experiences during the canning season. To meet the NPDES permit requirement of 80,000 pounds of BOD per month (36,360 kg BOD per month), the operator estimated the effluent BOD using daily COD tests. Figure 11.22 shows the cumulative plots of both the estimated BOD using the COD test results and the actual effluent BOD, which was obtained five days later. Use of the COD test results allows the operator to make changes immediately when a problem develops, rather than waiting five days. Using this approach also helps the operator meet NPDES permit requirements.

Under the overloaded conditions, the operator had to work very hard to meet the daily, seven-day average, and monthly NPDES permit requirements. Both the blowers and return sludge pumps worked at full capacity the entire month. Ferric chloride was added in the primary clarifier to reduce the solids and BOD loadings on the activated sludge process. Toward the end of the month, a portion of the effluent had to be diverted to storage oxidation ponds in order for the plant to meet permit requirements.

11.58 Recordkeeping

After a plant is started, the sampling and laboratory testing program must be continued to identify and correct operational problems whenever they start to develop. (See Lesson 6 for possible approaches to solving operational problems.)

Accurate daily plant and laboratory records on the following items can help the operator determine the best operating ranges for operational controls on the basis of plant performance. Records can also indicate when problems develop and help identify the source of the problem. Record the following data on a daily basis. (Also see Monthly Data Sheet in the Appendix at the end of this chapter.)

1. Suspended Solids and Volatile Content
   a. Primary effluent
   b. Aerator mixed liquor
   c. Return sludge and waste sludge
   d. Final clarifier or secondary sedimentation tank effluent

2. BOD, COD, or *TOC*[22]
   a. Plant influent
   b. Primary effluent
   c. Final clarifier or secondary sedimentation tank effluent

   NOTE: COD is recommended to determine the strength of influent wastewater because the results are available within four hours and can be used to control the activated sludge process. For many years, operators attempted to use the BOD test for operational control, but the test has the following disadvantages:

   (1) Procedural errors can cause a large variation in results.

   (2) Five days of waiting are required before results are available.

   (3) Only a portion of the load on the activated sludge process is measured by the test.

3. Dissolved Oxygen
   a. Aerator[23]
   b. Final clarifier or secondary sedimentation tank (inside the effluent weir)
   c. Final effluent

4. Settleable Solids
   a. Influent
   b. Mixed liquor settleability test
   c. Digester supernatant
   d. Final effluent

5. Temperature
   a. Influent
   b. Aerator
   c. Final effluent

6. pH
   a. Influent
   b. Primary effluent
   c. Aerator
   d. Final effluent

7. Clarity *(SECCHI DISK*[24]*)* or Turbidity (turbidimeter)
   a. Final clarifier

[22] *TOC* (pronounce as separate letters). Total Organic Carbon. TOC measures the amount of organic carbon in water.

[23] Measure aerator DO in the aerator or immediately after sample is collected because the DO can change very rapidly once the sample is out of the aerator. Do not take sample to the lab unless you follow procedures in Section 16.454 on pages 519 and 520.

[24] *Secchi* (SECK-key) *Disk.* A flat, white disk lowered into the water by a rope until it is just barely visible. At this point, the depth of the disk from the water surface is the recorded Secchi disk transparency.

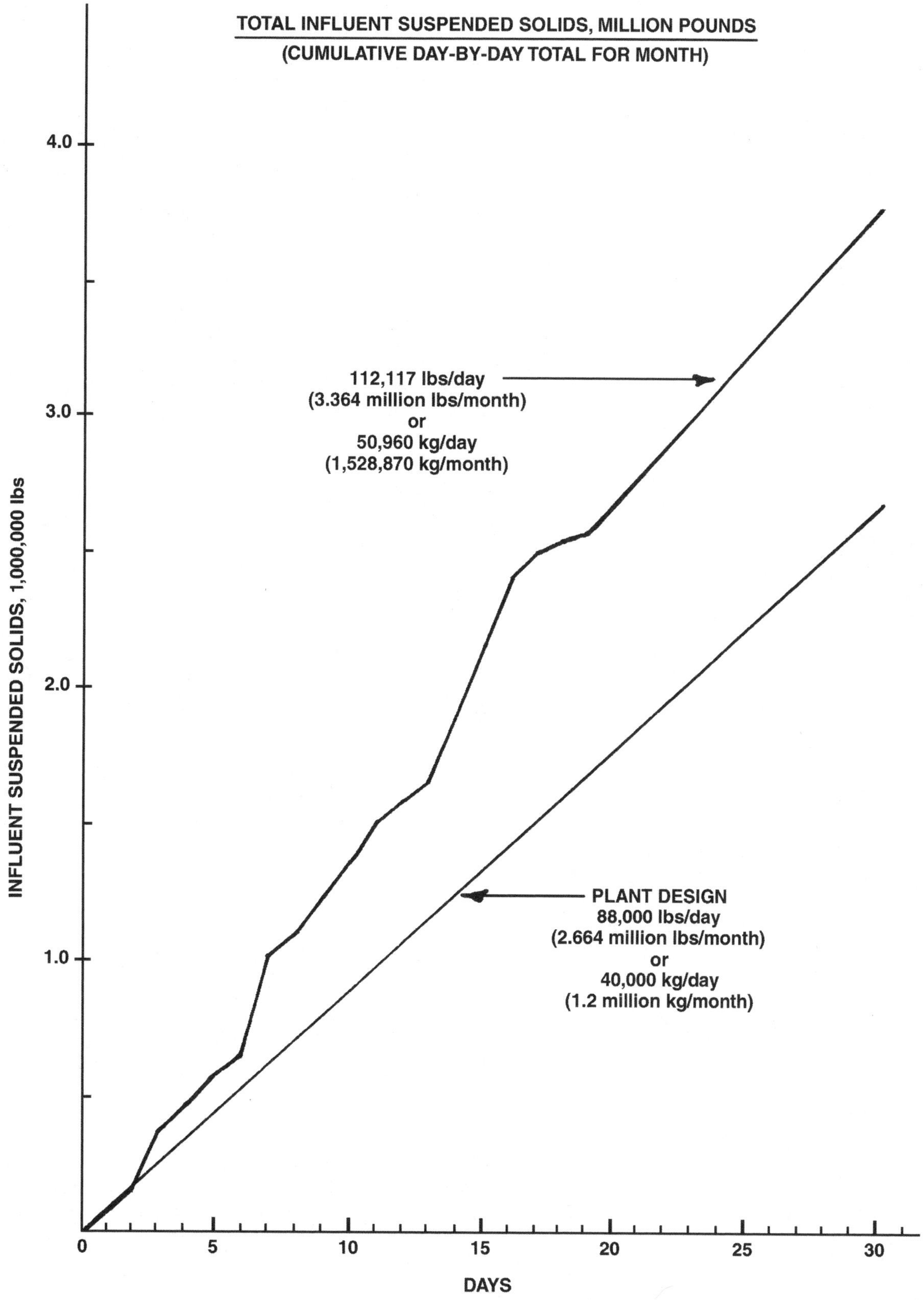

Fig. 11.21 Plant overload during canning season

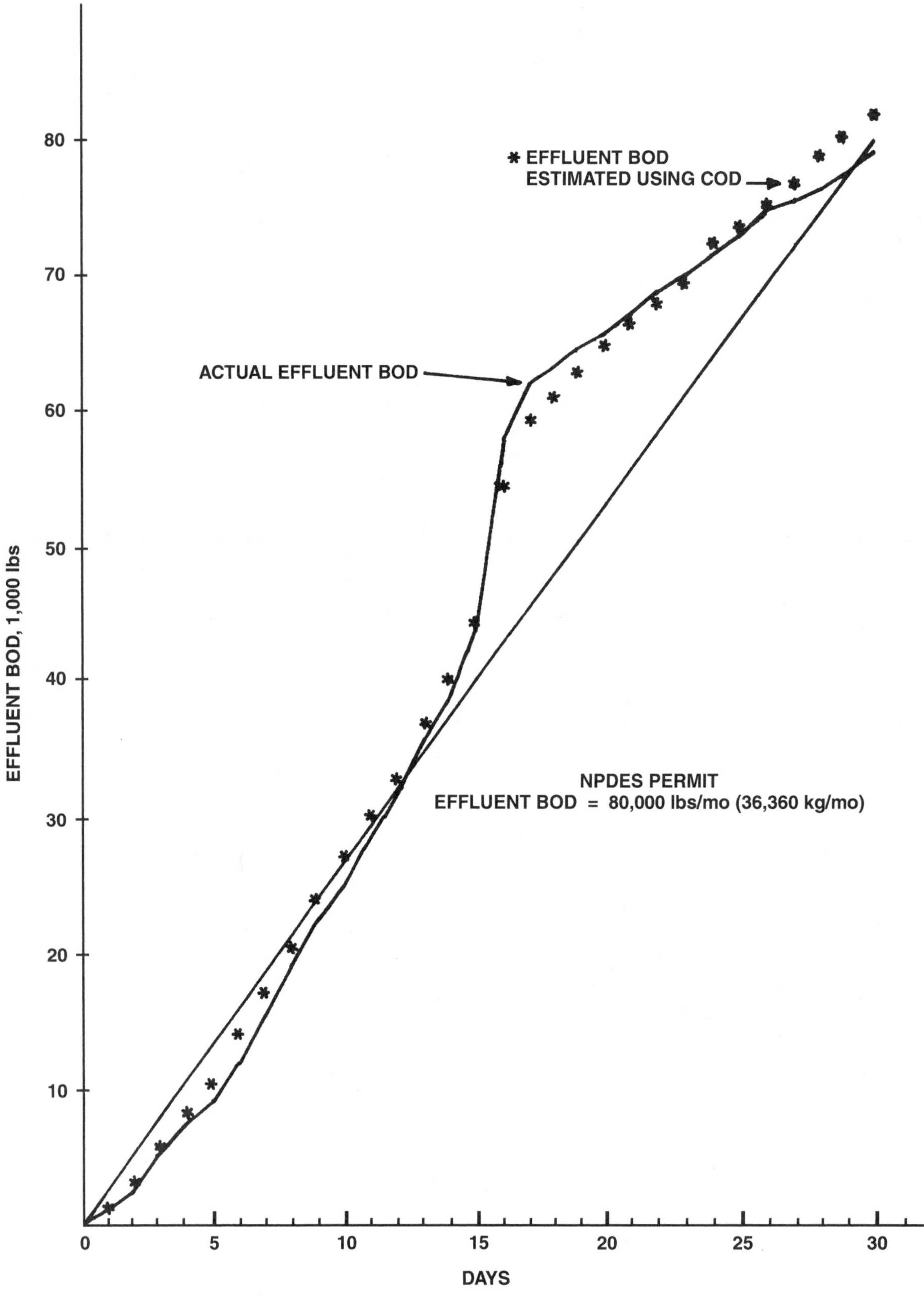

Fig. 11.22 Estimated and actual effluent BOD (cumulative day-by-day total for month)

8. Chlorine Demand
   a. Final clarifier effluent

9. Coliform Group Bacteria[25]
   a. Plant effluent

10. Meter Readings and Calculations
    a. Daily flow
    b. Pounds of solids under aeration
    c. Pounds per day of COD or BOD to aerators
    d. Pounds per day of solids in effluent
    e. Return sludge rate
    f. Waste sludge rate
    g. Air to aerators (diffused air system); hours operated at various speeds (mechanical aeration)
    h. Sludge age (other similar calculations include Food/Microorganism Ratio and Mean Cell Residence Time (see Sections 11.550 and 11.552)
    i. Pounds of solids in sludge to digester
    j. Pounds of solids in digester supernatant
    k. Power cost

11. Daily Observations
    a. Odors
    b. Influent—color and level of inflow
    c. Primary clarifier—scum and color
    d. Aeration tank—turbulence, color, and amount of surface foam and scum
    e. Secondary clarifier—effluent clear or turbid, type of solids on surface, and influent
    f. Return activated sludge—color and odor
    g. Equipment and motors—smooth operation, vibrations, noises, and temperature
    h. Condition of receiving water upstream and downstream from point of discharge

Accurate records will show you have established operating procedures that will produce the best possible effluent. This effluent will be low in COD (or BOD) and suspended solids, and the effluent clarity will be good. Waste loadings and operational procedures will change due to seasonal changes. This requires the operator to constantly review the plant records for changes and to make appropriate changes to maintain the best possible effluent quality. Process control consists not only of maintaining the equipment, but of making a constant daily review of process conditions to determine when adjustments must be made to compensate for the many variables that can influence effluent quality. Remember that the sight, smell, and touch observations often are your first indications that problems are developing and frequently offer indications of appropriate corrective action.

QUESTIONS

Write your answers in a notebook and then compare your answers with those on page 133.

11.5G List the three main types of inspections every shift operator responsible for an activated sludge process should conduct at the start of every shift.

11.5H Why should the strength or waste load of the influent to the activated sludge process be measured by the COD test instead of the BOD test?

11.5I Why should the aerator DO be measured in the aerator instead of the lab?

11.5J Sight and smell observations are often the operator's first indication that process problems are developing. True or False?

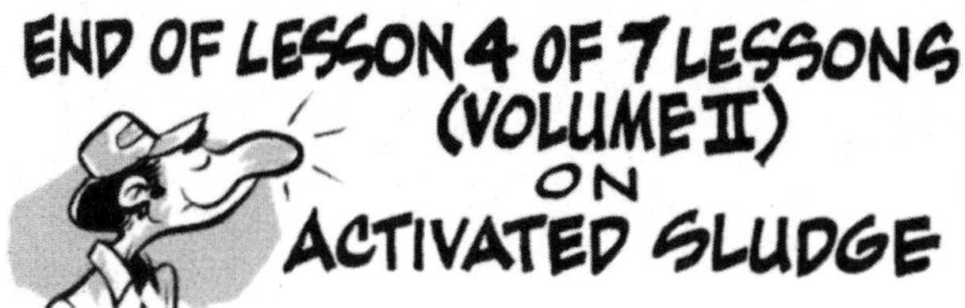

Please answer the discussion and review questions next.

[25] Check with your regulatory agency for test and procedures. Tests approved by agencies include MPN by multiple fermentation tubes, membrane filter, fecal coliform, and *E. coli.*

DISCUSSION AND REVIEW QUESTIONS

Chapter 11. ACTIVATED SLUDGE

(Lesson 4 of 7 Lessons)

Write the answers to these questions in your notebook. The question numbering continues from Lesson 3.

26. What do the letters MLSS stand for?
27. Estimate the pounds of solids under aeration in a 200,000-gallon tank when the suspended solids concentration is 1,600 mg/*L*. Show your work.
28. Calculate the desired pounds of mixed liquor suspended solids in an aeration tank if the primary effluent suspended solids are 1,800 pounds per day and the loading is based on a five-day sludge age. Show your work.
29. Why must some activated sludge be wasted?
30. Calculate the desired mixed liquor suspended solids concentration (mg/*L*) if 4,000 lbs are desired in a 200,000-gallon aeration tank. Show your work.
31. If sludge should be wasted at a rate of 0.05 MGD, what should be the waste pumping rate in GPM?
32. What is the formula for calculating the Mean Cell Residence Time?
33. Calculate the MCRT for an activated sludge process. Plant inflow is 4.0 MGD, MLSS is 2,700 mg/*L*, waste sludge flow is 0.05 MGD, aeration system volume is 2.0 MG, waste sludge suspended solids concentration is 6,500 mg/*L*, and final effluent suspended solids concentration is 15 mg/*L*.
34. What are some of the disadvantages of using the BOD test for operational control?
35. The operator of an activated sludge plant must constantly review plant records and make appropriate changes to account for seasonal changes. True or False?

CHAPTER 11. ACTIVATED SLUDGE

OPERATION OF CONVENTIONAL ACTIVATED SLUDGE PLANTS

(Lesson 5 of 7 Lessons)

11.6 ABNORMAL OPERATION (OPERATIONAL PROBLEMS)

11.60 Typical Problems

An activated sludge plant can accept quite a shock load now and then without adverse effects to the system, but it cannot survive a continuous series of shock loads.

Many factors may change that the operator cannot anticipate or control but must compensate for by adjusting the operational controls. For example, a conventional activated sludge plant has operated satisfactorily for several weeks. The secondary clarifier had good clarity of 68 inches (1.7 m) with a Secchi disc, and the effluent BOD and suspended solids were running from 5 to 18 mg/*L*. The aeration tanks had been maintained at 15,000 pounds (6,000 kg) of mixed liquor suspended solids with a volatile content of 78.5 percent, and sludge age of five days.

A minimum DO of 2.8 mg/*L* had been measured in the last two-thirds of the aerator. Sludge wasting had been at a rate of 2,000 lbs/day (900 kg/day) from the system.

This week the situation has changed; the clarity in the secondary tanks has dropped to 18 inches (0.5 m). The suspended solids in the secondary clarifier effluent have remained about the same, but the BOD test started five days ago came out at 38 mg/*L*. If a COD test had been run at the time the BOD was started, an operational correction could have been made at that time. Overall, the plant effluent has definitely deteriorated from the previous week.

Only you and your records can determine the cause and what corrective action should be taken. Has plant flow increased or decreased? Have air rates been maintained? Have you received some toxic or untreatable slug dose in the influent? Are your sludge return pump and lines clear? Has the BOD load to the aeration tank changed? Have mixed liquor solids been the same? These are just a few of the conditions that may change effluent quality.

The difficult decision after determining the cause or probable cause is—should a change be made? This is where an operator's thorough knowledge of plant processes pays off. If you know the situation is unusual and will only last a couple of days, minor changes may quickly improve the effluent quality. But if the condition occurred before and lasted several weeks according to past records, a process change may be necessary to compensate for it. This is where experience with your plant and records plays an important role in activated sludge operation.

By keeping accurate records, you can find the desirable operating range in terms of efficiency of waste removal and cost of operation. Usually each plant will have some mixed liquor suspended solids concentration where the plant will function best. This concentration should produce a clear final effluent, with low suspended solids and BOD of 8 to 20 mg/*L*. However, depending on plant design, type of waste, and season of year, the best mixed liquor suspended solids concentration might be found to be anywhere from 1,000 to 4,000 mg/*L*. When a satisfactory mixed liquor suspended solids concentration is found for a specific plant under certain conditions, the operator should attempt to maintain this level until something changes.

If the mixed liquor suspended solids are allowed to start building up, the final effluent will begin to deteriorate by becoming turbid. When the mixed liquor suspended solids are allowed to increase too high for the conventional activated sludge plant, other problems can develop. The previous return sludge rate for the plant flow would not be sufficient. Return rates may have to be increased considerably. If the return sludge rate was not increased, the activated sludge in the final clarifiers would build a higher blanket. The deep blanket in the final tank could cause solids to be swept over the weirs during peak flow.

Another limiting factor is aeration equipment. The amount of oxygen supplied to the aerator also limits the microorganism mass that can be maintained in an aerobic state. A high oxygen demand in the aerator can be created by a high solids content in the plant influent.

The other factor is the organisms themselves. If insufficient food is available, only a limited number of organisms will develop energy to multiply. This is where the struggle for survival begins. When food supply is low, the microorganisms begin to feed upon themselves (*ENDOGENOUS RESPIRATION*[26]). This is the period of most complete oxidation, and new sludge production is at a minimum. Extended aeration plants are designed to operate under these conditions that tend to increase solids in the plant effluent.

[26] *Endogenous* (en-DODGE-en-us) *Respiration.* A situation in which living organisms oxidize some of their own cellular mass instead of new organic matter they adsorb or absorb from their environment.

QUESTIONS

Write your answers in a notebook and then compare your answers with those on page 133.

11.6A Why are plant records of the activated sludge operation important?

11.6B What could happen if the mixed liquor suspended solids were allowed to increase beyond the best range in an activated sludge process?

11.61 Plant Changes

If the plant becomes upset, the first action before making any changes is to check the plant data for at least three previous weeks. The problem probably started last week or earlier. To look for the cause of the problem, ask yourself the following questions:

1. Have any changes been made to other plant units such as the digesters or primary clarifiers? Was a digester supernatant with excessive amounts of solids returned to the primary clarifiers? Return of supernatant should be slow and easy at low-load periods. Digester supernatant solids mixed with raw wastewater and waste activated sludge may create a light sludge that may be washed out of the primary clarifier. The solids washed out of the primary clarifier create undesirable recycle and loading problems.
2. Have plant daily flows and waste concentrations increased or decreased? Heavy rains following a dry spell, a new industrial plant, or a different process discharge from an existing industry can cause problems.
3. Has temperature of the influent changed a significant amount?
4. Has the sampling program been consistent?

Most of the time a plant upset is due to some in-plant problem and not the influent raw wastewater, unless your plant is frequently overloaded.

CONDITION 1: (High Solids in Digester Supernatant)

A high solids content in digester supernatant can throw a curve to the operator. The solids in digester supernatants are usually high in immediate oxygen demand and contain high colloidal and dissolved solids. If a large quantity of low volatile content solids escapes from the primary clarifier to the aeration system, several undesirable events occur. The supernatant solids are picked up by the activated sludge in the aerator and carried through the system. This creates extra oxygen demand, and air output must be increased. Digester solids make a good settling activated sludge, but the color of the floc will be darker. Total pounds of solids in the aerator will increase due to the supernatant solids, and the operator will normally increase waste sludge rates to hold the established range of solids or sludge age. Consequently, the effluent from the plant deteriorates. Why? Lab tests show that the solids in the aerator are at the desired level, and DO of the mixed liquor has been held at 2.0 mg/*L* (probably more air was required).

What has occurred is that by wasting apparently excess activated sludge, many of the microorganisms have been replaced in the aerator by digested or inert solids. They are sampled the same as mixed liquor suspended solids and are included as total pounds of solids under aeration. This is why many plants base aerator loadings on mixed liquor *volatile* suspended solids and not mixed liquor suspended solids. They are making the assumption that the volatile content of the mixed liquor suspended solids represents microorganisms.

Most activated sludge mixed liquor suspended solids fall into a range of 70 to 80 percent volatile content for municipal waste when the process is operating properly. This would mean that if you are striving to maintain a sludge age of five days, you are attempting to maintain a prescribed number of organisms for every pound of food applied to the aeration tanks. A five-day sludge age is equivalent to 20 pounds of food for every 100 pounds of organisms. When the supernatant was admitted into the aeration tank, the pounds of solids in the aerator was increased. When sludge was wasted to maintain the five-day sludge age, many of the organisms needed to treat the incoming wastes were replaced by the inert supernatant solids. This placed a higher food load on the remaining organisms of maybe 30 to 35 pounds per 100 pounds of organisms and reduced the effective sludge age from five days to possibly 3.5 days.

Storm flow may sweep excessive silt into the plant by infiltration into sewers or through combined sewer systems. Solids increase drastically but the percent of volatile solids may drop to 50 percent of the total solids. If you measure only total solids in this type of situation, it will appear that you have excess solids that need to be removed. However, wasting solids at this point may produce serious organism losses. If possible, change the plant to step-feed aeration to retain more mixed liquor solids in the aeration tank.

In the example plant, the supernatant solids may not appear to produce much of a change, but over a period of several days the system could become severely upset. When the total pounds of volatile solids in the aerator becomes too low due to excessive amounts of inert solids from digester supernatant or storm inflow, the solution to the problem is to reduce or stop wasting sludge for several days. This will provide the time necessary to rebuild the microorganism population to handle the incoming waste load. Another solution is to draw sludge from the digester to sludge handling facilities to reduce the supernatant solids load to the plant.

Try to hold solids in the digester a little longer and try to increase solids concentration in the sludge fed to the digester. It is possible that the poor supernatant was due to overloading the digester or insufficient seed sludge in the digester. In this case, the problem snowballs—first the digester is overloaded, then supernatant solids overload the aerator, which overloads the clarifier, and the problem keeps getting worse.

CONDITION 2: (Flow or Waste Changes)

Always be alert for the possibility of toxic dumps, accidental spills (particularly the midnight variety), storms, or other upsewer factors that may change the influent flow or waste characteristics.

A frequent problem is the increased flows from storm infiltration or other sources. These flows may create shorter aeration times or loss of activated sludge solids from the final clarifiers due to a hydraulic overload. To compensate for this condition, regulate return and waste sludge rates to hold as much of the solids as possible in the aerator. Or change plant operation to step-feed or contact stabilization mode of operation (see Sections 11.81, "Contact Stabilization" and 11.83, "Step-Feed Aeration").

Changes in waste characteristics may be caused by isolated dumps or spills, or changes may be seasonal. Become acquainted with plant managers whose activities may cause changes in the waste loadings on your plant and encourage these people to notify you whenever a problem discharge occurs. Try to convince them to release unusual dumps at a low discharge rate rather than all at once. Certain industries such as canneries create seasonal problems, which the operator should prepare for in advance.

CONDITION 3: (Temperature Changes)

The activated sludge system is influenced by temperature changes similar to the response of trickling filters to temperature changes in spring and fall. During the summer, the activated sludge plant may operate satisfactorily in a certain loading range and air rates, but in winter the best loading ranges and air rates change and the plant requires less air and more solids under aeration. Usually a temperature change is not significant unless it raises or lowers the wastewater temperature more than 10°F (6°C).

Temperature is an important factor in oxidation relative to sludge accumulation. A high temperature produces a rapid microorganism growth rate and more waste storage in the organism cell with less oxidation. Therefore, greater biological activity will result in more overall sludge production but the sludge may be thinner than usual.

During the colder winter months, operators increase the solids under aeration (MLSS) to provide more microorganisms to treat the biochemical oxygen demand. When the weather warms in the spring and summer, the microorganisms become more active. If there is poor settling of the activated sludge in the secondary clarifier, try increasing the wasting rate (by no more than ten percent per day) until you see an increase in settling and improved effluent quality.

CONDITION 4: (Changes in Sampling Program)

Data on system performance can be greatly affected by changes in a sampling program. If improper sampling locations or laboratory procedures are used, lab results could vary considerably. When the lab data varies widely from one day to the next, check sampling location, time, and lab procedures for errors.

When considering a major process change, first review the plant data. Next, make only one major change at a time. If two changes are made, you will not know whether one or both changes provided the corrective action. Allow one week for a plant to stabilize after a process change. An experienced operator who knows the plant may be able to determine if the proper changes have been made after several days, but some plants require up to two weeks to stabilize after a change.

QUESTIONS

Write your answers in a notebook and then compare your answers with those on page 133.

11.6C What is the first action that should be taken by an operator when the plant becomes upset?

11.6D What can happen when a large quantity of low volatile content solids from a digester supernatant escapes from the primary clarifier to the aeration system?

11.6E How would you correct or compensate for an upset created by high flows from stormwater infiltration?

11.6F How would you correct an upset apparently caused by an increase in temperature due to seasonal changes?

11.6G What would you do if a review of the lab data revealed considerable variation from day to day?

11.6H What would you do if an activated sludge plant became upset?

11.6I How long would you allow an activated sludge process to react and stabilize after a change?

11.62 Sludge Bulking[27]

Bulking is the term applied to the condition in which the mixed liquor solids tend to show a very slow settling rate and compact to a limited extent. The liquid that does separate from the solids usually has excellent clarity but generally there is not enough time for complete removal of the solids in the secondary clarifier. The sludge blanket in the clarifier becomes deeper and rises to overflow the weirs and is discharged with the effluent.

Bulking may be associated with production of a highly jellylike, waterlogged (hydrated) sludge that has a very low sludge density (*BOUND WATER*[28]). At other times, filamentous organisms may grow from one floc mass to another and act as stay rods to prevent compaction of the sludge particles and produce poor settling results.

Low pH, low DO, and low nutrient concentrations have been related to bulking. High food-to-microorganism loading rates (low sludge ages) are the major cause of repeated bulking. Organisms that grow rapidly tend to spread out and will not clump or form a floc mass until growth rates decrease. It is difficult to retain enough low-density (light) sludge to decrease the food-to-microorganism load ratio (or increase sludge age) without chemical flocculation or other techniques to increase the sludge density (weight). Rain may provide enough silt to favor increased sludge density. Low loads during weekends may help. Adding some preaerated digested sludge (Kraus process, Section 11.82) helps reduce bulking by the addition of nitrate, and the heavy solids in the digested sludge improve settleability. Some of the *POLYELECTROLYTE*[29] *FLOCCULANT*[30] aids

[27] *Bulking.* Clouds of billowing sludge that occur throughout secondary clarifiers and sludge thickeners when the sludge does not settle properly. In the activated sludge process, bulking is usually caused by filamentous bacteria or bound water.

[28] *Bound Water.* Water contained within the cell mass of sludges or strongly held on the surface of colloidal particles. One of the causes of bulking sludge in the activated sludge process.

[29] *Polyelectrolyte* (POLY-ee-LECK-tro-lite). A high-molecular-weight (relatively heavy) substance, having points of positive or negative electrical charges, that is formed by either natural or synthetic (manmade) processes. Natural polyelectrolytes may be of biological origin or obtained from starch products or cellulose derivatives. Synthetic polyelectrolytes consist of simple substances that have been made into complex, high-molecular-weight substances. Used with other chemical coagulants to aid in binding small suspended particles to larger chemical flocs for their removal from water. Often called a POLYMER.

[30] *Flocculation* (flock-yoo-LAY-shun). The gathering together of fine particles after coagulation to form larger particles by a process of gentle mixing. This clumping together makes it easier to separate the solids from the water by settling, skimming, draining, or filtering.

are very effective in controlling a bulking activated sludge. If it is possible, bulking may be reduced by decreasing the load to the aeration tanks until the sludge becomes sufficiently oxidized to flocculate. Addition of clay or bentonite has been used to control bulking.

The main objective of most bulking control procedures is to increase sludge age or decrease the ratio of waste (food) load added per day per unit of mixed liquor volatile solids in the aerator. Some good methods for holding solids under aeration are the addition of aluminum sulfate ($Al_2 (SO_4)_3 \bullet 14\ H_2O$) or ferric (iron) chloride ($FeCl_3$). Also, ferric sulfate ($Fe_2 (SO_4) \bullet 3\ H_2O$) may be used as a flocculant with alkaline (lime) addition to prevent the alkalinity from dropping below 50 to 100 mg/*L* as $CaCO_3$. The proper polyelectrolyte may cost more than other chemicals, but alkali addition may not be required to increase the alkalinity.

Chlorination of return sludge has been practiced quite extensively as a means of controlling bulking caused by filamentous growths. It has been ineffective when bulking is due to bound water. Chlorination of return sludge may be based on the dry solids content of the return sludge and a reasonable range is between 0.2 and 1.0 percent by weight. Therefore, the chlorine dose in pounds per day is equal to 0.002 to 0.010 times the return sludge dry solids in pounds per day. Effluent turbidity may increase until such time as the sludge is freed of the filamentous forms. The death of nitrifying bacteria from chlorine may also contribute to turbidity. Bulking is likely to return unless the cause is identified and permanently corrected.

When bulking occurs it will generally be associated with the load ratio or sludge age. Plant records should be reviewed in an attempt to locate the cause of the problem. Identification of the cause will not remedy the present bulking condition, but should be considered a valuable lesson, and measures should be taken to prevent the same conditions from occurring again.

To prevent sludge bulking from occurring, the following items should be carefully controlled in an activated sludge plant:

1. *SLUDGE AGE.* Carefully review plant records and maintain a sludge age that produces the best quality effluent. Watch influent solids loadings, maintain desired level of solids in the aerator, and carefully regulate waste sludge rates. Generally, bulking may be cured by increasing the sludge age.
2. *DO LEVEL.* Prevent low levels of DO from developing. Mixed liquor DO tests are a quick and simple test, or a DO probe installed in the aeration tank wall will give you a continuous reading. There is no valid excuse for low DO concentrations during normal conditions if sufficient oxygenation capacity is available, unless a slug of waste with an excessive oxygen demand is received.
3. *LENGTH OF AERATION PERIOD.* Bulking caused by the aeration period being too short is usually the result of a design problem, unless the operator has formed the habit of returning too high a volume of return sludge. To correct this problem, reduce the return sludge rate and thicken the return sludge solids concentration by *COAGULATION*[31] (if necessary). In this way, you still return the same number of organisms to meet the new food (waste) entering the aerator, but effectively reduce the total flow through the aerator and clarifier.
4. *FILAMENTOUS GROWTH.* The growth of filamentous organisms may be caused by incorrect sludge age or nutritional imbalances, such as a shortage or abundance of nitrogen, phosphorus, or carbon. If filamentous growths are allowed to become well established, they create a difficult problem to overcome. Control may be achieved by increasing MLSS (more microorganisms, which will increase sludge age), by maintaining higher DOs and, in special instances, supplementing a nutrient deficiency. Control by chlorination of return sludge, as discussed earlier, may also be used.

11.63 Septic Sludge

Septic sludge may be produced when any type of sludge remains too long in such places as hoppers and channels. It is likely to cause a foul odor, rises slowly, and sometimes rises in clumps. Even small amounts can upset an aerator.

Septic sludge may occur in poorly designed or constructed hoppers, wet wells, channels, or pipe systems. This occurs when activated sludge is allowed to be deposited and anaerobic decomposition starts. Septic sludge deposits also may develop on the floor of the aerator due to insufficient air rates that are not keeping the tank completely mixed. A high solids load also can cause septic problems.

To effectively control septic sludge, aerators must be thoroughly mixed and sludge must be pumped frequently. In channels and pipelines, a velocity over 1.5 feet per second (0.45 m/sec) will prevent the formation of sludge deposits that could become septic.

Sludge going septic in the secondary clarifier may develop from four causes:

1. Return sludge rate too low, thus holding the solids in the final clarifier too long and allowing them to become septic
2. Clarifier collection mechanism turned off, thus the sludge is not being moved to the draw-off hopper
3. Sludge draw-off lines plugged, obstructed, or used infrequently
4. Return sludge pump off or a valve closed

A good operator checks the system several times a day. In most new activated sludge plants the secondary clarifiers have air lift samplers or photocells to indicate sludge blanket level in the tank. Whenever the final clarifier sludge blanket level changes, an immediate investigation should be undertaken. In any of the cases above, the correction is quite obvious—restore suitable return sludge flow as soon as possible.

11.64 Toxic Substance

Toxicity causes a severe slowdown or death of working organisms and produces system and effluent upsets. The operator has limited control over toxic waste entering the plant. When this happens, however, sludge wasting should be

[31] *Coagulation* (ko-agg-yoo-LAY-shun). The clumping together of very fine particles into larger particles (floc) caused by the use of chemicals (coagulants). The chemicals neutralize the electrical charges of the fine particles, allowing them to come closer and form larger clumps.

stopped immediately and all available solids returned to the aerator. Toxic materials such as heavy metals, acids, insecticides, and pesticides should never be dumped into a sewer system without proper control.

11.65 Rising Sludge[32]

Rising sludge is not to be confused with bulking. The sludge settles and compacts satisfactorily on the bottom of the clarifier, but after settling it rises to the top of the secondary tank in patches or small particles the size of a pea. Rising sludge usually produces a fine scum or froth (brown in color) on the surface of the aeration and secondary tanks.

Rising sludge is caused by *DENITRIFICATION*[33] or *SEPTICITY*[34] and results from too long a detention time in the secondary clarifiers. The secondary clarifiers should be equipped with scum baffles and skimmers to prevent these solids from escaping in the plant effluent.

Denitrification is most common when the sludge age is high (extended aeration). When this type of activated sludge flows from the aerator to the secondary clarifier or becomes short of oxygen, the organisms first use the available dissolved oxygen, and then the oxygen in the nitrate compounds resulting in the release of nitrogen gas. Denitrification is an indication of good treatment, providing the sludge in the settleability test stays on the bottom of the cylinder for at least one hour, but floats to the surface in two hours. If it floats up too early in the settleability test, the sludge age should be reduced or the food-to-microorganism ratio should be increased. This solution will be successful if the nitrifying bacteria are washed out of the system. If the sludge stays down for an hour in the settleability test but problems are still present in the secondary clarifier, increase the return sludge rates to move the solids out of the clarifier at a faster rate. Under some circumstances this will not help and better results might be obtained by decreasing the return sludge rate, so be careful.

Rising sludge also may be controlled by increasing the load to the aerator by removing a primary clarifier from service if more than one is being used. During low flow periods, raw wastewater may be discharged directly to the aerator. Another option is to check the possibility of installing aeration diffusers in a tapered pattern. The tapered pattern is used to meet higher initial oxygen demands.

11.66 Foaming/Frothing

Aerator foaming or frothing has been a problem for some plants. There have been many theories presented on the cause, such as surfactants (detergents), polysaccharides, and overaeration. Whatever the cause, there is a definite relationship between froth buildup on the aerator and the amount of suspended solids in the mixed liquor and air supply to the aerator.

Operators may have to control different types of foam. The foam may be unstable and easy to control or the foam could be persistent and difficult to control.

Unstable foam may be caused by nutrient deficiencies or solids from dewatering processes (recycled solids). Polymer overdosing can be a cause of foam. Floating sludges and floating scum also are types of foams. These unstable foams are usually kept down using water sprays.

Persistent foams are often called filamentous or *Nocardia* foam and are difficult to control. These foams are brown, stable, viscous, and usually scum-like in appearance. These foams are often associated with high MCRT values. The higher the concentration of filaments, the greater the tendency for foaming. Also the higher the MLSS concentration, the more susceptible an aeration basin is to foaming. The aeration rate directly influences foaming and the height of foam.

Filamentous growth rates tend to increase with temperature. Many plants have experienced foaming problems during seasonal temperature changes in the spring or fall. Apparently, the optimum pH for filamentous growth is around a pH of 6.5.

For control of foaming:

1. Maintain higher mixed liquor suspended solids concentrations.
2. Reduce air supply during periods of low flow while still maintaining DO.
3. Return supernatant to the aeration tank during low flows (be cautious in this method—supernatant should be returned slowly and steadily because too much supernatant could cause an excess oxygen demand).

These solutions apply only to detergent foam. In some extended aeration systems or nitrification systems (see *ADVANCED WASTE TREATMENT* manual, Chapter 2, "Activated Sludge"), a froth builds up that can sometimes be controlled by higher sludge wasting rates (reduction of mixed liquor suspended solids, MLSS).

Frothing from filamentous organisms is usually present in aeration basins. When the number of filaments becomes excessive, the organisms may form a thick, dark brown scum or froth on the surface of the aeration basin. Microscopic examination of the foamy scum can confirm the presence of filamentous organisms (see Figure 11.45 on page 118).

---

[32] *Rising Sludge.* Rising sludge occurs in the secondary clarifiers of activated sludge plants when the sludge settles to the bottom of the clarifier, is compacted, and then starts to rise to the surface, usually as a result of denitrification, or anaerobic biological activity that produces carbon dioxide and/or methane.

[33] *Denitrification* (dee-NYE-truh-fuh-KAY-shun). (1) The anoxic biological reduction of nitrate nitrogen to nitrogen gas. (2) The removal of some nitrogen from a system. (3) An anoxic process that occurs when nitrite or nitrate ions are reduced to nitrogen gas and nitrogen bubbles are formed as a result of this process. The bubbles attach to the biological floc and float the floc to the surface of the secondary clarifiers. This condition is often the cause of rising sludge observed in secondary clarifiers or gravity thickeners. Also see NITRIFICATION.

[34] *Septic* (SEP-tick) or *Septicity.* A condition produced by bacteria when all oxygen supplies are depleted. If severe, the bottom deposits produce hydrogen sulfide, the deposits and water turn black, give off foul odors, and the water has a greatly increased oxygen and chlorine demand.

Filamentous foam has been controlled by MCRT control, RAS/MLSS chlorination, direct foam chlorination, selective foam wasting, use of water sprays, and selector technology. An MCRT of less than six days has been effective. MCRT can be reduced by slowly increasing the wasting rate with care to remain in compliance. Chlorination of RAS/MLSS or return activated sludge or both has been effective in controlling filamentous foam. If a stable foam has already formed, direct foam chlorination is the most effective method of killing foam-forming microorganisms in the foam. A foam trap is installed in the mixed liquor effluent or aeration tank and a highly concentrated chlorine spray is applied directly to the foam-forming microorganisms. Periodic chlorination of return sludge or MLSS can be useful as a preventive measure to control the number of filaments below the foaming threshold.

Aerobic, *ANOXIC,*[35] and anaerobic selectors have been used to prevent the growth of filamentous foam microorganisms by creating an environment in which they are at a competitive disadvantage to non-foam-forming organisms. Selectors are small reactors usually immediately upstream of the aeration tank. Aerobic selectors operate at high F/M ratios, at

11.67 Process Troubleshooting Guide (adapted from *PERFORMANCE EVALUATION AND TROUBLESHOOTING AT MUNICIPAL WASTEWATER TREATMENT FACILITIES,* Office of Water Program Operations, US EPA, Washington, DC)

INDICATOR/OBSERVATION	PROBABLE CAUSE	CHECK OR MONITOR	SOLUTION
1. Sludge floating to surface of secondary clarifiers.	1a. Filamentous organisms predominating in mixed liquor ("bulking sludge").	1a. SVI—if less than 100, 1(a) is not likely cause; microscopic examination can be used to determine presence of filamentous organisms.	1a. (1) Increase DO in aeration tank if less than 1.5 mg/*L* at effluent end of aerator. (2) Increase MCRT to greater than 6 days. (3) Increase sludge return rate and reduce or stop wasting. (4) Supplement deficiency of nutrients so that BOD to nutrient ratio is no more than 100 mg/*L* BOD to 5 mg/*L* total nitrogen, 1 mg/*L* phosphorus, and 0.5 mg/*L* iron. (5) Add 5-10 mg/*L* of chlorine to return sludge until SVI <150 (should be controlled within 2-3 days). Microscopically examine sludge to avoid destruction of beneficial organisms during chlorine application. (6) Increase pH to 7. (7) Add 50-200 mg/*L* of hydrogen peroxide to aeration tank until SVI <150.
	1b. Denitrification occurring in secondary clarifiers; nitrogen gas bubbles attaching to sludge particles; sludge rising in clumps.	1b. Nitrate concentration in clarifier influent; if no measurable NO_3^-, then 1(b) is not the cause.	1b. (1) Increase sludge return rate (will increase clarifier hydraulic load and reduce detention time). (2) Increase DO in aeration tank. (3) Reduce MCRT. (4) Reduce flow to offending unit if sludge return rate cannot be effectively increased.
2. Pin floc in secondary clarifier overflow—SVI is good but effluent is turbid.	2a. Excessive turbulence in aeration tanks.	2a. DO in aeration tank.	2a. Reduce aeration agitation (reduce blower CFM output or depth of submergence and RPM of mechanical aerator).
	2b. Overoxidized sludge.	2b. Sludge appearance.	2b. Increase sludge wasting to decrease MCRT.
	2c. Anaerobic conditions in aeration tank.	2c. DO in aeration tank.	2c. Increase DO in aeration tank to at least 1.0 to 1.5 mg/*L* in aerator effluent.
	2d. Toxic shock load.	2d. Microscopically examine sludge for inactive protozoa.	2d. (1) Re-seed sludge with sludge from another plant if possible; enforce industrial waste ordinances. (2) Stop wasting. (3) Return rate as high as possible to reestablish culture.

[35] *Anoxic* (an-OX-ick). A condition in which the aquatic (water) environment does not contain dissolved oxygen (DO), which is called an oxygen deficient condition. Generally refers to an environment in which chemically bound oxygen, such as in nitrate, is present. The term is similar to ANAEROBIC.

11.67 Process Troubleshooting Guide *(continued)* (adapted from *PERFORMANCE EVALUATION AND TROUBLESHOOTING AT MUNICIPAL WASTEWATER TREATMENT FACILITIES,* Office of Water Program Operations, US EPA, Washington, DC)

INDICATOR/OBSERVATION	PROBABLE CAUSE	CHECK OR MONITOR	SOLUTION
3. Very stable dark tan foam on aeration tanks which sprays cannot break up. *NOTE:* If not causing a problem, do nothing.	3a. MCRT is too long.	3a. If MCRT greater than 9 days, this is probable cause.	3a. Increase sludge wasting to reduce MCRT. Increases should be at a modest rate and trends watched carefully.
4. Thick billows of white sudsy foam on aeration tank.	4a. MLSS too low.	4a. MLSS.	4a. Decrease sludge wasting to increase MLSS and MCRT.
	4b. Presence of a non-biodegradable surface active material.	4b. If MLSS level is appropriate, surfactants are probable cause.	4b. Monitor industrial discharges.
5. Aerator contents turn dark—sludge blanket lost in secondary clarifier.	5a. Inadequate aeration, dead zones, and septic sludge.	5a. Aeration basin DO and aeration rates (blower CFM output or mechanical aerator speed and depth).	5a. (1) Increase aeration by placing another blower in service. (2) Decrease loading by placing another aeration basin in service. (3) Check aeration system piping for leaks. (4) Clean any plugged diffusers or add more diffusers if possible. (5) Increase blower CFM output or mechanical aerator speed and depth.
6. MLSS concentrations differ substantially from one aeration basin to another.	6a. Unequal flow distribution to aeration tanks.	6a. Flow to each basin.	6a. Adjust valves and/or inlet gates to equally distribute flow.
	6b. Return sludge distribution unequal to aeration basins.	6b. RAS flow to each basin.	6b. Check return sludge flows and discharge points.
7. Sludge blanket overflowing secondary clarifier weirs uniformly.	7a. Inadequate rate of sludge return.	7a. Sludge return pump output.	7a. (1) If return pump is malfunctioning, place another pump in service and repair. (2) If pump is in good condition increase rate of return and monitor sludge blanket depth routinely. Maintain 1-3 foot (0.3-0.9 m) deep blanket. When blanket increases in depth, increase rate of return. (3) Clean sludge return line if plugged.
	7b. Unequal flow distribution to clarifiers causing hydraulic overload.	7b. Flow to each clarifier.	7b. Adjust valves and/or inlet gates to equally distribute flow.
	7c. Peak flows are overloading clarifiers.	7c. Hydraulic overflow rates at peak flows if >1,000 GPD/sq ft, this is a likely cause.	7c. (1) Install flow equalization facilities or expand plant. (2) Switch aeration mode to step-feed.
	7d. Solids loadings are too high on clarifier.	7d. Loadings should not exceed 1.25 lbs/sq ft/hr.	7d. Reduce MCRT if plant design does not allow step-feed.
8. Sludge blanket overflowing secondary clarifier weirs in one portion of clarifier.	8a. Unequal flow distribution in clarifier.	8a. Effluent weir.	8a. (1) Level effluent weirs. (2) Dye test clarifier for short circuiting and install baffles or deflectors if a problem.
9. In diffused aeration basin, air rising in very large bubbles or clumps in some areas.	9a. Diffusers plugged or broken.	9a. Visually inspect diffusers.	9a. Clean or replace diffusers; check air supply; install air cleaners ahead of blowers to reduce plugging from dirty air.
10. pH of mixed liquor decreases to 6.7 or lower. Sludge becomes less dense.	10a. Nitrification occurring and wastewater alkalinity is low.	10a. Effluent NH_3; influent and effluent alkalinity.	10a. (1) Decrease sludge age by increased wasting if nitrification not required. (2) Add source of alkalinity—lime or sodium bicarbonate.
	10b. Acid wastewater entering system.	10b. Influent pH.	10b. Determine source and stop flow into system.

11.67 Process Troubleshooting Guide *(continued)* (adapted from *PERFORMANCE EVALUATION AND TROUBLESHOOTING AT MUNICIPAL WASTEWATER TREATMENT FACILITIES,* Office of Water Program Operations, US EPA, Washington, DC)

INDICATOR/OBSERVATION	PROBABLE CAUSE	CHECK OR MONITOR	SOLUTION
11. Sludge concentration in return sludge is low (<8,000 mg/*L*).	11a. Sludge return rate too high.	11a. Return sludge concentration, solids level (balance) around final clarifier, settleability test.	11a. Reduce sludge return rate.
	11b. Filamentous growth.	11b. Microscopic examination, DO, pH, nitrogen concentration.	11b. Raise DO, raise pH, supplement nitrogen, add chlorine (see item 1).
	11c. *Actinomycetes* predominate.	11c. Microscopic examination, dissolved iron content.	11c. Supplement iron feed if dissolved iron less than 5 mg/*L*.
	11d. Collector mechanism speed inadequate.	11d. Collector mechanism.	11d. Adjust speed as appropriate/turn collector on.
12. Dead spots in aeration tank.	12a. Diffusers plugged.	12a. Visually inspect diffusers.	12a. Clean or replace diffusers—check air supply—install air cleaners ahead of blowers to reduce plugging from dirty air.
	12b. Underaeration resulting in low DO.	12b. Check DO.	12b. Increase rate of aeration to bring DO concentration up to 1 to 3 mg/*L*.
	12c. Air supply valves improperly adjusted.	12c. Air supply valve settings.	12c. Adjust valving as appropriate.
	DIFFUSED AERATION		
	12d. Inadequate number of diffusers.	12d. (1) Check aeration rate CFM. (2) Check air back pressures in manifolds and headers.	12d. (1) Increase output CFM. (2) Add diffusers to air header to reduce back pressure on supply and allow more air to aeration tank.
	MECHANICAL AERATION		
	12e. Rotor or turbine speed too low.	12e. Speed of rotors or turbines.	12e. Increase speed of rotors or turbines.
	12f. Aeration basin water level too low.	12f. Water level in basin.	12f. Increase water level in basin.
	12g. Insufficient energy input.	12g. Dye test basin.	12g. Add rotors or turbines.

dissolved oxygen levels between 2 and 5 mg/*L* and at an MCRT of less than 5 days. Anoxic selectors contain no dissolved oxygen (less than 0.5 mg/*L*), contain recycled nitrate nitrogen, and operate at high MCRTs.

Most plants are equipped with water sprays along the aerator to dissipate the foam. If mixed liquor solids are allowed to be reduced, low water sprays will not be sufficient to hold the foam. When this occurs, two problems develop—maintenance and safety.

The froth from an aerator is an excellent vehicle for minute grease particles and, when deposited on Y-walls or walks, will leave a grease deposit that is very slippery. More than one operator has been injured by slipping on a walk or step previously coated with foam.

This deposit not only is unsafe, but unsightly, and it must be cleaned up immediately. The best way to remove this type of deposit is with water (preferably hot), trisodium phosphate (TSP), and a stiff bristle deck brush. Wet the area to be cleaned, lightly sprinkle TSP granules on the area, let the TSP dissolve for a few minutes, and then brush the area to spread the TSP and loosen the grease. Let it work for five minutes, rebrush, and then hose off.

QUESTIONS

Write your answers in a notebook and then compare your answers with those on page 133.

11.6J How would you tell the difference between bulking and rising sludge?

11.6K What would you do to correct a rising sludge problem?

11.6L How can foam be controlled in an aerator?

11.6M How can grease deposits (from froth) on walks be removed?

11.7 EQUIPMENT SHUTDOWN, ABNORMAL OPERATION, AND MAINTENANCE

11.70 Need for Understanding Equipment

Always read and understand manufacturers' literature before starting, operating, maintaining, or shutting down equipment. You may be very capable of operating the activated sludge process, but if your equipment does not perform, life can be very difficult. Proper shutdown procedures may be overlooked in manufacturers' literature and O & M manuals. If proper shutdown procedures are not followed, equipment can be damaged and not start again properly.

This section will identify important steps to follow when shutting down aeration system equipment, when attempting to handle abnormal conditions or troubleshooting problems, and when maintaining equipment. Remember that these steps are general and you must prepare your own detailed list if one is not available. There are many equipment manufacturers in business today and they are continually improving their products; therefore the lists in this section are presented to guide you in the preparation of your own lists for the equipment in your plant.

11.71 Shutdown

11.710 Surface Aerators

Aerator shutdown is required when any maintenance service is performed to prevent injury to maintenance personnel and possible damage to equipment.

1. Turn the ON-OFF-AUTO switch to OFF.
2. Turn the main power breaker to OFF.
3. Lock out and tag the main power breaker in the OFF position.

Maintenance service may be performed now. *REMEMBER*, wear an approved flotation device or fall arrest system, depending on the depth of water in the aeration basin.

11.711 Positive Displacement Blowers

Blower shutdown is required when any maintenance service is performed in order to prevent injury to maintenance personnel and possible damage to equipment.

1. Turn the ON-OFF-AUTO or ON-OFF switch to OFF.
2. Turn the main power breaker to OFF.
3. Lock out and tag the main power breaker in the OFF position.
4. Close the discharge and suction valves.

   When the blower is shut down, ensure that the check valve closes and seats. Many blowers run backward for a moment or two after shutdown and this reverse operation creates a vacuum condition that will pull liquid up the headers and into the air distribution system.

Maintenance service may be performed now.

11.712 Centrifugal Blowers

Blower shutdown is *REQUIRED* when any maintenance service is performed to prevent injury to maintenance personnel and possible damage to equipment.

1. Depress the stop button.
2. Open the bypass valve.
3. Close the discharge valve.
4. Let the auxiliary oil pump run for a 10-minute, post-lubrication period. This allows the bearings and gears to cool gradually.
5. Turn the control panel power to OFF.
6. Turn the main power breakers to OFF.
7. Lock out and tag the main power breakers in the OFF position.

Maintenance service may be performed now.

11.713 Air Distribution System

Periodically, the air distribution system may need to be shut down for repairs, modifications, or cleaning. If your distribution system is composed of different pipes serving different plant processes, and if regulating/isolation valves have been installed, the section to be serviced may be shut down effectively by closing and tagging the regulating/isolation valve. Keep in mind that when closing down a section of the distribution system, an increase of air flow, head loss, and system pressure in the active sections of the distribution system may occur.

Where only one or no regulating/isolation valve is installed or there is only one distribution line, the blower must be shut down before service on the distribution system begins. *CAUTION:* Before service begins on the distribution system, the regulating/isolation valve(s) or the blower must be locked out and tagged to prevent air from entering the system.

Do not attempt to repair even the smallest leaks or discrepancies in the distribution system unless the system is shut down. Although 6.0 to 10.0 psi (0.4 to 0.7 kg/sq cm) system pressure may not seem like much, *SERIOUS BODILY INJURY* may result from escaping air or foreign material.

11.714 Air Headers and Diffusers

Before shutting down the header(s) or diffusers, mark the valve position of the butterfly valve or count the number of turns that a gate valve is open and record it. This will provide a ready reference for positioning the valve when the header(s) and diffusers are returned to service.

Close the header(s) regulating/isolation valve.

QUESTIONS

Write your answers in a notebook and then compare your answers with those on page 133.

11.7A Why must both surface aerators and blowers be shut down before any maintenance service is performed?

11.7B List the steps you would follow to shut down an aerator.

11.7C Before service begins on the air distribution system, why should the regulating/isolation valve(s) or the blower be locked out and tagged?

11.7D Why should the air distribution system be shut down before attempting to repair even the smallest leaks?

11.72 Abnormal Equipment Operation

When making routine checks and inspections you may occasionally find some abnormal conditions. Serious damage may result to the equipment if these conditions are not corrected as soon as possible. Listed in the tables in this section are some abnormal conditions for activated sludge process equipment, possible causes, and operator response to the conditions that will aid you in the safe and efficient operation of the equipment in your plant. As you become familiar with your plant and discover abnormal conditions, add the causes and your response to these tables for use by other operators in your plant.

11.720 Surface Aerators

Table 11.3 lists abnormal surface aerator conditions, possible causes, and operator response. High oil temperature and motor amperage are not uncommon when breaking in a new or overhauled surface aerator.

11.721 Air Filters

The difference in pressure recorded by the manometers will be small when the filters are clean, but as they become dirtier, the manometer reading will increase and the fluid in each tube will move farther away from the manometer zero mark. When this difference reaches approximately two to three inches (50 to 75 mm) of water column more than the initial (clean filters) difference, it is time to clean the filters. If filters do not have a manometer, you can determine if filters need cleaning by removing them and observing the degree of dirt and dust build-up.

11.722 Blowers

Blower operation is essential for the efficient operation of the activated sludge process, and any other in-plant process that requires blower-supplied air. Loss of blower air for an extended period of time (longer than four hours) will turn the plant into an upset system and cause bulking of the activated sludge process. Table 11.4 lists abnormal blower conditions, possible causes, and operator response. High oil temperature, oil pressure, and motor amperage are not uncommon or abnormal when breaking in a new or overhauled blower unit. Realize that your response to an abnormal condition may depend on the type of blower and blower manufacturer.

11.723 Air Distribution System

Although there are only a few moving parts connected with the air distribution system, failures in the system can be expected. Table 11.5 lists some abnormal conditions, possible causes, and the operator's response to the conditions that will aid you in the safe and efficient operation and maintenance of the air distribution system.

11.724 Air Headers

The wet and corrosive environment that the header is exposed to, the upward water current drag forces exerted on the header, vibration, and normal deterioration all contribute to the abnormal conditions listed in Table 11.6 that should be anticipated.

11.725 Air Diffusers

Abnormal operation of the fine bubble diffusers is usually the result of the porous surfaces clogging due to dirt and biological growths. Abnormal operation of the coarse bubble diffusers, though infrequent, is usually the result of clogging due to biological growth or broken diffusers. Table 11.7 lists some abnormal conditions, possible causes, and the operator's response to the conditions that will aid you in the efficient operation and maintenance of the air diffusers.

QUESTIONS

Write your answers in a notebook and then compare your answers with those on page 134.

11.7E What factors might cause the motor on a surface aerator to draw a high or uneven amperage?

11.7F How can you determine if filters need cleaning if the filter chamber does not have a manometer?

11.7G What can clog fine bubble diffusers?

11.73 Maintenance

A comprehensive preventive maintenance program is an essential part of plant operations. Proper maintenance will ensure longer and better equipment performance than equipment that is given little, if any, care. This section should be used as a guideline in performing the required maintenance on activated sludge process equipment. For more detailed information on maintenance, see Chapter 15, "Maintenance," your treatment plant's operation and maintenance manual, and the manufacturers' manuals.

11.730 Surface Aerators

MOTORS

Motors should be greased after about 2,000 hours of operation. The motor must be stopped when greasing begins.

Remove filler and drain plugs, free the drain hole of any hardened grease, add new grease through the filler hole until it starts to come out of the drain hole. Start the motor and let it run for about 15 minutes to expel any excess grease. Stop the motor and install the filler and drain plugs.

After about five years of operation, the motor windings may tend to deteriorate due to moisture and heat. Have the motor inspected and removed from service for repair by an authorized motor repair facility.

GEAR REDUCER

Generally all new oil-lubricated equipment has a break-in period of about 400 hours. After this time the oil should be drained from the gear reducer, the reservoir flushed, and new oil added. This procedure removes fine metal particles that have worn off of the internal components as a result of the initial close tolerances during the break-in period. If large quantities of fine metal particles are found after the break-in period, the manufacturer should be consulted.

A high-quality, turbine-type oil is normally used in the gear reducer assembly. After an oil change has been completed and with the reducer assembly inspection plates removed, inspect the gears for proper operation and the oil for proper flow.

TABLE 11.3 ABNORMAL SURFACE AERATOR OPERATION

Item	Abnormal Condition	Possible Cause	Operator Response
Motor	High or uneven amperage	Moisture	Have electrician MEG[a] check motor. Have motor rewound.
		Winding breakdown	
		Degree of impeller submergence results in amperage draw in excess of motor amperage design	Adjust aerator.
		Excessive motor bearing or gear reducer friction	Inspect and lubricate bearings and gears. Overhaul if needed.
Gear Reducer	Bearing or gear noise	Lack of proper lubrication	Repair or replace oil pump. Change oil.
			Remove obstruction in oil line.
Shaft Coupling	Unusual noise and vibration	Cracked coupling	Replace coupling. Align impeller shaft.
		Loose coupling bolts/nuts as a result of vibration	Torque bolts.[b] Use "locking" nuts. Align impeller shaft.
Impeller	Unusual noise and vibration	Loose blades	Torque blade bolts.[b] Use lock-washers. Align impeller.
		Cracked blades	Replace. Torque bolts.[b] Align.

[a] Use instrument (megger) to check insulation resistance of motor.
[b] Tighten bolts to manufacturer's torque rating. Ratings are given in foot-pounds or kilogram-meters.

TABLE 11.4 ABNORMAL BLOWER OPERATION

Item	Abnormal Condition	Possible Cause	Operator Response
Unusual noise or vibration	Coupling misaligned	Incorrect installation	Align coupling with blower at operating temperature according to manufacturer.
	Loose nuts, bolts, or screws	Vibration	Tighten.
Air system pressure	Low pressure	Bypass valve open, leaks or breaks in distribution piping	Close valve, repair leaks or breaks.
		Diffusers came off air header	Replace diffusers.
	High pressure	Blockage or partially closed valve in distribution piping	Remove blockage or open valve.
		Plugged diffusers	Blow out or remove and clean.
Air flow	Low total flow	High ambient temperatures	Add more air if needed.
		Blower air control malfunction	Repair or replace control.
System oil pressure	Low pressure	Oil level too low	Add oil.
		Oil filter dirty	Replace.
		Check valve sticks open	Replace valve.
		Incorrect oil type	Drain and refill with proper oil type.
	High pressure	Incorrect oil type	Drain and refill with proper oil type.
Oil discharge pressure	Low pressure	Suction lift too high	Reduce lift.
		Air or vapor in oil	Purge air at filter.
		Coupling slipping on pump shaft	Secure coupling.
Oil temperature	Low temperature	Oil cooler water flow too high	Throttle water flow.
	High temperature	Oil cooler water flow too low	Increase water flow.
		Incorrect oil type	Drain and refill with proper oil type.
		Insufficient oil circulation	Replace oil filter, check oil lines for restrictions.
Bearings	Hot bearing(s)	Blower speed too high	Reduce speed to recommended RPM.
		Defective bearing(s)	Check bearing(s) for clearance, hot spots, cracks or other damage. Repair or replace.
		Oil cooler water flow too low	Increase water flow.
Motor	Will not start	Overload relay tripped	Correct and reset.
	Noisy	Noisy bearing	Check and lubricate.
	High temperature	Restricted ventilation	Check openings and duct work for obstructions.
		Electrical	Check for grounded or shorted coils and unbalanced voltages between phases.

TABLE 11.5 ABNORMAL AIR DISTRIBUTION SYSTEM OPERATION

Item	Abnormal Condition	Possible Cause	Operator Response
Meter(s)	High, low, or no indication	Loose movement	Tighten or replace.
		Out of calibration	Calibrate.
		Dirt in mechanism	Clean.
		Pointer dragging on scale plate	Adjust pointer.
		Bypass valve open or leaking	Close or repair.
		Meter piping leaks	Tighten or replace.
		Meter piping plugged	Clean piping.
Seals, gaskets, and flex connections	Leaking	Loose bolts or fittings	Tighten.
		Blown out	Replace.
	Worn	Usual deterioration	Replace.
Pipe	Corrosion	Condensate	Drain traps daily, install additional traps, flush pipe, paint pipe, and remove standing water from around pipe.
	Sludge inside pipe	Vacuum action by blower operating in reverse	Flush pipe, install check valve on blower, or repair check valve.
	Dirt	No or inefficient air filtration	Install filters. Clean filters more frequently.
Valves	Difficult to operate or frozen	Hardened grease	Remove old grease and apply seizing[a] inhibitor. Operate valves monthly.
		Corrosion	Drain condensate traps daily. Apply seizing inhibitor.

[a] Seizing. Seizing occurs when an engine overheats and a component expands to the point where the engine will not run. Also called freezing.

TABLE 11.6 ABNORMAL AIR HEADER OPERATION

Item	Abnormal Condition	Possible Cause	Operator Response
Valve	Valve leaks at stem	Loose stem packing nut	Tighten nut.
		Defective packing	Secure distribution system and replace packing.
	Valve will not seat closed	Corrosion	Secure distribution system and clean or replace valve.
		Butterfly rubber seat defective	Secure distribution system and replace rubber seat.
		Butterfly or gate has come off valve system	Secure distribution system and replace.
Swing header pivot joints	Air leaks from joint	Defective O-ring	Close header valve, pull header from tank with crane, and replace O-ring.
		Loose joint	Tighten.
		Insufficient grease in joint	Apply 3 to 5 shots of grease.
		Cracked joint	Replace.
Fixed header couplings or unions	Air leaks from couplings, unions, or end caps	PVC has defective glue bond	Remove, clean with PVC solvent, bond, and allow bond to cure.
		Pipe has leak through thread	Remove, apply Teflon tape, tighten.
Horizontal header	Uneven water motion (roll) in tank	Header not perfectly level, thus allowing more air to one side	Level header with surveyor's level (tank empty) or use Mason's level.
		O-rings or gaskets defective or connections loose	Replace O-rings or gaskets. Tighten connections.
Header pipe	Interior corrosion	Moisture	Use PVC or galvanized pipe.
	Exterior corrosion	Moisture	Use PVC, galvanized pipe, or paint pipe with an epoxy coating.
		Electrolysis	Use a sacrificial (magnesium) anode or coat surface.

Grease-lubricated bearings should be greased about every 500 hours of operation, depending on service conditions. *NOTE:* More damage is done to bearings by overgreasing than undergreasing.

Oil for gears and bearings should be changed after about 1,400 hours of operation under normal service and more frequently when required. *NOTE:* The proper type of oil is a necessity. If it is too thin or too thick, it will impede proper functioning of bearings and gears.

COUPLING AND IMPELLER

Every 6 to 12 months the aerator should be stopped and all bolts and nuts on the impeller and coupling should be retorqued according to the manufacturer's specifications. This is also a good time to inspect the metal surfaces for deterioration such as cracks or worn components. While the unit is off, check the impeller and shaft for proper alignment.

After performing routine maintenance, the unit and surrounding area must be wiped or washed. Be sure to remove any spilled oil or grease. Dispose of oil- and grease-soiled rags in a covered container to avoid a fire hazard.

11.731 Air Filters

When filter cleaning is scheduled, remove the filters from the filter chamber and check the inside of the chamber.

TABLE 11.7 ABNORMAL AIR DIFFUSER OPERATION

Item	Abnormal Condition	Possible Cause	Operator Response
Fine-bubble diffuser	Exterior clogged	Biological growth	Raise air header, remove diffuser, scrub and wash diffuser.
	Interior clogged	Dirt from distribution system	Raise air header, remove diffuser, scrub and wash diffuser. Install filters, clean filters more frequently.
Coarse-bubble diffuser	Exterior clogged	Biological growth	Raise air header, scrub and wash diffuser. Once a month increase air flow 2 to 3 times normal for 15 minutes to "blow out" diffuser orifices.
	Cracked	Overtightened when installed, structural failure	Replace.
Fine- and coarse-bubble diffusers	Accumulation of rags, hair, string	Inefficient pretreatment, normal conditions	Yearly, raise headers and clean diffusers.
	Insufficient diffusion pattern or oxygen transfer	Clogged diffusers	Clean the diffusers.
		Inadequate diffuser arrangement	Modify diffuser arrangement.
		Too few diffusers	Add diffusers or install a different type.

Remove any sand, dirt, paper, water, or other debris. Removing air filters can produce substantial quantities of airborne dust. Wear protective clothing and eye and respiratory protection if excessive dust exposure may occur.

Usually filters may be cleaned by using a relatively high-pressure stream of clean water or by steam cleaning. The filter manufacturer's equipment manual should be consulted for the recommended method of cleaning your particular type of filter.

Allow the filters to dry, securely install the filters in the filter chamber, and ensure that no tools or other items are left in the filter chamber that could be drawn into the blowers.

11.732 Blowers

Generally, all new oil-lubricated equipment has a break-in period of about 400 hours. After this time the oil should be drained from the blower, the reservoir and filter cleaned and flushed, and new oil added. This procedure removes the fine metal particles that have worn off of the internal components as a result of the initial close tolerances during the break-in period. If large quantities of fine metal particles are found after the break-in period, the equipment manufacturer should be consulted.

Grease-lubricated bearings should be greased about every 500 hours of operation, depending on service conditions. *NOTE:* More damage is done to bearings by overgreasing than undergreasing.

Oil for gears and bearings should be changed and the filters cleaned about every 1,400 hours of operation under normal service, more frequently when required. *NOTE:* The proper grade of oil is a necessity. If it is too thin or thick it will impede proper functioning of bearings and gears.

See Section 11.730, "Surface Aerators," *MOTORS*, for a discussion on blower motors. Maintenance on blower motors is similar to maintenance of surface aerator motors.

After performing routine maintenance, the units and surrounding area must be wiped or washed, removing any spilled oil or grease. Dispose of oil-soiled rags in a covered container.

A routine should be established whereby the blower, pressure relief valve, and blower suction and discharge valves are operated to prevent the equipment from *SEIZING*[36] up and becoming inoperable. During this operation the equipment should be checked for proper alignment. Blowers that are not routinely in service should be operated at least six hours per week on a given day.

Pressure relief valves should be checked at least once a month by manually lifting the valve off of the valve seat. Test the pressure relief valve for the correct setting by slowly closing the blower discharge valve slightly, with the blower operating, until the pressure relief valve opens. Observing the air system pressure gauge while noting when the relief valve opens will alert you to the pressure setting at that time. Instructions for making adjustments on the pressure relief valve

[36] *Seizing* or *Seize Up.* Seizing occurs when an engine overheats and a part expands to the point where the engine will not run. Also called freezing.

and the proper setting for your blower installation should be contained in the manufacturer's manual. *NOTE:* When testing the pressure relief valve, *do not close the blower discharge valve completely* or serious damage to the blower may result. Use this test procedure *only* if the pressure relief valve is located between the blower and the discharge valve.

Never use any oil or grease on the pressure relief valve. Due to the heat generated by compressing air, oil and grease lubricants will harden and may keep the pressure relief valve from opening. A metal seizing inhibitor may be applied and is generally available from most hardware stores.

Blower suction and discharge valves should be operated once a month, with the blower off, by fully closing and opening the valves. A metal seizing inhibitor should also be applied to the gear assembly of each valve.

The blower check valve should be checked once a month for free operation by opening and closing the valve while checking that the valve seats properly when it is in the closed position.

11.733 Air Distribution System

Depending on the harshness of the environment surrounding the distribution piping, an inspection schedule should be established for the entire piping system from at least once monthly to once every six months. During this inspection, look for loose pipe support clamps or cracked pipe support welds, shifting of pipe out of original position due to structural settling, loose nuts and bolts on fittings, stuck or difficult to operate valves, and damage from corrosion.

Exterior pipe, fittings, and valve surfaces should be painted with a primer coat and a finish coat. Painting should be done as frequently as needed to prevent and inhibit corrosion. Meters should be calibrated at least once a year to ensure meter accuracy.

All valves in the distribution system, including meter valves, should be fully opened and fully closed once a month (with blowers off) to ensure free operation.

For more detailed information on maintenance of your meters, fittings, valves, and pipes consult your manufacturer's manual, the plant operation and maintenance manual, and Chapter 15, "Maintenance," of this training manual.

11.734 Air Headers

Due to the severe environment that the headers are exposed to, a maintenance program should be adopted so that header failure and resultant lengthy downtime can be minimized or avoided. The following activities should be scheduled on a monthly and yearly basis.

1. Monthly
   a. Fully close and open all regulating/isolation valves to ensure free operation on coarse bubble diffusers, but *NOT* on porous media diffusers.
   b. Apply 3 to 5 shots of grease to the upper pivot swing joint O-ring cavity (swing header).
   c. Check for loose fittings, bolts, and nuts. Secure if loose.
2. Yearly
   a. Raise headers, clean and check for loose bolts, nuts, and fittings. Secure if loose.
   b. Apply 3 to 5 shots of grease to the pivot joint O-ring cavity (swing header).
   c. Check for corrosion. Properly prepare pipe and paint with epoxy coating where needed.

Hydraulic fluid in the swing header hoist should be changed at least once yearly.

A weatherproof tarpaulin should be provided to cover the hoist to protect it from the elements when it is not in use. Secure the tarpaulin to the hoist to prevent the wind from blowing it into the tanks.

11.735 Diffusers

Maintenance activities that should be scheduled on a monthly and yearly basis are listed in this section.

1. Monthly

   Increase the air flow to the diffusers 2 to 3 times normal for about 15 minutes to blow out the biological growths that have accumulated around the diffuser orifices.
2. Yearly, or as conditions warrant

   Raise the air header from the tank and clean diffusers, inspect for damage, and replace as needed.

Diffusers that are allowed to remain in a defective condition reduce the diffusers' effectiveness and will result in inefficient wastewater treatment.

QUESTIONS

Write your answers in a notebook and then compare your answers with those on page 134.

11.7H What causes motor windings to deteriorate?

11.7I How are air filters cleaned?

11.7J What causes equipment to seize up and become inoperable?

END OF LESSON 5 OF 7
(VOLUME II
ON
ACTIVATED

Please answer the discussion and review questions next.

DISCUSSION AND REVIEW QUESTIONS

Chapter 11. ACTIVATED SLUDGE

(Lesson 5 of 7 Lessons)

Write the answers to these questions in your notebook. The question numbering continues from Lesson 4.

36. When an activated sludge plant becomes upset, what should the operator do first?
37. Why do many activated sludge plants base aerator loadings on mixed liquor *volatile* suspended solids and not total mixed liquor suspended solids?
38. When attempting to correct an upset activated sludge process, why should only one major change be made at a time?
39. Is sludge bulking undesirable? Why?
40. What safety hazard can a frothing aerator cause?
41. Why must proper equipment shutdown procedures be followed?
42. Why should you develop your own lists of shutdown procedures for equipment in your plant?
43. How would you dispose of oil- and grease-soaked rags and why?

CHAPTER 11. ACTIVATED SLUDGE

OPERATION OF CONVENTIONAL ACTIVATED SLUDGE PLANTS

(Lesson 6 of 7 Lessons)

11.8 MODIFICATIONS OF THE ACTIVATED SLUDGE PROCESS

11.80 Reasons for Other Modes of Operation

Modifications of the conventional activated sludge process have been developed to improve operational results under certain circumstances. Some of these conditions may be:

1. Current or actual loadings are in excess of design loading for conventional operation.
2. Wastewater constituents require added nutrients to properly treat influent waste load.
3. Flow or strength of waste varies seasonally.

The remainder of Section 11.8 briefly describes several modes of operation, including contact stabilization, the Kraus process, step-feed aeration, complete mix aeration, modified aeration, and the Bardenpho process. Section 11.9 presents a more detailed explanation of one additional process, the sequencing batch reactor (SBR).

11.81 Contact Stabilization (Figure 11.23)

Operation of an activated sludge plant on the basis of contact stabilization requires two aeration tanks. One tank is for separate reaeration of the return sludge for at least four hours before it is permitted to flow into the other aeration tank to be mixed with the primary effluent requiring treatment. Overall loading factors are the same as for conventional activated sludge (see Tables 11.8 and 11.9 (metric)).

If the solids content in aeration tank "A" (mixed liquor aerator, Figure 11.23) and aeration tank "B" (return sludge reaeration only) are combined, the loading ratio of food/microorganisms is the same as conventional operation. However, if you only look at aeration tank "A" where the load is applied, the food/microorganism ratio is nearly double the usual load ratio for conventional activated sludge.

Contact stabilization attempts to have organisms assimilate (take in) and store large portions of the influent waste load in a short time (30 to 90 minutes). The activated sludge is separated from the treated wastewater in the secondary clarifier and returned to the reaeration tank "B." No new food is added to the reaeration tank and the organisms must use the waste material they collected and stored in the first aeration tank. When the stored food is used up, the organisms begin searching for more food and are ready to be returned to tank "A."

Process controls for a contact stabilization plant are the same as those described for a conventional plant in this chapter. The contact stabilization system with its off-stream

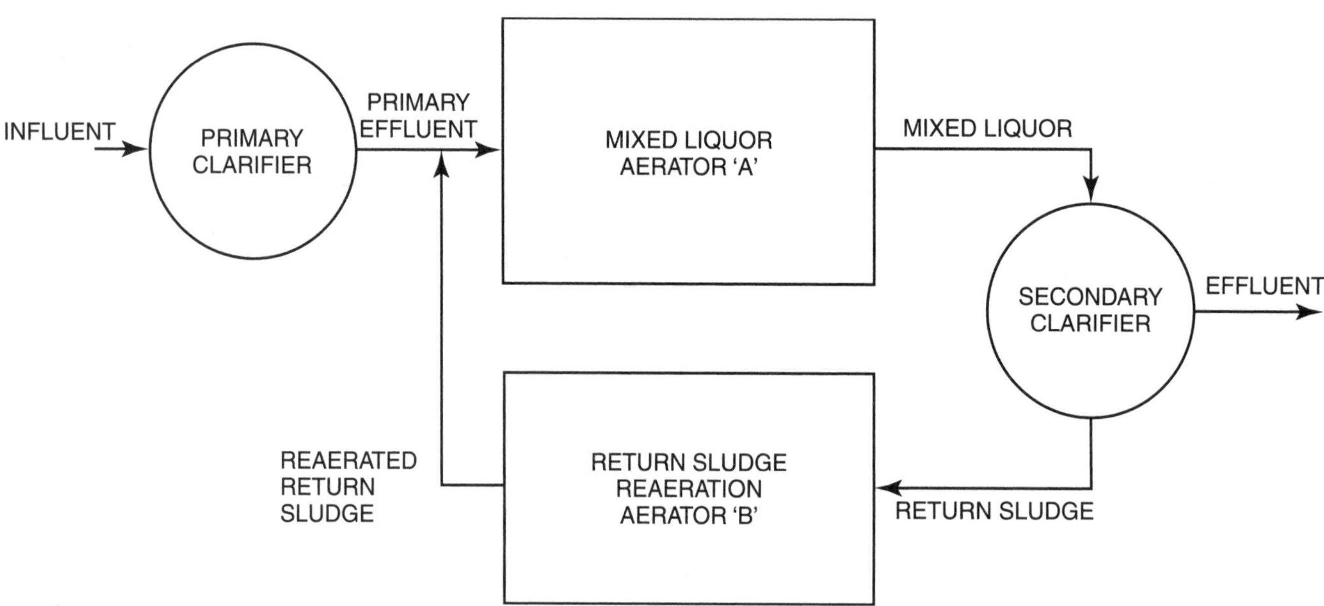

Fig. 11.23 Plan layout of contact stabilization plant

TABLE 11.8 AERATION TANK CAPACITIES AND PERMISSIBLE LOADINGS[a]

Process	Aeration Tank Organic Loading - lb BOD_5/day per 1,000 cu ft	F/M Ratio lb BOD_5/day per lb MLVSS	MLSS[b] mg/liter
Conventional Step Aeration Complete Mix	40	0.2 - 0.5	1,000 - 3,000
Contact Stabilization	50[c]	0.2 - 0.6	1,000 - 3,000
Extended Aeration Oxidation Ditch	15	0.05 - 0.1	3,000 - 5,000

[a] *RECOMMENDED STANDARDS FOR WASTEWATER FACILITIES (10 STATE STANDARDS),* Great Lakes-Upper Mississippi River Board of State and Provincial Public Health and Environmental Managers, 2004 Edition. Available from Health Education Services, PO Box 7126, Albany, NY 12224. Price, $17.00, includes cost of shipping and handling.
[b] MLSS values are dependent upon the surface area provided for sedimentation and the rate of sludge return as well as the aeration process.
[c] Total aeration capacity, includes both contact and reaeration capacities. Normally, the contact zone equals 30 to 35% of the total aeration capacity.

TABLE 11.9 AERATION TANK CAPACITIES AND PERMISSIBLE LOADINGS (METRIC)[a]

Process	Aeration Tank Organic Loading - kg BOD_5/day per 1,000 cu m	F/M Ratio kg BOD_5/day per kg MLVSS	MLSS[b] mg/liter
Conventional Step Aeration Complete Mix	640	0.2 - 0.5	1,000 - 3,000
Contact Stabilization	800[c]	0.2 - 0.6	1,000 - 3,000
Extended Aeration Oxidation Ditch	240	0.05 - 0.1	3,000 - 5,000

[a] Based on *RECOMMENDED STANDARDS FOR WASTEWATER FACILITIES (10 STATE STANDARDS),* Great Lakes-Upper Mississippi River Board of State and Provincial Public Health and Environmental Managers, 2004 Edition. Available from Health Education Services, PO Box 7126, Albany, NY 12224. Price, $17.00, includes cost of shipping and handling.
[b] MLSS values are dependent upon the surface area provided for sedimentation and the rate of sludge return as well as the aeration process.
[c] Total aeration capacity, includes both contact and reaeration capacities. Normally, the contact zone equals 30 to 35% of the total aeration capacity.

reservoir of organisms in aeration tank "B" avoids a complete solids wash-out when high flows occur or a kill of microorganisms when toxic wastes reach the plant. When calculating loading guidelines, the solids in the reaeration tank may be ignored. You must realize that the effluent quality will not be the same as for loadings on the conventional process, but the results from the calculations can be used to operate your plant and to reveal any trends.

11.82 Kraus Process (Figure 11.24)

The Kraus process is a modification of conventional activated sludge. This process is patented by its developer. The process is widely used when the wastewater contains a much greater ratio of carbonaceous to nitrogenous material than found in normal domestic wastewater and when activated sludge has poor settling characteristics.

This nutrient imbalance commonly occurs when wastes from canneries or dairies are treated. When the organisms use all of a limiting constituent (nitrogen, for example) they refuse to remove the remaining portions of the other constituents (carbon, for example). Normally, this nutrient deficiency is nitrogenous material, which is readily available in anaerobic digester supernatant and sludges. Feeding anaerobic digester supernatant or digester sludge to the aeration system will usually supply the proper nutrients to maintain the balance and also add inert suspended solids to assist with liquid-solid separation in the clarifier. The method of supernatant application is very important.

In the Kraus process, the return sludge is sent to the reaeration aerator ("B") to be mixed with the digested sludge from a completely mixed digester. In the reaeration tank ("B"), the digested sludge and the return sludge are mixed, reaerated,

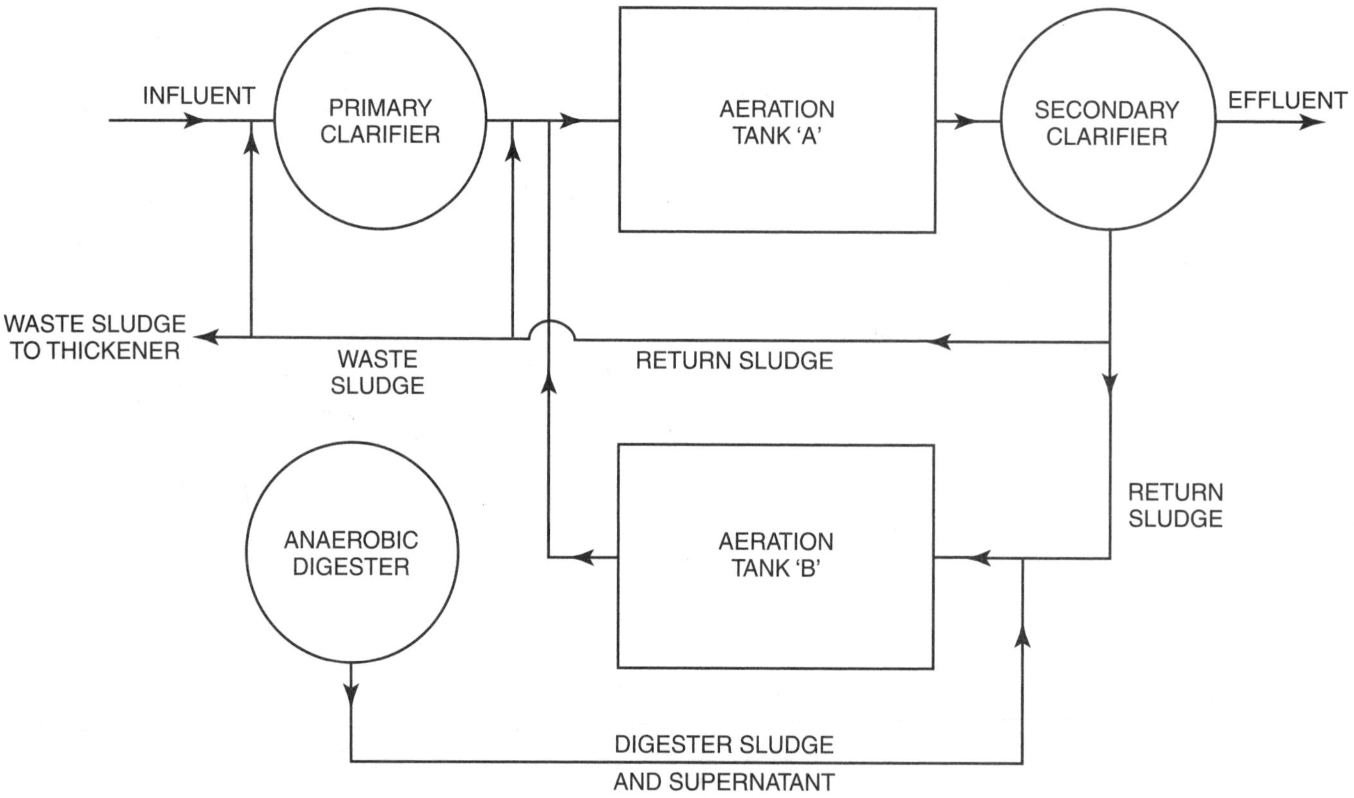

Fig. 11.24 Kraus process

and then sent to the mixed liquor aerator ("A"). The amount of digested sludge introduced to the system is determined by laboratory evaluation of the carbonaceous material removal through the system.

The same controls apply as described for controlling a conventional activated sludge plant. The main objective is to properly balance nutrients; however, other advantages include increased activated sludge settleability and (similar to contact stabilization) a reservoir of organisms off the main processing stream.

11.83 Step-Feed Aeration (Figure 11.25)

Step-feed aeration actually is a step-feed process based on conventional activated sludge loading guidelines. The difference between step-feed and conventional operation is that in conventional activated sludge, the primary effluent and return sludge are introduced at one point only, the entrance to the aeration tanks. In step-feed aeration, the return sludge is introduced separately and, in many cases, allowed a short reaeration period by itself at the entrance to the tank. The primary effluent is admitted to the aeration tanks at several different locations. These locations distribute the waste load over the aeration tank and reduce oxygen sags in an aerator. If you introduce the influent near the outlet end of the aeration tank, the process will become similar to contact stabilization.

Step-feed aeration distributes the oxygen demand from the wastewater over the entire aerator instead of concentrating it at the inlet end. Some of its advantages over conventional operation include less aeration volume to treat the same volumes of wastewater, better control in handling shock loads, and the potential for lower applied solids to the secondary clarifiers. When a conventional plant is operating above design loads or the secondary clarifiers cannot handle the solids load, switching to step-feed aeration or contact stabilization allows the operator to maintain more solids under aeration with a lower applied solids concentration to the secondary clarifiers. Successful operation requires transfer of wastes into the activated sludge solids in the short time interval before the waste reaches the effluent end of the aeration tank.

Step-feed aeration can be operated on a variable basis by using combinations of the modes shown in Figure 11.25. For example, use Mode 4 (contact stabilization) during peak flows resulting from storms or when treating strong industrial wastes in order to preserve the activated sludge solids in the system. Peak flows could wash out the activated sludge solids or the activated sludge culture could be killed by a strong industrial waste if you were operating in Mode 1 (conventional activated sludge process). Therefore, you would ideally operate in Mode 1 to treat ordinary domestic flows and switch to Modes 2, 3, and eventually 4 depending on the quantity of industrial waste flows, seasonal wastewaters, or temperature variations in relation to operating partially or fully nitrified (see Chapter 2, "Activated Sludge," in *ADVANCED WASTE TREATMENT* manual) or the ability of your existing facility to adequately process the existing flows.

Selection of the proper mode depends on influent characteristics (flow rate and waste strength) and the capabilities of the plant to handle these characteristics. Mode 1 is usually best for providing maximum treatment of normal domestic wastewaters because all of the wastewater is exposed to aeration and microorganisms for the longest possible time, which will produce the highest degree of treatment. If your activated

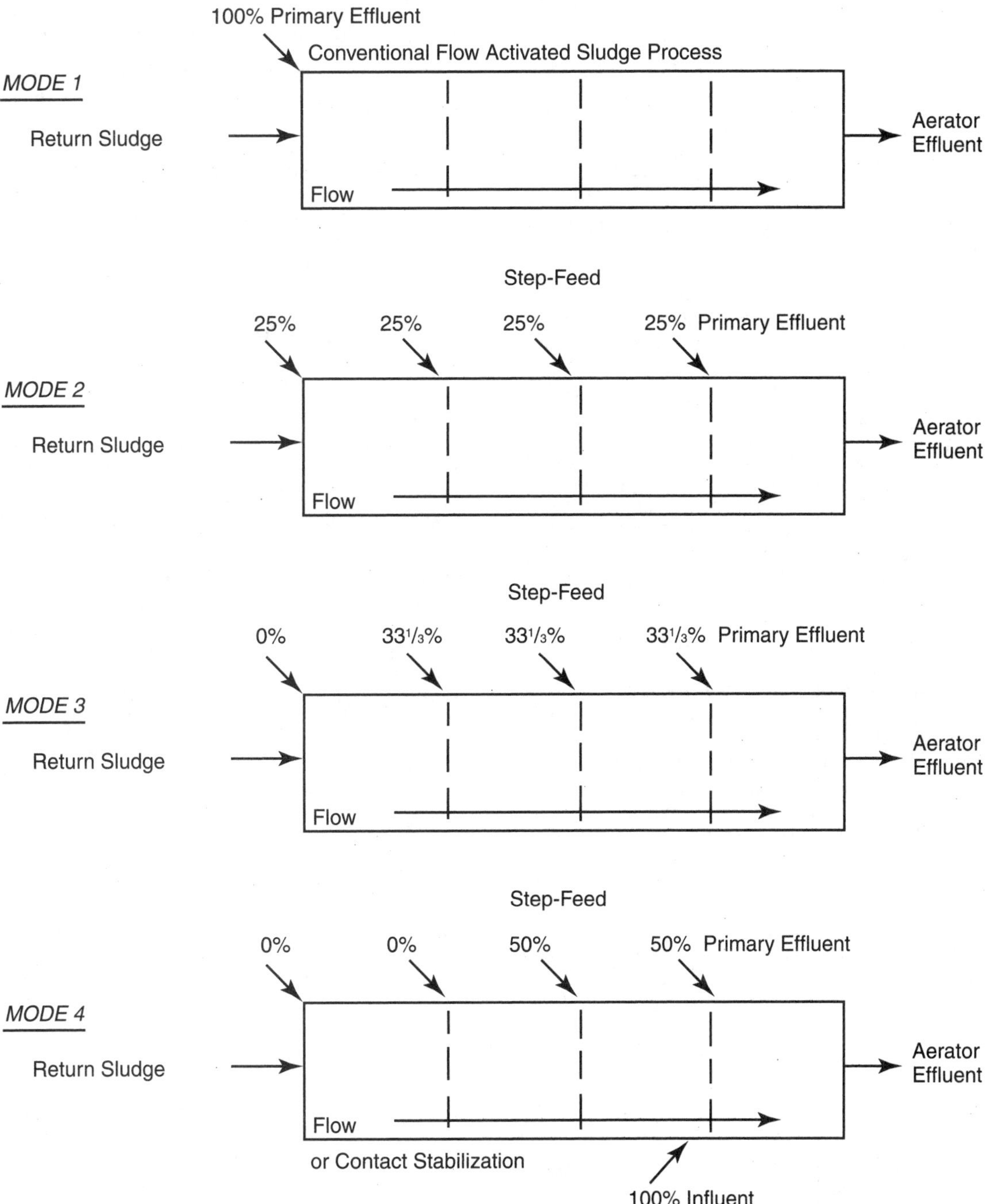

Several possible modes of feeding primary effluent to the aeration tanks. Some tanks may have more or fewer points of discharge into the tank.

Fig. 11.25 Modes of step aeration

sludge process develops a sludge bulking or rising sludge problem, a shift to Modes 2, 3, or 4 would help to cure the problem by: (1) allowing more solids to be maintained in the aeration system thus controlling the solids volume entering the secondary clarifiers (more in aeration and less in clarification), and (2) exposing the solids to a more evenly distributed oxygenated environment, which provides for healthy bugs throughout the aeration system. When dealing with a bulking sludge, the problem is usually caused by inadequate dissolved oxygen levels, which are conducive to the growth of filamentous organisms. Rising sludge usually results from high dissolved oxygen levels in the aerator and prolonged secondary clarifier detention times, which will cause denitrification.

When attempting to cure a bulking or rising sludge problem, an increase in return sludge flows may be helpful to reduce solids on a short-term basis. This procedure should not be relied on to solve the problem. Increased return sludge flows will cause the sludge blanket in the secondary clarifier to drop initially, yet the blanket could rise as the result of excessive clarifier underflow rates. Try to find the return rate that produces the lowest sludge blanket and seek to correct the bulking or rising problem by modifying the aeration system operational mode. Your job as an operator is to select the best mode that will do the best job for the normal wastewater characteristics and meet effluent requirements.

The step-feed mode of operation is controlled by many of the same procedures used for the conventional process. An exception is that the mixed liquor suspended solids determinations must be made at each point of wastewater addition. The purpose of this is to measure the waste content and dilution factor provided by the primary effluent in order to determine the total pounds of solids in the aeration tank.

11.84 Complete Mix (Figure 11.26, page 91)

The complete mix mode of operation is a design modification of tank mixing techniques that is made to ensure equal distribution of applied waste load, dissolved oxygen, and return sludge throughout the aeration tank. The theory of this modification is that all parts of the aeration tank should be similar in terms of amounts of food, organisms, and air. This is accomplished by providing diffuser location and application points of influent and return sludge to the aerator at several locations. Providing a similar condition throughout the entire aeration tank allows a food/organism ratio of 1/1 and still produces effluent qualities comparable to conventional operation. Generally, smaller aeration tanks are more completely mixed than larger ones. Usually aeration is more efficient in a complete mix facility, such as the one illustrated in Figure 11.26, because of the locations of the air headers.

11.85 Modified Aeration (Figure 11.27)

Modified aeration is also known as high-rate activated sludge. Frequently it is used as intermediate treatment where the discharge requirements demand higher treatment than primary, but not as high as conventional activated sludge, in terms of BOD and suspended solids removals.

Either raw wastewater or primary effluent is applied to an aeration tank with a detention time of 1.5 to 3 hours and a mixed liquor suspended solids concentration of less than 1,000 mg/*L*. Air requirements are lower because of fewer organisms (solids) under aeration as a result of a lower sludge age or a lower mean cell residence time. Effluent quality ranging from primary treatment to conventional activated sludge treatment can be achieved by the operator by controlling the air supply, aeration period, and the pounds of solids under aeration.

11.86 Bardenpho Process (Figure 11.28)

The Bardenpho process is used to remove between 90 and 95 percent of all the nitrogen present in the raw wastewater by recycling nitrate-rich mixed liquor from the aeration basin to an anoxic zone located ahead of the aeration basin. Denitrification takes place in the anoxic zone in the absence of dissolved oxygen. Further denitrification may be obtained by adding a second anoxic basin for the removal of nitrate remaining after recycling.

The degree of nitrate removal depends on the rate of recycling the mixed liquor from the aeration basin. Some plants have three recycle pumps that allow pumping of two, four, or six times the average dry weather flow back to the anoxic zone. Usually pumping four times the average dry weather flow is sufficient to achieve satisfactory nitrate removal. The correct recycle flow can be determined by monitoring the nitrate level in the effluent of the first anoxic basin. If the nitrate concentration in the effluent rises above about 1 mg/*L*, the recycle rate is too high because not enough detention time is provided in the anoxic zone for denitrification to occur.

If phosphorus removal is desired, a fermentation stage (tank) is added before the first anoxic zone. The return activated sludge is mixed with the influent to produce an organism stress condition in the absence of dissolved oxygen and nitrate. This stress condition allows phosphorus to be removed biologically in subsequent aeration basins.

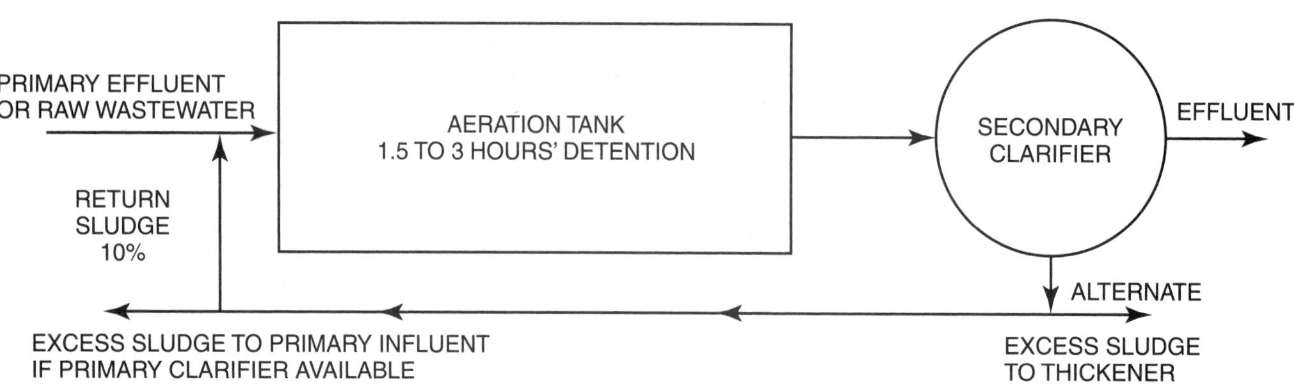

Fig. 11.27 Modified aeration

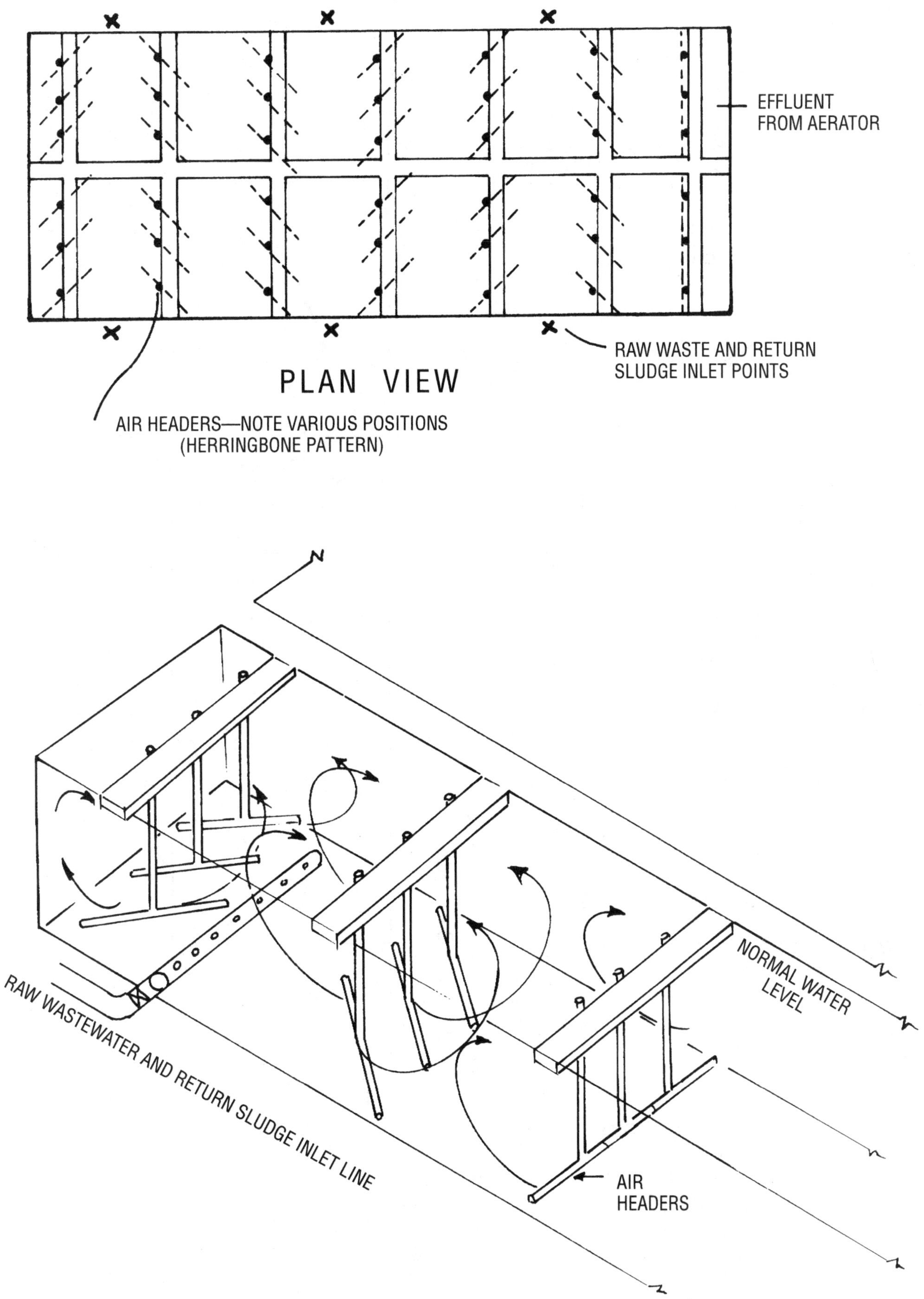

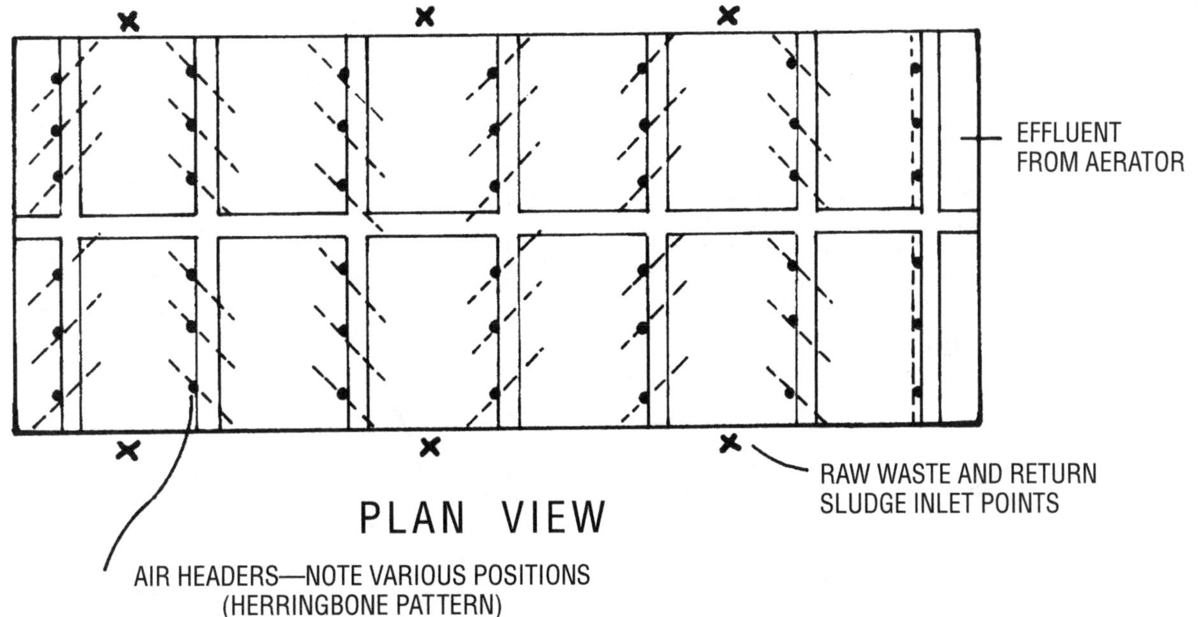

EFFLUENT
FROM AERATOR
RAW WASTE AND RETURN
SLUDGE INLET POINTS
PLAN VIEW
AIR HEADERS—NOTE VARIOUS POSITIONS
(HERRINGBONE PATTERN)
NORMAL WATER
LEVEL
RAW WASTEWATER AND RETURN SLUDGE INLET LINE
AIR
HEADERS

Fig. 11.26 Air header locations in complete mix

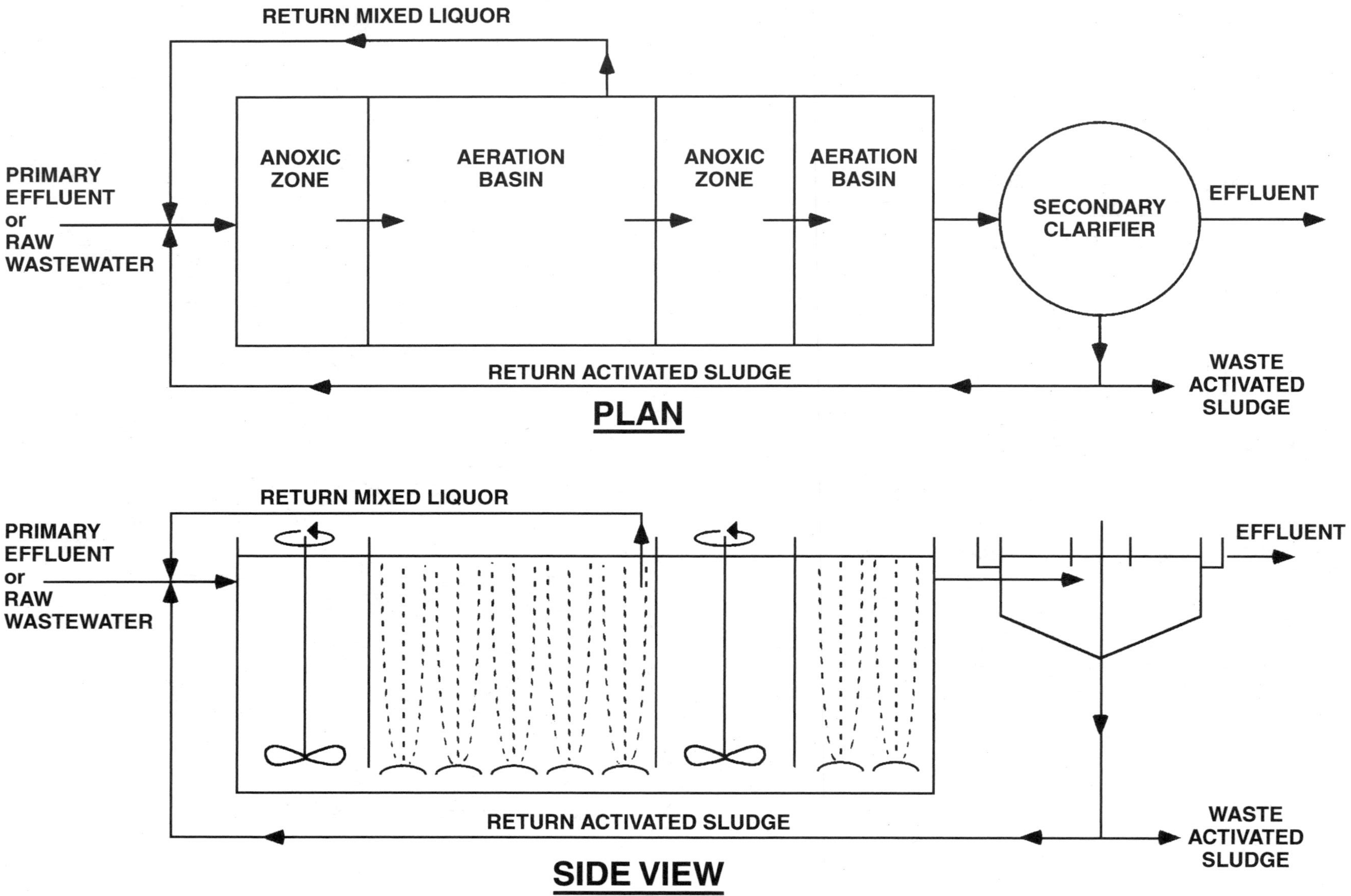

Fig. 11.28 Bardenpho process

This section on the Bardenpho process was prepared on the basis of information provided by Dr. James L. Barnard, Meiring & Barnard, Pretoria, South Africa. His assistance is greatly appreciated.

See *ADVANCED WASTE TREATMENT* in this series of manuals for additional information on nitrogen and phosphorus removal.

QUESTIONS

Write your answers in a notebook and then compare your answers with those on page 134.

11.8A Why does contact stabilization require two aeration tanks?

11.8B What is the difference between step-feed and conventional activated sludge aeration?

11.8C What type of modification is the complete mix mode of operation?

11.9 SEQUENCING BATCH REACTORS (SBRs)

11.90 Process Description

Sequencing batch reactors (SBRs) are especially well suited to the treatment needs of municipalities. They can handle fluctuating flows with ease. A skilled operator can quickly learn how to treat a batch of municipal wastewater to meet NPDES discharge permit requirements.

The activated sludge SBR process performs the same treatment steps as a conventional activated sludge process. Both systems accomplish aeration, sedimentation, and clarification. The difference between the two systems is that in conventional plants the process is a continuous flow process. All of the treatment steps are carried out at the same time in individual tanks. The SBR process of activated sludge treatment is accomplished in a single tank, through sequencing stages. This eliminates the need for additional tanks for settling and clarification, and does not require the conventional return sludge handling facilities. The sequencing series for treatment consists of the following process stages: *FILL, REACT* (aeration), *SETTLE, DECANT* (withdrawal of clarified effluent), and *IDLE.* This complete cycle generally occurs two to six times a day in each SBR tank.

Table 11.10 compares how a number of important factors affect the operation of activated sludge sequencing batch reactor (SBR) systems and continuous flow activated sludge (CFS) systems.

The intermittent cycle extended aeration system (ICEAS) is a modification of the SBR system. The inflow and outflow are periodic in the SBR (at the beginning and end of the treatment cycle). In contrast, the inflow to an ICEAS is continuous. An SBR system must have at least a storage tank and an SBR tank or a minimum of two SBR tanks to accommodate continuous inflow. In an ICEAS, a baffle wall may be installed in the treatment tank to buffer the continuous inflow. Apart from these differences, SBR and ICEAS layouts are very similar.

Figure 11.29 illustrates typical activated sludge sequencing batch reactor flow diagrams. This type of activated sludge plant (Figure 11.30) is becoming increasingly popular for municipal plants that must produce high-quality effluents and, in addition, may be required to achieve nutrient removal of both nitrogen and phosphorus.

Sequencing batch reactor systems are particularly practical for smaller municipal flows for three reasons:

1. The SBR plant's process controls are implemented by a *PROGRAMMABLE LOGIC CONTROLLER (PLC).*[37] Once the SBR system is put online and the start-up bugs have been worked out (normally performed by the contractor and equipment supplier), the PLC automatically controls routine process sequencing.
2. The SBR process is very stable due to the high sludge age associated with long sludge retention times (SRT). Since the treatment processes all take place in a single tank, high storm flows are less likely to cause sludge washouts.
3. Constructing an SBR plant usually costs less than a conventional activated sludge plant due to the elimination of secondary clarifiers and, in most instances, return sludge facilities. In some plants separate sludge digestion facilities are not needed due to the light organic loading and the high sludge age maintained in the SBR process. Sludge is wasted directly from the reactors to drying beds or composting processes.

[37] *Programmable Logic Controller (PLC).* A microcomputer-based control device containing programmable software; used to control process variables.

TABLE 11.10 COMPARISON OF BATCH AND CONTINUOUS ACTIVATED SLUDGE PROCESSES[a]

Feature	SBR System	Continuous Flow Activated Sludge System (CFS)[b]	Remarks
Concept	Time sequence in the same tank	Special sequence in different tanks	Time sequence can be varied in SBR; no such flexibility in CFS.
Inflow	Periodic—normal SBR Continuous—ICEAS	Continuous	
Discharge	Periodic	Continuous	Decant period can be easily changed in SBR. Further, somewhat possible to hold effluent until it meets specific requirements. CFS—inflexible.
Organic Load	Cyclic—normal SBR Continuous—ICEAS	Continuous	Several variations of organic loading are possible by changing durations of cycle periods. CFS—inflexible.
Aeration	Intermittent	Continuous	Increased flexibility in SBR. Both the aeration rate and the aeration duration can be varied. Aeration costs are often less than CFS. CFS—only aeration rate with respect to oxygen applied can be changed.
Mixed Liquor	Always in reactor; no recycle	Recycle through reactor and clarifier	No need for final clarifiers and RAS pumps in SBR. CFS requires above facilities.
Clarification	Ideal—normal SBR Not as ideal—ICEAS	Not ideal. Short-circuiting and density currents are common.	Several CFS systems are known to perform unsatisfactorily because of less than ideal settling conditions present in the clarifiers. SBR free from these problems.
Flow Pattern	Perfect plug flow	Complete mix or approaching plug flow	A perfect plug flow condition in SBR achieves rapid biodegradation of pollutants (shorter reaction time). CFS requires longer reaction time.
Equalization	Always occurs in single tank	None	SBR is an ideal reactor in situations with excessive diurnal (day/night) variations in flow and BOD. CFS can fail under above conditions.
Flexibility	Considerable	Limited	Operator can routinely change cycle durations, aeration/mixing strategies. CFS is somewhat limited in these areas.
Reactor Size	Could be larger than CFS because it must provide space for sludge blanket	Generally smaller than SBR	In spite of larger reactor size, SBR can be more compact and require less overall space because no separate clarifiers and RAS pumps are needed.
Operation (Process)	Relatively easy to operate—achieved by microprocessor technology	Same as SBR	SBR ideal for small plants. CFS may not be practical in small plants with excessive diurnal (day/night) flow variations.
Operation (Equipment)	Less mechanical equipment results in easier operation and maintenance	Same as SBR plus RAS pumps (except for very small plants)	
Effluent Quality	Excellent, in most cases	Excellent, in most cases	
Flexibility to meet changing effluent requirements (C, N, P removals)	Tremendous flexibility in SBR; achieved by changing operational strategy (cycle durations, cycle sequence, and aeration/mixing strategy). Somewhat less in ICEAS.	Limited as compared to SBR	

[a] Arora, Madan L. and Umphres, Peggy B., *TECHNICAL EVALUATION OF SEQUENCING BATCH REACTORS,* for US Environmental Protection Agency, Cincinnati, OH, September 1984.
[b] Includes aeration tank(s), clarifier(s), and return activated sludge (RAS) pumping.

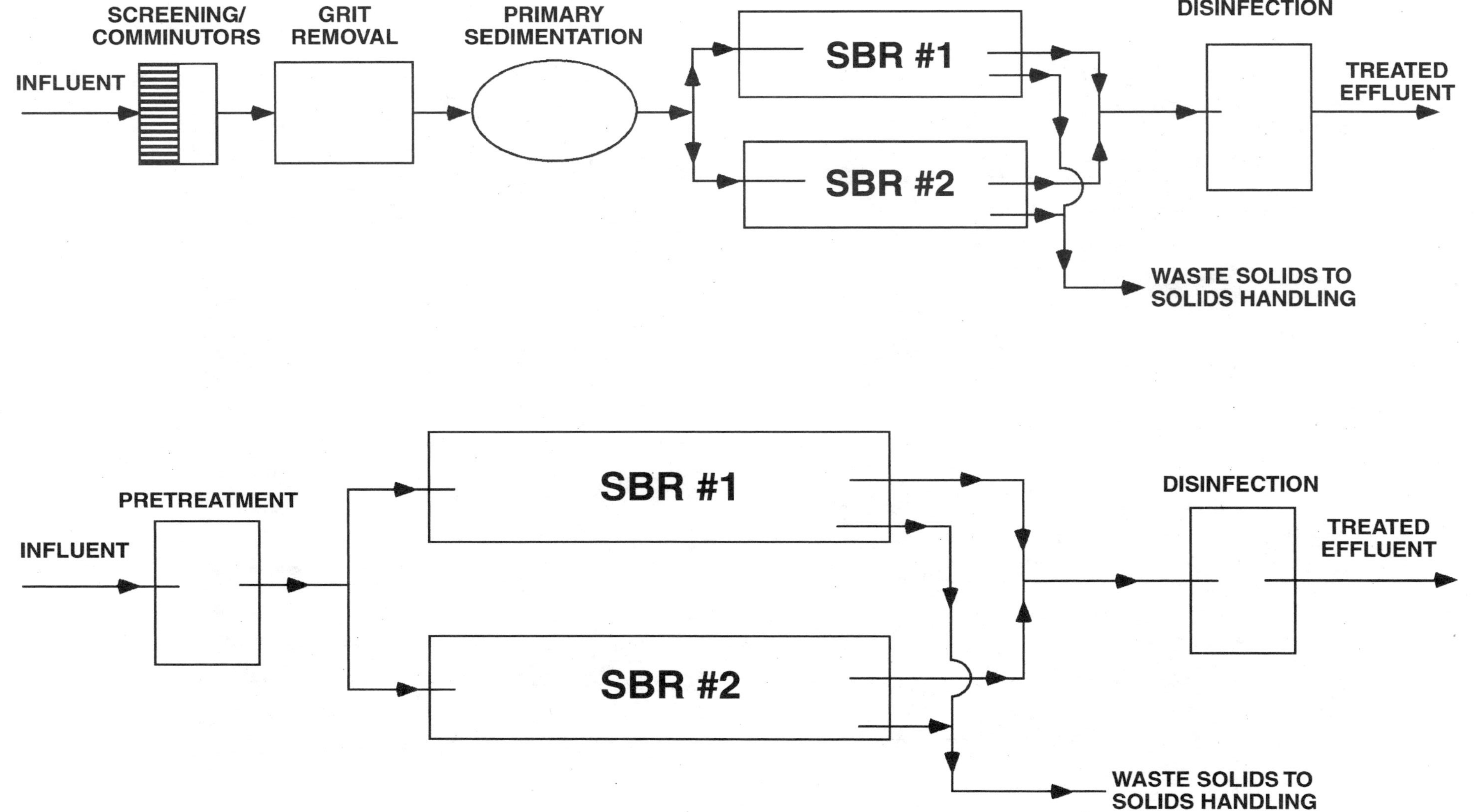

Fig. 11.29 Activated sludge sequencing batch reactor flow diagrams

Fig. 11.30 Activated sludge sequencing batch reactors

QUESTIONS

Write your answers in a notebook and then compare your answers with those on page 134.

11.9A List the sequence of treatment process stages for a sequencing batch reactor (SBR).

11.9B What is the difference between the mixed liquor in the SBR system and the mixed liquor in the continuous flow activated sludge system (CFS)?

11.9C Why are SBRs becoming practical for smaller municipal flows?

11.91 Components of a Sequencing Batch Reactor Activated Sludge Plant

11.910 Preliminary Treatment

Preliminary treatment requirements depend upon the makeup of the wastewater, the collection system type and length, and any other factor that may affect or occasionally degrade the quality of the plant's effluent. In most instances preliminary treatment at an SBR facility includes screening and removal of grit and grease.

1. Screens

Screening devices are frequently installed ahead of SBRs to remove debris from the wastewater stream. SBRs are vulnerable to sticks, rags, and plastic material that can jam pumps or block automatic valves from opening or closing. If problem-causing debris is not found in the wastestream, then screens would not be necessary. Where screening devices are required, automatic fine screens or hand-raked bar racks may be installed. The type of screening device needed depends on the type and quantity of debris to be removed from the wastewater.

2. Grit Removal

If grit is present in the wastewater, it is going to end up in the SBR's tanks. Grit removal is usually essential for wastestreams from combined sewers.

3. Grease Removal

Soaps, oils, and grease, if not removed prior to the SBR tanks, will accumulate into large floating balls and scum on the surface of the SBRs. This problem produces odors, hinders decant operations, degrades the effluent quality, and promotes insect breeding. Hand skimming of the reactors and chlorine contact chamber is a labor-intensive (and therefore expensive) process. The unsightly appearance and foul odors from floating grease and scum give visitors and regulators a bad impression of plant operation. Regardless of the type of collection system serving the SBR plant, some form of grease removal prior to the SBR tanks should be provided if grease is present in the wastestream.

11.911 Primary Treatment

Primary treatment is not provided in most SBR processes, but may be installed or kept in retrofitted activated sludge plants, or where excessive heavy solids loading is anticipated. Primary treatment would consist of primary sedimentation tanks or clarifiers.

11.912 Reactors

1. Number of Reactors

The SBR activated sludge process requires at least two reactors. One reactor is filling and reacting while the other is settling and decanting. Some operators believe that there should be at least three reactors so that two will always be available for operation even if one tank is out of service for extended maintenance periods.

2. Tank Configuration

Reactors have been constructed in various forms and shapes, including rectangular, circular, and oval. They may be equipped with straight or sloped side walls.

3. Reactor Depths

The wastewater level in an SBR reactor reaches its maximum height during the fill stage, and it drops to its minimum level after the decant (drawdown) stage. After drawdown, a large portion of the tank's contents still remain in the reactor. For this reason SBR reactors are quite deep. When the reactor has been filled to its highest working water level, this is called the *TOP WATER LEVEL (TWL)*. At the end of the decant cycle, after effluent has been drawn off to a prescribed depth, this depth is called the *BOTTOM WATER LEVEL (BWL)*. The water depth of most SBR reactors ranges from 11 to 20 feet (3.3 to 6.1 m) for TWL, and 7 to 14 feet (2.1 to 4.2 m) for BWL. The operating range (from BWL to TWL) of the reactors is from 3 to 6 feet (0.9 to 1.8 m) of water depth.

4. Decanters (Figure 11.31)

Decanters are installed in each reactor to remove the mixed liquor following treatment. Several types of decanting equipment are produced by several different manufacturers. Some decanters use fixed-elevation submersible pumps while others use a siphon. Other decanters may be elevated above the reactor's water surface in a so-called park mode. Decanters of the launder type convey the treated effluent from near the surface of the reactor after the settling period. Some decanters are designed to move downward with the fall of the water level in the reactor. Decanter working levels are from the TWL down to the prescribed set point of the tank water elevation for the BWL.

5. Aeration/Mixing

Mixing or aeration of the reactor may be accomplished by mechanical mixers or by blowers (Figure 11.32) supplying air to either jet or fine/coarse bubble diffusers. In some instances, a combination of mixing devices is used in the reactors to conserve energy. One disadvantage of using air diffusers in SBRs is the possibility of overaeration. At reactor BWLs, the lower head over air diffusers can greatly increase the air flow rates.

6. Wasting

The settled activated sludge blanket in the reactor ranges from 5 to 12 feet (1.5 to 3.6 m) in depth, depending on the depth of the reactor. Usually the sludge blanket measures 40 to 50 percent of the TWL.

Decanter out of water during react (aeration) cycle

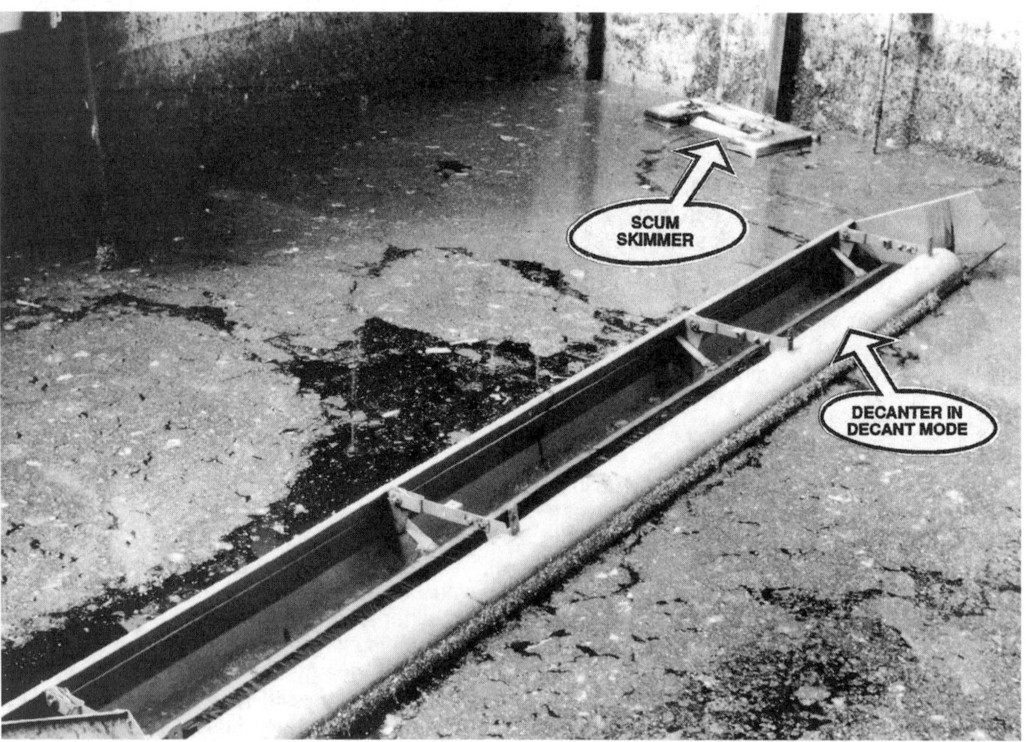

Decanter in water during settle cycle

NOTE: Grease and scum on surface and ineffectiveness of scum removal device at top of photo

Fig. 11.31 Decanters

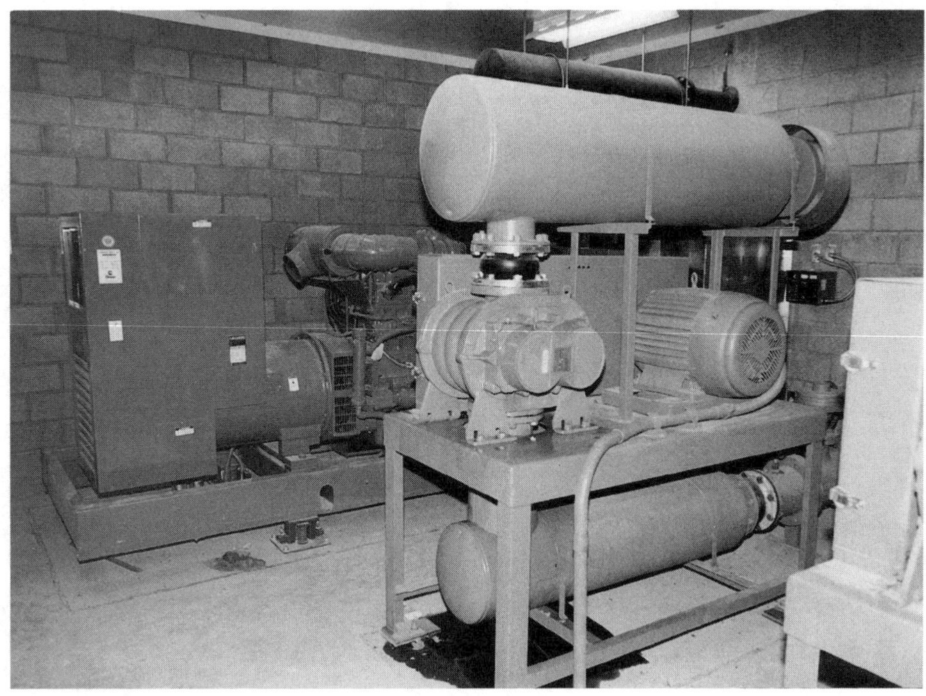

Fig. 11.32 Blowers

Wasting of excess activated sludge from each reactor may be accomplished in either of two ways: (1) gravity drawoff lines can be used to remove mixed liquor either during the aeration or settling periods, or (2) a submersible pump can be used. In some designs, the pump can be raised or lowered to various levels in the reactor. The waste activated sludge (WAS) is processed through other sludge handling facilities such as aerobic digesters or sludge drying beds (Figure 11.33).

11.913 Sequencing Control

SBRs are controlled by a microprocessor controller called a programmable logic controller (PLC) (Figure 11.34). The PLC controls the sequencing of reactor cycles for each stage of treatment by operating pneumatic, solenoid, or motorized valves. The PLC regulates aeration or mixing devices, and it controls the decanting equipment through level sensors, automatic timers, and flowmeters. The PLC may also be programmed to control other plant equipment such as screens, influent pumps, and chlorinators. These components of automated equipment and the computer *SOFTWARE*[38] are supplied by the manufacturers of the SBRs. The PLC hardware is built in modules and replacement of a faulty module is not difficult. An internal battery powers the PLC in case of a power failure. If both the battery and the main power supply fail, the software is backed up by a memory chip and can easily be reloaded.

11.914 Post Treatment

Usually the only post treatment of the SBR plant's effluent is disinfection before discharge. Chlorine or chlorine compounds are the most widely used disinfecting agent.

A significant factor affecting SBR effluent disinfection is the intermittent release of treated effluent from the reactors. After the settling period in the reactor, the decant cycle takes place. Depending on the decant equipment and the operation program in the PLC, a very rapid withdrawal of effluent from the reactor may occur. The effluent flow rate may be as high as four times normal plant inflow. Therefore, disinfection facilities must be sized to accommodate the high flow releases. In particular, tankage must be available to provide adequate contact time with the disinfection agent, and disinfection chemical addition equipment must be sized to provide the correct dosage rates to achieve adequate disinfection.

QUESTIONS

Write your answers in a notebook and then compare your answers with those on page 134.

11.9D What preliminary treatment processes may be used at a sequencing batch reactor treatment plant?

11.9E List the different types of decanters installed on SBRs.

11.9F How is the sequencing of reactor cycles in an SBR controlled?

11.92 Operation of Sequencing Batch Reactors

11.920 Sequencing Batch Reactor Operation Stages

SBRs automatically sequence through four or more process treatment stages depending on what treatment is needed to meet the plant's effluent NPDES permit requirements. Figure 11.35 shows a typical SBR operation for one cycle.

---

[38] *Software Program.* Computer program; the list of instructions that tell a computer how to perform a given task or tasks. Some software programs are designed and written to monitor and control treatment processes.

Fig. 11.33 Sludge drying bed

1. STAGE 1—FILL OR FEED

The fill or feed stage admits raw wastewater or primary effluent. The fill process allows the reactor water level to rise from the BWL, which is the level of the reactor after the decant or idle cycle. The reactor is filled to or nearly to capacity, which is the TWL.

The maximum level during fill ranges from 60 to 100 percent of the reactor volume. The same range applies for the draw (decant) stage. The BWL during the idle period ranges from 60 to 80 percent of the tank volume. The time (length) of the cycle is controlled by the PLC responding to a timer or water level sensors in the reactor. Fill cycles vary considerably due to the process control mode and flow rates at that particular time. For example, PLCs may be programmed to accelerate sequencing process cycles during wet weather periods. During the fill cycle, some degree of treatment is taking place within the reactor. The activated sludge microorganisms that remained in the reactor after the last draw cycle continue to break down and *METABOLIZE*[39] wastes.

2. STAGE 2—REACT OR AERATION

The start of the react or aeration cycle may depend on the type of aeration devices that are used to provide dissolved oxygen to the mixed liquor. It is desirable to immediately start mixing the reactor to resuspend the settled activated sludge

[39] *Metabolism.* All of the processes or chemical changes in an organism or a single cell by which food is built up (anabolism) into living protoplasm and by which protoplasm is broken down (catabolism) into simpler compounds with the exchange of energy.

Fig. 11.34 Programmable logic controller (PLC)

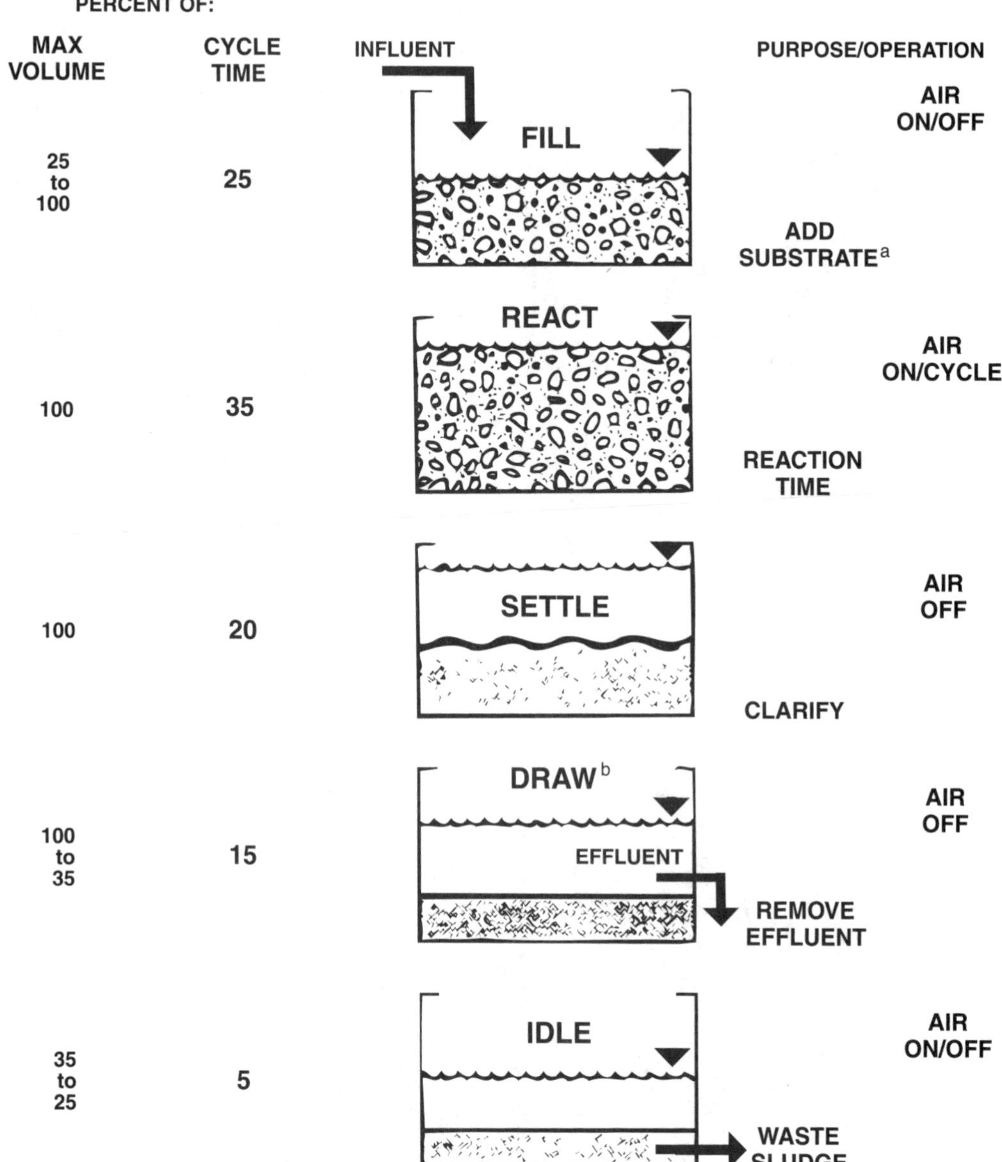

[a] Substrate—The base or food on which an organism lives.

[b] Draw—The decant or treated wastewater removal phase.

Fig. 11.35 Typical SBR operation for one cycle

(Source: Irvine, Robert L., *TECHNOLOGY ASSESSMENT OF SEQUENCING BATCH REACTORS*, US EPA, Cincinnati, OH)

and mix the influent with the mixed liquor. Mechanical aerators may start as soon as the fill cycle begins and continue operating until the end of the react cycle. Diffused air systems may be programmed not to start aeration until the reactor is almost to the TWL. This prevents overaeration and inefficient blower operation due to the low head over the diffusers. Some SBRs with diffused air aeration are designed to operate at reduced air flow rates during the fill cycle to facilitate mixing. Then, when the reactor reaches a predetermined water level, the air flow rates are increased to maintain the desired dissolved oxygen levels and mixing in the reactor. The treatment that occurs during the fill and react cycles is what would occur in a standard aeration tank where microorganisms convert the nutrients (organic material) to cell growth and provide some oxidation. The timing of the cycle, which is programmed into the PLC, is designed to achieve the particular plant's effluent quality requirements.

3. STAGE 3—SETTLE

All mixing or agitation of the reactor is stopped. This permits the activated sludge floc to form and settle to the bottom of the reactor, creating a solids blanket and leaving a clear supernatant above the settled solids. The rate of solids separation is determined by the condition of the activated sludge within the reactor. This may be observed by obtaining a sample of mixed liquor from the reactor near the end of the react cycle. The sample may be analyzed in the laboratory to determine the *SLUDGE VOLUME INDEX (SVI)*,[40] or a simple *JAR TEST*[41] may be used to measure sludge settling. The SBR frequently produces a better effluent more efficiently than a conventional system because in the settle cycle of the reactor the wastewater is completely quiescent (still).

The length of the settle cycle is based on the settling rates (SVI) usually found in the reactor. Cycle length may be controlled by elapsed time or by sludge blanket sensors. Sludge blanket sensors can be set to signal the PLC when the bottom of the supernatant layer or top of the sludge blanket reaches a predetermined level. SVIs may differ from reactor to reactor in the same wastewater treatment plant and from day to day in the same reactor. SVIs will be affected by such factors as influent wastewater strength, presence of filamentous bacteria, and presence of compounds that are toxic to the activated sludge microorganisms.

4. STAGE 4—DECANT (Withdrawal of Effluent)

The decant cycle provides the time to remove the upper layer of clarified wastewater from the reactor; this liquid is the plant effluent. The decant equipment is automatically actuated and starts the removal of effluent. Surface material (scum or floating debris) is prevented from leaving the reactor with the effluent by means of weirs or baffles. This prevents degrading the effluent due to an increase in suspended solids. The decant cycle lowers the reactor's wastewater liquid level from TWL down to the predetermined BWL. When the BWL is reached, the decant equipment is automatically taken out of service. Depending on the decant equipment, decant cycles may be short or long, but this cycle typically requires 45 minutes. Shorter or longer cycles may be programmed.

5. STAGE 5—IDLE

An idle cycle is used in SBR plants that have several reactors in service. This cycle lets one reactor completely fill during its fill cycle before switching to another reactor. Since idle is not a necessary stage, it is sometimes omitted from the process program. During the idle cycle or at the end of the decant cycle, waste sludge may be removed from the reactor.

11.921 Variations of Stages

SBRs may be programmed to accomplish various degrees of wastewater treatment. When nitrogen (N) or phosphorus (P) nutrient removals are needed, the stages are varied with different additions of fill, mix, aerate, quiescent periods, and combinations of cycles to accomplish the treatment desired. The PLCs provide the means to accomplish these programs in SBR plants. It would be much more difficult to achieve this process flexibility in a conventional continuous flow activated sludge plant because operators would be required to initiate each process change.

The flexibility of sequencing batch reactors is also shown by their ability to achieve nitrogen and phosphorus removal with simple operational modifications. Figure 11.36 shows a recommended strategy for accomplishing both nitrogen and phosphorus removal in an SBR system.

11.922 Sludge Wasting

Sludge wasting is normally required to control reactor mixed liquor solids. This, in turn, controls the F/M (food to microorganism ratio) or solids retention time (SRT)/sludge age for the activated sludge process.

An exception may be found in a few very small, low-flow plants with long sludge ages; these plants may function for long periods (several months) without the need of wasting solids. SBR plants with sludge ages of 30 to 45 days produce a fairly stable (oxidized) sludge that may not require additional oxidation. These plants generally waste directly to drying beds or compost systems for disposal of the activated sludge solids. Plants operating with lower sludge ages require additional oxidation of the wasted solids in separate facilities. The additional treatment may consist of aerobic or anaerobic digestion or wet combustion. Additional treatment is needed because waste sludges with lower sludge ages or solids retention times (SRT) contain a larger amount of residual volatile organic material, which could cause odors and other nuisances.

How the SBR plant is operated depends on the number of reactors that are available, the configuration or design of the

[40] *Sludge Volume Index (SVI).* A calculation that indicates the tendency of activated sludge solids (aerated solids) to thicken or to become concentrated during the sedimentation/thickening process. SVI is calculated in the following manner: (1) allow a mixed liquor sample from the aeration basin to settle for 30 minutes; (2) determine the suspended solids concentration for a sample of the same mixed liquor; (3) calculate SVI by dividing the measured (or observed) wet volume (m*L*/*L*) of the settled sludge by the dry weight concentration of MLSS in grams/*L*.

$$\text{SVI, m}L\text{/gm} = \frac{\text{Settled Sludge Volume/Sample Volume, m}L/L}{\text{Suspended Solids Concentration, mg/}L} \times \frac{1{,}000\text{ mg}}{\text{gram}}$$

[41] *Jar Test.* A laboratory procedure that simulates coagulation/flocculation with differing chemical doses. The purpose of the procedure is to estimate the minimum coagulant dose required to achieve certain water quality goals. Samples of water to be treated are placed in six jars. Various amounts of chemicals are added to each jar, stirred, and the settling of solids is observed. The lowest dose of chemicals that provides satisfactory settling is the dose used to treat the water.

Hours

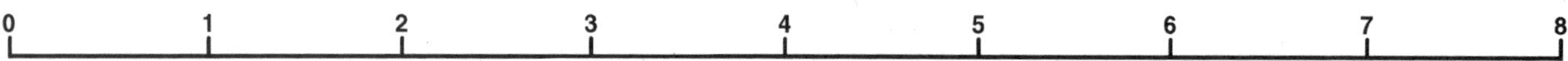

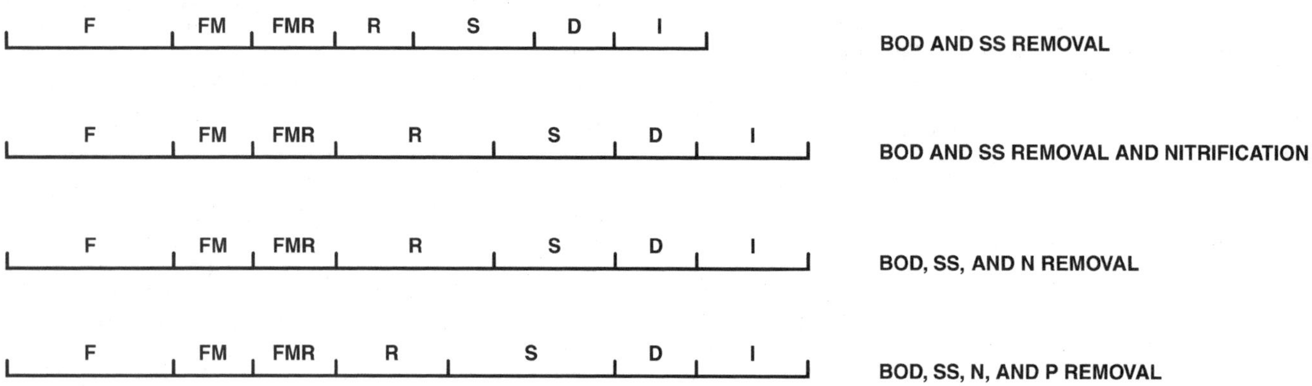

Fig. 11.36 Suggested operating strategies for SBRs for different water quality objectives

(Source: Arora, Madan L. and Umphres, Peggy B., *TECHNICAL EVALUATION OF SEQUENCING BATCH REACTORS,* US EPA, Cincinnati, OH).

reactors and, as previously mentioned, the quality of the effluent required to be produced.

11.923 Influent Variables

Important influent variables are pH, temperature, nutrients, and possible toxic substances. This section provides information on how SBR operators can prepare for and respond to these variations when necessary.

QUESTIONS

Write your answers in a notebook and then compare your answers with those on page 134.

11.9G What happens in the reactor during the react or aeration cycle?

11.9H Why is sludge wasted from an SBR?

11.93 Types of Sequencing Batch Reactors

11.930 Plants with a Single Reactor

This type of plant may be used in a municipality that has intermittent waste discharge flows from a lift station. For example, wastewater might be discharged a few times a day from a wet well or a holding tank to the SBR reactor.

11.931 Round Reactor Tanks

Round tanks limit the SBRs to the operational mode of *INTERMITTENT FILL, INTERMITTENT DECANT (IFID)*. The IFID reactor is considered the true SBR process. The round tank is a complete mix type of reactor because the tank's contents are homogeneous (similar throughout) during the react cycle. SBRs of this type can only accept intermittent flows because the reactor must have time to go through the entire treatment cycle (Figure 11.35) before more wastewater is added.

11.932 Plants with Rectangular Reactors

Rectangular reactors may be operated in a *PLUG FLOW*[42] mode if the width to length ratios are properly designed. In a properly designed tank, the flow of wastewater through the length of the tank is similar to flow in a pipe. The flow introduced at the inlet will basically stay together and come out the other end of the pipe without being mixed with liquid that entered the pipe either earlier or later.

[42] *Plug Flow.* A type of flow that occurs in tanks, basins, or reactors when a slug of water or wastewater moves through a tank without ever dispersing or mixing with the rest of the water or wastewater flowing through the tank.

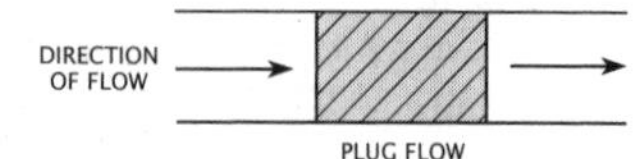

Rectangular reactors may also be operated in the IFID mode. If the plant has only one reactor, then raw wastewater storage must also be provided to accommodate intermittent feed to the reactor.

A rectangular or oval (oxidation ditch or Pasveer) reactor may also be operated in a *CONTINUOUS FILL, INTERMITTENT DECANT (CFID)* mode. Under the CFID mode of operation, a single reactor could treat the wastewater, even from a continuous flow collection system not provided with raw wastewater storage. Rectangular reactors designed for the CFID mode of operation normally have an additional partial-height wall across the tank at the inlet end. This wall creates a small filling zone where raw wastewater is mixed with activated sludge, a process known as contact stabilization, before being admitted to the main reactor compartment. The bottom of the wall has ports or slots to let the influent into the main chamber of the reactor. This wall is intended to limit mixing of the influent with the contents of the main reactor during the settle and decant cycles so that a high-quality reactor effluent is maintained. In SBR plants that require the removal of either nitrogen or phosphorus, the wall *must* be installed in the rectangular reactor to permit the operation of a treatment process that is slightly different for removal of those nutrients.

Most industrial wastewater treatment SBR plants are built with rectangular reactors. Usually two or more reactors are available to provide backup and treatment flexibility, although plants with oval (oxidation ditch) reactors can perform just as well. Round reactors are somewhat restricted in their ability to accomplish certain SBR process variations.

11.933 Sequencing Batch Reactor Operating Guidelines

The SBR plant may be operated to accommodate a range of flows from very small flows up to flows of 9.0 million gallons per day or more. Most SBR plants are operated in the extended aeration mode of the activated sludge process.

1. FLOW	Total SBR reactor volume should be equivalent to 1.2 to 2.0 times the average daily flow
2. NUMBER OF REACTORS	Typically two or more
3. REACTOR DEPTH	10 to 20 feet (3 to 6 m) working water depth plus 1.5 feet (0.5 m) *FREEBOARD*[43]
4. NUMBER OF CYCLES/DAY	Normally 2 to 6 cycles per day
5. F/M RATIO	0.02 to 0.05 lb BOD/day/lb MLVSS
6. MLSS CONCENTRATION	2,000 to 6,000 mg/*L*
7. SLUDGE AGE	25 to 45 days
8. DISSOLVED OXYGEN DURING REACT CYCLE	1.0 to 3.0 mg/*L*

NOTE: BOD and TSS removals of 85 to 98% can be expected. Nitrogen and phosphorus removal in SBRs is practiced but removal rates depend on several reactor operation variables. The SBR equipment supplier can provide you with information about the operating mode and sequencing stages you should use to achieve your desired removal rates.

11.94 New SBR Check and Start-Up

In a new plant the operator should have assistance from the designer, contractor, and equipment suppliers to put the plant on line. The operator has the responsibility to learn about the plant's physical components or structures, piping routes and sizes, and the installed equipment. A lot of this information can be obtained during construction of the facility. Being on site during construction enables you to learn about flow system routes, pipe locations and burial depths, valve placement, and locations of protective devices such as *THRUST BLOCKS*[44] or corrosion prevention systems. Watching the contractor install equipment will provide insight about the tools and meth-

ods used to lift or place equipment. You will find this information is invaluable when it becomes necessary to overhaul or repair the equipment. There are certain techniques for alignment of larger mechanical equipment (such as blowers, motors, and compressors) that are critical to their efficient operation and life expectancy. You should ask questions, read the equipment maintenance and operation manuals supplied by the manufacturers, and compare that material with the material in the plant O & M manual. Learn how to correctly start and stop various pieces of equipment, what is required for lubrication, and other factors involved in the operation of the machinery. If there are some items such as programming the PLC or aligning a blower and motor that you cannot do properly, try to learn how, or locate someone locally who can help you perform the task when it becomes necessary. Be sure to check the following items:

1. Every plant should have an up-to-date O & M manual and set of as-built plans (record drawings).
2. Just prior to placing the plant into service, thoroughly inspect the plant and equipment.
3. To the extent possible, operate and test all equipment before admitting wastewater for treatment.
4. Install protective safety guards on rotating equipment, open tanks, and channels.
5. Check and flush pipe systems and remove any refuse or debris left in them during construction.
6. Tanks must be clean. Be sure all debris, boards, ladders, and tools have been removed.

---

[43] *Freeboard.* (1) The vertical distance from the normal water surface to the top of the confining wall. (2) The vertical distance from the sand surface to the underside of a trough in a sand filter. This distance is also called AVAILABLE EXPANSION.

[44] *Thrust Block.* A mass of concrete or similar material appropriately placed around a pipe to prevent movement when the pipe is carrying water. Usually placed at bends and valve structures.

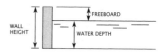

7. In plants equipped with diffused air aeration devices, check to see that all diffusers have been installed leaving no open bosses (where diffusers are connected) on the air header. It is critical that all air diffusers be at the same elevation on their manifold. If some are higher than others, more air will pass through the higher diffusers, which could reduce treatment efficiency.

The following procedures apply to the start-up of an SBR. As with any activated sludge plant, an activated sludge culture is needed. There are two ways to obtain or start a culture for a new SBR plant.

1. Obtain seed sludge from another nearby activated sludge plant. Approximately 1,500 to 2,000 gallons (5,680 to 7,570 liters) of settled activated sludge transported in a septic pumper tanker and discharged into the new reactor should be enough to get the process started. Seed sludge speeds up the growth of mixed liquor suspended solids; the plant is usually adequately treating the wastewater in one to two weeks.
2. Fill the reactor with plant influent wastewater and operate through the cycles to establish a new activated sludge culture. This procedure may require from one to four weeks or more to establish a mixed liquor solids concentration sufficient to properly treat the influent. Provisions must be made for the discharge of any partially treated wastewater during this start-up period.

An SBR is selected and the PLC program is activated to fill the reactor. Once the liquid level in the reactor reaches the bottom water level, the rest of the cycle should be automatic. Because the starting mixed liquor solids concentration is low whether the reactor was seeded or not, there will be excess dissolved oxygen levels and foam may appear on the reactor surface. If possible, aeration should be reduced during the react cycles of the start-up period. Foam buildups may be controlled by water sprays, hoses with nozzles, or chemical antifoaming agents. Once the desired mixed liquor suspended solids concentrations are obtained, the white foam problem should end.

If the plant has more than one SBR tank, all of the reactors should be started up. When establishing a new activated sludge culture, it is desirable to have long aeration periods and frequent additions of influent into the reactor as food to stimulate growth of the organisms. Operating two or more reactors will provide longer react (aeration) cycles in each reactor.

No wasting of activated sludge from the reactors should be necessary during the start-up period. If a laboratory is not available, the buildup of mixed liquor suspended solids can be monitored by sampling during the react cycle and allowing the sample to stand in a 1,000-m*L* graduated cylinder or a quart jar. The solids should come together forming a floc and then settle to the bottom of the container. The solids should have settled at least 50 percent by volume in 10 minutes, and 80 percent in 30 minutes. Each day you should be able to see an increase in the solids content that has settled out after 30 minutes. After the first three weeks of operation, analyze (in a laboratory) a sample of mixed liquor for suspended solids (SS) and volatile SS concentrations to determine the actual suspended solids content.

During the start-up period, the reactor will operate through the programmed cycles. Adjustments can be made to the react and decant cycles until the mixed liquor solids concentration has reached the desired level. The start-up period is usually sufficient to correct any deficiencies found in the various systems, and to alter PLC programs if needed.

QUESTIONS

Write your answers in a notebook and then compare your answers with those on page 134.

11.9I What is a plug flow?

11.9J Where can an operator find out how to remove nitrogen and phosphorus in an SBR?

11.9K Why should all air diffusers be at the same elevation on their manifold?

11.9L How can foam problems be controlled when starting a new SBR?

11.95 SBR Process Control

Process control of an SBR system is achieved by sampling and analyzing the reactor contents and determining the F/M ratio (or sludge age/SRT) and dissolved oxygen level. The same techniques and laboratory procedures are used as for all activated sludge plants. However, SBR operators must be alert during sampling of reactor contents for analysis and process control. SBRs have some unique requirements. Sampling of a reactor, for whatever constituent, must be performed:

1. At the same time in that reactor cycle
2. At the same reactor liquid level
3. At the same location in the reactor

Also remember that each SBR is an individual treatment plant.

The reactors operate through sequencing stages. Do not sample or sound for sludge blanket level or depth during the react cycle. There should not be any blanket during the react cycle. Measure sludge blanket depth at the end of the settle (quiescent) cycle just before the decant cycle begins. The mixed liquor suspended solids concentration will be much higher if sampled at the end of the quiescent or decant cycles, providing erroneous data for determinations of F/M or sludge age. Mixed liquor SS samples should be obtained halfway through or near the end of a react cycle, and when the reactor water level is at the same depth as when previous mixed liquor SS samples were obtained. Low reactor liquid levels concentrate SS solids; conversely, at high liquid levels, the reactor is full and SS solids concentrations are lower. Thus, when calculating F/M or sludge age/SRT, the answer could be in error unless the water level is the same each time samples are taken.

The SBR operator must be aware of a reactor's program and know which stage the reactor is at in a given cycle in order to obtain valid and consistent information for process control.

The PLC (Figure 11.34) controls the physical aspects of the reactor such as valve settings, cycle time lengths, starting and stopping of mixing or aeration equipment, and, in most instances, starting and stopping of other plant equipment. The PLC should be programmable by the operator, and you should learn how to change the program in order to improve plant efficiency. You should also know how to restore the PLC to service in case it is accidentally shut down due to power interruption or some other accident. (Some PLCs prevent the operator from making major program changes.) The equipment supplier can provide the necessary information and may be willing to teach you how to make program changes, if necessary. Remember, if you are not sure what you are doing, ask the supplier or a knowledgeable person to make the changes.

SBR treatment plants that are equipped with multiple reactors have a good chance of surviving a slug toxic waste dump. If the toxic waste is fed to only one reactor, it may destroy the culture in that reactor. The reactor contents will have to be removed and the reactor will have to be restarted. If the other reactors did not receive the toxic waste, seed sludge from them may be used to re-culture the reactor that was upset.

SBRs are less prone to the solids losses (washouts) due to high flows that commonly affect conventional continuous flow activated sludge plants. Since there are no continuous flows through an SBR, there is little opportunity for solids losses.

11.96 Maintenance

SBR plants are very similar to other activated sludge plants with regard to maintenance. There are several areas that should be monitored and may require some extra effort by the operators.

SBRs equipped with diffused air aeration usually use coarse bubble type diffusers (Figure 11.37). The diffusers are the disc or a sparger (sprinkler) type with a large orifice for air discharge to the mixed liquor. Fine bubble diffusers such as porous plate, sock, or Saran-wrapped tubes are rarely installed on SBR reactors due to possible fouling by the intermittent presence of a heavy sludge blanket. Several types of diffusers are shown in Figure 11.37.

During the cycle periods of *SETTLE, DECANT,* and *IDLE,* the air supply to the reactor is valved off by the PLC. During these periods, mixed liquor and settled sludge can flow back through the diffusers into the diffuser air manifold. These liquids and solids will not be displaced until the air is turned on again in the reactor for mixing and aeration during the *REACT* cycle. Deposits of solids in the air manifold and diffusers will continue to build up until the diffuser orifices are restricted, creating a back pressure on the aeration blowers. When back pressures approach the equipment manufacturer's guidelines, or mixing and DO levels become inadequate, the reactor should be taken out of service and dewatered so that the air manifold and diffusers can be cleaned. At this time any accumulated grit should also be removed from the reactor. If the reactor is equipped with exposed metal piping, it should be inspected for corrosion due to *ELECTROLYSIS*[45] and a protective coating should be applied or corrosion protection devices should be repaired or replaced.

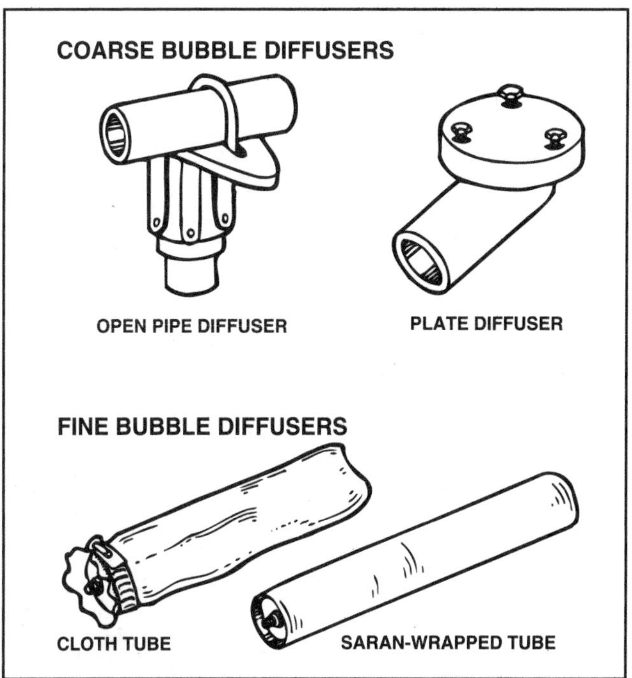

Fig. 11.37 Aeration diffusers
(Source: *PACKAGE TREATMENT PLANTS OPERATIONS MANUAL,* US EPA, Washington, DC)

Remember that entry into the reactor must be done in compliance with confined space regulations and that the application of protective coatings must be limited to avoid creating an atmospheric hazard.

Keep the surface of the reactors free of floating grease balls and scum or floating debris. Even with a preliminary treatment system for removing these constituents, some accumulation will occur on the surface of the reactors and may have to be removed by hand skimming with fine mesh net or a sieve (Figure 11.38). Take proper safety precautions to avoid falling into the reactor.

One last area that should be monitored is the disinfection system. The designer should have taken into consideration flow rates through disinfection structures to provide adequate treatment or contact times. Decant cycles may be very fast and the reactor may discharge high volumes of effluent in a short period of time. These flow rates may be three to five times higher than normal plant influent flow. The operator must make certain that adequate chemical feed is being delivered to the effluent and that sufficient contact time is provided under all possible flow conditions.

Even though the PLC performs the major portion of plant operation and relieves the operator of some duties, the SBR plant will require close attention to process conditions and vigorous performance of maintenance tasks to keep the system functioning properly.

[45] *Electrolysis* (ee-leck-TRAWL-uh-sis). The decomposition of material by an outside electric current.

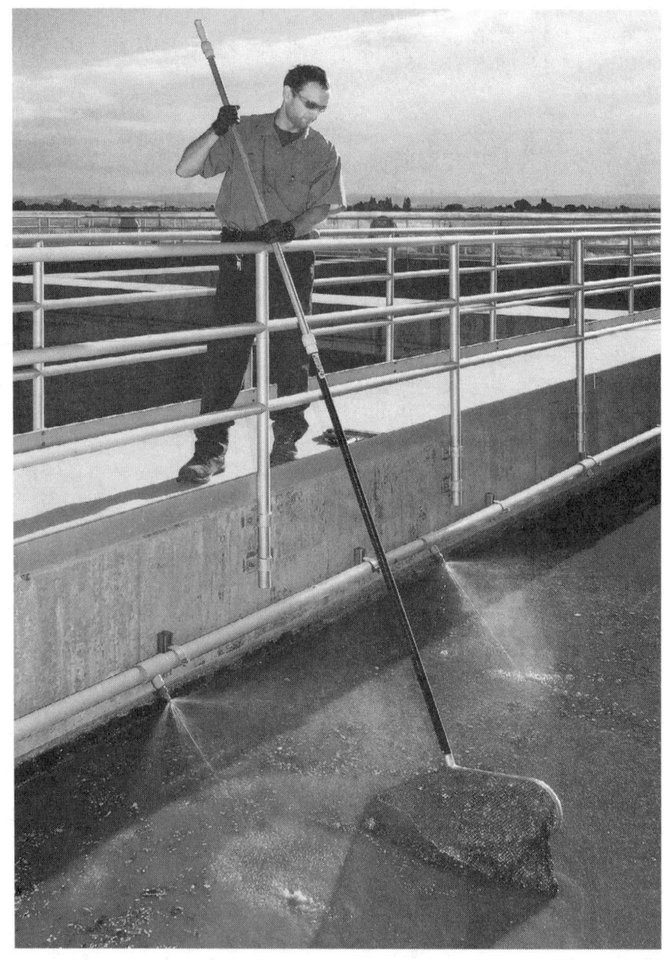

Fig. 11.38 Dip net or screen to remove floatables
(Source: Joe DiGiorgio, ECO:LOGIC Engineering)

QUESTIONS

Write your answers in a notebook and then compare your answers with those on page 134.

11.9M List the three important sampling requirements for an SBR.

11.9N When should the sludge blanket level or depth be measured?

11.9O Why do SBR treatment plants that are equipped with multiple reactors have a good chance of surviving a slug toxic waste dump?

11.9P How can floating grease balls and scum or floating debris be removed from the surface of an SBR?

11.97 Review of Plans and Specifications

11.970 Need for Operator Reviews

The operator who has the good fortune of being involved in the expansion of existing facilities or the design and construction of a new wastewater treatment facility will have a large advantage in operating and maintaining the facility. When reviewing plans and specifications, operators can be very helpful to design engineers by thinking of how they plan to operate and maintain each process or piece of equipment.

Advances in technology for equipment that can control processes and sensing devices to monitor dynamic process streams have led to the automation of proven wastewater treatment systems, with the SBRs being a good example. Automatic systems function 24 hours a day, 7 days a week, and with a greater reliability than was previously provided by human operators. The PLC and automatic monitoring equipment have taken over the routine tasks of adjusting valves, starting and stopping aeration equipment, measuring water levels, and keeping track of time periods to make appropriate adjustments. This leaves the SBR operator free to perform maintenance tasks and laboratory analyses and gives the operator time to think about what the system is accomplishing and how to improve its performance.

Since the SBR is a relatively new process, improved systems and new designs and equipment are continually being introduced. In the review of plans and specifications, the operator should carefully think about the items listed in Sections 11.971 and 11.972. Do not hesitate to ask questions and make suggestions for improving the layout or reducing safety hazards.

11.971 Physical Layout

1. Is there room to safely service and work on installed equipment such as blowers, motors, pumps, and instruments?
2. Is there access for heavy equipment, for example, portable pumps if needed to dewater a reactor cell by means other than installed equipment; cranes to lift out such equipment as decanters, skimming devices, surface aerators, or large slide gates for extensive maintenance without interfering with trees or overhead power and phone lines? Can installed heavy equipment such as blowers, pumps, and motors be removed easily from buildings through doors or access covers, and are lifting eyes installed in ceilings for lifting or hoisting equipment?
3. Is access to structures safe and well lighted for emergency night work? Are guardrails installed? Is safety and rescue equipment available and stored in a convenient location? Is the plant site fenced and are there adequate gates for easy equipment access? Are there signs that inform the public of potential health and safety hazards? Are fences grounded in case of downed power lines or lightning strikes?
4. Are there enough utility stations for cleaning and maintaining structures and equipment and are the stations properly located?
5. Are you aware of any site-specific conditions that the designer may not be totally aware of, such as flooding conditions, heavy snowfalls reducing ease of access to structures and components, vandalism, frequent power failures, or any condition that might interfere with the reliability or safety of the plant?

11.972 Specifications

Job specifications are the legal documents that list in detail the construction methods, equipment to be installed, and conditions for payments to the contractor and equipment suppliers. Specifications are sometimes difficult to interpret. Defining the true meaning of the terms often requires professional legal assistance. The SBR is a comprehensive system that relies on good, compatible, automated equipment such as variable-speed motors, actuators, instrumentation input-output devices and a programmable logic controller. To ensure that the industry receives an acceptable, fully operational system, the specifications should include the items listed below.

1. A performance bond large enough to readily correct any unsatisfactory or incomplete work by the contractor.
2. Due to the sophistication of an SBR system, the contractor should be responsible for purchasing and installing the equipment. As a precautionary measure, the contractor will usually withhold a portion of the final payment to the supplier until the system has been proven to be operational. An

industry or agency sometimes believes it can save a large portion of the investment by purchasing the equipment and then having the contractor install it. However, if the system fails to operate as intended, the industry or agency may have to pursue both the contractor and equipment suppliers to complete the job, and usually at a far greater cost, with a less satisfactory result.

3. Cooperation should be required between the contractor, equipment suppliers, designer, and the industry or agency during start-up to ensure that the plant is started up properly and is debugged. Ongoing assistance to the industry or agency should be required until the system is fully functional.

4. A comprehensive O & M manual should be written specifically for that plant. Three versions should be developed and delivered on the following schedule:

   - Available at start of construction:
     A rough draft of the manual, consisting of plans, process diagrams, control strategies, operational modes, laboratory testing, maintenance program, staffing requirements, and spare parts lists.
   - Available at the time of process start-up:
     Construction update of draft O & M manual reflecting any installed equipment changes or design changes made during construction; these are the as-built plans (record drawings).
   - Available after six months of operation:
     A final draft of the O & M manual revised and updated to include data accumulated during the first six months of operation.

QUESTIONS

Write your answers in a notebook and then compare your answers with those on page 135.

11.9Q Why should the operator of an SBR be involved in the design and construction of the unit?

11.9R What are some types of site-specific conditions that an SBR designer should know about but may not be aware of?

11.9S What information should be included in the final draft of the O & M manual?

Please answer the discussion and review questions next.

DISCUSSION AND REVIEW QUESTIONS

Chapter 11. ACTIVATED SLUDGE

(Lesson 6 of 7 Lessons)

Write the answers to these questions in your notebook. The question numbering continues from Lesson 5.

44. Why might an operator wish to use step-feed aeration?
45. How would you operate the step-feed aeration process during:
    1. Peak flows?
    2. Sludge bulking in the secondary clarifier?
46. What is the name of the treatment process that is capable of producing effluents with a quality (BOD and suspended solids) between primary treatment and conventional activated sludge?
47. What is the main difference between the sequencing batch reactor (SBR) process and the conventional activated sludge process?
48. What is the sequence of process stages that is used for wastewater treatment by the activated sludge sequencing batch reactor (SBR) process?
49. Why are activated sludge SBRs popular for smaller industrial plants?
50. Why should soaps, oils, and grease be removed from the wastewater prior to the SBR tanks?
51. Why does the SBR frequently produce a better effluent more efficiently than a conventional activated sludge flow system?
52. What can an operator learn during the construction of an SBR?
53. How can an operator obtain or start an activated sludge culture in a new SBR plant?

CHAPTER 11. ACTIVATED SLUDGE

OPERATION OF CONVENTIONAL ACTIVATED SLUDGE PLANTS

(Lesson 7 of 7 Lessons)

11.10 MICROBIOLOGY FOR ACTIVATED SLUDGE
by Paul V. Bohlier

11.100 Importance of Microbiology

The activated sludge process is a living biological process. Activated sludge is made up of many different types of microorganisms, referred to as a mixed culture. (A pure culture would be many microorganisms of the same type.) You should not only know what types of microorganisms are in your aeration tanks, but also how the microorganisms live in the environment you create for them. The more you learn about microbiology, the better you will understand how to control the microorganisms.

11.1000 Who Should Know Microbiology?

The person responsible for making process changes should have some knowledge of the microorganisms that are being controlled in the process. This person should teach the less experienced operators. If there is no one to teach you, this lesson will help you learn what you need to know to control biological treatment processes.

11.1001 Better Control of Process

Microbiology is another tool for you to use in controlling the activated sludge process. Because the process is so complex and the influent wastewater flow and characteristics are constantly changing, it is to your advantage to use all the tools available to help you meet your plant's effluent requirements.

1. Minimizes Guessing

There are times when you can successfully control the activated sludge process based on laboratory data, calculated process guidelines, and visual observation of aeration tanks and secondary clarifiers. However, there will be times when you will be trying to fit all this information together like pieces of a puzzle only to find that the pieces do not fit. There is conflicting data. You could make an educated guess as to which process change should be made and wait to see what happens. Or, you could use the same information with conflicting data along with the use of a microscope. Knowing the types and quantities of microorganisms present in the system will enable you to more accurately decide which process change is necessary.

2. Provides Signs of Trouble Ahead

There are many times that operators react to process problems instead of acting to prevent problems. Many times, effluent quality deteriorates to the point of violating effluent requirements or is close to violating before a process change is initiated.

If you use the microscope as recommended later in this lesson, you will see microbial changes while effluent quality is still good and before laboratory process data reflects any significant changes. This is the time to make a process change to maintain good effluent quality.

QUESTIONS

Write your answers in a notebook and then compare your answers with those on page 135.

11.10A The activated sludge process is a ________ biological process.

11.10B Why is activated sludge referred to as a "mixed culture"?

11.101 Collection of Samples

Like all other samples collected for laboratory analyses, it is very important to get a representative sample for microscopic observation. These samples should be grab samples from an aeration tank collected at the same time each day, preferably when the Sludge Volume Index (SVI) grab sample is taken. This will allow you to correlate your microscopic observations with changes in the SVI. When collecting grab samples, use proper safety precautions to avoid slipping or falling into the aeration tank. Never use composite samples or water from a sampling pipe for microscopic analyses. These samples will not give you an accurate count of the numbers and types of microorganisms in the aeration tanks.

11.1010 Equipment Needed

To collect the micro sample (micro sample is a sample taken for microscopic observation), you need a small plastic bottle that holds 100 to 300 m*L*, and a dipper pole made to hold the bottle (Figure 11.39). The pole can be a broom handle or a length of 1/2-inch (1.25-cm) aluminum conduit.

If you already have a dipper pole with a large bottle for the SVI grab sample, this may be used to grab a sample from the

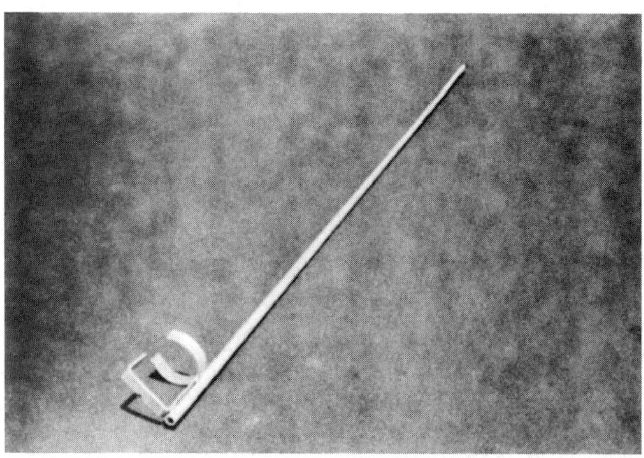

Fig. 11.39 Sampling pole with bottle holder

aeration tank. The small bottle may be filled from the large bottle. Samples collected in large bottles must be thoroughly mixed before filling small bottles.

11.1011 Sampling Location

The routine micro sample should always be taken from the same point in the aeration tanks. A point to remember is that you want to observe the condition of the activated sludge at a place in the aeration tanks where the microorganisms (bugs) should be hungry. From this observation you will know whether or not the bugs are healthy and ready to be fed again.

The proper sampling location in the aeration tanks depends on the mode of operation of your plant. Figure 11.40 is a diagram of four common modes of operation and the proper sampling location for each.

1. Conventional Mode

Regardless of whether you are operating rectangular aeration tanks in parallel or in series as shown, or you have circular tanks, the sample should be taken at the effluent end of the aeration system. This is the point where the food should be used up and the bugs are hungry and ready to be separated from the liquid in the secondary clarifiers and returned to the aeration system for more food.

2. Step-Feed Mode

The step-feed mode of operation usually has two separate zones, mixed liquor and reaeration. In the mixed liquor zone, the bugs absorb and adsorb soluble and colloidal organic matter. That is, some food is diffused through the cell wall to the inside of the cell (absorbed) while more food is attached to the outside of the cell wall (adsorbed) for future feeding. After the bugs are settled out in the secondary clarifiers, they are returned to the reaeration zone where the adsorbed food is used up. The bugs should be hungry at the end of the reaeration zone.

The best sampling location for the step-feed mode would be at the end of the reaeration zone, just before the bugs are fed. However, do not take the sample close to a feed gate because there is usually some back mixing of feed. Take the sample at a point upstream of the first feed gate where you know there is no contamination from back mixing.

3. Contact Stabilization Mode

The contact stabilization mode of operation is somewhat like the step-feed mode. The contact zone is similar to the mixed liquor zone in step-feed and the stabilization zone is similar to the reaeration zone in step-feed.

In contact stabilization, the contact time between the bugs and the food supply is very short. There is a high rate of adsorption and little absorption of organic matter by the bugs. After the bugs are settled out in the secondary clarifiers, they are returned to the stabilization zone to feed on the adsorbed matter picked up previously. All the food should be used up at the end of the stabilization zone, which is the point to collect the micro sample.

4. Extended Aeration

The best sample location for extended aeration is at the effluent end of the aeration system. The extended aeration process is usually used for wastewater flows of less than one MGD. The Mean Cell Residence Time (MCRT) and air rates are much higher than for the conventional mode.

QUESTIONS

Write your answers in a notebook and then compare your answers with those on page 135.

11.10C What precautions must be exercised when collecting micro samples?

11.10D What kind of sample would you collect for microscopic analysis or observation, and when would you collect the sample?

11.102 Sample Preparation

Once you have collected your sample, it should be taken to the laboratory and prepared for analysis within 15 minutes. Some bugs are sensitive to the lack of dissolved oxygen (DO) and will appear to be dead if the sample is not analyzed promptly.

11.1020 Laboratory Supplies Needed

The laboratory supplies you will need consist of: (1) flat glass slides (3 inches long × 1 inch wide × 0.96 to 1.06 mm thick) (7.5 cm long × 2.5 cm wide), (2) cover glass (25-mm (1-inch) square), (3) eyedropper or pipet, (4) lens cleaning paper, and (5) methylene blue solution. When handling the cover glass (also called cover slip), be very careful because they are thin and easily broken.

11.1021 Procedure for Preparing Sample

Technique and cleanliness are important aspects of the procedure for preparing a sample. The slide and cover glass should always be handled on the edges, not on the flat surfaces. Also, never set them on the countertop or table because they will pick up tiny dust particles. Although these particles are invisible to the naked eye, you will see them when looking

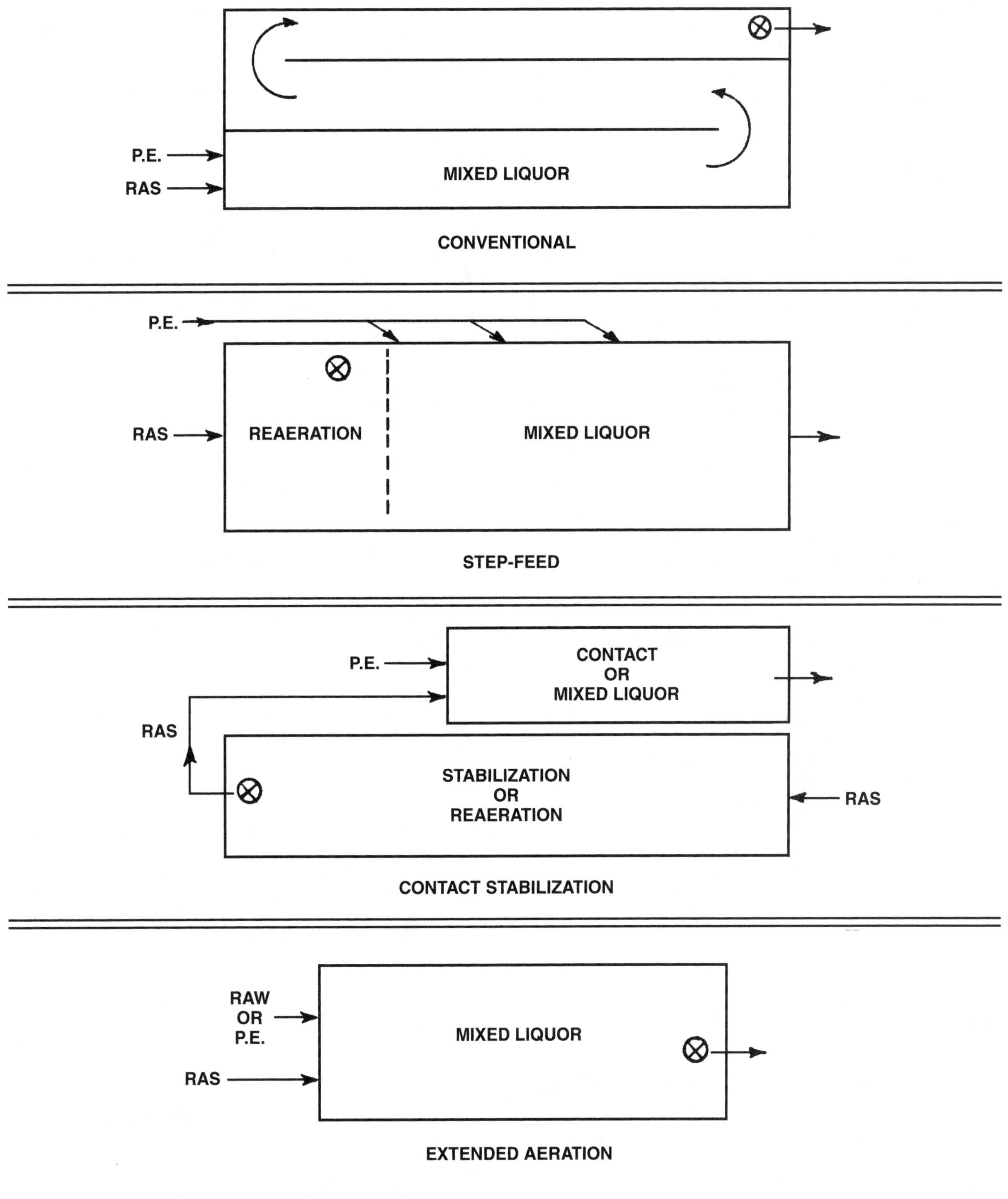

Fig. 11.40 Sampling locations for various modes of operation

through the microscope. These dust particles could be misinterpreted as "trash" in the activated sludge. If you must set a slide down, set it on a piece of lens cleaning paper.

There are two types of samples you will be preparing. The first, called a wet mount, will be used for observing live microorganisms; and the second, called a stained dried slide, will be used to observe stained filamentous (fill-uh-MEN-tuss) organisms. Stained slides should be saved for future reference.

1. Wet Mount (Live Sample)

To prepare a wet mount, take a slide from the box and hold it in one hand between the thumb and the index finger by grabbing the two long ends (see Figure 11.41). Wipe the slide clean (top and bottom) with lens cleaning paper. Now take a cover glass and hold it in the same hand between the index finger and the middle finger by grasping two edges. Wipe the cover glass clean (top and bottom) with lens cleaning paper. Practice this procedure over and over until it feels natural.

While holding the clean slide and cover glass, pick up the sample bottle with the other hand and gently rotate it 180 degrees and back (from right side up to upside down and back to right side up) several times to mix the sample. Do not shake the sample vigorously; this will break up the floc and may cause you to misinterpret the floc condition when you observe it through the microscope.

Next, remove the lid from the sample bottle, pick up the eyedropper, squeeze the bulb, insert tip of eyedropper about half way down in sample bottle and release bulb. This part of the procedure must be performed quickly before the activated sludge starts settling in the sample bottle. Pull the eyedropper out of the sample until the tip of the dropper is out of the liquid. Squeeze the bulb to release one or two drops from eyedropper back into the sample bottle.

At this point you are holding the slide and cover glass in one hand and the eyedropper with sample liquid in the other. While holding the slide in a level, horizontal position, place the tip of the eyedropper close to the surface and in the center of the slide. Squeeze the bulb to release one drop of sample onto the slide (Figure 11.41). Do not tilt the slide or the sample will run. Put the eyedropper down and, take the cover glass and gently place on the slide so the drop of sample is between the slide and cover glass. The sample is now ready to be observed.

2. Stained Dried Slide

To prepare a stained dried slide, place a drop of sample on a slide using the same procedure as for a wet mount but do not use a cover glass. Place the slide on a sheet of lens cleaning paper on the counter to air dry the sample, or place slide in a drying oven for rapid drying. Once the sample is dried, pass the bottom of the slide over a flame (two or three times just above the flame) to fix the solids on the slide. The flame from a match or cigarette lighter is sufficient for this. Do not expose the sample directly to the flame. With a pipet or another eyedropper, cover the entire dry sample with methylene blue and immediately rinse the slide under a gentle flowing stream of water to remove excess methylene blue. To dry the water from the slide, place the slide on a paper towel and gently blot dry the top of the slide. The sample is now ready to be observed through the microscope.

QUESTIONS

Write your answers in a notebook and then compare your answers with those on page 135.

11.10E What will happen if a micro sample is not analyzed promptly?

11.10F Wet mount slides and stained dried slides are used for what purposes?

11.103 Microscopic Observation

Like other laboratory equipment, the microscope is a delicate instrument and should never be handled in a rough manner. The microscope should be cleaned occasionally and should be covered when not in use. A terry cloth cover for a toaster or blender could be used.

11.1030 Microscope (Figure 11.42)

Microscopes were invented at the beginning of the 17th Century. These early microscopes were of two kinds, the simple microscope with a single lens, which was similar to the ordinary magnifying glass, and the compound microscope with a double lens system consisting of an ocular (eyepiece) and objective. Today's microscopes are of the compound type.

Microscopes fall into one of two broad categories, the light microscope and the electron microscope. The light microscope is divided into five different methods of microscopy, which are: (1) bright field, (2) dark field, (3) ultraviolet, (4) fluorescence, and (5) phase-contrast.

1. Identifying the Parts of the Microscope

The most common microscope used in treatment plants is the compound light, bright-field microscope. Figure 11.42 is a schematic of a compound light microscope with the names of the various parts. Light from the built-in illuminator or light source travels up through the condenser to give proper illumination in the objective. The condenser will usually have a diaphragm, which allows excess light in the objective to be reduced.

The prepared sample (specimen) rests on the flat portion of the microscope, which is called the stage. Most modern microscopes are available with a mechanical stage to allow easy and precise movement of the specimen forward, backward, left, or right.

The bright field microscope is usually limited to magnifications of 10× (enlarged 10 times), 43×, and 97× or 100×. You select one of these objectives by rotating the revolving nosepiece. You will be using only the two lower magnifications. The 97× or 100× objective requires the use of immersion oil between the objective lens and the cover glass.

The eyepiece should have a magnification of 10×. Other magnifications are available. The total magnification of the microscope is determined by multiplying the magnification of the eyepiece times that of the objective. Therefore, a 10× eyepiece with a 43× objective will give total magnification of 430×.

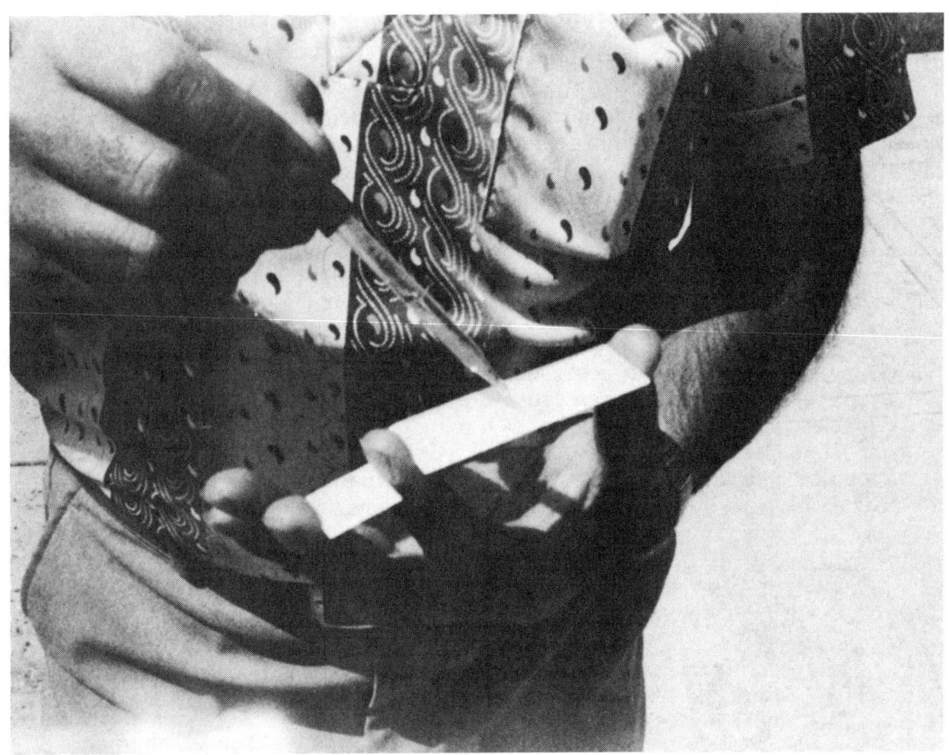

Fig. 11.41 Picture showing technique for holding slide and cover glass while depositing one drop of sample on the slide

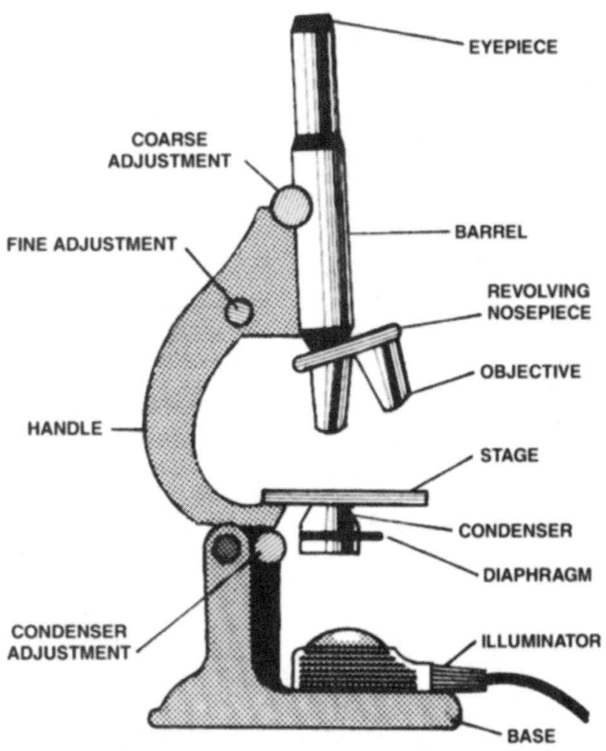

Fig. 11.42 Schematic of a compound light microscope with built-in light source

2. Type of Microscope Needed

If you are ready to purchase a microscope, you should shop around for the best combination of price and quality. The microscope should be a binocular (dual eyepiece), light microscope with (1) built-in light, (2) condenser and diaphragm, and (3) 10× and 43× objectives. One of the eyepieces should be fitted with a single hairline that will be used for counting filamentous organisms. A mechanical stage is not essential but is highly recommended. You may already have a microscope without a mechanical stage. If so, a mechanical stage may be purchased separately and adapted to your existing microscope.

The monocular (single eyepiece) microscope costs less than a binocular (dual eyepiece) microscope, but can cause eye fatigue. If money is a problem, you could purchase a monocular microscope, which will be less comfortable to work with. The monocular microscope should also be fitted with a single hairline in the eyepiece.

A specially made Polaroid camera can be purchased for either type of microscope. The camera mounts directly on top of the eyepiece of the monocular microscope. A more elaborate and expensive camera setup is available for the binocular microscope.

11.1031 Microorganisms of Importance

As mentioned previously, activated sludge is a mixed culture of microorganisms consisting of bacteria (back-TEER-e-uh), protozoans (pro-toe-ZOE-ans), rotifers (ROTE-uh-fers), and sometimes worms, fungi (FUN-ji), and algae (AL-jee).

All microorganisms used to be classified into either the plant kingdom or animal kingdom. Later, as scientists found more and more organisms that had some characteristics of both plants and animals, the taxonomy (classification of organisms) was revised. A third kingdom, protista, and a fourth kingdom, monera, were established (see Figure 11.43). Members of the third kingdom are called protists. Protozoa, algae, and fungi are protists. Bacteria and blue-green bacteria belong in the monera kingdom.

1. Bacteria

Bacteria are single-cell (unicellular) organisms and are the main workers in the activated sludge process. Individual cells have one of three shapes and are designated as: (1) coccus—round or spherical, (2) bacillus—cylindrical or rod-shaped, and (3) spirillum—spiral or corkscrew-shaped.

The unit of bacterial measurement is the micron. One micron is 1/1,000 mm or 1/25,400 inch (25,400 microns in one inch). Most bacteria measure approximately 0.5 to 1.0 micron wide and 2.0 to 5.0 microns long. A volume of one cubic inch (16.4 cu cm) could contain 9 trillion average-sized bacilli.

Bacteria reproduce by binary fission (a single cell divides into two cells). The time it takes a cell to divide is called the generation time. Different species of bacteria have different generation times ranging from several minutes to several days. Also, a changing environment will cause changes in the generation time of a particular bacterium. The *E. coli* bacteria found in the intestinal tract of warm-blooded animals and humans have a generation time of 17 minutes in a broth medium and a generation time of 12.5 minutes in milk. Each person eliminates about 200 billion *E. coli* per day as waste.

The typical growth pattern of bacteria is shown in Figure 11.44. The growth curve shows four distinct phases of growth. Not all cells go from one phase to the next at the same time (that is, growth is not synchronous). Because of this, there is a transition from one phase to the next in which the slower cells catch up to the faster ones. A brief and simplified description of the four phases is as follows:

1. Lag phase—cells grow in size, not in numbers
2. Log phase—cells grow in numbers
3. Stationary phase—population remains constant because there are as many cells dying as there are multiplying
4. Death phase—there are more cells dying than there are multiplying

This part of the lesson was intended to make you aware of the significance of bacteria in the activated sludge process. You will not be observing bacteria through the microscope, with the exception of filamentous (fill-uh-MEN-tuss) bacteria.

2. Filaments (fill-uh-MEN-ts)

Many filamentous growths in activated sludge are formed by individual bacterial cells attached end to end in multicellular chains called long-chain filaments.

a. Long Filaments

Many different types of long filaments can grow in activated sludge. You will usually find a small amount of filaments in the sludge floc. This "background" level of filaments aids in holding the floc particles together and can be called the backbone of the floc. Sludge bulking occurs when either the backbone filament or a different filament grows and extends beyond the floc. This prevents compaction in the secondary clarifiers and the sludge blanket increases in depth.

Long filaments usually grow quickly once they become established in the system. A non-bulking sludge can become a bulking sludge within two or three days. Routine observations of the sludge through the microscope will enable you to detect an increasing number of long filaments before there is a serious problem of sludge bulking.

There are many different types of long filaments and they grow in different environments. Undesirable environments that promote the growth of long filaments can be caused by: (1) process guideline(s) (such as low DO) out of adjustment, (2) nutrient deficiency (low nitrogen) in the influent wastewater, or (3) an undesirable substance in the influent wastewater or in a waste sidestream being

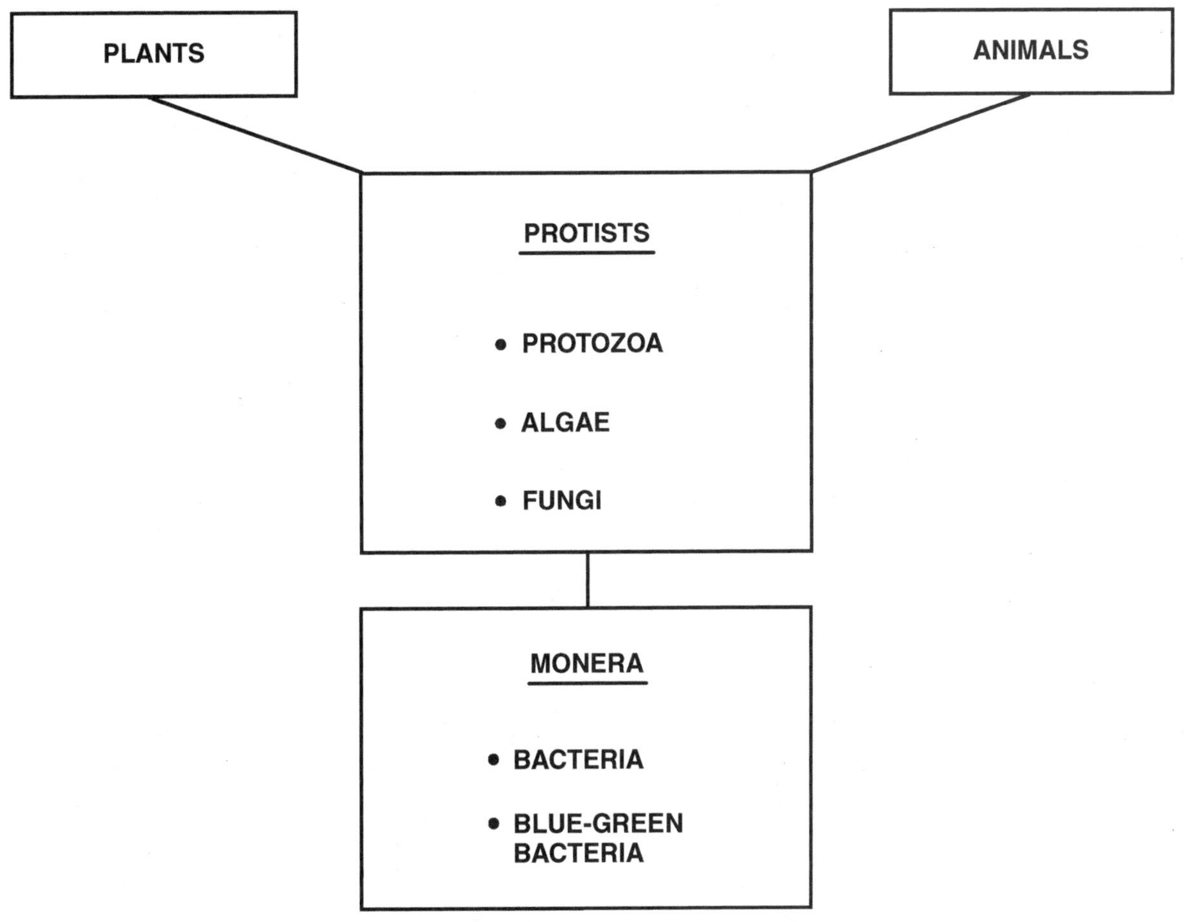

Fig. 11.43 Illustration showing placement of the kingdoms Protista *and* Monera *in relation to the plant and animal kingdoms*

returned to the plant (such as a toxic substance or hydrogen sulfide).

Most bulking sludge incidents occur during the warmer months in plants operating in the conventional mode. The problem is usually the result of one or more process guidelines being out of adjustment. One of the more common types of long filament, *Sphaerotilus* (sfer-AH-till-us), has been associated with low DO and high food/microorganism ratio (F/M ratio).

The sulfur bacterium, *Thiothrix* (THIGH-o-THR-ix), is a long filament that grows well in the presence of sulfide ions. High levels of sulfide ions may be present in the raw wastewater, may be generated in primary clarifiers with long detention times, may be present in waste side-streams returned to the aeration tanks or ahead of the aeration tanks, and may be generated in aeration tanks with low DO for excessive periods of time.

Operators are not expected to learn how to identify long filaments because there are many different types and the procedure for identifying filaments is very time consuming and requires special training.

b. Short Filaments

The most common short filament encountered in activated sludge plants is a type of actinomycete (ak-TIN-o-MY-see-tee) called *Nocardia* (no-CAR-dee-uh). These short filaments form branches that are weblike in appearance. Figure 11.45 shows a stained slide with *Nocardia. Nocardia* does not cause sludge bulking but is associated with foaming or frothing in the aeration tanks and excessive brown floating scum in the secondary clarifiers.

These short filaments are very slow growing and start growing inside the floc particles. Observation of a stained sample is necessary to detect *Nocardia* inside the floc during the early stages of growth. As the amount of *Nocardia* increases, you will start to see them in the open spaces between the floc particles. You will also see an increased amount of foam in the aeration tanks and more floating scum in the secondary clarifiers.

The presence of *Nocardia* in activated sludge is more common in plants operating in the step-feed mode, contact stabilization mode, and extended aeration plants. Short filaments are sometimes seen in conventional plants, although it is not common.

The growth of *Nocardia* has been associated with low DO and low F/M ratio. Foam or frothing is generated in the aeration tanks because *Nocardia* cells are poorly wetable and readily combine with activated sludge flocs and air bubbles to form a thick, dark brown scum or froth.

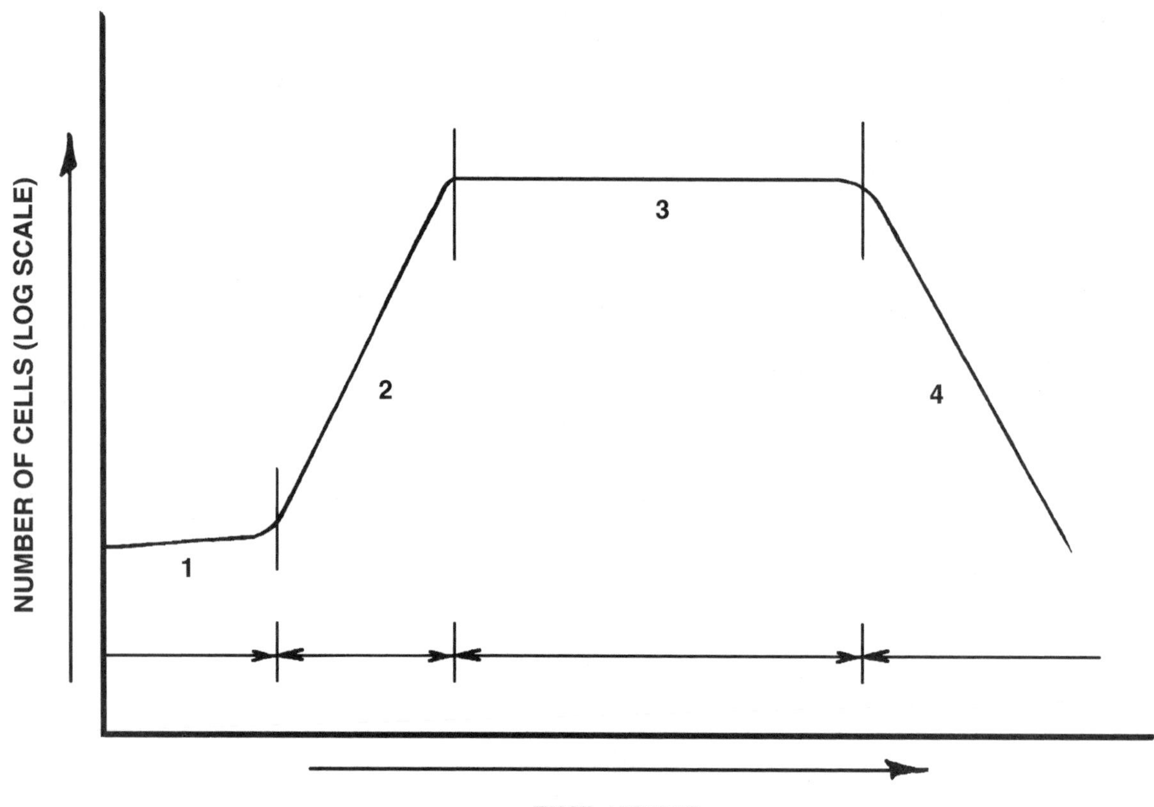

Fig. 11.44 Typical bacterial growth curve

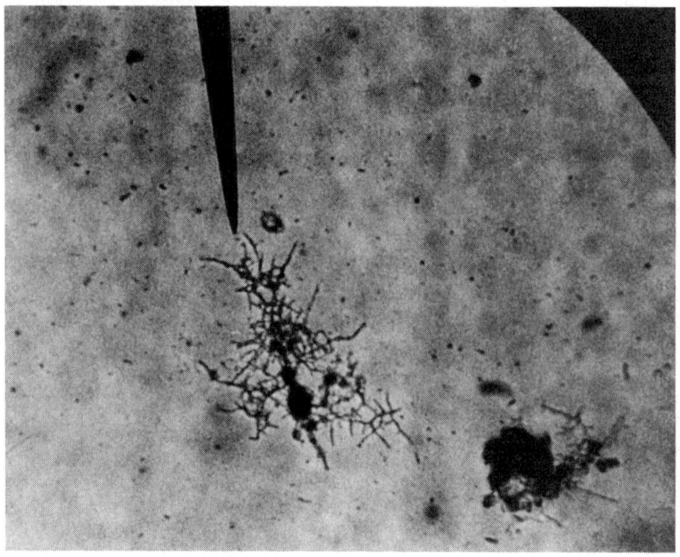

Fig. 11.45 Photomicrograph of Nocardia *at 430× magnification*

3. Protozoa

Protozoa are usually single-cell protists that range in size from 10 microns to over 300 microns and most are easily observed through the microscope at 100× magnification. Some people continue to classify protozoa in the animal kingdom.

Protozoa can be called "indicator organisms." Their presence or absence indicates the amount of bacteria in the activated sludge and the degree of treatment. The five types of protozoa (Figure 11.46) that you will be observing are:

a. Amoeba (uh-ME-buh);

b. *Mastigophora* (flagellate) (ma-STIGG-uh-FOUR-uh (FLAJ-eh-late))

c. Free-swimming ciliate (SILLY-ate)

d. Stalked ciliate

e. Suctoria (SUCK-tor-EE-uh)

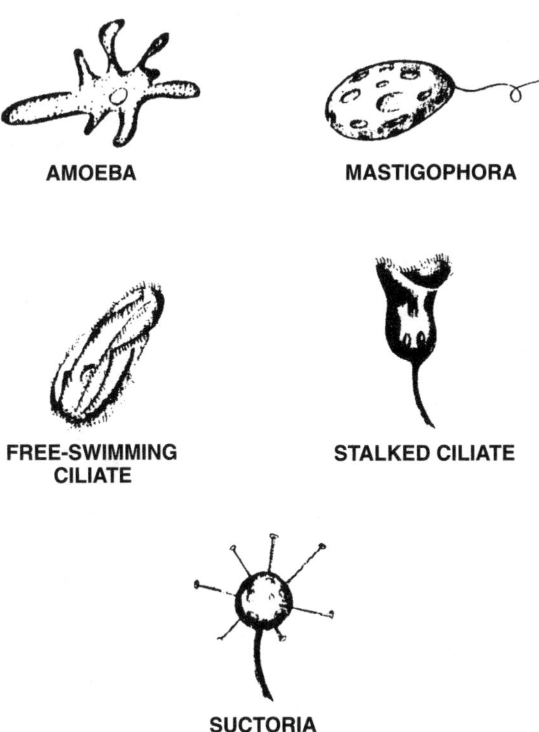

Fig. 11.46 Five types of protozoa

Amoeba

The amoeba has a flexible cell membrane that changes shape with the movement of protoplasm (living substance inside a cell). The amoeba moves very slowly in a particular direction by shifting the protoplasm to send out parts of the cell to form a pseudopod (SUE-doe-pod), or false foot.

Amoeboids range in size from 30 to 600 microns. They can absorb soluble food through the cell membrane or the pseudopods can be used to engulf solid particles and other microorganisms.

The amoeba is usually present in small numbers in most activated sludge processes. They can be seen through the microscope at 100× magnification but are frequently overlooked because of their slow movement and because they blend in with other solid particles. Amoeboids predominate in an activated sludge when a plant is undergoing start-up, when an established system is recovering from an upset condition, and when a system is operated at a high F/M ratio.

Mastigophora

Mastigophora are called flagellates because they possess one or more long, hair-like appendages called flagella. The cell propels itself in search of food by a whipping action of the flagella.

The most common *Mastigophora* are very small (10 to 20 microns) and use their flagella to attach themselves to floc particles. The cell whips back and forth very fast to absorb soluble food through the cell wall. Because of their small size, they are difficult to see at 100× magnification. These small *Mastigophora* predominate at high F/M ratios. If the F/M and MCRT are in normal ranges, their presence would indicate the DO is too low in (1) mixed liquor (conventional mode), or (2) reaeration zone (step-feed and contact stabilization modes).

The larger *Mastigophora* (40 to 70 microns) are easily seen at 100× magnification. They can absorb soluble food through the cell wall or ingest solid particles or other microorganisms. These flagellates also indicate a high F/M ratio and low MCRT.

Ciliate, Free-Swimming

The free-swimming ciliates have numerous short, hair-like extensions called cilia. The cilia beat rhythmically to propel the ciliate in search of food. Other cilia around the mouth pore direct solid food particles and bacteria through the mouth into the gullet. These protozoa range in size from 50 to 300 microns. The cilia can be seen at 430× magnification. Free-swimming ciliates expend a great deal of energy so they require large amounts of food. They are predominant in activated sludge with a large amount of bacteria. Their presence indicates the process is approaching stable operation.

Ciliate, Stalked

These ciliates grow on a flexible stalk that is attached to a solid particle. The most common of these ciliates, *Vorticella* (vor-ti-CELL-uh), are tulip-shaped or bell-shaped and have cilia around the outer edge of the bell. Bacteria are caught in a whirlpool created by the beating action of the cilia and enter the opening in the bell and into the gullet. When the gullet is full of food, the bell closes and the stalk recoils. When the food is ingested, the stalk straightens and the bell opens to catch more food. Usually when stalked ciliates first appear in the activated sludge, they occur as single organisms on single stalks. As the MCRT is increased, other species grow with a single stalk that branches out to accommodate many organisms, which is called a colony. The presence of stalked ciliates indicates a stable process that produces a low turbidity effluent.

Suctoria

Like the stalked ciliate, suctoria grow on a stalk. Suctoria have rigid tentacles that catch other protozoa swimming by. The tentacles are then used to suck the protoplasm out of the prey. Suctoria are about the same size as the stalked ciliates and can be seen through the microscope at 100× magnification. These microorganisms usually indicate an older sludge with a high MCRT and are most often found in extended aeration plants.

4. Rotifers

Rotifers are multicellular animals (classified in the animal kingdom) with rotating cilia on the head and a forked tail. The cilia are used to catch food in the same manner as the stalked ciliate. The rotifer also uses its cilia as a means of motility (to move about). The forked tail is used to attach the rotifer to a solid particle for feeding at a fixed location. The forked tail can be used for "walking" by pulling the tail under the main part of the body, grasping a solid particle with the fork, and pushing the main body forward. Rotifers consume enormous quantities of bacteria and can feed on solid particles such as pieces of floc. The most common rotifers in activated sludge range in size from 400 to 600 microns and appear very large in a microscope at 100× magnification. Rotifers are an indication of an old activated sludge with a high MCRT and are usually associated with a turbid effluent.

5. Worms

Little is known about the role of the worm, or nematode (NEM-uh-TOAD), in activated sludge. They are strict aerobes and can metabolize some solid organic matter that is not easily metabolized by other microorganisms. Worms are usually seen in sludges from extended aeration plants.

11.1032 Recording Observations

Operators must record everything they see through the microscope. These records can be useful in determining which microorganisms and how many of each type are present when a plant is producing good effluent quality, and also poor effluent quality. Although there are several methods for recording microscopic observations, one of the best ways is to use a worksheet like the one shown in Figure 11.47. Another method of documenting observations is to create files of pictures (photomicrographs) and stained slides.

1. Size and Nature of Floc Particles

The physical appearance of the floc is very important when evaluating the condition or health of activated sludge. Comments can be recorded on the worksheet such as: "floc is small, dark, and dispersed," or "floc is large, light, and stringy." Taking pictures with a camera is the best method of documenting the appearance of the floc.

2. Protozoa and Rotifer Count

Observing and counting every microorganism in the entire slide could be a very time-consuming chore. Do not try to count the tiny *Mastigophora* as they are difficult to see and there could be hundreds or thousands of them in a single drop. Make a comment on the worksheet about their presence and if the numbers appear light, moderate, or heavy. Stalked ciliates should be separated into two groups, singles (one or two bells on a stalk) and colonies (three or more bells on a single stalk).

Figure 11.48 shows a prepared slide and cover glass with circles that represent fields of view when looking through the microscope. Using the 25-mm (1-inch) square cover glass, there are about 14 fields across and 14 fields down (the number of fields will vary depending on the microscope). Using 100× total magnification and starting at one edge of the cover glass, observe consecutive fields across the entire length of the cover glass. Repeat the procedure going up or down the width of the cover glass.

3. Filament Index

Filament indexing, or filament counting, is a technique used for long filaments only. There is no known procedure for counting short filaments. If you are not experiencing bulking sludge or high SVI problems caused by long filaments, there is no need for routine counting of filaments. Filament counting should be initiated once the SVI shows a steadily increasing trend, or when you notice more filaments than normal while counting other microorganisms. To establish a normal or background level of filaments, you should perform the filament counting procedure when the activated sludge process is operating well at a normal SVI value.

When there is a need for filament counting, it should be done at the same time you are counting protozoa and rotifers, and using the same fields of view. Filament counting requires the eyepiece of the microscope to be fitted with a single hairline. The position of the hairline is not important, but it should not be moved while viewing a slide. Count the number of times that any filamentous organism intersects with the hairline (Figure 11.48). Add together the number of intersections for all fields examined. The sum of all 28 fields equals the filament count.

QUESTIONS

Write your answers in a notebook and then compare your answers with those on page 135.

11.10G What is the purpose of a mechanical stage on a microscope?

11.10H List the distinct growth phases in the typical growth pattern of bacteria?

11.10I What factors can cause filamentous growths in activated sludge?

11.104 Interpretation of Results

An operator using knowledge of microbiology as an aid for operating a treatment plant will generally be more successful than an operator who is not familiar with microbiology. However, the use of microbiology does not guarantee 100 percent success. There are still too many unknowns about the bugs in activated sludge.

The rest of this lesson is intended to give you some general guidelines to follow for operating your plant. No two plants can be operated identically; therefore, you are the best judge of your particular situation.

You may wish to use Figure 11.49 as a quick reference for determining where your process is or which direction it is headed in terms of SVI, F/M, and MCRT. For instance, if you see worms (nematodes) and a predominance of rotifers through the microscope, you can readily see that the F/M is too low (this would be normal for the extended aeration process).

11.1040 Desirable Microorganisms

When you have observed and recorded the microorganisms, you should see a predominance of free-swimming and

DATE ____________________

TIME ____________________

WORKSHEET FOR MICROORGANISM COUNTING

ENTER SUM OF ALL 28 FIELDS FOR TOTAL OF EACH TYPE OF BUG

FIELD		AMOEBA		LARGE FLAGELLATES		CILIATES						ROTIFIERS		LONG FILAMENTS	
						FREE SWIMMING		STALKED							
								SINGLES		COLONIES					
1	15														
2	16														
3	17														
4	18														
5	19														
6	20														
7	21														
8	22														
9	23														
10	24														
11	25														
12	26														
13	27														
14	28														
TOTAL															

HIGH ←	SVI	→ LOW
HIGH ←	F:M	→ LOW
LOW ←	MCRT	→ HIGH

TINY FLAGELLATES ____________________ NOCARDIA ____________________

WORMS ____________________ FLOC CONDITION ____________________

Fig. 11.47 Counting worksheet for microorganisms

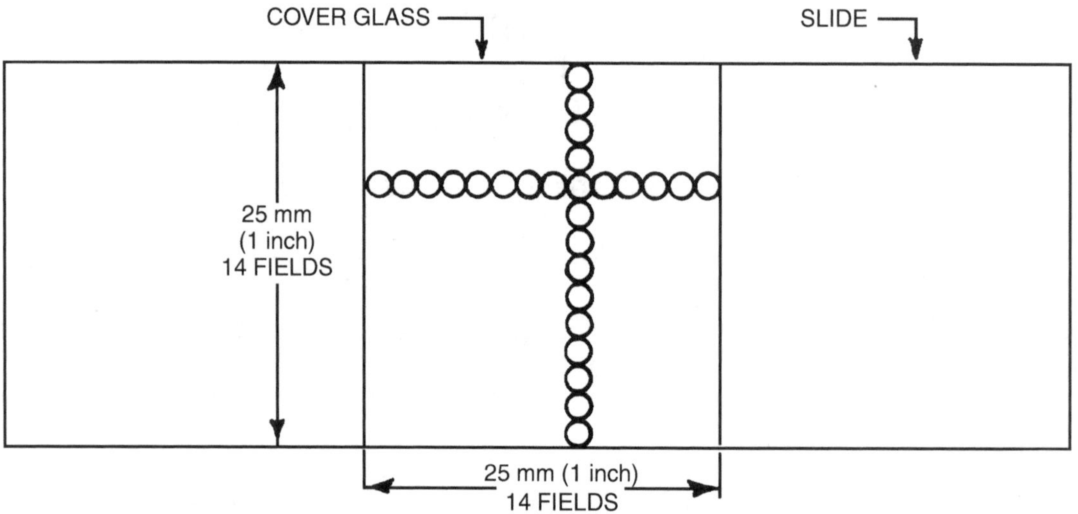

TECHNIQUE FOR COUNTING MICROORGANISMS

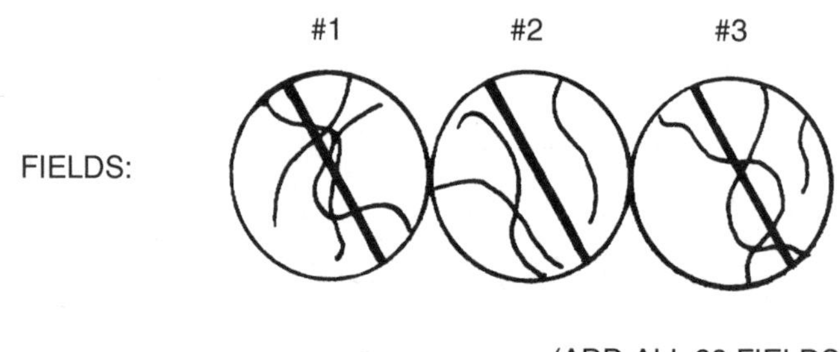

COUNT: 5 + 0 + 4 = (ADD ALL 28 FIELDS FOR TOTAL COUNT)

TECHNIQUE FOR COUNTING FILAMENTS*

**NOTE:* Techniques shown are modifications of original work by R. D. Beebe, City of San Jose, California, and D. Jenkins, University of California, Berkeley (1981). Paper presented at California Water Pollution Control Association meeting.

Fig. 11.48 Counting techniques

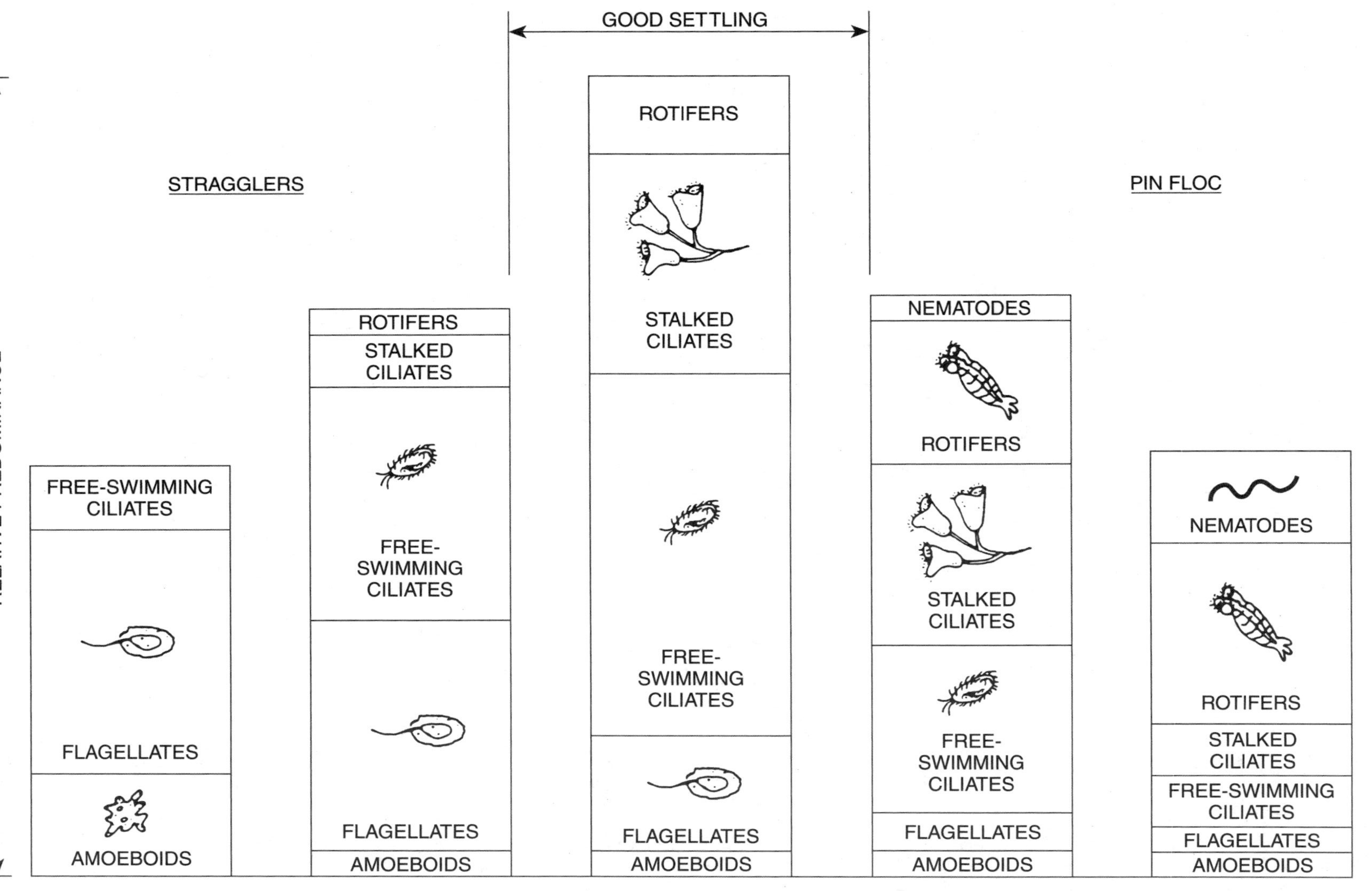

Fig. 11.49 *Population predominance versus operational guidelines (Bar graph taken from EPA PROCESS CONTROL MANUAL FOR AEROBIC BIOLOGICAL WASTEWATER TREATMENT FACILITIES, 1977)*

stalked ciliates with a few rotifers. There should also be a few long filaments.

11.1041 Undesirable Microorganisms

Flagellates and amoeboids are almost always present in activated sludge, but their numbers should be very low. Ideally, you should never see tiny flagellates *(Mastigophora)* nor short filaments *(Nocardia)*. Long filaments are not wanted in large numbers because they prevent good settling of the sludge in the secondary clarifiers. Some plants have established a maximum filament count of 15 for good operation and other plants can maintain good operation with 25 or more filaments.

When long filaments are present, check the aeration tanks and all flow streams entering the aeration tanks for hydrogen sulfide (H_2S). If you detect sulfide ions, it is possible that the long filaments are *Thiothrix*. If you eliminate the sulfide ions, the filaments (if they are *Thiothrix*) will die.

11.1042 Comparing Microscopic Results

The activated sludge process never reaches steady state conditions. This is because the influent wastewater flow, strength, and constituents are usually changing constantly throughout the day and night, from day to day, week to week, and so on.

When problems develop in the secondary process, you should first check the influent wastewater characteristics for changes from "normal." If nothing has changed, check the operation of the primary clarifiers. Is the detention time too long? Did the sludge blanket depth increase? Is the sludge going anaerobic before being removed? If chemicals are being added to the wastewater entering the primary clarifiers, the dosage could be too high, thus causing residual chemicals to be carried out in the primary effluent going to the aeration tanks. Waste sidestreams (digester supernatant, for example) being returned to the plant are often overlooked as a potential source of problems. If everything checked appears normal, the next step is to compare the results of your microscopic observation with laboratory process data and with past microscopic results.

The above comparisons should be made even when there are no problems. This will help you to analyze whether the activated sludge process will remain in a stable range of operation or if it is slowly wandering into an unstable range.

1. Laboratory Process Data

Laboratory process data, process control guidelines, and flow streams should be plotted on graphs, which will enable you to quickly observe upward and downward trends. At this point it is recommended that you plot graphs for the numbers of microorganisms that you observed and recorded on the worksheet.

Comparing microscopic results with laboratory process data is a check to support your evaluation and interpretation of the results from your microscopic observation. Example: If you see an upward trend in the numbers of rotifers and very few or no flagellates, the F/M should be decreasing and the MCRT increasing. This is confirmed when the laboratory process data show an upward trend in the suspended solids concentrations in the aeration tanks and a downward trend in the SVI.

On the other side of the comparison, if the microscopic results show a predominance of flagellates and no rotifers but the laboratory process data indicate there should be rotifers and few to no flagellates (low F/M, high MCRT), you should look for a dropping trend in the process air flow or DO. Another possible cause of the conflicting comparison could be from organic or toxic shock loadings.

2. Past Microscopic Results

When comparing present and past microscopic results, the comparison should be made with the immediate past (within the past month or so) or with the same time period last year. Note also that the wastewater temperature should be similar during both periods of time. If wastewater characteristics change (such as better industrial waste source control or a large industry shuts down), microscopic results after the change should not be compared with results before the change.

Reviewing other past records and comparing with past microscopic results may reveal events that lead to good or poor operation, which were overlooked at that time. Or, perhaps the review of past records may refresh your memory of the reason for the process going sour after many weeks of good operation.

11.1043 Changes in Numbers or Types of Microorganisms

Changes in the numbers or predominance of microorganisms in activated sludge is usually a gradual occurrence. The time required for a complete shift of predominance from one species to another species following a process change will normally be about two to three MCRTs. That is, if you are operating at a four-day MCRT, the shift of predominance would take 8 to 12 days. Of course, a drastic process change or several process changes made simultaneously will reduce the time.

With this in mind, imagine a stable process with an equal predominance of free-swimming and stalked ciliates. The numbers of each may bounce back and forth, but at some point in time a continuous trend will develop where one type of ciliate increases in numbers while the other type of ciliate decreases in numbers. A gradual shift in predominance can occur for several days before effluent quality is affected. Sooner or later a process change will be required to reverse the trend.

QUESTIONS

Write your answers in a notebook and then compare your answers with those on page 135.

11.10J List the expected *desirable* and *undesirable* microorganisms in the activated sludge process.

11.10K What changes would be expected in laboratory process data if you see an upward trend in the numbers of rotifers and very few or no flagellates?

11.105 Response to Results

After you have interpreted the results of your microscopic observations, you need to know whether something should be done about it or not.

11.1050 Decision: To Make or Not to Make a Process Change

Is effluent quality poor or starting to deteriorate? Is the SVI too high; too low? Were *Nocardia* observed through the microscope? Did you observe an increased number of flagellates or amoeboids? Are there fewer stalked ciliates? If the answer is yes to questions like these, you would probably want to make a process change.

Some people say, "If effluent quality is good, do not mess with anything." If you go by this philosophy and wait until effluent quality starts deteriorating before you implement a process change, you may see a period of poor effluent quality before it improves again. As mentioned earlier, most process changes require time to take effect. By basing most of the process decisions on your interpretation of microscopic observations, you will be able to maintain good effluent quality longer and more often than someone who waits to see the effluent quality deteriorate before doing something.

There may be times when effluent quality shows some deterioration or microbial populations shift and you decide not to make a process change because you know of some temporary condition that caused it. The following examples illustrate some such situations: (1) You have taken some secondary clarifiers out of service for maintenance. This would increase the hydraulic and solids loadings on the remaining clarifiers, thereby causing higher turbidity; (2) There may be a temporary increase in the flow or strength of a waste sidestream being returned to the plant; (3) You see indications that the process is recovering from a high COD shock loading. Whenever effluent quality deteriorates or populations change and you decide not to make a process change because you feel it is a temporary condition and the process will turn around on its own, you must watch everything very closely. A process change should be initiated immediately upon recognition of continued degradation.

11.1051 Waiting to Analyze More Data

If your plant is operating smoothly and effluent quality values (for example, BOD, turbidity) have been well below the limits of effluent requirements, you often will have time to wait for more data before making process changes in response to the first indications of a deterioration in the effluent quality. Frequently, some deterioration occurs but then levels off, with effluent values still well within the discharge requirements. In a case such as this, waiting for additional data before making any changes will help to ensure that the problem has been properly diagnosed and that an appropriate response has been selected.

Another time when you may need to wait for more data is when the existing data are conflicting and even the microscopic results are contradictory (for example, you see large numbers of rotifers and flagellates). You may feel lost and not sure of what is happening or what to do. If this happens to you, do not feel bad. There are times when even the most experienced and knowledgeable operator feels lost and not sure what to do. Usually, during times like these, most of the data point in one direction while the remaining data point in the other direction. If you are violating or close to violating effluent requirements, a process change must be made immediately. Go in the direction indicated by most of the data. Once you have a thorough knowledge of activated sludge with a basic understanding of the microbiology involved, there may be occasions when you allow your instincts to override conflicting data for making process changes.

11.1052 Making a Process Change

Assuming you made a decision to make a process change, which process change are you going to make and how much of a change is needed?

1. Deciding Which Change(s) to Make

When deciding which process change to make, evaluate the feasibility of changing each process control and how each change will affect other operating guidelines. The "Worksheet for Microorganism Counting" (Figure 11.47) was designed to help you see which direction the process is headed and at the bottom of the sheet you will see how some of the process guidelines are affected.

Example: If the numbers or types of microorganisms are shifting toward the left side of the sheet, you will see the F/M Ratio increasing, MCRT decreasing, and SVI increasing.

Process changes are used to control or change populations of microorganisms to achieve the desired level of treatment in a particular mode of operation. The physical process controls for making process changes in the conventional mode are:

1. Process air flow
2. Waste activated sludge (WAS) flow
3. Return activated sludge (RAS) flow
4. Number of tanks in service (primaries, aeration, secondaries)

The process controls listed above are ranked in order of priority. Process air flow was placed first on the list because the process air compressors or blowers are a major user of energy in most plants. If you decide a process change is needed and you conclude that you could either reduce the air flow or increase the WAS flow, the obvious choice is to reduce the air flow. This does not mean you should allow an upset to occur before increasing the air flow. Your first priority should be placed on meeting effluent requirements. If increasing the air flow is deemed necessary to comply with effluent requirements, then by all means, increase it.

Much mention has been made about air flow and very little about dissolved oxygen (DO). You should understand that DO in the aeration system is very important and that DO control for air requirements is one of the best alternatives available. However, when operating the activated sludge process by DO, you must always monitor the air flow as well. Two examples of why the air flow must be monitored are:

a. Assume the mixed liquor suspended solids (MLSS) have been slowly decreasing during the past week but you have been maintaining 2.0 mg/*L* DO at the end of the aeration tanks. During the time the MLSS have been slowly decreasing, you have been seeing fewer stalked ciliates and more free-swimming ciliates and flagellates. If you look at the air flow during the same time period, you will see that it has been decreasing even though you have been maintaining the same DO. Why? Fewer bugs require less air.

   In a situation like this you should increase the air flow significantly or increase the mixed liquor DO high enough to achieve a higher air flow. At the same time, you should reduce the WAS flow to increase the mixed liquor suspended solids. When everything looks normal again, the air flow/DO may be reduced.

b. Assume your plant is not required to produce a nitrified effluent. To determine the degree of nitrification, you need to analyze the 24-hour composite samples of primary effluent (PE) and secondary effluent (SE) for concentrations of ammonia nitrogen (NH_3). The degree of nitrification is expressed in percent using the removal efficiency formula.

$$\text{Nitrification, percent} = \frac{\text{PE } NH_3 - \text{SE } NH_3}{\text{PE } NH_3}(100\%)$$

A process can be non-nitrified (no change in NH_3), totally nitrified (100 percent nitrification), or partially nitrified (some change in NH_3 but less than 100 percent). In the process of nitrification, NH_3 is converted to nitrite (NO_2^-) by bacteria called *Nitrosomonas* (NYE-tro-so-MOAN-us), and NO_2^- is converted to nitrate (NO_3^-) by bacteria called *Nitrobacter* (NYE-tro-BACK-ter). Denitrification occurs when facultative microorganisms use the oxygen molecules from NO_3^-, leaving only nitrogen gas (N_2).

When a process totally nitrifies, the NO_2^- plus NO_3^- should almost equal the amount of NH_3 reduced or converted. If they are not almost equal (NO_2^- plus NO_3^- are significantly less than NH_3 converted), denitrification is occurring somewhere in the process.

In the second example, you assumed that your plant effluent was not required to be nitrified. Imagine your plant has been operating in the conventional mode and you have been maintaining the mixed liquor DO at 2.0 mg/*L*. One day you review your plant records and notice the process air flow has been increasing for the past week. You decide to do nothing about it. A few days later you have rising sludge in the secondary clarifiers. If you were observing the percent nitrification during this period of time, you would have seen it steadily increasing. As the percent nitrification increases, more air flow is required to maintain a given DO.

To correct a rising sludge problem, the air flow must be reduced immediately to achieve a mixed liquor DO of about 0.2 to 0.5 mg/*L* (at the end of the aeration system). The purpose of the low DO is to allow the sludge to denitrify in the aeration tanks. The rising sludge in the clarifiers will cease within a few hours following the reduction of DO, at which time the DO can be increased again but should be at a lower value than the previous value.

Many factors are involved for nitrification and denitrification to occur. Generally, if you stay below 50 percent nitrification, you should not experience rising sludge.

If you are forced into total nitrification because of effluent requirements, the opposite of the above action should be taken when you see rising sludge. This problem develops when the process is "slipping out" of nitrification. Therefore, you would increase the DO to stop denitrification.

Next on the list of physical process controls is the WAS flow. Changing the MCRT and F/M from one target value to another target value (by changing the WAS flow) would be called a process change. Routinely changing the WAS flow to *maintain* a specific MCRT or F/M value would be considered a fine-tune adjustment rather than a process change. Using the "Worksheet for Microorganism Counting" (Figure 11.47), you would want to reduce the WAS flow if the numbers of microorganisms are shifting toward the left side of the sheet.

The RAS flow is not normally used to control the process. However, there are times when an increased RAS flow will cause the SVI to increase, or reducing the RAS flow will cause the SVI to drop. Normally, the RAS flow is changed because of a problem. In the previously mentioned problem of rising sludge, the air flow was reduced. The RAS flow should have been increased to lower the sludge blankets until the reduced air flow takes effect.

There are times when taking process units (tanks) out of service could be considered a process change. If multiple primary clarifiers are used, one or more could be taken out of service to increase the F/M of the activated sludge.

If the flow through your plant is significantly less than what the plant was designed for, you may have frequent process problems due to excessive aeration detention times. This can cause a high degree of nitrification or turbid effluent because of pin floc from a low SVI. Taking an aeration tank or battery of tanks out of service can reduce the frequency and severity of process problems.

Secondary clarifiers can also be taken out of service during some of the above-mentioned problems. Taking one or more clarifiers out of service, along with other changes, can help in reducing or eliminating *Nocardia* and foaming in the aeration tanks. When clarifiers are operated with sludge blankets, the bugs in the sludge are usually in the clarifier much longer than the calculated hydraulic detention time. The bugs are without oxygen during this time.

The step-feed and contact stabilization modes have more physical process controls than the conventional mode. First, think of the reaeration zone and the mixed liquor zone in both of these modes as having separate functions. Process changes made in the reaeration zone of both modes are used to control the SVI and process changes made in the mixed liquor zone of both modes are used for controlling the degree of nitrification. Changes made in either zone are used to control the numbers and species of microorganisms and thereby control the effluent quality.

In either mode of operation, a low SVI can be increased by reducing the air flow in the reaeration zone, by increasing the RAS flow (which reduces the reaeration detention time), or by increasing the WAS flow to increase the F/M. In the step-feed mode, the primary effluent feed points would be changed to points upstream, which would reduce the reaeration detention time. To reduce a high SVI, the process controls would be changed in the opposite directions. Use the "Worksheet for Microorganism Counting" (Figure 11.47) to see which types of bugs are associated with a high or low SVI.

A major point to remember about the reaeration zone is that this is where the bugs assimilate the adsorbed food they picked up previously in the mixed liquor zone. If the reaeration time is too short, not all the adsorbed food will be used up before the bugs return to the mixed liquor for more food. This causes the bugs to become "fat," which results in a high SVI. A long reaeration time causes "skinny" bugs and a low SVI.

You may be fortunate to have a plant that was designed with a great deal of flexibility that enables you to operate the activated sludge process in any mode. When process problems occur in a particular mode of operation and process changes appear to be ineffective, or immediate results are needed, you should change the process to another mode. Following are a few examples of when this plan of action could be implemented.

If you are operating in the conventional mode and long filaments are increasing rapidly, thus causing the SVI and sludge blankets to increase, and process changes do not stop or reverse this trend, change the process to the step-feed or contact stabilization mode. For the first few days following the change, you will have to evaluate the effectiveness of the change by closely observing the SVI and sludge blanket depth. The sludge blanket depth should be reduced within three to six hours following the change, and the SVI should drop within 24 to 48 hours after the change. Remember, when counting filaments and other microorganisms in the conventional mode, your micro sample was being taken from mixed liquor. After changing the mode of operation, your micro sample will be taken from the reaeration zone. The microorganism counts will show higher values because of the higher concentration of suspended solids in the reaeration zone. Once the numbers of long filaments are reduced to low values, you may change back to the conventional mode.

Another example of when you should change the mode of operation is when you are in the step-feed or contact stabilization mode and large amounts of *Nocardia* are present causing severe foaming in the aeration tanks and an excessive amount of floating scum in the secondary clarifiers. If you are unable to control this problem by process changes, switch the mode of operation to the conventional mode. The height of foam in the aeration tanks should show a drop within a few hours following the change. Usually one to four weeks are required for the *Nocardia* and foam to completely disappear, at which time you may change back to the original mode of operation. Again remember that microorganism counts will be much different when changing from one mode of operation to another mode because of the change in location of the micro sample. *Nocardia* and foaming could be caused by industrial waste toxic shock loadings. If this is the case, do not change the mode of operation to conventional.

A last example of when you could change the mode of operation is when effluent quality is poor or close to violating effluent requirements while operating in step-feed or contact stabilization. If you are sure the problem is not caused by industrial wastes entering the plant, change to the conventional mode. After the change is made, calculate the mixed liquor detention time in the aeration tanks and the detention time in the secondary clarifiers. These combined times represent the amount of time before the secondary effluent turbidity should start dropping.

2. Deciding How Much Change Is Needed

The three previously mentioned examples of when to change the mode of operation are for serious problems that require drastic measures to obtain immediate results. During normal operation, if you are able to detect symptoms or first signs of a problem developing, a small degree of change should stop the process from deteriorating further. If the process is allowed to deteriorate beyond the first signs, a more significant change will be necessary to reverse the events.

You must know whether your process is fast or slow in responding to process changes. This knowledge can be obtained only from experience. To operate your plant successfully with few problems, get to know your plant intimately.

QUESTIONS

Write your answers in a notebook and then compare your answers with those on pages 135 and 136.

11.10L What conditions should be reviewed when considering making a process change?

11.10M List the physical process controls for making process changes in the conventional activated sludge mode.

11.10N When operating the contact stabilization mode of the activated sludge process, what process changes could be made to lower the SVI?

11.10O What can happen if the reaeration time is too short in the contact stabilization mode of the activated sludge process?

11.106 Monitoring

As mentioned earlier, monitoring the condition of the activated sludge, process laboratory data, operational guidelines, and visual observations in the aeration tanks and secondary clarifiers are important for you to see the complete picture of what is happening in the process.

11.1060 Frequency of Microscopic Observations

If you recently started using microbiology to help you operate your plant, you should make microscopic observations daily for several months. Also, you should observe your bugs during low flow and again at high flow times on the same day on as many days as possible.

1. Good Operation

Once you are familiar with the bugs in your process, you should be able to reduce the frequency of microscopic observations from daily to two or three times per week during good operation. You will be the best judge of the needed frequency by knowing how fast changes occur in your process.

2. Poor Operation

A person in charge of a treatment plant is under a lot of pressure and feels the tension or stress when the activated sludge process is operating poorly. There is a natural tendency to watch the process more closely during periods of poor operation. Microscopic observations should be performed daily or twice per day (morning and afternoon) during times like these.

3. Following a Process Change

The frequency of microscopic observation following a process change should also be daily or twice per day even if the process change was made during good operation for the purpose of maintaining good operation. This frequency is necessary to help you determine the effectiveness of the process change and whether you made the correct change. Remember that the most knowledgeable and experienced operator sometimes misinterprets process-related information or makes an incorrect process change. This can happen to anyone, even the best operator.

11.1061 Watching for Changes or Trends

Changes can occur in the activated sludge process from day to day or from hour to hour because of diurnal (daily) wastewater influent flow fluctuations, strength of the influent wastewater, waste sidestreams being returned to the plant, industrial wastes, chemical additions, or failure of mechanical equipment.

A trend is when something changes in the same direction in several successive periods of time. An example of trends can be seen of secondary effluent turbidity on a chart of a continuous recorder when you see the turbidity at a low value in the early morning hours and it goes to a higher value sometime later. You see an increasing trend when the turbidity is going up. The daily laboratory results for turbidity from secondary effluent composite samples for three consecutive days may show values of 2.3 TU, 2.6 TU, and 3.0 *TU*.[46] This is an increasing trend from day to day whereas the continuous recorder shows the hour-to-hour or even the minute-to-minute trend.

Be alert and respond immediately to the sudden appearance of undesirable bugs such as the tiny *Mastigophora,* which usually indicates that more air flow is needed.

There will be times when there are small or slow changes in the numbers and types of bugs and you want to wait to take action until you are sure of a continuing trend over a period of two, three, or four days. However, do not wait too long.

REMEMBER:
YOU SHOULD RUN THE PLANT,
DO NOT LET THE PLANT RUN YOU.

[46] *Turbidity* (ter-BID-it-tee) *Units (TU).* Turbidity units are a measure of the cloudiness of water. If measured by a nephelometric (deflected light) instrumental procedure, turbidity units are expressed in nephelometric turbidity units (NTU) or simply TU. Those turbidity units obtained by visual methods are expressed in Jackson turbidity units (JTU), which are a measure of the cloudiness of water; they are used to indicate the clarity of water. There is no real connection between NTUs and JTUs. The Jackson turbidimeter is a visual method and the nephelometer is an instrumental method based on deflected light.

QUESTIONS

Write your answers in a notebook and then compare your answers with those on page 136.

11.10P How frequently should microscopic observations be made?

11.10Q What is a trend?

11.107 Acknowledgment

Material in this section, "Microbiology for Activated Sludge," was reviewed by Virginia Alford and Anita Freudenthal.

11.11 ARITHMETIC ASSIGNMENT

Turn to the Arithmetic Appendix at the back of this manual. Read and work the problems in Section A.30, "Activated Sludge." Check the arithmetic in this section using an electronic calculator. You should be able to get the same answers. Section A.50 contains similar problems using metric units.

11.12 ACKNOWLEDGMENT

F. J. Ludzack, Chemist, National Training Center, US Environmental Protection Agency, provided many helpful comments during the original development of this chapter; his contributions are greatly appreciated.

11.13 ADDITIONAL READING

1. *MOP 11*, Chapter 20,* "Activated Sludge." New Edition of MOP 11, Volume II, *LIQUID PROCESSES*.
2. *NEW YORK MANUAL*, Chapter 5,* "Secondary Treatment, Activated Sludge."
3. *TEXAS MANUAL*, Chapter 13,* "The Activated Sludge Process."
4. *OPERATIONAL CONTROL PROCEDURES FOR THE ACTIVATED SLUDGE PROCESS*, by Alfred W. West. Obtain from National Technical Information Service (NTIS), 5285 Port Royal Road, Springfield, VA 22161. Order No. PB-286800. Price, $59.00, plus $5.00 shipping and handling per order.
5. *PROCESS CONTROL MANUAL FOR AEROBIC BIOLOGICAL WASTEWATER TREATMENT FACILITIES*, US Environmental Protection Agency. EPA No. 430-9-77-006. Obtain from National Technical Information Service (NTIS), 5285 Port Royal Road, Springfield, VA 22161. Order No. PB-279474. Price, $86.50, plus $5.00 shipping and handling per order.
6. Snelling, Donald P., "Move in the Right Direction to Control Your Sludge," *OPERATIONS FORUM*, March 1985, pages 10-13. An excellent and easy-to-read paper on how to determine whether or not to increase or decrease RAS and WAS pumping rates.

* Depends on edition.

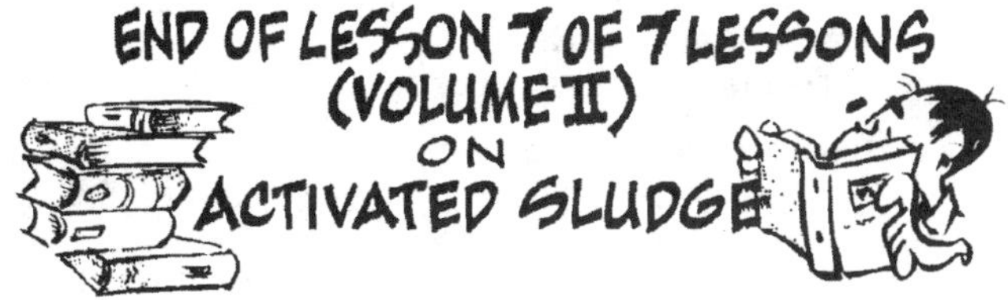

Please answer the discussion and review questions next.

DISCUSSION AND REVIEW QUESTIONS

Chapter 11. ACTIVATED SLUDGE

(Lesson 7 of 7 Lessons)

Write the answers to these questions in your notebook. The question numbering continues from Lesson 6.

54. Who should know microbiology?
55. Where should micro samples be collected in the various modes of operation of the activated sludge process?
56. What precautions should be taken with the slide and cover glass when preparing a micro sample for analysis and observation?
57. What causes sludge bulking?
58. What items should be checked when problems develop in a secondary process?
59. Under what circumstances could the effluent quality show some deterioration or the microbial populations shift in the wrong direction but still you decide not to make a process change?

SUGGESTED ANSWERS

Chapter 11. ACTIVATED SLUDGE

OPERATION OF CONVENTIONAL ACTIVATED SLUDGE PLANTS

ANSWERS TO QUESTIONS IN LESSON 1

Answers to questions on page 20.

11.0A The purpose of the activated sludge process in treating wastewater is to oxidize and remove soluble or finely divided suspended materials that were not removed by previous treatment.

11.0B A stabilized waste is a waste that has been treated or decomposed to the extent that, if discharged or released, its rate and state of decomposition would be such that the waste would not cause a nuisance or odors in the receiving water.

11.0C Air is added to the aeration tank in the activated sludge process to provide oxygen to sustain the living organisms as they oxidize wastes to obtain energy for growth. The application of air also encourages mixing in the aerator.

11.0D Air requirements increase when the strength (BOD) of the incoming wastewater increases because more food (wastes) encourages biological activity (reproduction and respiration).

11.0E Factors that could cause an unsuitable environment for the activated sludge process in an aeration tank include:

1. High concentrations of acids, bases, and other toxic substances
2. Uneven flows of wastewater that cause overfeeding or starvation
3. Failure to supply enough oxygen

11.0F Effluent quality requirements may be stated by regulatory agencies in terms of percentage removal of wastes or allowable quantities of wastes that may be discharged.

Answers to questions on page 22.

11.0G The two major variables that affect the operation of an activated sludge plant are: (1) changes in wastewater characteristics and collection system flows, and (2) in-plant operational variables.

11.0H Excessive stormwater can upset the activated sludge process by: (1) reducing treatment time of the wastewater, (2) increasing the amount of grit and silt, (3) increasing the organic loading, and (4) causing fluctuations in wastewater temperature and solids content.

11.0I Other variables in the collection system affecting the activated sludge plant include: (1) maintenance activities, and (2) the wastes discharged.

11.0J $$\text{Sludge Age, days} = \frac{\text{Suspended Solids Under Aeration, lbs or kg}}{\text{Suspended Solids Added, lbs/day or kg/day}}$$

ANSWERS TO QUESTIONS IN LESSON 2

Answers to questions on page 39.

11.1A Aeration serves the dual purposes of providing dissolved oxygen and mixing of the mixed liquor and wastewater in the aeration tank.

11.1B The two methods used to supply oxygen from air to bacteria in the activated sludge process are: (1) mechanical aeration, and (2) diffused aeration.

11.1C Diffusers are located near the bottom of the aeration tank to maximize the contact time of the air bubbles with the mixed liquor. Also, this location encourages mixing and discourages deposits on the tank bottom.

11.1D Guide vanes on centrifugal blowers are positioned manually by operating personnel or by plant instrumentation responding to dissolved oxygen levels in the aeration tanks or to plant influent flows.

11.1E Fine bubble diffusers are easily clogged because these diffusers are self-sealing from the inside if dirty air is pumped into the diffusers. They may clog from the outside due to biological growths.

Answers to questions on page 40.

11.2A Before attempting to maintain or repair a surface aerator, shut down the aerator and be sure the main power breaker is opened (shut off), locked out, and properly tagged.

11.2B A lockout tag must note the date of the lockout, the reason for the lockout, and the name of the person who locked out the equipment.

11.2C The following precautions should be taken when working with blowers:

1. Be sure all inlet and discharge valves are open before starting.
2. Remove all foreign matter that might enter blower.
3. All personnel must be clear of blower before starting.
4. Always wear hearing protection when working near an operating blower (make sure the hearing protectors reduce your noise exposure adequately).
5. When shutting down a blower, be sure the main power breaker is opened (shut off), locked out, and properly tagged.
6. Allow only qualified and authorized electricians to troubleshoot and repair electrical problems.

ANSWERS TO QUESTIONS IN LESSON 3

Answers to questions on page 44.

11.3A The main reasons an operator should completely check the equipment and structures before start-up are to: (1) familiarize the operator with plant equipment and locations of piping, and (2) to be reasonably sure that everything will function properly when the plant is put into service.

11.3B Usually, in the construction specifications, the equipment suppliers are requested to supply three or four copies of instructions for each piece of equipment. If they are lost, a new one may be obtained by writing to the manufacturer and giving the name, size, serial number, and the treatment plant location and contract number, if possible.

11.3C Most manufacturers normally supply a one-year warranty on their equipment. If the equipment is damaged due to improper operation, the warranty becomes void and your community or district must absorb the cost of repairs or replacement. Proper operation prolongs the life of equipment, reduces the cost of repairs and replacement, and helps prevent operators from being injured.

11.3D Proper starting, operating, and stopping procedures must be followed when running equipment to prolong the life of the equipment.

11.3E Lines and channels should be cleaned before start-up to remove any debris such as rocks, sand, timber, waste concrete, or other foreign material that will occupy space or damage equipment. Cleaning also will familiarize the operator with the flow routes.

11.3F Chips and scrapes on gates should be painted for the protection of the equipment against rust and corrosion. The cost of equipment is high, and it is part of the operator's responsibility to obtain as long a life as possible for plant equipment.

11.3G The main purpose of the stop-nut is to limit the length of travel. Without a stop-nut, the gate could fall into the aeration tank if it is opened too far or the stem could be bent if an operator attempts to close the gate farther than necessary.

Answers to questions on page 44.

11.3H The effluent weir should be level to prevent short-circuiting and an uneven distribution of solids in the effluent.

11.3I The water sprayed on a rail or walk may create a slipping hazard either from ice formation in the winter or algal growths in the summer.

Answers to questions on page 45.

11.3J Air is cleaned by the use of air filters.

11.3K When the difference in pressure recorded by the manometer reaches approximately 2 to 3 inches (5 to 8 cm) of water column more than the initial difference, it is probably time to clean or change the filters. See the manufacturer's manual for recommended pressure differences.

Answer to question on page 47.

11.3L The sharp edge of a metering orifice is usually on the same side as the orifice information, which is stamped on the orifice plate handle. The sharp edge should be placed so as to be facing the stream flow with the bevel on the discharge side. If the orifice is installed backward, it will give an incorrect reading.

11.3M If the hoist is not anchored properly and a header is being raised, the hoist could catapult into the aeration tank and could cause injury to operators and damage to equipment.

11.3N Important safety precautions that should be observed when working on a center Y-wall include the wearing of approved flotation devices or fall arrest systems, having help standing by, keeping the area clear and free of slippery materials, providing secure footing, and requiring adequate lighting.

11.3O If air diffuser headers, or the pipe containing them, are slanted so that one end of the header cross arm is higher than the other (by as little as ½ inch or 1.2 cm), the air will be unevenly released into the aerator. The diffusers on the high side at low air rates will allow most of the air to pass out of them through the path of least resistance rather than go through the diffusers that are located on the low end of the header and that have only ½ inch (1.2 cm) more water over the top of them. This creates an undesirable flow pattern in the tank.

Answers to questions on page 47.

11.3P A light application of grease should be applied to the threads of a diffuser to serve as a protective coating and to make it easier to remove the diffuser for cleaning.

11.3Q Whenever a blower is first put on the line, the bearing lubrication and temperature should be checked, along with driven equipment load, air flow rates, differential pressures across the filter system, and temperature.

Answer to question on page 48.

11.3R
1. Name of pump
2. Location of pump
3. Pump suction pressure
4. Pump discharge pressure
5. Water level in wet well
6. Motor amperage
7. Pump discharge (flow)
8. Date of test
9. Name of operator

Answers to questions on page 50.

11.4A Blowers should be started and air should be flowing out the diffusers before primary effluent is admitted to the aeration tanks to prevent diffuser clogging by waste solids. This is particularly important if fine bubble diffusers are used.

11.4B During start-up, return sludge pumping rates must be adjusted to rapidly return the solids (organisms) to the aeration tanks and to keep the sludge blanket in the secondary clarifier as low as possible.

11.4C Chlorination equipment should be put in service to disinfect the plant effluent and to protect the health of the receiving water users.

11.4D DO should be checked throughout the aerator to be sure DO is available for all organisms. If possible, use a DO probe and determine the DO along both sides of the aerator at both the water surface and the bottom of the tank.

Answers to questions on page 51.

11.4E To record the solids buildup in the aeration tank, the operator should collect a grab sample at the same time every day, preferably during peak flows. The sample of mixed liquor should be collected approximately 5 feet (1.5 m) from the effluent end of the aeration tank and 1.5 to 2 feet (0.4 to 0.6 m) below the water surface to ensure a good sample.

11.4F Calculate pounds of solids under aeration.

$$\text{Solids, lbs} = \text{MLSS, mg}/L \times \text{Aer Vol, MG} \times 8.34 \text{ lbs/gal}$$

$$= 640 \text{ mg}/L \times 0.25 \text{ MG} \times 8.34 \text{ lbs/gal}$$

$$= \frac{640 \text{ mg}}{\text{M mg}} \times 0.25 \text{ MG} \times 8.34 \text{ lbs/gal}$$

$$= 640 \times 2.1 \text{ lbs}$$

$$= 1{,}344 \text{ lbs}$$

11.4G Estimate return sludge pumping rate, GPM.

Flow to Aerator from Primary Clarifier	=	2.0 MGD
Return Sludge flow to Aerator	=	0.5 MGD
Total Flow Through Aerator	=	2.5 MGD

$$\text{Return Sludge Rate, MGD} = \text{Aerator Flow, MGD} \times \text{Return Sludge Ratio}$$

$$= 2.5 \text{ MGD} \times \left(\frac{170 \text{ m}L}{1{,}000 \text{ m}L - 170 \text{ m}L}\right)$$

$$= 0.512 \text{ MGD or } 512{,}000 \text{ GPD}$$

$$\text{Return Sludge Rate, GPM} = \frac{512{,}000 \text{ GPD}}{1{,}440 \text{ min/day}}$$

$$= 356 \text{ GPM}$$

ANSWERS TO QUESTIONS IN LESSON 4

Answers to questions on page 57.

11.5A

$$\text{Sludge Age, days} = \frac{(\text{Suspended Sol in Mixed Liq, mg}/L)(\text{Aerator Vol, MG})(8.34 \text{ lbs/gal})}{(\text{Suspended Sol in Primary Effl, mg}/L)(\text{Flow, MGD})(8.34 \text{ lbs/gal})}$$

$$= \frac{(2{,}100 \text{ mg}/L)(0.5 \text{ MG})(8.34 \text{ lbs/gal})}{(70 \text{ mg}/L)(4.0 \text{ MGD})(8.34 \text{ lbs/gal})}$$

$$= \frac{(30)(0.5)}{4}$$

$$= 3.75 \text{ days}$$

11.5B Some activated sludge must be wasted to prevent an excessive solids buildup in the aerator.

11.5C Excess activated sludge is normally wasted by diverting it to a primary clarifier, gravity thickener, gravity belt thickener, dissolved air flotation thickener (DAFT), aerobic digester, or anaerobic digester.

Answers to questions on page 60.

11.5D

$$\text{Solids Wasted, lbs} = \text{Solids in System, lbs} - \text{Desired Solids, lbs}$$

$$= 12{,}000 \text{ lbs} - 9{,}000 \text{ lbs}$$

$$= 3{,}000 \text{ lbs}$$

11.5E

$$\text{Waste Sludge Rate, MGD} = \frac{\text{Solids to Be Wasted, lbs/day}}{\text{Return Sludge Conc, mg}/L \times 8.34 \text{ lbs/gal}}$$

$$= \frac{3{,}000 \text{ lbs/day}}{6{,}000 \text{ mg}/L \times 8.34 \text{ lbs/gal}}$$

$$= \frac{3{,}000/\text{day}}{6{,}000} \times \frac{1}{8.34/\text{MG}} \quad \text{(Remember, one liter = 1 M mg)}$$

$$= \frac{3{,}000/\text{day}}{50{,}040/\text{MG}} \text{ or } \frac{3{,}000/\text{day}}{50{,}000/\text{MG}}$$

$$= 0.06 \text{ MGD}$$

$$\text{Waste Sludge Rate, GPM} = \frac{0.06 \text{ MGD}}{1{,}440 \text{ min/day}}$$

$$= \frac{60{,}000 \text{ gal}}{1{,}440 \text{ min}}$$

$$= 41.7 \text{ GPM Waste or 40 GPM}$$

11.5F

$$\text{Waste Sludge, lbs/day} = \frac{\text{Susp Sol in System, lbs}}{\text{MCRT, days}} - \text{Susp Sol in Effl, lbs/day}$$

$$= \frac{40{,}000 \text{ lbs}}{10 \text{ days}} - 10 \text{ mg}/L \times 5.0 \text{ MGD} \times 8.34 \text{ lbs/gal}$$

$$= 4{,}000 \text{ lbs/day} - 417 \text{ lbs/day}$$

$$= 3{,}583 \text{ lbs/day}$$

Answers to questions on page 67.

11.5G The three main types of inspections every shift operator responsible for an activated sludge process should conduct at the start of every shift are:

1. Review the log book.
2. Visually inspect the activated sludge process.
3. Review the lab results.

11.5H The COD test is recommended to measure the strength of the influent wastewater to the activated sludge process because the results are available within four hours and can be used to control the process.

11.5I The aerator DO should be measured in the aerator because the DO of the sample can change very rapidly outside the aerator.

11.5J True. Sight and smell observations are often the operator's first indication that process problems are developing.

ANSWERS TO QUESTIONS IN LESSON 5

Answers to questions on page 70.

11.6A Plant records of the activated sludge operation are important because they are helpful in identifying the cause of operational problems or upsets and indicating what corrective action should be taken.

11.6B If the mixed liquor suspended solids were allowed to build up too high, the quality of the plant effluent would deteriorate.

Answers to questions on page 71.

11.6C If the plant becomes upset, the first action before making any changes is to check the plant data for at least three previous weeks. The problem probably started last week or earlier.

11.6D When low volatile content solids from a digester escape the primary clarifier to the aeration system, the supernatant solids are picked up by the activated sludge in the aerator and carried through the system. This creates an extra oxygen demand, and air output must be increased.

11.6E If a plant becomes upset by high flows from stormwater infiltration, reduce wasting solids at this point to prevent organism losses.

11.6F If a plant becomes upset due to seasonal temperature changes, then the solids loading and air rates must be adjusted. During cold winter weather, the plant requires less air and more solids under aeration. During warmer weather, the plant requires more air and less solids under aeration.

11.6G When lab data varies considerably from day to day, check sampling location, time, and laboratory procedures for errors.

11.6H Check the plant records to determine the cause and then make adjustments to process. However, only one change at a time should be made to the activated sludge system, and one week should be allowed to observe the response.

11.6I Allow one week for a plant to stabilize after a process change. An experienced operator who knows the plant may be able to determine if the proper changes have been made after several days, but some plants require up to two weeks to stabilize after a change.

Answers to questions on page 76.

11.6J Bulking sludge will be indicated by clouds of billowing sludge or a tremendous amount of suspended solids being carried out of the final clarifier in the effluent. Rising sludge will be light flocculent particles collecting mainly on the surface of the final tanks and forming a thin surface scum.

11.6K Several adjustments may be attempted to correct a rising sludge problem, but only one should be undertaken at a time.

1. Increase return sludge rates, but control final clarifier sludge level.
2. Increase load to aerator by removing a primary clarifier from service if more than one is being used.
3. Admit raw wastewater directly to the aerator during low flows.
4. Check the possibility of changing the aeration diffuser placement to a tapered pattern.

11.6L Foam may be controlled by:

1. Maintaining higher mixed liquor suspended solids concentrations
2. Reducing air supply during periods of low flow while still maintaining DO
3. Returning supernatant to the aeration tank during low flows

11.6M Grease deposits can be removed from walks by using trisodium phosphate (TSP) and a stiff bristle deck brush and hose.

Answers to questions on page 77.

11.7A Both surface aerators and blowers must be shut down before any maintenance is performed to prevent injury to maintenance personnel and possible damage to equipment.

11.7B To shut down an aerator:

1. Turn the ON-OFF-AUTO switch to OFF.
2. Turn the main power breaker to OFF.
3. Lock out and tag the main power breaker in the OFF position.

11.7C The regulating/isolation valve(s) or the blower must be locked out and tagged before servicing the air distribution system to prevent air from entering the system.

11.7D The air distribution system must be shut down before attempting to repair even the smallest leak because serious bodily injury may result from escaping air or foreign material.

Answers to questions on page 78.

11.7E The motor on a surface aerator could draw a high or uneven amperage due to:

1. Moisture
2. Winding breakdown
3. Degree of impeller submergence
4. Excessive motor bearing or gear reducer friction

11.7F If filters do not have a manometer, you can determine if filters need cleaning by removing them and observing the degree of dirt and dust buildup.

11.7G Fine bubble diffusers can become clogged due to dirt and biological growths.

Answers to questions on page 84.

11.7H Motor windings deteriorate due to heat and moisture.

11.7I Air filters may be cleaned by using a relatively high-pressure stream of clean water or by steam cleaning.

11.7J Equipment may seize up and become inoperable when the equipment overheats and a component expands so the equipment will not work. This is also called freezing.

ANSWERS TO QUESTIONS IN LESSON 6

Answers to questions on page 93.

11.8A Contact stabilization requires two aeration tanks. One tank is for separate reaeration of the return sludge and the other tank is for conventional aeration of mixed liquor.

11.8B The difference between step-feed and conventional activated sludge operation is that in conventional activated sludge, the primary effluent and return sludge are introduced at one point only, the entrance to the aeration tanks. In step-feed aeration, the return sludge is introduced separately and, in many cases, allowed a short reaeration period by itself at the entrance to the tank.

11.8C The complete mix mode of operation is a design modification of tank mixing techniques to ensure equal distribution of applied waste load, dissolved oxygen, and return sludge throughout the aeration tank.

Answers to questions on page 97.

11.9A The sequence of treatment process stages for a sequencing batch reactor (SBR) is: *FILL, REACT* (aeration), *SETTLE, DECANT* (withdrawal of clarified effluent), and *IDLE.*

11.9B The difference between the mixed liquor in the SBR system and the continuous flow activated sludge system (CFS) is that in the SBR system the mixed liquor is always in the reactor and there is no recycle. In the CFS system, the mixed liquor is recycled through the reactor and clarifier.

11.9C SBRs are becoming practical for smaller municipal flows because: (1) the process controls are implemented by a programmable logic controller (PLC), (2) the process is very stable, and (3) construction of an SBR system usually costs less than a conventional activated sludge system.

Answers to questions on page 99.

11.9D Preliminary treatment processes that may be used at a sequencing batch reactor treatment plant include screening, grit removal, and grease removal.

11.9E The different types of decanters installed on SBRs include fixed-elevation submersible pumps, siphons, launders, and decanters designed to move downward with the fall of the water level in the reactor.

11.9F The sequencing of reactor cycles in an SBR is controlled by a programmable logic controller (PLC) (a microprocessor controller).

Answers to questions on page 104.

11.9G During the react or aeration cycle, the influent is mixed with the contents of the reactor and dissolved oxygen is added to the mixed liquor.

11.9H Sludge is wasted from an SBR to control reactor mixed liquor solids. This, in turn, controls the F/M (food to microorganism ratio) or solids retention time (SRT)/sludge age for the activated sludge process.

Answers to questions on page 106.

11.9I A plug flow is a type of flow that occurs in tanks, basins, or reactors when a slug of water or wastewater moves through a tank without ever dispersing or mixing with the rest of the water or wastewater flowing through the tank.

11.9J The SBR equipment supplier can provide information about the operating mode and sequencing stages that should be used to remove nitrogen and phosphorus.

11.9K All air diffusers should be at the same elevation on their manifold because if some diffusers are higher than others, more air will pass through the higher diffusers; this could reduce treatment efficiency.

11.9L When starting a new SBR, foam problems can be controlled by the use of water sprays, hoses with nozzles, or chemical anti-foaming agents.

Answers to questions on page 109.

11.9M The three important sampling requirements for an SBR are that sampling must be performed:

1. At the same time in the reactor cycle
2. At the same reactor liquid level
3. At the same location in the reactor

Also remember that each SBR is an individual treatment plant.

11.9N The sludge blanket level or depth should be measured at the end of the settle (quiescent) cycle just before the decant cycle begins.

11.9O If an SBR treatment plant has multiple reactors, then the plant has a good chance of surviving a slug toxic waste dump because only one reactor will probably receive the toxic waste.

11.9P Floating grease balls and scum or floating debris can be removed from the surface of an SBR by hand skimming with a fine mesh net or a sieve.

Answers to questions on page 110.

11.9Q SBR operators can be very helpful to design engineers by thinking of how they plan to operate and maintain each process or piece of equipment. An operator could make suggestions for improving the layout or reducing safety hazards and could provide site-specific information that the designers may not be aware of.

11.9R Types of site-specific conditions that the SBR designer should know about but may not be totally aware of include flooding conditions, heavy snowfalls reducing ease of access to structures and components, vandalism, frequent power failures, or any condition that might interfere with the reliability or safety of the plant.

11.9S The final draft of the O & M manual should contain as-built plans (record drawings), process diagrams, control strategies, operational modes, laboratory testing, maintenance program, staffing requirements, and spare parts lists, as well as data accumulated during the first six months of operation.

ANSWERS TO QUESTIONS IN LESSON 7

Answers to questions on page 111.

11.10A The activated sludge process is a living biological process.

11.10B Activated sludge is referred to as a "mixed culture" because it is made up of many different types of microorganisms rather than one single type.

Answers to questions on page 112.

11.10C When collecting micro samples, the following precautions must be exercised:

1. Collect a representative sample from the proper location.
2. Be sure that samples from large bottles are completely mixed before filling small bottles.
3. Use proper safety precautions to avoid slipping or falling into the aeration tank.

11.10D Grab samples should be taken at the same time of day every time a sample is collected for microscopic analysis or observation, preferably at the same time as the SVI grab sample.

Answers to questions on page 114.

11.10E If a micro sample is not analyzed promptly, some organisms will appear to be dead.

11.10F Wet mount slides are used for observing live microorganisms and stained dried slides are used to observe filamentous organisms.

Answers to questions on page 120.

11.10G The purpose of a mechanical stage on a microscope is to allow easy and precise movement of the specimen forward, backward, left, and right.

11.10H The distinct growth phases in the typical growth pattern of bacteria are: (1) lag phase, (2) log phase, (3) stationary phase, and (4) death phase.

11.10I Filamentous growths can be caused in activated sludge by undesirable environments. Undesirable environments that promote the growth of long filaments include: (1) process guideline(s) (such as low DO) out of adjustment, (2) nutrient deficiency (low nitrogen) in the influent wastewater, or (3) an undesirable substance in the influent wastewater or in a waste sidestream being returned to the plant (such as a toxic substance or hydrogen sulfide). The growth of short filaments has been associated with low DO and low F/M ratio.

Answers to questions on page 125.

11.10J *Desirable Microorganisms:* There should be a predominance of free-swimming and stalked ciliates with a few rotifers. There should also be a few long filaments.

Undesirable Microorganisms: Flagellates and amoeboids are almost always present in activated sludge, but their numbers should be very low. Ideally, you should never see tiny flagellates *(Mastigophora)* nor short filaments *(Nocardia)*.

11.10K If there is an upward trend in the numbers of rotifers and few or no flagellates, the F/M ratio should be decreasing and the MCRT increasing. Also, you would expect an upward trend in the suspended solids concentrations in the aeration tanks and a downward trend in the SVI.

Answers to questions on page 128.

11.10L When considering making a process change, review the following conditions:

1. Effluent quality and signs of deterioration
2. SVI high or low
3. *Nocardia* observed through microscope
4. Numbers (increased) of flagellates or amoeboids
5. Numbers (fewer) of stalked ciliates

11.10M Physical process controls for making process changes in the conventional activated sludge mode:

1. Process air flow
2. Waste activated sludge (WAS) flow
3. Return activated sludge (RAS) flow
4. Number of tanks in service (primaries, aeration, secondaries)

11.10N When operating the contact stabilization mode of the activated sludge process, high values of SVI can be lowered by increasing the air flow in the reaeration zone, by decreasing the RAS flow (which increases the reaeration detention time), or by decreasing the WAS flow to decrease the F/M.

11.10O If the reaeration time is too short in the contact stabilization mode of the activated sludge process, not all of the adsorbed food will be used up before the bugs return to the mixed liquor for more food. This causes the bugs to become "fat" and results in a high SVI.

Answers to questions on page 129.

11.10P If you recently started using microbiology to help operate your plant, microscopic observations should be made daily for several months. Also, the bugs should be observed during high and low flows on the same day. Once you are familiar with the bugs in your process, you should be able to reduce the frequency of microscopic observations from daily to two or three times per week during good operation.

11.10Q A trend occurs when something changes in the same direction in several successive periods of time.

APPENDIX

Monthly Data Sheet

CLEANWATER, U.S.A.

WATER POLLUTION CONTROL PLANT

MONTHLY RECORD ________ 20___ OPERATOR: ____________

DATE	DAY	WEATHER	FLOW – MGD	RAW WASTEWATER TEMP.	RAW pH	RAW SETT. SOLIDS	RAW B.O.D.	RAW SUSP. SOLIDS	PRIM. EFF. B.O.D.	PRIM. EFF. SUSP. SOLIDS	PRIM. EFF. D.O.	FINAL EFFLUENT pH	FINAL B.O.D.	FINAL SUSP. SOLIDS	FINAL D.O.	FINAL CL_2 RES.	AERATION LBS. VOL. SOLIDS	AERATION SUSP. SOLIDS	% VOL.	30 MIN. SETT.	S.V.I.	D.O.	RETURN SUSP. SOLIDS	RETURN-MGD	WASTE GAL. X 1000	WASTE LBS./DAY
1	S	CLEAR	1.782	75	7.2	14			118	84	0.6	6.9	19	18	0.9	2.7	6746	2036	78.9	150	73	2.5	5961	0.702	70	3480
2	M	CLEAR	2.347	74	7.3	13	218	150	156	84	0.3	6.8	19	15	1.0	2.8	6859	2078	78.6	150	72	1.4	4683	0.711	72	2812
3	T	CLEAR	2.165	74	7.3	8			109	66	0.8	6.8	14	9	1.2	8.8	7224	2211	77.8	170	76	2.1	6625	0.708	71	3922
4	W	CLEAR	2.012	74	7.3	12			135	74	0.5	6.8	16	14	0.8	4.4	7305	2213	78.6	180	81	2.0	6641	0.712	70	3877
5	T	CLEAR	2.483	74	7.2	13	189	138	134	62	0.3	6.8	9	11	1.7	5.2	7014	2106	79.3	170	80	2.6	6098	0.722	78	3966
6	F	CLEAR	2.386	74	7.2	13			112	60	0.4	6.8	18	6	2.6	6.0	6754	2069	79.0	160	77	3.5	5862	0.700	80	3911
7	S	CLEAR	2.131	75	7.3	13			89	66	0.7	6.9	14	7	1.2	4.4	6296	1905	78.7	150	78	2.6	5564	0.706	80	3712
8	S	CLEAR	1.867	76	7.4	12	174	134	84	74	0.9	6.9	9	15	0.8	4.2	7057	2138	78.6	180	84	0.9	6758	0.703	72	4058
9	M	CLEAR	2.634	75	7.3	14			117	68	0.3	6.9	11	11	1.6	3.5	6767	2037	79.1	160	78	2.5	6022	0.712	70	3515
10	T	CLEAR	2.307	76	7.3	18			120	66	0.3	7.1	8	8	1.5	6.6	6119	1861	78.3	170	91	2.8	6135	0.705	64	3274
11	W	CLEAR	2.198	76	7.3	18	192	142	111	68	0.4	7.0	9	10	1.1	6.6	7035	2123	78.9	200	94	1.7	6183	0.700	70	3609
12	T	CLEAR	2.202	76	7.3	11			99	72	0.4	7.0	11	9	2.0	3.8	6352	1954	77.4	190	97	3.0	6027	0.704	70	3518
13	F	CLEAR	2.178	77	7.3	11			81	58	0.6	7.0	15	18	3.5	4.0	6313	1937	77.6	170	87	4.8	5542	0.689	72	3327
14	S	CLEAR	2.006	78	7.3	12	155	156	105	76	0.3	6.9	12	8	3.1	3.8	6335	1929	78.2	160	82	4.3	4856	0.703	70	2834
15	S	CLEAR	1.942	78	7.2	12			113	74	0.4	6.9	9	9	1.3	4.4	6873	2090	78.3	180	86	2.2	5753	0.711	73	3502
16	M	CLEAR	2.464	78	7.2	11			128	74	0.3	6.9	10	10	1.8	3.0	7082	2162	78.0	200	92	2.5	6852	0.723	76	4343
17	T	CLEAR	2.321	78	7.1	8	168	144	110	64	0.4	6.8	10	11	1.9	6.6	6215	1937	76.4	190	98	2.8	6654	0.698	74	4106
18	W	CLEAR	2.611	78	7.3	15			105	64	0.2	6.9	11	7	2.2	4.4	6227	1923	77.1	190	98	3.3	5767	0.717	83	3992
19	T	CLEAR	2.457	78	7.3	12			87	72	0.5	6.9	10	7	2.9	6.2	4844	1534	75.2	170	110	4.5	4762	0.721	25	992
20	F	CLEAR	2.498	79	7.3	11	193	118	105	66	0.7	6.9	18	12	3.1	4.4	5846	1822	76.4	190	104	4.1	5123	0.719	0	0
21	S	CLEAR	2.213	76	7.1	12			109	76	0.6	7.1	10	9	1.2	4.2	6892	2096	78.3	200	95	2.6	5928	0.706	35	1730
22	S	CLEAR	1.878	76	7.3	12			131	78	0.4	6.9	14	10	0.5	4.2	7518	2263	79.1	260	114	1.8	3894	0.703	35	1136
23	M	CLEAR	2.901	77	7.3	12	187	142	133	89	0.2	6.9	13	13	0.3	2.5	8388	2541	78.6	310	121	1.9	8396	0.741	70	4901
24	T	N.M. CLEAR	2.346	78	7.3	13			114	56	0.3	6.9	14	10	2.2	4.2	7962	2409	78.7	230	95	3.6	8824	0.700	71	5225
25	W	CLEAR	2.421	78	7.3	13			89	56	0.4	7.0	12	8	2.8	4.0	7688	2332	78.5	230	98	4.1	7382	0.713	72	4432
26	T	CLEAR	2.562	79	7.3	12	212	170	143	87	0.6	7.0	10	6	1.7	4.3	6697	2047	77.9	210	102	2.6	6867	0.698	70	4008
27	F	CLEAR	2.428	79	7.3	10			128	84	0.5	6.8	15	10	0.5	3.8	6923	2103	78.2	200	94	1.2	7436	0.702	64	3969
28	S	CLEAR	2.149	78	7.3	12			84	66	0.9	6.9	16	5	0.6	3.5	7169	2180	78.3	200	91	1.7	8412	0.706	68	4770
29	S	CLEAR	1.862	79	7.3	7	176	102	117	60	0.5	6.9	14	8	0.5	3.9	7852	2397	78.0	230	95	1.0	7117	0.700	66	3917
30	M	CLEAR	2.746	79	7.3	13			107	73	0.2	6.8	8	8	1.6	3.5	7688	2335	78.4	220	94	2.9	7735	0.713	70	4515
31																										
MAX			2.901	79	7.4	18	218	170	156	89	0.9	7.1	19	18	3.5	8.8	8388	2541	79.3	310	121	4.8	8824	0.741	83	5225
MIN			1.782	74	7.1	7	155	102	84	56	0.2	6.8	8	5	0.5	2.5	4844	1534	76.4	150	72	0.9	4683	0.698	0	0
AVG			2.283	77	7.3	12	186	139	112	70		6.9	12	10	1.6	4.4	6868	2092	78.1	192	91	2.6	6328	0.708	65	3511

SUMMARY DATA

% REMOVAL	B.O.D.	S.S.
INF – PRI	39.7	49.6
INF – EFF	93.5	92.8

SLUDGE DATA

% SOLIDS — AVG.	5.6
LBS. DRY SOLIDS / DAY	5579
% VOL. SOLIDS — AVG.	79.8
LBS. VOL. SOLIDS / DAY	4452
LBS. VOL. SOL./1000 FT^3/DAY	67.5
GALS. SLUDGE TO BEDS	28,000
CU. YDS. CAKE REMOVED	63
FT^3 GAS / LB. VOL. SOLIDS	6.8
FT^3 GAS / LB. MG FLOW	14,286

COST DATA

MAN DAYS 63 PAYROLL	2,325.78
POWER PURCHASED	520.32
OTHER UTILITIES (GAS, H_2O)	NONE
GASOLINE, OIL, GREASE	108.56
CHEMICALS AND SUPPLIES	547.25
MAINTENANCE	238.48
VEHICLE COSTS	NONE
OTHER	NONE
TOTAL	$ 3,740.39
OPER. COST/MG TREATED	$ 54.62
OPER. COST/CAPITA/MO.	$ 0.158

FLOW METER: LAST 222046; 1st 153549; TOTAL 68.497 MG

ELECTRIC METER: LAST 7838; 1st 5670; TOTAL 2168; MULT 40 X 2168 = 86,720 KWH

RAW SLUDGE: LAST 798324; 1st 432984; STROKES 365340; TOTAL 365340 X 1.0 = 365,340 GALS

GAS METER: LAST 2181110; 1st 1265230; TOTAL 915880 FT^3

RETURN SLUDGE: LAST 67635048; 1st 67613800; TOTAL 21.248 MG

WASTE SLUDGE: LAST 134251; 1st 132560; TOTAL 1961 x 1000 MG

CHAPTER 12

SLUDGE DIGESTION AND SOLIDS HANDLING

by

John Brady

(With a Special Section by William Garber)

Revised by

James F. Stahl

TABLE OF CONTENTS

Chapter 12. SLUDGE DIGESTION AND SOLIDS HANDLING

OBJECTIVES

Chapter 12. SLUDGE DIGESTION AND SOLIDS HANDLING

Following completion of Chapter 12, you should be able to:

1. Explain how a sludge digester works and what factors influence and control the digestion process.
2. Inspect new sludge digesters and solids handling facilities for proper installation.
3. Place a new sludge digester into operation.
4. Schedule and conduct operation and maintenance duties.
5. Collect samples, interpret lab results, and make appropriate adjustments in the sludge digestion and solids handling processes.
6. Recognize factors that indicate the sludge digestion and solids handling processes are not performing properly, identify the source of the problem, and take corrective action.
7. Discuss the various methods of solids handling and know how to operate and maintain these processes.
8. Determine loadings on sludge digesters and solids handling facilities.
9. Conduct your duties in a safe fashion.
10. Keep records for the sludge digestion and solids disposal processes.
11. Develop an operational strategy for a sludge digester.
12. Review plans and specifications for a sludge digester.

WORDS

Chapter 12. SLUDGE DIGESTION AND SOLIDS HANDLING

ACID REGRESSION STAGE | ACID REGRESSION STAGE

A stage of anaerobic digestion during which the production of volatile acids is reduced and acetate and ammonia compounds form, causing the pH to increase.

AEROBIC (air-O-bick) DIGESTION | AEROBIC DIGESTION

The breakdown of wastes by microorganisms in the presence of dissolved oxygen. This digestion process may be used to treat only waste activated sludge, or trickling filter sludge and primary (raw) sludge, or waste sludge from activated sludge treatment plants designed without primary settling. The sludge to be treated is placed in a large aerated tank where aerobic microorganisms decompose the organic matter in the sludge. This is an extension of the activated sludge process.

ALKALINITY (AL-kuh-LIN-it-tee) | ALKALINITY

The capacity of water or wastewater to neutralize acids. This capacity is caused by the water's content of carbonate, bicarbonate, hydroxide, and occasionally borate, silicate, and phosphate. Alkalinity is expressed in milligrams per liter of equivalent calcium carbonate. Alkalinity is not the same as pH because water does not have to be strongly basic (high pH) to have a high alkalinity. Alkalinity is a measure of how much acid must be added to a liquid to lower the pH to 4.5.

ANAEROBIC (AN-air-O-bick) DIGESTION | ANAEROBIC DIGESTION

A treatment process in which wastewater solids and water (about 5 percent solids, 95 percent water) are placed in a large tank (the digester) where bacteria decompose the solids in the absence of dissolved oxygen. At least two general groups of bacteria act in balance: (1) saprophytic bacteria break down complex solids to volatile acids, the most common of which are acetic and propionic acids; and (2) methane fermenters break down the acids to methane, carbon dioxide, and water.

BTU (pronounce as separate letters) | BTU

British Thermal Unit. The amount of heat required to raise the temperature of one pound of water one degree Fahrenheit. Also see CALORIE.

BIODEGRADATION (BUY-o-deh-grah-DAY-shun) | BIODEGRADATION

The breakdown of organic matter by bacteria to more stable forms that will not create a nuisance or give off foul odors.

BIOSOLIDS | BIOSOLIDS

A primarily organic solid product produced by wastewater treatment processes that can be beneficially recycled. The word biosolids is replacing the word sludge when referring to treated waste.

BUFFER | BUFFER

A solution or liquid whose chemical makeup neutralizes acids or bases without a great change in pH.

BUFFER CAPACITY | BUFFER CAPACITY

A measure of the capacity of a solution or liquid to neutralize acids or bases. This is a measure of the capacity of water or wastewater for offering a resistance to changes in pH.

CENTRATE | CENTRATE

The water leaving a centrifuge after most of the solids have been removed.

CHEMICAL OXYGEN DEMAND (COD) | CHEMICAL OXYGEN DEMAND (COD)

A measure of the oxygen-consuming capacity of organic matter present in wastewater. COD is expressed as the amount of oxygen consumed from a chemical oxidant in mg/*L* during a specific test. Results are not necessarily related to the biochemical oxygen demand (BOD) because the chemical oxidant may react with substances that bacteria do not stabilize.

CONING | CONING

Development of a cone-shaped flow of liquid, like a whirlpool, through sludge. This can occur in a sludge hopper during sludge withdrawal when the sludge becomes too thick. Part of the sludge remains in place while liquid rather than sludge flows out of the hopper. Also called coring.

DECOMPOSITION or DECAY | DECOMPOSITION or DECAY

The conversion of chemically unstable materials to more stable forms by chemical or biological action.

DEWATERABLE | DEWATERABLE

This is a property of sludge related to the ability to separate the liquid portion from the solid, with or without chemical conditioning. A material is considered dewaterable if water will readily drain from it.

ELUTRIATION (e-LOO-tree-A-shun) | ELUTRIATION

The washing of digested sludge with either fresh water, plant effluent, or other wastewater. The objective is to remove (wash out) fine particulates and/or the alkalinity in sludge. This process reduces the demand for conditioning chemicals and improves settling or filtering characteristics of the solids.

ENDOGENOUS (en-DODGE-en-us) RESPIRATION | ENDOGENOUS RESPIRATION

A situation in which living organisms oxidize some of their own cellular mass instead of new organic matter they adsorb or absorb from their environment.

ENZYMES (EN-zimes) | ENZYMES

Organic or biochemical substances that cause or speed up chemical reactions.

HYDROLYSIS (hi-DROLL-uh-sis) | HYDROLYSIS

(1) A chemical reaction in which a compound is converted into another compound by taking up water.

(2) Usually a chemical degradation of organic matter.

INOCULATE (in-NOCK-yoo-late) | INOCULATE

To introduce a seed culture into a system.

JOULE (JOOL) | JOULE

A measure of energy, work, or quantity of heat. One joule is the work done when the point of application of a force of one newton is displaced a distance of one meter in the direction of the force. Approximately equal to 0.7375 ft-lbs (0.1022 m-kg).

LIQUEFACTION (lick-we-FACK-shun) | LIQUEFACTION

The conversion of large, solid particles of sludge into very fine particles that either dissolve or remain suspended in wastewater.

LYSIMETER (lie-SIM-uh-ter) | LYSIMETER

A device containing a mass of soil and designed to permit the measurement of water draining through the soil.

MESOPHILIC (MESS-o-FILL-ick) BACTERIA | MESOPHILIC BACTERIA

Medium temperature bacteria. A group of bacteria that grow and thrive in a moderate temperature range between 68°F (20°C) and 113°F (45°C). The optimum temperature range for these bacteria in anaerobic digestion is 85°F (30°C) to 100°F (38°C).

POSITIVE PRESSURE | POSITIVE PRESSURE

A positive pressure is a pressure greater than atmospheric. It is measured as pounds per square inch (psi) or as inches of water column. A negative pressure (vacuum) is less than atmospheric and is sometimes measured in inches of mercury. In the metric system, pressures are measured in kg/sq m, kg/sq cm, or pascals (1 psi = 6,895 Pa = 6.895 kN/sq m).

PSYCHROPHILIC (sy-kro-FILL-ick) BACTERIA | PSYCHROPHILIC BACTERIA

Cold temperature bacteria. A group of bacteria that grow and thrive in temperatures below 68°F (20°C).

PUTREFACTION (PYOO-truh-FACK-shun) | PUTREFACTION

Biological decomposition of organic matter, with the production of foul-smelling and -tasting products, associated with anaerobic (no oxygen present) conditions.

SAPROPHYTES (SAP-row-fights) | SAPROPHYTES

Organisms living on dead or decaying organic matter. They help natural decomposition of organic matter in water or wastewater.

SEED SLUDGE | SEED SLUDGE

In wastewater treatment, seed, seed culture, or seed sludge refer to a mass of sludge that contains populations of microorganisms. When a seed sludge is mixed with wastewater or sludge being treated, the process of biological decomposition takes place more rapidly.

STASIS (STAY-sis) | STASIS

Stagnation or inactivity of the life processes within organisms.

STRUVITE (STREW-vite) | STRUVITE

A deposit or precipitate of magnesium ammonium phosphate hexahydrate found on the rotating components of centrifuges and centrate discharge lines. Struvite can be formed when anaerobic sludge comes in contact with spinning centrifuge components rich in oxygen in the presence of microbial activity. Struvite can also be formed in digested sludge lines and valves in the presence of oxygen and microbial activity. Struvite can form when the pH level is between 5 and 9.

STUCK DIGESTER | STUCK DIGESTER

A stuck digester does not decompose organic matter properly. The digester is characterized by low gas production, high volatile acid/alkalinity relationship, and poor liquid-solids separation. A digester in a stuck condition is sometimes called a sour or UPSET DIGESTER.

SUPERNATANT (soo-per-NAY-tent) | SUPERNATANT

Liquid removed from settled sludge. Supernatant commonly refers to the liquid between the sludge on the bottom and the scum on the surface.

THERMOPHILIC (thur-moe-FILL-ick) BACTERIA | THERMOPHILIC BACTERIA

A group of bacteria that grow and thrive in temperatures above 113°F (45°C). The optimum temperature range for these bacteria in anaerobic decomposition is 120°F (49°C) to 135°F (57°C). Aerobic thermophilic bacteria thrive between 120°F (49°C) and 158°F (70°C).

UPSET DIGESTER | UPSET DIGESTER

An upset digester does not decompose organic matter properly. The digester is characterized by low gas production, high volatile acid/alkalinity relationship, and poor liquid-solids separation. A digester in an upset condition is sometimes called a sour or STUCK DIGESTER.

CHAPTER 12. SLUDGE DIGESTION AND SOLIDS HANDLING

(Lesson 1 of 6 Lessons)

12.0 NEED FOR SLUDGE DIGESTION

In primary treatment, the settled sludge solids that are removed from the bottom and the floating scum removed from the top of the clarifiers or sedimentation tanks are a watery, odorous mixture called raw sludge and scum. The raw sludge and sludges from secondary trickling filter and activated sludge processes are most commonly pumped to a sludge digester for treatment. The sludge digester can either be an aerobic or anaerobic treatment system. In either treatment system, bacteria decompose the sludges to simpler forms prior to ultimate disposal of the sludges or reuse of the biosolids. (The word "biosolids" refers to a primarily organic solid product produced by wastewater treatment processes that can be beneficially recycled. The word biosolids is replacing the word sludge when referring to treated waste. After sludge is treated, the resulting solids are ready for disposal or recycling and are called "biosolids.") This chapter discusses the operation of both anaerobic and aerobic sludge digesters.

The basics of operation of an anaerobic digester are discussed in Sections 12.1 through 12.5. Aerobic digesters are discussed in Section 12.6.

Figure 12.1 shows the location (solids handling) of either digestion system in a typical plant. Figures 5.2, 6.2, and 8.4 in Volume I also show plan views of the location of sludge digestion and solids handling facilities in relation to other treatment processes.

12.00 Purpose of Anaerobic Sludge Digestion

ANAEROBIC DIGESTION[1] reduces wastewater solids from a sticky, smelly mixture to a mixture that is relatively odor free, *DEWATERABLE,*[2] and capable of being disposed of without causing a nuisance. In this process, organic solids in the sludge are liquified, the solids volume is reduced, and valuable methane gas is produced in the digester by the action of two different groups of bacteria living together in the same environment. One group consists of *SAPROPHYTIC ORGANISMS,*[3] commonly referred to as "acid formers." The second group, which uses the acid produced by the saprophytes, are the "methane fermenters." Initially, the organic or volatile matter is converted to volatile acids by acid formers, and then the volatile acids are converted to methane gas by methane fermenters. The methane fermenters are not as abundant in raw wastewater as are the acid formers. The methane fermenters desire a pH range of 6.6 to 7.6 and will reproduce only in that range.

The object of good digester operation is to maintain suitable conditions in the digester for a growing (reproducing) population of both acid formers and methane fermenters. You must do this by controlling the loading rate or food supply (organic solids), volatile acid/alkalinity relationship, mixing, and temperature. Generally, you have done your job properly if the digester reduces the volatile (organic) solids content by between 50 and 60 percent of what they were in the raw sludge.

To obtain the desired degree of organic solids reduction may require from 5 to 120 days of digestion time. The time required depends on how good a job you are required to do on digesting the sludge, and on the adequacy of mixing, the organic loading rate, and the temperature at which the bacterial culture is maintained. Most digesters require a minimum of 15 days digestion time.

12.01 How Anaerobic Sludge Digestion Works
by William Garber

The equations shown in Figure 12.2 illustrate one way of outlining what happens in a digester. These equations indicate two general types of reactions:

1. Acid forming reactions, which proceed at a rate dependent on temperature, pH, and food conditions
2. Methane fermentation reactions, which proceed at a rate dependent on temperature, pH, and food conditions

You must try to operate an anaerobic sludge digester so that the rate of acid formation and methane formation are approximately equal; otherwise the reaction will get out of balance. The most common condition of imbalance is excess acid formation (fermentation). This occurs when the methane fermenters fail to keep pace with the acid formers. As the rate at which methane fermenters convert acids slows down, the digester becomes more and more acid and the level of imbalance rises. Efficient operation depends on keeping both types of organisms in proper balance. The literature has been full of terms such as "standard-rate" and "high-rate" digestion. These terms refer to digester loading and not to the rates of

---

[1] *Anaerobic* (AN-air-O-bick) *Digestion.* A treatment process in which wastewater solids and water (about 5 percent solids, 95 percent water) are placed in a large tank (the digester) where bacteria decompose the solids in the absence of dissolved oxygen. At least two general groups of bacteria act in balance: (1) saprophytic bacteria break down complex solids to volatile acids, the most common of which are acetic and propionic acids; and (2) methane fermenters break down the acids to methane, carbon dioxide, and water.

[2] *Dewaterable.* This is a property of sludge related to the ability to separate the liquid portion from the solid, with or without chemical conditioning. A material is considered dewaterable if water will readily drain from it.

[3] *Saprophytes* (SAP-row-fights). Organisms living on dead or decaying organic matter. They help natural decomposition of organic matter in water or wastewater.

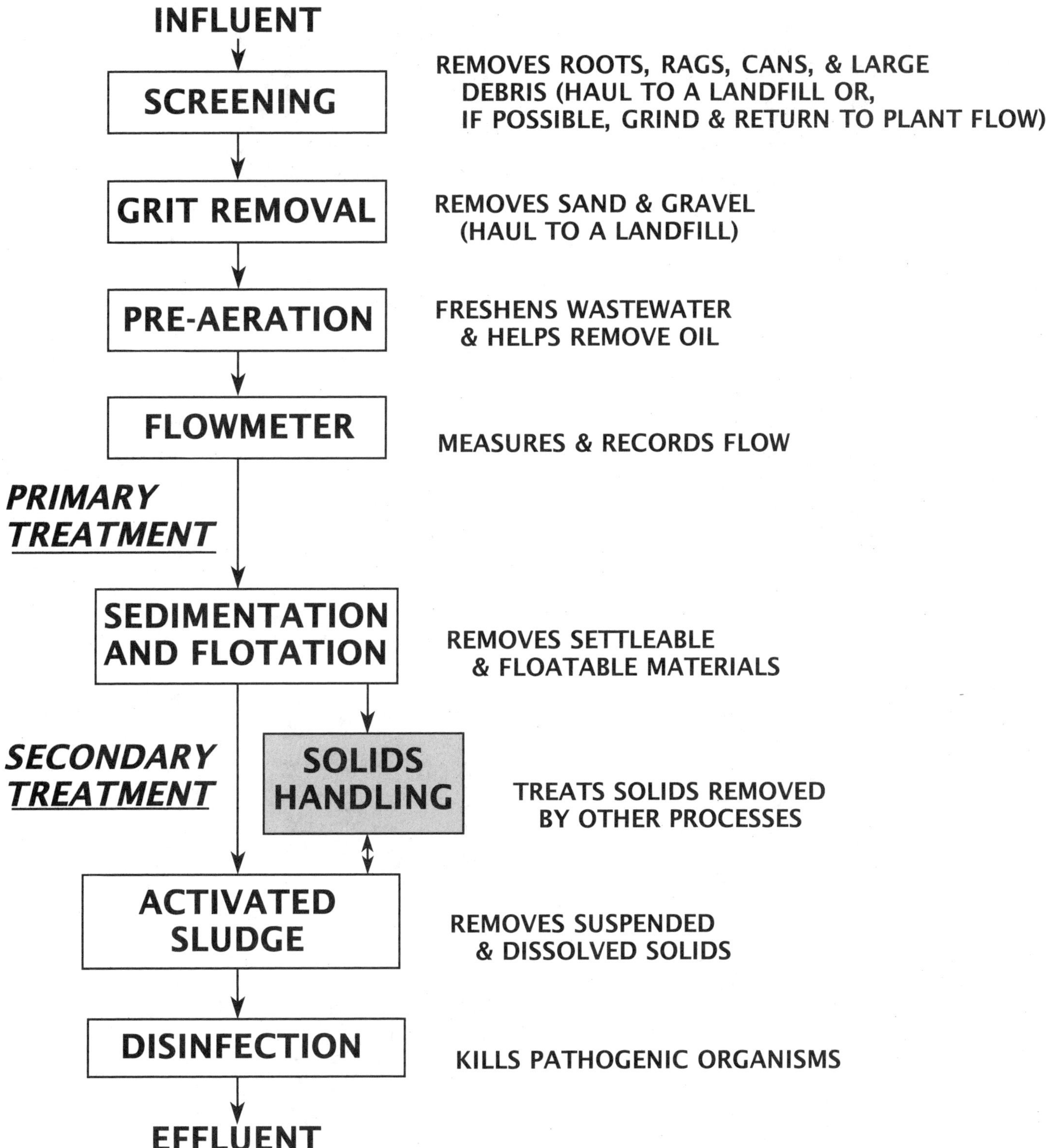

Fig. 12.1 Flow diagram of a typical plant

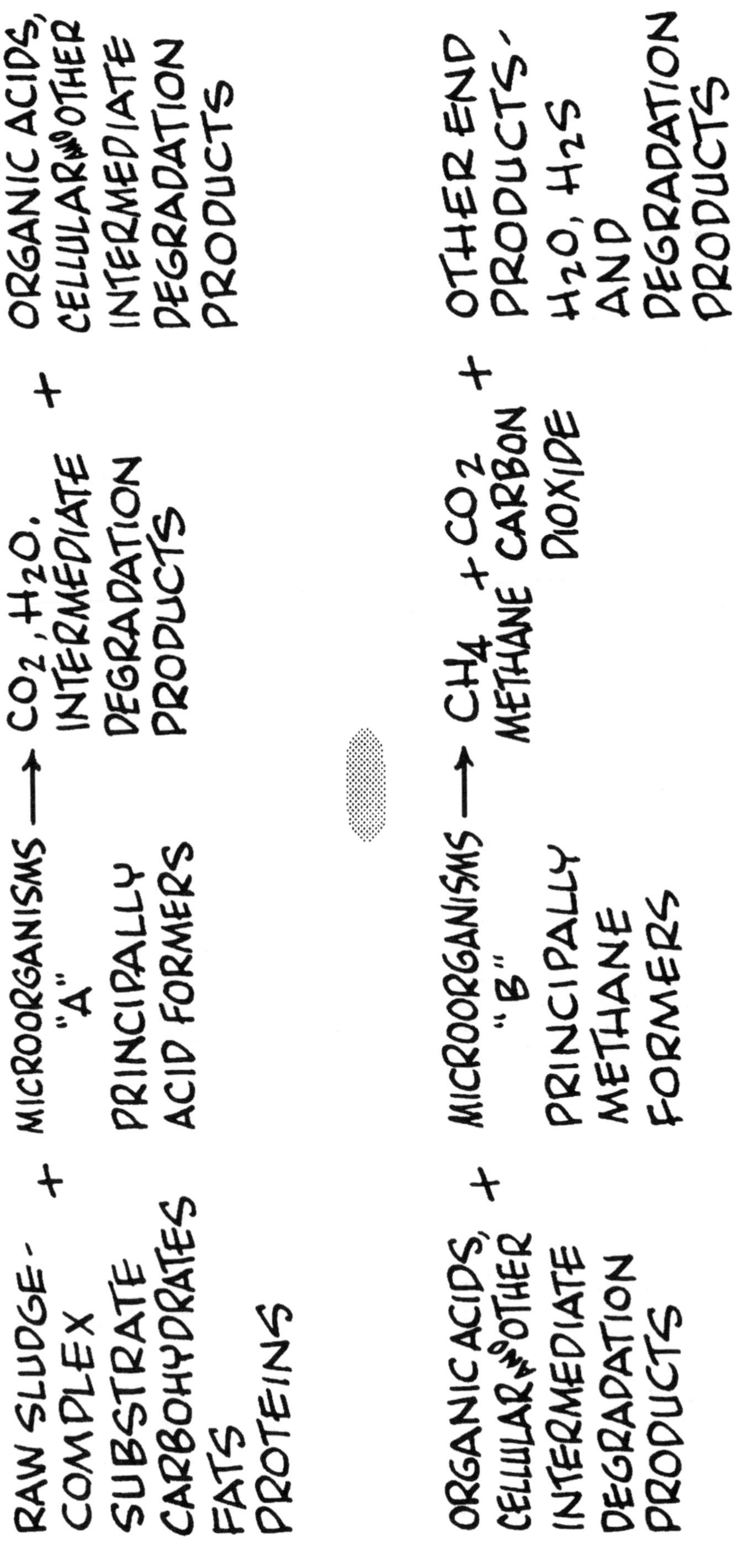

RAW SLUDGE- COMPLEX SUBSTRATE CARBOHYDRATES FATS PROTEINS	+	MICROORGANISMS "A" PRINCIPALLY ACID FORMERS	⟶	CO_2, H_2O. INTERMEDIATE DEGRADATION PRODUCTS	+ ORGANIC ACIDS, CELLULAR AND OTHER INTERMEDIATE DEGRADATION PRODUCTS
ORGANIC ACIDS, CELLULAR AND OTHER INTERMEDIATE DEGRADATION PRODUCTS	+	MICROORGANISMS "B" PRINCIPALLY METHANE FORMERS	⟶	CH_4 METHANE + CO_2 CARBON DIOXIDE	+ OTHER END PRODUCTS- H_2O, H_2S AND DEGRADATION PRODUCTS

Fig. 12.2 Reactions in a digester

bacterial action. In high-rate systems, mixing is used to obtain the best possible distribution of the substrate (food) and seed (organisms) so that more bacterial reaction can occur.

Mixing is the most important factor in the high-rate processes, but is very important in the operation of any digester. Mixing accomplishes the following:

1. Uses as much of the total content of a digester as possible
2. Quickly distributes the raw sludge (food) throughout the volume of sludge in the tank
3. Puts the microorganisms in contact with the food
4. Dilutes the inhibitory by-products of microbiological reactions throughout the sludge mass
5. Achieves good pH control by distributing *BUFFERING*[4] *ALKALINITY*[5] throughout the digestion tank
6. Obtains the best possible distribution of heat through the tank
7. Minimizes the separation of grit and inert solids to the bottom or floating scum material to the top

An anaerobic digester may be operated in one of three temperature zones or ranges, each of which has its own particular type of bacteria. The lowest range (in an unheated digester) uses *PSYCHROPHILIC* (cold temperature loving) *BACTERIA*.[6]

The organisms in the sludge within an unheated digester tend to adjust to the outside temperature. However, below 50°F (10°C) little or no bacterial activity occurs and the necessary reduction in sludge volatiles (organic matter) will not occur. When the temperature increases above 50°F (10°C), bacterial activity increases to measurable rates and digestion starts again. The bacteria appear to be able to survive temperatures well below freezing with little or no harm. The psychrophilic upper range is around 68°F (20°C). Digestion in this range requires from 50 to 180 days, depending on the degree of treatment (solids reduction) required. Few digesters are designed today to operate in this range, but there are a few still in use, including most Imhoff tanks and similar, unheated digesters with no mixing devices. Generally, these digesters are not very effective in digesting sludge.

Organisms in the middle temperature range are called the *MESOPHILIC* (medium temperature loving) *BACTERIA*[7]; they thrive between about 68°F (20°C) and 113°F (45°C). The optimum temperature range is 85°F (30°C) to 100°F (38°C), with temperatures being maintained at about 95°F (35°C) in most anaerobic digesters. Digestion at 95°F (35°C) may take from 5 to 50 days or more (normally around 25 to 30 days), depending on the required degree of volatile solids reduction and adequacy of mixing. The high-rate processes are usually operated within the mesophilic temperature range. These are nothing more than procedures to obtain good mixing so that the organisms and the food can be brought together to allow the digestion processes to proceed as rapidly as possible. With the most favorable conditions, the time may be no more than five days for an intermediate level of digestion.

Organisms in the third temperature range are called *THERMOPHILIC* (hot temperature loving) *BACTERIA*[8] and they thrive above 113°F (45°C). The optimum temperature range is considered 120°F (49°C) to 135°F (57°C). The time required for digestion in this range falls between 5 and 12 days, depending on operational conditions and degree of volatile solids reduction required. However, the problems of maintaining temperature, sensitivity of the organisms to temperature change, and some reported problems of poor solids-liquid separation are reasons why only a few plants have actually been operated in the thermophilic range.

You cannot merely raise the temperature of the digesters and have a successful operation in another range. The bacteria must have time to adjust to the new temperature zone and to develop a balanced culture before continuing to work. An excellent rule for digestion is never change the temperature more than one degree per day to allow the bacterial culture to become acclimated (adjust to the temperature changes).

Secondary digestion tanks are sometimes used to allow liquids *(SUPERNATANT*[9]*)* to separate from the solids, to provide a small amount of additional digestion, and to act as a *SEED SLUDGE*[10] source (the settled, digested sludge). However, digestion tanks generally have too small a surface area-to-depth ratio to be good sedimentation tanks. Separation of solids from liquids is more efficient in other processes such as centrifuges, vacuum filters, filter presses, lagoons, or in tanks designed for separation. If a significant amount of digestion occurs in the secondary tank, the result may be poor separation of solids. Secondary digesters should be used for solids concentration and for a reservoir of alkalinity and seed sludge, which may be returned to the primary digester when needed.

You have certain other items you can use for control in addition to mixing and temperature selection. These include:

1. Varying the sludge concentration or water added to the system
2. Varying the rate and frequency of feeding, with continuous feed the most desirable

[4] *Buffer.* A solution or liquid whose chemical makeup neutralizes acids or bases without a great change in pH.

[5] *Alkalinity* (AL-kuh-LIN-it-tee). The capacity of water or wastewater to neutralize acids. This capacity is caused by the water's content of carbonate, bicarbonate, hydroxide, and occasionally borate, silicate, and phosphate. Alkalinity is expressed in milligrams per liter of equivalent calcium carbonate. Alkalinity is not the same as pH because water does not have to be strongly basic (high pH) to have a high alkalinity. Alkalinity is a measure of how much acid must be added to a liquid to lower the pH to 4.5.

[6] *Psychrophilic* (sy-kro-FILL-ick) *Bacteria.* Cold temperature bacteria. A group of bacteria that grow and thrive in temperatures below 68°F (20°C).

[7] *Mesophilic* (MESS-o-FILL-ick) *Bacteria.* Medium temperature bacteria. A group of bacteria that grow and thrive in a moderate temperature range between 68°F (20°C) and 113°F (45°C). The optimum temperature range for these bacteria in anaerobic digestion is 85°F (30°C) to 100°F (38°C).

[8] *Thermophilic* (thur-moe-FILL-ick) *Bacteria.* A group of bacteria that grow and thrive in temperatures above 113°F (45°C). The optimum temperature range for these bacteria in anaerobic decomposition is 120°F (49°C) to 135°F (57°C). Aerobic thermophilic bacteria thrive between 120°F (49°C) and 158°F (70°C).

[9] *Supernatant* (soo-per-NAY-tent). Liquid removed from settled sludge. Supernatant commonly refers to the liquid between the sludge on the bottom and the scum on the surface.

[10] *Seed Sludge.* In wastewater treatment, seed, seed culture, or seed sludge refer to a mass of sludge that contains populations of microorganisms. When a seed sludge is mixed with wastewater or sludge being treated, the process of biological decomposition takes place more rapidly.

3. Closely controlling grit and skimming so that the capacity of the tank is reduced as little as possible by these materials
4. Cleaning regularly to maintain capacity
5. Setting up a good maintenance program to maintain the maximum degree of flexibility
6. Maintaining records and laboratory control so that the process condition is known at all times

Although anaerobic digestion is a complex process and only a portion of its theory is understood, enough is known to allow you to exercise good operational control. For anaerobic sludge digestion, as for any of the wastewater processes, remember that for the most successful operation you need to do the following:

1. Understand the theory of the process so you know what you are basically trying to do.
2. Know your facilities thoroughly so that you can attain maximum flexibility of operation.
3. Keep careful records and use laboratory analyses to follow the process continually.
4. Maintain your facilities in the best possible condition at all times.

QUESTIONS

Write your answers in a notebook and then compare your answers with those on page 231.

12.00A Why must raw sludge be digested?

12.00B What happens during anaerobic digestion?

12.00C What are some of the important factors in controlling the rate of reproduction of acid forming and methane fermenting bacteria in an anaerobic digester?

12.1 COMPONENTS IN THE ANAEROBIC SLUDGE DIGESTION PROCESS

To understand and operate an anaerobic sludge digester, the operator must be familiar with the location and function of the various components of the digestion facility.

12.10 Pipelines and Valves

Raw sludge pipelines are usually constructed of cast iron or steel to withstand pumping pressures. In recent years, glass-lined or epoxy-lined sludge lines have been used to reduce the problem of grease deposits. These deposits cut capacity and may cause stoppages. To clean lines, some plants use steam, "go-devil" type cleaners, or hot chemical solutions such as TSP.

The valves used in sludge and scum lines are mostly of the plug type. Plug-type valves give positive control where a gate or butterfly valve may become blocked by rags or other material that will not allow the valve to seat. In some cases, a gate or butterfly valve is preferred because a quick-closing plug valve could result in water hammer and damage the pipeline.

CAUTION

Never start a positive displacement pump against a closed valve because excessive pressure may result in damaging the line or the pump. All pumps should be equipped with pressure cut-off devices, but sometimes these devices fail or allow the pressure to build up high enough to cause damage. A sludge line should never be isolated by closing the valves on each end for several days because gas production can generate sufficient pressures to cause the line to fail. Also, the solids will form an almost immovable mass in the line.

QUESTIONS

Write your answers in a notebook and then compare your answers with those on page 231.

12.10A Why are plug-type valves used in sludge lines?

12.10B Why should a positive displacement pump never be started against a closed valve?

12.10C Why should a sludge line never be closed at both ends?

12.11 The Anaerobic Digester

Anaerobic digestion tanks may be cylindrical or cubical in shape. Most tanks constructed today are cylindrical. The floor of the tank is sloped so that sand, grit, and heavy sludge will tend to be removed from the tank. Most digesters constructed today have either fixed or floating covers. The parts of a digester and their purposes are summarized in Table 12.1.

12.110 Fixed Cover Tanks

A fixed cover tank is constructed of concrete or steel and has a stationary roof, which may be flat, conical, or dome-shaped. A typical fixed cover digester is shown in Figure 12.3. Both flat and conical covers are normally designed to maintain no more than an eight-inch (20-cm) water column of gas pressure on the tank roof (Figure 12.4), but some are designed for a pressure of 25 inches (64 cm) or more. If the pressure is exceeded, the water seal could be broken, allowing air to enter the tank and form an explosive mixture of gases. High gas pressures may cause structural damage to the tank in severe cases. The domed cover is designed to hold a larger volume of gas. Any type of mixing device may be used with a fixed cover tank, and the tank must be equipped with pressure and vacuum relief valves to break a vacuum or bleed off excessive pressure to protect the digester from structural damage.

A fixed cover digester may develop an explosive mixture in the tank when sludge is withdrawn if proper precautions are not taken to prevent air from being drawn into the tank. Each time a new charge of raw sludge is added, an equal amount of supernatant is displaced because the tank is maintained at a fixed level.

TABLE 12.1 PURPOSE OF ANAEROBIC SLUDGE DIGESTER PARTS

Part[a]	Purpose
1. Sludge Feed	Conveys sludge to digester for treatment.
2. Draft Tube	Directs mixture of raw and digested sludge to heating region.
3. Gas Injectors	Inject gas into sludge to mix sludge and increase heat transfer.
4. Hot Water Jacket	Contains hot water that heats sludge.
5. Deflection Shield	Deflects heated sludge throughout digester.
6. Gas Mixer	Pumps gas from gas draw-off to gas injectors.
7. Gas Draw-Off	Draws gas from top of digester to gas system.
8. Hot Water Piping	Conveys hot water from heat exchanger to hot water jacket.
9. Sludge Draw-Off	Removes digested sludge from bottom of digester.
10. Supernate (supernatant) Draw-Off Tubes	Draw off supernatant from various levels in digester.
11. Supernate (supernatant) Box	Collects supernatant for transport to headworks or primary clarifier influent.
12. Water Seal	Prevents air from entering digester or digester gas from escaping. Prevents development of explosive conditions.
13. Vacuum Relief Valve	Relieves excessive vacuums so digester cover will not collapse.
14. Pressure Relief Valve	Relieves excessive pressures in digester so water seal will not be blown out.
15. Flame Arrester	Prevents spark or flame from entering digester.
16. Sediment Trap	Traps sediment in digester gas.
17. Condensate Drain	Drains condensate removed from digester gas.
18. Gas Piping	Conveys digester gas from digester to heaters, mixers, or waste gas burners (see Section 12.12, "Gas System").

[a] Numbers 1-11 correspond with parts on Figure 12.3 and numbers 12-18 correspond with parts on Figure 12.4.

12.111 Floating Cover

A floating cover moves up and down with the tank level and gas pressure. Normally, the vertical travel of the cover is about eight feet (2.5 m), with stops (corbels) or landing edges for down (lowering) control and maximum water level for upward travel. Maximum water level is controlled by an overflow pipe that must be kept clear to prevent damage to the floating cover by overfilling. Gas pressure depends on the weight of the cover. The advantages of a floating cover include less danger of explosive mixtures forming in the digester, better control of supernatant withdrawal, and better control of scum blankets. Disadvantages include higher construction and maintenance costs. For more details on floating cover digesters, see Section 12.16, "Floating Cover Digesters."

12.112 Digester Depth

A typical operation depth for digesters is around 20 feet (6 m) (sidewall water level depth). The bottom slopes downward to the center of the tank. A gas space of two to three feet (0.6 to 1.0 m) is usually provided above normal liquid sludge level, but some floating covers allow more room for gas storage.

12.113 Raw Sludge Inlet

Typically the raw sludge feed is piped to the top of the primary digester and admitted to the side opposite where the supernatant overflow (Figures 12.3 and 12.4) is conveyed to the secondary digester. The raw sludge feed line also carries any recirculated digested sludge in the system so that the raw sludge is immediately seeded with anaerobic bacteria as it enters the tank.

12.114 Supernatant Tubes (Figures 12.3 and 12.5)

On a fixed cover digester there may be three to five supernatant tubes set at different levels for supernatant removal and to obtain samples of digester contents. Normally only one tube is used at a time. The tube used is selected to return the supernatant liquor with the lowest quantity of solids from the primary digester to the secondary digester or from the secondary digester back to the primary clarifier, or to the sludge drying beds, provided space is available.

A single adjustable tube is also used at some plants. On the floating cover digester there is usually only one supernatant tube. This may be adjusted to pull supernatant liquor from various levels of the tank by raising or lowering the tube. In smaller plants the supernatant withdrawal may be done only once or twice a day, because the floating cover allows the tank to handle volume changes. An adjustable tube usually allows a supernatant with the least solids content to be selected. The digester should be visually checked a minimum of once per day for liquor levels to prevent overfilling and structural damage to the tank.

12.115 Sludge Draw-Off Lines (Figures 12.3 and 12.4)

The sludge draw-off lines are typically placed on blocks along the sloping floor of the digester. Sludge is withdrawn from the center of the tank. Very seldom are they placed under the floor of the digester because they would not be accessible in case of blockages. These lines are normally six inches (150 mm) in diameter and equipped with plug valves. The lines are used to transfer the digested sludge periodically to a sludge disposal system of either drying beds or some type of dewatering

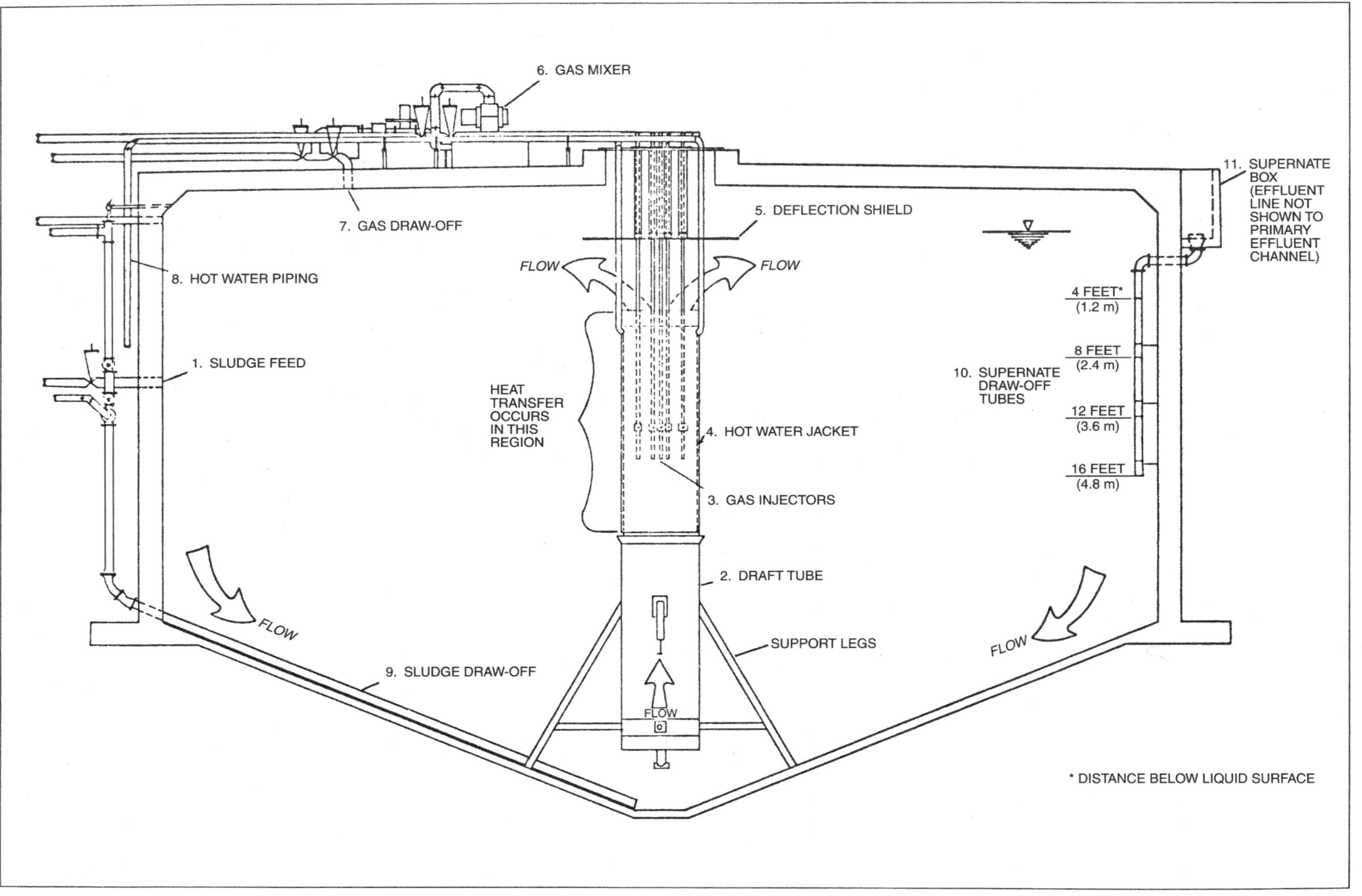

Fig. 12.3 Fixed cover digester
(Permission of County Sanitation Districts of Los Angeles County, California)

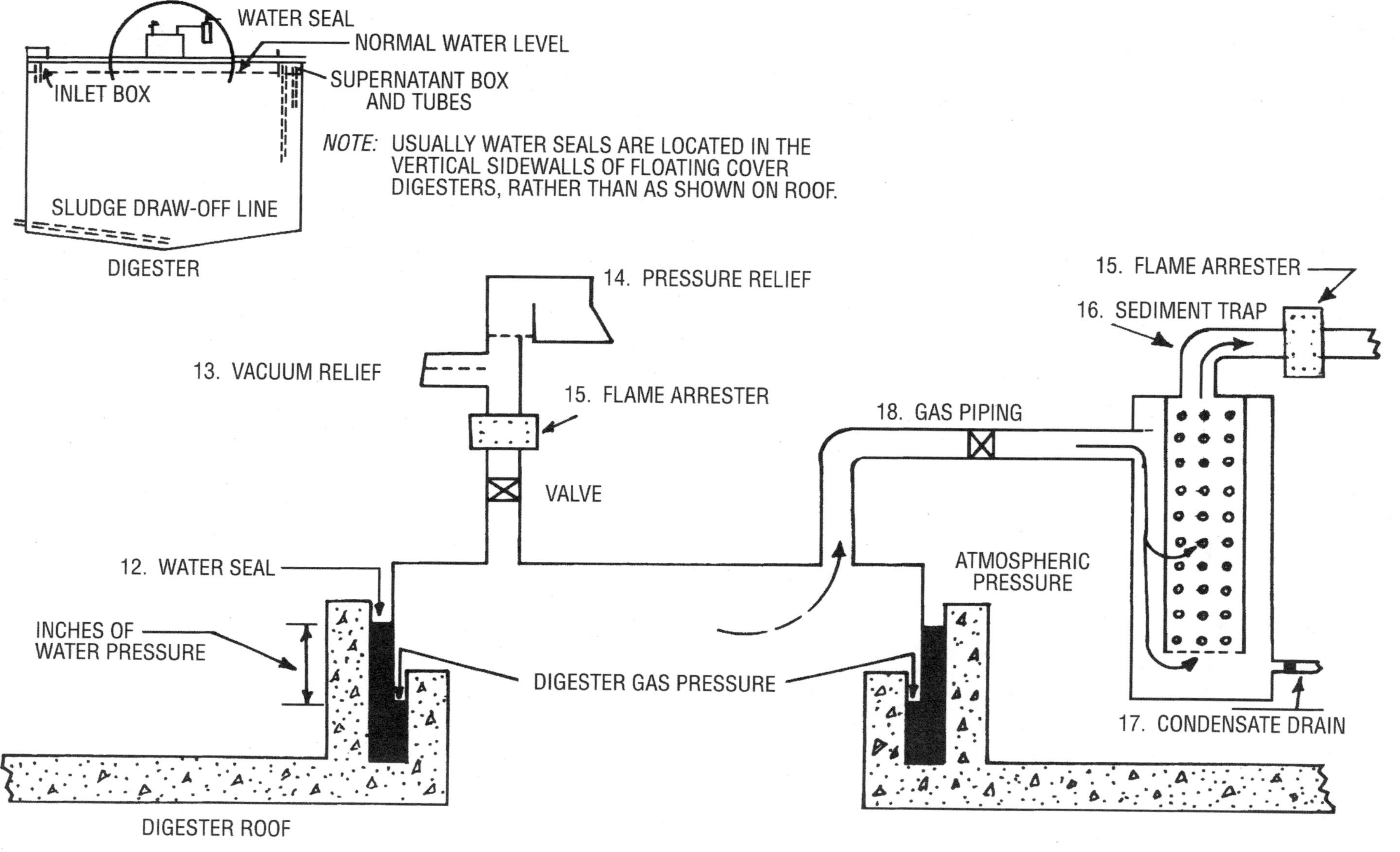

Fig. 12.4 Water seal on digester

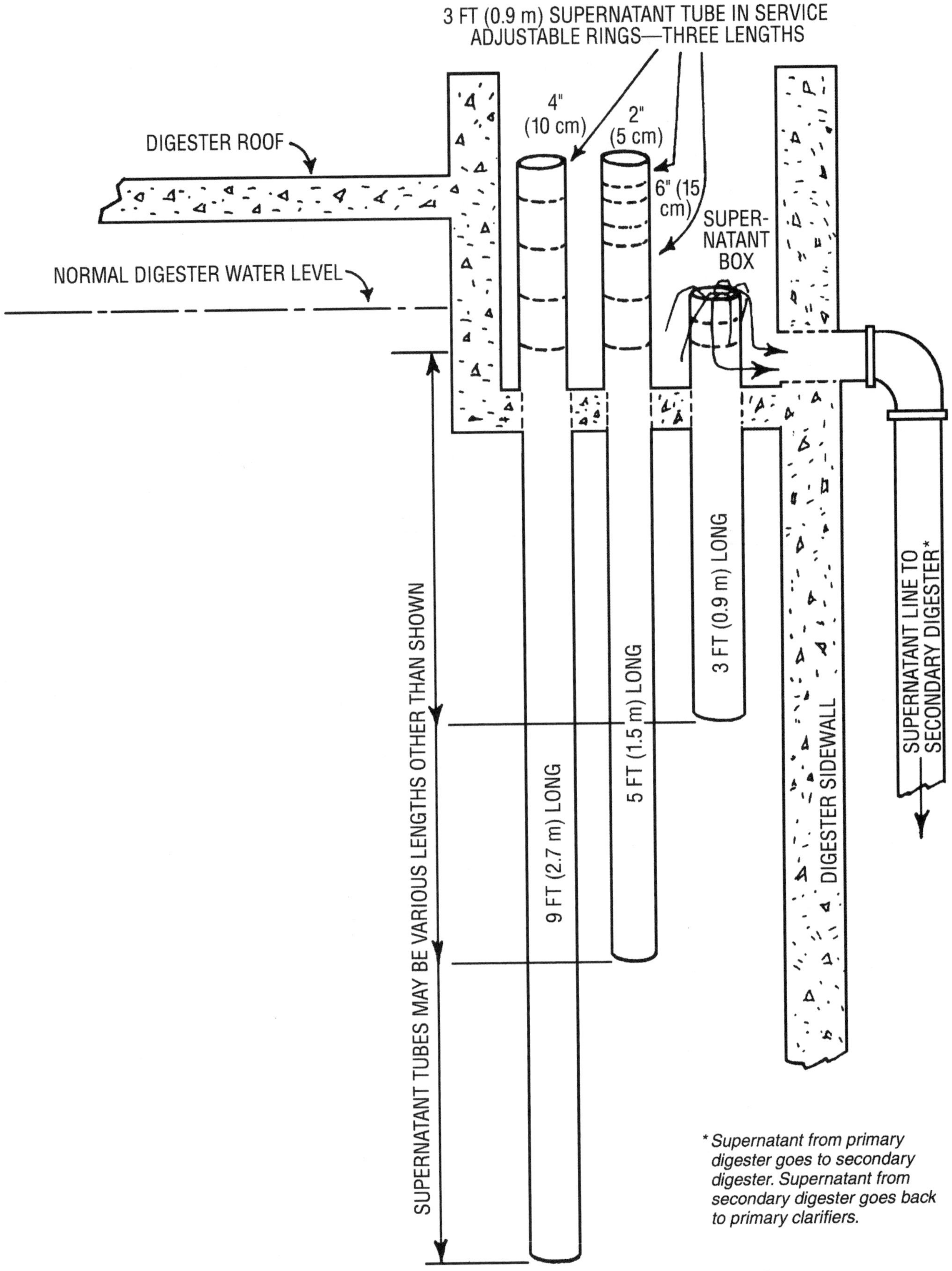

3 FT (0.9 m) SUPERNATANT TUBE IN SERVICE
ADJUSTABLE RINGS—THREE LENGTHS

DIGESTER ROOF

4" (10 cm)

2" (5 cm)

6" (15 cm)

SUPERNATANT BOX

NORMAL DIGESTER WATER LEVEL

SUPERNATANT TUBES MAY BE VARIOUS LENGTHS OTHER THAN SHOWN

9 FT (2.7 m) LONG

5 FT (1.5 m) LONG

3 FT (0.9 m) LONG

DIGESTER SIDEWALL

SUPERNATANT LINE TO SECONDARY DIGESTER*

Supernatant from primary digester goes to secondary digester. Supernatant from secondary digester goes back to primary clarifiers.

Fig. 12.5 Supernatant tubes and box for fixed cover digester

system. These lines also transfer seed sludge from the secondary digester to the primary digester and recirculate bottom sludge to seed and break up a scum blanket.

QUESTIONS

Write your answers in a notebook and then compare your answers with those on page 231.

12.11A Why should you maintain no more than an eight-inch (20-cm) water column of gas pressure on the roof of a fixed cover digester?

12.11B Why must a fixed cover digester be equipped with pressure and vacuum relief valves?

12.11C What are the advantages of a floating cover in comparison with a fixed cover digester?

12.11D Why is it desirable to mix recirculated digester sludge with raw sludge?

END OF LESSON 1 OF 6 LESSONS
on
SLUDGE DIGESTION AND SOLIDS HANDLING

Please answer the discussion and review questions next.

DISCUSSION AND REVIEW QUESTIONS

Chapter 12. SLUDGE DIGESTION AND SOLIDS HANDLING

(Lesson 1 of 6 Lessons)

At the end of each lesson in this chapter, you will find some discussion and review questions. The purpose of these questions is to indicate to you how well you understand the material in the lesson. Write the answers to these questions in your notebook.

1. Briefly explain what happens when sludge is added to an anaerobic digester.
2. Why is it important to keep the contents of a digester well mixed?
3. Why should the floor of a digester be sloped?
4. Why do digesters have supernatant tubes?

CHAPTER 12. SLUDGE DIGESTION AND SOLIDS HANDLING

(Lesson 2 of 6 Lessons)

12.12 Gas System (Figure 12.6)[11]

The anaerobic digestion process produces 8 to 12 cubic feet of gas for every pound (0.44 to 0.75 cu m/kg) of volatile matter added, and from 12 to 18 cubic feet of gas for every pound (0.75 to 1.2 cu m/kg) of volatile matter destroyed, depending on the characteristics of the sludge. The gas consists mainly of methane (CH_4) and carbon dioxide (CO_2). The methane content of the gas in a properly functioning digester will vary from 65 to 70 percent, with carbon dioxide running around 30 to 35 percent by volume. One or two percent of the digester gas is composed of hydrogen, nitrogen, hydrogen sulfide, and other gases. The characteristics of these gases are described in Chapter 14, Table 14.1, page 268.

WARNING

Digester gas can be extremely dangerous in two ways. When mixed with oxygen in certain proportions, it can form explosive mixtures, and it also can cause asphyxiation or oxygen starvation. Smoking, open flames, or sparks must not be tolerated around the digesters or sludge pumping facilities.

Digester gas (due to the methane) possesses a heat value of approximately 500 to 600 *BTU*[12] per cubic foot (18,630 to 22,350 k *J*[13]/cu m), whereas natural gas with a higher methane content may range from 900 to 1,200 BTU per cubic foot (33,530 to 44,700 k J/cu m).

Digester gas is used in plants in various ways: for heating and mixing the digesters, for heating the plant buildings, for running engines, for air blowers for the activated sludge process, and for electrical power for the plant.

The gas system removes the gas from the digester to a point of storage, use, or to be burned in the waste gas burner as excess. The following items are components of the gas system.

12.120 Gas Dome

This is a point in the digester roof where the gas from the tank is removed. On fixed cover tanks there may also be a water seal (Figure 12.4) to protect the tank structure from excess *POSITIVE PRESSURE*,[14] or vacuum (negative pressure) created by the addition or withdrawal of sludge or gas too rapidly.

If gas pressure is allowed to build up to 11 inches (28 cm) of water column pressure, it will escape around the water seal to the atmosphere without lifting the roof. If sludge or gas is withdrawn too rapidly, the vacuum could exceed eight inches (20 cm) and break the water seal, thus allowing air to enter the tank. Without the water seal, the vacuum could become great enough to collapse the tank. Air in the digester between 85% and 95% by volume (5% to 15% methane) creates an explosive condition (see Figure 12.7). In addition, sulfuric acid corrosion is often found where air is consistently in contact with the gas.

QUESTIONS

Write your answers in a notebook and then compare your answers with those on page 231.

12.12A What are the two main gaseous components of digester gas after gas production has become well established?

12.12B Why must the digester gas be handled with extreme caution?

12.12C What are some uses of digester gas?

12.121 Pressure Relief and Vacuum Relief Valves

The pressure relief valve (Figure 12.8) will operate when the waste gas burner cannot handle excessive gas pressures; valve operation prevents the breaking of the water seal. The vacuum relief valve will operate if sludge or gas is withdrawn from the digester too quickly and a vacuum in the digester develops. The tank could collapse if the vacuum relief valve failed.

---

[11] Many figures in this section were made available courtesy of Whessoe Varec, Inc. (now part of Endress + Hauser Systems & Gauging, Inc.). Mention of commercial products or manufacturers is for illustrative purposes and does not imply endorsement by California State University, Sacramento, EPA/WQO, or any state or federal agency.

[12] *BTU* (pronounce as separate letters). British Thermal Unit. The amount of heat required to raise the temperature of one pound of water one degree Fahrenheit. Also see CALORIE.

[13] *Joule* (JOOL). A measure of energy, work, or quantity of heat. One joule is the work done when the point of application of a force of one newton is displaced a distance of one meter in the direction of the force. Approximately equal to 0.7375 ft-lbs (0.1022 m-kg).

[14] *Positive Pressure.* A positive pressure is a pressure greater than atmospheric. It is measured as pounds per square inch (psi) or as inches of water column. A negative pressure (vacuum) is less than atmospheric and is sometimes measured in inches of mercury. In the metric system, pressures are measured in kg/sq m, kg/sq cm, or pascals (1 psi = 6,895 Pa = 6.895 kN/sq m).

WASTE GAS SYSTEMS

• Anaerobic Digesters
• Municipal Landfills
• Biogas Utilization

GENERAL

VAREC's waste gas equipment provides a safe and efficient method of controlling the gases produced during the anaerobic digestion of organic solids. The gas generated by this process is an energy source which can be collected and utilized. Waste gas is typically a highly moist mixture of gases. It consists of approximately 55 to 70% methane, 25 to 35% carbon dioxide and trace amounts of nitrogen and hydrogen sulfide. This gas can be used to heat the digester and to supply much of the power needed by the treatment plant.

The solids in municipal and industrial wastewater streams may be treated by anaerobic digestion to meet environmental requirements. The anaerobic process occurs in closed vessels (digesters), covered ponds or lagoons. Soluble organics are converted to methane and CO_2 in a two-step process. Additionally, the BOD and COD of the wastewater is lowered, and the sludge is stabilized and reduced in volume. Landfills generate waste gas naturally as buried organic refuse biodegrades.

DIGESTER GAS HANDLING SYSTEM

Gas collection and utilization is an important part of the anaerobic digestion process. The low Btu gas is saturated, and contains elements harmful to personnel and corrosive to piping and equipment. Waste gas must be properly handled to ensure a protected environment.

The key aspect of design is to recognize that gas collection equipment operates as a system. This bulletin is intended to provide general guidelines for the design and selection of a complete gas handling system. With over five decades of experience and service to the industry, Varec is proud of its high quality world-wide performance record.

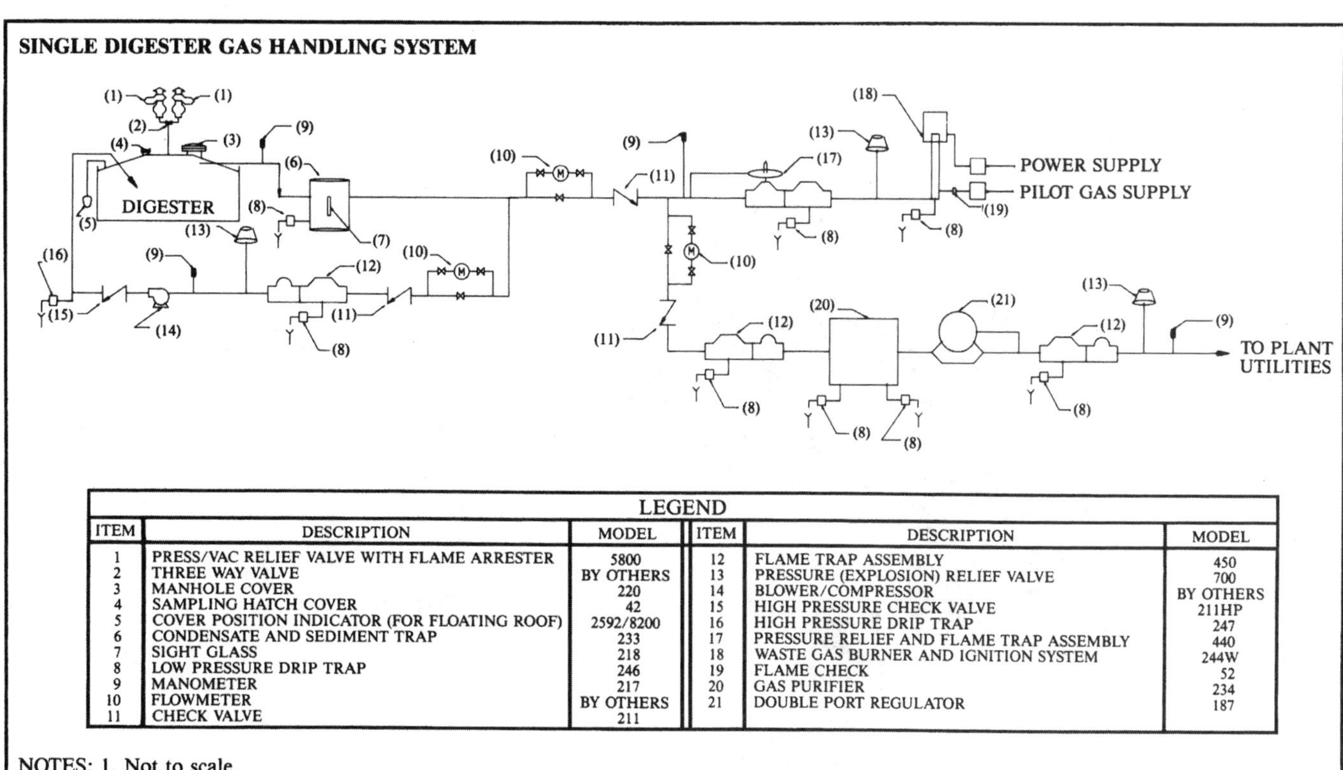

LEGEND					
ITEM	DESCRIPTION	MODEL	ITEM	DESCRIPTION	MODEL
1	PRESS/VAC RELIEF VALVE WITH FLAME ARRESTER	5800	12	FLAME TRAP ASSEMBLY	450
2	THREE WAY VALVE	BY OTHERS	13	PRESSURE (EXPLOSION) RELIEF VALVE	700
3	MANHOLE COVER	220	14	BLOWER/COMPRESSOR	BY OTHERS
4	SAMPLING HATCH COVER	42	15	HIGH PRESSURE CHECK VALVE	211HP
5	COVER POSITION INDICATOR (FOR FLOATING ROOF)	2592/8200	16	HIGH PRESSURE DRIP TRAP	247
6	CONDENSATE AND SEDIMENT TRAP	233	17	PRESSURE RELIEF AND FLAME TRAP ASSEMBLY	440
7	SIGHT GLASS	218	18	WASTE GAS BURNER AND IGNITION SYSTEM	244W
8	LOW PRESSURE DRIP TRAP	246	19	FLAME CHECK	52
9	MANOMETER	217	20	GAS PURIFIER	234
10	FLOWMETER	BY OTHERS	21	DOUBLE PORT REGULATOR	187
11	CHECK VALVE	211			

NOTES: 1. Not to scale.
2. This schematic is for general guidance purposes only and is not intended to represent a specific design.

Fig. 12.6 Single digester gas handling system
(Permission of Whessoe Varec)

An explosive gas mixture was accidentally ignited.

The digester cover blew off …

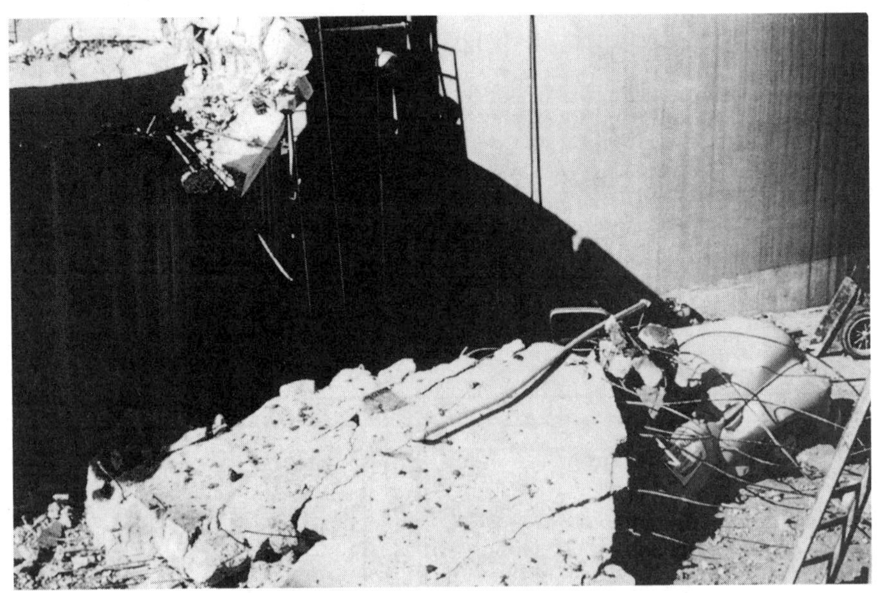

… and landed on top of a pickup truck.

Fig. 12.7 Results of digester explosion

PRESSURE RELIEF AND FLAME TRAP ASSEMBLY

- Spring Loaded Diaphragm With Visual Indicator
- Double Acting Needle Valve
- Positive Tight Shut-off
- Extensible Bank Assembly
- Three to Four Times Net Free Area Through Bank

INTRODUCTION

The Whessoe Varec 440 Series Pressure Relief and Flame Trap Assembly is a combination of the Whessoe Varec 386 Series Back Pressure Regulator, Model 5010 Flame Arrester, and a thermal by-pass shut-off valve. This unit combines the sensitive Whessoe Varec Single Port Regulator with the easily serviced Whessoe Varec extensible bank flame arrester and adds a by-pass valve for maximum protection and reliable operation.

APPLICATION

The Whessoe Varec 440 Series Pressure Relief and Flame Trap Assembly is designed to control upstream pressure while protecting the waste gas piping system from flashback fires. Waste gas is typically a highly moist mixture of gases. It consists of approximately 55 to 70% methane, 25 to 35% carbon dioxide and trace amounts of nitrogen and hydrogen sulfide. This assembly is generally installed in the waste gas piping just upstream of the waste gas burner.

The regulator valve will remain closed until the line pressure exceeds the set-point, thereby maintaining a predetermined back pressure throughout the gas control system. By selecting a regulator set-point higher than the required supply pressure to waste gas utilization equipment (boiler, engine generator, etc.), waste gas is first supplied to these devices. When gas production exceeds the capacity of this equipment, the regulator opens and relieves surplus gas to the waste gas burner. While gas is flaring, the flame arrester and thermal by-pass shut-off valve serve to inhibit a possible flashback of the flare into the gas control system. The 440 Series Assembly is designed for installation in low pressure horizontal gas lines. The flame arrester portion is fitted with a drain plug at the low point. A Whessoe Varec Drip Trap should be specified for field installation at this connection to provide safe removal of condensate.

The Model 5010 Flame Arrester is designed to stop the propagation of flame from external sources. It can be mounted up to 15 feet upstream of the ignition source when used in accordance with UL Approval..

For further details on the components of this assembly, see Bulletin PDS 386WT for the Back Pressure Regulator, and Bulletin PDS 5000WT for the Model 5010 Flame Arrester.

OPERATION

Operation of the Whessoe Varec 440 Pressure Relief and Flame Trap Assembly is automatic. The regulator constantly senses the upstream waste gas pressure beneath the diaphragm, while the upper diaphragm chamber is vented to the atmosphere through the by-pass valve. As the line pressure increases above the set-point, the diaphragm lifts upward and the connecting valve opens to relieve the over pressure. As the upstream pressure falls below the set-point, the diaphragm lowers and the valve closes.

If a flashback occurs downstream of the 440 assembly, the flame arrester prevents propagation of the flame through dissipation of heat. The bank assembly sheets absorb heat from the ignited gas faster than it is developed. The temperature of the gas is lowered below its ignition point, quenching the flame.

Should a major flashback occur, the fusible element is contacted by the flame and melts, changing the position of the needle valve. This closes the vent to the atmosphere and applies full upstream waste gas pressure from beneath the diaphragm to the upper diaphragm chamber. Pressure is equalized on both sides of the diaphragm. Both gravity and the compression spring force the regulator valve closed, shutting off the gas flow and extinguishing the flame.

A 1/2" NPT connection is provided in the regulator diaphragm housing for field installation of the sensing line. This line should be connected to the waste gas piping at least 10 feet (3 m) upstream of the regulator to avoid improper operation caused by turbulence which may occur as the valve opens and closes.

If the unit is installed indoors, the vent in the by-pass valve should be extended outdoors with tubing. This will prevent gas from escaping to the room should the diaphragm develop a leak. A Whessoe Varec Model 52 Flame Check should be installed near the open end of the vent line for maximum protection.

Fig. 12.8 Pressure relief valve and flame trap assembly

(Permission of Whessoe Varec)

The pressure relief valve and the vacuum relief valve both are attached to a common pipe, but each works independently. The pressure relief valve is equipped with a seat, or pressure relief pallet, which is weighted with lead washer weights. Each weight is stamped with its equivalent *WATER COLUMN HEIGHT*[15] such as 1 inch H_2O or 3 inches H_2O (2.5 or 7.5 cm). There should be sufficient weights, combined with the weight of the pallet, to equal the designed holding pressure of the tank. The gas pressure is normally established between 6 inches and 8 inches (15 and 20 cm) of water column pressure. Check the pressure with a water manometer and adjust the lead weights as needed to maintain correct pressure. If the gas pressure in the tank exceeds the pop-off setting, then the valve will open and vent to the atmosphere through the pressure relief valve. This should occur before the water seal blows out. The water seal can be broken when a tank is overpumped or gas removal is too slow.

The vacuum relief valve operates like the pressure relief valve except that it relieves negative pressures to prevent the tank from collapsing. Operation of either one of these valves is undesirable because this allows the mixing of digester gas with air and can create an explosive atmosphere outside the tank if the pressure relief valve opens, or inside the tank if the vacuum relief valve opens.

These two valves should be checked at least every six months for proper operation and daily under freezing conditions.

QUESTIONS

Write your answers in a notebook and then compare your answers with those on page 231.

12.12D How would you adjust the pressure relief valve to prevent pressures within the digester from exceeding the design pressure?

12.12E How could the water seal be broken in a digester?

12.12F Why is the operation of either the pressure relief valve or the vacuum relief valve undesirable?

12.122 Flame Arresters *(Figure 12.9)*

A typical flame arrester is a rectangular box holding approximately 50 to 100 corrugated aluminum plates. If a flame should develop in the gas line, it would be cooled below the ignition point as it attempted to pass through the baffles, but gas could continue to flow through the line with little loss in pressure.

To prevent explosions, flame arresters should be installed:

1. Between vacuum and pressure relief valves and the digester dome
2. After sediment traps on gas lines from digester
3. At waste gas burners
4. Before every boiler, furnace, or flame

Flame arresters should be serviced every three months. Before servicing a flame arrester, valve off the gas flow to the line and put out all flames and pilot lights in the area. Remove the end housing and slide out the cartridge containing the flame arrester baffles. The cartridge is designed to slide open to expose the baffle plates and may be cleaned without complete dismantling. Separate and wash the baffle plates in a solvent; then dry them. A buildup of scale, salts from condensate, and residue buildup on the plates restricts gas flow. Reinstall the cartridge and replace the end plate. When the unit is reassembled, it should be tested for leaks by swabbing a soapsuds solution over potentially leaky areas and inspecting for bubbles.

12.123 Thermal Valves *(Figure 12.10)*

Another protective device installed near a flame source and near the gas dome is the thermal valve. The valve is round, with a weighted or spring-loaded seat attached to a stem. The stem rests on a fusible disk holding the seat up. If enough heat is generated by a flame, the fusible element melts and drops the stem and valve seat to cut off gas flow. Most valves are equipped with a wing nut on top of the valve body. If the wing nut is removed, it uncovers a glass tube that shows visually if the stem is up. If the stem cannot be seen, then the valve is closed and no gas can flow. If this occurs, the valve is removed and heated in boiling water to remove the melted fusible slug. A new slug is installed (slightly larger than an aspirin tablet), the stem replaced on top of it, and the valve is ready for service. These valves should be dismantled at least once a year to ensure that the stem is free to fall and is not gummed up with residue or scale from the gas.

12.124 Flame Trap Assembly *(Figure 12.11)*

The flame trap assembly is a combination of a thermally operated shut-off valve and a flame arrester. The thermal valve will shut down the flow of gas if a flame melts the valve's fuse. The flame arrester prevents the propagation of the flame.

QUESTIONS

Write your answers in a notebook and then compare your answers with those on pages 231 and 232.

12.12G How would you service a flame arrester?

12.12H Why should you check the thermal valves at least once a year?

12.125 Sediment Traps

The sediment trap is usually located on top of the digester near the gas dome. The inlet gas line is near the top of the

[15] *Water Column Height.* When pressure builds up in a digester, the gas pressure would force water up a tube of water connected to the outside of the digester. The higher the water column height, the greater the gas pressure.

RELIEF VALVE AND FLAME ARRESTER

- Vent to Atmosphere or Pipe Away Model
- Oversized Pressure and Vacuum Ports
- Extensible Bank Assembly
- Replaceable Pressure and Vacuum Seat Rings
- Optional "All-Weather" Coating of Valve Seats and Guides

5810 SERIES

INTRODUCTION

The VAREC 5810/5820 Series Relief Valve and Flame Arrester is a combination of the VAREC 2010/2020 Series Pressure and Vacuum Relief Valve and the VAREC 5000 Series Flame Arrester. This unit combines the high flow capacity of the VAREC relief valve with the easily serviced VAREC extensible bank flame arrester for maximum protection and reliable operation.

APPLICATION

VAREC 5810/5820 Series Relief Valve/Flame Arresters are installed on anaerobic digester covers or low pressure gas holder roofs. The relief valve protects the cover from excessive pressure and vacuum within the tank. In addition, it maintains system operating pressure so waste gas is not routinely vented to the atmosphere. The flame arrester protects from accidental ignition of sludge gas within the digester or gas holder. It is designed to stop the propagation of flame from external sources. For further information on these devices, see Bulletin PDS 2010WT and PDS 5000.

Several material combinations are available to suit various climates. The standard aluminum construction is suitable for moderate climates. In climates with extremes of hot, humid and freezing weather, the "All-Weather" 5811 series is recommended. This design incorporated added features which reduce potential malfunctions from these extreme conditions.

For further protection from cold weather, VAREC offers the 5820 Series. This model incorporates an enclosed pressure port on the relief valve. This allows gas to be vented away from the valve. It is especially suited for field installation of insulation jackets or insulated shelters. The 5821 Series includes the pipe-away outlet and the "All-Weather" features for maximum cold weather protection.

Where corrosion from H_2S is of concern, the relief valve can be supplied with optional 316 SS trim. The flame arrester bank sheets are also available in 316 SS material.

OPERATION

Whenever the system gas pressure or vacuum exceeds the valve setting, the pallets lift. Only excess pressure is vented to the atmosphere. Air is drawn into the digester only to relieve an excess vacuum condition. The valve remains closed while the gas utilization system remains within normal operating pressure.

The flame arrester stops the propagation of a flame by absorbing and dissipating heat through the surface area of the bank sheets. Heat is absorbed as ignited gas attempts to pass through the small passages within the bank assembly. This action lowers the temperature of the gas below its ignition point and quenches the flame.

Fig. 12.9 Relief valve and flame arrester

(Permission of Whessoe Varec)

THERMAL OPERATED SHUT-OFF VALVE

INTRODUCTION

The Whessoe Varec 430 Series Thermal Operated Shut-off Valve is designed to protect low pressure waste gas systems by shutting off the gas flow if a fire is burning in the piping. Once gas is no longer supplied at the source of combustion, the fire is extinguished.

The 430 Series Thermal Valve is typically installed in lines leading to boilers, blowers, compressors, or other possible sources of combustion. The valve may be installed in either horizontal or vertical piping and is suitable for working pressures up to 5 psig (34.5 kPa). For higher pressure ratings, consult factory.

For maximum protection from flame fronts caused by flashbacks in the waste gas piping, a Whessoe Varec Model 5010 Flame Arrester is recommended in addition to the thermal valve. This combination unit is specified as Model 450 Flame Trap Assembly and is described in detail in Bulletin PDS 450WT.

OPERATIONS AND FEATURES

The Model 430 Valve is a fusible element released, spring operated pallet type shut-off valve. During normal operation, the pallet is positioned above the valve seat. The pallet is connected to a rod which sits atop a fusible element. In this position, the valve is open, permitting a constant flow of gas through the piping system. Should the valve be subjected to flashback temperatures, the fusible element in the valve will melt within 15 seconds upon reaching 260°F (127°C) for valves, sizes 2"–4" and within 22–45 seconds for valves, sizes 6"–12". The valve compression spring immediately forces the pallet down against the valve seat, closing the valve, thereby shutting off the gas flow.

The thermal valve is designed for ease of operation and maintenance. A sight glass is located beneath the winged inspection cap and an indicator rod is connected to the pallet. The pallet position may be determined simply by viewing the indicator rod through the sight glass, without having to remove the valve from service. This allows the operator to quickly determine whether the valve is open or closed without having to use a manometer or pressure gauge when trying to locate a line blockage. The gas tight fuse plug allows the fusible element to easily be replaced without having to remove the valve cover or disassemble the valve.

AT A GLANCE

- ***Fusible Element and Spring Provide Quick Shut-Off***
- ***Sight Glass With Indicator Rod***
- ***Gas Tight Fuse Plug Allows Replacement Without Disassembling Valve***

CONSTRUCTION

The valve body and cover are constructed of heavy wall 356 HT cast aluminum. The pallet assembly is of low copper aluminum and the compression spring is of 304 stainless steel. The sight glass is acrylic, and is isolated by Neoprene gaskets. Fusible element is antimony lead. Inlet and outlet flanges are drilled to 125 PSI ANSI flat face drilling.

Fig. 12.10 Thermal operated shut-off valve
(Permission of Whessoe Varec)

FLAME TRAP ASSEMBLY

- Spring Actuated Thermal Valve
- Isolated Sight Glass and Indicator Rod
- Removable Fusewell
- Extensible Bank Assembly
- Three to Four Times Net Free Area Through Bank

INTRODUCTION

The Whessoe Varec 450 Series Flame Trap Assembly is a combination of the Whessoe Varec 430 Series Thermal Operated Shut-off Valve and the Whessoe Varec 5010 Series Flame Arrester. This unit combines the time-tested thermal valve with the easily serviced extensible bank flame arrester for maximum protection and reliable operation.

APPLICATION

The 450 Series Flame Traps are designed to protect the waste gas utilization system from flashbacks in the piping. Waste gas is typically a highly moist mixture of gases. It consists of approximately 55 to 70% methane, 25 to 35% carbon dioxide and trace amounts of nitrogen and hydrogen sulfide. Flame traps are installed in waste gas systems just upstream of boilers, blowers, compressors, stationary gas engines, burners or other equipment.

The Model 5010 Flame Arrester is designed to stop the propagation of flame from external sources. It can be mounted up to 15 feet upstream of the ignition source when used in accordance with UL Approval. The Model 430 Thermal Valve provides added protection. It is designed to shut off the gas flow if a fire is burning in the line.

The 450 Flame Trap Assembly may be installed in the vertical or horizontal position. For horizontal installation, the flame arrester portion is fitted with a drain plug at the low point. A Whessoe Varec Drip Trap should be specified for field installation at this connection to provide for safe removal of condensate.

The 450 Series Flame Trap should be used in application where back pressure control is not required. For service requiring back pressure regulation in combination with flame trap protection, the Whessoe Varec 440 Series Pressure Relief and Flame Trap Assembly as described in PDS 440WT.

OPERATION

The 450 Series Flame Trap provides two types of protection from flashback fires. If a flashback occurs downstream of the 450 Assembly, the flame arrester prevents propagation of the flame through dissipation of heat. The bank assembly sheets absorb heat from the ignited gas faster than it is developed. The temperature of the gas is lowered below its ignition point, quenching the flame.

The thermal shut-off valve is normally open, permitting a constant flow of gas through the piping system. The pallet is held above the valve seat by a fusible element. Should a major flashback occur, the fuse is contacted by the flame and melts. The valve compression spring immediately forces the pallet against seat, closing the valve. The gas flow is shut off until the fusible element is replaced.

Fig. 12.11 Flame trap assembly

(Permission of Whessoe Varec)

tank and on the side. The outlet line comes directly from the top of the sediment tank. The sediment trap is also equipped with a perforated inner baffle, and a condensate drain near the bottom. The gas enters the side at the top of the tank, passes down and through the baffle, then up and out the side at the top opposite the inlet. Moisture is collected from the gas in the trap, and any large pieces of scale are trapped before entering the gas system. The trap should be drained of condensate frequently, but may have to be drained twice a day during cold weather because greater amounts of water will be condensed.

12.126 *Drip Traps—Condensate Traps* (*Figures 12.12 and 12.13*)

Digester gas is quite wet and, in traveling from the heated tank to a cooler temperature, the water condenses. The water must be trapped at low points in the system and removed or it will impede gas flow and cause damage to equipment, such as compressors, and interfere with gas utilization. Traps are usually constructed to have a storage space of one to two quarts (one or two liters) of water. All drip traps on gas lines should be located in the open air and be of the manual operation type. Traps should be drained at least once a day and possibly more often in cold weather. Actual required frequency of draining traps depends on the location of the trap in the gas system, temperature changes, and the type of digester mixing system. Automatic drip traps are not recommended because many automatic traps are equipped with a float and needle valve orifice and corrosion, sediment, or scale in the gas system can keep the needle from seating. The resulting leaks may create gas concentrations with a potential hazard to life and equipment.

12.127 *Gas Meters*

Gas meters may be of various types, such as bellows, diaphragm, shunt flow, propeller, and orifice plate or differential pressure. They are described in detail in the metering section of Chapter 15, "Maintenance."

12.128 *Manometers*

Manometers are installed at several locations to indicate gas pressure within the system in inches (or centimeters) of water column.

12.129 *Pressure Regulators* (*Figures 12.8, 12.14, and 12.15*)

Pressure regulators are typically installed next to and before the waste gas burner, at various points in the system, to regulate the gas pressure to boilers, heaters, and engines. Regulators on the waste gas burner are usually of the diaphragm type and control the gas pressure on the whole digester gas system. They are normally set at eight inches (20 cm) of water column by adjusting the spring tension on the diaphragm. Whenever an adjustment of a pressure setting is made, check the gas system pressure with a manometer for the proper range. If the gas pressure in the system is below eight inches (20 cm) of water column, no gas flows to the burner. If gas pressure reaches eight inches (20 cm) of water column, the regulator opens slightly, allowing gas to flow to the burner. If the pressure continues to increase, the regulator opens further to compensate. The only maintenance this unit requires is on the thermal valve on the discharge side, which protects the system from backflashes. This unit is spring-loaded and controlled by a fusible element that vents one side of the diaphragm, thus stopping the gas flow when heated. Maintenance includes checking for proper operation of the regulator and of the fusible element. The condition of the diaphragms in the regulators should also be checked at periodic intervals.

12.1210 *Waste Gas Burner* (*Figures 12.16 and 12.17*)

Waste gas burners are used to burn the excess gas from the digestion system. The waste gas burner is equipped with a continuously burning pilot flame so that any excess gas will pass through the gas regulator and be burned. The pilot flame should be checked daily to be sure that it has not been blown out by wind. If the pilot is out, gas will be vented to the atmosphere creating an odorous and potentially explosive condition.

WARNING

A gas mixture ratio of 5.3 to 19.3% digester gas to air is *explosive* (lower and upper limits). (Figure 12.7)

QUESTIONS

Write your answers in a notebook and then compare your answers with those on page 232.

12.12I How frequently should you drain a sediment trap?

12.12J Why must drip or condensate traps be installed in gas lines?

12.12K What is a deficiency in automatic drip and condensate traps?

12.12L How would you adjust the gas pressure of the digester gas system?

12.12M Why should the pilot flame in the waste gas burner be checked daily?

12.13 Sampling Well (Thief Hole) (Figure 12.18)

The sampling well consists of 2- to 10-inch pipe (5- to 25-cm) (with a hinged-seal cap) that goes into the digestion tank, through the gas zone, and is always submerged a foot (0.3 m) or so into the digester sludge. This permits the sampling of the digester sludge without loss of digester gas pressure, or the creation of dangerous conditions caused by the mixing of air and digester gas. However, caution must be used not to breathe gas, which will always be present in the sample well and will be released when first opened. A sampling well is sometimes referred to as a "thief hole."

12.14 Digester Heating

Digesters can be heated in several ways. Some facilities heat digesters by recirculating the digester sludge through an external hot water heat exchanger (Figures 12.19 and 12.20). Digester gas is used to fire the boiler, which is best maintained between 140° and 180°F (60° to 82°C) for proper operation. The hot water is then pumped from the boiler to the heat exchanger where it passes through the jacket system, while the recirculating sludge passes through an adjacent jacket, picking up heat from the hot water. In some units the boiler and exchanger are combined and the sludge also is passed through the unit or to the draft tube of the gas mixer.

Circulation of 130°F (54°C) water through pipes or heating coils attached to the inside wall of the digester is another method of heating digesters. This approach creates problems of cooking sludge on the pipes and insulating them, thus reducing the amount of heat transferred. Some facilities use

LOW-PRESSURE MANUAL DRIP TRAP

- Positive Seal Against Gas Escape
- 2.5 Quart and 6 Quart Reservoir Capacity
- Simple Operation
- 5 psig Working Pressure

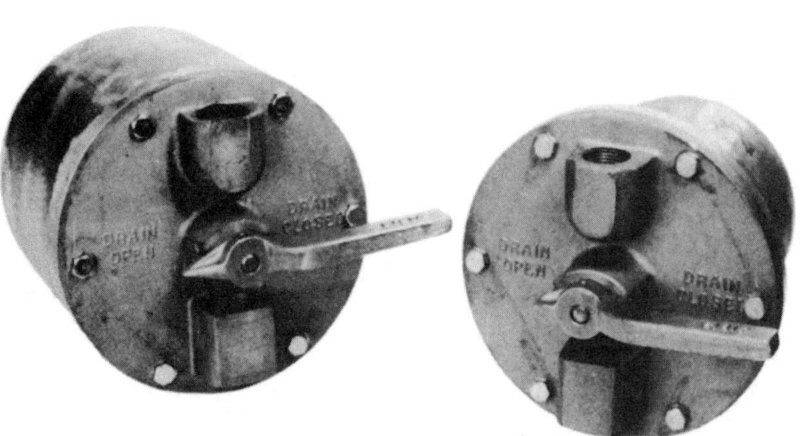

INTRODUCTION

VAREC Series 246 Manually Operated Drip Traps are designed for collection and safe removal of condensate from waste gas. Diligent removal of condensate from the waste gas piping system is necessary to protect piping and equipment from possible damage caused by corrosion or water hammer. In addition, water lying in low spots in piping can restrict gas flow resulting in increased pressure drop within this low-pressure system.

Because the waste gas is saturated and hot when it leaves the digester, it continues to cool and drop out water throughout the piping system. In addition to all low spots, VAREC drip traps should be installed frequently along horizontal piping runs. A drip trap should also be installed on each VAREC Model 232 or 233 Sediment Trap for convenient removal of accumulated condensate.

The VAREC Series 246 Drip Trap is suitable for working pressures up to 5 psig (34.5 kPa). For drip traps with higher pressure ratings, see PDS 245WT and PDS 247WT.

OPERATION AND FEATURES

Condensate accumulates in the drip trap reservoir and is drained manually by simply rotating the handle which is connected to an internal ported rotating disc. When all accumulated condensate has been drained, the handle is returned to the "fill" position.

The VAREC 246 Series Drip Trap is designed to prevent gas from escaping while draining. The disc, ports, and "O" ring seals are positioned to block the waste gas before opening the drain outlet ensuring that gas cannot escape regardless of disc position. The air inlet port located at the outlet connection allows condensate to flow freely from the reservoir when draining.

CONSTRUCTION

The Model 246 Drip Trap is constructed of 356 HT low copper cast aluminum body and handle, anodized cast aluminum cover plate and disc, neoprene "O" rings, and stainless steel internal working parts and hardware. Inlet and outlet connections are 1-inch NPT. Condensate reservoir capacity is available in 2.5 quarts (2.4 liters) or 6.0 quarts (5.7 liters), whichever is specified.

Fig. 12.12 Low-pressure drip trap

(Permission of Whessoe Varec)

HIGH-PRESSURE MANUAL DRIP TRAP

- Gas Isolated Before Draining
- Double-Seal Ball Plug Valves
- 100 psig Working Pressure
- 4 Quart Capacity
- Locking Lever Secures Handle Position

INTRODUCTION

VAREC Series 247 Manually Operated Drip Traps are designed for collection and safe removal of condensate from high-pressure waste gas piping. Removal of condensate from the waste gas piping system is important to protect piping and equipment from possible damage caused by corrosion or water hammer. In addition, water lying in low spots in piping can restrict gas flow resulting in increased pressure drop within the system.

Because the waste gas is saturated and hot when it leaves the digester, it continues to cool and drop out water throughout the piping system. In addition to all low spots, VAREC Model 247 drip traps should be installed frequently along horizontal high-pressure pipe runs. When a VAREC Model 233 Sediment Trap is installed at working pressures greater than 5 psig (34.5 kPa) and less than 25 psig (172 kPa), a Model 247 Drip Trap is recommended for convenient removal of accumulated condensate.

The VAREC Series 247 Drip Trap is suitable for working pressures up to 100 psig (688 kPa). For low-pressure manually operated drip traps or automatic drip traps, see PDS 246WT and PDS 245WT.

OPERATION AND FEATURES

Condensate accumulates in the drip trap reservoir and is drained manually by operating the valve handles. When all accumulated condensate has been drained, the handles are returned to the "fill" position, and locked in place. The VAREC Model 247 Drip Trap is designed to prevent gas from escaping while operating. The double-seal ball plug valves and locking lever isolate the gas line connection before opening the drain port.

CONSTRUCTION

The VAREC Model 247 Drip Trap body is fabricated of boiler plate carbon steel. The plug valves include a brass body with teflon seats and 316 SS ball and stem. Fill and drain pipes are of galvanized steel. Inlet and outlet connections are 1-inch NPT. Condensate reservoir capacity is 4.0 quarts (3.8 liters).

Fig. 12.13 High-pressure drip trap

(Permission of Whessoe Varec)

SINGLE PORT BACK PRESSURE REGULATOR

- Positive Shut-off
- Diaphragm Operated
- Spring Loaded for Easy Setting Adjustment
- Controls Upstream Pressure

INTRODUCTION

The Whessoe Varec 386 Series Single Port Back Pressure Regulator is designed to control upstream pressure in low pressure gas piping systems. The 386 Series Regulator is installed in piping leading to gas utilization equipment providing a reliable means of controlling line pressure and sequentially directing gas flow to boilers, engines, burners, etc.

The Model 386 Regulator should be used wherever a positive "tight" shut-off is necessary. This single port regulator is less sensitive than the Whessoe Varec 186/187 Series Double Port Regulator, yet will maintain a back pressure within approximately 10% of the setting.

Note — for maximum safety, the regulator located just upstream of the waste gas burner should be specified as a Whessoe Varec 440 Series Pressure Relief and Flame Trap Assembly. The Model 440 combination unit includes a Model 386 Regulator to control back pressure, Model 5010 Flame Arrester to provide protection from possible flashback fires, and a thermal by-pass shut-off valve. See Bulletin PDS 440WT for details on this combination unit.

OPERATION AND FEATURES

The Whessoe Varec 386 Series Pressure Regulator is a single port, diaphragm operated, spring loaded valve. The large diaphragm area provides sensitive operation while the compression spring allows field adjustment of relief setting.

The ratio of diaphragm area to valve port area of the Whessoe Varec spring loaded regulator is greater than that of typical weight loaded regulators. This results in greater sensitivity of operation which minimizes upstream pressure fluctuations that might otherwise create nuisance tripping of pressure switch controls.

A spring adjusting screw is easily accessed by unscrewing the spring barrel cap, providing a simple means of varying the pressure setting when necessary. This adjustable spring feature eliminates the need to unbolt and remove a large diaphragm housing cover to facilitate setting changes as is necessary with weight loaded regulators.

A glass enclosed pointer and scale provide the operator with a quick visual indication of the set-point. Standard setting range is 2" to 12" (50 mm to 300 mm) W.C. (sizes 2-inch through 4-inch), and 2" to 10" (50 mm to 250 mm) W.C. (sizes 6-inch and 8-inch). Standard scale markings are in 1/2" W.C. increments. Special spring and scale sets are available for higher setting ranges.

The valve is designed to remain closed until the upstream gas pressure, as sensed beneath the diaphragm, begins to exceed the regulator setting. The diaphragm plate lifts, opening the valve to relieve the over pressure, than lowers to close the valve as the upstream pressure decreases to below the regulator set pressure.

A 1/2" NPT connection is provided in the lower diaphragm housing for field installation of the sensing line. This line should be connected to the waste gas piping at least 10 feet (3 m) upstream of the regulator to avoid improper operation caused by turbulence which may occur as the valve opens and closes.

If the regulator is installed indoors, the vent in the upper diaphragm housing should be extended outdoors with tubing. This will prevent gas from escaping to the room should the diaphragm develop a leak. A Whessoe Varec Model 52 Flame Check should be installed near the open end of the vent line for maximum protection.

CONSTRUCTION

The 386 Series Regulator is constructed of 356 HT cast aluminum body, diaphragm and spring housings, and diaphragm plate. The pallet is low copper aluminum, with 304 stainless steel stem and bushings. Diaphragm is molded Buna-N rubber with nylon reinforcement. Aluminum indicator scale is protected by a gasketed glass window. Setting spring is zinc plated steel, and adjusting screw is 304 SS. Inlet and outlet flanges are drilled to 125 PSI ANSI flat face drilling. Maximum working pressure is 5 psig (34 kPa).

Fig. 12.14 Single port pressure regulator

(Permission of Whessoe Varec)

DOUBLE PORT REGULATOR

- Back Pressure or Pressure Reducing Models
- Rotary Linkage Action
- Double Port/Balanced Plugs
- Large Sensing Diaphragm
- Weighted Lever Field Adjustable Setting

INTRODUCTION

Vapor recovery and gas blanketing systems require that gas be removed or added as the pressure within the tank changes. This pressure fluctuation is proportional to the thermal expansion and contraction of the stored product as well the change in volume due to filling and emptying. Vapor recovery systems typically operate at atmospheric pressure, requiring regulators which provide very sensitive operation. The VAREC 180/181 Series regulators provide highly sensitive control so that the operating pressure within the tank remains stable. They are ideal for use in petroleum or chemical industry tank farms.

APPLICATION

MODEL 180 is primarily a BACK PRESSURE regulator. It is used to relieve excess pressure inside the tank. The Model 180 is installed in the line leading to the suction side of the vapor recovery compressor. By relieving the pressure the Model 180 protects the tank from over pressure which may otherwise cause the tank to rupture.

MODEL 181 is primarily a PRESSURE REDUCING regulator. It is used to provide make-up or blanketing gas to the tank. The Model 181 is installed in the line between the make-up gas supply and the tank. By providing gas during a vacuum condition the Model 181 protects the tank from under pressure which may otherwise cause the tank to collapse.

Correct regulator valve sizing is important. The smallest size valve capable of handling the expected flow should be selected. An oversized valve can oscillate, reducing performance and subjecting the seat and other components to excessive wear.

The field installed control line should be connected to the piping at least 10 feet (3 m) from the regulator. This is necessary so valve operation is not affected by turbulence in the piping from the throttling valve.

OPERATION

VAREC 180/181 Series Regulators are double port, diaphragm operated valves. The VAREC regulator is self contained—no outside power source is required. Operation is completely automatic. Set pressure is controlled by a weighted lever for settings below 1″ w.c. and a combination weighted lever and spring for settings between 1″ and 1.5″ w.c.

Tank pressure as sensed through control line piping is applied to one side of the diaphragm. This pressure acts against the weighted lever arm, moving the diaphragm linkage. The movement positions the valve to throttle (regulate) the flow of gas.

The Model 180 control line terminates upstream. The weights act to hold the valve closed. As the upstream (tank) pressure increases beyond the set point of the regulator, the diaphragm moves and the valve throttles open. Gas flows from the tank to the compressor suction, relieving the excess pressure as required to maintain a constant operating pressure within the tank.

The Model 181 control line terminates downstream. The weights act to open the valve as downstream pressure decreases. Gas flows from the make up gas supply to the tank to maintain a constant pressure within the tank.

Fig. 12.15 Double port pressure regulator (back pressure or pressure reducing models)

(Permission of Whessoe Varec)

WASTE GAS BURNER & IGNITION SYSTEM

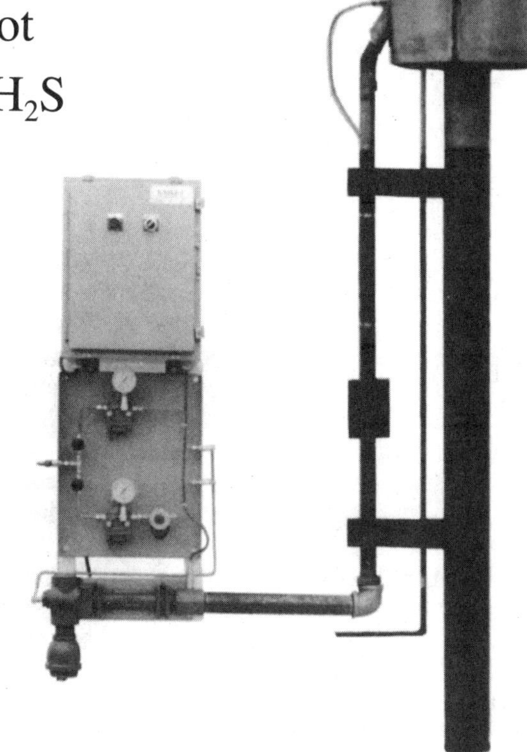

- Highly Reliable Stoichiometric Pilot
- High Temperature Conversion of H_2S
- Simplified Maintenance — All Performed at Ground Level
- 110 mph Wind Survival Without Secondary Stack
- Long Life Stainless Steel Tip
- Solid State Controller Provides Automatic Re-Ignition

INTRODUCTION

The VAREC 244W Series Waste Gas Burner is a highly reliable flare and ignition system. It is developed from systems used extensively in the petroleum industry. The pilot has proven reliable, even in the extremes of Alaska's climate. The Model 244W is ideal for use in the Waste Treatment Industry to burn excess waste gas.

APPLICATION

Excess waste gas must be disposed of safely. The gas is flared to avoid an odor nuisance or an explosion hazard. Waste gas is generated by the anaerobic digestion of organic solids. It is produced in municipal or industrial anaerobic digesters, lagoons, and municipal landfills. The waste gas is typically a highly moist mixture of gases. It consists of approximately 55 to 70% methane, 25 to 35% carbon dioxide and trace amounts of nitrogen and hydrogen sulfide. The waste gas often has a fluctuating flow and BTU value. The VAREC Model 244W is designed to operate reliably at low and high flow rates, and is not affected by changes in the waste gas BTU value.

OPERATION

The VAREC 244W Burner utilizes a patented pilot ignition system. Pilot gas and air are mixed and ignited at ground level, remote from the burner stack. This controlled method results in a stable pilot flame with an ideal gas-to-air burning ratio. The pilot burns a true stoichiometric, non-smoking flame. It is not affected by changes in the waste gas flow rate or BTU content.

The electronics package controls pilot ignition and monitoring. During the ignition cycle, pilot gas is directed to the flame retention nozzle. Additionally, gas is directed to the venturi where air is inspirated. The air/gas mixture is ignited at the venturi outlet. This generates a flame front which exits the continuous flame nozzle. The flame retention nozzle is ignited by this flame front. Once the continuous flame nozzle piping purges, it is automatically re-ignited by the flame retention pilot.

A thermocouple is installed in the continuous flame nozzle. When it reaches the temperature setting, the controller shuts off the flame retention pilot to conserve fuel. The pilot in the larger nozzle burns continuously. It ignites the waste gas as it is relieved through the burner. Should the pilot go out, an alarm is energized and the controller cycles to relight the pilot. If unsuccessful after several attempts, a second alarm is activated. The system continues to cycle for re-ignition until the operator changes the controller mode.

Fig. 12.16 Waste gas burner and ignition system

(Permission of Whessoe Varec)

ENCLOSED FLARE SYSTEM

- Low Operating Pressure
- 99.9% Destruction Efficiency
- Low NO_X Emissions
- High Turn Down Available
- No Visible Flame
- Low Operating Noise Level

INTRODUCTION

The Whessoe Varec 249 Series Enclosed Flare System is designed for burning excess waste gas generated in anaerobic digesters. The Whessoe Varec 249 is specifically designed for low pressure digester gas systems. It is used in areas where flare stack emissions are regulated or where it is desirable to hide the flame from the neighbors view.

OPERATION

An enclosed flare differs from a candle type (244W) flare in a number of ways. Whereas the combustion in a candle type flare takes place in the open atmosphere, the combustion in an enclosed flare takes place inside a combustion chamber or stack. These stacks can be anywhere from 4 to 15 feet in diameter and 30 to 50 feet tall. They can come in round or square configurations. Sizing of the flare is based on gas flow rate and destruction efficiency required.

By controlling the combustion atmosphere, an enclosed flare is capable of greatly improving the destruction efficiency under all operating conditions when compared to a candle type flare. The main factors used in calculating destruction efficiency rate are the retention time, controlled by the height of the stack, and the temperature, which is maintained by controlling the combustion air available. The combustion air is controlled via a set of motorized shutters at the base of the stack. The modulation shutters, controlled by a thermocouple located near the top of the stack, maintains the desired operating temperatures. In this way, a minimum temperature can be maintained.

FEATURES

The Whessoe Varec 249 Series Enclosed Flare System has a number of unique features that make it ideal for digester gas. It has burners which have been specifically designed for gas with a pressure as low as 4" WC. The unique air mixing design of the burner provides for low NOx emissions, as much as 50% lower than conventional enclosed flares.

The 249 Series uses a unique manifolding system to provide turn down ratios of 15:1 or higher. The turn down ratio is used to determine the minimum flow rate at which the flare will maintain efficiency standards, compared to the maximum flow rate. A high turn down ratio is important since most plants are designed for a much greater gas flow than they actually experience.

The unit is designed with a dual pilot fuel system. In normal operation the unit will use the waste gas as a pilot. Make-up gas (either LPG or natural gas) will be switched to automatically should the waste gas quality become too poor to sustain the pilot flame. Whessoe Varec 386 Series Back Pressure Control Valves are supplied as standard to allow the burner to operate only when the pressure in the gas system shows that there is an excess.

Fig. 12.17 Enclosed flare system
(Permission of Whessoe Varec)

SAMPLING AND GAUGING HATCH COVER

- Non-Sparking (Buna-N Insert)
- Gas-Tight (Buna-N Insert)
- Corrosion Resistant 356 HT Aluminum Construction
- Quick-Opening
- Safety Foot Pedal
- Handwheel with Padlock Feature
- Sizes 4" Through 10"

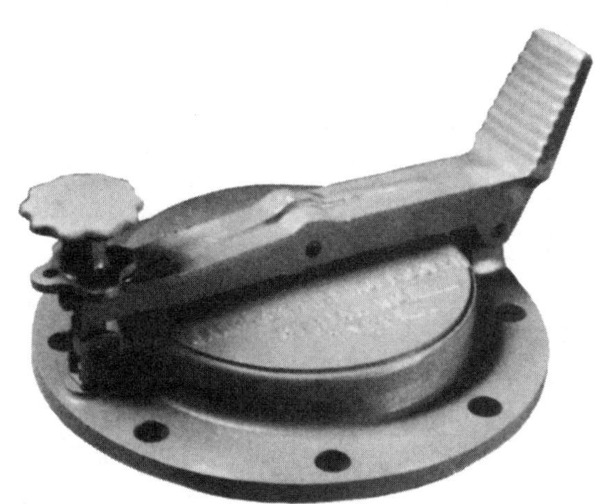

INTRODUCTION

Varec 42 Series Sampling and Gauging Hatch Covers are designed to provide quick access for product gauging, temperature measurement or sampling. They are installed on tank roofs or roof flanges.

Model 42 hatch covers are secured by tightening the handwheel. A padlock feature is incorporated to prevent unauthorized tank access through the hatch. The Model 42 hatch cover maintains a gas-tight seal at working pressures of up to 3 PSIG (20.7 kPa), when supplied with a Buna-N insert.

Varec Model 42 covers are designed with a foot pedal which provides quick tank access for sampling. This pedal is depressed to hold the cover open, enabling the operator free use of both hands to gauge or sample the tank contents. The foot pedal is designed with inclined thread for added safety. The cover closes automatically when the foot pedal is released.

CONSTRUCTION

Varec 42 Series hatch covers are available in 356 HT aluminum construction as standard. The aluminum flange is flat face, drilled per ANSI Class 125 dimensions. An optional steel base is available for welding to the tank roof. A Buna-N rubber insert is supplied for non-sparking operation. The gauging hatch covers are available in 4" through 8" sizes.

Fig. 12.18 Sampling and gauging hatch cover
(Permission of Whessoe Varec)

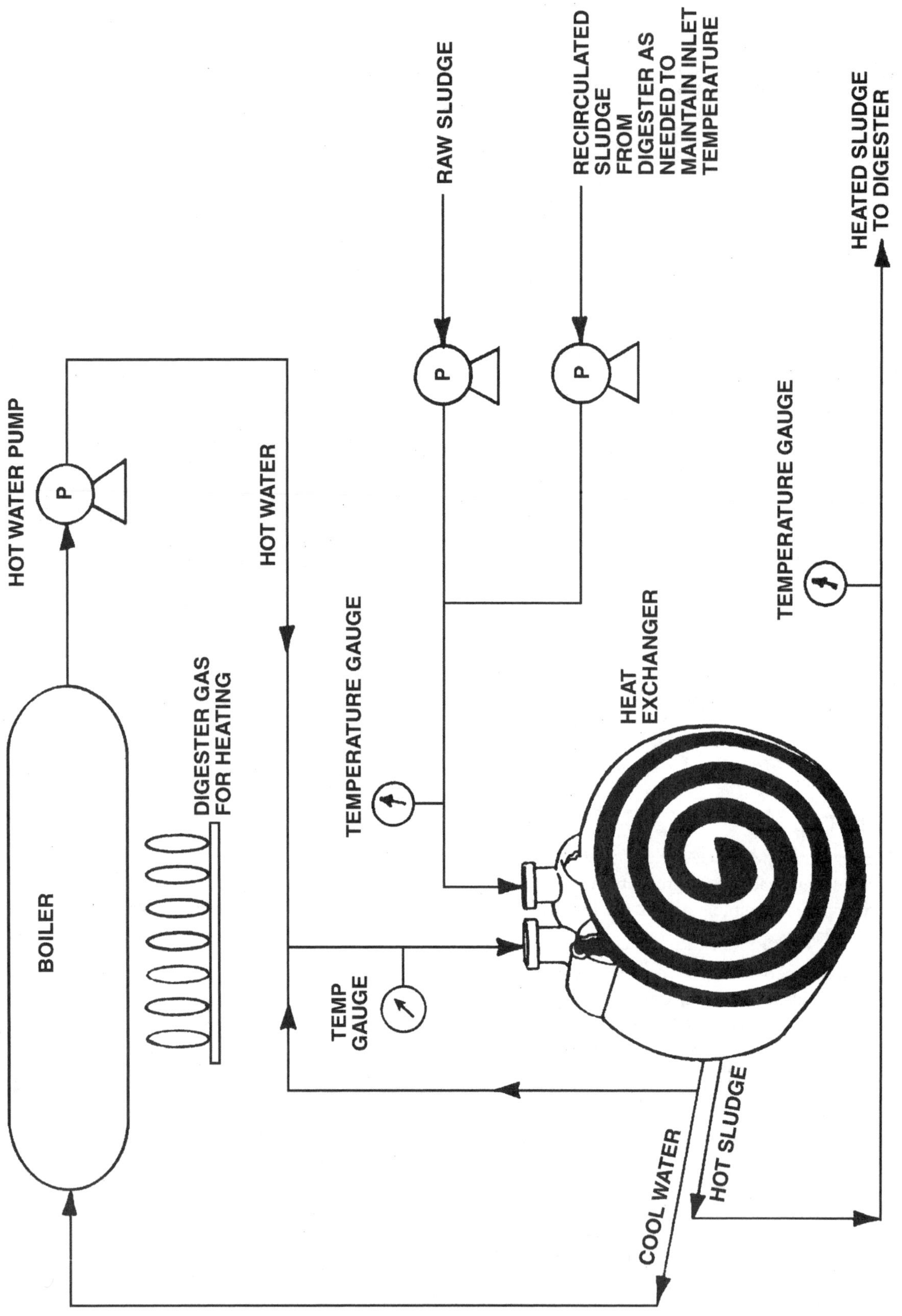

Fig. 12.19 Gas-fired external hot water heat exchanger

Fig. 12.20 Hot water heat exchanger
(Permission of Dorr-Oliver Incorporated)

submerged combustion of the gas with heat exchange between the hot, gaseous products evolved and the liquid sludge.

Other plants inject steam directly into the digesters for heating. The steam is produced in separate boilers or is recovered in connection with vapor phase cooling of engines. Careful treatment of the evaporated water to prevent scaling of the system is necessary so the practice is generally confined to plants with good laboratory control.

QUESTIONS

Write your answers in a notebook and then compare your answers with those on page 232.

12.13A Why should a digester have a special sampling well?

12.14A What causes a reduction in the amount of heat transferred from coils within the digester?

12.15 Digester Mixing

Mixing is very important in a digester because it greatly speeds up the digestion rate. Digesters are mixed by gas or by mechanical means, such as propellers, pumps, or draft tube propeller mixers.

Several important objectives are accomplished in a well-mixed digester.

a. *INOCULATION*[16] of the raw sludge immediately with microorganisms

b. Prevention of scum blanket formation

c. Maintenance of homogeneous (similar) contents throughout the tank, including even distribution of food, organisms, alkalinity, heat, and waste bacterial products

d. Use of as much of the total contents of the digester as possible and minimization of the buildup of grit and inert solids on the bottom

Some plants are designed with egg-shaped digesters. These digesters cost more to construct than other digester shapes, but the egg shape allows easy mixing of the digester contents.

12.150 Gas Mixing

Gas mixing is the method most widely used in recent years, and various approaches have been patented by manufacturers. Gas is pulled from the tank, compressed, and discharged through gas outlets (Figure 12.3) or orifices within the digester, or at some point several feet below the sludge surface. The gas rising to the surface through the digesting sludge carries sludge with it, creating a gas lift with a rolling action of the tank contents. The gas mixer may be operated on either a start and stop or a continuous basis, depending on tank conditions. The parts required for gas mixing include inlet and discharge gas lines, a positive displacement compressor, and a stainless-steel gas line header in the digester. The gas header is equipped with a cross arm to hold a specified number of gas outlets, and may be mounted in a draft tube. The gas compressor is sized for the digester and may range from 30 to 200 CFM (cubic feet per minute) (1 to 6 cu m/min) of gas.

Work with "natural gas evolution" mixing at the Los Angeles County Sanitation Districts' plants has indicated that loadings of over 0.4 pound of volatile solids per cubic foot per day (6.4 kg/cu m/day) were possible, but that if the loading dropped below 0.3 pound (4.8 kg/cu m/day) immediate stratification (layering of contents) occurred. In terms of gas recirculation, adequate mixing has been calculated from this study to require about 500 CFM of gas per 100,000 cubic feet (500 cu m/min of gas per 100,000 cubic meters) of tank capacity if released at about a 15-foot (4.5-m) depth. If released at a 30-foot (9-m) depth, about 250 CFM per 100,000 cubic feet (250 cu m/min per 100,000 cubic meters) of tank capacity should be satisfactory. If hydraulic processes are used, either by recirculation or by draft tubes and propellers, then something like 30 HP per 10,000 cubic feet (4.8 M m-kg/min per 10,000 cu m) of tank capacity is required. The latter figure is highly dependent on the type of mixer and the geometry (shape) of the tank.

Maintenance requires that the condensate be drained from the lines at least twice a day, that the diffusers be cleaned to prevent high discharge pressures, and that the compressor unit be properly lubricated and cooled.

QUESTIONS

Write your answers in a notebook and then compare your answers with those on page 232.

12.15A Why should a digester be kept well mixed?

12.15B What maintenance is necessary for the proper operation of a gas-type mixing system in a digester?

12.151 Mechanical Mixing

PROPELLER MIXERS (Figures 12.21 and 12.22) are found mainly on fixed cover digesters. Normally, two or three of these units are supported from the roof of the tank with the propeller blades submerged 10 to 12 feet (3 to 3.5 m) in the sludge. An electric motor drives the propeller stirring the sludge. Digested sludge normally contains a great deal of grit and debris. As this material comes in contact with the mixer shaft and impeller, the surface of the impeller wears down or

[16] *Inoculate* (in-NOCK-yoo-late). To introduce a seed culture into a system.

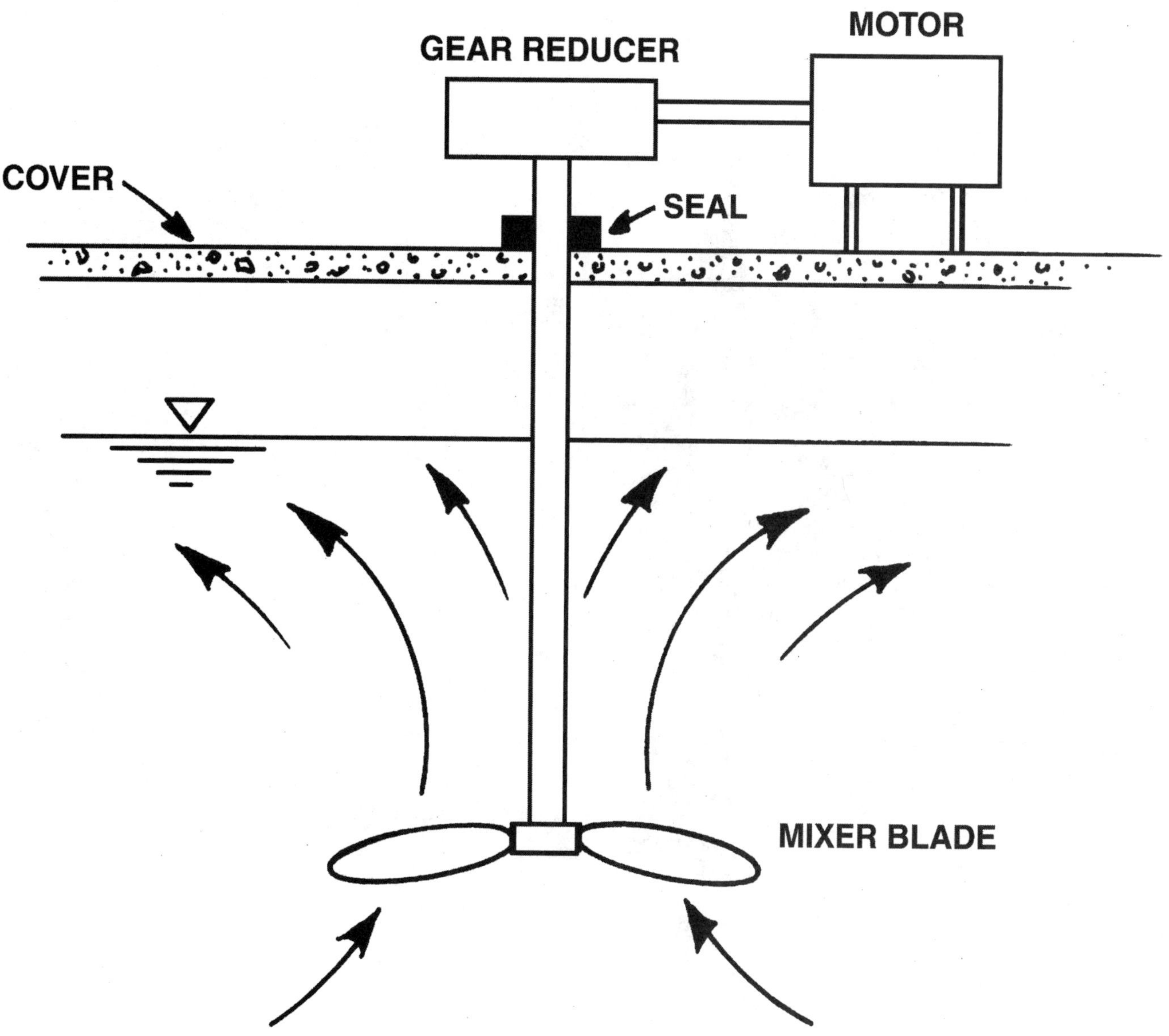

Fig. 12.21 Propeller mixer

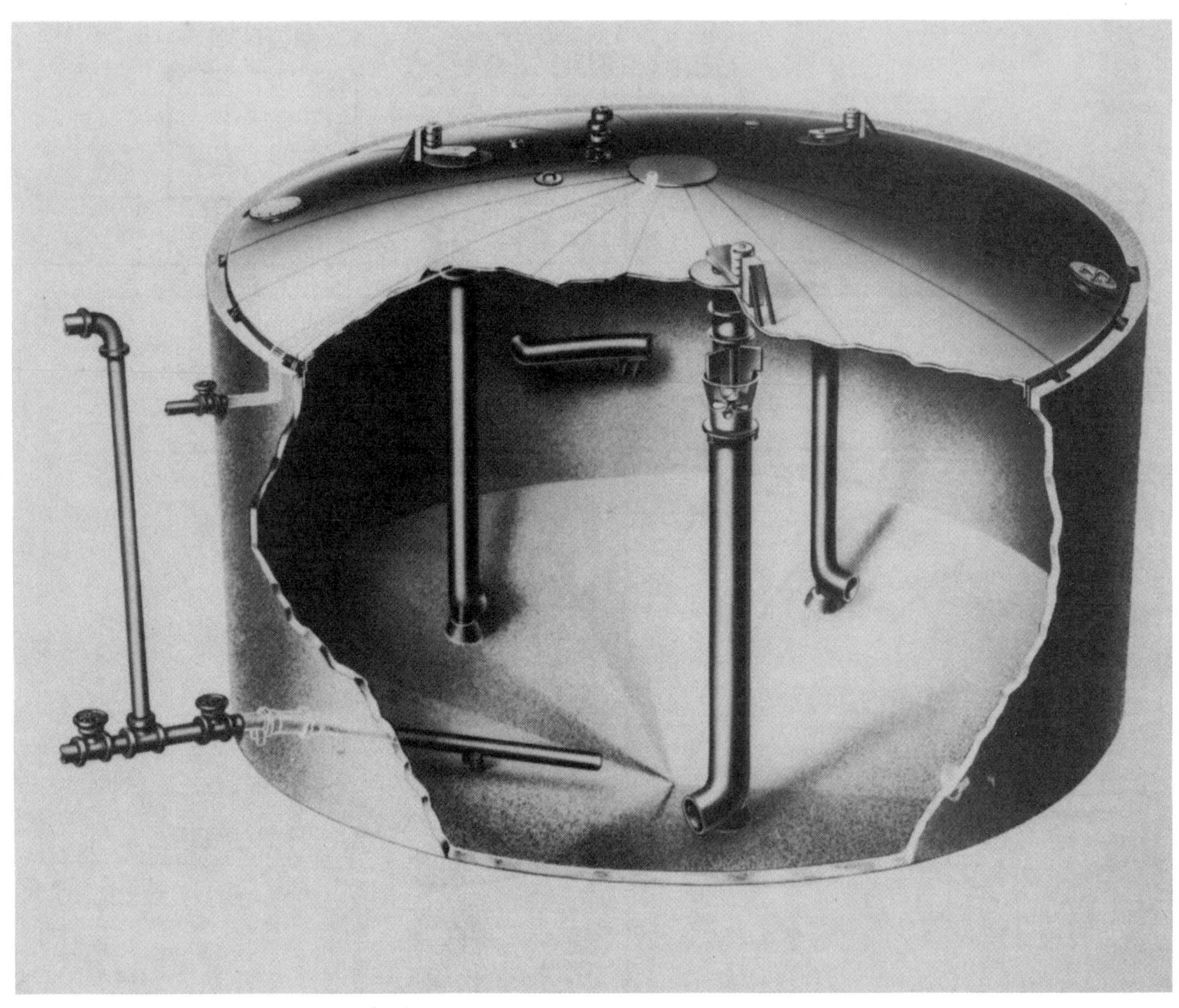

Fig. 12.22 Propeller mixer, draft tube, and digester
(Permission of Dorr-Oliver Incorporated)

the shaft becomes unbalanced and vibrates. When vibrations develop, the mixer shaft and impeller must first be cleaned and then replaced if excessive vibration continues.

DRAFT TUBE PROPELLER MIXERS (Figure 12.22) are either single or multiple unit installations. The tubes are of steel and range from 18 to 24 inches (45 to 60 cm) in diameter. The top of the draft tube has a rolled lip and is located approximately 18 inches (45 cm) below the normal water level of the tank. The bottom of the draft tube may be straight or equipped with a 90-degree elbow. The 90-degree elbow type is placed so that the discharge is along the outside wall of the tank to create a vortex (whirlpool) action.

The electric motor-driven propeller is located about 2 feet (0.6 m) below the top of the draft tube. This type unit usually has reversible motors so the propeller may rotate in either direction. In one direction the contents are pulled from the top of the digester and forced down the draft tube to be discharged at the bottom. By operating the motor in the opposite direction, the digested sludge is pulled from the bottom of the tank and discharged over the top of the draft tube to the surface. Reversible motors also assist in minimizing accumulations of rags on the propeller.

If two units are installed in the same tank, an effective way to break up a scum blanket is operating one unit in one direction and the other unit in the opposite direction, thereby creating a push-pull effect. The direction of flow in the tubes should be reversed each day. The draft tube units are subject to shaft bearing failure due to the abrasiveness of sludge and corrosion by hydrogen sulfide (H_2S) in the digester gas. Maintenance consists of lubrication and, if belt-driven, adjustment of belt tension.

A limitation of draft tube-type mixers is the potential formation of a scum blanket. If the water level is maintained at a constant elevation, a scum blanket forms on the surface. The scum blanket may be a thick layer and the draft will only pull liquid sludge from under the blanket, not disturbing it. Lowering the level of the digester to just three or four inches (7 to 10 cm) over the top of the draft tube forces the scum to move over and down the draft tube. This applies mainly to single-direction mixers.

PUMPS are sometimes used to mix digesters. This method is common in smaller tanks. When external heat exchangers are used, a larger centrifugal pump recirculates the sludge and discharges it back into the digester through one or two directional nozzles at the rate of 200 to 1,000 GPM (1,000 to 5,000 cu m/day).

The tank may or may not be equipped with a draft tube positioned in such a way that the pump suction may be from the top or valved from the bottom of the digester. Control of scum blankets with this method of mixing depends on how the operator maintains the sludge level and where the pump is pulling from and discharging to the digester.

Maintenance of the pump requires normal lubrication and a good pump shaft sealing water system. The digested sludge is abrasive and pump packing, shafts, wearing rings, and impellers are rapidly worn. Another problem associated with pump mixing is the clogging of the pump impeller with rags, rubber goods, and plastic material. A pump may run for days not pumping due to clogging because the operator was not checking the equipment for proper operation.

Pressure gauges should be installed on the pump suction and discharge pipes. A change in pressures could indicate that the pump is not functioning properly and the desired mixing may not be taking place in the digester.

QUESTIONS

Write your answers in a notebook and then compare your answers with those on page 232.

12.15C How would you break up a scum blanket in a digester with two or more draft tube propeller mixers?

12.15D Why should pressure gauges be installed on mixing pump suction and discharge lines?

12.16 Floating Cover Digesters

12.160 Purpose of Floating Covers

Floating covers on anaerobic digesters provide a flexible space for digester gas storage. A properly operated floating cover can effectively control the scum blanket in a digester. A floating cover can keep the scum blanket submerged in the digesting sludge. Submerged scum blankets are wet and easier to digest than a dry scum blanket floating on the surface of digesting sludge.

Explosive gas-air mixtures caused by too rapid withdrawal of digested sludge or supernatant are less apt to develop in anaerobic digesters with floating covers than in digesters with fixed covers. If sludge or supernatant withdrawal is too rapid, fixed covers may suffer structural damage due to the development of negative pressures.

12.161 Types of Floating Covers

Digester floating covers may be called flat, dome, gas holder, convex, or wigwam. These names describe the shape of the inside bottom of the floating cover. Otherwise they are essentially all the same.

12.162 Parts of Floating Covers (*Figure 12.23*)

CORBELS are spaced about 10 feet (3 m) apart on the inside of the digester wall. They stick out from the wall and provide support for the floating cover at its lowest position. A floating cover resting on the corbels can drop no lower. Corbels support the floating cover when the digester is emptied for maintenance and cleaning. They are usually placed from 10 to 16 feet (3 to 5 m) from the top of the digester wall. This location permits the floating cover to move from 6 to 10 feet (2 to 3 m) depending on the design of the digester.

ROLLER GUIDES are located on the inner wall of the digester. Roller guides are usually made of channel steel and

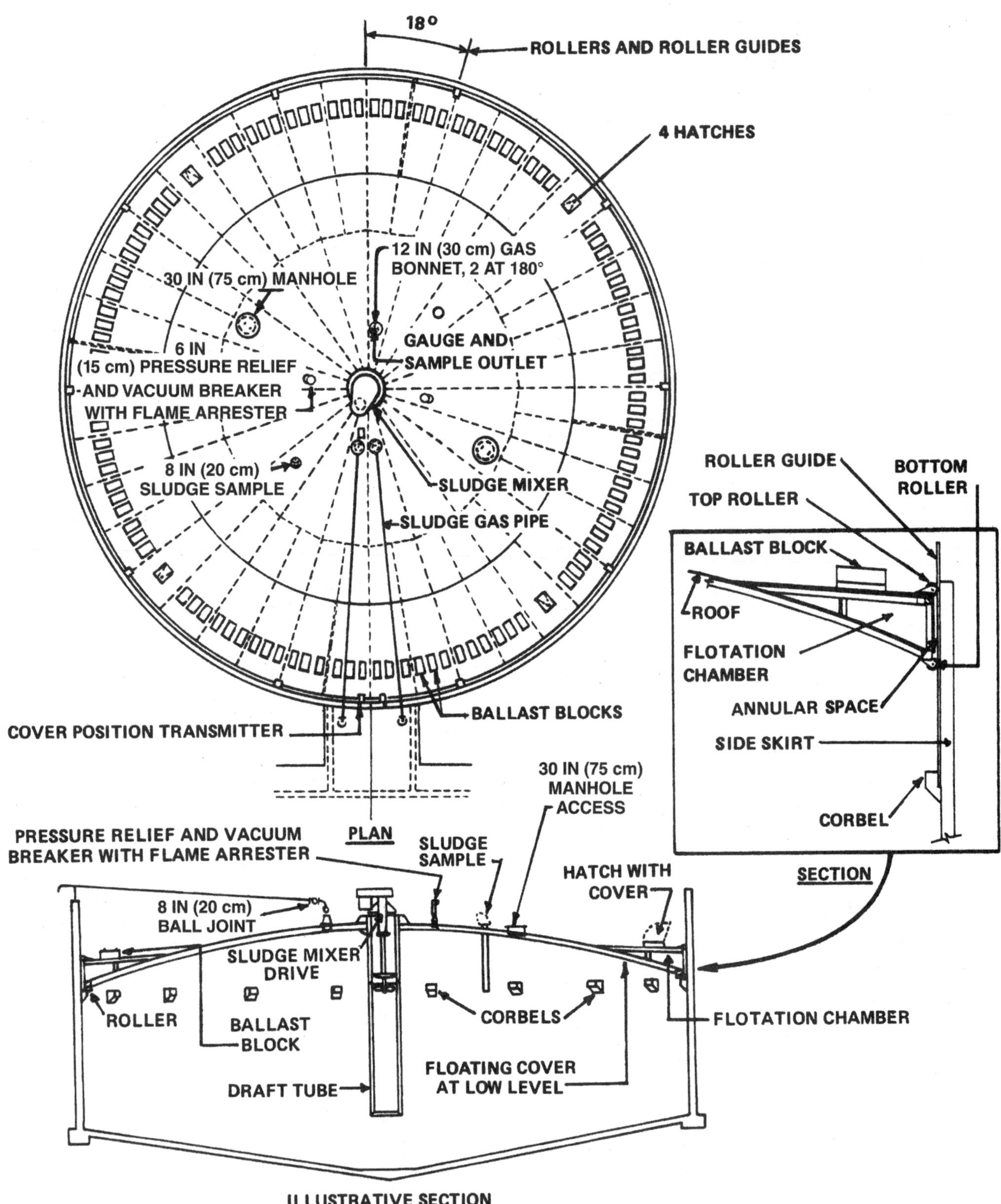
18°
ROLLERS AND ROLLER GUIDES
4 HATCHES
12 IN (30 cm) GAS
BONNET, 2 AT 180°
30 IN (75 cm) MANHOLE
GAUGE AND
SAMPLE OUTLET
6 IN
(15 cm) PRESSURE RELIEF
AND VACUUM BREAKER
WITH FLAME ARRESTER
8 IN (20 cm)
SLUDGE SAMPLE
SLUDGE MIXER
SLUDGE GAS PIPE
ROLLER GUIDE
BOTTOM
ROLLER
TOP ROLLER
BALLAST BLOCK
ROOF
FLOTATION
CHAMBER
ANNULAR SPACE
SIDE SKIRT
CORBEL
SECTION
BALLAST BLOCKS
COVER POSITION TRANSMITTER
30 IN (75 cm)
MANHOLE
ACCESS
PLAN
PRESSURE RELIEF AND VACUUM
BREAKER WITH FLAME ARRESTER
SLUDGE
SAMPLE
HATCH WITH
COVER
8 IN (20 cm)
BALL JOINT
SLUDGE MIXER
DRIVE
ROLLER
BALLAST
BLOCK
CORBELS
FLOTATION CHAMBER
FLOATING COVER
AT LOW LEVEL
DRAFT TUBE
ILLUSTRATIVE SECTION

Fig. 12.23 Digester floating cover

attached vertically to the wall of the digester. Rollers are attached to the outer edge of the floating cover at equal distances to prevent the cover from scraping the sidewall or rotating in the tank. Also, this keeps the cover floating on top of the sludge.

BALLAST BLOCKS are large concrete blocks positioned along the outside top edge of the floating cover to provide stability and the proper cover buoyancy on top of the digesting sludge and gas. Some covers have a thin layer (3 to 6 inches or 7 to 15 cm) of concrete on the cover surface to reduce, or eliminate, the ballast blocks.

ANNULAR SPACE is the space full of digesting sludge between the floating cover and digester sidewall. The annular space provides the digester gas seal until the digester contents are lowered below the corbels. If the digester contents drop below the corbels, the gas seal is broken and air can enter the digester through the space (annular space) between the outside edge of the floating cover and the digester wall. *When air enters the digester an explosive atmosphere can develop.* Extreme caution must be exercised to prevent an explosion whenever an anaerobic digester is being emptied or dewatered for maintenance or cleaning.

ACCESS HATCHES are usually installed on opposite sides of the floating cover. Smaller covers have two hatches and larger covers have four hatches or one in each quarter (quadrant) of the cover. They provide for cover access and maintenance.

FLOTATION CHAMBER is used to prevent the cover from sinking into the digesting sludge and resting submerged on the corbels. Flotation chambers are located in the roof of the cover. Digester covers have sunk as a result of leaks in the bottom of the flotation chamber, cracks on the surface or roof that allow rainfall to enter the chamber, or of an absent-minded person leaving an access hatch open. Flotation chambers should be inspected frequently for leaks from rain or washdown water. Also look for signs of corrosion on steel or other metal parts that could develop into leaks.

Hazardous atmospheric conditions could be encountered inside the flotation chamber in a floating cover digester. Before entering, treat the chamber like any other confined space. Use the proper instruments to test the atmosphere for sufficient oxygen (at least 19.5 percent oxygen), explosive conditions, and toxic gases (hydrogen sulfide). See Chapter 14, Section 14.12, "Confined Spaces."

Many flotation chambers have sumps to aid in removing any rainwater or hosedown water that enters the chamber. Small hand pumps or portable submersible pumps are used to empty the sump. If a cover leaks in one quadrant or section, it may tilt the cover. A tipped cover either jams at an angle in the tank or sinks to the corbels. Under these conditions the cover must be freed by either pumping in or withdrawing sludge from the digester until the cover is in a normal position. If the cover has sunk to the corbels, the digester contents must be lowered to the level of the bottom of the flotation chamber, pumped out, and the cover restored to service.

SKIRTS are found on most floating covers. The skirts are approximately 18 to 24 inches (45 to 60 cm) wide, go around the outside of the cover and extend into the digester parallel to the digester walls. Skirts are used to prevent foaming in the annular space. Foaming can get as high as 2 to 4 feet (0.6 to 1.2 m) above the top of a floating cover. Foaming may be caused by excessive methane gas production or high-energy digester mixing. Offsets must be provided in the skirts to allow the main structural components of the floating covers to rest on the corbels rather than the bottom of the skirt resting on the corbels.

COVER INDICATOR DIALS show at a glance the location of the cover relative to the amount of cover travel. If the floating cover can move a distance of 8 feet (2.4 m) and the dial reads from 0 to 8, then a dial reading of 2 indicates that the cover is 2 feet (0.6 m) from the top. If the indicator goes past the zero mark, sludge could flow over the digester walls or there could be structural damage to the cover or the digester. Cover indicators may be equipped with an alarm to warn of high or low cover positions.

GAS PIPING consists of flexible hoses (usually rubber) or mechanical pivoting joints connecting pipes on the digester cover to pipes on the digester wall. Flexible connections allow digester gas to be removed from the digester regardless of the level of the cover. Hoses and connections must be inspected monthly for wear, deterioration, cracks, and leaks.

ACCESS LADDERS provide the operator with convenient access to the roof of the cover regardless of the level or position of the floating cover.

Floating cover digesters are equipped with other standard components found on fixed cover digesters such as sample holes (thief holes), supernatant withdrawal tubes, gas domes, and mixing equipment (propeller or gas mixers). Safety devices include pressure and vacuum relief valves, flame arresters, sediment traps, and isolation valves on sludge and gas lines.

12.163 Safety

Never allow the floating cover to be raised above the maximum level because sludge could flow over the digester walls or the rollers could leave the roller guides. Do not permit the level of digesting sludge to drop so low that the cover rests on the corbels, except when cleaning the digester. If the gas seal breaks and air enters the digester, *an explosive atmosphere can exist in the digester.* Inspect the cover frequently (daily) to prevent the cover from becoming crooked, tipped, stuck, or jammed. Carefully control digester sludge feed and withdrawal rates. An additional inspection of the cover every time the level of the digester contents is changed is a good idea.

QUESTIONS

Write your answers in a notebook and then compare your answers with those on page 232.

12.16A What is the purpose of floating covers on anaerobic sludge digesters?

12.16B Why are explosive conditions less apt to develop in an anaerobic digester with a floating cover?

12.16C What is the annular space in the digester?

12.16D What kinds of hazardous atmospheric conditions could be encountered inside the flotation chamber of a floating cover digester?

12.16E What could happen if the floating cover rises too high or drops too low?

END OF LESSON 2 OF 6 LESSONS
on
SLUDGE DIGESTION AND SOLIDS HANDLING

Please answer the discussion and review questions next.

DISCUSSION AND REVIEW QUESTIONS

Chapter 12. SLUDGE DIGESTION AND SOLIDS HANDLING

(Lesson 2 of 6 Lessons)

Write the answers to these questions in your notebook. The question numbering continues from Lesson 1.

5. Why is digester gas considered dangerous?
6. What is the function of the digester gas system?
7. Under what types of circumstances will the pressure relief valve and vacuum relief valve operate?
8. Where should flame arresters be installed in the digester gas system?
9. How would you test for gas leaks around a flame arrester after it has been serviced?
10. Why must drip and condensate traps be drained regularly?
11. What means are used to mix the contents of digesters?

CHAPTER 12. SLUDGE DIGESTION AND SOLIDS HANDLING

(Lesson 3 of 6 Lessons)

12.2 OPERATION OF ANAEROBIC DIGESTERS

A digester can be compared with your own body. Both require food; but if fed too much, both become upset. Excess acid will upset both. Both like to be warm, with a body temperature of 98.6°F (37°C) near optimum. Both have digestive processes that are similar. Both discharge liquid and solid waste. Both use food for cell reproduction and energy. If something causes upsets in a digester, just think how you would react if it happened to you and recall what would be the proper remedy. The remedies for curing upset digesters will be discussed throughout this chapter.

12.20 Raw Sludge, Scum, Waste Activated Sludge

Raw sludge is normally composed of solids settled and removed from the primary clarifiers. Raw sludge contains carbohydrates, proteins, and fats, plus organic and inorganic chemicals that are added by domestic and industrial uses of water.

Solids are composed of organic (volatile) and inorganic material with the volatile content ranging from about 60 to 80 percent of the total, by weight. Some plants do not have grit removal equipment, so the bulk of the inert (inorganic) material such as sand, eggshells, and other debris will end up on the bottom of the digester occupying active digestion space. The rate of debris accumulation is predictable so that the amount is a function of the period of time between digester cleanings. Where cleaning has been neglected, a substantial portion of the active volume of the digester becomes filled with inert debris. Scum-forming products, such as kitchen grease, soaps, oils, cellulose, and other floatable debris, are generally all organic in nature but may create problems if the scum blanket in the digester is not controlled. Control is provided by adequate mixing and heat.

Several products end up in the digester that are not desirable because the bacteria cannot effectively use or digest them, and they cannot be readily removed by the normal process. These products include:

1. Petroleum products and mineral oils
2. Rubber goods
3. Plastics (back sheets to diapers)
4. Filter tips from cigarettes
5. Hair
6. Grit (sand and other inorganics)

Consequently, these items tend to accumulate in the digester and, without adequate mixing, may form a hard, floating mat and a substantial bottom deposit. On the other hand, a well-mixed tank may also present operational problems. For example, the material shredded by a comminutor or barminutor may become balled together by the mixing action and plug the digester supernatant lines.

Scum from the primary clarifiers consists mainly of grease and other floatable material. It may be collected and held in a scum box and then pumped to the digester once a day, or it may be added continuously or at a frequency necessary to maintain the proper removal of scum from the raw wastewater flow. Many operators prefer not to pump scum to the digesters, but to dispose of it by burning or burial. Scum may also refer to the floating and gas-buoyed material found on the surface of poorly mixed digesters. This material may contain much cellulose, rubber particles, mineral oil, plastic, and other debris. It may become 5 to 15 feet (2 to 5 m) thick in a digester, but should not occur in a properly operating digester. A thick scum layer will reduce the active digestion capacity of a digester.

Waste activated sludge is the sludge intentionally removed from the activated sludge process and consists almost entirely of the microorganisms from the activated sludge aeration tanks. The volatile solids content of this material typically ranges from 75 to 85 percent. The density of the sludge will depend on whether it is taken directly from the aeration tank, the return sludge line from the secondary clarifier, or if it is additionally thickened, whether by gravity, gravity belt, dissolved air flotation, or centrifugation. Under these different circumstances the solids concentrations can vary from 2,000 mg/*L* if taken from the aeration tank to 50,000 to 150,000 mg/*L* if thickened by a centrifuge. The waste activated sludge can be combined with the raw sludge and this mixture digested, or the waste activated sludge can be digested separately. In either case it must be recognized that the activated sludge will not be destroyed in the digester to the same extent as the raw sludge. This occurs because of the complex organic nature of the microorganisms and their resistance to digestion.

12.21 Starting a Digester

When wastewater solids are first added to a new digester, naturally occurring bacteria attack the most easily digestible food available, such as sugar, starches, and soluble nitrogen. The anaerobic acid producers change these foods into organic acids, alcohols, and carbon dioxide, along with some hydrogen sulfide. The pH of the sludge drops from 7.0 to about 6.0 or lower. An *ACID REGRESSION STAGE*[17] then starts

[17] *Acid Regression Stage.* A stage of anaerobic digestion during which the production of volatile acids is reduced and acetate and ammonia compounds form, causing the pH to increase.

and lasts as long as six to eight weeks. During this time, ammonia and bicarbonate compounds are formed, and the pH gradually increases to around 6.8 again, establishing an environment for the methane fermentation or alkaline fermentation phase. Organic acids are available to feed the methane fermenters. Larger quantities of methane gas are produced as well as carbon dioxide, and the pH increases to 7.0 to 7.2. Once alkaline fermentation is well established, strive to keep the digesting sludge in the 7.0 to 7.2 pH range.

If too much raw sludge is added to the digester, the acid fermenters will predominate, driving the pH down and creating an undesirable condition for the methane fermenters. The digester will go sour or acid again. When a digester recovers from a sour or acid condition, the breakdown of the volatile acids and formation of methane and carbon dioxide is usually very rapid. The digester may then foam or froth, forcing sludge solids through water (which forms the gas seals) and the gas lines and causing a fairly serious operational problem. A sour digester usually requires 30 to 60 days to recover.

As noted at the beginning of this section, the first group of organisms must do its part before food is available to the next group. Once the balance is upset, so is the food cycle to the next group. When the tank reaches the methane fermentation phase, there is sufficient alkaline material to buffer the acid stage and maintain the process. Operational actions such as poor mixing, addition of excess food, excess water supplied, which dilutes the alkaline buffer, overdrawing digested sludge, or improper temperature changes can cause souring again.

The simplest way to start a digester is with seed sludge (actively digesting material) from another digester. Seed sludge should be added to a new digester to provide methane fermenters and alkalinity so the digestion process will start and continue in balance. The amount of seed to use depends on factors such as mixing processes, digester sizes, and sludge characteristics, but amounts between 10 and 50 percent of the digester capacity have been used.

EXAMPLE 1 (seed volume based on tank capacity):

Calculate the volume of seed sludge needed for a 40-foot diameter digester with a normal water depth of 20 feet, if the seed required is 25 percent of the tank volume. Most digesters have sloping bottoms, but assume the normal sidewall water depth represents the average digester depth:

$$\text{Tank Diameter, D} = 40 \text{ feet}$$

$$\text{Depth, H} = 20 \text{ feet}$$

$$\frac{\pi}{4} = 0.785$$

$$\begin{aligned}\text{Tank Volume, cu ft} &= \text{Area, sq ft} \times \text{Depth, ft}\\ &= \frac{\pi}{4} D^2 \times H\\ &= 0.785(40 \text{ ft})^2 \times 20 \text{ ft}\\ &= 25{,}120 \text{ cu ft}\end{aligned}$$

$$\begin{aligned}\text{Tank Volume, gal} &= (25{,}120 \text{ cu ft})(7.5 \text{ gal/cu ft})\\ &= 188{,}400 \text{ gal}\end{aligned}$$

Seed required assumed to be 25 percent or 1/4 of the digester tank volume:

$$\begin{aligned}\text{Seed Volume, gal} &= \frac{\text{Tank Volume, gal}}{4}\\ &= \frac{188{,}400 \text{ gal}}{4}\\ &= 47{,}100 \text{ gal (180 cu m)}\end{aligned}$$

Therefore, 47,100 gallons (180 cu m) of seed sludge would be needed. If seed sludge is not available, the tank may be started by filling the digester with raw wastewater and heating the tank to 85-95°F (30-35°C) with natural gas or other fuel. Allow the bacteria to take the natural course of decomposition as earlier described. The time required for a start of this nature ranges from 45 to 180 days.

Rather than estimate the volume of seed sludge on the basis of digester capacity, a better approach is to determine the volume of seed necessary to maintain digestion under the expected initial loading. To use this approach, allow 0.03 to 0.10 pound of new volatile solids to be added per day per pound of volatile solids under digestion (0.03 to 0.1 kg added per day per kg under digestion).

EXAMPLE 2 (seed volume based on raw sludge to be added):

Initially, a new plant expects to pump 500 gallons of raw sludge per day to the digester. The raw sludge is estimated to contain 6 percent solids with a volatile content of 68 percent. Estimate the pounds of volatile solids needed by the digester and the gallons of seed sludge, assuming the seed sludge contains 10 percent solids with 50 percent volatile solids and weighs 9 pounds per gallon. (Digested sludge containing 10 percent solids weighs more than water (8.34 lbs/gal) without any solids.)

Find pounds of volatile matter pumped to digester per day.

$$\begin{aligned}\text{Volatile Matter Pumped, lbs/day} &= (\text{Vol of Sl, GPD})(\text{Solids, \%})(\text{Volatile, \%})(8.34 \text{ lbs/gal})\\ &= (500 \text{ gal/day})(0.06)(0.68)(8.34 \text{ lbs/gal})\\ &= 170 \text{ lbs/day}\end{aligned}$$

Select a digester loading between 0.03 and 0.10 pound of new volatile solids added per day per pound of volatile solids in digester. Try 0.05 lb VM per day per pound under digestion for a starting loading.

Find pounds of seed volatile matter needed.

$$\frac{0.05 \text{ lb VM Added/day}}{1 \text{ lb VM in Digester}} = \frac{170 \text{ lbs VM Pumped/day}}{\text{Seed, lbs VM}}$$

$$\begin{aligned}\text{Seed, lbs VM} &= \frac{170 \text{ lbs VM Pumped/day}}{0.05 \text{ lb VM Added/day}} (\text{lbs VM})\\ &= 3{,}400 \text{ lbs VM}\end{aligned}$$

Find gallons of seed sludge needed.

$$\text{Seed Sludge, gal} = \frac{\text{Seed, lbs VM}}{(9\text{ lbs/gal})(\text{Solids, \%})(\text{VM, \%})}$$

$$= \frac{3{,}400\text{ lbs VM}}{(9\text{ lbs/gal})(0.10)(0.50\text{ VM})}$$

$$= 7{,}560\text{ gal (29 cu m)}$$

To start a digester, add seed sludge (7,500 to 8,000 gallons (28 to 30 cu m) of digested sludge) and then add enough wastewater (primary or secondary effluent) to create a water seal on the supernatant lines or any other possible vent lines. Now start adding 500 gallons (1.9 cu m) of raw sludge per day to the digester. Some operators do not completely fill a digester during start-up but this practice is not recommended. It is hazardous to start a digester when it is only partially full due to explosive conditions created by a mixture of air and methane in partially full digesters.

During the start-up of a digester, once production of a good, burnable gas is obtained, the raw sludge feed rate can be gradually increased until the system is handling the total load.

QUESTIONS

Write your answers in a notebook and then compare your answers with those on pages 232 and 233.

12.20A What is (1) raw sludge? (2) scum? (3) waste activated sludge?

12.21A What happens if you add too much raw sludge to the digester?

12.21B When starting a digester, what could cause a digester to foam and froth?

12.21C Calculate the recommended volume of seed sludge to start a digester 50 feet in diameter and 25 feet deep (average). Assume 700 gallons per day of raw sludge will be added, containing 6.5% solids and 70% volatile matter. Assume seed sludge contains 10% solids with 50% volatile solids and weighs 9 pounds per gallon. Use a digester loading of 0.05 lb VM added per day per lb VM under digestion.

12.21D Why is it hazardous to start a digester when it is only partially full?

12.21E How could you determine when a new digester is ready for the raw sludge feed rate to be gradually increased to the full plant load?

12.22 Feeding

Food for the bacteria in the digester is the raw sludge from the primary clarifier and/or waste activated sludge or trickling filter "humus" from the secondary clarifiers. Make every effort to pump as thick a sludge to the digester as possible. To obtain a thick sludge, hold a blanket of sludge as long as possible in the primary clarifier, long enough to allow sludge concentration, but not long enough for sludge to start gassing or rising. In some plants, concentration is accomplished in separate sludge thickening or flotation tanks.

Better operational performance occurs when the digester is fed several times a day, rather than once a day, because this strategy avoids temporary overloads on the digester and uses available space more effectively. Several pumpings a day not only helps the digestion process, but maintains better conditions in the clarifiers, permits thicker sludge pumping, and prevents *CONING*[18] in the primary clarifier hopper. On fixed cover digesters, frequent feeding spreads the return of digester supernatant over the entire day instead of a return in one slug with possible upset of the secondary treatment system. Sludge is usually concentrated by holding a thick blanket on the bottom of the clarifier; but if sludge sits for a prolonged period, the lowest layers may stick to the bottom and will no longer flow with the liquid. When pumping is attempted, liquid flows but solids remain in the hopper in a cone around the outlet. As thick a sludge as possible may be pumped to the digester by operating the sludge pump for several minutes each hour (at a rate not to exceed 50 GPM (3 *L*/sec)) to clear the sludge hopper.

Never pump thin sludge or water to a digester. A sludge is considered thin if it contains less than 3.5 percent solids (too much water). Reasons for not pumping a thin sludge include:

1. Excess water requires more heat than may be available.
2. Excess water reduces the holding time of the sludge in the digester.
3. Excess water forces seed and alkalinity from the digester, jeopardizing the system due to insufficient *BUFFER CAPACITY*[19] for the acids produced by digestion of the raw sludge.
4. Excess water imposes a heavier supernatant load on the plant.

Sludge concentrations above about 10 to 12 percent solids will usually not digest well in conventional digestion tanks since adequate mixing cannot be obtained. This, in turn, leads to improper distribution of food, seed, heat, and metabolic

[18] *Coning.* Development of a cone-shaped flow of liquid, like a whirlpool, through sludge. This can occur in a sludge hopper during sludge withdrawal when the sludge becomes too thick. Part of the sludge remains in place while liquid rather than sludge flows out of the hopper. Also called coring.

[19] *Buffer Capacity.* A measure of the capacity of a solution or liquid to neutralize acids or bases. This is a measure of the capacity of water or wastewater for offering a resistance to changes in pH.

products so that souring and a *STUCK*[20] digester results. However, most plants have difficulty in obtaining a raw sludge of 8 percent solids. Where a trickling filter or activated sludge process is used as the secondary system, sludges may have a solids range from 1 to 3 percent. If additional thickening of waste activated sludge is used, such as dissolved air flotation or centrifuges, the solids content may range from 4.5 to 6 percent solids.

Feeding a digester must be regulated on the basis of laboratory test results in order to ensure that the volatile acid/alkalinity relationship does not start to increase and become too high. See Section 12.3B, "Volatile Acid/Alkalinity Relationship."

QUESTIONS

Write your answers in a notebook and then compare your answers with those on page 233.

12.22A Why should sludge be pumped occasionally throughout the day rather than as one slug each day?

12.22B How would you attempt to pump as thick a sludge as possible to a digester?

12.22C Why should the pumping of thin sludge be avoided?

12.23 Neutralizing a Sour Digester

The recovery of a sour digester can be accelerated by neutralizing the acids with a caustic material such as anhydrous ammonia, soda ash, or lime, or by transferring alkalinity in the form of digested sludge from the secondary digester. Such neutralization reduces the volatile acid/alkalinity to a level suitable for growth of the methane fermenters and provides buffering material that will help maintain the required volatile acid/alkalinity relationship and increase the pH to 7.0. When ammonia is added to a digester, an added load is eventually placed on the receiving waters. The application of lime will increase the solids handling problems. Soda ash is more expensive than lime, but does not add as much to the solids deposits. Transferring secondary digester sludge has the advantage of not adding anything extra to the system that was not there at an earlier time and, if used properly, will reduce both the effluent load and the solids handling problem.

If digestion capacity and available recovery time are great enough, it is probably preferable to simply reduce loading while heating and mixing so that natural recovery occurs. However, there are often conditions in which such neutralization is necessary.

When neutralizing a digester, the prescribed dose must be carefully calculated. Too little will be ineffective, and too much is both toxic and wasteful. In considering dosage with lime, the small plant without laboratory facilities could use as a rough guide a dosage of about one pound of lime added for every 1,000 gallons (120 gm/1,000 *L*) of sludge to be treated. Thus, a 188,000-gallon (711-cu m) digester full of sludge would receive 188 pounds (85 kg) of lime. A more accurate method is to add sufficient lime to neutralize 100 percent of the volatile acids in the digester liquor. (See Volatile Acids Test in Chapter 16, "Laboratory Procedures and Chemistry.")

You must realize that neutralizing a sour digester will only bring the pH to a suitable level, it will not cure the cause of the upset.

EXAMPLE 3:

Volatile acids in digester sludge = 230 mg/*L*. We should add lime equivalent to 230 mg/*L*.

$$\text{Lime Required, lbs} = \text{Volatile Acids, mg/}L \times \text{Tank Volume, MG} \times 8.34 \text{ lbs/gal}$$

$$= 230 \text{ mg/}L \times 0.188 \text{ MG} \times 8.34 \text{ lbs/gal}$$

$$= 230 \times 1.57 \text{ lbs}$$

$$= 361 \text{ lbs}$$

Assume the acids are reported as acetic acid and the lime is calcium hydroxide ($Ca(OH)_2$). Another procedure would be to take a sludge sample of known volume into the lab and add lime until the desired pH is obtained. Use this result to determine the amount of lime to be added to the digester.

The lime must be mixed into a solution before being added to the digester because dry lime would settle to the bottom in lumps that are not only ineffective but take up digester capacity and are difficult to remove when cleaning the digester. Use all of the mixing energy available while liming and thereafter in digester mixing. The easiest application point is through the scum box, if one is available. Add small quantities of lime daily until the pH and volatile acid/alkalinity relationship (Section 12.3B) of the tank are restored to desired levels and gas production is normal. In any case, use lime only if recovery by natural methods cannot be accomplished within the time available.

Caution must be used when adding any basic chemical to decrease the volatile acid/alkalinity relationship to the desired level and to keep it there. Lime, for example, can combine with carbon dioxide in the digester gas and form a dangerous vacuum in the digester. To prevent the formation of a vacuum, add the chemical very carefully and take all possible steps to ensure continuous, thorough mixing in the digester.

Although it is seldom used because of higher costs, sodium bicarbonate is a good substitute for lime. Besides requiring smaller quantities, use of sodium bicarbonate presents none of the problems described in this section associated with lime.

[20] *Stuck Digester.* A stuck digester does not decompose organic matter properly. The digester is characterized by low gas production, high volatile acid/alkalinity relationship, and poor liquid-solids separation. A digester in a stuck condition is sometimes called a sour or UPSET DIGESTER.

QUESTIONS

Write your answers in a notebook and then compare your answers with those on page 233.

12.23A Why is lime added to a sour digester?

12.23B How much lime should be added to a 100,000-gallon digester, using the "rough guide" dosage in the previous section?

12.23C Why should lime be added in solution to the digester rather than in dry form?

12.23D For how long a time should you add lime to a digester?

12.24 Enzymes[21]

Bacteria secrete enzymes that help break down compounds that the bacteria in the digester may use as food. In recent years several products containing "commercial" enzymes or other biocatalysts (BUY-o-CAT-uh-lists) have been marketed for starting digesters, controlling scum, or simply to maintain operation. Such biocatalysts or enzymes have rarely been shown to be effective in controlled tests and could, in fact, cause as much harm as good. A biological system such as found in the digesters develops a balanced enzyme and biocatalyst system for the conditions under which it is operating. The quantities of natural enzymes developed within the digesting sludge are many, many times greater than any amount you could either add or afford to purchase.

12.25 Foaming

Large amounts of foam may be generated during start-up by the almost explosive generation of gas during the time of acid recovery. Foaming is the result of active gas production while solids separation has not progressed far enough (insufficient digestion). Overfeeding during start-up aggravates foaming problems. Foaming can be prevented by adequate mixing of the digester contents before foaming starts.

Bacteria can go to work very quickly when they have the proper environment. Almost overnight they can generate enough gas to create a terrible mess of black foam and sludge that must be controlled. The foam not only plugs gas piping systems, but can exert excess pressures on digester covers, cause odor problems, and ruin the paint on tanks and buildings.

Foaming also may occur when a thick sludge blanket is broken up, temperature changes radically, or the sludge feeding to the digester is increased. Avoid any conditions that give the acid formers the opportunity to produce more food than the methane fermenters can handle, because when the methane fermenters are ready, they may work too fast. Steps you can take to prevent foaming from recurring include:

1. Maintain a constant temperature in digester.
2. Feed sludge at regular, short intervals.
3. Exercise caution when breaking up scum blankets.
4. Do not overdrain sludge from digester.
5. Keep contents of digester well mixed from bottom to top at all times.

To control the foaming, the best method is to stir the tank gently to release as much of the trapped gas from the foam as possible. Some operators even stop mechanical mixing equipment and stir with long wooden poles. During stirring with poles, air will get into the digester and create an explosive condition. (Be sure to take appropriate safety precautions.) This explosive condition will persist until the digesting sludge produces enough gas to carry the oxygen from the air out of the digester by venting to the atmosphere or by way of the waste gas burner. Try not to add too much water from the cleaning hoses as this reduces the temperature and dilutes the tank, which could create conditions for more foaming later. Do not feed the tank heavily, preferably not at all, until the foaming has subsided.

To clean up the mess created by a foaming problem, first drop the level of the digester a couple of feet by withdrawing some supernatant. Next, cut off the gas system and flush it with water. Then hose down the outside of the digester as soon as possible or the paint will be stained a permanent gray. Drain and refill the water seal to remove the water fouled by the foaming. Use a strainer-type skimming device to remove any rubber goods and plastic materials that have entered the water seal.

With adequate mixing, foaming problems will not develop. Start mixing from bottom to top of the tank before foaming starts, not afterward.

12.26 Gas Production

When a digester is first started, extremely odorous gases are produced, including a number of nitrogen and sulfur compounds such as skatole, indole, mercaptans, and hydrogen sulfide. Many of these are also produced during normal digestion phases, but they are generally so diluted by carbon dioxide and methane that they are hardly noticeable. Their presence can be determined by testing, if so desired.

During the first phases of digester start-up, most of the gas is carbon dioxide (CO_2) and hydrogen sulfide (H_2S). This combination will not burn and therefore is usually vented to the atmosphere. In such venting, the operator should be aware of the potential odor problem that can result. When methane fermentation starts and the methane content reaches around 60 percent, the gas will be capable of burning. Methane production eventually should predominate, generating a gas with 65 to 70 percent methane and 30 to 35 percent CO_2 by volume. Digester gas will burn when it contains 56 percent methane, but is not usable as a fuel until the methane content approaches 62 percent. When the gas produced is burnable, it may be

[21] *Enzymes* (EN-zimes). Organic or biochemical substances that cause or speed up chemical reactions.

used to heat the digester as well as for powering engines and for providing building heating.

QUESTIONS

Write your answers in a notebook and then compare your answers with those on page 233.

12.24A What is the function of enzymes in digestion?

12.25A What causes foaming in a digester?

12.25B What steps would you take to prevent foaming from recurring?

12.26A Why is the gas initially produced in a digester not burnable?

12.27 Supernatant and Solids

Some plants are constructed with two separate digestion tanks (Figure 12.24) or one tank with two divided sections. One tank is called the primary digester and is used for heating, mixing, and breakdown of raw sludge. The second tank, or secondary digester, is used as a holding tank for separation of the solids from the liquor. To accomplish such separation, the secondary tank must be operated with alternating quiescent (kwee-ES-sent) (without mixing) periods and shorter periods of mixing.

Most of the sludge stabilization work is accomplished in the primary digester, and 90 percent of the gas production occurs there. The primary tank must be very thoroughly mixed, but it is undesirable to return the digested mixture to the plant as a supernatant. Therefore, when raw sludge is pumped to the primary digester, an equal volume is transferred to the secondary digester, and settled supernatant from the secondary digester is returned to the plant.

In the primary digester, the binding property of the sludge is broken, allowing the water to be released. In the secondary digester, the digested sludge is allowed to settle and compact, with some digestion continuing. When the solids settle, they leave a light, amber-colored liquor zone between the top of the settled sludge and the surface of the digester. By adjusting or selecting the supernatant tube, the liquor with the least solids is returned to the plant. On a fixed cover digester, select the supernatant tube that reaches the clearest supernatant zone in the digester. On floating cover tanks, raise or lower the supernatant draw-off line to the clearest supernatant zone.

The settled solids in the secondary digester are allowed to compact so that a minimal amount of water will be handled in the sludge dewatering system. These solids are excellent seed or buffer sludge in case the primary digester becomes upset. A reserve of 30 to 100 thousand gallons (100 to 400 cu m) or 25 percent of the volume undergoing active digestion should always be held in the secondary digester. This represents a natural enzyme reserve and may save the system during a shock load. Primary and secondary sludge digesters should be operated as a complement to each other. If you need more seed or buffering capacity in the primary digester, it should be taken from the secondary digester.

The secondary tanks should be mixed frequently, preferably after sludge has been withdrawn and when supernatant will not be returned to the plant. Usually, secondary digesters are provided with mixers or recirculating pumps, preferably arranged for vertical mixing. This periodic mixing prevents coning of solids on the bottom of the tank and the formation of a scum blanket on the top. Mixing also helps the release of slowly produced gas that may float solids or scum.

If your plant has only one digester, stop mixing for one day before withdrawing digested sludge to drying beds.

Although this section and other textbooks describe supernatant as a clear liquid with an amber color, this condition is rarely observed. In some plants, the secondary digester can sit for two days without mixing and no significant difference can be observed between the digested sludge on the bottom and the digested sludge on the top. In other words, the liquid did not separate from the digested sludge. This occurrence is not uncommon, especially when treating waste activated sludge. For this reason, the trend appears to be away from the use of secondary digesters for liquid-solids separation. Chapter 3, *ADVANCED WASTE TREATMENT* manual, describes liquid-solids separation processes, including vacuum filters, pressure filters, centrifuges, dryers, and multiple hearth furnaces.

12.28 Rate of Sludge Withdrawal

The withdrawal rate of sludge from either digester should be no faster than a rate at which the gas production from the system is able to maintain a positive pressure in the digester (at least two inches (5 cm) of water column).

WARNING

If the draw-off rate is too fast, the gas pressure drops due to volume expansion. If continued, a negative pressure develops on the system (vacuum). This may create an explosive hazard by drawing air into the digester. If the primary digester has a floating cover, the sludge may be drawn down to where the cover rests on the corbels without danger of losing gas pressure.

Some operators prefer to pump raw sludge or wastewater to a digester during digested sludge draw-off to maintain a positive pressure. If gas storage lines permit, gas should be returned to the digester to maintain pressure in the digester.

12.29 Struvite Control

Struvite (magnesium ammonium phosphate ($MgNH_4PO_4 \cdot 6H_2O$)) scale sometimes forms in anaerobic digesters and also in downstream digested sludge concentration and handling equipment. Anaerobic digestion of sludges converts ammonia (NH_3-N), phosphate (PO_4-P), and magnesium (Mg) to

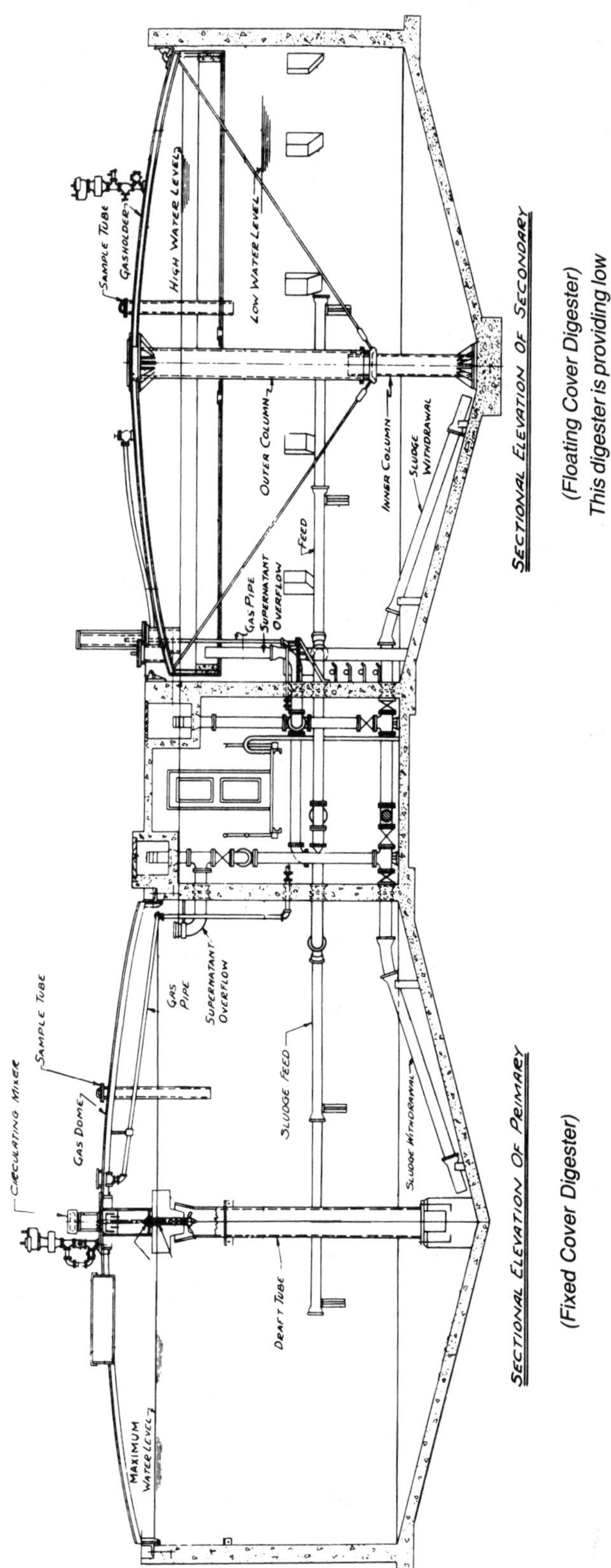

(Fixed Cover Digester)

(Floating Cover Digester)
This digester is providing low pressure (less than 8 inches (20 cm) of water column pressure) gas storage for both digesters.

Fig. 12.24 Primary and secondary digester
(Permission of Dorr-Oliver Incorporated)

soluble forms. When the concentrations of these constituents of struvite exceed the saturation level, struvite scale deposits will form on piping and equipment. One method used to control struvite is the removal (precipitation) of phosphate (PO_4-P) by the addition of an iron salt (ferric chloride ($FeCl_3$)), or ferrous chloride ($FeCl_2$)). These iron salts are added either to the wastewater flow or to the digesting sludge.

Studies conducted at the City of San Francisco's Southeast Water Pollution Control Plant determined that the optimum dose of an iron salt for struvite control depends on the amount of soluble phosphorus (P) and soluble magnesium (Mg) available for precipitation as well as the ratio (0.37 for San Francisco) of phosphate (PO_4-P) removed per iron (Fe) added. The optimum doses determined at San Francisco were 2,200 mg/*L* ferric chloride or 100 kg ferric chloride per ton of total solids under digestion. Specific values can be determined for a particular plant by conducting a series of continuous-flow anaerobic digester experiments at various doses of ferric chloride. For additional information, see Pitt, P. *et al.,* "Struvite Control Using Ferric Chloride at the Southeast Plant, San Francisco, CA." This paper was presented at the 65th Annual Conference and Exposition of Water Environment Federation, New Orleans, Louisiana, September 20–24, 1992.

The Office of Water Programs has developed a struvite tool designed to calculate the struvite precipitation potential for a facility based on water quality indicators input by the user. It allows the user to vary input parameters to determine "what-if" scenarios when conditions are changed to control struvite precipitation. This tool is a must for struvite control planning. It runs within Microsoft Office Excel and comes with complete operating instructions. The "Struvite Precipitation Potential Calculation Tool" is available from the Office of Water Programs at California State University, Sacramento. Price, $75.00.

QUESTIONS

Write your answers in a notebook and then compare your answers with those on page 233.

12.27A What is the purpose of the secondary digester?

12.27B When raw sludge is pumped to the primary digester, what happens in the secondary digester?

12.27C How is the level of supernatant withdrawal selected?

12.28A How would you determine the rate of sludge withdrawal?

END OF LESSON 3 OF 6 LESSONS
on
SLUDGE DIGESTION AND SOLIDS HANDLING

Please answer the discussion and review questions next.

DISCUSSION AND REVIEW QUESTIONS

Chapter 12. SLUDGE DIGESTION AND SOLIDS HANDLING

(Lesson 3 of 6 Lessons)

Write the answers to these questions in your notebook. The question numbering continues from Lesson 2.

12. What kinds of material or products frequently end up in digesters that are not desirable because bacteria cannot effectively use or digest them?
13. Why should seed sludge be added to a new digester?
14. Why should a digester be fed at regular intervals during the day, rather than once a day?
15. What are enzymes?
16. How can an operator attempt to prevent a digester from starting to foam?
17. Why should secondary digesters be mixed, if at all?

CHAPTER 12. SLUDGE DIGESTION AND SOLIDS HANDLING

(Lesson 4 of 6 Lessons)

12.3 ANAEROBIC DIGESTION CONTROLS AND TEST INTERPRETATION

NOTE: See Chapter 16, "Laboratory Procedures and Chemistry," for testing procedures.

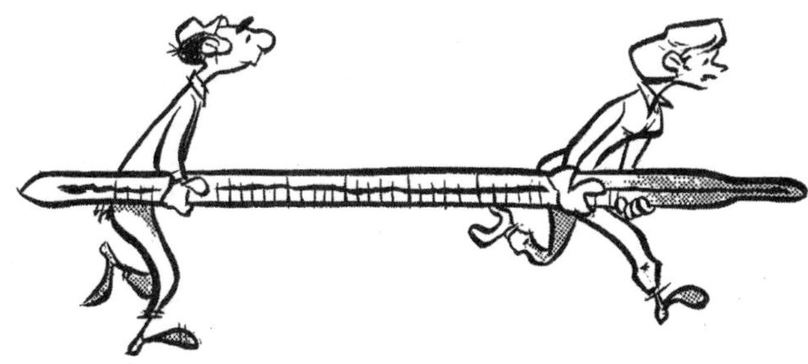

A. *TEMPERATURE*

Temperature is measured in a digester to ensure the proper temperature environment is maintained for the organisms. A thermometer is usually installed in the recirculated sludge line from the digester to the heat exchanger. This thermometer will accurately measure the temperature of the digester contents when circulation is from bottom to top. Accurate temperature readings also may be taken from the flowing supernatant tube or from the heat exchanger sludge inlet line. The same temperature should be maintained at all levels of the tank. The temperature from the digester is recorded and should be maintained between about 95° and 98°F (35° and 37°C) for mesophilic digestion. The temperature in a digester should not be changed by more than 1°F or 0.5°C per day to allow the organisms time to adjust to the temperature change.

B. *VOLATILE ACID/ALKALINITY RELATIONSHIP*

The volatile acid/alkalinity relationship is the key to successful digester operation. As long as the volatile acids remain low and the alkalinity stays high, anaerobic sludge digestion will occur in a digester. Each treatment plant will have its own characteristic ratio for proper sludge digestion (generally less than 0.1). When the ratio starts to increase, corrective action must be taken immediately. This is the first warning that trouble is starting in a digester. If corrective action is not taken immediately or is not effective, eventually the CO_2 content of the digester gas will increase, the pH of the sludge in the digester will drop, and the digester will become sour or upset.

A good procedure is to measure the volatile acid/alkalinity relationship at least twice a week, plot the volatile acid/alkalinity relationship against time, and watch for any adverse trends to develop. Whenever something unusual happens, such as an increased solids load from increased waste discharges or a storm, the volatile acid/alkalinity relationship should be watched closely. Chapter 16, "Laboratory Procedures and Chemistry," contains a procedure for measuring volatile acids by titration that gives satisfactory results for operational control.

The volatile acid/alkalinity relationship is an indication of the buffer capacity of the digester contents. A high buffer capacity is desirable and is achieved by a low ratio, which exists when volatile acids are low and the alkalinity is high (120 mg/*L* volatile acids/2,400 mg/*L* alkalinity). Excessive feeding of raw sludge to the digester, removal of digested sludge, or a shock load such as produced by a storm flushing out the collection system may unbalance the volatile acid/alkalinity relationship.

A definite problem is developing when the volatile acid/alkalinity relationship starts increasing. Once the relationship reaches the vicinity of 0.5/1.0 (1,000 mg/*L* volatile acids/2,000 mg/*L* alkalinity), serious decreases in the alkalinity usually occur. At a relationship of 0.5/1.0 the concentration of CO_2 in digester gas will start to increase. When the relationship reaches 0.8 or higher, the pH of the digester contents will begin to drop. When the relationship first starts to increase, ample warning is given for corrective action to be taken before problems develop and digester control is lost.

RESPONSE TO AN INCREASE IN THE VOLATILE ACID/ALKALINITY RELATIONSHIP

When the volatile acid/alkalinity relationship starts to increase, extend mixing time of digester contents, control heat more evenly, decrease raw sludge feed rates, and decrease digested sludge withdrawal rates from digesters. Pumping a thicker sludge to the digester can help prevent a loss of alkalinity. Mixing should be vertical mixing from the bottom of the tank to the top of any scum blanket. If possible, some of the concentrated sludge in the secondary digester should be pumped back to help correct the ratio. In addition, the primary digester should not be operated as a continuous overflow unit when raw sludge is added, but it should be drawn down to provide room for some sludge from the secondary digester too. During heavy rains when extra solids are flushed into the plant, it may be necessary to add some digested sludge to the primary digester. Use the volatile acid/alkalinity ratio as a guide to determine the amount of digested sludge that should be returned to the primary digester for control purposes.

C. *DIGESTER GAS* (CO_2 and Gas Production)

This is a useful test to record. The change of CO_2 in the gas is an indicator of the condition of the digester. Good digester gas will have a CO_2 content of 30 to 35 percent. The volatile acid/alkalinity relationship will start to increase *BEFORE* the carbon dioxide (CO_2) content begins to climb. If the CO_2 content exceeds 42 percent, the digester is considered in poor condition and the gas is close to the burnable limit (44 to 45 percent CO_2).

Gas production in a properly operating digester should be constant if feed is reasonably constant. If the volume produced gradually starts falling, trouble of some sort is indicated.

D. *pH*

pH is normally monitored on raw sludge, recirculated sludge, and supernatant. This information is strictly for the record and not for plant control. The raw sludge, if stale, will be acid and run in the range of 5.5 to 6.8. Digester liquors should stay around 7.0 or higher. pH is a poor indicator of approaching trouble because it is usually the last indicator to change and, by the time it changes significantly, the digester is already in serious trouble.

QUESTIONS

Write your answers in a notebook and then compare your answers with those on pages 233 and 234.

12.3A Where would you obtain the temperature of a digester?

12.3B Why is the volatile acid/alkalinity relationship very useful in digester control?

12.3C What should be done when the volatile acid/alkalinity relationship starts to increase?

12.3D Why is pH a poor indicator of approaching trouble in a digester?

E. *SOLIDS TEST*

Samples are collected of the raw sludge, recirculated sludge, and supernatant. Each sample is tested for total solids and volatile solids.

The information from these tests is used to determine the pounds of sludge, the pounds of volatile sludge available to the bacteria, the pounds of volatile sludge destroyed or reduced, the digester loading rates, reductions, and the amount of solids handled through the system. All of these tests are necessary for the maintenance of efficient digester operation.

F. *VOLUME OF SLUDGE*

The volume of sludge pumped per day is needed to determine gallons pumped per day, pounds of dry solids per day to the digester, estimated gallons of supernatant returned to the plant, and volume of sludge for ultimate disposal.

Some plants use a magnetic flowmeter to measure the volume of sludge pumped. In smaller plants that use a positive displacement pump, the volume of raw sludge is determined by the volume the pump displaces during each revolution. For instance, a 10-inch (25-cm) diameter piston pump with a 3-inch (7.5-cm) stroke will discharge one gallon per revolution. These pumps are equipped with a counter on the end of the shaft and are seldom operated faster than 50 GPM (3 *L*/sec).

EXAMPLE 4:

Calculate the volume pumped per stroke (revolution) by a piston pump with a 10-inch diameter piston and the stroke set at 3 inches.

Convert inches to feet.

$$\frac{10 \text{ in}}{12 \text{ in/ft}} = 0.833 \text{ or } 0.83 \text{ ft}$$

$$\frac{3 \text{ in}}{12 \text{ in/ft}} = 0.25 \text{ ft}$$

$$\begin{aligned} \text{Volume of Cylinder, cu ft} &= \text{Area, sq ft} \times \text{Depth, ft} \\ &= 0.785 \text{ D}^2\text{H} \\ &= 0.785 \times (0.83 \text{ ft})^2 \times 0.25 \text{ ft} \\ &= 0.785 \times 0.69 \text{ ft} \times 0.25 \text{ ft} \\ &= 0.785 \times 0.17 \\ &= 0.133 \text{ cu ft} \end{aligned}$$

$$\begin{aligned} \text{Volume of Cylinder, gal} &= 0.133 \text{ cu ft} \times 7.48 \text{ gal/cu ft} \\ &= 0.995 \text{ gal/stroke} \\ &= 1.0 \text{ gal/stroke (3.8 liters) (approximately)} \end{aligned}$$

This is the maximum volume that can be pumped per three-inch stroke with this unit. Slow or incomplete valve closures are likely to reduce this amount. You may check it by taking the delivery volume and dividing it by the number of strokes to fill a drying bed or tank.

UNITS:

The piston travels the depth of the cylinder each stroke. We could have written our original equation in volume per stroke by indicating depth as distance per stroke.

$$\text{Volume, cu ft/stroke} = \text{Area, sq ft} \times \text{Depth, ft/stroke}$$

Therefore, if the pump counter recorded 2,800 revolutions, ideally the pump handled a total of 2,800 gallons for that period of time, which is normally a 24-hour period.

If a centrifugal pump is provided, it would be necessary to determine the volume pumped within the system. Thus, by determining how long it took to pump one foot of sludge to the digester, the volume per minute could be determined. The quantity of sludge pumped per day is an important variable, and the operator should make a real effort to determine the quantity. The volume of sludge pumped to the digester should be approximately the same each day.

QUESTIONS

Write your answers in a notebook and then compare your answers with those on page 234.

12.3E Why would you run a solids test on digester sludge?

12.3F Why would you want to know the volume of sludge pumped per day?

12.3G Calculate the volume of sludge pumped per stroke by a 12-inch diameter piston pump with the stroke set at 4 inches.

12.3H If a piston pump discharges 1.2 gallons of sludge per stroke and the counter indicates 2,000 revolutions during a 24-hour period, estimate how many gallons were pumped during that day.

G. *RAW SLUDGE*

During a 24-hour period, 2,800 gallons of 6.5 percent total solids sludge with a volatile content of 68 percent was pumped to a digester.

1. How many pounds of dry sludge were handled?
2. What part is subject to digestion (volatile solids)?

EXAMPLE 5:

Sludge Pumped = 2,800 gallons

Solids = 6.5%

Volatile = 68%

Dry Solids, lbs = Gal Pumped × % Solids as decimal × 8.34 lbs Water/gal

= 2,800 gal × 0.065 × 8.34 lbs/gal

= 182 gal × 8.34 lbs/gal of Solids

= 1,518 lbs Dry Solids

Volatile Solids, lbs = % Volatile as decimal × Total Solids, lbs

= 0.68 × 1,518 lbs

= 1,032 lbs of Volatile Solids

H. *RECIRCULATED SLUDGE*

Laboratory tests indicate that the dry solids in a recirculated digested sludge sample were 4.5 percent and contained 54.2 percent volatile content. This indicates that the process is reducing the volatile content of the sludge because the 4.5 percent solids is lower than that of the raw sludge being pumped to the digester. The reduction is a result of the conversion of a substantial portion of the volatile solids in the raw sludge to methane, carbon dioxide, and water. Therefore, the reduction in solids comes from some of the solids being converted to gas and some of the solids being washed out in the supernatant.

The reduction of volatile solids that has occurred in the primary digester is arrived at mathematically by the following formula:

$$P = \frac{R - D}{R - (R \times D)} \times 100\% = \frac{\text{In} - \text{Out}}{\text{In} - (\text{In} \times \text{Out})} \times 100\%$$

P = Percent Reduction of Volatile Matter

In R = Percent Volatile Matter in Raw Sludge

Out D = Percent Volatile Matter in Digested Sludge

This formula assumes that the digester is completely mixed, and that the pounds of fixed solids entering the digester equals the pounds of fixed solids leaving. If these assumptions do not apply to the situation, the answer given by the formula may be in error.

EXAMPLE 6:

In = 68% Volatile Matter in Raw Sludge

Out = 54% Volatile Matter in Digested Sludge

$$P = \frac{\text{In} - \text{Out}}{\text{In} - (\text{In} \times \text{Out})} \times 100\%$$

$$= \frac{0.68 - 0.54}{0.68 - (0.68 \times 0.54)} \times 100\%$$

$$= \frac{0.14}{0.68 - 0.37} \times 100\%$$

$$= \frac{0.14}{0.31} \times 100\%$$

$$= 0.45 \times 100\%$$

$$= 45\%$$

What is actually happening in the calculation of the percent reduction of volatile matter may be visualized by the following example. Start with 100,000 pounds of raw sludge solids consisting of 75 percent volatile (organic) solids and 25 percent fixed (inorganic) solids. After digestion, 50,000 pounds of volatile matter has been converted to methane, carbon dioxide, and supernatant water containing recycle solids, nitrogen, and a chemical oxygen demand (COD). The remaining digested sludge consists of 25,000 pounds volatile matter and 25,000 pounds of fixed solids.

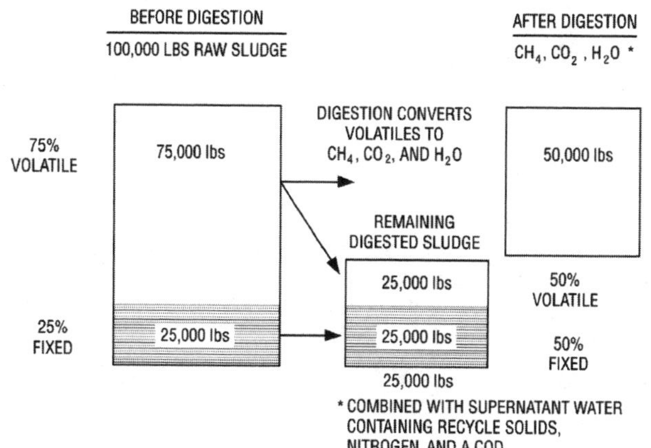

CHECK 1:

$$\text{Percent Reduction of Volatile Matter} = \frac{(\text{Reduction of Vol Solids, lbs}) \times 100\%}{\text{Starting Amt of Vol Solids, lbs}}$$

$$= \frac{(75{,}000 \text{ lbs} - 25{,}000 \text{ lbs}) \times 100\%}{75{,}000 \text{ lbs}}$$

$$= \frac{50{,}000 \times 100\%}{75{,}000}$$

$$= 66.7\%$$

CHECK 2:

$$P = \frac{\text{In} - \text{Out}}{\text{In} - (\text{In} \times \text{Out})} \times 100\%$$

$$= \frac{0.75 - 0.50}{0.75 - (0.75 \times 0.50)} \times 100\%$$

$$= \frac{0.25}{0.375} \times 100\%$$

$$= 66.7\%$$

As previously mentioned it is important to make a mass balance, or take an accounting of the fixed solids when analyzing the destruction of volatile matter in a digester. For example, in the previous analysis the pounds of fixed solids before digestion and after digestion were equal, and thus it would seem that the sampling and digester mixing were good. However, what if the pounds of fixed solids before and after digestion were not approximately equal? It would suggest that either there was bad sampling, inadequate digester mixing, or a combination of both. If this condition occurs, the previous calculations would reflect the process problems.

I. *SECONDARY DIGESTER SLUDGE*

Laboratory results indicate that a total digested sludge solids sample was 9.6 percent solids and 42.8 percent volatile content. The raw sludge solids volatile content was 68 percent. The overall percent reduction, P, could then be arrived at by using the formula,

$$P = \frac{\text{In} - \text{Out}}{\text{In} - (\text{In} \times \text{Out})} \times 100\%$$

$$= \frac{(0.68 - 0.43) \times 100\%}{0.68 - (0.68)(0.43)}$$

$$= \frac{0.25 \times 100\%}{0.68 - 0.29}$$

$$= \frac{0.25}{0.39} \times 100\%$$

$$= 0.64 \times 100\%$$

$$= 64\%$$

If sludge is drawn from the secondary digester, the total pounds of dry solids may be calculated by using the 9.6 percent solids results for that example and the volume of the withdrawn sludge. Thus,

Vol Sec Dig Sludge, gal × Solids, % × Weight of Sludge, lbs/gal[22]

= Total Solids, lbs

Total Solids, lbs × Volatile Solids, % = Volatile Solids, lbs

By subtracting from the raw sludge figures, the pounds reduction of total and volatile solids can be found.

QUESTIONS

Write your answers in a notebook and then compare your answers with those on page 234.

12.3I During a 24-hour period, 3,000 gallons of 5 percent total solids sludge with a volatile content of 70 percent was pumped to a digester.

1. How many pounds of dry sludge were handled?
2. What part is subject to digestion (volatile solids)?

12.3J Calculate the reduction in volatile solids if the percent volatile entering the digester is 70 percent and the percent leaving is 45 percent.

$$P = \frac{\text{In} - \text{Out}}{\text{In} - (\text{In} \times \text{Out})} \times 100\%$$

J. *DIGESTER SUPERNATANT*

The total solids test is run on the secondary digester supernatant to determine the solids load returned to the plant. The total solids in the digester supernatant should be kept below 0.5 percent (0.005 or 5,000 mg/*L*). High solids content in the supernatant usually indicates that too much seed or digested sludge is being withdrawn or "lost" from the digester.

Another simple method for checking supernatant is to draw a sample into a 1,000 m*L* graduate and let it stand for four or five hours. The sludge on the bottom of the graduate should be below 50 m*L*, with an amber-colored liquor above it. If supernatant solids are allowed to build too high, an excessive solids and BOD load is placed on the secondary system and primary clarifier. When the supernatant load becomes too heavy on the plant, the point of supernatant withdrawal from the secondary digester should be changed or the supernatant removal tubes should be moved to a different level in the digester where the liquor contains the least amount of solids.

Plants should be designed with the capacity to divert all sludge solids and liquids to a lagoon or some such system for final or ultimate disposal, rather than returning them to the plant.

K. *COMPUTING DIGESTER LOADINGS*

Digester loadings are reported as pounds of volatile matter (VM) per cubic foot or 1,000 cubic feet of digester volume per day. The loading rate should be around 0.15 to 0.35 pound of volatile solids per cubic foot (2.4 to 5.6 kg/cu m) in a heated and mixed high-rate digester. For an unmixed or cold digester, the loading rate should not exceed 0.05 pound of volatile matter per cubic foot (0.8 kg/cu m), assuming that each cubic foot contains approximately 0.5 pound of predigested solids (each cubic meter contains 8 kg of solids).

EXAMPLE 7:

A raw sludge volume of 2,800 gallons was pumped to a digester having a 40-foot diameter and a 20-foot water depth.

[22] A gallon of water weighs 8.34 pounds. A gallon of digested sludge will weigh slightly more due to solids. The best way to find the weight is to weigh a gallon of the sludge.

The raw sludge contained 6.5 percent solids and 68 percent volatile matter (VM). With these figures, it can be determined that 1,032 pounds of volatile solids were added to the digester per day. (See Section 12.21 for the formula to calculate Volatile Matter Pumped.)

Calculate the digester loading in lbs VM/cu ft/day.

$$\text{Digester Loading, lbs VM/cu ft/day} = \frac{\text{VM Added, lbs/day}}{\text{Volume of Digester, cu ft}}$$

$$= \frac{\text{1,032 lbs of VM/day}}{\text{25,120 cu ft}}$$

$$= \text{0.041 lb Volatile Matter/cu ft/day}$$

This would be a light loading, but it is not uncommon in small, new plants.

The pounds of solids that should remain in the digester to maintain a suitable environment must be determined too. To retain a favorable volatile acid/alkalinity relationship of around 0.1, at least 10 pounds of digested sludge should be retained in the digester for every pound of volatile matter added to the digester. Therefore, if 1,032 lbs VM are added, 10,320 lbs of digested sludge should be retained (lbs VM added times 10).

$$\text{Digested Sludge in Storage, lbs} = \frac{\text{VM Added, lbs/day} \times \text{10 lbs Dig Sludge in Storage}}{\text{1 lb VM Added per day}}$$

$$= \text{1,032 lbs/day} \times \frac{\text{10 lbs Storage}}{\text{1 lb Added/day}}$$

$$= \text{10,320 lbs Old Sludge in Storage on a Dry Solids Basis}$$

The actual amount of sludge retained will depend on digester conditions and the volatile acid/alkalinity relationship.

EXAMPLE 8:

Sometimes data are reported as pounds of volatile matter destroyed per cubic foot or 1,000 cubic feet of digester capacity per day. Using the data from Example 7 and assuming a volatile solids reduction of 50 percent, calculate the VM destroyed in lbs/day/cu ft and lbs/day/1,000 cu ft.

Volatile Matter Destroyed, lbs/day/cu ft

$$= \frac{\text{Volume of Sludge Pumped, gal/day} \times \text{\% Solids} \times \text{\% Volatile} \times \text{\% Reduction} \times \text{8.34 lbs/gal}}{\text{Volume of Digester, cu ft}}$$

$$= \frac{\overbrace{\text{2,800 gal/day} \times 0.065}^{\text{Volume of Solids}} \times 0.68 \times 0.50 \times \text{8.34 lbs/gal}}{\text{25,120 cu ft}}$$

$$= \frac{\text{2,800 GPD} \times 0.065 \times 0.34 \times \text{8.34 lbs/gal}}{\text{25,120 cu ft}}$$

$$= \frac{\text{516 lbs/day}}{\text{25,120 cu ft}}$$

$$= \text{0.0205 lb/day/cu ft}$$

or

$$= \text{20.5 lbs/day/1,000 cu ft}$$

L. *COMPUTING GAS PRODUCTION*

Digester gas data should be recorded in cubic feet produced per day by the digestion system, as recorded daily from the gas meter. The carbon dioxide (CO_2) content should normally be tested once or twice a week. (See Chapter 16, "Laboratory Procedures and Chemistry.") Gas production should range between 12 and 18 cubic feet for each pound of volatile matter destroyed (0.75 and 1.12 cu m/kg) in the digesters.

EXAMPLE 9:

Assume that the gas meter readings have averaged 6,000 cubic feet of gas per day. Using the data from Example 8 (516 pounds of volatile matter destroyed per day), compute gas produced per pound of volatile matter destroyed.

Gas Produced, cu ft/lb of VM destroyed

$$= \frac{\text{Gas Produced, cu ft/day}}{\text{Volatile Matter Destroyed, lbs/day}}$$

$$= \frac{\text{6,000 cu ft/day}}{\text{516 lbs/day}}$$

$$= \text{11.6 cu ft Gas/lb VM Destroyed (0.75 cu m Gas/kg VM Destroyed)}$$

QUESTIONS

Write your answers in a notebook and then compare your answers with those on page 234.

12.3K Why would you run a total solids test on the digester supernatant?

12.3L What would you do if the total solids were too high in the digester supernatant?

M. *SOLIDS BALANCE*
by F. Ludzack

What comes into a treatment plant must go out. This is the basis of the solids balance concept. If you measure what comes into your plant and can account for at least 90 percent of this material leaving your plant as a solid (sludge), liquid (effluent), or gas (digester gas), then you have control of your plant and know what is going on in the treatment processes. This approach provides a good check on your metering devices, sampling procedures, and analytical techniques. It is an eye-opener when tried for the first time and advanced operators are urged to calculate the solids balance for their plant.

Using the data from Example 5, page 193, the following series of calculations will illustrate the solids balance concept on a digester and drying bed.

INPUT TO DIGESTER

2,800 gallons of raw sludge with
solids content, 6.5% and
volatile solids content, 68%

DIGESTER OUTPUT or INPUT TO DRYING BED

Digested solids with
solids content, 4.5% and
volatile solids content, 54%

Calculate the pounds of total solids, water, volatile solids, and inorganic solids pumped into the digester.

Total solids to digester.

$$\text{Dry Solids, lbs} = \text{Gal Pumped} \times \text{\% Solids as decimal} \times 8.34 \text{ lbs Water/gal}$$

$$= 2{,}800 \text{ gal} \times 0.065 \times 8.34 \text{ lbs/gal}$$

$$= 1{,}518 \text{ lbs Solids}$$

Total water and solids to digester.

$$\text{Water and Solids, lbs} = \frac{\text{Total Solids, lbs}}{\text{\% Solids as decimal}}$$

$$= \frac{1{,}518 \text{ lbs}}{0.065}$$

$$= 23{,}400 \text{ lbs}$$

Water to digester.

$$\text{Water, lbs} = \text{Water and Solids, lbs} - \text{Solids, lbs}$$

$$= 23{,}400 \text{ lbs} - 1{,}518 \text{ lbs}$$

$$= 21{,}882 \text{ lbs}$$

or

$$= 21{,}900 \text{ lbs}$$

Volatile solids to digester.

$$\text{Volatile Solids, lbs} = \text{Total Solids, lbs} \times \text{\% Volatile Solids as decimal}$$

$$= 1{,}518 \text{ lbs} \times 0.68$$

$$= 1{,}032 \text{ lbs}$$

Inorganic solids to digester.

$$\text{Inorganic Solids, lbs} = \text{Total Solids, lbs} - \text{Volatile Solids, lbs}$$

$$= 1{,}518 \text{ lbs} - 1{,}032 \text{ lbs}$$

$$= 486 \text{ lbs}$$

Calculate the percent reduction in volatile matter in the digester to find the pounds of gas produced during digestion.

Percent reduction of volatile matter.

$$P = \frac{\text{In} - \text{Out}}{\text{In} - (\text{In} \times \text{Out})} \times 100\%$$

$$= \frac{0.68 - 0.54}{0.68 - (0.68 \times 0.54)} \times 100\%$$

$$= 45\%$$

Gas out of digester.

$$\text{Gas, lbs} = \text{Volatile Solids, lbs} \times \text{\% reduction as a decimal}$$

$$= 1{,}032 \text{ lbs} \times 0.45$$

$$= 465 \text{ lbs}$$

Determine the pounds of total, volatile, and inorganic solids removed from the digester to the drying bed as digested sludge.

Volatile solids to drying bed.

$$\text{Volatile Solids, lbs} = \text{Volatile Solids to Digester, lbs} - \text{Volatile Solids out as Gas, lbs}$$

$$= 1{,}032 \text{ lbs} - 465 \text{ lbs}$$

$$= 567 \text{ lbs}$$

Total solids to drying beds.

$$\text{Total Solids, lbs} = \frac{\text{Volatile Solids, lbs}}{\text{\% Volatile Solids as a decimal}}$$

$$= \frac{567 \text{ lbs}}{0.54}$$

$$= 1{,}050 \text{ lbs}$$

Inorganic solids to drying beds.

$$\text{Inorganic Solids, lbs} = \text{Total Solids, lbs} - \text{Volatile Solids, lbs}$$

$$= 1{,}050 \text{ lbs} - 567 \text{ lbs}$$

$$= 483 \text{ lbs}$$

NOTE: Almost same as 486 lbs to digester.

Total solids and water to drying bed.

$$\text{Water and Solids, lbs} = \frac{\text{Total Solids, lbs}}{\text{\% Solids as decimal}}$$

$$= \frac{1{,}050 \text{ lbs}}{0.045}$$

$$= 23{,}400 \text{ lbs}$$

NOTE: Same volume as put into digester because of thinner sludge going out.

Find total pounds of water to drying bed and compare amounts of water into and out of digester.

Water to drying bed.

$$\text{Water, lbs} = \text{Water and Solids, lbs} - \text{Solids, lbs}$$

$$= 23{,}400 \text{ lbs} - 1{,}050 \text{ lbs}$$

$$= 22{,}350 \text{ lbs}$$

or say

$$= 22{,}400 \text{ lbs}$$

Compare amounts of water in and out of digester.

$$\text{Water Change, lbs} = \text{Water Out, lbs} - \text{Water In, lbs}$$

$$= 22{,}400 \text{ lbs} - 21{,}900 \text{ lbs}$$

$$= 500 \text{ lbs Drawdown in Digester}$$

In this case, more water was withdrawn in the thin sludge than was added with the thick sludge. No supernatant was withdrawn from the digester or recycled. All of the recycle material must come from the dewatering operation.

DRYING BED OUTPUT

Dried residue removed, 2 lbs water/1 lb solids or 33% solids

Determine the pounds of water removed with the dried solids and the pounds of drainage water recycled to the plant.

Water in solids.

$$\text{Water in Solids, lbs} = \text{Total Solids, lbs} \times 2 \text{ lbs Water/lb Solids}$$

$$= 1{,}050 \text{ lbs} \times 2 \text{ lbs Water/lb}$$

$$= 2{,}100 \text{ lbs}$$

Drainage water from drying bed recycled to plant.

$$\text{Recycle Water, lbs} = \text{Water to Drying Bed, lbs} - \text{Water in Solids, lbs}$$

$$= 22{,}400 \text{ lbs} - 2{,}100 \text{ lbs}$$

$$= 20{,}300 \text{ lbs} \quad \text{(less evaporation)}$$

Summary

Constituent	Digester In	DIGESTER OUT To Drying Bed	DIGESTER OUT To Plant Recycle
Total Solids, lbs	1,518	1,050	
Volatile Solids, lbs	1,032	567	
Inorganic Solids, lbs	486	483	
Water, lbs	21,900	22,400	20,300
Gas Out, lbs		465	

To complete the solids balance, the quantity of water actually recycled and its solids content should be compared with the calculated values. Another helpful solids balance is to compare calculated and actual digester inputs and outputs on an annual basis.

N. *OTHER COMPUTATIONS*

Determine the minutes per hour that a primary sludge pump should operate if the influent flow to the clarifier is 1.8 MGD, the influent suspended solids 240 mg/*L* and the effluent suspended solids 100 mg/*L*. The raw sludge density is 4 percent solids and the pumping rate is 40 GPM.

Known		Unknown
Flow, MGD	= 1.8 MGD	Pump Operating Time, minutes/hour
Infl SS, mg/*L*	= 240 mg/*L*	
Effl SS, mg/*L*	= 100 mg/*L*	
Sludge, %	= 4%	
Pump, GPM	= 40 GPM	

Calculate the pump operating time in minutes per hour.

$$\text{Operating Time, min/hr} = \frac{(\text{Flow, MGD})(\text{Infl SS, mg}/L - \text{Effl SS, mg}/L)(100\%)}{(\text{Pump, GPM})(\text{Sludge Solids, \%})(24 \text{ hr/day})}$$

$$= \frac{(1.8 \text{ MGD})(240 \text{ mg}/L - 100 \text{ mg}/L)(100\%)}{(40 \text{ GPM})(4\%)(24 \text{ hr/day})}$$

$$= 6.6 \text{ min/hr}$$

Calculate the solids concentration fed to a digester that consists of the mixing of a 4 percent primary sludge flowing at 6,500 GPD and a 6 percent thickened secondary sludge flowing at 4,800 GPD.

Known		Unknown
Primary Solids, %	= 4%	Mixture Solids, %
Primary Flow, GPD	= 6,500 GPD	
Secondary Solids, %	= 6%	
Secondary Flow, GPD	= 4,800 GPD	

Determine the solids concentration of the mixture.

$$\text{Mixture Solids, \%} = \frac{(\text{Prim Sol, \%})(\text{Prim Flow, GPD}) + (\text{Sec Sol, \%})(\text{Sec Flow, GPD})}{\text{Prim Flow, GPD} + \text{Sec Flow, GPD}}$$

$$= \frac{(4\%)(6{,}500 \text{ GPD}) + (6\%)(4{,}800 \text{ GPD})}{6{,}500 \text{ GPD} + 4{,}800 \text{ GPD}}$$

$$= 4.85\%$$

Estimate the organic loading on an anaerobic digester from the primary clarifiers in pounds of volatile solids applied per day per pound of volatile solids under digestion. The plant inflow is 1.8 MGD. The influent suspended solids are 240 mg/*L* with a 4 percent sludge density and 77 percent volatile solids content. The effluent suspended solids are 100 mg/*L*. The digester contains 100,000 gallons of digesting sludge solids with a sludge density of 6 percent solids and 55 percent volatile solids.

Known		Unknown
Plant Flow, MGD	= 1.8 MGD	Loading, $\frac{\text{lbs VS/day}}{\text{lb VS dig}}$
Infl SS, mg/*L*	= 240 mg/*L*	
Effl SS, mg/*L*	= 100 mg/*L*	
Raw Sludge, %	= 4%	
Raw Sl VS, %	= 77%	
Digester Sl Vol, gal	= 100,000 gal	
Dig Sludge, %	= 6%	
Dig Sl VS, %	= 55%	

Calculate the sludge added in pounds of volatile solids per day.

$$\text{Sludge Added, lbs VS/day} = \frac{(\text{Flow, MGD})(\text{Infl SS, mg}/L - \text{Effl SS, mg}/L)(8.34 \text{ lbs/gal})(\text{RS VS, \%})}{100\%}$$

$$= \frac{(1.8 \text{ MGD})(240 \text{ mg}/L - 100 \text{ mg}/L)(8.34 \text{ lbs/gal})(77\%)}{100\%}$$

$$= 1{,}618 \text{ lbs VS/day}$$

Determine the pounds of volatile solids under digestion.

$$\text{Digestion VS, lbs} = \frac{(\text{Dig Sl Vol, gal})(8.34\text{ lbs/gal})(\text{Dig Sl, \%})(\text{DS VS, \%})}{(100\%)(100\%)}$$

$$= \frac{(100{,}000\text{ gal})(8.34\text{ lbs/gal})(6\%)(55\%)}{(100\%)(100\%)}$$

$$= 27{,}522\text{ lbs VS Under Digestion}$$

Estimate the organic loading in pounds of volatile solids added per day per pound of volatile solids under digestion.

$$\text{Loading, } \frac{\text{lbs VS/day}}{\text{lb VS dig}} = \frac{\text{Sludge Added, lbs VS/day}}{\text{Digestion VS, lbs}}$$

$$= \frac{1{,}618\text{ lbs VS/day}}{27{,}522\text{ lbs VS Under Digestion}}$$

$$= 0.059\text{ lb VS/day/lb VS dig}$$

END OF LESSON 4 OF 6 LESSONS
on
SLUDGE DIGESTION AND SOLIDS HANDLING

Please answer the discussion and review questions next.

DISCUSSION AND REVIEW QUESTIONS

Chapter 12. SLUDGE DIGESTION AND SOLIDS HANDLING

(Lesson 4 of 6 Lessons)

Write the answers to these questions in your notebook. The question numbering continues from Lesson 3.

18. Why is temperature measured in a digester?
19. Why should the temperature in the digester not be changed by more than one degree Fahrenheit per day?
20. What is the first warning that trouble is developing in an anaerobic digester?
21. How often should the volatile acid/alkalinity relationship in a digester be checked.
22. How would you try to stop and reverse an increasing volatile acid/alkalinity relationship?
23. What is the percent reduction in volatile matter in a primary digester if the volatile content of the raw sludge is 69% and the volatile content of the digested sludge is 51%?

$$P = \frac{\text{In} - \text{Out}}{\text{In} - (\text{In} \times \text{Out})} \times 100\%$$

CHAPTER 12. SLUDGE DIGESTION AND SOLIDS HANDLING

(Lesson 5 of 6 Lessons)

12.4 OPERATIONAL STRATEGY

All previous discussions and problem assignments were intended to provide you with the basic working principles of anaerobic sludge digestion. Their successful application in the operation of a digestion system requires sound and thorough daily operational checks in combination with adequate sampling and neat, well-organized records of the resulting data. Many operators find that plotting certain operational data in a graphical form is very helpful to recognize changes or trends in digester performance. Informative operational data that could be plotted against time include:

1. Digester loading
   a. Volatile solids added, lbs/day per cubic foot of digester capacity, or
   b. Volatile solids added, lbs/day per volatile solids under digestion, lbs
2. Volatile acid/alkalinity relationship

   Volatile acids, mg/*L* to alkalinity, mg/*L*
3. Gas production

   1,000 cubic feet of gas produced per day
4. Carbon dioxide content of digester gas

   Percent carbon dioxide
5. Temperature

   Degrees Fahrenheit or Degrees Celsius

Problems with anaerobic digesters can be anticipated by daily inspection of the digester data, plotting the data, and being alert for changes or trends in the wrong direction. Corrective action should be taken whenever trends or changes start in the wrong direction.

12.40 Operation and Maintenance Checklist

The checklist on page 200 is intended to help the operator remain on top of the system. This list is general in nature, and does not cover all situations, but serves as an example of the checklist that should be made for each plant. You should prepare a similar checklist for the anaerobic sludge digesters at your treatment plant. As you make your rounds inspecting each item, be alert. Investigate and record anything that looks different or unusual, smells different, feels different (hotter or vibrating more), and sounds different. If problems appear to be developing, correct them now or alert your supervisor to the changes.

12.41 Sampling and Data Checklist (See page 202)

Results and interpretation of lab tests tell you what you are feeding a digester and how the digester is treating the sludge. Graphically recording lab results helps to interpret what is happening in a digester. If undesirable trends start to develop, refer to the appropriate section in this manual for the proper corrective action.

12.42 Normal Operation

In this chapter we have discussed the following important topics regarding digester operation:

1. Section 12.1, "Components in the Anaerobic Sludge Digestion Process"
2. Section 12.2, "Operation of Anaerobic Digesters"
3. Section 12.3, "Anaerobic Digestion Controls and Test Interpretation"

This section combines the highlights of those portions of the previous sections that are critical to the actual day-to-day operation of an anaerobic sludge digester. For details, refer to the actual section. The normal operation of a digester involves the following activities:

1. Feeding sludge to the digester (Section 12.22, "Feeding")
2. Maintaining the proper temperature (Section 12.14, "Digester Heating" and Section 12.3, A. "Temperature")
3. Keeping the contents of the digester mixed (Section 12.15, "Digester Mixing")
4. Removing supernatant (Section 12.27, "Supernatant and Solids")
5. Withdrawing sludge (Section 12.28, "Rate of Sludge Withdrawal")

Let us study each one of these activities.

1. Feeding Sludge to the Digester (See page 202, A. "Raw Sludge")
   a. Pump as thick a sludge as possible to the digester. Watch sludge being pumped, listen to sound of sludge pump, and observe any instruments that indicate thickness of sludge.
   b. Pump small amounts of sludge at regular intervals to prevent adding too much raw sludge too fast for the organisms or for the temperature controls to maintain a constant temperature.
   c. Calculations
      (1) Try not to add more than one pound of volatile matter per day for every 10 pounds of digested sludge in storage (1 kg VM/day per 10 kg digested sludge). This ratio may vary from digester to digester and from season to season.
      (2) Calculate the volatile acid/alkalinity relationship and plot the results. If the relationship starts to increase, try to pump a thicker sludge or reduce the amount of volatile matter added per day. Also reduce the pumping rate of digested sludge.

OPERATION AND MAINTENANCE CHECKLIST	SCHEDULE				
ITEM	DAILY	WEEKLY	MONTHLY	TWICE/ YEAR	AS REQUIRED
A. Raw Sludge Pumping					
1. Total sludge volume pumped in 24 hours or individual feed periods. Record pump stroke counter or meter reading.	X				
2. Proper operation of pump(s). Check oil level. While operating check motor, pump, packing (leaks), suction, and discharge pressure.	X				
3. If density meter is used, check for proper operation during pump run.	X				
4. Instrumentation, especially pump time clock operation.	X				
5. Sludge line valve positions.	X				
6. Visual observation of raw sludge being pumped. Note consistency (thick or thin), color, and odor (septic).	X				
7. Automatic sampler operation.	X				
8. Exercise all sludge valves by opening and closing.		X			
9. Lubricate all valve stems. Inspect and grease pump motor bearings according to manufacturer's recommendations.				X	
B. Boiler and Heat Exchanger					
1. Temperature of the recirculated sludge.	X				
2. Temperature of the recirculated hot water.	X				
3. Boiler and heat exchanger temperature and pressure.	X				
4. Water level in sight glass of day-water tank.	X				
5. Boiler and heat exchanger operation.	X				
a. Gas pressure	X				
b. Makeup water valve	X				
c. Pressure relief (pop-off) valve	X				
d. Power failure or low gas pressure shutdown	X				
e. Safety devices			X		
6. Boiler firing (flame-air mixture).			X		
7. Recirculated sludge pump operation. Check oil level. While pump is operating check motor, pump, packing (leaks), suction, and discharge pressures.	X				
8. Inspect and grease pump motor bearings according to manufacturer's recommendations.				X	
C. Digesters					
1. Record gas meter reading.	X				
2. Check gas manometers (digester gas pressure).	X				
3. Record digester gas pressure or floating cover position and indicator level reading.	X				
4. Drain gas line condensate traps and sedimentation traps (from one to four times per day depending on location of trap in gas system, temperature changes, and digester mixing systems).					X
5. Check liquid level in the digester.	X				
6. Check supernatant tubes for operation and wash down supernatant box.	X				
7. Check digester gas safety analyzer (LEL) and recorder.	X				

OPERATION AND MAINTENANCE CHECKLIST
(continued)

ITEM	DAILY	WEEKLY	MONTHLY	TWICE/ YEAR	AS REQUIRED
8. Check and record level of water seal (located on center dome of fixed cover digesters).	X				
9. Check operation of mixing equipment.					
GAS					
a. Flow rate, CFM	X				
b. Pressure, psi	X				
c. Compressor operation	X				
MECHANICAL					
a. Motor operation	X				
b. Drive belts or gear reducers	X				
c. Vibrations	X				
d. Direction of mixing (down/up)	X				
10. Examine waste gas burner for proper operation.					
a. Pilot on	X				
b. Number of burners on		X			
c. Digester gas pressure (wasting or excess)		X			
11. Exercise all sludge and gas system valves by opening and closing.		X			
12. Check all supernatant tubes for operation and sample each for clearest liquor for supernatant removal from digester.		X			
13. Check digester for scum blanket buildup.		X			
14. Examine the digester structure and piping system for possible gas leaks. Examine the digester structure for cracks.			X		
15. Clean, inspect, and calibrate the digester gas safety analyzer and recorder.			X		
16. Lubricate all valve stems and rotating equipment as required by the manufacturer.				X	
17. Clean and refill gas manometers with proper fluids to levels specified by manufacturers.				X	
18. Flush and refill water seals (from 2 to 6 months).					X
19. For floating cover digesters, inspect flotation compartment for leakage or excessive condensation buildup (pump out) and look for corrosion of cover interior.		X			
20. Dewater digester and clean out, repair, and paint. Normal cleanout schedules are three (3) to eight (8) years.					X

SAMPLING AND DATA CHECKLIST

ITEM	SCHEDULE: DAILY	TWICE/ WEEK	WEEKLY	MONTHLY
A. Raw Sludge				
1. Composite raw sludge sample. If grab is taken instead, then prepare a composite twice a week.	X	(X)		
2. Total and volatile solids.	X	(X)		
3. pH.		X		
B. Supernatant				
1. Solids (total and volatile) and COD. Graphically record the data and be alert to long-term decreasing quality (increased levels of solids and COD) of supernatant.				X
C. Digested Sludge				
1. Grab sample.	X			
2. Temperature.	X			
3. pH.	X			
4. Cubic feet of total gas and CO_2 content.	X			
5. Calculate and graphically record gas production and CO_2 content.	X			
6. Calculate and graphically record loading rate (solids and hydraulic).	X			
7. Volatile acids.	X			
8. Alkalinity.		X		
9. Calculate and graphically record volatile acid/alkalinity relationship.		X		
10. Digested sludge total solids and volatile solids.		X		
11. Solids (total and volatile) and temperature profile at five-foot (1.5-m) intervals from the digester bottom up to the surface. If scum blanket present, try to break it up.				X
D. Solids Balance				
1. Calculate the solids balance on the digesters (see Section 12.3, M. "Solids Balance"). This calculation helps indicate to you how well you are controlling the digester operation.				X

In item C-11 above, as regards the profile sampling of the digester, the solids and temperature data should be carefully examined for indications of poor mixing in the digester or grit accumulation at the bottom of the digester. The operator should use the data to calculate the useful volume of the digester (total volume minus the grit volume). Such data can be graphically plotted against time to show the rate of grit buildup and the date for digester cleaning. An example of such a plot is illustrated in Figure 12.25, although actual data may not plot a straight line. Digester cleaning, for example, will substantially increase the active volume of the unit.

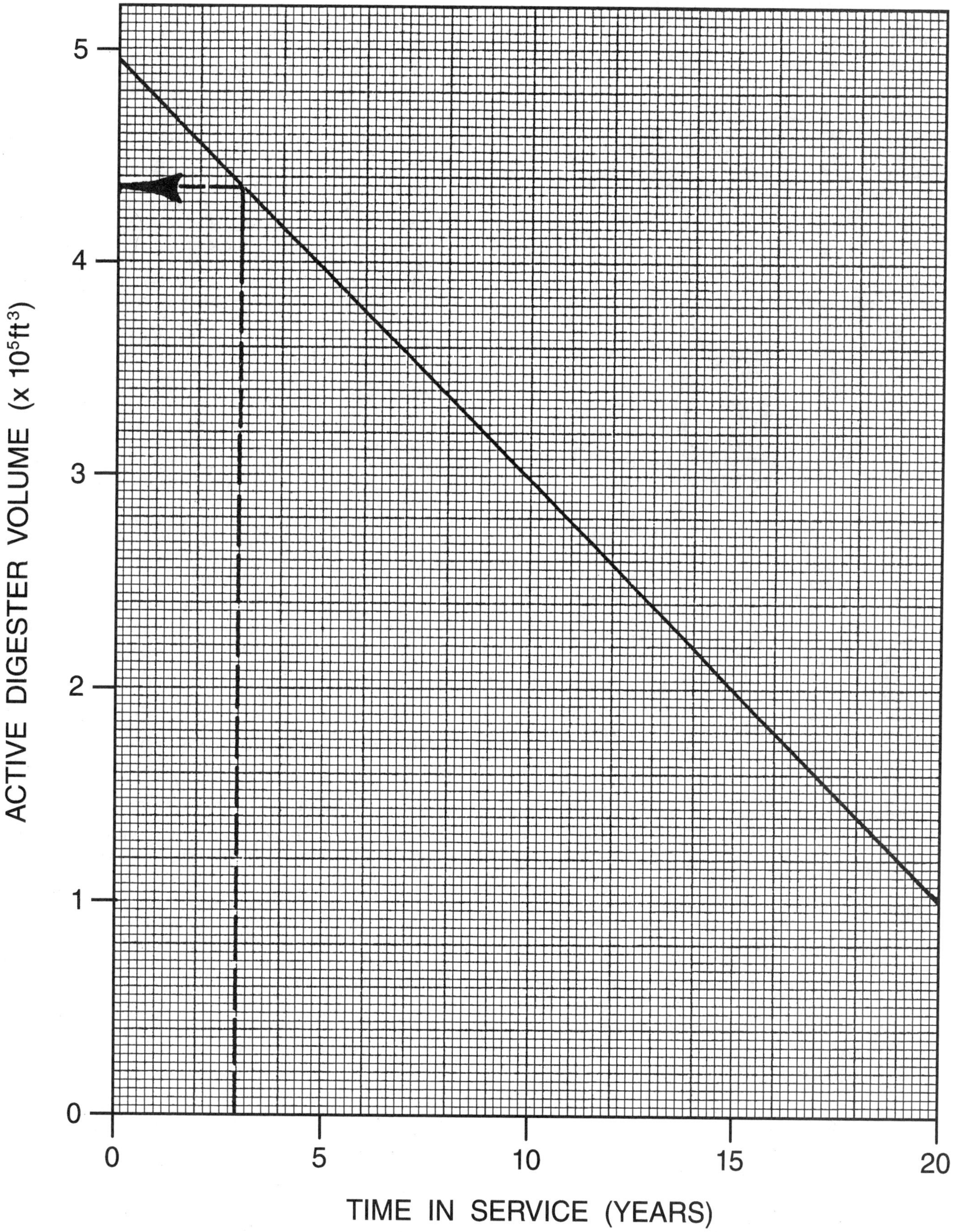

Fig. 12.25 Active volume of a digester tank

(Permission of Sanitation Districts of Los Angeles County)

See Section 12.3, "Anaerobic Digestion Controls and Test Interpretation," B. "Volatile Acid/Alkalinity Relationship," and K. "Computing Digester Loadings" for details.

2. Maintaining the Proper Temperature (See page 200, B. "Boiler and Heat Exchanger")

   Record the temperature of the recirculated sludge every day. If the temperature changes from the desired level, adjust the temperature controls. Do not allow the temperature to change more than 1°F (0.5°C) per day. Determine the temperature (usually between 95° and 98°F or 35° and 37°C) that best suits your digester.

3. Keeping the Contents of the Digester Mixed

   How a digester is mixed depends on the mixing equipment and whether you have a single-stage or two-stage digestion process. Digester contents must be well mixed to provide an even distribution of food (raw sludge), organisms, alkalinity, heat, and waste bacterial products. Good mixing should prevent the buildup of a scum blanket and the deposition of grit on the bottom of the digester. If mixing is inadequate, try increasing the mixing time or looking for equipment problems. Primary digesters should be mixed continuously when you have secondary digesters.

4. Removing Supernatant

   Supernatant should be removed from the digesters on a daily basis. If you have a single-stage digestion process, mixing should be stopped for 6 to 12 hours before supernatant removal to allow the supernatant to separate from the digested sludge. If you have a two-stage digestion process, the primary digesters should be mixed continuously and supernatant is transferred to the secondary digester automatically whenever raw sludge is added to the primary digester. Mix these secondary digesters whenever primary supernatant is added. Also mix the secondary digesters occasionally to keep the scum blanket down. Adjust or select the supernatant tube that produces the least solids to remove supernatant from the digester back to the plant. Carefully observe your other treatment processes to be sure the supernatant does not cause a solids or BOD overload on the other processes. Remove supernatant and digested sludge until sufficient space is obtained in the digesters for the incoming raw sludge.

5. Withdrawing Sludge

   Before withdrawing sludge, stop mixing for 6 to (preferably) 12 hours to allow the digested sludge to separate from the supernatant. The digester contents must be well mixed before stopping mixing so a lot of raw sludge will not be removed with the digested sludge. Good mixing also prevents the buildup of a scum blanket and the development of coning during the removal of digested sludge. The withdrawal rate of sludge from either digester should be no faster than a rate at which the gas production from the system is able to maintain a positive pressure in the digester (at least 2 inches (5 cm) of water column).

12.43 Troubleshooting

Using the information obtained from the analysis of the samples and the daily rounds, the knowledgeable and alert operator can note changes from normal operation. The first step is to realize that there is a problem, and the second step is to take the appropriate corrective action. Table 12.2 is an example of a logical sequence that can be followed to identify and correct an impending or actual digester upset. The four indicators of a problem tell you to look for one or more of the problem areas listed that need correcting.

Toxicity can be a very difficult problem to identify and solve. Heavy metals can gradually creep up in concentrations until toxic levels are reached. Also, as the pH decreases the concentrations of dissolved metals tend to increase and become toxic to bacteria in the digester.

Possible methods of controlling toxic materials include:

1. Remove toxic material from waste at source.
2. Dilute toxic material below its toxic level.
3. Add a chemical that will neutralize the toxic material.
4. Add a chemical that will cause the toxic material to precipitate out of solution or form an insoluble compound.

If soluble, toxic, heavy metals are present, *SODIUM SULFIDE* (Na_2S)[23] can be added, which will cause the formation of nontoxic, insoluble, heavy metal sulfide compounds. Digesters are similar to people in many ways. A small amount of something may be very good for a digester, but too much may be toxic as shown in Table 12.3.

[23] *Sodium Sulfide (Na_2S).* Use extreme care when handling sodium sulfide. Anhydrous sodium sulfide may ignite spontaneously on exposure to air. When dissolved in water it is a strong base, it reacts violently with acid and is corrosive toward aluminum and zinc. Handle sodium sulfide only after you are aware of and comply with all related safety requirements. Read the Material Safety Data Sheet (MSDS) very carefully.

TABLE 12.2 DIGESTER OPERATION TROUBLESHOOTING

Indication from Data	Problem Area	Possible Cause
• Rise in VA/Alk Ratio • Gas Production Decrease or Increase in CO_2 • Decrease in VS Reduction • High Solids in Supernatant	Digester Loading →	1. Change in raw sludge pumping. 2. Raw sludge density or VS changed. 3. Raw sludge pH change. 4. Decrease in effective volume of the digester.
	Digester Heating →	1. Heat exchangers plugged. 2. Recirculated sludge pump not working. 3. Boiler malfunction. 4. Unsteady sludge temperatures—more than 1°F/day or 0.5°C/day.
	Digester Mixing →	1. Fouled draft tube. 2. Mechanical or electrical failure. 3. In case of gas mixing, inadequate recirculation.
	Gas System →	1. Gas meter failure. 2. Leaking gas. 3. Abnormal pressure. 4. Plugged gas line.
	Toxicity →	1. Slug of toxic material. 2. Constant feed that has reached toxic limit.

TABLE 12.3 BENEFICIAL AND TOXIC CONCENTRATIONS OF MATERIALS ON DIGESTION PROCESS

Material	Beneficial	Moderately Inhibitory	Toxic
Ammonia Nitrogen, mg/*L*	50-200	1,500-3,000[a]	3,000
Calcium, mg/*L*	100-200	2,500-4,500	8,000
Magnesium, mg/*L*	75-150	1,000-1,500	3,000
Potassium, mg/*L*	200-400	2,500-4,500	12,000
Sodium, mg/*L*	100-200	3,500-5,500	8,000

[a] Toxic at higher pH values

12.44 Troubleshooting Guide (adapted from *PERFORMANCE EVALUATION AND TROUBLESHOOTING AT MUNICIPAL WASTEWATER TREATMENT FACILITIES,* Office of Water Program Operations, US EPA, Washington, DC)

INDICATOR/OBSERVATION	PROBABLE CAUSE	CHECK OR MONITOR	SOLUTION
1. A rise in the volatile acid/alkalinity (VA/Alk) ratio.	1a. Pumping thin (<3%) sludge, accidental overpumping, withdrawing too much sludge.	1a. Monitor the following twice daily until problem is corrected: —pump feed rate and % solids —volatile acids —alkalinity —temperature	1a. If ratio increases to 0.3: (1) thicken raw sludge. (2) decrease sludge withdrawal rate to keep seed sludge in digester (and/or) (3) extend mixing time. (4) check sludge temperatures closely and control heating, if needed. (5) add seed sludge from secondary digester, if available.
	1b. Organic overload.	1b. Monitor sludge pumping volume, amount of volatile solids in feed sludge; check for increase in septic tank sludge discharged to plant or industrial wastes.	1b. See 1a.
	1c. Discharge of toxic materials to digesters such as heavy metals, sulfide, ammonia.	1c. Volatile acids, pH, gas production; check industrial wastes at source; check for inadequate sludge pumping generating sulfide.	1c. Use any or combination of the following: (1) solids recycle. (2) liquid dilution. (3) decrease feed concentration. (4) precipitate heavy metals with sulfur compound. Be sure pH in digester is greater than 7.0. (5) use iron salts to precipitate sulfide. (6) institute source control program for industrial wastes.
2. CO_2 in gas starts to increase.	2. VA/Alk ratio has increased to 0.5.	2a. Percent CO_2 in gas. 2b. Gas analyzer. 2c. Waste gas burner.	2. (1) Check 1a, sludge feed, mixing, and temperature. (2) See Item 1 and start adding alkalinity (lime) using the volatile acids to calculate the amount.
3. pH starts to drop and CO_2 increases to the point (42-45%) that no burnable gas is obtained.	3. VA/Alk ratio has increased to 0.8.	3a. Monitor as indicated above.	3a. Decrease loading to less than 0.01 lb vol. solids/cu ft/day until ratio drops to 0.5 or below. Continue mixing and maintain temperature.
		3b. Hydrogen sulfide (rotten egg) odor.	3b. Add alkalinity (lime).
		3c. Rancid butter odor.	
4. The supernatant quality returning to process is poor, causing plant upsets.	4a. Supernatant draw-off point not at same level as supernatant layer.	4a. Locate depth of supernatant by sampling at different depths.	4a. Adjust tank operating level or draw-off pipe.
	4b. Excessive mixing and not enough settling time.	4b. Withdraw sample and observe separation pattern.	4b. Allow longer periods for settling before withdrawing supernatant.
	4c. Raw sludge feed point too close to supernatant draw-off line.	4c. Determine volatile solids content. Should be close to value found in well-mixed sludge and much lower than raw sludge.	4c. Schedule pipe revision for soonest possible time when digester can be dewatered.
	4d. Not withdrawing enough digested sludge.	4d. Compare feed and withdrawal rates—check volatile solids to see if sludge is well digested.	4d. Increase digested sludge withdrawal rates. Withdrawal should not exceed 5% of digester volume per day.
			4e. Review feasibility of adding powdered carbon to digesters with consultant or regulatory agency. Also look for another way to treat supernatant (lagooning, filtering) rather than returning to plant for biological (secondary) treatment.

12.44 Troubleshooting Guide *(continued)* (adapted from *PERFORMANCE EVALUATION AND TROUBLESHOOTING AT MUNICIPAL WASTEWATER TREATMENT FACILITIES,* Office of Water Program Operations, US EPA, Washington, DC)

INDICATOR/OBSERVATION	PROBABLE CAUSE	CHECK OR MONITOR	SOLUTION
5. Supernatant has a sour odor from either primary or secondary digester.	5a. Overloaded digester (rotten egg odor).	5a. See Item 3.	5a. See Item 3.
	5b. The pH of digester is too low.	5b. See Item 3.	5b. See Item 3.
	5c. Toxic load (rancid butter odor).	5c. See Item 1c.	5c. See Item 1c.
6. Foam observed in supernatant from single stage or primary tank.	6a. Organic overload.	6a. Volatile solids loading ratio.	6a. Reduce feeding rate.
	6b. Scum blanket breaking up.	6b. Check condition of scum blanket.	6b. Normal condition but should stop withdrawing supernatant if possible.
	6c. Excessive gas recirculation.	6c. 20 CFM/1,000 cu ft is adequate.	6c. Throttle compressor output.
7. Bottom sludge too watery or sludge at disposal point too thin.	7a. Short-circuiting.	7a. Draw-off line open to supernatant zone.	7a. Change to bottom draw-off line and drain or pump out slowly to prevent coning.
	7b. Excessive mixing.	7b. Take sample and check how it concentrates in settling vessel.	7b. Shut off mixing for 24-48 hours before drawing sludge.
	7c. Sludge coning, allowing lighter solids to be pulled into pump suction.	7c. Total solids test or visual observation. Measure solids in digester at various levels.	7c. (1) "Bump" the pump 2 or 3 times by starting and stopping. (2) Mix digester thoroughly and then allow to settle.
8. Sludge temperature is falling and cannot be maintained at normal level.	8a. Plugging of external heat exchanger.	8a. Check inlet and outlet pressure of exchanger.	8a. Open heat exchanger and clean.
	8b. Sludge recirculation line is partially or completely plugged.	8b. Check pump inlet and outlet pressure.	8b. (1) Backflush the line with heated digester sludge. (2) Use mechanical cleaner. (3) Apply water pressure. Do not exceed working line pressure. (4) Add approx. 3 lbs/100 gal (3.6 kg/1,000 *L*) water of trisodium phosphate (TSP) or commercial degreasers. (Most convenient method is to fill scum pit to a volume equal to the line, add TSP or other chemical, then admit to the line and let stand for an hour.)
	8c. Inadequate mixing.	8c. Check temperature profile in digester.	8c. Increase mixing.
	8d. Hydraulic overload.	8d. Incoming sludge concentration and pumping rate.	8d. See Item 1a.
	8e. Low water feed rate in internal coils used for heat exchange.	8e. (1) Air lock in line. (2) Valve partially closed.	8e. (1) Bleed air relief valve. (2) Upstream valve may be partially closed.
	8f. Boiler burner not firing on digester gas.	8f. (1) Low gas pressure. (2) Unburnable gas due to process upset.	8f. (1) Locate and repair leak. (2) See Item 3. (3) Switch boiler to another fuel (natural gas or LP gas).
	8g. Heating coils inside digester have coating.	8g. Temperature of inlet and outlet water is about the same.	8g. (1) Remove coating, may require draining tank. (2) Control water temperature to 130°F (55°C) maximum.
	8h. Heat exchanger water temperature too low.	8h. Inlet water temperature to heat exchanger.	8h. Increase temperature as appropriate.

12.44 Troubleshooting Guide *(continued)* (adapted from *PERFORMANCE EVALUATION AND TROUBLESHOOTING AT MUNICIPAL WASTEWATER TREATMENT FACILITIES,* Office of Water Program Operations, US EPA, Washington, DC)

INDICATOR/OBSERVATION	PROBABLE CAUSE	CHECK OR MONITOR	SOLUTION
9. Sludge temperature is rising.	9. Temperature controller is not working properly.	9. Check water temperature and controller setting.	9. If over 130°F (55°C), reduce temperature. Repair or replace controller.
10. Recirculation pump not running; power circuits OK.	10. Temperature override in circuit to prevent pumping too hot water through tubes.	10. Visual check, no pressure on sludge line.	10a. Allow system to cool off. 10b. Check temperature control circuits.
11. Gas mixer feed lines plugging.	11a. Lack of flow through gas line. 11b. Debris in gas lines.	11. Identify location of low pressure (gas feed pipes or manometer).	11a. Flush out with water. 11b. Ensure proper operation/ pressure of compressor. 11c. Clean feed lines and/or valves. 11d. Give thorough service when tank is drained for inspection.
12. Gear reducer wear on mechanical mixers.	12a. Lack of proper lubrication.	12a. Excessive motor amperage, excessive noise and vibration, evidence of shaft wear.	12a. Verify correct type and amount of lubrication from manufacturer's literature.
	12b. Rags.	12b. Check impeller for rag buildup.	12b. Correct imbalances caused by accumulation of material on the internal moving parts.
	12c. Tank level.	12c. Check tank level for proper range—not too low or too high.	12c. Maintain proper tank level.
	12d. Poor alignment of equipment.	12d. See Item 15.	12d. Realign. Replace worn or broken guide bushings or bearings.
13. Shaft seal leaking on mechanical mixer.	13. Packing or seal dried out or worn.	13. Evidence of gas leakage (evident odor of gas).	13a. Follow manufacturer's instructions for repacking. 13b. Replace packing or mechanical seal any time the tank is empty if it is not possible when unit is operating.
14. Wear on internal parts of mechanical mixer.	14. Grit or misalignment.	14. Visual observation when tank is empty, compare with manufacturer's drawings for original size. Motor amperage will also go down as moving parts are worn away and get smaller.	14. Replace or rebuild—experience will determine the frequency of this operation.
15. Imbalance of internal parts because of accumulation of debris on the moving parts of mechanical mixers (large-diameter impellers or turbines would be affected most).	15. Poor comminution and/ or screening.	15. Vibration, heating of motor, excessive amperage, noise.	15a. Reverse direction of mixer if it has this feature. 15b. Stop and start alternately. 15c. Open inspection hole and visually inspect. 15d. Draw down tank and clean moving parts.
16. Rolling movement of scum blanket is slight or absent.	16a. Mixer is off.	16a. Mixer switch or timer.	16a. May be normal if mixers are set on a timer. If not and mixers should be operating, check for malfunction.
	16b. Inadequate mixing.	16b. Amount of mixing.	16b. Increase mixing.
	16c. Scum blanket is too thick.	16c. Measure blanket thickness.	16c. See Items 18 and 19.
17. Scum blanket is too high.	17. Supernatant overflow is plugged.	17. Check gas pressure, it may be above normal or relief valve may be venting to atmosphere.	17. Lower contents through bottom draw-off then rod supernatant line to clear plugging.

12.44 Troubleshooting Guide *(continued)* (adapted from *PERFORMANCE EVALUATION AND TROUBLESHOOTING AT MUNICIPAL WASTEWATER TREATMENT FACILITIES,* Office of Water Program Operations, US EPA, Washington, DC)

INDICATOR/OBSERVATION	PROBABLE CAUSE	CHECK OR MONITOR	SOLUTION
18. Scum blanket is too thick.	18. (1) Lack of mixing, high grease content. (2) Low temperature.	18. Probe blanket for thickness through thief hole or in gap beside floating cover.	18a. Break up blanket by using mixers. 18b. Use sludge recirculation pumps and discharge above the blanket. 18c. Raise digester temperature to 99-100°F (37-38°C). 18d. Break up blanket physically with pole. 18e. Use chemicals to soften blanket. 18f. Modify tank.
19. Draft tube mixers not moving surface adequately.	19. Scum blanket too high and allowing thin sludge to travel under it.	19. Breakup of scum blanket to obtain rolling movement on sludge surface.	19. Lower sludge level to 3-4" (7.5-10 cm) above top of tube allowing thick material to be pulled into tube—continue for 24-48 hours. Reverse direction (if possible) of mixers.
20. Gas is leaking through pressure relief valve (PRV) on roof.	20a. Plugged gas line or closed valve. 20b. Valve not seating properly or is stuck open.	20. Check the manometer to see if digester gas pressure is normal. If pressure is high, check 20a; if low, check 20b.	20a. Check gas system valves, regulators, safety thermal valves, water traps, and drains. 20b. Remove PRV cover and move weight holder until it seats properly. Install new ring, if needed. Rotate a few times for good seating. Adjust gas pressure to normal level.
21. Manometer shows digester gas pressure is above normal.	21a. Obstruction or water in main gas line.	21a. If all use points are operating and normal, then check for a waste gas line restriction or a plugged or stuck safety device.	21a. Purge with air, drain condensate traps, check for low spots. Care must be taken not to force air into digester.
	21b. Waste gas burner line pressure control valve is closed.	21b. Gas meters show excess gas is being produced, but not going to waste gas burner.	21b. Open/adjust burner line pressure control valve accordingly.
	21c. Digester PRV is stuck shut.	21c. Gas is not escaping as it should.	21c. Remove PRV cover and manually open valve, clean valve seat.
22. Manometer shows digester gas pressure below normal.	22a. Too fast withdrawal causing a vacuum inside digester.	22a. Check vacuum breaker to be sure it is operating properly.	22a. Stop supernatant discharge or sludge withdrawal and close off all gas outlets from digester until pressure returns to normal.
	22b. Adding too much lime.	22b. Sudden increase in CO_2 in digester gas.	22b. Stop addition of lime and increase mixing.
23. Pressure regulating valve not opening as pressure increases.	23a. Inflexible diaphragm.	23a. Isolate valve and open cover.	23a. If no leaks are found (using soap solution) diaphragm may be lubricated and softened using neat's-foot oil.
	23b. Ruptured diaphragm.	23b. Visual inspection.	23b. Ruptured diaphragm would require replacement.
	23c. Internal mechanism of regulator may be frozen (seized) (that is, the spring).	23c. Visual inspection.	23c. Disassemble and free mechanism.
24. Yellow gas flame from waste gas burner.	24. Poor quality gas with a high CO_2 content or air shutter closed on burner.	24. Check CO_2—content will be higher than normal.	24. Check concentration of sludge feed—may be too dilute. If so, increase sludge concentration. See Item 1a. Open air shutter, if appropriate.
25. Gas meter failure (propeller or lobe type).	25a. Water or debris in line.	25a. Condition of gas line.	25a. Flush with water, isolating digester and working from digester toward points of usage.
	25b. Mechanical failure.	25b. Fouled or worn parts.	25b. Wash with kerosene or replace worn parts.

12.44 Troubleshooting Guide *(continued)* (adapted from *PERFORMANCE EVALUATION AND TROUBLESHOOTING AT MUNICIPAL WASTEWATER TREATMENT FACILITIES,* Office of Water Program Operations, US EPA, Washington, DC)

INDICATOR/OBSERVATION	PROBABLE CAUSE	CHECK OR MONITOR	SOLUTION
26. Gas meter failure (bellows type).	26a. Inflexible diaphragm.	26a. Isolate valve and open cover.	26a. If no leaks are found (using soap solution) diaphragm may be lubricated and softened using neat's-foot oil.
	26b. Ruptured diaphragm.	26b. Visual inspection.	26b. Replace diaphragm.
			26c. Metal guides may need to be replaced if corroded.
27. Gas pressure higher than normal during freezing weather.	27a. Supernatant line plugged.	27a. Supernatant overflow lines.	27a. Check every two hours during freezing conditions, inject steam, protect line from weather by covering and insulating overflow box.
	27b. Pressure relief stuck or closed.	27b. Weights on pressure relief valves.	27b. If freezing is a problem, apply light grease layer impregnated with rock salt.
			27c. Moisture in regulator lines may become frozen, rendering diaphragm inoperable. Remove moisture.
28. Gas pressure lower than normal.	28a. Pressure relief valve or other pressure control devices stuck open.	28a. Pressure relief valve and devices.	28a. Manually operate vacuum relief and remove corrosion if present and interfering with operation.
	28b. Gas line or hose leaking.	28b. Gas line and/or hose.	28b. Repair as needed.
29. Leaks around metal covers.	29. Anchor bolts pulled loose and/or sealing material moved or cracking.	29. Concrete broken around anchors, tie-downs bent, sealing materials displaced.	29. Repair concrete with fast-sealing, concrete repair material. New tie-downs may have to be welded (be careful with open flames) onto old ones and redrilled. Tanks should be drained and well ventilated for this procedure. New sealant material should be applied to leaking area. Confined space procedures must be implemented if tank entry is required to accomplish repair.
30. Suspected gas leaking through concrete cover.	30. Freezing and thawing causing widening of construction cracks.	30. Apply soap solutions to suspected area and check for bubbles.	30. If this is a serious problem, drain tank, clean cracks, and repair with concrete sealers. Tanks should be drained and well ventilated for this procedure. Confined space procedures must be implemented if tank entry is required to accomplish repair.
31. Floating cover tilting, little or no scum around the edges.	31a. Water in floating cover compartment.	31a. Visual inspection of flotation chamber for water or sludge.	31a. Siphon or pump out any water or sludge. Repair leak in either cover, top or bottom.
	31b. Water from condensation or rain water collecting on top of metal cover in one location.	31b. Check around the edges of the metal cover. (Some covers with insulating wooden roofs have inspection holes for this purpose.)	31b. Use siphon or other means to remove the water. Repair roof if leaks in the roof are contributing to the water problem.
	31c. Weight distributed unevenly.	31c. (1) Location of weights. (2) Check buoyancy chamber for leakage.	31c. (1) If movable ballast or weights are provided, move them around until the cover is level. If no weights are provided, use a minimal number of sandbags to cause cover to level up. (Note: pressure relief valves may need to be reset if significant amounts of weight are added.) (2) Lower digester level to corbels to support and level cover. Inspect and repair internal leakage. Confined space procedures must be used for cover entry.

12.44 Troubleshooting Guide *(continued)* (adapted from *PERFORMANCE EVALUATION AND TROUBLESHOOTING AT MUNICIPAL WASTEWATER TREATMENT FACILITIES,* Office of Water Program Operations, US EPA, Washington, DC)

INDICATOR/OBSERVATION	PROBABLE CAUSE	CHECK OR MONITOR	SOLUTION
32. Floating cover tilting, heavy thick scum accumulating around edges.	32a. Excess scum in one area, causing excess drag.	32a. Probe with a stick or some other method to determine the condition of the scum.	32a. Use chemicals or degreasing agents such as Digest-aid or Sanfax to soften the scum, then hose down with water. Continue on regular basis every two to three months or more frequently, if needed.
	32b. Guides or rollers out of adjustment. Rollers "frozen."	32b. Distance between guides or rollers and the wall.	32b. Soften up the scum (as in 32a) and readjust rollers or guides so that skirt does not rub on the walls. Free rollers and ensure adequate lubrication.
	32c. Rollers or guides broken.	32c. Determine the normal position if the suspected broken part is covered by sludge. Verify correct location using manufacturer's information and/or prints, if necessary.	32c. Drain tank, if necessary, taking care as cover lowers to corbels not to allow it to bind or come down unevenly. It may be necessary to use a crane or jacks in order to prevent structural damage with this case.
33. Cover binding even though rollers and guides are free.	33. Internal guide or guy wires are binding or damaged (some covers are built like umbrellas with guides attached to the center column).	33. Lower down to corbels. Open hatch and, following confined space procedures, inspect from the top. If cover will not go all the way down, it may be necessary to secure in one position with a crane or by other means to prevent skirt damage to sidewalls.	33. Drain and repair, holding the cover in a fixed position, if necessary.

12.45 Actual Digester Operation

By using the procedures outlined in this section, digesters can be operated successfully without any problems. The data plotted in Figure 12.26 shows some of the information used to operate four digesters with a total capacity of 6.9 million gallons (26,000 cu m). This activated sludge plant treats an average daily flow of approximately 18 MGD (68,130 cu m/day) with flows averaging over 24 MGD (90,840 cu m/day) during the canning season. Under adverse conditions, the digesters have provided only 8 days of detention time, yet the digesters have never become upset.

Raw sludge from the primary clarifiers and gravity thickened waste activated sludge are fed on a regular basis throughout the day to each digester. Every 2 hours the operators read and record the gauges, pump meters, and temperature readings. Temperatures are controlled by adjusting the heat exchanger.

Digester contents are continuously mixed through draft tubes. Every day the flows through the draft tubes are reversed for 2 hours to knock off rags accumulated on the draft tubes. Additional mixing is available using digested sludge recirculation pumps, if necessary. The operator reviews the lab data and, if problems appear to be developing, additional mixing is applied, if appropriate. If everything is satisfactory and mixing is greater than usual, mixing is reduced.

The following information is recorded with regard to the digesters:

1. Raw Sludge and Thickened Waste Activated Sludge to Digesters
   a. Volume, gallons per day
   b. pH
   c. Total solids, %
   d. Volatile solids, %
2. Digester Gas
   a. Total production, cubic feet per day
   b. Carbon dioxide, %
3. Digested Sludge (mixed digester contents)
   a. Volatile acids, mg/*L*
   b. Alkalinity, mg/*L*
   c. Total solids, %
   d. Volatile acids, %
   e. pH
4. Sludge Removed (mixed digester contents)
   a. Volume, gallons per day
   b. Total solids, %
   c. Volatile solids, %

The volatile acid/alkalinity relationship has been the key to successful digester operation over the last nine years without any of the four digesters becoming upset. Volatile acids and alkalinity tests are normally run three times per week on each digester. If one high volatile acid reading is observed, the volatile

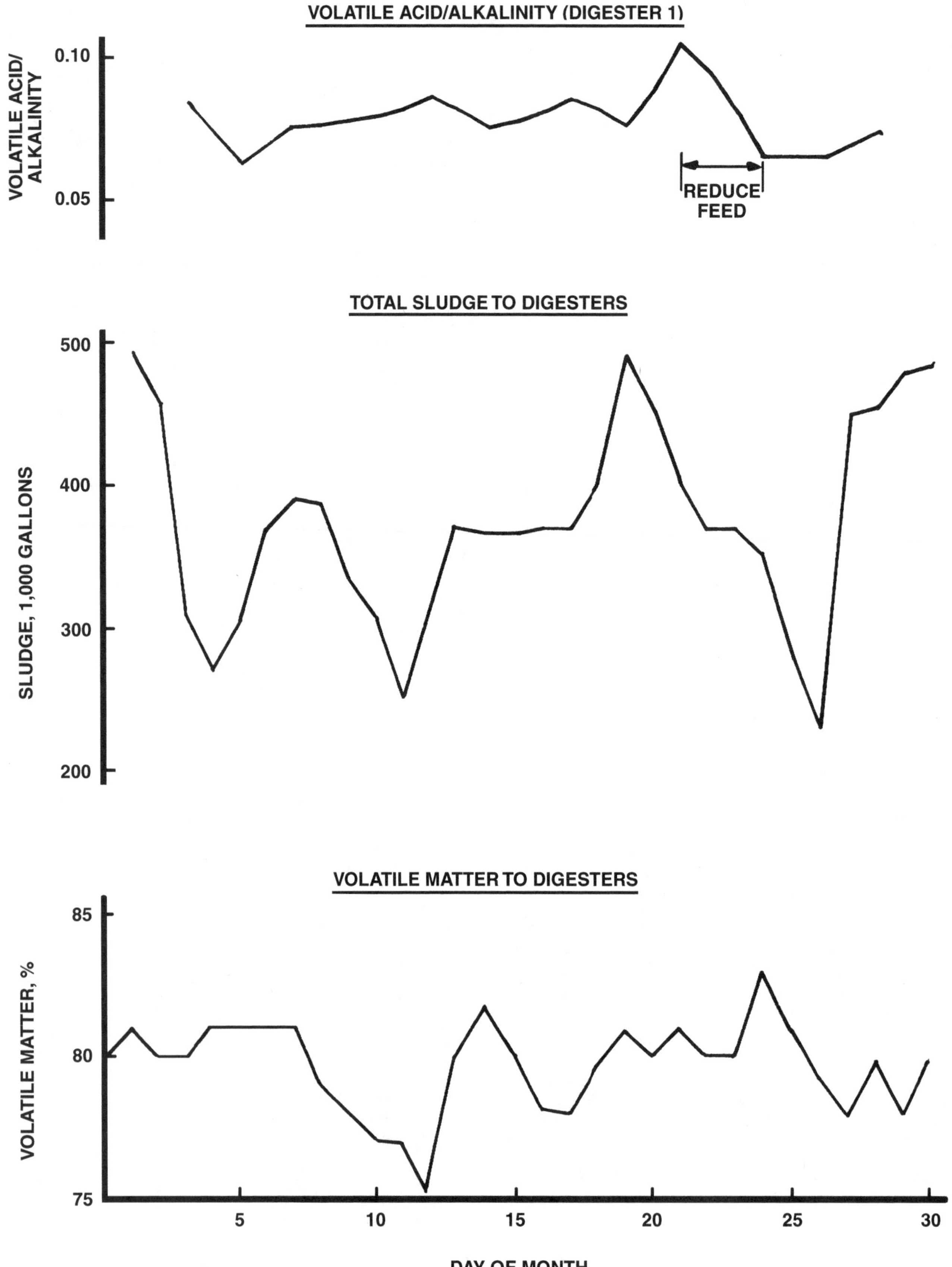

Fig. 12.26 Actual digester operating data

acid test is repeated the next day. Usually the volatile acid value is back down to the normal range the next day. If the volatile acid value is high, the raw sludge pumped to the digester is cut in half or stopped until the volatile acid reading is normal again. Usually this requires only one or two days.

These digesters are not used for liquid-solids separation. Therefore, no information is collected on the supernatant.

QUESTIONS

Write your answers in a notebook and then compare your answers with those on page 234.

12.4A List the activities involved in the normal operation of a digester.

12.4B How often should gas line condensate traps and sedimentation traps be drained?

12.4C How can you tell if the quality of the digester supernatant is decreasing?

12.4D What are the possible causes of a floating cover tilting with thick scum showing around the edges?

12.5 DIGESTER CLEANING

12.50 Need for Digester Cleaning

Under the ideal conditions of perfect mixing and no grease and grit accumulation, the total design volume of a digester will always be available for operation. However, operators are well aware that ideal conditions seldom exist in the field, and that grease and grit do accumulate in digesters even with the best mixing devices. As the grease and grit volume increases, the digester volume available for good digestion (active volume) decreases. This action is graphically shown in Figure 12.25 for a typical digester. As shown by that curve, at the end of eight years of service, only 67 percent or $^{2}/_{3}$ of the design digester volume is available for anaerobic digestion. Thus if the design digester detention time was calculated to be 30 days, at the end of eight years of service it would be 20 days ($^{2}/_{3} \times 30$ days = 20 days). A continued decrease in digester volume will result in further decreases in volatile solids destruction, lower gas production, and, ultimately, failure of the digestion process if the digester is not cleaned. Figure 12.25 is meant only as an example of the decrease in the active volume of a digestion tank with time. For the particular situation that an operator may face, how rapidly the digester volume decreases and when to clean the tank will depend on many factors. Among these factors are:

1. Rate of grit and grease accumulation (this should be periodically measured as indicated in the checklist on page 202)
2. Type and extent of mixing
3. Shape of the digestion tank
4. Operating condition of the equipment inside the digester
5. Type of sludge treated

The normal cleaning interval of anaerobic digesters is between three and eight years, but this will vary greatly with the factors listed above. Between every six months to one year, the depth of heavy solids and scum should be measured. The important thing for the operator is to record the condition of the digester volume with time and thus have a planned or lead time for the cleaning operation. When approximately one-third of the capacity is filled or when the process starts to fail, clean the digester.

12.51 Pre-Cleaning Decisions

When it has been decided that a digester must be cleaned, there are certain questions that must be answered before actual cleaning begins.

1. While the digester is being cleaned, what will be done with the raw sludge?

   If the plant has only one digester, then the answers to this question are limited: (1) haul the sludge to a nearby treatment plant for processing, (2) temporarily convert an activated sludge aeration tank to an aerobic digester, (3) construct a temporary anaerobic lagoon, or (4) hold the sludge in the primary clarifiers but keep the collector mechanism running. The final two alternatives could present significant odor problems and also seriously affect surface or groundwaters. Do not attempt these methods without the permission of the appropriate regulatory agencies. If more than one digester is available at the plant, then the digester remaining in service can be used to digest all of the raw sludge for a short time (usually not more than 10 days).

2. Where will the cleanings be disposed of?

   The digested sludge obtained from the digester while it is being pumped down can be disposed of in the normal plant manner. However, the digester cleanings, which are largely composed of grit, grease, and other heavy solids and fibrous material, will normally require additional considerations for disposal. This can vary from hauling directly to a landfill, which can be very expensive, to ponding the material for solar drying and subsequent disposal in a landfill or on the land. Remember that the material has a large amount of grit and grease and this may require special equipment for the sludge hauling vehicles to control odor.

3. Will the cleaning be accomplished by contract or with the use of plant personnel?

   Normally, this decision is based on a comparison of the cost of hiring a contractor to clean the digester with the cost of using plant personnel and equipment. However, if you do not have the necessary equipment or experienced operators to do the job, your decision will be based on competitive bids from qualified contractors.

12.52 Cleaning Methods and Necessary Equipment

The methods used in cleaning a digester can vary from simply opening a valve and draining the digester contents to a large drying bed with the aid of a washdown hose or fire hose, to using an extensive system such as shown in Figure 12.27. In this system, the digested sludge is first drained from the digester to the normal sludge processing station. Then, with the aid of several turret nozzle mechanisms mounted over access holes on the top of the digester, the remaining grease, grit, and other accumulated solids are washed from the digester and pumped to the digester cleaning system. The cleaning system consists of an inclined screen for the removal of all large solids such as hair, grease, seeds, and cigarette butts. A centrifugal separator or degritter is used for the removal of all remaining suspended and settleable material.

The above-described system could be temporarily put together with leased equipment, but certainly there are other ways to clean digesters. In summary, these can be broken

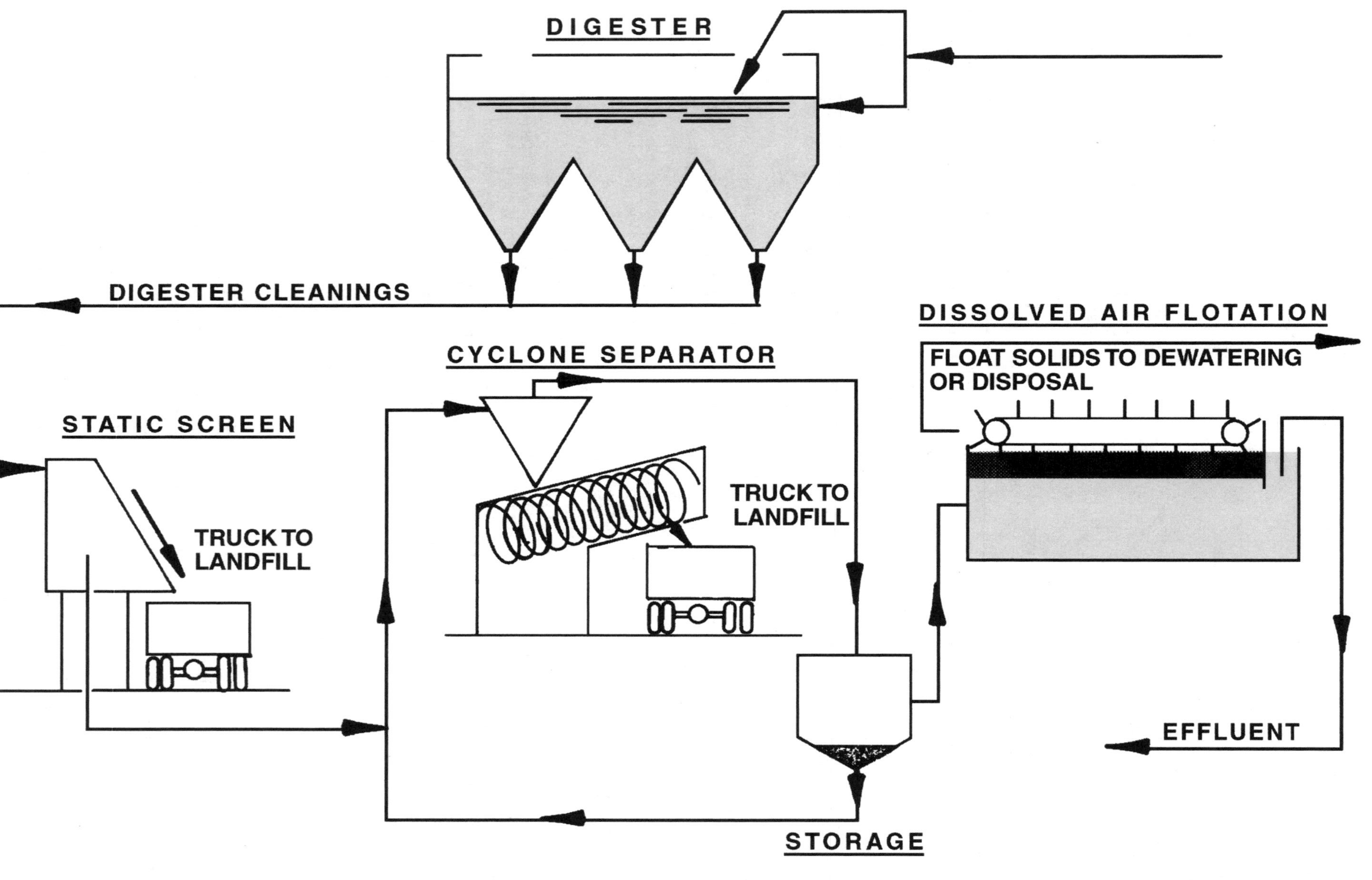

Fig. 12.27 Flow diagram of digester cleanings treatment system

down into three approaches, depending on how the digester cleanings are to be handled:

1. Gravity flow to the disposal area
2. Pump to the disposal area
3. Pump to a tank truck for distant disposal

A general list of the equipment necessary to clean a digester under any of the above methods is summarized in Table 12.4.

TABLE 12.4 DIGESTER CLEANING EQUIPMENT SUMMARY[a]

Equipment Item	Digester	Cleanings	Disposal Option
	1	**2**	**3**
1. Sludge line valves	x[b]	x	x
2. Sludge line (permanent)	x	x	x
3. Sludge line (temporary)	x	x	x
4. Digester access	x	x	x
5. Explosion-proof vent fan	x	x	x
6. Atmospheric monitor	x	x	x
7. Safe ladder	x	x	x
8. Self-contained breathing apparatus	x	x	x
9. Safety harness	x	x	x
10. Explosion-proof lights	x	x	x
11. Water source	x	x	x
12. Washdown hose	x	x	x
13. Nozzle with shutoff	x	x	x
14. Wash water pump	o	x	o[c]
15. Fixed sludge pump		x	o[c]
16. Portable sludge pump		x	o[c]
17. Turret nozzle			o[c]
18. Tripod or hoist			o[c]
19. Tank truck	o	x	o[c]
20. Crane[c]			x

[a] Adapted from *OPERATIONS MANUAL—ANAEROBIC SLUDGE DIGESTION,* US Environmental Protection Agency. EPA No. 430-9-76-001. Obtain from National Technical Information Service (NTIS), 5285 Port Royal Road, Springfield, VA 22161. Order No. PB-250129. Price, $59.50, plus $5.00 shipping and handling per order.

[b] "x" indicates definitely needed equipment.
"o" indicates possibly needed equipment.

[c] This equipment will be definitely needed if there is an abnormally heavy accumulation of solidified solids, such as grease and grit.

12.53 Safety

The efficient cleaning of a digester demands that the operator follow appropriate safety rules. Some of the more important safety precautions are listed below.

1. When the digester is taken out of service, make sure the gas collection system and sludge system have been effectively isolated and provide adequate ventilation through the access holes with the use of explosion-proof vent fans.
2. Follow the procedures outlined in Chapter 14, Section 14.12, "Confined Spaces," prior to entry and during occupancy.
3. Test for explosive conditions with an approved atmospheric monitor.
4. Before entering a digester, always test the atmosphere for oxygen content, flammable/explosive gases, and toxic gases (hydrogen sulfide). Record all atmosphere test results.
5. Provide adequate ventilation at all times.
6. Always use explosion-proof motors and electrical equipment when working near openings in the digester.
7. When cleaning a digester, rope off or barricade the area to prevent unauthorized personnel from entering the area or the digester.
8. Always use a bucket and rope to lower tools and equipment.

QUESTIONS

Write your answers in a notebook and then compare your answers with those on page 234.

12.5A What kinds of materials accumulate in digesters and reduce the active volume for digesting sludge?

12.5B What happens in a digester when the active volume is reduced?

12.5C What could be done with the raw sludge while the digester is being cleaned if a plant has only one digester?

12.5D What kinds of tests should be performed on the digester atmosphere before entry?

END OF LESSON 5 OF 6 LESSONS
on
SLUDGE DIGESTION AND SOLIDS HANDLING

Please answer the discussion and review questions next.

DISCUSSION AND REVIEW QUESTIONS

Chapter 12. SLUDGE DIGESTION AND SOLIDS HANDLING

(Lesson 5 of 6 Lessons)

Write the answers to these questions in your notebook. The question numbering continues from Lesson 4.

24. How can you anticipate problems with anaerobic digesters and correct them before a problem becomes serious?
25. When observing raw sludge being pumped, what items would you observe?
26. How would you feed sludge to a digester?
27. How would you determine if an anaerobic digester was not operating properly?
28. How would you determine when a digester needs cleaning?

CHAPTER 12. SLUDGE DIGESTION AND SOLIDS HANDLING

(Lesson 6 of 6 Lessons)

12.6 AEROBIC SLUDGE DIGESTION

12.60 Comparisons Between Anaerobic and Aerobic Digestion (also see Chapter 3, "Residual Solids Management," in the *ADVANCED WASTE TREATMENT* manual)

Aerobic digestion of solids occurs, whether intentional or not, in any of the conventional secondary treatment processes. In the extended aeration process, the aerobic digestion process is continued almost to the maximum obtainable limit of volatile matter reduction. A separate aerobic digester is intended mainly to ensure that residual solids from aerobic biological treatment processes are digested to the extent that they will not cause objectionable odors during disposal. An aerobic digester is commonly used to avoid the problems encountered when a waste aerobic activated sludge with low solids content is placed in an anaerobic digester. An aerobic digester may be used to treat mixtures of waste activated sludge or trickling filter sludge and primary sludge, or only waste activated sludge, or waste activated sludge from plants with no primary sedimentation tanks. The aerobic digester is a separate operation following other processes. Its purpose is to extend decomposition of solids and regrowth of organisms to a point where available energy in active cells and storage of waste materials are low enough and the material is stable enough for ultimate disposal. Neither aerobic nor anaerobic sludge digestion completes the oxidation of volatile materials in the digester.

Important comparisons between aerobic and anaerobic sludge digestion are summarized in the following sections.

ANAEROBIC SLUDGE DIGESTION

1. Does not use aeration as part of the process.
2. Works best on fresh wastes that have not been treated by prior stabilization processes.
3. Uses *PUTREFACTION*[24] as a basic part of the process.
4. Tends to concentrate sludge and improves drainability.
5. Produces methane gas that provides energy for other operations.
6. Generates major digestion products consisting of solids, carbon dioxide, water, methane, and ammonia.
7. Produces liquids that may be difficult to treat when returned to the plant.
8. Generates sludges that need additional stabilization before ultimate disposal.

AEROBIC SLUDGE DIGESTION

1. Has lower equipment costs, but operating costs are higher, mainly because of energy requirements.
2. Tends to produce fewer noxious odors.
3. Produces liquids that usually are easier to treat when returned to the plant.
4. Generates major digestion products consisting of residual solids, carbon dioxide, water, sulfate, and nitrate compounds. Most of these products are close to the final stabilization stage.
5. May achieve nitrogen removal by stopping aeration long enough to allow the conversion of nitrate to nitrogen gas. Aeration must be restarted before sulfate compounds are converted to sulfide (H_2S).
6. Tends to work better on partially stabilized solids from secondary processes that are difficult to treat by the anaerobic digestion process.
7. Produces a sludge that has a higher water content. Aerobic sludges are difficult to concentrate higher than 4 percent solids.
8. Uses oxygenation and mixing provided by aeration process equipment.
9. Has fewer hazardous cleaning and repairing tasks.
10. Works by aerobic decay, which produces fewer odors when operated properly.

12.61 Process Description

Aerobic digestion tanks may be either round or rectangular, 18 to 20 feet (5.4 to 6 m) deep, with or without covers, depending on geographical location and climatic conditions. The tanks use aeration equipment (mechanical or diffused air) to

[24] *Putrefaction* (PYOO-truh-FACK-shun). Biological decomposition of organic matter, with the production of foul-smelling and -tasting products, associated with anaerobic (no oxygen present) conditions.

maintain aerobic conditions. Each tank has a sludge feed line above the high water level of the tank, a sludge draw-off line at the bottom of the tank, and a flexible, multilevel supernatant draw-off line to remove liquor from the upper half of the tank.

Covers are used in colder climates to help maintain the temperature of the waste being treated. Covers should not be used if they reduce evaporative cooling too much and the liquid contents become too warm. When the liquid becomes too warm, offensive odors may develop and the process effluent will have a very poor quality.

Detention time depends on the origin of the sludge being treated. Twenty days will provide sufficient digestion time for sludges from an extended aeration process where the sludges are already well digested. Sludges from a contact stabilization process require more than 20 days. When temperatures are very low, the sludge may have to be held until the weather warms in the spring. Maximum volatile acids reduction ranges between 45 and 70 percent, depending on temperature.

12.62 Operation

Aerobic digesters are operated under the principle of extended aeration from the activated sludge process, relying on the mode or region called *ENDOGENOUS RESPIRATION.*[25] Aerobic digestion consists of continuously aerating the sludge without the addition of new food, other than the sludge itself, so the sludge is always in the endogenous region. Aeration continues until the volatile suspended solids are reduced to a level where the sludge is reasonably stable, does not create a nuisance or odors, and will readily dewater.

To place aerobic digesters (assume this plant has three aerobic digesters) in series into service, fill the first digester with primary effluent to within three feet (1 m) of the normal water level and start the aeration equipment. Pump to the aerobic digestion process whenever sludge is pumped. Waste aerobic sludge from the secondary clarifier will provide the seed to start the process. Maintain a dissolved oxygen level near 1.0 mg/*L* in the aerobic digester.

Pump raw primary and secondary sludges to the aerobic digester in the same manner sludge is pumped to an anaerobic digester, except sludge concentrations in the range of 1.5 to 4 percent are commonly pumped to the aerobic digester.

When the aerobic digester has filled to normal water level, turn off the aeration equipment and allow the solids to settle to the bottom of the tank. This will leave a supernatant above the solids. Do not leave the aeration equipment off too long because odors will start to develop.

After the solids have settled, adjust the flexible, multilevel supernatant line to draw off a foot or two (0.3 to 0.6 m) of water from the upper portion of the tank. Sufficient water is removed from the digester to accommodate another 24-hour flow of sludges from the primary and secondary clarifiers. Restart the aeration equipment when sufficient water has been removed.

Water withdrawn from the aerobic digester may be discharged to a pond or returned to the primary clarifier. If the water is returned to the primary clarifier, the clarifier should be capable of handling the extra flow. Primary clarifier effluents frequently have undesirably high solids levels.

On the next day, repeat the process of stopping aeration, allowing settling, and removing a portion of the supernatant liquor to make room for another day's pumping of sludge. After a week or two, the solids level will build up to occupy approximately 50 percent of the tank volume during the settling period with a suspended solids concentration of 10,000 to 15,000 mg/*L*.

Place the second aerobic digestion tank in service at this time. Fill the second aerobic digester with primary or secondary effluent to within three feet (1 m) of the normal water level. Transfer a foot (0.3 m) or so of sludge from the bottom of the first digester to the second one, leaving sufficient room in the first to accept another 24-hour period of sludge pumping. Start the aeration equipment in both digesters.

On the next day, supernatant should be removed only from the second digester to the primary clarifier. Transfer enough supernatant from the first aerobic digester to allow enough room for one day's sludge pumping. When the second aerobic digestion tank attains the desirable solids level, place the third aerobic digester into service.

After all three tanks are in operation, the aeration equipment is seldom stopped in the first tank. Remove supernatant from the second and third tanks only. Withdraw solids from the third tank for disposal to drying beds or mechanical dewatering, as required. The water levels in the tanks should be kept equal when the tanks are operated in a series.

New sludge is introduced into the first tank. All of the tanks receive organisms and their stored materials as food. When starting with new cell mass containing negligible silt, up to about 40 percent of the volatile material can be digested. By the time the sludge reaches the third tank, most of the food has been used by the organisms, but they still require energy. Under these conditions they use their own cell material to the extent that only their empty shells remain.

The greatest oxygen demand is exerted in the first tank, and the demand decreases as the sludge is moved to the second and third tanks. Usually, sufficient oxygen is being supplied in the third tank if the sludge is kept mixed and not allowed to settle to the bottom of the digester. Dissolved oxygen levels in aerobic digesters should be maintained between 1.0 and 2.0 mg/*L* throughout the tank.

12.63 Operational Records

Successful operation requires the operator to measure and record the following information:

DAILY

1. Volume of raw and secondary sludges transferred to the aerobic digesters

---

[25] *Endogenous* (en-DODGE-en-us) *Respiration.* A situation in which living organisms oxidize some of their own cellular mass instead of new organic matter they adsorb or absorb from their environment.

2. Pounds of solids transferred and volatile content
3. Volume of supernatant liquor withdrawn from last digestion tank

WEEKLY

1. Supernatant solids and volatile solids content in digesters

WHEN SLUDGE IS WITHDRAWN

1. Volume of sludge withdrawn for dewatering
2. Pounds of solids dewatered and volatile content
3. Pounds of volatile solids destroyed during digestion

12.64 Operational Problems

12.640 Scum

The aerobic digesters will have to be skimmed periodically to remove floating grease and other material that will not digest. This material should be disposed of by incineration or burial with the scum collected from the primary clarifier.

12.641 Odors

Odors should not be a problem in aerobic digestion unless insufficient oxygen is supplied or a shock load reaches the aerobic digestion tanks. If an odor problem does occur, a very effective cure is to recycle sludge from the bottom of the second or third tank back to the first tank. This is also good practice in activated sludge plants that have bulking problems because sludge from the last aerobic digester responds very quickly when returned to an aerator.

12.642 Floating Sludge

Floating sludge may become quite thick in the second and third tanks when aeration is stopped during removal of the supernatant. To avoid clogging, the supernatant draw-off line should be installed so the withdrawal point is from two to six feet (0.5 to 1.8 m) below the water surface. The floating sludge is a problem only during supernatant removal. Scum and solids must be removed from the supernatant to prevent interference with other treatment processes and degradation of the plant effluent.

12.65 Maintenance Problems

Usually, this process requires very little maintenance. Routinely hose the side walls of open tanks for appearance and fly control.

12.650 Diffuser Maintenance

If diffused air is used for aeration, only open orifice or nozzle-type diffusers should be installed because the daily stopping of air flow during supernatant removal tends to increase plugging of other types of diffusers.

12.651 Aeration Equipment

Aeration equipment should be operated continuously except when settling is needed for supernatant removal. Both settling and supernatant removal should be accomplished in 0.5 to 1.5 hours.

QUESTIONS

Write your answers in a notebook and then compare your answers with those on page 235.

12.6A Why do some plants have aerobic digesters?

12.6B What dissolved oxygen levels should be maintained in aerobic digesters?

12.6C What operational problems could occur in aerobic digesters?

12.7 DIGESTED SLUDGE HANDLING

After sludge has passed through an aerobic or anaerobic digestion system, it must be dewatered and disposed of. Small treatment plants are usually provided with sludge drying beds, while larger plants use mechanical dewatering and drying systems. These processes will be briefly discussed in this section. For a more detailed and complete discussion, refer to Chapter 3, "Residual Solids Management," *ADVANCED WASTE TREATMENT.*

12.70 Sludge Drying Beds (Figure 12.28)

The drying process is accomplished through evaporation and percolation of the water from the sludge after it is spread on a drying bed. The drying bed is constructed with an underdrain system covered with coarse crushed rock. Over the rock is a layer of gravel, and then a layer of pea gravel covered with six to eight inches (15 to 20 cm) of sand.

Before sludge is applied, loosen the compacted sand layer by using a sludge fork with tines 8 to 12 inches (20 to 30 cm) long. Insert the tines of the fork into the sand bed and rock the fork back and forth several times. This is to loosen the sand only, and care should be taken that the gravel and sand layers are not mixed. After the whole surface of the bed is loosened, rake it with a garden rake to break up the sand clods. Then level the bed by raking or dragging a 4" × 6" or 2" × 12" (10 cm × 15 cm or 5 cm × 30 cm) board on ropes to smooth the surface.

Sludge is then drawn to the bed from the bottom of the secondary anaerobic digester, or the aerobic digester. Draw the sludge slowly so as not to create a negative pressure in the anaerobic digester and to prevent coning of sludge in the bottom of the digester. A thick sludge of 8 percent solids travels slowly, and if the draw-off rate is too fast, the sludge around the pipe flows out and the thicker sludge on the bottom moves too slowly to fill the void. Consequently, the thinner sludge above the draw-off pipe moves in; and when it does, the supernatant level is reached, thus allowing almost nothing but water to go to the drying bed. The thin sludge and supernatant flowing down to the draw-off pipe washes a hole (shaped like a cone) in the bottom sludge. When this occurs, it sometimes may be remedied by "bumping." This is accomplished by quickly closing and opening the draw-off valve on a gravity flow system, which creates a minor shock wave and sometimes washes the heavier sludge into the cone. If the digested sludge is pumped to the drying bed, quickly start and

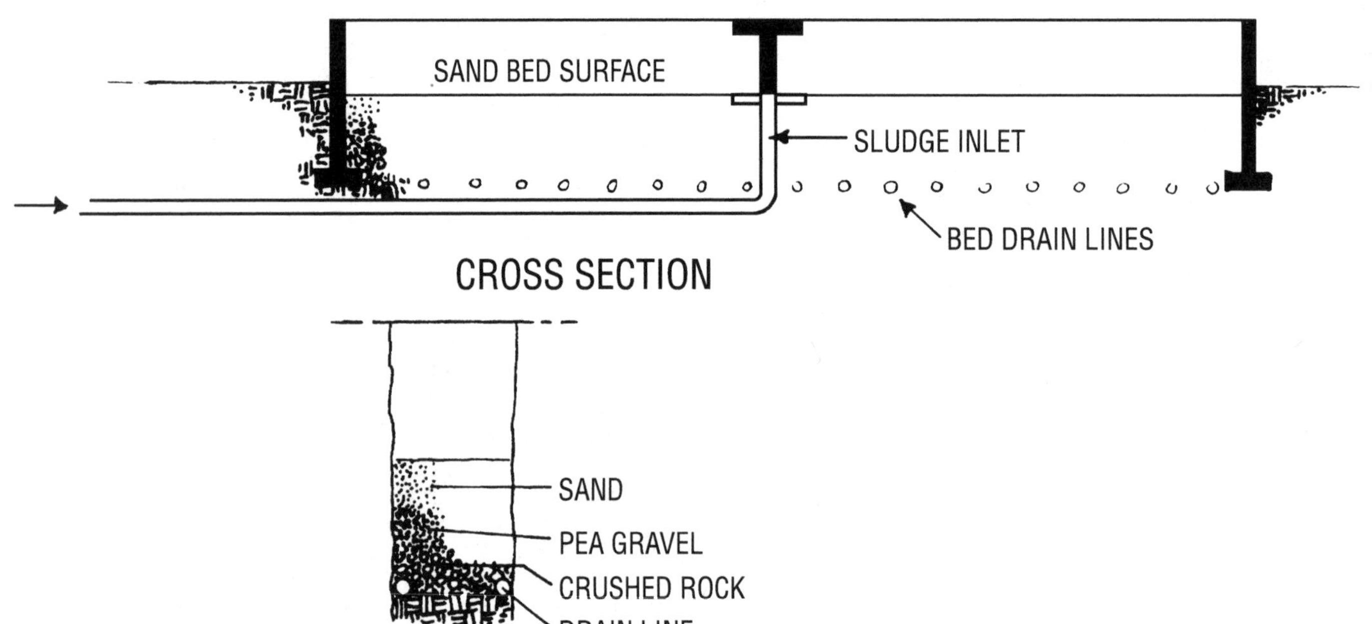

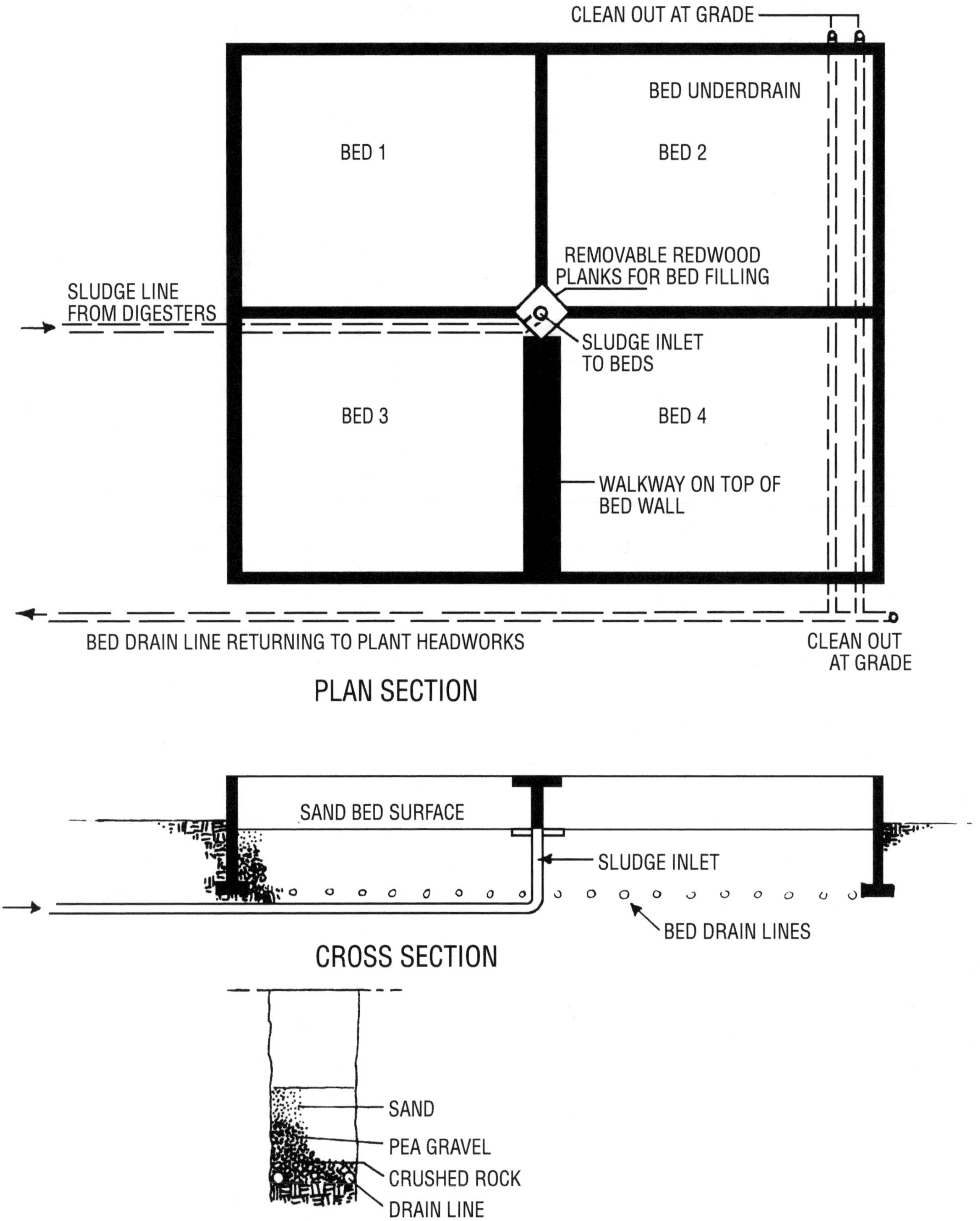

Fig. 12.28 Sludge drying bed

stop the pump using the power switch to create the bumping action.

Withdrawing sludge slowly from a digester is time consuming and requires frequent checks to be sure it does not thicken and stop flowing completely or cone and run too fast.

The sludge being drawn to the bed is sampled at the beginning of the fill, when the bed is half full, and just before the bed is filled to the desired level. The samples may be mixed together or analyzed separately for total and volatile solids.

The depth to which the sludge is applied is normally around 12 inches (30 cm), but sometimes it is as deep as 18 inches (45 cm) in arid regions. If it is deeper, the time required for drying is too long. A bed filled with 20 inches (50 cm) of sludge would require approximately the same time to dry as a bed loaded with 14 inches (35 cm), dried and removed and filled with another load 14 inches (35 cm) deep. Drying anaerobically digested sludge requires special precautions.

WARNING

No smoking or open flames should be allowed in the vicinity where sludge is being drawn for drying. The sludge still contains some methane gas. This is shown on a fresh bed of sludge by the number of small holes and bubbles on the surface of the sludge. There have been cases of explosions and fires caused by an operator throwing a lighted match or cigarette onto a drying bed of sludge from a digester.

After a bed of sludge is drawn, the sludge draw-off line should be flushed and cleared with water so the solids will not cement in the line. One end of the line should be left open so any gas produced will not rupture the line. The line should be drained if freezing is a problem.

In warm weather, a good sand bed will have the sludge dry enough for removal within four weeks. The water separates from the sludge and drains down through the sand. Evaporation also dries the sludge and will cause it to crack.

When the sludge has formed cracks clear to the sand, it may then be removed by hand with forks. The one major drawback of sand beds is that heavy equipment, such as a skip loader, cannot be used because the weight could damage the underdrain system. Also, the scraping action could mix the sand with the gravel or remove some of the sand with the dried sludge. The sand would then have to be replaced.

Some operators lay 2" × 12" (5 cm × 30 cm) boards across the sand for wheelbarrows or light trucks and fork the sludge cake into them to haul to a disposal site. The dried sludge cake is normally three to six inches (8 to 15 cm) thick and is not heavy unless a large amount of grit was present in the sludge. The operator calculates the amount of cake in cubic feet by the depth of the dry sludge cake and surface area of the bed. The total dry pounds is arrived at from the total solids in the sludge samples when the sludge was drawn.

Dried sludge makes an excellent soil conditioner and a low-grade fertilizer. However, in many states air-dried digested sludge may only be used on lawns, shrub beds, and orchards. It cannot be used on root crop vegetables unless heat dried (at 1,450°F or 790°C), or unless it has been in the ground that the crop is to be planted in for over one year. Always check with the state or local health department before dried sludge or "biosolids" is used on a food crop.

If a bed of "green" (partially digested) sludge is accidentally drawn, it will require special attention. The water will not drain rapidly, odors will be produced, and the water held will provide an excellent breeding ground for nuisance insects. Flies, rat-tail maggots, psychoda flies, and mosquitoes will breed profusely in this environment. An application of dry lime, spread over the bed by shovel, and a spraying of a pesticide or insecticide are beneficial. The use of pesticides and insecticides may require special training and licensing for the operator. Check with your local health department or safety agency. Use appropriate safety gear (eye/face protection, respiratory protection, and rubber gloves) when handling lime, pesticides, or insecticides. The sludge from such a bed should never be used for fertilizer.

Dry sludge cake will burn at a slow smoldering pace, producing quite an offensive odor; therefore, do not allow it to catch fire.

WARNING

If a sludge powder or dust is present and kicked into the air, it can explode, similar to a dust explosion in a flour mill. Once sludge begins burning, it is extremely difficult to extinguish. Water sprays are best used in extinguishing a sludge fire.

Write your answers in a notebook and then compare your answers with those on page 235.

12.70A How would you prepare a drying bed prior to applying sludge?

12.70B Why should sludge be drawn slowly from the digester?

12.70C What would you do if thin sludge suddenly started flowing onto the drying bed on a gravity flow system, indicating that a sludge cone had formed in the bottom of the digester?

12.70D Why should no smoking or open flames be allowed in the vicinity where the anaerobically digested sludge is being drawn?

12.70E What should you do to the sludge draw-off line after sludge is applied to the drying bed?

12.70F Why should heavy equipment such as skip loaders not be used to remove dried sludge from a sand drying bed?

12.70G What is the volume of dry sludge in a bed 100 feet long and 25 feet wide if the dried sludge is six inches thick? How many two-cubic-yard dump truckloads would be required to haul away this sludge?

12.70H If green (partially digested) sludge accidentally was applied to a drying bed, how would you handle this situation?

12.71 Blacktop Drying Beds (Figure 12.29)

This type of bed has become prevalent and has worked well with aerobically digested sludges, if designed properly. The bed is made of blacktop or asphalt with both sides sloping gradually to the center to a one-foot (0.3-m) wide drain channel. The drain channel runs the full length of the bed with a three- or four-inch (75- or 100-mm) drain line on the bottom. The drain line is covered with rock, gravel, and sand as in a sand bed. The drain line usually has a cleanout at the upper end, and a control valve on the discharge end.

When the bed is to be used, the cleanout on the drain line is removed, the line is flushed with clear water, and the cleanout cover replaced. The drain line valve should be closed and the drain line and drain channel should be filled with water to the top of the sand so that the sand will not become sealed with sludge. Sludge is then applied to the bed until it is 18 to 24 inches (45 to 60 cm) deep. Some plants have operated successfully without pre-filling the collection system with water.

To reduce sludge drying time, sample the sludge in the same way as when using a sand bed, except take one additional sample in a glass jar or beaker and set aside. By watching the jar of sludge, you can observe at some time during the first 24 to 36 hours that the sludge will rise to the top, leaving liquor on the bottom. (This is primarily caused by the gas in the sludge. Later, the sludge will again settle to the bottom and the liquor will be on the surface.) When the sludge rises to the surface of the sample, open the drain line and slowly bleed off the lower liquor. Return the drained liquor to the primary clarifiers.

After the sludge has started to crack and has a crust, drying time may be reduced by driving a vehicle through the bed to mix the sludge and expose the wet sludge. When the cake is dry, a skip loader is used to clean the bed.

Blacktop beds may be able to handle two to three times as much sludge as sand beds in a given period of time because mechanical equipment can be used to mix the drying sludge, expose the sludge to the atmosphere, and clear the bed of dried sludge.

12.72 Sludge Lagoons

Sludge lagoons are deep ponds that normally hold anaerobically digested sludge and, in some instances, supernatant. Digested sludge is drawn to the lagoon periodically and may require a year or two to fill. When the lagoon is full, sludge is discharged into another lagoon while the first one dries. This drying period can require a year or two before the sludge is removed. Some large cities have used lagoons for many years, avoiding the use of covered secondary anaerobic digestion tanks. Sludge depths usually range from 2.5 to 4 feet (0.7 to 1.2 m). Typically sludge is pumped to the lagoon for 18 months and then the lagoon is rested for 6 months.

QUESTIONS

Write your answers in a notebook and then compare your answers with those on page 235.

12.71A How would you attempt to reduce sludge drying time in blacktop beds?

12.72A How does a sludge lagoon operate?

12.73 Withdrawal to Land

Wet sludge (either aerobically or anaerobically digested) can be spread on land to reclaim the land or on farm land and ploughed in as a soil conditioner and fertilizer. Used with lagoons, this gives a flexible system. This is an excellent method of sludge disposal, wherever applicable, because it returns the nutrients to the land and completes the cycle as intended by nature.

Transporting sludge to the disposal site is accomplished by tank truck or pipeline. The application of wet sludge to the land depends on the topography and the crop to be raised on that land. When applied to grass or low ground cover crops, application may be by spraying from the back of the tank truck while driving over the land, by the use of irrigation piping, or by shallow flooding.

The best method, but most costly, is leveling the land, constructing ridges and furrows, and then pumping the sludge down the furrows similar to irrigation practices used in arid regions. This method is not only capable of reclaiming land unsuitable for growing plants and trees, but may yield crops equal to or greater than those raised with commercial fertilizers.

Some precautions that must be practiced with this method of sludge disposal include:

1. Never apply partially digested (green) sludge or scum.
2. Residential areas must not be located near land disposal sites.
3. Land disposal sites must not be located on a flood plain where the sludge may be washed into the receiving waters during flooding.
4. Domestic water wells must not be located on the land receiving the sludge.
5. Root crop vegetables must not be grown on the land.
6. Cooperation with the landowner as to application time, drying, and covering must be guaranteed.
7. Access to the land during wet weather must be provided.

12.74 Mechanical Dewatering

In plants where large volumes of sludge are handled and drying beds are not feasible, mechanical dewatering may be used. Mechanical dewatering methods available are vacuum filtration, pressure filtration, centrifugation, and filter presses. Two methods, vacuum filters and centrifuges, will be discussed in this section. Each is capable of reducing the moisture content of sludge by 60 to 80 percent leaving a wet, pasty cake containing 20 to 40 percent solids. This cake may then be disposed of in a sanitary landfill, dried in furnaces for fertilizer, incinerated to ash in furnaces or wet oxidation units, or aerobically composted and used as a fertilizer.

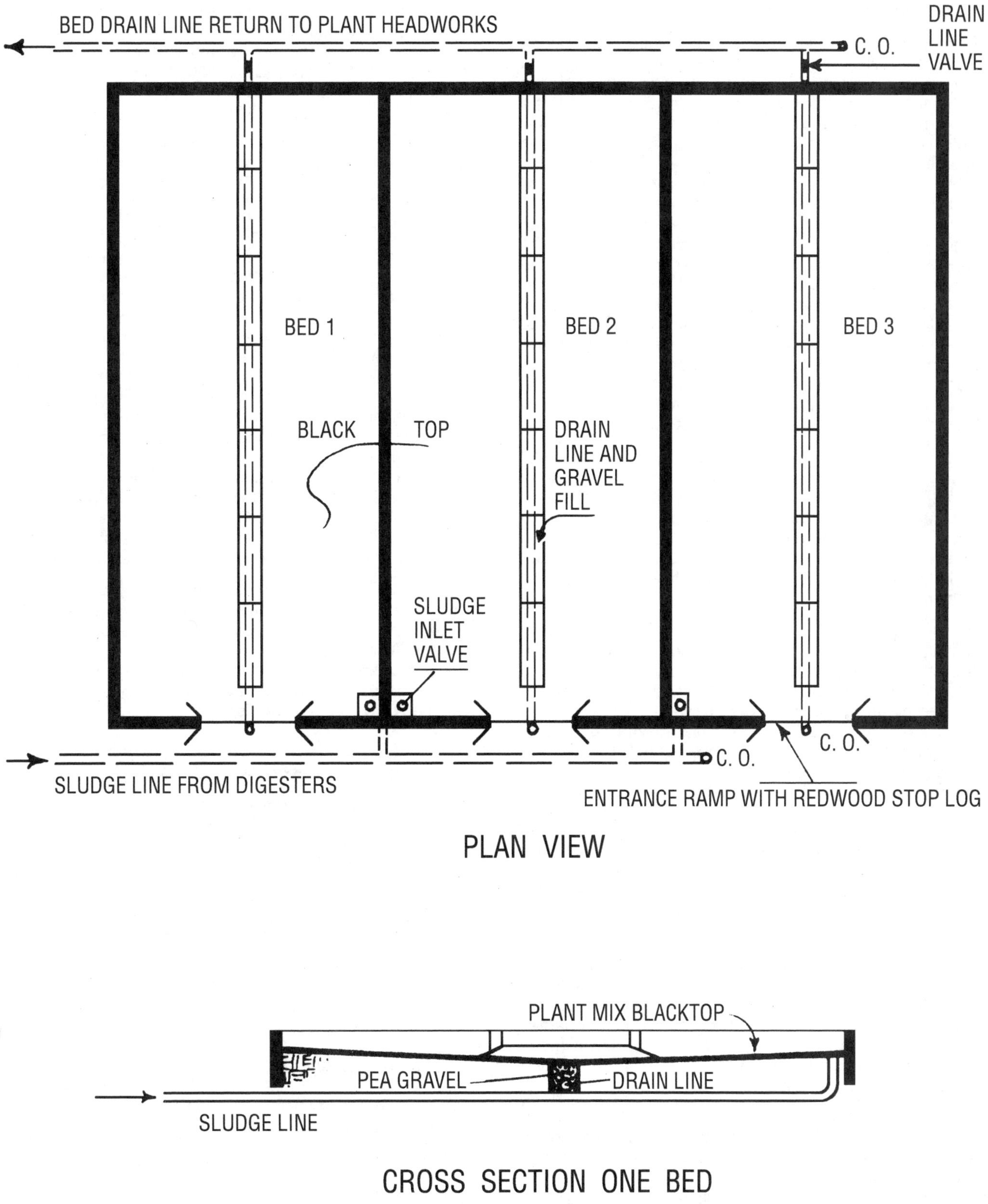

BED DRAIN LINE RETURN TO PLANT HEADWORKS
C. O.
DRAIN LINE VALVE
BED 1
BED 2
BED 3
BLACK TOP
DRAIN LINE AND GRAVEL FILL
SLUDGE INLET VALVE
C. O.
C. O.
SLUDGE LINE FROM DIGESTERS
ENTRANCE RAMP WITH REDWOOD STOP LOG
PLAN VIEW
PLANT MIX BLACKTOP
PEA GRAVEL
DRAIN LINE
SLUDGE LINE
CROSS SECTION ONE BED

Fig. 12.29 Blacktop drying bed

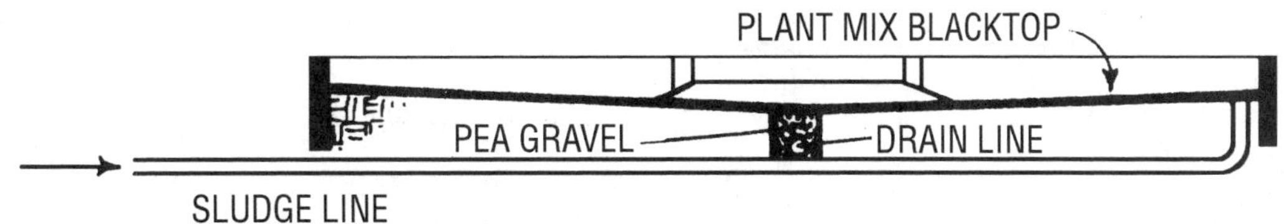

12.740 Vacuum Filters (Figures 12.30 and 12.31)

Dewatering aerobically or anaerobically digested sludge with a vacuum filter usually requires washing the sludge first and then conditioning it with chemicals. Elutriation (e-LOO-tree-A-shun) is the washing of digested sludge in water at a suitable ratio of sludge to water. Plant effluent is commonly used as the water source. Elutriation may be accomplished in from one to three separate tanks, similar to small rectangular clarifiers. The sludge is pumped to the elutriation tank and mixed with water (plant effluent). Next this mixture is admitted to the other tanks to establish a countercurrent wash. The sludge is then allowed to settle and is collected by flights and pumped to the next elutriation tank. After one to three washings it is then pumped to the conditioning tanks. The main purpose of elutriation is to remove (wash out) fine particulates or the alkalinity in sludge. It also removes amino acids and salts, which may have a small coagulant demand. After elutriation, the sludge will react better with the chemicals and produce better cake. The elutriate (effluent from elutriation tanks) is returned to the primary clarifiers and may result in a very heavy recirculating load since it is chiefly fine solids.

Many treatment plants have discontinued the practice of elutriation. Although the process may save approximately $2 per ton ($2.20 per 1,000 kg) of dry solids handled on chemical costs, the costs are excessive for treating the elutriate (wash water) in the biological treatment processes. Other problems from elutriation include odors and the adverse impact of the recycled suspended solids on the other treatment processes and the plant effluent.

Sludge conditioning is accomplished by the addition of various coagulants or flocculating agents such as ferric chloride, alum, lime, and polymers. The amount of chemical solution added to the conditioning tank is normally established by laboratory testing of sludge grab samples. Various chemical concentrations are added to the grab samples to obtain a practical filtration rate by vacuum with a Buchner funnel. This test establishes the operating rate for the chemical feed pumps or rotameters from the chemical head tanks. The rate is normally less than 10 percent of the dry sludge solids rate to the conditioning tank. (Both rates could be in pounds (kg) per 24 hours.) In this tank, the chemical is mixed into the sludge by gentle agitation for several minutes. The conditioned sludge then flows to the filter bath where it is continuously and gently agitated. After operation has started, chemical feed is regulated according to cake appearance and behavior.

Filter drums are 10 to 18 feet (3 to 5.5 m) in diameter, and 12 to 20 feet (3.5 to 6 m) in length. They may use cloth blankets of Dacron, nylon, or wool, or use steel coil springs in a double layer, to form the outer drum covering and filter media. The drum inside is a maze of pipes running from a metal screen and wood surface skin, and connecting to a rotating valve port at each end of the drum.

Cloth blankets are stretched and caulked to the surface of the filter drums with short sections of 1/4-inch (6-mm) cotton rope at every screen section. The sides of the blanket are also stretched and stapled to the end of the drums. The *NAP*[26] of the blanket should be out. After the blanket is stretched completely around the drum, it is then wrapped with two strands of 1/8-inch (3-mm) stainless-steel wire, approximately 2 inches (5 cm) apart for the full length of the drum.

The installation of a blanket may require several days' work. A blanket will usually last from 200 to 20,000 hours, but this depends greatly on the blanket material, conditioning chemical, backwash frequency, and acid bath frequency. An improper adjustment of the scraper blade, or accidental tear in the blanket, will usually require its replacement.

Both cloth blankets and coil springs filters require a high-pressure wash after 12 to 24 hours of operation and, in some instances, an acid bath after 1,000 to 5,000 operating hours.

The filter drum is equipped with a variable-speed drive to turn the drum from 1/8 to 1 RPM. Normally, the lower RPM range is used to give the filter time to pick up sufficient sludge as it passes through the conditioned sludge tub under the filter. Normally less than 1/5 of the filter surface is submerged in the tub and pulling sludge to the blanket or springs by vacuum to form the cake mat. As that area passes through the conditioned sludge, the vacuum holds a layer 1/8- to 1/2-inch (0.3- to 1.2-cm) thick of sludge to the media, and continues to pull the water from the sludge to approximately 210 degrees from the bottom point of the filter after it leaves the vat. This is the drying cycle. At this point the vacuum is released and a light air pressure (3.0 psi or 0.2 kg/sq cm) is applied to the inside of the blanket, lifting the sludge so that it falls from the blanket into a hopper or conveyor belt. The drum then rotates past a scraper blade to remove sludge that did not fall. The applied air is then phased out as that section starts into the filter tub and vacuum is applied to pick up another coating of sludge.

The thickness of the sludge cake and moisture content depend on the sludge, chemical feed rate, drum rotation speed, mixing time, and condition of the blanket or coil springs. A filter may blank out (lose sludge cake) for any of the above reasons or due to the loss of vacuum or filtrate pumps. Filtrate is the liquor separated from the sludge by the filter; it is returned to the primary clarifiers.

QUESTIONS

Write your answers in a notebook and then compare your answers with those on page 235.

12.73A What are some of the advantages of applying sludge to land?

12.74A How is sludge treated for disposal in many large plants or areas where drying beds are not feasible?

12.74B How would you prepare digested sludge for drying by vacuum filtration?

12.74C How would you determine the chemical feed rate to condition sludge?

12.74D What factors influence the life of a filter blanket?

***12.741 Centrifuge** (also see Chapter 3, "Residual Solids Management," in ADVANCED WASTE TREATMENT)*

Centrifuges are used for dewatering raw or primary sludges for disposal in furnaces or incineration units. Their use on digested sludge is becoming more widespread. Most digested sludges are conditioned with polymers before being fed to a centrifuge.

[26] *Nap.* The soft, fuzzy surface of the fabric.

Fig. 12.30 Vacuum filter
(Permission of Komline-Sanderson Engineering Corporation)

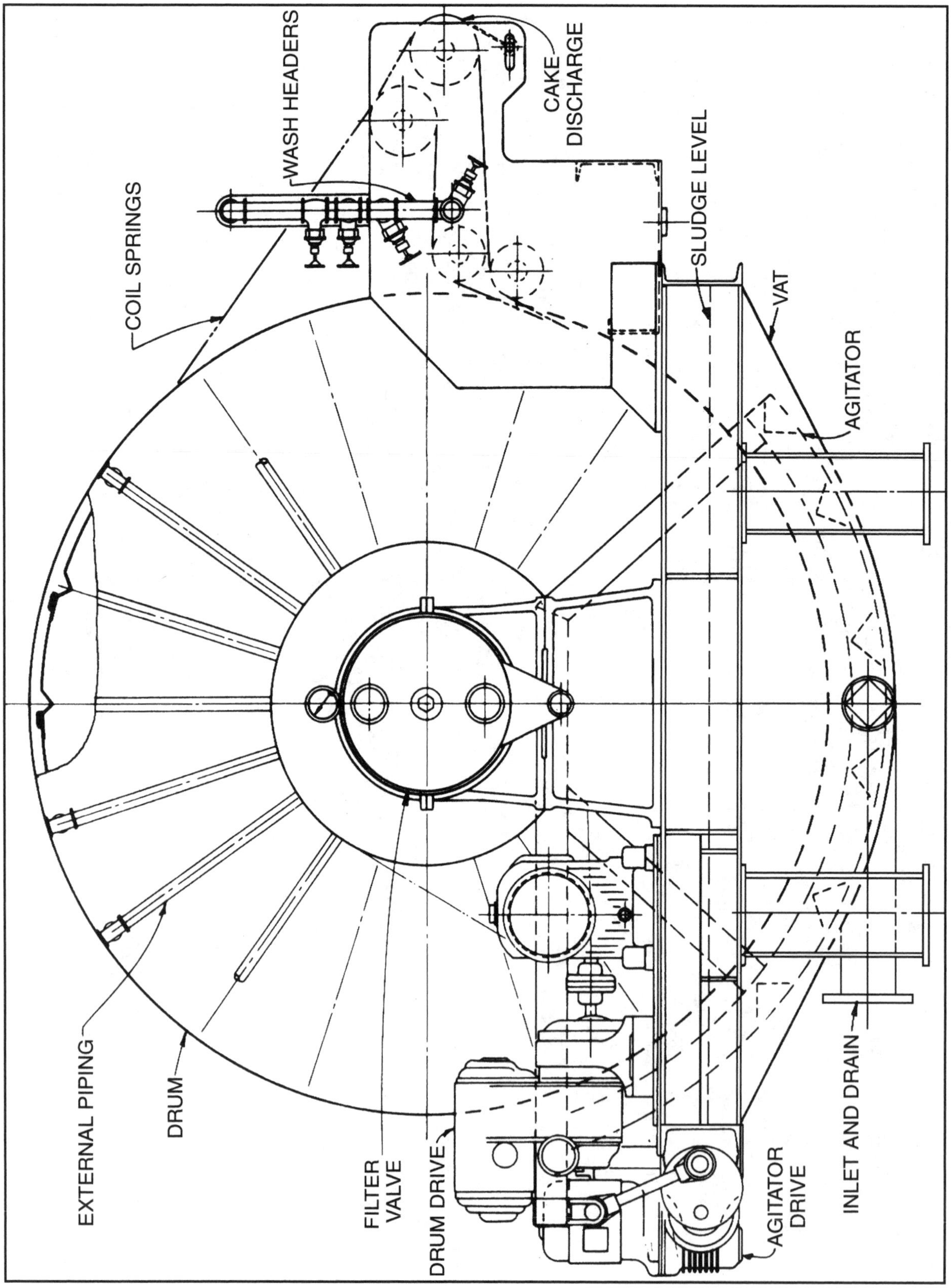

Fig. 12.31 Coil filter elevation
(Permission of Komline-Sanderson Engineering Corporation)

Centrifuges are various sized cylinders that rotate at high speeds. The sludge is pumped to the center of the bowl where centrifugal force established by the rotating unit separates the lighter liquid from the denser solids. The *CENTRATE*[27] is returned to the primary clarifiers, and the sludge cake is removed to a hopper or to a conveyor for disposal.

The condition of the sludge cake or slurry and the quality of the centrate from a centrifuge are regulated by the sludge feed rate, bowl speed, and, if chemical conditioners are used, by dosage rates and pool depth. The centrate usually contains a high amount of suspended solids that become difficult to handle in the primary clarifiers and digesters. A large amount of grit in the sludge greatly increases the wear rate on the centrifuge. Similar to the wash water from the elutriation process, the suspended solids in the centrate from centrifuges also exert a difficult load on biological treatment processes.

QUESTIONS

Write your answers in a notebook and then compare your answers with those on page 235.

12.74E Centrifuges are commonly used to dewater what types of sludges?

12.74F How would you regulate the condition of the sludge cake or slurry and the quality of the centrate from a centrifuge?

12.8 SLUDGE DISPOSAL

The US Environmental Protection Agency (EPA) and state regulatory agencies are continually developing and enforcing stricter rules regarding the ultimate disposal of municipal wastewater sludge. The goal of all regulations governing the ultimate disposal of sludge or reuse of *BIOSOLIDS*[28] is the protection of public health and the environment. Major areas of concern are the emissions to the atmosphere from furnaces, groundwater and surface water contamination, and the potential health consequences of applying biosolids to land used for food production. Keep in contact with your state regulatory agency for the current rules and regulations applicable to the management and disposal of sludge from your treatment plant. Regulations apply to the disposal of sludge produced in the treatment of domestic wastewater using the following ultimate disposal practices:

1. Land application
2. Distribution and marketing (D & M)
3. Monofilling (sludge-only landfills)
4. Surface disposal
5. Incineration

Land application is the disposal of domestic wastewater sludge on agricultural lands or in a sanitary landfill at the treatment plant site or offsite. Regulations describe the requirements for stabilization of sludge to reduce the levels of pathogens and the attraction of vectors (an insect or rodent capable of spreading germs or other agents of disease). Limits are also placed on the concentrations of metals, especially zinc and copper, in the sludge applied to land. Surface runoff from the disposal site must be prevented and percolation of leachate to groundwater must be carefully controlled or eliminated.

Beneficial reuse of biosolids for agricultural purposes is encouraged, and is safe when accomplished in accordance with federal and state guidelines. If biosolids are to be reused for agricultural purposes in areas where groundwater contamination is a concern, monitoring wells should be installed. Where groundwater is not a problem, *LYSIMETERS*[29] placed at several intervals can indicate if nitrate is moving through the soil. Use of biosolids on agricultural land requires close monitoring of nitrogen and heavy metals; cadmium is one metal of concern. Monitoring for toxic substances and pathogens must also be conducted.

Distribution and marketing refers to treated sludge products (biosolids) sold to the general public. This category includes composted sludge and other products produced by high-temperature drying and inactivation processes. Rules are generally more restrictive than the land application regulations.

Monofilling is the practice of burying sludge in a dedicated landfill. At dedicated land disposal (DLD) sites, stabilized sludge is applied to the land and then ploughed under. At sites of this type, sludge must be covered the same day it is applied. Public access must be avoided because pathogens or parasites may not have been removed. Regulations are very restrictive and may result in the ultimate elimination of this method.

Surface disposal includes those applications in which only wastewater sludge is applied to an area of land. The sites do not include a vegetative cover. This is a disposal practice with no intent of beneficial use. The regulations are similar to monofills.

Incineration is the burning (combustion) of sludge at a high temperature in a furnace. Incineration reduces the volume of sludge while producing heat, dry inorganic ash, and gaseous emissions. The rules for incineration deal mainly with gas emissions. Of concern is the fate of metals and certain organics as they pass through the incineration process and are either destroyed, removed with the ash, or pass through the gaseous emission scrubbing system and exit with the exhaust gases. Scrubbing equipment removes particulates and undesirable gases from gaseous emissions by passing dirty air through water sprays.

---

[27] *Centrate.* The water leaving a centrifuge after most of the solids have been removed.

[28] *Biosolids.* A primarily organic solid product produced by wastewater treatment processes that can be beneficially recycled. The word biosolids is replacing the word sludge when referring to treated waste.

[29] *Lysimeter* (lie-SIM-uh-ter). A device containing a mass of soil and designed to permit the measurement of water draining through the soil.

The federal, state, and local regulations concerning ultimate disposal or reuse of sludge-based products from wastewater treatment plants have been developed to protect the environment and public health. Complying with these regulations and securing necessary approvals and permits for sludge management projects has become increasingly challenging, and may be more difficult for municipal wastewater facilities treating industrial wastes because of the potential for the sludge-based products to be classified as a hazardous material. A listing and discussion of all of the applicable regulations relative to sludge management practices is beyond the scope of this operator training manual. However, the planners and operators of wastewater treatment plants should be familiar with the general procedures that should be followed for implementing a sludge management project, and with the self-implementing EPA sewage sludge regulations discussed below. Some general sludge management procedures are:

Step 1 - Analyze the sludge or otherwise predict the concentration of pollutants identified by the appropriate state health department and the Environmental Protection Agency (EPA) that could cause the sludge to be classified as a hazardous waste.

Step 2 - If the concentrations of pollutants exceed established limits, the sludge will be classified as hazardous and can only be disposed of at an approved Class I landfill.

Step 3 - If the concentrations of all pollutants are lower than the established limits, the sludge may be classified as nonhazardous and may fall under applicable regulations that govern the disposal or reuse of municipal treatment plant sludges.

Step 4 - If the sludge is classified as nonhazardous by the federal, state, and local regulatory authorities, then it should be analyzed and compared with the requirements of the EPA 40 CFR 257 or 503 regulations and the appropriate state health department guidelines. These regulatory requirements will establish the acceptable alternatives for disposing of sludges.

Step 5 - Keep in contact with your state regulatory agency for current rules and regulations concerning the management and disposal of sludge from your treatment plant.

The Sewage Sludge Use and Disposal Regulation (40 CFR Part 503) sets national standards for pathogens and 10 heavy metals in sewage sludge.[30] It also defines standards (or management practices) for the safe handling and use of sewage sludge. This rule is designed to protect human health and the environment when sewage sludge is beneficially applied to the land, placed in a surface disposal site, or incinerated.

The scientific research used to develop this rule shows that most sewage sludge can be safely and beneficially used in a wide variety of ways. It can be applied safely to agricultural land, lawns and gardens, golf courses, forests and parks, and is a valuable resource for land reclamation projects. This rule is designed to protect human health and the environment with a margin of safety equal to any of the unregulated use or disposal practices. It sets standards for pathogens and limits for 12 pollutants that have the potential for adverse effects, and explains why limits are not needed for 61 other pollutants that were considered. Additionally, it contains a comprehensive set of management practices to ensure that sewage sludge is beneficially used or disposed of properly. Table 12.5 summarizes the numerical criteria for Part 503 Sewage Sludge Regulations.

To convert the information in Table 12.5 from metric units to English units, use the following procedures:

1. Milligrams per kilogram (mg/kg) is the same as parts per million or pounds per million pounds. Therefore a cadmium limit of 39 mg/kg means 39 milligrams of cadmium per kilogram of dry sludge solids or 39 pounds of cadmium per million pounds of dry sludge solids.

$$\text{Limit, } \frac{\text{lbs Cadmium}}{\text{ton Dry Solids}} = \frac{\text{39 mg Cadmium}}{\text{1 kg Dry Solids}}$$

$$= \frac{\text{39 lbs Cadmium}}{\text{1,000,000 lbs Dry Sludge Solids}}$$

$$= \frac{\text{(39 lbs Cadmium)(2,000 lbs/ton)}}{\text{1,000,000 lbs Dry Sludge Solids}}$$

$$= \text{0.078 lbs Cadmium/ton Dry Solids}$$

2. Kilograms per hectare per year. This value is often converted to tons per acre per year. Therefore a cadmium **A**nnual **P**ollutant **L**oading **R**ate of 1.9 kg/ha/yr would be converted to pounds or tons per acre per year by the following calculation.

$$\text{Cadmium APLR, lbs/acre/yr} = \text{1.9 kg/ha/yr}$$

$$= \text{(1.9 kg/ha/yr)(2.2 lbs/kg)(2.47 ha/ac)}$$

$$= \text{10.3 lbs Cadmium/acre/yr}$$

$$\text{Cadmium APLR, tons/acre/yr} = \frac{\text{10.3 lbs Cadmium/acre/yr}}{\text{2,000 lbs/ton}}$$

$$= \text{0.0052 tons Cadmium/acre/yr}$$

For additional information, see Chapter 3, "Residual Solids Management," in *ADVANCED WASTE TREATMENT* in this series of operator manuals.

12.9 REVIEW OF PLANS AND SPECIFICATIONS

Operators should be given the opportunity to review the contract plans and specifications to become knowledgeable about proposed treatment processes. However, it is just as important that operators participate in the design of the system

[30] The Water Environment Federation (WEF) refers to sludge produced at wastewater treatment plants as "wastewater solids." After these solids receive further treatment and meet US federal criteria for beneficial use, the solids are called "biosolids."

TABLE 12.5 PART 503 SEWAGE SLUDGE REGULATIONS LIMITS SUMMARY

	Land Application				Surface Disposal		Incineration
	Ceiling (mg/kg)	Cumulative Load (kg/ha)	APL[a] (clean sludge) (mg/kg)	APLR[b] (kg/ha/yr)	Criterion Unlined (mg/kg)	Criterion Lined (mg/kg)	
Arsenic	75	41	41	2.0	73	—	c
Beryllium	—	—	—	—	—	—	10 gm/24 hr
Cadmium	85	39	39	1.9	—	—	c
Chromium	3,000	—	—	—	600	—	c
Copper	4,300	1,500	1,500	75	—	—	—
Lead	840	300	300	15	—	—	c
Mercury	57	17	17	0.85	—	—	3,200 gm/24 hr
Molybdenum	75	18	18	0.90	—	—	—
Nickel	420	420	420	21	420	—	c
Selenium	100	100	100	5.0	—	—	—
Zinc	7,500	2,800	2,800	140	—	—	—
Total Hydrocarbons	—	—	—	—	—	—	100 ppm

[a] APL. Monthly **A**verage **P**ollutant **L**oad
[b] APLR. **A**nnual **P**ollutant **L**oading **R**ate
[c] Allowable concentration in sewage sludge is determined site-specifically for each incinerator through performing incinerator testing and emissions dispersion modeling.

whenever possible. Operators can supply information about specific requirements for effective and safe operation and maintenance of treatment processes.

In a sludge digestion system, the following areas are among those the operator should carefully examine when reviewing plans and specifications. The operator should look for those items that are important for proper operation and maintenance and ask questions on unclear or confusing items. You must realize that many items may not be shown in particular detail on the plans, but will be described fully in the specifications.

A. Generally review the specifications with regard to design loading rates (peak and average) and check with any plant operational history. Check:

    1. Equipment details regarding sizes, capacities, flow rates, pressure, horsepower, efficiencies
    2. Performance requirements or capabilities
    3. Paints and protective coatings
    4. Instrumentation—remote and local control board items and recorders provided
    5. Equipment warranties and responsibility for acceptance testing
    6. Adequate supply of equipment operation and maintenance manuals
    7. Adequate number of operator and maintenance personnel training hours

B. In examining the plans, emphasize those areas having a direct influence on plant operation and maintenance. For a digestion system, the following items are recommended for review.

    1. Examine the general site layout and make sure of adequate access for maintenance equipment and personnel. If overhead electrical power and telephone lines are involved, note location and make sure of proper clearance for any boom crane.
    2. Note location and sufficiency of power outlets and wash water faucets.
    3. Carefully trace the piping flow scheme and examine all valve placements and ability to route flow. Examine operation under automatic and manual modes.
    4. Check for adequate access to all valves and piping for maintenance and repairs. Give particular attention to pipeline cleaning.
    5. If pump or other heavy equipment is located in underground galleries, check for provisions for removal.
    6. Check for adequate backup or standby raw sludge pumps.
    7. Examine the digester hopper configuration and suitability for cleaning. Note provisions for adequate ventilation and access during cleaning. Determine how sludge is transferred from the hoppers. Note provisions for line cleaning.
    8. Determine how the raw sludge flow is to be measured. Check if the sludge pump is to be controlled by timers, sludge density, clarifier blanket level, or some combination thereof.
    9. Note how the digester gas is to be measured and check for adequate liquid traps and provisions for cleaning.
    10. Check for location, size, and type of sampling ports.
    11. Examine the digester heating system. If heat exchangers are used, check for cleaning access. Note the boiler operation and provision for temperature control.
    12. Examine the mixing system. If a propeller mixer is used, note provisions for removing the gear box or entire unit.
    13. Examine the supernatant line and its control. Determine the number of sludge draw-off points.

14. Check for combustible gas analyzers and their location in galleries or closed areas.
15. Carefully study the digester gas system (Figure 12.6, page 159) for your plant and be sure the sediment and drip traps and flame traps are properly located and that none are missing.

QUESTIONS

Write your answers in a notebook and then compare your answers with those on page 235.

12.8A What is the goal of the regulations governing the ultimate disposal of sludge or reuse of biosolids?

12.8B List four methods used for the ultimate disposal of domestic wastewater sludge.

12.9A Why is it important that operators participate in the design of a treatment system?

12.10 ARITHMETIC ASSIGNMENT

Turn to the Arithmetic Appendix at the back of this manual. Read and work the problems in Section A.31, "Sludge Digestion." Check the arithmetic in this section using an electronic calculator. You should be able to get the same answers. Section A.51 contains similar problems using metric units.

12.11 ADDITIONAL READING

1. *MOP 11,* Chapter 30,* "Anaerobic Digestion," and Chapter 31,* "Aerobic Digestion." New edition of MOP 11, Volume III, *SOLIDS PROCESSES.*
2. *NEW YORK MANUAL,* Chapter 7,* "Solids Handling and Disposal."
3. *TEXAS MANUAL,* Chapter 17,* "Sludge Conditioning and Thickening," Chapter 18,* "Separate Sludge Digestion," Chapter 19,* "Sludge Drying Beds," and Chapter 20,* "Sludge Disposal."
4. Chapter 3, "Residual Solids Management," in *ADVANCED WASTE TREATMENT* manual in this series. This chapter covers gravity thickening, dissolved air flotation thickeners (DAFT), centrifuges, aerobic digestion, chemical stabilization, conditioning, filter presses, vacuum filters, drying, composting, incineration, and land disposal. Available from Office of Water Programs, California State University, Sacramento, 6000 J Street, Sacramento, CA 95819-6025. Price, $49.00.
5. *ANAEROBIC SLUDGE DIGESTION* (MOP 16). Obtain from Water Environment Federation (WEF), Publications Order Department, 601 Wythe Street, Alexandria, VA 22314-1994. Order No. M0024. Price to members, $20.00; nonmembers, $30.00; plus shipping and handling.
6. *OPERATIONS MANUAL—ANAEROBIC SLUDGE DIGESTION,* US Environmental Protection Agency. EPA No. 430-9-76-001. Obtain from National Technical Information Service (NTIS), 5285 Port Royal Road, Springfield, VA 22161. Order No. PB-250129. Price, $59.50, plus $5.00 shipping and handling per order.

* Depends on edition.

END OF LESSON 6 OF 6 LESSONS
on
SLUDGE DIGESTION AND SOLIDS HANDLING

Please answer the discussion and review questions next.

DISCUSSION AND REVIEW QUESTIONS

Chapter 12. SLUDGE DIGESTION AND SOLIDS HANDLING

(Lesson 6 of 6 Lessons)

Write the answers to these questions in your notebook. The question numbering continues from Lesson 5.

29. What types of sludges are treated by aerobic digesters?
30. What kind of sludge should be placed on a sand drying bed?
31. What precautions should be taken when applying sludge to a drying bed?
32. What are the advantages of a blacktop drying bed over a sand drying bed?
33. Why have some plants discontinued the practice of elutriation?
34. What are some of the operational problems encountered in using a centrifuge to dewater sludge?

SUGGESTED ANSWERS

Chapter 12. SLUDGE DIGESTION AND SOLIDS HANDLING

ANSWERS TO QUESTIONS IN LESSON 1

Answers to questions on page 152.

12.00A Raw sludge must be digested so wastewater solids may be disposed of without creating a nuisance.

12.00B During anaerobic digestion, organic solids in the sludge are liquified, the solids volume is reduced, and valuable methane gas is produced in the digester by the action of two different groups of bacteria living together in the same environment.

12.00C Some of the important factors in controlling the rate of reproduction of acid forming and methane fermenting bacteria in an anaerobic digester include regulation of food supply (organic solids), volatile acid/alkalinity relationship, mixing, and temperature.

Answers to questions on page 152.

12.10A Plug-type valves give positive control where a gate or butterfly valve may become blocked by rags or other material that will not allow the valve to seat.

12.10B Never start a positive displacement pump against a closed valve because excessive pressure may result in damaging the line or the pump.

12.10C A sludge line should never be isolated by closing the valves on each end for several days because gas production can generate sufficient pressures to cause the line to fail. Also, the solids will form an almost immovable mass in the line.

Answers to questions on page 157.

12.11A Normally, the digester roof is designed to contain a maximum operating gas pressure. If the pressure is exceeded, the water seal could be broken, allowing air to enter the tank and form an explosive mixture of gases. High gas pressures may cause structural damage to the tank in severe cases.

12.11B Fixed cover digesters must be equipped with pressure and vacuum relief valves to break a vacuum or bleed off excessive pressure to protect the digester from structural damage.

12.11C The advantages of a floating cover include less danger of explosive mixtures forming in the digester, better control of supernatant withdrawal, and better control of scum blankets.

12.11D Mixing recirculated digester sludge with raw sludge provides immediate seeding of the raw sludge with anaerobic bacteria from the digester.

ANSWERS TO QUESTIONS IN LESSON 2

Answers to questions on page 158.

12.12A The two main gaseous components of digester gas are methane and carbon dioxide.

12.12B Digester gas can be extremely dangerous in two ways. When mixed with oxygen in certain proportions, it can form explosive mixtures, and it also can cause asphyxiation or oxygen starvation. Smoking, open flames, or sparks must not be tolerated around the digesters or sludge pumping facilities.

12.12C Digester gas is used in plants in various ways: for heating and mixing the digesters, for heating the plant buildings, for running engines, for air blowers for the activated sludge process, and for electrical power for the plant.

Answers to questions on page 162.

12.12D The pressure relief valve is adjusted by placing the correct amount of lead weights on the pressure relief pallet and then checking the digester pressure with a water manometer to ensure the proper setting.

12.12E The water seal in a digester can be broken when a tank is overpumped or gas removal is too slow.

12.12F The operation of either the pressure or vacuum relief valves can create an explosive atmosphere when the digester gas mixes with air.

Answers to questions on page 162.

12.12G Flame arresters should be serviced every three months.

1. Before servicing, valve off the gas flow to the line and put out all flames and pilot lights in the area.
2. Remove end housing and slide out cartridge containing flame arrester baffles.
3. Clean the baffles in solvent and dry.
4. Reinstall cartridge and replace end plate.
5. Return pressure to gas line and soap test for gas leaks.

12.12H Thermal valves should be dismantled at least once a year to ensure that the stem is free to fall and is not gummed up with residue or scale from the gas.

Answers to questions on page 166.

12.12I A sediment trap should be drained of condensate frequently, but may have to be drained twice a day during cold weather because greater amounts of water will be condensed.

12.12J Drip or condensate traps should be installed to keep water out of the gas lines where it would restrict gas flows and damage equipment.

12.12K Automatic drip and condensate traps may stick open, venting gas to the atmosphere and creating a hazardous condition.

12.12L The gas pressure of the digester gas system may be adjusted by connecting a manometer to the gas system and adjusting the regulator to hold eight inches (20 cm) of water column of gas pressure on the digester.

12.12M The pilot flame in the waste gas burner should be checked daily to prevent unburned waste gas from being vented to the atmosphere and creating an odorous and potentially hazardous condition.

Answers to questions on page 176.

12.13A A digester should have a special sampling well so that the digester sludge may be sampled without loss of digester gas pressure, or the creation of dangerous conditions caused by the mixing of air and digester gas.

12.14A If the temperature of the hot water in the heating coils of this type of heating system is maintained too high, it will cook the sludge onto the coils, thus acting as insulation and reducing the heat transferred.

Answers to questions on page 176.

12.15A In addition to speeding up the digestion rate, several important objectives are accomplished in a well-mixed digester: (1) inoculation of the raw sludge immediately with microorganisms, (2) prevention of scum blanket formation, (3) maintenance of homogeneous contents throughout the tank, including even distribution of food, organisms, alkalinity, heat, and waste bacterial products, and (4) use of as much of the total contents of the digester as possible and minimization of the buildup of grit and inert solids on the bottom.

12.15B Maintenance of a gas-type mixing system in a digester requires that the condensate be drained from the lines at least twice a day, that the diffusers be cleaned to prevent high discharge pressures, and that the compressor unit be properly lubricated and cooled.

Answers to questions on page 179.

12.15C An effective operation for breaking up scum blankets in digesters with draft tube propeller mixers is to operate one unit (tube) as a top suction and the other unit as a bottom suction. The direction of flow in the tubes should be reversed each day.

12.15D Pressure gauges should be installed on mixing pump suction and discharge lines. A change in pressures could indicate that the pump is not functioning properly and the desired mixing may not be taking place in the digester.

Answers to questions on page 182.

12.16A The purpose of floating covers on anaerobic sludge digesters is to provide a flexible space for digester gas storage. Floating covers also can keep the scum blanket submerged in the digesting sludge, which helps to digest the scum.

12.16B Explosive conditions are less apt to develop in digesters with floating covers because the cover can drop during sludge or supernatant withdrawal and avoid vacuum conditions that pull air into the digester.

12.16C The annular space is the space full of digesting sludge between the floating cover and digester sidewall.

12.16D Hazardous atmospheric conditions that could be encountered inside the flotation chamber of a floating cover digester include oxygen deficiency, explosive gas-air mixtures, and toxic gases (hydrogen sulfide).

12.16E If a floating cover rises too high, sludge can overflow the digester walls or the rollers could leave the roller guides. If the cover drops too low and rests on the corbels, the gas seal formed by the digesting sludge could break and explosive conditions could develop inside the digester.

ANSWERS TO QUESTIONS IN LESSON 3

Answers to questions on page 185.

12.20A Raw sludge is normally composed of solids settled and removed from the primary clarifiers. Raw sludge contains carbohydrates, proteins, and fats, plus organic and inorganic chemicals that are added by domestic and industrial uses of water. Scum is composed mainly of grease and other floatable material. Waste activated sludge is the sludge intentionally removed from the activated sludge process and consists almost entirely of the microorganisms from the activated sludge aeration tanks.

12.21A If too much raw sludge is added to the digester, the acid fermenters will predominate, driving the pH down and creating an undesirable condition for the methane fermenters.

12.21B When starting a digester, if too much raw sludge is added, the digester will go sour or acid. When a digester recovers from a sour or acid condition, the breakdown of the volatile acids and formation of methane and carbon dioxide is usually very rapid. The digester may then foam or froth.

12.21C Find pounds of volatile matter pumped to digester per day.

$$\text{Volatile Matter Pumped, lbs/day} = (\text{Vol Sludge, GPD})(\text{Solids, \%})(\text{Volatile, \%})(8.34\text{ lbs/gal})$$

$$= (700\text{ gal/day})(0.065)(0.70)(8.34\text{ lbs/gal})$$

$$= 265\text{ lbs/day}$$

Find pounds of seed volatile matter needed.

$$\frac{0.05\text{ lb VM Added/day}}{1\text{ lb VM in Digester}} = \frac{265\text{ lbs VM Pumped/day}}{\text{Seed, lbs VM}}$$

$$\text{Seed, lbs VM} = \frac{(265\text{ lbs VM Pumped/day})\text{ lbs VM}}{0.05\text{ lb VM Added/day}}$$

$$= 5{,}300\text{ lbs VM}$$

Find gallons of seed sludge needed.

$$\text{Seed Sludge, gal} = \frac{\text{Seed, lbs VM}}{(9\text{ lbs/gal})(\text{Solids, \%})(\text{VM, \%})}$$

$$= \frac{5{,}300\text{ lbs VM}}{(9\text{ lbs/gal})(0.10)(0.50\text{ VM})}$$

$$= 11{,}800\text{ gal}$$

12.21D It is hazardous to start a digester when it is only partially full due to explosive conditions created by a mixture of air and methane in partially full digesters.

12.21E A new digester is ready for the raw sludge feed rate to be gradually increased to the full plant load when it is producing a burnable gas.

Answers to questions on page 186.

12.22A Better operational performance occurs when the digester is fed several times a day, rather than once a day, because this strategy avoids temporary overloads on the digester and uses available space more effectively. Several pumpings a day not only helps the digestion process, but maintains better conditions in the clarifiers, permits thicker sludge pumping, and prevents coning in the primary clarifier hopper.

12.22B As thick a sludge as possible may be pumped to the digester by operating the sludge pump for several minutes each hour (at a rate not to exceed 50 GPM) to clear the sludge hopper.

12.22C The pumping of thin sludge should be avoided because too much water pumped to the digester increases heating requirements, reduces digester holding time, washes buffer and seed sludge out of the digester, and imposes a heavier supernatant load on the plant.

Answers to questions on page 187.

12.23A Lime is added to a sour digester in an attempt to neutralize the acids and increase the pH to 7.0.

12.23B Assume a dosage of one pound of lime per 1,000 gallons of digester sludge.

$$\text{Lime Dose, lbs} = \frac{\text{Digester Sludge Volume, gal}}{1{,}000\text{ gal/lb Lime}}$$

$$= \frac{100{,}000\text{ gal}}{1{,}000\text{ gal/lb Lime}}$$

$$= 100\text{ lbs of Lime}$$

12.23C Lime must be mixed into a solution before being added to a digester because dry lime would settle to the bottom in lumps that are not only ineffective but take up digester capacity and are difficult to remove when cleaning the digester.

12.23D Lime should be added daily until the volatile acid/alkalinity relationship, gas production, and pH levels are restored.

Answers to questions on page 188.

12.24A Bacteria secrete enzymes that help break down compounds that the bacteria in the digester may use as food.

12.25A Foaming in a digester is caused by an imbalance between the acid-forming bacteria and the methane-fermenting bacteria. This imbalance results in active gas production while solids separation has not progressed far enough (insufficient digestion).

12.25B To prevent foam from recurring:

1. Maintain a constant temperature in digester.
2. Feed sludge at regular, short intervals.
3. Exercise caution when breaking up scum blankets.
4. Do not overdrain sludge from digester.
5. Keep contents of digester well mixed from bottom to top at all times.

12.26A The gas initially produced in a digester is not burnable because it contains mostly CO_2 and H_2S. Generally, digester gas will burn when the methane content reaches 56 percent but, for use as a fuel, the methane content should be at least 62 percent.

Answers to questions on page 190.

12.27A The purpose of the secondary digester is to allow separation of the sludge from the liquor, to store digested sludge, and to allow more complete digestion. This reserve of digested sludge is needed to act as seed sludge or buffer sludge to be transferred to the primary digester if it becomes upset.

12.27B When raw sludge is pumped to the primary digester, usually an equal volume of sludge from the primary digester is transferred to the secondary digester and supernatant is displaced from the secondary digester back to the plant.

12.27C The level of supernatant withdrawal is selected on a fixed cover digester by selecting the supernatant tube that reaches the clearest supernatant zone in the digester. On floating cover tanks, the supernatant draw-off line is raised or lowered to the clearest supernatant zone in the digester.

12.28A The withdrawal rate of sludge from either digester should be no faster than a rate at which the gas production from the system is able to maintain a positive pressure in the digester (at least two inches (5 cm) of water column).

ANSWERS TO QUESTIONS IN LESSON 4

Answers to questions on page 192.

12.3A A thermometer is usually installed in the recirculated sludge line from the digester to the heat exchanger. Accurate temperature readings also may be taken from the flowing supernatant tube or from the heat exchanger sludge inlet line.

12.3B The volatile acid/alkalinity relationship is useful in digester control because it is the first indicator that the digestion process is starting to get out of balance and that corrective action is necessary.

12.3C When the volatile acid/alkalinity relationship starts to increase, extend mixing time of digester contents, control heat more evenly, decrease raw sludge feed rates, and decrease digested sludge withdrawal rates from digesters. Mixing should be vertical mixing from the bottom of the tank to the top of any scum blanket. If possible, some of the concentrated sludge in the secondary digester should be pumped back to help correct the ratio.

12.3D pH is a poor indicator of approaching trouble because it is usually the last indicator to change and, by the time it changes significantly, the digester is already in serious trouble.

Answers to questions on pages 192 and 193.

12.3E Solids tests are run on digester sludge to determine the pounds of sludge, the pounds of volatile sludge available to the bacteria, the pounds of volatile sludge destroyed or reduced, the digester loading rates, reductions, and the amount of solids handled through the system. All of these tests are necessary for the maintenance of efficient digester operation.

12.3F The volume of sludge pumped per day is needed to determine:

1. Gallons pumped per day
2. Pounds of dry solids per day to the digester
3. Estimated gallons of supernatant returned to the plant
4. Volume of sludge for ultimate disposal

12.3 G

$$\text{Volume, cu ft} = \text{Area, sq ft} \times \text{Depth, ft}$$

$$= \frac{\pi D^2}{4} \times \text{Depth, ft}$$

$$= 0.785 \times \left[\frac{12 \text{ in}}{12 \text{ in/ft}}\right]^2 \times \frac{4 \text{ in}}{12 \text{ in/ft}}$$

$$= 0.259 \text{ cu ft}$$

$$\text{Volume, gal} = 0.259 \text{ cu ft} \times 7.48 \text{ gal/cu ft}$$

$$= 1.937 \text{ or } 1.9 \text{ gal/stroke}$$

12.3H

$$\text{Sludge Pumped, gal/day} = 1.2 \text{ gal/rev} \times 2{,}000 \text{ rev/day}$$

$$= 1.2 \text{ gal} \times 2{,}000\text{/day}$$

$$= 2{,}400 \text{ gal/day}$$

Answers to questions on page 194.

12.3I

$$\text{Dry Solids, lbs} = \text{Gal Pumped} \times \% \text{ Solids} \times 8.34 \text{ lbs/gal}$$

$$= 3{,}000 \text{ gal} \times 0.05 \times 8.34 \text{ lbs/gal}$$

$$= 1{,}251 \text{ lbs Dry Solids}$$

$$\text{Volatile Solids, lbs} = \% \text{ Volatile} \times \text{Total Solids, lbs}$$

$$= 0.70 \times 1{,}251 \text{ lbs}$$

$$= 875.70 \text{ lbs of Volatile Solids}$$

12.3J

$$P = \frac{\text{In} - \text{Out}}{\text{In} - (\text{In} \times \text{Out})} \times 100\%$$

$$= \frac{0.70 - 0.45}{0.70 - (0.70 \times 0.45)} \times 100\%$$

$$= \frac{0.25}{0.70 - 0.315} \times 100\%$$

$$= \frac{0.25}{0.385} \times 100\%$$

$$= 0.65 \times 100\%$$

$$= 65\% \text{ Reduction of Volatile Matter}$$

Answers to questions on page 195.

12.3K Total solids tests are run on the digester supernatant to estimate the solids load being returned to the plant.

12.3L If the total solids reached 0.5 percent, the point of supernatant withdrawal from the secondary digester should be changed or the supernatant removal tubes should be moved to a different level in the digester where the liquor contains the least amount of solids.

Answers to questions on page 213.

12.4A The normal operation of a digester involves the following activities:

1. Feeding sludge to the digester
2. Maintaining the proper temperature
3. Keeping the contents of the digester mixed
4. Removing supernatant
5. Withdrawing sludge

12.4B Gas line condensate traps and sedimentation traps should be drained as required (from one to four times per day).

12.4C The quality of the digester supernatant is decreasing when the solids level and COD are increasing.

12.4D Possible causes of a floating cover tilting include excess scum in one area, causing excess drag; guides or rollers out of adjustment; and rollers or guides broken.

Answers to questions on page 215.

12.5A Grease (scum) and grit (sand, seeds, cigarette butts) accumulate in digesters and reduce the active volume for digesting sludge.

12.5B When the active volume in a digester is reduced, the amount of volatile solids destruction is decreased, gas production is lowered, and, ultimately, the digestion process will fail if the digester is not cleaned.

12.5C While a plant's only digester is being cleaned: (1) haul the raw sludge to a nearby treatment plant for processing, (2) temporarily convert an activated sludge aeration tank to an aerobic digester, (3) construct a temporary anaerobic lagoon, or (4) hold the sludge in the primary clarifiers.

12.5D Before entering a digester, test the digester atmosphere for oxygen content, flammable/explosive gases, and toxic gases (hydrogen sulfide). Follow confined space testing requirements prior to entry and during occupancy.

ANSWERS TO QUESTIONS IN LESSON 6

Answers to questions on page 219.

12.6A Aerobic digestion is commonly used to handle waste activated sludge because of the operational problems encountered when a waste aerobic activated sludge with a low solids content is placed in an anaerobic digester.

12.6B Dissolved oxygen levels in aerobic digesters should be maintained between 1.0 and 2.0 mg/*L* throughout the tank.

12.6C Operational problems that could occur in aerobic digesters include scum accumulation, odors, and floating sludge.

Answers to questions on pages 221 and 222.

12.70A Prior to applying sludge to a drying bed, the operator should loosen the sand, break the clods, and level the sand bed.

12.70B Sludge should be drawn slowly from the digester so as not to create a negative pressure in the anaerobic digester and to prevent coning of sludge in the bottom of the digester.

12.70C To eliminate a sludge cone, quickly close and open the draw-off valve on a gravity flow system, which creates a minor shock wave and sometimes washes the heavier sludge into the cone. If the digested sludge is pumped to the drying bed, quickly start and stop the pump using the power switch to create a bumping action.

12.70D Flames and smoking should not be allowed due to the presence of methane gas, which is flammable.

12.70E After sludge has been applied to the drying bed, the draw-off line should be flushed with water so the solids will not cement in the line. One end of the line should be left open so any gas produced will not rupture the line. The line should be drained if freezing is a problem.

12.70F Heavy equipment should not be used to remove dried sludge from a sand drying bed because the equipment may damage the underdrain system, mix the sand and gravel in the bed, or remove sand that will have to be replaced.

12.70G

$$\text{Volume, cu ft} = \text{Length, ft} \times \text{Width, ft} \times \text{Depth, ft}$$

$$= 100 \text{ ft} \times 25 \text{ ft} \times 0.5 \text{ ft}$$

$$= 1{,}250 \text{ cu ft}$$

$$\text{Volume, cu yd} = \frac{1{,}250 \text{ cu ft}}{27 \text{ cu ft/cu yd}}$$

$$= 46.3 \text{ cubic yards of Dry Sludge}$$

$$\text{No of Truckloads} = \frac{\text{Volume, cu yd}}{\text{Truck Capacity, cu yd/Truckload}}$$

$$= \frac{46.3 \text{ cu yd}}{2 \text{ cu yd/Truckload}}$$

$$= 23.15 \text{ Truckloads}$$

$$= 23 \text{ or } 24 \text{ Truckloads}$$

12.70H If green sludge were accidentally applied to a drying bed, then the operator should apply dry lime and, if allowable, a pesticide or insecticide to control insects. Use appropriate safety gear (eye/face protection, respiratory protection, and rubber gloves) when handling lime, pesticides, or insecticides. The sludge should never be used for fertilizer.

Answers to questions on page 222.

12.71A To reduce sludge drying time in a blacktop drying bed, obtain a separate sample of the sludge applied to the bed. When the sludge rises to the surface of the sample, open the drain line and slowly bleed off the lower liquor. When the sludge begins to dry and crack, mix the bed, thereby exposing wet sludge.

12.72A A sludge lagoon is filled with anaerobically digested sludge (and sometimes supernatant) and then allowed to dry by evaporation. When dry, the sludge is removed and disposed of.

Answers to questions on page 224.

12.73A Applying sludge to land improves the condition of the soil and returns nutrients to the soil.

12.74A In plants where large volumes of sludge are handled and drying beds are not feasible, mechanical dewatering may be used. Mechanical dewatering methods available are vacuum filtration, pressure filtration, centrifugation, and filter presses.

12.74B Digested sludge can be prepared for vacuum filtration by elutriation (washing) and conditioning with a coagulant.

12.74C The chemical feed rate to condition sludge is determined by sampling and adding various dosages of flocculant to the samples and running filterability tests with a Buchner funnel.

12.74D The life of a filter blanket is influenced by care, maintenance, and the type of material.

Answers to questions on page 227.

12.74E Centrifuges are used to dewater raw or primary sludges and digested sludges for disposal in incinerators or furnaces.

12.74F The condition of the sludge cake or slurry and the quality of the centrate from a centrifuge are regulated by the sludge feed rate, bowl speed, and, if chemical conditioners are used, by dosage rates and pool depth.

Answers to questions on page 230.

12.8A Protection of public health and the environment is the goal of the regulations governing the ultimate disposal of sludge or reuse of biosolids.

12.8B Ultimate sludge disposal methods include land application, distribution and marketing, monofilling, surface disposal, and incineration.

12.9A It is important that operators participate in the design of a treatment system because operators can supply information about specific requirements for effective and safe operation and maintenance of treatment processes.

APPENDIX

Monthly Data Sheet

CLEANWATER, USA

WATER POLLUTION CONTROL PLANT

MONTHLY RECORD __________ 20__ DIGESTION OPERATOR: __________

DATE	DAY	RAW SLUDGE: GALLONS PER DAY	RAW SLUDGE: % SOLIDS	RAW SLUDGE: % VOLATILE	RAW SLUDGE: pH	RAW SLUDGE: SCUM GAL./DAY	RECIRCULATED SLUDGE: TEMP. °F	RECIRCULATED SLUDGE: % SOLIDS	RECIRCULATED SLUDGE: pH	RECIRCULATED SLUDGE: VOL. ACIDS MG/L	RECIRCULATED SLUDGE: ALKALINITY MG/L	RECIRCULATED SLUDGE: MIXING HRS.	GAS: GAS PROD. FT^3/DAY	GAS: FT^3/M.G. FLOW	GAS: % CO_2	SLUDGE DISPOSAL: GALS. TO BED	SLUDGE DISPOSAL: BED NO.	SLUDGE DISPOSAL: % SOLIDS	SLUDGE DISPOSAL: % VOL. SOL.	SLUDGE DISPOSAL: LBS. DRY SOLIDS	SLUDGE DISPOSAL: CU. YDS. CAKE REM.	REMARKS
1	S	12850	2.4	80.7	5.9		91	1.3	6.9	103	1450	24	31080									
2	M	12280				1090	91					24	37170									
3	T	11990					91					24	26800									
4	W	12020					91					24	28900									
5	T	11460	3.1	73.5	6.0	1700	91	2.8	6.8	137	1400	24	28320		41							
6	F	11720					91					24	28710									
7	S	11990					91					24	25580									
8	S	13090	3.4	73.1	5.9		91	1.6	6.9	86	1690	24	24310		39							
9	M	11170				1190	92					24	24550									
10	T	11000					92					24	24550									
11	W	12000					92					24	24550									
12	T	11490	3.5	83.4	5.7	720	92	1.6	6.9	69	1700	24	25640									
13	F	12000					92					24	25940									
14	S	11760				740	93					24	28750		39							
15	S	12460	3.8	79.2	5.9		93	1.6	6.8	86	1790	24	31450									
16	M	11820				540	93					24	34150									
17	T	12280					93					24	32550									
18	W	11440					94					24	28400									
19	T	11560	1.3	81.7	6.3	580	94	1.4	6.9	120	1610	24	27900		39							
20	F	12070					94					24	31000									
21	S	11960					94					24	32380									
22	S	12510				850	94					24	34420		32							
23	M	12180	1.9	75.3	—		95	1.3	—	86	1280	24	37810									
24	T	10560					94					24	37430									
25	W	11870					94					24	37560									
26	T	11430	2.9	70.6	—		94	1.4	—	—	—	24	37130									
27	F	11510				1150	94					24	35320									
28	S	11840	2.9	73.2	6.1		94	1.6	6.8	120	1670	24	31150		33							
29	S	12120				650	94					24	29520									
30	M	11700	4.6	77.7	6.2		94	1.5	6.9	120	1400	24	32860									
31																						
MAX.		13090	4.6	83.4	6.3	1700	95	2.8	6.9			24	37820		41							
MIN.		10560	1.3	70.6	5.7	540	91	1.3	6.8			24	24310		32							
AVG.		11871	2.9	76.8	6.0	307/DAY	92	1.6	6.9	92	1554	24	30529		36							

CHAPTER 13

EFFLUENT DISCHARGE, RECLAMATION, AND REUSE

by

Bill B. Dendy

Revised by

Daniel J. Hinrichs

TABLE OF CONTENTS

Chapter 13. EFFLUENT DISCHARGE, RECLAMATION, AND REUSE

OBJECTIVES

Chapter 13. EFFLUENT DISCHARGE, RECLAMATION, AND REUSE

Following completion of Chapter 13, you should be able to:

1. Properly reclaim and reuse plant effluents for beneficial uses.
2. Develop an operational strategy for effluent discharge, reclamation, and reuse.
3. Troubleshoot an effluent discharge, reclamation, and reuse system.
4. Develop a receiving water monitoring program.
5. Select the proper locations to collect samples.
6. Determine when and how often samples should be collected.
7. Collect representative samples.
8. Conduct an effluent monitoring program in a safe fashion.
9. Review plans and specifications for an effluent discharge, reclamation, and reuse system.

WORDS

Chapter 13. EFFLUENT DISCHARGE, RECLAMATION, AND REUSE

AESTHETIC (es-THET-ick) AESTHETIC

Attractive or appealing.

BIOASSAY (BUY-o-AS-say) BIOASSAY

(1) A way of showing or measuring the effect of biological treatment on a particular substance or waste.

(2) A method of determining the relative toxicity of a test sample of industrial wastes or other wastes by using live test organisms, such as fish.

COMPOSITE (PROPORTIONAL) SAMPLE COMPOSITE (PROPORTIONAL) SAMPLE

A composite sample is a collection of individual samples obtained at regular intervals, usually every one or two hours during a 24-hour time span. Each individual sample is combined with the others in proportion to the rate of flow when the sample was collected. Equal volume individual samples also may be collected at intervals after a specific volume of flow passes the sampling point or after equal time intervals and still be referred to as a composite sample. The resulting mixture (composite sample) forms a representative sample and is analyzed to determine the average conditions during the sampling period.

ESTUARIES (ES-chew-wear-eez) ESTUARIES

Bodies of water that are located at the lower end of a river and are subject to tidal fluctuations.

FIXED SAMPLE FIXED SAMPLE

A sample is fixed in the field by adding chemicals that prevent the water quality indicators of interest in the sample from changing before final measurements are performed later in the laboratory.

GRAB SAMPLE GRAB SAMPLE

A single sample of water collected at a particular time and place that represents the composition of the water only at that time and place.

HYDRAULIC JUMP HYDRAULIC JUMP

The sudden and usually turbulent abrupt rise in water surface in an open channel when water flowing at high velocity is suddenly retarded to a slow velocity.

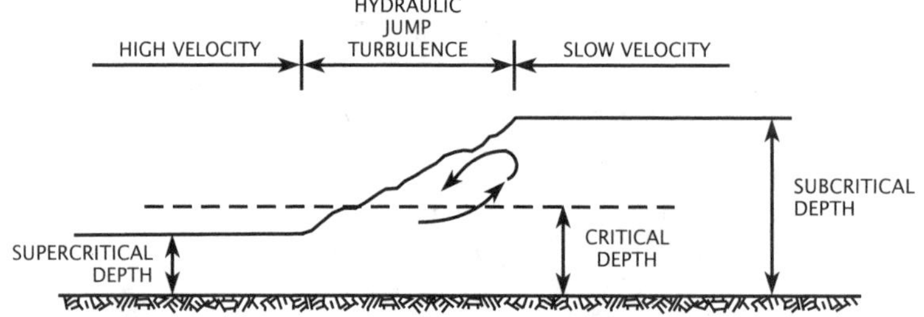

NPDES PERMIT NPDES PERMIT

National Pollutant Discharge Elimination System permit is the regulatory agency document issued by either a federal or state agency that is designed to control all discharges of potential pollutants from point sources and stormwater runoff into US waterways. NPDES permits regulate discharges into US waterways from all point sources of pollution, including industries, municipal wastewater treatment plants, sanitary landfills, large animal feedlots, and return irrigation flows.

OUTFALL OUTFALL

(1) The point, location, or structure where wastewater or drainage discharges from a sewer, drain, or other conduit.

(2) The conduit leading to the final discharge point or area. Also see OUTFALL SEWER.

REPRESENTATIVE SAMPLE REPRESENTATIVE SAMPLE

A sample portion of material, water, or wastestream that is as nearly identical in content and consistency as possible to that in the larger body being sampled.

RESPIRATION RESPIRATION

The process in which an organism takes in oxygen for its life processes and gives off carbon dioxide.

CHAPTER 13. EFFLUENT DISCHARGE, RECLAMATION, AND REUSE

13.0 IMPORTANCE OF EFFLUENT DISCHARGE, RECLAMATION, AND REUSE

Proper discharge of plant effluent is the final process for the operator. The job you have done operating and maintaining all of the processes contributes to the quality of the final effluent. There are several means of effluent discharge, reclamation, and reuse. This chapter emphasizes the most common method, discharge into surface waters (Figure 13.1). Reuse on land and direct or indirect reuse are covered in Chapter 8, "Wastewater Reclamation and Reuse," *ADVANCED WASTE TREATMENT.* Included there you will find a discussion of discharge to land by the use of irrigation, groundwater recharge basins, and underground discharge. However, the most common reuse method is direct discharge to surface waters. Surface waters include rivers, streams, lakes, *ESTUARIES,*[1] and oceans. Discharge requirements vary depending on the probable use of the receiving waters. Typical water quality indicators for various water uses are listed in Table 13.1.

In addition to its contribution to successful operation, effluent control is also extremely important in legal matters. For example, there may be several effluents being discharged to the receiving water. If a problem develops, then all discharges would be investigated by the regulatory agency. If proper sampling and analysis techniques were carefully followed and results recorded, then the regulatory agency can look for problem causes elsewhere if your discharge is not causing the problem. Another type of legal issue may be a change in the discharge standards. If accurate sampling and analysis techniques show that a certain constituent level is not causing a problem in the receiving stream, then the required constituent level can be changed in the discharge permit. For example, a chlorine residual of 1.0 mg/*L* may be required in the effluent. If sampling and analysis showed this level was not necessary or was excessive in controlling coliform bacteria, then a formal request can be made for a change in the standard. The result could mean a substantial savings in chlorine costs.

Due to water shortages in many areas and extremely stringent requirements for direct discharge in some locations, reclamation and reuse are becoming popular. Disposal systems that once discharged to land are now more often land reclamation systems. Usually, secondary treatment systems remove suspended solids and BOD, but not nutrients such as nitrogen and phosphorus. The nitrogen and phosphorus are extremely valuable to farmers who irrigate with treated effluent. In some instances, they do not need to purchase and apply chemical fertilizers. In all instances they can reduce their fertilizer costs. Land discharge systems are beneficial to the municipality in that the soil layer and plant material serve as a "living filter" to remove unwanted materials from the effluent. Groundwater recharge basins or infiltration–percolation basins are extremely simple to operate and maintain. An underground discharge system can consist of a deep well injection system, a subsurface irrigation system, or subsurface leach lines. The deep well injection system may be used to displace oil or gas that is being pumped out of the ground or to provide a barrier between underground salt water and fresh groundwater. The subsurface irrigation system is popular for landscaping in urban areas. Leach lines are usually limited to small systems and provide a means of groundwater recharge.

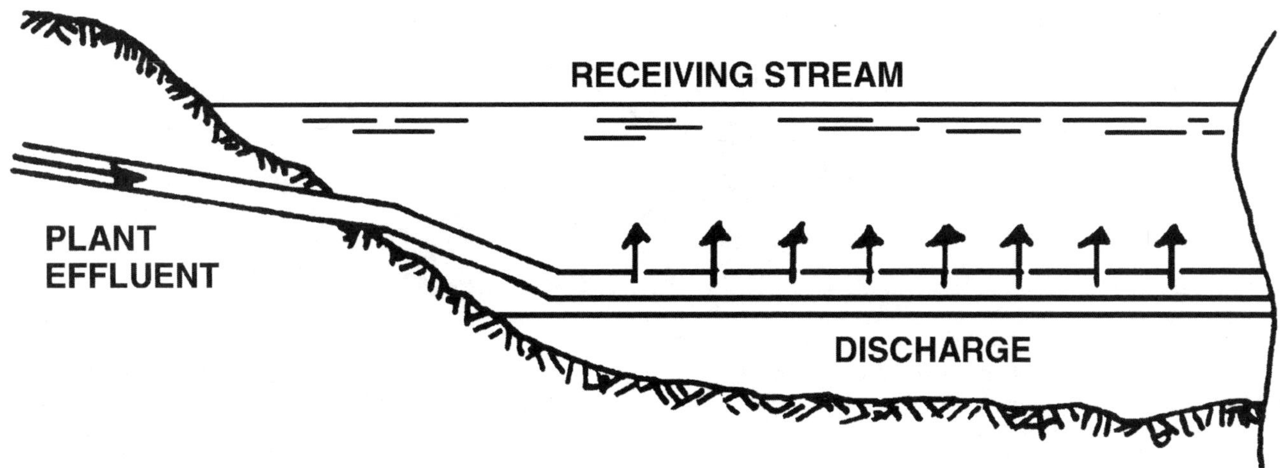

Fig. 13.1 Effluent discharge into surface waters

[1] *Estuaries* (ES-chew-wear-eez). Bodies of water that are located at the lower end of a river and are subject to tidal fluctuations.

TABLE 13.1 WATER QUALITY CRITERIA [a]

(All units in mg/*L* unless otherwise noted)

Water Quality Indicator	Water Use					
	Non-Contact Recreation	Contact Recreation	Fish Propagation	Drinking Water Supply	Irrigation	Livestock and Wildlife
Aquatic Growth	Virtually Free	Virtually Free	—	—	—	Avoid blue-green algae
Coliform Bacteria	200 MPN/ 100 m*L*	100 MPN/ 100 m*L*	—	20,000 MPN/ 100 m*L* [b]	1,000 MPN/ 100 m*L*	5,000/100 m*L*
COD	60	30	—	—	—	—
Floating Debris and Scum	Virtually Free	Virtually Free	—	—	—	—
Odor, Units	Virtually Free	Virtually Free	—	Virtually Free	—	—
Oxygen, Dissolved	Aerobic	Aerobic	5.0	—	—	Aerobic
pH, Units	6.5-8.3	6.5-8.3	6.5-9.0	—	6.0-9.0	7.0-9.0
Settleable Solids	Virtually Free	Virtually Free	—	—	—	—
Nitrogen, Total	—	10	—	—	—	—
Nitrite as N	—	—	—	1.0	—	10
Nitrite and Nitrate as N	—	—	—	10	—	100
Ammonia as N	—	—	0.02 (un-ionized)	0.05	—	—
Phosphate	—	0.2	—	—	—	—
Suspended Solids	—	5	25	—	—	—

NOTE: There are other elements that interfere with use of effluent for these purposes; however, these other elements cannot be controlled by conventional wastewater treatment processes.

[a] *WATER REUSE—VOLUME 2, EVALUATION OF TREATMENT TECHNOLOGY* (OWRT/RU-79/2) by Culp, Wesner, Culp; report for US Department of Interior, Office of Water Research and Technology, Washington, DC, 1979.

[b] Raw water source

Direct and indirect reuse of treated effluent is limited only by the costs to treat the effluent to the necessary requirements for a particular use. Normally, reuse systems supply wash-down water or cooling water.

Why might the effluent from your wastewater treatment plant be reclaimed and reused directly by someone? The main reason is that someone needs water and the effluent from your wastewater treatment plant is of an acceptable quality to meet their needs. Effluent reuse is considered when: (1) the volume of municipal water needed is not available, (2) the cost of purchasing available treated water is too expensive, (3) surface waters are not available or the cost of treatment is excessive, and (4) groundwaters are either not available or the costs of pumping and any treatment are prohibitive. In the future, as discharge requirements become more and more stringent, treatment plant effluents become more and more attractive as the best available source of water. For these reasons, you must be able to produce an effluent that can be used either directly or reclaimed for beneficial uses. In many regions, after treated effluent is discharged to receiving waters, they are diluted and used indirectly for many beneficial uses.

QUESTIONS

Write your answers in a notebook and then compare your answers with those on page 256.

13.0A What is the most common method of effluent discharge, reclamation, and reuse?

13.0B What are other common methods of wastewater discharge, reclamation, and reuse?

13.1 EFFLUENT DISCHARGE TO SURFACE WATERS

13.10 Treatment Requirements

Prior to discharge, the wastewater is treated to protect the health of the people who may come in contact with it, to prevent nuisances due to odors or unsightliness, and to prevent the wastewater from interfering with the many uses of the surface waters. Governmental regulatory agencies set discharge requirements to ensure that surface waters receiving effluent are protected. The requirements will vary depending on local conditions but may consist of both *TREATMENT EFFICIENCY*[2] standards and effluent quality standards. For example, a plant may be required to remove 85 percent of the influent suspended solids and to maintain an effluent suspended solids concentration of less than 30 mg/*L*. The operator measures both the influent and effluent concentrations and from these determines the percent reduction. These measurements are called "in-plant measurements." Those in-plant measurements that show the results of an individual unit process or the entire plant are "performance evaluation" tests. In-plant measurements that assist the operator in deciding how to operate a unit are called "process control" tests. For example, the dissolved oxygen test in an aeration basin is a process control test. The results indicate to the operator whether the aeration system is providing sufficient oxygen to the organisms treating the wastes.

A second type of measurement consists of "receiving water measurements." These tests are used to determine the effect of the waste discharge on the receiving waters and on the beneficial uses (water supply, recreation, fishery) of the receiving waters after the effluent has mixed with the receiving waters. A plant could be operating according to its design and be removing 90 percent of the suspended solids, yet still be causing a bad impact on the water quality in the water that receives the effluent. This could happen because the plant was not designed to meet more demanding effluent standards or there has been a change in the receiving water, such as a reduction in flow or new industrial discharges since the plant was built.

Receiving water measurements are as important as the in-plant measurements because the ultimate purpose of the plant is to protect the receiving waters. However, plants should always be operated as efficiently as possible, therefore in-plant measurements must not be slighted. Also, the plant effluent must be tested or measured so that it can be related to effects in the receiving waters.

The usual approach in measuring the impact of a discharge on receiving waters is to take a measurement in an area that is not affected (upstream) and in an area that is affected (downstream) and compare the two results. This comparison shows how much of an impact the discharge has on the receiving waters. Analysis of the differences shows whether the discharge is causing a violation of the water quality objectives or standards that have been set by the regulatory agency.

Successful operation of a surface discharge system depends on conscientious operators using proper techniques for sampling and analysis.

13.11 Sampling and Analysis Equipment Variations

Equipment requirements include sampling and metering equipment. The sampling equipment can vary from sample bottles for manual *GRAB SAMPLES*[3] to complex devices consisting of timers and flow proportioning equipment for obtaining composite samples. As the name grab sample suggests, these samples are taken at no regularly scheduled place or time. A *COMPOSITE SAMPLE*[4] is one that consists of a series of samples taken over a given time period at a particular location. For example, a 24-hour composite sample may consist of 24 individual samples taken each hour or the sample may consist of a continuous composite obtained by continually pumping a sample. Some composite samplers are more complex than others. Each of the samples in the composite may be equal in size or proportional to the flow at the time of the sampling.

Dissolved oxygen (DO), temperature, and pH may be monitored or determined from in-stream measurements taken from the actual flow stream (wastewater being treated). Sampling equipment is usually required, however, to collect field samples that will be analyzed in the lab for such water quality indicators as biochemical oxygen demand (BOD) and suspended solids (SS). Total coliform and fecal coliform determinations are performed in the lab only on carefully collected grab samples. Each area of surface water will have certain special requirements for additional data or additional tests that must be conducted. For example, in areas with slowly moving streams or rivers, fish may be endangered by ammonia (NH_3) resulting from the ammonium ion (NH_4^+) in the effluent.

---

[2] $\text{Treatment Efficiency, \%} = \frac{(\text{Concentration In, mg}/L - \text{Concentration Out, mg}/L)100\%}{\text{Concentration In, mg}/L}$

[3] *Grab Sample.* A single sample of water collected at a particular time and place that represents the composition of the water only at that time and place.

[4] *Composite (Proportional) Sample.* A composite sample is a collection of individual samples obtained at regular intervals, usually every one or two hours during a 24-hour time span. Each individual sample is combined with the others in proportion to the rate of flow when the sample was collected. Equal volume individual samples also may be collected at intervals after a specific volume of flow passes the sampling point or after equal time intervals and still be referred to as a composite sample. The resulting mixture (composite sample) forms a representative sample and is analyzed to determine the average conditions during the sampling period.

Flow-metering equipment is usually located at the plant headworks. In most situations, the actual effluent flow is not critical. However, some discharge requirements include a limit on the total pounds per day of a constituent (suspended solids) in addition to the concentration. The total pounds/day is equal to the flow (MGD) times concentration (mg/*L*) times the constant 8.34 pounds per gallon.[5] Metering devices include weirs, Parshall flumes, Venturi meters, propeller meters, and magnetic meters.

13.12 Control Considerations

Physical control of effluent for discharge to receiving waters is usually limited to discharging all flow that enters the plant. However, some systems may have emergency effluent storage lagoons available. Physical control is generally not needed unless some special problem has arisen, such as a toxic waste spill into the collection system. In this situation, there are two solutions. The affected flow may be stored for future treatment or treated chemically prior to entering the plant. Usually, storage is not possible, however, and a neutralizing chemical may be added prior to discharge. If the treatment processes fail, or if water quality levels in receiving waters become critical, the emergency effluent storage facilities can be used.

Operators can detect adverse impacts on receiving waters caused by the treatment plant effluent by noticing odors, visible floating substances, grease, or abnormal colors. Odors may indicate failure in a treatment process in the plant or the discharge of an industrial waste into the collection system or to the receiving water. Visible floating substances may be indicative of a malfunctioning skimmer or a final sedimentation tank.

QUESTIONS

Write your answers in a notebook and then compare your answers with those on page 256.

13.1A What water quality indicators may be monitored or determined by in-stream measurements?

13.1B What water quality indicators are usually determined by samples collected in the field and analyzed in the lab?

13.2 OPERATING PROCEDURES

The operating procedures for effluent discharge to surface streams are similar to other treatment processes and are fairly straightforward. Before starting your plant, operating procedures covering initial start-up, routine operation, shutdown, and abnormal conditions should be developed. These procedures can help avoid forgetting a critical item that could result in the violation of your *NPDES PERMIT*[6] or cause a fish kill in the receiving waters. The following procedures apply to most plants, and you should use them as a guide for developing procedures to be used at your own plant.

13.20 Start-Up

1. Inspect the discharge line, where possible.
2. Open all appropriate valves.
3. Start the pumps (unless it is a gravity system).
4. Observe the flow entering the discharge system for visible pollutants.
5. Look at the water surface over the diffuser (if your plant has a diffuser) or *OUTFALL*[7] pipe for unusual turbulence caused by a break in the line (or conversely—a lack of turbulence and possible clogged line). Also, look for foam, oil, grease, or discoloration of the water, which would be undesirable from an *AESTHETIC*[8] viewpoint.

13.21 Normal Operation

1. Conduct a monitoring program. See Section 13.3, "Receiving Water Monitoring," for detailed procedures.
2. Inspect the plant operation if a sudden change in the visual appearance of the effluent is observed.

13.22 Shutdown

1. Stop the pumps (unless it is a gravity system).
2. Close the valve preceding the discharge line.
3. Flush the line and diffuser with fresh water, if possible.

13.23 Operational Strategy

An operational strategy consists of doing everything possible to minimize the discharge of pollutants to the receiving

[5] Loading on Receiving Waters, lbs/day = (Flow, MGD)(Concentration, mg/*L*)(8.34 lbs/gal)
or metric
Loading on Receiving Waters, kg/day = (Flow, cu m/day)(Concentration, mg/*L*)(1,000 *L*/cu m)(1,000,000 mg/kg)
Remember that mg/*L* = $\frac{\text{lbs}}{\text{M lbs}} = \frac{\text{mg}}{\text{M mg}}$

[6] *NPDES Permit.* National Pollutant Discharge Elimination System permit is the regulatory agency document issued by either a federal or state agency that is designed to control all discharges of potential pollutants from point sources and stormwater runoff into US waterways. NPDES permits regulate discharges into US waterways from all point sources of pollution, including industries, municipal wastewater treatment plants, sanitary landfills, large animal feedlots, and return irrigation flows.

[7] *Outfall.* (1) The point, location, or structure where wastewater or drainage discharges from a sewer, drain, or other conduit. (2) The conduit leading to the final discharge point or area. Also see OUTFALL SEWER.

[8] *Aesthetic* (es-THET-ick). Attractive or appealing.

waters. This may involve the use of emergency holding tanks or reservoirs during times when treatment processes have failed or toxic wastes reach the plant. After the failure has been corrected, the stored effluent must be properly discharged. Discharge can be accomplished by any of three methods. (1) The stored water may be slowly added to the treated effluent (discharge by dilution in treated effluent). Another procedure (2) consists of returning the stored water to the plant for treatment. This would be done during periods of low flow into the plant so the treatment processes would not become overloaded. (3) If the waste is extremely toxic, treat it chemically or allow it to become concentrated by evaporation in the storage pond. After concentration, dispose of toxic waste in an approved manner.

The decision of which method to use depends on the condition of the stored water. If the wastes can be diluted with treated effluent and the discharge requirements met, then this is the preferred method. Otherwise, treatment is required. This treatment should be done during times when the plant inflow is low and returned at a low flow rate. A high rate of flow returned during the day when flows coming to the plant are high could cause the treatment processes to be upset.

13.24 Emergency Operating Procedures

The impact on the discharge system due to loss of power is critical only with pumped discharge systems. These systems should have standby generators, which can be started to provide power to run pump motors.

Loss of upstream treatment units will result in poorer effluent quality. If available, emergency storage reservoirs should be used and discharge to the receiving stream stopped until upstream units are working properly again.

13.25 Troubleshooting

If your plant is not meeting National Pollutant Discharge Elimination System permit requirements, try to identify the cause of the problem and to select the proper corrective action. Solutions to the problems listed in this section (Tables 13.2, 13.3, and 13.4) have been covered in more detail in previous chapters.

QUESTIONS

Write your answers in a notebook and then compare your answers with those on page 256.

13.2A Operating procedures should be developed for what conditions in an effluent discharge, reclamation, and reuse program?

13.2B What is the main consideration for an operational strategy?

13.3 RECEIVING WATER MONITORING

13.30 Types of Monitoring Programs

There are two types of monitoring programs. The first is the stream or water quality survey. To plan a water quality survey, you must understand the reasons for, or the objectives of, the survey. A typical objective could be to detect any adverse impact of the plant effluent on the receiving waters. An effluent sampling program is the other type of monitoring program. The overall objectives of each survey greatly influence the location of sampling stations, types of samples, frequency and time of day of collecting samples, and any other critical factors. When developing sampling programs, survey planners also must realize that water quality characteristics vary from one body of water to another, from place to place in a given body of water, and from time to time at a fixed location in a given body of water.

A sampling program must be prepared in a manner that will produce accurate and useful results. The collection, handling, and testing of each sample should be scheduled and conducted in such a manner as to ensure that the results will be descriptive of the sources of the individual samples at the time and place of collection. Select locations for sampling stations and collect samples during times of the day and/or night that will provide the data needed to meet the objectives of your survey. Collect enough data over a period of time to adequately describe the condition or quality of the water at each sampling station.

To illustrate these important items, consider a simplified example of a waste discharge into a flowing river. Assume the river flows at 500 cubic feet per second (CFS) (323 MGD or 14 cu m/sec) and the treatment plant discharges at 10 CFS (6.46 MGD or 0.3 cu m/sec). Assume that it is desirable to find out what is the impact on the river. To determine the impact, the following questions must be answered:

1. What are the characteristics of the river upstream from the discharge?
2. What are the characteristics downstream?
3. If upstream and downstream river characteristics are different, does the discharge cause the difference?
4. Are the downstream river characteristics in violation of established standards or objectives?
5. If the river downstream is in violation of standards or objectives, did the discharge cause it?

Start with Question 1. Assume that you wish to measure:

a. Temperature
b. Dissolved Oxygen

13.31 Temperature

You must develop a temperature measurement program that will accurately describe the river temperature upstream from the discharge. How can this be done so that the changes from hour to hour during the day and from month to month during the year are known? Also, how can the temperature be measured so that the average for the river cross section (see Figure 13.2, page 248) can be found, as well as the variations from the average in the cross section?

In some rivers that are deep and slow-moving, the temperature may be several degrees cooler on the bottom than on the top. Thus, if the temperature measurement upriver from the discharge was taken near the bottom and the one downriver

TABLE 13.2 TROUBLESHOOTING POND EFFLUENTS

Indicator/Observation	Probable Cause	Check or Monitor	Solution
PONDS			*REVIEW CHAPTER 9, VOLUME I*
1. Floatables in effluent	1a. Outlet baffle not at proper location	1a. Visually inspect outlet baffle.	1a. Adjust outlet baffle.
	1b. Excessive floatables and scum on surface	1b. Visually inspect pond surface.	1b. Remove floatables from pond surface using hand rakes or skimmers. Scum can be broken up using jets of water or a motor boat. Broken scum often sinks.
	1c. Excessive velocity or insufficient detention time	1c. Visually inspect pond effluent.	1c. Reduce flow from upstream units.
	1d. Excessive solids accumulated on bottom have gasified and floated to surface	1d. Sound pond to determine depth of sludge accumulation.	1d. Remove accumulated sludge.
2. Excessive algae in effluent	2. Temperature or weather conditions may favor a particular species of algae	2. Visually observe effluent or run suspended solids test.	2. Operate ponds in series. Draw off effluent from below pond surface by use of a good baffle arrangement.
3. Excessive BOD in effluent	3. Detention time too short, hydraulic or organic overload, poor inlet and/or outlet arrangements and possible toxic discharges	3a. Influent flows. 3b. Calculate organic loading. 3c. Observe flow through inlet and outlet. 3d. Dead algae in effluent.	3a. Inspect collection system. 3b. Use pumps to recirculate pond contents. 3c. Rearrange inlets and outlets or install additional ones. 3d. Prevent toxic discharges.

TABLE 13.3 TROUBLESHOOTING BIOLOGICAL PROCESS EFFLUENTS

Indicator/Observation	Probable Cause	Check or Monitor	Solution
SECONDARY CLARIFIERS FOR TRICKLING FILTERS, ROTATING BIOLOGICAL CONTACTORS, OR ACTIVATED SLUDGE			*REVIEW CHAPTER 6, VOLUME I* *REVIEW CHAPTER 7, VOLUME I* *REVIEW CHAPTER 8, VOLUME I, and CHAPTER 11*
1. Floatables in effluent	1a. Clarifiers hydraulically overloaded	1a. Visually observe effluent or calculate hydraulic loadings.	1a. Install hardware cloth or similar screening device in effluent channels. Review Chapter 5, Volume I.
	1b. Skimmers not operating properly	1b. Observe skimmer movement.	1b. Lower skimmer arm, replace neoprene, or adjust arm.
2. Excessive solids in effluent	2a. Clarifiers hydraulically overloaded	2a. See 1a above.	2a. See 1a above and review operation of biological treatment process.
	2b. Biological treatment process organically overloaded	2b. Calculate BOD or organic loading.	2b. Review operation of biological treatment process.
	2c. Excessive solids in system	2c. Sludge blanket level.	2c. Increase sludge wasting rate.
3. High BOD, organic overload	3. See 2b above	3. See 2b above.	3. See 2b above.

TABLE 13.4 TROUBLESHOOTING DISINFECTED EFFLUENTS

Indicator/Observation	Probable Cause	Check or Monitor	Solution
DISINFECTION			*REVIEW CHAPTER 10, VOLUME I*
1. Unable to maintain chlorine residual	1a. Chlorinator not working properly	1a. Inspect chlorinator.	1a. Repair chlorinator.
	1b. Increase in chlorine demand	1b. Run chlorine demand tests.	1b. Increase chlorine dose and/or identify and correct cause of increase in demand.
2. Unable to meet coliform requirements	2a. Chlorine residual too low	2a. See 1a and b above.	2a. See 1a and b above.
	2b. Chlorine contact time too short	2b. Measure time for dye to pass through contact basin.	2b. Improve baffling arrangement.
	2c. Solids in effluent	2c. Observe solids or run suspended solids test.	2c. Install hardware cloth or similar screening device in effluent channels. Review operation of biological treatment process.
	2d. Sludge in chlorine contact basin	2d. Look for sludge deposits in contact basin.	2d. Drain and clean contact basin.
	2e. Diffuser not properly discharging chlorine	2e. Lower tank water level and inspect.	2e. Clean diffuser.
	2f. Mixing inadequate	2f. Add dye to diffuser.	2f. Add mechanical mixer or move diffuser.
	2g. Nitrate and/or nitrite combining with chlorine and decreasing effectiveness of disinfection	2g. Nitrite concentration.	2g. Adjust biological processes.

near the top, it might appear that the discharge had caused the stream to warm up when it actually had little impact.

The first thing to do is locate the river cross section (line across the stream) to be sampled. This may be located at a bridge or near a boat dock or some other accessible place. Then measure the temperature at several points across the stream and at several depths at each point (see Figure 13.2). Measurements also should be taken near shorelines, in backwater areas, and near the stream bottom. These are locations where problems first develop. If the temperatures are all about the same (within about 1°C or 2°F),[9] you can assume the stream is well mixed with a uniform temperature.

The next thing to consider is the time of day. Most streams will be cooler at night than during the day. Usually mid-channel temperature measurements vary less than those in shallow stretches. The minimum (lowest) temperature usually occurs about dawn and the highest in the late afternoon. The best way to measure these variations is to use a 24-hour recorder. If no recorder is available, take a measurement each hour for 24 hours (a little night work never hurt anybody!) to get an average temperature for the day: add up all the numbers and divide by 24. Then see what time of day the average value and the maximum value usually occur. (This may vary with the season.) Usually, it is accurate enough for most streams to measure the temperature at those times and use the values for an average daily and maximum daily. For example, assume the following measurements were recorded:

Time	Temperature, °C
12 noon	12.4
1 pm	12.8
2 pm	13.2
3 pm	13.6
4 pm	14.0
5 pm	13.8
6 pm	13.6
7 pm	13.3
8 pm	13.0
9 pm	12.7
10 pm	12.4
11 pm	12.0
12 midnight	11.6
1 am	11.2
2 am	10.8
3 am	10.5
4 am	10.3
5 am	10.1
6 am	10.0
7 am	10.2
8 am	10.4
9 am	10.8
10 am	11.2
11 am	11.8
TOTAL =	285.7°C

[9] °C means "degrees Centigrade or degrees Celsius," both of which refer to a particular temperature scale; °F means "degrees Fahrenheit," a different temperature scale.

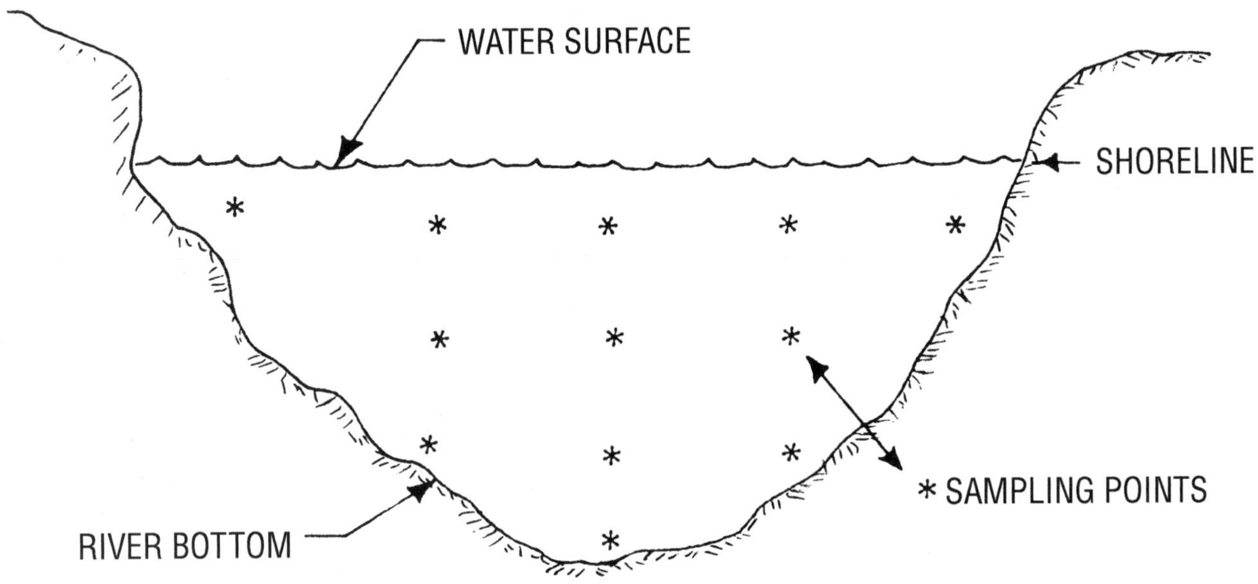

NOTE: Sampling points should be located near the shoreline and approximately one foot (0.3 m) below the water surface and one foot (0.3 m) above the bottom. The number of sampling points between the water surface and the bottom will depend on the depth of the water, and the number of vertical sampling sections will depend on the width of the stream. Total number of sampling points also will depend on budget, time, water quality indicators being sampled, and desired accuracy.

Fig. 13.2 River cross section showing typical sampling points

$$\text{Average Temperature, °C} = \frac{\text{Sum of Measurements, °C}}{\text{Number of Measurements}}$$

$$= \frac{285.7\text{°C}}{24}$$

$$= 11.9\text{°C}$$

The average temperature of 11.9°C occurs at about 11 am and again about 11 pm. If every day is like this, a measurement taken at 11 am each day will give a fairly accurate record of the daily average temperature. Periodic rechecks of the hourly variation should be made. The same goes for the maximum value, which occurs at 4 pm.

The next thing to consider is the seasonal variation in temperature. Streams normally warm up in summer and cool off in winter. Obviously, if a measurement is taken daily (as explained above) this record will show all variations throughout the year. But, usually the daily average temperature does not change very much from day to day.

Assume that the daily values for each month have been used to calculate monthly averages and that the following numbers have been obtained.

Month	Monthly Average Temperature, °C
January	7.0
February	6.0
March	8.0
April	10.0
May	12.0
June	14.0
July	16.0
August	19.0
September	18.0
October	14.0
November	10.0
December	8.0
TOTAL =	142.0°C

$$\text{Yearly Average, °C} = \frac{\text{Sum of Measurements, °C}}{\text{Number of Measurements}}$$

$$= \frac{142.0\text{°C}}{12}$$

$$= 11.8\text{°C}$$

Minimum Monthly Average, °C = 6.0°C

Maximum Monthly Average, °C = 19.0°C

The month-to-month changes in temperature are not very predictable because some years are colder than others, or summer lasts longer, or something else unusual can happen. The minimum and maximum monthly averages indicate the extent of the monthly changes for the observed year.

This discussion of temperature measurement does not mean that temperature is the most important characteristic to measure, although it is important from the standpoint of protecting fish. What is important is for you to be careful when selecting a sample for measuring *any* characteristic of wastewater or the receiving waters. Be sure to plan ahead so that each sample will indicate or represent the actual conditions of the river. If this is accomplished, you have obtained a *REPRESENTATIVE SAMPLE.*[10] Going out and blindly taking measurements (sometimes known as "flailing the water") can yield a lot of numbers which do not mean very much, but only the person who takes the measurement knows that. Others who use the numbers may assume they are meaningful and act accordingly. Always plan ahead to get maximum benefit from your receiving water sampling program.

Now go back to Question 2. What are the characteristics (in this case, temperature) downriver from the discharge? The same procedure for sample selection downstream should be used as previously explained. There is an additional consideration, however. The cross section selected for a sampling station should be far enough downstream for the waste discharge to have become well mixed. If the stream is very sluggish and deep, the discharge may not mix thoroughly for a mile or more, so it will be necessary to sample in such a way that the unmixed condition can be described properly (Figure 13.3). The higher the stream velocity, shallower the water, and the sharper the bends in the stream, the greater the turbulence and thus the quicker the discharge becomes mixed with the receiving waters.

Before going out in the field to measure water quality, obtain a range of expected values for guidance in sampling and interpreting results. Try various sampling locations to find high and low values for different times of the day and season.

Look now at the problem of measuring the temperature (or other characteristic) of the treatment plant effluent. Wastewater flow from a municipal discharge normally has a variable flow rate and variable characteristics. Fortunately, this variation follows similar patterns from day to day, week to week, year to year, so a logical sampling program can be set up to keep track of the characteristics of the plant effluent.

[10] *Representative Sample.* A sample portion of material, water, or wastestream that is as nearly identical in content and consistency as possible to that in the larger body being sampled.

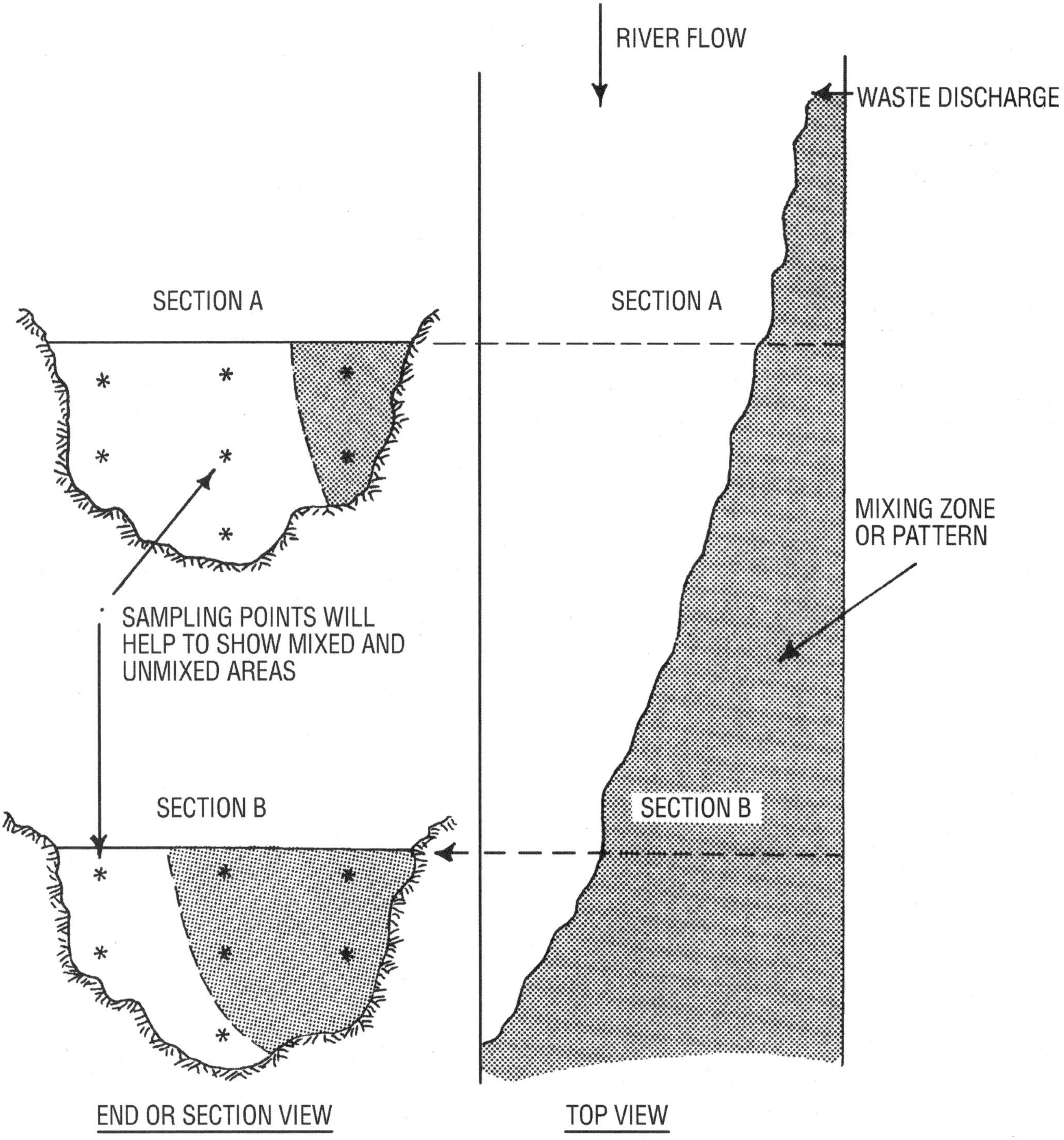

RIVER FLOW

WASTE DISCHARGE

SECTION A

SECTION A

MIXING ZONE OR PATTERN

SAMPLING POINTS WILL HELP TO SHOW MIXED AND UNMIXED AREAS

SECTION B

SECTION B

END OR SECTION VIEW

TOP VIEW

Fig. 13.3 Waste mixing in a river

Remember: The effluent measurement program must be designed to tell the operator how much volume of flow and what quality of constituents are entering the receiving waters hourly, daily, weekly, monthly, and yearly.

Try to select a convenient, accessible location where the effluent can be sampled easily. A remote sampling location or one that is difficult to reach will discourage regular sampling. Wherever the sampling station is located, it must provide meaningful samples.

13.32 Dissolved Oxygen

Another measurable characteristic of the stream is dissolved oxygen. The principles for collecting samples for measuring dissolved oxygen are the same as for measuring temperature. In fact, the amount of oxygen that can be dissolved in water depends on temperature, among other things.

Cold water will hold more dissolved oxygen than hot water. This does not mean that cold water always will contain more oxygen. Cold waters tend to slip under warmer waters because they have a greater density. In this position as the lower layers of a body of water, the cold waters are farther from the sources of oxygen created by surface aeration and algal activity. Bottom waters also may be close to deposits of organic materials containing organisms that use oxygen during *RESPIRATION*.[11] The net result could be lower oxygen concentrations in colder waters if they remain near the bottom too long.

In measuring the dissolved oxygen downstream from a treatment plant discharge, it is important to remember that a decrease in dissolved oxygen may not be noticeable immediately downstream, even if the effluent is well mixed with the stream. Many hours of flow time may be required for the oxygen to be reduced due to organic material in the discharge. Therefore, it is necessary to make an "oxygen profile" of the stream to get a good measure of the effect of the effluent.

Making a profile means merely measuring the dissolved oxygen at several different cross sections downstream from the discharge to find out where the lowest dissolved oxygen level occurs. For this example, assume the same waste discharge and river used in the previous example.

Number the cross sections to be sampled as follows:

Cross Section No.	Location
1	1 mile above discharge
2	1 mile below discharge
3	3 miles below discharge
4	5 miles below discharge
5	7 miles below discharge
6	9 miles below discharge
7	11 miles below discharge

Identify the location of any additional waste discharges or points of inflow from tributary streams. Selection of the number and location of sampling cross sections depends on stream characteristics, accessibility, and information desired. Normally, locations are selected to show critical conditions and changes in the receiving waters.

At each cross section, be sure representative samples are being selected. Always remember that a gallon of sample is supposed to be nearly identical to the millions of gallons of water that flow past the sampling point.

Now, assume that you have checked and found that only one properly located sample was required to represent each cross section and that the following measurements were obtained:

Cross Section No.	Temperature, °C	Dissolved Oxygen, mg/*L*
1	13.5	10.5
2	13.5	10.0
3	13.5	9.0
4	13.5	7.5
5	13.5	6.0
6	13.5	7.1
7	13.5	8.9

(Note that the temperature is constant for all cross sections. This is to simplify the example. If the temperature increased downstream from the discharge, some of the drop in dissolved oxygen would be due to the temperature increase and some would be due to the organic material. You also should be sure to notice any effects due to tributaries or other waste discharges.)

Figure 13.4 shows a plot or graph of the measurements listed above. This is a good way to show the dissolved oxygen profile. Profiles for different days or months can be plotted on the same sheet in different colors to show how the profile changes from season to season or from year to year. The location of the low point may move up or down the stream, depending on the amount of flow in the stream and other factors. Therefore, several points must be obtained for each profile to

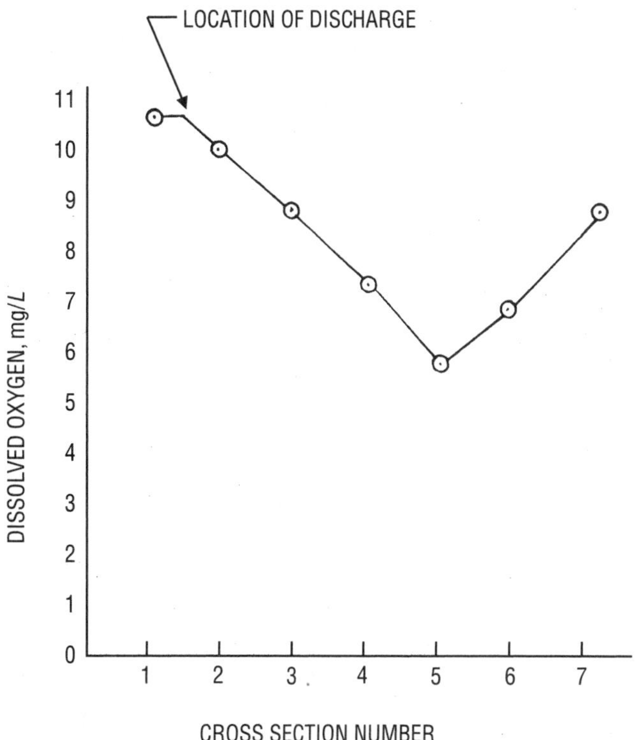

Fig. 13.4 Dissolved oxygen profile

[11] *Respiration.* The process in which an organism takes in oxygen for its life processes and gives off carbon dioxide.

be sure the amount and location of the low point can be determined. Additional discharges will complicate the profile.

13.33 Review of Sampling Results

To determine if sampling stations are in the proper location and producing meaningful results, the results from the testing program must be carefully reviewed. If the results do not appear correct, try to determine why they appear strange. Look for sampling errors, testing errors, and recording errors. Attempt to verify each step in your sampling program. Remember that you sample because something unusual can happen. Do not reject strange results because they are unusual, but investigate and attempt to identify the reasons for the results. Establish additional sampling locations when necessary and eliminate or relocate stations that are not producing meaningful results.

QUESTIONS

Write your answers in a notebook and then compare your answers with those on page 256.

13.3A An assistant plant operator collected samples and measured the temperature at one stream cross section each hour for 24 hours. Following are the results reported:

Time	Temperature, °C	Time	Temperature, °C
1 pm	13.5	1 am	12.4
2 pm	13.4	2 am	11.0
3 pm	14.3	3 am	11.0
4 pm	14.7	4 am	10.8
5 pm	14.5	5 am	10.6
6 pm	13.8	6 am	10.5
7 pm	14.0	7 am	10.7
8 pm	13.7	8 am	11.6
9 pm	13.4	9 am	11.4
10 pm	22.7	10 am	11.8
11 pm	12.7	11 am	12.4
12 midnight	12.2	12 noon	13.0
		TOTAL =	310.1°C

$$\text{Average Temperature, °C} = \frac{\text{Sum of Measurements, °C}}{\text{Number of Measurements}}$$

$$= \frac{310.1\text{°C}}{24}$$

$$= 12.9\text{°C}$$

You are the supervisor. What would be your response to these results?

13.3B Which measurement would you consider doubtful?

a. 12 midnight
b. 4 am
c. 10 pm
d. 6 am

13.3C Assume that you tell the operator to collect a new set of measurements two days later. The operator measures (not copies) all the same numbers except for the one you questioned earlier, and the new reading for that hour is 13.0. What are the new total and average values?

a. 296.3 and 13.2
b. 316.9 and 12.6
c. 300.4 and 12.5
d. 306.1 and 12.7
e. 298.9 and 13.0

13.3D a. What are the new maximum and minimum values?
b. At what times did they occur?
c. Which samples most nearly represented the average value?

13.34 Interpretation of Test Results and Follow-Up Corrections

A sudden drop in the receiving stream dissolved oxygen (DO) below the discharge line without a similar drop upstream from the line indicates that the plant biochemical oxygen demand (BOD) removal efficiency has decreased and should be corrected. The sudden drop in DO might have been caused by an increase in temperature resulting from an industrial waste discharge or stormwater discharge. The source should be determined and corrective action taken.

Sudden changes in effluent constituent concentrations may be due to process failure, a sudden increase in flow quantity, or a change in influent characteristics such as industrial waste discharges into the system. A review of influent data will determine whether the problem is in-plant or due to a change in the influent. If the problem is in the influent, the source will take time to trace. Until the source can be found and corrective action taken, adjustments should be made in the plant to minimize the detrimental effect on the receiving stream.

13.4 SAMPLING AND ANALYSIS

13.40 Collection

The proper technique to follow and precautions to observe to be sure a good sample is obtained are adequately covered in several publications. Complete information about proper sampling techniques can be found in the publications listed in Section 13.10, "Additional Reading." Also see Chapter 16, Section 16.3, "Sampling," in this manual. All of these books emphasize having the proper equipment to do a good job. Remember that behind all the instructions on techniques is the basic idea that the sample must be collected and preserved in such a manner that it does not change significantly from the time it is first obtained in the field until the final analysis is completed.

For example, if a gallon of water is selected from a stream for a dissolved oxygen measurement, be sure that the amount of dissolved oxygen in the sample does not change before it is measured. The same goes for any other characteristic.

Some characteristics change so rapidly that they should be measured immediately. This is true of temperature, pH, and dissolved oxygen. Some of the dissolved gases can be *FIXED*[12] for a while to allow transporting the sample to the laboratory for measurement. Procedures for "fixing" can be found

[12] *Fixed Sample.* A sample is fixed in the field by adding chemicals that prevent the water quality indicators of interest in the sample from changing before final measurements are performed later in the laboratory.

in Chapter 16, Section 16.36, "Preservation of Samples," *STANDARD METHODS,* and other publications on analysis. Usually, it is not a good idea to try to fix a sample containing a significant amount of organic matter, such as a plant effluent, because it tends to change anyway. Take a field kit to the sampling location and test at the site. Be sure the field kit equipment is capable of producing the desired accuracy.

13.41 Frequency of Sampling

Regular sampling intervals should be developed in cooperation with the regulatory agency having authority over the plant and the receiving waters. When the treatment plant effluent is not meeting discharge requirements or the receiving waters are not meeting established water quality standards, the frequency of sampling may be increased.

13.42 Size of Sample

When samples are tested in the field, the size of sample should be sufficient to perform the desired tests. If samples are preserved and transported to a lab for analysis, the size of sample should be at least twice the amount needed to perform the desired tests to allow for backup or repeat tests.

13.43 Labeling of Samples

A record must be made of every sample collected. Every sample bottle must be identified and should have attached to it a label or tag indicating the exact location where the sample was collected, date, hour, air and water temperatures, and name of collector. Other pertinent data such as water level or river flow and weather conditions should be recorded. Precipitation, cloud conditions, and prevailing winds during the previous few days should be noted when collecting samples. The weather during the three or four days before sampling may be entirely different from the weather on the day of sampling and may significantly affect the character of the sample. Sampling points should be identified on maps, including a detailed description, and identified in the field by easily located markers or landmarks.

13.44 What to Measure

There are hundreds, possibly thousands, of characteristics that could be measured in receiving waters and plant effluents. Many of them are not important to the operation of a treatment plant; however, it is possible to list a minimum number of characteristics that should enable the operator to measure the effect of the plant's effluent and find out if it meets water quality objectives. Measurements required for plant effluents are specified in NPDES permits and vary with location and use of receiving waters.

Table 13.5 contains a list of water quality characteristics that should be observed or measured in the receiving waters and in the plant effluent.

Many of these effluent tests are for the record rather than for plant control purposes. The operator can do nothing in adjustment of treatment plant processes to affect the characteristics. However, the operator will often be asked to measure them because they are listed in the plant's NPDES discharge requirements or receiving waters standards. Some of them, such as toxicity, can be controlled by ordinances that prevent toxic substances from being put in the wastewater collection system. Others, such as total dissolved solids, may require the city to find a new water supply.

TABLE 13.5 EFFLUENT AND RECEIVING WATERS CHARACTERISTICS

Effluent	Receiving Waters
1. Visual Appearance (color, floating materials)	Visual Appearance (color, floating materials)
2. Coliform Group Bacteria and Chlorine Residual	Coliform Group Bacteria
3. Biochemical Oxygen Demand	Dissolved Oxygen
4. Suspended Solids and Settleable Solids	Suspended Solids and Clarity
5. Temperature	Temperature
6. pH	pH
7. Odor	Odor
8. Grease	Grease

The list above is basic. Some additional characteristics that may be important in various situations are:

1. Total Dissolved Solids	Total Dissolved Solids
2. Chlorine	Chlorine
3. Hardness	Hardness
4. Toxicity (*BIOASSAY*[13])	Health of Aquatic Animals
5. Biostimulants (such as nitrogen, phosphorus)	Algae and other Aquatic Plants
6. Iron and Manganese	Iron and Manganese
7. Chlorinated Hydrocarbons (pesticides)	Chlorinated Hydrocarbons (pesticides)
8. Fluoride	Fluoride
9. MBAS	MBAS
10. Phenols	Phenols

[13] *Bioassay* (BUY-o-AS-say). (1) A way of showing or measuring the effect of biological treatment on a particular substance or waste. (2) A method of determining the relative toxicity of a test sample of industrial wastes or other wastes by using live test organisms, such as fish.

13.5 SAFETY

Take adequate precautions to prevent falling or slipping into the water when sampling. Not only can this save your life, it also will prevent muddying the sample. Choose sampling cross sections or "stations" carefully so that safe access is possible during all seasons of the year. Use snap-on safety belts when leaning over bridge railings or stream banks. When sampling on a bridge, be sure there is enough room for the person collecting the samples and traffic. Wear a flagging vest (bright red/orange) to alert vehicular traffic of your presence. Wear a Coast Guard-approved life jacket when sampling from a boat; and be sure the boat is well marked with lights, reflectors, and flags to prevent collisions. At least two people should be in a sampling boat. When the sampler's attention is focused on collecting samples, the other person must watch for other watercraft.

13.6 MAINTENANCE

Effluent discharge, reclamation, and reuse involves very few maintenance considerations except for pumps. These are discussed in Chapter 15, "Maintenance." Metering system maintenance consists mainly of cleaning and visual inspections. Otherwise, maintenance requires good housekeeping at all times.

13.7 REVIEW OF PLANS AND SPECIFICATIONS

A careful review of plans and specifications by an operator can provide the design engineer with helpful suggestions. These suggestions can produce a plant that is easier to operate and maintain.

1. Flowmeters

Plant effluent flow rates may be measured in either open channels or pipelines. Parshall flumes, weirs, or similar devices are used to measure flows in open channels and pipes flowing partially full. Magnetic flowmeters and Venturi meters are used to measure flows in pipes flowing completely full or under pressure. Magnetic flowmeters require an auxiliary power source, but Venturi meters do not require any power.

2. Outfall

The plant effluent should be discharged below the surface of the receiving waters at all times to prevent complaints about foaming and discoloration of the receiving waters. Inspect the actual location in the field. Be sure the outfall is below low water levels and the discharge conditions (such as location of river channel) will not change in the near future. In areas near oceans, the river level may fluctuate due to tidal action.

If the submerged outfall has a number of outlet ports or diffusers, determine if any maintenance will be necessary. If so, decide how to do the maintenance job and determine if there might be an easier way. Be sure a cleanout is located near the end of the outfall and is easily accessible. Often a minimum outlet flow or velocity must be maintained to prevent diffusers from clogging. Determine if minimum plant flows during start-up or low-flow periods will be adequate to prevent clogging. Maintenance consists of clearing outlet ports of any debris that gets into the effluent channels and removing any silt deposits or debris left by the river.

Sometimes receiving waters become very high during flood conditions. Determine if the plant effluent pumps have sufficient capacity and discharge head to perform as intended if the plant receives high inflows when flood conditions exist in the receiving waters.

3. Emergency Storage

What happens if your plant's chlorination facilities do not work or other treatment processes are unable to provide adequate treatment? Newer plants are being constructed with emergency storage basins to hold peak storm runoff or inadequately treated wastewater until this wastewater can be properly treated by the plant. If receiving waters serve as a source for a downstream domestic water supply or other use requiring high-quality water, construction of an emergency storage basin can be very important.

13.8 OTHER TYPES OF RECEIVING WATERS

This chapter on effluent reclamation and reuse by discharge to surface waters has emphasized very simple examples of flowing streams or rivers. Some other types of receiving waters you may encounter are:

1. Oceans
2. Estuaries
3. Groundwaters
4. Lakes

These receiving waters usually require a more sophisticated approach to sampling and measurement of characteristics. The basic rules are the same, however, for sample selection and collection techniques. The best answer to sampling these types of receiving waters is to seek advice from a consultant, a regulatory agency, or other experts on where and how to sample and on how to evaluate the results.

13.9 ARITHMETIC ASSIGNMENT

Turn to the Arithmetic Appendix at the back of this manual. Read and work the problems in Section A.32, "Effluent Discharge, Reclamation, and Reuse." Check the arithmetic in this section using an electronic calculator. You should be able to get the same answers. Section A.52 contains similar problems using metric units.

13.10 ADDITIONAL READING

1. *MOP 11,* Chapter 26,* "Effluent Disinfection," New Edition of MOP 11, Volume II, *LIQUID PROCESSES.*
2. *NEW YORK MANUAL*, Chapter 10,* "Sampling and Testing Procedures."
3. *STANDARD METHODS FOR THE EXAMINATION OF WATER AND WASTEWATER,* 21st Edition, 2005. Obtain from Water Environment Federation (WEF), Publications Order Department, 601 Wythe Street, Alexandria, VA 22314-1994. Order No. S82011. Price to members, $194.75; nonmembers, $259.75; price includes cost of shipping and handling.
4. *GOOD UNTIL THE LAST DROP: A PRACTITIONER'S GUIDE TO WATER REUSE.* Obtain from American Public Works Association (APWA), Bookstore, PO Box 802296, Kansas City, MO 64180-2296. Order No. PB.A515. Price to members, $54.00; nonmembers, $65.00; price includes cost of shipping and handling. Phone (800) 848-2792 or visit www.apwa.net/bookstore/.

* Depends on edition.

QUESTIONS

Write your answers in a notebook and then compare your answers with those on page 256.

13.4A What is a fixed sample?

13.5A What safety precautions should be taken when sampling?

13.6A What are the most important maintenance considerations for effluent discharge, reclamation, and reuse systems?

13.7A What items should be examined when reviewing the plans and specifications for an effluent discharge, reclamation, and reuse system?

Please answer the discussion and review questions next.

DISCUSSION AND REVIEW QUESTIONS

Chapter 13. EFFLUENT DISCHARGE, RECLAMATION, AND REUSE

The purpose of these questions is to indicate to you how well you understand the material in the chapter. Write the answers to these questions in your notebook.

1. Why should receiving waters be sampled and tested?
2. Can a treatment plant operate as effectively as possible and still have a bad effect on the receiving waters? Why?
3. How would you determine the impact of a plant effluent on receiving waters?
4. What would you do if your plant is not meeting NPDES permit requirements?
5. Where should samples be collected in the cross section of a stream at a particular sampling location? Assume you are trying to find the point in the section that will give you a representative measurement of the entire section.
6. What does the term "representative sample" mean?
7. Why should more than one sampling station be established downstream from the point of wastewater discharge?
8. What is a major concern regarding sample water quality indicators after the sample has been collected?
9. Why are some water quality indicators measured instream and others measured in the lab?
10. How large a sample should be collected?
11. What information should be included on a sample label?
12. What safety precautions should be taken when sampling from a bridge?

SUGGESTED ANSWERS

Chapter 13. EFFLUENT DISCHARGE, RECLAMATION, AND REUSE

Answers to questions on page 243.

13.0A The most common method of effluent discharge, reclamation, and reuse is discharge into surface waters.

13.0B Other common methods of wastewater discharge, reclamation, and reuse include discharge to land by the use of irrigation, groundwater recharge basins, and underground discharge.

Answers to questions on page 244.

13.1A Water quality indicators that may be monitored or determined by in-stream measurements include dissolved oxygen (DO), pH, and temperature.

13.1B Water quality indicators usually determined by samples collected in the field and analyzed in the lab include biochemical oxygen demand (BOD), suspended solids (SS), and total coliform and fecal coliform bacteria.

Answers to questions on page 245.

13.2A Operating procedures should be developed for the following conditions in an effluent discharge, reclamation, and reuse program:

1. Start-up
2. Routine operation
3. Shutdown
4. Abnormal conditions

13.2B An operational strategy should attempt to do everything possible to minimize the discharge of pollutants to the receiving waters.

Answers to questions on page 252.

13.3A I would contact the assistant plant operator and attempt to verify the results recorded. Also, I would ask the operator if anything unusual was observed while collecting the sample.

13.3B (c) 10 pm

13.3C (c) 300.4 and 12.5

13.3D a. Maximum 14.7°C and Minimum 10.5°C
b. Time of maximum temperature, 4 pm; minimum temperature, 6 am
c. 11 am and 1 am

Answers to questions on page 255.

13.4A A fixed sample is a sample that has chemicals added to prevent a particular water quality indicator from changing before the sample can be analyzed.

13.5A Adequate safety precautions should be taken when sampling to prevent falling or slipping into the water. Choose sampling cross sections or "stations" carefully so that safe access is possible during all seasons of the year. Use snap-on safety belts when leaning over bridge railings or stream banks. When sampling on a bridge, be sure there is enough room for the person collecting the samples and traffic. Wear a flagging vest (bright red/orange) to alert vehicular traffic of your presence. Wear a Coast Guard-approved life jacket when sampling from a boat; and be sure the boat is well marked with lights, reflectors, and flags to prevent collisions. At least two people should be in a sampling boat.

13.6A Important maintenance considerations for effluent discharge, reclamation, and reuse systems include:

1. Pumps
2. Meters
3. Good housekeeping at all times

13.7A When reviewing plans and specifications for an effluent discharge, reclamation, and reuse system, consider:

1. Flowmeters
2. Outfall
3. Emergency storage

CHAPTER 14

PLANT SAFETY

by

Robert Reed

Revised by

Russ Armstrong

TABLE OF CONTENTS

Chapter 14. PLANT SAFETY

SAFETY ASSIGNMENT: DO AS I SAY... NOT AS I DO!!

OBJECTIVES

Chapter 14. PLANT SAFETY

Following completion of Chapter 14, you should be able to:

1. Identify the types of hazards you may encounter operating a wastewater treatment plant.
2. Recognize unsafe conditions and correct them whenever they develop.
3. Organize regular "tailgate" safety meetings.
4. Develop the habit of always "thinking safety and working safely."

NOTE: Special safety information is given in other chapters because of the importance of safety considerations at all times.

S SAFETY FIRST
A ACCIDENTS COST LIVES
F FASTER IS NOT ALWAYS BETTER
E EXPECT THE UNEXPECTED
T THINK BEFORE YOU ACT
Y YOU CAN MAKE THE DIFFERENCE

ACCIDENTS DO NOT JUST HAPPEN...
THEY ARE CAUSED!

WORDS

Chapter 14. PLANT SAFETY

ACUTE HEALTH EFFECT | ACUTE HEALTH EFFECT

An adverse effect on a human or animal body, with symptoms developing rapidly.

AIR GAP | AIR GAP

An open, vertical drop, or vertical empty space, between a drinking (potable) water supply and potentially contaminated water. This gap prevents the contamination of drinking water by backsiphonage because there is no way potentially contaminated water can reach the drinking water supply.

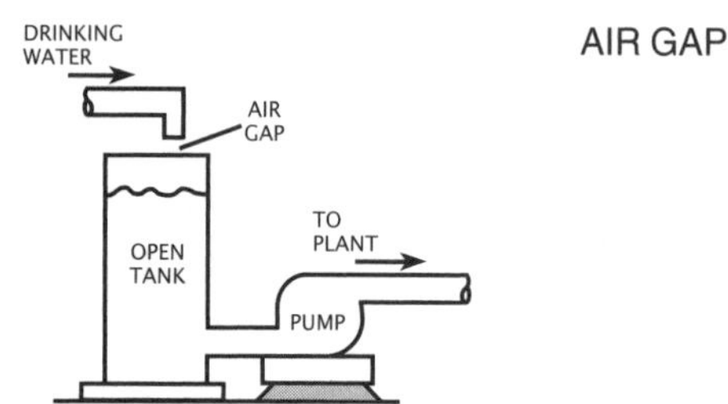

CFR | CFR

Code of Federal Regulations. A publication of the US government that contains all of the proposed and finalized federal regulations, including safety and environmental regulations.

CARCINOGEN (kar-SIN-o-jen) | CARCINOGEN

Any substance that tends to produce cancer in an organism.

CHRONIC HEALTH EFFECT | CHRONIC HEALTH EFFECT

An adverse effect on a human or animal body with symptoms that develop slowly over a long period of time or that recur frequently.

COMPETENT PERSON | COMPETENT PERSON

A competent person is defined by OSHA as a person capable of identifying existing and predictable hazards in the surroundings, or working conditions that are unsanitary, hazardous, or dangerous to employees, and who has authorization to take prompt corrective measures to eliminate the hazards.

CONFINED SPACE | CONFINED SPACE

Confined space means a space that:

(1) Is large enough and so configured that an employee can bodily enter and perform assigned work; and

(2) Has limited or restricted means for entry or exit (for example, manholes, tanks, vessels, silos, storage bins, hoppers, vaults, and pits are spaces that may have limited means of entry); and

(3) Is not designed for continuous employee occupancy.

Also see DANGEROUS AIR CONTAMINATION and OXYGEN DEFICIENCY.

CONFINED SPACE, NON-PERMIT | CONFINED SPACE, NON-PERMIT

A non-permit confined space is a confined space that does not contain or, with respect to atmospheric hazards, have the potential to contain any hazard capable of causing death or serious physical harm.

CONFINED SPACE, PERMIT-REQUIRED (PERMIT SPACE) | CONFINED SPACE, PERMIT-REQUIRED (PERMIT SPACE)

A confined space that has one or more of the following characteristics:

(1) Contains or has a potential to contain a hazardous atmosphere,

(2) Contains a material that has the potential for engulfing an entrant,

(3) Has an internal configuration such that an entrant could be trapped or asphyxiated by inwardly converging walls or by a floor that slopes downward and tapers to a smaller cross section, or

(4) Contains any other recognized serious safety or health hazard.

DANGEROUS AIR CONTAMINATION DANGEROUS AIR CONTAMINATION

An atmosphere presenting a threat of causing death, injury, acute illness, or disablement due to the presence of flammable and/or explosive, toxic, or otherwise injurious or incapacitating substances.

(1) Dangerous air contamination due to the flammability of a gas, vapor, or mist is defined as an atmosphere containing the gas, vapor, or mist at a concentration greater than 10 percent of its lower explosive (lower flammable) limit (LEL).

(2) Dangerous air contamination due to a combustible particulate is defined as a concentration that meets or exceeds the particulate's lower explosive limit (LEL).

(3) Dangerous air contamination due to the toxicity of a substance is defined as the atmospheric concentration that could result in employee exposure in excess of the substance's permissible exposure limit (PEL).

NOTE: A dangerous situation also occurs when the oxygen level is less than 19.5 percent by volume (OXYGEN DEFICIENCY) or more than 23.5 percent by volume (OXYGEN ENRICHMENT).

DECIBEL (DES-uh-bull) DECIBEL

A unit for expressing the relative intensity of sounds on a scale from zero for the average least perceptible sound to about 130 for the average level at which sound causes pain to humans. Abbreviated dB.

ENGULFMENT ENGULFMENT

Engulfment means the surrounding and effective capture of a person by a liquid or finely divided (flowable) solid substance that can be aspirated to cause death by filling or plugging the respiratory system or that can exert enough force on the body to cause death by strangulation, constriction, or crushing.

ENTRAIN ENTRAIN

To trap bubbles in water either mechanically through turbulence or chemically through a reaction.

FLAME POLISHED FLAME POLISHED

Melted by a flame to smooth out irregularities. Sharp or broken edges of glass (such as the end of a glass tube) are rotated in a flame until the edge melts slightly and becomes smooth.

HYDRAULIC JUMP HYDRAULIC JUMP

The sudden and usually turbulent abrupt rise in water surface in an open channel when water flowing at high velocity is suddenly retarded to a slow velocity.

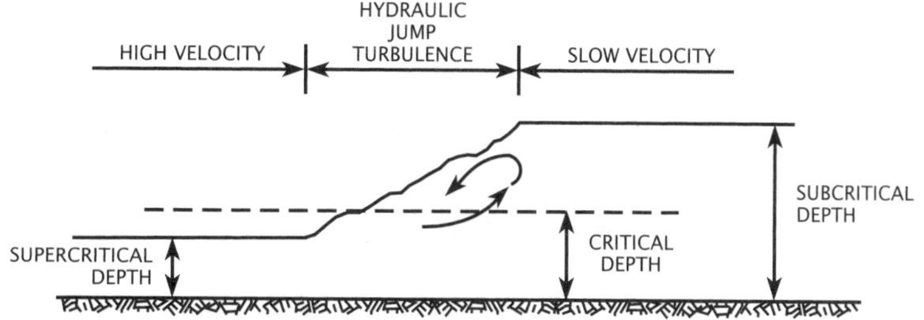

IDLH IDLH

Immediately Dangerous to Life or Health. The atmospheric concentration of any toxic, corrosive, or asphyxiant substance that poses an immediate threat to life or would cause irreversible or delayed adverse health effects or would interfere with an individual's ability to escape from a dangerous atmosphere.

LOWER EXPLOSIVE LIMIT (LEL) LOWER EXPLOSIVE LIMIT (LEL)

The lowest concentration of a gas or vapor (percent by volume in air) that explodes if an ignition source is present at ambient temperature. At temperatures above 250°F (121°C) the LEL decreases because explosibility increases with higher temperature.

MATERIAL SAFETY DATA SHEET (MSDS) MATERIAL SAFETY DATA SHEET (MSDS)

A document that provides pertinent information and a profile of a particular hazardous substance or mixture. An MSDS is normally developed by the manufacturer or formulator of the hazardous substance or mixture. The MSDS is required to be made available to employees and operators or inspectors whenever there is the likelihood of the hazardous substance or mixture being introduced into the workplace. Some manufacturers are preparing MSDSs for products that are not considered to be hazardous to show that the product or substance is not hazardous.

NONSPARKING TOOLS NONSPARKING TOOLS

These tools will not produce a spark during use. They are made of a nonferrous material, usually a copper-beryllium alloy.

OSHA (O-shuh) OSHA

The Williams-Steiger Occupational Safety and Health Act of 1970 (OSHA) is a federal law designed to protect the health and safety of workers, including the operators of water supply and treatment systems and wastewater collection and treatment systems. The Act regulates the design, construction, operation, and maintenance of water and wastewater systems. OSHA regulations require employers to obtain and make available to workers the Material Safety Data Sheets (MSDSs) for chemicals used at industrial facilities and treatment plants. OSHA also refers to the federal and state agencies that administer the OSHA regulations.

OLFACTORY (all-FAK-tore-ee) FATIGUE OLFACTORY FATIGUE

A condition in which a person's nose, after exposure to certain odors, is no longer able to detect the odor.

OXYGEN DEFICIENCY OXYGEN DEFICIENCY

An atmosphere containing oxygen at a concentration of less than 19.5 percent by volume.

OXYGEN ENRICHMENT OXYGEN ENRICHMENT

An atmosphere containing oxygen at a concentration of more than 23.5 percent by volume.

SET POINT SET POINT

The position at which the control or controller is set. This is the same as the desired value of the process variable. For example, a thermostat is set to maintain a desired temperature.

SEWER GAS SEWER GAS

(1) Gas in collection lines (sewers) that results from the decomposition of organic matter in the wastewater. When testing for gases found in sewers, test for oxygen deficiency, oxygen enrichment, and also for explosive and toxic gases.

(2) Any gas present in the wastewater collection system, even though it is from such sources as gas mains, gasoline, and cleaning fluid.

SPECIFIC GRAVITY SPECIFIC GRAVITY

(1) Weight of a particle, substance, or chemical solution in relation to the weight of an equal volume of water. Water has a specific gravity of 1.000 at 4°C (39°F). Particulates with specific gravity less than 1.0 float to the surface and particulates with specific gravity greater than 1.0 sink.

(2) Weight of a particular gas in relation to the weight of an equal volume of air at the same temperature and pressure (air has a specific gravity of 1.0). Chlorine gas has a specific gravity of 2.5.

SPOIL SPOIL

Excavated material, such as soil, from the trench of a water main or sewer.

SURFACE-ACTIVE AGENT SURFACE-ACTIVE AGENT

The active agent in detergents that possesses a high cleaning ability. Also called a SURFACTANT.

TAILGATE SAFETY MEETING TAILGATE SAFETY MEETING

Brief (10 to 20 minutes) safety meetings held every 7 to 10 working days. The term comes from the safety meetings regularly held by the construction industry around the tailgate of a truck.

TIME-WEIGHTED AVERAGE (TWA) TIME-WEIGHTED AVERAGE (TWA)

The average concentration of a pollutant based on the times and levels of concentrations of the pollutant. The time-weighted average is equal to the sum of the portion of each time period (as a decimal, such as 0.25 hour) multiplied by the pollutant concentration during the time period divided by the hours in the workday (usually 8 hours). 8TWA PEL is the time-weighted average permissible exposure limit, in parts per million, for a normal 8-hour workday and a 40-hour workweek to which nearly all workers may be repeatedly exposed, day after day, without adverse effect.

VISCOSITY (vis-KOSS-uh-tee) VISCOSITY

A property of water, or any other fluid, that resists efforts to change its shape or flow. Syrup is more viscous (has a higher viscosity) than water. The viscosity of water increases significantly as temperatures decrease. Motor oil is rated by how thick (viscous) it is; 20 weight oil is considered relatively thin while 50 weight oil is relatively thick or viscous.

CHAPTER 14. PLANT SAFETY

(Lesson 1 of 3 Lessons)

14.0 WHY SAFETY?

A cat may have nine lives, but you have only one. Protect it. Others may try, but only your efforts in thinking about safety and acting safely can ensure you the opportunity of continuing to live your one life.

You are working at an occupation that has an accident frequency rate second only to that of the mining industry. This is not a very desirable record. This chapter is intended to make you aware of the many hazards that may be encountered at your plant. Guidelines for working safely are provided. Precise requirements for your plant and work may vary depending on the specific design of your plant and the equipment and processes being used. Site-specific safety procedures should be confirmed with your local safety regulatory agency.

Your employer has the responsibility of providing you with a safe place to work. But you, the operator, who has overall responsibility for your treatment plant, must accept the task of seeing to it that your plant is maintained in such a manner as to continually provide a safe place to work. This can only be done by constantly "thinking safety and working safely."

You have the responsibility of protecting yourself and other plant personnel or visitors by establishing safety procedures for your plant and then by seeing that they are followed. Train yourself to analyze jobs, work areas, and procedures from a safety standpoint. Learn to recognize potentially hazardous actions or conditions. When you do recognize a hazard, take immediate steps to eliminate it by corrective action. Corrective actions can range from temporary isolation of the hazard through the placement of barricades or warning signs to engineering redesign and physical modifications to permanently eliminate the hazard(s). As an individual, you can be held liable for injuries or property damage that results from an unsafe condition, act, or situation that you knew of, or through reasonable diligence, could have known of.

REMEMBER: Accidents do not just happen—they are caused! Behind every accident there was a chain of events that led to an unsafe act, unsafe condition, or a combination of both. THINK SAFETY!

Accidents may be prevented by using good common sense, applying a few basic safety rules, and, particularly, by acquiring a good knowledge of the hazards unique to your job as a plant operator.

The Bell System had one of the best safety records of any industry. A variation of their successful policy statement is:

> "There is no job so important nor emergency so great that we cannot take time to do our work safely."

Although this chapter is intended primarily for the wastewater treatment plant operator, the operators of many small plants also have the responsibility of sewer maintenance. Therefore, the safety aspects of both sewer maintenance and plant operation will be discussed.

14.1 TYPES OF HAZARDS

You are equally exposed to accidents whether working on the collection system or working in a treatment plant. As an operator, you may be exposed to:

1. Physical injuries
2. Infections and infectious diseases
3. Confined spaces
4. Oxygen deficiency or enrichment
5. Toxic or suffocating gases or vapors
6. Toxic and harmful chemicals
7. Radiological hazards
8. Explosive gas mixtures
9. Fires
10. Electric shock/stored energy
11. Noise
12. Dusts, fumes, mists, gases, and vapors

14.10 Physical Injuries

The most common physical injuries are cuts, bruises, strains, and sprains. Injuries can be caused by many things including moving machinery, improper lifting techniques, or slippery surfaces. Falls from or into tanks, wet wells, catwalks, or conveyors can also be disabling or fatal.

If you work in an area six feet (1.8 meters) or more above a lower level and you are not protected by guardrails or other fall protection, you must use a personal fall arrest system. The fall arrest system may consist of a combination of anchorage points, connectors, a body harness, and a lanyard, deceleration device, or a lifeline. Connectors used in the fall restraint system must be the locking type with a self-closing, self-locking keeper that stays closed and locked until physically unlocked and pushed open for connecting and disconnecting. Lanyards, lifelines, and anchorage points must be rated at 5,000 pounds (2,273 kilograms) breaking and tensile strength. The fall arrest system must be rigged to prevent you from contacting the lower level or free falling more than six feet (1.8 meters). Fall arrest systems must be inspected prior to each use to detect damage, deterioration, or wear that could compromise the user's safety. Never use any suspect equipment. If the system is subjected to impact loading (someone falls), it must be removed from service until a *COMPETENT PERSON*[1] inspects it and approves it for reuse. Additional information and training on fall arrest systems can be obtained from your local safety regulatory agency and safety equipment vendor.

Working with ladders can also be very hazardous. Do not overreach when on a ladder and make sure that the ladder is positioned properly, is secured, and is appropriate for the job. Portable ladders are classified according to the weight they can sustain and should be labeled to identify the manufacturers' rated capacity. Make sure the ladder you use is rated for your weight and the weight of any tools or equipment it may have to support. Do not use a conductive (metal) ladder when working around electrical equipment.

Most injuries can be avoided by the proper use of ladders, hand tools, and safety equipment, and by following established safety procedures.

14.11 Infections and Infectious Diseases

Although treatment plants and plant personnel are not expected to be pristine, personal hygiene is the best protection against the risk of infections and infectious diseases such as typhoid fever, dysentery, hepatitis, and tetanus. Immunization shots for protection against tetanus, polio, and for both hepatitis A and hepatitis B are often available free or for a minor charge from your local health department. *REMEMBER,* many pathogenic organisms can be found in wastewater. Some diseases that may be transmitted by wastewater are anthrax, tuberculosis, paratyphoid fever, cholera, and polio. Tapeworms and the organisms associated with food poisoning may also be present.

The possibility that Acquired Immune Deficiency Syndrome (AIDS), which is caused by a virus, can be contracted from exposure to raw wastewater has been discounted by researchers who have found that although the AIDS virus is present in the wastes from AIDS victims, the raw wastewater environment is hostile to the virus itself and has not been identified as a mode of transmission to date. Needle sticks from potentially contaminated syringes should remain a concern to operators and maintenance personnel. Fluids in or on syringes may provide a less severe environment than raw wastewater where dilution and chlorination significantly reduce infection potential.

Make it a habit to thoroughly wash your hands before eating or smoking, as well as before and after using the restroom. *ALWAYS* wear proper protective gloves when you may contact wastewater or sludge in any form. Bandages covering wounds should be changed frequently.

Do not wear your work clothes home because diseases may be transmitted to your family. Provisions should be made in your plant for a locker room where each employee has a locker. Work clothes should be placed in lockers and not thrown on the floor. Your work clothes should be cleaned as often as necessary. If you are required to wear protective clothing because of the possibility of contamination with toxic materials, you should store your street clothes and your protective clothing in separate lockers. If your employer does not supply you with uniforms and laundry service, investigate the availability of disposable clothing for "dirty" jobs. If you must take your work clothes home, launder them separately from your regular family wash. All of these precautions will reduce the possibility of you or your family becoming ill because of your contact with wastewater.

What is wrong with the above sketch? *NEVER* stick objects in your mouth that you do not intend to eat.

14.12 Confined Spaces

This section outlines procedures for preventing personal exposure to dangerous air contamination and/or oxygen deficiency/enrichment when working within such spaces as tanks, channels, boilers, sewers, or manholes. If you enter confined spaces, you must develop and implement written, understandable procedures in compliance with OSHA standards and you must provide training in the use of these procedures for all persons whose duties may involve confined space entry. *The procedures presented here are intended as guidelines. Exact procedures for work in confined spaces may vary with different agencies and geographical locations and must be confirmed with the appropriate regulatory safety agency.*

A confined space may be defined as any space that: (1) is large enough and so configured that an employee can bodily enter and perform assigned work; and (2) has limited or restricted means for entry or exit (for example, manholes, tanks, vessels, silos, storage bins, hoppers, vaults, and pits are spaces that may have limited means of entry); and (3) is not designed for continuous employee occupancy. One easy way to identify a confined space is by whether or not you can enter it by simply walking while standing fully upright. In general, if you must duck, crawl, climb, or squeeze into the space, it is considered a confined space.

[1] *Competent Person.* A competent person is defined by OSHA as a person capable of identifying existing and predictable hazards in the surroundings, or working conditions that are unsanitary, hazardous, or dangerous to employees, and who has authorization to take prompt corrective measures to eliminate the hazards.

A major concern in confined spaces is whether the existing ventilation is capable of removing *DANGEROUS AIR CONTAMINATION*[2] and/or oxygen deficiency/enrichment that may exist or develop. In wastewater treatment, we are concerned primarily with oxygen deficiency (less than 19.5 percent oxygen by volume), oxygen enrichment (greater than 23.5 percent by volume), methane (explosive), hydrogen sulfide (toxic), and other gases as identified in Table 14.1.

The potential for buildup of toxic or explosive gas mixtures and/or oxygen deficiency/enrichment exists in all confined spaces. The atmosphere must be checked with reliable, calibrated instruments before every entry. When testing the atmosphere, first test for oxygen deficiency/enrichment, then combustible gases and vapors, and then toxic gases and vapors. The oxygen concentration in normal breathing air is 20.9 percent. The atmosphere in the confined space must not fall below 19.5 percent or exceed 23.5 percent oxygen. Engineering controls are required to prevent low or high oxygen levels. However, personal protective equipment is necessary if engineering controls are not possible. In atmospheres where the oxygen content is less than 19.5 percent, supplied-air or self-contained breathing apparatus (SCBA) is required. SCBAs are sometimes referred to as scuba gear because they look and work much like the air tanks used by divers.

Entry into confined spaces is never permitted until the space has been properly ventilated using specially designed forced-air ventilators. These blowers force all the existing air out of the space, replacing it with fresh air from outside. This crucial step must *ALWAYS* be taken even if atmospheric monitoring instruments show the atmosphere to be safe. Because some of the gases likely to be encountered in a confined space are combustible or explosive, the blowers must be specially designed so that the blower itself will not create a source of ignition that could cause an explosion.

There are two general classifications of confined spaces: (1) non-permit confined spaces, and (2) permit-required confined spaces (permit spaces).

A *NON-PERMIT CONFINED SPACE* is a confined space that does not contain or, with respect to atmospheric hazards, have the potential to contain any hazard capable of causing death or serious physical harm. The following steps are recommended *PRIOR* to entry into *ANY* confined space:

1. Ensure that all employees involved in confined space work have been effectively trained.
2. Identify and close off or reroute any lines that may carry harmful substance(s) to, or through, the work area.
3. Empty, flush, or purge the space of any harmful substance(s) to the extent possible.
4. Monitor the atmosphere at the work site and within the space to determine if dangerous air contamination and/or oxygen deficiency/enrichment exists.
5. Record the atmospheric test results and keep them at the site throughout the work period.
6. If the space is interconnected with another space, each space must be tested and the most hazardous conditions found must govern subsequent steps for entry into the space.

7. If an atmospheric hazard is noted, use portable blowers to further ventilate the area; retest the atmosphere after a suitable period of time. Do not place the blowers inside the confined space.
8. If the *ONLY* hazard posed by the space is an actual or potential hazardous atmosphere and the preliminary ventilation has eliminated the atmospheric hazard or continuous forced ventilation *ALONE* can maintain the space safe for entry, entry into the area may proceed.

A *PERMIT-REQUIRED CONFINED SPACE* (permit space) is a confined space that has one or more of the following characteristics:

1. Contains or has the potential to contain a hazardous atmosphere
2. Contains a material that has the potential for engulfing an entrant
3. Has an internal configuration such that an entrant could be trapped or asphyxiated by inwardly converging walls or by a floor that slopes downward and tapers to a smaller cross section
4. Contains any other recognized serious safety or health hazard

OSHA regulations require that a confined space entry permit be completed for each permit-required confined space entry (Figure 14.1). The permit must be renewed each time the space is left and re-entered, even if only for a break or lunch, or to go get a tool. The confined space entry permit is "an authorization and approval in writing that specifies the location and type of work to be done, certifies that all existing

[2] *Dangerous Air Contamination.* An atmosphere presenting a threat of causing death, injury, acute illness, or disablement due to the presence of flammable and/or explosive, toxic, or otherwise injurious or incapacitating substances.

(1) Dangerous air contamination due to the flammability of a gas, vapor, or mist is defined as an atmosphere containing the gas, vapor, or mist at a concentration greater than 10 percent of its lower explosive (lower flammable) limit (LEL).

(2) Dangerous air contamination due to a combustible particulate is defined as a concentration that meets or exceeds the particulate's lower explosive limit (LEL).

(3) Dangerous air contamination due to the toxicity of a substance is defined as the atmospheric concentration that could result in employee exposure in excess of the substance's permissible exposure limit (PEL).

NOTE: A dangerous situation also occurs when the oxygen level is less than 19.5 percent by volume (OXYGEN DEFICIENCY) or more than 23.5 percent by volume (OXYGEN ENRICHMENT).

TABLE 14.1 COMMON DANGEROUS GASES ENCOUNTERED IN WASTEWATER COLLECTION SYSTEMS AND AT WASTEWATER TREATMENT PLANTS[a]

Name of Gas and Chemical Formula	8TWA PEL[b]	Specific Gravity or Vapor Density[c] (Air = 1)	Explosive Range (% by volume in air) Lower Limit	Upper Limit	Common Properties (Percentages below are percent in air by volume)	Physiological Effects (Percentages below are percent in air by volume)	Most Common Sources in Sewers	Method of Testing[d]
Oxygen, O_2 (in Air)		1.11	Not flammable		Colorless, odorless, tasteless, non-poisonous gas. Supports combustion.	Normal air contains 20.93% of O_2. If O_2 is less than 19.5%, do not enter space without respiratory protection.	Oxygen depletion from poor ventilation and absorption or chemical consumption of available O_2.	Oxygen monitor.
Gasoline Vapor, C_5H_{12} to C_9H_{20}	300	3.0 to 4.0	1.3	7.0	Colorless, odor noticeable in 0.03%. Flammable. Explosive.	Anesthetic effects when inhaled. 2.43% rapidly fatal. 1.1% to 2.2% dangerous for even short exposure.	Leaking storage tanks, discharges from garages, and commercial or home dry-cleaning operations.	Combustible gas monitor.
Carbon Monoxide, CO	50	0.97	12.5	74.2	Colorless, odorless, nonirritating. Tasteless. Flammable. Explosive.	Hemoglobin of blood has strong affinity for gas causing oxygen starvation. 0.2 to 0.25% causes unconsciousness in 30 minutes.	Manufactured fuel gas.	1. CO monitor. 2. CO tubes.
Hydrogen, H_2		0.07	4.0	74.2	Colorless, odorless, tasteless, non-poisonous, flammable. Explosive. Propagates flame rapidly; very dangerous.	Acts mechanically to deprive tissues of oxygen. Does not support life. A simple asphyxiant.	Manufactured fuel gas.	Combustible gas monitor.
Methane, CH_4		0.55	5.0	15.0	Colorless, tasteless, odorless, non-poisonous. Flammable. Explosive.	See hydrogen.	Natural gas, marsh gas, manufactured fuel gas, gas found in sewers.	Combustible gas monitor.
Hydrogen Sulfide, H_2S	10	1.19	4.3	46.0	Rotten egg odor in small concentrations, but sense of smell rapidly impaired. Odor not evident at high concentrations. Colorless. Flammable. Explosive. Poisonous.	Death in a few minutes at 0.2%. Paralyzes respiratory center.	Petroleum fumes, from blasting, gas found in sewers.	1. H_2S monitor. 2. H_2S tubes.
Carbon Dioxide, CO_2	5,000	1.53	Not flammable		Colorless, odorless, nonflammable. Not generally present in dangerous amounts unless there is already a deficiency of oxygen.	10% cannot be endured for more than a few minutes. Acts on nerves of respiration.	Issues from carbonaceous strata. Gas found in sewers.	Carbon dioxide monitor.
Ethane, C_2H_4		1.05	3.1	15.0	Colorless, tasteless, odorless, non-poisonous. Flammable. Explosive.	See hydrogen.	Natural gas.	Combustible gas monitor.
Chlorine, Cl_2	0.5	2.5	Not flammable Not explosive		Greenish yellow gas, or amber color liquid under pressure. Highly irritating and penetrating odor. Highly corrosive in presence of moisture.	Respiratory irritant, irritating to eyes and mucous membranes. 30 ppm causes coughing. 40–60 ppm dangerous in 30 minutes. 1,000 ppm apt to be fatal in a few breaths.	Leaking pipe connections. Overdosage.	1. Chlorine monitor. 2. Strong ammonia on swab gives off white fumes.
Sulfur Dioxide, SO_2	2	2.3	Not flammable Not explosive		Colorless compressed liquified gas with a pungent odor. Highly corrosive in presence of moisture.	Respiratory irritant, irritating to eyes, skin, and mucous membranes.	Leaking pipes and connections.	1. Sulfur dioxide monitor. 2. Strong ammonia on swab gives off white fumes.

[a] Originally printed in Water and Sewage Works, August 1953. Adapted from "Manual of Instruction for Sewage Treatment Plant Operators," State of New York.
[b] 8TWA PEL is the Time Weighted Average permissible exposure limit, in parts per million, for a normal 8-hour workday and a 40-hour workweek to which nearly all workers may be repeatedly exposed, day after day, without adverse effect.
[c] Gases with a specific gravity less than 1.0 are lighter than air; those more than 1.0, heavier than air.
[d] The first method given is the preferable testing procedure.

Confined Space Pre-Entry Checklist/Confined Space Entry Permit

Date and Time Issued: ____________ Date and Time Expires: ____________ Job Site/Space I.D.: ____________

Job Supervisor: ____________ Equipment to be worked on: ____________ Work to be performed: ____________

Standby personnel: ____________ ____________ ____________

1. Atmospheric Checks: Time ______ Oxygen ______ % Toxic ______ ppm
   Explosive ______ % LEL Carbon Monoxide ______ ppm

2. Tester's signature: ____________

3. Source isolation: (No Entry)

	N/A	Yes	No
Pumps or lines blinded, disconnected, or blocked	( )	( )	( )

4. Ventilation Modification:

	N/A	Yes	No
Mechanical	( )	( )	( )
Natural ventilation only	( )	( )	( )

5. Atmospheric check after isolation and ventilation: Time ______

Oxygen	______	% > 19.5%	< 23.5%	Toxic	______	ppm	< 10 ppm H_2S
Explosive	______	% LEL	< 10%	Carbon Monoxide	______	ppm	< 35 ppm CO

Tester's signature: ____________

6. Communication procedures: ____________

7. Rescue procedures: ____________
____

8. Entry, standby, and backup persons

	Yes	No
Successfully completed required training?	( )	( )
Is training current?	( )	( )

9. Equipment:

	N/A	Yes	No
Direct reading gas monitor tested	( )	( )	( )
Safety harnesses and lifelines for entry and standby persons	( )	( )	( )
Hoisting equipment	( )	( )	( )
Powered communications	( )	( )	( )
SCBAs for entry and standby persons	( )	( )	( )
Protective clothing	( )	( )	( )
All electric equipment listed for Class I, Division I, Groups A, B, C, and D, and nonsparking tools	( )	( )	( )

10. Periodic atmospheric tests:

Oxygen:	____% Time ____;	____% Time ____;	____% Time ____;	____% Time ____;
Explosive:	____% Time ____;	____% Time ____;	____% Time ____;	____% Time ____;
Toxic:	____ppm Time ____;	____ppm Time ____;	____ppm Time ____;	____ppm Time ____;
Carbon Monoxide:	____ppm Time ____;	____ppm Time ____;	____ppm Time ____;	____ppm Time ____;

We have reviewed the work authorized by this permit and the information contained herein. Written instructions and safety procedures have been received and are understood. Entry cannot be approved if any brackets ( ) are marked in the "No" column. This permit is not valid unless all appropriate items are completed.

Permit Prepared By: (Supervisor) ____________ Approved By: (Unit Supervisor) ____________

Reviewed By: (CS Operations Personnel) ____________
(Entrant) (Attendant) (Entry Supervisor)

This permit to be kept at job site. Return job site copy to Safety Office following job completion.

Fig. 14.1 Confined space pre-entry checklist/confined space entry permit

hazards have been evaluated by a competent person, and that necessary protective measures have been taken to ensure the safety of each worker." A competent person, in this case, is a person designated in writing as capable, either through education or specialized training, of anticipating, recognizing, and evaluating employee exposure to hazardous substances or other unsafe conditions in a confined space. This person is authorized to specify control procedures and protective actions necessary to ensure worker safety.

The following procedures must be observed before entry into a permit-required confined space:

1. Ensure that personnel are effectively trained.
2. If the confined space has both side and top openings, enter through the side opening if it is within 3½ feet (1.1 meters) of the bottom.
3. Wear appropriate, approved respiratory protective equipment.
4. Ensure that written operating and rescue procedures are at the entry site.
5. Wear an approved harness with an attached line. The free end of the line must be secured outside the entry point.
6. Test for atmospheric hazards as often as necessary to determine that acceptable entry conditions are being maintained.
7. Station at least one person to stand by on the outside of the confined space and at least one additional person within sight or call of the standby person.
8. Maintain effective communication between the standby person and the entry person.
9. The standby person, equipped with appropriate respiratory protection, should only enter the confined space in case of emergency.
10. If the entry is made through a top opening, use a hoisting device with a harness that suspends a person in an upright position. A mechanical device must be available to retrieve personnel from vertical spaces more than five feet (1.5 meters) deep.
11. If the space contains, or is likely to develop, flammable or explosive atmospheric conditions, do not use any tools or equipment (including electrical) that may provide a source of ignition.
12. Wear appropriate protective clothing when entering a confined space that contains corrosive substances or other substances harmful to the skin.
13. At least one person trained in first aid and cardiopulmonary resuscitation (CPR) should be immediately available during any confined space job.

    Individuals designated to provide first aid or CPR should be included in a Bloodborne Pathogens (BBP) program. These employees may be exposed to contact with blood or other potentially infectious materials from the performance of their duties. The BBP program includes training in exposure potential determination, engineering and work practice controls, personal protective equipment (PPE), and the availability of the hepatitis B vaccination series (29 CFR 1910.1030). If the operator(s) must enter confined spaces to perform rescue services, they must be trained specifically to perform the assigned rescue duty and to use required personal protective equipment (PPE) and rescue equipment. Rescue practice sessions must be held at least once every 12 months.

If you arrange to have a contractor perform work in confined spaces at your facility or within your collection system, you must inform the contractor:

- That the contractor must comply with confined space regulations.
- Of hazards that you have identified and your experience with the space(s).
- Of precautions or procedures you have implemented for the protection of employees in or near the space where the contractor's personnel will be working.
- That a debriefing must occur at the conclusion of the entry operations regarding the confined space program followed and any hazards encountered or created during the entry operations.

To enhance safety, communications, and coordination of confined space activities, the contractor is also required to obtain available information from you and to inform you of the confined space program the contractor will follow. This exchange of information must occur before the confined space is entered by any operator.

Confined space work can present serious hazards if you are uninformed or untrained. The procedures presented are only guidelines and exact requirements for confined space work for your locale may vary. Contact your local regulatory safety agency for specific requirements.

14.13 Oxygen Deficiency or Enrichment

Low oxygen levels may exist in any poorly ventilated, low-lying structure where gases such as hydrogen sulfide, gasoline vapor, carbon dioxide, or chlorine may be produced or may accumulate (see Table 14.1, "Common Dangerous Gases Encountered in Wastewater Collection Systems and at Wastewater Treatment Plants"). Oxygen in a concentration above 23.5 percent (oxygen enrichment) also can be dangerous because it speeds up combustion.

Oxygen deficiency is most likely to occur when structures or channels are installed below grade (ground level). Several gases (including hydrogen sulfide and chlorine) have a tendency to collect in low places because they are heavier than air. The specific gravity of a gas indicates its weight as compared to an equal volume of air. Since air has a specific gravity of exactly 1.0, any gas with a specific gravity greater than 1.0 may sink to low-lying areas and displace the air from that area or structure. (On the other hand, methane may rise out of a manhole because it has a specific gravity of less than 1.0, which means that it is lighter than air.) You should never rely solely on the specific gravity of a gas to tell you where it is. Air movement or temperature differences within a confined space may affect the location of atmospheric hazards. The only effective way of ensuring safe atmospheric conditions prior to entering a confined space is to test the atmosphere with an appropriate monitor(s) at various levels and locations throughout the space.

When oxygen deficiency or enrichment is discovered, the area should be ventilated with fans or blowers and checked again for oxygen deficiency/enrichment before anyone enters the area to work. Ventilation may continue to be provided by fans or blowers. You should follow confined space procedures before entering and during occupancy of any suspect area. *ALWAYS* get air into the confined space *BEFORE* you enter to work and maintain the ventilation until you have left the space.

Equipment is available to measure oxygen concentration as well as toxic and combustible atmospheric conditions. You must use this equipment whenever you encounter a potential confined space situation. Ask your local safety regulatory agency or wastewater association about sources of this type of equipment in your area.

> **Never enter an enclosed, poorly ventilated area, whether a manhole, sump, or other structure, without first following confined space entry procedures.**

14.14 Toxic or Suffocating Gases or Vapors

The most common toxic gas in wastewater treatment is hydrogen sulfide, which is produced during the anaerobic decomposition of certain materials containing sulfur compounds. Hydrogen sulfide will tend to accumulate in the lower voids of sewers, tanks, channels, and manholes because it is heavier than air, with a specific gravity of 1.19. The only reliable method of detecting hydrogen sulfide is by atmospheric monitoring since it has the unique ability to affect your sense of smell. You will lose the ability to smell hydrogen sulfide's "rotten egg" odor after only a short exposure. Your loss of ability to smell hydrogen sulfide is known as *OLFACTORY FATIGUE*.[3]

Other toxics, such as carbon monoxide, chlorinated solvents, and industrial toxins, may enter your plant as a result of industrial discharges, accidental spills, or illegal disposal of hazardous materials. You must become familiar with the waste discharges into your system. Table 14.1, "Common Dangerous Gases Encountered in Wastewater Collection Systems and at Wastewater Treatment Plants," contains information on methods of testing for several gases.

14.15 Toxic and Harmful Chemicals

Strong acids, bases, and chlorine are examples of toxic and harmful chemicals that operators may encounter working in and around treatment plants and laboratories. Be very careful when handling and using these chemicals. Chemical hazards may be present in many forms—vapors, dusts, mists, liquids, gases, and particles. The seriousness of a chemical hazard depends on exposure time and the concentration of the chemical, as well as which chemical you are exposed to. To avoid injury, be sure all hazardous chemicals are clearly labeled; obtain and read the health and safety data about the chemical *BEFORE* using the chemical; and learn and practice safe handling procedures and precautions.

A *MATERIAL SAFETY DATA SHEET*[4] (MSDS) (see Section 14.9, Figure 14.18) is your best source of information about hazardous chemicals and can be provided by your chemical supplier. The MSDS will provide at least the following information:

1. Identification of composition, formula, and synonyms—chemicals that are classified as health hazards must be identified if they are present in concentrations at 1.0 percent or greater. Carcinogens (cancer-causing chemicals) must also be identified if they are present at concentrations of 0.1 percent or greater.
2. Physical and chemical properties, such as specific gravity.
3. Incompatible substances and decomposition products.
4. *ACUTE*[5] and *CHRONIC*[6] health hazards.
5. Environmental impacts.
6. Personal protective measures and engineering/administrative controls.
7. Safe handling, storage, disposal, and cleanup procedures.

No chemical should be received, stored, or handled without essential safety information being provided to those who may come into contact with the substance.

Containers of hazardous chemicals at your plant must be labeled, tagged, or marked to identify the hazardous contents and appropriate hazard warnings (Figure 14.2). Exceptions are that you are not required to label portable containers that you transfer hazardous chemicals into for your personal use during your shift and individual stationary process containers can be identified by using signs or other written materials instead of attaching labels.

Operator exposure to chemical contaminants must be kept to a minimum. OSHA requires that engineering controls be initiated in preference to administrative controls or the use of personal protective equipment to minimize employee exposure. Engineering controls are considered the most effective means of protecting operators. Frequently, a combination of controls, training, safe work practices, and the use of personal protective equipment may be required to adequately reduce the potential for operator exposure.

Employers must develop, implement, and maintain a written hazard communication program describing how requirements for labeling and other forms of warning, material safety data sheets, and employee information and training will be met. The program must also include a list of the hazardous chemicals at your plant.

Training must include: (1) methods and observations that can be used to detect the presence or release of a hazardous chemical, (2) the physical and health hazards of the chemicals, (3) measures that operators can take to protect themselves from those hazards, and (4) the details of the plant's specific hazard communication program, including labeling, material safety data sheets, and how employees can obtain and use the hazard information. Additional information concerning the Hazard Communication Standard can be found in Section 14.9.

Remember, do not work with a chemical unless you understand the hazards involved and are using the protective equipment necessary to protect yourself. Contact your local safety regulatory agency about specific chemicals you may deal with if there is any doubt in your mind about safe procedures.

---

[3] *Olfactory* (all-FAK-tore-ee) *Fatigue.* A condition in which a person's nose, after exposure to certain odors, is no longer able to detect the odor.

[4] *Material Safety Data Sheet (MSDS).* A document that provides pertinent information and a profile of a particular hazardous substance or mixture. An MSDS is normally developed by the manufacturer or formulator of the hazardous substance or mixture. The MSDS is required to be made available to employees and operators or inspectors whenever there is the likelihood of the hazardous substance or mixture being introduced into the workplace. Some manufacturers are preparing MSDSs for products that are not considered to be hazardous to show that the product or substance is not hazardous.

[5] *Acute Health Effect.* An adverse effect on a human or animal body, with symptoms developing rapidly.

[6] *Chronic Health Effect.* An adverse effect on a human or animal body with symptoms that develop slowly over a long period of time or that recur frequently.

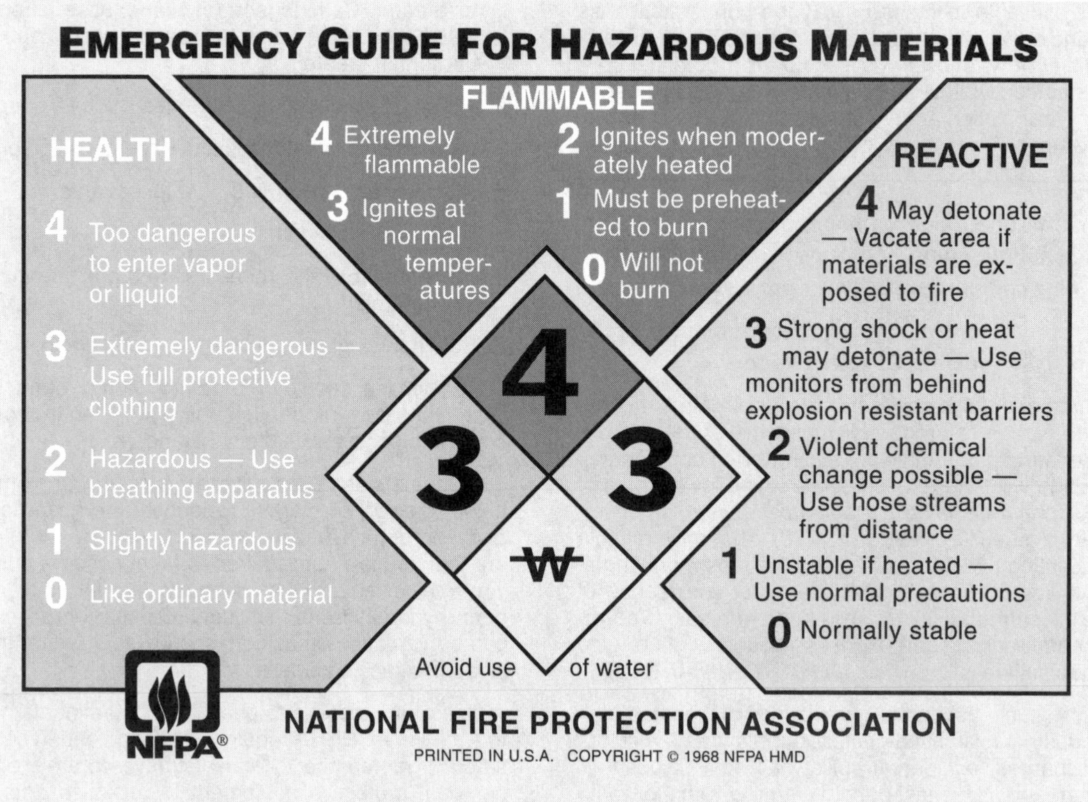

Fig. 14.2 NFPA hazard warning label
(Permission of National Fire Protection Association)

14.16 Radiological Hazards

Creation of another hazard to plant operators is a result of the use of radioactive isotopes in hospitals, research labs, and various industries. Check your sewer service area for the possible use of these materials. The routine handling and disposal of radioactive materials is stringently controlled by the US Nuclear Regulatory Commission (NRC). If you are receiving a discharge that may contain a radioactive substance, the NRC should be contacted. The only legal sources of radioactive wastes to a treatment plant are those specifically licensed by the NRC. Any personnel dealing with radioactive wastes must be monitored for exposure levels by pocket dosimeters or film badges. Special protective clothing must be worn and all work must be done under the direction of qualified staff.

Some plants use radioactive substances in control elements (for example, density elements or level indicators) or in various laboratory equipment. If your plant uses radioactive isotopes in these or other applications, you may be required to comply with NRC regulations and Title 10 of the Code of Federal Regulations, Parts 19 and 20. The requirements include, but are not limited to, licensing, training, recordkeeping, testing source integrity, and performing radiation surveys to determine potential exposure levels for personnel in the area of each source. You may also be required to have a "Radiation Safety Officer" on site. Consult the Title 10 regulations or contact the NRC to confirm the requirements for your radiation sources and to obtain information on how to safely deal with maintenance activities and accidents involving radioactive substances.

14.17 Explosive Gas Mixtures

Explosive gas mixtures may develop in many areas of a treatment plant from mixtures of air and methane, natural gas, manufactured fuel gas, hydrogen, or gasoline vapors. Table 14.1 lists the common dangerous gases that may be encountered in a treatment plant and identifies their explosive range where appropriate. The upper explosive limit (UEL) and lower explosive limit (LEL) indicate the range of concentrations at which combustible or explosive gases will ignite when an ignition source is present at ambient temperature. No explosion or ignition occurs when the concentration is outside these ranges. Gas concentrations below the LEL are too lean to ignite; there is not enough flammable gas or vapor to support combustion. Gas concentrations higher than the UEL are too rich to ignite; there is too much flammable gas or vapor and not enough oxygen to support combustion (Figure 14.3).

Explosive ranges can be measured by using a combustible gas detector calibrated for the gas of concern. Do not rely on your nose to detect gases. The sense of smell is absolutely unreliable for evaluating the presence of dangerous gases. Some gases have no smell and hydrogen sulfide can paralyze the sense of smell.

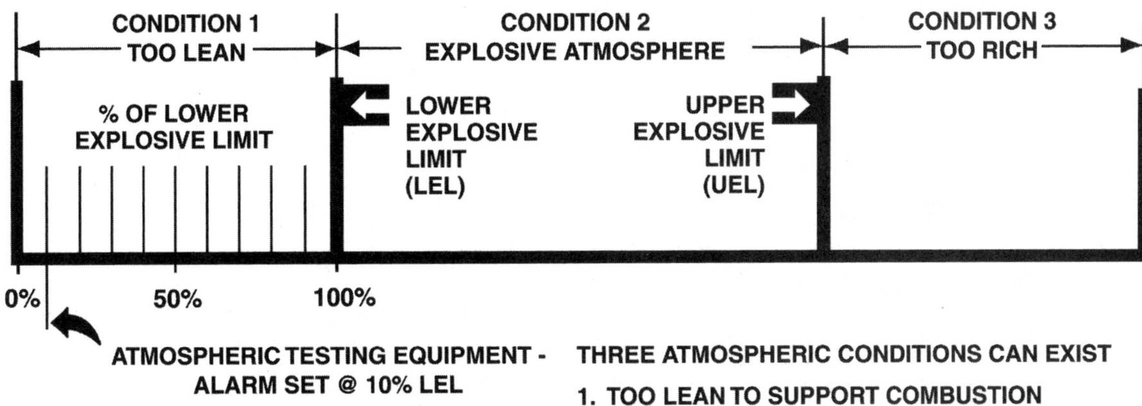

Fig. 14.3 Relationship between the lower explosive limit (LEL) and the upper explosive limit (UEL) of a mixture of air and gas

Avoid explosions by eliminating all sources of ignition in areas potentially capable of developing explosive mixtures. Only explosion-proof electrical equipment and fixtures should be used in these areas (influent/bar screen rooms, gas compressor areas, digesters, battery charging stations). Provide adequate ventilation in all areas that have the potential to develop an explosive atmosphere.

The National Fire Protection Association Standard 820 (NFPA 820), *FIRE PROTECTION IN WASTEWATER TREATMENT AND COLLECTION FACILITIES,* lists requirements for electrical classifications, ventilation, gas detection, and fire control methods in various wastewater treatment and collection system areas. (For ordering information, see Section 14.11, "Additional Reading," item 7.) Comparing these requirements with your plant's existing design and equipment may indicate deficiencies, which should be remedied to minimize potential hazards.

14.18 Fires

Burns can be very serious and cause painful injuries. Structural damage from fires can be very costly. Every facility should develop a fire prevention plan, with input from the local fire marshall, fire chief, and insurance company. The plan may be very simple or very complex, depending on the specific facility needs. Some items that may be included in any plan are:

1. Regulate the use, storage, and disposal of all combustible materials/substances.
2. Provide periodic cleanup of weeds or other vegetation in and around the plant.
3. Develop written response procedures for reacting to a fire situation, to include evacuation.
4. Provide required service on all fire detection and response equipment (inspection, service, hydrostatic testing).
5. Routinely inspect fire doors to ensure proper operation and unobstructed access.
6. Immediately repair, remove, or replace any defective wiring.
7. Restrict the use of any equipment that may provide a source of ignition in areas where combustible gases may exist.
8. Maintain clear access to fire prevention equipment at all times.
9. Develop a written hot work procedure and permit to provide written authorization to perform operations (for example, welding, cutting, burning, and heating) that involve a source of ignition.

Regardless of the size of the facility, each operator should know the location of fire protection equipment in their work area and must be trained in the proper use of fire extinguishing equipment and methods of extinguishing fires. Training must be provided upon initial employment and at least annually thereafter.

A portable extinguisher must also be visually inspected monthly and must receive an annual maintenance check. Maintenance checks must be documented and the records must be retained for one year. Hydrostatic testing of extinguishers is also required every 5 to 12 years, depending on the type of extinguisher. Remember, always have extinguishers serviced promptly after use so they will be ready if you need them again.

QUESTIONS

Write your answers in a notebook and then compare your answers with those on page 319.

14.1A How can you help prevent the spread of infectious diseases from your job to you and your family?

14.1B When testing the atmosphere before entry in any confined space, what procedure should be used?

14.1C According to OSHA regulations, when is a confined space entry permit required?

14.1D How do toxic and suffocating gases or vapors enter the wastewater treatment plant?

14.1E Why is your sense of smell not a reliable method of detecting hydrogen sulfide?

14.1F What do the initials UEL and LEL stand for?

14.19 Other Hazards

14.190 Electric Shock/Stored Energy

Electric shock frequently causes serious injury. Do not attempt to repair electrical equipment unless you know what you are doing. You must be qualified and authorized to work on electrical equipment before you attempt any troubleshooting or repairs. Ordinary 120 volt electricity may be fatal; 12 volts may, on good contact, cause injury. Any electrical system, regardless of voltage, should be considered dangerous unless you know positively that it is de-energized. Remember these basic safety rules when working around electrical equipment:

1. Keep your mind on the potential hazard at all times.
2. Always lock out and tag out any electrical equipment being serviced. *NEVER* remove anyone else's lock or tag.
3. Do not use portable ladders with conductive side rails.
4. Never override any electrical safety device.
5. Inspect extension cords for abrasion, insulation failure, and evidence of possible internal damage.
6. Use only grounded or insulated (Underwriter's Laboratory (UL) approved) electrical equipment.
7. Take care not to accidentally ground yourself when in contact with electrical equipment or wiring.
8. Do not alter or connect attachment plugs and receptacles in a manner that could prevent proper grounding.
9. Do not use flexible electrical cords connected to equipment to raise or lower the equipment.
10. Wear nonconductive head protection if there is a danger of head injury from contact with exposed energized parts.
11. Use a ground-fault circuit interrupter in damp locations.
12. Do not wear conductive articles of jewelry or clothing if they might contact exposed energized parts (unless they are covered, wrapped, or otherwise insulated).

EMERGENCY PROCEDURES

In the event of electric shock, the following steps should be taken:

1. Survey the scene and see if it is safe to enter.
2. If necessary, free the victim from a live power source by shutting power off at a nearby disconnect, or by using a dry stick or some other nonconducting object to move the victim.
3. Send for help, calling 911 or whatever the emergency number is in your community. Check for breathing and pulse. Begin CPR (cardiopulmonary resuscitation) immediately, if needed.

Remember, only trained and qualified individuals working in pairs should be allowed to service, repair, or troubleshoot electrical equipment and systems.

Whenever replacement, repair, renovation, or modification of equipment is performed, OSHA laws require that all equipment that could unexpectedly start up or release stored energy must be locked out or tagged out to protect against accidental injury to personnel. Some of the most common forms of stored energy are electrical and hydraulic energy. The energy isolating devices (switches, valves) for the equipment should be designed to accept a lockout device. A lockout device (Figure 14.4) uses a positive means such as a lock to hold a switch or valve in a safe position and prevent the equipment from becoming energized or moving. In addition, prominent warnings, such as the tags illustrated in Figure 14.5, must be securely fastened to the energy isolating device and the equipment (in accordance with an established written procedure) to indicate that both it and the equipment being controlled may not be operated until the tag and lockout device are removed by the person who installed them.

For the safety of all personnel, each plant should develop a standard operating procedure that must be followed whenever equipment must be shut down or turned off for repairs. If every operator follows the same procedures, the chances of an accidental start-up injuring someone will be greatly reduced. The following procedures are intended as guidelines and can be used as a model for developing your own written standard operating procedure for lockout/tagout.

Training must be provided to ensure that the purpose and function of the lockout/tagout program are understood and that the knowledge and skill required for the safe use of energy controls are gained. Each employee using lockout/tagout must be aware of applicable energy sources and the methods necessary for their effective isolation and control.

If you hire a contractor, you must inform each other of your respective lockout/tagout procedures. You must ensure that you and your employees or co-workers comply with the restrictions and prohibitions of the contractor's program.

Periodic inspections of the lockout/tagout program are also required (at least annually) to ensure that the requirements of the program are being followed. The inspection(s) must be done by someone other than the one(s) using the procedure.

BASIC LOCKOUT/ TAGOUT PROCEDURES

1. Notify all affected employees that a lockout or tagged system is going to be used and the reason why. The authorized employee shall know the type and magnitude of energy that the equipment uses and shall understand the hazard thereof.
2. If the equipment is operating, shut it down by the normal stopping procedure.
3. Operate the switch, valve, or other energy isolating device(s) so that the equipment is isolated from its energy source(s). Stored energy such as that in springs; elevated machine members; rotating flywheels; hydraulic systems; and systems using air, gas, steam, or water pressure must be dissipated or restrained by methods such as repositioning, blocking, or bleeding down.
4. Lock out or tag out the energy isolating device with assigned individual lock or tag. If a tagout device is used, it must be substantial enough to prevent accidental removal. The attachment means for a tagout device must be a nonreusable type, attachable by hand, self-locking, and nonreleasable with a minimum unlocking strength of no less than 50 pounds (22.7 kg).
5. After ensuring that no personnel are exposed, and as a check that the energy source is disconnected, operate the push button or other normal operating controls to make certain the equipment will not operate. *Caution: Return operating controls to the neutral or off position after the test.*

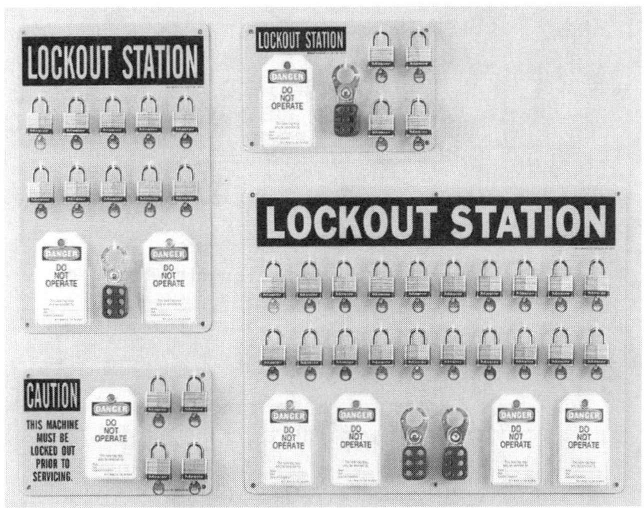

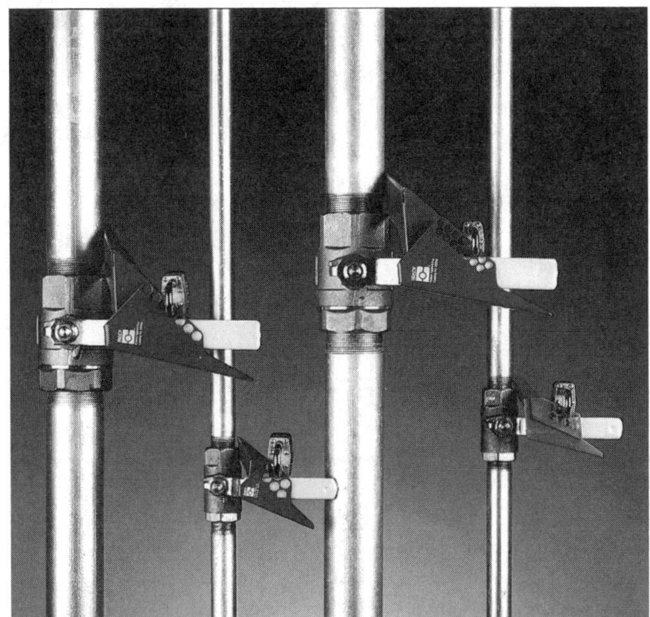

Fig. 14.4 Typical lockout devices

(Courtesy of Brady Worldwide)

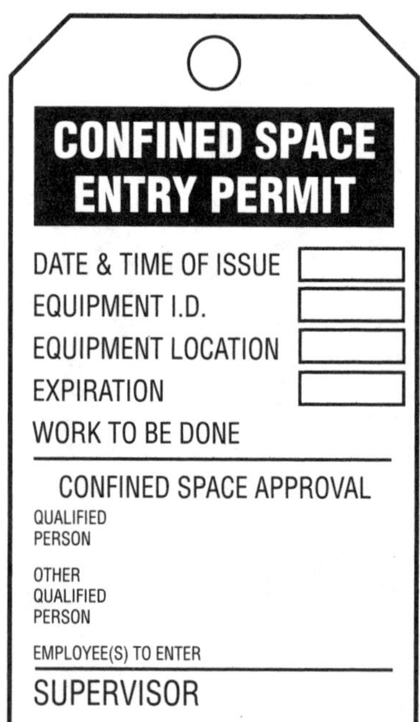

Fig. 14.5 Typical lockout warning tags

6. The equipment is now locked out or tagged out and work on the equipment may begin.

7. After the work on the equipment is complete, all tools have been removed, guards have been reinstalled, and employees are in the clear, remove all lockout or tagout devices. Operate the energy isolating devices to restore energy to the equipment.

14.191 Noise

Wastewater treatment facilities contain some equipment that produces high noise levels, intermittently or continuously. Operators must be aware of this and use safeguards such as hearing protectors that eliminate or reduce noise to acceptable levels. In general, if you have to shout or cannot hear someone talking to you in a normal tone of voice, the noise level is excessive. Prolonged or regular daily exposure to high noise levels can produce at least two harmful, measurable effects: hearing damage and masking of desired sounds such as speech or warning signals.

Noise source monitoring should be conducted to measure noise levels in all suspect areas of the treatment plant and to identify excessive tool or equipment noise generation. Follow-up testing should be performed when any equipment or process is placed into service or modified.

The ideal method of dealing with any high-noise environment is the elimination or reduction of all sources through feasible engineering or administrative controls. This approach is frequently not possible; therefore, employers are required to identify and monitor operators whose normal noise exposure might equal or exceed an 8-hour *TIME-WEIGHTED AVERAGE*[7] (TWA) of 85 *DECIBELS*[8] (A-scale).

To ensure that the welfare of operators is not compromised and to comply with federal regulations (29 CFR 1910.95), a comprehensive hearing conservation program should be implemented. All individuals whose normal noise exposure equals or exceeds the 8-hour TWA of 85 dBA must be included in this program. The primary elements of the program are monitoring, audiometric testing, hearing protection, training in the use of protective equipment and procedures, access to noise level information, and recordkeeping. The purpose of the conservation program is to prevent hearing loss that might affect an operator's ability to hear and understand normal speech.

A selection of hearing protection devices must be available to operators; however, a certain degree of confusion can arise concerning the adequacy of a particular protector. To estimate the adequacy of a hearing protector, use the noise reduction rating (NRR) shown on the hearing protector package. Subtract 7 dB from the NRR and subtract the remainder from the individual's A-weighted TWA noise environment to obtain the estimated A-weighted TWA under the ear protector. To provide adequate protection, the value under the ear protector should be 85 dB or less, the lower the better.

It is essential that individuals use properly rated protective devices in high-noise areas or during high-noise activities. Employee training must include information on (1) the effects of noise on hearing; (2) the purpose of hearing protectors and the advantages, limitations, and effectiveness of various types, and instruction on selection, fitting, use, and care; and (3) the purpose of the audiometric testing and an explanation of the test procedures. The training must be repeated annually. Audiometric test records must be retained for as long as the affected employee works at the plant. Contact your local health department or safety regulatory agency for assistance in the development of a program for your specific treatment plant.

14.192 Dusts, Fumes, Mists, Gases, and Vapors

The ideal way to control occupational diseases caused by breathing air contaminated with harmful dusts, fumes, mists, gases, and vapors is to prevent atmospheric contamination from occurring. This can sometimes be accomplished through engineering control measures. Remember, OSHA requires that engineering controls be implemented whenever feasible to eliminate or reduce operator exposure to a hazard. When effective engineering controls are not feasible, however, appropriate respirators must be used.

Respirators must be provided by the employer when they are necessary to protect the health of the operator. The respirators must be appropriate and suitable for the purpose intended. The four most common respirators are (1) Self-Contained Breathing Apparatus (SCBA), (2) Supplied-Air Respirators (SAR), (3) Powered Air-Purifying Respirators (PAPR), and (4) Air-Purifying Respirators (APR).

SARs and SCBAs, known as air-supplying respirators, supply clean air to the wearer from an independent source and provide the highest levels of protection. The air source may either be remotely located tanks or a tank carried by the user.

APRs and PAPRs take air from the immediate surroundings and purify it by passing it through filters, cartridges, or canisters. APRs and PAPRs provide lower levels of protection than air-supplying types of respirators. These respirators (APRs and PAPRs) *ARE NOT* suitable for potentially oxygen-deficient atmospheres.

Positive-pressure respirators, in which the pressure inside the facepiece during inhalation (breathing in) remains higher than the pressure outside, help prevent contaminants from entering and offer the greatest protection. Conversely, negative-pressure respirators, which allow the interior facepiece pressure to drop below the outside pressure during the inhalation cycle, may not prevent contaminants from leaking into an improperly sealed facepiece.

Positive-pressure SCBAs and full-facepiece positive-pressure SARs with an escape SCBA can be used in oxygen-deficient atmospheres (containing less than 19.5 percent oxygen) and atmospheres that are considered immediately dangerous to life or health (IDLH) as defined in OSHA regulations.

Negative-pressure respirators (PAPRs and APRs) are approved only for atmospheres that are not immediately dangerous to life or health.

Selection of a respirator is based on the type of hazard and the contaminant concentration. Each respirator type is given

---

[7] *Time-Weighted Average (TWA).* The average concentration of a pollutant based on the times and levels of concentrations of the pollutant. The time-weighted average is equal to the sum of the portion of each time period (as a decimal, such as 0.25 hour) multiplied by the pollutant concentration during the time period divided by the hours in the workday (usually 8 hours). 8TWA PEL is the time-weighted average permissible exposure limit, in parts per million, for a normal 8-hour workday and a 40-hour workweek to which nearly all workers may be repeatedly exposed, day after day, without adverse effect.

[8] *Decibel* (DES-uh-bull). A unit for expressing the relative intensity of sounds on a scale from zero for the average least perceptible sound to about 130 for the average level at which sound causes pain to humans. Abbreviated dB.

an assigned protection factor (APF) by OSHA indicating the maximum contaminant level for which the respirator can be used. OSHA rates contaminants according to their permissible exposure limit (PEL). A contaminant's PEL is the legally established maximum time-weighted average level of contaminant to which an operator can be exposed during a work shift. The proper respirator is, therefore, chosen according to its APF and the PEL of the contaminant. For example, a respirator with an APF of 100, approved for a given contaminant, can be used in atmospheres containing 100 times the PEL of the contaminant.

Remember, you, the operator, must use the provided respiratory protection in accordance with instructions and training provided to you.

Employers are also responsible for establishing and maintaining a respiratory protection program. Some of the basic elements of a respiratory protection program are:

1. Written standard operating procedures (SOPs) governing the selection and use of respirators.
2. Instruction and training in the proper use of respirators and their limitations (to include annual fit testing).
3. Physical assessment of individuals assigned tasks requiring the use of respirators.

These are only a few requirements for the safe use of respiratory protection. Specific requirements for a respiratory protection program for your application must be confirmed with your local regulatory safety agency.

QUESTIONS

Write your answers in a notebook and then compare your answers with those on page 319.

14.1G What is a lockout device and when is it used?

14.1H What types of equipment or systems are potentially hazardous for operators due to their stored energy?

14.1I Identify the primary elements of the hearing conservation program.

14.1J What types of respirators must be used in oxygen-deficient atmospheres?

14.2 SPECIFIC HAZARDS

The remainder of this chapter will acquaint you with the specific hazards, by location or type of work, that you may encounter in the field of wastewater collection and treatment.

14.20 Collection Systems

Good design and the use of safety equipment will not prevent physical injuries in sewer work unless safety practices are understood by the *entire crew* and are enforced.

Never attempt to do a job unless you have sufficient training, assistance, the proper tools, and the necessary safety equipment. *There are no shortcuts to safety.*

14.200 Traffic Hazards

Before starting any job in a street or other traffic area, even if you are just going to open a manhole, study the work area and plan your work. Your task must be regulated to provide maximum safety.

Working in a roadway represents a significant hazard to an operator as well as pedestrians and drivers. Drivers can be seen applying makeup, shaving, talking on cellular phones, and changing tapes, CDs, or radio stations rather than concentrating on driving. The control of traffic is necessary to reduce the risk of injury or death while working in this hazardous area. The purpose of traffic control is to provide safe and effective work areas and to warn, control, protect, and expedite vehicular and pedestrian traffic. This can be accomplished by appropriate and prudent use of traffic control devices (Figures 14.6 and 14.7).

The traffic control zone is the area between the first advance warning sign and the point beyond the work area where traffic is no longer affected. Most traffic control zones can be divided into these specific areas (Figure 14.8):

1. Advance warning area
2. Transition area
3. Buffer space
4. Work area
5. Termination area

Upon arrival at the job site, look for a safe place to park your vehicle. If it must be parked in the street to do the job, route traffic around the job site before parking your vehicle in the street. If practical, park your vehicle between oncoming traffic and the job site, outside the buffer zone, to serve as a warning barricade and to discourage reckless drivers from plowing into you. The use of flashing/revolving warning lights is an excellent method of alerting traffic to your presence. Remember, you need protection from the drivers as well.

Traffic must be warned of your presence in the street. BE PREPARED TO STOP, SHOULDER WORK, and UTILITY WORK AHEAD are signs that are effective. Signs with flashing warning lights and vehicles with yellow rotating beacons are used to warn other motorists. Vehicle-mounted traffic guides are also helpful. Use trained flaggers to alert drivers and to direct traffic around the work site. Warning signs and flaggers must be located far enough in advance of the work area to allow motorists time to realize they must slow down, be alert for activity, and safely change lanes or follow a detour. Exact distances and the nature of the advance warning depend on traffic speed, congestion, roadway conditions, and local regulations. Figure 14.9 provides an example of appropriate placement of traffic control devices and a key (legend) for interpreting the diagrams.

For additional information about traffic control and working safely in streets, see *OPERATION AND MAINTENANCE OF WASTEWATER COLLECTION SYSTEMS,* Volume I, Chapter 4, "Safe Procedures," in this series of operator training manuals, and *MANUAL ON UNIFORM TRAFFIC CONTROL DEVICES,* Federal Highway Administration (FHWA), 2003 edition, available online at www.fhwa.dot.gov.

14.201 Manholes[9]

Manholes are confined spaces and the requirements for entry into a confined space and the atmospheric hazards that may be encountered in a confined space have been discussed earlier in this chapter in Section 14.12. The following items are

[9] Also see OSHA standards regarding safe procedures for entry into confined spaces.

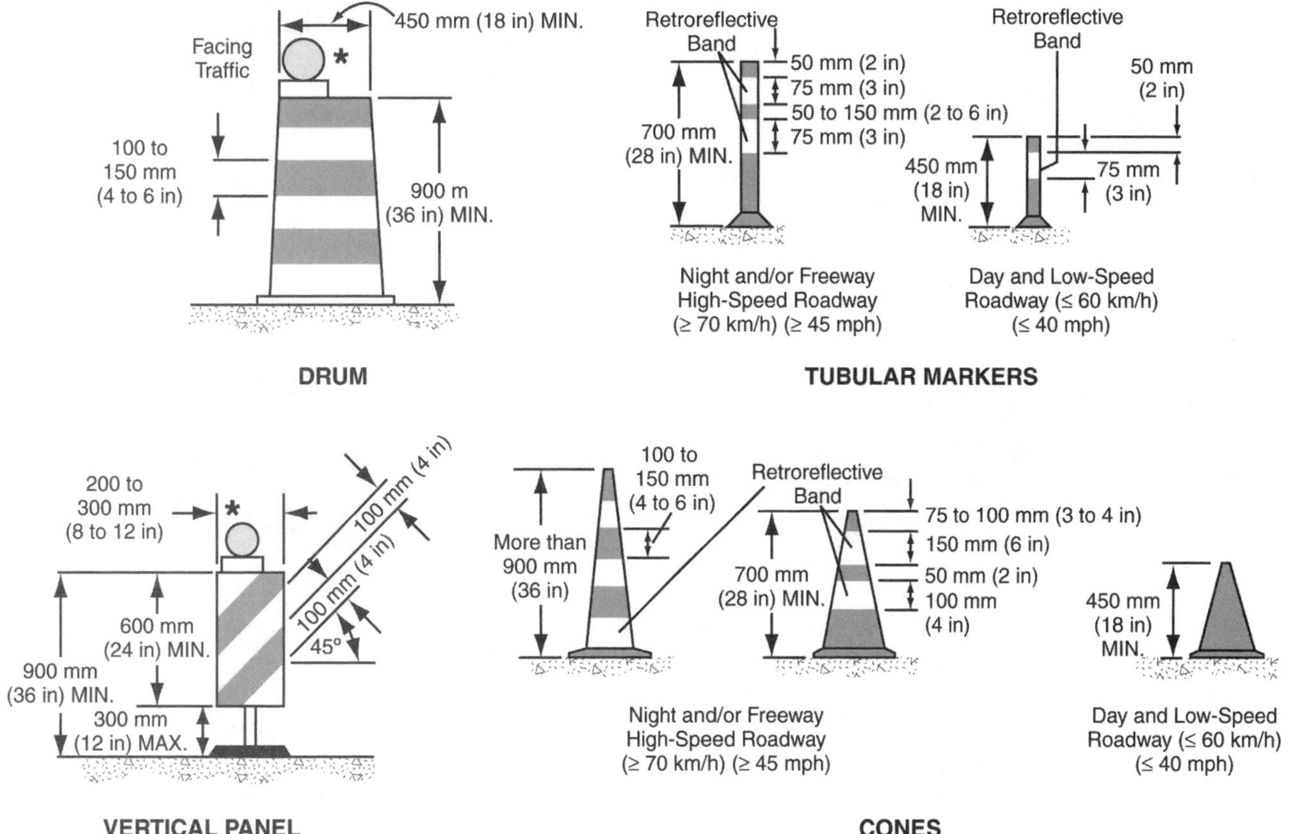

* Warning lights (optional)

Note: If drums, cones, or tubular markers are used to channelize pedestrians, they shall be located such that there are no gaps between the bases of the devices, in order to create a continuous bottom, and the height of each individual drum, cone, or tubular marker shall be no less than 900 mm (36 in) to be detectable to users of long canes.

Fig. 14.6 Delineating and channelizing devices

(Source: Manual on Uniform Traffic Control Devices, 2003)

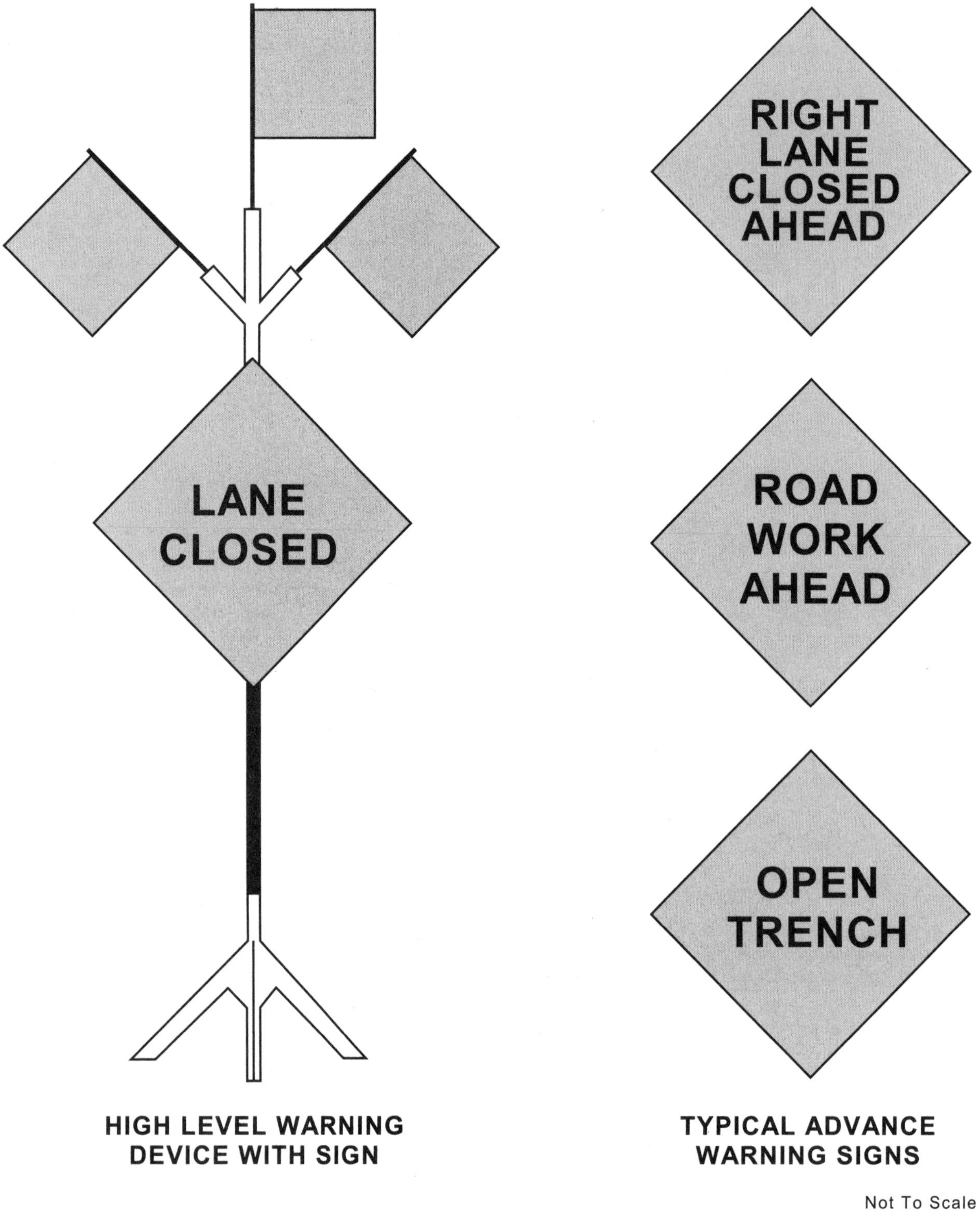

LANE CLOSED

RIGHT LANE CLOSED AHEAD

ROAD WORK AHEAD

OPEN TRENCH

HIGH LEVEL WARNING DEVICE WITH SIGN

TYPICAL ADVANCE WARNING SIGNS

Not To Scale

Fig. 14.7 Warning devices and signs

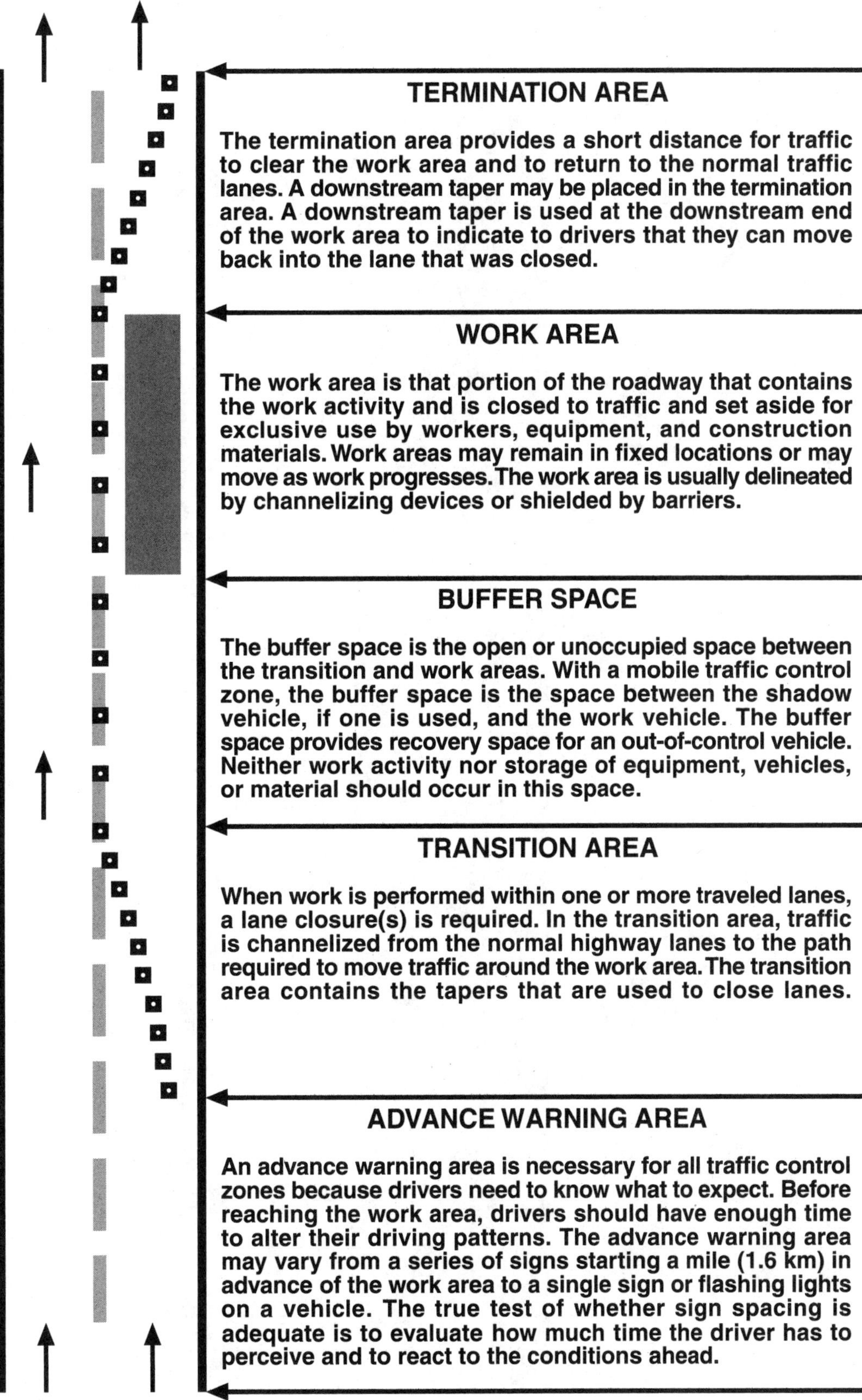

TERMINATION AREA
The termination area provides a short distance for traffic to clear the work area and to return to the normal traffic lanes. A downstream taper may be placed in the termination area. A downstream taper is used at the downstream end of the work area to indicate to drivers that they can move back into the lane that was closed.
WORK AREA
The work area is that portion of the roadway that contains the work activity and is closed to traffic and set aside for exclusive use by workers, equipment, and construction materials. Work areas may remain in fixed locations or may move as work progresses. The work area is usually delineated by channelizing devices or shielded by barriers.
BUFFER SPACE
The buffer space is the open or unoccupied space between the transition and work areas. With a mobile traffic control zone, the buffer space is the space between the shadow vehicle, if one is used, and the work vehicle. The buffer space provides recovery space for an out-of-control vehicle. Neither work activity nor storage of equipment, vehicles, or material should occur in this space.
TRANSITION AREA
When work is performed within one or more traveled lanes, a lane closure(s) is required. In the transition area, traffic is channelized from the normal highway lanes to the path required to move traffic around the work area. The transition area contains the tapers that are used to close lanes.
ADVANCE WARNING AREA
An advance warning area is necessary for all traffic control zones because drivers need to know what to expect. Before reaching the work area, drivers should have enough time to alter their driving patterns. The advance warning area may vary from a series of signs starting a mile (1.6 km) in advance of the work area to a single sign or flashing lights on a vehicle. The true test of whether sign spacing is adequate is to evaluate how much time the driver has to perceive and to react to the conditions ahead.

Fig. 14.8 Parts of a traffic control zone

Lane Closure on Low-Volume, Two-Lane Road

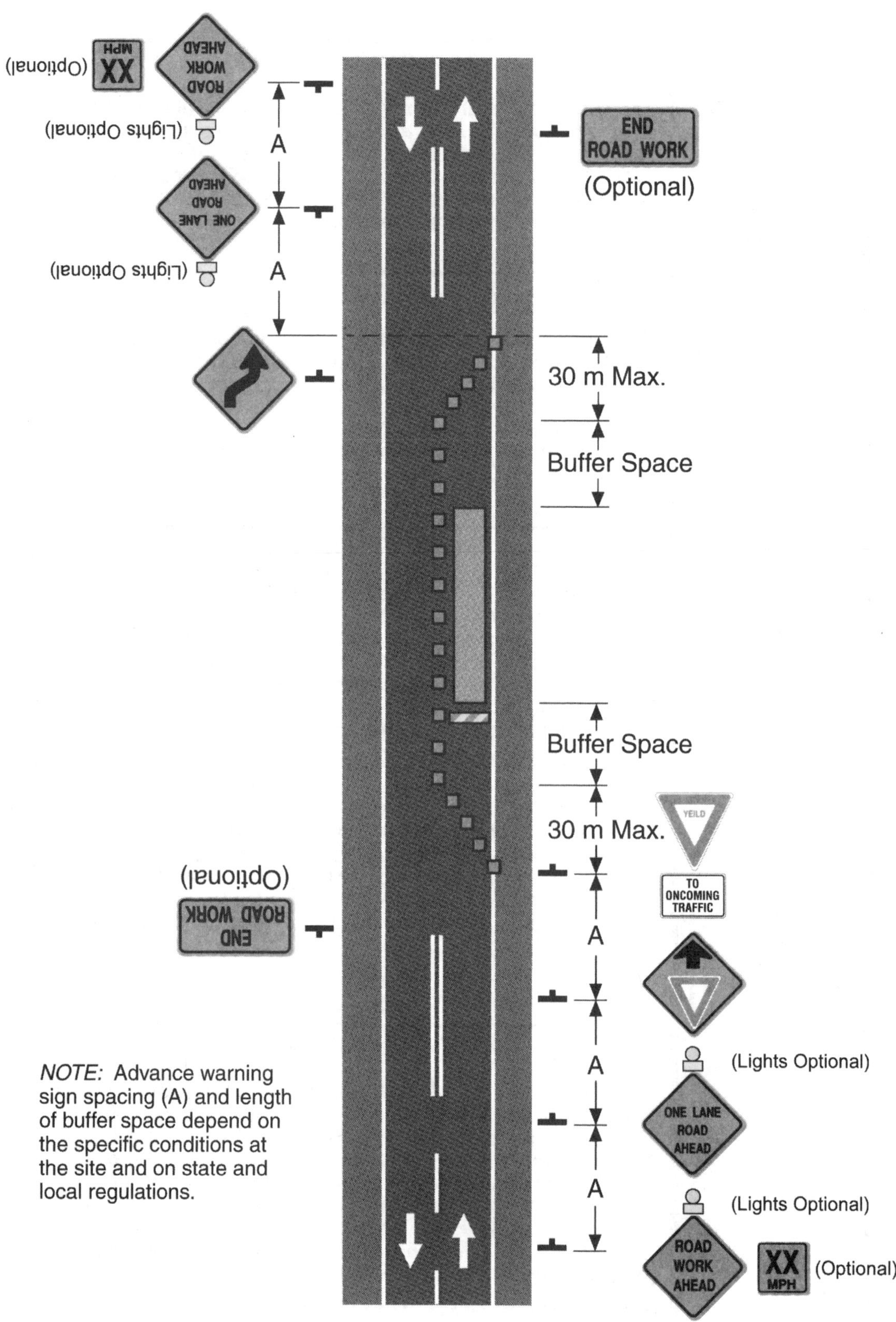

Fig. 14.9 Typical placement of traffic control devices

(Source: California Department of Transportation)

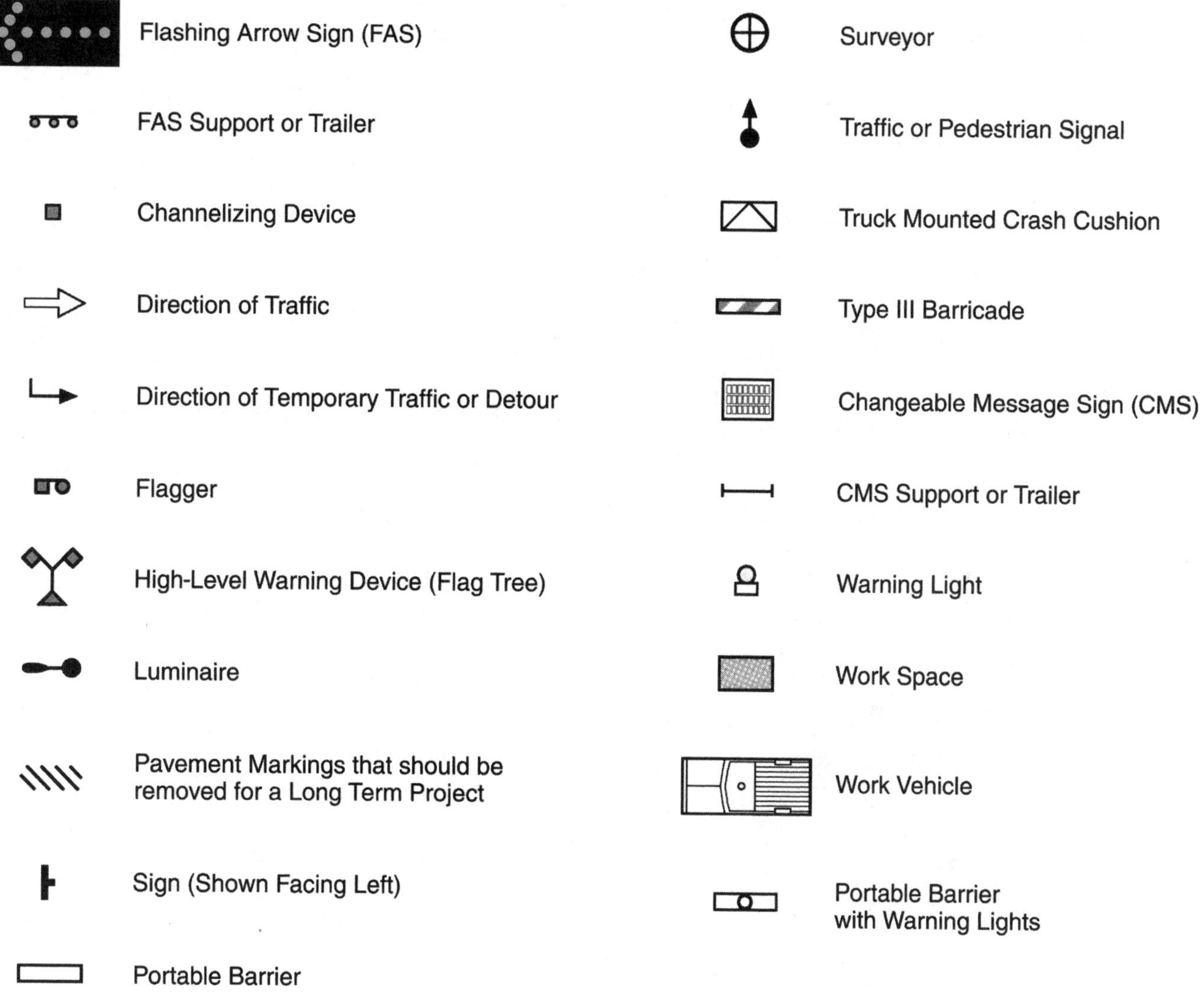

Fig. 14.9 Typical placement of traffic control devices (continued)

(Source: California Department of Transportation)

examples of other hazards that may exist in manholes, depending on the design and use of the manhole:

- Mechanical hazards, such as hot or cold surfaces, steam leaks, or rotating equipment.
- Electrical hazards involving conduit, energized circuits, lights, portable power tools, or moisture/water accumulation.
- Engulfment hazards, such as flooding from wastewater, sludge, or chemicals.
- Physical injury, such as sprains, strains, scrapes, and cuts due to uneven footing, inadequate working room, poor balance, or awkward positioning.
- Infections and diseases from bacteria, parasites, and viruses found in the wastewater stream.
- Bites from insects and rodents.
- Toxic exposure from substances illegally or accidentally discharged into the collection system or plant drainage system.

Manhole work will usually require job site protection, either a manhole safety enclosure by itself or in conjunction with traffic controls. Job site protection when working in manholes should include barricades and traffic warning devices for the safety of vehicles, bikes, pedestrians, and workers. In addition to following confined space procedures for entry and occupancy of the manhole, the atmosphere within the manhole should be tested prior to removing the manhole cover. A spark created by removing the cover could cause an explosion if a combustible atmosphere exists within the manhole. Also, it may be helpful to evaluate the conditions within the manhole before any ventilation occurs.

Remove the manhole cover with a tool specifically designed for the purpose, not your fingers. You have only 10 fingers. Protect them. When lifting a lid, the use of the rule "Lift with your legs, not with your back" will help eliminate back strains (Figure 14.10). Once the lid is removed, leave it flat on the ground and far enough away from the manhole to provide adequate room for a working area. This is usually at least two to three feet (0.6 to 1 m).

REMEMBER: Never enter a manhole without fully complying with the requirements for work in confined spaces.

BACK INJURIES ARE A FREQUENT INJURY

BE CAREFUL...

THE BACK YOU SAVE MAY BE YOUR OWN!

Think about the following steps when lifting:

1. Size up the load.
2. Bend your knees.
3. Get a firm grip.
4. Lift with your legs... gradually.
5. Keep the load close to you.

IT IS HOW YOU LIFT as well as WHAT YOU LIFT.

Fig. 14.10 Lifting guidelines

Be very cautious when using ladder rungs or steps installed in the side of the manhole. Be alert for loose or corroded steps. Always test each step individually before placing your weight on it. If possible, it is much safer to use a portable ladder as a means of entering a manhole. Be certain, how ever, that the bottom feet are properly placed so tha ladder will not slip or twist when your weight is placed The top of the ladder should extend three feet (1 m) abov ground level to facilitate getting on and off the ladder. A chanical lifting device is the safest possible way to be lowered into a manhole, wet well, or other below-grade work area or to be lifted to an elevated work area.

If you are working in wastewater, be sure to wear properly fitted rubber gloves and boots, or approved substitutes that will provide protection from infection. Be aware of possible needle sticks and other puncture wounds.

Tools and equipment should be lowered into a manhole by means of a bucket or basket. Do not drop them into the manhole for a person to catch. Attempting to carry tools in one hand while climbing up or down a ladder is an unsafe practice.

14.202 Excavations[10]

If it becomes necessary for you to excavate a sewer line, remember to contact utility companies to locate underground

[10] Also see *OPERATION AND MAINTENANCE OF WASTEWATER COLLECTION SYSTEMS,* Volume I, Chapter 4, in this series of manuals.

telephone, gas, fuel, electric, cable, and water lines *before* opening the excavation. If you cannot establish the exact location of the underground utilities, you must proceed with caution and you should use detection equipment, if possible, to locate the service lines. These lines can present a very significant safety hazard to the operator(s) and to the public during excavation activities.

Become familiar with the fundamentals of excavating and the proper, safe approach for shoring, shielding, and sloping and benching before excavating. Without a proper protective system the bank (wall) of a trench or excavation can cave in and kill you. It is strongly recommended that some type of adequate cave-in protection be provided when the trench or excavation is four feet (1.2 meters) deep or deeper. OSHA requirements state that adequate protection is absolutely required if the trench is five feet (1.5 meters) or more in depth. Types of adequate protection include shoring, shielding, and sloping and benching.

SHORING is a complete framework of wood or metal that is designed to support the walls of a trench (see Figure 14.11). Sheeting is the solid material placed directly against the side of the trench. Either wooden sheets or metal plates might be used. Uprights are used to support the sheeting. They are usually placed vertically along the face of the trench wall. Spacing between the uprights varies depending on the stability of the soil. Stringers (or walers) are placed horizontally along the uprights. Trench braces are attached to the stringers and run across the excavation. The trench braces must be adequate to support the weight of the wall to prevent a cave-in. Examples of different types of trench braces include solid wood or steel, screw jacks, or hydraulic jacks.

The space between the shoring and the sides of the excavation should be filled in and compacted in order to prevent a cave-in from getting started. If properly done, shoring may be the operator's best choice for cave-in protection because it actually prevents a cave-in from starting and does not require additional space.

SHIELDING is accomplished by using a two-sided, braced steel box that is open on the top, bottom, and ends (Figure 14.12). This "drag shield," as it is sometimes called, is pulled through the excavation as the trench is dug out in front and filled in behind. Operators using a drag shield must always work only within the walls of the shield and are not allowed in the shield when it is being installed, removed, or moved. If the trench is left open behind or in front of the shield, the temptation could be high to wander outside of the shield's protection. Shielding does not actually prevent a cave-in, as the space between the trench wall and the drag shield is left open, allowing a cave-in to start. There have been cases where a drag shield was literally crushed by the weight of a collapsing trench wall.

SLOPING and BENCHING are practices that simply remove the trench wall itself (Figure 14.13). The amount of soil needed to be removed will vary, depending on the stability of the soil. A good rule of thumb is to always slope or bench at least one foot back for every one foot (0.3 m) of depth on *both* sides of the excavation. Exact sloping angles and benching dimensions largely depend on the type of soil that is being excavated. OSHA has established three types of soil classifications, Type A, Type B, and Type C, in decreasing order of stability. The type of soil dictates sloping/benching requirements. A competent person must examine the work site and determine the soil classification in order to determine requirements for a specific site.

Certain soil conditions can contribute to the chances of a cave-in. These conditions include low cohesion, high moisture content, freezing conditions, or a recent excavation at the same site. Other factors to be considered are the depth of the trench, the soil weight, the weight of nearby equipment, and vibration from equipment or traffic. It is worth repeating that regardless of the presence or absence of any or all of the above factors, the trench must still have proper cave-in protection if it is five feet (1.5 meters) or more deep. Excavations less than five feet (1.5 meters) deep may also require protection if a competent person determines that there are indications of a potential cave-in. The spoil (dirt removed from the trench) must be placed at least two feet (0.6 meter) back from the trench and should be placed on one side of the trench only. A stairway, ramp, or ladder is required in the trench if it is four feet (1.2 meters) or more deep. The means of leaving the trench must be placed so that no more than 25 feet (7.5 meters) of travel is required to exit the trench.

Atmospheric monitoring prior to entering excavations more than four feet (1.2 meters) deep is required if a hazardous atmosphere can reasonably be expected to exist. Excavations in landfill areas, in areas where hazardous substances are stored nearby, and in which a connection(s) is being made to an in-service manhole are examples of situations where a hazardous atmosphere could exist or develop. Emergency rescue equipment such as breathing apparatus, safety harness, and lifeline must also be available *and* attended.

Excavations and adjacent areas must be inspected on a daily basis by a competent person for evidence of potential cave-ins, protective system failures, hazardous atmospheres, or other hazardous conditions. The inspections are only required if an employee exposure is anticipated. Walkways must also be provided where personnel must cross the excavation. If the excavation is six feet (1.8 meters) or more in depth at the crossing point, guardrails must also be provided on the walkways.

Accidents at the site of trenching and shoring activities are all too common. In addition to protecting workers from the danger of a cave-in, safety precautions must also be taken to protect them from traffic hazards if the work is performed in a street. Check with your local safety regulatory agency. They can provide you with the appropriate regulations. Do not wait until an emergency arises to obtain the information.

14.203 Sewer Cleaning

Never use a tool or piece of equipment unless you have received training in its proper use or operation. Require the equipment vendor to provide you with this training. Know your limitations and the limitations and capabilities of your tools and equipment. Do not use tools or equipment improperly; you could be seriously injured.

If you use chemicals of any kind for root or grease control in your system, become thoroughly familiar with their use and, specifically, with any hazards involved. See Section 14.15, "Toxic and Harmful Chemicals."

14.204 Acknowledgment

Major portions of Section 14.12, "Confined Spaces," Section 14.190, "Electric Shock/Stored Energy," and Section 14.202, "Excavations," were adapted from "Wastewater System Operations Certification Study Guide," developed by the State of Oklahoma. Permission to use this material is greatly appreciated.

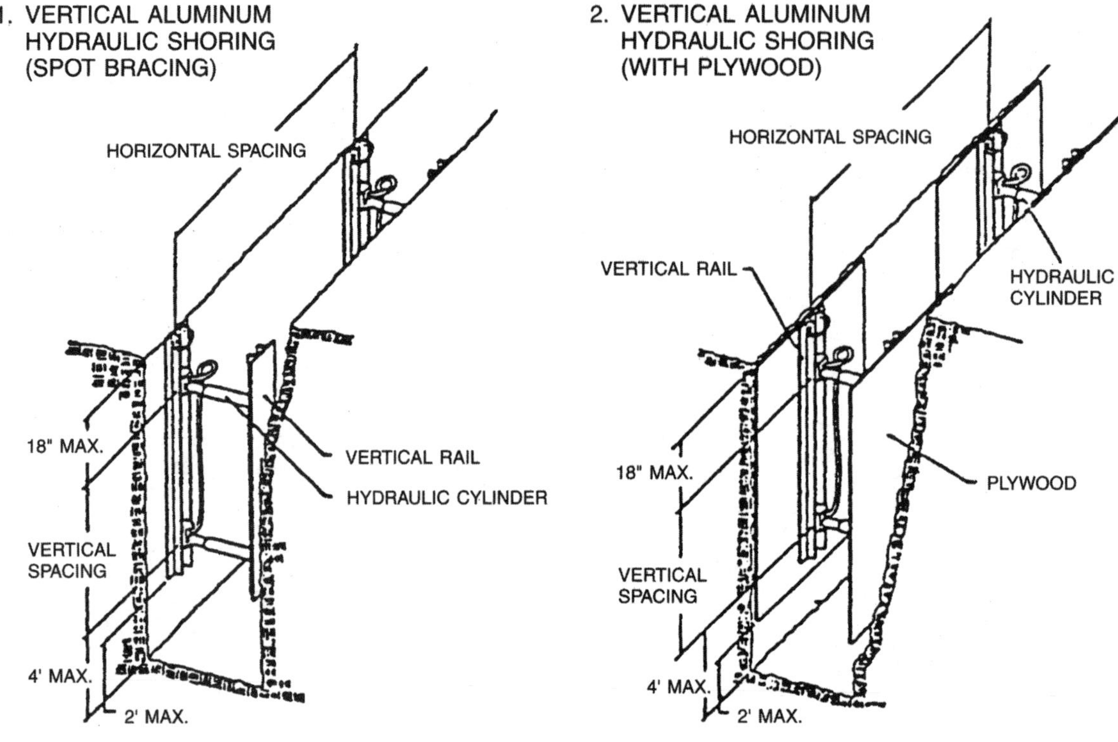

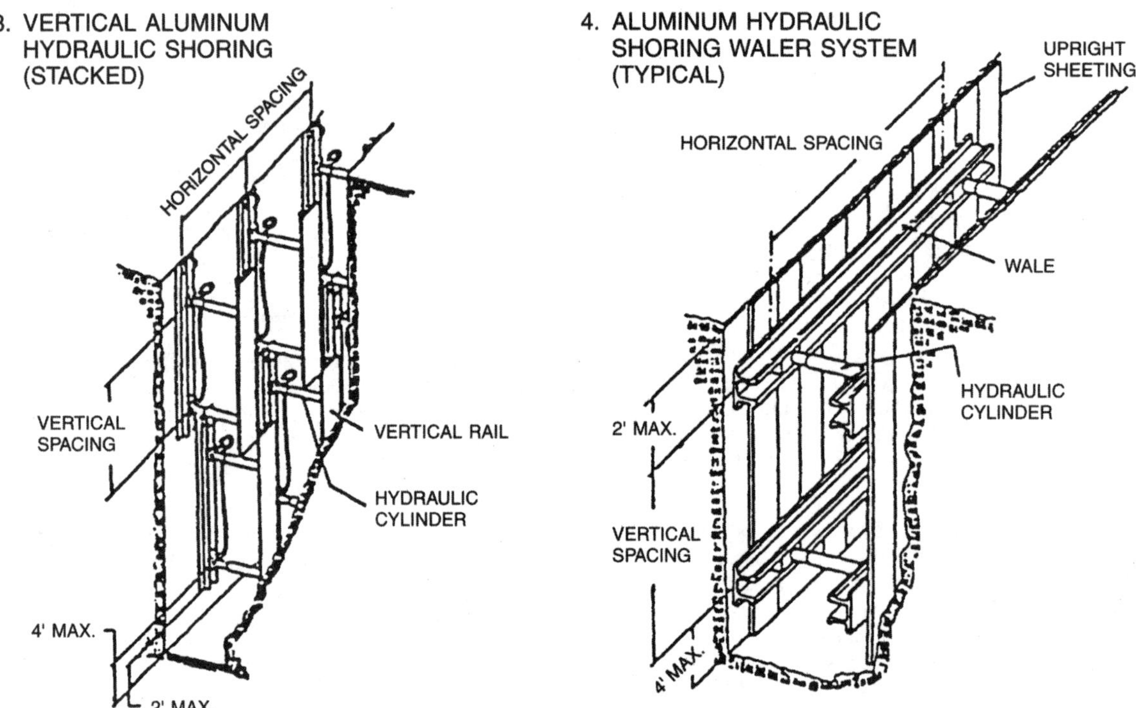

NOTE: Multiply inches × 2.5 to obtain centimeters; multiply feet × 0.3 to obtain meters.

Fig. 14.11 Typical installations of aluminum hydraulic shoring

(Source: 29 Code of Federal Regulations)

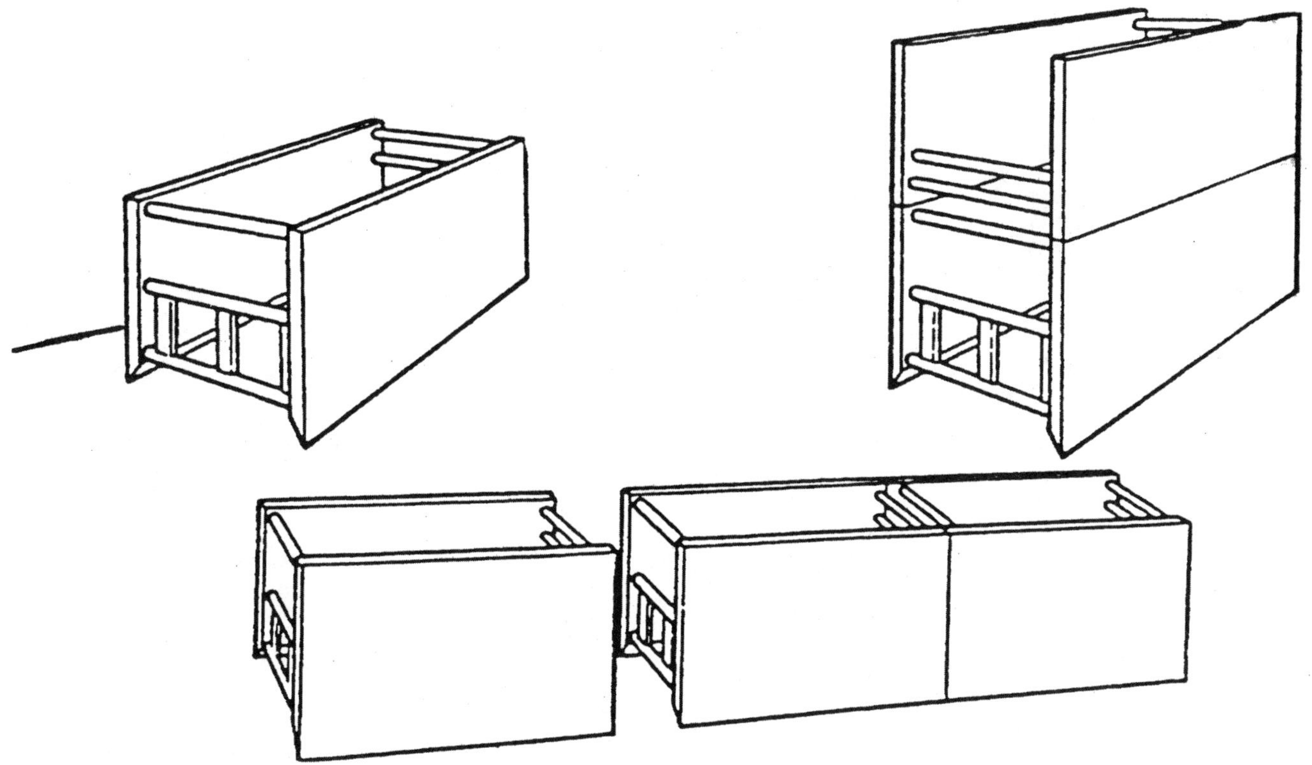

Fig. 14.12 Trench shields
(Source: 29 Code of Federal Regulations)

EXCAVATIONS MADE IN TYPE B SOIL

1. All simple slope excavations 20 feet or less in depth shall have a maximum allowable slope of 1:1.

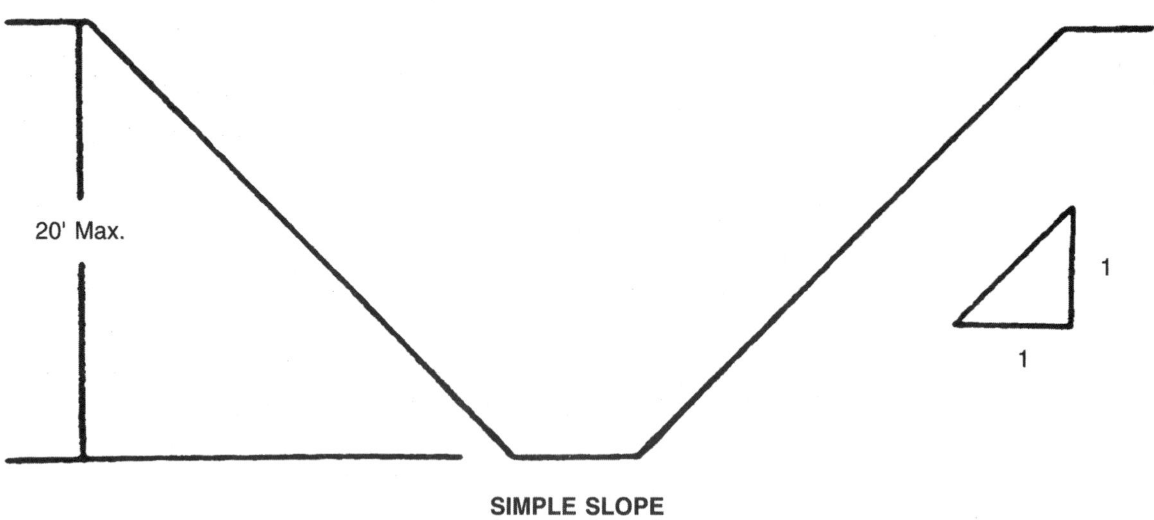

SIMPLE SLOPE

2. All benched excavations 20 feet or less in depth shall have a maximum allowable slope of 1:1 and maximum bench dimensions as follows:

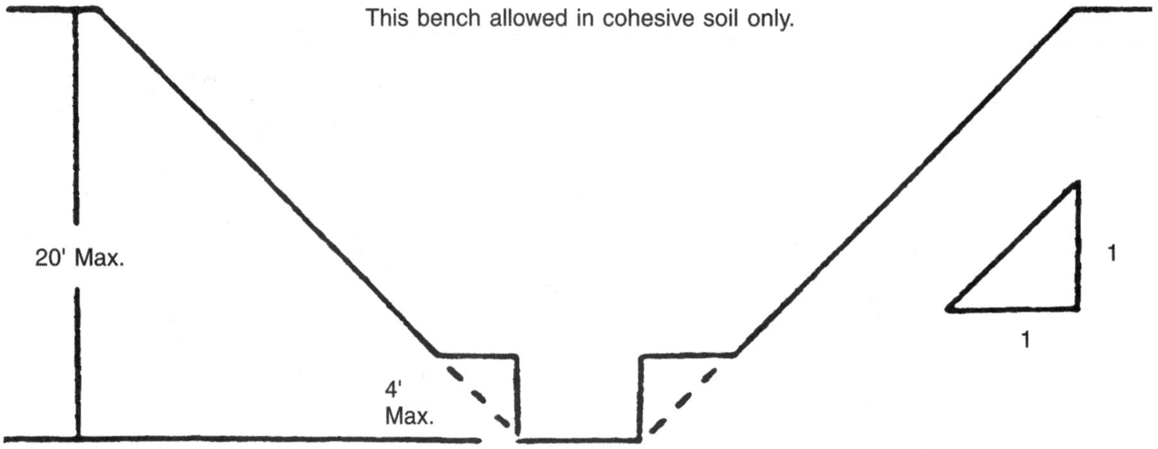

This bench allowed in cohesive soil only.

20' Max.

1

1

4' Max.

SINGLE BENCH

This bench allowed in cohesive soil only.

20' Max.

4' Max.

4' Max.

1

1

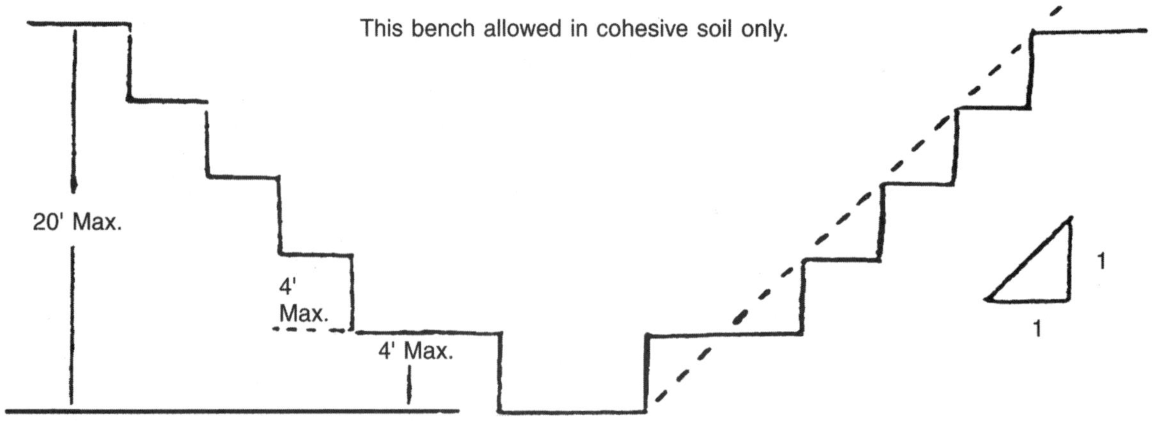

MULTIPLE BENCH

NOTE: Multiply inches × 2.5 to obtain centimeters; multiply feet × 0.3 to obtain meters.

Fig. 14.13 Typical sloping/benching schemes

(Source: 29 Code of Federal Regulations)

QUESTIONS

Write your answers in a notebook and then compare your answers with those on page 319.

14.2A List three ways to warn traffic that you are working in a street or traffic area.

14.2B List at least four hazards that may be encountered when entering a manhole, excluding atmospheric hazards.

14.2C List three protective systems for excavations.

14.2D At what excavation depth does OSHA require a protective system?

14.2E From whom should you receive training regarding the proper use or operation of sewer cleaning tools and equipment?

END OF LESSON 1 OF 3 LESSONS
on
PLANT SAFETY

Please answer the discussion and review questions next.

DISCUSSION AND REVIEW QUESTIONS

Chapter 14. PLANT SAFETY

(Lesson 1 of 3 Lessons)

At the end of each lesson in this chapter, you will find some discussion and review questions. The purpose of these questions is to indicate to you how well you understand the material in the lesson. Write the answers to these questions in your notebook.

1. What is the operator's responsibility with regard to safety?
2. Accidents do not just happen—they are _________ !
3. How can an operator avoid physical injuries?
4. Immunization shots protect against which infectious diseases?
5. What precautions should you take to avoid transmitting disease to your family?
6. What should you do when you discover an area with an oxygen deficiency or enrichment?
7. What kind of job site protection is usually required when you are working in a manhole?
8. Lift with your legs, not with your _________.
9. How should tools and equipment be transported to the bottom of a manhole?

CHAPTER 14. PLANT SAFETY

(Lesson 2 of 3 Lessons)

14.21 Pumping Stations

Pumping stations may vary from small, telemetered lift stations that are visited monthly to large pumping stations that have operators on duty 24 hours a day. Regardless of the size, type, or complexity of the pumping station, safe procedures must be followed at all times. Safety precautions discussed in Section 14.12, "Confined Spaces," are very important for work in any potential confined space, including both wet wells and dry wells.

Always be aware of the possibility of oxygen deficiency/enrichment, explosive or flammable gases, and toxic gases that could be generated in, or be discharged into, the sewers. A properly operating ventilation system is essential, particularly in below-grade pumping stations where heavier-than-air gases may collect. Before entering a potential confined space in an unattended lift station, test the atmosphere for oxygen deficiency/enrichment, explosive or flammable conditions (lower explosive limit), and toxic gases (hydrogen sulfide).

Do not work on electrical systems or controls unless you are qualified and authorized to do so. Even if you are qualified and authorized, use caution when operating and maintaining electrical controls, circuits, and equipment. Operate only those switches and electrical controls installed for the purpose of your job. Do not open or work inside electrical cabinets or switch boxes unless you are an authorized and qualified electrician.

Be aware of moving equipment, especially reciprocating equipment and rotating shafts. Moving parts that create a contact hazard to employees must be guarded. Do not wear loose clothing, rings, or other jewelry around machinery. Long hair must be secured. Wear gloves and use appropriate tools when cleaning pump casings to protect your hands from dangerous sharp objects.

When starting equipment, everyone should stand away from rotating parts. Dust and oil or loose metal may be thrown from shafts and couplings, or sections of a long, vertical shaft could come loose and whip around, especially during start-up of equipment.

If stairs are installed in a pumping station, they should have handrails and nonslip treads. Where space limitations prevent the installation of stairs, a spiral stairway, ship's ladder, or vertical ladder should be provided. Vertical ladders must be provided with a cage enclosure if they exceed 20 feet (6.1 meters) in unbroken length. If the ladder is not provided with a safety device or cage, an intermediate platform or landing should be provided for each 20 feet (6.1 meters) of height or fraction thereof.

Lighting should be sufficient to avoid glare and shadows.

Fire extinguishers must be provided in the station, properly located and maintained. They should be of a type that may be used on electrical equipment as well as on solid material or power overload-type fires. The use of liquid-type fire extinguishers should be avoided. All-purpose, ABC chemical-type fire extinguishers are recommended.

Good housekeeping is a necessity in a pumping station. Common housekeeping problems include water and oil on the floor, dirty or oily rags from cleanup operations, dirty lighting fixtures leading to poor visibility, and stairways with dirt or grease carried in by maintenance or repair crews.

Always be sure you have properly secured the pumping station when you leave. This procedure is necessary to prevent injury to neighborhood children and possible vandalism to the station. Both of these problems can be very costly to your employer if proper precautions are not taken.

QUESTIONS

Write your answers in a notebook and then compare your answers with those on pages 319 and 320.

14.2F What types of moving equipment should you be aware of in a pumping station and what safety precautions should be taken to protect yourself from injury?

14.2G What type of fire extinguisher should be available in a pumping station?

14.22 Treatment Plants

Because hazards found in pumping stations are similar to those found in treatment plants, the items discussed below may be applied to both situations. Safety precautions outlined in the previous section apply both to pumping stations and treatment plants.

14.220 Headworks

Structures and equipment in this category may consist of bar screens, racks, comminuting or grinding equipment, pump rooms, wet pits, and chlorination facilities.

1. *BAR SCREENS OR RACKS.* These may be either manually or mechanically cleaned. When manually cleaning screens or racks, be certain that you have a clean, firm surface to stand on. Remove all slimes, rags, greases, or other material that may create slip or trip hazards. *Good housekeeping in these areas is absolutely necessary.*

   When raking screens, leave plenty of room for the length of your rake handle so you will not be thrown off balance if the handle strikes a wall, railing, or light fixture. Wear gloves to avoid slivers from the rake handle or scraping your knuckles on concrete. Injury may allow an infection to enter your body.

Place all material in a container that may be easily removed from the structure. Be careful lifting containers full of heavy material such as grit. Use of containers sized to prevent excessive weight is recommended. It is better to make more than one trip than to risk injury. Do not allow material to build up on the working surface.

If your rack area is provided with railings, check to see that they are properly anchored before you lean against them. If removable safety chains are provided, never use these to lean against or as a means of providing extra leverage for removing large amounts of material.

A hanging or mounting bracket should be used to hold the rake when not in use. Do not leave it lying on the deck.

If mechanically raked screens or racks are installed, never work on the electrical or mechanical part of this equipment without locking out the unit. Always open, lock out, and tag the main circuit breaker before you begin repairs.

The tag (Figure 14.5) should be securely fastened to the breaker handle or to the lockout device to notify others that you are working on the equipment and that it must remain de-energized.

The time and date the unit was locked out should be noted on the tag, as well as the reason it was locked out. The tag should be signed by the person who locked out the unit. No one should then close the main breaker and start the unit until the tag and lock have been removed by the person who placed them there, or until specific instructions are received from the person who tagged the breaker. Your local safety equipment supplier can obtain tags and lockout devices for you.

2. *COMMINUTING OR GRINDING EQUIPMENT.* This equipment may consist of barminutors, comminutors, grinders, or disintegrators.

   NEVER work on the mechanical or electrical parts of the unit without first locking out the unit at its main circuit breaker. Be certain the breaker is properly tagged as explained in the previous section.

   Good housekeeping is essential in the area of comminuting equipment. Keep all walking areas clean and free of slimes, oils, greases, or other materials. Clean up all spills immediately. Provide a storage place for equipment and tools used in this area.

   See that proper guards are installed and kept in place around moving parts that create a contact hazard to employees. De-energize, lock out, and tag equipment before removing guards to make adjustments. Replace all of the guards prior to returning the equipment to service. Use extension tools to make adjustments if the equipment must be operated unguarded. Do not expose yourself to the hazard of moving parts.

3. *PUMP ROOMS.* The same basic precautions apply here as to any type of enclosed room or pit where wastewater or gases may enter and accumulate.

   Always provide adequate ventilation to remove atmospheric hazards. If the room is below ground level and provided with forced-air ventilation, be certain the fan is operating properly before entering the area. Use confined space procedures (refer to Section 14.12) when entering pits, wet wells, or tanks.

   The tops of all stairwells or ladders should be protected by a removable safety chain. Keep this chain in place when the stairwell or ladder is not being used.

   Guards should be installed around all moving parts that create a contact hazard to employees. Never remove equipment guards without first locking out and tagging the equipment at the main breaker. Always replace all guards before starting units.

   Maintain good housekeeping in pump rooms. Remove all oil and grease, and clean up spills immediately.

   If you have a multilevel pump building, never remove and leave off equipment access hatches unless you are actually removing or replacing equipment. Be sure to provide barricades or posts with safety chains around the opening to prevent falls. Be extremely cautious when working around openings that have raised edges. These are hazardous because you can easily stumble over them.

   Never start a positive displacement pump against a closed valve. Positive displacement pumps can create extreme pressures that can cause equipment or piping failures and endanger operators in the vicinity. Positive displacement pumps should be equipped with a pressure switch to shut the unit off before dangerous pressures are encountered.

   Lighting fixtures, ventilation units, electric motors, and other electrical devices in pump rooms may be required to be rated for use in hazardous atmospheres (explosion proof), depending on the design and contents of the specific area. The integrity of electrical devices in potentially hazardous atmospheres must be maintained. Any defective or inoperative equipment must be repaired or replaced immediately by a qualified person. Only qualified electricians should work on an electrical control panel because unqualified persons may be seriously injured by electric shock and they could damage the equipment. Errors in reassembly of explosion-proof equipment can create a significant hazard potential. Sources of ignition must be controlled if the pump room has the potential to contain an explosive/flammable atmosphere. Hot work such as welding or soldering and work involving standard portable electrical tools should only be allowed in the area if atmospheric conditions can be maintained in a safe state while the ignition source(s) exist.

> CAUTION
>
> **Unless you are a qualified electrician, stay out of the inside of all electrical panels. If you are not trained, authorized, or qualified to work with the equipment, leave it alone!**

4. *WET PITS—SUMPS.* Covered wet pits or sumps are potential death traps. They are confined spaces. Always strictly follow confined space procedures (refer to Section 14.12) when you work in these areas. Requirements for work in confined spaces include, but are not limited to, written procedures and permits, monitoring, rescue and respiratory protective equipment, training, atmospheric monitoring, and ventilation. In addition to the atmospheric, engulfment, and entrapment hazards common to many confined spaces, many other hazards may also exist.

Extreme care must be used when climbing up and down access ladders to pit areas. The application of a nonslip coating on ladder rungs is helpful. If available, a hoisting mechanism is safer than a ladder for entering pit and sump areas. Watch your footing on the floor of pits and sumps. They are very slippery.

Never attempt to carry tools or equipment up or down ladders into pits or sumps. Always use a bucket and handline or a sling for this purpose.

Only explosion-proof lights and equipment should be used in these areas if the potential for a flammable/explosive atmosphere exists.

A good safety practice is to effectively block the entry of potentially hazardous contaminants into the wet pit or sump. Isolate sources such as wastewater or sludge flows, chlorination, or plant drainage, whether they discharge upstream, downstream, or directly into the wet pit or sump. This exercise, coupled with forced ventilation, will minimize the potential for an atmospheric hazard to develop while the space is occupied.

Chlorination safety is discussed in Volume I, Chapter 10, "Disinfection and Chlorination."

QUESTIONS

Write your answers in a notebook and then compare your answers with those on page 320.

14.2H Why should slimes, rags, or greases be removed from around bar screens or racks?

14.2I What precautions would you take when working on electrical or mechanical parts of equipment?

14.2J What parts of equipment should have guards installed around them?

14.2K How would you transport tools or equipment into or out of pits or sumps?

14.221 Grit Channels

Grit channels may be of various designs, sizes, and shapes, but they all have two things in common: they are confined spaces and they get dirty. Good housekeeping is needed. Keep walking surfaces free of grit, grease, oil, slimes, or other material that may create a slip or trip hazard.

Before working on mechanical or electrical equipment, be certain that it is locked out and properly tagged (Figure 14.5). Install and maintain guards over moving parts that could be contacted accidentally.

If it becomes necessary to enter the channel for cleaning or other work, do so with extreme caution. Follow all confined space procedures (refer to Section 14.12). Always provide sufficient ventilation to maintain safe atmospheric conditions within the channel. Before any entry occurs, use a high-pressure water hose to agitate any material that may have accumulated on the bottom of the channel. Agitate the material sufficiently to release any *ENTRAINED*[11] gases. Test the atmosphere first for oxygen deficiency/enrichment, then combustible gases, and then toxic gases.

Be sure of your footing when working in these structures. Rubber boots with steel safety toes and a nonskid, cleat-type sole should be worn. Use safety sole inserts in your boots to prevent puncture wounds. Step slowly and cautiously as there is usually an accumulation of slippery material or slimes on the bottom. Use hand holds and railings; if none are available, install them now.

Use ladders, whether vertical or ships' ladders, cautiously. If possible, apply nonslip material or coatings to ladder rungs. Keep handrails free of grease and other slippery substances.

When using a portable ladder, the horizontal offset should not be less than one-quarter of the ladder distance between supports. The ladder should extend three feet (one meter) beyond the access point to facilitate easy access to and from the ladder at the top.

14.222 Clarifiers or Sedimentation Basins

The greatest hazard involved in working on or in a clarifier is the danger of slipping. If possible, maintain a good, nonskid surface on all stairs, ladders, and catwalks. This may be done by using nonskid strips or coating. Be extremely cautious during freezing weather. A small amount of ice can be very dangerous. Be careful and do not fall in.

Your housekeeping program should include the brushing or cleaning of effluent weirs and launders (effluent troughs). Effluent weirs and launders on clarifiers should be brushed or cleaned to avoid the accumulation of slippery slimes and solids. Uneven flow over an effluent weir can encourage short-circuiting. When it is necessary to actually climb down into the launder, always use a fall arrest system as discussed in Section 14.10, page 265. A fall may result in a very serious injury. If you work in areas where the danger of drowning exists, you must wear a US Coast Guard-approved life jacket or buoyant work vest.

Remember that a clarifier is a confined space and confined space procedures must be followed for entry and occupancy (see Section 14.12). Be cautious when working on the bottom of a clarifier. When hosing down, always hose a clean path to walk upon. Avoid walking on the remaining sludge whenever possible.

Always turn off, lock out, and tag the clarifier breaker before working on the drive unit. Adjustments should not be made on flights or scrapers while the unit is in operation. Keep in mind that, although these are moving quite slowly, there is tremendous power behind their movement. Stay clear of any situation where your body or the tools you are using may get caught under one of the flights or scrapers.

Guards must be installed over moving parts that could be contacted accidentally. Keep these in place whenever the unit is in operation.

Railings must be installed along the tank side of all normal walkways. If the unit is elevated above ground, railings should

[11] *Entrain.* To trap bubbles in water either mechanically through turbulence or chemically through a reaction.

be installed along the *outside* of all walkways, also. Standard guardrails usually consist of a top rail, an intermediate rail, and posts or uprights. The top rail should be 42 inches (107 cm) high, measured from the top of the rail to the surface of the walkway. The guardrail system is typically designed to support a load of at least 200 pounds (91 kilograms) applied in any direction at any point on the rail. Check with your state safety office for requirements on railing installation.

14.223 Digesters and Digestion Equipment

Digesters and their related equipment include many hazardous areas and potential dangers. The National Fire Protection Association Standard 820 (NFPA 820), *FIRE PROTECTION IN WASTEWATER TREATMENT AND COLLECTION FACILITIES,* lists requirements for electrical classifications, ventilation, gas detection, and fire control methods in sludge and gas handling areas. For ordering information, see Section 14.11, "Additional Reading," item 7.

NO SMOKING AND NO OPEN FLAMES should be allowed in the vicinity of digesters, in digestion control buildings, or in any other areas or structures used in the sludge digestion system. This includes pipe galleries, compressor or heat exchanger rooms, and others. All these areas should be posted with noticeable signs that forbid smoking and open flames. *Methane gas produced by anaerobic conditions is explosive when mixed with the proper proportion of air. For this reason, NFPA 820 prohibits the use of standard electrical systems on anaerobic digesters.*

Electrical systems on or within five feet (1.5 meters) of an anaerobic digester must be rated for use in Class I, Division I locations. Class I, Division I locations are those areas where a hazardous concentration of digester gas may exist (1) under normal operating conditions, or (2) frequently because of leakage repair/maintenance activities, or (3) because of a breakdown or faulty operation that might release hazardous concentrations of digester gas.

All rooms, galleries, and tunnels in the sludge and gas-handling areas should be provided with continuous forced ventilation. Required air changes per hour are specified in NFPA 820. Certain areas such as gas compressor rooms are also required to be fitted with classified electrical devices and gas detection systems.

Digesters are confined spaces. Follow all confined space procedures if it becomes necessary to work in a digester. In addition to confined space procedures, several additional measures can enhance the safety of digester entry. Stopping sludge addition several days before removing the digester from service will drastically reduce gas production. Piping that could introduce hazards to the digester such as gas and sludge lines should be mechanically separated from the digester (or pancaked by inserting a solid disc between flanges) to eliminate potential hazards from valve leakage. Purging of gas lines and equipment is also recommended. Purging is the displacement of hazardous gas with an inert substance such as nitrogen, carbon dioxide, or water. Purged gases are typically discharged at a controlled rate to a safe outside area where no ignition source is present and where no employee exposure will occur.

Explosion-proof ventilation, lighting, and "nonsparking" tools must also be used when working around digesters and gas-handling equipment unless safe atmospheric conditions can be maintained. Continuous atmospheric monitoring at the immediate work site should be performed throughout the work period.

Never allow smoking, open flames, or other sources of ignition within an empty digester unless safe atmospheric conditions can be ensured and maintained! (Figure 14.14)

Guardrails should be installed along the edges of the digester roof or cover in areas where it is necessary to work close to the edge. As previously discussed in Section 14.10, page 265, a fall arrest system must be used if you work in an unguarded elevated area six feet (1.8 meters) or more above a lower level. A fall from the top of a digester could be fatal. Toe boards should also be installed to prevent objects/tools from being kicked over the edge.

When working on equipment such as draft tube mixers, compressors, and diffusers, be certain that the unit that operates or supplies gas to the equipment is properly valved or locked out and appropriately tagged (Figure 14.5).

If you have a heated digester, read and heed the manufacturer's instructions before working on the boiler or heat exchanger. Know that the main gas valve is turned off before attempting to light the pilot. Be certain that the fire box has been ventilated according to the manufacturer's instructions before lighting the pilot. Do not work on the boiler unless you are authorized and qualified to do so.

When it becomes necessary to clean tubes or coils in a heat exchanger, turn off the hot water supply far enough in advance to allow the heat exchanger to cool. Never open the unit without ensuring that sludge and water flows have been effectively isolated and that water and sludge temperatures have cooled down to body temperature or lower.

CAUTION

Waste gas burners are noted for blowing out in a moderate wind. Before you attempt to relight the unit, be certain that the main valve has been turned off and the stack allowed to vent itself for a few minutes. These units have been known to "backflash" when re-ignited.

Before working on any sludge pump, whether it is centrifugal or positive displacement, be certain that the unit is effectively isolated from process piping and that it is locked out and properly tagged (Figure 14.5).

Positive displacement pumps should be equipped with a pressure switch to shut the unit off at a preset pressure. Never start a positive displacement pump against a closed discharge valve because pressure could build up and burst a line or damage the pump. If you have closed this valve to inspect or clean the pump, double check to be sure that it is open before starting the unit.

Sludge pump rooms should be well ventilated to remove any gases that might accumulate from leakage, spillage, or from a normal pump cleaning. If you spill digesting sludge, clean it up immediately to prevent the possible accumulation of gases or a slippery walkway.

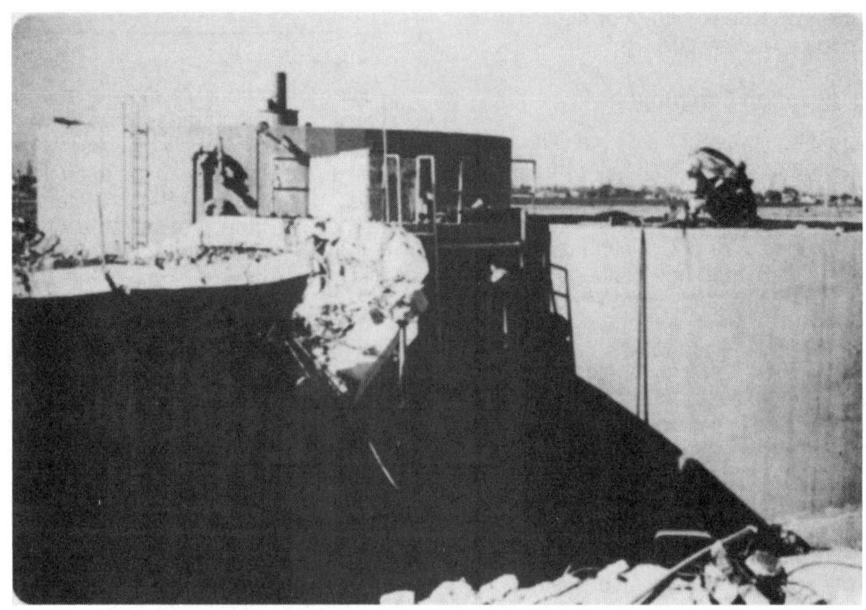

An explosion blew off the top of a digester ...

... and it landed on top of a pickup truck.

Fig. 14.14 Blown-up digester

Provide thorough, regularly scheduled inspections and maintenance of your gas collection system. Inspect drip traps regularly. The so-called "automatic" drip trap is known to jam open frequently, allowing gas to escape, and, therefore, should not be installed in any area where leakage could create a hazard.

Routine effective maintenance of safety equipment in gas handling systems is essential to maintain a safe working environment. Flame arresters, vacuum and pressure relief valves, and thermal valves in gas piping cannot prevent injury or property damage if they do not function properly. Equipment interlocks, automatic isolation valves, and gas detection and alarm systems all require regular testing and calibration to ensure that they will operate properly if and when they are needed.

Additional safety information for working in and around digesters is also contained in Chapter 12, "Sludge Digestion and Solids Handling."

QUESTIONS

Write your answers in a notebook and then compare your answers with those on page 320.

14.2L How can the danger of slipping be reduced on slippery surfaces?

14.2M Why should no smoking or open flames be allowed in the vicinity of digesters?

14.2N What safety precautions would you take before entering a recently emptied digester?

14.2O What would you do before relighting a waste gas burner?

14.2P Why should you never start a positive displacement pump against a closed discharge valve?

14.224 Trickling Filters

When it becomes necessary to inspect or service a rotating distributor, stop the flow of wastewater to the unit and allow it to come to rest. Lock out and tag the influent valve or pump supplying flow and secure the distributor arm so no one can start the distributor while you are working.

Never stand or walk on the filter media while the rotating distributor is in motion.

Provide an approved ladder or stairway for access to the media surface. Be positive this is free from obstructions such as hose bibs and valve stems.

Use extreme caution when walking on the filter media. The biological slimes make the media very slippery. Move cautiously and be certain of your footing.

Never allow anyone to ride a rotating distributor.

Although a rotating distributor moves fairly slowly, the force behind it is powerful. If you fall off and are dragged by a distributor, you will be fortunate if you can walk away under your own power.

When inspecting underdrains, remember that these are dangerous confined spaces and, if entry is necessary, confined space procedures must be used. Check to determine that the channels or conduits are adequately ventilated. Gases are not normally a problem here, but may be if there is a buildup of solids that have become septic.

If it becomes necessary to jack up a distributor mechanism for inspection or repair, always try to provide a firm base off the media or drainage system for the jack plate. A firm base may be provided by wooden planks, which will spread the weight over a large area. However, sometimes the only way to obtain firm support is to remove the media and use the drainage system as a firm base. Again, wooden planks should be placed over the drainage system to distribute the weight better. Remember, you are lifting a heavy weight. Do not attempt inspection or repair work until the distributor has been adequately and properly blocked in its raised position.

14.225 Aerators

Guardrails must be installed on the tank side of usual work areas or walkways. If the tank is elevated above ground, guardrails should also be installed on the ground side of the tank as discussed in Section 14.222.

When working on Y-walls, or other unguarded areas where work is done infrequently, at least two people should do the work. US Coast Guard-approved life preservers with permanently attached handlines should be accessible at strategic locations around the aerator. The life preservers must be provided with at least 90 feet (27 meters) of handline and be located not more than 200 feet (61 meters) apart. If you work in areas where the danger of drowning exists, you must wear a US Coast Guard-approved life jacket or buoyant work vest.

An experiment in England found that if an operator fell into a diffused aeration tank, the operator should be able to survive because air will collect in the clothing and tend to help keep the operator afloat.[12] Drownings apparently occur when a person is overcome by the initial shock or there is nothing to grab hold of to keep afloat or to pull oneself out of the aerator. In aerators where diffused air is supplied only along the walls, strong currents develop, which could pull anyone but a very strong swimmer under water. *ALWAYS wear a Coast Guard-approved flotation device when working unguarded over water.*

When removing or installing diffusers, be aware of the limitations of your working area. Inspect and properly position hoists and other equipment used in servicing swing diffusers.

Remember, an aerator is a confined space. When it is necessary to work in an empty aerator, follow confined space procedures and comply with all relevant safety requirements. Portable ladders can be awkward but they are safe if positioned and used properly. A good practice is to use a fall arrest system when climbing up or down a ladder and to secure the top of the ladder so that it cannot slip. Be extremely careful when walking in an aerator; the floor of the aerator can be very slippery.

---

[12] M. A. Kershaw, "Buoyancy of Aeration Tank Liquid," *JOURNAL WATER POLLUTION CONTROL FEDERATION,* Vol. 33, No. 11, p. 1151 (Nov. 1961) and Patrick L. Stevens, "I Fell into an Aeration Tank," *OPERATIONS FORUM,* Vol. 3, No. 5, p. 21 (May 1986).

If your plant is in an area subject to freezing weather, be aware of possible ice conditions around these units and use caution accordingly.

Additional safety information for working in and around aerators is contained in Chapter 11, "Activated Sludge."

14.226 Ponds

Ponds of any kind present the same hazards. Therefore, the following safety measures will apply to ponds in general.

If it is necessary to drive a vehicle on top of the pond levees, maintain the roadway in good driving condition by surfacing it with gravel or asphalt. Do not allow chuckholes or the formation of ruts. Be extremely cautious in wet weather. The material used in the construction of most levees becomes very slippery when wet. Slippery conditions should be corrected using crushed rock or other suitable material.

Never go out on the pond for sampling or other purposes when you are by yourself. Someone should be standing by on the bank in case you get into trouble. Always wear a US Coast Guard-approved life jacket when working from a boat or raft. And, as in any boating activity, do not stand up in the boat while performing work.

QUESTIONS

Write your answers in a notebook and then compare your answers with those on page 320.

14.2Q How would you stop the rotating distributor on a trickling filter?

14.2R Why should you never work alone on the center Y-wall of an aerator?

14.2S What precautions should you take when using a boat to collect samples from a pond?

14.227 Chemical Treatment

14.2270 CHLORINE AND SULFUR DIOXIDE

Chlorine and sulfur dioxide are very similar when considering safety and health aspects of the two chemicals. Because of this similarity, the following information is applicable to both substances. Additional safety considerations and information are contained in Volume I, Chapter 10, "Disinfection and Chlorination."

Chlorine and sulfur dioxide are usually purchased on a bid contract. Contracts should specify the conditions under which containers delivered to your plant will be accepted. The container should be stamped with the last date when it was pressure tested. Do not accept containers that have not been pressure tested within five years of the delivery date. Cover cap should be securely in place over the valve mechanism. All threaded connections should be clean and not worn or cross threaded. New, approved gaskets should be provided with each container. Do not accept containers not meeting these standards. A few minutes spent inspecting containers when they are delivered can prevent serious problems in the future.

> ***Chlorine gas and sulfur dioxide gas are highly irritating and corrosive gases. Danger...handle with caution!***

The most common causes of accidents involving chlorine and sulfur dioxide are leaking pipe connections and excessive dosage rates.

Containers should be stored in a cool, dry place, away from direct sunlight or from heating units. Some heat is needed to cause desired evaporation and to control moisture condensation on containers. Containers should never be dropped or allowed to strike each other with any force. Cylinders should be stored in an upright position and secured with a chain, wire rope, or clamp. They should be moved only by hand truck and should be well secured during moving. One-ton containers should be blocked so that they cannot move. They should be lifted only by an approved lifting bar with hooks over the ends of the containers. Never lift a container with an improvised sling.

Always wear a faceshield when changing chlorine or sulfur dioxide containers. Escape supplied-air respirators are also recommended. Connections to containers should be made only with approved fittings. Always inspect all surfaces and threads of the connector before connecting. If you are in doubt as to their condition, do not use the connector. Always use a *new,* approved-type gasket when making a connection. The reuse of gaskets very often will result in a leak. Check for leaks as soon as the connection is completed. Use aqua ammonia vapor to check for chlorine or sulfur dioxide leaks. The ammonia vapor will combine with any chlorine or sulfur dioxide present to form a white cloud. Do not apply ammonia solution directly to piping or connections; corrosion will result. Never wait until you smell chlorine or sulfur dioxide. If you discover even the slightest leak, correct it immediately, as leaks will get worse rather than better. Like accidents, leaks generally are caused by faulty procedure or carelessness.

Obtain from your supplier and post in a conspicuous place (*outside* the chlorination and sulfonation room and by the telephone) the name and telephone number of the nearest emergency service in case of a severe leak.

Storage and equipment rooms must be provided with forced ventilation. As chlorine and sulfur dioxide are approximately two and a half times heavier than air, vents should be provided at floor level. Some installations have a blower mounted on the roof to blow air into the room and are vented at the floor level to allow escaped gas to be blown out of the building. Other installations may have the ventilation system and duct work interlocked to a leak detection system to "bottle up" the room when a leak occurs. This prevents uncontrolled dispersion of chlorine or sulfur dioxide to the atmosphere.

The Uniform Fire Code, Article 80, Hazardous Materials, requires that exhaust ventilation for chlorine and sulfur dioxide areas must be taken from a point within 12 inches (30.5 centimeters) of the floor. Mechanical ventilation should be at a rate of not less than one cubic foot of air per minute per square foot (0.00508 cubic meter of air per second per square meter) of floor area of storage handling area. Normally, ventilation from these areas is discharged to the atmosphere, but when a leak occurs, the ventilated air containing the chlorine or sulfur dioxide should be routed to a treatment system to remove the chemical. A caustic scrubbing system can be used to treat the air from a leak. Treatment systems must be designed to reduce the maximum allowable discharge concentration of chlorine or sulfur dioxide to one-half the IDLH (Immediately Dangerous to Life or Health[13]) at the point of discharge to the atmosphere. The IDLH for chlorine is 10 ppm. For sulfur dioxide the IDLH is 100 ppm. A secondary, standby source of power is required for the detection, alarm, ventilation, and treatment systems.

Always enter storage or equipment rooms with caution. If you smell chlorine or sulfur dioxide when opening the door to the area, immediately close the door, leave the ventilation on, and seek assistance.

Never attempt to enter an atmosphere containing chlorine or sulfur dioxide when you are by yourself. Operators engaged in emergency response to a hazardous materials release must be effectively trained in hazard recognition, repair techniques, use of personal protective equipment such as clothing and respiratory protective equipment, decontamination procedures, if required, and use of the buddy system. In a buddy system, each operator in the work group is designated to observe another operator in the group, with the purpose of providing assistance in an emergency. Responses to chlorine or sulfur dioxide releases should be coordinated with your local fire department and hazardous materials (HAZMAT) response agency.

Excellent booklets may also be obtained from PPG Industries, Inc. (*CHLORINE SAFE HANDLING*[14]) and from The Chlorine Institute, Inc. (*CHLORINE MANUAL*[15]). Safety information on chlorine handling is also contained in Volume I, Chapter 10, "Disinfection and Chlorination." Your chlorine and sulfur dioxide suppliers will probably provide you with all the information you need to handle and use chlorine and sulfur dioxide safely. It is your responsibility to obtain, read, and understand safety information and to practice safety.

If you are the manager of a wastewater treatment facility that uses large quantities of chlorine or sulfur dioxide, you may be required to develop and implement written safety plans for dealing with the effects of both on-site and off-site chemical releases. See Chapter 20, Section 20.10, "Emergency Response," for information about Process Safety Management (PSM) programs and Risk Management Programs (RMPs).

14.2271 POLYMERS

The primary use for polymers in wastewater treatment is the conditioning of sludge to facilitate removal of water in subsequent treatment processes such as belt filter presses, centrifuges, gravity belt thickeners, and dissolved air flotation thickeners. Polymers are available in either liquid or dry forms. The type of polymer you use depends on the application, product performance, volume of use, space availability, and equipment capabilities.

Dry polymers come in various types of powder, crystals, and beads. Most dry polymers have a carrier in them and are about 94 percent to 96 percent active. The powder form is the most economical, but it may release a fine dust when handled, which can present a safety hazard.

Liquid polymers are available in either solution or emulsion/dispersion types. Both types consist of a polymer material in a medium, either water or oil, respectively. Emulsion/dispersion polymers range from 25 percent to 50 percent concentration and must be mixed with a large volume of water to achieve full activation before use. When water is added to emulsion/dispersion polymers, there is a dramatic increase in *VISCOSITY*.[16] Solution polymers range from 4 percent to 50 percent actual polymer and are extremely viscous and difficult to pump if not mixed with water. A widely used solution polymer is the ***mannich reaction*** polymer, which typically contains 4 percent to 6 percent active polymer. Mannich reaction polymers are produced by using formaldehyde as a catalyst. Since vapors from these polymers pose a safety hazard (formaldehyde is a *CARCINOGEN*[17]), they should be stored and used only by trained personnel in well-ventilated areas.

The use of polymers exposes operators to a number of safety hazards that require appropriate precautions to prevent injury or illness. The hazards generally fall into two areas, slipping hazards and personal exposure hazards from contact or inhalation:

- Slipping hazards—polymers have a moisture attracting property. Even a thin film of dry polymer can combine with moisture and become extremely slippery. Liquid polymers are inherently slippery. Polymer spills must be cleaned up immediately. Gather up as much of the spilled material as possible by using an appropriate method, such as gentle sweeping, vacuuming, soaking it up with rags, or using kitty litter as an absorbent. Water flushing and the use of household bleach can remove remaining polymer material. The proper disposal of cleaned-up material should be confirmed with your local environmental regulatory agency.
- Personal exposure hazards—irritation of skin, eyes, or lungs can result from contact with polymer, polymer dust, or polymer fumes. Personal protective equipment (PPE), such as chemical-resistant gloves, splash-proof goggles, and an apron, may be required. If exposure to polymer dust or fumes from mannich reaction polymer may occur, appropriate respiratory protection should be used. Environmental monitoring should be performed to determine the extent of exposure as well as the required level of protection.

Areas where polymers are handled or stored should be equipped with sufficient continuous ventilation and emergency eye wash stations and deluge showers. Remember, read and heed the safety requirements identified on your polymer's Material Safety Data Sheet (MSDS).

[13] *IDLH.* Immediately Dangerous to Life or Health. The atmospheric concentration of any toxic, corrosive, or asphyxiant substance that poses an immediate threat to life or would cause irreversible or delayed adverse health effects or would interfere with an individual's ability to escape from a dangerous atmosphere.

[14] PPG Industries, Inc., Monroeville Chemical Center, 440 College Park Drive, Monroeville, PA 15146. No charge.

[15] The Chlorine Institute, Inc., Bookstore, PO Box 1020, Sewickley, PA 15143-1020. Pamphlet 1. Price to members, $28.00; nonmembers, $70.00; plus $6.95 shipping and handling.

[16] *Viscosity* (vis-KOSS-uh-tee). A property of water, or any other fluid, that resists efforts to change its shape or flow. Syrup is more viscous (has a higher viscosity) than water. The viscosity of water increases significantly as temperatures decrease. Motor oil is rated by how thick (viscous) it is; 20 weight oil is considered relatively thin while 50 weight oil is relatively thick or viscous.

[17] *Carcinogen* (kar-SIN-o-jen). Any substance that tends to produce cancer in an organism.

14.228 Applying Protective Coatings

CAUTION: When applying protective coatings in a clarifier or any other tank, channel, or pit, whether enclosed or open topped, you can create an atmospheric hazard. These areas are confined spaces and the coating system used can create additional hazards by generating flammable or toxic vapors during application and curing periods. Whenever possible, use nonhazardous coating materials. The control of application rates and methods such as brushing or rolling rather than spraying can also minimize potential hazards. Always attempt to find an appropriate atmospheric monitor to ensure that a hazard is not being created; however, with the sophisticated coating products available now, this may not be possible. Industrial hygiene calculations can be made to control hazards if atmospheric monitors are not available for the specific hazardous component(s) in the coating material. These calculations consider characteristics of the coating material, application rates, ventilation rates, and a safety factor that can be increased as desired. Consult your local safety regulatory agency for assistance.

In addition to the atmospheric hazards encountered during coatings operations, some coatings (asphaltic or bitumastic) can also cause skin burns if you are exposed to higher vapor concentrations. Engineering controls such as ventilation and a combination of personal protective equipment (protective clothing and creams) may be required to prevent exposure and injury or illness.

Check with your paint suppliers for any hazards involved in using their products. Study the appropriate Material Safety Data Sheet (MSDS) and follow the safety guidelines provided.

14.229 Housekeeping

Good housekeeping can and has prevented many accidents. Housekeeping tasks that will keep a treatment plant a safer, cleaner place to work include:

1. Have a place for your tools and equipment. When they are not being used, see that they are kept in their proper place.
2. Clean up all spills of oil, grease, chemicals, polymers, wastewater, and sludge. Keep walkways and work areas clean. A clean plant will reduce the possibility of physical injuries and infections.
3. Provide proper containers for wastes, oily rags, and papers. Empty these frequently.
4. Store flammable substances in an approved storage cabinet.
5. Remove snow and ice in areas where a person may slip and fall.

QUESTIONS

Write your answers in a notebook and then compare your answers with those on page 320.

14.2T How should one-ton chlorine containers be lifted?

14.2U Why are chlorine vents placed at floor level?

14.2V What should you do if you open a door and smell chlorine or sulfur dioxide?

14.2W List four housekeeping tasks that will keep a treatment plant a safer, cleaner place to work.

14.23 Industrial Waste Treatment

If your wastewater treatment plant treats only industrial wastes or a mixture of both industrial and municipal wastes, you must be aware of the industries and the types of wastes that are discharged. Also, in spite of an effective sewer-use ordinance, you must be alert for accidental spills and other unanticipated discharges. These discharges can be toxic to you and the organisms in your treatment processes and corrosive to your equipment. From a safety viewpoint, if you know how to identify the types of wastes that may reach your plant, you can be prepared to take the proper action and safety precautions. This section will discuss some of the hazardous or toxic substances that could reach your treatment plant.

14.230 Fuels

Fuels may be dumped or drained into storm sewers connected to sanitary sewers or they may enter the collection system as a result of a leaking underground fuel line or tank. Fuel oil and gasoline usually float on the surface of wastewater and are not diluted by mixing. Therefore, most floating fuels tend to collect in wet wells and can create both explosive and toxic conditions or atmospheres.

To reduce the chances of an explosion in a wet well and downstream enclosed structures, a combustible-gas detector should be installed in wet wells to sound an alarm and transmit a signal to the main control panel before explosive conditions are reached. Explosion-proof wiring, equipment, and fixtures will help to prevent fires and explosions in hazardous areas such as wet wells. Adequate ventilation also is essential. Oil skimmers should be installed in wet wells to remove floating fuel oil and gasoline. This equipment may be rarely used, but can remove explosive fuels without exposing operators to hazardous conditions.

Fuels may be detected by permanent or portable devices that measure either hydrocarbons or the lower explosive limit (LEL). These devices must be installed in treatment plants using pure oxygen treatment processes (activated sludge). They are located in the collection system upstream from the plant, in the headworks, and at the oxidation reactors (aerators). If hydrocarbons or explosive conditions reach a given *SET POINT,*[18] an alarm is usually generated. If the concentrations increase beyond this point, the oxygen gas flow to the processes is automatically shut off. The system is purged with air to prevent a possible explosion. These detection devices must be properly located and maintained at frequent intervals to provide reliable service.

If gasoline reaches the wet well in the headworks of your treatment plant:

1. Try to remove the gasoline from the surface of the wet well with skimmers (if available) or with a portable pump.
2. Apply as much ventilation as possible to prevent an explosive atmosphere from developing.
3. Monitor the atmosphere for toxic and explosive conditions.
4. Keep personnel away from the area.
5. Do not allow any flames, sparks, or other sources of ignition in the area (use nonsparking tools and explosion-proof equipment and wiring).

---

[18] *Set Point.* The position at which the control or controller is set. This is the same as the desired value of the process variable. For example, a thermostat is set to maintain a desired temperature.

14.231 Toxic Gases

Hydrogen sulfide (H_2S) is the most common toxic gas encountered by operators because it is produced in collection systems by decomposing organic matter under anaerobic conditions. Hydrogen sulfide and other toxic gases can be discharged into sewers or produced by chemical and biological reactions in sewers, in pretreatment facilities, or at the wastewater treatment plant. A serious hydrogen sulfide problem can develop when a discharge of sodium sulfide (Na_2S) from one industry mixes with the discharge of an acid waste from another industry. This mixture can produce extremely hazardous concentrations of hydrogen sulfide that are not only toxic, but explosive, flammable, and very odorous. Sodium sulfide also reacts violently with oxidants to form sulfur dioxide, which is toxic and may accumulate in below-grade areas.

Other toxic gases include phosgene ("war or mustard gas"), chlorine, and tear-producing substances (lacrimators). See Table 14.1, page 268, "Common Dangerous Gases Encountered in Wastewater Collection Systems and at Wastewater Treatment Plants."

Phosgene is produced in sewers when discharges of alcohol saturated with phosgene-wasted chloroformates is back-hydrolyzed (reverse chemical reaction) to phosgene. If an industry or laundry quickly dumps a few hundred to one thousand gallons of bleach, a slug of chlorine can occur in the plant's influent. Naphtha is used as both a fuel and a solvent and can create hazardous conditions.

Tear-gas type substances (lacrimators) occasionally may reach treatment plants. Certain organic insecticide wastes, when only partially chlorinated at an industrial wastewater pretreatment plant, can form tear-gas type substances.

Toxic gases may be detected by probes that measure the concentration of a particular gas, such as hydrogen sulfide, chlorine, or sulfur dioxide. Most instruments are capable of detecting the lower explosive limit (LEL), an oxygen deficiency/enrichment, and hydrogen sulfide. Portable sensors and monitors for toxic gases usually require daily calibration. Permanent systems usually require weekly calibration. Both systems require regular maintenance.

14.232 Amines

Amines are compounds formed from ammonia. Some of these compounds may react with other substances in wastewater to form nitrosamines, some of which are considered carcinogens (capable of causing cancer in humans). A remote possibility exists that nitrosamines from industrial dumps or chemical-biological reactions in sewers could contaminate the air space around treatment plants. If this problem is discovered at a treatment plant, all treatment process structures (wet wells, clarifiers, aeration tanks) could have to be covered. Exhaust air from these sources would have to be treated to remove harmful contaminants and offensive odors. Operators must test the atmosphere for harmful contaminants before entering the area. The type of testing equipment needed will depend on the contaminants that are expected to be present.

14.233 Surface-Active Agents

Concentrated industrial *SURFACE-ACTIVE AGENTS,*[19] either accidentally spilled or dumped into a collection system, can upset wastewater treatment processes. Certain industrial wastes can serve as super floc agents that produce a much denser sludge than usually pumped. This sludge can be too thick to pump, which then requires special handling to reslurry the sludge so it can be pumped. The opposite can occur when antifloc agents lower the capture of suspended solids in primary clarifiers. Excessive solids can reach aeration tanks and even flow out in the plant effluent. Rejected batches of detergents can contain antifloc agents.

Foaming agents often cause treatment plant operators very serious problems. Foam in wet wells can prevent inspection and operation of screens and grinders. Foam on aeration tanks that blows into neighbors' yards and on the surface of receiving water is objectionable to the public. If foam is formed in a chlorination chamber by a *HYDRAULIC JUMP,*[20] the foam bubbles could contain chlorine gas. Detection devices are available that are capable of detecting excessive levels of foaming agents. These foam detection devices also can be programmed to automatically feed an antifoaming agent to keep foaming under control.

14.234 Biocides

Poisons or biocides from industries can be harmful to operators as well as toxic to organisms purifying the wastewater in treatment processes. Unfortunately, biocides frequently cannot be detected until after the organisms in the treatment processes have been killed. Not only can the activated sludge process and the digesters be put out of action, but the sludge can be so contaminated that it cannot be disposed of on land or in landfills.

14.235 High or Low pH

Highly acidic or alkaline wastes can be very hazardous. They are dangerous to personnel, treatment processes, and equipment. pH probes installed in the headworks can detect abnormal pH levels. A low pH caused by an acid can be increased by the addition of sodium hydroxide (NaOH). Sulfuric acid (H_2SO_4) can be added at the headworks to lower the pH of an alkaline waste. Study the appropriate Material Safety Data Sheet(s) (MSDS) and follow the safety guidelines provided.

---

[19] *Surface-Active Agent.* The active agent in detergents that possesses a high cleaning ability. Also called a SURFACTANT.

[20] *Hydraulic Jump.* The sudden and usually turbulent abrupt rise in water surface in an open channel when water flowing at high velocity is suddenly retarded to a slow velocity.

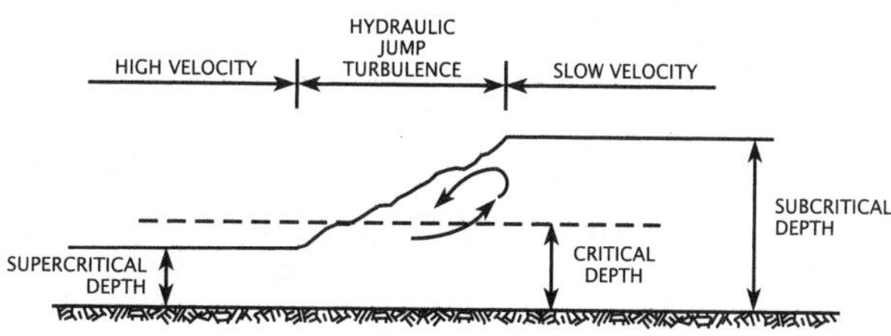

14.236 Summary

Your main concerns as an operator are your own personal survival and that of your co-workers, and the preservation of your plant and the organisms in the treatment processes. Industrial dumps can produce especially serious hazards. Effective sewer-use ordinances and industrial pretreatment facilities can help greatly to reduce the frequency and severity of industrial dumps. This section only covers a few of the many potential hazards that can be created by industrial wastes.

QUESTIONS

Write your answers in a notebook and then compare your answers with those on page 320.

14.2X List the major types of discharges from industrial plants that could create safety hazards.

14.2Y What would you do if a gasoline truck was in an accident and the fire department washed the spilled gasoline into a storm drain? The storm drain conveys wastewater to the wet well in the headworks of your treatment plant.

14.2Z What types of problems can be caused by surface-active agents?

END OF LESSON 2 OF 3 LESSONS
on
PLANT SAFETY

Please answer the discussion and review questions next.

DISCUSSION AND REVIEW QUESTIONS

Chapter 14. PLANT SAFETY

(Lesson 2 of 3 Lessons)

Write the answers to these questions in your notebook. The question numbering continues from Lesson 1.

10. When cleaning racks or screens, on what kind of surface should the operator stand?
11. Never lean against a removable safety chain. True or False?
12. Why should only qualified electricians work on an electrical control panel?
13. Why should effluent weirs and launders on clarifiers be brushed or cleaned?
14. Why should no smoking or open flames be allowed in the vicinity of the digester or sludge digestion system?
15. Why should you never go out on a pond for sampling or other purposes by yourself?
16. Where should the name and telephone number of the nearest emergency chlorine leak repair service be posted?
17. What safety precautions should be taken when applying protective coatings?
18. Why are discharges of amines into collection systems considered dangerous?

CHAPTER 14. PLANT SAFETY

(Lesson 3 of 3 Lessons)

14.3 SAFETY IN THE LABORATORY [21]

In addition to all safety practices and procedures mentioned in the previous sections of this chapter, the collecting of samples and the performance of laboratory tests require that you be aware of the specific hazards involved in this type of work.

Laboratories use many hazardous chemicals. These chemicals should be kept in limited amounts and used with respect. Your chemical supplier may be able to supply you with a safety manual. Everyone who may handle, use, or dispose of hazardous chemicals must be trained in appropriate methods to ensure their safety. See Section 14.15, "Toxic and Harmful Chemicals."

14.30 Sampling Techniques

Wear disposable, impervious gloves if your hands may come in contact with wastewater or sludge. When you have finished sampling, dispose of the gloves and wash your hands thoroughly, using a disinfectant-type soap, to prevent the spread of disease.

> *Never collect any samples with your bare hands. If you have any broken skin areas, such as cuts or scratches, you may easily become infected.*

Do not climb over or go beyond guardrails or chains when collecting samples. Do not lean on safety chains. Use sample poles, ropes, and other devices as necessary to collect samples.

14.31 Equipment Use and Testing Procedures

The following are some basic procedures to keep in mind when working in the laboratory:

> *Never look into the open end of a container during a reaction or when heating a container.*

1. Use proper safety goggles or a faceshield in all tests where there is danger to the eyes.
2. Use care in making rubber-to-glass connections. Lengths of glass tubing should be supported while they are being inserted into rubber. The ends of the glass should be *FLAME POLISHED*[22] to smooth them out, and a lubricant such as water should be used. Never use grease or oil. Wear gloves or some other form of protection for the hands when making such connections. Hold the tubing as close to the end being inserted as possible to prevent bending or breaking. Never try to force rubber tubing or stoppers from glassware. Cut the rubber as necessary to remove it.
3. Always check labels on bottles to make sure that the proper chemical is selected. Never permit unlabeled or undated containers to accumulate around or in the laboratory. Unlabeled containers are only allowed if the person using the chemical dispenses and uses the chemical during the shift the chemical was placed in the container. Keep storage areas well organized to prevent mistakes when selecting chemicals for use. Clean out old or excess chemicals. Separate flammable, explosive, or special hazard items for storage in an approved manner. See Section 14.11, "Additional Reading," Reference 11.

> *All chemical containers should be clearly labeled, indicating contents and date opened or solution prepared. All chemicals must be labeled with approved warning labels.*

4. Never handle chemicals with your bare hands. Use a spoon or spatula for this purpose.
5. Be sure that your laboratory is adequately ventilated.

> *Always work in a fume hood if working with chemicals or samples having toxic fumes.*

Even mild concentrations of fumes or gases can be dangerous.

There are minimum requirements for air velocity through laboratory hoods. A "normal hood" must have an average "face" velocity of 100 fpm (feet per minute) (30 mpm (meters per minute)) with a minimum of 70 fpm (21 mpm) at any point. If carcinogens are handled in the hood, a face velocity of 150 fpm (39 mpm) with a minimum of 125 fpm (38 mpm) is required. You must also provide a

[21] Also see *FISHER SAFETY CATALOG.* Obtain from Fisher Scientific Company, Safety Division, 4500 Turnberry Drive, Suite A (Customer Service), Hanover Park, IL 60103 or phone (800) 772-6733.

[22] *Flame Polished.* Melted by a flame to smooth out irregularities. Sharp or broken edges of glass (such as the end of a glass tube) are rotated in a flame until the edge melts slightly and becomes smooth.

method of indication that the air flow is active at each hood. Contact your local safety regulatory agency to verify specific requirements.

6. Never use laboratory glassware for a cup or food dish. This is particularly dangerous when dealing with wastewaters.
7. When handling hot equipment of any kind, always use tongs, insulated gloves, or other suitable tools. Burns can be painful and can cause more problems (encourage spills, fire, and shock).
8. When working in the lab, do not smoke or eat except in prescribed coffee break areas or in the lunch areas.

> Always thoroughly wash your hands before smoking or eating to prevent the spread of disease.

9. Do not pipet chemicals or wastewater samples by mouth. Always use a suction bulb or an automatic buret.
10. Handle all chemicals and reagents with care to protect your body from serious injuries and possible poisoning. Read and become familiar with all precautions or warnings on labels. Know and have available the antidote for all chemicals in your lab.
11. A short section of flexible tube on each water outlet is an excellent water flusher to wash away harmful chemicals from the eyes and skin. It is easy to reach and can quickly be directed on the exposed area. Eyes and skin can be saved if dangerous materials are washed away quickly. Provide an emergency eye wash and deluge shower in any area where corrosive, caustic, or otherwise hazardous substances are handled and stored. This equipment requires monthly operation and flushing to ensure that it will operate when needed.
12. Unsafe glassware is the largest single cause of accidents in the laboratory. Chipped glassware may still be used if it is possible to fire polish the chip in order to eliminate the sharp edges. This may be done by slowly heating the chipped area until it reaches a temperature at which the glass will begin to melt. At this point remove from flame and allow to cool.

> *Never* hold any piece of glassware or equipment in your bare hands while heating. Always use a suitable glove or tool.

13. Dispose of all broken or cracked glassware immediately. A special receptacle for broken glass should be available and well identified.
14. Wear a protective smock or apron when working in the lab. This may save you the cost of replacing your work clothes or uniform and prevent serious injury to yourself. Protective gloves and goggles should be worn when working with corrosive/dangerous chemicals.

> Remember to add acid to water, but *never* the reverse.

15. Electrical equipment must be properly grounded and safeguards provided to prevent insertion of improper plugs into the equipment.
16. Do not keep your lunch in a refrigerator that is used for samples or chemical storage.
17. Where cylinders of compressed gas are used or stored, the area must be well ventilated and heat sources eliminated. Ensure that the cylinders are properly secured.
18. Carbon dioxide (CO_2) or all-purpose, ABC chemical-type extinguishers should be mounted in readily accessible locations throughout the laboratory. A D-type extinguisher should be available where burning metals might be encountered.
19. Maintain spill control stations in sufficient quantity and locations so that chemical spills can be cleaned up in a safe and timely manner.
20. Provide appropriate first-aid kits at accessible and strategic locations within the laboratory.
21. Properly store and dispose of hazardous chemicals and wastes. Requirements for hazardous waste storage include, but are not limited to:
    - Do not place incompatible wastes in the same container.
    - Store hazardous waste containers in a secondary container to prevent uncontrolled leaks in the event the primary container fails.
    - Use a container made of a material that will not react with the hazardous material being stored. Containers should be DOT (Department of Transportation) approved for ultimate transport off site.
    - Keep the hazardous waste container closed unless it is necessary to add or remove waste.
    - Do not open, handle, or store a hazardous waste container in a manner that may cause it to rupture or leak.
    - At least weekly, inspect areas where hazardous waste containers are stored; look for evidence of leaks and container deterioration. Take corrective action as required.
    - Label containers with a hazardous waste label describing the waste contained and the date that accumulation started.

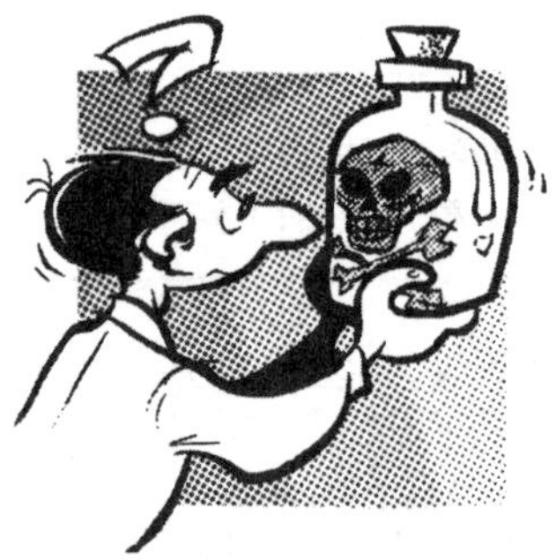

There are many more requirements for dealing with hazardous waste generation and disposal. Most areas limit the amount of time that a hazardous waste can be stored on site to 90 days, however, there are exceptions for certain operations. Your plant may also be required to obtain a federal EPA identification number that must be used on hazardous waste manifests that are required when you transport, or arrange transport of, hazardous wastes from your plant. Contact your local or state Department of Environmental Management for guidance and requirements for your particular plant and wastestreams. Remember, items such as lead-acid batteries, used oil and oil filters, spent or discarded laboratory chemicals, contaminated containers, and asbestos gaskets can all be classified as hazardous wastes.

Treatment plants that have a laboratory must develop and implement a written chemical hygiene plan. Some of the items that must be included in the plan are:

- Standard operating procedures to be followed when using hazardous chemicals.
- Criteria that will be used to determine and implement control measures to reduce employee exposure, including engineering controls, personal protective equipment, and hygiene practices.
- A requirement that the operation of fume hoods will comply with regulatory requirements and that specific measures will be taken to ensure that all protective equipment will function properly.
- Provisions for employee training that includes, but is not limited to, hazard communication (see Section 14.9).
- The circumstances under which a particular laboratory operation or activity will require prior approval.
- Provisions for medical consultation and medical examinations, if required.
- Assignment of a chemical hygiene officer.
- Provisions for additional employee protection for work with particularly hazardous substances such as select carcinogens and reproductive toxins.

This listing does not include all of the detailed requirements of a chemical hygiene plan. You should contact your local safety regulatory agency if you need assistance in the development of your program.

QUESTIONS

Write your answers in a notebook and then compare your answers with those on page 321.

14.3A What safety precautions would you take when collecting laboratory samples from a plant influent?

14.3B Why should you always wash your hands before eating?

14.3C Why should chemicals and reagents be handled with care?

14.4 FIRE PREVENTION

Fires are a serious threat to the health and safety of the operator and to the buildings and equipment in a treatment plant. Fires may injure or cause the death of an operator. Equipment damaged by fire may no longer function properly, and your treatment plant may have difficulty adequately treating the influent wastewater.

Good safety practices with respect to fire prevention require a knowledge of:

1. Ingredients necessary for a fire
2. Fire control methods
3. Fire prevention practices

14.40 Ingredients Necessary for a Fire

The three essential ingredients of all ordinary fires are:

1. FUEL—paper, wood, oil, solvents, and gas.
2. HEAT—the degree necessary to vaporize fuel according to its nature.
3. OXYGEN—normally at least 15 percent of oxygen in the air is necessary to sustain a fire. The greater the concentration, the brighter the blaze and more rapid the combustion.

14.41 Fire Control Methods

To extinguish a fire, it is necessary to remove only one of the essentials by:

1. Cooling (temperature and heat control)
2. Smothering (oxygen control)
3. Isolating (fuel control)

Fire classifications are important for determining the type of fire extinguisher needed to control the fire. Classifications also aid in recordkeeping. Fires are classified as A, B, C, or D fires based on the type of material being consumed: A, ordinary combustibles; B, flammable liquids and vapors; C, energized electrical equipment; and D, combustible metals. Fire extinguishers are also classified as A, B, C, or D to correspond with the class of fire each will extinguish.

Class A fires: ordinary combustibles, such as wood, paper, cloth, rubber, many plastics, dried grass, hay, and stubble. Use foam, water, soda-acid, carbon dioxide gas, or almost any type of extinguisher.

Class B fires: flammable and combustible liquids, such as gasoline, oil, grease, tar, oil-based paint, lacquer, and solvents, and also flammable gases. Use foam, carbon dioxide, or dry chemical extinguishers.

Class C fires: energized electrical equipment, such as starters, breakers, and motors. Use carbon dioxide or dry chemical extinguishers to smother the fire; both types are nonconductors of electricity.

Class D fires: combustible metals, such as magnesium, sodium, zinc, and potassium. Operators rarely encounter this type of fire. Use a Class D extinguisher or use fine dry soda ash, sand, or graphite to smother the fire. Consult with your local fire department about the best methods to use for specific hazards that exist at your facility.

Multipurpose extinguishers are also available, such as a Class BC carbon dioxide extinguisher that can be used to smother Class B and Class C fires. A multipurpose ABC carbon dioxide extinguisher will handle most laboratory fire situations. (When using carbon dioxide extinguishers, remember that the carbon dioxide can displace oxygen—take appropriate precautions.)

There is no single type of fire extinguisher that is effective for all fires so it is important that you understand the class of fire you are trying to control. You must be trained in the use of the different types of extinguishers, and the proper type should be located near the area where that class of fire may occur.

14.42 Fire Prevention Practices

You can prevent fires by:

1. Maintaining a neat and clean work area, preventing accumulation of debris and combustible materials.
2. Putting oil- and paint-soaked rags in covered metal containers and regularly disposing of them in a safe manner.
3. Observing all "no smoking" signs.
4. Keeping fire doors, exits, stairs, fire lanes, and firefighting equipment clear of obstructions.
5. Keeping all combustible materials away from furnaces and other sources of ignition.
6. Reporting any fire hazards you see, especially electrical hazards, which are sources of many fires.

Finally, here again are the things to remember:

1. Prevent fires by good housekeeping and proper handling of flammables.
2. Make sure that everyone obeys "no smoking" signs in all classified areas of your facility.
3. In case of fire, turn in the alarm immediately and make sure that the fire department is properly directed to the place of the fire.
4. Action during the first few seconds of ignition generally means the difference between destruction and control. Use the available portable firefighting equipment to control the fire until help arrives only if you have been trained and are qualified to do so.
5. Use the proper extinguisher for that fire.
6. Learn how to operate the extinguishers *before* an emergency arises.

If it is necessary, evacuate the building; do not stop to get anything—just leave!

Can you prevent fires? You can if you try, so let us see what we can do to preserve our well-being and the water pollution control system.

If you guard against fires, you will be protecting your lives and your community.

Additional recommendations can be found in Section 14.18.

QUESTIONS

Write your answers in a notebook and then compare your answers with those on page 321.

14.4A What are the necessary ingredients of a fire?

14.4B How should oil- and paint-soaked rags be handled?

14.5 WATER SUPPLIES

Inspect your plant to see if there are any cross connections between your potable (drinking) water and items such as water seals on pumps, feed water to boilers, hose bibs below grade where they may be subject to flooding with wastewater or sludges, or any other location where wastewater could contaminate a domestic water supply.

If any of these or other existing or potential cross connections are found, be certain that your drinking water supply source is properly protected by the installation of an approved backflow prevention device. Many treatment plants use an *AIR GAP DEVICE*[23] (Figure 14.15) to protect their drinking water supply.

It is a good practice to have your drinking water tested at least monthly for coliform group organisms. Sometimes the best of backflow prevention devices do fail.

Never drink from outside water connections such as faucets and hoses. The hose you drink from may have been used to carry effluent or sludge.

You may find in your plant that it will be more reliable and more economical to use bottled drinking water. If so, be sure to post conspicuous signs that your plant water is not potable at all outlets. This also applies to all hose bibs in the plant from which you may obtain water other than a potable source. This is a must in order to inform visitors or absent-minded or thirsty employees that the water from each marked location is not for drinking purposes. This practice reduces the possibility of the spread of disease from unknown cross connections or defective devices installed to prevent contamination by backflows.

14.6 SAFETY EQUIPMENT AND INFORMATION

Conspicuously post on your bulletin board the location and types of safety equipment available at your plant (such as first-aid kits, respiratory protective devices, and atmospheric monitors). You, as the plant operator, should be thoroughly familiar with the operation and maintenance of each piece of equipment. You should review these at fixed intervals to be certain that you can safely use the piece of equipment as well as to be sure that it is in good operating condition.

[23] *Air Gap.* An open, vertical drop, or vertical empty space, between a drinking (potable) water supply and potentially contaminated water. This gap prevents the contamination of drinking water by backsiphonage because there is no way potentially contaminated water can reach the drinking water supply.

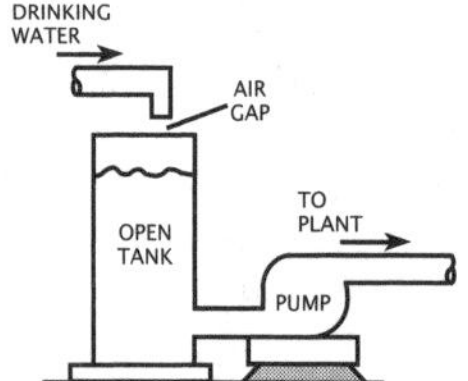

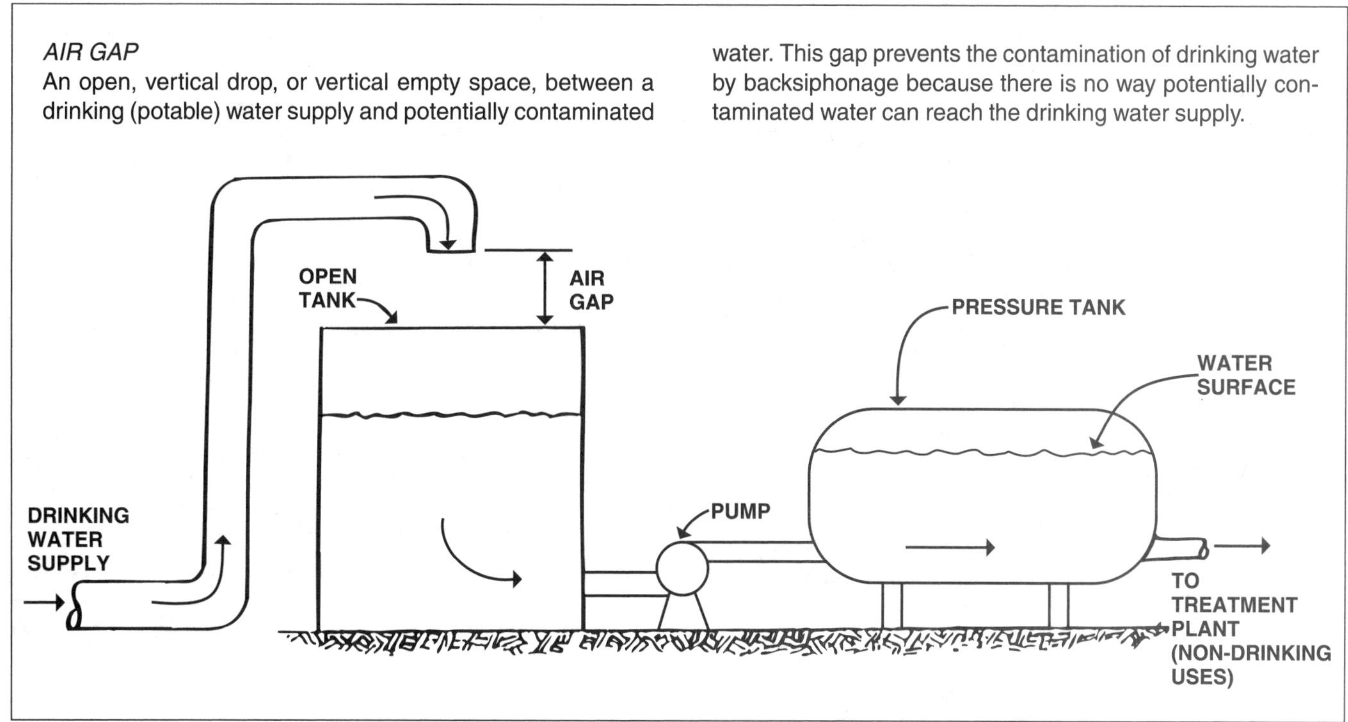
AIR GAP
An open, vertical drop, or vertical empty space, between a drinking (potable) water supply and potentially contaminated water. This gap prevents the contamination of drinking water by backsiphonage because there is no way potentially contaminated water can reach the drinking water supply.
OPEN TANK
AIR GAP
PRESSURE TANK
WATER SURFACE
DRINKING WATER SUPPLY
PUMP
TO TREATMENT PLANT (NON-DRINKING USES)

Fig. 14.15 Air gap device

Contacts should be made with your local fire and police departments to acquaint them with hazards at your plant as well as to inform them of the safety equipment that is necessary to cope with problems that may arise. Arrange a joint training session with these people in the use of safety equipment and the handling of emergencies. They also should know access routes to and around the treatment plant.

If you have any specific problems of a safety nature, do not hesitate to contact officials in your state safety agency. They can be of great assistance to you. And do not forget your equipment manufacturers; their familiarity with your equipment will be of great value to you.

Also, posted in conspicuous places in your plant should be such information as the phone numbers of your fire and police departments, ambulance service, chemical supplier or repair service, and the nearest doctor who has agreed to be available on call. Having these immediately available at telephone sites may save your or a fellow worker's life. Check and make sure these numbers are listed at your plant. If they are not listed, *add them now.* Also see Section 15.042, page 337, "Emergencies," for a list of emergency response agencies.

Prepare an emergency medical information sheet for each operator. Keep all of these sheets together in one binder. Send a copy of the appropriate sheet with the ambulance that takes an injured operator to the hospital.

QUESTIONS

Write your answers in a notebook and then compare your answers with those on page 321.

14.5A Why do some wastewater treatment plants use bottled water for drinking purposes?

14.6A What emergency phone numbers should be listed in a conspicuous place in your plant?

14.7 "TAILGATE" SAFETY MEETINGS [24]

Safety is crucial. Accidents cost money. No one can afford to lose time from their job due to injury. The purposes of tailgate safety meetings are to remind operators of the need for safety, to explain and discuss safe procedures and safe conditions, and to review potential hazards and how to correct or avoid dangerous situations.

In some states, you are required by law to conduct safety meetings at fixed intervals with employees. Whether this is required or not, it certainly is a good practice. Invite police and fire personnel to participate from time to time so you get to know them and they become acquainted with you and your facilities. Once every 7 to 10 working days is a good frequency. These meetings should usually be confined to one topic, and should be from 10 to 20 minutes long. It will be worthwhile to

[24] *Tailgate Safety Meeting.* Brief (10 to 20 minutes) safety meetings held every 7 to 10 working days. The term comes from the safety meetings regularly held by the construction industry around the tailgate of a truck.

review any accidents that have occurred since the last meeting. Do not use this meeting to fix blame. Try to determine the cause and what can be or has been done to prevent a similar accident in the future.

To help you conduct tailgate safety meetings, this chapter was arranged to discuss the safety aspects of different plant operations. The material in some sections was deliberately repeated to cover the topic and to remind you of dangers. Some plants select topics for their tailgate safety meetings from a "safety goof box." The box is placed in a convenient location. Whenever anyone sees an unsafe situation or sees someone perform a hazardous act without proper safety precautions, this person places a note in the box identifying the situation or act. The box is opened at each safety meeting, and the cause of the "goof" and the steps that can be taken to correct and prevent it from happening again are discussed.

Your state safety agency, your insurance company, equipment and material suppliers, and the Water Environment Federation are all excellent sources of literature and visual aids that may help you in conducting tailgate safety meetings. Some of these agencies may be able to supply you with posters, signs, and slogans that are very effective safety reminders. You may wish to create some reminders of your own.

QUESTIONS

Write your answers in a notebook and then compare your answers with those on page 321.

14.7A What is the purpose of tailgate safety meetings?

14.7B How frequently should safety meetings be held for treatment plant operators?

14.8 HOW TO DEVELOP SAFETY TRAINING PROGRAMS

14.80 Conditions for an Effective Safety Program

Effective safety programs rely on many techniques to help workers recognize hazards and learn safe procedures. These safety programs can range from highly organized meetings to tailgate safety sessions to informal get-togethers. Safety programs of all types have proven very effective and usually stress the following points:

1. Basic safety concepts and practices are thoroughly understood by all.
2. Everyone participates and accepts personal responsibility for their own safety and that of their fellow workers (participation at all levels is absolutely essential to the continued success of any safety program).
3. Adequate safety equipment is available and its capabilities are thoroughly understood. Responsible individuals must regularly review and drill in the actual use of the equipment in order to safely respond to an emergency and hazardous conditions.
4. Everyone realizes that safety is a continuing learning and re-learning process—a way of life that must become habit.
5. Accidents are studied step by step and thoroughly reviewed with the attitude that "they are caused and do not just happen." Every reasonable step will be taken to reduce the chance of an accident happening again to as near zero as is practical.
6. Every detail of work is a subject for discussion to the extent it will improve safety.
7. Operators realize they should stop anyone performing an unsafe act and remind the person that they are not following safe procedures, why, and how the job can be done safely.
8. Before starting a job, operators ensure that they can do the job without injury. If assigned work they are not qualified to perform, they do not just blindly do it, but bring it to the attention of their supervisor.
9. Before starting a job, operators thoroughly understand the work to be done, the job, and the safety rules that apply. Tailgate safety sessions or pre-job discussions will help promote safe operations.
10. Management actively supports the safety training program and demands that safe equipment and procedures be used at all times.

14.81 Start at the Top

An effective safety training program must start at the top. The person who controls the purse strings and makes final decisions must not only support safety but must promote it from the start and continuously promote safety on a day-to-day basis. The safety director must have direct access to this person. Without this type of organization, a safety program may be put off, watered down, or even eliminated in the name of urgency, time, and cost.

Top management's attitude and approach toward safety will probably be reflected in the attitude of the supervisors, and the operator's attitude will most likely be the same as that of the supervisors. Thus, if management is not committed to safety, no one else is likely to be. Management must issue a safety policy statement that makes it clear that safety will take precedence over expediency or shortcuts and that the safety of employees, the public, and plant operations is a top priority.

14.82 The Supervisor's Role

Supervisors are the key people in a safety program because they are in constant contact with operators. Supervisors can actively support the plant's safety policy. This sets an example for their operators and gives safety the emphasis it must have. First-line supervisors should be responsible to see that operators:

- Understand the hazards of the various processes at the plant.
- Observe necessary precautions when performing operations, maintenance, or laboratory tasks, including the use of safe work techniques and appropriate personal protective equipment.
- Understand and properly follow established work procedures for their safety and the safety of others.

Supervisors are the plant's safety representatives in the field and their actions and attitudes will directly affect the entire safety program.

14.83 Plan for Emergencies

Start where you are. Nothing is going to stop while you get your safety program organized. Emergencies, accidents, and injuries can happen at any time and usually at the wrong time. Try to minimize the impact of accidents while trying to develop or improve your plan for prevention. The first step is to prepare emergency procedures for your treatment plant, collection system, and vehicles. These plans should include:

1. What to do and what not to do for the injured.
2. How to contact the nearest fire department, rescue squad, or ambulance service.
3. Identification of the injured and notification of relatives.
4. Directions for emergency vehicles to reach the scene and to locate the victim.
5. Prevention of further damage to people and property.
6. Names of persons and authorities to be notified after the emergency.

All employees must be interested and trained in these procedures and copies should be posted in prominent places in all plant areas, pumping stations, and vehicles. The fact that employees are preparing for emergencies will have a positive effect in reducing accidents.

14.84 Promote Safety

Start early with your promotion of safety. Be proactive, not reactive. Make safety a part of discussions and work procedures. All types of on-the-job training should emphasize the importance of learning and practicing safe work procedures. Develop a written Standard Operating Practice (SOP) for routine duties or equipment operation and have regular training sessions over each SOP. This will not only point out safety aspects of the job, but will also be a way to train people in the most efficient way to work. Follow safe practices yourself—this applies especially to supervisory personnel. Example is a powerful incentive.

14.85 Hold Safety Drills and Train for Safety

No matter how well safety is engineered into your plant or a job, much of the safety of operators depends on their own conduct. Some people work safely in dangerous surroundings while others have accidents on jobs that seem very safe.

The training of operators begins the day they start work. Whether or not your plant has a formal safety orientation program, the operators start to learn about their jobs and to form attitudes about many things, including safety, the first day.

Use every opportunity to give safety instruction from 10-minute, on-the-spot chats to supervisory safety meetings. Vary the techniques and timing with chats, meetings, drills, exercises, workshops, and seminars. Cover all the subjects. Match discussions to incidents, such as slips and falls during the slippery season, defensive driving if a bad accident has occurred in the area, and chlorine safety if there have been problems with leaks. Make your point about safety while details of specific situations are fresh in everyone's mind.

Remember those fire drills in school? Try drilling on how employees should respond during emergencies, including evacuation of facilities. All facilities should have the necessary safety equipment (fire extinguishers, self-contained breathing apparatus, and atmospheric monitors), written procedures, and appropriate personnel trained for emergency response. Do not wait for an emergency before trying to learn how to use this equipment. Get proper instructions and conduct practice sessions.

First aid and chemical safety are important steps in organizing others to assist in your safety program. You can use the Red Cross multimedia program to train first-aid teams and instructors. Chemical manufacturers or distributors provide excellent instruction in chemical hazards and safety precautions.

When developing your training courses, try to emphasize the most hazardous tasks that are likely to cause accidents. Studies have shown that injuries most often occur when doing activities that are not routine. In your course discussions try to identify how hazards can cause injuries, how bad the injuries can be, and ways to avoid injury.

14.86 Purchase the Obvious Safety Equipment First

Hard hats, safety footwear, eye protection, and hearing protection apply to all personnel in designated areas and specific jobs. Purchase this equipment and post the areas where it must be used. The purchase of more specific and expensive equipment such as atmospheric monitors, light meters, and noise monitoring devices, is also very important. As your safety program develops, the benefits of more specific and more sophisticated safety equipment will become clear; the need and time to purchase them will be obvious.

14.87 Safety Is Important for Everyone

As your safety program develops, you will realize that safety is the responsibility of everyone, from managers to workers. Everyone must be involved. Organize safety committees and meetings from top to bottom, as well as from bottom to top. If you do it well, then safety practices will progress from top to bottom. Ideas and suggestions will come if they are recognized and implemented.

14.88 Necessary Paperwork

When you start to develop your safety program, concentrate your efforts on programs that apply generally to all employees. Paperwork can be helpful to identify the causes of accidents and to develop corrective procedures.

1. Accident report forms (Figure 14.16). Use these forms to record and analyze the causes of accidents and to prevent future accidents.
2. Safety policy. The plant manager must establish a safety policy and repeatedly demonstrate support of the policy.
3. Safety rules. Safety rules are as important as work rules and they should be implemented and enforced in the same manner. Most people perform better and with more confidence if they know the rules of the game. These rules must apply to everyone. Supervisors should serve as examples to the operators.
4. Supervisors' guidelines. Supervisors must have guidelines on how to promote and implement a safety program and enforce the rules.
5. Facility plans and specifications; plant inspection reports. State and OSHA regulations must be used when reviewing plans and specifications. Checklists are a tremendous aid during plant inspections.

CITY OF ________ WASTEWATER TREATMENT PLANT ACCIDENT REPORT

Date of this report ______ Name of person injured ______

Date of injury ______ Time ______ a.m. p.m. Occupation ______

Home address ______ Age ______ Sex ______

Check ___ First aid case, or ___ disabling (lost time) injury

___ Employee ___ on duty, or ___ off duty

___ Visitor injury

Date last worked ______ Date returned to work ______

Person reporting ______

DESCRIPTION OF ACCIDENT

1. Description of Accident ______ (Describe in detail what happened) (Name machine, tool, appliance, ______ gas or liquid involved—if machine or vehicle—name part, gears, pulley, etc.)
2. Accident occurred where? ______ If vehicle accident, make simple sketch of scene of accident.
3. Describe nature of injury and part of body affected ______ (Amputation of finger, laceration of leg, back strain, etc.)
4. Were other persons involved? ______ (If yes, give names and addresses.) ______
5. Names and addresses of witnesses ______
6. If property damage involved, give brief description ______
7. If hospitalized, name of hospital ______
8. Name and address of physician ______
9. Treatment given for injuries ______

Fig. 14.16 Typical accident report form

14.89 Summary

All types of safety programs are helpful. If variety is the spice of life, let variety add spice to your safety program. Informal chats on safety do not replace formal safety meetings or vice versa. Every type of safety meeting can help you develop a more effective safety program.

Your safety program should include the following items:

1. Get your top official to support and promote safety (have the official issue a written safety policy statement).
2. Give your safety officer direct access to the plant manager.
3. Direct your program from general topics to the more specific.
4. Organize from top to bottom.
5. Establish rules and implement and enforce them.
6. Train at all levels from employment to retirement.
7. *MAKE SAFETY A HABIT.*

QUESTIONS

Write your answers in a notebook and then compare your answers with those on page 321.

14.8A List three types of safety meetings.

14.8B What is the role of management in an effective safety training program?

14.8C Why should safety drills be held regularly?

14.8D What types of paperwork are necessary for an effective safety training program?

14.9 HAZARD COMMUNICATION (WORKER RIGHT-TO-KNOW LAWS)

Operators must be concerned with the increased emphasis nationally on hazardous materials and wastes. Much of this attention has focused on hazardous and toxic waste dumps, and the efforts to clean them up after the long-term effects on

human health were recognized. Each year, thousands of new chemical compounds are produced for industrial, commercial, and household use. Frequently, the long-term effects of these chemicals are unknown. Exposure to the wastewater treatment plant operator can occur from one or more of the following:

1. Use of the collection system as an intentional or accidental disposal method for hazardous materials, for example:
   a. Industrial solvents
   b. Acids
   c. Flammable/explosive compounds
   d. Caustics
   e. Toxics
2. Chemicals and chemical compounds that we use every day for operation and maintenance in the treatment plant, for example:
   a. Solvents
   b. Degreasers
   c. Acids
   d. Chlorine
   e. Industrial cleaners

As a result, federal and state laws have been enacted to control all aspects of hazardous materials handling and use. These laws are more commonly known as Worker Right-To-Know (R-T-K) laws. Every state is covered by one or more laws regarding Worker Right-To-Know. The Federal Occupational Safety and Health Administration (OSHA) Standard 29 CFR 1910.1200—Hazard Communication forms the basis of most laws. Although the federal standards were originally directed at the manufacturing sector, they now include all industries including the public sector and, therefore, collection systems and treatment plants.

In many cases, the individual states have the authority under the OSHA standard to develop their own state Worker Right-To-Know laws and most states have adopted their own laws. Unfortunately, state laws vary significantly from state to state. The state laws that have been passed are at least as stringent as the federal standard and, in most cases, are even more stringent and already apply to treatment plant operators.

State laws are also under continuous revision and, because a strong emphasis is being placed on hazardous materials and worker exposure, state laws can be expected to be amended in the near future to apply to virtually everybody in the workplace.

The purpose of this section is to familiarize you with general requirements of Worker RTK elements so that you are better prepared as a treatment plant operator to minimize risk to yourself and your co-workers from hazardous materials, and to comply with state and federal laws as well. Because of the wide diversity of existing laws, these guidelines may or may not meet your state's requirements. This section will give you an overview of the basic elements of a hazard communication program (see Table 14.2), which can be particularly useful if your agency currently does not have one. An extremely effective program can be developed in house through your safety program and committee. By cutting through the bureaucratic/legal language and applying some common sense to what the law is trying to accomplish (protection of the worker in the workplace from hazardous materials), an effective program can be developed.

Basically, the different elements of the program are as follows:

1. Identify Hazardous Materials

While there are thousands and thousands of chemical compounds that would fall under this definition in a technical sense, treatment plant operators should be concerned, first of all, with the materials they use in their everyday operation and maintenance activities; and also with materials that can be introduced to the collection system through intentional or accidental spills. Information on materials that could be introduced into the collection system can be obtained from industrial pretreatment inspectors. (Also see the manual, *PRETREATMENT FACILITY INSPECTION,* in this series of manuals.)

In particular, you should be familiar with the industries in your area and be aware of the materials they routinely discharge into the collection system as well as the types of hazardous materials they might accidentally discharge. In some cases, special precautions may have to be taken. For example, when your collection system operators are performing routine maintenance with a high-velocity cleaner (jet machine) in a section of line, you may wish to draw up an agreement with an industrial discharger upstream to halt discharges on the day the sewer maintenance is being performed. Hazardous materials can be broken down into general categories as follows:

1. Corrosives
2. Toxics
3. Flammables and explosives
4. Asphyxiants
5. Harmful physical agents
6. Infectious agents

A complete inventory of materials in use will produce a list similar to the following:

1. Corrosives
   a. Sodium hydroxide
   b. Calcium oxide (lime)
   c. Hydrochloric acid
   d. Ferric chloride
2. Toxics
   a. Hydrogen sulfide
   b. Chlorine
   c. Carbon monoxide
3. Flammables and Explosives
   a. Methane
   b. Acetylene
   c. Gasoline
   d. Solvents
4. Asphyxiants
   a. Carbon Dioxide
   b. Nitrogen
5. Harmful Physical Agents
   a. Noise
   b. Temperature
   c. Radiation
6. Infectious Agents

2. Obtain Chemical Information and Define Hazardous Conditions

Once the inventory is complete, the next step is to obtain specific information on each of the chemicals and hazardous conditions. This information is generally incorporated into a standard format form called the "Material Safety Data Sheet" (MSDS). Figure 14.18 (page 313) is the Material Safety Data Sheet (OSHA 174, September, 1985) produced by the US Department of Labor, Occupational Safety and Health Administration. This information is commonly available from manufacturers. Many agencies request an MSDS when the purchase order is generated and will refuse to accept delivery of the shipment if the MSDS is not included.

The purpose of the MSDS is to have a readily available reference document that includes complete information on common names, safe exposure level, effects of exposure, symptoms of exposure, flammability rating, type of first-aid procedures, and other information about each hazardous substance.

Operators must be trained to read and understand the MSDS forms. The forms themselves must be stored in a convenient location where they are readily available for reference.

3. Properly Label Hazards

Once the physical, chemical, and health hazards have been identified and listed, a labeling and training program must be implemented. To meet labeling requirements on hazardous materials, specialized labeling is available from a number of sources, including commercial label manufacturers. Exemptions to labeling requirements do exist, so consult your local safety regulatory agency for specific details.

TABLE 14.2 ELEMENTS OF A HAZARD COMMUNICATION PROGRAM

I. *WRITTEN HAZARD COMMUNICATION PROGRAM*

A. Hazard Determination (Chemical manufacturers and importers only)

1. Person(s) responsible for evaluating the chemical(s)
2. List of sources to be consulted
3. Criteria to be used to evaluate studies
4. A plan for reviewing information to update MSDSs if new and significant information is found

B. Labels and Other Forms of Warning

1. Person(s) responsible for labeling in-plant containers, if used
2. Person(s) responsible for labeling shipped containers
3. Description of labeling system
4. Description of written alternatives to labeling of in-plant containers, if used
5. Procedure to review and update label information

C. Material Safety Data Sheets (MSDSs)

1. Person(s) responsible for obtaining or maintaining the MSDSs
2. Description of how the MSDSs will be made available to employees
3. Procedure to follow when the MSDS is not received at time of first shipment
4. Procedure to review and update MSDS information
5. Description of alternatives to actual data sheets in the workplace, if used

D. Training

1. Person(s) responsible for conducting training
2. Format of the program to be used
3. Elements of the training program

   a. Requirements of the OSHA standard
   b. Operations, where hazardous chemicals are present in routine tasks, non-routine tasks, and foreseeable emergencies
   c. Location and availability of

      i. Written hazard communication program
      ii. List of hazardous chemicals
      iii. MSDSs

   d. Methods and observations that may be used to detect the presence or release of hazardous chemicals in the work area
   e. Physical and health hazards of chemicals in the work area
   f. Measures employees can take to protect themselves from hazards
   g. Details for the Hazard Communication Program developed

4. Procedure to train new employees, as well as current employees, when a new chemical hazard is introduced into the workplace

E. List of hazardous chemicals

F. Procedure to inform employees of the hazards of chemicals in unlabeled pipes

G. Procedure to inform on-site contractors

II. *LABELS AND OTHER FORMS OF WARNING*

A. Labels or other markings on each container of hazardous chemicals

1. Process vessels
2. Storage tanks
3. Compressed gas cylinders
4. Product containers
5. Tank truck and tank car labels

III. *MATERIAL SAFETY DATA SHEETS (MSDSs)*

A. MSDSs developed or obtained for all hazardous chemicals

B. Employees have access on each shift

C. MSDS completed appropriately

D. When no MSDS is available, the documentation requesting an MSDS from supplier is maintained

TABLE 14.2 ELEMENTS OF A HAZARD COMMUNICATION PROGRAM *(continued)*

IV. *TRAINING*

A. Employee training files

B. Employee questioning:

1. Are they aware of the Hazard Communication Program (HCP) and its requirements?
2. Have they received training?
3. Are they able to locate the MSDSs?
4. Do they have a general familiarity with the hazardous properties of the chemicals in their workplace?

Each hazardous substance container in the workplace must be labeled, tagged, or marked with the name of the hazardous substance and appropriate warning labels. There are a number of acceptable labeling systems available, including private labeling systems such as the one illustrated in Figure 14.17 produced by the J. T. Baker Chemical Company. The second illustration shows a label produced by a commercial label maker (Figure 14.19), and the third, in Figure 14.20, is one designed by the National Fire Protection Association (NFPA).

These are standardized formats that use a combination of pictographs, a numbering system, and colors to indicate various levels of conditions. In some cases, the MSDS sheet can be incorporated into the labeling requirements by locating the appropriate MSDS in close proximity to drums or storage areas. The labeling requirements offer the treatment plant operator virtually instant recognition of the hazards in dealing with specific substances, protective equipment required, and other information.[25]

4. Train Operators

The last element in the Hazard Communication Worker Right-To-Know Program is training and making information available to the collection system and treatment plant operators. A common-sense approach eliminates the confusing issue of which of the thousands of substances operators should be trained for, and concentrates on those that they will be exposed to or use in everyday maintenance routines. Obviously, the protection from hazardous materials is tied in with the confined space policy, since average domestic wastewater found in collection systems and treatment plants does not contain sufficient concentrations of hazardous materials to address each of the compounds that may be found. The use of ventilating equipment, atmospheric testing instrumentation, and the other precautions defined in the confined space procedure assist in protecting the collection system and treatment plant operators from hazardous materials.

Industrial dischargers can and do discharge significant concentrations of hazardous materials, either intentionally or unintentionally. A treatment plant operator should be familiar with each industry that discharges to the collection system, since some special precautions may need to be taken to minimize exposure to specific hazardous materials.

The last element is a formal training program provided by qualified personnel designed to accomplish the following:

1. Make employees aware of the hazard communication standard and its requirements.
2. Familiarize employees with potentially dangerous operations and hazardous substances that are present in their routine and non-routine operation and maintenance tasks and how to deal with emergency situations involving those materials and conditions.
3. Inform employees of the location and availability of the written hazard communication program.
4. Train employees in the methods and observations that they may use to detect the presence or release of a hazardous substance in the work area such as visual appearance, odor, and monitoring.
5. Inform employees of the measures they can take to protect themselves from the physical and health hazards of the chemicals in the work area.
6. Explain how to read and interpret MSDS forms and have an MSDS file readily available for reference.
7. Train employees to use the labeling format.
8. Document what training has been performed.

The hazard communication standard and the individual state requirements are a very complex set of regulations. Remember, however, the ultimate goal of these regulations and other treatment plant procedures is to provide additional employee protection. These standards and regulations, once the intent is understood, are relatively easy to implement and can certainly be accomplished in house by a safety-conscious organization.

When all is said and done, common sense, knowledge, awareness, and commitment are the keys to complying with the hazardous material and the Worker Right-To-Know regulations.

Depending on what state you are in, your agency could be in violation of federal and state regulations if it does not currently have a Worker Right-To-Know policy. As more states that are covered by the federal standard revise their laws, it is anticipated that in the near future all employees will be covered by Worker Right-To-Know laws.

[25] The United States and Europe are working on an internationally acceptable labeling system, which ultimately could lead to standardized labeling for 575,000 chemical mixtures.

J.T.Baker®

BAKER SAF-T-DATA™ Guide

An easy-to-understand hazard classification appears on J. T. Baker labels. It will help increase your awareness of vital occupational health and safety practices. Using the BAKER SAF-T-DATA™ System as a guide you can quickly learn the hazards each substance presents to your health and safety, personal laboratory protective equipment that should be used for handling, and the recommended storage of compatible products by color code.

A NUMERICAL HAZARD CODE

Substances are rated on a scale of 0 (non-hazardous) to 4 (extremely hazardous) in each of four hazard categories:

- **Health hazard** - the danger or toxic effect a substance presents if inhaled, ingested, or absorbed.
- **Flammable hazard** - the tendency of the substance to burn.
- **Reactivity hazard** - the potential of a substance to explode or react violently with air, water or other substances.
- **Contact hazard** - the danger a substance presents when exposed to skin, eyes, and mucous membranes.

Rating Scale

4	3	2	1	0
Extreme	Severe	Moderate	Slight	None*

*No scientific data in the standard references that suggests the substance is hazardous.

HAZARD SYMBOL

A substance rated 3 or 4 in any hazard category will also display a hazard symbol. These easy-to-understand pictograms emphasize the serious hazards related to a substance:

HEALTH | FLAMMABILITY | CONTACT

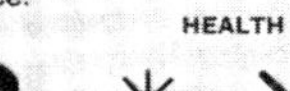

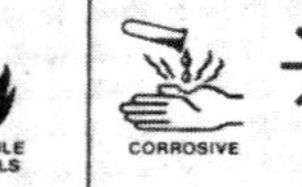

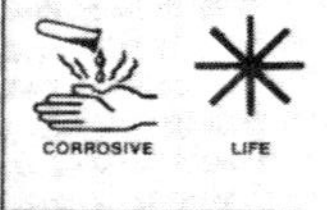
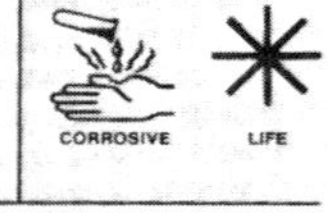

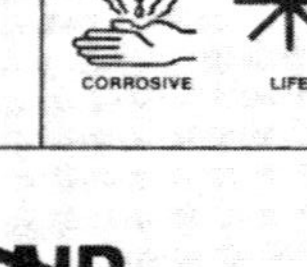

REACTIVITY

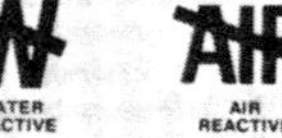

Sample label:

BAKER SAF-T-DATA™ System

HEALTH	FLAMMABILITY	REACTIVITY	CONTACT
1	3	2	1
SLIGHT	SEVERE	MODERATE	SLIGHT

LABORATORY PROTECTIVE EQUIPMENT: SAFETY GLASSES, LAB COAT, VENT HOOD, PROPER GLOVES, EXTINGUISHER

STORAGE COLOR: RED

4 L — 9006-3

Acetone

$(CH_3)_2CO$ — FW 58.1

'BAKER ANALYZED' Reagent

ACTUAL ANALYSIS, LOT **238093** — MEETS A.C.S. SPECIFICATIONS

Assay $[(CH_3)_2CO]$(by GC, corrected for H_2O)		99.8	%
Color (APHA)	<	5	
Density (g/ml) at 25°C		0.7852	
Residue after Evaporation		0.00004	%
Solubility in Water		Passes Test	
Titrable acid		0.0002	meq/g
Titrable base		0.00008	meq/g
Aldehyde (as HCHO)	<	0.002	%
Isopropyl Alcohol $(CH_3CHOHCH_3)$(by GC)	<	0.01	%
Methanol (CH_3OH)(by GC)	<	0.01	%
Substances Reducing Permanganate		Passes Test	
Water (H_2O)(by Karl Fischer titrn)		0.2	%
Trace Impurities (in ppm):			
Copper (Cu)	<	0.05	
Heavy Metals (as Pb)		0.3	
Iron (Fe)		0.06	
Nickel (Ni)	<	0.1	

DANGER!

EXTREMELY FLAMMABLE
HARMFUL IF SWALLOWED OR INHALED

Keep away from heat, sparks, flame. Keep in tightly closed container. Avoid breathing vapor. Avoid contact with eyes, skin, clothing. Use with adequate ventilation. Wash thoroughly after handling. In case of fire, use water spray, alcohol foam, dry chemical, CO_2. Flush spill area with water spray.

FIRST AID: If swallowed, if conscious, immediately induce vomiting and call a physician. **If inhaled,** remove to fresh air. If not breathing, give artificial respiration. If breathing is difficult, give oxygen. Call a physician.

FLASH POINT: ~0°F(SCT)
CAS 67-64-1 — 402-2275
DOT Description: Ⓡ **Acetone UN 1090**
IMO Description: Ⓡ **Acetone**
EPA-HW: Ignitables
Solusorb* solvent absorbent is recommended for spills of Acetone.

NFPA: 3, 1, 0

© J.T. Baker Chemical Co. Phillipsburg, NJ 08865

B LABORATORY PROTECTIVE EQUIPMENT

This series of pictograms suggests the personal protective clothing and equipment recommended for use when handling the substance in a laboratory situation. The pictograms relate to the combination of hazards presented by the substance.

SAFETY GLASSES — GOGGLES — GOGGLES & SHIELD — LAB COAT — LAB COAT & APRON

PROPER GLOVES — VENT HOOD — EXTINGUISHER — EXTINGUISHER — EXTINGUISHER

The stop sign indicates the substance represents a special extreme hazard and the MSDS and other references should be consulted before handling.

C STORAGE COLOR CODING

The SAF-T-DATA label suggests a unique method for setting up your chemical storage area. Compatible products are labelled with the same color. Simply group these colors together and follow the recommendations for appropriate storage:

BLUE - health hazard. Store in a secure poison area.

RED – flammable hazard. Store in a flammable liquid storage area.

YELLOW - reactivity hazard. Store separately and away from flammable or combustible materials.

WHITE – contact hazard. Store in a corrosion-proof area.

ORANGE - substances with no rating higher than 2 in any hazard category. Store in a general chemical storage area.

STRIPED LABELS - incompatible materials of the same color class have striped labels. These approximately 40 products should not be stored adjacent to substances with the same colored labels. Proper storage must be individually assessed.

D SPILL CONTROL CODE

This statement indicates which J.T. Baker spill control kit is recommended for use with the substance.

E NFPA SYSTEM*

This system was adopted by the NFPA in 1975 to safeguard the lives of firefighters. It is based on the hazards created by a substance in a fire situation. For this reason, the hazard ratings in the Baker SAF-T-DATA™ System, which are based on substances in a laboratory situation, will not always correspond with the NFPA ratings.

*National Fire Protection Association

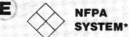

Fig. 14.17 J. T. Baker hazardous substance labeling system

(Permission of J. T. Baker Chemical Company)

Material Safety Data Sheet
May be used to comply with
OSHA's Hazard Communication Standard,
29 CFR 1910.1200 Standard must be
consulted for specific requirements.

U.S. Department of Labor
Occupational Safety and Health Administration
(Non-Mandatory Form)
Form Approved
OMB No. 1218-0072

IDENTITY *(As Used on Label and List)*	*Note: Blank spaces are not permitted. If any item is not applicable, or no information is available, the space must be marked to indicate that.*

Section I

Manufacturer's Name	Emergency Telephone Number
Address *(Number, Street, City, State, and ZIP Code)*	Telephone Number for Information
	Date Prepared
	Signature of Preparer *(optional)*

Section II—Hazardous Ingredients/Identity Information

Hazardous Components (Specific Chemical Identity: Common Name(s))	OSHA PEL	ACGIH TLV	Other Limits Recommended	%*(optional)*

Section III—Physical/Chemical Characteristics

Boiling Point		Specific Gravity (H_2O = 1)	
Vapor Pressure (mm Hg.)		Melting Point	
Vapor Density (AIR = 1)		Evaporation Rate (Butyl Acetate = 1)	

Solubility in Water

Appearance and Odor

Section IV—Fire and Explosion Hazard Data

Flash Point (Method Used)	Flammable Limits	LEL	UEL

Extinguishing Media

Special Fire Fighting Procedures

Unusual Fire and Explosion Hazards

(Reproduce locally) OSHA 174, Sept. 1985

Fig. 14.18 Material Safety Data Sheet

Material Safety Data Sheet
May be used to comply with
OSHA's Hazard Communication Standard,
29 CFR 1910.1200 Standard must be
consulted for specific requirements.

U.S. Department of Labor
Occupational Safety and Health Administration
(Non-Mandatory Form)
Form Approved
OMB No. 1218-0072

IDENTITY *(As Used on Label and List)*	*Note: Blank spaces are not permitted. If any item is not applicable, or no information is available, the space must be marked to indicate that.*

Section I

Manufacturer's Name	Emergency Telephone Number
Address *(Number, Street, City, State, and ZIP Code)*	Telephone Number for Information
	Date Prepared
	Signature of Preparer *(optional)*

Section II—Hazardous Ingredients/Identity Information

Hazardous Components (Specific Chemical Identity: Common Name(s))	OSHA PEL	ACGIH TLV	Other Limits Recommended	%*(optional)*

Section III—Physical/Chemical Characteristics

Boiling Point		Specific Gravity (H_2O = 1)	
Vapor Pressure (mm Hg.)		Melting Point	
Vapor Density (AIR = 1)		Evaporation Rate (Butyl Acetate = 1)	
Solubility in Water			
Appearance and Odor			

Section IV—Fire and Explosion Hazard Data

Flash Point (Method Used)	Flammable Limits	LEL	UEL
Extinguishing Media			
Special Fire Fighting Procedures			
Unusual Fire and Explosion Hazards			

(Reproduce locally) OSHA 174, Sept. 1985

Fig. 14.18 Material Safety Data Sheet (continued)

CHECK
DANGER
EYE PROTECTION REQUIRED HERE
DANGER
HAND PROTECTION REQUIRED
DANGER
RESPIRATOR REQUIRED
DANGER
FOOT PROTECTION REQUIRED
CAUTION
USE APRON
CAUTION
USE PROTECTIVE CLOTHING
CAUTION
USE FACE SHIELD
CAUTION
USE CHEMICAL GOGGLES
DANGER
"NO SMOKING OR OPEN FLAME"
CAUTION
GROUND BEFORE POURING
CHECK
DANGER
TOXIC
ACID
ALKALINE
FLAMMABLE
COMBUSTIBLE
USE VENTILATION
CHEMICAL NAME ______
USE ______

Fig. 14.19 Commercial warning labels
(Permission of the SIGNMARK Division, W. H. Brady Co.)

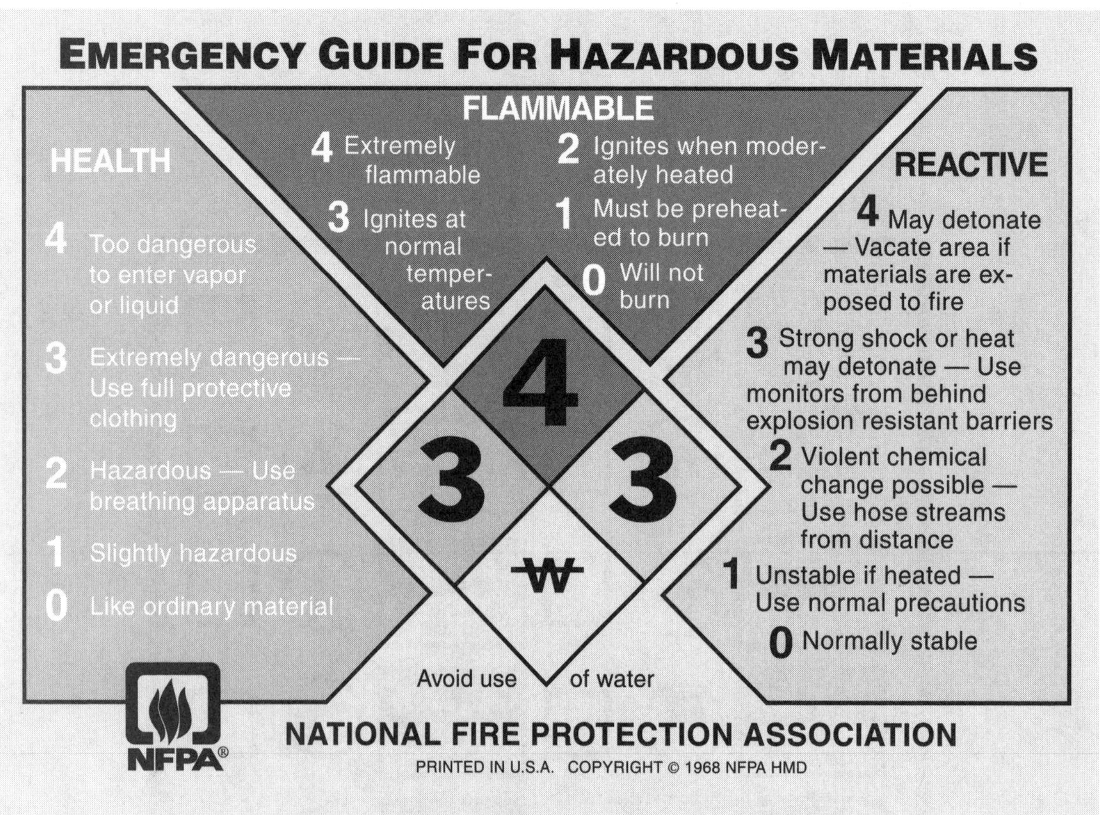

Fig. 14.20 NFPA hazard warning label
(Permission of National Fire Protection Association)

QUESTIONS

Write your answers in a notebook and then compare your answers with those on page 321.

14.9A List the four basic elements of a hazard communication program.

14.9B List the general categories of hazardous materials.

14.9C What is the purpose of the Material Safety Data Sheet (MSDS)?

14.9D What information must be on the label of a hazardous material container?

14.10 SAFETY SUMMARY

Following is a summary of the safety precautions that have been discussed in the previous sections.

1. Good design without proper safety precautions will not prevent accidents. All personnel must be involved in a safety program and provided with frequent safety reminders.
2. Never attempt to do a job unless you have sufficient help, adequate training and skills, the proper tools, and necessary safety equipment.
3. Remove a manhole cover or heavy grate using the proper tool, not your fingers.
4. "Lift with your legs, not with your back" to prevent back strains.
5. Use ladders of any kind with caution. Be certain that portable ladders are positioned so they will not slip or twist and that they are placed at the appropriate angle. Whenever possible, secure the top of a ladder used to enter below-grade structures. Do not use metal ladders near electrical boards or appliances.
6. Never enter a manhole, pit, sump, or other enclosed area without complying with all confined space requirements.
7. Always test manholes, pits, sumps, and other enclosed areas for atmospheric hazards. When testing the atmosphere, first test for oxygen deficiency/enrichment, then combustible gases and vapors, and then toxic gases and vapors. Before entering, thoroughly ventilate with a forced air blower.
8. Wear or use personal protective equipment (PPE) such as safety harnesses, gas detectors, and rubber gloves to prevent infections and injuries.
9. Never use a tool or piece of equipment unless you have been trained and are thoroughly familiar with its use or operation and know its limitations.
10. When working in traffic areas, always provide:
    a. Adequate advance warning to traffic by signs and flags.
    b. Traffic cones, barricades, or other approved items for channeling the flow of traffic around your work area.
    c. Protection to workers by placing your vehicle between traffic and the job area, if practical, or by use of flashing or revolving lights, or other devices.
    d. Flaggers, when necessary, to direct and control flow of traffic.

11. Before starting a job, be certain that the work area is of adequate size. If not, make allowances for this. Keep all working surfaces free of material that may cause slip or trip hazards.
12. See to it that all guardrails and chains are properly installed and maintained.
13. Provide and maintain guards on all chains, sprockets, gears, shafts, and other similar moving pieces of equipment that can be accidentally contacted.
14. Before working on mechanical or electrical equipment, properly lock out and tag breakers and other sources of energy to prevent the accidental starting or movement of the equipment while you are working on it. Wear approved, insulated gloves and boots wherever you may contact live electric circuits.
15. Never enter a launder, channel, conduit, or other slippery area when by yourself. Use a fall arrest system if working in an unguarded, elevated area six feet (1.8 meters) or more above a lower level.
16. Do not allow smoking, open flames, or other sources of ignition in the area of, on top of, or in any structure in your digestion and gas-handling system. Post all these areas with warning signs in conspicuous places.
17. Never enter a chlorine or sulfur dioxide atmosphere by yourself or without proper protective equipment. Seek the cooperation of your local fire department in supplying support when responding to a release.
18. Obtain and post in a conspicuous location the name and telephone number of the nearest chlorine and sulfur dioxide emergency service. Acquaint your police and fire department with this service.
19. Inspect all chlorine and sulfur dioxide connectors and lines before using. Replace any of these that appear defective.
20. Keep all chlorine and sulfur dioxide containers secure to prevent falling or rolling. Use only approved methods of moving and lifting containers.
21. Maintain a good housekeeping program. This is a proven method of preventing many accidents.
22. Conduct an effective safety awareness and training program.

These are the highlights of what has been previously discussed. Whenever in doubt about the safety of any piece of equipment, structure, operation, or procedure, contact the equipment manufacturer, your city or county safety officer, or your state safety office. One of these should be able to supply you with an answer to your questions.

ACCIDENTS DO NOT JUST HAPPEN... THEY ARE CAUSED!

You can be held personally liable for injuries or damages caused by an accident as a result of your negligence.

Can you afford the price of one?

Can you afford the loss of one or more operators?

Can your family afford to lose *YOU*?

REMEMBER: SAFETY IS NO ACCIDENT!

14.11 ADDITIONAL READING

1. *MOP 11*, Chapter 5,* "Occupational Safety and Health," New Edition of MOP 11, Volume I, *MANAGEMENT AND SUPPORT SYSTEMS.*
2. *NEW YORK MANUAL,* Chapter 13,* "Safety."
3. *TEXAS MANUAL,* pages 689–706.*
4. *CHLORINE SAFE HANDLING BOOKLET.* Obtain from PPG Industries, Inc., Monroeville Chemical Center, 440 College Park Drive, Monroeville, PA 15146. No charge.
5. *SAFETY AND HEALTH IN WASTEWATER SYSTEMS* (MOP 1). Obtain from Water Environment Federation (WEF), Publications Order Department, 601 Wythe Street, Alexandria, VA 22314-1994. Order No. MO2001. Price to members, $20.00; nonmembers, $30.00; plus shipping and handling.
6. *CHLORINE MANUAL,* Sixth Edition. Obtain from the Chlorine Institute, Inc., Bookstore, PO Box 1020, Sewickley, PA 15143-1020. Pamphlet 1. Price to members, $28.00; nonmembers, $70.00; plus $6.95 shipping and handling.
7. National Fire Protection Association Standard 820 (NFPA 820), *FIRE PROTECTION IN WASTEWATER TREATMENT AND COLLECTION FACILITIES.* Obtain from National Fire Protection Association (NFPA), 11 Tracy Drive, Avon, MA 02322-9908. Item No. 82099. Price to members, $32.85; nonmembers, $36.50; plus $7.95 shipping and handling.
8. *SAFETY IN THE CHEMICAL LABORATORY,* edited by Norman V. Steere. Out of print.
9. *GENERAL INDUSTRY, OSHA SAFETY AND HEALTH STANDARDS* (CFR, Title 29, Labor Pt. 1900–1910. (most recent edition)). Obtain from the US Government Printing Office, Superintendent of Documents, PO Box 371954, Pittsburgh, PA 15250-7954. Order No. 869-060-00108-5. Price, $61.00.
10. *FISHER SAFETY CATALOG.* Obtain from Fisher Scientific Company, 4500 Turnberry Drive, Suite A (Customer Service), Hanover Park, IL 60103 or phone (800) 772-6733.
11. *SAFETY RULES,* published by National Environmental Training Association (NETA), 5320 N. 16th Street, Suite 114, Phoenix, AZ 85016-3241. Out of print.
12. Safety-related training products are available from Communication Arts Multimedia, Inc., 226 Scenic View Lane, Ligonier, PA 15658. The titles are:
    a. Confined Space Safety/Manhole Entry. Price for video or DVD, $155.00.
    b. Shoring and Trenching. Price for video or DVD, $155.00.

    The cost of shipping and handling is $5.00 per order.

* Depends on edition.

END OF LESSON 3 OF 3 LESSONS
on
PLANT SAFETY

Please answer the discussion and review questions next.

DISCUSSION AND REVIEW QUESTIONS

Chapter 14. PLANT SAFETY

(Lesson 3 of 3 Lessons)

Write the answers to these questions in your notebook. The question numbering continues from Lesson 2.

19. How can samples for lab tests be collected without going beyond guardrails or chains?
20. What should be done with the jagged ends of glass tubes?
21. How should hot lab equipment be handled?
22. How can a fire be extinguished?
23. Fires can be prevented by good housekeeping and proper handling of flammables. True or False?
24. Why should plant water supplies be checked monthly for coliform group bacteria?
25. Why should safety equipment be checked periodically?
26. Where could you obtain safety literature and visual aids to help you in conducting tailgate safety meetings?
27. Carefully study the illustration below. List the safety hazards and indicate how each one can be corrected.

SUGGESTED ANSWERS

Chapter 14. PLANT SAFETY

ANSWERS TO QUESTIONS IN LESSON 1

Answers to questions on pages 273 and 274.

14.1A You can help prevent the spread of infectious diseases from your job to you and your family by immunization shots, thoroughly washing your hands before eating or smoking, wearing proper protective gloves when you may contact wastewater or sludge in any form, not wearing your work clothes home, keeping your work clothes clean, storing your street clothes and your work clothes in separate lockers, and, if your employer does not supply you with uniforms and laundry service, washing your work clothes separately from your regular family wash.

14.1B When testing the atmosphere before entry in any confined space, first test for oxygen deficiency/enrichment, then combustible gases and vapors, and then toxic gases and vapors.

14.1C OSHA regulations require that a confined space entry permit be completed for each permit-required confined space entry. The permit must be renewed each time the space is left and re-entered, even if only for a break or lunch, or to go get a tool.

14.1D The most common toxic gas in wastewater treatment is hydrogen sulfide, which is produced during the anaerobic decomposition of certain materials containing sulfur compounds. Other toxics, such as carbon monoxide, chlorinated solvents, and industrial toxins, may enter your plant as a result of industrial discharges, accidental spills, or illegal disposal of hazardous materials.

14.1E Hydrogen sulfide has the unique ability to affect your sense of smell. You will lose the ability to smell hydrogen sulfide's "rotten egg" odor after only a short exposure. Your loss of ability to smell hydrogen sulfide, and certain other odors, is known as olfactory fatigue. *NEVER* depend on your nose for detection, regardless of the situation.

14.1F UEL stands for upper explosive limit and LEL stands for lower explosive limit. UEL and LEL indicate the range of concentrations at which combustible or explosive gases will ignite when an ignition source is present at ambient temperature.

Answers to questions on page 278.

14.1G A lockout device uses a positive means such as a lock to hold a switch or valve in a safe position and prevent a piece of equipment from becoming energized or moving. Lockout devices are used whenever equipment must be repaired or replaced to ensure that the equipment will not start up or move unexpectedly and possibly injure workers.

14.1H Examples of equipment or systems that are potentially dangerous for operators due to stored energy include: springs; elevated machine members; rotating flywheels; hydraulic systems; and systems using air, gas, steam, or water pressure.

14.1I The primary elements of the hearing conservation program are monitoring, audiometric testing, hearing protection, training, access to information, and recordkeeping.

14.1J Positive-pressure self-contained breathing apparatus (SCBAs) and full-facepiece positive-pressure supplied-air respirators (SARs) with an escape SCBA can be used in oxygen-deficient atmospheres.

Answers to questions on page 289.

14.2A Traffic must be warned that you are working in a street or traffic area. BE PREPARED TO STOP, SHOULDER WORK, and UTILITY WORK AHEAD are signs that are effective. Signs with flashing warning lights and vehicles with yellow rotating beacons are used to warn other motorists. Vehicle-mounted traffic guides are also helpful. Use trained flaggers to alert drivers and to direct traffic around the work site. Warning signs and flaggers must be located far enough in advance of the work area to allow motorists time to realize they must slow down, be alert for activity, and safely change lanes or follow a detour.

14.2B Mechanical, electrical, engulfment, and toxic exposure hazards may be encountered when entering a manhole. Physical injury, infections and diseases, and insect and rodent bites are other potential hazards.

14.2C Three protective systems for excavations are shoring, shielding, and benching and sloping.

14.2D OSHA requires a protective system when excavations are five feet (1.5 m) or more in depth.

14.2E You should receive training regarding the proper use or operation of sewer cleaning tools and equipment from the equipment vendor.

ANSWERS TO QUESTIONS IN LESSON 2

Answers to questions on page 290.

14.2F Be aware of moving equipment in pumping stations, especially reciprocating equipment and rotating shafts. Moving parts that create a contact hazard to employees must be guarded. Do not wear loose clothing, rings, or other jewelry around machinery. Long hair must be secured.

14.2G Fire extinguishers in pumping stations should be of a type that may be used on electrical equipment as well as on solid material or power overload-type fires. The use of liquid-type fire extinguishers should be avoided. All-purpose, ABC chemical-type fire extinguishers are recommended.

Answers to questions on page 292.

14.2H Slimes, rags, or greases should be removed from any area because they create slip and trip hazards.

14.2I When working on mechanical or electrical parts of equipment, you should lock out the main circuit breaker and fasten a tag to the handle or to the lock-out device to notify others that you are working on the equipment and that it must remain de-energized.

14.2J Guards should be placed around all moving parts that create a contact hazard to employees.

14.2K Tools and equipment should not be carried, but should be transported in and out of pits and sumps by the use of buckets and a handline or sling.

Answers to questions on page 295.

14.2L Slippery surfaces, such as walkways, stairs, ladders, and catwalks, can be made less dangerous by keeping them free of grease, oil, slimes, or other materials and by applying nonslip strips or coatings. To reduce slipping risks, wear safety shoes with nonslip soles and install hand holds and railings.

14.2M Smoking and open flames should not be allowed in the vicinity of digesters because methane may be present and when methane gas is mixed with the proper proportion of air it forms an explosive mixture.

14.2N Follow all confined space procedures before entering *AND* during occupancy of a recently emptied digester. *DO NOT* enter unless you are certain that sources of contaminants, such as sludge and gas lines, have been effectively isolated and that safe atmospheric conditions exist and can be maintained before entry and during occupancy.

14.2O Before relighting a waste gas burner, the main gas valve should be turned off and the stack allowed to vent itself for a few minutes.

14.2P If a positive displacement pump is started against a closed discharge valve, pressures could build up and break a pipe or damage the pump.

Answers to questions on page 296.

14.2Q The rotating distributor should be stopped by turning off the flow of wastewater. Extreme care must be taken because of the powerful force behind the distributor.

14.2R You should never work alone on the center Y-wall of an aerator because you could fall into the aerator and need help getting out.

14.2S When using a boat to collect samples from a pond, have someone standing by in case you get into trouble, wear a Coast Guard-approved life jacket, and do not stand up in the boat while performing work.

Answers to questions on page 298.

14.2T One-ton chlorine containers should only be lifted by an approved lifting bar with hooks over the ends of the container.

14.2U Chlorine gas is two and a half times heavier than air and is best removed when leaks occur by blowing the gas out of the room at floor level.

14.2V If you open a door and smell chlorine or sulfur dioxide, immediately close the door, leave ventilation on, and seek help.

14.2W Housekeeping tasks that will keep a treatment plant a safer, cleaner place to work include:

1. Designate a place for tools and equipment and, when they are not being used, keep them in their proper places.
2. Clean up spills of oils, grease, chemicals, polymers, wastewater, and sludge. Keep walkways and work areas clean.
3. Provide proper containers for wastes, oily rags, and papers and empty the containers frequently.
4. Store flammable substances in an approved storage cabinet.
5. Remove snow and ice in areas where a person might slip and fall.

Answers to questions on page 300.

14.2X Major types of discharges from industrial plants that could create safety hazards include:

1. Fuels
2. Toxic gases
3. Amines
4. Surface-active agents
5. Biocides
6. Highly acidic or alkaline wastes

14.2Y If gasoline reaches the wet well in the headworks of your treatment plant:

1. Try to remove the gasoline from the surface of the wet well with skimmers (if available) or with a portable pump.
2. Apply as much ventilation as possible to prevent an explosive atmosphere from developing.
3. Monitor the atmosphere for toxic and explosive conditions.
4. Keep personnel away from the area.
5. Do not allow any flames, sparks, or other sources of ignition in the area (use nonsparking tools and explosion-proof equipment and wiring).

14.2Z Surface-active agents can cause three types of problems:

1. Super floc agents can produce a much denser sludge than usually pumped. This sludge can be too thick to pump, which then requires special handling to reslurry the sludge so it can be pumped.
2. Antifloc agents can lower the capture of suspended solids in primary clarifiers and cause a solids overload on downstream treatment processes.
3. Foaming agents can cause foam to cover treatment processes and prevent inspection and maintenance. Foam that blows into neighbors' yards or forms on the surface of receiving waters is objectionable to the public.

ANSWERS TO QUESTIONS IN LESSON 3

Answers to questions on page 303.

14.3A When collecting influent samples, disposable, impervious gloves should be worn to protect the operator's hands if there is any chance of direct contact with the wastewater. If possible, sample poles or other similar types of samplers should be used.

14.3B Hands should always be washed before eating to prevent the spread of disease.

14.3C Chemicals and reagents should be handled with care to protect your body from serious injuries and possible poisoning.

Answers to questions on page 304.

14.4A The necessary ingredients of a fire are fuel, heat, and oxygen.

14.4B Oil- and paint-soaked rags should be placed in covered metal containers and regularly disposed of in a safe manner.

Answers to questions on page 305.

14.5A Some treatment plants use bottled drinking water because it is an economical and reliable source of potable water. This practice reduces the possibility of the spread of disease from unknown cross connections or defective devices installed to prevent contamination by backflows.

14.6A The following phone numbers should be conspicuously posted in your plant: fire department, police department, ambulance service, chemical supplier or repair service, and physician. Check your list to be sure they are all listed and the numbers are correct.

Answers to questions on page 306.

14.7A The purposes of tailgate safety meetings are to remind operators of the need for safety, to explain and discuss safe procedures and safe conditions, and to review potential hazards and how to correct or avoid dangerous situations.

14.7B Safety meetings should be held every 7 to 10 working days.

Answers to questions on page 308.

14.8A Safety meetings could be:

1. Formal, organized meetings
2. Tailgate safety sessions
3. Informal get-togethers

14.8B Management must be committed to safety and actively support a safety training program. An effective safety training program must start at the top.

14.8C Safety drills should be held regularly so everyone knows how to respond during emergencies. Do not wait for an emergency before trying to learn how to use safety equipment. Get proper instructions and conduct practice sessions.

14.8D Paperwork necessary for an effective safety training program includes:

1. Accident report forms
2. Safety policy
3. Safety rules
4. Supervisors' guidelines
5. Checklists for review of facility plans and specifications and also for plant inspections

Answers to questions on page 316.

14.9A The four basic elements of a hazard communication program include:

1. Identify hazardous materials
2. Obtain chemical information and define hazardous conditions
3. Properly label hazards
4. Train operators

14.9B The general categories of hazardous materials include:

1. Corrosives
2. Toxics
3. Flammables and explosives
4. Asphyxiants
5. Harmful physical agents
6. Infectious agents

14.9C The purpose of the Material Safety Data Sheet (MSDS) is to have a readily available document to be used for training and as an immediate reference, since it includes information on common names, safe exposure level, the effects of exposure, the symptoms of exposure, whether it is flammable, what type of first aid should be administered, and other information.

14.9D The label on a hazardous material container must indicate the name of the hazardous substance and appropriate warnings.

APPENDIX

REGULATORY INFORMATION
APPLICABLE OSHA REGULATIONS

REGULATORY INFORMATION

Safety regulations are generally developed and enforced under the overall jurisdiction of the Federal Government's Department of Labor Occupational Safety and Health Administration (OSHA). In general, coverage of the OSHA Act extends to all employers and their employees in the 50 states, the District of Columbia, Puerto Rico, and all other territories under Federal Government jurisdiction. Coverage is provided either directly by federal OSHA or through an OSHA-approved state program.

OSHA has developed a comprehensive set of safety regulations, many of them with specific applications to wastewater collection and treatment systems. Where OSHA has not developed specific standards, employers are responsible for following the Act's general duty clause, which states that each employer "shall furnish...a place of employment which is free from recognized hazards that are causing or are likely to cause death or serious physical harm to his employees."

States with OSHA-approved occupational safety and health programs must set standards that are at least as effective as the federal standards. Many states adopt standards identical to the federal standards. Where states adopt and enforce their own standards under state law, copies of state standards may be obtained from the individual states.

The Federal Register is one of the best sources of information on standards, since all OSHA standards are published there when adopted, as are all amendments, corrections, insertions, or deletions. The Federal Register is available in many public libraries. Annual subscriptions are available from the US Government Printing Office, Superintendent of Documents, PO Box 371954, Pittsburgh, PA 15250-7954, phone (866) 512-1800.

Each year the Office of the Federal Register publishes all current regulations and standards in the Code of Federal Regulations (CFR), available at many libraries and from the Government Printing Office. OSHA's regulations are collected in Title 29 of the Code of Federal Regulations (CFR), Part 1900-1999. Copies of the CFR may also be purchased from the Superintendent of Documents at the address above.

APPLICABLE REGULATIONS

OSHA regulations fill several hundred pages of text, which cannot be printed in this manual. Of the regulations, Part 1910, OCCUPATIONAL SAFETY AND HEALTH STANDARDS, contains the regulations most applicable to the work we do in wastewater collection and treatment systems. If your state is one of the 25 that has a state program, there will be a comparable set of regulations similar to Part 1910. Listed next are some of the OSHA regulations that apply.

PART 1910 — OCCUPATIONAL SAFETY AND HEALTH STANDARDS

Subpart D — Walking/Working Surfaces

1910.21 Definitions.

1910.22 General requirements.

1910.23 Guarding floor and wall openings and holes.

1910.24 Fixed industrial stairs.

1910.25 Portable wood ladders.

1910.26 Portable metal ladders.

1910.27 Fixed ladders.

1910.28 Safety requirements for scaffolding.

1910.29 Manually propelled mobile ladder stands and scaffolds (towers).

1910.30 Other working surfaces.

1910.31 Sources of standards.

1910.32 Standards organizations.

Subpart E — Means of Egress

1910.35 Definitions.

1910.36 General requirements.

1910.37 Means of egress, general.

1910.38 Employee emergency plans and fire prevention plans.

1910.39 Sources of standards.

1910.40 Standards organizations.

1910.40A Appendix to Subpart E — Means of Egress.

Subpart F — Powered Platforms, Manlifts, and Vehicle-Mounted Work Platforms

1910.66 Power platforms for exterior building maintenance.

1910.66A Appendix A to §1910.66 — Guideline (Advisory)

1910.66B	Appendix B to §1910.66 — Exhibits (Advisory)
1910.66C	Appendix C to §1910.66 — Personal Fall Arrest System (Section I — Mandatory; Sections II and III — Nonmandatory)
1910.66D	Appendix D to §1910.66 — Existing Installations (Mandatory)
1910.67	Vehicle-mounted elevating and rotating work platforms.
1910.68	Manlifts.
1910.69	Sources of standards.
1910.70	Standards organizations.

Subpart G — Occupational Health and Environmental Control

1910.94	Ventilation.
1910.95	Occupational noise exposure.
1910.96	Ionizing radiation.
1910.97	Nonionizing radiation.
1910.98	Effective dates.
1910.99	Sources of standards.
1910.100	Standards organizations.

Subpart H — Hazardous Materials

1910.101	Compressed gases (general requirements).
1910.102	Acetylene.
1910.103	Hydrogen.
1910.104	Oxygen.
1910.106	Flammable and combustible liquids.
1910.107	Spray finishing using flammable and combustible materials.
1910.108	Dip tanks containing flammable or combustible liquids.
1910.110	Storage and handling of liquefied petroleum gases.
1910.115	Sources of standards.
1910.116	Standards organizations.
1910.119	Process safety management of highly hazardous chemicals.
1910.119A	Appendix A to §1910.119 — List of Highly Hazardous Chemicals, Toxics and Reactives (Mandatory).
1910.119B	Appendix B to §1910.119 — Block Flow Diagram and Simplified Process Flow Diagram (Nonmandatory).
1910.119C	Appendix C to §1910.119 — Compliance Guidelines and Recommendations for Process Safety Management (Nonmandatory).
1910.119D	Appendix D to §1910.119 — Sources of Further Information (Nonmandatory).
1910.120	Hazardous waste operations and emergency response.
1910.120A	Appendix A to §1910.120 — Personal Protective Equipment Test Methods.
1910.120B	Appendix B to §1910.120 — General Description and Discussion of the Levels of Protection and Protective Gear.
1910.120C	Appendix C to §1910.120 — Compliance Guidelines.
1910.120D	References.

Subpart I — Personal Protection Equipment

1910.132	General requirements.
1910.133	Eye and face protection.
1910.134	Respiratory protection.
1910.135	Occupational head protection.
1910.136	Occupational foot protection.
1910.137	Electrical protective devices.

Subpart J — General Environmental Controls

1910.141	Sanitation.
1910.144	Safety color code for marking physical hazards.
1910.145	Specifications for accident prevention signs and tags.
1910.145A	Appendix A to §1910.145(f) — Recommended Color Coding.
1910.145B	Appendix B to §1910.145(f) — References for further information.
1910.146	Permit-required confined spaces.
1910.146A	Appendix A to §1910.146 — Permit-required Confined Space Decision Flow Chart.
1910.146B	Appendix B to §1910.146 — Procedures for Atmospheric Testing.
1910.146C	Appendix C to §1910.146 — Examples of Permit-required Confined Space Programs.
1910.146D	Appendix D to §1910.146 — Sample Permits.
1910.146E	Appendix E to §1910.146 — Sewer System Entry.
1910.147	The control of hazardous energy (lockout/tagout).
1910.147A	Appendix A to §1910.147 — Typical Minimal Lockout Procedure.
1910.148	Standards organizations.
1910.149	Effective dates.

Subpart K — Medical and First Aid

1910.150	Sources of Standards.
1910.151	Medical services and first aid.
1910.153	Sources of standards.

Subpart L — Fire Protection

PORTABLE FIRE SUPPRESSION EQUIPMENT

1910.157	Portable fire extinguishers.
1910.158	Standpipe and hose systems.

FIXED FIRE SUPPRESSION EQUIPMENT

1910.159	Automatic sprinkler systems.
1910.160	Fixed extinguishing systems, general.
1910.161	Fixed extinguishing systems, dry chemical.
1910.162	Fixed extinguishing systems, gaseous agent.
1910.163	Fixed extinguishing systems, water spray and foam.

OTHER FIRE PROTECTION SYSTEMS

1910.164 Fire detection systems.

1910.165 Employee alarm systems.

Subpart M — Compressed Gas and Compressed Air Equipment

1910.169 Air receivers.

1910.170 Sources of standards.

1910.171 Standards organizations.

Subpart N — Materials Handling and Storage

1910.176 Handling materials — general.

1910.177 Servicing multi-piece and single piece rim wheels.

1910.178 Powered industrial trucks.

1910.179 Overhead and gantry cranes.

1910.180 Crawler locomotive and truck cranes.

1910.181 Derricks.

1910.182 Effective dates.

1910.183 Helicopters.

1910.184 Slings.

1910.189 Sources of standards.

1910.190 Standards organizations.

Subpart O — Machinery and Machine Guarding

1910.211 Definitions.

1910.212 General requirements for all machines.

1910.213 Woodworking machinery requirements.

1910.214 Cooperage machinery.

1910.215 Abrasive wheel machinery.

1910.216 Mills and calenders in the rubber and plastics industries.

1910.217 Mechanical power presses.

1910.218 Forging machines.

1910.219 Mechanical power-transmission apparatus.

1910.220 Effective dates.

1910.221 Sources of standards.

1910.222 Standards organizations.

Subpart P — Hand and Portable Powered Tools and Other Hand-Held Equipment

1910.241 Definitions.

1910.242 Hand and portable powered tools and equipment, general.

1910.243 Guarding of portable powered tools.

1910.244 Other portable tools and equipment.

1910.245 Effective dates.

1910.246 Sources of standards.

1910.247 Standards organizations.

Subpart Q — Welding, Cutting, and Brazing

1910.251 Definitions.

1910.252 General requirements.

1910.253 Oxygen-fuel gas welding and cutting.

1910.254 Arc welding and cutting.

1910.255 Resistance welding.

1910.256 Sources of standards.

1910.257 Standards organizations.

1910.269 Electric power generation, transmission, and distribution.

1910.269A Appendix A to §1910.269 — Flow Charts.

1910.269A-1 Appendix A-1 to §1910.269 — Application of Section 1910.269 and Subpart S of this Part to Electrical Installations.

1910.269A-2 Appendix A-2 to §1910.269 — Application of Section 1910.269 and Subpart S of this Part to Electrical Safety-Related Work Practices.

1910.269A-3 Appendix A-3 to §1910.269 — Application of Section 1910.269 and Subpart S of this Part to Tree-Trimming Operations.

1910.269A-4 Appendix A-4 to §1910.269 — Application of Section 1910.147, Section 1910.269, and Section 1910.333 to Hazardous Energy Control Procedures (Lockout/Tagout).

1910.269A-5 Appendix A-5 to §1910.269 — Application of Section 1910.146 and Section 1910.269 to Permit-Required Confined Spaces.

1910.269B Appendix B to §1910.269 — Working on Exposed Energized Parts.

1910.269C Appendix C to §1910.269 — Protection from Step and Touch Potentials.

1910.269D Appendix D to §1910.269 — Methods of Inspecting and Testing Wood Poles.

1910.269E Appendix E to §1910.269 — Reference Documents.

Subpart S — Electrical

GENERAL

1910.301 Introduction.

DESIGN SAFETY STANDARDS FOR ELECTRICAL SYSTEMS

1910.302 Electric utilization systems.

1910.303 General requirements.

1910.304 Wiring design and protection.

1910.305 Wiring methods, components, and equipment for general use.

1910.306 Specific purpose equipment and installations.

1910.307 Hazardous (classified) locations.

1910.308 Special systems.

SAFETY-RELATED WORK PRACTICES

1910.331 Scope.

1910.332 Training.

1910.333 Selection and use of work practices.

1910.334 Use of equipment.

In addition to the regulations listed above, 29 CFR Part 1926, SAFETY AND HEALTH FOR CONSTRUCTION, defines the regulations for Excavations and Trenches under Subpart P. There are other federal regulations that may also apply to wastewater collection and treatment systems operation and maintenance, such as Department of Transportation (DOT), as well as state and local regulations for traffic control and protection of utilities during excavations using "one-call" notification systems.

CHAPTER 15

MAINTENANCE

GENERAL PROGRAM
by
Norman Farnum

MECHANICAL MAINTENANCE
by
Stan Walton

UNPLUGGING PIPES, PUMPS, AND VALVES
by
John Brady

FLOW MEASUREMENT
by
Roger Peterson

Revised by
Malcolm Carpenter and
Rick Arbour

TABLE OF CONTENTS

Chapter 15. MAINTENANCE

Page

The format of this section differs slightly from the others. The arrangement of procedures was designed specifically to assist you in planning an effective preventive maintenance program. The contents are at the beginning of the section, and the paragraphs are numbered for easy reference on equipment service record cards.

Paragraph

OBJECTIVES

Chapter 15. MAINTENANCE

Following completion of Chapter 15, you should be able to:

1. Develop a maintenance program for your plant, including equipment, buildings, grounds, channels, and tanks.
2. Start a maintenance recordkeeping system that will provide you with information to protect equipment warranties, to prepare budgets, and to satisfy regulatory agencies.
3. Schedule maintenance of equipment at proper time intervals.
4. Perform maintenance as directed by manufacturers.
5. Recognize symptoms that indicate equipment is not performing properly, identify the source of the problem, and take corrective action.
6. Start and stop pumps.
7. Unplug pipes, pumps, and valves.
8. Explain the operation and maintenance of sensors, transmitters, receivers, and controllers.
9. Determine when you need assistance to correct a problem.

NOTE: Special maintenance information is given in the previous chapters on treatment processes where appropriate.

WORDS

Chapter 15. MAINTENANCE

AIR GAP AIR GAP

An open, vertical drop, or vertical empty space, between a drinking (potable) water supply and potentially contaminated water. This gap prevents the contamination of drinking water by backsiphonage because there is no way potentially contaminated water can reach the drinking water supply.

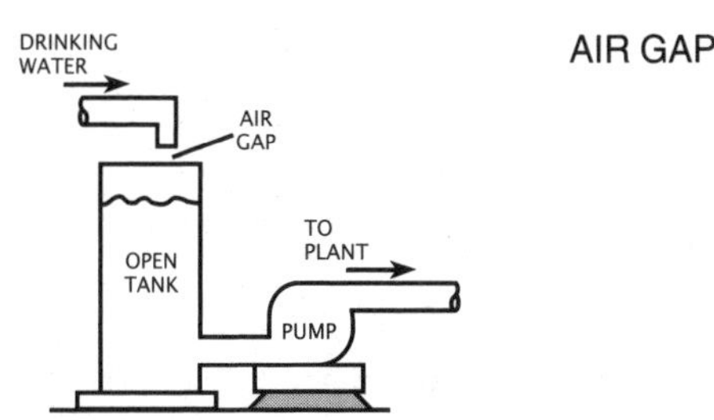

ANALOG READOUT ANALOG READOUT

The readout of an instrument by a pointer (or other indicating means) against a dial or scale. Also see DIGITAL READOUT.

AXIAL TO IMPELLER AXIAL TO IMPELLER

The direction in which material being pumped flows around the impeller or flows parallel to the impeller shaft.

AXIS OF IMPELLER AXIS OF IMPELLER

An imaginary line running along the center of a shaft (such as an impeller shaft).

BRINELLING (bruh-NEL-ing) BRINELLING

Tiny indentations (dents) high on the shoulder of the bearing race or bearing. A type of bearing failure.

CAVITATION (kav-uh-TAY-shun) CAVITATION

The formation and collapse of a gas pocket or bubble on the blade of an impeller or the gate of a valve. The collapse of this gas pocket or bubble drives water into the impeller or gate with a terrific force that can knock metal particles off and cause pitting on the impeller or gate surface. Cavitation is accompanied by loud noises that sound like someone is pounding on the impeller or gate with a hammer.

CROSS CONNECTION CROSS CONNECTION

(1) A connection between drinking (potable) water and an unapproved water supply.
(2) A connection between a storm drain system and a sanitary collection system.
(3) Less frequently used to mean a connection between two sections of a collection system to handle anticipated overloads of one system.

DATEOMETER (day-TOM-uh-ter) DATEOMETER

A small calendar disk attached to motors and equipment to indicate the year in which the last maintenance service was performed.

DIGITAL READOUT DIGITAL READOUT

The readout of an instrument by a direct, numerical reading of the measured value or variable.

JOGGING JOGGING

The frequent starting and stopping of an electric motor.

MEGOHM (MEG-ome) MEGOHM

Millions of ohms. Mega- is a prefix meaning one million, so 5 megohms means 5 million ohms.

MULTISTAGE PUMP MULTISTAGE PUMP

A pump that has more than one impeller. A single-stage pump has one impeller.

NAMEPLATE NAMEPLATE

A durable, metal plate found on equipment that lists critical operating conditions for the equipment.

POLE SHADER POLE SHADER

A copper bar circling the laminated iron core inside the coil of a magnetic starter.

PROGRAMMABLE LOGIC CONTROLLER (PLC) PROGRAMMABLE LOGIC CONTROLLER (PLC)

A microcomputer-based control device containing programmable software; used to control process variables.

PRUSSIAN BLUE PRUSSIAN BLUE

A blue paste or liquid (often on a paper like carbon paper) used to show a contact area. Used to determine if gate valve seats fit properly.

RADIAL TO IMPELLER RADIAL TO IMPELLER

Perpendicular to the impeller shaft. Material being pumped flows at a right angle to the impeller.

SINGLE-STAGE PUMP SINGLE-STAGE PUMP

A pump that has only one impeller. A multistage pump has more than one impeller.

STATOR STATOR

That portion of a machine that contains the stationary (nonmoving) parts that surround the moving parts (rotor).

STETHOSCOPE STETHOSCOPE

An instrument used to magnify sounds and carry them to the ear.

CHAPTER 15. MAINTENANCE

(Lesson 1 of 7 Lessons)

15.0 TREATMENT PLANT MAINTENANCE—GENERAL PROGRAM

A treatment plant operator has many duties. Most of them have to do with the efficient operation of the plant. An operator has the responsibility to discharge an effluent that will meet all the requirements established for the plant. By doing this, the operator develops a good working relationship with the regulatory agencies, water recreationists, water users, and plant neighbors.

Another duty an operator has is that of plant maintenance. A good maintenance program is essential for a wastewater treatment plant to operate continuously at peak design efficiency. A successful maintenance program will cover everything from mechanical equipment, such as pumps, valves, scrapers, and other moving equipment, to the care of the plant grounds, buildings, and structures.

Mechanical maintenance is of prime importance as the equipment must be kept in good operating condition for the plant to maintain peak performance. Manufacturers provide information on the mechanical maintenance of their equipment. You should thoroughly read their literature on your plant equipment and understand the procedures. Contact the manufacturer or the local representative if you have any questions. Follow the instructions very carefully when performing maintenance on equipment. You also must recognize tasks that may be beyond your capabilities or repair facilities, and you should request assistance when needed.

For a successful maintenance program, your supervisors must understand the need for and benefits from equipment that operates continuously as intended. Disabled or improperly working equipment is a threat to the quality of the plant effluent, and repair costs for poorly maintained equipment usually exceed the cost of maintenance.

15.00 Preventive Maintenance Records

Preventive maintenance programs help operating personnel keep equipment in satisfactory operating condition and aid in detecting and correcting malfunctions before they develop into major problems.

A frequent occurrence in a preventive maintenance program is the failure of the operator to record the work after it is completed. When this happens the operator must rely on memory to know when to perform each preventive maintenance function. As days pass into weeks and months, the preventive maintenance program is lost in the turmoil of everyday operation.

The only way an operator can keep track of a preventive maintenance program is by good recordkeeping. A good recordkeeping system tells when maintenance is due and also provides a record of equipment performance. Poor performance is a good justification for replacement or new equipment. Good records also help keep your warranty in force. Whatever recordkeeping system is used, it should be kept up to date on a daily basis and not left to memory for some other time. Equipment service cards and service record cards (Figure 15.1) are easy to set up and require little time to keep up to date.

An *EQUIPMENT SERVICE CARD* (master card) should be filled out for each piece of equipment in the plant. Each card should have the equipment name on it, such as Sludge Pump No. 1, Primary Clarifier.

1. List each required maintenance service with an item number.
2. List maintenance services in order of frequency of performance. For instance, show daily service as items 1, 2, and 3 on the card; weekly items as 4 and 5; monthly items as 6, 7, 8, and 9; and so on.
3. Describe each type of service under work to be done.

Make sure all necessary inspections and services are shown. For reference data, list paragraph or section numbers as shown in the mechanical maintenance section of this chapter (Section 15.7, page 413). Also, list frequency of service as shown in the time schedule columns of the same section. Under time, enter day or month service is due. Service card information may be changed to fit the needs of your plant or particular equipment as recommended by the equipment manufacturer. Be sure the information on the cards is complete and correct.

The *SERVICE RECORD CARD* should have the date and work done, listed by item number and signed by the operator who performed the service. Some operators prefer to keep both cards clipped together, while others place the service record card near the equipment.

When the service record card is filled, it should be filed for future reference and a new card attached to the master card. The equipment service card tells what should be done and when, while the service record card is a record of what you did and when you did it. Many plants keep this information on a computer.

EQUIPMENT SERVICE CARD				
EQUIPMENT: #1 Raw Wastewater Lift Pump				
Item No.	**Work To Be Done**	**Reference**	**Frequency**	**Time**
1	Check water seal and packing gland	Par. 1	Daily	
2	Operate pump alternately	Par. 1	Weekly	Monday
3	Inspect pump assembly	Par. 1	Weekly	Wed.
4	Inspect and lube bearings	Par. 1	Quarterly	1-4-7-10*
5	Check operating temperature of bearings	Par. 1	Quarterly	1-4-7-10
6	Check alignment of pump and motor	Par. 1	Semiannually	4 & 10
7	Inspect and service pump	Par. 1	Semiannually	4 & 10
8	Drain pump before shutdown	Par. 1		

* 1-4-7-10 represent the months of the year when the equipment should be serviced—1. January, 4. April, 7. July, and 10. October.

SERVICE RECORD CARD					
EQUIPMENT: #1 Raw Wastewater Lift Pump					
Date	**Work Done (Item No.)**	**Signed**	**Date**	**Work Done (Item No.)**	**Signed**
1-5-07	1 & 2	J.B.			
1-6-07	1	J.B.			
1-7-07	1-3-4-5-	R.W.			

Fig. 15.1 Equipment service card and service record card

QUESTIONS

Write your answers in a notebook and then compare your answers with those on page 460.

15.0A Why should you plan a good maintenance program for your treatment plant?

15.0B What general items would you include in your maintenance program?

15.0C Why should your maintenance program be accompanied by a good recordkeeping system?

15.0D What is the difference between an equipment service card and a service record card?

15.01 Building Maintenance

Building maintenance is another program that should be maintained on a regular schedule. Buildings in a treatment plant are usually built of sturdy materials to last for many years. Buildings must be kept in good repair. In selecting paint for a treatment plant, it is always a good idea to have a painting expert help the operator select the types of paint needed to protect the buildings from deterioration. The expert also will have some good ideas as to color schemes to help blend the plant in with the surrounding area. Consideration should also be given to the quality of paint. A good quality, more expensive material will usually give better service over a longer period of time than the economy-type products.

Building maintenance programs depend on the age, type, and use of a building. New buildings require a thorough check to be certain essential items are available and working properly. Older buildings require careful watching and prompt attention to keep ahead of leaks, breakdowns, replacements when needed, and changing uses of the building. Attention must be given to the maintenance requirements of many items in all plant buildings, such as electrical systems, plumbing, heating, cooling, ventilating, floors, windows, roofs, and drainage around the buildings. Regularly scheduled examinations and necessary maintenance of these items can prevent many costly and time-consuming problems in the future.

In each plant building, periodically check all stairways, ladders, catwalks, and platforms for adequate lighting, head clearance, and sturdy and convenient guardrails. Protective devices should be around all moving equipment. Whenever any repairs, alterations, or additions are built, avoid building accident traps such as pipes laid on top of floors or hung from the ceiling at head height, which could create serious safety hazards.

Organized storage areas should be provided and maintained in an accessible and neat manner.

Keep all buildings clean and orderly. Janitorial work should be done on a regular schedule. All tools and plant equipment should be kept clean and in their proper place. Floors, walls, and windows should be cleaned at regular intervals to maintain a neat appearance. A treatment plant kept in a clean, orderly condition makes a safer place to work and aids in building good public and employee relations.

15.02 Plant Tanks and Channels

Wastewater channels, wet wells, and tanks such as clarifiers and grit tanks should be drained and inspected at least once a year. Be sure that the groundwater level is down far enough so the tanks will not float on the groundwater when empty or develop cracks from groundwater pressure. Most of the tanks in recently constructed facilities contain relief valves in the floor of the tank to prevent tank flotation if it is dewatered under high groundwater conditions.

Schedule inspections and maintenance of tanks and channels during periods of low inflow to minimize the load on other plant treatment units. Route flows through alternate units, if available; otherwise provide the best possible treatment with remaining units not being inspected or repaired.

All metal and concrete surfaces that come in contact with wastewater and covered surfaces exposed to fumes should have a good protective coating. The coating should be reapplied where necessary at each inspection.

Digesters should also be drained and cleaned on a regular basis. Once every five years (actual times range from three to eight years) has been accepted as an approximate interval for this operation. Most digesters have a sludge inlet box on one side and a supernatant box on the opposite side. A sludge sampler can be lowered through the pipes into both of these boxes to check for sand and grit buildup. To determine the amount of grit buildup, you must know the sidewall depth of the digester. If the sludge sampler will only drop to within four feet (1.2 m) of the bottom, you can assume that you have a four-foot (1.2-m) buildup of sand and grit. By measuring the depth of sand and grit at periodic intervals, you can determine how fast the buildup is accumulating. In digesters, all metal and concrete surfaces should be inspected for deterioration.

On surfaces where the protective coatings are dead and flake off, it is necessary to sandblast the entire surface before new coatings are applied. Usually, two or more coats are needed for proper protection. Be aware that sandblasting old coatings may result in the production of a hazardous waste. Many coating materials used in wastewater facilities contain lead and the resulting spent blast material may contain sufficient lead levels to be classified as a hazardous waste. Test kits are available to determine if the coating is lead based. Very stringent regulations exist that deal with abrasive blasting and disposal of lead-based coating materials. Check with your local environmental health and safety agencies.

The protective coatings used on these types of tanks and channels can also be black asphaltic-type paint. These coatings should be used wherever practical. In areas where fumes and moisture are not severe, aluminum coating or a color scheme may be desirable. In these areas, a rubber-base paint or some similar material may be used. Follow the recommendations of a paint expert.

CAUTION

Periodic drainage, inspection, and repair of tanks and channels is essential. Failure to do so may result in complete disruption of operations during the critical receiving water low-flow/high-temperature season. Select a time for maintenance (removing obstructions and repairing gates, concrete pipes, and pumps) when you can minimize the discharge of harmful wastes to receiving waters. Schedule as many concurrent events as possible during a shutdown to minimize the time the plant or parts of the plant must be taken out of service.

15.03 Plant Grounds

Plant grounds that are well groomed and kept in a neat condition will greatly add to the overall appearance of the plant area. Well-groomed and neat grounds are important because many people judge the ability of the operator and the plant performance on the basis of the appearance of the plant. Management, also, tends to view well-kept grounds as evidence of an operator's ability and competence.

If the plant grounds have not been landscaped, it is sometimes the responsibility of the operator to do so. This may consist of planting shrubs and lawns or just keeping the grounds neat and weed free. Some plant grounds may be entirely paved. In any case, they should be kept clean and orderly at all times.

Control rodents and insects so they will not spread diseases or cause nuisances.

For the convenience of visitors and new operators, signs directing people to the plant, pointing the way to different plant facilities, identifying plant buildings, and indicating the direction of flow and contents flowing in a pipe can all be very helpful. Well-lighted and well-maintained walks and roadways are very important. Plant grounds should be fenced to prevent unauthorized persons and animals from entering the area. Keep seldom-used items and old, discarded equipment neatly stored to avoid the appearance of a cluttered junkyard. Take pride in your plant grounds and you will be amazed at the favorable impression your facility will convey to the public and administrators.

QUESTIONS

Write your answers in a notebook and then compare your answers with those on page 460.

15.0E What items should be included in a building maintenance program?

15.0F When plant tanks and channels are drained, what items would you inspect?

15.0G Why are neat and well-groomed grounds important?

15.04 Chlorinators

15.040 Maintenance

Chlorine gas leaks around chlorinators or containers of chlorine will cause corrosion of equipment. Check every day for leaks. Large leaks will be detected by odor; small leaks may go unnoticed until damage results. A green or reddish deposit on metal indicates a chlorine leak. Any chlorine gas leakage in the presence of moisture will cause corrosion. Always plug the ends of any open connection to prevent moisture from entering the lines. Never pour water on a chlorine leak because this will only create a bigger problem by enlarging the leak. Chlorine gas reacts with water to form hydrochloric acid.

WARNING

The most important reason for preventing chlorine leaks is that the gas is toxic to humans.

Ammonia water will detect any chlorine leak. Use a concentrated ammonia solution containing 28 to 30 percent ammonia as NH_3 (this is the same as 58 percent ammonium hydroxide, NH_4OH, or commercial 26° Baumé). A small piece of cloth, soaked with ammonia water and wrapped around the end of a short stick, makes a good leak detector. Wave this stick in the general area of the suspected leak (do not touch the equipment with it). If chlorine gas leakage is occurring, a white cloud of ammonium chloride will form. You should make this test at all gas pipe joints, both inside and outside the chlorinators, at regular intervals. Bottles of ammonia water should be kept tightly capped to avoid loss of strength. All pipe fittings must be kept tight to avoid leaks. New gaskets should be used for each new connection.

CAUTION

To locate a chlorine leak, *do not spray or swab equipment with ammonia water!* Wave an ammonia-soaked rag or paint brush in the general area and you can detect the presence of many leaks. Some operators prefer to wave a stick with an ammonia-soaked cloth on the end in front of them when looking for chlorine leaks.

A squeeze bottle containing a piece of sponge soaked with ammonia may be used to look for chlorine leaks around connections and cylinders. However, do not use a squeeze bottle in a room where large amounts of chlorine gas have already leaked into the air. After one squeeze, the entire area may be full of white smoke and you will have trouble locating the leak. Under these conditions, use a cloth soaked in ammonia water to look for leaks.

The exterior casing of chlorinators should be painted as required; however, most chlorinators have plastic cases that do not require protective coatings. A clean machine is a better operating machine. Parts of a chlorinator handling chlorine gas must be kept dry to prevent the chlorine and moisture from forming hydrochloric acid. Some parts may be cleaned, when required, first with water to remove water-soluble material, then with wood alcohol, followed by drying. The above chemicals leave no moisture residue. Another method would be to wash the parts with water and dry them in a pan or use a heater to remove all traces of moisture. Pipelines should be purged with dry air (–40°F dew point) or nitrogen to remove moisture.

Water strainers on chlorinators frequently clog and require attention. They may be cleaned by flushing with water or, if badly fouled, they may be cleaned with dilute hydrochloric acid, followed with a water rinse.

The atmosphere vent lines from chlorinators must be open and free. These vent lines evacuate the chlorine to the outside atmosphere when the chlorinator is being shut down. Place a screen over the end of the pipe to keep insects from building a nest in the pipe and clogging it up.

When chlorinators are removed from service, as much chlorine gas as possible should be removed from the supply lines and machines. The chlorine valves at the containers are shut off and the chlorinator injector is operated for a period to remove the chlorine gas. In V-notch chlorinators (Volume I, Chapter 10), the rotameter goes to the bottom of the manometer tube when the chlorine gas has been expelled.

All chlorinators will give continuous, trouble-free operation if properly maintained and operated. Each chlorinator manufacturer provides with each machine a maintenance and operations instruction booklet with line diagrams showing the operation of the component parts of the machine. Manufacturer's instructions should be followed for maintenance and lubrication of your particular chlorinator. If you do not have an instruction booklet, you may obtain one by contacting the manufacturer's representative in your area.

15.041 Chlorine and Sulfur Dioxide Safety

For information on chlorine and sulfur dioxide safety, see Volume I, Chapter 10, "Disinfection and Chlorination," and Chapter 14, "Plant Safety." Read these chapters before attempting maintenance on chlorinators, sulfonators, lines, or cylinders.

Sulfur dioxide (SO_2) is commonly used today for dechlorination purposes. Most of the principles that apply to chlorine also apply to sulfur dioxide. Leaks are detected with ammonia water. Sulfonators (the machine through which the chemical is fed) are identical in appearance and function to chlorinators. Strainers should be kept clear to avoid plugging. Sulfonators should be completely emptied before performing maintenance. You must always remember that sulfur dioxide (SO_2) is just as toxic to humans as chlorine. One person should never attempt to repair a chlorine or sulfur dioxide leak alone because the person could be overcome by fumes. Valuable time, needed to repair and correct a serious emergency, could be lost rescuing a foolish person.

15.042 Emergencies

If your plant has not developed procedures for handling potential emergencies, do it now. Emergency procedures must be established for operators to follow when emergencies are caused by the release of chlorine, sulfur dioxide, or other hazardous chemicals. These procedures should include a list of emergency phone numbers located near a telephone that is unlikely to be affected by the emergency.

1. Police
2. Fire
3. Hospital or Physician
4. Responsible Plant Officials
5. Local Emergency Disaster Office
6. *CHEMTREC,* (800) 424-9300
7. Emergency Team (if your plant has one)

The *CHEMTREC* toll-free number may be called at any time. Personnel at this number will give information on how to handle emergencies created by hazardous materials and will notify appropriate emergency personnel.

An emergency team for your plant may be trained and assigned the task of responding to specific emergencies such as chlorine or sulfur dioxide leaks. An emergency team to repair chlorine leaks is important because chlorine is a hazardous chemical. An emergency team is specially trained and qualified to control and repair emergency chlorine leaks. Unqualified and untrained personnel may injure themselves and create hazards for others. This emergency team must meet the following strict specifications at all times.

1. Team personnel must be physically and mentally qualified.
2. Proper equipment must be available at all times, including:
   a. Protective equipment, including self-contained breathing apparatus
   b. Repair kits
   c. Repair tools
3. Proper training must take place on a regular basis and include instruction about:
   a. Properties and detection of hazardous chemicals
   b. Safe procedures for handling and storage of chemicals
   c. Types of containers, safe procedures for shipping containers, and container safety devices
   d. Installation of repair devices
4. Team members must be exposed regularly to simulated field emergencies or practice drills. Team response must be carefully evaluated and any errors or weaknesses corrected.
5. Emergency team performance must be reviewed annually on a specified date. Review must include:
   a. Training program
   b. Response to actual emergencies
   c. Team physical and mental examinations

WARNING: One person must never be permitted to attempt an emergency repair alone. Always wait for trained assistance. Valuable time, needed to correct a serious emergency, could be lost rescuing a foolish individual.

Chlorine or sulfur dioxide releases that exceed their reportable quantity (RQ), 10 pounds (4.5 kg) and 1 pound (0.45 kg) respectively, must be reported to certain regulatory agencies, including the National Response Center (NRC). It should be noted that the United States Environmental Protection Agency (EPA) is revising the RQ for several chemicals. Contact your local or state environmental management agency for current requirements for your plant.

15.05 Library

A plant library can contain helpful information to assist in plant operation. Material in the library should be cataloged and filed for easy use. Items in the library should include:

1. Plant operation and maintenance manual
2. Plant plans and specifications
3. Manufacturers' equipment instructions

4. Reference books on wastewater treatment, including manuals in this series
5. Professional journals and publications
6. Manuals of Practice and safety literature published by the Water Environment Federation (WEF), 601 Wythe Street, Alexandria, VA 22314-1994
7. First-aid book
8. Reports from other plants
9. A dictionary
10. Local, state, and federal safety regulations

QUESTIONS

Write your answers in a notebook and then compare your answers with those on page 460.

15.0H How would you search for chlorine and sulfur dioxide leaks?

15.0I Why should chlorine and sulfur dioxide leaks be detected and repaired?

15.0J Prepare a list of emergency phone numbers for your treatment plant.

15.0K What items should be included in the training program for an emergency team?

Please answer the discussion and review questions next.

DISCUSSION AND REVIEW QUESTIONS

Chapter 15. MAINTENANCE

(Lesson 1 of 7 Lessons)

At the end of each lesson in this chapter, you will find some discussion and review questions. The purpose of these questions is to indicate to you how well you understand the material in the lesson. Write the answers to these questions in your notebook.

1. Why should the operator thoroughly read and understand manufacturers' literature before attempting to maintain plant equipment?
2. Why must administrators or supervisors be made aware of the need for an adequate maintenance program?
3. What is the purpose of a maintenance recordkeeping program?
4. What kinds of maintenance checks should be made periodically of stairways, ladders, catwalks, and platforms?
5. When should inspection and maintenance of the underwater portions of plant structures such as clarifiers and digesters be scheduled?
6. Why should rodents and insects be controlled?
7. Why should one person never be permitted to repair a chlorine or sulfur dioxide leak alone?
8. Why should your plant have an emergency team to repair chlorine leaks?
9. What items should be included in a plant library?

CHAPTER 15. MAINTENANCE

(Lesson 2 of 7 Lessons)

15.1 MECHANICAL EQUIPMENT

Mechanical equipment commonly used in treatment plants is described and discussed in this section. Equipment used with specific treatment processes such as clarifiers or aeration basins are not discussed. You must be familiar with equipment and understand what it is intended to do before developing a preventive maintenance program and maintaining equipment.

15.10 Repair Shop

Many large plants have fully equipped machine shops staffed with competent mechanics. But for smaller plants, adequate machine shop facilities often can be found in the community. In addition, most pump manufacturers maintain pump repair departments where pumps can be fully reconditioned.

The pump repair shop in a large plant commonly includes such items as welding equipment, lathes, drill press and drills, power hacksaw, flame-cutting equipment, micrometers, calipers, gauges, portable electric tools, grinders, a hydraulic press, metal-spray equipment, and sandblasting equipment. You must determine what repair work you can and should do and when you need to request assistance from an expert. Help may be obtained from the manufacturer, the local representative, a consulting engineer, or another operator.

15.11 Pumps

Pumps serve many purposes in wastewater collection systems and treatment plants. They may be classified by the character of the material handled: raw wastewater, grit, effluent, activated sludge, raw sludge, or digested sludge. Or, they may relate to the conditions of pumping: high lift, low lift, recirculation, or high capacity. They may be further classified by principle of operation, such as centrifugal, propeller, reciprocating, and turbine.

The type of material to be handled and the function or required performance of the pump vary so widely that the design engineer must use great care in preparing specifications for the pump and its controls. Similarly, the operator must conduct a maintenance and management program adapted to the specific characteristics of the equipment.

15.110 Centrifugal Pumps

A centrifugal pump is basically a very simple device: an impeller rotating in a casing. The impeller is supported on a shaft, which is, in turn, supported by bearings. Liquid coming in at the center (eye) of the impeller (Figure 15.2) is picked up by the vanes and by the rotation of the impeller and then is thrown out by centrifugal force into the discharge.

To help you understand how pumps work and the purpose of the various parts, a section titled "Let's Build a Pump" has been included on the following pages. This material, reprinted with the permission of Allis-Chalmers Corporation, has been edited for style. Originally the material was printed in Allis-Chalmers Bulletin No. OBX62568.

Let's Build a Pump!

A student of medicine spends long years learning exactly how the human body is built before attempting to prescribe for its care. Knowledge of pump anatomy is equally basic in caring for centrifugal pumps.

But, whereas the medical student must take a body apart to learn its secrets, it will be far more instructive to us if we put a pump together (on paper, of course). Then we can start at the beginning—adding each new part as we need it in logical sequence.

As we see what each part does, how it does it... we will see how it must be cared for.

Another analogy between medicine and maintenance: there are various types of human bodies, but if you know basic anatomy, you understand them all. The same is true of centrifugal pumps. In building one basic type, we will learn about all types.

Part of this will be elementary to some maintenance people... but they will find it a valuable "refresher" course, and, after all, maintenance just cannot be too good.

So with a side glance at the centrifugal principle on page 341, let us get on with building our pump...

FIRST WE REQUIRE A DEVICE TO SPIN LIQUID AT HIGH SPEED...

That paddle-wheel device is called the "impeller" (Figure 15.2)... and it is the heart of our pump.

Note that the blades curve out from its hub. As the impeller spins, liquid between the blades is impelled outward by centrifugal force.

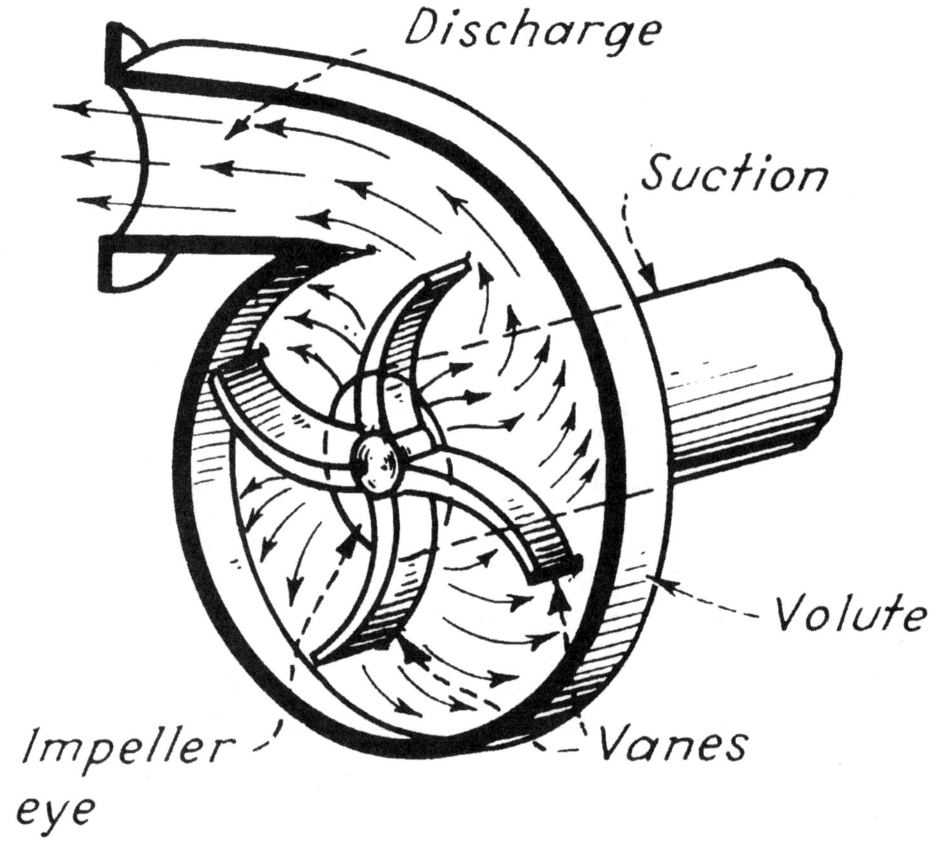

Refer to Figure 15.3 (page 347) for location of impeller in pump.

Fig. 15.2 Diagram showing details of centrifugal pump impeller
(Source: *CENTRIFUGAL PUMPS* by Karassik and Carter of Worthington Corporation)

Note, too, that our impeller is open at the center—the "eye." As liquid in the impeller moves outward, it will suck more liquid in behind it through this eye... provided it is not clogged.

Maintenance Rule No. 1: If there is any danger that foreign matter (sticks, refuse, etc.) may be sucked into the pump—clogging or wearing the impeller unduly—provide the intake end of the suction piping with a suitable screen.

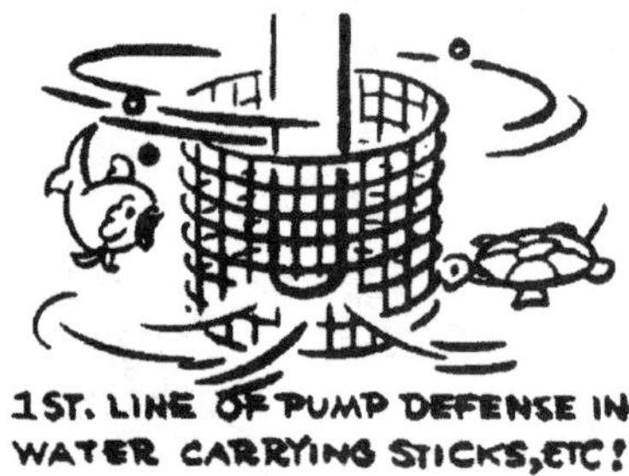

1ST. LINE OF PUMP DEFENSE IN WATER CARRYING STICKS, ETC!

NOW WE NEED A SHAFT TO SUPPORT AND TURN THE IMPELLER . . .

Our shaft looks heavy—and it is. It must maintain the impeller in precisely the right place.

But that ruggedness does not protect the shaft from the corrosive or abrasive effects of the liquid pumped... so we must protect it with sleeves slid on from either end.

CENTRIFUGAL FORCE IN ACTION--

ALL MOVING BODIES TEND TO TRAVEL IN A STRAIGHT LINE. WHEN FORCED TO TRAVEL IN A CURVE, THEY CONSTANTLY *TRY* TO TRAVEL ON A TANGENT...

...IN AN "AIRPLANE RIDE"

Centrifugal force pushes dummy planes swung in a circle *away* from center of rotation.

...IN A WHIRLPOOL

Centrifugal force tends to push swirling water *outward*... forming vortex in center.

What these sleeves—and the impeller, too—are made of depends on the nature of the liquid we are to pump. Generally, they are bronze, but various other alloys, ceramics, glass, or even rubber-coating are sometimes required.

Maintenance Rule No. 2: Never pump a liquid for which the pump was not designed.

Whenever a change in pump application is contemplated and there is any doubt as to the pump's ability to resist the different liquid, check with your pump manufacturer.

WE MOUNT THE SHAFT ON SLEEVE, BALL, OR ROLLER BEARINGS...

As we will see later, clearances between moving parts of our pump are quite small.

If bearings supporting the turning shaft and impeller are allowed to wear excessively and lower the turning units within a pump's closely fitted mechanism, the life and efficiency of that pump will be seriously threatened.

Maintenance Rule No. 3: Keep the right amount of the right lubricant in bearings at all times. Follow your pump manufacturer's lubrication instructions to the letter.

Main points to keep in mind are...

1. Although too much oil will not harm sleeve bearings, too much grease in antifriction type bearings (ball or roller) will promote friction and heat. The main job of grease in antifriction bearings is to protect steel elements against corrosion, not friction.
2. Operating conditions vary so widely that no one rule as to frequency of changing lubricant will fit all pumps. So play it safe: if anything, change lubricant *before* it is too worn or too dirty.

TO CONNECT WITH THE MOTOR, WE ADD A COUPLING FLANGE...

Some pumps are built with pump and motor on one shaft and, of course, offer no alignment problem.

But our pump is to be driven by a separate motor...and we attach a flange to one end of the shaft through which bolts will connect with the motor flange.

Use a straightedge or a dial indicator to ensure shaft alignment (see pages 431 and 432, Figures 15.53 and 15.54).

Maintenance Rule No. 4: See that pump and motor flanges are parallel vertically and axially...and that they are kept that way.

If shafts are eccentric or meet at an angle, every revolution throws tremendous extra load on bearings of both pump and motor. Flexible couplings will not correct this condition, if excessive.

Checking alignment should be a regular procedure in pump maintenance. Foundations can settle unevenly, piping can change pump position, bolts can loosen. Misalignment is a major cause of pump and coupling wear.

NOW WE NEED A "STRAW" THROUGH WHICH LIQUID CAN BE SUCKED...

Notice two things about the suction piping: (1) the horizontal piping slopes upward toward the pump; (2) any reducer that connects between the pipe and pump intake nozzle should be horizontal at the top—(eccentric, not concentric).

This up-sloping prevents air pocketing in the top of the pipe...where trapped air might be drawn into the pump and cause loss of suction.

Maintenance Rule No. 5: Any down-sloping toward the pump in suction piping (as exaggerated in the previous diagrams) should be corrected.

This rule is *VERY* important. Loss of suction greatly endangers a pump... as we will see shortly.

WE CONTAIN AND DIRECT THE SPINNING LIQUID WITH A CASING . . .

We got a little ahead of our story in the previous paragraphs... because we did not yet have the casing to which the suction piping bolts. And the manner in which it is attached is of great importance.

Maintenance Rule No. 6: See that piping puts absolutely no strain on the pump casing.

THE WEIGHT OF PIPING CAN
EASILY RUIN A PUMP!

When the original installation is made, all piping should be in place and self-supporting before connection. Openings should meet with no force. Otherwise, the casing is apt to be cracked... or sprung enough to allow closely fitted pump parts to rub.

It is good practice to check the piping supports regularly to see that loosening, or settling of the building, has not put strains on the casing.

NOW OUR PUMP IS ALMOST COMPLETE, BUT IT WOULD LEAK LIKE A SIEVE...

We are far enough along now to trace the flow of water through our pump. It is not easy to show suction piping in the cross-section view above, so imagine it stretching from your eye to the lower center of the pump.

Our pump happens to be a "double suction" pump, which means that water flow is divided inside the pump casing... reaching the eye of the impeller from either side.

LIQUID GOES IN HERE (UNDER SUCTION)
AND IT COMES OUT HERE (UNDER PRESSURE)
BUT SOME OF IT LEAKS BACK FROM PRESSURE TO SUCTION!

As water is drawn into the spinning impeller, centrifugal force causes it to flow outward... building up high pressure at the outside of the pump (which will force water out) and creating low pressure at the center of the pump (which will draw water in). This situation is diagrammed in the upper half of the pump, above.

So far so good... except that water tends to be drawn back from pressure to suction through the space between impeller and casing—as diagrammed in the lower half of the pump, above—and our next step must be to plug this leak, if our pump is to be very efficient.

SO WE ADD WEARING RINGS TO PLUG INTERNAL LIQUID LEAKAGE...

You might ask why we did not build our parts closer fitting in the first place—instead of narrowing the gap between them by inserting wearing rings (Figure 15.3, page 347).

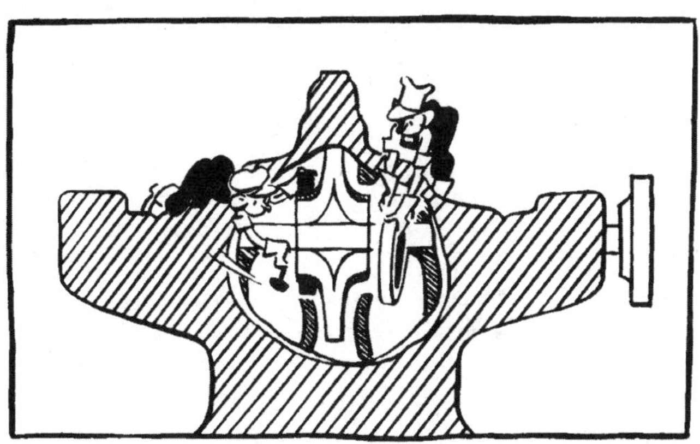

The answer is that those rings are removable and replaceable... when wear enlarges the tiny gap between them and the impeller. (Sometimes rings are attached to impeller rather than casing—or rings are attached to both so they face each other.)

PRIME
DESTRUCTIVE FRICTION AND HEAT
YOUR PUMP
A LOT DEPENDS ON MAINTAINING PRIME!

Maintenance Rule No. 7: Never allow a pump to run dry (either through lack of proper priming when starting or through loss of suction when operating). Water is a lubricant between rings and impeller.

Maintenance Rule No. 8: Examine wearing rings at regular intervals. When seriously worn, their replacement will greatly improve pump efficiency.

TO KEEP AIR FROM BEING DRAWN IN, WE USE STUFFING BOXES...

We have two good reasons for wanting to keep air out of our pump: (1) we want to pump water, not air; (2) air leakage is apt to cause our pump to lose suction.

Each stuffing box we use consists of a casing, rings of packing, and a gland at the outside end. A mechanical seal may be used instead.

Maintenance Rule No. 9: Packing should be replaced periodically—depending on conditions—using the packing recommended by your pump manufacturer. Forcing in a ring or two of new packing instead of replacing worn packing is bad practice. It is apt to displace the seal cage.

Put each ring of packing in separately, seating it firmly before adding the next. Stagger adjacent rings so the points where their ends meet do not coincide. See Figure 15.48, "How to pack a pump," on pages 420 and 421.

Maintenance Rule No. 10: Never tighten a gland more than necessary... as excessive pressure will wear shaft sleeves unduly.

Maintenance Rule No. 11: If shaft sleeves are badly scored, replace them immediately... or packing life will be entirely too short.

TO MAKE PACKING MORE AIRTIGHT, WE ADD WATER SEAL PIPING...

In the center of each stuffing box is a "seal cage." By connecting it with piping to a point near the impeller rim, we bring liquid under pressure to the stuffing box.

This liquid acts both to block out air intake and to lubricate the packing. It makes both packing and shaft sleeves wear longer... providing it is clean liquid.

WATER IS A LUBRICANT!

Maintenance Rule No. 12: If the liquid being pumped contains grit, a separate source of sealing liquid should be obtained (for example, it may be possible to direct some of the pumped liquid into a container and allow the grit to settle out).

To control liquid flow, draw up the gland just tight enough so a thin stream (approximately one drop per second) flows from the stuffing box during pump operation.

DISCHARGE PIPING COMPLETES THE PUMP INSTALLATION—AND NOW WE CAN ANALYZE THE VARIOUS FORCES WE ARE DEALING WITH...

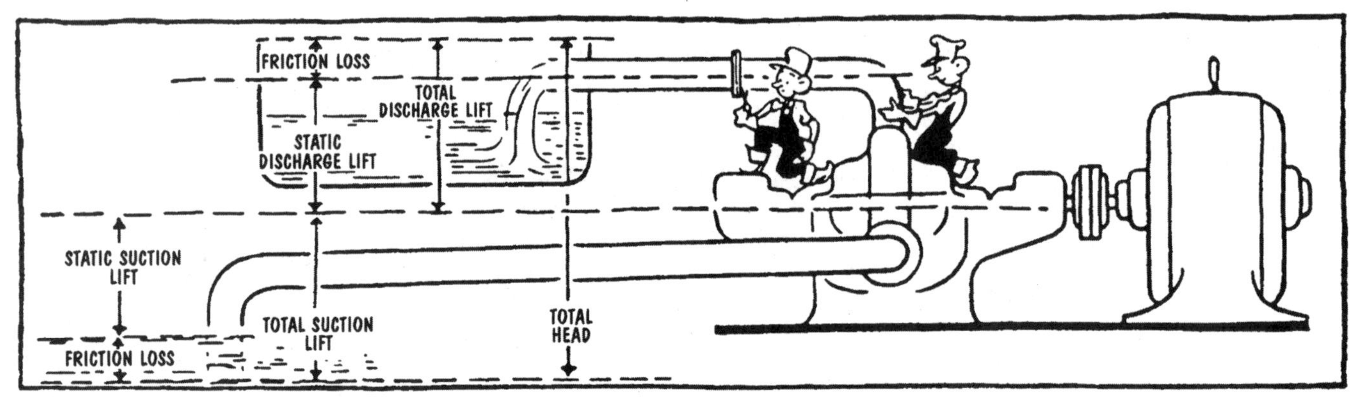

SUCTION. At least 75 percent of centrifugal pump troubles trace to the suction side. To minimize them...

1. Total suction lift (distance between centerline of pump and liquid level when pumping, plus friction losses) generally should not exceed 15 feet (4.5 meters).
2. Piping should be at least a size (diameter) larger than pump suction nozzle.
3. Friction in piping should be minimized...use as few and as easy bends as possible...avoid scaled or corroded pipe.

DISCHARGE lift, plus suction lift, plus friction in the piping from the point where liquid enters the suction piping to the end of the discharge piping equals total head.

Pumps should be operated near their rated heads. Otherwise, the pump is apt to operate under unsatisfactory and unstable conditions, which reduce efficiency and operating life of the unit, or "cavitation" could occur. Note the description of "cavitation" on page 346. Cavitation can seriously damage your pump.

PUMP CAPACITY generally is measured in gallons per minute (liters per second or cu m per second). A new pump is guaranteed to deliver its rating in capacity and head. But whether a pump retains its actual capacity depends to a great extent on its maintenance.

Wearing rings must be replaced when necessary—to keep internal leakage losses down.

Friction must be minimized in bearings and stuffing boxes by proper lubrication...and misalignment must not be allowed to force scraping between closely fitted pump parts.

POWER of the driving motor, like capacity of the pump, will not remain at a constant level without proper motor maintenance.

Starting load on motors can be reduced by throttling or closing the pump discharge valve (*NEVER* the suction valve!)...but the pump must not be operated for long with the discharge valve closed. Power then is converted into friction, overheating the water with serious consequences.

Centrifugal pumps designed for pumping wastewater (Figures 15.3 and 15.4) usually have smooth channels and impellers with large openings to prevent clogging.

Impellers may be of the open or closed type (Figure 15.5). Submersible pumps (Figure 15.6) usually have open impellers and are frequently used to pump wastewater from wet wells in lift stations.

15.111 Propeller Pumps

There are two basic types of propeller pumps (Figure 15.7), axial-flow and mixed-flow impellers. The axial-flow propeller pump is one having a flow parallel to the *AXIS*[1] of the impeller (Figure 15.8). The mixed-flow propeller pump is one having a flow that is both *AXIAL*[2] and *RADIAL*[3] to the impeller (Figure 15.8).

15.112 Vertical Wet Well Pumps

A vertical wet well pump is a vertical shaft, diffuser-type centrifugal pump with the pumping element suspended from the discharge piping (Figure 15.9). The needs of a given installation determine the length of discharge column. The pumping bowl assembly may connect directly to the discharge head for shallow sumps, or may be suspended several hundred feet for raising water from wells. Vertical turbine pumps are used to pump water from deep wells, and may be of the *SINGLE-STAGE* or *MULTISTAGE TYPE*.[4]

15.113 Reciprocating or Piston Pumps

The word "reciprocating" means moving back and forth, so a reciprocating pump is one that moves water or sludge by a piston that moves back and forth. A simple reciprocating pump is shown in Figure 15.10. If the piston is pulled to the left, Check Valve A will be open and sludge will enter the pump and fill the casing.

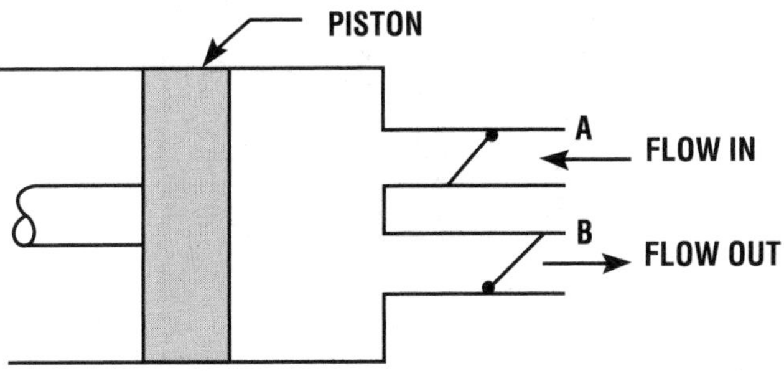

Fig. 15.10 Simple reciprocating pump
(See page 422 for pump details)

When the piston reaches the end of its travel to the left and is pushed back to the right, Check Valve A will close, Check Valve B will open, and wastewater will be forced out the exit line.

A piston pump is a positive-displacement pump. Never operate it against a closed discharge valve or the pump, valve, or pipe could be damaged by excessive pressures. Also, the suction valve should be open when the pump is started. Otherwise, an excessive suction or vacuum could develop and cause problems.

15.114 Incline Screw Pumps

Incline screw pumps (Figure 15.11) consist of a screw operating at a constant speed within a housing or trough. When the screw rotates, it moves the wastewater up the trough to a discharge point. The screw is supported by two bearings, one at the top and one at the bottom.

15.115 Progressive Cavity (Screw-Flow) Pumps (Figure 15.12)

Operation of a progressive cavity pump is similar to that of a precision incline screw pump (Figure 15.11). The progressive

[1] *Axis of Impeller.* An imaginary line running along the center of a shaft (such as an impeller shaft).
[2] *Axial to Impeller.* The direction in which material being pumped flows around the impeller or flows parallel to the impeller shaft.
[3] *Radial to Impeller.* Perpendicular to the impeller shaft. Material being pumped flows at a right angle to the impeller.
[4] Single-stage type pumps have only one impeller while multistage type pumps have more than one impeller.

HAVE A HEALTHY RESPECT FOR CAVITATION!

IF PUMP CAPACITY, SPEED, HEAD, AND SUCTION LIFT ARE NOT FIGURED PROPERLY, CAVITATION CAN EAT AN IMPELLER AWAY *FAST!* A LABORATORY WATER HAMMER INDICATES ITS EROSIVE FORCE . . .

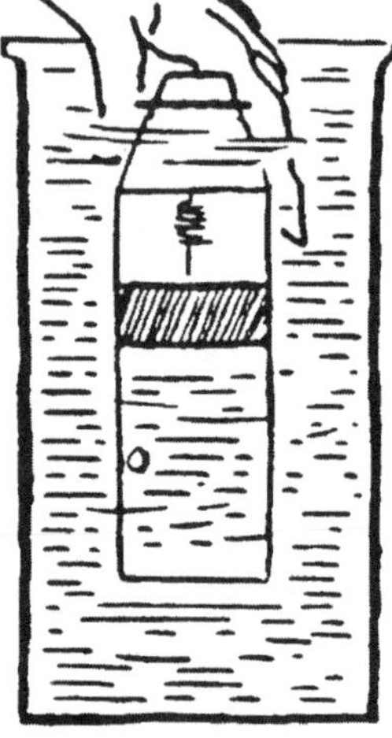

1 VESSEL FILLED WITH WATER DROPPED TO BOTTOM OF TANK

BRASS PLATE

2 MOMENTUM OF WATER & WEIGHT PRODUCE CAVITY BENEATH INSERTED BRASS PLATE

3 PRESSURE CLOSES CAVITY - WATER PUNCHES HOLE IN BRASS PLATE!

1 FAST MOVEMENT OF IMPELLER BLADE THROUGH WATER ...

2 PRODUCES CAVITY BEHIND BLADE ...

3 LOCAL PRESSURE INCREASE DRIVES WATER INTO METAL WITH TERRIFIC FORCE.

MORAL: BE SURE YOUR HEAD IS RIGHT FOR YOUR PUMP!

Cavitation is a condition that can cause a drop in pump efficiency, vibration, noise, and rapid damage to the impeller of a pump. Cavitation occurs due to unusually low pressures within a pump. These low pressures can develop when pump inlet pressures drop below the design inlet pressures or when the pump is operated at flow rates considerably higher than design flows. When the pressure within the flowing water drops very low, the water starts to boil and vapor bubbles form. These bubbles then collapse with great force, which knocks metal particles off the pump impeller. This same action can and does occur on pressure-reducing valves and partially closed gate and butterfly valves.

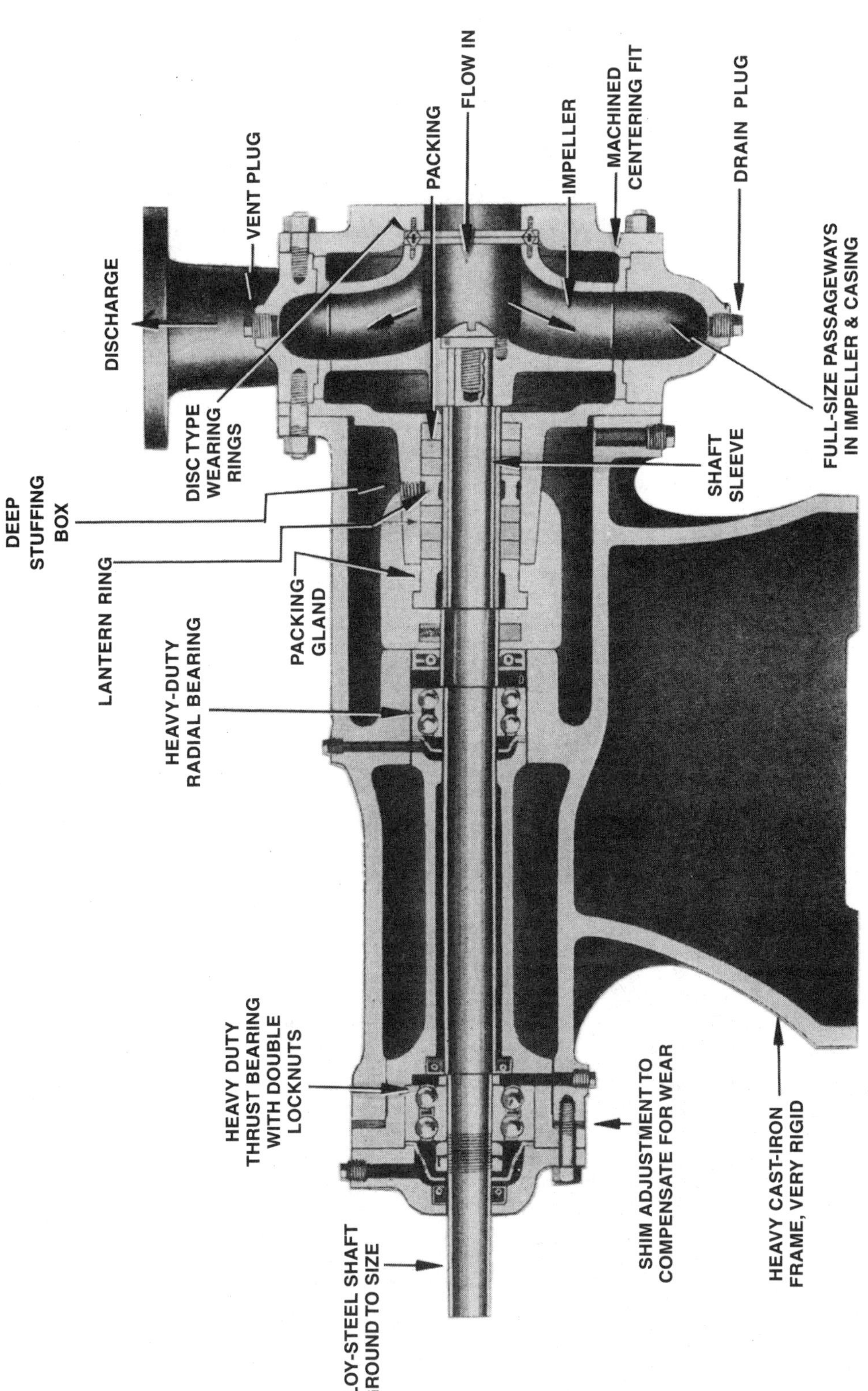

Fig. 15.3 Horizontal nonclog wastewater pump with open impeller
(Source: War Department Technical Manual TM5-666)

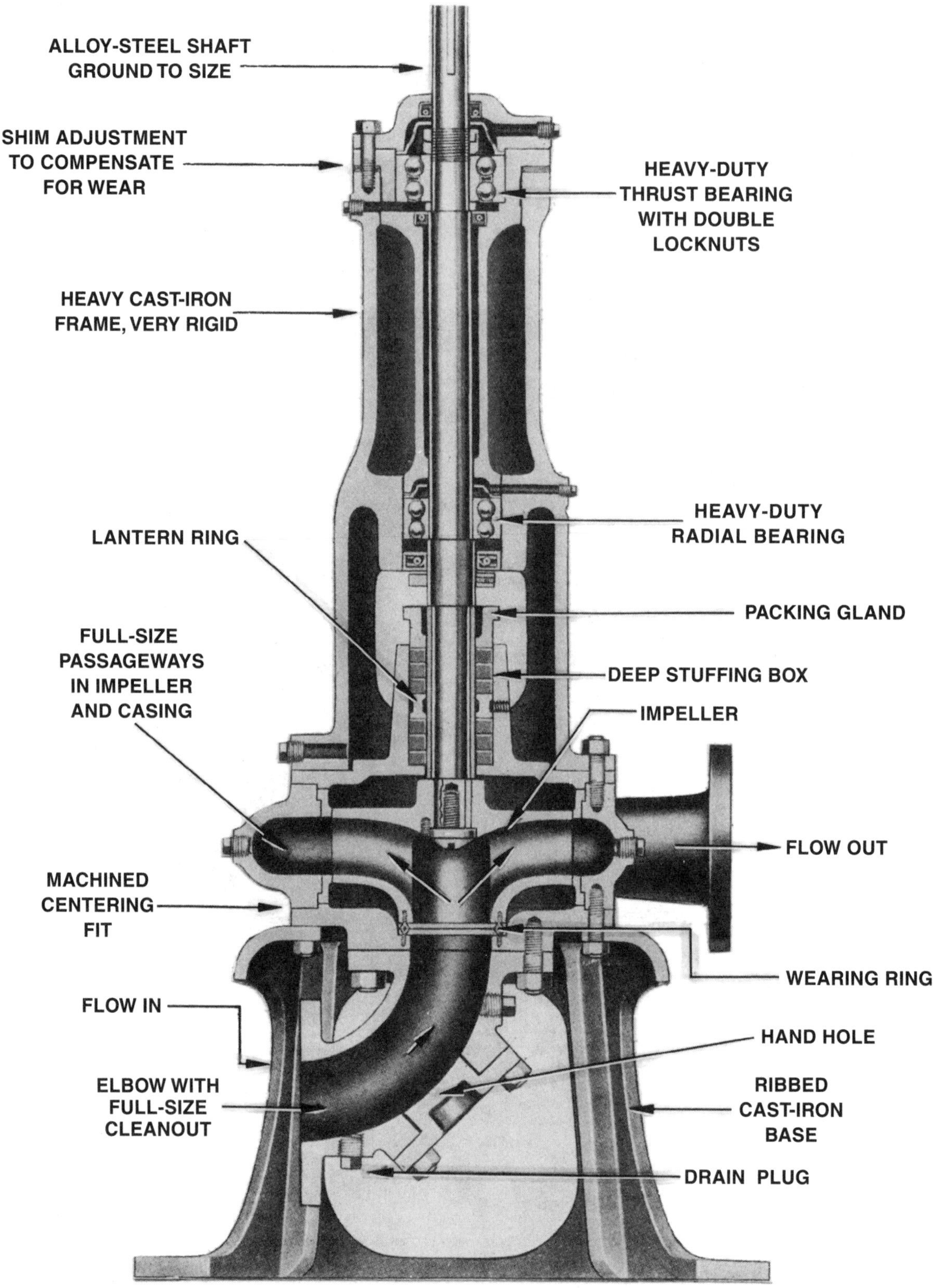

Fig. 15.4 Vertical ball bearing-type wastewater pump

(Source: War Department Technical Manual TM5-666)

Closed Radial
(Closed radial impellers are used in wastewater treatment plants.)

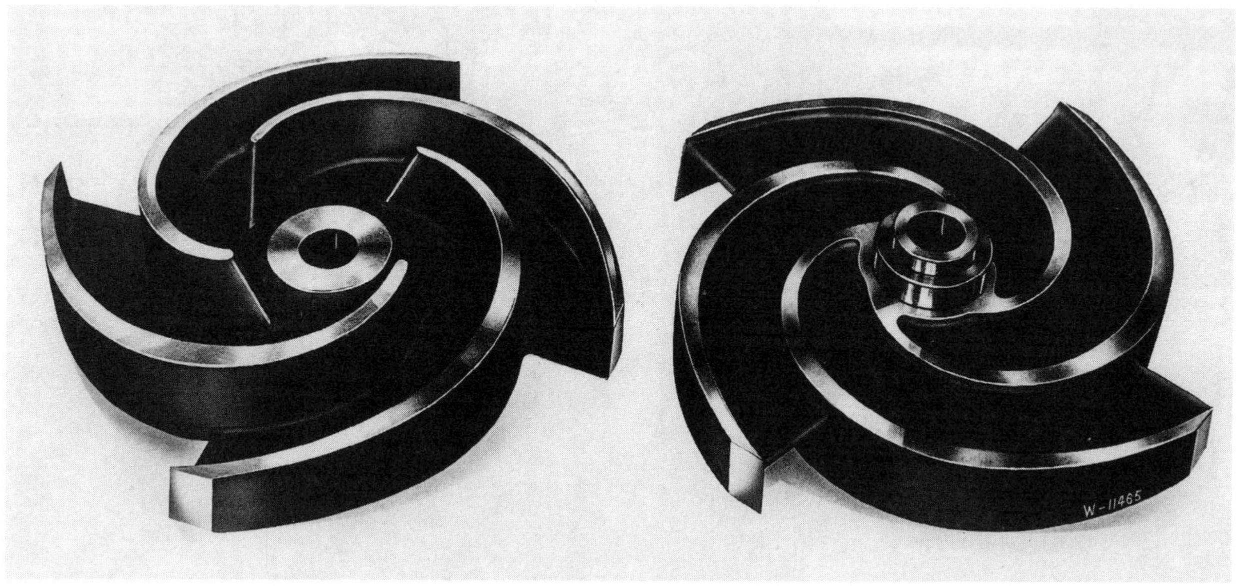

Open Radial

Fig. 15.5 Impellers
(Source: *CENTRIFUGAL PUMPS* by Karassik and Carter of Worthington Corporation)

1. LIFTING HANDLE
2. JUNCTION CHAMBER WITH WATERTIGHT CABLE ENTRIES
3. ANTIFRICTION BEARINGS
4. SHAFT
5. STATOR WITH TEMPERATURE SENSING THERMISTORS
6. ROTOR
7. STATOR HOUSING LEAKAGE SENSOR
8. BEARING TEMPERATURE THERMISTOR
9. SHAFT SEAL
10. OIL CHAMBER
11. VOLUTE
12. NONCLOG IMPELLER
13. COOLING JACKET
14. SLIDING BRACKET
15. AUTOMATIC DISCHARGE CONNECTION

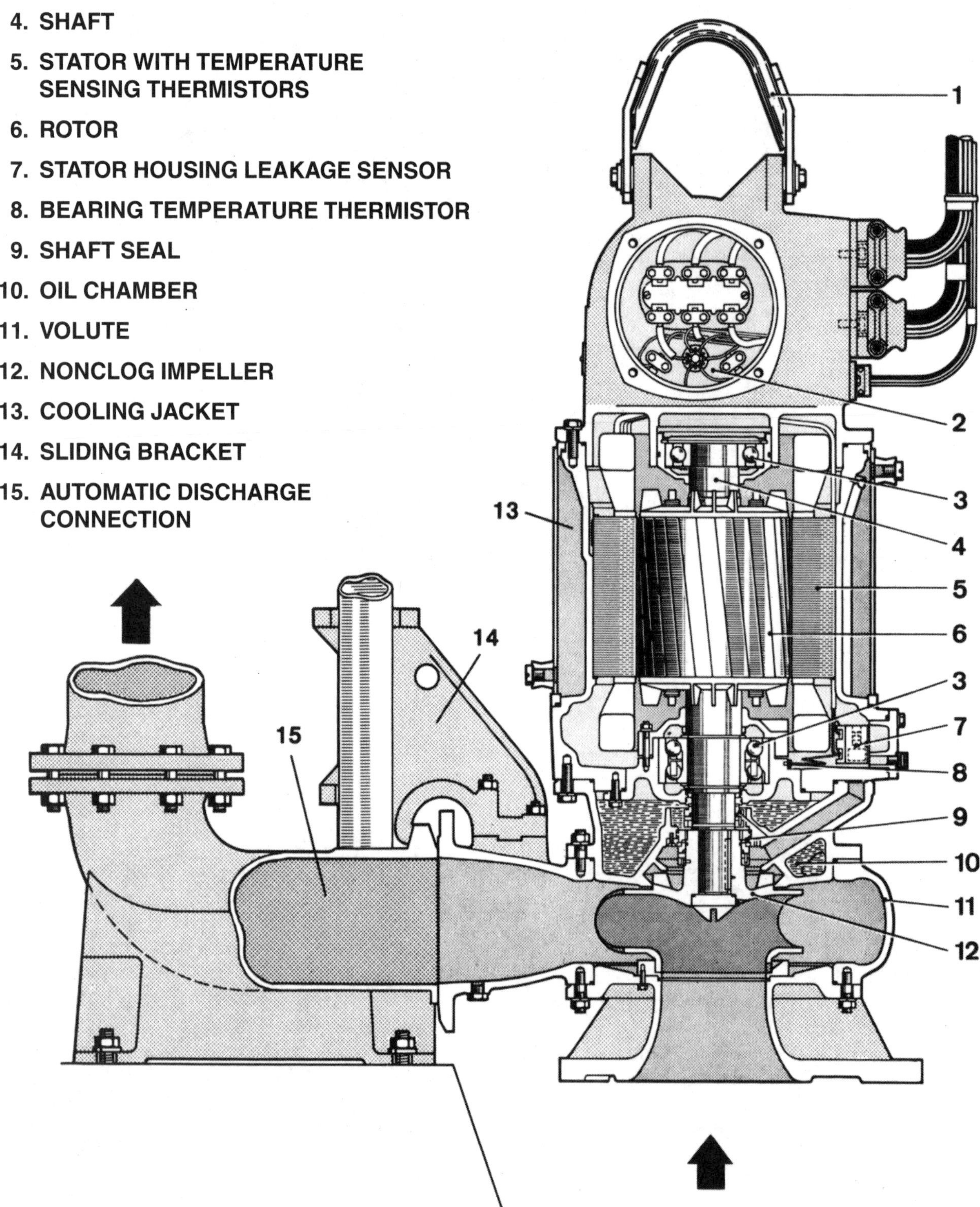

Fig. 15.6 Submersible wastewater pump

(Courtesy of Flygt Corporation)

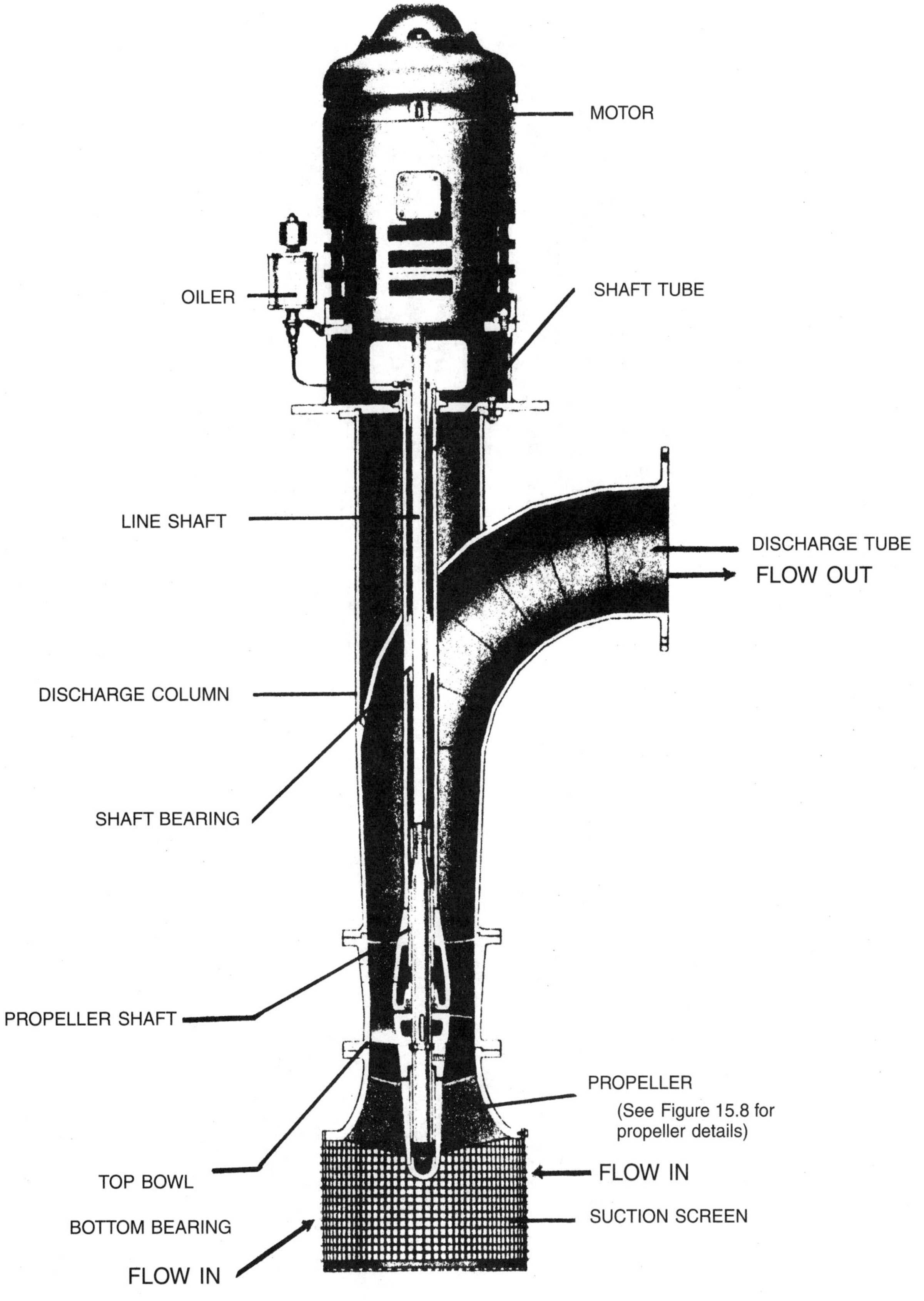

Fig. 15.7 Propeller pump

Axial-Flow

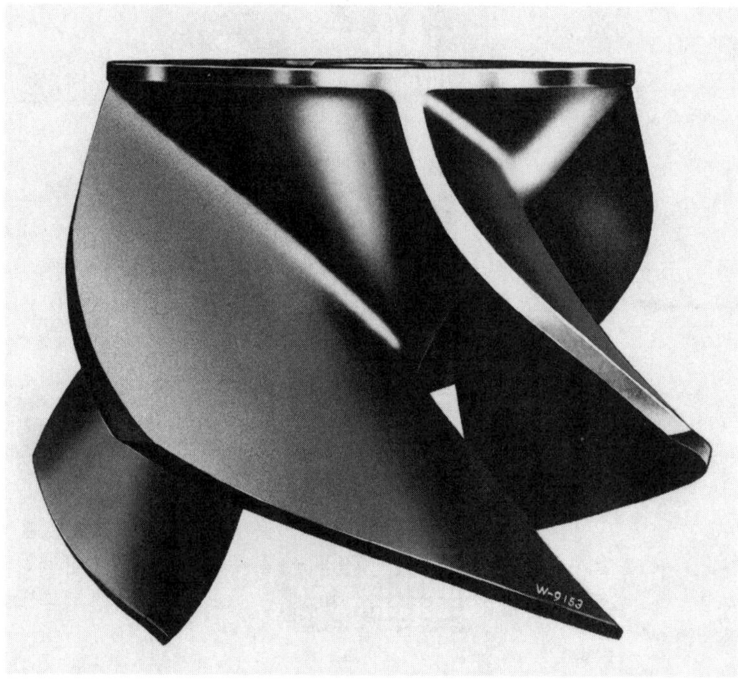

Mixed-Flow

Fig. 15.8 Propeller-type impellers

(Source: *CENTRIFUGAL PUMPS* by Karassik and Carter of Worthington Corporation)

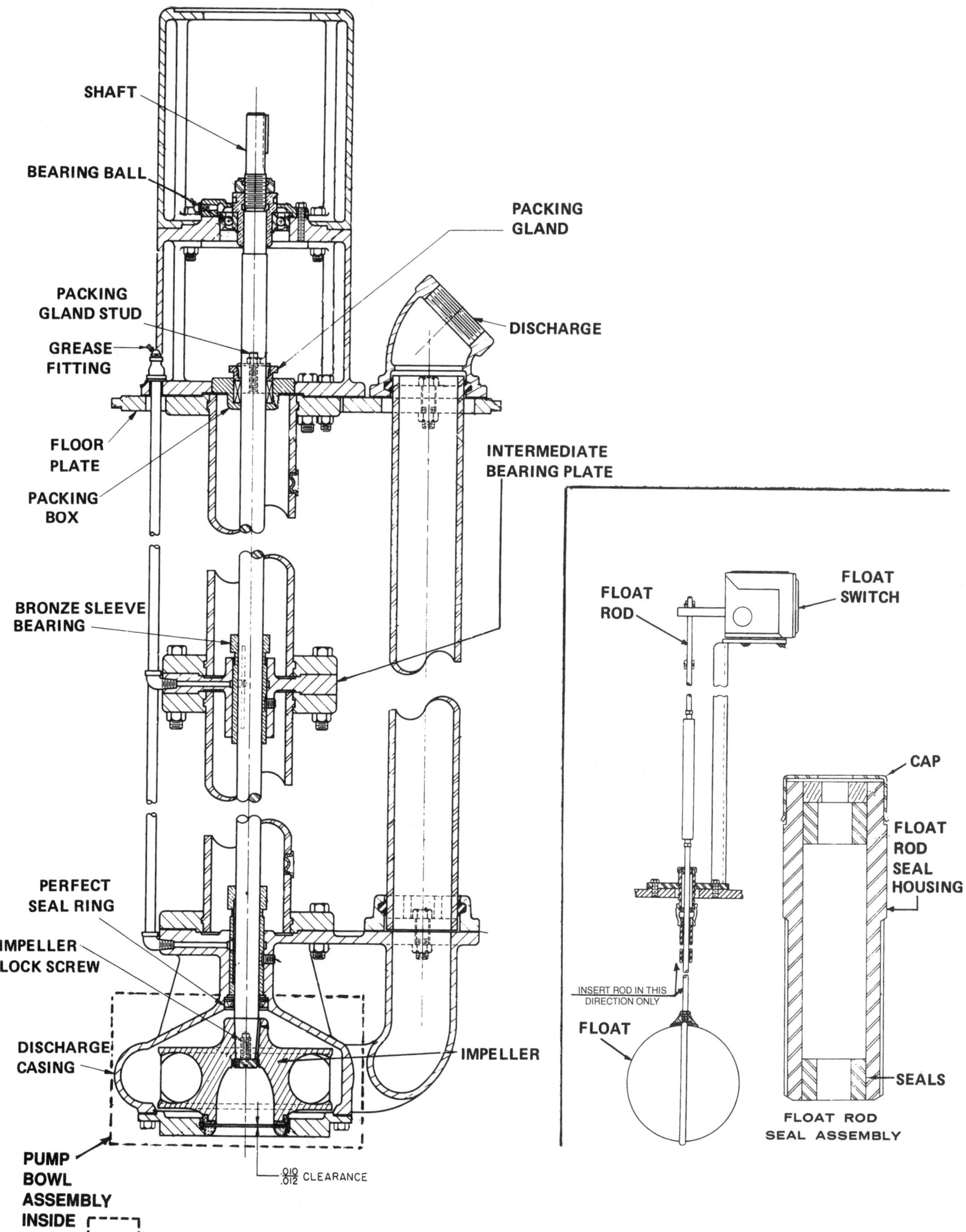

Fig. 15.9 Vertical wet well pump (motor not shown)

(Courtesy Chicago Pump)

Note: Figure 15.10 is on page 345

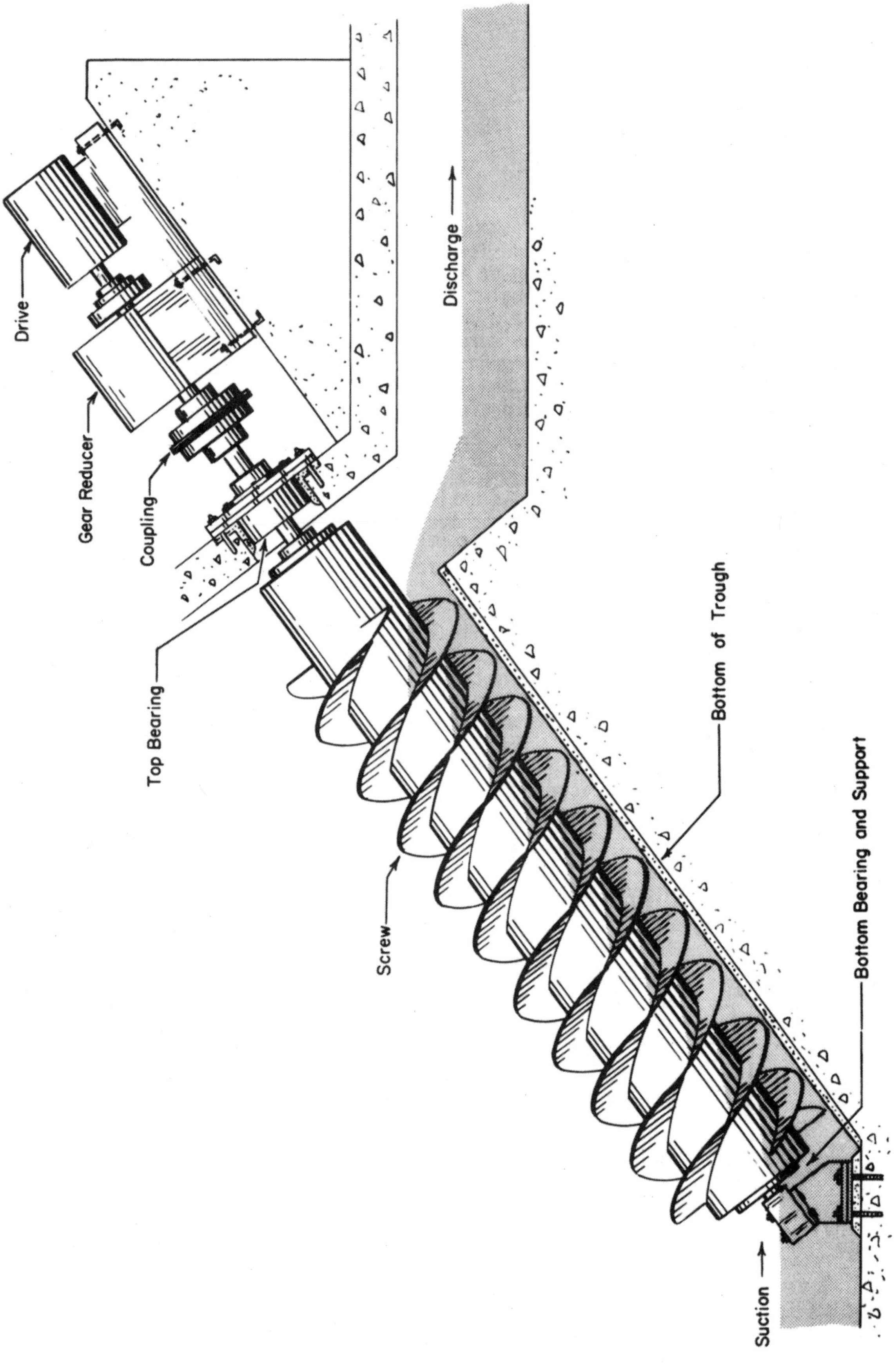

Fig. 15.11 Incline screw pump
(Courtesy of FMC Corporation, Environmental Equipment Division)

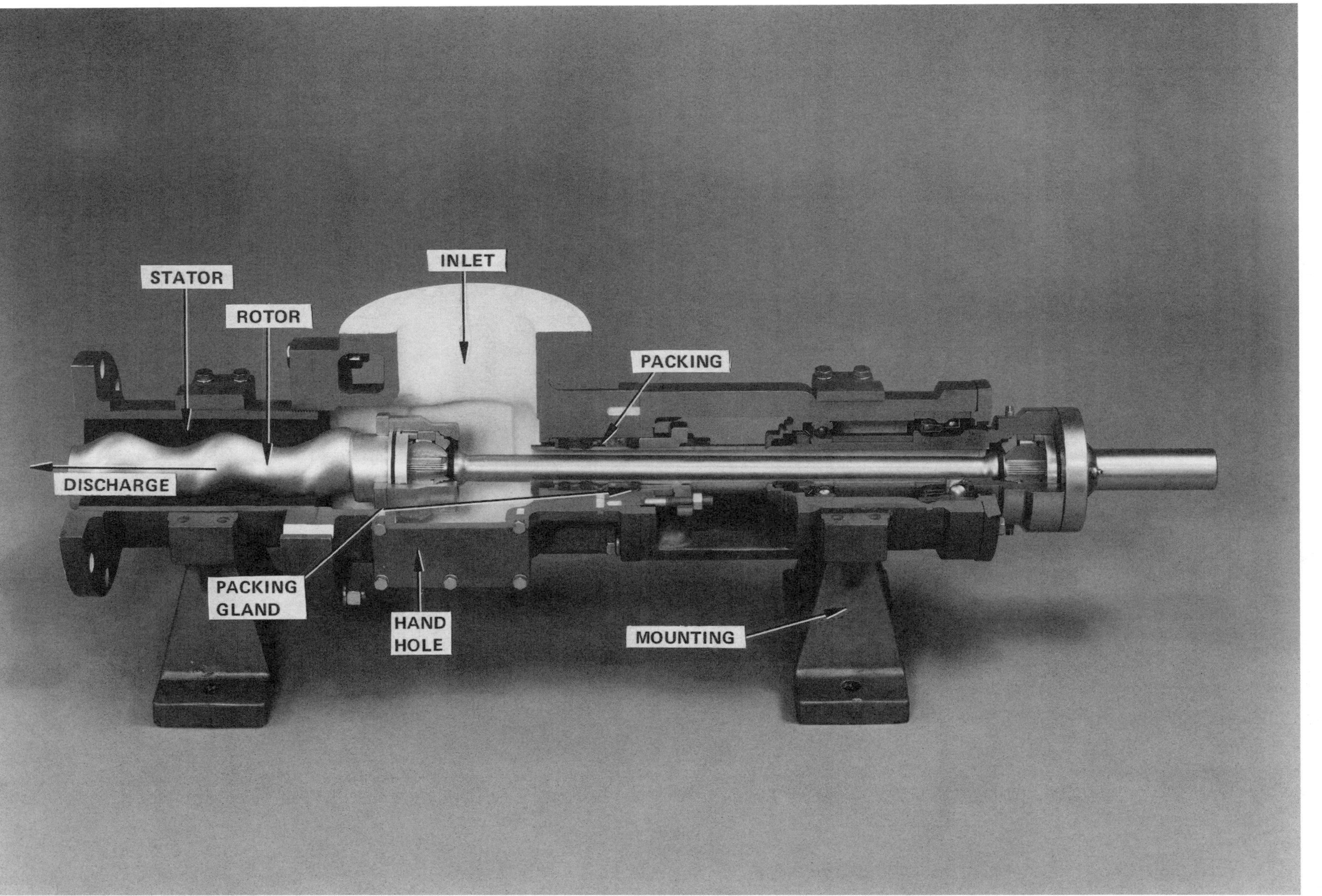

Fig. 15.12 Progressive cavity (screw-flow) pump
(Permission of Moyno Pump Division, Robbins & Meyer, Inc.)

cavity pump consists of a screw-shaped rotor snugly enclosed in a nonmoving *STATOR*[5] or housing (Figure 15.13). The threads of the screw-like rotor (commonly manufactured of chromed steel) make contact along the walls of the stator (usually made of synthetic rubber). The gaps between the rotor threads are called "cavities." When wastewater is pumped through an inlet valve, it enters the cavity. As the rotor turns, the waste material is moved along until it leaves the conveyor (rotor) at the discharge end of the pump. The size of the cavities along the rotor determines the capacity of the pump.

All progressive cavity pumps operate on the basic principle described above. To further increase capacity, some models have a shaped inside surface of the stator (housing) with a similarly shaped rotor. In addition, some models use a rotor that moves up and down inside the stator as well as turning on its axis (Figure 15.12). This allows a further increase in the capacity of the pump.

These pumps are recommended for materials that contain higher concentrations of suspended solids. They are commonly used to pump sludges. Progressive cavity pumps should never be operated dry (without liquid in the cavities), nor should they be run against a closed discharge valve.

15.116 Pneumatic Ejectors (*Figure 15.14*)

Pneumatic ejectors are used when it is necessary to handle limited flows. Centrifugal pumps are highly efficient for pumping large flows; however, when scaled down for lower flows, they tend to plug easily. Unstrained solids will tend to block the small impeller opening (if less than two inches (5.1 cm)) of a centrifugal pump and will quickly reduce the flow. Pneumatic ejectors, on the other hand, are capable of passing solids up to the size of the inlet and discharge valves, and there is nothing on the inside of the ejector-receiver to restrict the flow. The ejector may be useless, however, if a stick or another object gets stuck under the inlet or discharge check valve preventing it from closing.

15.12 Pump Lubrication, Seals, and Bearings

15.120 Lubrication

Pumps, motors, and drives should be oiled and greased in strict accordance with the recommendations of the manufacturer. Cheap lubricants may often be the most expensive in the end. Oil should not be put in the housing while the pump shaft is rotating because the rotary action of the ball bearings will pick up and retain a considerable amount of oil. When the unit comes to rest, an overflow of oil around the shaft or out of the oil cup will result.

15.121 Mechanical Seals (*Figure 15.15*)

Many pumps use mechanical seals in place of packing. Mechanical seals serve the same purpose as packing; that is, they prevent leakage between the pump casing and shaft. Like packing, they are located in the stuffing box where the shaft goes through the volute; however, they should not leak.

Mechanical seals have two faces that mate tightly and prevent water from passing through them. One half of the seal is mounted in the pump or gland with an O-ring or gasket, thus providing sealing between the housing and seal face. This prevents water from going around the seal face and housing. The other half of the mechanical seal is installed on the pump shaft. This part also has an O-ring or gasket between the shaft and seal to prevent water from leaking between the seal part and shaft. There is a spring located behind one of the seal parts that applies pressure to hold the two faces of the seal together and keeps any water from leaking out. One half of the seal is stationary and the other half is revolving with the shaft.

Materials used in the manufacture of mechanical seal parts are carbon, stainless steel, ceramic, tungsten carbide, brass, and many others. The different type materials are selected for their best application. Some of the variables are:

1. Liquid and solids being pumped
2. Shaft speed
3. Temperature
4. Corrosion resistance
5. Abrasives

Initially, mechanical seals are more expensive than packing to install in a pump. This cost is gained back in maintenance savings during a period of time.

Some of the advantages of mechanical seals are as follows:

1. They last from three to four years without having to touch them, resulting in labor savings.
2. Usually, there is not any damage to the shaft sleeve when they need replacing (expensive machine work is not needed).
3. Continual adjusting, cleaning, or repacking is not required.
4. The possibility of flooding a lift station because a pump has thrown its packing is eliminated; however mechanical seals can fail and flood a lift station too.

Some of the limitations of mechanical seals are as follows:

1. High initial cost.
2. Competent mechanic required for installation.
3. When they fail, the pump must be shut down.
4. Pump must be dismantled to repair.

Mechanical seals are always flushed in some manner to lubricate the seal faces and minimize wear. This may be the liquid being pumped and is referred to as "source lubrication." If fresh water is used, it is connected from the high-pressure side and back to the stuffing box low-pressure side. In some lift stations where fresh water is not available, the wastewater is used in this manner but it must be filtered. Another method is to use fresh water from an air gap system. Still another way of lubricating the seal is to use a spring-loaded grease cup.

Whatever method is used, the mechanical seal must be inspected frequently. The seal water is adjusted to five psi (0.35 kg/sq cm) above maximum discharge pressure to keep the

[5] *Stator.* That portion of a machine that contains the stationary (nonmoving) parts that surround the moving parts (rotor).

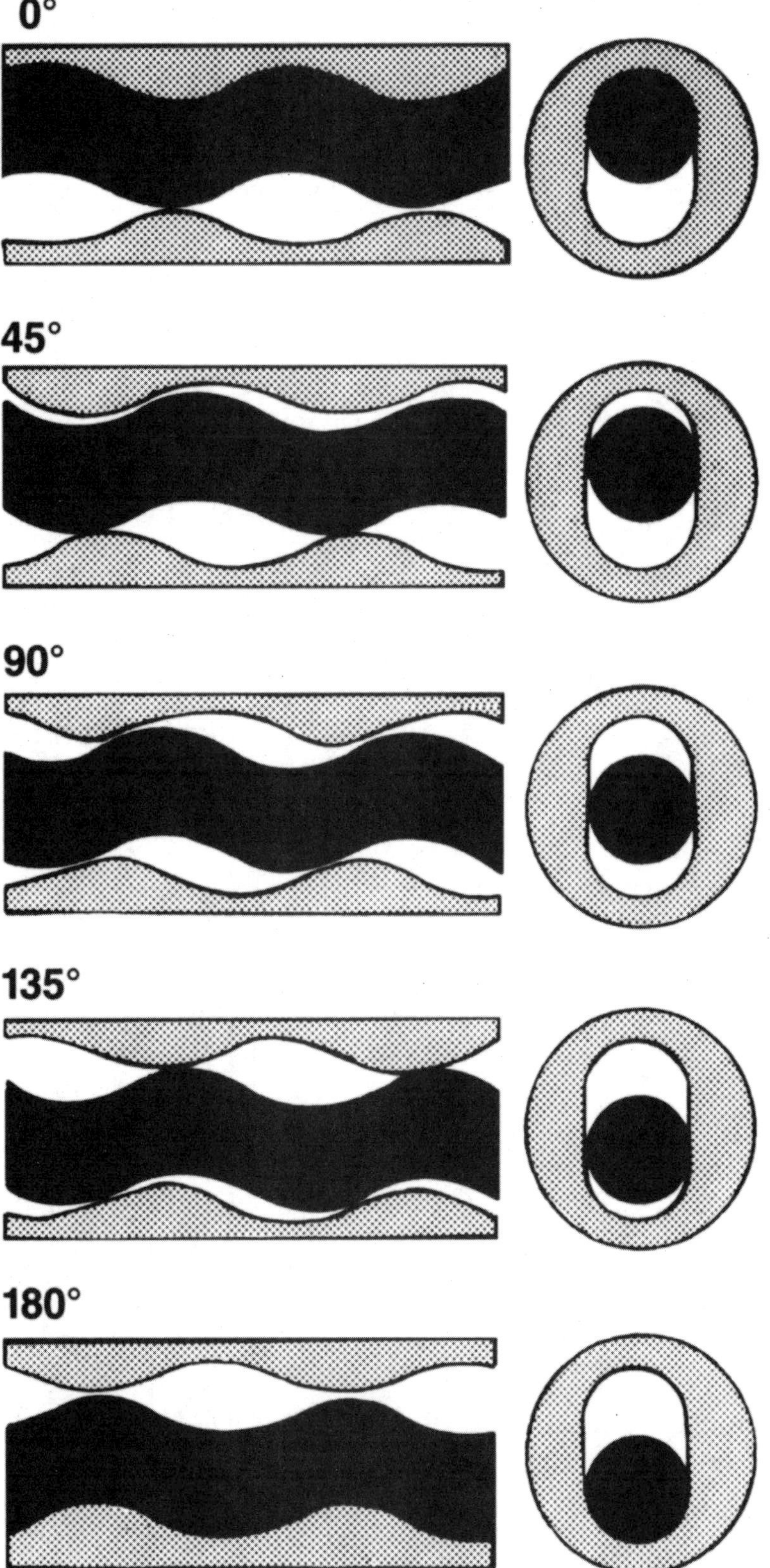

Fig. 15.13 Pumping principle of a progressive cavity pump

(Permission of Allweiler Pumps, Inc.)

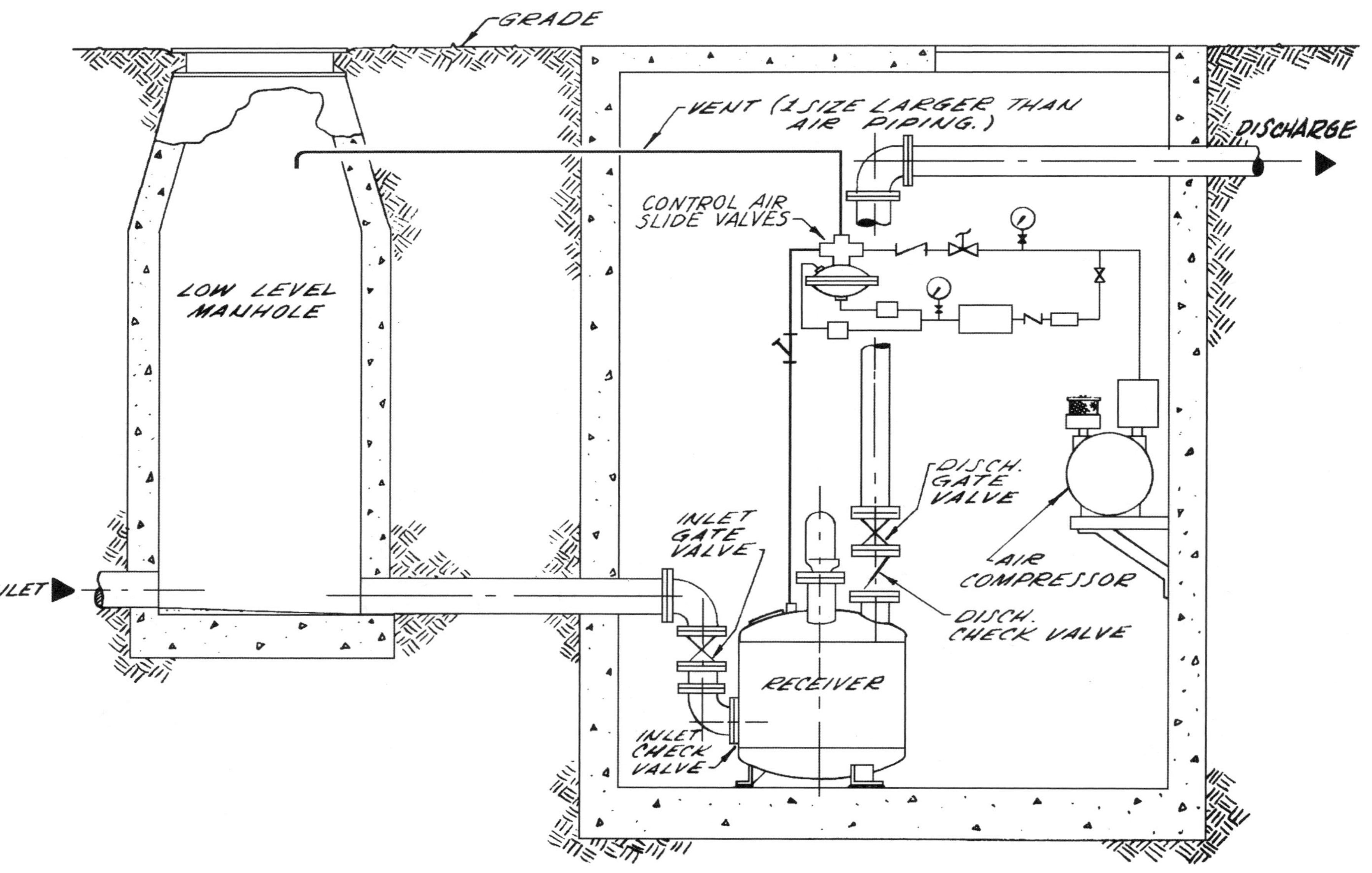

Fig. 15.14 Pneumatic ejector system
(Courtesy of James Equipment and Manufacturing Company)

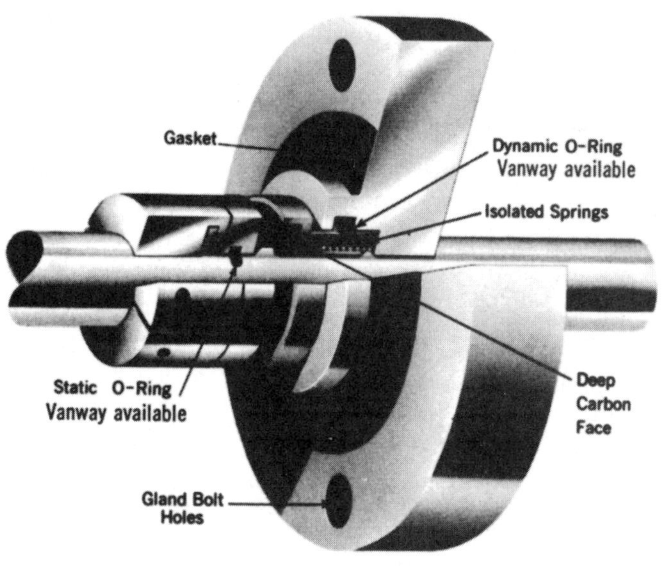

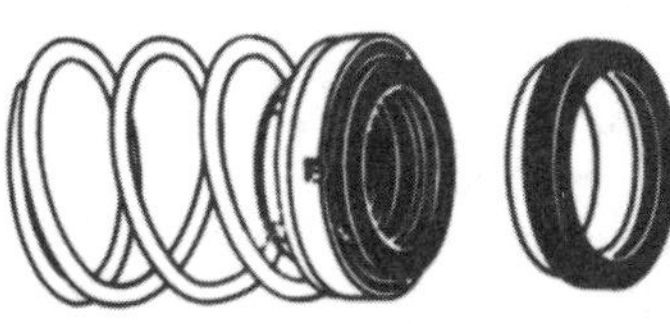

Fig. 15.15 Mechanical seals

(Courtesy A. W. Chesterton Co.)

wastewater and grit from entering the seal housing and contaminating the seal faces. Grease cups must be kept full at all times and inspected to make sure they are operating properly.

When a pump is fitted with a mechanical seal, it must never run dry or the seal faces will be burned and ruined. Mechanical seals are not supposed to have any leakage from the gland. If a leak develops, the seal may require resurfacing or it may have to be replaced.

Repair or replacement of mechanical seals requires the pump to be removed and dismantled. Seals are quite delicate and special care must be taken when installing them. Mechanical seals differ widely in their construction and installation, and individual manufacturer's instructions must be followed. Due to the complexity of installing mechanical seals, special tools and equipment are needed as well as a qualified pump mechanic.

The following is a checklist for cost-effective installation and use of mechanical seals:

1. Use cartridge-mounted seals. They are easy to install and do not damage either the seal or the pump.
2. Purchase standard, off-the-shelf seals to ensure seals will be available from suppliers, even in an emergency.
3. Return seals to the original manufacturer for repair.
4. Obtain bearing isolators (labyrinths) or magnetic seals to protect the bearing from intrusion of moisture or debris.
5. Use synthetic lubricants in all pumps, motors, bearings, and compressors. Synthetic lubricants are very cost-effective because they lower the operating temperature, tolerate water intrusion better, lower high-frequency vibration, and last two or three times longer than the low-priced mineral oils.
6. Every time the motor or pump is worked on, be sure to perform precision coupling alignment by using a dial indicator or laser to ensure long-term service life of bearings and mechanical seals.

15.122 Bearings

Pump bearings usually should last for years if serviced properly and used in their proper application. There are several types of bearings used in pumps such as ball bearings, roller bearings, and sleeve bearings. Each bearing has a special purpose, such as thrust load, radial load, and speed. The type of bearing used in each pump depends on the manufacturer's design and application. Whenever a bearing failure occurs, the bearing should be examined to determine the cause and, if possible, to eliminate the problem. Many bearings are ruined during installation or start-up. Bearing failures may be caused by:

1. Fatigue failure
2. Contamination
3. *BRINELLING*[6]
4. False brinelling
5. Thrust failures
6. Misalignment
7. Electric arcing
8. Lubrication failure
9. Cam failure

15.13 Starting a New Pump

The initial start-up work described in this paragraph should be done by a qualified and trained person, such as a manufacturer's representative, consulting engineer, or an experienced operator. The operator can learn a lot about pumps and motors by accompanying and helping an equipment representative put new equipment into operation.

Before starting a pump, lubricate it according to the lubrication instructions. Turn the shaft by hand to see that it rotates freely. Then check to see that the shafts of the pump and motor are aligned and the flexible coupling adjusted. (Refer to Section 15.7, Paragraph 11, "Couplings," page 429.) If the unit is belt driven, sheave (pulley) alignment and belt adjustment should be checked. (Refer to Paragraph 8, "Belt Drives.")

Check the electric current characteristics with the motor characteristics and inspect the wiring. See that thermal units in the starter are set properly. Turn on the motor just long enough to see that it turns the pump in the direction indicated by the rotational arrows marked on the pump. If separate water seal units or vacuum primer systems are used, these should be started. Finally, make sure lines are open. Sometimes, there is an exception (see following paragraph) in the case of the discharge valve.

A pump should not be run without first having been primed. To prime a pump, the pump must be completely filled with water or wastewater. In some cases, automatic primers are provided. If they are not, it is necessary to vent the casing. Most pumps are provided with a valve to accomplish this. Allow the trapped air to escape until water or wastewater flows from the vent; then replace the vent cap. In the case of suction lift applications, the pump must be filled with water unless a self-primer is provided. In nearly every case, you may start a pump with the discharge valve open. Exceptions to this, however, are where water hammer or velocity disturbances might result, or where the motor does not have sufficient margin of safety or power. Sometimes, there are no check valves in the discharge line. In this case (with the exception of positive displacement pumps), it is necessary to start the pump and then open the discharge lines. Where there are common discharge headers, it is essential to start the pump and then open the discharge valve. A positive displacement pump (reciprocating or progressive cavity types) should never be operated against a closed discharge line.

After starting the pump, again check to see that the direction of rotation is correct. Packing gland boxes (stuffing boxes) should be observed for slight leakage (approximately 60 drops per minute), as described in Paragraph 1, "Pumps, General (Including Packing)." Check to see that the bearings do not overheat from over- or underlubrication. The flexible coupling should not be noisy; if it is, the noise may be caused by misalignment or improper clearance or adjustment. Check to be sure pump anchorage is tight. To find out if a new pump is delivering design flows and pressures, measure the flows and

---

[6] *Brinelling* (bruh-NEL-ing). Tiny indentations (dents) high on the shoulder of the bearing race or bearing. A type of bearing failure.

pressures and compare them with the pump performance curves supplied by the manufacturer. (See Volume I, Appendix, "How to Solve Wastewater Treatment Plant Arithmetic Problems," Section A.8, *PUMPS*.) If pump delivery falls below performance curves, look for obstructions in the pipelines. Inspect piping for leaks.

15.14 Pump Shutdown

When shutting down a pump for a long period, the motor disconnect switch should be opened, locked out, and tagged with the reason for the shutdown noted on the tag. All valves on the suction, discharge, and water seal lines should be tightly shut. Completely drain the pump by removing the vent and drain plugs. Do not permit sludge to remain in pumps or piping for any length of time; there are cases on record in which the gas produced has created sufficient pressure to rupture pipes and sludge pumps.

Inspect the pump and bearings thoroughly so that all necessary servicing may be done during the inactive period. Drain the bearing housing and then add fresh lubricant. Follow any additional manufacturer's recommendations.

15.15 Pump-Driving Equipment

Driving equipment used to operate pumps includes electric motors and internal combustion engines. In rare instances, pumps are driven with steam turbines, steam engines, air, and hydraulic motors.

In all except the large installations, electric motors are used almost exclusively, with synchronous and induction types being the most commonly used. Synchronous motors operate at constant speeds and are used chiefly in large sizes. Three-phase, squirrel-cage induction motors are most often used in treatment plants. These motors require little attention and, under average operating conditions, the factory lubrication of the bearing will last approximately one year. (Check with the manufacturer for average number of operating hours for bearings.) When lubricating motors, remember that too much grease may cause bearing trouble or damage the winding.

Clean and dry all electrical contacts. Inspect for loose electrical contacts. Only qualified and authorized electricians should be permitted to work on electrical equipment and circuitry. Make sure that hold-down bolts on motors are secure. Check voltage while the motor is starting and running. Examine bearings and couplings.

15.16 Electrical Controls

A variety of electrical equipment is used to control the operation of wastewater pumps or to protect electric motors. The simplest type of control unit consists of a counterweighted float that triggers a switch. When the float is raised by the wastewater to a predetermined level, a switch is tripped to start the pump. When the wastewater level falls to the cutoff level, the float switch stops the pump. The time required for each cycle and the length of time between cycles depend on the pumping rate and the quantity of wastewater flow.

If starters, disconnect switches, and cutouts are used, they should be installed in accordance with the local regulations (city or county codes) regarding this equipment. In the case of larger motors, the power company often requires starters that do not overload the power lines.

The electrode-type, bubbler-type, and diaphragm-type water level control systems are all similar in effect to the float-switch system. Scum is a problem with most water level controls that operate pumps and it must be removed on a regular basis.

QUESTIONS

Write your answers in a notebook and then compare your answers with those on page 460.

15.1A Where would you find out how to lubricate a pump?

15.1B Why should a pump with a mechanical seal never be allowed to run dry?

15.1C What should be done when a mechanical seal develops a leak?

15.1D What problems can develop if too much grease is used in lubricating a motor?

15.17 Variable-Speed, Alternating Current (AC) Motors
by Ken Peschel

15.170 Description

Many of today's collection system pumping stations incorporate a wet well design that uses variable-speed pumping equipment. Lower building costs and the smaller (or nonexistent) wet well both contribute to the growing popularity of this type of installation. A variable-speed pump facility can maintain a predetermined level of flow in the incoming line during minimal flows in dry weather as well as during maximum flows in the peak wet weather season. Still, all components of this type of facility require a set preventive maintenance program to avoid costly breakdowns. Experience has shown that the motors themselves are most often overlooked during preventive maintenance.

A standard three-phase, single-speed, synchronous AC motor can be satisfactorily maintained with a standard preventive maintenance program consisting primarily of lubrication, cleaning, and testing of electric circuits. In contrast, a standard variable-speed, three-phase AC motor incorporates three copper or copper alloy slip rings or "collectors" attached to the shaft of a wound rotor. These three slip rings receive a secondary electrical current through sets of carbon brushes from the rotor windings. The slip rings and brushes on such a motor require additional preventive maintenance procedures.

To properly maintain a standard three-phase, variable-speed, synchronous AC motor, you must have some idea of what to look for when examining the slip rings and brushes. Examine these components before initial start-up, if you possibly can. You must also keep a close watch during the "filming" or seating period immediately after start-up; this may take anywhere from a few hours to several hundred hours, but proper formation of the film is crucial to the operation of the motor.

The film that forms on the slip rings during operation is the conducting medium by which the electric current is carried from the carbon brush to the metal slip ring. Without it, satisfactory operation of the sliding contact is impossible. This film is a form of corrosion caused by the chemical reaction between the metals and the atmosphere. The first layers protect the metal and the rate of film growth will decrease with time. This film is actually an oxide that starts forming when the ring surface is machined. Initial oxidation takes place in a matter of seconds and will have a thickness of about 20 molecules of oxide. This film must be deposited uniformly on the ring by some means throughout the life of the ring and brush. Film formation is accelerated by the heat of machining the ring surface, seating of brushes, friction while running, electrical load, and by mechanical burnishing during operation.

A uniform film of oxidation is essential to the flow of electricity, but it serves another purpose as well by protecting the metal surfaces from wear. The powerful forces of attraction cause rapid wear of the surfaces. In some cases, these forces may cause seizure and result in motor failure. Fortunately, these forces are effective only over short distances, and a boundary film separating and lubricating the components is sufficient to prevent seizure and wear.

The brush is the other main component of the variable-speed motor that needs periodic care and attention. The four basic types of brush composition are: carbongraphite, electrographite, graphite, and metal graphite. Complete details of these various types together with characteristics, shunts, connections, and shunt insulation, may be obtained from brush manufacturers' publications.

It is extremely important to maintain a regular examination and cleaning schedule for all types of graphite brushes. The normal wear of brushes may cause a buildup of conductive particles along the creepage paths of the winding, the slip ring assembly, and the brushholder assembly. Some particles will be carried to various parts of the machine by ventilating air. They may clog the vent ducts, or settle between the brush and the pocket causing the machine to overheat or the brushes to bind. The buildup of particles could contribute to ring threading or grooving, as well as sparking, burning, and flashovers if they are allowed to accumulate. The importance of keeping these areas clean to avoid flashovers and breakdown cannot be overemphasized.

Satisfactory operation of brushes on AC slip rings requires that mechanical conditions be as perfect as possible. Intimate contact between the two surfaces must be maintained; the brushholder design and assembly are vital in this respect. The brushholder assembly normally provides for two or more brushes, which are mounted radially with respect to the ring surface. When mounting the brushholder assembly, position the brushes on the center of the slip ring. The holder should clear the ring by approximately $^1/_{16}$ to $^1/_8$ inch (1.6 to 3.2 mm) when the rings are new, and not over $^1/_4$ inch (6.4 mm) as the rings wear. The brushholder must have solid, rigid mountings to minimize vibration.

The brushholder assembly also provides a means of maintaining the proper pressure between the ring and brush surface—a most important factor in brush and ring performance. The rate of brush wear varies with the brush pressure. At light pressures, electrical wear is dominant because the brush can jump off the rings, sparking occurs, and the filming action on the ring becomes erratic. At higher pressures, mechanical wear is dominant because of high friction losses, needless heating, and needless abrasion. Check the machine manufacturer's recommended pressure for the type and grade of brush you are using.

Pressures of all brushes on a given ring must be as equal as possible. Check the brush pressures at regular intervals as the brushes wear to be sure that pressures stay equal at the proper value. If constant-pressure springs are used, adjustment may not be necessary. However, springs sometimes lose their temper (resilience) and require replacing—another reason to schedule regular maintenance.

It is also important to remember that the brushholders respond to machine vibration caused by any unbalanced condition. Brushholder mountings cannot eliminate the vibration that disrupts the contact surface between brush and ring. The causes of machine vibration are beyond the scope of this section, but everything possible should be done to eliminate vibration that disrupts brush contact.

It is highly recommended that you check slip ring runout with a dial indicator. Excessive runout of 0.003 of an inch (0.076 mm) or more on any slip ring will result in poor motor performance, rapid brush wear, and eventual motor failure. As Murphy's Law dictates, motors fail when they are needed most. A motor that has slip rings with excessive runout must be taken out of service as soon as possible. Send it to a reliable motor service shop to have the rings resurfaced to tolerances as close to zero runout as possible.

15.171 Troubleshooting

Table 15.1 lists the indications and sources of unsatisfactory performance of AC slip rings and brushes. Indications of unsatisfactory performance are divided into those that appear at the brushes, at the ring surface, and as heating. The immediate causes of a particular indication of unsatisfactory performance are listed in the middle column. The primary fault, responsible for a specific immediate cause, is listed by a number in the third column.

Table 15.2 lists the primary sources of unsatisfactory AC brush and ring performance. Sources discussed in this section are listed together with others that are beyond the scope of this section. They are numbered for use with Table 15.1 and, in addition, are broken down into basic groups of association as follows:

A. Preparation and care of machine

B. Machine adjustment

C. Mechanical fault in machine

D. Electrical fault in machine

E. Machine design

F. Load or service condition

G. Disturbing external condition

H. Wrong brush grade

I. Wrong brush shunting

By using the two summary tables, an indication of unsatisfactory AC slip ring and brush performance can be traced to the primary source of the trouble. For example, from Table 15.1, one immediate cause of flashover at the brushes is "lack of attention." The primary fault would be items 3 and 7 of Table 15.2. Looking now at Table 15.2, these items indicate a need for periodic cleaning and for cleaning after seating brushes (item 3) or incorrect spring tension (item 7). The basic fault is improper care of the machine (Part A of Table 15.2) or incorrect machine adjustment (Part B).

The two summary tables list only the general troubles most frequently experienced with the wound-rotor machines.

TABLE 15.1 INDICATIONS AND SOURCES OF UNSATISFACTORY PERFORMANCE OF AC SLIP RINGS AND BRUSHES

Indications	Immediate Causes	Primary Faults (See Table 15.2)
APPEARING AT BRUSHES:		
Sparking	Ring surface condition	1, 2, 26, 27, 28, 29, 32, 40, 41
	Faulty machine adjustment	5, 7
	Mechanical fault in machine	8, 9, 10, 11, 12, 14
	Electrical fault in machine	18, 19
	Bad load conditions	23, 24, 25
	Vibration	34, 35
	Chattering of brushes	See "Chattering and noisy brushes"
	Wrong brush grade	38, 40
Pitting of brush face	Glowing	See "Glowing at brush face"
	Embedded copper	See "Copper in brush face"
Rapid brush wear	Ring surface condition	See specific indication under "APPEARING AT RING SURFACE"
	Severe sparking	See "Sparking"
	Imperfect contact with ring	7, 8, 9, 10, 34, 35
	Wrong brush grade	39
Glowing at brush face	Embedded copper	See "Copper in brush face"
	Severe load condition	23, 24, 25
	Bad service condition	29, 30
	Wrong brush grade	38, 40, 42
Copper in brush face	Ring surface condition	2
	Bad service condition	26, 29, 30, 31, 32
	Wrong brush grade	40, 42
Flashover at brushes	Machine condition	3, 8
	Bad load condition	23, 24, 36, 41
	Lack of attention	3, 7
Chattering and noisy brushes	Ring surface condition	See specific indication under "APPEARING AT RING SURFACE"
	Looseness in machine	9, 10, 11
	Faulty machine adjustment	6, 7
	High friction	26, 28, 32, 35, 39, 40
	Wrong brush grade	39, 40
Brush chipping or breakage	Ring surface condition	See specific indication under "APPEARING AT RING SURFACE"
	Looseness in machine	9, 10, 11
	Vibration	35
	Sluggish brush movement	8
Burned shunts		See "Heating at shunts"
APPEARING AT RING SURFACE:		
Rough or uneven surface		1, 2, 11
Dull or dirty surface		3, 27, 33, 41
Eccentric surface		1, 12, 14, 35
Grooving or threading of surface	Sparking	26, 27, 28, 29, 32, 40
	Copper or foreign material in brush face	2, 29, 30, 31, 42
	Glowing	See "Glowing at brush face"
Ring burning	Sparking	2, 3, 7, 8, 23, 24, 25, 36, 41
	Flashover	3, 7, 8, 23, 24, 25, 36, 41
Brush outline appearing on ring at brush spacing	Sparking	33

TABLE 15.1 INDICATIONS AND SOURCES OF UNSATISFACTORY PERFORMANCE OF AC SLIP RINGS AND BRUSHES *(continued)*

APPEARING AT RING SURFACE: *(continued)*		
Flat spot	Sparking	1, 2, 36
	Flashover	3, 7, 8, 23, 24, 25, 36
	Lack of attention	1, 3, 7
Discoloration of surface	High temperature	See "Heating at ring"
	Atmospheric condition	27, 29
	Wrong brush grade	41
Raw-material surface	Embedded copper	See "Copper in brush face"
	Bad surface condition	26, 28, 30, 32
	Wrong brush grade	40, 42
Rapid ring wear with blackened surface	Burning	2, 7, 8
	Severe sparking	See "Sparking"
Rapid ring wear with bright surface	Foreign material in brush face	26, 28, 30, 32
	Wrong brush grade	42
APPEARING AS HEATING:		
Heating in windings	Severe load condition	23, 25, 36
	Unbalanced magnetic field	12, 18, 19, 20
	Unbalanced currents in the windings	12, 14, 16, 18, 19, 20, 22
	Lack of ventilation	4
Heating at ring	Severe load condition	23, 25
	Severe sparking	5, 13, 28, 38
	High friction	6, 7, 21, 26, 28, 32, 39, 40
	Poor ring surface	See specific indication under "APPEARING AT RING SURFACE"
	Degradation of ring material	15
	High contact resistance	37
Heating at brushes	Severe load condition	23, 25
	Faulty machine adjustment	6, 7, 17
	Severe sparking	See "Sparking"
	Raw streaks on ring surface	See "Grooving or threading of surface"
	Embedded copper	See "Copper in brush face"
	Wrong brush grade	38, 39, 40, 42, 43
Heating at shunts	Severe load condition	36, 38
	Selective action	8, 17, 43, 45, 46, 49
	Inadequate heat transfer	4, 17, 44, 46, 47, 48
	Corroded shunts	27, 28

TABLE 15.2 PRIMARY SOURCES OF UNSATISFACTORY AC BRUSH AND RING PERFORMANCE

A. POOR PREPARATION AND CARE OF MACHINE
   1. Poor preparation of ring surface
   2. Porous ring material (inclusions, blowholes)
   3. Need for periodic cleaning and for cleaning after seating brushes
   4. Clogged ventilating ducts, poor ventilation in general

B. INCORRECT MACHINE ADJUSTMENT
   5. Poor alignment of brushholders
   6. Incorrect brush angle
   7. Incorrect spring tension

C. MECHANICAL FAULT IN MACHINE
   8. Brushes tight in holders
   9. Brushes too loose in holders
   10. Brushholders loose at mountings
   11. Ring(s) loose
   12. Loose or worn bearings
   13. Unequal air gap around stator
   14. Dynamic unbalance
   15. Ring diameter too small

D. ELECTRICAL FAULT IN MACHINE
   16. Open or high-resistance connection at ring
   17. Poor connection of brush shunt terminal
   18. Short circuit in primary or secondary winding
   19. Ground in primary or secondary winding
   20. Reversed coil(s) in primary or secondary winding

E. POOR MACHINE DESIGN
   21. High ratio of brush contact to ring surface area
   22. Insufficient cross section of parallel rings in winding

F. IMPROPER LOAD OR SERVICE CONDITION
   23. Overload
   24. Rapid change in load
   25. Plugging
   26. Low average current density in brushes
   27. Contaminated atmosphere
   28. "Contact poisons"
   29. Oil on ring or oil mist in air
   30. Abrasive dust in air
   31. Humidity too high
   32. Humidity too low
   33. Down for extended time periods

G. DISTURBING EXTERNAL CONDITION
   34. Loose or unstable foundation
   35. External source of vibration
   36. External short circuit or very heavy load surge

H. WRONG BRUSH GRADE
   37. Contact drop of brushes too high
   38. Contact drop of brushes too low
   39. Coefficient of friction too high
   40. Lack of film-forming properties in brush
   41. Lack of polishing action in brush
   42. Brushes too abrasive
   43. Lack of carrying capacity
   44. Wrong thermal conductivity

I. WRONG BRUSH SHUNTING
   45. Cross section too small
   46. Poor shunt connection at brush
   47. Insufficient dissipating surface area
   48. Wrong shunt insulation
   49. Improper temperature-resistance relationship

15.172 Preventive Maintenance

Accumulation of dry or oily dirt containing carbon from the brushes should be kept to a minimum by a regular preventive maintenance schedule as often as operating conditions may require. Flashover at the rings is apt to occur at start-up because the voltage between rings is highest at the moment when power is first applied. This secondary voltage is reduced to nearly zero as the motor reaches full speed. A dirty, low-resistance leakage path may withstand the lower secondary voltage during normal running operation, but may flash over at the next start-up. Therefore, it is recommended that the critical areas shown in Figure 15.16 be cleaned after the seating of new brushes, or after prolonged periods of operation or shutdown.

Critical leakage paths (refer to Figure 15.16) between phases and from phase to ground along the insulated surfaces are:

A. *Support-ring* surfaces and edges, between coils, and between coils and metal surfaces.

B. *Slip ring sleeve* surfaces and edges, between rings, and from rings to shafts.

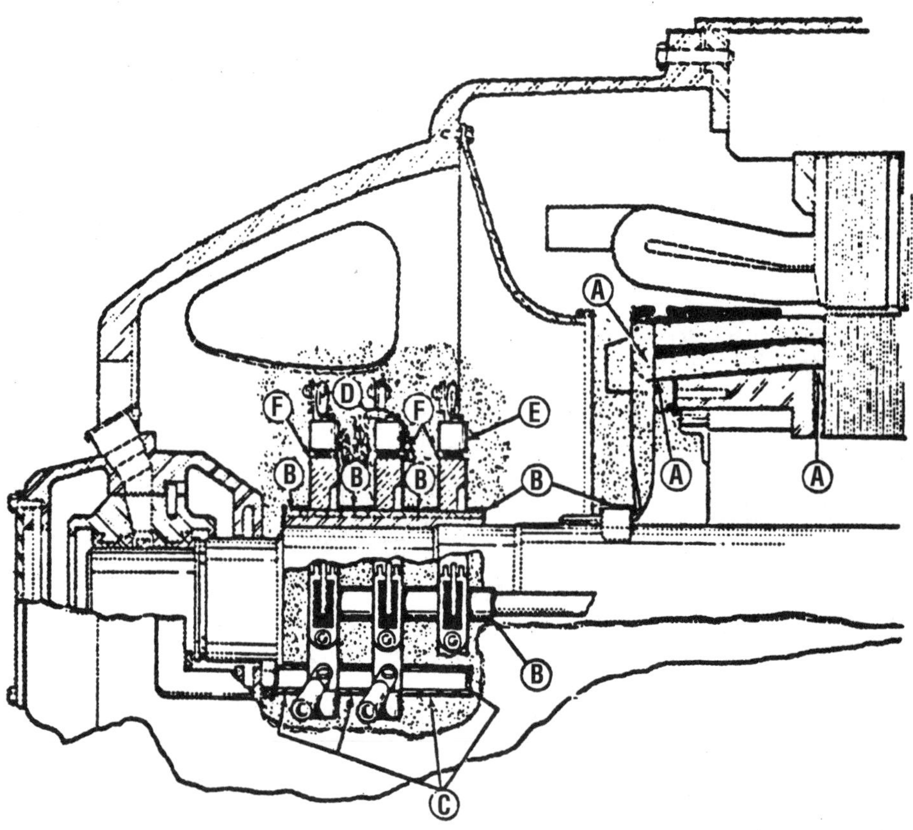

Fig. 15.16 Critical leakage paths

C. *Brushholder-stud* surfaces and ends, between holders, and from holder to bracket.

D. *Through the air* separating the coils, the rings, and the brushholders at A, B, and C, but especially in the area of the brush shunts D where misplaced shunt reduces creepage through the air.

Recommended methods for cleaning the critical leakage paths are:

A. *For areas designated A.* Blow out the dirt with clean, dry, compressed air at 30 to 50 psi (2.1 to 3.5 kg/sq cm). Excessive oil or oil accumulations may require wiping with a clean, dry cloth or washing with an approved insulation cleaner (such as VM & P Naphtha, Stoddard's Solvent, or chloroethylene) or dry steam. Washing, blowing, or wiping may have to be repeated to ensure a clean leakage path of high resistance. *CAUTION: Use eye protection when cleaning with compressed air. Study appropriate Material Safety Data Sheets (MSDSs) and follow all safety requirements.*

B. *For areas designated B.* Wipe thoroughly with a clean, dry cloth and repeat as required to ensure a clean leakage path. Avoid accumulations of dirt containing carbon at the end of slip ring sleeve, along leads, and at junctions of a lead and its ring.

C. *For areas designated C.* Wipe surface as for B above.

D. *For areas designated D.* Here it is most important to maintain maximum creepage through the air. Do not allow the shunts to droop over. Keep them in line with the holder.

The critical leakage paths A, B, C, and D should be inspected for cleanliness before each start-up. If cracks appear in the insulated surfaces, check the insulation value with a megger. The megohm reading should exceed kV+1 megohms, where kV is the secondary voltage (given on the nameplate) divided by 1,000. If the megohm reading is low, clean the cracks, fill and seal with an insulating varnish such as glyptol; retest the insulation resistance value. In some instances, replacement with new insulation may be advantageous.

The other areas of preventive maintenance are:

E. *Brushes.* Brushes may be lifted off the rings if the machine is to be idle for a period extending beyond one week, especially if the absolute humidity is above one grain per cubic foot. An alternative way to protect brushes during a period of shutdown involves carefully placing a sheet of polyethylene film between the brushes and the ring. This will minimize chances of chipping brushes, changing brush pressure setting, and disturbing the film on the ring. Be sure to remove the polyethylene before the next start-up and check for freedom of brush movement in the brush pocket, for proper seating of brushes, and for correct, uniform brush pressure.

   The brushes are centered on the rings at the factory for operation with the rotor on magnetic center. If the rotor is pulled off magnetic center in normal operation, determine if the brushes are positioned on the rings so that they will not override the ring at any positions of axial "float" during start-up or shutdown. It may be necessary to relocate the brushholders on the studs.

F. *Slip ring surfaces.* Unless the rings are refinished, the film that develops in the brush track during normal operation should not be removed or disturbed. This is the film that is vital and necessary for satisfactory operation. Remove any loose foreign material by blowing or wiping lightly with a clean, dry cloth. If the machine is to be shut down for several months or placed in storage, it is desirable to cover the rings with polyethylene film to prevent contamination of the ring surfaces. Do not paint, grease, or oil the ring surface.

QUESTIONS

Write your answers in a notebook and then compare your answers with those on pages 460 and 461.

15.1E Why do wet wells use variable-speed pumping equipment?

15.1F What preventive maintenance is required by a standard three-phase, single-speed, synchronous AC motor and what additional maintenance is required for a variable-speed, three-phase AC motor?

15.1G The rate of brush wear is influenced by what factors?

15.1H Indications of unsatisfactory performance of AC slip rings and brushes appear at or as what items?

15.18 Operating Troubles

The following list of operating troubles includes most of the causes of failure or reduced operating efficiency. The remedy or cure is either obvious or may be identified from the description of the causes. The lists of causes are not arranged in priority.

SYMPTOM A—Pump Will Not Start

Causes:

1. Blown fuses or tripped circuit breakers due to:
   a. Rating of fuses or circuit breakers not correct
   b. Switch (breakers) contacts corroded or shorted
   c. Terminal connections loose or broken somewhere in the circuit
   d. Automatic control mechanism not functioning properly
   e. Motor shorted or burned out
   f. Wiring hookup or service not correct
   g. Switches not set for operation
   h. Contacts of the control relays dirty and arcing
   i. Fuses or thermal units too warm
   j. Wiring short-circuited
   k. Shaft binding or sticking due to rubbing impeller, tight packing glands, or clogging of pump
2. Loose connection, fuse, or thermal unit

SYMPTOM B—Reduced Rate of Discharge

Causes:

1. Pump not primed
2. Mixture of air in the wastewater
3. Speed of motor too low
4. Improper wiring
5. Defective motor
6. Discharge head too high
7. Suction lift higher than anticipated
8. Impeller clogged
9. Discharge line clogged
10. Pump rotating in wrong direction
11. Air leaks in suction line or packing box
12. Inlet to suction line too high, permitting air to enter
13. Valves partially or entirely closed
14. Check valves stuck or clogged
15. Incorrect impeller adjustment
16. Impeller damaged or worn
17. Packing worn or defective
18. Impeller turning on shaft because of broken key
19. Flexible coupling broken
20. Loss of suction during pumping may be caused by leaky suction line, ineffective water or grease seal
21. Belts slipping
22. Worn wearing ring

SYMPTOM C—High Power Requirements

Causes:

1. Speed of rotation too high
2. Operating heads lower than rating for which pump was designed, resulting in excess pumping rates
3. Check valves open, draining long force main back into wet well

4. Specific gravity or viscosity of liquid pumped too high
5. Clogged pump
6. Sheaves on belt drive misaligned or maladjusted
7. Pump shaft bent
8. Rotating elements binding
9. Packing too tight
10. Wearing rings worn or binding
11. Impeller rubbing

SYMPTOM D—Noisy Pump

Causes:

1. Pump not completely primed
2. Clogged inlet
3. Inlet not submerged
4. Pump not lubricated properly
5. Worn impellers
6. Strain on pumps caused by unsupported piping fastened to the pump
7. Insecure foundation
8. Mechanical defects in pump
9. Misalignment of motor and pump where connected by flexible shaft
10. Rags or sticks bound (wrapped) around impeller

QUESTIONS

Write your answers in a notebook and then compare your answers with those on page 461.

15.1I What items would you check if a pump will not start?

15.1J How would you attempt to increase the discharge from a pump if the flow rate is lower than expected?

15.19 Starting and Stopping Pumps

The operator must determine what treatment processes will be affected by either starting or stopping a pump. The pump discharge point must be known and valves either opened or closed to direct flows as desired by the operator when a pump is started or stopped.

15.190 Centrifugal Pumps

Figure 15.17 illustrates a typical wet well and pumping system. The purpose of each part is explained in Table 15.3. Basic rules for the operation of centrifugal pumps include the following items.

1. Do not operate the pump when safety guards are not installed over or around moving parts.
2. Do not start a pump that is locked or tagged out for maintenance or repairs; serious injury to the operator or damage to the equipment could result.
3. Never run a centrifugal pump when the impeller is dry. Always be sure the pump is primed.
4. Never attempt to start a centrifugal pump whose impeller or shaft is spinning backward.
5. Do not operate a centrifugal pump that is vibrating excessively after start-up. Shut unit down and isolate pump from system by closing the pump suction and discharge valves. Look for a blockage in the suction line and the pump impeller.

There are several situations in which it may be necessary to start a centrifugal pump against a closed discharge valve until the pump has picked up its prime and developed a satisfactory discharge head for that operating system. Once the pump is primed, slowly open the pump discharge valve until the pump is fully on line. This procedure is used with treatment processes or piping systems with vacuums or pressures that cannot be dropped or allowed to fluctuate greatly while an alternate pump is put on the line.

TABLE 15.3 PURPOSE OF WET WELL AND PUMP SYSTEM PARTS

Part	Purpose
1. Inlet	Carries wastewater to wet well.
2. Wet Well	Stores wastewater for removal by pump.
3. Suction Bell	Guides wastewater into pump suction pipe and reduces pipe entrance energy losses.
4. Suction Valve	Isolates pump and piping from wet well.
5. Suction Gauge	Indicates suction head or lift on suction side of pump.
6. Volute (not shown in Figure 15.17)	Collects wastewater discharged by pump impeller and directs flow to pump discharge.
7. Volute Bleed Line	Keeps pump primed for automatic operation by allowing entrapped gases (air) to escape from the pump volute.
8. Discharge Gauge	Indicates discharge head (energy imparted to wastewater by pump).
9. Discharge Valve	Isolates pump from discharge system.
10. Discharge Check Valve	Prevents discharge pipe and treatment process tanks from draining back through pump and into wet well.
11. Process Valves	Direct flow.

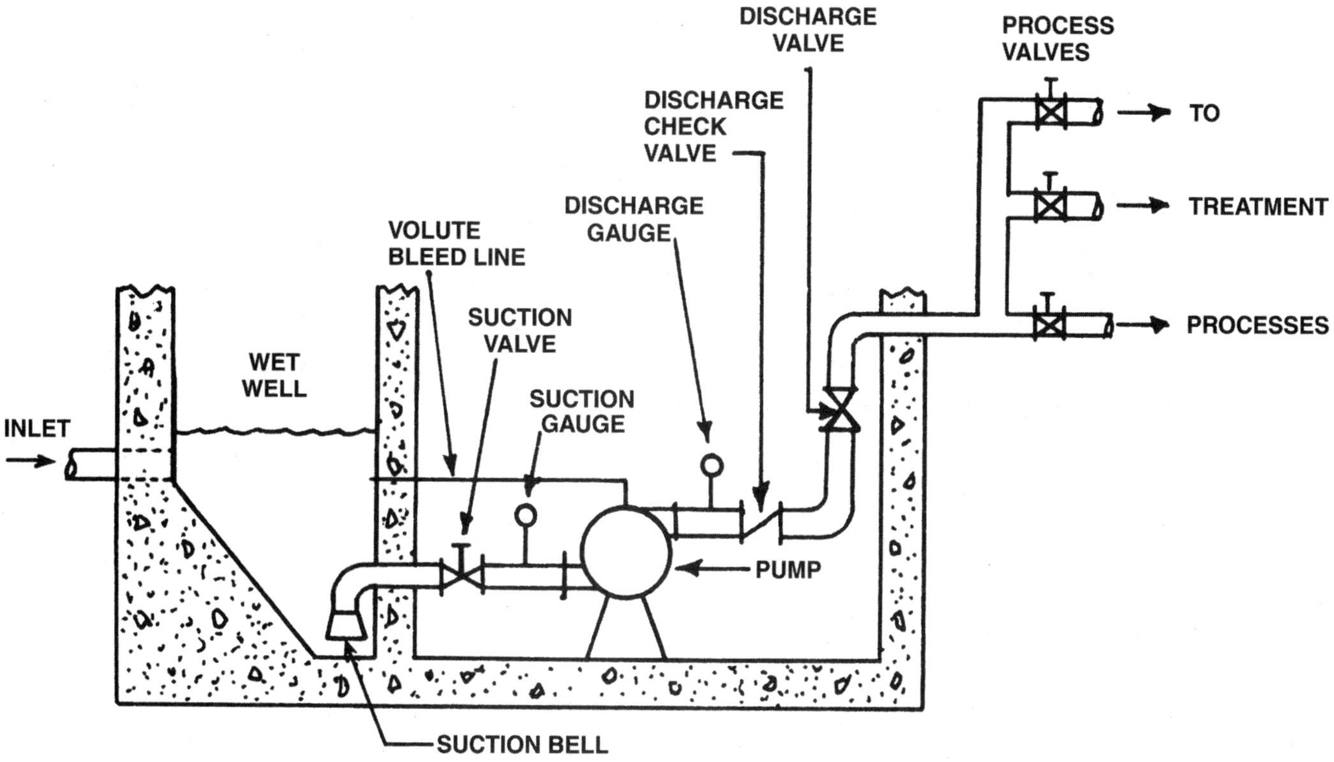

Fig. 15.17 Wet well and pump system

Most centrifugal pumps used in wastewater treatment plants are designed so that they can be easily started even if they have not been primed. This is accomplished with a positive static suction head or a low suction lift. On most of these arrangements, the pump will not require priming as long as the pump and the piping system do not leak. Leaks would allow the water to drain out of the pump volute. When pumps in wastewater systems lose their prime, the cause is often a faulty check valve on the pump discharge line. When the pump stops, the discharge check valve will not seal (close) properly. Wastewater previously pumped then flows back through the check valve, down through the pump rotating the pump impeller and shaft backward, and back into the wet well. The pump is drained and has lost its prime.

The other danger associated with a faulty check valve is that if the pump wet well has a high inflow, plus inflow from the water running back through the pump, the wet well water level will rise to the elevation set to turn on the pump. If the pump attempts to start while the pump, motor, and shafting are rotating in the opposite direction, very serious damage can occur to the pumping equipment. Many pumps are equipped with antirotational devices that minimize this potential problem.

About 95 percent of the time, the centrifugal pumps in wastewater treatment plants are ready to operate with suction and discharge valves open and seal water turned on. When the automatic start or stop command is received by the pump from an air or electronic controller, the pump is ready to respond properly.

When the pumping equipment must be serviced, take it off the line by locking and tagging out the pump main breaker until all service work is completed.

QUESTIONS

Write your answers in a notebook and then compare your answers with those on page 461.

15.1K Why should a pump that has been locked or tagged out for maintenance or repairs not be started?

15.1L Under what conditions might a centrifugal pump be started against a closed discharge valve?

STOPPING PROCEDURES

This section contains a typical sequence of procedures to follow to stop a centrifugal pump.

1. Inspect process system affected by pump, start alternate pump if required, and notify supervisor or log action.
2. Before stopping the operating pump, check its operation. This will give an indication of any developing problems, required adjustments, or problem conditions of the unit. This procedure only requires a few minutes. Items to be inspected include:

   a. Pump packing gland

      (1) Seal water pressure

      (2) Seal leakage (too much, sufficient, or too little leakage)

      (3) Seal leakage drain flowing clear

      (4) Mechanical seal leakage (if equipped)

b. Pump operating pressures

(1) Pump suction (Pressure-Vacuum)

A higher vacuum than normal may indicate a partially plugged or restricted suction line. A lower vacuum may indicate a higher wet well level or a worn pump impeller or wearing rings.

(2) Pump discharge pressure

System pressure is indicated by the pump discharge pressure. Lower than normal discharge pressures can be caused by:

(a) Worn impeller or wearing rings in the pump

(b) A different point of discharge can change discharge pressure conditions

(c) A broken discharge pipe can change the discharge head

NOTE: To determine the maximum head a centrifugal pump can develop, slowly close the discharge valve at the pump. Read the pressure gauge between the pump and the discharge valve when the valve is fully closed. This is the maximum pressure the pump is capable of developing. Do not operate the pump longer than a few minutes with the discharge valve closed completely because the energy from the pump is converted to heat and water in the pump can become hot enough to damage the pump.

c. Motor temperature and pump bearing temperature

If motor or bearings are too hot to touch, further checking is necessary to determine if a problem has developed or if the temperature is normal. High temperatures may be measured with a thermometer.

d. Unusual noises, vibrations, or conditions about the equipment

If any of the above items indicate a change from the pump's previous operating condition, additional service or maintenance may be required during shutdown.

3. Actuate stop switch for pump motor and lockout switch. If possible, use switch next to equipment so that you may observe the equipment stop. Observe the following items:

   a. Check valve closes and seats.

      Valve should not slam shut, or discharge piping will jump or move in their supports. There should not be any leakage around the check valve shaft. If check valve is operated automatically, it should close smoothly and firmly to the fully closed position.

NOTE: If the pump is not equipped with a check valve, close discharge valve before stopping pump.

   b. Motor and pump should wind down slowly and not make sudden stops or noises during shutdown.

   c. After equipment has completely stopped, pump shaft and motor should not start backspinning. If backspinning is observed, close the pump discharge valve slowly. Repair faulty check or foot valve.

4. Go to power control panel containing the pump motor starters just shut down and open motor breaker switch, lock out, and tag.

5. Return to pump and close:

   a. Discharge valve

   b. Suction valve

   c. Seal water supply valve

   d. Pump volute bleed line (if so equipped)

6. If pump is to be left out of service more than two days, drain pump volute and leave volute empty. Indicate "volute empty" on lockout tag.

7. If required, close and open appropriate valves along piping system through which pump was discharging.

STARTING PROCEDURES

This section contains a typical sequence of procedures to follow to start a centrifugal pump.

1. Check motor control panel for lock and tags. Examine tags to be sure that no item is preventing start-up of equipment.

2. Inspect equipment.

   a. Be sure stop switch is locked out at equipment location.

   b. Guards over moving parts must be in place.

   c. Cleanout on pump volute and drain plugs should be installed and secure.

   d. Valves should be in closed position.

   e. Pump shaft must rotate freely.

   f. Pump motor should be clean and air vents clear.

   g. Pump, motor, and auxiliary equipment lubricant level must be at proper elevations.

   h. Determine if any special considerations or precautions are to be taken during start-up.

3. Follow pump discharge piping route. Be sure all valves are in the proper position and that the pump flow will discharge where intended.

4. Return to motor control panel.

   a. Remove tag.

   b. Remove padlock.

   c. Close motor main breaker.

   d. Place selector switch to manual (if you have automatic equipment).

5. Return to pump equipment.

   a. Open seal water supply line to packing gland. Be sure seal water supply pressure is adequate.

   b. Open pump suction valve slowly.

   c. Bleed air out of top of pump volute to prime pump. Some pumps are equipped with air relief valves or bleed lines back to the wet well for this purpose.

   d. When pump is primed, slowly open pump discharge valve and recheck prime of pump. Be sure no air is escaping from volute.

   e. Unlock stop switch and actuate start switch. Pump should start.

6. Inspect equipment.

   a. Motor should come up to speed promptly. If ammeter is installed on instrument panel, then read or watch for excessive draw of power (amps) during start-up and normal operation. If ammeter is not installed on panel and power leads are exposed from conduit or motor connection box, then a portable ammeter may be used to observe draw of power.

   b. No unusual noise or vibrations should be observed during start-up.

   c. Check valve should be open and no chatter or pulsation should be observed.

   d. Pump suction and discharge pressure readings should be within normal operating range for this pump.

   e. Packing gland leakage should be normal.

   f. If a flowmeter is on the pump discharge, record pump output.

7. If the unit is operating properly, return to the motor control panel and place the motor mode of operation selector in the proper operation position (manual-auto-off).

8. The pump and auxiliary equipment should be inspected routinely after it has been placed back into service.

QUESTIONS

Write your answers in a notebook and then compare your answers with those on page 461.

15.1M What should be done before stopping an operating pump?

15.1N What could cause a pump shaft or motor to spin backward?

15.1O Why should the position (open or closed) of all valves be checked before starting the pump?

15.191 Positive Displacement Pumps

Steps for starting and stopping positive displacement pumps are outlined in this section. There are two basic differences in the operation of positive displacement pumps as compared with centrifugal pumps. Centrifugal pumps (due to their design) will permit an operator error, but a positive displacement pump will not and someone will have to pay for correcting the damages.

Important rules for operating positive displacement pumps include:

1. **Never operate a positive displacement pump against a closed valve, especially a discharge valve.** The pipe, valve, or pump could rupture from excessive pressure. The rupture will damage equipment and possibly seriously injure or kill someone standing nearby.

2. Positive displacement pumps are used to pump solids (sludge) and certain precautions must be taken to prevent injury or damage. If the valves on both ends of a sludge line are closed tightly, the line becomes a closed vessel. Gas from decomposition can build up to a pressure that will rupture pipes or valves.

3. Positive displacement pumps also are used to meter and pump chemicals. Care must be exercised to avoid venting chemicals to the atmosphere.

4. Never operate a positive displacement pump when it is dry or empty, especially the progressive cavity types that use rubber stators. A small amount of liquid is needed for lubrication in the pump cavity between the rotor and the stator.

In addition to never closing a discharge valve on an operating positive displacement pump, the only other difference (when compared with a centrifugal pump) may be that the positive displacement pump system may or may not have a check valve in the discharge piping after the pump. Installation of a check valve depends on the designer and the material being pumped.

Other than the specific differences mentioned in this section, the starting and stopping procedures for positive displacement pumps are similar to the procedures for centrifugal pumps.

QUESTIONS

Write your answers in a notebook and then compare your answers with those on page 461.

15.1P What is the most important rule regarding the operation of positive displacement pumps?

15.1Q What could happen if a positive displacement pump is started against a closed discharge valve?

15.1R Why should both ends of a sludge line never be closed tight?

Please answer the discussion and review questions next.

DISCUSSION AND REVIEW QUESTIONS

Chapter 15. MAINTENANCE

(Lesson 2 of 7 Lessons)

Write the answers to these questions in your notebook. The question numbering continues from Lesson 1.

10. What should you do if you cannot understand the manufacturer's instructions?

Select the correct word:

11. Cheap lubricants may be the (1) *MOST,* or (2) *LEAST* expensive in the end.
12. Start-up of a new pump should be done by (1) *A NEW OPERATOR,* or (2) *A TRAINED PERSON.*
13. How can you determine if a new pump will turn in the direction intended?
14. How can you tell if a new pump is delivering design flows and pressures?
15. When shutting down a pump for a long period, what precautions should be taken with the motor disconnect switch?
16. What is a maintenance problem with water level float controls?

CHAPTER 15. MAINTENANCE

(Lesson 3 of 7 Lessons)

15.2 BEWARE OF ELECTRICITY

RECOGNIZE YOUR LIMITATIONS

In the wastewater collection and treatment maintenance departments of all cities, there is a need for maintenance operators to know something about electricity. Duties could range from repairing a taillight on a trailer or vehicle to repairing complex pump controls and motors. Very few maintenance operators do the actual electrical repairs or troubleshooting because this is a highly specialized field and unqualified people can seriously injure themselves and damage costly equipment. For these reasons, you must be familiar with electricity, know the hazards, and recognize your own limitations when you must work with electrical equipment.

Most municipalities employ electricians or contract with a commercial electrical company that they call when major problems occur. However, the maintenance operator should be able to explain how the equipment is supposed to work and what it is doing or is not doing when it fails.

After this lesson, you should be able to tell an electrician what appears to be the problem with electrical panels, controls, circuits, and equipment. Even though operators may only be assigned maintenance tasks on electrical systems, the more operators know about electricity, the better equipped they are to diagnose electrical problems, especially during emergency situations.

When an outside electrical contractor performs maintenance on your system, ask the contractor to assign the same person each time. This allows one person to become familiar with the pump station controls and electrical systems and, therefore, troubleshoot and repair much more rapidly during failures. This person should be accompanied by an assistant so a backup person will be available with some knowledge of your system when the regular person is not available.

The need for safety should be apparent. If proper safe procedures are not followed in operating and maintaining the various electrical equipment used in wastewater collection and treatment facilities, accidents can happen that cause injuries, permanent disability, or loss of life. Some of the serious accidents that have happened, and could have been avoided, occurred when machinery was not shut off, locked out, and tagged properly (Figure 15.18) as required by OSHA. Possible accidents include:

1. Maintenance operator could be cleaning pump and have it start, thus losing an arm, hand, or finger.
2. Electrical motors or controls not properly grounded could lead to possible severe shock, paralysis, or death.
3. Improper circuits created by mistakes, such as wrong connections, bypassed safety devices, wrong fuses, or improper wire, can cause fires or injuries due to incorrect operation of machinery.

Another consideration for having a basic working knowledge of electricity is to prevent financial losses resulting from motors burning out and from damage to equipment, machinery, and control circuits. Additional costs result when damages have to be repaired, including payments for outside labor.

WARNING

Never work in electrical panels or on electrical controls, circuits, wiring, or equipment unless you are qualified and authorized. By the time you find out what you do not know, you could find yourself too dead to use the knowledge.

QUESTIONS

Write your answers in a notebook and then compare your answers with those on page 461.

15.2A Why must unqualified or inexperienced people be extremely careful when attempting to troubleshoot or repair electrical equipment?

15.2B What could happen when machinery is not shut off, locked out, and tagged properly?

15.3 ELECTRICAL EQUIPMENT MAINTENANCE

15.30 Introduction

This section contains a basic introduction to electrical terms and information plus directions on how to troubleshoot problems with electrical equipment.

Most electrical equipment used in wastewater collection systems and treatment plants is labeled with *NAMEPLATE*[7] information indicating the proper voltage and allowable current in amps.

[7] *Nameplate.* A durable, metal plate found on equipment that lists critical operating conditions for the equipment.

DANGER

OPERATOR WORKING ON LINE

DO NOT CLOSE THIS SWITCH WHILE THIS TAG IS DISPLAYED

TIME OF F: ____________________

DATE: ____________________

SIGNATURE: ____________________

This is the ONLY person authorized to remove this tag.

INDUSTRIAL INDEMNITY/INDUSTRIAL UNDERWRITERS/
INSURANCE COMPANIES

4E210—R66

Fig. 15.18 Typical warning tag

(Source: Industrial Indemnity/Industrial Underwriters/Insurance Companies)

15.31 Volts, Amps, Watts, and Power Requirements

VOLTS

Voltage (E) is also known as Electromotive Force (EMF), and is the electrical pressure available to cause a flow of current (amperage) when an electric circuit is closed.[8] This force can be compared with the pressure or force that causes water to flow in a pipe. Some pressure in a water pipe is required to make the water move. The same is true of electricity. A force is necessary to push electricity or electric current through a wire. This force is called voltage.

There are two types of voltage: Direct Current (DC) and Alternating Current (AC).

DIRECT CURRENT

Direct current (DC) is flowing in one direction only and is essentially free from pulsation. Direct current is seldom used in wastewater treatment plants except in motor-generator sets, some control components of pump drives, and standby lighting. Direct current is used exclusively in automotive equipment, certain types of welding equipment, and a variety of portable equipment. Direct current is found in various voltages, such as 6 volts, 12 volts, 24 volts, 48 volts, and 110 volts. All batteries are direct current. DC is tested by holding the multimeter leads on the positive and negative poles on a battery. These poles are usually marked Positive (+) and Negative (–). Direct current usually is not found in higher voltages (over 24 volts) around plants and lift stations unless in motor-generator sets. Care must be taken when installing battery cables and wiring that Positive (+) and Negative (–) poles are connected properly to wires marked (+) and (–). If not properly connected, you could get an arc across the unit that could cause an explosion.

ALTERNATING CURRENT

Alternating current (AC) is periodic current that has alternating positive and negative values. In other words, it goes from zero to maximum strength, back to zero, and to the same strength in the opposite direction, which comprises a cycle. Our AC voltage is 60-cycle frequency, or "Hertz," which means that this happens 60 times per second. Alternating current is classified as:

a. Single Phase

b. Two Phase

c. Three Phase or Polyphase

The most common of these are single phase and three phase. The various voltages you probably will find on your job are 110 volts, 120 volts, 208 volts, 220 volts, 240 volts, 277 volts, 440 volts, 460 volts, 480 volts, and 550 volts.

Single-phase power is found in lighting systems, small pump motors, various portable tools, and throughout our homes. It is usually 120 volts or 240 volts. Single phase means that only one phase of power is supplied to the main electrical panel at 240 volts and the power supply has three wires or leads. Two of these leads have 120 volts each, the other lead is neutral and usually is coded white. The neutral lead is grounded. Many appliances and power tools have an extra ground (commonly a green wire) on the case for additional protection.

Three-phase power is generally used with motors and transformers found in lift stations and wastewater treatment plants. This power generally is 208, 220, 240 volts, or 440, 460, 480, and 550 volts. Higher voltages are used in some lift stations. Three phase is used when higher power requirements or larger motors are used because efficiency is usually higher and motors require less maintenance. Generally, all motors above two horsepower are three phase unless there is a problem with the power company getting three phase to the installations. Quite a few residential lift stations are on single-phase power due to their remote locations. Three-phase power usually is brought in to the point of use with three leads and there is power in all three leads.

When taking a voltage check between any two of the three leads, you measure 208, 220, 240 volts, or 440, 460, 480 volts depending on the supply voltage. There are some instances where you might measure a difference between the leads due to the use of different transformers. If there is power in three leads and if a fourth lead is brought in, it is a neutral lead.

Two-phase and polyphase systems will not be discussed because they generally are not found in wastewater collection and treatment facilities.

AMPS

An ampere (I) is the practical unit of electric current. This is the current produced by a pressure of one volt in a circuit having a resistance of one ohm. Amperage is the measurement of current or electron flow and is an indication of work being done or "how hard the electricity is working."

In order to understand amperage, one more term must be explained. The ohm is the practical unit of electrical resistance (R). "Ohm's Law" states that in a given electrical circuit the amount of current in amperes (I) is equal to the pressure in volts (E) divided by the resistance (R) in ohms. The following three formulas are given to provide you with an indication of the relationships among current, resistance, and EMF (electromotive force).

$$\text{Current, amps} = \frac{\text{EMF, volts}}{\text{Resistance, ohms}} \qquad I = \frac{E}{R}$$

$$\text{EMF, volts} = (\text{Current, amps})(\text{Resistance, ohms}) \qquad E = IR$$

$$\text{Resistance, ohms} = \frac{\text{EMF, volts}}{\text{Current, amps}} \qquad R = \frac{E}{I}$$

These equations are used by electrical engineers for calculating circuit characteristics. If you memorize the following relationship, you can always figure out the correct formula.

[8] Electricians often talk about closing an electric circuit. This means they are closing a switch that actually connects circuits together so electricity can flow through the circuit. Closing an electric circuit is like opening a valve on a water pipe.

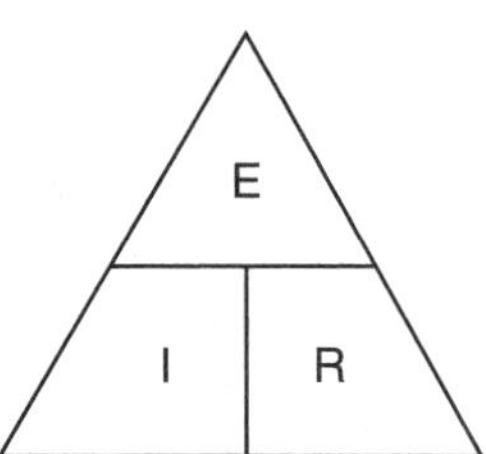

To use the above triangle you cover up with your finger the term you do not know or are trying to find out. The relationship between the other two known terms will indicate how to calculate the unknown. For example, if you are trying to calculate the current, cover up I. The two knowns (E and R) are shown in the triangle as E/R. Therefore, I = E/R. The same procedure can be used to find E when I and R are known or to find R when E and I are known.

WATTS

The watt and kilowatt (one thousand watts) are measures of power units used to rate electrical machines or motors or to indicate power in an electric circuit. The watts or power (P) required by a machine is obtained by multiplying the amps (I) required (or pulled) times the potential drop in volts (E) or the electromotive force (EMF). The amount of watts in a circuit is equal to the voltage (E) times the amperage (I). The total power in a circuit, measured in watts, at any given moment is similar to the power produced by a motor, measured in horsepower, at any given moment.

Power, watts = (Electromotive Force, volts)(Current, amps)

or P, watts = (E, volts)(I, amps)

For the sake of comparisons,

1 horsepower = 746 watts = 0.746 kilowatt = 550 ft-lbs/sec

The actual horsepower output of a motor also depends on the power factor (a number less than one) and the efficiency of the motor.

$$\text{Output, HP} = \frac{(\text{P, watts})(\text{Power Factor})(\text{Efficiency, \%})}{(746\ \text{watts/horsepower})(100\%)}$$

POWER REQUIREMENTS

Power requirements (PR) are expressed in kilowatt hours: 500 watts for two hours or one watt for 1,000 hours equals one kilowatt hour (kW-hr or kWh). The power company charges so many cents per kilowatt hour.

Power Req, kW-hr = (Power, kilowatts)(Time, hours)

PR, kW-hr = (P, kW)(T, hr)

CONDUCTORS AND INSULATORS

A material, like copper, that permits the flow of electric current is called a conductor. Material that will not permit the flow of electricity, like rubber, is called an insulator. When such a material is wrapped or cast around a wire, it is called insulation. Insulation is commonly used to prevent the loss of electrical flow by two conductors coming into contact with each other.

QUESTIONS

Write your answers in a notebook and then compare your answers with those on page 461.

15.3A How can you determine the proper voltage and allowable current in amps for a piece of equipment?

15.3B What are two types of voltage?

15.3C Amperage is a measurement of what?

15.32 Tools, Meters, and Testers

> **WARNING**
>
> **Never enter any electrical panel or attempt to troubleshoot or repair any piece of electrical equipment or any electric circuit unless you are qualified and authorized.**

A wide variety of instruments are used to maintain lift station electrical systems. These instruments measure current, voltage, and resistance. They are used not only for troubleshooting, but for preventive maintenance as well. These instruments may have either an *ANALOG READOUT,*[9] which uses a pointer and scale, or a *DIGITAL READOUT,*[10] which gives a numerical reading of the measured value.

To check for voltage, a *MULTIMETER* is needed. There are several types on the market and all of them work. They are designed to be used on energized circuits and care must be

[9] *Analog Readout.* The readout of an instrument by a pointer (or other indicating means) against a dial or scale. Also see DIGITAL READOUT.

[10] *Digital Readout.* The readout of an instrument by a direct, numerical reading of the measured value or variable.

exercised when testing. By holding one lead on ground and the other on a power lead, you can determine if power is available. You also can tell if it is AC or DC and the intensity or voltage (110, 220, 480, or whatever) by testing the different leads.

A multimeter can also be used to measure voltage, current, and resistance. A digital multimeter is shown in Figure 15.19 and an analog clamp-on multimeter is shown in Figure 15.20.

Fig. 15.19 Digital multimeter

(Reproduced with permission of Fluke Corporation)

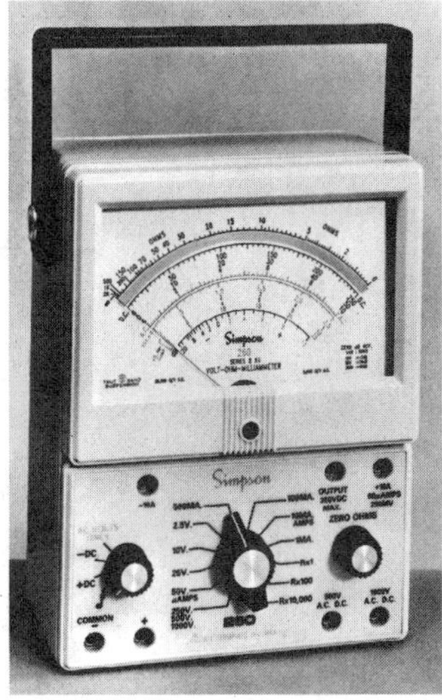

Fig. 15.20 Analog clamp-on multimeter

(Permission of Simpson Electric)

Do not work on any electric circuits unless you are qualified and authorized. (Refer also to Chapter 14 for more information on electrical safety and OSHA requirements.) Use a multimeter and other circuit testers to determine if the circuit is energized, or if all voltage is off. This should be done after the main switch is turned off to make sure it is safe to work inside the electrical panel. Always be aware of the possibility that even if the disconnect to the unit you are working on is off, the control circuit may still be energized if the circuit originates at a different distribution panel. Check with a multimeter before and during the time the main switch is turned off to have a double-check. This procedure ensures that the multimeter is working and that you have good continuity to your tester. Use circuit testers to measure voltage or current characteristics to a given piece of equipment for making sure that you have or do not have a "live" circuit. *WARNING: Switches can fail and the only way to ensure that a circuit is dead is to test the circuit.*

In addition to checking for power, a multimeter can be used to test for open circuits, blown fuses, single phasing of motors, grounds, and many other uses. Some examples are illustrated in the following paragraphs.

In the circuit shown in Figure 15.21, test for power by holding one lead of the multimeter on point "A" and the other at point "B." If no power is indicated, the switch is open or faulty. The sketch shows the switch in the "open" position.

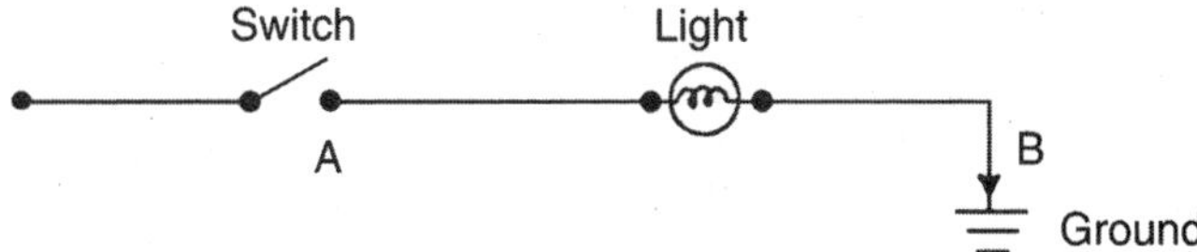

Fig. 15.21 Single-phase circuit

To test for power at Point "A" and Point "B" in Figure 15.22, open the switches as shown. Using a multimeter with clamp-on leads, clamp a lead on L1 and a lead on L2 between the fuses and the load. Bring the multimeter and leads out of the panel and close the panel door as far as possible without cutting or damaging the meter leads. Some switches cannot be closed if the panel door is open. The panel door is closed when testing because hot copper sparks could seriously injure you when the circuit is energized and the voltage is high. Close the switches.

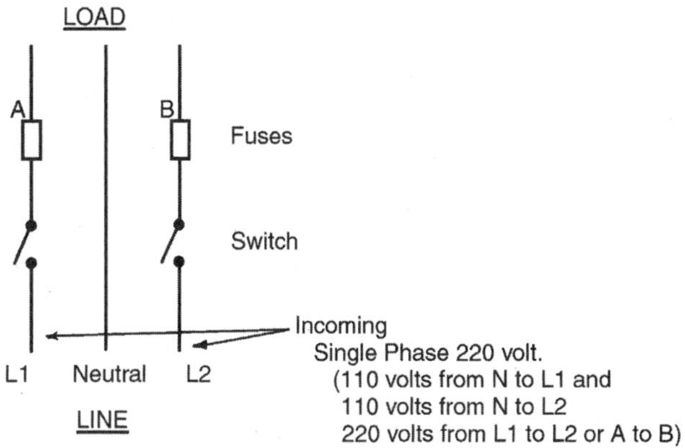

Fig. 15.22 Single-phase, three-lead circuit

1. Multimeter should register at 220 volts. If there is no reading at points "A" and "B," the fuse or fuses could be "blown."
2. Move multimeter down below fuses to line 1 and line 2 (L1, L2). If there is still no reading on the multimeter, check for an open switch in another location, or call the power company to find out if power is out.
3. If a 220-volt reading is registered at line 1 and line 2, move the test leads to point "A" and the neutral lead. If a reading of 110 volts is observed, the fuse on line "A" is OK. If there is no voltage reading, the fuse on line "A" is blown. Move the lead from line "A" to line "B." Observe the reading. If 110-volt power is recorded, the fuse on line "B" is OK. If there is no voltage reading, the fuse on line "B" is blown. Another possibility to consider is that the neutral line could be broken.

> **WARNING**
>
> Turn off power and be sure that there is no voltage in either power line before changing fuses. Use a fuse puller. Test circuit again in the same manner to make sure fuses or circuit breakers are OK. 220 volts power or voltage should be present between points "A" and "B." If fuse or circuit breaker trips again, shut off and determine the source of the problem.

Test for power at points "A," "B," and "C" in a three-phase circuit (Figure 15.23). Place multimeter leads on lines "A" and "B." Close all switches. 220 volts should register on multimeter. Check between lines "A" and "C," and between lines "B" and "C." 220 volts should be recorded between all of these points. If voltage is not present, one or all of the fuses are blown or the circuit breaker has been tripped. First, check for voltage above the fuses at all of these points, "A" to "B," "A" to "C," and "B" to "C," to make sure power is available (see 220 readings in Figure 15.23). If voltage is recorded, move leads back down to bottom of fuses. If voltage is present from "A" to "B," but not at "A" to "C" and "B" to "C," the fuse on line "C" is blown. If there were no voltage readings at any of the test points, all the fuses could be blown.

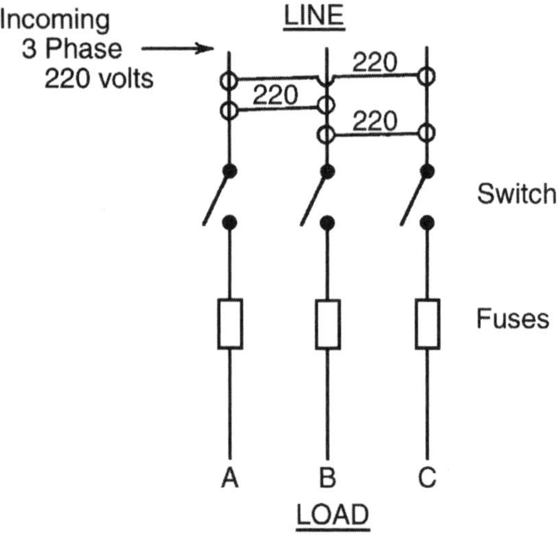

Fig. 15.23 Three-phase circuit, 220 volts

Another way of checking the fuses on this three-phase circuit would be to take your multimeter and place one lead on the bottom, and one lead on the top of each fuse. You should not get a voltage reading on the multimeter. This is because electricity takes the path of least resistance. If you get a reading across any of the fuses (top to bottom), that fuse is bad.

Always make sure that when you use a multimeter it is set for the proper voltage. If voltage is unknown and the meter has different scales that are manually set, always start with the highest voltage range and work down. Otherwise, the multimeter could be damaged. Look at the equipment instruction manual or nameplate for the expected voltage. Actual voltage should not be much higher than given unless someone goofed when the equipment was wired and inspected.

Voltage readings are important because they determine how you connect motor leads, relays, and transformers. Low or high voltages can drastically affect motors. Operators of small wastewater systems should also be aware of the common but little understood problem of unbalanced current. Operating a pumping unit with unbalanced current can seriously damage three-phase motors and cause early motor failure. Unbalanced current reduces the starting torque of the motor and can cause overload tripping, excessive heat, vibrations, and overall poor performance. (Section 15.431, Problem 7, describes how to test circuits for unbalanced current.)

Another meter used in electrical maintenance and testing is the *AMMETER.* The ammeter records the current or amps in the circuit. There are several types of ammeters, but only two will be discussed in this chapter. The ammeter generally used for testing is called a clamp-on type (Figures 15.24 and 15.25). The term "clamp-on" means that it can be clamped around a lead or each lead supplying a motor lead, and no direct electrical connection needs to be made. These are used by clamping the meter over only one of the power leads to the motor or other apparatus and taking a direct reading. Each "leg" or lead on a three-phase motor must be checked by itself.

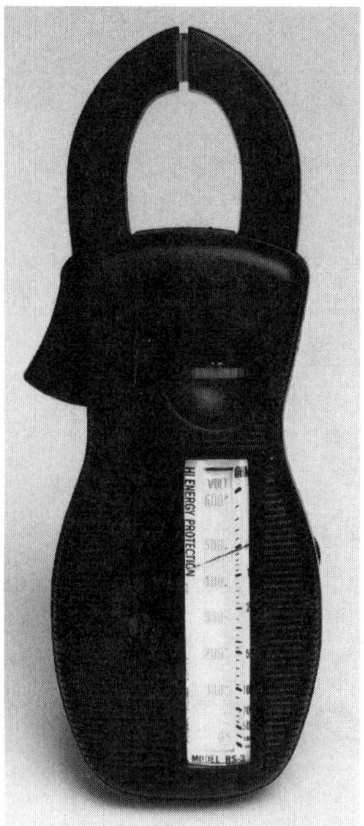

Fig. 15.24 Analog clamp-on type ammeter

(Permission of Amprobe)

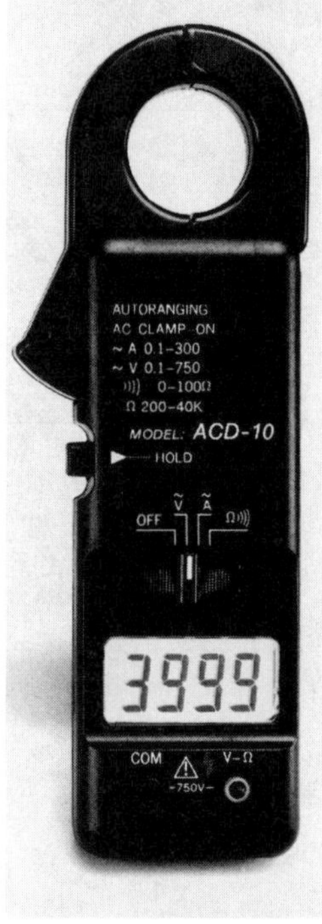

Fig. 15.25 Digital clamp-on type ammeter

(Permission of Amprobe)

The first step should be to read the motor nameplate data and find what the amperage reading should be for the particular motor or device you are testing. After you have this information, set the ammeter to the proper scale. Set it on a higher scale than necessary if the expected reading is close to the top of the meter scale. Place the clamp around one lead at a time. Record each reading and compare with the nameplate rating. If the readings do not compare with the nameplate rating, find the cause, such as low voltage, bad bearings, poor connections, plugging, or excessive load. If the ammeter readings are higher than expected, the high current could produce overheating and damage the equipment. Try to find the problem and correct it.

When using a clamp-on ammeter, be sure to set the meter on a high enough range or scale for the starting current if you are testing during start-up. Starting currents range from 500 to 700 percent higher than running currents and using too low a range can ruin an expensive and delicate instrument. Newer clamp-on ammeters automatically adjust to the proper range and can measure both starting or peak current and normal running current.

Another type ammeter is one that is connected in line with the power lead or leads. Generally, they are not portable and are usually installed in a panel or piece of equipment. They require physical connections to put them in series with the motor or apparatus being tested. These ammeters are usually more accurate than the clamp-on type and are used in motor control centers and pump panels.

A *MEGGER* or *MEGOHMMETER* is used for checking the insulation resistance on motors, generators, feeders, bus bar systems, grounds, and branch circuit wiring. This device actually applies a DC test voltage, which can be as high as 5,000 volts DC, depending on the megohmmeter selected. The one shown in Figure 15.26 is a hand-held, hand-cranked system that applies 500 volts DC and is particularly useful for testing motor insulation. Battery-operated and instrument-style meggers are also available in both analog and digital models.

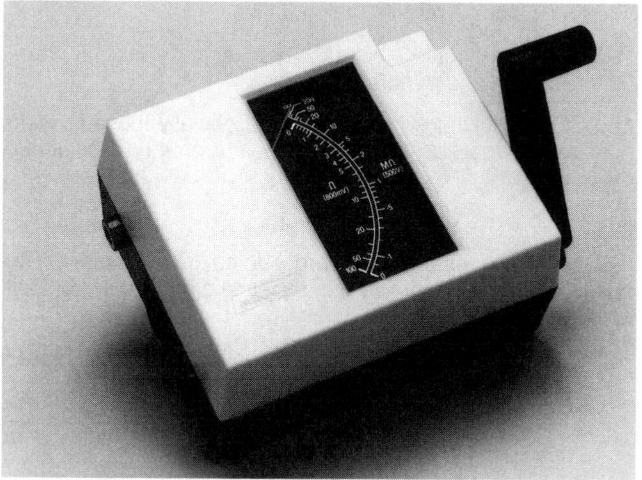

Fig. 15.26 Hand-cranked megohmmeter

(Permission of AVO International)

WARNING

Turn off the circuit breaker when using a megger.

To use a megger, there are two leads to connect. One lead is clamped to a ground lead and the other to the lead you are testing. The readings on the megger will range from "0" (ground) to infinity (perfect), depending on the condition of your circuit.

The megger is usually connected on the motor terminals, one at a time, at the starter, and the other lead to the ground lead. Results of this test indicate if the insulation is deteriorating or cut.

If a low reading is obtained, disconnect motor leads from power or line leads. Meg motor and if a low reading is observed, the motor winding insulation is breaking down. If a good reading is obtained, meg the circuit or branch wiring. If this reading is low, the wiring to the motor is bad. A rule of thumb is not to run a motor if it is less than one *MEGOHM*[11] per horsepower. This means a 5-horsepower motor should have a 5-megohm reading; 10 horsepower, a 10-megohm reading; 20 horsepower, a 20-megohm reading; and so forth.

Motors and wirings should be megged at least once a year, and twice a year, if possible. The readings taken should be recorded and plotted to determine when insulation is breaking down. Meg motors and wirings after a pump station has been flooded. If insulation is wet, excessive current could be drawn and cause pump motors to "kick out" (shut off).

OHMMETERS, sometimes called circuit testers, are valuable tools used for checking electric circuits. An ohmmeter is used only when the electric circuit is OFF, or de-energized. The ohmmeter supplies its own power by using batteries. An ohmmeter is used to measure the resistance (ohms) in a circuit. These are most often used in testing the control circuit components, such as coils, fuses, relays, resistors, and switches. They are used also to check for continuity. An ohmmeter has several scales that can be used. Typical scales are: R × 1, R × 10, R × 1,000, and R × 10,000. Each scale has a level of sensitivity for measuring different resistances. To use an ohmmeter, set the scale, start at the low point (R × 1), and put the two leads across the part of the circuit to be tested, such as a coil or resistor, and read the resistance in ohms. A reading of infinity would indicate an open circuit, and a "0" would indicate no resistance. Ohmmeters usually would be used only by skilled technicians because they are very delicate instruments.

The motor rotation indicator illustrated in Figure 15.27 is another specialized instrument used in electrical maintenance. This device is useful for determining the phase rotation of utility power when connecting 3-phase motors to ensure that the motor is connected properly for correct rotation.

All meters should be kept in good working order and calibrated periodically. They are very delicate, susceptible to damage, and should be well protected during transportation. When readings are taken, they should always be recorded on a machinery history card for future reference. Meters are a good way to determine pump and equipment performance.

[11] *Megohm* (MEG-ome). Millions of ohms. Mega- is a prefix meaning one million, so 5 megohms means 5 million ohms. For additional information, see *A STITCH IN TIME: THE COMPLETE GUIDE TO ELECTRICAL INSULATION TESTING.* Available from AVO Training Institute, Technical Resource Center, Electrical Training Services, 4271 Bronze Way, Dallas, TX 75237-1019. Order No. AVOB001. Price, $7.00.

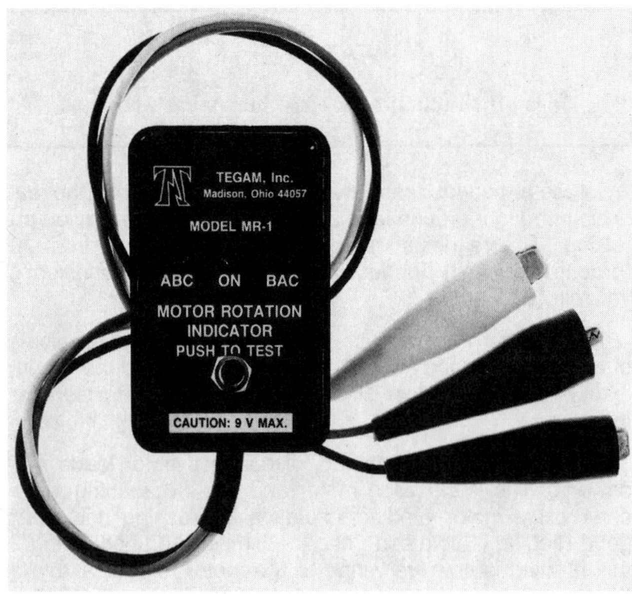

Fig. 15.27 Motor rotation indicator
(Permission of Tegam Inc.)

CAUTION: Never use a meter unless you are qualified and authorized. The risk of electric shock or electrocution is very high if meters are used by unqualified personnel.

QUESTIONS

Write your answers in a notebook and then compare your answers with those on page 461.

15.3D How can you determine if there is voltage in a circuit?

15.3E What are some of the uses of a multimeter?

15.3F What precautions should be taken before attempting to change fuses?

15.3G How do you test for voltage with a multimeter when the voltage is unknown?

15.3H What could be the cause of amp readings different from the nameplate rating?

15.3I How often should motors and wirings be megged?

15.3J An ohmmeter is used to check the ohms of resistance in what control circuit components?

15.33 Electrical System Equipment

15.330 Need for Maintenance

Electrical system equipment is frequently the least understood and, therefore, most neglected equipment in a pump station. Usually, we do not think of it as a system that requires frequent inspection or maintenance; however, the reverse of this is actually true. Electrical equipment can be damaged more readily by operating conditions than almost any other kind of equipment. Water, dust, heat, cold, humidity, corrosive atmospheres, and vibration are all common pump station conditions that can affect the performance and the life of electrical equipment.

This section will discuss various elements in the electrical system of a pump station including motor control center components such as circuit breakers, contactors, protective devices, transformers, and control relays. The section will also explain in detail important maintenance and operating aspects of the AC induction motor, the most common pump driver used in pump stations.

Electrical equipment should be inspected and maintained on at least an annual basis or more frequently depending on the equipment and the application. Inspection should include a thorough examination, replacement of worn and expendable parts, and operational checks and tests.

Listed below are examples of electrical equipment maintenance tasks. Check and inspect each of these items annually.

1. All switch gear and distribution equipment. Look for worn parts and note general condition.
2. Wiring integrity.
3. Terminal connections. Tighten, if necessary (often needed with aluminum wire conductors).
4. Interlocking devices to prevent unauthorized entry.
5. Control circuits. Check operation and verify sequencing, including actual sequencing of pump motors as a function of wet well level.
6. All panel instruments. Clean and check for accuracy. Permanently installed, panel-mounted current and multimeters can be added to existing electrical systems and should be specified on new projects. This eliminates the need for routine access to the inside of the electrical system to obtain voltage current readings and exposure to hazardous voltages. Select a multimeter that measures both phase-to-phase and phase-to-neutral voltage. Current meters should also allow switching to measure current in each of the three phases.
7. Mechanical disconnect switches. Service and lubricate all switches, fuses, disconnects, and transfer switches.
8. Fuses. Verify proper application, size, and general condition. Check inventory of spare fuses.
9. Circuit breakers. Cycle each breaker and check for proper response and performance.
10. Contacts. Check for response (especially those in motor starters that carry high switching current); replace, if necessary.
11. Enclosures. Clean and vacuum.

If these basic maintenance procedures are done at least annually, it will help minimize pump station electrical system failures including:

1. Current imbalances or unbalances, which ultimately result in motor failure
2. Loose contacts or terminals in control and power circuits causing high-resistance contacts
3. Overheating resulting in arcing, fire, and electrical system damage
4. Dirty enclosures and components, which allow a conductive path to build up between incoming phases causing phase-to-phase shorts
5. Corrosion that causes high-resistance contacts and heating

15.331 Equipment Protective Devices

Electricity, by its very nature, is extremely hazardous and safety devices are needed to protect operators and equipment. Water and wastewater systems have pressure valves, pop-offs, and different safety equipment to protect the pipes and equipment. So must electricity have safety devices to contain the voltage and amperage that come in contact with the wiring and equipment.

Pump station electrical equipment protective devices may consist of fuses, circuit breakers, motor and circuit overload devices, and grounds. These devices usually are found in combination so that all three are present. For example, a fused disconnect switch (see Figure 15.28) may protect the entire electrical distribution system within the pump station; circuit breakers may be used to protect branch motor circuits from short circuits; and overload elements would be installed to protect the motor from overloads.

Figure 15.29 is a block diagram of motor circuit elements showing a typical arrangement of control and protective devices used to apply 3-phase line voltage to the motor. The first component in the circuit is some means of disconnecting the motor. This can be either a safety switch, a fused disconnect switch, or a circuit breaker. When using a fused disconnect switch or circuit breaker (the second element in the block diagram), the motor branch circuit overcurrent protection is incorporated. The next element in the block diagram is the motor controller, also referred to as a contactor. This is a 3-phase switch that connects 3-phase line voltage from the supply to the motor terminals. The final element in the diagram is the motor running overcurrent protection device. This device can be one that uses thermal elements that sense motor running current and trip under overload conditions or one of the newer solid-state motor control devices that also sense motor running current and provide overload protection. When the motor controller, or contactor, and the motor overload device are combined in a single unit, they are referred to as a motor starter.

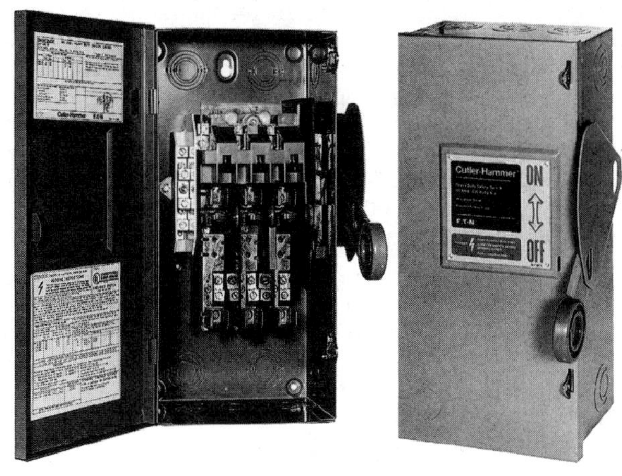

Fig. 15.28 Fused disconnect switch

(Permission of Eaton)

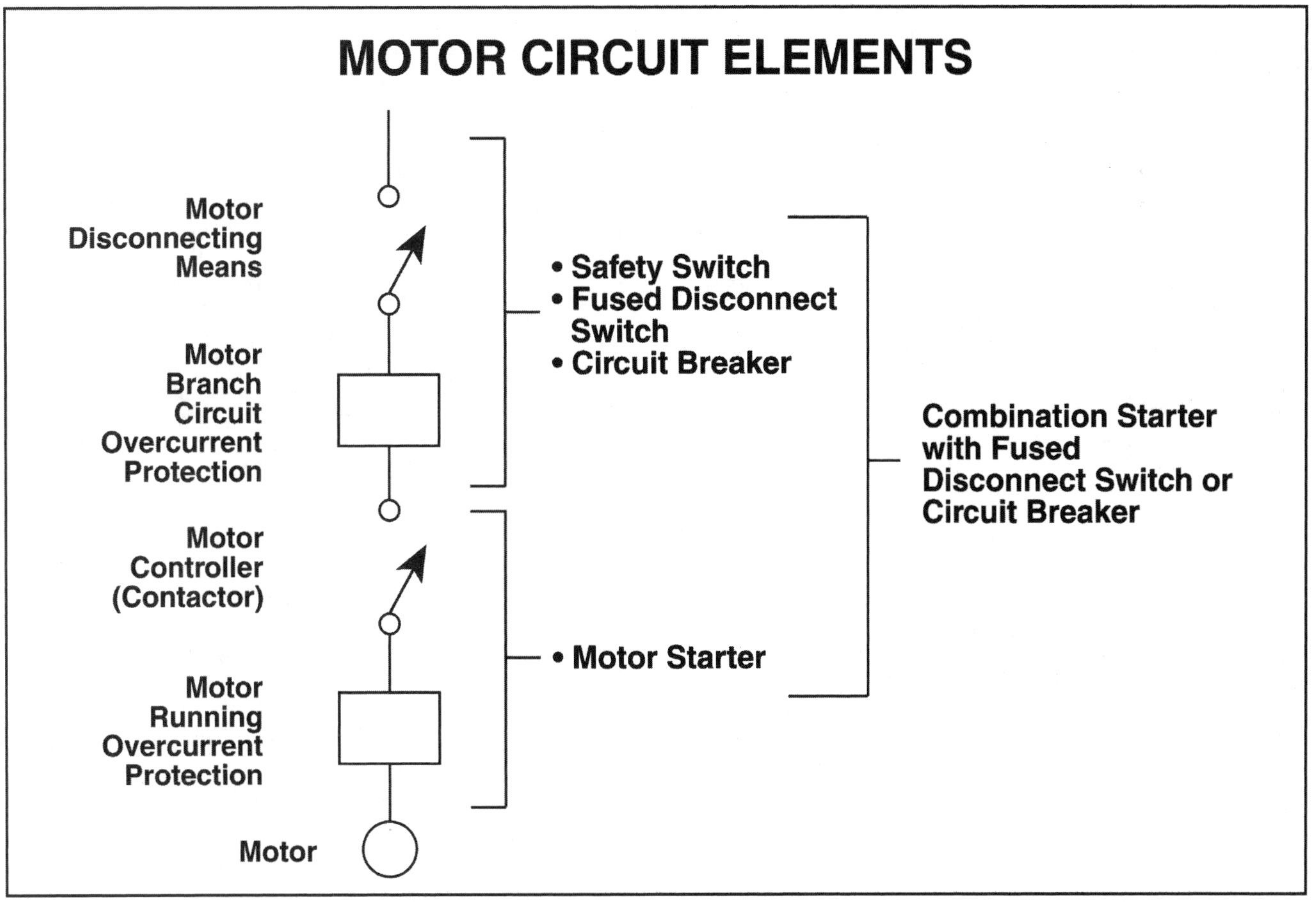

MOTOR CIRCUIT ELEMENTS

Motor Disconnecting Means

Motor Branch Circuit Overcurrent Protection

- **Safety Switch**
- **Fused Disconnect Switch**
- **Circuit Breaker**

Motor Controller (Contactor)

Motor Running Overcurrent Protection

- **Motor Starter**

Combination Starter with Fused Disconnect Switch or Circuit Breaker

Motor

Fig. 15.29 Block diagram of motor circuit elements

The motor branch circuit breaker may provide only short-circuit fault current protection if it is a magnetic only breaker or, in the case of a thermal magnetic circuit breaker, it may provide overload protection as well. Typically the motor branch circuit breaker is a molded-case circuit breaker (described later in Section 15.333).

Regardless of which type of protective device you are working with, if it blows or trips, the source of the problem should be investigated, identified, and corrected. All too frequently, a fuse is simply replaced or a circuit breaker simply reset without any investigation into the problem. If the protective device has operated reliably for long periods of time (for example, no nuisance tripping), then tripping is almost a sure indication that a problem exists somewhere in the circuit and the device is trying to protect the circuit. Simply resetting the circuit breaker or replacing a fuse and then restarting the system may result in damage or more severe damage to the equipment that is being protected.

Once again, if you are not qualified to perform electrical troubleshooting, you should not attempt to proceed any further than providing diagnostic analysis. Maintenance must always be performed by trained personnel. It does not take high voltage or extreme currents to seriously injure an operator or cause a fatality.

15.332 Fuses (Figure 15.30)

The power company installs fuses on their power poles to protect their equipment from damage. We also must install something to protect the main control panel and wiring from damage due to excessive voltage or amperage.

A fuse is a protective device having a strip or wire of fusible metal which, when placed in a circuit, will melt and break the electric circuit when subjected to an excessive temperature. This temperature will develop in the fuse when a current flows through the fuse in excess of what the circuit will carry safely. This means that the fuse must be capable of de-energizing the circuit before any damage is done to the wiring it is safely protecting. Fuses are used to protect operators, wiring, main circuits, branch circuits, heaters, motors, and various other electrical equipment. A fuse must never be bypassed or jumped because the fuse is the only protection the circuit has. Without it, serious damage to equipment and possible injury to operators can occur.

There are several types of fuses, each being used for a certain type of protection. Some of these are:

1. Current-Limiting Fuses: Used to protect against current in circuits.
2. Dual-Element Fuses: Used for motor protection.
3. Time-Delay Fuses: Used in electronic and motor starting circuits.
4. Sand-Filled Fuses: Used on high voltage.
5. Phase Fuses: Used to protect phase sequence.
6. Voltage-Sensitive Fuses: Used where close voltage control is needed.

Since fuses are one-time-use devices, little can be done to check them during maintenance procedures; however, the following tasks should be performed at least annually:

1. Inspect bolted connections at the fuse clip or fuse holder for signs of looseness.
2. Check connections for any evidence of corrosion from moisture or atmosphere (air pollution).
3. Tighten connections.
4. Check fuse for obvious overheating.
5. Inspect insulation on the conductors coming into the fuses on the line side and out of the fuses on the load side for evidence of discoloration or bubbling, which would indicate overheating of the conductors.

Definition Line Side/Load Side

Line side/load side are terms frequently used to describe the incoming and outgoing conductors of circuit breakers, motor starters, and other devices. The line side of the device is where incoming power is fed into the device. The load side is the terminal where power is fed to the load, for example, a motor.

Usually, when a fuse blows it is not visibly apparent and it is necessary to check the fuse with a meter. The following procedure should be followed to determine whether or not a fuse has blown:

1. Ensure the main disconnect or circuit breaker is in the OFF position and locked out.
2. Check for live voltage in the panel (power may feed into a control system from other sources).
3. Remove fuse with a fuse-pulling device.
4. Test resistance of fuse using an ohmmeter. An open circuit (infinite resistance) indicates a blown fuse, whereas, a short circuit indicates a good fuse).
5. Only in the event it is not possible to disconnect incoming power should the fuse be checked with power applied. In this case, the procedure as outlined in Section 15.32, "Tools, Meters, and Testers," may be used.

15.333 Circuit Breakers (Figure 15.31)

A circuit breaker is a switch that is opened automatically when the current or the voltage exceeds or falls below a certain limit. Unlike a fuse that has to be replaced each time it blows, a circuit breaker can be reset after a short delay to allow time for cooling. Circuit breakers are used as disconnecting devices and to protect electrical systems primarily from short-circuit conditions that may occur on the load side of the circuit breaker. In some cases, circuit breakers can also be used in conjunction with motor starters to provide motor overload protection. In most cases, however, they are sized to protect the entire system as opposed to specific components in the circuit from short-circuit fault currents.

Molded-Case Circuit Breakers (Figure 15.32)

Pump station motor control center circuit breakers are typically housed in a molded case unless the motor or other electrical equipment is high horsepower. Molded-case circuit breakers can be thermal-magnetic trip or magnetic only.

Thermal-magnetic trip circuit breakers provide both overload protection and short-circuit protection. If the circuit is overloaded, a thermal sensing element will detect this and cause the circuit breaker to open or trip. In addition, the magnetic sensing element of the circuit breaker rapidly senses extremely high currents, which flow under a short-circuit condition, and will also trip the circuit breaker.

The amount of current that can flow during a short-circuit condition is a function of many things. For this reason, circuit breaker replacement should be done very carefully and by personnel who are qualified to evaluate circuit characteristics.

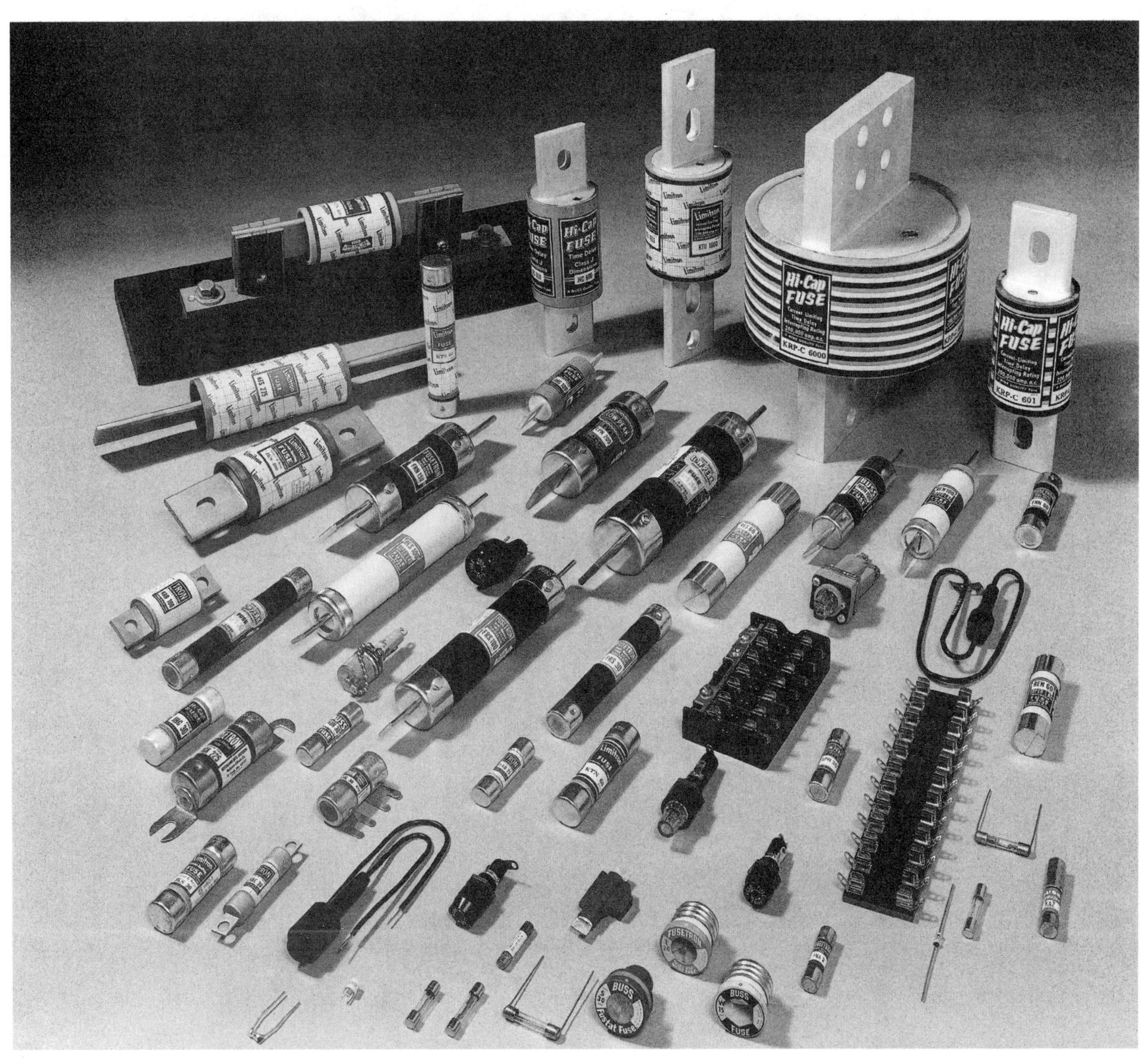

Fig. 15.30 Fuses
(Courtesy of Bussman Manufacturing, McGraw-Edison Company Division)

Fig. 15.31 Circuit breaker panelboard
(Permission of Square D Company)

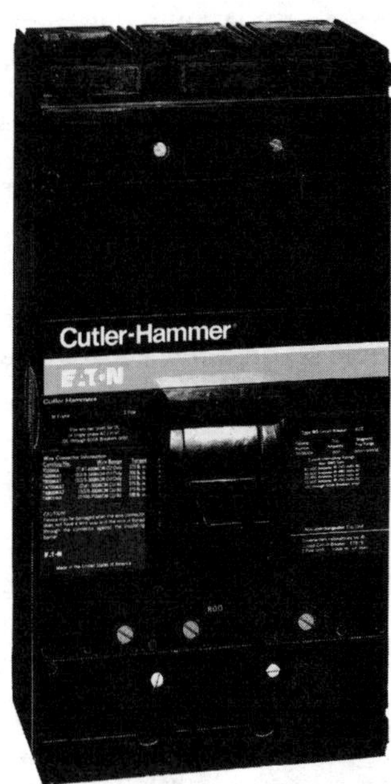

Fig. 15.32 Molded-case circuit breaker

(Permission of Eaton Corporation)

Molded-case circuit breakers require little maintenance other than:

1. Manually trip and operate the mechanism.
2. Check connections for tightness.
3. Inspect for evidence of overheating on the line and load side conductors.
4. Inspect the circuit breaker case for evidence of overheating.

The following additional tests can be performed on molded-case circuit breakers, but require specialized equipment.

1. Insulation resistance test. This test uses a high-voltage DC megger device (described previously in this chapter) to check the internal resistance of the circuit breaker.
   a. Phase-to-phase check. Apply high voltage between the three phases of the circuit breaker, which ultimately could lead to a phase-to-phase short. This check is performed by applying the high voltage to each combination of two phases of the three phases (1 and 2, 1 and 3, and 2 and 3) one at a time. At least one megohm of resistance in the insulation is adequate. Inadequate insulation resistance could lead to a phase-to-phase short.
   b. Phase-to-ground. To check the insulation resistance between each individual phase to ground, connect one phase on the circuit breaker at a time to ground, apply a high voltage, and measure the resistance. At least one megohm of resistance in the insulation is desirable.
2. Contact resistances. With a DC power supply, measure the voltage drop from the line side to the load side of the circuit breaker with the circuit breaker closed, but power disconnected. An excessive voltage drop indicates a high-resistance condition between the contacts, and the circuit breaker must be replaced.
3. Overload trip test. This test requires a specialized circuit breaker tester that is capable of generating currents in the range of the trip devices of the circuit breaker. Connect the test device between the line and the load and run a current equal to the circuit breaker capacity through the circuit breaker. A current in the range of the trip devices that causes them to trip will indicate that the circuit breaker is functioning properly.

Because the interior components of molded-case circuit breakers are not accessible, repair or replacement of internal parts is not possible. The entire circuit breaker must be replaced when it fails.

Motor Protection Devices

Since circuit breakers are normally used as a disconnecting device and as a protective device for the entire circuit, including the conductors that are connected to the breaker, the circuit breaker must be sized for the current carrying capacity of all components or equipment connected to it. For example, a thermal-magnetic circuit breaker must have a continuous current rating of at least 115 percent of motor full-load amps and the rating may be as great as 250 percent. This means that the thermal overload sensing device will not trip the breaker unless there is 115 percent overload present. In most cases, by the time this occurs, the motor will have been seriously damaged.

To overcome this problem and to provide further protection for motors, a device called a motor starter is usually installed in pump station motor control centers. The device operates through the supervisory control system responding automatically to the level in the wet well. A typical configuration is referred to as a magnetic motor starter and can be operated manually as well as automatically. Section 15.44 of this chapter contains a complete discussion of motor starters.

CAUTION: Never increase the rating of the overload heaters because of tripping. You should find the problem and repair it.

Internal thermal protection devices are another form of motor and insulation protection. Such devices are especially desirable on motors that are likely to experience occasional overload conditions or voltage imbalances. Thermal protection devices may include the following features:

1. Built-in thermal switches that automatically open when the motor overheats. This type of device will not reset until the motor cools, at which time the switch resets automatically. (In some cases, there may be a reset button on the motor enclosure.) This type of protection is normally limited to small, single-phase motors.
2. Three-phase motors with internal thermal protection have thermostats embedded in each phase of the three-phase windings. The thermostats are connected in series and brought out into the motor conduit box and labeled P1 and

P2. The switches are connected into the motor control circuit, which will open the magnetic starter contacts when an over-temperature condition occurs in the windings.

Two circuit configurations are available:

a. One in which the control circuit automatically restarts when the thermal switches and the motor reset

b. A circuit that is locked out and must be manually reset in order to restart

Figure 15.33 illustrates a typical control circuit in which normally closed thermostats are connected in series with the motor starter coil in the control circuit. *NOTE:* This is the type of control circuit that will automatically reset when the motor thermostat's switches are reset.

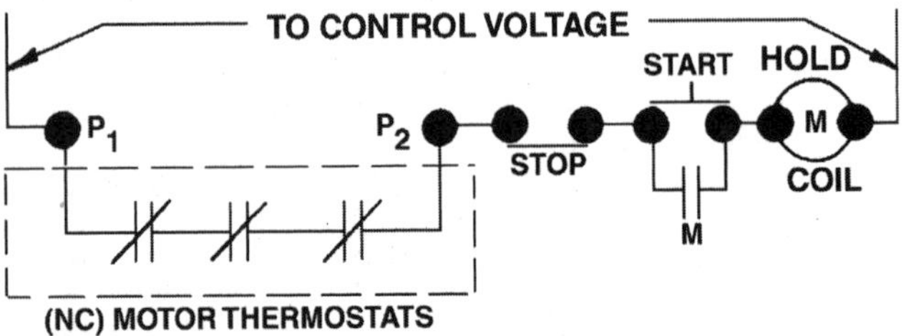

Fig. 15.33 Closed thermostats connected in series with the motor starter coil

There are many other protective devices for electricity, such as motor winding thermostats, phase protectors, low-voltage protectors, and ground-fault circuit interrupters. Each has its own special applications and should never be tampered with or jumped (bypassed).

15.334 Ground

"Ground" is an expression representing an electrical connection to earth or a large conductor that is at the earth's potential or neutral voltage. Motor frames and all electrical tools and equipment enclosures should be connected to ground. This is generally referred to simply as grounding, or equipment ground.

Connecting motor frames, tools, and electrical equipment to ground is a safety precaution that protects you and the motors, tools, and equipment. If one of the conductors opens up and is not connected to ground, then *you* become the ground, the current could flow through you, and you could receive a severe or fatal electric shock. If the current flows through motors, tools, or equipment, severe damage could occur.

The third prong on cords from electric hand tools is the equipment ground and must never be removed. When an adapter is used with a two-prong receptacle, the green wire on the adapter should be connected under the center screw on the receptacle cover plate. Many times equipment grounding, especially at home, is achieved by connecting onto a metal water pipe or drain rather than a rod driven into the ground. This practice is not recommended when plastic pipes and other nonconducting pipe materials are used. Also, corrosion can be accelerated if pipes of different metals are used. A rod driven into dry ground is not very effective as a ground.

QUESTIONS

Write your answers in a notebook and then compare your answers with those on pages 461 and 462.

15.3K What is the most common pump driver used in pump stations?

15.3L Basic pump station maintenance procedures performed at least annually will help minimize what types of pump station failures?

15.3M What should be done when a fuse or circuit breaker blows or trips?

15.3N What are two types of safety devices found in main electrical panels or control units?

15.3O What are fuses used to protect?

15.3P Why must a fuse never be bypassed or jumped?

15.3Q What types of annual maintenance should be performed with regard to fuses?

15.3R How does a circuit breaker work?

15.34 Motor Control/Supervisory Control and Electrical System

The motor and supervisory control systems are composed of the auxiliary electrical equipment, such as relays, transformers, lighting panels, pump control logic, alarms, and other electrical equipment typically found in a pump station electrical system, over and above the protective devices and the motor starters.

In general, annual maintenance should be performed on all these systems as follows:

1. Control Transformers
   a. Check primary, secondary, and ground connections.
   b. Check for loose windings/coils.
   c. Inspect insulation for signs of overheating as a result of overloading.
   d. Check mounting for tightness.
   e. Check primary/secondary fusing and fuse clips for tightness.
2. Motor Control Centers
   a. Check panel lights for operation.
   b. Check control knobs/switches for freedom of movement and contact condition.

   c. Check horizontal and vertical bus and supports for evidence of heating or arcing and tighten. (Bus refers to the copper or aluminum bars that run horizontally and vertically in the motor control center; they feed the three-phase power to the branch circuits.)

3. Control Relays

   a. Check mounting for looseness.

   b. Tighten all screw terminal connections.

   c. Check for evidence of overheating or arcing indicated by carbon buildup or discoloration of plastic housing.

4. Clean and vacuum enclosure

The use of aluminum wire as a conductor has become very common because of its economic advantage over copper. Copper traditionally has been used for virtually all wiring applications in wastewater lift stations including the conductors feeding the station from the utility transformers, bus bar, motor control centers, control wiring, as well as power wiring to motors.

The greatest advantage of copper is that it oxidizes very slowly. Even though this reaction with air or moisture results in a surface layer of impurities, the surface is soft and is easily penetrated by the connecting device. With the exception of the annual tightening of the terminals, copper requires very little maintenance.

Aluminum is much softer and reacts much more rapidly with the air so aluminum begins oxidizing almost immediately when exposed to air. This oxide, as opposed to that found on copper, tends to form an insulating layer over the aluminum wire. Aluminum also expands and contracts 36 percent more than copper, which will result in loose connections, high-resistance contacts, heat formation, and failure of the connection. When using aluminum conductors, the following rules must be observed:

1. The connecting terminal must be specifically designed to accommodate aluminum conductors (connectors that are rated for this use are stamped with the letters CU/AL indicating that they are suitable for use with copper (CU) or aluminum (AL) conductors).

2. The termination point of the aluminum conductor must be coated with a compound to prevent the formation of the insulating oxides and should be wire-pressure scraped before the compound is applied.

Because of these limitations, aluminum conductors and bus bars should be inspected and maintained more frequently than the traditional copper conductors. Conduct a visual inspection of the terminal, conductor, and insulation for any evidence of discoloration and overheating.

Three basic factors contribute to the reliable operation of electrical systems found in pump stations. They are:

1. An adequate preventive maintenance program must be implemented and maintained.

2. A knowledge of the collection system by the pump station operator, even though the pump operator does not perform the actual collection system maintenance.

3. Adherence to three principles: KEEP IT CLEAN! KEEP IT DRY! KEEP IT TIGHT!

QUESTIONS

Write your answers in a notebook and then compare your answers with those on page 462.

15.3S Motor and supervisory control systems are composed of what types of auxiliary electrical equipment?

15.3T What are the three basic factors that contribute to the reliable operation of electrical systems found in pump stations?

15.4 MOTORS

15.40 Types

Electric motors are the machines most commonly used to convert electrical energy into mechanical energy. Although a multitude of different types of motors are produced today, the most common pump motor is the AC (Alternating Current) induction motor.

Two types of induction motor construction are typically encountered when dealing with AC induction motors:

1. Squirrel Cage Induction Motor (SCIM, by far the most numerous)

2. Wound Rotor Induction Motor (WRIM, used for variable-speed applications)

Figure 15.34 illustrates a horizontal squirrel cage induction motor with a drip-proof enclosure. Major components of the motor are:

1. Stator winding with connection to the three-phase power supply

2. Shaft end and opposite end ball bearings

3. Rotor assembly, including rotor, shaft, and fans on the rotor cage

4. Enclosure

Stator construction is generally the same in both types of motors. The primary difference is that the squirrel cage induction motor has no electrical connections to the rotor circuit. In contrast, a wound rotor motor has a slip ring assembly and brushes that connect the rotor circuit to an external electric circuit. This arrangement varies the resistance, thus causing the speed characteristics of the wound rotor motor to change. Therefore, a wound rotor is usually found in a variable-speed

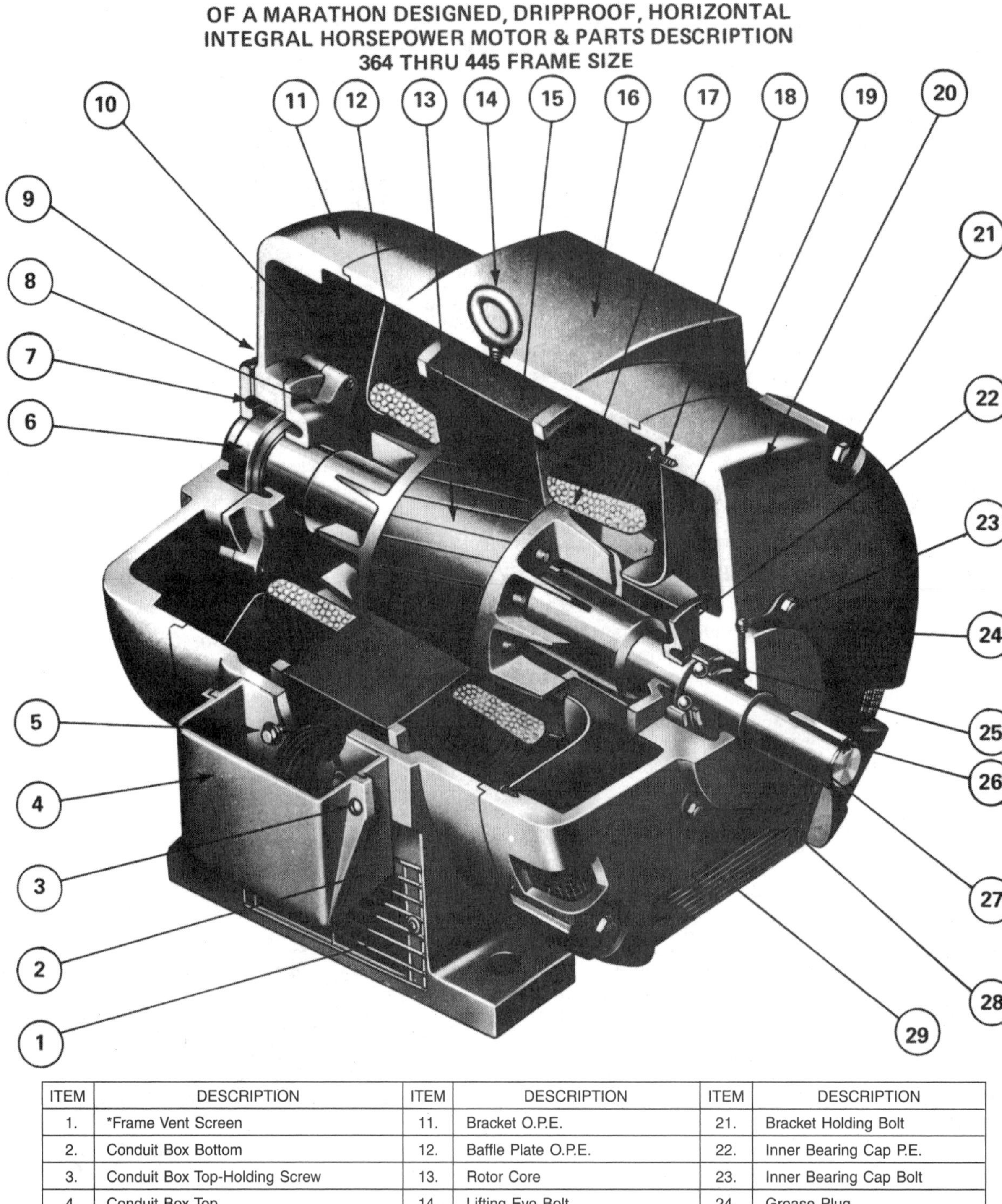

ITEM	DESCRIPTION	ITEM	DESCRIPTION	ITEM	DESCRIPTION
1.	*Frame Vent Screen	11.	Bracket O.P.E.	21.	Bracket Holding Bolt
2.	Conduit Box Bottom	12.	Baffle Plate O.P.E.	22.	Inner Bearing Cap P.E.
3.	Conduit Box Top-Holding Screw	13.	Rotor Core	23.	Inner Bearing Cap Bolt
4.	Conduit Box Top	14.	Lifting Eye Bolt	24.	Grease Plug
5.	Conduit Box Bottom Holding Bolt	15.	Stator Core	25.	**Ball Bearing P.E.
6.	**Ball Bearing O.P.E.	16.	Frame	26.	Shaft Extension Key
7.	Pre-Loading Spring	17.	Stator Winding	27.	Shaft
8.	Inner Bearing Cap O.P.E.	18.	Baffle Plate Holding Screw	28.	Drain Plug (grease)
9.	Grease Plug	19.	Baffle Plate P.E.	29.	*Bracket Screen
10.	Inner Bearing Cap Bolt	20.	Bracket P.E.		

P.E. = Pulley End
O.P.E. = Opposite Pulley End
* = Bracket and frame screens are optional.
** = Bearing Numbers are shown on motor nameplate. When requesting information or parts, always give complete motor description, model, and serial numbers

Fig. 15.34 Horizontal squirrel cage induction motor

(Courtesy of Marathon Electric)

pump station application. This section will deal specifically with the squirrel cage induction motor since this is the type you will encounter most often.

15.41 Nameplate Data (Figure 15.35)

As part of your preventive maintenance program, record the motor nameplate information in the file for each motor or piece of equipment as recommended in NEMA (National Electrical Manufacturers Association) publication MG1, Section 10.3A. Motor nameplate data must be recorded and filed so the information is available when needed to repair the motor or to obtain replacement parts.

Following is a brief description of the information you could find on a nameplate and how it relates to motor performance.

Serial Number. This is a unique number assigned by the manufacturer based on the manufacturer's numbering system to identify that specific motor. (This number should be available and used whenever it is necessary to communicate with the manufacturer.)

Type. This may be a combination of letters and numbers, established by the manufacturer to identify the type of enclosure and any modifications to it.

Model Number. The manufacturer's model number.

Horsepower. Rated horsepower is the horsepower that the motor is designed to produce at the shaft when power is applied at the rated frequency and voltage and the motor is operating at a service factor of 1.0.

Frame. This identifies the frame size in accordance with established NEMA Standards and does not vary from manufacturer to manufacturer. Therefore, the motor from one manufacturer with a NEMA frame is dimensionally identical to that same frame size from another manufacturer.

Service Factor. Service factors of 1.0 and 1.15 are commonly found on pump station motors. A service factor of 1.0 indicates that the motor may be run continuously at its rated horsepower without causing damage to the insulation system. A 1.15 service factor indicates that the motor may occasionally be run at a horsepower equal to the rated horsepower times the service factor without serious injury to the insulation system. This allows for intermittent variations in voltage, which will cause some internal heating on the motor windings but not enough to cause damage. A service factor above 1.0 should never be relied upon to accommodate continuous loads because such use will quickly degrade the insulation system.

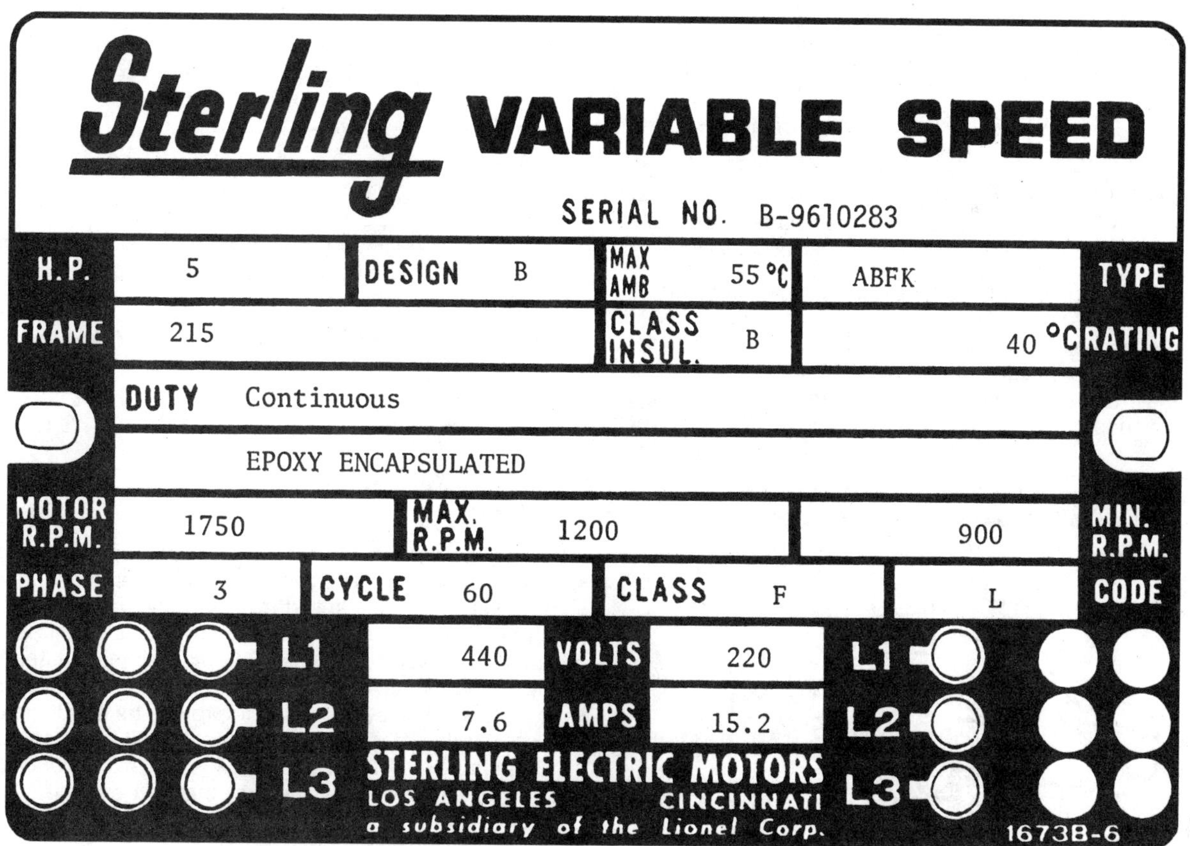

NOTE:

1. The motor for this unit is rated at 1,750 RPM and the maximum speed for the variable drive unit is 1,200 RPM.
2. The 40°C rating is the allowable operating temperature above ambient temperature.

Fig. 15.35 Typical nameplate

(Courtesy of Sterling Power Systems. Inc.)

Amps. The current drawn by the motor at rated voltage and frequency at rated horsepower.

Volts. This is the voltage that would be measured at the terminals of the motor as opposed to the voltage of the supply.

Class of Insulation. Various classes of insulation material (described in Section 15.43) are available and the listed class will determine the operating temperatures at which the motor can safely operate.

RPM. Speed of the motor shaft, in revolutions per minute, at rated horsepower with rated voltage and frequency being supplied.

Hertz. Frequency, in cycles per second, of the utility company power.

Duty. Normally stamped "continuous," which means the motor can operate 24 hours a day, 365 days a year. In some cases, the motor nameplate will indicate "intermittent duty" for a specified time interval. This means the motor can operate at full load for the time interval specified, and then must be shut down and allowed to cool before restarting.

Ambient Temperature. Specifies the maximum ambient (surrounding) temperature at which the motor can safely operate. If the ambient temperature is exceeded, a corresponding increase of operating temperatures in the winding will degrade the insulation system and cause premature failure of the motor.

Phase. This indicates the number of phases; for example, 3 or three-phase, which indicates the phase for which the motor is designed.

KVA Code. Starting inrush current, which relates KiloVolt Amps to HorsePower (KVA/HP).

Design. This letter indicates the electrical design characteristics and, therefore, torque, speeds, inrush current, and slip values (these values and codes are also specified by NEMA).

Bearings. This number and letter sequence is designated by the Anti-Friction Bearing Manufacturers' Association (AFBMA) standards. It specifies looseness of the bearing fit, type of retainer, degree of protection, and dimension of the bearings. In most cases, unless the bearing is an extremely specialized type, the standard allows the use of bearings of different manufacturers.

Efficiency. This indicates the nominal operating efficiency of the motor at full load (Power Out/Power In)(100%).

15.42 Causes of Failure

Over the years, a considerable amount of research has been (and continues to be) conducted by various associations and manufacturers of electrical equipment to identify types of motor failures and the reasons for failures in AC induction motors. This section will deal with some of the common causes of motor failure and will suggest what you can do to minimize motor failures in pump stations.

In a survey of 9,000 motor failures by the Electrical Research Association, Letterhead, England, the following causes and percentages of motor failures were identified:

Causes of Motor Malfunction	Frequency, %
1. Overload (Thermal)	30
2. Contaminants	19
3. Single phasing	14
4. Bearing failures	13
5. Old age	10
6. Miscellaneous	9
7. Rotor failures	5

A study conducted by the Dymac Division of Scientific Atlanta identifies the frequency with which specific components failed.

Failed Component	Frequency, %
1. Stator windings	70
2. Rotor windings	10
3. Bearings	10
4. Other	10

This information suggests that a large percentage of motor failures could be prevented. Sixty-three percent of the failures examined were primarily due to overload, single phasing, and contaminants, and all of these factors can be controlled by the operator. In addition, since 70 percent of the failures occur in the stator, we can assume that many of the failures are related to the insulation system used to protect the motor stator windings. For this reason, the next section will focus on the motor insulation system and show how a variety of conditions can lead to excessive heat buildup that will eventually result in motor failure.

15.43 Insulation

15.430 Types and Specifications

Four classes or levels of insulation systems are available:

1. Class A
2. Class B

3. Class F

4. Class H

Each class of insulation is defined by the temperature limitation of the insulation itself. If the temperature limitation is exceeded, the insulation will deteriorate and, ultimately, cause a premature motor failure. The temperature rating consists of three components:

1. Ambient operating temperature—the air temperature where the motor is operating, for example, air temperature inside the pump station

2. Temperature rise—the maximum inside temperature of the windings during normal operation

3. Hot-spot allowance—because of inconsistencies in the manufacturing process and the winding materials, a 10°C hot-spot allowance is included

Older motors, those built before 1964, are likely to have Class A insulation systems. A standard motor today has a Class B system or, in some cases, Class F. In extreme environments, a Class H insulation system may be used.

Table 15.4, AC Motor Temperature Limits, lists the temperature limitations for the four classes of insulation as they relate to the type of motor construction. For example, for all motors with a 1.15 or higher service factor, the total temperature that the insulation can withstand is 140°C for Class F insulation, or 165°C with Class H insulation. Table 15.5 lists typical materials used for motor insulation.

15.431 Causes of Failure

Obviously, induction motors are a critical and integral part of the operating ability, efficiency, and reliability of the lift station. When failures occur, they can be extremely serious because the loss of pumping capacity creates critical operating problems. In addition, repair or replacement of a motor is an expensive, labor-intensive job.

If a motor is operated in a clean, dry environment and within its specified nameplate load and operating characteristics, there is no reason why the motor should not operate for years and years without major maintenance. Unfortunately, a number of different factors can cause a motor to operate outside its specified ranges. When this occurs, the life of the motor is shortened significantly.

In the paragraphs that follow, these conditions will be discussed in terms of problems/solutions. You will see that, in one way or another, failure to adhere to motor nameplate guidelines frequently results in excessive heat buildup, damaged motor insulation, and eventual motor failure.

1. *PROBLEM:* Contaminants. Insulation failure can occur when deposits of dust, grease, or other foreign material accumulate on the windings and prevent the dissipation of the heat generated in the motor winding during normal operation. This causes local hot spots in the winding and, when the insulation breaks down completely, it will cause a phase-to-phase short characterized by arcing and melting between the phases' windings.

Vertical motors are susceptible to improper greasing methods and materials. Frequently, grease escapes from the bearing and the bearing housing and then contaminates the upper end turns on the stator winding, resulting in winding failure.

1. *SOLUTION:* Keep motors clean and free of dirt or grease accumulations. Follow the manufacturer's recommended methods and materials for greasing the equipment.

TABLE 15.4 AC MOTOR TEMPERATURE LIMITS[a]

	Temperature (Degrees C)				
		Class A	Class B	Class F	Class H
1.0 Service Factor					
Drip-Proof					
Ambient Temperature	40	40	40	40	40
Rise by Thermometer	40	..	..	..	..
Rise by Resistance	..	50	80	105	125
Service-Factor Margin	10	10	..	..	..
Hot-Spot Allowance	15	5	10	10	15
Total Temperature	105	105	130	155	180
TEFC					
Ambient Temperature	40	40	40	40	40
Rise by Thermometer	55	..	..	..	..
Rise by Resistance	..	60	80	105	125
Hot-Spot Allowance	10	5	10	10	15
Total Temperature	105	105	130	155	180
TENV[b]					
Ambient Temperature	40	40	40	40	40
Rise by Thermometer	55	..	85	..	..
Rise by Resistance	..	65	..	110	135
Hot-Spot Allowance	10	0	5	5	5
Total Temperature	105	105	130	155	180
Encapsulated[c]					
Ambient Temperature				40	40
Rise by Thermometer				85	110
Hot-Spot Allowance				5	5
Total Temperature				130	155
1.15 or Higher Service Factor					
All Motors					
Ambient Temperature				40	40
Rise by Thermometer[d]				90	115
Hot-Spot Allowance				10	10
Total Temperature				140	165

[a] Adapted from National Electrical Manufacturers Association (NEMA) publication MG 1-12.39 and 12.40.
[b] Including all fractional-horsepower totally enclosed motors and fractional-horsepower motors smaller than frame 42.
[c] Enclosed.
[d] At service-factor load.

2. *PROBLEM:* Short cycling or excessive starts. Short cycling occurs when the automatic control system triggers frequent starts and stops of the pump and motor in response to fluctuating wet well elevations or because of a failure in the control system.

When induction motors start, the current (called locked rotor amps) required to magnetize the windings and start the rotor can be five to eight times the normal running current. For example, with a motor rated at 100 amps for full-load current, the starting sequence will require anywhere from 500 to 800 amps to start the motor rotating.

When frequent starts occur, the heat generated by the locked rotor currents never has a chance to dissipate and the internal winding temperature increases with each successive start of the motor. Typically, motors up to 100 horsepower (75 kilowatts) should not exceed more than four or five starts per hour and as the horsepower increases, motor starting limitations may reduce the number to as few as one start per hour.

2. *SOLUTION:* To overcome excessive heat buildup from frequent starts and stops of a motor, use a reduced-voltage method of starting the motor. Table 15.6 compares different reduced-voltage starting methods.

TABLE 15.5 MOTOR INSULATION MATERIALS

	Class A Systems-105°C	Class B Systems-130°C	Class F Systems-155°C	Class H Systems-180°C
Varnish	Modified phenolic Modified asphalt Alkyd polyester	Unmodified polyester Epoxy Modified phenolic	Modified polyester Epoxy	Silicone Polyimide
Wire Insulation	Vinyl acetal enameled	Modified polyester enameled Epoxy enameled Enamel plug glass yarn	Modified polyester enameled Epoxy enameled	Glass yarn, silicone varnish covered Polyimide
Other	Rag paper Kraft paper Polyester film Acetate film Varnished cambric Wood Fiber Cotton cord	Polyester film Polyester mat Varnished glass Mica flake or mica paper Polyester glass Laminated glass Asbestos Melamine Glass cord	Varnished glass Laminated glass Mica flake or mica paper Glass cord Polyester glass Tetrafluoro-ethylene resin	Glass cord Mica flake or mica paper Silicone varnished glass Laminated glass Polyimide film Polyimide varnished glass Polyimide filament

TABLE 15.6 REDUCED-VOLTAGE STARTING[a]

Type of Starting	Relative Starting Current	Relative Starting Torque
1. Across-the-Line	100%	100%
2. Resistors/reactor (at 65% voltage)	65%	65%
3. Auto Transformer (at 65% voltage)	42%	42%
4. Wye-Delta Winding	33%	33%
5. Two-Part Winding	50%	50%

[a] Compared with full-voltage, across-the-line starting, which typically draws 6.5 times the full-load current.

3. *PROBLEM:* High ambient operating temperature. If the ambient operating temperature exceeds that specified on the nameplate, it will contribute to a higher internal operating temperature. For all classes of insulation, the maximum ambient temperature is specified as 40°C (approximately 104°F).

3. *SOLUTION:* Provide adequate ventilation in lift stations and outdoor motor installations, particularly in southern climates where temperatures can be expected to exceed 104°F.

4. *PROBLEM:* Obstructed enclosure vents. All motor enclosures are designed for maximum dissipation of internally generated heat.

4. *SOLUTION:* In open, drip-proof motors, do not allow obstructions of the ventilation openings. Similarly, in totally enclosed, fan-cooled motors, a buildup of dirt, grease, or dust on the ribs of the enclosure will decrease the ability of the enclosure to dissipate heat. Keep the enclosure clean.

In outside installations, construct the enclosure to prevent entry by rodents. Frequently, check rodent screens, if present, to ensure they are not clogged with foreign material. In areas of high humidity, treat the enclosure with a fungicide to prevent formation of fungus on the insulation surface.

5. *PROBLEM:* Single phasing. Single phasing refers to the condition that occurs when one phase of the power source to the motor is lost, either from the utility company or from a fuse blowing in one phase in the motor control center. Under a single-phase condition, an induction motor that is already rotating will continue to rotate; however, it will be characterized by increased noise and vibration. If the motor is not rotating, a single-phase condition will not start the motor, will cause excessive noise and vibration, and will continue to do so until the protective device senses the condition and trips the overloads. Single-phase condition causes unbalanced currents to circulate in the rotor causing increases in internal motor heating.

5. *SOLUTION:* To correct a single-phasing problem, determine why one phase is missing. Did the utility company lose a phase (a common problem) or is a fuse blown or a circuit breaker tripped in the motor control center? Once the source of the problem is identified, then it can be corrected by notifying the utility company, replacing the blown fuse, or resetting the tripped circuit breaker.

6. *PROBLEM:* Motor overloading. This is operation of the motor in a way that causes it to draw current in excess of the motor nameplate current value. Overloading can happen inadvertently through improper operation on the pump curve by changing impeller diameters or through a change in the dynamic operating conditions of the pump, which changes the total dynamic head (TDH). Other conditions that can cause motor overloading include bearing problems and jamming of material between the rotating impeller and the stationary pump housing. More energy is required to operate the pump when rags, rocks, or timbers interfere with free rotation of the impeller.

The heat generated by continuous operation of a motor above its design rating is extremely damaging to the insulation. For example, a relatively minor overload of 6 percent will cause a 10°C increase in temperature; continuous operation at this elevated temperature will reduce the insulation life by 50 percent. A continuous 12-percent overload cuts insulation life to one-quarter of its design life.

6. *SOLUTION:* A thorough understanding of hydraulics and the existing operating conditions is required before changes in the pump can be made since improper changes can have disastrous effects on the motor. Know what you are doing before making any changes. Also, avoid continuous operation of a motor above its design rating. For additional protection, consider installing an overload protective device (described in Section 15.331, "Equipment Protective Devices").

7. *PROBLEM:* Voltage imbalance. A common problem found in pump stations with a high rate of motor failure is voltage imbalance or unbalance. Unlike a single-phase condition, all three phases are present but the phase-to-phase voltage is not equal in each phase.

Voltage imbalance can occur in either the utility side or the pump station electrical system. For example, the utility company may have large, single-phase loads (such as residential services) that reduce the voltage on a single phase. This same condition can occur in the pump station if a large number of 120/220 volt loads are present. Slight differences in voltage can cause disproportional current imbalance; this may be six to ten times as large as the voltage imbalance. For example, a two-percent voltage imbalance can result in a 20-percent current imbalance. A 4.5-percent voltage imbalance will reduce the insulation life to 50 percent of the normal life. This is the reason a dependable voltage supply at the motor terminals is critical. Even relatively slight variations can greatly increase the motor operating temperatures and burn out the insulation.

It is common practice for electrical utility companies to furnish power to three-phase customers in open delta or wye configurations. An open delta or wye system is a two-transformer bank that is a suitable configuration where lighting loads are *large* and three-phase loads are *light.* This is the exact opposite of the configuration needed by most pumping facilities where three-phase loads are *large.* (Examples of three-transformer banks include Y-delta, delta-Y, and Y-Y.) In most cases, three-phase motors should be fed from three-transformer banks for proper balance. The capacity of a two-transformer bank is only 57 percent of the capacity of a three-transformer bank. The two-transformer configuration can cause one leg of the three-phase current to furnish higher amperage to one leg of the motor, which will greatly shorten its life.

Operators should acquaint themselves with the configuration of their electric power supply. When an open delta or wye configuration is used, operators should calculate the degree of current imbalance existing between legs of their polyphase motors. If you are unsure about how to determine the configuration of your system or how to calculate the percentage of current imbalance, always consult a qualified electrician. *Current imbalance between legs should never exceed 5 percent under normal operating conditions* (NEMA Standards MGI-14.35).

Loose connections will also cause voltage imbalance as will high-resistance contacts, circuit breakers, or motor starters.

Another serious consideration for operators is voltage fluctuation caused by neighborhood demands. A pump motor in near perfect balance (for example, 3 percent unbalance) at 9:00 am could be as much as 17 percent unbalanced by 4:00 pm on a hot day due to the use of air conditioners by customers on the same grid. Also, the hookup of a small market or a new home to the power grid can cause a significant change in the degree of current unbalance in other parts of the power grid. Because energy demands are constantly changing, wastewater system operators should have a qualified electrician check the current balances between legs of their three-phase motors at least once a year.

7. *SOLUTION:* Motor connections at the circuit box should be checked frequently (semiannually or annually) to ensure that the connections are tight and that vibration has not caused the insulation on the conductors to wear away. Measure the voltage at the motor terminals and calculate the percentage imbalance (if any) using the procedures below.

Do not rely entirely on the power company to detect unbalanced current. Complaints of suspected power problems are frequently met with the explanation that all voltages are within the percentages allowed by law and no mention is made of the percentage of current unbalance, which can be a major source of problems with three-phase motors. A little research of your own can pay large benefits. For example, a small water company in Central California configured with an open delta system (and running three-phase unbalances as high as 17 percent as a result) was routinely spending $14,000 a year for energy and burning out a 10-HP motor on the average of every 1.5 years (six 10-HP motors in 9 years). After consultation, the local power utility agreed to add a third transformer to each power board to bring the system into better balance. Pump drop leads were then rotated, bringing overall current unbalances down to an average of 3 percent; heavy-duty, three-phase capacitors were added to absorb the prevalent voltage surges in the area; and computerized controls were added to the pumps to shut them off when pumping volumes got too low. These modifications resulted in a saving in energy costs the first year alone of $5,500.

FORMULAS

Percentage of current unbalance can be calculated by using the following formulas and procedures:

$$\text{Average Current} = \frac{\text{Total of Current Value Measured on Each Leg}}{3}$$

$$\text{\% Current Unbalance} = \frac{\text{Greatest Amp Difference from the Average}}{\text{Average Current}} \times 100\%$$

PROCEDURES

1. Measure and record current readings in amps for each leg (Hookup 1). Disconnect power.
2. Shift or roll the motor leads from left to right so the drop cable lead that was on terminal 1 is now on 2, lead on 2 is now on 3, and lead on 3 is now on 1 (Hookup 2). Rolling the motor leads in this manner will not reverse the motor rotation. Start the motor, measure and record current reading on each leg. Disconnect power.
3. Again shift drop cable leads from left to right so the lead on terminal 1 goes to 2, 2 goes to 3, and 3 to 1 (Hookup 3). Start pump, measure and record current reading on each leg. Disconnect power.
4. Add the values for each hookup.
5. Divide the total by 3 to obtain the average.
6. Compare each single leg reading to the average current amount to obtain the greatest amp difference from the average.
7. Divide this difference by the average to obtain the percentage of unbalance.
8. Use the wiring hookup that provides the lowest percentage of unbalance.

CORRECTING THE THREE-PHASE POWER UNBALANCE

Example: Check for current unbalance for a 230-volt, 3-phase, 60-Hz submersible pump motor, 18.6 full load amps.

Solution: Steps 1 to 3 measure and record amps on each motor drop lead for Hookups 1, 2, and 3 (Figure 15.36).

	Step 1 (Hookup 1)	Step 2 (Hookup 2)	Step 3 (Hookup 3)
(T_1)	DL_1 = 25.5 amps	DL_3 = 25 amps	DL_2 = 25.0 amps
(T_2)	DL_2 = 23.0 amps	DL_1 = 24 amps	DL_3 = 24.5 amps
(T_3)	DL_3 = 26.5 amps	DL_2 = 26 amps	DL_1 = 25.5 amps
Step 4	Total = 75 amps	Total = 75 amps	Total = 75 amps
Step 5	Average Current =	$\frac{\text{Total Current}}{\text{3 readings}}$ =	$\frac{75}{3}$ = 25 amps
Step 6	Greatest amp difference from the average:	(Hookup 1) = 25 − 23 = 2 (Hookup 2) = 26 − 25 = 1 (Hookup 3) = 25.5 − 25 = .5	
Step 7	% Unbalance	(Hookup 1) = 2/25 × 100 = 8 (Hookup 2) = 1/25 × 100 = 4 (Hookup 3) = 0.5/25 × 100 = 2	

As can be seen, Hookup 3 should be used since it shows the least amount of current unbalance. Therefore, the motor will operate at maximum efficiency and reliability on Hookup 3.

By comparing the current values recorded on each leg, you will note the highest value was always on the same leg, L_3. This indicates the unbalance is in the power source. If the high current values were on a different leg each time the leads were changed, the unbalance would be caused by the motor or a poor connection.

If the current unbalance is greater than 5 percent, contact your power company for help.

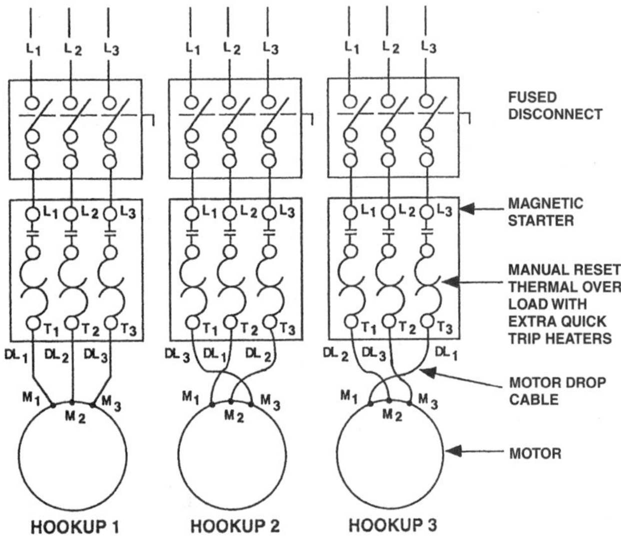

Fig. 15.36 Three hookups used to check for current unbalance

Acknowledgment

Material on unbalanced current was provided by James W. Cannell, President, Canyon Meadows Mutual Water Company, Inc., Bodfish, California. His contribution is greatly appreciated.

15.432 Increasing Resistance Value

The minimum insulation value in megohms is calculated by the following procedures. Divide the rated voltage by 1,000 and add 1 to the result. For example, for a motor operating at 460 volts, the minimum resistance value is 1.46 megohms.

$$\frac{460}{1{,}000} + 1 = 1.46$$

Measure the actual insulation resistance using a megger and compare the measured megger value with the calculated value; then any one of the three following procedures can be used to increase the actual (measured) insulation resistance value. The time required to increase the insulation resistance value depends on the wetness of the insulation and the size of the motor.

1. Remove the motor and bake in an oven at a temperature of not more than 194°F (90°C) until the resistance reaches an acceptable level.

2. Cover motor with a canvas or tarp and insert heating units or lamps. Heat until the resistance reaches an acceptable level. The heating time depends on the number and size of the heating units.
3. Provide a low-voltage current at the motor terminals that will generate heat within the windings. Heat until the resistance reaches an acceptable level.

15.44 Starters

A motor starter is a device or group of devices that are used to connect the electrical power to a motor. These starters can be either manually or automatically controlled.

Manual and magnetic starters range in complexity from a single ON/OFF switch to a sophisticated automatic device using timers and coils. The simplest motor starter is used on single-phase motors where a circuit breaker is turned on and the motor starts. This type of starter also is used on three-phase motors of smaller horsepower as well as on fan motors, machinery motors, and other motors, usually where it is not necessary to have automatic control.

Magnetic starters (Figure 15.37) are usually used to start pumps, compressors, blowers, and anything where automatic or remote control is desired. They permit low-voltage circuits to energize the starter of equipment at a remote location or to start larger starters (Figure 15.38). A magnetic starter is operated by electromagnetic action. It has contactors and these operate by energizing a coil that closes the contact, thus starting the motor. The circuit that energizes the starter is called the control circuit and it is usually operated on a lower voltage (115 volts) than the motor. Whenever a starter is used as a part of an integrated circuit (such as for flow, pressure, or temperature control), a magnetic starter or controller is necessary.

Magnetic starters are sized for their voltage and horsepower ratings. Additional information can be found in electrical catalogs, manuals, and manufacturers' brochures.

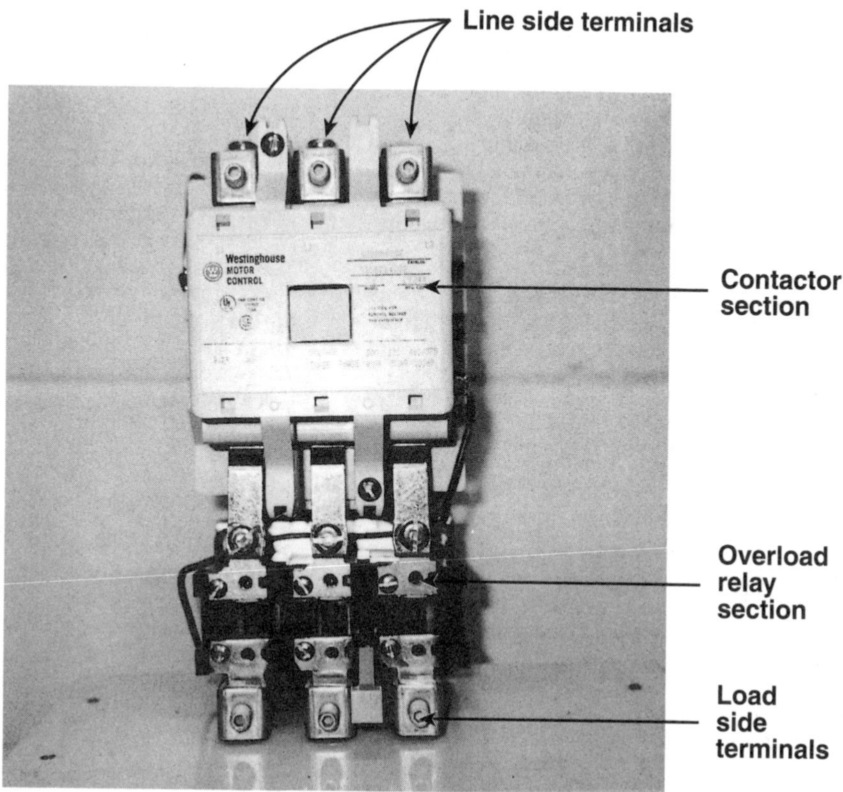

Fig. 15.37 Magnetic starter

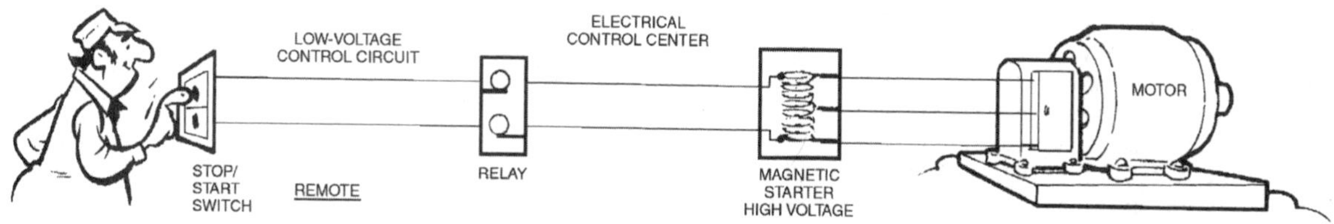

Fig. 15.38 Application of magnetic starter

A magnetic starter actually consists of two distinct sections:

1. The contacts, which connect and disconnect the power to the motor
2. Overload protection

Figure 15.37 illustrates a three-phase magnetic starter. The replaceable contacts (in the upper portion of the illustration) close when the motor is required to start, thus closing the electric circuit to the motor. Similarly, when the wet well level gets to a point where the pump is no longer required, a signal is sent to the motor starter and the contacts open, thus breaking the electric circuit to the motor. Each time this occurs, an arc takes place between the movable contact and the stationary contact and pitting occurs. This is why contacts must be replaced as a regular part of your preventive maintenance program.

The control coil usually uses a lower voltage than the line voltage to the motors. The magnetic coil is the device that actually causes the contacts to energize and de-energize.

In addition, each phase has an overload protection device that operates as a function of the length of the overload and the amount of the overload. Two types of devices are used:

1. A bimetallic strip that is precisely calibrated to open under higher temperature conditions to de-energize the coil
2. A small solder pot within the coil that melts because of the heat and will de-energize the system

A more common term for the overload protection devices is "heater elements." They are replaceable and can be selected and changed to correspond to the desired trip setting. Typically, the overload heater is selected for a trip setting that will de-energize the contactor when a 10-percent overload occurs.

Figure 15.39 shows a typical schematic diagram (referred to as a ladder diagram) for the control of one pump. This diagram is intended only for illustrative purposes. It does not include required grounding. Electrical systems must conform with the National Electrical Code in all cases.

Three-phase, 480-volt power is fed into terminals L1, L2, and L3 on the line side of the main circuit breaker. The main circuit breaker is normally located in the motor control center and provides circuit protection for all electrical equipment on the load side of the circuit breaker. On the load side of the circuit breaker, connections are made for a branch circuit breaker for the motor circuit. The load side of the branch circuit breaker feeds into motor starter 1M, which is a combination contactor and thermal overload. The load side of the motor starter 1M is then connected to the motor terminals.

Figure 15.39 shows the circuit components needed to get the 3-phase, 480-volt power to the motor terminals. However, some means must be provided for turning the motor on and off in response to wet well levels. The 24-volt AC, low-voltage control circuit illustrates how this is accomplished. Two phases of the 480-volt power are fed to a low-voltage control circuit breaker. The load side of the circuit breaker is connected to transformer T1, which reduces the line voltage to 24 volts AC Power is then supplied to the low-voltage control circuit through fuse F1. Under a rising wet well condition, float switch FS1 is energized (as shown in Figure 15.39). This means that the float has tipped and the contacts are made in the float switch.

FS1 is the low-level or pump shutoff switch. As the level rises in the wet well, the pump start float switch FS2 tips. The contacts close and the circuit is completed through the hand-off-auto (HOA) switch; this energizes control relay CR1 and the motor starter contactor coil 1M. In a nonoverload condition, thermal overload relay contacts OL1 are in the closed position allowing the circuit to be completed. When CR1 energizes, the contacts in parallel with FS2 close. This is necessary because as the wet well is pumping down, float switch FS2 will de-energize; however, we want the pump to continue operating until float switch FS1 de-energizes. When 1M energizes, this closes the contacts in the motor starter connecting the 3-phase power to the motor terminals. In the event of an overload condition, which would be sensed by the thermal overloads in motor starter 1M, the overload relay contact OL1 in the low-voltage control circuit will open, thereby shutting off the motor. As the wet well continues to drop, float switch FS1 will return to the normal position when the water level drops below the float. This opens the control circuit and de-energizes relay coil 1M, which, in turn, opens the contacts in motor starter 1M.

If more than one motor is installed in the pump station, which is usually the case, additional branch circuit breakers, motor starters, float switches, and control devices are required. Additional control functions are easily added, such as indicating lights, alarms, and other control elements. As previously discussed, a *PROGRAMMABLE LOGIC CONTROLLER (PLC)*[12] could be installed to control pump functions. The PLC would essentially replace the control circuit elements, thus reducing the number of relays and the wiring required in pump motor control circuits.

Figure 15.40 illustrates how a motor contactor operates. The coil, when energized from the low-voltage control circuit, pulls the armature up in a vertical movement. The armature is attached to a bell crank lever that translates the vertical motion into horizontal motion, moving the contacts to the right where they make contact with the stationary contacts. Three-phase voltage is brought in on the line side connection and, in the closed position, the contacts allow that voltage to be applied through the load side connection to the motor. Both the stationary contacts and the moving contacts are replaceable. *NOTE:* This diagram does not illustrate the thermal overload protection.

Starters are also available without the overload function. They provide only the motor disconnecting means and are referred to as contactors rather than motor starters. Manual contactors are available as well as magnetic contactors, which are capable of automatic control.

Magnetic starter maintenance consists of at least an annual inspection of the equipment, including the following tasks:

1. Inspect line and load conductors for evidence of high temperatures as indicated by bubbling or discoloring of the wire insulation.
2. Tighten all line and load connection terminals, including all low-voltage terminals (control, auxiliary switches).

[12] *Programmable Logic Controller (PLC).* A microcomputer-based control device containing programmable software; used to control process variables.

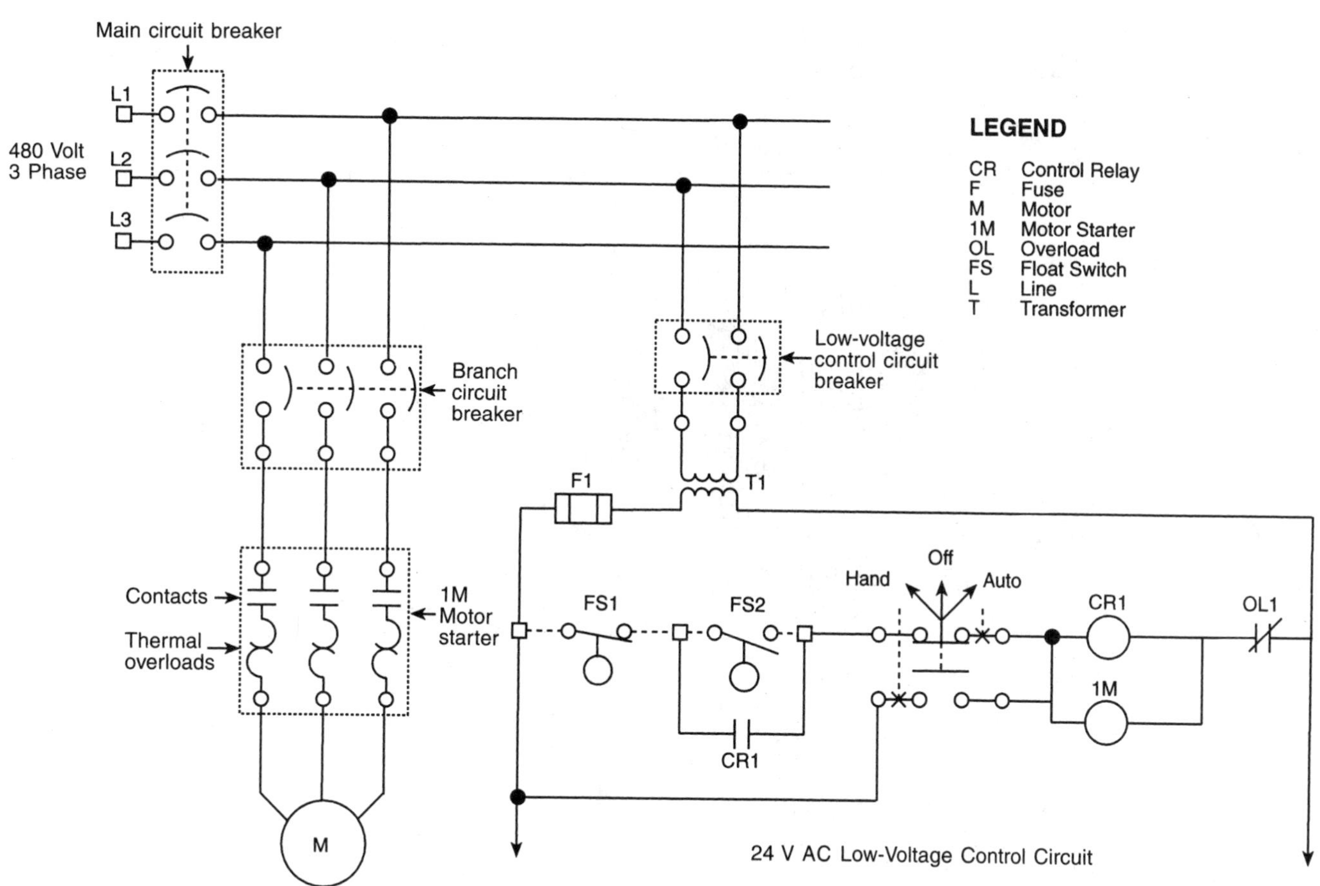

Fig. 15.39 Schematic (ladder) diagram for control of one pump

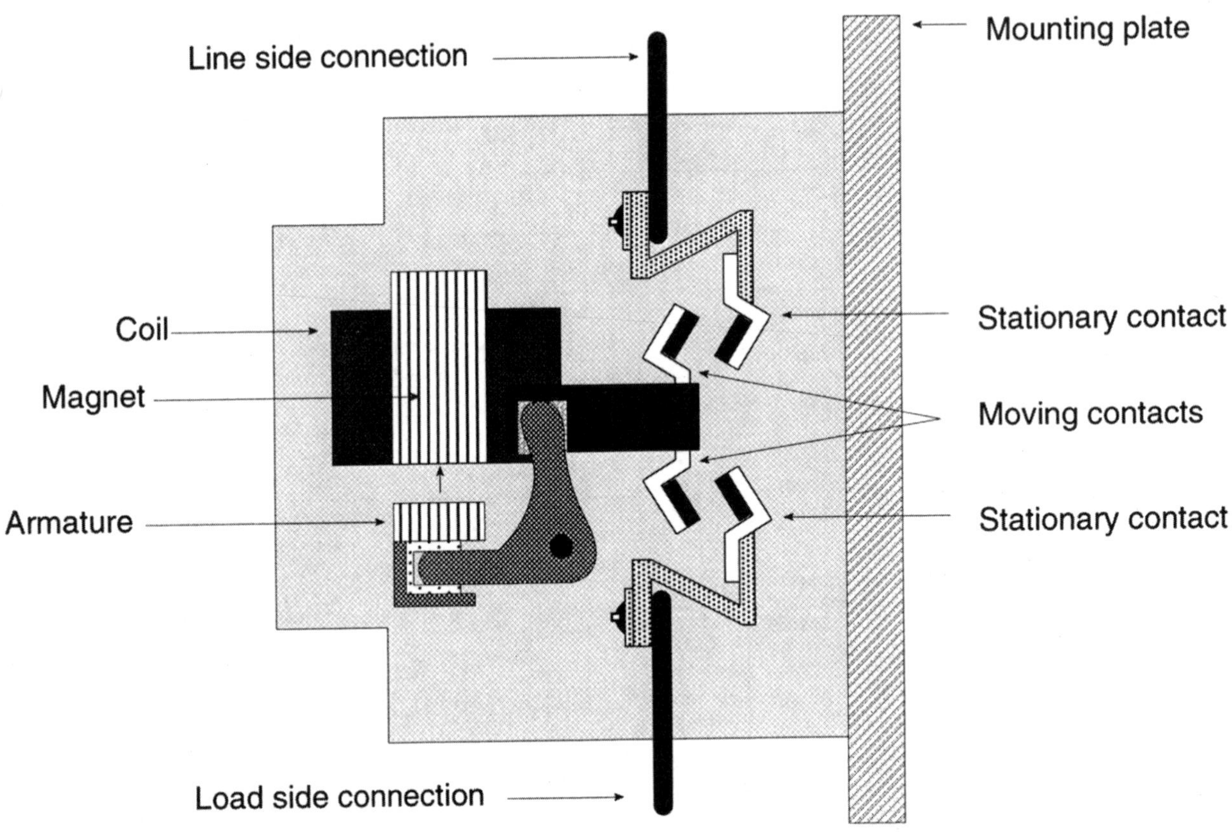

Fig. 15.40 How a motor contactor operates

3. Inspect and replace, if necessary, the stationary and movable contacts. See Figure 15.41, which illustrates the appearance of new contacts, contacts that are used but still suitable for use, and contacts that are used and should be replaced. A troubleshooting guide for magnetic starters is presented in Section 15.472. *CAUTION: No attempt should be made to file down contacts to restore the surface to a new condition, since this will result in an uneven surface and uneven distribution of electric energy across the face of the contacts.*

15.45 Safety

1. The eye bolt used for lifting and moving motors is designed for the weight of the motor alone without other equipment attached.
2. Whenever physically working on rotating equipment, including the pump and the motor, always open, lock out, and tag the electrical disconnect switch. This will prevent accidental energization of the motor, either by a careless operator or as a result of motor thermal switches automatically resetting and restarting the motor.
3. Ground all motors in accordance with requirements of National Electrical Code, Article 340.
4. Always keep hands and clothing away from moving parts.
5. Discharge all capacitors including power factor direction capacitors before servicing motor or motor controller.
6. Be sure that required safety guards are always in place.

15.46 Other Motor Considerations

15.460 Alignment

Horizontal motors should be mounted so that all four mounting feet are aligned to within 0.010 inch (0.25 mm) of each other for NEMA Frame 56 to NEMA Frame 210 and within 0.015 inch (0.38 mm) for NEMA Frame 250 to NEMA Frame 680. This alignment ensures a good, rigid foundation and also will make alignment of the motor and pump easier.

Whenever two pieces of rotating equipment, such as a pump and a motor, are used, there must be some means of transmitting the torque from the motor to the pump. Couplings are designed to do this. To function as intended, the equipment must be properly aligned at the couplings. Misalignment of the pump and the motor, or any two pieces of rotating equipment, can seriously damage the equipment and shorten the life of both the pump and the motor. Misalignment can cause excessive bearing loading as well as shaft bending, which will cause premature bearing failure, excessive vibration, or permanent damage to the shaft. Remember that the

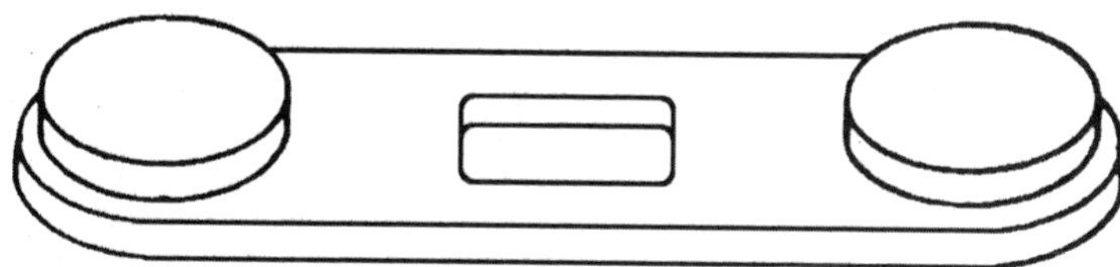

NEW

Smooth surface. May be bright or dull and somewhat discolored due to oxidation or tarnishing.

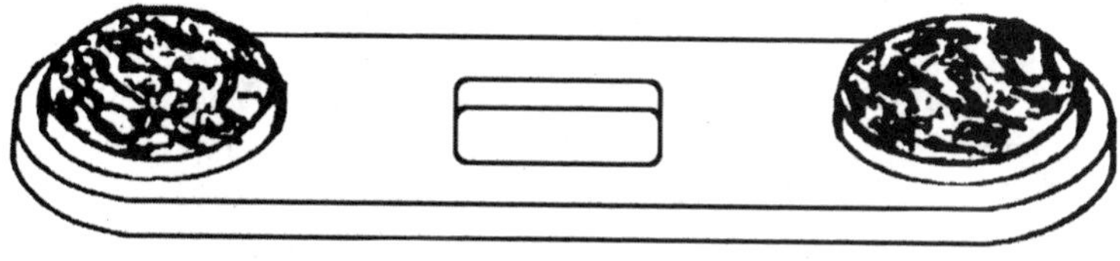

USED

Surface may be pitted and have discolored areas of black, brown, or may have blue (heat) tint. If half of the thickness (mass) of the silver points is still intact they are usable. This is the time to order a backup set.

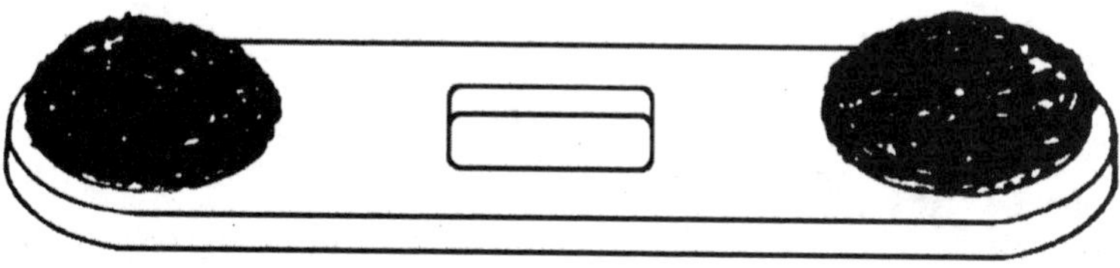

SEVERE OR LONG-TIME USE

Surface badly pitted and eroded with badly feathered and lifting edges. Replace entire contact set.

Fig. 15.41 Visual inspection of contact points

purpose of the coupling is to transmit power and, unless the coupling is of special design, it is not to be used to compensate for misalignment between the motor and the pump.

When connecting a pump and a motor, there are two important types of misalignment: (1) parallel, and (2) angular. Parallel misalignment occurs when the centerlines of the pump shaft and the motor shaft are offset. Figure 15.42 illustrates parallel misalignment. The pump and the motor shafts remain parallel to each other but are offset by some amount. Parallel misalignment can be detected very easily by holding a straightedge on one hub of the coupling and measuring the gap between the straightedge and the other hub of the coupling. Feeler gauges can be used to measure the amount of offset misalignment.

The second type of alignment problem is angular misalignment, also shown in Figure 15.42. In this case, the shaft centerlines are not parallel but instead form an angle, which represents the amount of angular misalignment. This type of misalignment can also be detected with a feeler gauge, calipers, or more sophisticated laser alignment equipment by measuring the distance between the coupling hubs at the point of maximum and minimum openings in the hubs, which would be 180° from each other. When using a dial indicator to measure angular misalignment, a general rule of thumb is that angular misalignment of the shafts must not exceed a total indicator reading of 0.002 inch (0.05 mm) for each inch (mm) of diameter of the coupling hub. To check for angular misalignment, mount a dial indicator on one coupling hub as shown in Diagram 1, Figure 15.43, with the finger or the button of the indicator against the finished face of the other hub and the dial set at zero. While rotating the shaft, note the reading on the indicator dial at each revolution.

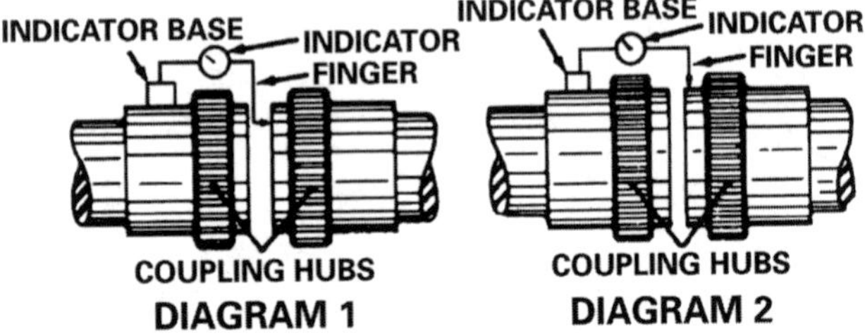

Fig. 15.43 Use of dial indicator to check for shaft angular alignment and trueness

In reality, misalignment usually includes both parallel and angular misalignment. The goal when aligning machines is to reduce both types of misalignment to a minimum. The purpose of this is two-fold: (1) couplings are not designed to accommodate large differences in parallel or angular misalignment, and (2) most alignment takes place when the machines being aligned are cold. However, all metal has a coefficient of expansion, which means that the metal expands as it heats up during operation. Since various types of metals expand at different rates and to different degrees, some additional misalignment may occur during operation. Couplings are designed to accommodate this type of misalignment.

In addition to misalignment, we are also concerned with end float in the shafts on the pump and the motor and with runout. End float is an in-and-out movement of the shaft along the axis of the shaft (see Figure 15.42).

Runout should also be checked. This checks the trueness or straightness of the shaft. Diagram 2, Figure 15.43, illustrates how the trueness is checked with the dial indicator mounted, again, on one coupling hub and the finger or button mounted on the outside surface of the second hub. As the shaft is rotated, note the indicator reading. The reading should not exceed 0.002 inch (0.05 mm). A reading in excess of 0.002 inch (0.05 mm) indicates a bent shaft or a shaft that is eccentrically machined.

Misalignment is one of the most frequent causes of vibration problems in lift station motors and pumps. It frequently causes premature failure of bearings, mechanical seals, and packing. While these failures in and of themselves are expensive and affect equipment availability and reliability, the failures can also cause more catastrophic related failures, such as broken shafts, destruction of the rotor element in the motor, or damage to motor windings.

15.461 Changing Rotation Direction

Changing the rotation direction of a three-phase motor is accomplished simply by changing any two of the power leads. Generally, this is done on the load side of a magnetic starter. The direction of motor rotation must be changed if a pump is rotating in the wrong direction.

15.462 Allowable Voltage and Frequency Deviations

1. Voltage 10 percent above or below the value stamped in the nameplate.
2. Frequency 5 percent above or below the value stamped in the nameplate.
3. Voltage and frequency together within 10 percent providing frequency is less than 5 percent above or below the value stamped in the nameplate.

As mentioned previously, both voltage and frequency have a direct effect on the performance and life of the motor, as illustrated in Tables 15.7 and 15.8. A 10-percent increase in the rated voltage, for example, results in a 0 to 17 percent increase in temperature, which will have a significant effect on the insulation life of the motor (see Table 15.8, Voltage DP, 110% of Rated Voltage, 1-200 HP, Temperature Rise, Full-Load, 0 to 17%).

15.463 Maximum Vibration Levels Allowed by NEMA

Speed, RPM	Maximum Vibration Amplitude, Inches
3,000–4,000	0.001
1,500–2,999	0.002
1,000–1,499	0.0025
0–999	0.003

15.464 Lubrication

The correct procedure for lubricating motors is as follows:

1. Stop motor, lock out, and tag.
2. Wipe all grease fittings.
3. Remove filler and drain plugs. *CAUTION: Zerk fittings should not be installed in both the filler and the drain holes.*
4. Free drain hole of any hard grease using a piece of wire, if necessary.

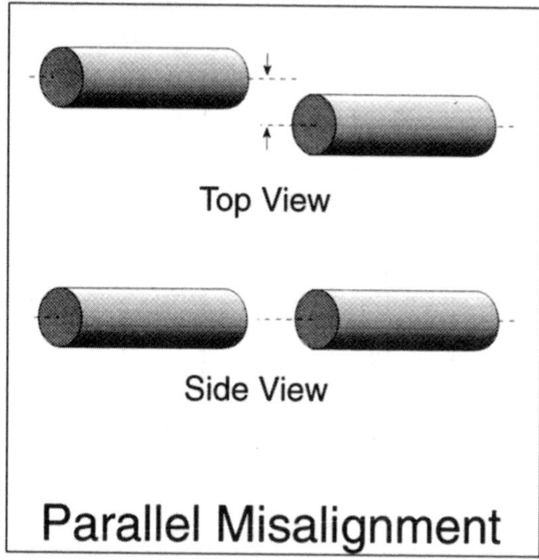

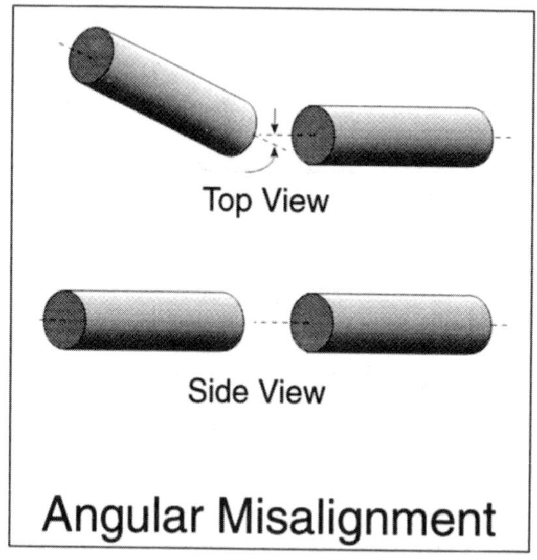

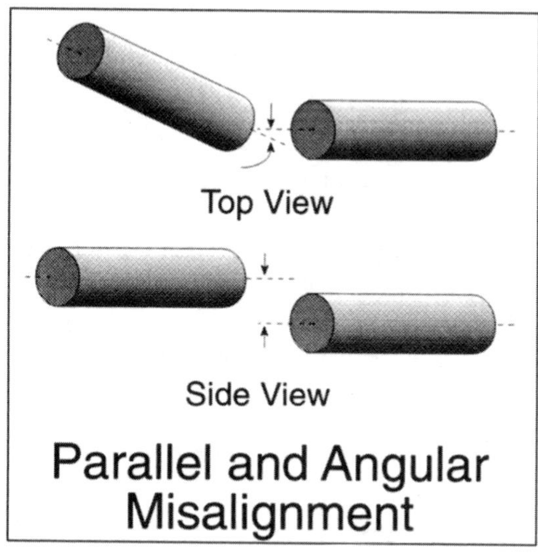

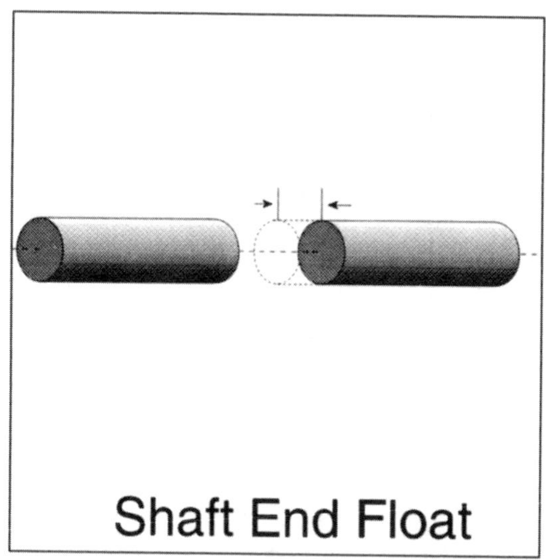

Fig. 15.42 Types of shaft misalignment and end float

TABLE 15.7 TYPICAL MOTOR PERFORMANCE VARIATIONS DUE TO POWER SUPPLY VARIATIONS

POLYPHASE - INTEGRAL HORSEPOWER

				TORQUE					
				Starting (Locked Rotor)	DIP	Break-Down	Full-Load	Speed Full-Load	Power Factor Full-Load
		1% Unbalance		Slight –	Slight –	Slight –	Slight +	Slight –	–5.5%
		2% Unbalance		Slight –	Slight –	Slight –	Slight +	Slight –	–7.1%
VOLTAGE	DP	110% of Rated Volt.	1-10 HP[a] 15-30 HP 40-75 HP 100-200 HP	+21 to 23%	+21 to 23%	+21 to 23%	–1.0% –0.6% –0.5% –0.3%	+1.0% +0.6% +0.5% +0.3%	–13 to 10% –9 to 8% –8 to 6% –6 to 4%
VOLTAGE	DP	90% of Rated Volt.	1-10 HP 15-30 HP 40-75 HP 100-200 HP	–17 to 19%	–17 to 19%	–17 to 19%	+1.5% +1.0% +0.6% +0.3%	–1.5% –1.0% –0.6% –0.3%	+11 to 7% +6 to 3% +3 to 2% +2 to 3%
VOLTAGE	TEFC	110% of Rated Volt.	1-10 HP 15-30 HP 40-75 HP 100-200 HP	+21 to 23%	+21 to 23%	+21 to 23%	–1.0% –0.6% –0.5% –0.3%	+1.0% +0.6% +0.5% +0.3%	–13 to 6% –5 to 3% –3 to 2% +2 to 0%
VOLTAGE	TEFC	90% of Rated Volt.	1-10 HP 15-30 HP 40-75 HP 100-200 HP	–17 to 19%	–17 to 19%	–17 to 19%	+1.5% +1.0% +0.6% +0.3%	–1.5% –1.0% –0.6% –0.3%	+11 to 4% +2 to 0% +1 to 0% +1 to 0%
FREQ	DP & TEFC	105% of Rated Freq.	1 to 200 HP	–10%	–10%	–10%	–5%	+5%	Slight +
FREQ	DP & TEFC	95% of Rated Freq.	1 to 200 HP	+11%	+11%	+11%	+5%	–5%	Slight –

[a] Multiply HP x 0.746 to obtain kilowatts.

+ Increase
– Decrease

TABLE 15.8 TYPICAL MOTOR PERFORMANCE VARIATIONS DUE TO POWER SUPPLY VARIATIONS

				AMPS				
				Starting (Locked Rotor)	Full-Load	No-Load	Effic. Full-Load	Temp. Rise Full-Load
		1% Unbalance		+1.5%	+8%	+13%	−2%	+2%
		2% Unbalance		+3%	+17%	+27%	−8%	+8%
VOLTAGE	DP	110% of Rated Volt.	1-10 HP[a]	+10 to 12%	+8 to 4%	+25 to 37%	−6 to 1%	+17 to 9%
			15-30 HP		+4 to 1%	+32 to 37%	−1 to 0%	+7 to 3%
			40-75 HP		−0 to 2%	+37%	No Change	+2 to 0%
			100-200 HP		−3 to 5%	+36 to 30%	+0 to 0.3%	−0 to 4%
		90% of Rated Volt.	1-10 HP	−10 to 12%	−3 to +5%	−20 to 19%	+1 to 0%	−6 to +12%
			15-30 HP		+6 to 9%	−19 to 18%	−0 to 0.4%	+15 to 19%
			40-75 HP		+9 to 10%	−18 to 17%	−0.4 to 0.1%	+19 to 20%
			100-200 HP		+9 to 10%	−17 to 16%	−0.1%	+19 to 20%
	TEFC	110% of Rated Volt.	1-10 HP	+10 to 12%	+8 to −4%	+37 to 27%	−6 to 1%	+17 to −5%
			15-30 HP		−5 to 6%	+26 to 25%	−1 to 0%	−7 to 9%
			40-75 HP		−6 to 7%	+24 to 21%	No Change	−9 to 10%
			100-200 HP		−7 to 9%	+20 to 13%	+0 to 0.3%	−10 to 11%
		90% of Rated Volt.	1-10 HP	−10 to 12%	−3 to +5%	−20 to 16%	+1 to 0%	−6 to +15%
			15-30 HP		+6 to 11%	−14 to 12%	−0 to 0.4%	+19 to 23%
			40-75 HP		+12%	−12 to 10%	−0.4 to 0.1%	+24 to 23%
			100-200 HP		+11 to 9%	−10%	−0.1%	+24 to 23%
FREQ	DP & TEFC	105% of Rated Freq.	1 to 200 HP	−5 to 6%	Slight −	−5 to 6%	Slight +	Slight −
		95% of Rated Freq.	1 to 200 HP	+5 to 6%	Slight +	+5 to 6%	Slight −	Slight +

[a] Multiply HP x 0.746 to obtain kilowatts.

+ Increase
− Decrease

5. Add appropriate amount and type of grease using low-pressure grease gun.
6. Start motor and let run for approximately 30 minutes. (This allows excess grease to drain out. If this is not done, pressure buildup in the bearing can blow out the bottom seal, allowing grease to run down the shaft onto the top of the motor windings.)
7. Stop motor, wipe off any drained grease, and replace filler and drain plugs.

Figure 15.44 shows a cross section of the bearing cap and filler/drain plugs.

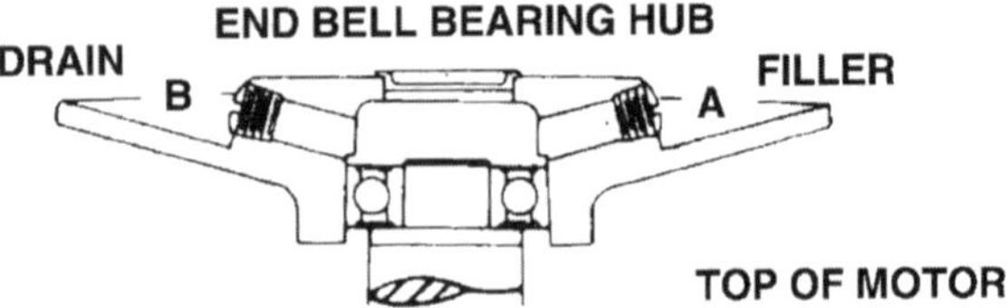

Fig. 15.44 Cross section of bearing cap and filler/drain plugs

Excessive greasing can cause as much damage to motor bearings as undergreasing. Therefore, manufacturer's recommendations should be followed without exception. The amount and frequency of greasing depends on a number of factors, including RPM, operating temperature, duty cycle, and environmental conditions.

QUESTIONS

Write your answers in a notebook and then compare your answers with those on page 462.

15.4A What are the two types of induction motor construction typically encountered when dealing with AC induction motors?

15.4B List the five most common causes of electric motor failure.

15.4C List the three components of an insulation temperature rating.

15.4D How are motor starters controlled?

15.4E When are magnetic starters used?

15.4F What two types of overload protection devices are used on magnetic starters?

15.4G How can the direction of rotation of a three-phase motor be changed?

15.47 Troubleshooting

15.470 Step-By-Step Procedures

The key to effective troubleshooting is the use of practical, step-by-step procedures combined with a common-sense approach.

"NEVER TAKE ANYTHING FOR GRANTED"

1. Gather preliminary information. The first step in troubleshooting any motor control that has developed trouble is to understand the circuit operation and other related functions. In other words, what is supposed to happen, operate, and so forth when it is working properly? Also, what is it doing now? The qualified maintenance operator should be able to do the following:
   a. *KNOW WHAT SHOULD HAPPEN WHEN A SWITCH IS PUSHED:* When switches are pushed or tripped, what coils go in, contacts close, relays operate, and motors run?
   b. *EXAMINE ALL UNUSUAL FACTORS:* What unusual things are happening in the pump station (facility) now that this circuit does not work properly? Lights dimmed, other pumps ran faster, lights went out when it broke, everything was flooded, operators were hosing down area, and many other possible factors.
   c. *ANALYZE WHAT YOU KNOW:* What part of it is working correctly? Is switch arm tripped? Everything but this is all right, except pump gets plugged with rags frequently. Is it a mechanical failure or an electrical problem caused by a mechanical failure?
   d. *SELECT SIMPLE PROCEDURES:* To localize the problem, select logical ways that can be simply and quickly accomplished.
   e. *MAKE A VISUAL INSPECTION:* Look for burned wires, loose wires, area full of water, coil burned, contacts loose, or strange smells.
   f. *CONVERGE ON SOURCE OF TROUBLE:* Mechanical or electrical. Motor or control, whatever it might be. Electrical problems result from some type of mechanical failure.
   g. *PINPOINT THE PROBLEM:* Exactly where is the problem and what do you need for repair?
   h. *FIND THE CAUSE:* What caused the problem? Moisture, wear, poor design, voltage, or overloading.
   i. *REPAIR THE PROBLEM AND ELIMINATE THE CAUSE, IF POSSIBLE:* If the problem is inside switch gear or motors, call an electrician. Give the electrician the information you have regarding the equipment. Do not attempt electrical repairs unless you are qualified and authorized, otherwise you could cause excessive damage to yourself and to the equipment.
2. Some of the things to look for when troubleshooting are given in the remainder of this section.

15.471 Troubleshooting Guide for Electric Motors

Symptom	Cause	Result[a]	Remedy
1. Motor does not start (switch is on and not defective)	a. Incorrectly connected	a. Burnout	a. Connect correctly per diagram on motor.
	b. Incorrect power supply	b. Burnout	b. Use only with correctly rated power supply.
	c. Fuse out, loose or open connection	c. Burnout	c. Correct open circuit condition.
	d. Rotating parts of motor may be jammed mechanically	d. Burnout	d. Check and correct: 1. Bent shaft 2. Broken housing 3. Damaged bearing 4. Foreign material in motor
	e. Driven machine may be jammed	e. Burnout	e. Correct jammed condition.
	f. No power supply	f. None	f. Check for voltage at motor and work back to power supply.
	g. Internal circuitry open	g. Burnout	g. Correct open circuit condition.
2. Motor starts but does not come up to speed	a. Same as 1-a, b, c above	a. Burnout	a. Same as 1-a, b, c above.
	b. Overload	b. Burnout	b. Reduce load to bring current to rated limit. Use proper fuses and overload protection.
	c. One or more phases out on a 3-phase motor	c. Burnout	c. Look for open circuits.
3. Motor noisy (electrically)	a. Same as 1-a, b, c above	a. Burnout	a. Same as 1-a, b, c above.
4. Motor runs hot (exceeds rating)	a. Same as 1-a, b, c above	a. Burnout	a. Same as 1-a, b, c above.
	b. Overload	b. Burnout	b. Reduce load.
	c. Impaired ventilation	c. Burnout	c. Remove obstruction.
	d. Frequent starts or stops	d. Burnout	d. 1. Reduce number of starts or reversals. 2. Secure proper motor for this duty.
	e. Misalignment between rotor and stator laminations	e. Burnout	e. Realign.
5. Motor noisy (mechanically)	a. Misalignment of coupling or sprocket	a. Bearing failure, broken shaft, stator burnout due to motor drag	a. Correct misalignment.
	b. Mechanical unbalance of rotating parts	b. Same as 5-a	b. Find unbalanced part, then balance.
	c. Lack of or improper lubricant	c. Bearing failure	c. Use correct lubricant, replace parts as necessary.
	d. Foreign material in lubricant	d. Bearing failure	d. Clean out and replace bearings.

15.471 Troubleshooting Guide for Electric Motors *(continued)*

Symptom	Cause	Result[a]	Remedy
5. Motor noisy (mechanically) *(continued)*	e. Overload	e. Bearing failure	e. Remove overload condition. Replace damaged parts.
	f. Shock loading	f. Bearing failure	f. Correct causes and replace damaged parts.
	g. Mounting acts as amplifier of normal noise	g. Annoying	g. Isolate motor from base.
	h. Rotor dragging due to worn bearings, shaft, or bracket	h. Burnout	h. Replace bearings, shaft, or bracket as needed.
6. Bearing failure	a. Same as 5-a, b, c, d, e	a. Burnout, damaged shaft, damaged housing	a. Replace bearings and follow 5-a, b, c, d, e.
	b. Entry of water or foreign material into bearing housing	b. Burnout, damaged shaft, damaged housing	b. Replace bearings and seals and shield against entry of foreign material (water, dust, etc.). Use proper motor.

Symptom	Caused By	Appearance
1. Shorted motor winding	a. Moisture, chemicals, foreign material in motor, damaged winding	a. Black or burned coil with remainder of winding good.
2. All windings completely burned	a. Overload	a. Burned equally all around winding.
	b. Stalling	b. Burned equally all around winding.
	c. Impaired ventilation	c. Burned equally all around winding.
	d. Frequent reversal or starting	d. Burned equally all around winding.
	e. Incorrect power	e. Burned equally all around winding.
3. Single-phase condition	a. Open circuit in one line. The most common causes are loose connection, one fuse out, loose contact in switch.	a. If 1,800 RPM motor—four equally burned groups at 90° intervals.
		b. If 1,200 RPM motor—six equally burned groups at 60° intervals.
		c. If 3,600 RPM motor—two equally burned groups at 180° intervals.
		NOTE: If Y connected, each burned group will consist of two adjacent phase groups. If delta connected, each burned group will consist of one-phase group.
4. Other	a. Improper connection	a. Irregularly burned groups or spot burns.
	b. Ground	

[a] Many of these conditions should trip protective devices rather than burn out motors. Also, many burnouts occur within a short period of time after motor is started up. This does not necessarily indicate that the motor was defective, but usually is due to one or more of the above-mentioned causes. The most common causes of failure shortly after start-up are improper connections, open circuits in one line, incorrect power supply, or overload.

15.472 Troubleshooting Guide for Magnetic Starters

Trouble	Possible Cause	Remedy
CONTACTS		
Contact chatter	1. Broken *POLE SHADER*[13]	1. Replace.
	2. Poor contact in control circuit	2. Improve contact or use holding circuit interlock.
	3. Low voltage	3. Correct voltage condition. Check momentary voltage drop.
Welding or freezing	1. Abnormal surge of current	1. Use larger contactor and check for grounds, shorts, or excessive motor load current.
	2. Frequent *JOGGING*[14]	2. Install larger device rated for jogging service or caution operators.
	3. Insufficient contact pressure	3. Replace contact spring; check contact carrier for damage.
	4. Contacts not positioning properly	4. Check for voltage drop during start-up.
	5. Foreign matter preventing magnet from seating	5. Clean contacts.
	6. Short circuit	6. Remove short fault and check that fuse and breaker are right.
Short contact life or overheating of tips	1. Contacts poorly aligned, poorly spaced, or damaged	1. Do not file silver-faced contacts. Rough spots or discoloration will not harm contacts. Replace.
	2. Excessively high currents	2. Install larger device. Check for grounds, shorts, or excessive motor currents.
	3. Excessive starting and stopping of motor	3. Caution operators. Check operating controls.
	4. Weak contact pressure	4. Adjust or replace contact springs.
	5. Dirty contacts	5. Clean with approved solvent.
	6. Loose connections	6. Check terminals and tighten.
Coil overheated	1. Starting coil may not kick out	1. Repair coil.
	2. Overload will not let motor reach minimum speed	2. Remove overload.
	3. Overvoltage or high ambient temperature	3. Check application and circuit.
	4. Incorrect coil	4. Check rating; if incorrect, replace with proper coil.
	5. Shorted turns caused by mechanical damage or corrosion	5. Replace coil.
	6. Undervoltage, failure of magnet to seal it	6. Correct system voltage.
	7. Dirt or rust on pole faces increasing air gap	7. Clean pole faces.

[13] *Pole Shader.* A copper bar circling the laminated iron core inside the coil of a magnetic starter.
[14] *Jogging.* The frequent starting and stopping of an electric motor.

15.472 Troubleshooting Guide for Magnetic Starters *(continued)*

Trouble	Possible Cause	Remedy
CONTACTS (continued)		
Overload relays tripping	1. Sustained overload	1. Check for grounds, shorts, or excessive motor currents. Mechanical overload.
	2. Loose connection on all or any load wires	2. Check, clean, and tighten.
	3. Incorrect heater	3. Replace with correct size heater unit.
	4. Fatigued heater blocks	4. Inspect and replace.
Failure to trip	1. Mechanical binding, dirt, or corrosion	1. Clean or replace.
	2. Wrong heater, or heaters omitted and jumper wires used	2. Check ratings. Apply heaters of proper rating.
	3. Motor and relay in different temperatures	3. Adjust relay rating accordingly, or install temperature compensating relays.
MAGNETIC AND MECHANICAL PARTS		
Noisy magnet (humming)	1. Broken shading coil	1. Replace shading coil.
	2. Magnet faces not mating	2. Replace magnet assembly or realign.
	3. Dirt or rust on magnet faces	3. Clean and realign.
	4. Low voltage	4. Inspect system voltage and voltage dips or drops during start-up.
Failure to pick up and seal	1. Low voltage	1. Inspect system voltage and correct.
	2. Coil open or shorted	2. Replace coil.
	3. Wrong coil	3. Check coil number and voltage rating.
	4. Mechanical obstruction	4. With power off, check for free movement of contact and armature assembly. Repair.
Failure to drop out	1. Gummy substance on pole	1. Clean with solvent.
	2. Voltage not removed from coil	2. Check coil circuit.
	3. Worn or rusted parts causing binding	3. Replace or clean parts as necessary.
	4. Residual magnetism due to lack of air gap in magnet path	4. Replace worn magnet parts or align, if possible.
	5. Welded contacts	5. Shorted circuit, grounded, overloaded.

15.473 Trouble/Remedy Procedures for Induction Motors

1. Motor will not start.

   Overload control tripped. Wait for overload to cool, then try to start again. If motor still does not start, check for the causes outlined below.

   a. Open fuses: test fuses.

   b. Low voltage: check nameplate values against power supply characteristics. Also, check voltage at motor terminals when starting motor under load to check for allowable voltage drop.

   c. Wrong control connections: check connections with control wiring diagram.

   d. Loose terminal-lead connection: turn power off and tighten connections.

   e. Drive machine locked: disconnect motor from load. If motor starts satisfactorily, check driven machine.

   f. Open circuit in stator or rotor winding: check for open circuits.

   g. Short circuit in stator winding: check for short.

   h. Winding grounded: test for grounded wiring.

   i. Bearing stiff: free bearing or replace.

   j. Overload: reduce load.

2. Motor noisy.

   a. Three-phase motor running on single phase: stop motor, then try to start. It will not start on single phase. Check for open circuit in one of the lines.

   b. Electrical load unbalanced: check current balance.

   c. Shaft bumping (sleeve-bearing motor): check alignment and conditions of belt. On pedestal-mounted bearing, check for play and axial centering of rotor.

   d. Vibration: driven machine may be unbalanced. Remove motor from load. If motor is still noisy, rebalance.

   e. Air gap not uniform: center the rotor and, if necessary, replace bearings.

   f. Noisy ball bearing: check lubrication. Replace bearings if noise is excessive and persistent.

   g. Rotor rubbing on stator: center the rotor and replace bearings, if necessary.

   h. Motor loose on foundation: tighten hold-down bolts. Motor may possibly have to be realigned.

   i. Coupling loose: insert feelers at four places in coupling joint before pulling up bolts to check alignment. Tighten coupling bolts securely.

3. Motor at higher than normal temperature or smoking. (Measure temperature with thermometer or thermister and compare with nameplate value.)

   a. Overload: measure motor loading with ammeter. Reduce load.

   b. Electrical load imbalance: check for voltage imbalance or single-phasing.

   c. Restricted ventilation: clean air passage and windings.

   d. Incorrect voltage and frequency: check nameplate values with power supply. Also check voltage at motor terminals with motor under full load.

   e. Motor stalled by driven tight bearings: remove power from motor. Check machine for cause of stalling.

   f. Stator winding shorted or grounded: test windings by standard method.

   g. Rotor winding with loose connection: tighten, if possible, or replace with another rotor.

   h. Belt too tight: remove excessive pressure on bearings.

   i. Motor used for rapid reversing service: replace with motor designed for this service.

4. Bearings hot.

   a. End shields loose or not replaced properly: make sure end shields fit squarely and are properly tightened.

   b. Excessive belt tension or excessive gear side thrust: reduce belt tension or gear pressure and realign shafts. See that thrust is not being transferred to motor bearing.

   c. Bent shaft: straighten shaft or send to motor repair shop.

5. Sleeve bearings.

   a. Insufficient oil: add oil—if supply is very low, drain, flush, and refill.

   b. Foreign material in oil or poor grade of oil: drain oil, flush, and relubricate using industrial lubricant recommended by a reliable oil manufacturer.

   c. Oil rings rotating slowly or not rotating at all: oil too heavy; drain and replace. If oil ring has worn spot, replace with new ring.

   d. Motor tilted too far: level motor or reduce tilt and realign, if necessary.

   e. Rings bent or otherwise damaged in reassembling: replace rings.

   f. Rings out of slot (oil-ring retaining clip out of place): adjust or replace retaining clip.

   g. Defective bearings or rough shaft: replace bearings. Resurface shaft.

6. Ball bearings.

   a. Too much grease: remove relief plug and let motor run. If excess grease does not come out, flush and relubricate.

   b. Wrong grade of grease: flush bearing and relubricate with correct amount of proper grease.

   c. Insufficient grease: remove relief plug and grease bearing.

   d. Foreign material in grease: flush bearing, relubricate; make sure grease supply is clean (keep can covered when not in use).

15.5 RECORDS

Records are a very important part of electrical maintenance. They must be accurate and complete. Pages 411 and 412 are examples of typical record sheets. Most of the information you will need to complete these forms can be found on the manufacturer's data sheet or in the instruction manual.

Whenever a piece of equipment is changed, repaired, or tested, the work performed should be recorded on an equipment history card of some type. Complete, up-to-date equipment records will enable you to evaluate the reliability of your equipment and will provide the basis for a realistic preventive maintenance program.

15.6 ADDITIONAL READING

1. *BASIC ELECTRICITY* by Van Valkenburgh, Nooger & Neville, Inc. Published by Delmar, an imprint of Thomson Learning. Obtain from Thomson Learning, Attn.: Order Fulfillment, PO Box 6904, Florence, KY 41022-6904. ISBN 978-0-7906-1041-2. Price, $49.95, plus shipping and handling.
2. "Instrumentation and Control Systems" by Leonard Ainsworth, Chapter 9 in *ADVANCED WASTE TREATMENT.* Obtain from the Office of Water Programs, California State University, Sacramento, 6000 J Street, Sacramento, CA 95819-6025. Price, $49.00.
3. "Maintenance" by Parker Robinson, Chapter 18 in *WATER TREATMENT PLANT OPERATION,* Volume II. Obtain from the Office of Water Programs, California State University, Sacramento, 6000 J Street, Sacramento, CA 95819-6025. Price, $49.00.
4. *MAINTENANCE ENGINEERING HANDBOOK* by L. Higgins and K. Mobley. Obtain from the McGraw-Hill Companies, Order Services, PO Box 182604, Columbus, OH 43272-3031. ISBN 978-0-07-028819-5. Price, $150.00, plus nine percent of order total for shipping and handling.

QUESTIONS

Write your answers in a notebook and then compare your answers with those on page 462.

15.4H What is the key to effective troubleshooting?

15.4I What are some of the steps that should be taken when troubleshooting magnetic starters?

15.5A What kind of information should be recorded regarding electrical equipment?

Please answer the discussion and review questions next.

DISCUSSION AND REVIEW QUESTIONS

Chapter 15. MAINTENANCE

(Lesson 3 of 7 Lessons)

Write the answers to these questions in your notebook. The question numbering continues from Lesson 2.

17. Why should inexperienced, unqualified, or unauthorized persons and even qualified and authorized persons be extremely careful around electrical panels, circuits, wiring, and equipment?
18. What is the difference between direct current (DC) and alternating current (AC)?
19. What meters and testers are used to maintain, repair, and troubleshoot electric circuits and equipment? Discuss the use of each meter and tester.
20. What protective or safety devices are used to protect operators and equipment from being harmed by electricity?
21. Why must motor nameplate data be recorded and filed?
22. How would you attempt to find the cause when a pump motor will not start?

PUMP RECORD CARD

NAME________ MAKE________ MODEL________

TYPE________ SIZE________ SERIAL #________

ORDER NUMBER________ SUPPLIER________ DATE PURCHASED________

DATE INSTALLED________ APPLICATION________ PLANT #________

Nameplate Data and Pump Info		Stuffing Box Data	Motor Data
GPM ____		Diameter____ Depth____	Name____ Serial #____
TDH ____		Pack. Size____ Type____	H.P.____ Speed____
RPM ____		Length____ No. Rings____	Ambient°____
Gage Press Disc ____ ____		Lantern Ring____ Flushed____	RPM____ Frame____
Gage Press Suc ____ ____		Mech. Seal Name____ Size____	Volts____ Amps____
Shut off Press ____ ____		Type____	Phase____ Cycle____
Suction Head ____ ____		Pump Materials	Shaft Size____ Key____
Rotation ____ ____		Casing ____	Bearing Front____
Impeller Type____		Shaft____	Rear____
Impeller Dia.____		Wearing Rings Casing ____	Code____ Type____
Impeller Clear____		Wearing Rings Impeller____	Amps @ Max Speed____
Coupl Type & Size____		Shaft Sleeve____	Amps @ Shut Off____
Front Brg #____		Slinger____	Control Data Info
Rear Brg #____		Shims____	Starter____
Lub Interval____		Gaskets____	NEMA Size____
Lubricant____		"O" Rings____	Cat. #____
Wearing Rings____		Brg. Seals Front____	Heater Size____
Shaft Sleeve Size____		Rear____	Rated @____
Pump Shaft Size____		Casing Wear Ring Size ID____	Control Voltage____
Pump Keyway____		OD____	Variable Speed Type____
		Width____	Speed Max____
Other Related Information:		Impeller Wear Ring ID____	Speed Min____
		OD____	
		Width____	

MOTOR STARTERS Number________

Title:________

Mfg.:________ Address________

Style:________ Class________ Size________

Type:________ ________ ________

O.L. HEATERS

Style________ Code________

Amps________ ________

________ ________

O.L. TRIP UNITS

Mfg:________ Style:________

Type:________ ________

Amps Range:________ ________

CIRCUIT BREAKER

Mfg:________ Address________

Style:________ Frame:________ Volts________ Amps Setting________

Cat. No.________ ________ ________ ________

MOTOR Number________

TITLE________

Mfg:________ Address________

HP:________ Volts:________ Ser. No.________ Duty:________

Phase:________ Amps:________ Frame:________ Temp:________

Cycles:________ RPM:________ Type________ Class:________

Code:________ S.F.:________ Model________ Spec.:________

SO#________ S#________ Style:________ CSA App:________

Form________ Spec.________ Shft. Brg.________ Rear Brg.________

50 Cycle Data________

Suitable for 208V Network:________

Additional data________

____

____

____

Connection Diagram

(6) (5) (4)   (6) (5) (4)

(7) (8) (9)   (7) (8) (9)

(1) (2) (3)   (1) (2) (3)

The format of this section differs from the other chapters. The table of contents is outlined below and the paragraphs are numbered for easy reference when you use the Equipment Service Cards and Service Record Cards mentioned in Section 15.00, pages 333 and 334.

15.7 MECHANICAL MAINTENANCE

CHAPTER 15. MAINTENANCE

(Lesson 4 of 7 Lessons)

The format of this section differs from the other chapters. This format was designed specifically to assist you in planning an effective preventive maintenance program. The table of contents is outlined on the preceding page and the paragraphs are numbered for easy reference when you use the Equipment Service Cards and Service Record Cards mentioned in Section 15.00, pages 333 and 334. You can also use this paragraph numbering system when your maintenance program is on a computer.

An entire book could be written on the topics covered in this section. Step-by-step details for maintaining equipment are not provided because manufacturers are continually improving their products and these details could soon be out of date. You are assumed to have some familiarity with the equipment being discussed. For details concerning a particular piece of equipment, you should contact the manufacturer. This section indicates to you the kinds of maintenance you should include in your program and how you could schedule your work. Carefully read the manufacturer's instructions and be sure you clearly understand the material before attempting to maintain and repair equipment. If you have any questions or need any help, do not hesitate to contact the manufacturer or your local representative.

A glossary is not provided in this section because of the large number of technical words that require familiarization with the equipment being discussed. The best way to learn the meaning of these new words is from manufacturers' literature or from their representatives. Some new words are described in the lessons where necessary.

Preventive Maintenance

The following paragraphs list some general preventive maintenance services and indicate frequency of performance.

There are many makes and types of equipment and the wide variation of functions cannot be included; therefore, you will have to use some judgment as to whether the services and frequencies will apply to your equipment. If something goes wrong or breaks in your plant, you may have to disregard your maintenance schedule and fix the problem now.

NOTE: If you need to shut a unit down, make sure it is also locked out and tagged properly.

CODE
D MEANS DAILY; W, WEEKLY; M, MONTHLY;
Q, QUARTERLY; S, SEMIANNUALLY; A, ANNUALLY

Paragraph 1: Pumps, General (Including Packing)

This paragraph lists some general preventive maintenance services and indicates frequency of performance. Typical centrifugal pump sections are shown in Figures 15.3 and 15.4, pages 347 and 348.

Frequency of Service	
D	1. CHECK WATER-SEAL PACKING GLANDS FOR LEAKAGE. See that the packing box is protected with a clear-water supply from an outside source; make sure that water seal pressure is at least 5 psi (0.35 kg/sq cm) greater than maximum pump discharge pressure. See that there are no *CROSS CONNECTIONS.*[15] Check packing glands for leakage during operation. Allow a slight seal leakage when pumps are running to keep packing cool and in good condition. The proper amount of leakage depends on equipment and operating conditions. Sixty drops of water per minute is a good rule of thumb. If excessive leakage is found, hand tighten gland nuts evenly, but not too tight. After adjusting packing glands, be sure shaft turns freely by hand. If serious leakage continues, renew packing, shaft, or shaft sleeve.

[15] *Cross Connection.* (1) A connection between drinking (potable) water and an unapproved water supply. (2) A connection between a storm drain system and a sanitary collection system. (3) Less frequently used to mean a connection between two sections of a collection system to handle anticipated overloads of one system.

Frequency of Service	
D	2. CHECK GREASE-SEALED PACKING GLANDS. When grease is used as a packing gland seal, maintain constant grease pressure on packing during operation. When a spring-loaded grease cup is used, keep it loaded with grease. Force grease through packing at a rate of about one ounce (30 gm) per day. Never allow the seal to run dry.
W	3. OPERATE PUMPS ALTERNATELY. If two or more pumps of the same size are installed, alternate their use to equalize wear, keep motor windings dry, and distribute lubricant in bearings.
W	4. INSPECT PUMP ASSEMBLY. Check float controls noting how they respond to rising water level. See that unit starts when float switch makes contact and that pump empties basin at a normal rate. Apply light oil to moving parts.
D	5. CHECK MOTOR CONDITION. See Paragraph 7.

W	6. CLEAN PUMP. First lock out power and tag switch (Figure 15.45). Cleanout handholes are provided on the pump volute. To clean pump, close all valves, drain pump, remove handhole cover, and remove all solids. Wear gloves to protect your hands from sharp objects.
W	7. CHECK PACKING GLAND ASSEMBLY. Check packing gland, the unit's most abused and troublesome part. If stuffing box leaks excessively when gland is pulled up with mild pressure, remove packing and examine shaft sleeve carefully. Replace grooved or scored shaft sleeve because packing cannot be held in stuffing box with roughened shaft or shaft sleeve. Replace the packing a strip at a time, tamping each strip thoroughly and staggering joints. (See Figure 15.46.) Position lantern ring (water-seal ring) properly. If grease sealing is used, completely fill lantern ring with grease before putting remaining rings of packing in place. The type of packing used (Figure 15.47) is less important than the manner in which packing is placed. Both types of packing wrap around and score the shaft sleeve or are thrown out against outer wall of stuffing box, allowing wastewater to leak through and score the shaft. The proper size of packing should be available in your plant's equipment files. See pages 420 and 421, Figure 15.48, for illustrated steps on how to pack a pump.
W	8. CHECK MECHANICAL SEALS. Mechanical seals usually consist of two subassemblies: (1) a rotating ring assembly, and (2) a stationary assembly (Figure 15.15, page 359). Inspect seal for leakage and excessive heat. If any part of the seal needs replacing, replace the entire seal (both subassemblies) with a new seal that has been provided by the manufacturer. Before installing a new seal, be sure that there are no chips or cracks on the sealing surface. Keep a new mechanical seal clean at all times. Always be sure that a mechanical seal is surrounded with water before starting and running the pump.
Q	9. INSPECT AND LUBRICATE BEARINGS. Unless otherwise specifically directed for a particular pump model, drain lubricant and wash out oil wells and bearing with solvent. Check sleeve bearings to see that oil rings turn freely with the shaft. Repair or replace, if defective. Refill with proper lubricant. Check bearings by feeling for rough spots and looking for signs of binding and excessive movement up, down, back and forth when rotated. Replace those bearings that are worn excessively. Generally, allow clearance of 0.002 inch plus 0.001 inch for each inch or fraction of inch of shaft-journal diameter (0.05 mm plus 0.025 mm for each 25 mm or fraction of 25 mm of shaft-journal diameter).
Q	10. CHECK OPERATING TEMPERATURE OF BEARINGS. Check bearing temperature with thermometer, not by hand. If antifriction bearings are running hot, check for overlubrication and relieve, if necessary. If sleeve bearings run too hot, check for lack of lubricant. If proper lubrication does not correct condition, disassemble and inspect bearing. Check alignment of pump and motor if high temperatures continue.
S	11. CHECK ALIGNMENT OF PUMP AND MOTOR. For method of aligning pump and motor, see Paragraph 11. If misalignment recurs frequently, inspect entire piping system. Unbolt piping at suction and discharge nozzles to see if it springs away, indicating strain on casing. Check all piping supports for soundness and effective support of load. Vertical pumps usually have flexible shafting, which permits slight angular misalignment; however, if solid shafting is used, align exactly. If beams carrying intermediate bearings are too light or are subject to contraction or expansion, replace beams and realign intermediate bearings carefully.

DANGER

OPERATOR
WORKING
ON LINE

DO NOT CLOSE THIS
SWITCH WHILE THIS
TAG IS DISPLAYED

TIME OF OFF: ______________________

DATE: ______________________

SIGNATURE: ______________________

This is the ONLY person authorized to remove this tag.

INDUSTRIAL INDEMNITY/INDUSTRIAL UNDERWRITERS/
INSURANCE COMPANIES

4E210—R66

Fig. 15.45 Typical warning tag

(Source: Industrial Indemnity/Industrial Underwriters/Insurance Companies)

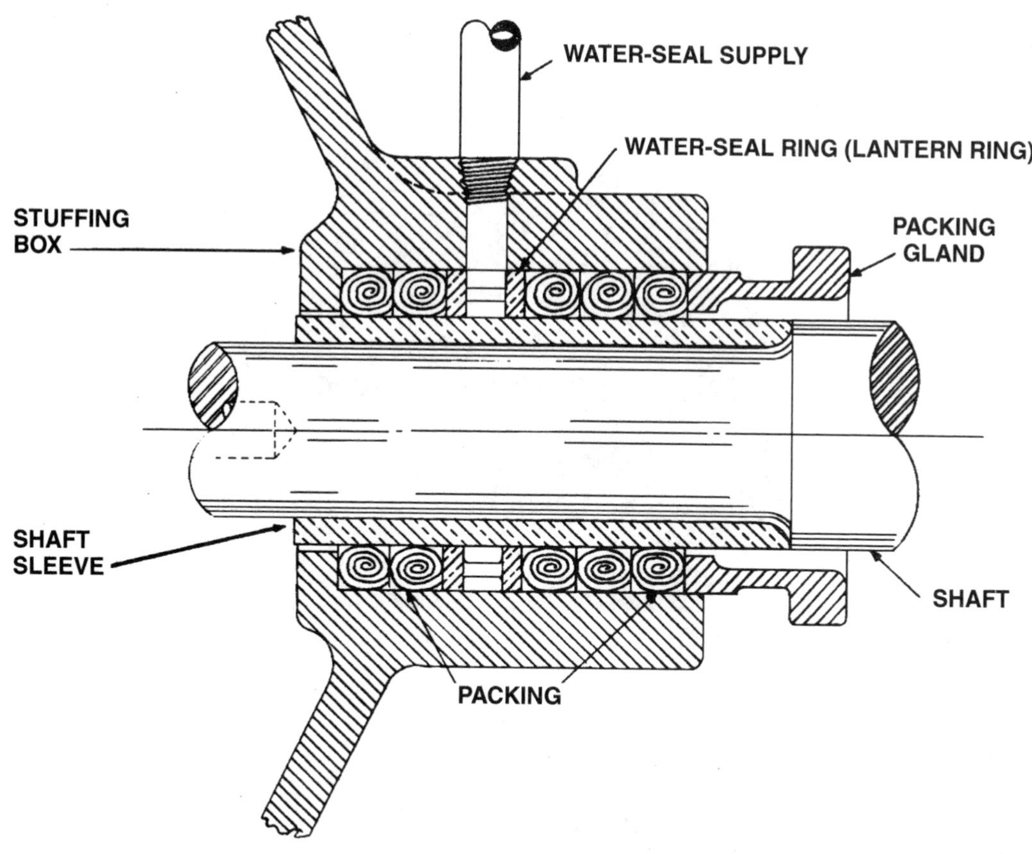

Fig. 15.46 Method of packing shaft
(Source: War Department Technical Manual TM5-666)

Frequency of Service

S 12. INSPECT AND SERVICE PUMPS.

a. Remove rotating element of pump and inspect thoroughly for wear. Order replacement parts, where necessary. Check impeller clearance between volute.

b. Remove any deposit or scaling. Clean out water-seal piping.

c. Determine pump capacity by pumping into empty tank of known size or by timing the draining of pit or sump.

$$\text{Pump Capacity, GPM} = \frac{\text{Volume, gallons}}{\text{Time, minutes}}$$

or

$$\text{Pump Capacity, } \frac{\text{liters}}{\text{sec}} = \frac{\text{Volume, liters}}{\text{Time, seconds}}$$

See Arithmetic Appendix, Section A.33, "Maintenance" for a detailed example.

d. Test pump efficiency. Refer to pump manufacturer's instructions on how to collect data and perform calculations.

e. Measure total dynamic suction lift and discharge head to test pump and pipe condition. Record figures for comparison with later tests.

f. Inspect foot and check valves, paying particular attention to check valves, which can cause water hammer when pump stops. (See Paragraph 14 also.) Foot valves are used when pumping fresh water or plant effluent. Wet wells must be dewatered before foot valves can be inspected.

g. Examine wearing rings. Replace seriously worn wearing rings to improve efficiency. Check wearing ring clearances, which generally should be no more than 0.003 inch per inch of wearing diameter (0.003 mm per mm of wearing diameter). *CAUTION: To protect rings and casings, never allow pump to run dry through lack of proper priming when starting or loss of suction when operating.*

Teflon Packing

Graphite Packing

Fig. 15.47 Packing
(Courtesy A. W. Chesterton Co.)

Frequency of Service	
A	13. DRAIN PUMP FOR LONG-TERM SHUTDOWN. When shutting down pump for a long period, open motor disconnect switch; shut all valves on suction, discharge, water-seal, and priming lines; drain pump completely by removing vent and drain plugs. This procedure protects pump against corrosion, sedimentation, and freezing. Inspect pump and bearings thoroughly and perform all necessary servicing. Drain bearing housings and replenish with fresh oil, purge old grease and replace. When a pump is out of service, run it monthly to warm it up and to distribute lubrication so the packing will not "freeze" to the shaft. Resume periodic checks after pump is put back in service.

QUESTIONS

Write your answers in a notebook and then compare your answers with those on pages 462 and 463.

15.7A What is a cross connection?

15.7B Is a slight water-seal leakage desirable when a pump is running? If so, why?

15.7C How would you measure the capacity of a pump?

15.7D Estimate the capacity of a pump (in GPM) if it lowers the water in a 10-foot wide × 15-foot long wet well 1.7 feet in five minutes.

15.7E What should be done to a pump before it is shut down for a long time, and why?

Paragraph 2: Reciprocating Pumps, General (See Figure 15.49, page 422)

The general procedures in this paragraph apply to all reciprocating sludge pumps described in this section.

Frequency of Service	
W	1. CHECK SHEAR PIN ADJUSTMENT. Set eccentric by placing shear pin through proper hole in eccentric flanges to give required stroke. Tighten the two 5/8- or 7/8-inch (1.5- or 2.2-cm) hexagonal nuts on connecting rods just enough to take spring out of lock washers. (See Paragraph 12.) When a shear pin fails, eccentric moves toward neutral position, preventing damage to the pump. Remove cause of obstruction and insert new shear pin. Shear pins fail because of one of three common causes: a. Solid object lodged under piston b. Clogged discharge line c. Stuck or wedged valve
D	2. CHECK PACKING ADJUSTMENT. Give special attention to packing adjustment. If packing is too tight, it reduces efficiency and scores piston walls. Keep packing just tight enough to keep sludge from leaking through gland. Before pump is installed or after it has been idle for a time, loosen all nuts on packing gland. Run pump with sludge suction line closed and valve covers open for a few minutes to break in the packing. Turn down gland nuts no more than necessary to prevent sludge from getting past packing. Tighten all packing nuts uniformly. When packing gland bolts cannot be taken up farther, remove packing. Remove old packing and thoroughly clean cylinder and piston walls. Place new packing into cylinder, staggering packing-ring joints, and tamp each ring into place. Break in and adjust packing as explained above. When chevron type packing is used, tighten gland nuts only finger tight because excessive pressure ruins packing and scores plunger.
Q	3. CHECK BALL VALVES. When valve balls are so worn that diameter is 5/8-inch (1.5-cm) smaller than original size, they may jam into guides in valve chamber. Check size of valve balls and replace if badly worn.
Q	4. CHECK VALVE CHAMBER GASKETS. Valve chamber gaskets on most pumps serve as a safety device and blow out under excessive pressure. Check gaskets and replace, if necessary. Keep additional gaskets on hand for replacement.
A	5. CHECK ECCENTRIC ADJUSTMENT. To take up babbitt bearing, remove brass shims provided on connecting rod. After removing shims, operate pump for at least one hour and check to see that eccentric does not run hot.
D	6. NOTE UNUSUAL NOISES. Check for noticeable water hammer when pump is operating. This noise is most pronounced when pumping water or very thin sludge; it decreases or disappears when pumping heavy sludge. Eliminate noise by opening the 1/4-inch (0.6-cm) petcock on the discharge air chamber slightly; this draws in a small amount of air, keeping the discharge air chamber full of air at all times. This must be done with the pump locked out, discharge line isolated, and line pressure relieved to allow air to enter the chamber.
D	7. CHECK CONTROL VALVE POSITIONS. Because any plunger pump may be damaged if operated against closed valves in the pipeline, especially the discharge line, make all valve setting changes with pump shut down; otherwise pumps that are installed to pump from two sources or to deliver to separate tanks at different times may be broken if all discharge line valves are closed simultaneously for a few seconds or discharge valve directly above pump is closed.
W	8. GEAR REDUCER. Check oil level by removing plug on the side of the gear case. Unit should not be in operation.
Q	9. CHANGE OIL AND CLEAN MAGNETIC DRAIN PLUG.

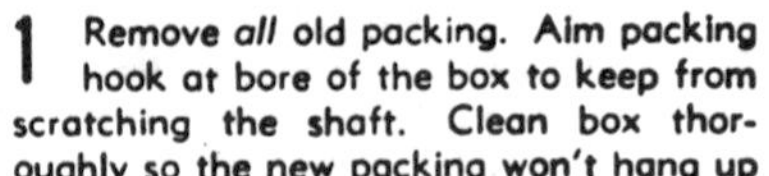

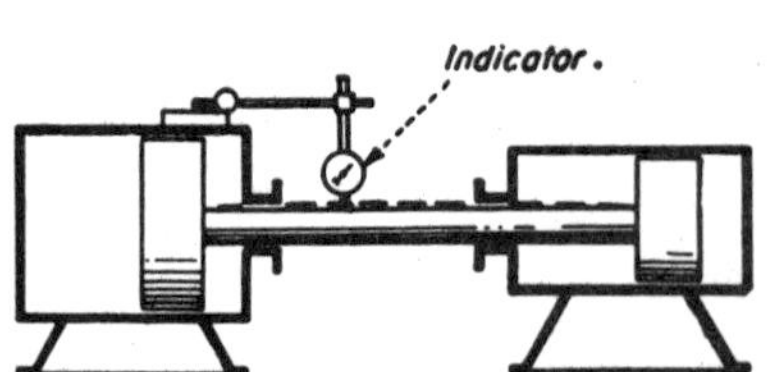

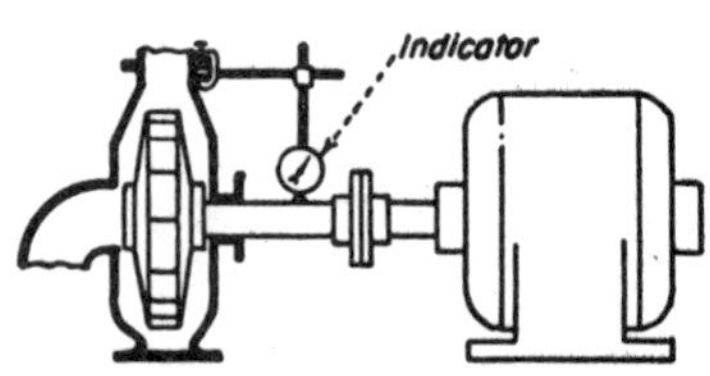

1 Remove *all* old packing. Aim packing hook at bore of the box to keep from scratching the shaft. Clean box thoroughly so the new packing won't hang up

2 Check for bent shaft, grooves or shoulders. If the neck bushing clearance in bottom of box is great, use stiffer bottom ring or replace the neck bushing

3 Revolve rotary shaft. If the indicator runs out over 0.003-in., straighten shaft, or check bearings, or balance rotor. Gyrating shaft beats out packing

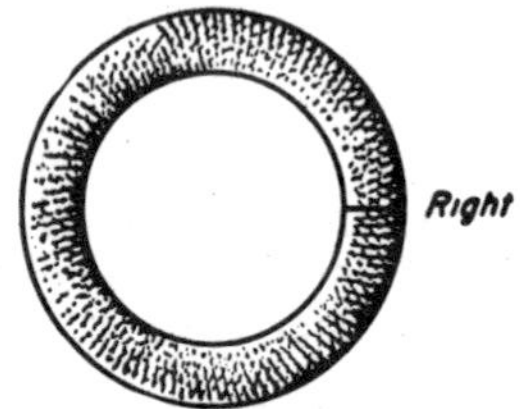

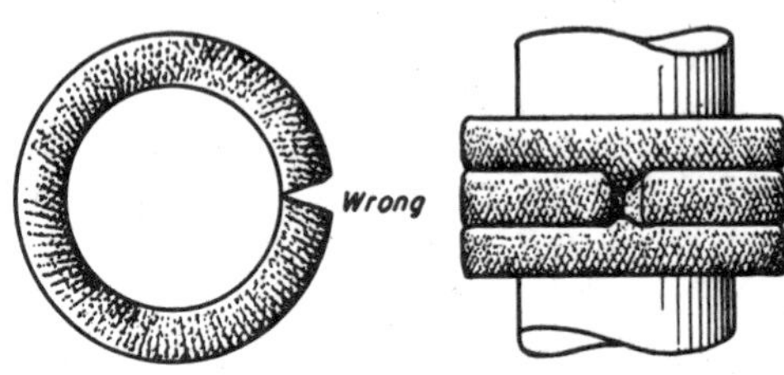

6 Cutting off rings while packing is wrapped around shaft will give you rings with parallel ends. This is very important if packing is to do job

7 If you cut packing while stretched out straight, the ends will be at an angle. With gap at angle, packing on either side squeezes into top of gap and ring cannot close. This brings up the question about gap for expansion. Most packings need none. Channel-type packing with lead core may need slight gap for expansion

HOW TO PACK A PUMP

(*Editor's Note:* This step-by-step illustration of a basic maintenance duty was brought to our attention by Anthony J. Zigment, Director, Municipal Training Division, Department of Community Affairs.)

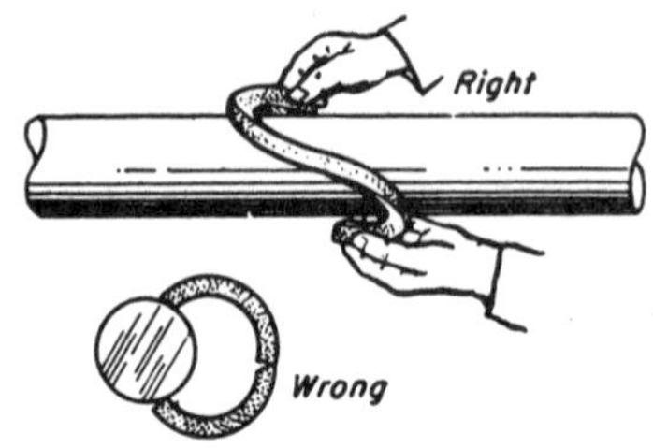

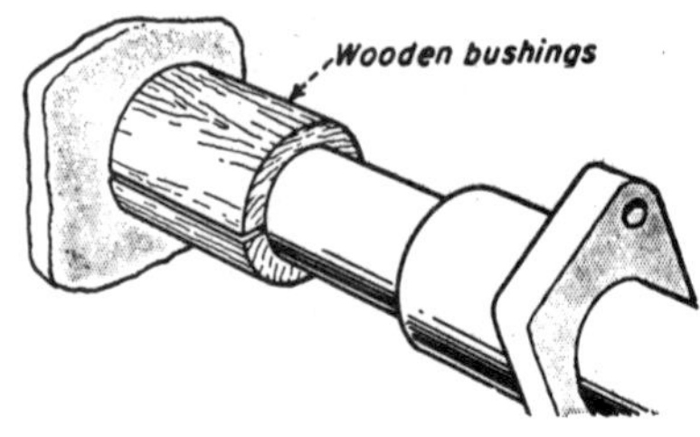

11 Open ring joint sidewise, especially lead-filled and metallic types. This prevents distorting molded circumference—breaking the ring opposite gap

12 Use split wooden bushing. Install first turn of packing, then force into bottom of box by tightening gland against bushing. Seat each turn this way

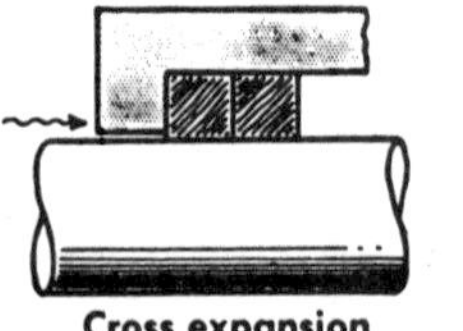

Cross expansion

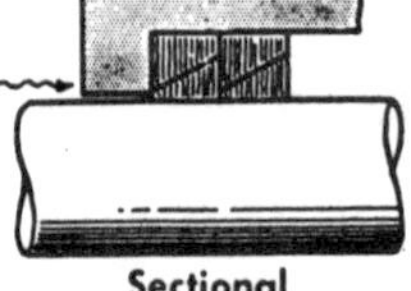

Sectional

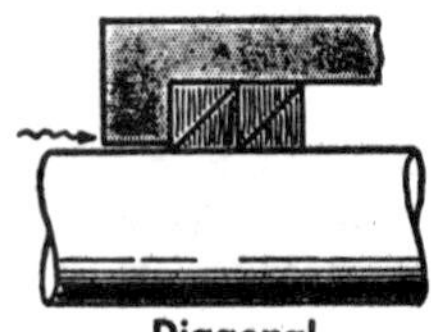

Diagonal

15 Always install cross-expansion packing so plies slope toward the fluid pressure from housing. Place sectional rings so slope between inside and outside ring is toward the pressure. Diagonal rings must also have slope toward the fluid pressure. Watch these details for best results when installing new packing in a box

Fig. 15.48 How to pack a pump

(Source: Water Pollution Control Association of Pennsylvania Magazine. January-February, 1976)

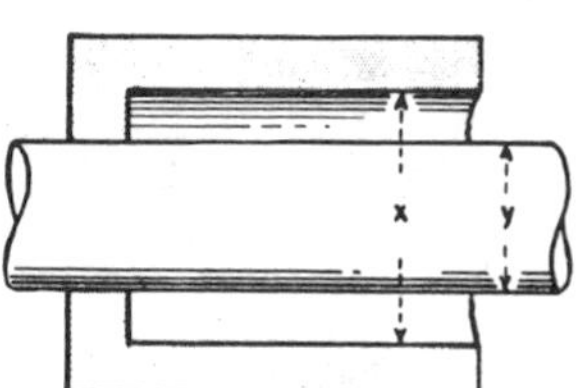

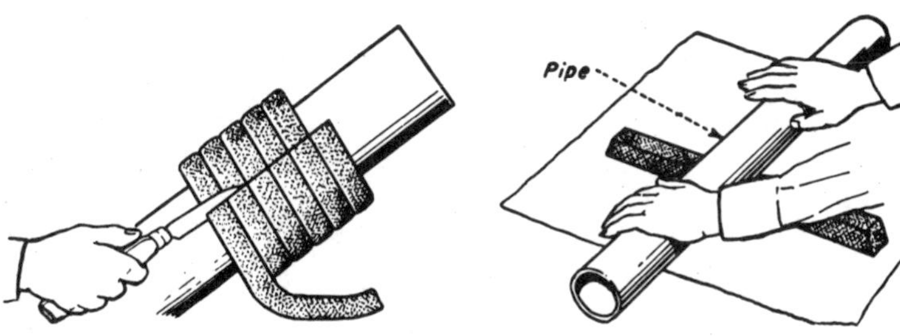

4 To find the right size of packing to install, measure stuffing-box bore and subtract shaft diameter, divide by 2. Packing is too critical for guesswork.

5 Wind packing, needed for filling stuffing box, snugly around shaft (or same size shaft held in vise) and cut through each turn while coiled, as shown. If the packing is slightly too large, never flatten with a hammer. Place each turn on a clean newspaper and then roll out with pipe as you would with a rolling pin

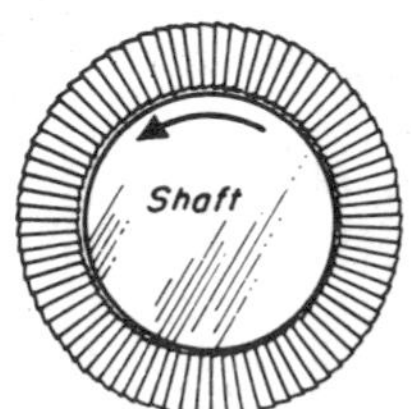

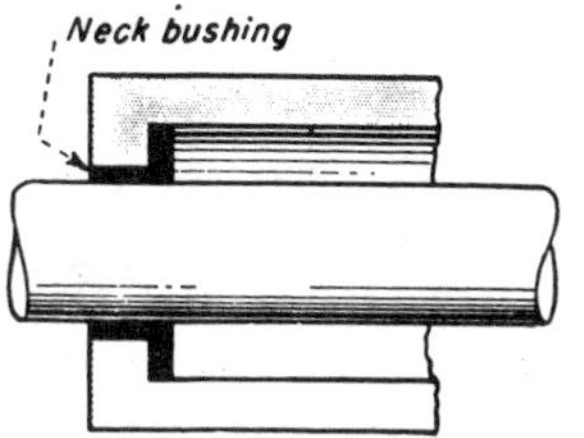

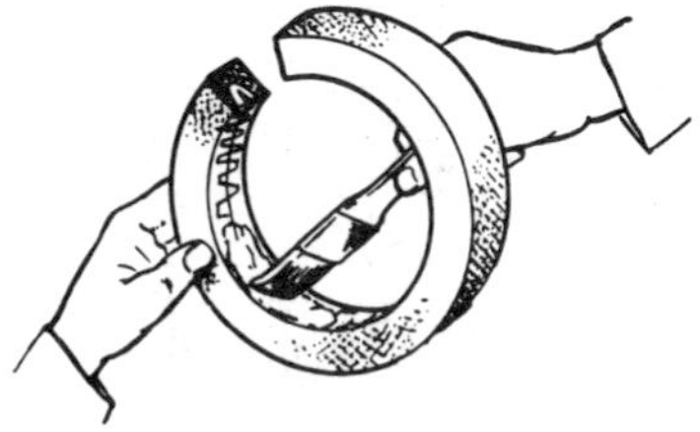

8 Install foil-wrapped packing so edges on inside will face direction of shaft rotation. This is a must; otherwise, thin edges flake off, reduce packing life

9 Neck bushing slides into stuffing box. Quick way to make it is to pour soft bearing metal into tin can, turn and bore for sliding fit into place

10 Swabbing new metallic packings with lubricant supplied by packing maker is OK. These include foil types, leadcore, etc. If the shaft is oily, don't swab it

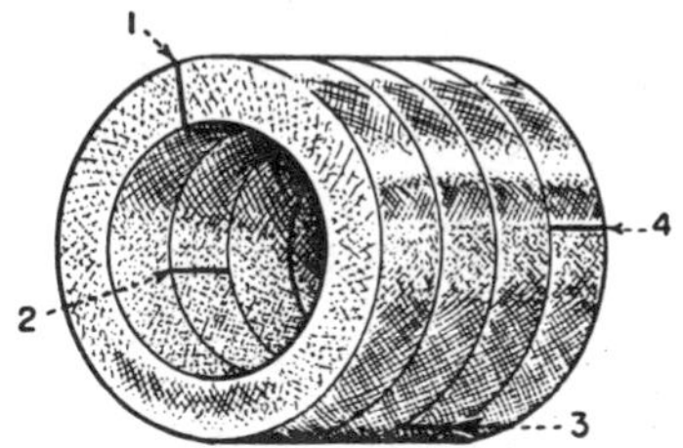

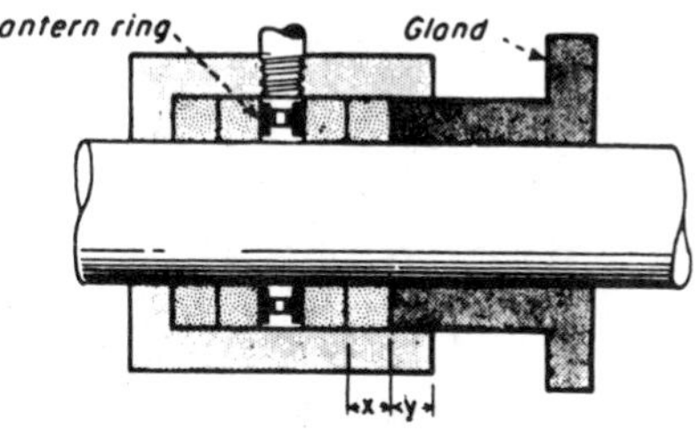

13 Stagger joints 180 degrees if only two rings are in stuffing box. Space at 120 degrees for three rings, or 90 degrees if four rings or more are in set

14 Install packing so lantern ring lines up with cooling-liquid opening. Also, remember that this ring moves back into box as packing is compressed. Leave space for gland to enter as shown. Tighten gland with wrench—back off finger-tight. Allow the packing to leak until it seats itself, then allow a slight operating leakage.

Hydraulic-packing pointers

First, clean stuffing box, examine ram or shaft. Next, measure stuffing box depth and packing set—find difference. Place 1/8-in. washers over gland studs as shown. Lubricate ram and packing set (if for water). If you can use them, endless rings give about 17% more wear than cut rings. Place male adapter in bottom, then carefully slide each packing turn home—don't harm lips. Stagger joints for cut rings. Measure from top of packing to top of washers, then compare with gland. Never tighten down new packing set until all air has chance to work out. As packing wears, remove one set of washers after more wear, remove other washer.

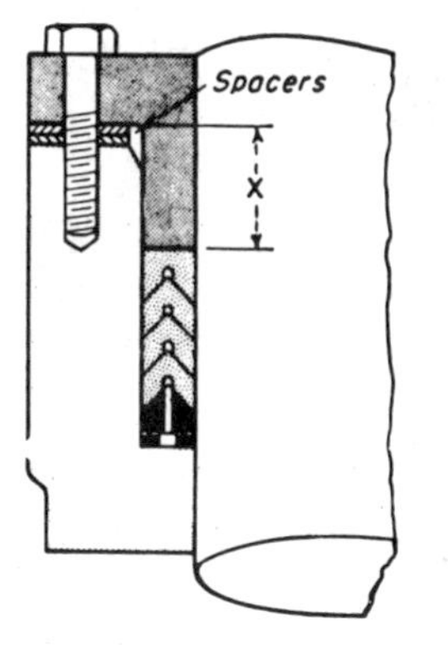

Fig. 15.48 How to pack a pump (continued)

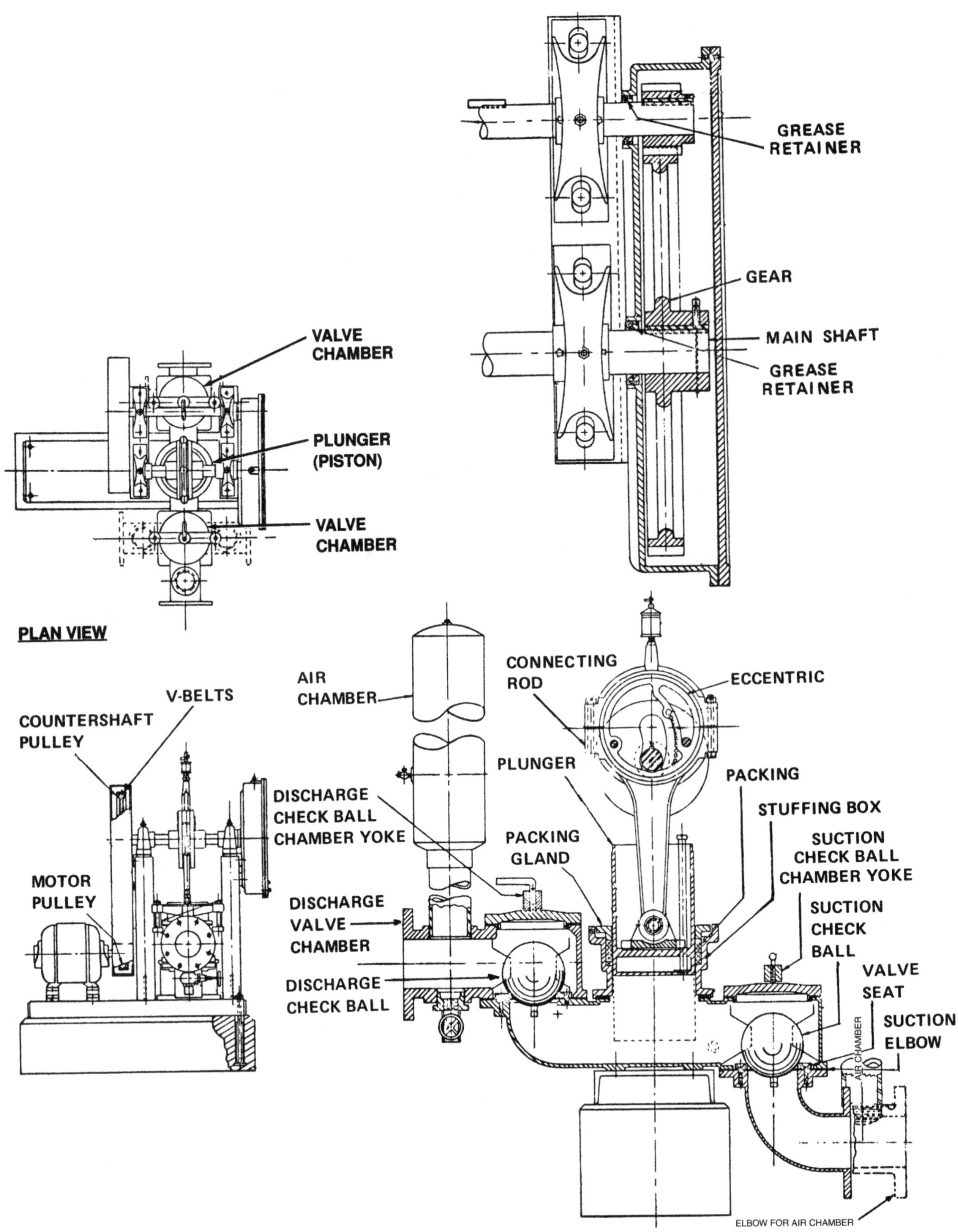

Fig. 15.49 Reciprocating pump

(Courtesy ITT Marlow, a Unit of International Telephone Corp.)

Frequency of Service	
W	10. CONNECTING RODS. Set oilers to dispense two drops per minute.
W	11. PLUNGER CROSSHEAD. Fill plunger as required to half cover the wrist pin with oil.
D	12. PLUNGER TROUGH. Keep small quantity of oil in trough to lubricate the plunger.
M	13. MAIN SHAFT BEARING. Grease bearings monthly. Pump should be in operation when lubricating to avoid excessive pressure on seals.
	14. CHECK ELECTRIC MOTOR. See Paragraph 7.

Paragraph 3: Propeller Pumps, General (Figure 15.7)

D	1. CHECK MOTOR CONDITION. See Paragraphs 7.1 and 7.2.
D	2. CHECK PACKING GLAND ASSEMBLY. See Paragraph 1.7.
W	3. INSPECT PUMP ASSEMBLY. See Paragraph 1.4.
W	4. LUBE LINE SHAFT AND DISCHARGE BOWL BEARING. Maintain oil in oiler at all times. Adjust feed rate to approximately four drops per minute.
W	5. LUBE SUCTION BOWL BEARING. Lube through pressure fitting. Usually, three or four strokes of gun are enough.
W	6. OPERATE PUMPS ALTERNATELY. See Paragraph 1.3.
A	7. LUBE MOTOR BEARINGS. See Paragraph 7.3.

Paragraph 4: Progressive Cavity Pumps, General (Figure 15.12)

D	1. CHECK MOTOR CONDITION. See Paragraphs 7.1 and 7.2.
D	2. CHECK PACKING GLAND ASSEMBLY. See Paragraph 1.7.
D	3. CHECK DISCHARGE PRESSURE. A higher than normal discharge pressure may indicate a line blockage or a closed valve downstream. An abnormally low discharge pressure can mean reduced rate of discharge.
S	4. INSPECT AND LUBRICATE BEARINGS—GREASE. If possible, remove bearing cover and visually inspect grease. When greasing, remove relief plug and cautiously add 5 or 6 strokes of the grease gun. Afterward, check bearing temperature with thermometer. If over 220°F (104°C), remove some grease.
S	5. LUBEFLUSH MOTOR BEARINGS. See Paragraph 7.3.
S	6. CHECK PUMP OUTPUT. Check how long it takes to fill a vessel of known volume or quantity; or check performance against a meter, if available.
A	7. SCOPE MOTOR BEARINGS. See Paragraph 7.4.
A	8. SCOPE PUMP BEARINGS. See Paragraph 7.4.

Paragraph 5: Pneumatic Ejectors, General (Figure 15.14)

D	1. INSPECT UNIT. Check unit through a complete cycle. Look for air or water leaks. Keep units clean.
W	2. BLOW DOWN RECEIVERS.
M	3. CHECK VALVES. Keep check valve packing tight enough to prevent air or water leaks.
S	4. CLEAN AIR STRAINERS. Isolate air inlet line to ejector pot. Remove strainer and clean with water and wire brush ensuring free strainer openings. In case of different types of filters, consult particular manufacturer's literature.
A	5. CLEAN RECEIVER. After completely isolating ejector and power to it, remove and clean electrodes. Open ejector pot inspection plate and scrape inside walls of the pot.
A	6. INSPECT CHECK VALVES. Check operation by closing discharge valve, and actuating ejector on "hand" until pressure relief valve operates continuously for several minutes. Non-operation, or intermittent operation of relief valve in this mode indicates inlet check valve not holding. Discharge check valve can be checked by automatic ejection of pot, followed by locking out controls, closing inlet and discharge valves, removing top inspection plate, and checking for fluid leaking by the check valve. Some check valves can be checked by manually operating and feeling if they seat. At times, gauges in strategic locations can indicate proper or improper operation. Inspect all check valves for leakage around packing.
A	7. INSPECT MAIN AIR VALVE. Check packing nut for lubrication. Check for smooth operation.
A	8. INSPECT PILOT VALVE. Time the ejection and check pressure gauge for maximum steady pressure.

Paragraph 6: Float and Electrode Switches

To ensure the best operation of the pump, a systematic inspection of the water-level controls should be made at least once a week. Check to see that:

Frequency of Service	
W	1. CHECK CONTROLS. Controls respond to a rising water level in the wet well.
W	2. START-UP. The unit starts when the float switch or electrode system makes contact, and the pump stops at the prescribed level in the wet well.
W	3. MOTOR SPEED. The motor speed comes up quickly and is maintained.
W	4. SPARKING. A brush-type motor does not spark profusely in starting or running.
W	5. INTERFERENCE WITH CONTROLS. Grease and trash are not interfering with controls. Be sure to remove scum from water-level float controls.
W	6. ADJUSTMENTS. Any necessary adjustments are properly completed.

QUESTIONS

Write your answers in a notebook and then compare your answers with those on page 463.

15.7F What are some of the common causes of shear pin failure in reciprocating pumps?

15.7G What may happen when water or a thin sludge is being pumped by a reciprocating pump?

END OF LESSON 4 OF 7 LESSONS ON MAINTENANCE

Please answer the discussion and review questions next.

DISCUSSION AND REVIEW QUESTIONS

Chapter 15. MAINTENANCE

(Lesson 4 of 7 Lessons)

Write the answers to these questions in your notebook. The question numbering continues from Lesson 3.

23. What would you do if considerable water was leaking from the water-seal of a pump?
24. When two or more pumps of the same size are installed, why should they be operated alternately?
25. What should be checked if pump bearings are running hot?
26. What happens when the packing is too tight on a reciprocating pump?
27. Why should changes in control valve settings for reciprocating pumps be adjusted when the pump is shut down?

CHAPTER 15. MAINTENANCE

(Lesson 5 of 7 Lessons)

Paragraph 7: Electric Motors (Figure 15.50)

To ensure the proper and continuous function of electric motors, the items listed in this paragraph must be performed at the designated intervals. If operational checks indicate a motor is not functioning properly, these items will have to be checked to locate the problem.

Frequency of Service

D 1. CHECK MOTOR CONDITIONS.

a. Keep motors free from dirt, dust, and moisture.

b. Keep operating space free from articles that may obstruct air circulation.

c. Check for excessive grease leakage from bearings.

W 2. NOTE ALL UNUSUAL CONDITIONS.

a. Unusual noises in operation.

b. Motor failing to start or come to speed normally, sluggish operation.

c. Motor or bearings that feel or smell hot.

d. Continuous or excessive sparking commutator or brushes. Blackened commutator.

e. Intermittent sparking at brushes.

f. Fine dust under couplings having rubber bushings or pins.

g. Smoke, charred insulation, or solder whiskers extending from armature.

h. Excessive humming.

i. Regular clicking.

j. Rapid knocking.

k. Brush chatter.

l. Vibration.

m. Hot commutator.

A 3. LUBRICATE BEARINGS (Figure 15.51).

a. Check grease in ball bearings and replenish when necessary.

Follow instructions below when preparing bearings for grease.

b. Wipe pressure gun fitting, bearing housing, and relief plug to make sure that no dirt gets into bearing with grease.

c. Before using grease gun, always remove relief plug from bottom of bearing to prevent excessive pressure in housing, which might rupture bearing seals.

d. Use clean screwdriver or similar tool to remove hardened grease from relief hole and permit excess grease to run freely from bearing.

e. While motor is running, add grease with hand-operated pressure gun until it flows from relief hole, purging housing of old grease. If there is no bottom or relief plug on bearing housing, insert grease cautiously through upper plug. Usually, four or five strokes of gun are enough. If bearing is overlubricated, seal may be ruptured. If lubricating a running motor is dangerous, follow above procedure with motor at a standstill. Lubricating a running motor is dangerous if you must remove protective gear such as guards over moving parts in order to lubricate the motor. Extension tubes and appropriate fittings should be installed wherever practical to facilitate lubrication with the guards in place.

f. Allow motor to run for five minutes or until all excess grease has drained from bearing house.

g. Stop motor and replace relief plug tightly with wrench.

A 4. USING A *STETHOSCOPE,*[16] CHECK BOTH BEARINGS. Listen for whines, gratings, or uneven noises. Listen all around the bearing and as near as possible to the bearing. Listen while the motor is being started and shut off. If unusual noises are heard, pinpoint the location.

5. IF YOU THINK THE MOTOR is running unusually hot, check with a thermometer. Place the thermometer in the casing near the bearing, holding it there with putty or clay. Magnetic thermometers are also available for this purpose.

[16] *Stethoscope.* An instrument used to magnify sounds and carry them to the ear.

DRIP PROOF

ITEM NO.	PART NAME
1	Wound Stator w/ Frame
2	Rotor Assembly
3	Rotor Core
4	Shaft
5	Bracket
6	Bearing Cap
7	Bearings
8	Seal, Labyrinth
9	Thru Bolts/Caps
10	Seal, Lead Wire
11	Terminal Box
12	Terminal Box Cover
13	Fan
14	Deflector
15	Lifting Lug

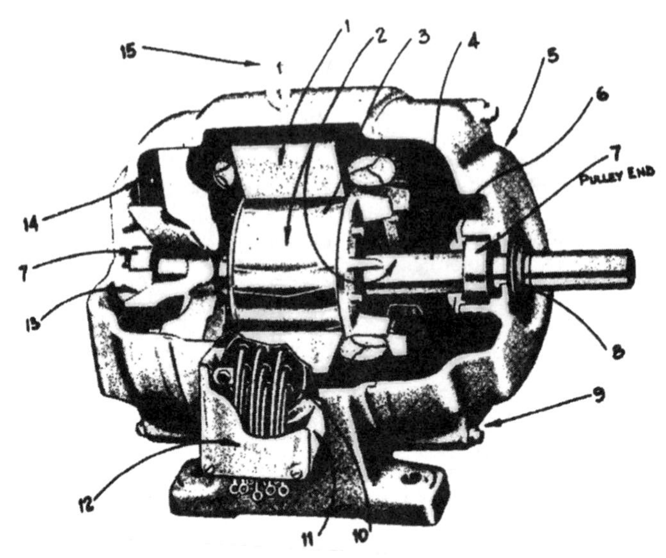

TOTALLY ENCLOSED FAN COOLED

ITEM NO.	PART NAME
1	Wound Stator w/ Frame
2	Rotor Assembly
3	Rotor Core
4	Shaft
5	Brackets
6	Bearings
7	Seal, Labyrinth
8	Thru Bolts/Caps
9	Seal, Lead Wire
10	Terminal Box
11	Terminal Box Cover
12	Fan, Inside
13	Fan, Outside
14	Fan Grill
15	Fan Cover
16	Fan Cover Bolts
17	Lifting Lug

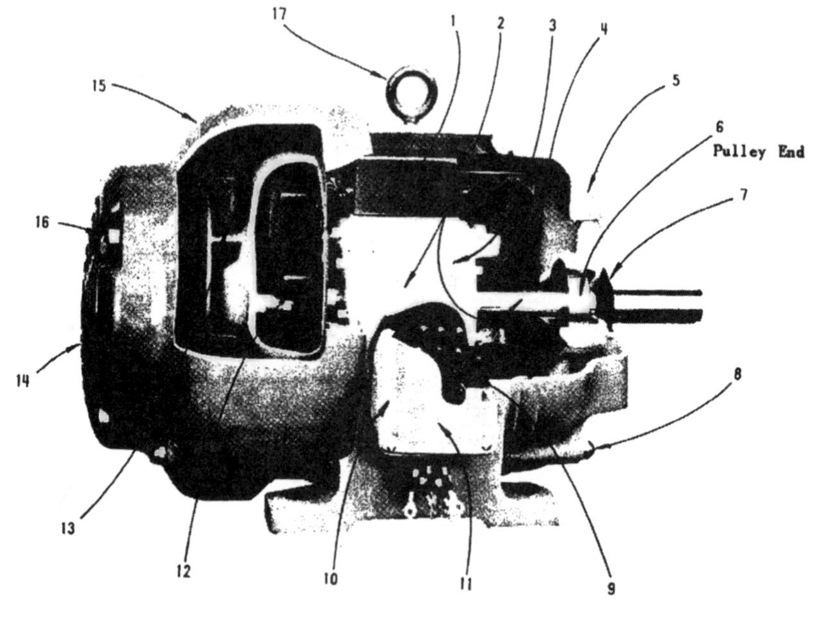

Fig. 15.50 Typical motors
(Courtesy of Sterling Power Systems, Inc.)

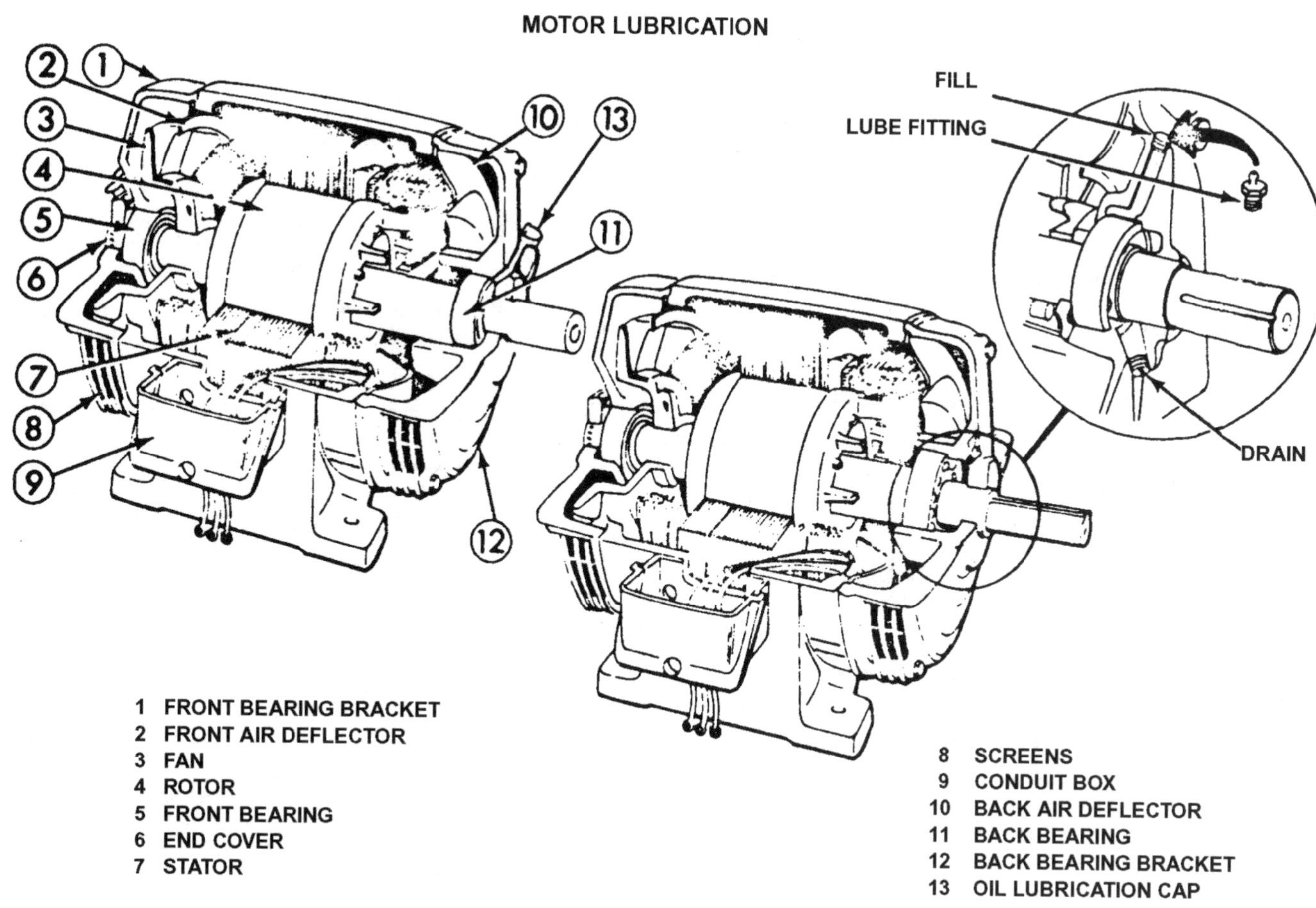

Fig. 15.51 Electric motor lubrication

A 6. *DATEOMETER.*[17] If there is a dateometer on the motor, after changing the oil in the motor, loosen the dateometer screw and set to the corresponding year.

QUESTION

Write your answer in a notebook and then compare your answer with the one on page 463.

15.7H What are the major items you would include when checking an electric motor?

Paragraph 8: Belt Drives

1. GENERAL. Maintaining a proper tension and alignment of belt drives ensures long life of belts and sheaves. Incorrect alignment causes poor operation and excessive belt wear. Inadequate tension reduces the belt grip, causes high belt loads, snapping, and unusual wear.

Frequency of Service

a. Cleaning belts. Keep belts and sheaves (pulleys) clean and free of oil, which causes belts to deteriorate. To remove oil, take belts off sheaves and wipe belts and sheaves with a rag moistened in a non-oil-base solvent. Follow the requirements on the solvent's MSDS concerning ventilation and personal protective equipment (PPE).

b. Installing belts. Before installing belts, replace worn or damaged sheaves, then slack off on adjustments. Do not try to force belts into position. Never use a screwdriver or similar lever to get belts onto sheaves. After belts are installed, adjust tension; recheck tension after eight hours of operation. (See Table 15.9.)

c. Replacing belts. Replace belts as soon as they become frayed, worn, or cracked. *Never replace only one V-belt on a multiple*

[17] *Dateometer* (day-TOM-uh-ter). A small calendar disk attached to motors and equipment to indicate the year in which the last maintenance service was performed.

Frequency of Service

drive. Replace the complete set with a set of matched belts, which can be obtained from any supplier. All belts in a matched set are machine-checked to ensure equal size and tension.

d. Storing spare belts. Store spare belts in a cool, dark place. Tag all belts in storage to identify them with the equipment on which they can be used.

2. V-BELTS. A properly adjusted V-belt has a slight bow in the slack side when running; when idle it has an alive springiness when thumped with the hand. An improperly tightened belt feels dead when thumped.

If the slack side of the drive is less than 45° from the horizontal, the vertical sag at the center of the span may be adjusted in accordance with Table 15.9 below:

TABLE 15.9 HORIZONTAL BELT TENSION

Span (inches)		10	20	50	100	150	200
Vertical Sag (inches)	From	.01	.03	.20	.80	1.80	3.30
	To	.03	.09	.58	2.30	4.90	8.60
Span (millimeters)		250	500	1,250	2,500	3,750	5,000
Vertical Sag (millimeters)	From	0.25	0.75	5.00	20.0	45.0	82.5
	To	0.75	2.25	14.50	57.5	122.5	215.0

M a. Check tension. If tightening belt to proper tension does not correct slipping, check for overload, oil on belts, or other possible causes. Never use belt dressing to stop belt slippage. Rubber wearings near the drive are a sign of improper tension, incorrect alignment, or damaged sheaves.

M b. Check sheave (pulley) alignment. Lay a long straightedge or string across outside faces of pulley and allow for differences in dimensions from centerlines of grooves to outside faces of the pulleys being aligned. Be especially careful in aligning drives with more than one V-belt on a sheave, as misalignment can cause unequal tension.

Paragraph 9: Chain Drives

1. GENERAL. Chain drives may be designated for slow, medium, or high speeds.

a. Slow-speed drives. Because slow-speed drives are usually enclosed, adequate lubrication is difficult. Heavy oil applied to the outside of the chain seldom reaches the working parts; in addition, the oil catches dirt and grit and becomes abrasive. For lubricating and cleaning methods, see 5 and 6 below.

b. Medium- and high-speed drives. Medium-speed drives should be continuously lubricated with a device similar to a sight-feed oiler. High-speed drives should be completely enclosed in an oil-tight case and the oil maintained at proper level.

D 2. CHECK OPERATION. Check general operating condition during regular tours of duty.

Q 3. CHECK CHAIN SLACK. The correct amount of slack is essential to proper operation of chain drives. Unlike other belts, chain belts should not be tight around the sprocket; when chains are tight, working parts carry a much heavier load than necessary. Too much slack is also harmful; on long centers particularly, too much slack causes vibrations and chain whip, reducing life of both chain and sprocket. A properly installed chain has a slight sag or looseness on the return run.

S 4. CHECK ALIGNMENT. If sprockets are not in line or if shafts are not parallel, excessive sprocket and chain wear and early chain failure result. Wear on inside of chain, side walls, and sides of sprocket teeth are signs of misalignment. To check alignment, remove chain and place a straightedge against sides of sprocket teeth.

S 5. CLEAN. On enclosed types, flush chain and enclosure with a petroleum solvent (kerosene). On exposed types, remove chain and soak and wash it in solvent. Clean sprockets, install chain, and adjust tension.

S 6. CHECK LUBRICATION. Soak exposed-type chains in oil to restore lubricating film. Remove excess lubricant by hanging chains up to drain.

Do not lubricate underwater chains that operate in contact with considerable grit. If water is clean, lubricate by applying waterproof grease with brush while chain is running.

Do not lubricate chains on elevators or on conveyors of feeders that handle dirty or gritty materials. Dust and grit combine with lubricants to form a cutting compound that reduces chain life.

S 7. CHANGE OIL. On enclosed types only, drain oil and refill case to proper level.

S 8. INSPECT. Note and correct abnormal conditions before serious damage results. Do not put a new chain on worn sprockets. Always replace worn sprockets when replacing a chain because out-of-pitch sprockets cause as much chain wear in a few hours as years of normal operation.

Frequency of Service

9. TROUBLESHOOTING. Some common symptoms of improper chain drive operation and their remedies follow:

   a. Excessive noise. Correct alignment, if misaligned. Adjust centers for proper chain slack. Lubricate in accordance with aforementioned methods. Be sure all bolts are tight. If chain or sprockets are worn, reverse or renew, if necessary.

   b. Wear on chain, side walls, and sides of teeth. Remove chain and correct alignment.

   c. Chain climbs sprockets. Check for poorly fitting sprockets and replace, if necessary. Make sure tightener is installed on drive chain.

   d. Broken pins and rollers. Check for chain speed that may be too high for the pitch, and substitute chain and sprockets with shorter pitch, if necessary. Breakage also may be caused by shock loads.

   e. Chain clings to sprockets. Check for incorrect or worn sprockets or heavy, tacky lubricants. Replace sprockets or lubricants, if necessary.

   f. Chain whip. Check for too-long centers or high, pulsating loads and correct cause.

   g. Chains get stiff. Check for misalignment, improper lubrication, or excessive overloads. Make necessary corrections or adjustments.

Paragraph 10: Variable-Speed Belt Drives (See Figure 15.52)

W 1. CLEAN DISCS. Remove grease, acid, and water from disc faces.

D 2. CHECK SPEED-CHANGE MECHANISM. Shift drive through entire speed range to make sure shafts and bearings are lubricated and discs move freely in lateral direction on shafts.

W 3. CHECK V-BELT. Make sure it runs level and true. If one side rides high, a disc is sticking on shaft because of insufficient lubrication or wrong lubricant. In this case, stop the drive at once, remove V-belt, and clean disc hub and shaft thoroughly with petroleum solvent until disc moves freely. Relubricate with soft, ball bearing grease and replace V-belt in opposite direction from that in which it formerly ran.

M If drive is not operated for 30 days or more, shift unit to minimum speed position, placing spring on variable-speed shaft at minimum tension and relieving belt of excessive pressure.

4. LUBRICATE DRIVE. Make sure to apply lubricant at all the six force-feed lubrication fittings (Figure 15.52: A, B, D, E, G, and H) and the one cup-type fitting (C). *NOTE: If the drive is used with a reducer, fitting E is not provided.*

W a. Once every ten days to two weeks, use two or three strokes of a grease gun through fittings A and B at ends of shifting screw and variable-speed shaft, respectively, to lubricate bearings of movable discs. Then, with unit running, shift drive from one extreme speed position to the other to ensure thorough distribution of lubricant over disc-hub bearings.

Q b. Add two or three shots of grease through fittings D and E to lubricate frame bearing on variable-speed shaft.

Q c. Every 90 days, add two or three cupfuls of grease to Cup C, which lubricates thrust bearing on constant-speed shaft.

Q d. Every 90 days, use two or three strokes of grease gun through fittings G and H to lubricate motor frame bearings.

CAUTION: Be sure to follow manufacturer's recommendation on type of grease. After lubricating, wipe excess grease from sheaves and belt.

QUESTIONS

Write your answers in a notebook and then compare your answers with those on page 463.

15.7I How can you tell if a belt on belt-drive equipment has proper tension and alignment?

15.7J Why should worn sprockets be replaced when replacing a chain in a chain-drive unit?

Paragraph 11: Couplings

1. GENERAL. Unless couplings between the driving and driven elements of a pump or any other piece of equipment are kept in proper alignment, breaking and excessive wear results in either or both the driven machinery and the driver. Burned-out bearings, sprung or broken shaft, and excessively worn or ruined gears are some of the damages caused by misalignment. To prevent outages and the expense of installing replacement parts, check the alignment of all equipment before damage occurs.

   a. Improper original installation of the equipment may not necessarily be the cause of the trouble. Settling of foundations, heavy floor loadings, warping of bases, excessive bearing wear, and many other factors cause misalignment. A rigid base is not always security against misalignment. The base may have been mounted off level, which could cause it to warp.

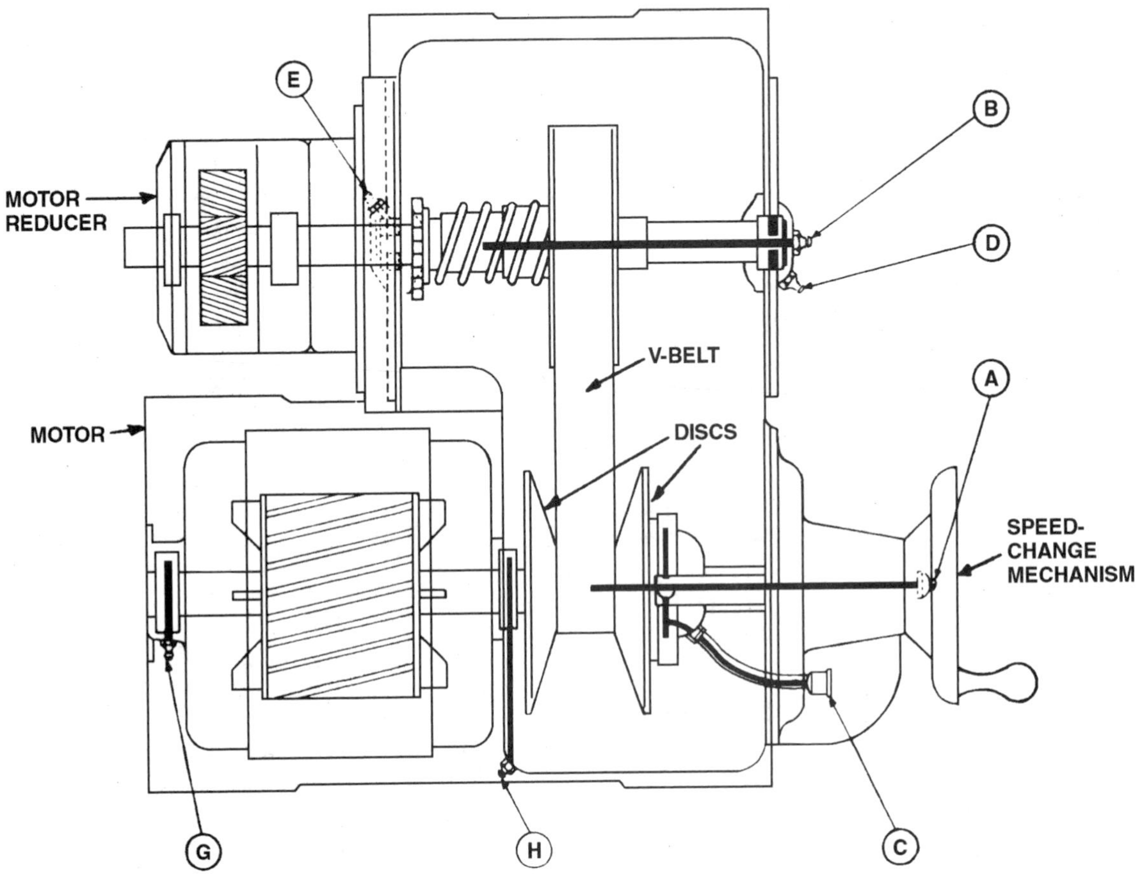

NOTE: A, B, D, E, G, and H are force-feed lubrication fittings. C is a cup-type lubrication fitting.

Fig. 15.52 Reeves varidrive (variable-speed belt drive)

(Source: War Department Technical Manual TM5-666)

Frequency of Service

b. Flexible couplings permit easy assembly of equipment, but they must be aligned as exactly as flanged couplings if maintenance and repair are to be kept to a minimum. Rubber-bushed types cannot function properly if the bolts cannot move in their bushings.

S 2. CHECK COUPLING ALIGNMENT (straightedge method). Excessive bearing and motor temperatures caused by overload, noticeable vibration, or unusual noises may all be warnings of misalignment. Realign when necessary (Figure 15.53) using a straightedge and thickness gauge or wedge. To ensure satisfactory operation, level up to within 0.005 inch (0.13 mm) as follows:

a. Remove coupling pins.

b. Rigidly tighten driven equipment; slightly tighten bolts holding drive.

c. To correct horizontal and vertical misalignment, shift or shim drive to bring coupling halves into position so no light can be seen under a straightedge laid across them. Place straightedge in four positions, holding a light in back of straightedge to help ensure accuracy.

d. Check for angular misalignment with a thickness or feeler gauge inserted at four places to make certain space between coupling halves is equal.

e. If proper alignment has been secured, coupling pins can be put in place easily using only finger pressure. Never hammer pins into place.

f. If equipment is still out of alignment, repeat the procedure.

S 3. CHECK COUPLING ALIGNMENT (dial indicator method). Dial indicators also are used to measure coupling alignment. This method produces better results than the straightedge method. The dial indicates very small movements or distances, which are measured in mils (one mil equals $^{1}/_{1000}$ of an inch). The indicator consists of a dial with a graduated face (with "plus" and "minus" readings), a pedestal, and a rigid indicator bar (or "fixture") as shown in Figure 15.54.

The dial indicator is attached to one coupling at the fixture and adjusted to the zero position or reading. When the shaft of the machine is rotated, misalignment will cause the pedestal to compress (a "plus" reading), or extend (a "minus" reading). Literature provided by the manufacturer of machinery usually will indicate maximum allowable tolerances or movement.

Carefully study the manufacturer's literature provided with your dial indicator before attempting to use the device.

A 4. CHANGE OIL IN FAST COUPLINGS. Drain out old oil and add gear oil to proper level. Correct quantity is given on instruction card supplied with each coupling.

Paragraph 12: Shear Pins

Many wastewater treatment units use shear pins as protective devices to prevent damage in case of sudden overloads. To serve this purpose, these devices must be in operational condition at all times. Under some operating conditions, shearing surfaces of a shear pin device may freeze together so solidly that an overload fails to break them.

Manufacturers' drawings for particular installations usually specify shear pin material and size. If this information is not available, obtain the information from the manufacturer, giving the model, serial number, and load conditions of unit. When necessary to determine shear pin size, select the lowest strength that does not break under the unit's usual loads. When proper size is determined, never use a pin of greater strength, such as a bolt or a nail.

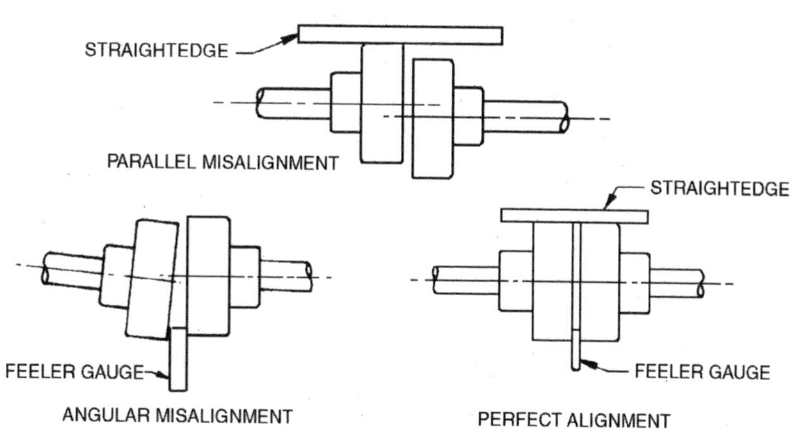

Fig. 15.53 Testing alignment, straightedge

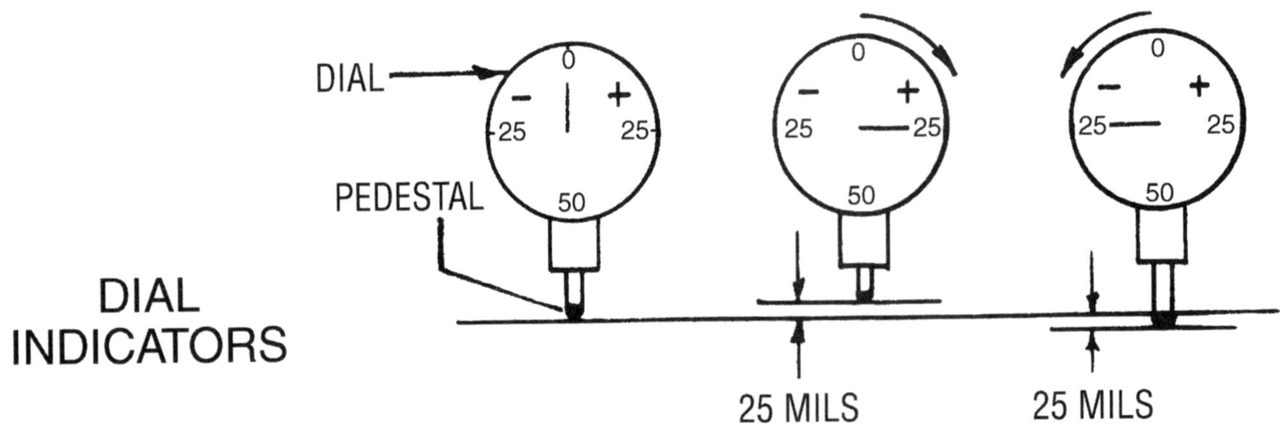

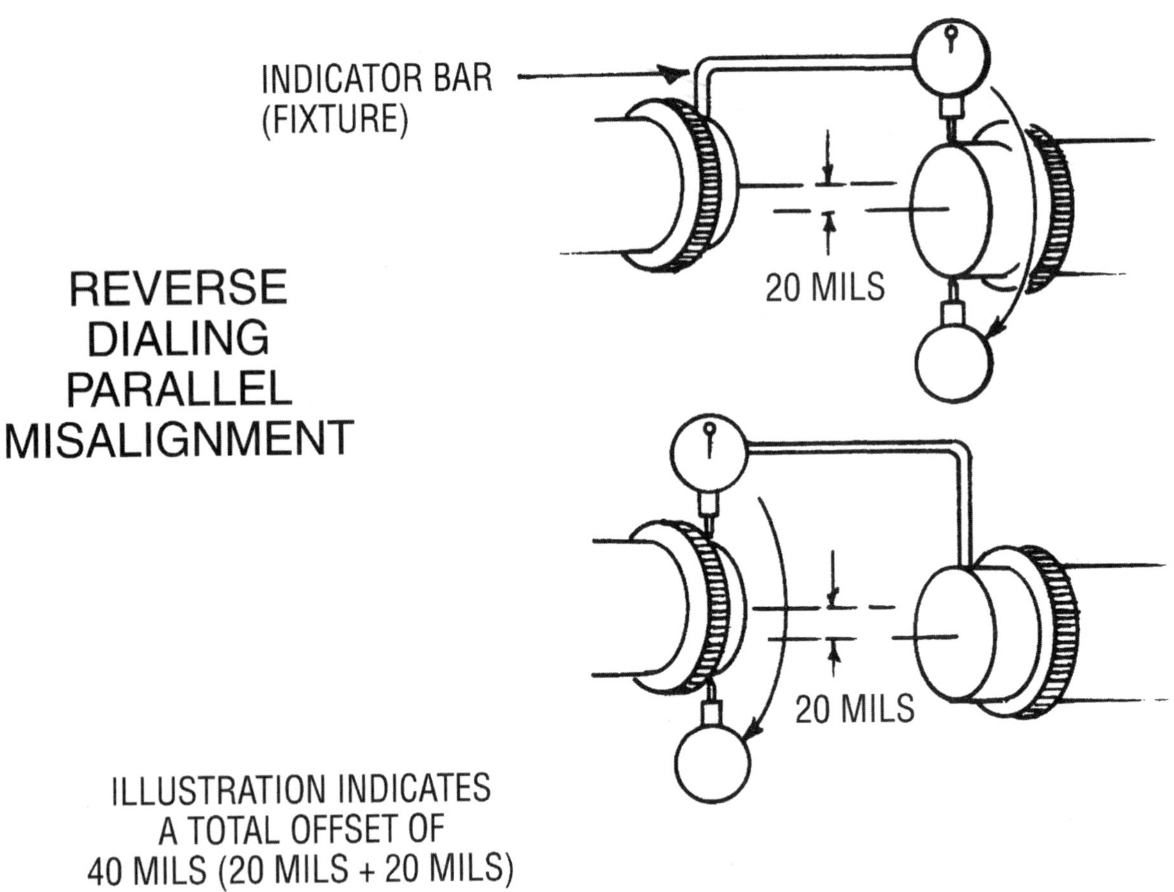

Fig. 15.54 Use of a dial indicator

(Permission of DYMAC, a Division of Spectral Dynamics Corporation)

If necked pins are used, be sure the necked-down portion is properly positioned with respect to shearing surfaces. When a shear pin breaks, determine and remedy the cause of failure before inserting new pin and starting drive in operation.

Frequency of Service	
M	1. GREASE SHEARING SURFACES.
Q	2. REMOVE SHEAR PIN. Operate motor for a short time to smooth out any corroded spots.
A	3. CHECK SPARE INVENTORY. Make sure an adequate supply is on hand, properly identified and with record of proper pin size, necked diameter, and longitudinal dimensions.

QUESTIONS

Write your answers in a notebook and then compare your answers with those on page 463.

15.7K What factors could cause couplings to become out of alignment?

15.7L What is the purpose of shear pins?

END OF LESSON 5
OF 7 LESSONS ON
MAINTENANCE

Please answer the discussion and review questions next.

DISCUSSION AND REVIEW QUESTIONS

Chapter 15. MAINTENANCE

(Lesson 5 of 7 Lessons)

Write the answers to these questions in your notebook. The question numbering continues from Lesson 4.

28. Why would you use a stethoscope to check an electric motor?
29. How would you determine if a motor is running unusually hot?
30. How would you clean belts and sheaves (pulleys) on a belt drive?
31. Why should you never replace only one belt on a multiple-drive unit?
32. What do rubber wearings near a belt drive indicate?
33. How can you determine if a chain in a chain-drive unit has the proper slack?
34. What happens when couplings are not in proper alignment?

CHAPTER 15. MAINTENANCE

(Lesson 6 of 7 Lessons)

Paragraph 13: Gate Valves (Figures 15.55 and 15.56)

Gate valves require the following maintenance: oiling, tightening, and replacing the stem stuffing box packing.

Frequency of Service	
A	1. REPLACE PACKING. Modern gate valves can be repacked without removing them from service. Before repacking, open valve wide. This prevents excessive leakage when the packing or the entire stuffing box is removed. It draws the stem collar tightly against the bonnet on a non-rising stem valve, and tightly against the bonnet bushing on a rising stem valve. a. Stuffing box. Remove all old packing from stuffing box with a packing hook. Clean valve stem of all adhering particles and polish it with fine emery cloth. After polishing, remove the fine grit with a clean cloth to which a few drops of oil have been added. b. Insert packing. Insert new split-ring packing in stuffing box and tamp it into place with packing gland. Stagger ring splits. After stuffing box is filled, place a few drops of oil on stem, assemble gland, and tighten it down on packing.
S	2. OPERATE VALVE. Operate inactive gate valves to prevent sticking.
A	3. LUBRICATE GEARING. Lubricate gate valves as recommended by manufacturer. Lubricate thoroughly any gearing in large gate valves. Wash open gears with solvent and lubricate with grease.
S	4. LUBRICATE RISING STEM THREADS. Clean threads on rising stem gate valves and lubricate with grease.
A	5. LUBRICATE BURIED VALVES. If a buried valve works hard, lubricate it by pouring oil down through a pipe that is bent at the end to permit oiling the packing follower below the valve nut.
A	6. REFACE LEAKY GATE VALVE SEATS. If gate valve seats leak, reface them immediately, using the method discussed below. A solid wedge disc valve is used for illustration, but the general method also applies to other types of repairable gate valves. Proceed as follows: a. Remove bonnet and clean and examine disc and body thoroughly. Carefully determine extent of damage to body rings and disc. If corrosion has caused excessive pitting or eating away of metal, as in guide ribs in body, repairs may be impractical. b. Check and service all parts of valve completely. Remove stem from bonnet and examine it for scoring and pitting where packing makes contact. Polish lightly with fine emery cloth to put stem in good condition. Use soft jaws if stem is put in vise. c. Remove all old packing and clean out stuffing box. Clean all dirt, scale, and corrosion from inside of valve bonnet and other parts. d. Do not salvage an old gasket. Remove it completely and replace with one of proper quality and size. e. After cleaning and examining all parts, determine whether valve can be repaired by removing cuts from disc and body seat faces or by replacement of body seats. If repair can be made, set disc in vise with face leveled, wrap fine emery cloth around a flat tool, and rub or lap off entire bearing surface on both sides to a smooth, even finish. Remove as little metal as possible. f. Repair cuts and scratches on body rings, lapping with an emery block small enough to permit convenient rubbing all around rings. Work carefully to avoid removing so much metal that disc will seat too low. When seating surfaces of disc and seat rings are properly lapped in, coat faces of disc with *PRUSSIAN BLUE*[18] and drop disc in body to check contact. When good, continuous contact is obtained, the valve is tight and ready for assembly. Insert stem in bonnet, install new packing, assemble

[18] *Prussian Blue.* A blue paste or liquid (often on a paper like carbon paper) used to show a contact area. Used to determine if gate valve seats fit properly.

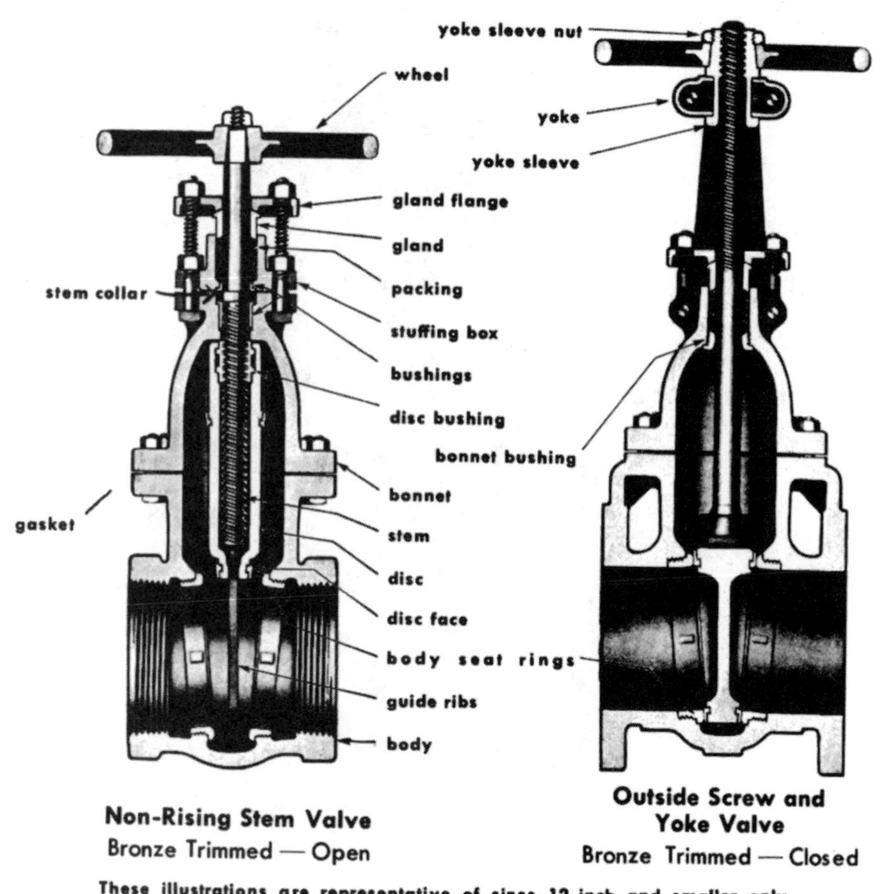

Fig. 15.55 Wedge gate valve

(Source: Crane Co.)

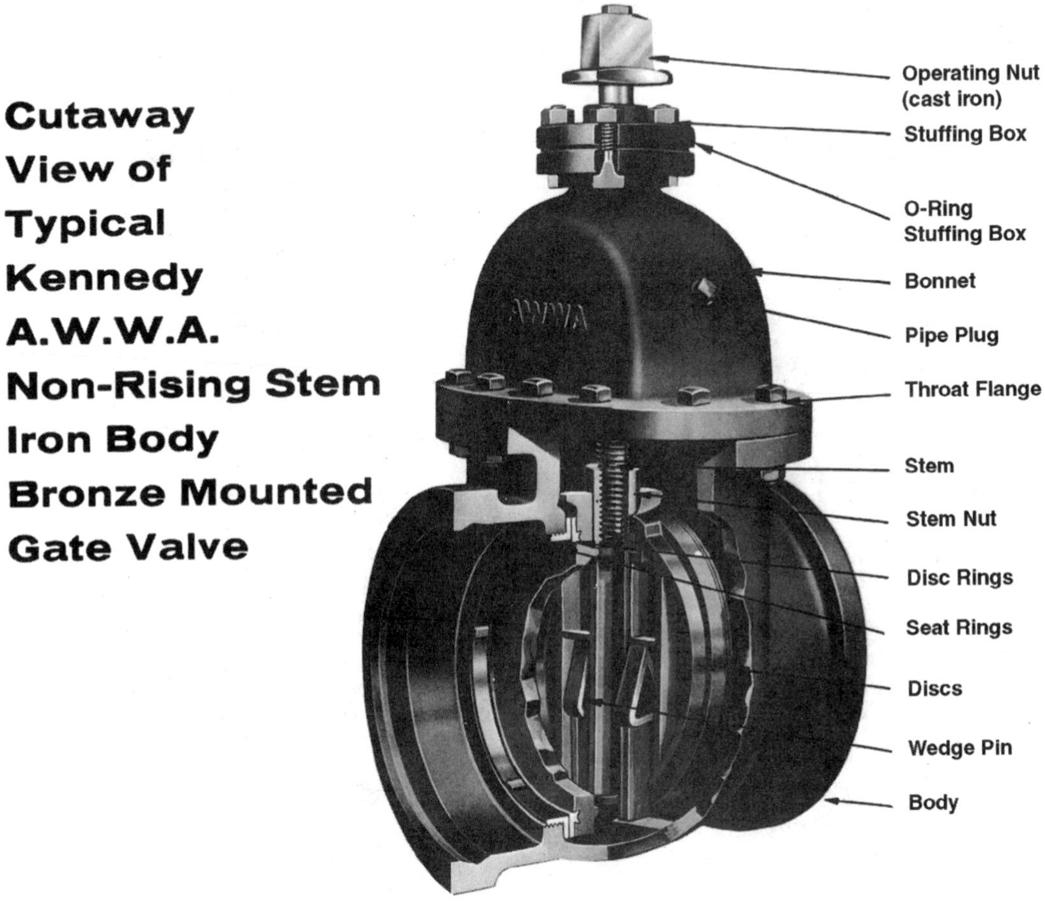

Cutaway View of Typical Kennedy A.W.W.A. Non-Rising Stem Iron Body Bronze Mounted Gate Valve

Operating Nut (cast iron)
Stuffing Box
O-Ring Stuffing Box
Bonnet
Pipe Plug
Throat Flange
Stem
Stem Nut
Disc Rings
Seat Rings
Discs
Wedge Pin
Body

Fig. 15.56 Non-rising stem gate valve

(Permission of Kennedy Valve, Division of ITT Grinnell Valve Co., Inc.)

Frequency of Service

other parts, attach disc to stem, and place assembly in body. Raise disc to prevent contact with seats so bonnet can be properly seated on body before tightening the joint.

g. Test repaired valve before putting it back in line to ensure that repairs have been properly made.

h. If leaky gate valve seats cannot be refaced, remove and replace seat rings with a power lathe. Chuck up body with rings vertical to lathe and use a strong steel bar across ring lugs to unscrew them. They can be removed by hand with a diamond point chisel if care is taken to avoid damaging threads. Drive new rings home tightly. Use a wrench on a steel bar across lugs when putting in rings by hand. Always coat threads with a good lubricant before putting threads into the valve body. This helps to make the threads easier to remove the next time the seats have to be replaced. Lap in rings to fit disc perfectly.

Paragraph 14: Check Valves (Figure 15.57)

A 1. INSPECT DISC FACING. Open valves to observe condition of facing on swing check valves equipped with leather or rubber seats on disc. If metal seat ring is scarred, dress it with a fine file and lap with fine emery paper wrapped around a flat tool.

A 2. CHECK PIN WEAR. Check pin wear on balanced disc check valve, since disc must be accurately positioned in seat to prevent leakage.

Paragraph 15: Plug Valves (Figures 15.58, 15.59, 15.60, 15.61, and 15.62)

M 1. ADJUST GLAND. The adjustable gland holds the plug against its seats in body and acts through compressible packing, which functions as a thrust cushion. Keep gland tight enough at all times to hold plug in contact with its seat. If this is not done, the lubricant system cannot function properly and solid particles may enter between the body and plug and cause damage.

M 2. LUBRICATE ALL VALVES. Apply lubricant by removing lubricant screw and inserting stick of plug valve lubricant for stated temperature conditions. The check valve fitting within the shank prevents line pressure from blowing out when lubricant screw is removed. Inject lubricant into valve by turning screw down to keep valve in proper operating condition. If lubrication has been neglected, several sticks of lubricant may be needed before lubricant system is refilled to operating condition. Be sure to lubricate valves that are not used often to ensure that they are always in operating condition. Leave lubricant chamber nearly full so extra supply is available by turning screw down. Use lubricant regularly to increase valve efficiency and service, promote easy operation, reduce wear and corrosion, and seal valve against internal leakage.

Paragraph 16: Sluice Gates (Figure 15.63)

There are two general types of light-duty sluice gates: those that seat with the pressure, and those that seat against the pressure. Both are maintained similarly. Heavy-duty sluice gates (Figure 15.63) can seat under seating and unseating pressure.

M 1. TEST FOR PROPER OPERATION. Operate inactive sluice gates. Oil or grease stem threads.

A 2. CLEAN AND PAINT. Clean sluice gate with wire brush and paint with proper corrosion-resistant paint.

A 3. ADJUST FOR PROPER CLEARANCE. For gates seating against pressure, check and adjust top, bottom, and side wedges until in closed position each wedge applies nearly uniform pressure against gate (Figure 15.64).

QUESTION

Write your answer in a notebook and then compare your answer with the one on page 463.

15.7M What maintenance is required by:

1. Gate valves?
2. Sluice gates?

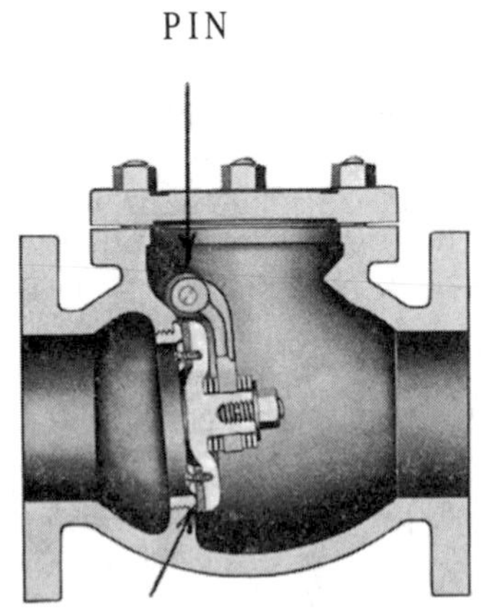

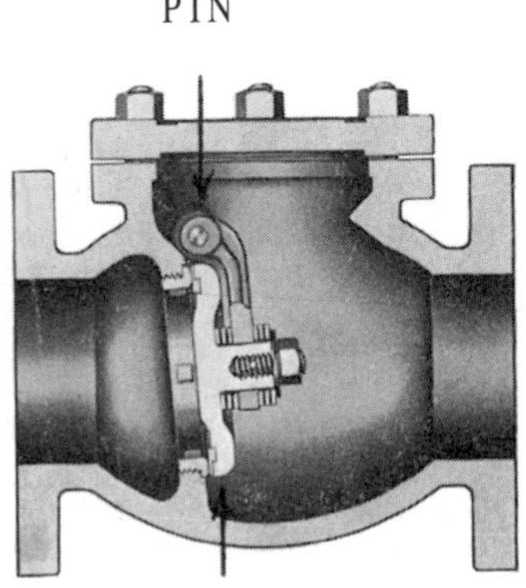

Fig. 15.57 Check valves
(Source: Crane Company)

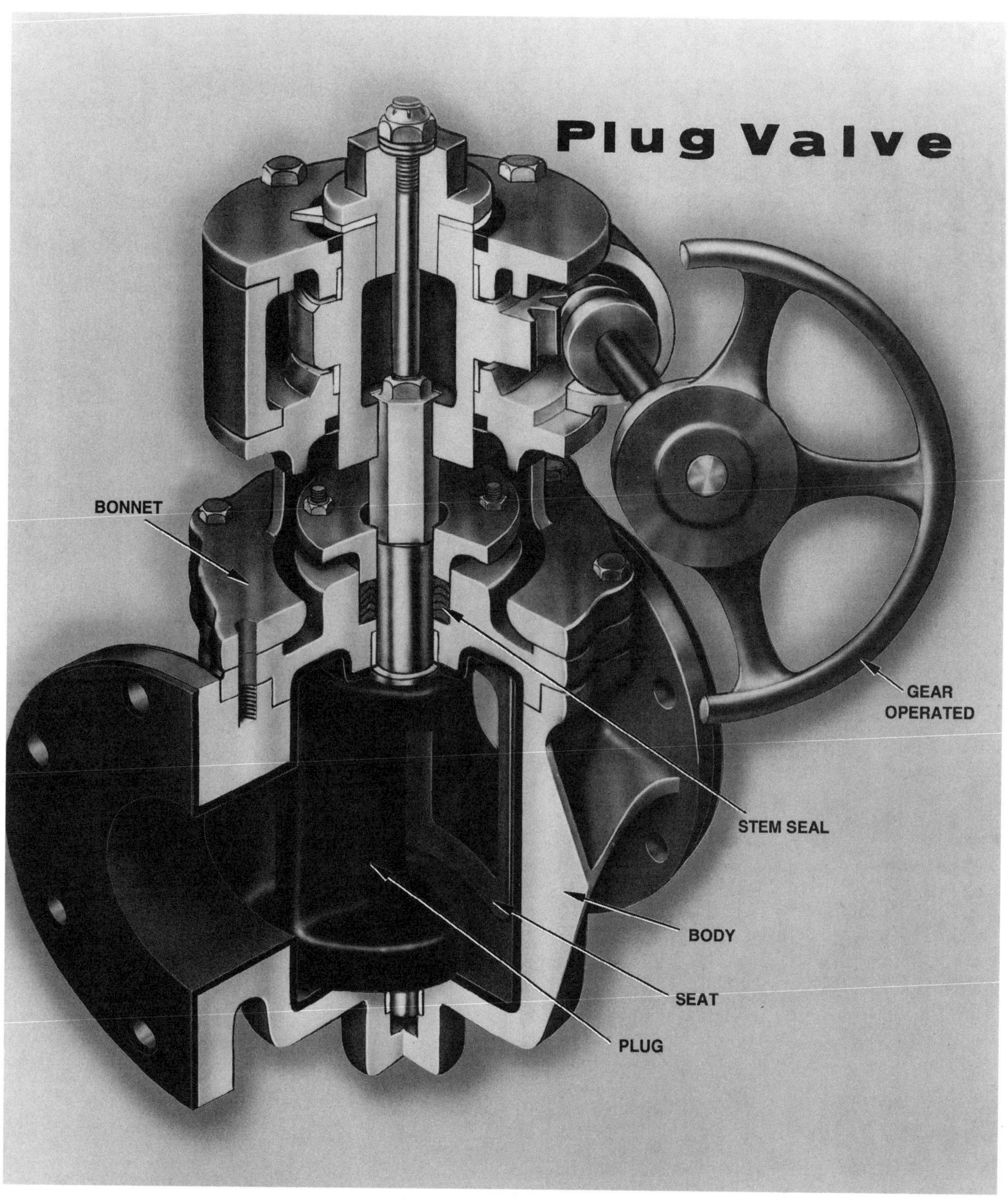

Fig. 15.58 Eccentric plug valve
(Permission of DeZurik, a Unit of General Signal)

Fig. 15.59 Plug valve, lever operated
(Permission of DeZurik, a Unit of General Signal)

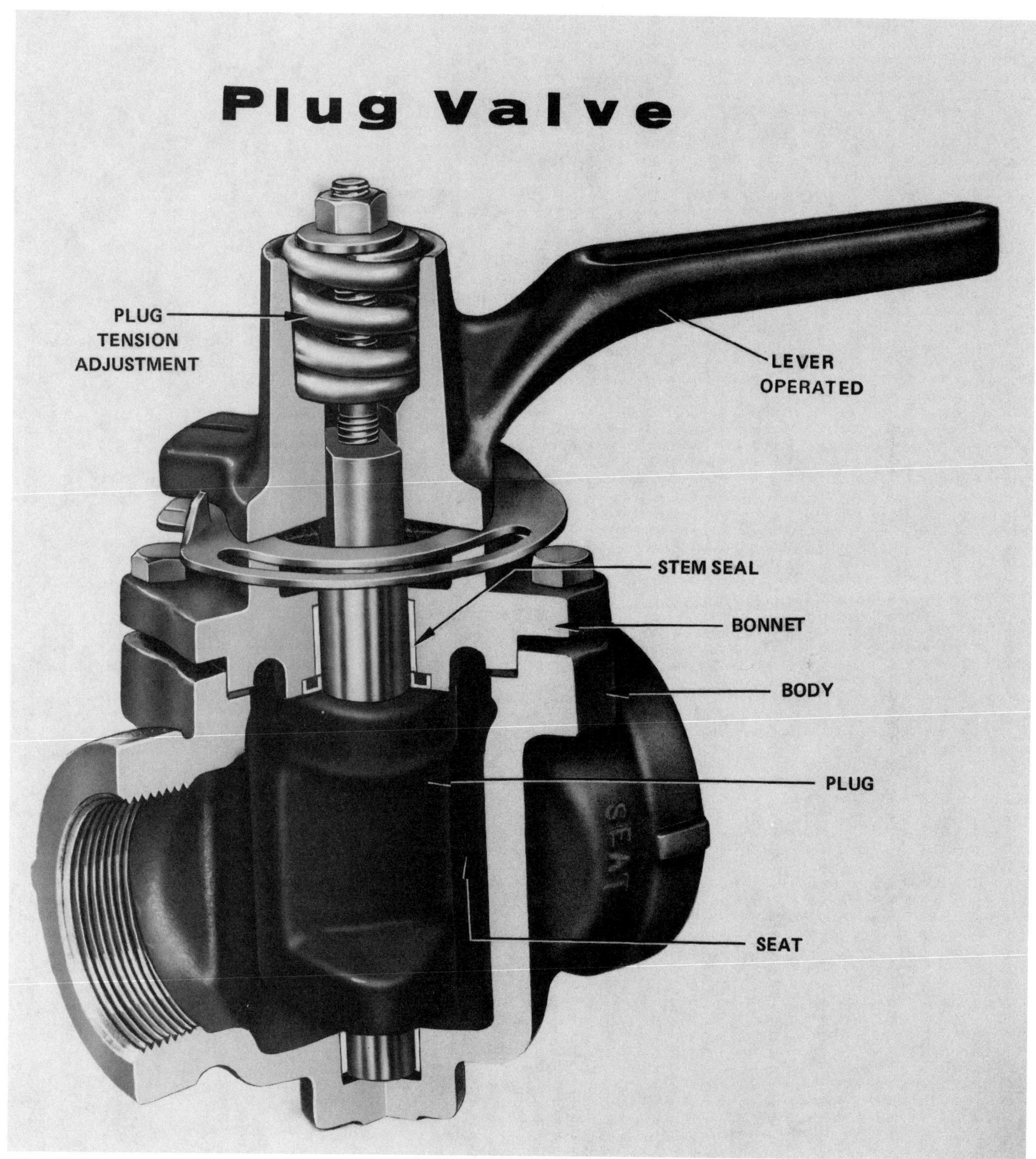

Fig. 15.60 Plug valve, lever operated
(Permission of DeZurik, a Unit of General Signal)

Fig. 15.61 Plug valve, gear operated
(Permission of DeZurik, a Unit of General Signal)

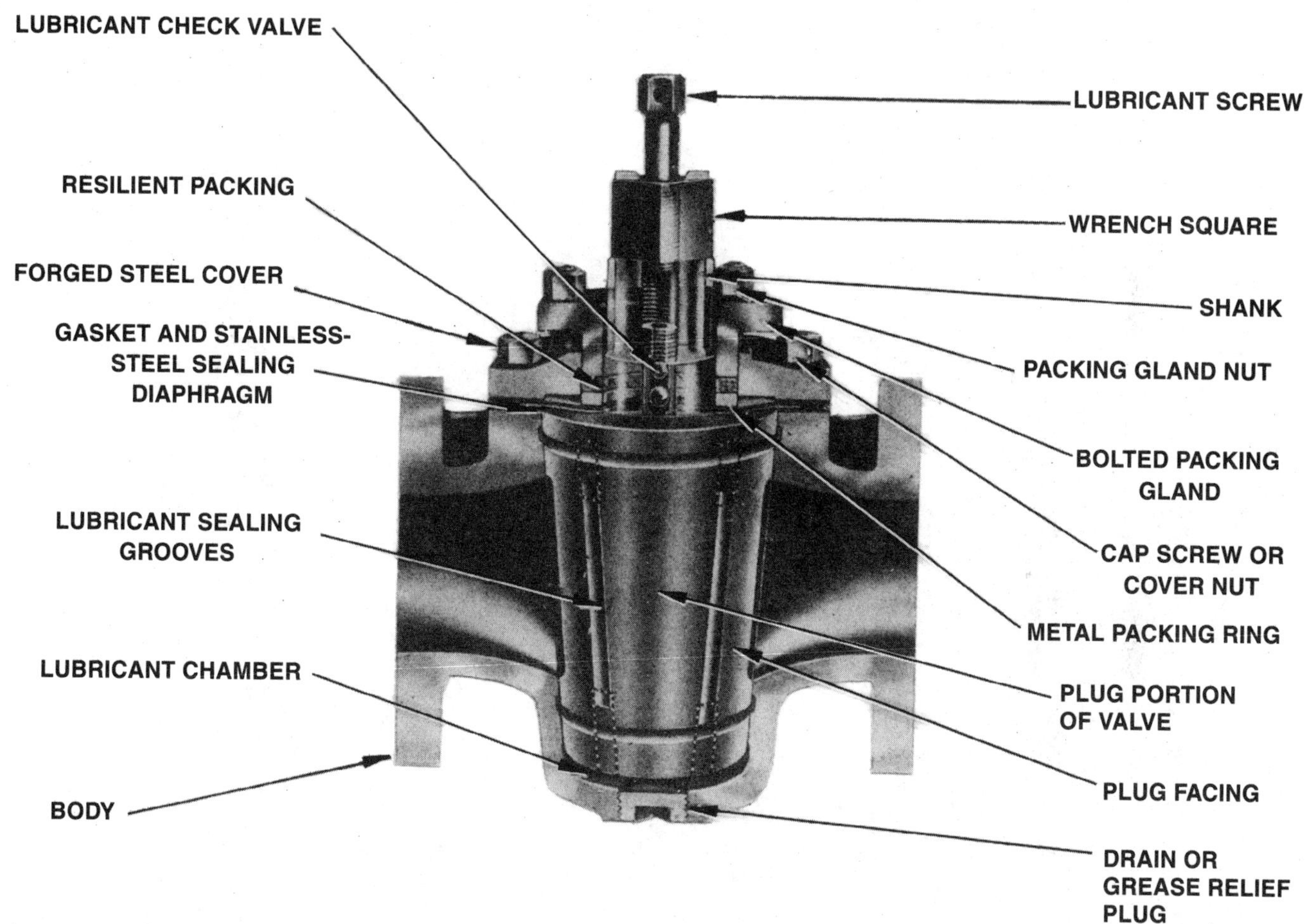

Fig. 15.62 Plug valve

(Source: War Department Technical Manual TM5-666)

Fig. 15.63 Heavy-duty sluice gate
(Permission of ARMCO)

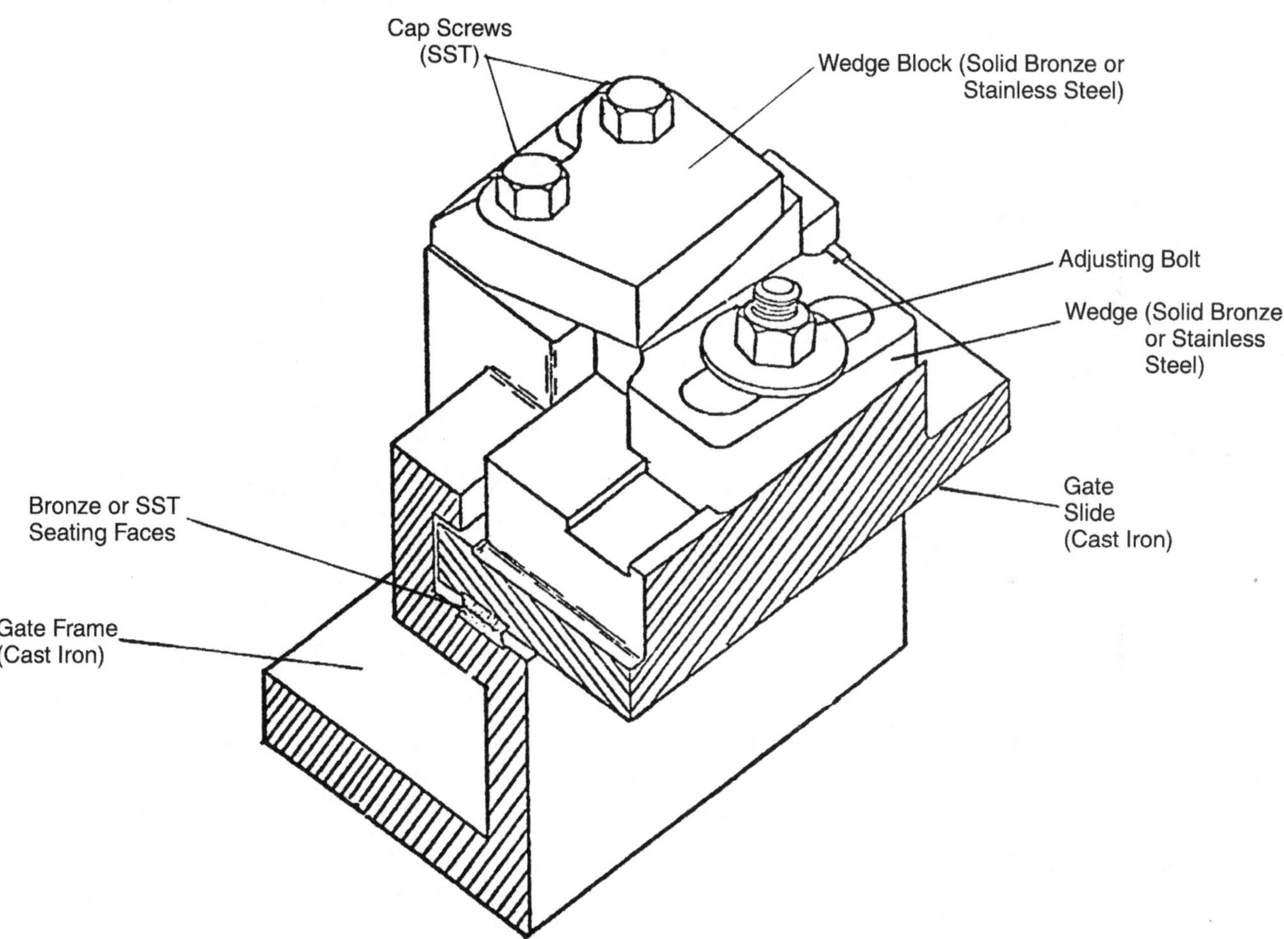

Fig. 15.64 Adjustment of sluice gate side wedges
(Permission of ARMCO)

Paragraph 17: Dehumidifiers

The job of the dehumidifier is to remove moisture from the surrounding atmosphere. Moisture can accumulate in locations that are below grade since the surrounding air temperature will remain relatively constant. This is especially true of pump stations. In the summertime, warm, humid air will enter the pump stations and, unless the air is dehumidified, moisture will form on pump controls, circuits, and other essential equipment. This moisture will cause malfunctions if allowed to exist, due to corrosion and oxidation.

Frequency of Service

D 1. Check dehumidifier performance by noting condensation on the walls of the station. If condensation continues, set the dehumidifier to a drier setting. Inspect dehumidifier coils. If frost is on coils, turn unit off until it defrosts.

S 2. To lube dehumidifier, disassemble the unit and clean it thoroughly, including the drip pan and drain hose. Lube the fan with a few drops of oil. After the unit is reassembled and installed, check condensation buildup daily.

Q 3. Be sure that the drain remains clear, since buildup of water will make the unit operate less efficiently.

Paragraph 18: Air Gap Separation Systems

An air gap separation system provides a physical break between the wastewater treatment plant's municipal or freshwater supply (well) and the plant's treatment process systems. The purpose of this system is to protect the potable (drinking) water supply in case wastewater backs up from a treatment process. For example, wastewater could travel back through a pump seal and cause contamination of a drinking water system.

A 1. Installation of air gap systems is controlled by health regulations and periodically should be inspected by the local health department or the public water supply agency.

W 2. Precautions must be taken to ensure that the air gap separation from the discharge of the water supply pipe is at least two pipe diameters above the rim of the air gap tank (see Figure 15.65). This separation prevents water from being drawn back down into the main water supply line under any circumstances because the water can never reach the elevation of the discharge pipe above the rim of the tank. Wind guards may be necessary around the rim to prevent water loss or damage by wind spray. The device must be easily accessible for weekly inspections.

An examination of Figure 15.65 shows how the plant receives its water supply in the receiving tank. If a plant power outage or other occurrence causes negative pressures in the plant water lines, it would be physically impossible for any contaminated water to enter the drinking water system.

3. Preventive maintenance should include the following regularly scheduled items:

D a. Pump and motor maintenance

W b. Servicing of float and control valve

A c. Periodic draining and cleaning of air gap tank

W d. Routine operational inspections should be conducted at least once a week

Paragraph 19: Plant Safety Equipment

Plant safety equipment must be maintained on a regular basis so it will always be ready for use when needed.

After use or W 1. AIR BLOWERS (for ventilation of confined space). Examine carefully each time used. Inspect hose carefully. If used infrequently, check weekly.

After use or W 2. SAFETY HARNESSES. Examine carefully each time used. Inspect stitching and rings.

W 3. PROTECTIVE CLOTHING. Inspect for rips or holes.

W 4. GAS DETECTORS (PORTABLE). Examine carefully before each use. Ensure battery is charged. Calibrate using known standard gas concentrations obtained from detector manufacturer.

M 5. FIRST-AID KITS. Inventory contents. Replace contents whenever used or contents discovered missing during inventory.

M 6. FIRE EXTINGUISHERS. Inspect and inventory. Check extinguishers for adequate pressure, pin retainers, and service dates.

A An authorized service representative must check/recharge and recertify portable extinguishers.

M 7. Inspect all emergency respiratory protective equipment.

M 8. Inspect and flush all emergency eye wash stations and deluge showers.

M 9. Check face velocities of laboratory fume hoods.

Acknowledgment

Major portions and basic concepts in this section on mechanical maintenance are from the War Department Technical Manual, TM5-666, "Inspections and Preventive Maintenance Services, Sewage Treatment Plants and Sewer Systems at Fixed Installations," War Department.

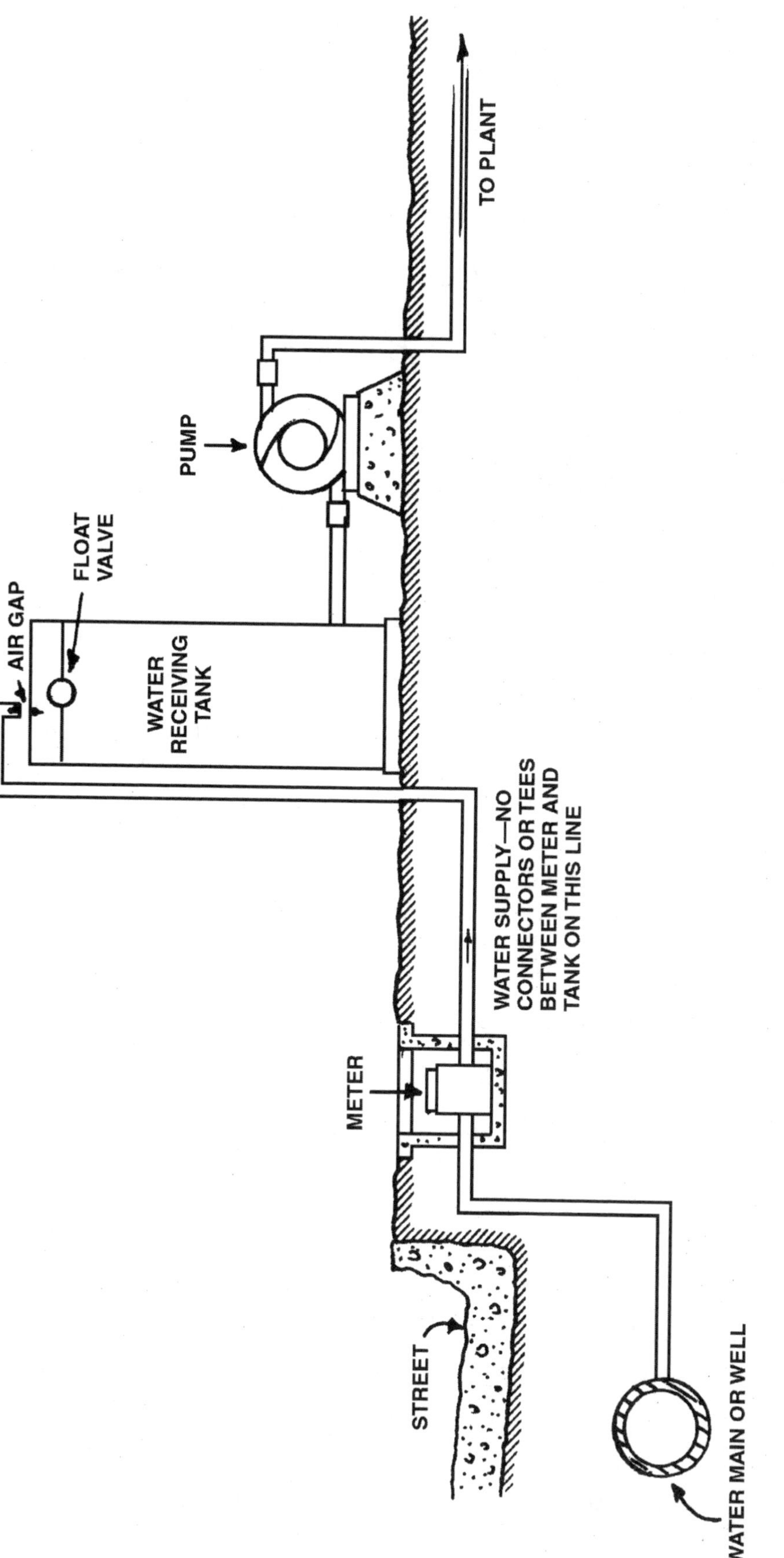

Fig. 15.65 Air gap separation method

QUESTIONS

Write your answers in a notebook and then compare your answers with those on page 463.

15.7N What is the job of the dehumidifier?

15.7O What happens if the dehumidifier does not do its job?

15.7P What is the purpose of an air gap separation system?

15.7Q How would you maintain a portable gas detector?

END OF LESSON 6 OF 7 LESSONS
ON
MAINTENANCE

Please answer the discussion and review questions next.

DISCUSSION AND REVIEW QUESTIONS

Chapter 15. MAINTENANCE

(Lesson 6 of 7 Lessons)

Write the answers to these questions in your notebook. The question numbering continues from Lesson 5.

35. Why should inactive gate valves be operated periodically?
36. Why should plug valves that are not used very often be lubricated regularly?
37. Why is regular maintenance of plant safety equipment necessary?

CHAPTER 15. MAINTENANCE

(Lesson 7 of 7 Lessons)

15.8 UNPLUGGING PIPES, PUMPS, AND VALVES

15.80 Plugged Pipelines

Plugged pipelines are encountered in lines transporting scum, raw sludge, digested sludge, or grit. The frequency of a particular line plugging depends on the type of material passing through the line, the construction material of the line, the type of pumps or system used to move the material, and the routine maintenance performed on the line. This section outlines the preventive maintenance measures to reduce plugging problems in the different lines in a wastewater treatment plant and the methods of unplugging pipes, pumps, and valves.

15.81 Scum Lines

Scum will cause more problems in pipelines than any other substance pumped in a wastewater treatment plant. Problems are more frequent and more severe in colder weather when grease tends to harden quickly.

Preventive maintenance includes:

1. Hose down scum troughs, hoppers, and flush lines to scum box at least every two hours when an operator is on duty and problems are occurring.
2. Clean lines monthly using:
   a. Rods equipped with cutters
   b. High-pressure hydraulic pipe cleaning units
   c. Steam cleaning units
   d. Chemicals such as strong commercial hydroxides. This method is least desirable because of costs and the possibility that the chemicals could be harmful to biological treatment processes. *CAUTION: Study appropriate Material Safety Data Sheets (MSDSs) and follow all safety requirements.*

15.82 Sludge Lines

Sludge lines will plug more often when scum and raw sludge are pumped through the same line, or when stormwaters carry in grit and silt that are not effectively removed by the grit removal facilities.

Preventive maintenance includes:

1. Flush lines monthly with plant effluent or wastewater.
2. If possible, recirculate warm, digested sludge through the line for an hour each week if grease tends to build up on pipe walls.
3. Rod or high-pressure clean lines monthly or quarterly, depending on severity of problem.
4. If possible, force cleaning tool (pig) through line using pressures produced by pump. Line must be equipped with valves and wyes to insert and remove pig. Pumps must be located to allow pig to be forced through the line. A plastic bag full of ice cubes makes an excellent cleaning tool or pig. Force the bag down the line with hot water. If the line plugs, the ice will melt to the point where the bag will continue down the line.

15.83 Digested Sludge Lines

Problems develop in digested sludge lines of small plants from infrequent use, ineffective grit removal, and failure to remove sludge from the line after withdrawing sludge to a drying bed.

Preventive maintenance includes checking:

1. Condition of pipeline for wear or obstructions, such as sticks and rags.
2. Pump impellers for wear. A worn impeller will not maintain desired velocity and pressure in the line.

15.84 Unplugging Pipelines

Selection of a method to unplug a pipe depends on the location of the blockage and access to the plugged line. Pressure methods and cutting tools are the most common techniques used to clear stopped lines.

15.840 Pressure Methods

REQUIREMENTS:

1. Must be able to valve off or plug one end of pipeline in order to move obstruction or blockage down the line and out other end to a free discharge.
2. Pressure may be developed using water or air pressure. Maximum available pressures are usually less than 80 psi (5.6 kg/sq cm).
3. Pipeline must have tap and control valves to control applied water or air pressure.

PRECAUTIONS:

1. Never use water connected to a domestic water supply because you may contaminate the water supply.

2. Do not exceed pipeline working design pressures, usually 125 psi (8.8 kg/sq cm).
3. Never attempt to use a positive displacement pump by overriding the safety cut-out pressure switches. This practice may damage the pump.

PROCEDURE:

1. Plug or valve off one end of pipe, but leave other end open. For example: (1) close valve to digester but open line to the drying beds, or a raw sludge line, or (2) close suction valve on raw sludge pump, and open pipe back to primary clarifier hopper.
2. Connect hose from pressure supply to tap and valve on pipeline as close as possible to the plugged or valved-off end.
3. Apply pressure to supply hose and then slowly open control tap valve and allow pressure to build up until obstruction is moved.

Do not exceed pipeline working design pressure.

15.841 Cutting Tools

Cutting tools are usually available from sewer maintenance crews and may consist of hand rods, power rods, or snakes, which are capable of cutting or breaking up material causing a stoppage.

REQUIREMENTS:

1. One end of the line must be open and reasonably accessible.
2. Cutting tools should be able to remove material causing stoppage when line is cleared.

LIMITATIONS:

1. Most of these units cannot clean lines with sharp bends or pass through some of the common types of plug valves used in sludge lines.
2. A 4-inch (10-cm) cutter tool may have to be used on a 6-inch (15-cm) line due to 90-degree bends.
3. A part of the line may have to be dismantled to use a cutting tool.
4. Rods are difficult to hand push over 300 feet (90 m). The operator must have firm footing and room to work.

HAND RODS:

1. Use sufficient sections to clean full length of line.
2. Insert cutter in the open end of the pipeline and twist rods as they are pushed up the line.
3. If rods start to twist up due to torque, pull back and let rod unwind.

POWER RODS:

1. Power drive unit must be located over plugged line. Do not attempt to run 40 feet (12 m) across a clarifier and then into sludge line.
2. Do not run rods into line too fast. You may hit an obstruction or valve and break the cutter off of the rods. They will be very difficult to recover.

15.842 High-Velocity Pressure Units

This unit is very good for removing grease, sludge, or grit from pipelines.

PROCEDURE:

1. Insert nozzle and hose 3 feet (1 m) into line.
2. Increase pressure in cleaning system to 600 to 1,000 psi (42 to 70 kg/sq cm) and slowly unreel the hose into the pipeline.
3. Keep track of how much hose is in the line to prevent the nozzle from attempting to go through an open valve. The nozzle and hose may catch on the valve and require taking apart the valve to free the nozzle.
4. Run water through nozzle while reeling in hose.

15.843 Last Resort

If the methods described in this section fail, the only solution is to attempt to locate the position of the stoppage, drain the line, take apart the plugged section of pipe, and remove the obstruction.

15.85 Plugged Pumps and Valves

Isolate the plugged pump or valve from the remainder of the treatment plant by valving off the plugged section and tagging and locking out the power supply to the pump. Remove the pump inspection plate or dismantle the valve and remove the material causing the blockage. When removing the pump inspection plate, loosen the bolts and allow the pump to drain *before* removing the bolts in case any of the pipes have not been properly valved off. Exercise caution when removing materials to avoid damaging the pump or valve.

QUESTIONS

Write your answers in a notebook and then compare your answers with those on page 463.

15.8A What methods are available for clearing plugged pipelines?

15.8B How would you clear a plugged pump?

15.9 FLOW MEASUREMENTS—METERS AND MAINTENANCE (Also see Chapter 9, "Instrumentation," in *ADVANCED WASTE TREATMENT.*)

15.90 Flow Measurements, Use and Maintenance

Flow measurement is the determination of the quantity of a mass in movement within a known length of time (Fig. 15.66). The mass may be solid, liquid, or gas and is usually contained within physical boundaries such as tanks, pipelines, and open channels or flumes. The limits of such physical or mechanical boundaries provide a measurable dimensional *AREA* that the mass is passing through. The speed at which the mass passes through these boundaries is related to dimensional distance and units of time; it is referred to as *VELOCITY*. Therefore, we have the basic flow formula:

Quantity = Area × Velocity

$Q = AV$

or

Q, cu ft/sec = (Area, sq ft)(V, ft/sec)

or

Q, cu m/sec = (Area, sq m)(V, m/sec)

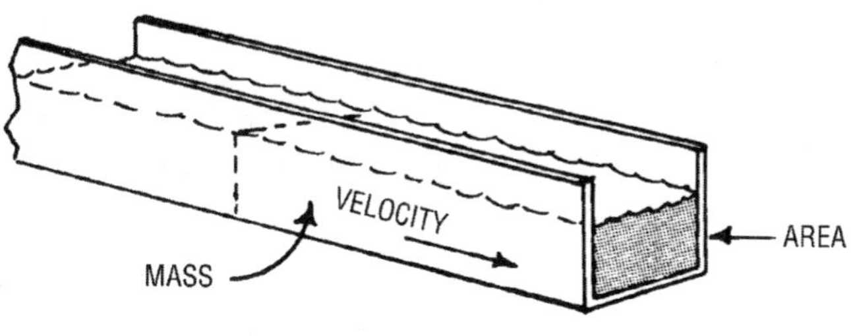

Fig. 15.66 Flow mass

The performance of a treatment facility cannot be evaluated or compared with other plants without flow measurement. Individual treatment units or processes in a treatment plant must be observed in terms of flow to determine their efficiency and loadings. Flow measurement is important to plant operation as well as to records of operation. The devices used for such measurement must be understood, be used properly, and, most important, be maintained so that information obtained is accurate and dependable.

15.91 Operators' Responsibilities

Instrumentation and flow measurement devices are often fragile mechanisms. Rough handling will damage the units in as serious a manner as does neglect. Treat the devices with care, keep them clean, and they will perform their designed functions with accuracy and dependability.

15.92 Various Devices for Flow Measurement

The selection of a type of flow metering device, and its location, is made by the designer in the case of new plant construction. It is also possible that a metering device will have to be added to an existing facility. In both cases, the various types available, their limitations, and criteria for installation should be known. Often the criteria for installation must be understood for the proper use and maintenance of a fluid flowmeter. Table 15.10 provides a summary of the types of flow metering devices commonly used in treatment facilities, their common names, and applications. A discussion of each type follows. Note that each meter type is actually defined by a description of how it works.

TABLE 15.10 SUMMARY OF FLOW METERING DEVICES

Type	Common Name	Application
Constant Differential	Rotameters	Liquids and Gases a. Chlorination
Velocity Meters	Propeller	Liquids—channel flow, clean water piped flow
	Magnetic	Liquids and sludge in closed pipes a. Influent b. Basin control c. Sludge recirculation d. Distribution
	Shuntflow	Gases—closed pipes a. Digester gas
Differential Producers	Venturi Tube Flow Nozzle Orifice	Gases and liquids in closed pipes a. Influent b. Basin control c. Effluent d. Digester gas e. Distribution
Head Area	Weirs Rectangular Cipolletti V-Notch Proportional	Liquids—partially filled channels, basins, or clarifiers a. Influent b. Basin control c. Effluent d. Distribution
	Flumes Parshall Palmer-Bowlus Nozzles	Liquids—partially filled pipes and channels a. Influent b. Basin control c. Effluent d. Distribution
Displacement Units	Piston Diaphragm	Gases and liquids in closed pipes a. Plant water b. Digester gas

CONSTANT DIFFERENTIAL—A mechanical device called the "float" is placed in a tapered tube in the flow line (Figure 15.67). The difference in pressures above and below the float causes the float to move with flow variations. Instantaneous rate of flow is read out directly on a calibrated scale attached to the tube. Read scale behind top of float to obtain flow rate.

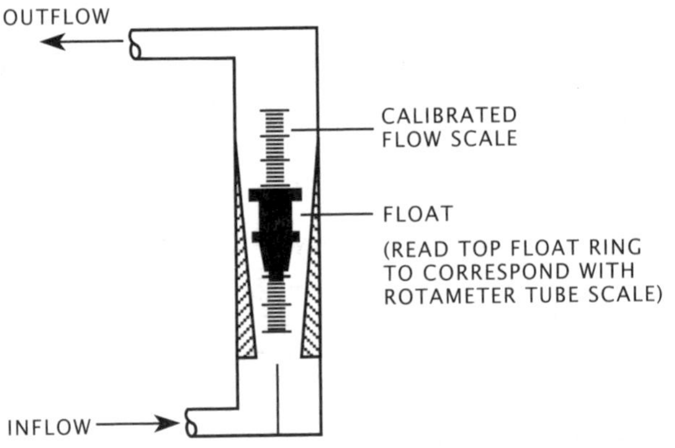

Fig. 15.67 Rotameter

VELOCITY METERS—The velocity of the liquid flowing past the measurement point through a given area gives a direct relation to flow rate. The propeller-type meter is turned by fluid flow past propeller vanes that move gear trains. These gear trains are used to indicate the fluid velocity or flow rate. The velocity of liquid flow past the probes of a magnetic meter is related to flow of electric current between the probes and is read out as the flow rate through secondary instrumentation. (See Section 15.94.) Pitot tubes are used to measure the velocity head (H) in flowing water to give the flow velocity ($V = \sqrt{2gH}$) (Figure 15.68).

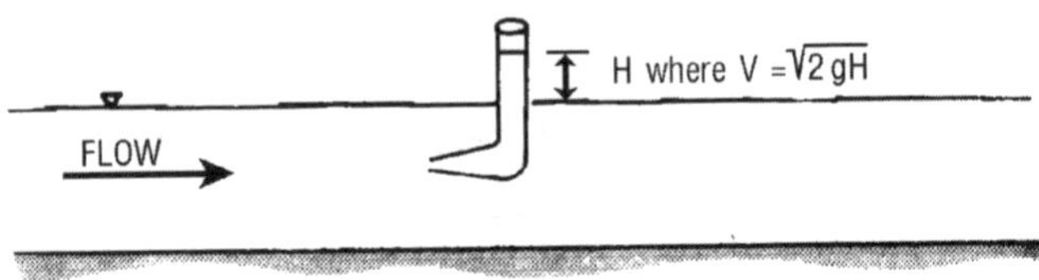

Fig. 15.68 Pitot tube

DIFFERENTIAL PRODUCERS—A mechanical constriction (Figure 15.69) in pipe diameter (reduction in pipe diameter) is placed in the flow line shaped to cause the velocity of flow to increase through the restriction. When the velocity increases, a pressure drop is created at the restriction. The difference between line pressure at the meter inlet and reduced pressure at the throat section is used to determine the flow rate, which is indicated by a secondary instrument.

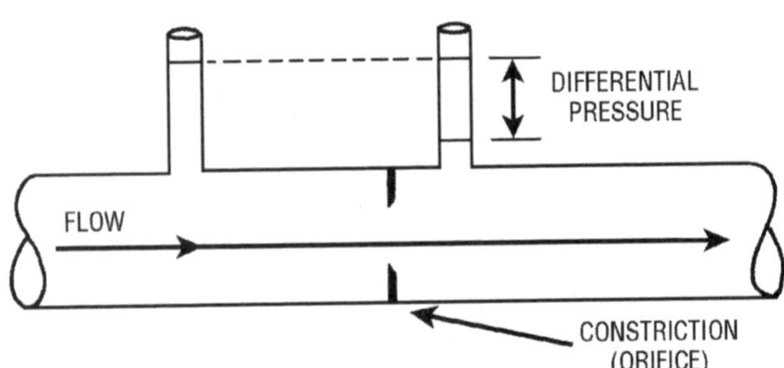

Fig. 15.69 Differential producer

HEAD AREA—A mechanical constriction or barrier is placed in the open flow line causing an upstream rise in liquid level (Figures 15.70, 15.71, and 15.72). The rise or "head" (H) is mathematically related to velocity (speed) of the flow. The head measurement can be used in a formula to calculate flow rate.

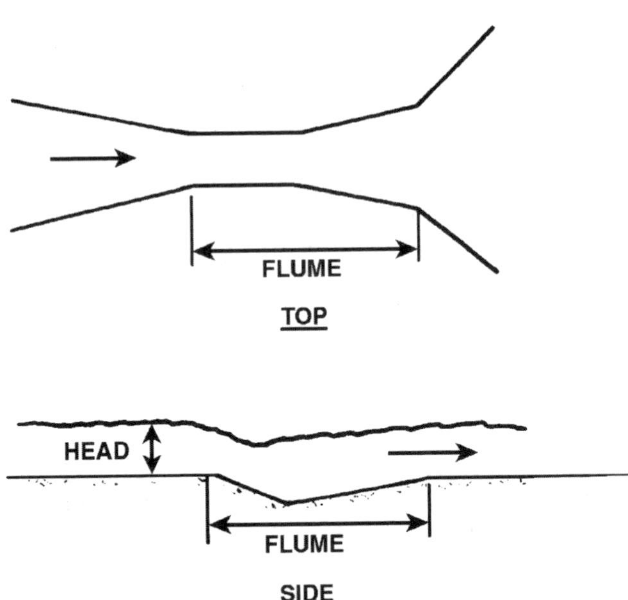

Fig. 15.70 Parshall flume

DISPLACEMENT UNITS—Liquid or gas enters, fills a gas tank or chamber of known dimensions, activates a mechanical counter, and empties the tank in readiness for another filling. As the chamber fills and empties, mechanical gearing activates a counter, which measures the amount of time the cycle takes. Flow rate can then be calculated using the size of the tank and the time factor.

15.93 Location of Measuring Devices

The selection of a particular type of meter or measuring device and its location in a particular flow line or treatment facility is usually a decision made by the plant designer. Ideally, the flow should be in a straight section before the flowmeter. The device must be accessible for servicing. In open channels, the flow should not be changing directions, nor should waves be present in the metering section above the measuring device. Valves, elbows, and other items that could disrupt the flow ahead of a meter can upset the accuracy and reliability of a flowmeter. Flowmeters in pipes will produce accurate flowmeter readings when the meter is located at least five pipe diameters distance downstream from any pipe bends, elbows, or valves and also at least two pipe diameters distance upstream from any pipe bends, elbows, or valves. Flowmeters also should be calibrated (checked for accuracy) in place to ensure accurate flow measurements. Most flowmeters are calibrated in the factory, but they also should be checked in their actual field installation as a check against the factory calibration and to be sure the indicated results are not distorted by field conditions. When a properly installed and field-calibrated meter starts to give strange results, check for obstructions in the flow channel and the flow metering device.

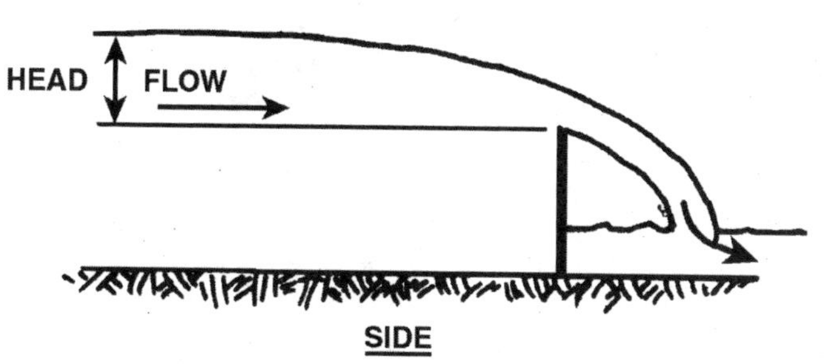

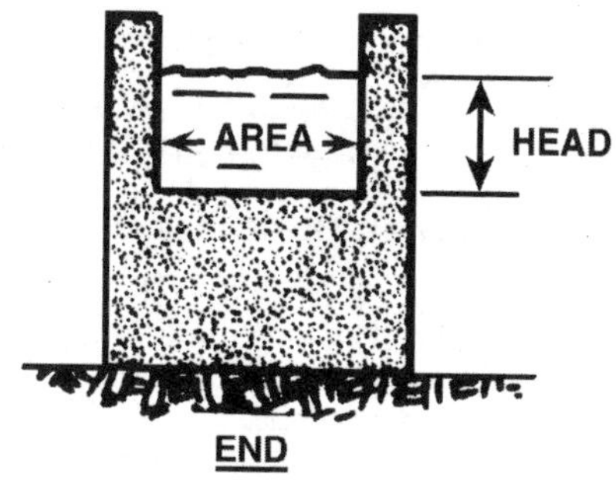

Fig. 15.71 Rectangular weir

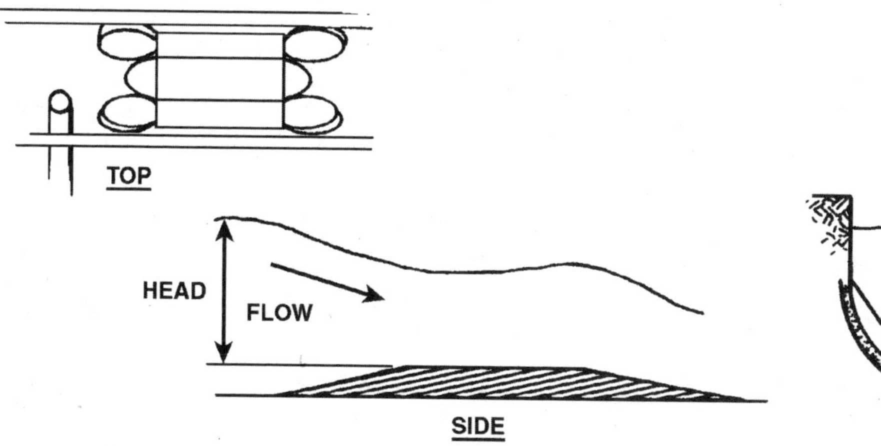

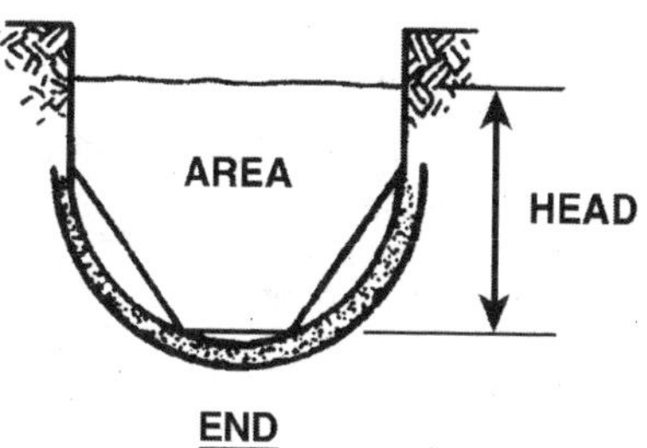

Fig. 15.72 Palmer-Bowlus flume

QUESTIONS

Write your answers in a notebook and then compare your answers with those on pages 463 and 464.

15.9A What is flow measurement?

15.9B Write the basic flow formula.

15.9C Why should flow be measured?

15.9D List several types of flow measuring devices.

15.9E If a flowmeter does not read properly, what items should be checked as potential causes of error?

15.94 Conversion and Readout Instruments and Controls

Conversion and readout instrumentation is used to convert the initial measurement (for example, depth of water) to a more commonly used number or value (depth of water in a Parshall flume to flow of water in MGD (M*L*D)). The type of device depends on what the primary element or sensor measures and what kind of results are desired. Often, the conversion device will only transmit the signal (depth of water) to another meter, which will interpret the signal and convert it to a usable number (flow in MGD (or M*L*D)). Instruments used with flow measurement equipment are classified as transmitters, receivers, recorders, controllers, and integrators or totalizers. All of the different devices available are too numerous to list. Most devices used today will fall into the classifications outlined in the following paragraphs.

15.940 Mechanical Meters

Mechanical meters are devices that measure the variable flow indicator and convert this value into a usable number. Conversion of the flow variable to a scale or meter giving the usable number may be by gear trains, hydraulic connections, magnetic sensing, electrical connections, and many other devices.

15.941 Transmitters

Transmitters send the flow variable, as measured by the measuring device, to another device for conversion to a usable number. Variables are transmitted mechanically, electrically, and pneumatically.

A type of transmitter that has come into great use is the magnetic flow transmitter. This in-line unit contains two sensors (electrodes) within an open tube. A third electrode serves as a ground. When liquid flows in the tube, voltage is conducted between the two sensors. This offers a measurable signal for flow measurement.

15.942 Receivers

Receivers pick up the transmitted signal and convert it to a usable number. There are four types of receivers (receiving meters):

1. *INDICATING* (Figure 15.73): Indicators give a visual representation of a variable's present value, either as an analog or digital display. The analog display uses some type of pointer or graphical display against a scale. A digital display is a direct numerical readout.

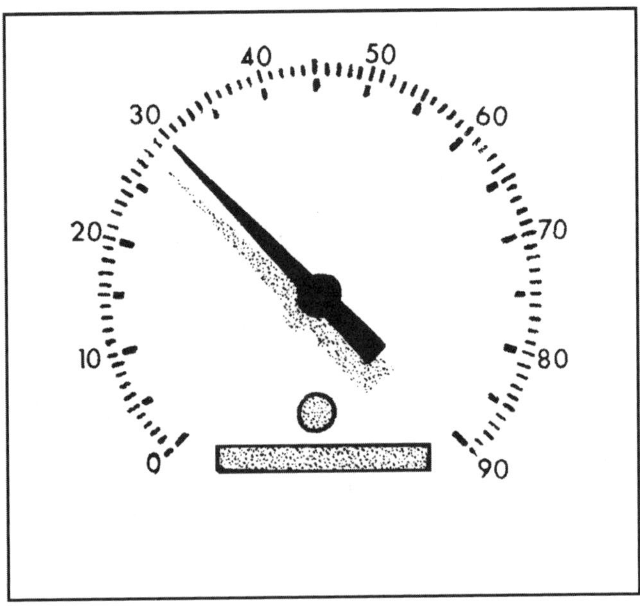

Fig. 15.73 Indicating receiver

2. *RECORDING*: Recorders are indicators designed to show how the value of the variable has changed with time (Figure 15.74). Usually, this is done by attaching a pen (or stylus) to an indicator's arm, which then marks or scribes the value of the variable onto a continuously moving chart. The chart is marked on a horizontal, vertical, or circular scale in time units. Chart records are also stored on a computer disk or in memory for download from the recorders.
3. *TOTALIZING:* Totalizing functions are commonly incorporated into a computer control system, which performs the calculations and stores the historical record of the totals. The historical data may be displayed on the operator interface in trend displays, which are graphical displays similar to strip-chart recorder outputs. The data can be displayed on the computer screen, printed, or sent electronically for analysis by charting and graphing programs.
4. *MULTIPURPOSE:* Logically, this type of receiver has features of two or more of the previously mentioned receivers. Common units have an indicator, a recording chart, and a totalizer.

15.943 Controllers

Controllers are similar to receivers except they are capable of comparing received signals with other values and sending corrective or adjusting signals when necessary. The compared value may be manually set or it may be based on another received signal. The correction or adjustment may be proportional to the size of the deviation of the compared values, may be a

Fig. 15.74 Chart recorder with digital indicator

gradual adjustment, or may provide a predetermined correction based on the size of the deviation and your objectives.

Selection and adjustment of controllers should be done by a specialist in the field or the manufacturer's representative. Maintenance must be done according to manufacturer's instructions.

15.95 Sensor Maintenance

Each individual sensing meter will have its own maintenance requirements. In any instrument, the sensor is the most common source of problems.

Common preventive maintenance suggestions include:

Meter Type	Suggested Maintenance
Constant Differential	
Rotameters	Disassemble and clean tube and float when deposits are observed.
Head Area	
Weirs: Rectangular Cipolletti V-Notch Proportional	Flow formula is based on square, clean edges of the weir in contact with water and with free fall over the weir (no downstream water backing up and flooding the weir). Clean and brush off deposits that accumulate in the weir. Keep weir and upstream and downstream channels clear of foreign bodies and interference.
Flumes: Parshall Palmer-Bowlus Nozzles	Normally used with float wells, keep sensor line between well and flume clean; clean off deposits.
Velocity Meters	
Propeller	Should not be used on anything but clear water. Grease and inspect yearly.
Shuntflow	Keep dampening chamber fluid level up to indicating line; periodically drain to remove collected sediment.
Magnetic	Manufacturers are providing various cleaning mechanisms to clean the internal parts regularly. If you, as an operator, manually operate, be sure to perform maintenance on schedule; if automatically, check action frequently. Provide for periodic meter removal from line and physically clean meter. Provide for periodic back-flushing of unit.
Differential Producers	Venturi, nozzle, and orifice hydraulic connections should be back-flushed regularly. Installation should be arranged for internal surface cleaning on a reasonable schedule.
Displacement Units	Periodically drain and flush. Keep greased as necessary; check frequently on operation.

The single most important item to be considered in sensor maintenance is good housekeeping. Always keep sensors and all instrumentation very clean. Good housekeeping, the act of providing preventive maintenance for each of the various sensors, includes being sure that foreign bodies are not interfering with the measuring device. Check for and remove deposits that will build up from normal use. Repair the sensor or measuring device whenever it is damaged.

External connections between the sensing and conversion and readout devices should be checked to ensure such connections are clean and connections are firm. Be sure no foreign obstruction will interfere or promote wear. On mechanical connections, grease as directed; on hydraulic or pneumatic connections, disconnect and ensure free flow in the internal passage.

15.96 Conversion and Readout Instrument Maintenance

Both the mechanically actuated unit and the transmitters will have direct sensor connections. Cleaning and checking on a regular schedule is essential to avoid problems with the usual accumulation of foreign material. Maintenance for the internal parts of either device is minimized when the sensor connections are clean and operable. Normal wear will occur and is increased when sediments and deposits are not removed regularly as directed. Lubricate mechanical components as directed by the equipment manufacturers' instrument manuals. Do not overlubricate because it causes other difficulties equally as troublesome as underlubrication.

Receiver maintenance is limited to periodic checking of mechanical parts, proper lubrication, and good housekeeping within the unit. When inking systems are used, ensure an adequate supply of ink and a free flow through the capillary tube. Moisture should be eliminated by heat, if required. Pneumatic instruments should be watched carefully to ensure that foreign particles that might be introduced by the air supply do not cause clogging in the actuating elements.

Pneumatic systems are usually protected by air filters or traps at the supply source and individual units at the instrument. Filters should be cleaned and blown down on a regular schedule to ensure their efficient operation in cleaning the air supply. Double the frequency of blowdowns during wet weather and periods of high humidity. In the case of clogging of small orifices and devices of the pneumatic system, *do not attempt to pressurize the system at higher than normal operating pressure for cleaning.* Such action will damage internal parts. Check for air leaks, which will cause decreased efficiency. Follow procedures as outlined by the manufacturer and as shown in the instrumentation manuals.

15.97 Manufacturers' Responsibilities

Most reputable manufacturers are equipped to provide repair service in the case of worn parts or mechanical failure.

It is recommended that major service and repair of instrumentation be left to trained employees of the manufacturer. They have the training, experience, tools, and replacement parts. It is preferred that manufacturers have field service available for repair on the plant premises; however, if such service is not available, the device should be returned to the factory for servicing.

Many manufacturers have a maintenance contract service available in which a trained service employee periodically, on a prescribed schedule, checks the instrument in all ways including accuracy and wear factors. Such periodic checking allows for replacement of parts prior to a complete breakdown. Parts that would normally wear over a time period are replaced by this service representative who will anticipate such need based on experience.

Do not attempt instrument service, parts replacement, or repair work unless you have read the instruction manual thoroughly, or have been trained, and you understand what you are doing. Follow the procedures as set forth in the instruction manual carefully.

All instruments are connected to a power supply from some source. That power supply is potentially dangerous unless handled properly. Be sure all electrical power is shut off and secured so that others cannot unintentionally switch on the source. On electrical and electronic devices, the electrical power used or generated within the device is exceptionally dangerous, both to the operator and to the other component equipment. Do not attempt service unless you are qualified and authorized to do so.

Recording charts often seem to accumulate at a rapid rate, and a decision must be made whether to store or destroy old records. Inconvenient as it may be, records should be retained. They are the backbone of reference information needed for future planning and plant expansion, when necessary. Above all, if properly used, they are a valuable source of information for checking efficiency of the plant processes. Today, these records are usually stored electronically in the computer control system.

15.98 Calibration and Cross-Checking Meter Performance

Meters should be calibrated and performance cross-checked regularly to ensure accurate and reliable meter readings under all conditions at all times. Ensuring proper meter performance protects the plant equipment and contributes to a smooth-running treatment facility. Properly calibrated meters reduce the need for many time-consuming tasks (for example, laboratory tests). Since instrumentation is the "nervous system" of the plant, field calibration of meters should be done daily.

Recording meters are often difficult to calibrate; this task is usually reserved for service instrumentation technicians.

Indicating meters often have a switch or knob which, when placed in "calibration" mode, will cause the indicating hand to move to a preset location on the scale. These meters also may be compared with another meter or checked against a substance of known value; for example, calibrating a pH meter with a buffer solution of known value. The meter also may be checked for performance. If an indicating flowmeter attached to a pump registers 50 GPM (3.15 *L*/sec), check to see if it takes a minute to fill a 55-gallon (200-liter) drum to the 50-gallon (190-liter) mark or another mark in a vessel of known volume.

Totalizing meters can be checked by adaptation of the same principle. Compare the total flow shown by the meter with the flow into a vessel with a known volume.

Multipurpose meters can usually be checked for calibration by comparing one portion of the meter (for example, the indicating meter) to another portion (the value displayed on the recording meter strip chart).

15.99 Troubleshooting Meters

If plant maintenance personnel are responsible for meter troubleshooting, they will need:

1. Adequate spare parts
2. Necessary test equipment
3. Proper tools
4. Proper shop area (clean and well lighted)
5. A thorough knowledge of operating principles of the plant's instruments
6. A complete set of service manuals and parts catalogs

When a flowmeter does not read properly, check the primary element, transmitter, receiver, and power supply. Troubleshooting information on specific pieces of instrumentation will usually be covered in the instrument's service maintenance manual. Unfortunately, the completeness of this information often leaves much to be desired.

QUESTIONS

Write your answers in a notebook and then compare your answers with those on page 464.

15.9F What is the purpose of transmitting instruments?

15.9G What is the most important item in sensor maintenance?

15.9H What should you do with old recording chart records?

15.9I Why should you calibrate and cross-check meter performance regularly?

15.9J If a flowmeter does not read properly, what instrumentation items should be checked as potential causes of error?

15.10 REVIEW OF PLANS AND SPECIFICATIONS

This section covers pumps and lift stations. Many large municipal agencies do their own design work for pumps and lift stations. Smaller agencies usually rely on a consulting engineering firm for design. In either case, operators should be given the opportunity to review the prints and specifications of a new lift station or pumping facility before the award of a contract for construction and installation. This review is very important to be sure adequate provisions have been made for the station pumps, equipment, and instrumentation to be easily and properly operated and maintained.

15.100 Examining Prints

When examining the prints, operators should look for accessibility not only for equipment, but for operators to get to the facilities. Is there sufficient space for vehicles to park and not restrict vehicles passing on streets or pedestrians on sidewalks? Is there room to use hydrolifts, cranes, and high-velocity cleaners, as needed, at the lift station? Are overhead clearances of power lines, trees, and roofs adequate for a crane to remove large equipment? Is there sufficient room to set up portable pumping units or other necessary equipment in cases of major station failures or disasters? Are floors sloped to provide drainage, where needed, and are drains located in low spots? Are station doors and access hatches large enough to remove the largest piece of equipment? Are lifting eyes or overhead rails available, where needed, in the structure? Has sufficient overhead and work room been provided around equipment and control panels to work safely? Is lighting inside and outside the facility provided and is it adequate? Does the alarm system signal high water levels in the wet well and water on the floor of the dry well? Is there sufficient water at a high enough pressure to adequately wash down the wet well? Are there any low-hanging projections or pipes, or other hazards such as unprotected holes, or unsafe stairs or platforms? Is there access to the wet well? If you have to clean out incoming lines, it may be necessary to put a temporary pump in the wet well. All of these questions must be answered satisfactorily if the lift station or pumping facility is to be easily operated and maintained.

Equipment should be laid out orderly with sufficient work room and access to valves and other station equipment, controls, wiring, pipes, and valves. If there is any possibility of future growth and the station may be enlarged, be sure provisions are made to allow for pumping units to be changed or for the installation of additional pumping units. If additional pumping units will be necessary, be sure spools and valves are built now for ease of expansion. Be sure there is sufficient room to add electrical switchgear for future units. If stationary standby power units are not provided, make certain there are external connections and transfer switches for a portable generator.

15.101 Reading Specifications

Plans and specifications should be carefully reviewed before construction and installation because changes are easily made on paper but are much more difficult in the field.

Review the specifications for the acceptability of the equipment, piping, electrical system, instrumentation, and auxiliary equipment. Determine if the equipment is familiar to your agency and if its reliability has been proven. Find out what warranties, guarantees, and operation and maintenance aids will be provided with the equipment and the lift station. Require a list of names, addresses, e-mail addresses, and phone numbers of persons to contact in case help is needed regarding supplies or equipment during start-up and shakedown runs. Be sure that the equipment brochures and other information apply to the equipment supplied. Sometimes new models are installed and you are provided with old brochures.

Be sure the painting and the color coding on pipes and electric circuits meet with your agency's practice. Try to standardize electrical equipment and components as much as possible so one manufacturer cannot blame the other when problems develop. Standardization also can help to reduce the inventory of spare parts necessary for replacement. A few hours spent reviewing plans and specifications will save many days of hard and discouraging labor in the future when it is a major job to make a change.

QUESTIONS

Write your answers in a notebook and then compare your answers with those on page 464.

15.10A Why should operators be given an opportunity to review the prints and specifications of a new lift station before the award of a construction contract?

15.10B Why is accessibility to pumping facilities important?

15.10C Why should plans and specifications be carefully reviewed before construction and installation?

15.10D What information should be determined for equipment to be installed in a pumping facility?

15.11 SUMMARY

1. Establish and follow a regular maintenance program.
2. Thoroughly read and understand manufacturers' maintenance instructions. Ask for assistance if you do not understand them. Follow the manufacturers' instructions in your maintenance program.
3. Critically evaluate the maintenance and repair capabilities of yourself and your facilities. Request the help of an expert, when necessary.

15.12 ARITHMETIC ASSIGNMENT

Turn to the Arithmetic Appendix at the back of this manual. Read and work the problems in Section A.33, "Maintenance." Check the arithmetic in this section using an electronic calculator. You should be able to get the same answers. Section A.53 contains similar problems using metric units.

15.13 ADDITIONAL READING

1. *MOP 11,* Chapter 12,* "Maintenance." New edition of MOP 11, Volume I, *MANAGEMENT AND SUPPORT SYSTEMS.*
2. *NEW YORK MANUAL,* Chapter 12,* "Maintenance of Plant and Equipment."

3. *TEXAS MANUAL,* Chapter 7,* "Lift Stations and Sewage Pumps," Chapter 8,* "Measurement of Wastewater Flow," and Chapter 24,* "Plant Maintenance."

4. *OPERATION AND MAINTENANCE OF WASTEWATER COLLECTION SYSTEMS,* Kenneth D. Kerri and John Brady. Obtain from the Office of Water Programs, California State University, Sacramento, 6000 J Street, Sacramento, CA 95819-6025. See Chapter 6, "Pipeline Cleaning and Maintenance Methods," Chapter 8, "Lift Stations," and Chapter 9, "Equipment Maintenance," Volumes I and II, $49.00 each.

5. "Planned Management Systems for Municipal Wastewater Treatment Plants," US Environmental Protection Agency. EPA No. 600-2-73-004. Obtain from National Technical Information Service (NTIS), 5285 Port Royal Road, Springfield, VA 22161. Order No. PB-233111. Price, $47.50, plus $5.00 shipping and handling per order.

6. "Maintenance Management Systems for Municipal Wastewater Facilities," US Environmental Protection Agency. EPA No. 430-9-74-004. Obtain from National Technical Information Service (NTIS), 5285 Port Royal Road, Springfield, VA 22161. Order No. PB-256611. Price, $52.00, plus $5.00 shipping and handling per order.

7. "Blue Plains Manages Maintenance" (implementation of a computerized system), *OPERATIONS FORUM,* November 1987.

8. *PUMP HANDBOOK,* Third Edition, edited by Igor Karassik, Joseph Messina, Paul Cooper, and Charles Heald. Obtain from the McGraw-Hill Companies, Order Services, PO Box 182604, Columbus, OH 43272-3031. ISBN 978-0-07-034032-9. Price, $135.00, plus nine percent of order total for shipping and handling.

* Depends on edition.

END OF LESSON 7
OF 7 LESSONS
ON
MAINTENANCE

Please answer the discussion and review questions next.

DISCUSSION AND REVIEW QUESTIONS

Chapter 15. MAINTENANCE

(Lesson 7 of 7 Lessons)

Write the answers to these questions in your notebook. The question numbering continues from Lesson 6.

38. How can scum lines be kept from plugging?
39. Which methods can be used by operators to unplug pipelines?
40. Calculate the quantity of flow in cubic feet per second when wastewater flows through an area of 2.5 square feet at a velocity of 1.5 feet per second.
41. What type of flowmeter is used to measure the flow of chlorine gas?
42. What does a pitot tube measure?
43. Why should a flowmeter be calibrated in its field installation?
44. What is the most common source of problems in any instrument?
45. Why should major service and repair of instrumentation be conducted by trained employees of the manufacturer?
46. Why should operators review plans and specifications for a new facility?

SUGGESTED ANSWERS

Chapter 15. MAINTENANCE

ANSWERS TO QUESTIONS IN LESSON 1

Answers to questions on page 335.

15.0A A good maintenance program is essential for a wastewater treatment plant to operate continuously at peak design efficiency.

15.0B The most important item is maintenance of the mechanical equipment—pumps, valves, scrapers, and other moving equipment. Other items include plant grounds, buildings, and structures.

15.0C A good recordkeeping system tells when maintenance is due and also provides a record of equipment performance. Poor performance is a good justification for replacement or new equipment. Good records also help keep your warranty in force.

15.0D An equipment service card tells what service or inspection work should be done on a piece of equipment and when, while the service record card is a record of what was done and when.

Answers to questions on page 336.

15.0E A building maintenance program will keep the building in good shape and includes painting, when necessary. Attention also must be given to electrical systems, plumbing, heating, cooling, ventilating, floors, windows, roofs, and drainage around the buildings. The building should be kept clean, tools should be stored in their proper place, and essential storage should be available. In each plant building, periodically check all stairways, ladders, catwalks, and platforms for adequate lighting, head clearance, and sturdy and convenient guardrails.

15.0F When plant tanks and channels are drained, the operator should check surfaces for wear and deterioration from wastewater or fumes. Protective coatings should be applied where necessary to prevent further damage.

15.0G Well-groomed and neat grounds are important because many people judge the ability of the operator and the plant performance on the basis of the appearance of the plant.

Answers to questions on page 338.

15.0H Chlorine and sulfur dioxide leaks can be detected by smell or by releasing ammonia vapors near areas where leaks might develop. A white cloud will indicate the presence of a leak.

15.0I Chlorine and sulfur dioxide are toxic to humans and will cause corrosion damage to equipment.

15.0J Emergency phone numbers for a treatment plant should include the phone numbers for police, fire, hospital or physician, responsible plant officials, local emergency disaster office, emergency team, and *CHEMTREC,* (800) 424-9300.

15.0K A training program for an emergency team should include:

1. Use of proper equipment (self-contained breathing apparatus, repair kits, and repair tools)
2. Properties and detection of hazardous chemicals
3. Safe procedures for handling and storage of chemicals
4. Types of containers, safe procedures for shipping containers, and container safety devices
5. Installation of repair devices
6. Simulated field emergencies

ANSWERS TO QUESTIONS IN LESSON 2

Answers to questions on page 361.

15.1A Pumps must be lubricated in accordance with manufacturers' recommendations. Quality lubricants should be used.

15.1B A pump with a mechanical seal (or any other type of seal) should never be run dry because the seal faces will be burned and ruined.

15.1C Mechanical seals are not supposed to have any leakage from the gland; if a leak develops, the seal may require resurfacing or it may have to be replaced.

15.1D In lubricating motors, too much grease may cause bearing trouble or damage the winding.

Answers to questions on page 367.

15.1E Wet wells use variable-speed pumping equipment because of lower building costs and the smaller (or nonexistent) wet well. A variable-speed pump facility can maintain a predetermined level of flow in the incoming line during minimal flows in dry weather as well as during maximum flows in the peak wet weather season.

15.1F A standard three-phase, single-speed, synchronous AC motor requires a preventive maintenance program consisting primarily of lubrication, cleaning, and testing of electric circuits. The slip rings and brushes of a variable-speed motor require additional preventive maintenance procedures.

15.1G The rate of brush wear varies with the brush pressure. At light pressures, electrical wear is dominant because the brush can jump off the rings, sparking occurs, and the filming action on the ring becomes erratic. At higher pressures, mechanical wear is dominant because of high friction losses, needless heating, and needless abrasion.

15.1H Indications of unsatisfactory performance of AC slip rings and brushes are those indicators that appear at the brushes, at the ring surface, and as heating.

Answers to questions on page 368.

15.1I If a pump will not start, check for blown fuses or tripped circuit breakers and the cause. Also, check for a loose connection, fuse, or thermal unit.

15.1J To increase the rate of discharge from a pump, you should look for something causing the reduced rate of discharge, such as pumping air, motor malfunction, plugged lines or valves, impeller problems, or other factors.

Answers to questions on page 369.

15.1K If a pump that has been locked or tagged out for maintenance or repairs is started, an operator working on the pump could be seriously injured and also equipment could be damaged.

15.1L Normally, a centrifugal pump should be started after the discharge valve is opened. Exceptions are treatment processes or piping systems with vacuums or pressures that cannot be dropped or allowed to fluctuate greatly while an alternate pump is put on the line.

Answers to questions on page 371.

15.1M Before stopping an operating pump:

1. Start another pump (if appropriate).
2. Inspect the operating pump by looking for developing problems, required adjustments, and problem conditions of the unit.

15.1N A pump shaft or motor will spin backward if wastewater being pumped flows back through the pump when the pump is shut off. This will occur if there is a faulty check valve or foot valve in the system.

15.1O The position of all valves should be checked before starting a pump to ensure that the wastewater being pumped will go where intended.

Answers to questions on page 371.

15.1P The most important rule regarding the operation of positive displacement pumps is to *NEVER* operate the pump against a closed valve, especially a discharge valve.

15.1Q If a positive displacement pump is started against a closed discharge valve, the pipe, valve, or pump could rupture from excessive pressure. The rupture will damage equipment and possibly seriously injure or kill someone standing nearby.

15.1R Both ends of a sludge line should never be closed tight because gas from decomposition can build up and rupture pipes or valves.

ANSWERS TO QUESTIONS IN LESSON 3

Answers to questions on page 373.

15.2A Unqualified or inexperienced people must be extremely careful when attempting to troubleshoot or repair electrical equipment because they can be seriously injured and damage costly equipment if a mistake is made.

15.2B When machinery is not shut off, locked out, and tagged properly, the following accidents could occur:

1. Maintenance operator could be cleaning pump and have it start, thus losing an arm, hand, or finger.
2. Electrical motors or controls not properly grounded could lead to possible severe shock, paralysis, or death.
3. Improper circuits created by mistakes, such as wrong connections, bypassed safety devices, wrong fuses, or improper wire, can cause fires or injuries due to incorrect operation of machinery.

Answers to questions on page 376.

15.3A The proper voltage and allowable current in amps for a piece of equipment can be determined by reading the nameplate information or the instruction manual for the equipment.

15.3B The two types of voltage are Direct Current (DC) and Alternating Current (AC).

15.3C Amperage is a measurement of current or electron flow and is an indication of work being done or "how hard the electricity is working."

Answers to questions on page 380.

15.3D You test for voltage by using a multimeter.

15.3E A multimeter can be used to test for voltage, open circuits, blown fuses, single phasing of motors, and grounds.

15.3F Before attempting to change fuses, turn off power and check both power lines for voltage. Use a fuse puller.

15.3G If the voltage is unknown and the multimeter has different scales that are manually set, always start with the highest voltage range and work down. Otherwise, the multimeter could be damaged.

15.3H Amp readings different from the nameplate rating could be caused by low voltage, bad bearings, poor connections, plugging, or excessive load.

15.3I Motors and wirings should be megged at least once a year, and twice a year, if possible.

15.3J An ohmmeter is used to test the control circuit components, such as coils, fuses, relays, resistors, and switches.

Answers to questions on page 386.

15.3K The most common pump driver used in pump stations is the AC induction motor.

15.3L Types of pump station failures that can occur due to lack of proper maintenance include:

1. Current unbalances, which ultimately result in motor failure
2. Loose contacts or terminals in control and power circuits causing high-resistance contacts
3. Overheating resulting in arcing, fire, and electrical system damage
4. Dirty enclosures and components, which allow a conductive path to build up between incoming phases causing phase-to-phase shorts
5. Corrosion that causes high-resistance contacts and heating.

15.3M If a fuse or circuit breaker blows or trips, the source of the problem should be investigated, identified, and corrected.

15.3N The two types of safety devices in main electrical panels or control units are fuses or circuit breakers.

15.3O Fuses are used to protect operators, wiring, circuits, heaters, motors, and various other electrical equipment.

15.3P A fuse must never be bypassed or jumped because the fuse is the only protection the circuit has. Without it, serious damage to equipment and possible injury to operators can occur.

15.3Q Annual maintenance that should be performed with regard to fuses includes:

1. Inspect bolted connections at the fuse clip or fuse holder for signs of looseness.
2. Check connections for any evidence of corrosion from moisture or atmosphere (air pollution).
3. Tighten connections.
4. Check fuse for obvious overheating.
5. Inspect insulation on the conductors coming into the fuses on the line side and out of the fuses on the load side for evidence of discoloration or bubbling, which would indicate overheating of the conductors.

15.3R A circuit breaker is a switch that is opened automatically when the current or the voltage exceeds or falls below a certain limit. Unlike a fuse that has to be replaced each time it blows, a circuit breaker can be reset after a short delay to allow time for cooling.

Answers to questions on page 387.

15.3S The motor and supervisory control systems are composed of the auxiliary electrical equipment, such as relays, transformers, lighting panels, pump control logic, alarms, and other electrical equipment typically found in a pump station electrical system.

15.3T The three basic factors that contribute to the reliable operation of electrical systems found in pump stations include:

1. An adequate preventive maintenance program must be implemented and maintained.
2. A knowledge of the collection system by the pump station operator, even though the pump operator does not perform the actual collection system maintenance.
3. Adherence to three principles: KEEP IT CLEAN! KEEP IT DRY! KEEP IT TIGHT!

Answers to questions on page 404.

15.4A The two types of induction motor construction that are typically encountered when dealing with AC induction motors are:

1. Squirrel cage induction motor
2. Wound rotor induction motor

15.4B The five most common causes of electric motor failure are: (1) overload (thermal), (2) contaminants, (3) single phasing, (4) bearing failures, and (5) old age.

15.4C The three components of an insulation temperature rating consist of:

1. Ambient operating temperature
2. Temperature rise
3. Hot-spot allowance

15.4D Motor starters can be either manually or automatically controlled.

15.4E Magnetic starters are usually used to start pumps, compressors, blowers, and anything where automatic or remote control is desired.

15.4F The two types of overload protection devices used with magnetic starters are:

1. A bimetallic strip that is precisely calibrated to open under higher temperature conditions to de-energize the coil
2. A small solder pot within the coil that melts because of the heat and will de-energize the system

15.4G The direction of rotation of a three-phase motor can be changed by changing any two of the power leads.

Answers to questions on page 410.

15.4H The key to effective troubleshooting is practical, step-by-step procedures combined with a common-sense approach.

15.4I When troubleshooting magnetic starters:

1. Gather preliminary information
2. Inspect:
   a. Contacts
   b. Mechanical parts
   c. Magnetic parts

15.5A Information that should be recorded regarding electrical equipment includes every:

1. Change
2. Repair
3. Test

ANSWERS TO QUESTIONS IN LESSON 4

Answers to questions on page 419.

15.7A A cross connection is a connection between a drinking (potable) water system and an unapproved water supply. For example, if you have a pump moving nonpotable water and hook into the drinking water system to supply water for the pump seal, a cross connection or mixing between the two water systems can occur. This mixing may lead to contamination of the drinking water.

15.7B Yes. A slight water-seal leakage is desirable when the pumps are running to keep the packing cool and in good condition.

15.7C To measure the capacity of a pump, measure the volume pumped during a specific time period.

$$\text{Capacity, GPM} = \frac{\text{Volume, gallons}}{\text{Time, minutes}}$$

or

$$\text{Capacity, } \frac{\text{liters}}{\text{sec}} = \frac{\text{Volume, liters}}{\text{Time, sec}}$$

15.7D
$$\text{Capacity, GPM} = \frac{\text{Volume, gallons}}{\text{Time, minutes}}$$

$$= \frac{10 \text{ ft} \times 15 \text{ ft} \times 1.7 \text{ ft} \times 7.5 \text{ gal/cu ft}}{5 \text{ minutes}}$$

$$= 382.5 \text{ GPM}$$

or

$$\text{Capacity, } \frac{\text{liters}}{\text{sec}} = \frac{\text{Volume, liters}}{\text{Time, sec}}$$

$$= \frac{3 \text{ m} \times 5 \text{ m} \times 0.5 \text{ m} \times 1{,}000 \text{ } L\text{/cu m}}{5 \text{ minutes} \times 60 \text{ sec/min}}$$

$$= 25 \text{ liters/sec}$$

15.7E When shutting down a pump for a long period, open the motor disconnect switch; shut all valves on suction, discharge, water-seal, and priming lines; drain the pump completely by removing the vent and drain plugs. This procedure protects the pump against corrosion, sedimentation, and freezing.

Answers to questions on page 424.

15.7F Shear pins commonly fail in reciprocating pumps because of: (1) a solid object lodged under the piston, (2) a clogged discharge line, or (3) a stuck or wedged valve.

15.7G A noise may develop when pumping thin sludge, due to water hammer, but will disappear when heavy sludge is pumped.

ANSWERS TO QUESTIONS IN LESSON 5

Answer to question on page 427.

15.7H When checking an electric motor, the following items should be checked periodically, as well as when trouble develops:

1. Motor condition
2. Note all unusual conditions
3. Lubrication of bearings
4. Listen to motor
5. Check temperature
6. Set dateometer

Answers to questions on page 429.

15.7I A properly adjusted horizontal belt has a slight bow in the slack side when running. When idle, it has an alive springiness when thumped with the hand. To check for proper alignment, place a straightedge or string against the pulley face or faces and allow for differences in dimensions from centerlines of grooves to outside faces of the pulleys being aligned.

15.7J Always replace worn sprockets when replacing a chain because out-of-pitch sprockets cause as much chain wear in a few hours as years of normal operation.

Answers to questions on page 433.

15.7K Improper original installation of equipment, settling of foundations, heavy floor loadings, warping of bases, and excessive bearing wear could cause couplings to become out of alignment.

15.7L Shear pins are designed to fail if a sudden overload occurs that could damage expensive equipment.

ANSWERS TO QUESTIONS IN LESSON 6

Answer to question on page 437.

15.7M Maintenance required by (1) gate valves is oiling, tightening, and replacing the stem stuffing box packing. Maintenance required by (2) sluice gates is testing for proper operation, cleaning and painting, and adjusting for proper clearance.

Answers to questions on page 448.

15.7N The job of the dehumidifier is to remove moisture from the atmosphere.

15.7O If the dehumidifier does not do its job, moisture will form on pump controls, circuits, and other essential equipment and cause malfunctions due to corrosion and oxidation.

15.7P The purpose of an air gap separation system is to protect the potable (drinking) water supply in case wastewater backs up from a treatment process.

15.7Q To maintain a portable gas detector:

1. Be sure battery is charged
2. Calibrate using known standard gas concentrations obtained from detector manufacturer

ANSWERS TO QUESTIONS IN LESSON 7

Answers to questions on page 450.

15.8A Plugged pipelines may be cleared by the use of pressure methods, cutting tools, high-velocity pressure units and, as a last resort, dismantling the plugged section and removing the obstruction.

15.8B To clear a plugged pump, isolate the pump from the remainder of the plant by valving off the plugged section and tagging and locking out the power supply to the pump. Loosen the pump inspection plate, allow the pump to drain, and remove the material causing the blockage.

Answers to questions on page 454.

15.9A Flow measurement is the determination of the quantity of a mass in movement within a known length of time. The mass may be solid, liquid, or gas and is usually contained within physical boundaries such as tanks, pipelines, and open channels or flumes.

15.9B Quantity = Area × Velocity, or Q = AV.

15.9C Flow should be measured in order to determine wastewater treatment plant loadings and efficiency.

15.9D Different types of flow measuring devices include constant differential, velocity meters, differential producers, head area, and displacement units.

15.9E Potential causes of flowmeter errors include foreign objects fouling the system or the meter may not be installed in the intended location. Liquids should flow smoothly through the meter and flow should not be changing directions, nor should waves be present on the liquid surface above the measuring device.

Answers to questions on page 457.

15.9F Transmitters send the flow variable, as measured by the measuring device, to another device for conversion to a usable number.

15.9G The most important item in sensor maintenance is good housekeeping. Your instruments must be kept clean and in good working condition.

15.9H Old recording charts should be stored for future reference, such as checks on plant performance, budget justifications, and information needed for future planning. Today, these records are usually stored electronically in the computer control system.

15.9I Meters should be calibrated and performance cross-checked regularly to ensure accurate and reliable meter readings under all conditions at all times.

15.9J When a flowmeter does not read properly, check the primary element, transmitter, receiver, and power supply.

Answers to questions on page 458.

15.10A Operators should review prints and specifications of a new lift station before construction to be sure adequate provisions have been made for the facility to be easily and properly operated and maintained.

15.10B Accessibility to a pumping facility is important not only for equipment but also because operators need to be able to easily reach the facility and have room to park vehicles and work.

15.10C Plans and specifications should be carefully reviewed before construction and installation because changes are easily made on paper but are much more difficult in the field.

15.10D Before equipment is installed in a pumping facility, determine if the equipment is familiar to your agency and if its reliability has been proven. Find out what warranties, guarantees, and operation and maintenance aids will be provided with the equipment and the lift station.

CHAPTER 16

LABORATORY PROCEDURES AND CHEMISTRY

by

James Paterson

(With a Special Section by Joe Nagano)

Revised by

James Sequeira

TABLE OF CONTENTS

Chapter 16. LABORATORY PROCEDURES AND CHEMISTRY

OBJECTIVES

Chapter 16. LABORATORY PROCEDURES AND CHEMISTRY

Following completion of Chapter 16, you should be able to:

1. Work safely in a laboratory.
2. Operate laboratory equipment.
3. Collect representative samples of influents to and effluents from a treatment process as well as sample the process.
4. Prepare samples for analysis.
5. Perform plant control tests.
6. Analyze plant effluent in accordance with NPDES permit requirements.
7. Recognize shortcomings or precautions for the plant control and NPDES tests.
8. Record laboratory test results.

CHAPTER 16. LABORATORY PROCEDURES AND CHEMISTRY

16.0 IMPORTANCE OF LABORATORY PROCEDURES

Laboratory control tests are the means by which we collect the data we use to control the efficiency of our wastewater treatment processes. By relating laboratory results to operations, the plant operator can select the most effective operational procedures, determine the efficiency of treatment unit processes, and identify developing treatment problems before they seriously affect effluent quality. For these reasons, a clear understanding of laboratory procedures is a must to any operator.

16.00 Material in This Lesson

A few of the lab procedures outlined in this chapter are not "Standard Methods" (5),[1] but are used by many operators because they are simple and easy to perform. Some of these procedures are not accurate enough for scientific investigations, but are satisfactory for successful plant control and operation. See Section 16.4, "Laboratory Procedures for Plant Control." When lab data must be submitted to regulatory agencies for monitoring and enforcement purposes, you must use approved test procedures. See Section 16.5, "Laboratory Procedures for NPDES Monitoring."

Each test section contains the following information.

1. Discussion of test
2. What is tested
3. Apparatus
4. Reagents
5. Procedures
6. Precautions
7. Examples
8. Calculations

If you would like to read an introductory discussion on laboratory equipment and analysis, the Water Environment Federation has a good publication entitled "Basic Laboratory Procedures for Wastewater Examination" (4).

16.01 References

1. *LABORATORY PROCEDURES AND CHEMISTRY FOR OPERATORS OF WATER POLLUTION CONTROL PLANTS.* This publication is a reproduction of *OPERATION OF WASTEWATER TREATMENT PLANTS,* Volume II, Chapter 16, "Laboratory Procedures and Chemistry." Obtain from California Water Environment Association (CWEA), 7677 Oakport Street, Suite 525, Oakland, CA 94621. Price to members, $13.00; nonmembers, $17.85; plus $9.50 shipping and handling.
2. "Methods for Chemical Analysis of Water and Wastes," March 1983, US Environmental Protection Agency. EPA No. 600-4-79-020. Obtain from National Technical Information Service (NTIS), 5285 Port Royal Road, Springfield, VA 22161. Order No. PB84-128677. Price, $117.00, plus $5.00 shipping and handling per order.
3. *HANDBOOK FOR ANALYTICAL QUALITY CONTROL IN WATER AND WASTEWATER LABORATORIES,* US Environmental Protection Agency. EPA No. 600-4-79-019. Obtain from National Technical Information Service (NTIS), 5285 Port Royal Road, Springfield, VA 22161. Order No. PB-297451. Price, $52.00, plus $5.00 shipping and handling per order.
4. *BASIC LABORATORY PROCEDURES FOR WASTEWATER EXAMINATION.* Obtain from Water Environment Federation (WEF), Publications Order Department, 601 Wythe Street, Alexandria, VA 22314-1994. Order No. P02404. Price to members, $31.95; nonmembers, $39.95; plus shipping and handling.
5. *STANDARD METHODS FOR THE EXAMINATION OF WATER AND WASTEWATER,* 21st Edition, 2005. Obtain from Water Environment Federation (WEF), Publications Order Department, 601 Wythe Street, Alexandria, VA 22314-1994. Order No. S82011. Price to members, $194.75; nonmembers, $259.75; price includes cost of shipping and handling.
6. *HANDBOOK FOR EVALUATING WATER BACTERIOLOGICAL LABORATORIES,* US Environmental Protection Agency. EPA No. 670-9-75-006. Obtain from National Technical Information Service (NTIS), 5285 Port Royal Road, Springfield, VA 22161. Order No. PB-247145. Price, $59.50, plus $5.00 shipping and handling per order.
7. *40 CFR 136,* July 1, 2006, "Guidelines Establishing Test Procedures for the Analysis of Pollutants." US Government Printing Office, Superintendent of Documents, PO Box 371954, Pittsburgh, PA 15250-7954.
8. *WATER ANALYSIS HANDBOOK* (CD version). Obtain from HACH Company, PO Box 389, Loveland, CO 80539-0389 or www.hach.com. Request Literature No. WA02. No Charge.

16.02 Acknowledgments

Many of the illustrated laboratory procedures were provided by Joe Nagano, Laboratory Director, Hyperion Treatment Plant, City of Los Angeles, CA. These procedures originally appeared in *LABORATORY PROCEDURES FOR OPERATORS OF WATER POLLUTION CONTROL PLANTS,* prepared by Mr. Nagano and published by the California Water Pollution Control Association.

[1] Numbers in parentheses refer to references in Section 16.01.

CHAPTER 16. LABORATORY PROCEDURES AND CHEMISTRY

(Lesson 1 of 9 Lessons)

16.1 BASIC LABORATORY WORDS, EQUIPMENT, AND TECHNIQUES

16.10 Laboratory Words

>GREATER THAN

DO >5 mg/*L* would be read as DO GREATER THAN 5 mg/*L*.

<LESS THAN

DO <5 mg/*L* would be read as DO LESS THAN 5 mg/*L*.

ALIQUOT (AL-uh-kwot)

Representative portion of a sample. Often, an equally divided portion of a sample.

AMBIENT (AM-bee-ent) TEMPERATURE

Temperature of the surroundings.

AMPEROMETRIC (am-purr-o-MET-rick) AMPEROMETRIC

A method of measurement that records electric current flowing or generated, rather than recording voltage. Amperometric titration is a means of measuring concentrations of certain substances in water.

ANAEROBIC (AN-air-O-bick) ANAEROBIC

A condition in which atmospheric or dissolved oxygen (DO) is *NOT* present in the aquatic (water) environment.

ASEPTIC (a-SEP-tick) ASEPTIC

Free from the living germs of disease, fermentation, or putrefaction. Sterile.

BIOASSAY (BUY-o-AS-say) BIOASSAY

(1) A way of showing or measuring the effect of biological treatment on a particular substance or waste.

(2) A method of determining the relative toxicity of a test sample of industrial wastes or other wastes by using live test organisms, such as fish.

BIOMONITORING BIOMONITORING

A term used to describe methods of evaluating or measuring the effects of toxic substances in effluents on aquatic organisms in receiving waters. There are two types of biomonitoring, the BIOASSAY and the BIOSURVEY.

BIOSURVEY BIOSURVEY

A survey of the types and numbers of organisms naturally present in the receiving waters upstream and downstream from plant effluents. Comparisons are made between the aquatic organisms upstream and those organisms downstream of the discharge.

BLANK BLANK

A bottle containing only dilution water or distilled water; the sample being tested is not added. Tests are frequently run on a sample and a blank and the differences are compared. The procedure helps to eliminate or reduce test result errors that could be caused when the dilution water or distilled water used is contaminated.

BUFFER BUFFER

A solution or liquid whose chemical makeup neutralizes acids or bases without a great change in pH.

BUFFER CAPACITY | BUFFER CAPACITY

A measure of the capacity of a solution or liquid to neutralize acids or bases. This is a measure of the capacity of water or wastewater for offering a resistance to changes in pH.

CHAIN OF CUSTODY | CHAIN OF CUSTODY

A record of each person involved in the handling and possession of a sample from the person who collected the sample to the person who analyzed the sample in the laboratory and to the person who witnessed disposal of the sample.

COLORIMETRIC MEASUREMENT | COLORIMETRIC MEASUREMENT

A means of measuring unknown chemical concentrations in water by measuring a sample's color intensity. The specific color of the sample, developed by addition of chemical reagents, is measured with a photoelectric colorimeter or is compared with color standards using, or corresponding with, known concentrations of the chemical.

COMPOSITE (PROPORTIONAL) SAMPLE | COMPOSITE (PROPORTIONAL) SAMPLE

A composite sample is a collection of individual samples obtained at regular intervals, usually every one or two hours during a 24-hour time span. Each individual sample is combined with the others in proportion to the rate of flow when the sample was collected. Equal volume individual samples also may be collected at intervals after a specific volume of flow passes the sampling point or after equal time intervals and still be referred to as a composite sample. The resulting mixture (composite sample) forms a representative sample and is analyzed to determine the average conditions during the sampling period.

COMPOUND | COMPOUND

A pure substance composed of two or more elements whose composition is constant. For example, table salt (sodium chloride, NaCl) is a compound.

DESICCATOR (DESS-uh-kay-tor) | DESICCATOR

A closed container into which heated weighing or drying dishes are placed to cool in a dry environment in preparation for weighing. The dishes may be empty or they may contain a sample. Desiccators contain a substance (DESICCANT), such as anhydrous calcium chloride, that absorbs moisture and keeps the relative humidity near zero so that the dish or sample will not gain weight from absorbed moisture.

DISTILLATE (DIS-tuh-late) | DISTILLATE

In the distillation of a sample, a portion is collected by evaporation and recondensation; the part that is recondensed is the distillate.

ELEMENT | ELEMENT

A substance that cannot be separated into its constituent parts and still retain its chemical identity. For example, sodium (Na) is an element.

END POINT | END POINT

The completion of a desired chemical reaction. Samples of water or wastewater are titrated to the end point. This means that a chemical is added, drop by drop, to a sample until a certain color change (blue to clear, for example) occurs. This is called the end point of the titration. In addition to a color change, an end point may be reached by the formation of a precipitate or the reaching of a specified pH. An end point may be detected by the use of an electronic device, such as a pH meter.

FACULTATIVE (FACK-ul-tay-tive) BACTERIA | FACULTATIVE BACTERIA

Facultative bacteria can use either dissolved oxygen or oxygen obtained from food materials such as sulfate or nitrate ions. In other words, facultative bacteria can live under aerobic, anoxic, or anaerobic conditions.

FLAME POLISHED | FLAME POLISHED

Melted by a flame to smooth out irregularities. Sharp or broken edges of glass (such as the end of a glass tube) are rotated in a flame until the edge melts slightly and becomes smooth.

GRAB SAMPLE | GRAB SAMPLE

A single sample of water collected at a particular time and place that represents the composition of the water only at that time and place.

GRAVIMETRIC | GRAVIMETRIC

A means of measuring unknown concentrations of water quality indicators in a sample by weighing a precipitate or residue of the sample.

INDICATOR INDICATOR

(1) (Chemical indicator) A substance that gives a visible change, usually of color, at a desired point in a chemical reaction, generally at a specified end point.

(2) (Instrument indicator) A device that indicates the result of a measurement, usually using either a fixed scale and movable indicator (pointer), such as a pressure gauge, or a moving chart with a movable pen like those used on a circular flow-recording chart. Also called a RECEIVER.

M or MOLAR *M* or MOLAR

A molar solution consists of one gram molecular weight of a compound dissolved in enough water to make one liter of solution. A gram molecular weight is the molecular weight of a compound in grams. For example, the molecular weight of sulfuric acid (H_2SO_4) is 98. A one *M* solution of sulfuric acid would consist of 98 grams of H_2SO_4 dissolved in enough distilled water to make one liter of solution.

MPN MPN

MPN is the Most Probable Number of coliform-group organisms per unit volume of sample water. Expressed as a density or population of organisms per 100 m*L* of sample water.

MSDS MSDS

See MATERIAL SAFETY DATA SHEET.

MATERIAL SAFETY DATA SHEET (MSDS) MATERIAL SAFETY DATA SHEET (MSDS)

A document that provides pertinent information and a profile of a particular hazardous substance or mixture. An MSDS is normally developed by the manufacturer or formulator of the hazardous substance or mixture. The MSDS is required to be made available to employees and operators or inspectors whenever there is the likelihood of the hazardous substance or mixture being introduced into the workplace. Some manufacturers are preparing MSDSs for products that are not considered to be hazardous to show that the product or substance is not hazardous.

MENISCUS (meh-NIS-cuss) MENISCUS

The curved surface of a column of liquid (water, oil, mercury) in a small tube. When the liquid wets the sides of the container (as with water), the curve forms a valley. When the confining sides are not wetted (as with mercury), the curve forms a hill or upward bulge.

MOLECULAR WEIGHT MOLECULAR WEIGHT

The molecular weight of a compound in grams per mole is the sum of the atomic weights of the elements in the compound. The molecular weight of sulfuric acid (H_2SO_4) in grams is 98.

Element	Atomic Weight	Number of Atoms	Molecular Weight
H	1	2	2
S	32	1	32
O	16	4	64
			98

MOLECULE MOLECULE

The smallest division of a compound that still retains or exhibits all the properties of the substance.

N or NORMAL *N* or NORMAL

A normal solution contains one gram equivalent weight of reactant (compound) per liter of solution. The equivalent weight of an acid is that weight that contains one gram atom of ionizable hydrogen or its chemical equivalent. For example, the equivalent weight of sulfuric acid (H_2SO_4) is 49 (98 divided by 2 because there are two replaceable hydrogen ions). A one *N* solution of sulfuric acid would consist of 49 grams of H_2SO_4 dissolved in enough water to make one liter of solution.

NIOSH (NYE-osh) NIOSH

The National Institute of Occupational Safety and Health is an organization that tests and approves safety equipment for particular applications. NIOSH is the primary federal agency engaged in research in the national effort to eliminate on-the-job hazards to the health and safety of working people. The NIOSH Publications Catalog, Seventh Edition, NIOSH Pub. No. 87-115, lists the NIOSH publications concerning industrial hygiene and occupational health. To obtain a copy of the catalog, write to National Technical Information Service (NTIS), 5285 Port Royal Road, Springfield, VA 22161. NTIS Stock No. PB88-175013.

OXIDATION — OXIDATION

Oxidation is the addition of oxygen, removal of hydrogen, or the removal of electrons from an element or compound; in the environment and in wastewater treatment processes, organic matter is oxidized to more stable substances. The opposite of REDUCTION.

OXIDATION STATE/OXIDATION NUMBER — OXIDATION STATE/OXIDATION NUMBER

In a chemical formula, a number accompanied by a polarity indication (+ or –) that together indicate the charge of an ion as well as the extent to which the ion has been oxidized or reduced in a REDOX REACTION.

Due to the loss of electrons, the charge of an ion that has been oxidized would go from negative toward or to neutral, from neutral to positive, or from positive to more positive. As an example, an oxidation number of 2+ would indicate that an ion has lost two electrons and that its charge has become positive (that it now has an excess of two protons).

Due to the gain of electrons, the charge of the ion that has been reduced would go from positive toward or to neutral, from neutral to negative, or from negative to more negative. As an example, an oxidation number of 2– would indicate that an ion has gained two electrons and that its charge has become negative (that it now has an excess of two electrons). As an ion gains electrons, its oxidation state (or the extent to which it is oxidized) lowers; that is, its oxidation state is reduced. Also see REDOX REACTION.

OXIDATION-REDUCTION POTENTIAL (ORP) — OXIDATION-REDUCTION POTENTIAL (ORP)

The electrical potential required to transfer electrons from one compound or element (the oxidant) to another compound or element (the reductant); used as a qualitative measure of the state of oxidation in water and wastewater treatment systems. ORP is measured in millivolts, with negative values indicating a tendency to reduce compounds or elements and positive values indicating a tendency to oxidize compounds or elements.

OXIDATION-REDUCTION (REDOX) REACTION — OXIDATION-REDUCTION (REDOX) REACTION

See REDOX REACTION.

PERCENT SATURATION — PERCENT SATURATION

The amount of a substance that is dissolved in a solution compared with the amount dissolved in the solution at saturation, expressed as a percent.

$$\text{Percent Saturation, \%} = \frac{\text{Amount of Substance That Is Dissolved} \times 100\%}{\text{Amount Dissolved in Solution at Saturation}}$$

pH (pronounce as separate letters) — pH

pH is an expression of the intensity of the basic or acidic condition of a liquid. Mathematically, pH is the logarithm (base 10) of the reciprocal of the hydrogen ion activity.

$$\text{pH} = \text{Log}\,\frac{1}{\{H^+\}}$$

If $\{H^+\} = 10^{-6.5}$, then pH = 6.5. The pH may range from 0 to 14, where 0 is most acidic, 14 most basic, and 7 neutral.

PILLOWS — PILLOWS

Plastic tubes shaped like pillows that contain exact amounts of chemicals or reagents. Cut open the pillow, pour the reagents into the sample being tested, mix thoroughly, and follow test procedures.

PLUG FLOW — PLUG FLOW

A type of flow that occurs in tanks, basins, or reactors when a slug of water or wastewater moves through a tank without ever dispersing or mixing with the rest of the water or wastewater flowing through the tank.

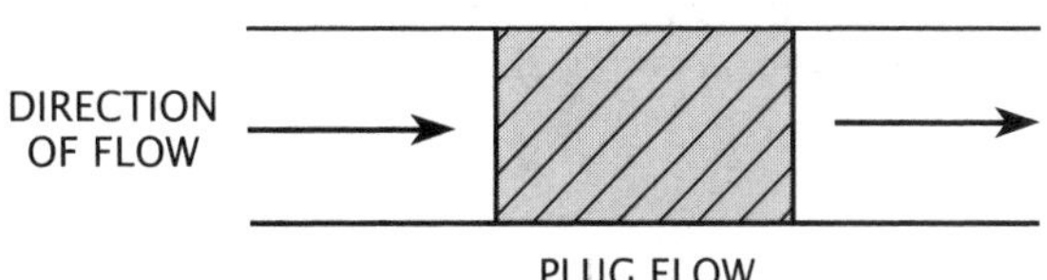

PLUG FLOW

PRECIPITATE (pre-SIP-uh-TATE) — PRECIPITATE

(1) An insoluble, finely divided substance that is a product of a chemical reaction within a liquid.

(2) The separation from solution of an insoluble substance.

REAGENT (re-A-gent) — REAGENT

A pure, chemical substance that is used to make new products or is used in chemical tests to measure, detect, or examine other substances.

REDOX (REE-docks) REACTION | REDOX REACTION

A two-part reaction between two ions involving a transfer of electrons from one ion to the other. Oxidation is the loss of electrons by one ion, and reduction is the acceptance of electrons by the other ion. Reduction refers to the lowering of the OXIDATION STATE/ OXIDATION NUMBER of the ion accepting the electrons.

In a redox reaction, the ion that gives up the electrons (that is oxidized) is called the reductant because it causes a reduction in the oxidation state or number of the ion that accepts the transferred electrons. The ion that receives the electrons (that is reduced) is called the oxidant because it causes oxidation of the other ion. Oxidation and reduction always occur simultaneously.

REDUCTION (re-DUCK-shun) | REDUCTION

Reduction is the addition of hydrogen, removal of oxygen, or the addition of electrons to an element or compound. Under anaerobic conditions (no dissolved oxygen present), sulfur compounds are reduced to odor-producing hydrogen sulfide (H_2S) and other compounds. In the treatment of metal finishing wastewaters, hexavalent chromium (Cr^{6+}) is reduced to the trivalent form (Cr^{3+}). The opposite of OXIDATION.

REFLUX | REFLUX

Flow back. A sample is heated, evaporates, cools, condenses, and flows back to the flask.

REPRESENTATIVE SAMPLE | REPRESENTATIVE SAMPLE

A sample portion of material, water, or wastestream that is as nearly identical in content and consistency as possible to that in the larger body being sampled.

SOLUTION | SOLUTION

A liquid mixture of dissolved substances. In a solution it is impossible to see all the separate parts.

SPECIFIC GRAVITY | SPECIFIC GRAVITY

(1) Weight of a particle, substance, or chemical solution in relation to the weight of an equal volume of water. Water has a specific gravity of 1.000 at 4°C (39°F). Particulates with specific gravity less than 1.0 float to the surface and particulates with specific gravity greater than 1.0 sink.

(2) Weight of a particular gas in relation to the weight of an equal volume of air at the same temperature and pressure (air has a specific gravity of 1.0). Chlorine gas has a specific gravity of 2.5.

STANDARD METHODS | STANDARD METHODS

STANDARD METHODS FOR THE EXAMINATION OF WATER AND WASTEWATER, 21st Edition. A joint publication of the American Public Health Association (APHA), American Water Works Association (AWWA), and the Water Environment Federation (WEF) that outlines the accepted laboratory procedures used to analyze the impurities in water and wastewater. Available from: American Water Works Association, Bookstore, 6666 West Quincy Avenue, Denver, CO 80235. Order No. 10084. Price to members, $198.50; nonmembers, $266.00; price includes cost of shipping and handling. Also available from Water Environment Federation, Publications Order Department, 601 Wythe Street, Alexandria, VA 22314-1994. Order No. S82011. Price to members, $194.75; nonmembers, $259.75; price includes cost of shipping and handling.

STANDARD SOLUTION | STANDARD SOLUTION

A solution in which the exact concentration of a chemical or compound is known.

STANDARDIZE | STANDARDIZE

To compare with a standard.

(1) In wet chemistry, to find out the exact strength of a solution by comparing it with a standard of known strength. This information is used to adjust the strength by adding more water or more of the substance dissolved.

(2) To set up an instrument or device to read a standard. This allows you to adjust the instrument so that it reads accurately, or enables you to apply a correction factor to the readings.

SURFACTANT (sir-FAC-tent) | SURFACTANT

Abbreviation for surface-active agent. The active agent in detergents that possesses a high cleaning ability.

THIEF HOLE | THIEF HOLE

A digester sampling well that allows sampling of the digester contents without venting digester gas.

TITRATE (TIE-trate) | TITRATE

To titrate a sample, a chemical solution of known strength is added drop by drop until a certain color change, precipitate, or pH change in the sample is observed (end point). Titration is the process of adding the chemical reagent in small increments (0.1–1.0 milliliter) until completion of the reaction, as signaled by the end point.

TOTALIZER — TOTALIZER

A device or meter that continuously measures and sums a process rate variable in cumulative fashion over a given time period. For example, total flows displayed in gallons per minute, million gallons per day, cubic feet per second, or some other unit of volume per time period. Also called an INTEGRATOR.

TURBIDITY (ter-BID-it-tee) UNITS (TU) — TURBIDITY UNITS (TU)

Turbidity units are a measure of the cloudiness of water. If measured by a nephelometric (deflected light) instrumental procedure, turbidity units are expressed in nephelometric turbidity units (NTU) or simply TU. Those turbidity units obtained by visual methods are expressed in Jackson turbidity units (JTU), which are a measure of the cloudiness of water; they are used to indicate the clarity of water. There is no real connection between NTUs and JTUs. The Jackson turbidimeter is a visual method and the nephelometer is an instrumental method based on deflected light.

VOLATILE (VOL-uh-tull) — VOLATILE

(1) A volatile substance is one that is capable of being evaporated or changed to a vapor at relatively low temperatures. Volatile substances can be partially removed from water or wastewater by the air stripping process.

(2) In terms of solids analysis, volatile refers to materials lost (including most organic matter) upon ignition in a muffle furnace for 60 minutes at 550°C (1,022°F). Natural volatile materials are chemical substances usually of animal or plant origin. Manufactured or synthetic volatile materials, such as plastics, ether, acetone, and carbon tetrachloride, are highly volatile and not of plant or animal origin. Also see NONVOLATILE MATTER.

VOLATILE ACIDS — VOLATILE ACIDS

Fatty acids produced during digestion that are soluble in water and can be steam-distilled at atmospheric pressure. Also called organic acids. Volatile acids are commonly reported as equivalent to acetic acid.

VOLATILE LIQUIDS — VOLATILE LIQUIDS

Liquids that easily vaporize or evaporate at room temperature.

VOLATILE SOLIDS — VOLATILE SOLIDS

Those solids in water, wastewater, or other liquids that are lost on ignition of the dry solids at 550°C (1,022°F). Also called organic solids and volatile matter.

VOLUMETRIC — VOLUMETRIC

A measurement based on the volume of some factor. Volumetric titration is a means of measuring unknown concentrations of water quality indicators in a sample by determining the volume of titrant or liquid reagent needed to complete particular reactions.

16.11 The Metric System

The metric system is based on the decimal system. All units of length, volume, and weight (mass) use factors of 10 to express larger or smaller quantities of these units. The metric system is used exclusively in the wastewater plant laboratory. Below is a summary of metric and English unit names and abbreviations.

Type of Measurement	English System	Metric Name	Metric Abbreviation
Length	inch feet yard	meter	m
Volume	quart gallon	liter	*L*
Weight	ounce pound	gram	gm
Temperature	Fahrenheit	Celsius	°C
Time	second	second	sec

In the laboratory we sometimes use smaller amounts than a meter, a liter, or a gram. To express these smaller amounts, prefixes are added to the names of the metric units. There are many prefixes in use; however, we commonly use two prefixes more than any others in the laboratory.

Prefix	Abbreviation	Meaning
milli-	m	1/1,000 of, *OR* 0.001 times
centi-	c	1/100 of, *OR* 0.01 times

One milliliter (m*L*) is 1/1,000 of a liter and likewise one centimeter (cm) is 1/100 of a meter.

EXAMPLE:

(1) Convert 2 grams into milligrams.

1 milligram = 1 mg = 1/1,000 gm

therefore, 1 gram = 1,000 milligrams

2 grams × 1,000 mg/gram = 2,000 mg

(2) Convert 500 m*L* to liters.

1 m*L* = 1/1,000 liter

therefore, 1 liter = 1,000 m*L*

500 m*L* × 1 liter/1,000 m*L* = 0.500 liters

The Celsius (or centigrade) temperature scale is used in the laboratory rather than the more familiar Fahrenheit scale.

	Fahrenheit (°F)	Celsius (°C)
Boiling point of water	212	100
Freezing point of water	32	0

To convert Fahrenheit to Celsius you can use the following formula:

$$\text{Temperature, °C} = \frac{5}{9}(\text{°F} - 32°)$$

EXAMPLE: Convert 68°F to °C.

$$\begin{aligned}\text{Temperature, °C} &= \frac{5}{9}(\text{°F} - 32°)\\ &= \frac{5}{9}(68\text{°F} - 32°)\\ &= \frac{5}{9}(36)\\ &= 20\text{°C}\end{aligned}$$

To convert Celsius to Fahrenheit you can use the following formula:

$$\text{Temperature, °F} = \left(\text{°C} \times \frac{9}{5}\right) + 32°$$

EXAMPLE: Convert 35°C to °F.

$$\begin{aligned}\text{Temperature, °F} &= \left(\text{°C} \times \frac{9}{5}\right) + 32°\\ &= \left(35\text{°C} \times \frac{9}{5}\right) + 32°\\ &= 63 + 32\\ &= 95\text{°F}\end{aligned}$$

16.12 Chemical Names and Formulas

Chemical symbols are "shorthand" for the names of the elements. Names and symbols for some of these elements are listed below.

Chemical Name	Symbol
Calcium	Ca
Carbon	C
Chlorine	Cl
Hydrogen	H
Iron	Fe
Oxygen	O
Potassium	K
Sodium	Na
Sulfur	S

Many different compounds can be made from the same two or three elements, therefore you must read the formula and name carefully to prevent errors and accidents. A chemical formula is a "shorthand" or abbreviated way to write the name of a compound. For example, the name sodium chloride (table salt) can be written "NaCl." Table 16.1 lists commonly used chemicals found in the wastewater treatment plant laboratory.

The following procedures are given to show the use of some of the chemicals whose names and formulas you will see in wastewater tests:

Dissolved Oxygen Procedure

"...To 300 milliliter (m*L*) sample placed in a BOD bottle, add 1 m*L* of manganous sulfate solution. Now add 1 m*L* of alkaline iodine-sodium azide solution. Shake well and add 1 m*L* of concentrated sulfuric acid. Titrate with sodium thiosulfate solution until..."

Preparation of Manganous Sulfate Solution

"...Weigh 480 grams (gm) $MnSO_4 \cdot H_2O$ and dissolve in distilled water. Dilute to 1 liter..."

TABLE 16.1 NAMES AND FORMULAS OF CHEMICALS COMMONLY USED IN WASTEWATER ANALYSES

Chemical Name	Chemical Formula
Ammonium chloride	NH_4Cl
Calcium chloride (heptahydrate)*	$CaCl_2 \cdot 7\ H_2O$
Dipotassium hydrogen phosphate	K_2HPO_4
Disodium hydrogen phosphate (heptahydrate)*	$Na_2HPO_4 \cdot 7\ H_2O$
Ferric chloride	$FeCl_3$
Magnesium sulfate (heptahydrate)*	$MgSO_4 \cdot 7\ H_2O$
Manganous sulfate (tetrahydrate)*	$MnSO_4 \cdot 4\ H_2O$
Phenylarsine oxide	C_6H_5AsO
Potassium iodide	KI
Sodium azide	NaN_3
Sodium chloride	NaCl
Sodium hydroxide	NaOH
Sodium iodide	NaI
Sodium thiosulfate (pentahydrate)*	$Na_2S_2O_3 \cdot 5\ H_2O$
Sulfuric acid	H_2SO_4

* Note that: tetra = 4, penta = 5, and hepta = 7, thus heptahydrate = 7 H_2O

Poor results and safety hazards are often caused by using a chemical from the shelf that is not the same chemical called for in the procedure. This mistake usually occurs when the chemicals have similar names or formulas. This problem can be eliminated if you use both the chemical name and formula as a double check. As you can see in the table of chemical names, the spellings of many chemicals are very similar. These slight differences are critical because the chemicals do not behave alike. For example, the chemicals potassium nitr*A*te (KNO_3) and potassium nitr*I*te (KNO_2) are just as different in meaning chemically as the words f*A*t and f*I*t are to your doctor.

16.13 Laboratory Equipment

The equipment in a wastewater laboratory is the lab analyst's "tools of the trade." In any laboratory there are certain pieces of equipment that are used routinely to perform chemical tests such as those in wastewater analyses. The pictures and names of these are given with a statement concerning the use of each piece.

Volumetric glassware (graduated cylinders and pipets) is calibrated either "to contain" (TC) or "to deliver" (TD). Glassware designed "to deliver" will do so accurately only when the inner surface is so scrupulously clean that water wets the surface immediately and forms a uniform film on the surface upon emptying.

BEAKERS. Beakers are the most common pieces of laboratory equipment. They come in sizes from 1 m*L* to 4,000 m*L*. They are used mainly for mixing chemicals and to measure approximate volumes.

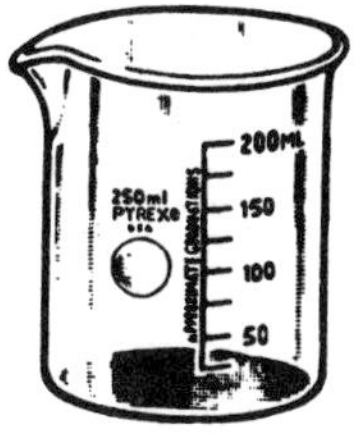

Beaker

GRADUATED CYLINDERS. Graduated cylinders also are basic to any laboratory and come in sizes from 5 m*L* to 4,000 m*L*. They are used to measure volumes more accurately than beakers.

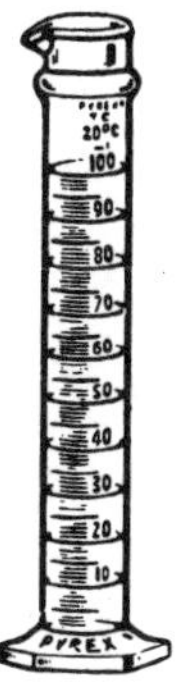

Cylinder, Graduated

PIPETS. Pipets are used to deliver accurate volumes and range in size from 0.1 m*L* to 100 m*L*.

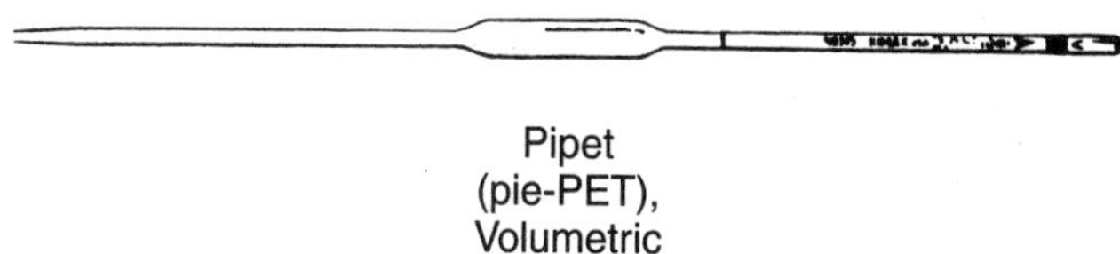

Pipet (pie-PET), Volumetric

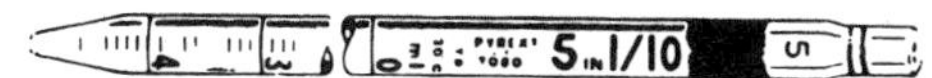

Pipet, Serological

BURETS. Burets are also used to deliver accurate volumes. They are especially useful in a procedure called "titration." Burets come in sizes from 10 to 1,000 m*L*.

FLASKS. Flasks are used for containing and mixing chemicals. There are many different sizes and shapes.

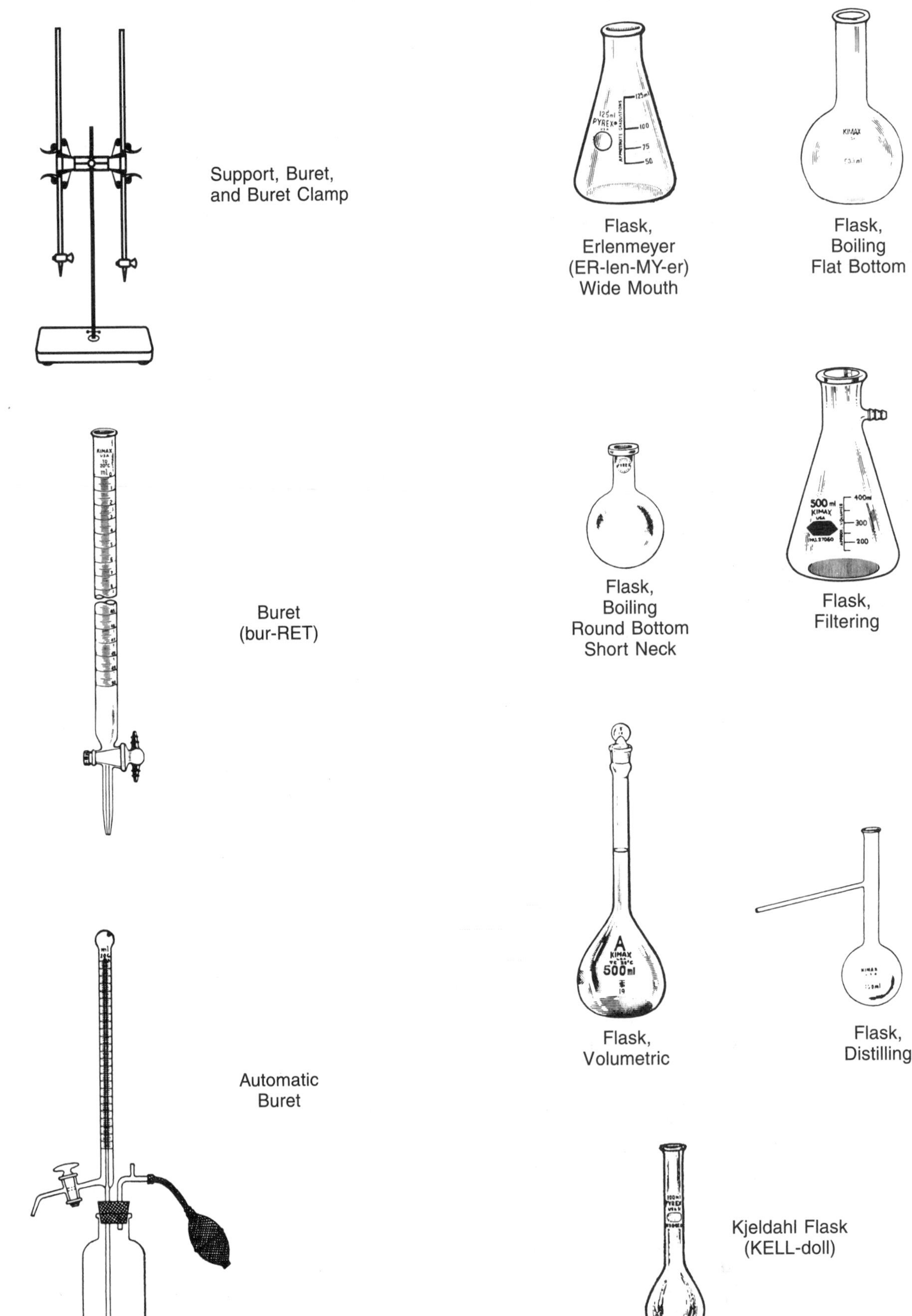

BOTTLES. Bottles are used to store chemicals, to collect samples for testing purposes, and to dispense liquids.

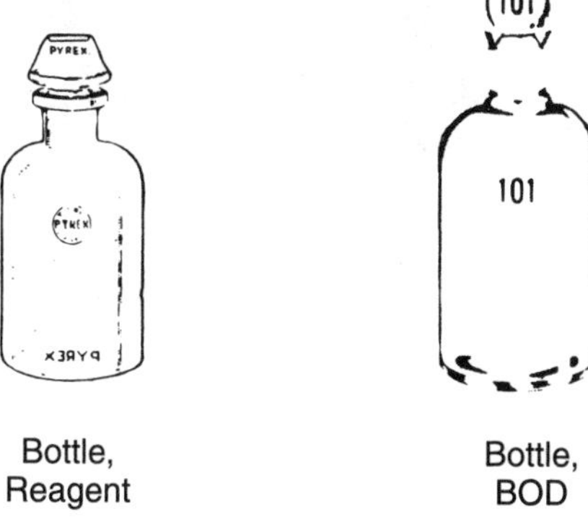

Bottle, Reagent

Bottle, BOD

FUNNELS. A funnel is used for pouring solutions or transferring solids chemicals. This funnel also can be used with filter paper to remove solids from a solution.

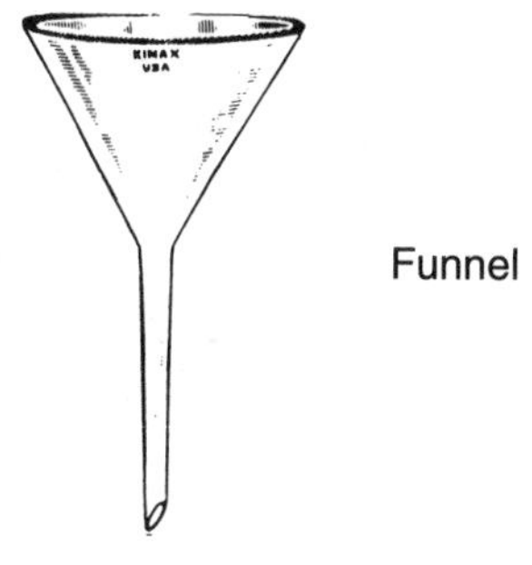
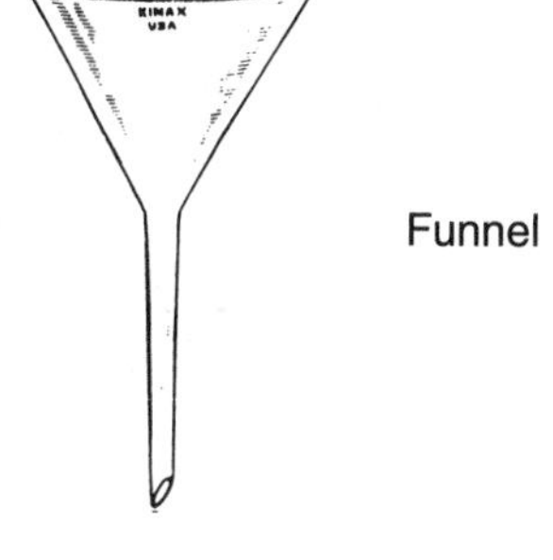

Funnel

A Buchner funnel is used to separate solids from a mixture. It is used with a filter flask and a vacuum.

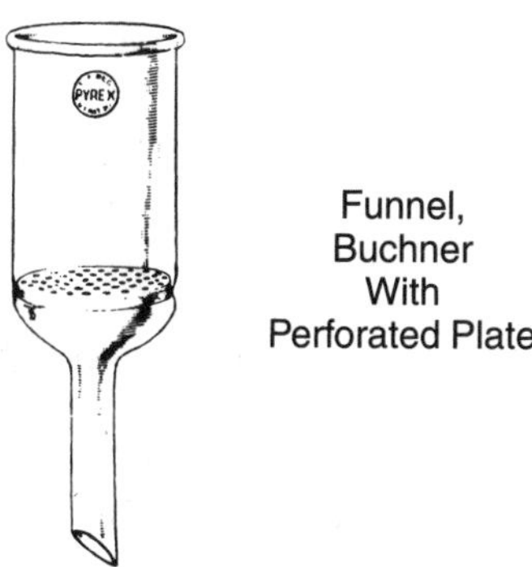

Funnel, Buchner With Perforated Plate

Separatory funnels are used to separate one chemical mixture from another. The separated chemical usually is dissolved in one or two layers of liquid.

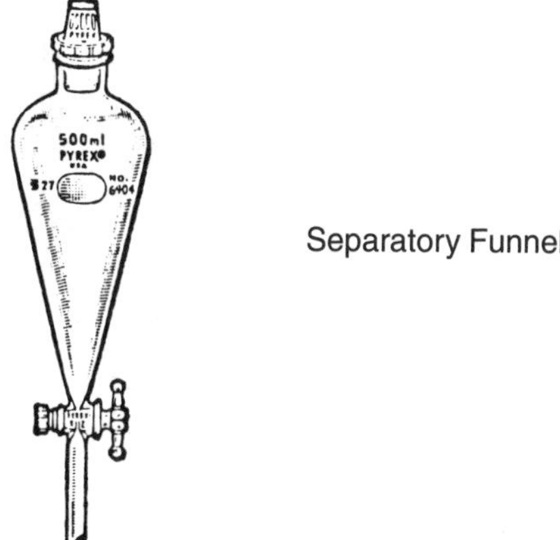

Separatory Funnel

TUBES. Test tubes are used for mixing small quantities of chemicals. They are also used as containers for bacterial testing (culture tubes).

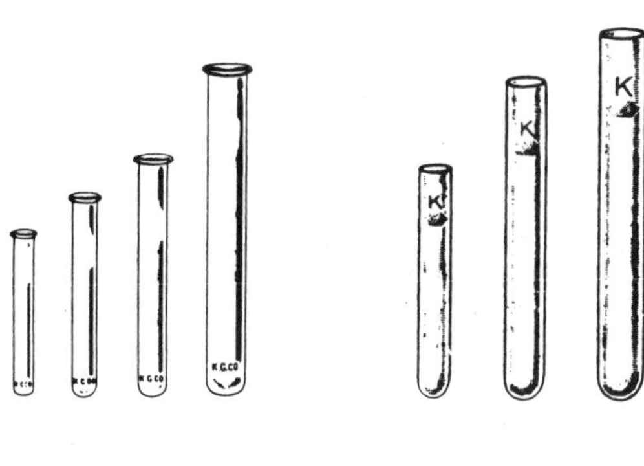

Test Tube

Culture Tube Without Lip

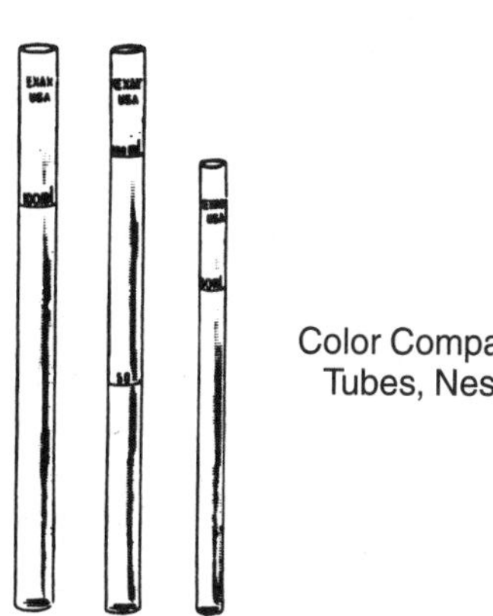

Color Comparison Tubes, Nessler

OTHER LABWARE AND EQUIPMENT.

Cone,
Imhoff
(IM-hoff)

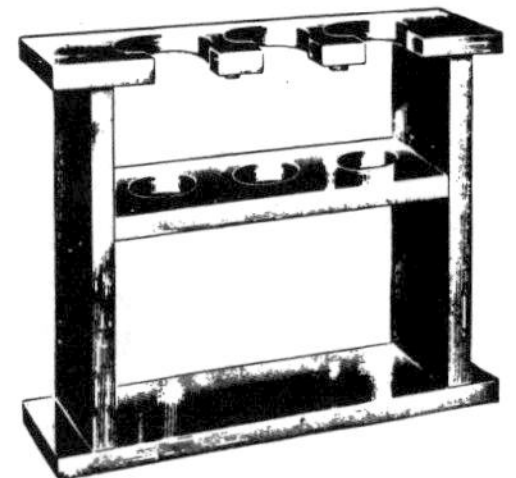

Cone Support

Hot Plate

Condenser

Dish, Petri

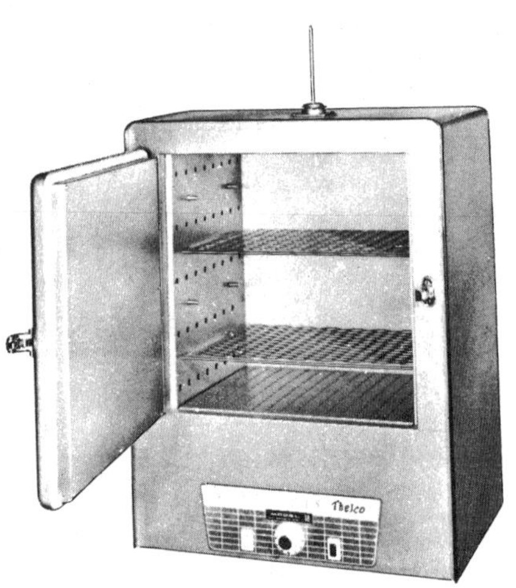

Oven, Mechanical Convection

Desiccator
(DES-uh-KAY-tor)

Thermometer, Dial

Muffle Furnace, Electric

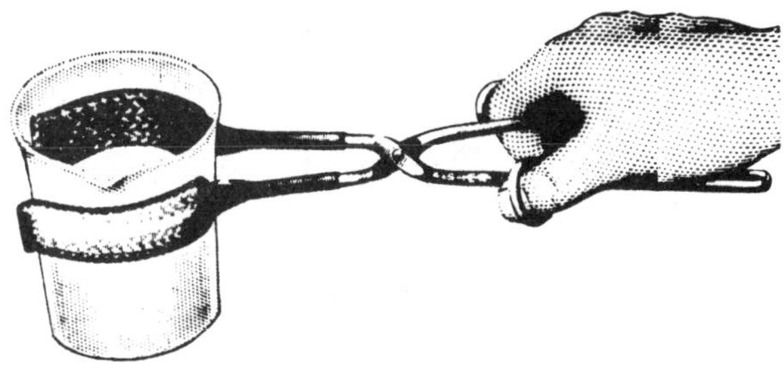

Clamp, Beaker,
Safety Tongs

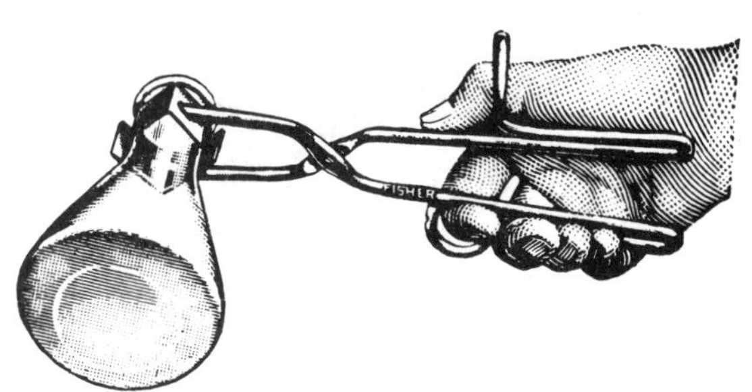

Clamp, Dish,
Safety Tongs

Clamp, Flask,
Safety Tongs

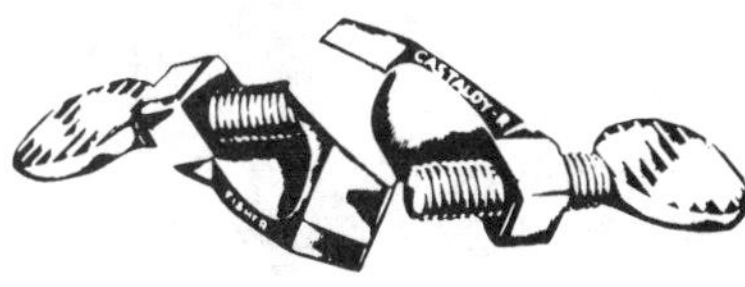

Clamp, Test Tube

Clamp Holder

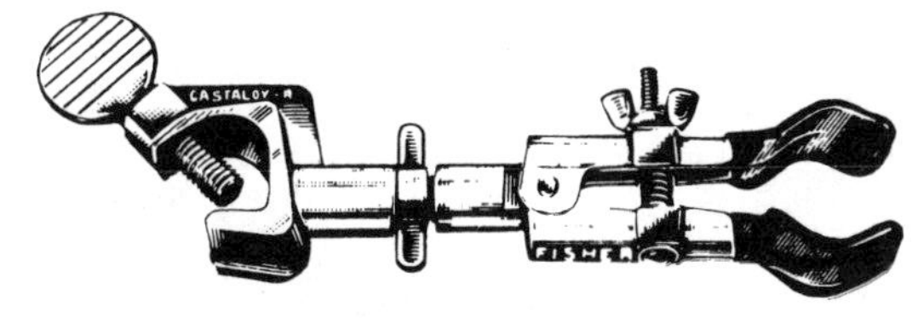

Clamp, Utility

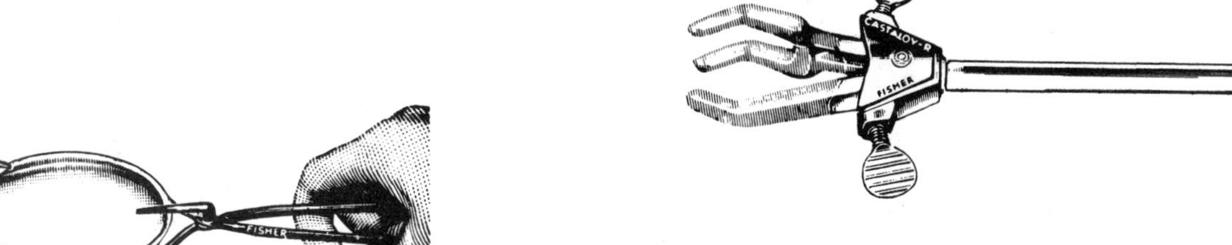

Clamp

Tripod, Concentric
Ring

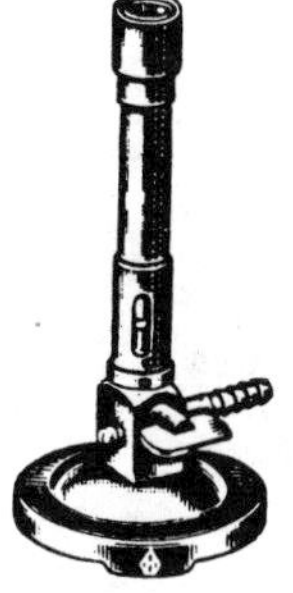

Burner, Bunsen

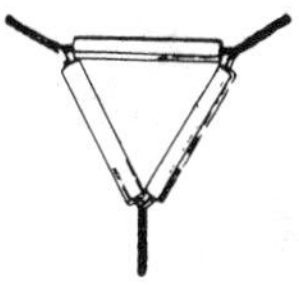

Triangle,
Fused

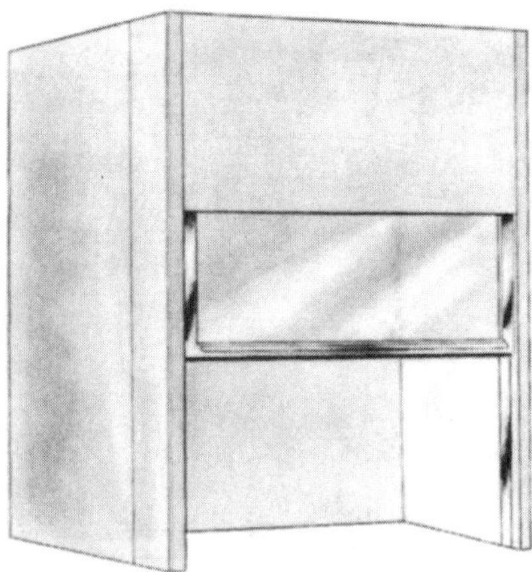

Fume Hood

Portable Dissolved Oxygen Meter
(with computer docking station)
(Courtesy of HACH Company)

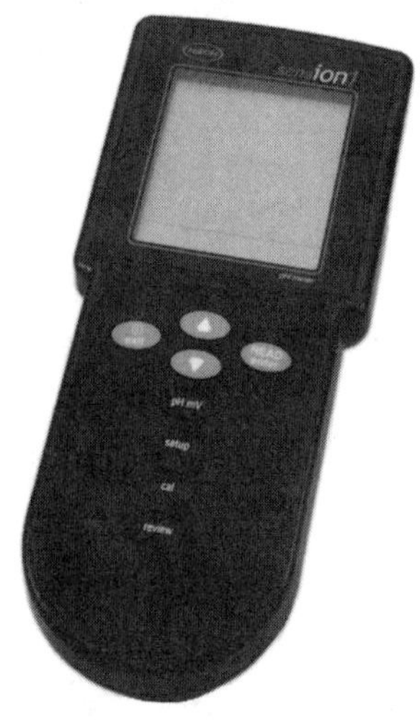

Portable pH Meter
(Courtesy of HACH Company)

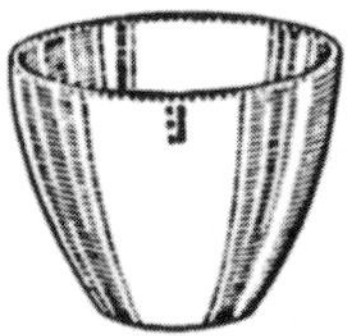

Crucible,
Porcelain

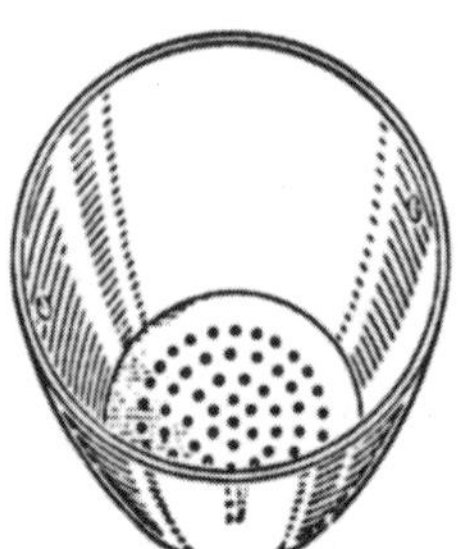

Crucible,
Gooch
(GOO-ch)
Porcelain

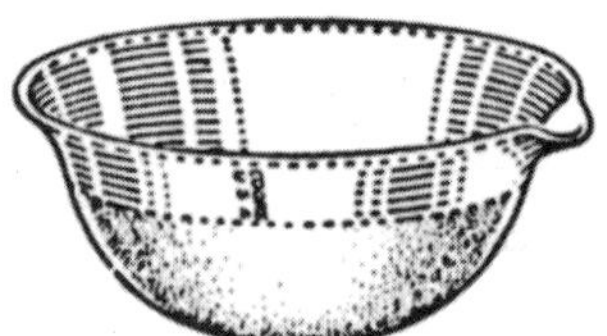

Dish,
Evaporating

Test Paper, pH 1-11

Pump, Air Pressure and Vacuum

BOD Incubator

Weight = 95.5580 gm.

Balance, Analytical
(Permission of Mettler)

16.14 Use of Laboratory Glassware

BURETS

A buret is used to give accurate measurements of liquid volumes. The stopcock controls the amount of liquid that will flow from the buret. A glass stopcock must be lubricated (stopcock grease) and should not be used with alkaline solutions. A Teflon stopcock never needs to be lubricated.

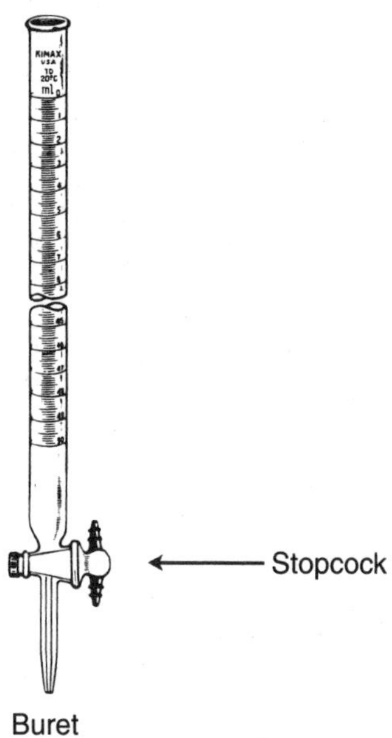

Buret

Burets come in several sizes, with those holding 10 or 25 milliliters used most frequently.

When a buret is filled with liquid, the surface of the liquid is curved. This curve of the surface is called the meniscus (meh-NIS-cuss). Depending on the liquid, the curve forms a valley, as with water, or forms a hill, as with mercury. Since most solutions used in the laboratory are water-based, always read the bottom of the meniscus with your eye at the same level (Figure 16.1). If you have the meniscus at eye level, the closest marks that go all the way around the buret will appear as straight lines, not circles.

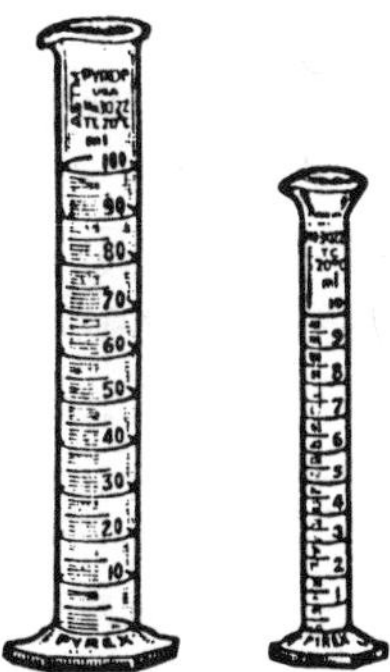

GRADUATED CYLINDERS

The graduated cylinder or "graduate" is one of the most often used pieces of laboratory equipment. This cylinder is made either of glass or of plastic and ranges in size from 10 m*L* to 4 liters. The graduate is used to measure volumes of liquid with an accuracy *less* than burets but *greater* than beakers or flasks. Graduated cylinders should never be heated in an open flame because they will break.

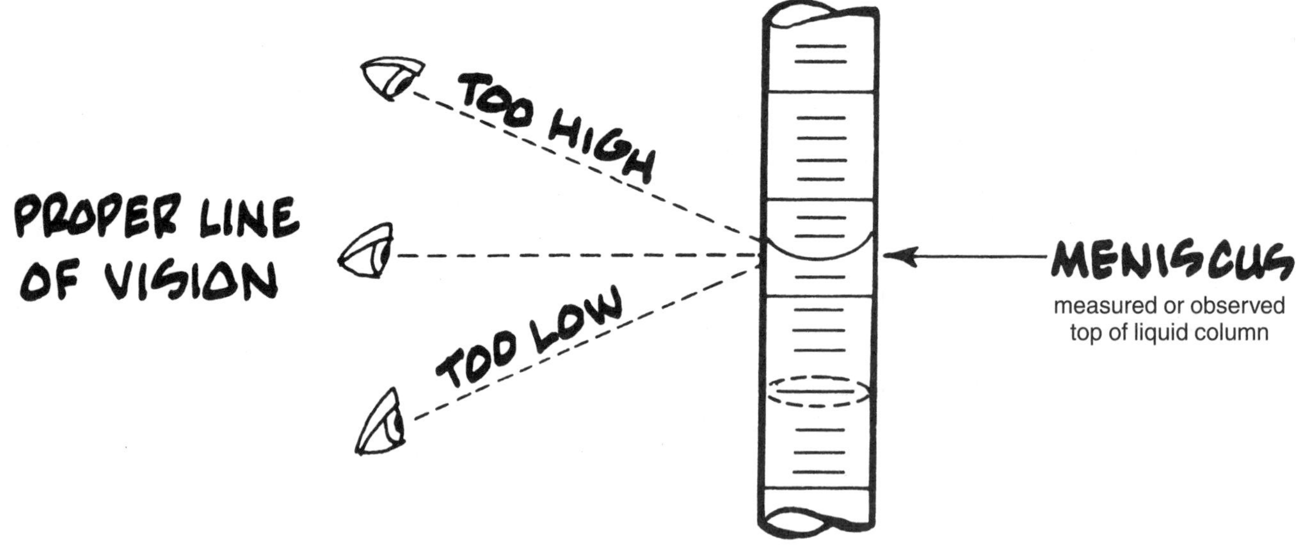

Fig. 16.1 How to read meniscus

FLASKS AND BEAKERS

Beakers and flasks are used for mixing, heating, and weighing chemicals. Most beakers and flasks are not calibrated with exact volume lines; however, they are sometimes marked with approximate volumes and can be used to estimate volumes.

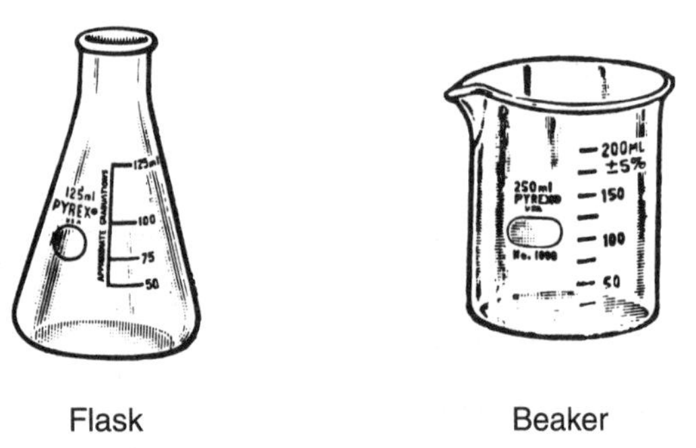

Flask Beaker

VOLUMETRIC FLASKS

Volumetric flasks are used to prepare solutions and come in sizes from 10 to 2,000 m*L*. Volumetric flasks should never be heated. Rather than store liquid chemicals in volumetric flasks, the chemicals should be transferred to a storage bottle. Volumetric flasks are more accurate than graduated cylinders.

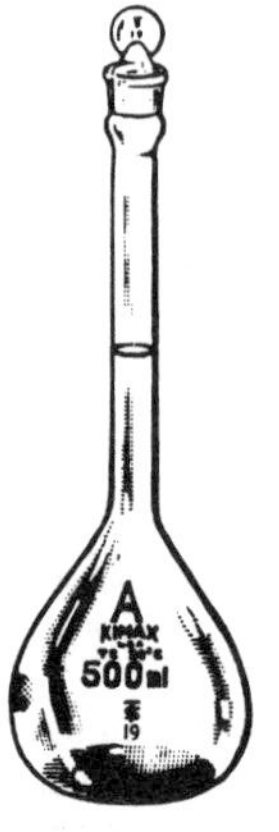

PIPETS

Pipets are used for accurate volume measurements and transfer. There are three types of pipets commonly used in the laboratory—volumetric pipets, graduated (measuring) or Mohr pipets, and serological pipets.

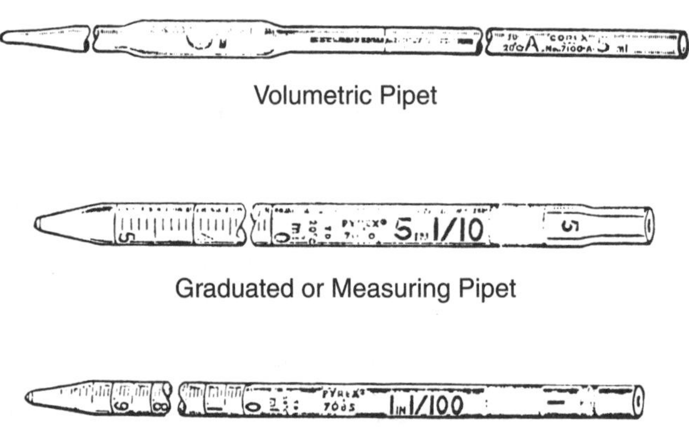

Volumetric Pipet

Graduated or Measuring Pipet

Serological Pipet

Volumetric pipets are available in sizes such as 1, 10, 25, 50, and 100 m*L*. They are used to deliver a single volume. Measuring (graduated) and serological pipets, however, will deliver fractions of the total volume indicated on the pipet.

Volumetric pipets should be held in a vertical position when emptying and the outflow should be unrestricted. The tip should be touched to the wet surface of the receiving vessel and kept in contact with it until the emptying is complete. Under no circumstance should the small amount remaining in the tip be blown out.

Measuring (graduated) and serological pipets should be held in the vertical position. After outflow has stopped, the tip should be touched to the wet surface of the receiving vessel. No drainage period is allowed. Where the small amount remaining in the tip is to be blown out and added, indication is made by a frosted band near the top of the pipet.

16.15 Solutions

Many laboratory procedures do not give the concentrations of standard solutions in grams/Liter or milligrams/liter. Instead, the concentrations are usually given as normality (*N*), which is the standard designation for solution strengths in chemistry.

EXAMPLES:

0.025 *N* H_2SO_4 means a 0.025 normal solution of sulfuric acid

2 *N* NaOH means that the normality of the sodium hydroxide solution is 2

The *larger* the number in front of the *N*, the *more* concentrated the solution. For example, 1 *N* NaOH solution is more concentrated than a 0.2 *N* NaOH solution.

When the exact concentration of a chemical or compound in a solution is known, it is referred to as a "standard solution." Many times standard solutions can be ordered already prepared. Once a standard has been prepared, it can then be used to standardize other solutions. To standardize a solution means to determine its concentration accurately, thereby making it a standard solution. "Standardization" is the process of using one solution of known concentration to determine the concentration of another solution. This action often involves a procedure called a "titration."

16.16 Titrations

A titration involves the addition of one solution, which is generally in a buret, to another solution in a flask or beaker. The solution in the buret is referred to as the "titrant" and is added to the other solution until there is a measurable change in the solution in the flask or beaker. This change is frequently a color change as a result of the addition of a special chemical called an "indicator" to the solution in the flask before the titration begins. The solution in the buret is added slowly to the flask until the change, which is called the "end point," is reached. The entire process is a "titration."

16.17 Use of a Spectrophotometer

In the field of wastewater analysis, many determinations such as phosphorus, nitrite and nitrate are based on measuring the intensity of color at a particular wavelength. The color is formed in the sample by adding a specific developing reagent to it. The intensity of the color formed is directly related to the amount of material (such as phosphorus) in the sample. For the analysis of phosphorus present in wastewater, for example, ammonium molybdate reagent is added as the developing reagent; if phosphorus is present, a blue color develops. The more phosphorus there is, the deeper and darker the blue color.

The human eye can detect some differences in color intensity; however, for very precise measurements an instrument called a spectrophotometer is used.

THE SPECTROPHOTOMETER. A spectrophotometer is an instrument generally used to measure the color intensity of a chemical solution. A spectrophotometer in its simplest form consists of a light source that is focused on a prism or other suitable light dispersion device to separate the light into its separate bands of energy. Each different wavelength or color may be selectively focused through a narrow slit. This beam of light then passes through the sample to be measured. The sample is usually contained in a glass tube called a cuvette (que-VET). Most cuvettes are standardized to have a 1.0-cm light path length, however many other sizes are available.

After the selected beam of light has passed through the sample, it emerges and strikes a photoelectric cell. If the solution in the sample cell has absorbed any of the light, the total energy content will be reduced. If the solution in the sample cell does not absorb the light, then there will be no change in energy. When the transmitted light beam strikes the photoelectric tube, it generates an electric current that is proportional to the intensity of light energy striking it. By connecting the photoelectric tube to a galvanometer (a device for measuring electric current) with a graduated scale, a means of measuring the intensity of the transmitted beam is achieved.

The diagram below illustrates the working parts of a spectrophotometer.

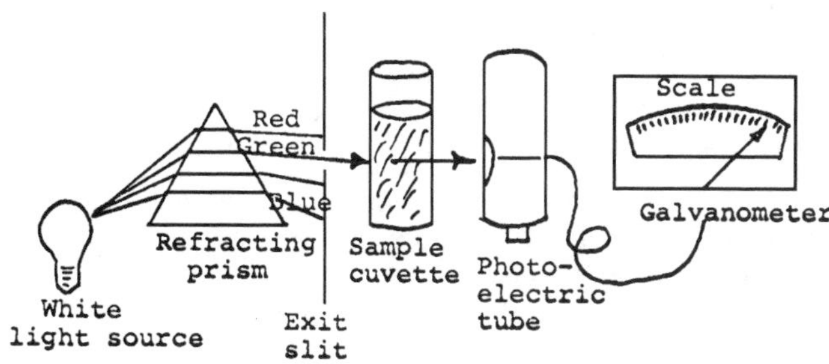

The operator should always follow the working instructions provided with the instrument.

UNITS OF SPECTROSCOPIC MEASUREMENT. The scale on spectrophotometers is generally graduated in two ways:

(1) In units of percent transmittance (%T), an arithmetic scale with units graded from 0 to 100%

(2) In units of absorbance (A), a logarithmic scale of non-equal divisions graduated from 0.0 to 2.0

Some specialized spectrophotometers also contain a scale that directly reads the concentration of one chemical constituent in milligrams per liter.

Both the units percent transmittance and absorbance are associated with the words color intensity. That is, a sample that has a low color intensity will have a high percent transmittance but a low absorbance.

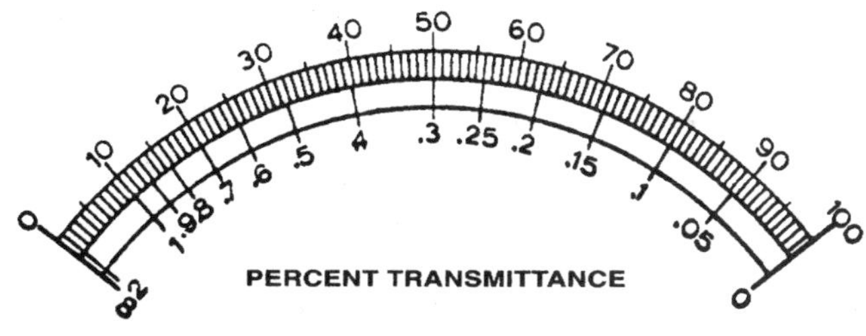

Absorbance

As illustrated above, the absorbance scale is ordinarily calibrated on the same scale as percent transmittance on spectrophotometers. The chief usefulness of absorbance lies in the fact that it is a logarithmic function rather than linear (arithmetic) and a law known as Beer's Law states that the concentration of a light absorbing colored solution is directly proportional to absorbance over a given range of concentrations. If one were to plot a graph showing %T or percent transmittance versus concentration on straight graph or line paper and another showing absorbance versus concentration on the same paper, the following curves (graphs) would result:

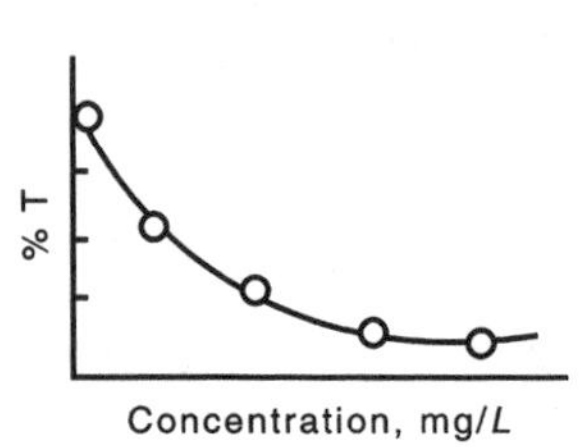

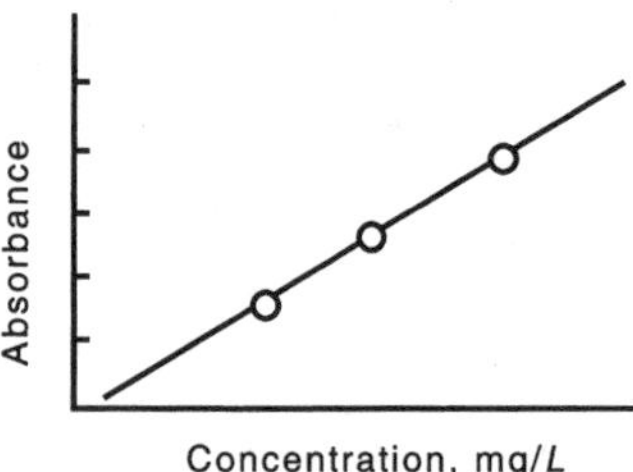

CALIBRATION CURVES. The calibration curve is used to determine the concentration of the water quality indicator (phosphorus, nitrite) contained in a sample. Three steps must be completed in order to prepare a calibration graph.

First, a series of standards must be prepared. A standard is a solution that contains a known amount of the same chemical constituent that is being determined in the sample.

Second, these standard solutions and a sample containing none of the constituent being tested for (usually distilled water and generally referred to as a blank) must be treated with the developing reagent in the same manner as the sample would be treated.

Third, using a spectrophotometer the absorbance or transmittance at the specified wavelength of the standards and blank must be determined. From the values obtained, a calibration curve of absorbance versus concentration can be plotted. Once these several points have been plotted, you can then extend the plotted points by connecting the known points with a straight line. For example, with the data given below one could construct the following calibration curve.

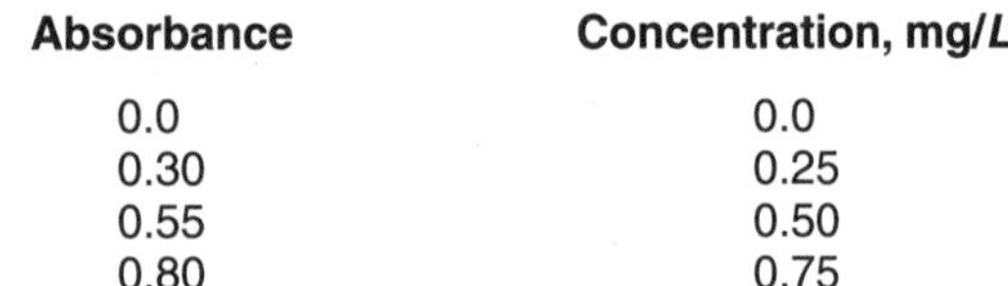

Absorbance	Concentration, mg/*L*
0.0	0.0
0.30	0.25
0.55	0.50
0.80	0.75

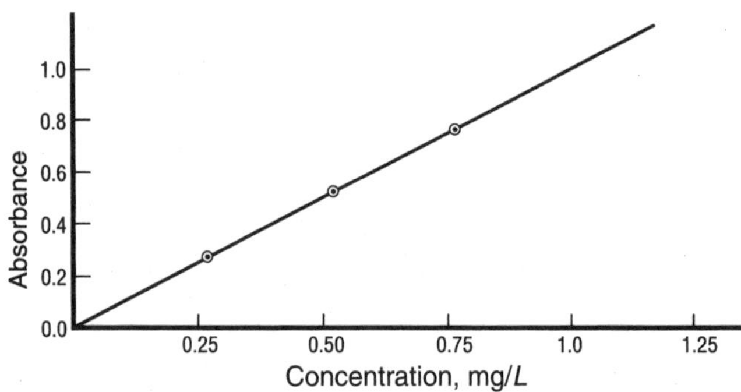

Once you have established a calibration curve for the water quality indicator in question, you can easily determine the amount of that substance contained in a solution of unknown concentration. You merely take an absorbance reading on the color developed by the unknown and locate it on the vertical axis. Then a straight line is drawn to the right on the graph until it intersects with the experimental standard curve. A line is then dropped to the horizontal axis and this value identifies the concentration of your unknown water quality indicator.

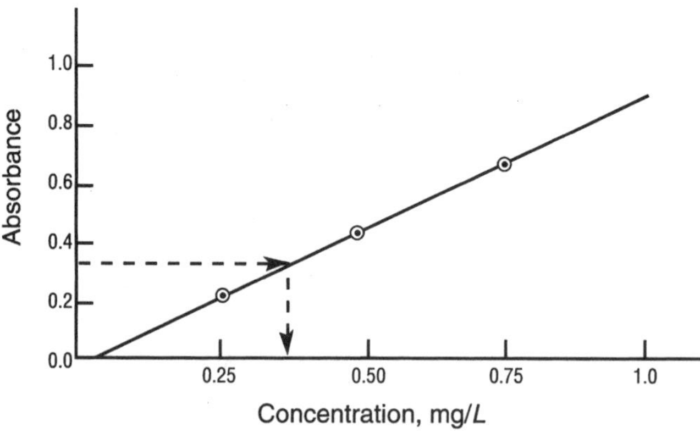

In this example, an absorbance reading of 0.32 was read on the unknown solution or sample, which indicates a concentration of about 0.37 mg/*L*.

16.18 Indicators

Indicators used in water testing are chemicals that will change color in the presence of specific chemical ions. Even though the indicator changes color as the test is run, it does not get involved or interfere with the reaction that takes place during the test.

There are many different indicators that are used in the testing of water samples. The person who sets up the test procedure will select the indicator that will give the most recognizable and reproducible results. Generally the test kits and procedures that are used in small plants will specify or include the best indicator for each test.

Indicators are used in three different levels of testing.

1. Qualitative Analysis. In this type of test the indicator is added to an unmeasured amount of sample. The only purpose of the test is to find out if a specific chemical, one that will cause the color change, is present in the sample.
2. Quantitative Titration. In this test a reagent solution that has a very accurately known concentration is added to an accurately measured volume of a test sample that contains the selected indicator. The reagent can be added by using a standardized dropper bottle or a laboratory apparatus called a buret. The purpose of the test is to measure how much reagent it takes to react with all of the test chemical that is in the sample. When the indicator changes color, it signals that the reaction is complete. Then, by using the magic of chemistry-mathematics, the amount of reagent used in the test is converted to the amount of test chemical present in the sample. Charts provided in the test kits or

laboratory manuals show the equivalent concentrations of the chemical being tested in milligrams per liter.

3. Quantitative Colorimetric Analysis. In this test an accurately measured volume of sample is placed in a glass tube called a cuvet and the indicator is added to the sample to develop the color. The intensity of the color indicates the concentration of the test chemical that is in the sample. The color intensity is measured in one of two ways:

    a. Color comparator: This method enables the tester to match the color of the test sample to the color of one of the standards provided with the test kit.

    b. Photoelectric cell: The prepared test sample is placed in a device or meter that directs a light source through the sample. A photocell then measures the amount of light that passes through the sample and, using the magic of photo electronics, converts the amount of light to the concentration of the test chemical in the sample. The test devices either report the concentration directly on a dial on the meter, or they report a value that can be used to calculate the concentration in milligrams per liter.

16.19 Laboratory Work Sheets and Notebooks

The plant operator has two goals in using a laboratory notebook and work sheets: (1) to record data, and (2) to arrange data in an orderly manner. Often, days of work can be wasted if data are written down on a scrap of paper, which is usually misplaced or thrown away. Work sheets and notebooks help prevent error and provide a record of the work. The routine use of work sheets and notebooks is the only way an operator can be sure that all important information for a test is properly recorded.

There is no standard laboratory form. Most treatment plants usually develop their own data sheets for recording laboratory results and other important plant data. The data sheets are prepared in a manner that makes it easy for you to record the data, review it, and recover it when necessary. Each plant may have different needs for collecting and recording data and many plants may use from five to eight different data sheets. Figures 16.2 and 16.3 illustrate typical laboratory work sheets (sometimes called bench sheets).

Laboratory work sheets provide an organized method for recording data. They are used to review effluent quality, identify problems, and search for the cause of problems. These sheets provide the information needed to complete NPDES permit reporting forms.

16.110 Acknowledgment

Pictures of laboratory glassware and equipment in this manual are reproduced with the permission of VWR Scientific, San Francisco, California.

QUESTIONS

Write your answers in a notebook and then compare your answers with those on page 609.

16.1A For each of the items listed below, describe the item and its use or purpose.

1. Beakers
2. Graduated Cylinders
3. Pipets
4. Burets

16.1B Why should graduated cylinders never be heated in an open flame?

16.1C What is a bench sheet?

END OF LESSON 1 OF 9 LESSONS
ON
LABORATORY PROCEDURES AND CHEMISTRY

Please answer the discussion and review questions next.

DISCUSSION AND REVIEW QUESTIONS

Chapter 16. LABORATORY PROCEDURES AND CHEMISTRY

(Lesson 1 of 9 Lessons)

At the end of each lesson in this chapter, you will find some discussion and review questions. The purpose of these questions is to indicate to you how well you understand the material in the lesson. Write the answers to these questions in your notebook.

1. Why must chemicals be properly labeled?
2. How are pipets emptied or drained? Why is this procedure important?
3. How would you titrate a solution?

PLANT ____________________

DATE ____________________

SUSPENDED SOLIDS AND DISSOLVED SOLIDS

SAMPLE							
Crucible							
Sample, m*L*							
Wt Dry + Dish, gm Wt Dish, gm							
Wt Dry, gm							
$mg/L = \frac{\text{Wt Dry, gm} \times 1{,}000{,}000}{\text{Sample, m}L}$							
Wt Dish + Dry, gm Wt Dish + Ash, gm							
Wt Volatile, gm							
$\% = \frac{\text{Wt Vol}}{\text{Wt Dry}} \times 100\%$							

BOD

\# Blank ____________________

SAMPLE										
DO Sample										
Bottle #										
% Sample										
Blank or adjusted blank DO after incubation										
Depletion, 5 days										
Depletion, %										

SETTLEABLE SOLIDS

Sample, m*L* ____________________ ____________________

Direct m*L*/*L* ____________________ ____________________

COD

Sample ____________________ ____________________

Blank Titration ____________________ ____________________

Sample Titration ____________________ ____________________

Depletion ____________________ ____________________

$mg/L = \frac{\text{Dep} \times N \text{ FAS} \times 8{,}000}{\text{Sample, m}L}$ ____________________ ____________________

Fig. 16.2 Typical laboratory work sheet

TOTAL SOLIDS

SAMPLE					
Dish No.					
Wt Dish + Wet, gm Wt Dish, gm					
Wt Wet, gm					
Wt Dish + Dry, gm Wt Dish, gm					
Wt Dry, gm					
% Solids = $\frac{\text{Wt Dry}}{\text{Wt Wet}} \times 100\%$					
Wt Dish + Dry, gm Wt Dish + Ash, gm					
Wt Volatile, gm					
% Volatile = $\frac{\text{Wt Vol} \times 100\%}{\text{Wt Dry}}$					
pH					
Volatile Acid, mg/*L*					
Alkalinity as $CaCO_3$, mg/*L*					

Grease

Sample ____________________

Sample, m*L* ____________________

Wt Flask + Grease, mg ____________________

Wt Flask, mg ____________________

Wt Grease, mg ____________________

$$\text{mg}/L = \frac{\text{Wt Grease, mg} \times 1{,}000}{\text{Sample, m}L}$$ ____________________

Fig. 16.3 Typical laboratory work sheet

CHAPTER 16. LABORATORY PROCEDURES AND CHEMISTRY

(Lesson 2 of 9 Lessons)

16.2 SAFETY AND HYGIENE IN THE LABORATORY

Safety is just as important in the laboratory as in the rest of the treatment plant. State laws and the Occupational Safety and Health Act (OSHA) demand proper safety procedures to be exercised in the laboratory at all times. OSHA specifically deals with "safety at the place of work." The act requires that "each employer has the general duty to furnish all employees, employment free from recognized hazards causing, or likely to cause death or serious physical harm."

Personnel working in a wastewater treatment plant laboratory must realize that a number of hazardous materials and conditions exist. Be alert and careful. Be aware of potential dangers at all times. Safe practice in the laboratory requires hardly any more effort than unsafe practice, and the important results are prevention of injury or bodily damage.

Each laboratory must develop and implement a written hazard communication program to ensure that personnel dealing with chemicals are appropriately informed of any health and safety considerations and that they are effectively trained in methods to safely deal with the chemicals. As a minimum, the written program must describe how criteria are met concerning labels and other forms of warnings, Material Safety Data Sheets (MSDSs), and employee information and training. Plants that have a laboratory must also develop and implement a written Chemical Hygiene Plan (CHP). Required elements of the plan are described in Chapter 14, "Plant Safety."

DO NOT PIPET HAZARDOUS LIQUIDS BY MOUTH.

16.20 Laboratory Hazards

Working with chemicals and other materials in the wastewater treatment plant laboratory can be dangerous. Dangers include:

Infectious Materials	Electric Shock
Poisons	Toxic Fumes
Explosions	Fire
Cuts and Bruises	Burns

The above dangers to yourself and others can be minimized, however, by using proper techniques and equipment.

16.200 Infectious Materials

Wastewater and sludge contain millions of biological organisms. Some of these are infectious and dangerous and can cause diseases such as tetanus, typhoid, dysentery, and hepatitis. Personnel handling these materials should thoroughly wash their hands with soap and water, particularly before handling food or smoking.

Do not pipet *anything* by mouth. Pipet all samples and reagents by mechanical means (rubber pipet bulbs) to avoid taking a chance on severe illness or death.

Never drink from a beaker or other laboratory glassware. A beaker left "specifically" for drinking is a menace to the laboratory.

Though not mandatory, inoculations by your local Health Department are recommended for each employee to reduce the possibility of contracting diseases.

16.201 Corrosive Chemicals

Acids

1. Examples: Sulfuric acid (H_2SO_4), hydrochloric (or muriatic) acid (HCl), nitric acid (HNO_3), glacial acetic acid (CH_3COOH), and chromic acid (H_2CrO_4).
2. Acids are extremely corrosive to human tissue, metals, clothing, wood, cement, stone, and concrete.
3. Commercially available spill cleanup materials should be kept on hand to neutralize acid in the event of an accidental spill.

Bases

1. Examples: Sodium hydroxide (lye or caustic soda—NaOH), potassium hydroxide (KOH), and alkaline iodide–azide reagent (used in dissolved oxygen test)
2. Bases are extremely corrosive to skin, clothing, and leather.
3. Commercially available spill cleanup materials should be kept on hand for use in the event of an accidental spill.

Miscellaneous

1. Examples: Chlorine, ferric salts (ferric chloride), and other strong oxidants.

16.202 Toxic Materials

1. Solids: Cyanide, chromium, orthotolidine, cadmium, and other heavy metals.
2. Liquids: Carbon tetrachloride, chloroform, ammonium hydroxide, nitric acid, bromine, chlorine water, aniline dyes, formaldehyde, and carbon disulfide. Carbon tetrachloride is absorbed into the skin on contact; its vapors will damage the lungs; and it will build up in your body to a dangerous level.
3. Gases: Hydrogen sulfide, chlorine, ammonia, and sulfur dioxide.

16.203 Explosive or Flammable Materials

1. Liquids: Carbon disulfide, benzene, ethyl ether, petroleum ether, acetone, and gasoline.
2. Gases: Acetylene and hydrogen.

16.21 Personal Safety and Hygiene

16.210 Laboratory Safety

Laboratory work can be dangerous if proper precautions and techniques are not taken. *Always* follow these basic rules:

1. *Never* work alone in the laboratory. Someone should always be available to help in case you should have an accident that blinds you, leaves you unconscious, or starts a fire you cannot handle. If necessary, have someone check on you regularly to be sure you are OK.
2. Wear protective goggles or safety glasses at all times in the laboratory. Contact lenses may be worn under safety goggles, but contact lenses are not eye protective devices, and wearing them does not reduce the requirement for eye and face protection. Because fumes can seep between the lens and the eyeball and irritate the eye, the National Institute of Occupational Safety and Health (NIOSH) recommends an eye injury hazard evaluation be conducted in accordance with the safety guidelines listed in Current Intelligence Bulletin (CIB) 59 2005-139 (http://www.cdc.gov/niosh/docs/2005-139/).

Safety Glasses

3. Wear a face shield if there is any danger of a hot liquid erupting from a container or flying pieces of glassware from an exploded apparatus. If in doubt as to its need—wear it.
4. Wear protective or insulated gloves when handling hot equipment or very cold objects, or when handling liquids or solids that are skin irritants.
5. Always wear a lab coat or apron in the laboratory to protect your skin and clothes.
6. Never pipet hazardous materials by your mouth.
7. Never eat or smoke in the laboratory. Never use laboratory glassware for serving food or drinks.
8. Do not keep food in a refrigerator that is used for chemical or sample storage.
9. Use ventilated laboratory fume hoods whenever handling toxic chemicals.
10. Maintain clear access to emergency eye wash stations and deluge showers.
11. Practice good housekeeping as an effective way to prevent accidents.

16.211 Personal Hygiene

Although it is highly unlikely that personnel will contract diseases by working in wastewater treatment plants, such a possibility does exist with certain diseases.

1. Some diseases are contracted through breaks in skin, cuts, or puncture wounds. In such cases the bacteria causing the disease may be covered over and trapped by flesh, creating a suitable *ANAEROBIC*[2] environment in which the bacteria may thrive and produce toxins that may spread throughout the body.

   For protection against diseases contracted through breaks in the skin, cuts, or puncture wounds, everyone working in or around wastewater should receive immunization from tetanus. Immunization must be received *before* the infection occurs. To prevent diseases from entering open wounds, care must be taken to keep wounds protected either with band-aids or, if necessary, with rubber gloves or waterproof protective clothing.
2. Diseases that may be contracted through the gastrointestinal system or through the mouth are typhoid, cholera, dysentery, amebiasis, worms, salmonella, infectious hepatitis, and polio virus. These diseases are transmitted when the infected wastewater materials are ingested or swallowed by careless persons. The best protection against those diseases is furnished by thorough cleansing. Hands, face, and body should be thoroughly washed with soap and water, particularly the hands, in order to prevent the transfer of any unsanitary materials or germs to the mouth while eating.

   A change of working clothes into street clothes before leaving work is highly recommended to prevent carrying unsanitary materials to your home. Personal hygiene, thorough cleansing, and washing of the hands are effective means of protection. Store your work clothes and street clothing in separate lockers.

[2] *Anaerobic* (AN-air-O-bick). A condition in which atmospheric or dissolved oxygen (DO) is *NOT* present in the aquatic (water) environment.

Immunization is available for typhoid, polio, and hepatitis A and B. Hepatitis A and B can be transmitted by wastewater. Hepatitis is frequently associated with gross wastewater pollution.

3. Diseases that may be contracted by breathing contaminated air include tuberculosis, infectious hepatitis, and San Joaquin fever. There has been no past evidence to indicate the transmission of tuberculosis through the air at wastewater treatment plants. However, there was one case of tuberculosis being contracted by an employee who fell into wastewater and, while swimming, inhaled wastewater into his lungs. San Joaquin fever is caused by a fungus that may be present in wastewater or dried sludge solids. However, there is no record of operators contracting the disease while on the job.

The best insurance against these diseases is proper personal hygiene and immunization. Your plant should have an immunization program against: (1) tetanus, (2) polio, and (3) hepatitis A and B. Immunization against typhoid should be considered if the possibility exists that typhoid could be found in your area. The immunizations should be provided to protect you. Check with your local or state health department for recommendations regarding immunization.

In the washing of hands, the kind of soap is less important than the thorough use of the soap. (Special disinfectant soaps are not essential.)

The use of protective clothing is very important, particularly gloves and boots. The protection of wounds and cuts is also important. Report injuries and take care of them.

The responsibility rests upon you.

There is no absolute insurance against contraction of disease in a wastewater treatment plant. However, the likelihood of transmission is practically negligible. There appears to be no special risk in working at treatment plants. In fact, operators may develop a natural immunity by working in this environment.

The possibility that Acquired Immune Deficiency Syndrome (AIDS), which is caused by a virus, can be contracted from exposure to raw wastewater has been discounted by researchers. Researchers have found that, although the AIDS virus is present in the wastes from AIDS victims, the raw wastewater environment is hostile to the virus itself and has not been identified as a mode of transmission to date. Needle sticks from potentially contaminated syringes should remain a concern to operators, maintenance personnel, and laboratory analysts. Fluids in or on syringes may provide a less severe environment for pathogens than raw wastewater where dilution and chlorination significantly reduce infection potential.

16.22 Prevention of Laboratory Accidents

16.220 Chemical Storage

An adequate storeroom is essential for safety in the wastewater laboratory. The storeroom should be properly ventilated and lighted and be laid out to segregate incompatible chemicals. Flammable liquids, acids, bases, and oxidizing agents should be separated from each other by distances, partitions, or other means so as to prevent accidental contact between them. Order and cleanliness must be maintained. All chemicals and bottles or reagents must be clearly labeled and dated. Never handle chemicals with bare hands. Use a spoon, spatula, or tongs.

Heavy items should be stored on or as near to the floor as possible. *VOLATILE LIQUIDS,*[3] which may escape as a gas, such as ether, must be kept away from heat sources, sunlight, and electrical switches. Do not store ether in the refrigerator. Several laboratory explosions have occurred because the ether gas will ignite when there is a small area and when the refrigerator light comes on.

CLAMPS, RAISED SHELF EDGES, AND PROPER ARRANGEMENT PREVENT STOCKROOM FALLOUT.

Cylinders of gas in storage should be capped and secured to prevent rolling or tipping. They should also be placed away from any possible sources of heat or open flames.

Flammable gases must be stored separately. The storage room should be fitted with explosion-proof wiring and lighting fixtures. Appropriate warning signs prohibiting sources of ignition must be posted in conspicuous locations.

The usual common sense rules of storage should be followed. Good housekeeping is a most significant contribution toward an active safety campaign.

16.221 Movement of Chemicals

The next area of concern is the transfer of chemicals, apparatus, gases, or other hazardous materials from the storeroom to the laboratory for use. Use cradles or tilters to facilitate handling carboys or other larger chemical vessels.

Drum Tilter

[3] *Volatile Liquids.* Liquids that easily vaporize or evaporate at room temperature.

In transporting cylinders of compressed gases, use a trussed hand truck. Never move or transport a cylinder without the valve protection hood in place. Never roll a cylinder by its valve. Immediately after they are positioned for use, cylinders should be clamped securely into place to prevent shifting or toppling.

Carry flammable liquids in safety cans or, in the case of reagent-grade chemicals, protect the bottle with a carrier. Always wear protective gloves, safety shoes, and rubber aprons in case of accidental spilling of chemical containers.

16.222 Proper Laboratory Techniques

Faulty technique is one of the chief causes of accidents and, because it involves the human element, is one of the most difficult to correct.

Because of their nature and prevalence in the laboratory, acids and other corrosive materials constitute a series of hazards ranging from poisoning, burning, and gassing through explosion. Always flush the outsides of acid bottles before opening them. Do not lay the stopper down on the countertop where a person might lay a hand or rest an arm on it. Keep all acids tightly stoppered when not in use and make sure no spilled acid remains on the floor, table, or bottle after use. To avoid splashing, do not pour water into acid; *always pour acid into water,* like you would when adding acid to a swimming pool.

Mercury requires special care. Even a small amount in the bottom of a drawer can poison the atmosphere in a room. After an accident involving mercury, go over the entire area carefully until there are no globules remaining. Follow all of the safety requirements specified on the MSDS. Keep all mercury containers tightly stoppered.

16.223 Accident Prevention

ELECTRIC SHOCK. Whenever there are electrical outlets, plugs, and wiring connections, there is a danger of electric shock. The usual "do's" and "don'ts" of protection against shock in the home are equally applicable in the laboratory. Do not use worn or frayed wires. Replace connections when there is any sign of thinning insulation. Ground all apparatus using plugs or pigtail adapters. Ground-fault circuit interrupters (GFIs) should be installed on all electrical circuits near laboratory sinks or liquid operations or activities. Do not continue to run a motor after liquid has spilled on it. Turn it off immediately and clean and dry the inside thoroughly before attempting to use it again.

Electrical units that are operated in an area exposed to flammable vapors should be explosion-proof. All permanent wiring should be installed by an electrician with proper conduit or BX cable to eliminate any danger of circuit overloading.

CUTS. Some of the pieces of glass used in the laboratory, such as glass tubing, thermometers, and funnels, must be inserted through rubber stoppers. If the glass is forced through the hole in the stopper by applying a lot of pressure, the glass usually breaks. This is one of the most common sources of cuts in the laboratory.

Use care in making rubber-to-glass connections. Lengths of glass tubing should be supported while they are being inserted into rubber. The ends of the glass should be *FLAME POLISHED*[4] and either wetted or covered with a lubricating jelly for ease in joining connections. Never use oil or grease. Gloves should be worn when making such connections, and the tubing should be held as close to the end being inserted as possible to prevent bending or breaking. Never try to force rubber tubing or stoppers from glassware. Cut off the rubber or material.

A FIRST-AID kit must be available in the laboratory.

BURNS. All glassware and porcelain look cold after the red from heating has disappeared. The red is gone in seconds but the glass is hot enough to burn for several minutes. After heating a piece of glass, put it out of the way until cool.

Spattering from acids, caustic materials, and strong oxidizing solutions should be washed off immediately with large quantities of water. Neutralize an acid with sodium carbonate or bicarbonate. Every worker in the wastewater laboratory should have access to a sink and an emergency deluge shower and eye wash station.

Many safeguards against burns are available. Special gloves, safety tongs, aprons, and emergency deluge showers are but a few examples. Never decide it is too much trouble to put on a pair of gloves or use a pair of tongs to handle a dish or flask that has been heated.

USE TONGS—DO NOT JUGGLE HOT CONTAINERS.

Perhaps the most harmful and painful chemical burn occurs when small objects, chemicals, or fumes get into your eyes. Immediately flood your eyes with water or a special "eyewash" solution from a safety kit or from an eye wash station or fountain. Washing with large amounts of water for at least 15 minutes is recommended.

---

[4] *Flame Polished.* Melted by a flame to smooth out irregularities. Sharp or broken edges of glass (such as the end of a glass tube) are rotated in a flame until the edge melts slightly and becomes smooth.

TOXIC FUMES. Certain chemicals are dangerous to breathe, therefore, use a ventilated laboratory fume hood for routine reagent preparation or whenever handling substances that may generate harmful atmospheric contaminants. Select a hood that has adequate air displacement and expels harmful vapors and gases at their source. The face velocity of laboratory fume hoods must average from 100 to 150 FPM (feet per minute) (30 to 45 meters per minute), depending on the substance handled. A method to continuously indicate that air is flowing should also be provided. Check with your local safety regulatory agency for specific requirements for your facility. An annual check should be made of the entire laboratory building. Sometimes noxious fumes are spread by the heating and cooling system of the building.

WASTE DISPOSAL. A good safety program requires constant care in disposal of laboratory waste. Hazardous chemicals/substances must be disposed of by methods that comply with local environmental regulations. Confirm the local requirements *before* disposal.

To protect maintenance personnel, use separate, clearly marked, covered containers to dispose of broken glass.

DO NOT POUR VOLATILE LIQUIDS INTO THE SINK.

FIRE. The laboratory should be equipped with a fire blanket. The fire blanket is used to smother clothing fires. Small fires that occur in an evaporating dish or beaker may be put out by covering the container with a glass plate, wet towel, or wet blanket. For larger fires, or ones that may spread rapidly, promptly use a fire extinguisher. Do not use a fire extinguisher on small beaker fires because the force of the spray will knock over the beaker and spread the fire. Become familiar with the operation and use of your fire extinguisher *before* an emergency arises.

Fire classifications are important for determining the type of fire extinguisher needed to control the fire. Classifications also aid in recordkeeping. Fires are classified as A, B, C, or D fires based on the type of material being consumed: A, ordinary combustibles; B, flammable liquids and vapors; C, energized electrical equipment; and D, combustible metals. Fire extinguishers are also classified as A, B, C, or D to correspond with the class of fire each will extinguish.

CHOOSE AN EXTINGUISHER BY CLASS OF FIRE—DO NOT GUESS.

Class A fires: ordinary combustibles such as wood, paper, cloth, rubber, many plastics, dried grass, hay, and stubble. Use foam, water, soda-acid, carbon dioxide gas, or almost any type of extinguisher.

Class B fires: flammable and combustible liquids such as gasoline, oil, grease, tar, oil-based paint, lacquer, and solvents, and also flammable gases. Use foam, carbon dioxide, or dry chemical extinguishers.

Class C fires: energized electrical equipment such as starters, breakers, and motors. Use carbon dioxide or dry chemical extinguishers to smother the fire; both types are nonconductors of electricity.

Class D fires: combustible metals such as magnesium, sodium, zinc, and potassium. Operators rarely encounter this type of fire. Use a Class D extinguisher or use fine dry soda ash, sand, or graphite to smother the fire. Consult with your local fire department about the best methods to use for specific hazards that exist at your facility.

Multipurpose extinguishers are also available, such as a Class BC carbon dioxide extinguisher that can be used to smother Class B and Class C fires. A multipurpose ABC carbon dioxide extinguisher will handle most laboratory fire situations. (When using carbon dioxide extinguishers, remember that the carbon dioxide can displace oxygen—take appropriate precautions.)

There is no single type of fire extinguisher that is effective for all fires so it is important that you understand the class of fire you are trying to control. You must be trained in the use of the different types of extinguishers, and the proper type should be located near the area where that class of fire may occur.

16.23 Acknowledgments

Portions of this section were taken from material written by A. E. Greenberg, "Safety and Hygiene," which appeared in the California Water Pollution Control Association's *OPERATORS' LABORATORY MANUAL.* Some of the ideas and material also came from the *FISHER SAFETY MANUAL.*

16.24 Additional Reading

1. *FISHER SAFETY CATALOG.* Obtain from Fisher Scientific Company, Safety Division, 4500 Turnberry Drive, Suite A (Customer Service), Hanover Park, IL 60103 or phone (800) 772-6733.
2. *GENERAL INDUSTRY, OSHA SAFETY AND HEALTH STANDARDS* (CFR, Title 29, Labor Pt. 1900-1910. (most recent edition)). Obtain from the US Government Printing Office, Superintendent of Documents, PO Box 371954, Pittsburgh, PA 15250-7954. Order No. 869-060-00108-5. Price, $61.00.

QUESTIONS

Write your answers in a notebook and then compare your answers with those on page 609.

16.2A Why should you always use a rubber bulb to pipet wastewater or polluted water?

16.2B Why are inoculations against disease recommended for people working around wastewater?

16.2C True or False? You may add acid to water, but never water to acid.

16.2D What would you do if you spilled a concentrated acid on your hand?

END OF LESSON 2 OF 9 LESSONS
ON
LABORATORY PROCEDURES AND CHEMISTRY

Please answer the discussion and review questions next.

DISCUSSION AND REVIEW QUESTIONS

Chapter 16. LABORATORY PROCEDURES AND CHEMISTRY

(Lesson 2 of 9 Lessons)

Write the answers to these questions in your notebook. The question numbering continues from Lesson 1.

4. What precautions should you take to protect yourself from diseases when working in a wastewater treatment plant?
5. Discuss the basic rules for working in a laboratory.
6. Why should work with certain chemicals be conducted under a ventilated laboratory fume hood?

CHAPTER 16. LABORATORY PROCEDURES AND CHEMISTRY

(Lesson 3 of 9 Lessons)

16.3 SAMPLING

16.30 Importance

The basis for any plant monitoring program rests on information obtained by sampling. Decisions based on incorrect data may be made if sampling is performed in a careless and thoughtless manner. Obtaining good results will depend to a great extent on the following factors:

1. Ensuring that the sample taken is truly representative of the wastestream
2. Using proper sampling techniques
3. Protecting and preserving the samples until they are analyzed
4. Following a proper *CHAIN OF CUSTODY*[5] procedure if samples are sent off site for analysis

The greatest errors produced in laboratory tests are usually caused by improper sampling, poor preservation, or lack of enough mixing during compositing[6] and testing.

16.31 Representative Sampling

You must always remember that wastewater flows can vary widely in quantity and composition over a 24-hour period. Also, composition can vary within a given stream at any single time due to partial settling of solids or floating of light materials. Samples should therefore be taken from the wastestream where it is well mixed. Obtaining a representative sample should be of major concern in any sampling and monitoring program.

Laboratory equipment, in itself, is generally quite accurate. Analytical balances weigh to 0.1 milligram. Graduated cylinders, pipets, and burets usually measure to 1 percent accuracy, so that the errors introduced by these items should total less than 5 percent, and under the worst possible conditions only 10 percent. Under ideal conditions let us assume that a test of raw wastewater for suspended solids should run about 300 mg/*L*. Because of the previously mentioned equipment or apparatus variables, the value may actually range from 270 to 330 mg/*L*. Results in this range are reasonable for operation. Other less obvious factors are usually present that make it quite possible to obtain results that are 25, 50, or even 100 percent in error, unless certain precautions are taken. The following example will illustrate how these errors are produced.

The Dumpmore Wastewater Treatment Plant is a secondary treatment facility with a flow of 8 million gallons per day (30 M*L*/day). The plant has an aerated grit basin, two circular primary clarifiers of 750,000-gallon (2.8-M*L*) capacity, two digesters, two aeration basins, two secondary clarifiers, four chlorinators, and two chlorine contact basins.

Monthly summary calculations based on the suspended solids test showed that about 8,000 pounds (3,640 kg) of suspended solids were being captured per day during primary sedimentation, assuming 200 mg/*L* for the influent and 100 mg/*L* for the effluent. However, it also appeared that 12,000 pounds (5,450 kg) per day of raw sludge solids were being pumped out of the primary clarifiers and to the digester. Obviously, if sampling and analyses had been perfect, these weights would have been balanced, provided the waste activated sludge was not returned to the primary clarifiers. The capture should equal the removal of solids. A study was made to determine why the variance in these values was so great. It would seem logical to expect that the problem could be due to: (1) incorrect testing procedures, (2) poor sampling, (3) incorrect metering of the wastewater or sludge flow, or (4) any combination of the three or all of them.

In the first case, the equipment was in excellent condition. The operator was a conscientious and able person who was found to have carried out the laboratory procedures carefully and who had previously run successful tests on comparative samples. It was concluded that the equipment and test procedures were completely satisfactory.

A survey was then made to determine if sampling stations were in need of relocation. By using Imhoff cones and running settleable solids tests along the influent channel and the aerated grit chamber, one could quickly recognize that the best mixed and most representative samples were to be taken from the aerated grit chamber rather than the influent channel.

The settleable solids ran 13 m*L*/*L* in the aerated grit chamber against 10 m*L*/*L* in the channel. By the simple process of determining the best sampling station, the suspended solids value in the influent was corrected from 200 mg/*L* to the more representative 300 mg/*L*. Calculations, using the correct figures, changed the solids capture from 8,000 pounds to 12,000 pounds per day (3,640 kg to 5,450 kg/day) and a balance was obtained.

This example clearly illustrates the importance of selecting a good sampling point in securing a truly representative sample.

---

[5] *Chain of Custody.* A record of each person involved in the handling and possession of a sample from the person who collected the sample to the person who analyzed the sample in the laboratory and to the person who witnessed disposal of the sample.

[6] *Composite (Proportional) Sample.* A composite sample is a collection of individual samples obtained at regular intervals, usually every one or two hours during a 24-hour time span. Each individual sample is combined with the others in proportion to the rate of flow when the sample was collected. Equal volume individual samples also may be collected at intervals after a specific volume of flow passes the sampling point or after equal time intervals and still be referred to as a composite sample. The resulting mixture (composite sample) forms a representative sample and is analyzed to determine the average conditions during the sampling period.

It emphasizes the point that even though a test is accurately performed, the results may be entirely erroneous and meaningless insofar as use for process control is concerned, unless a good representative sample is taken. Furthermore, a good sample is highly dependent on the sampling station. Whenever possible, select a place where mixing is thorough and the wastewater quality is uniform. As the solids concentration increases, above about 200 mg/*L,* mixing becomes even more significant because the wastewater solids will tend to separate rapidly with the heavier solids settling toward the bottom, the lighter solids in the middle, and the floatables rising toward the surface. If, as is usual, a one-gallon portion is taken as representative of a million gallon flow, the job of sample location and sampling must be taken seriously.

16.32 Time of Sampling

Let us consider next the time and frequency of sampling. In carrying out a testing program, particularly where personnel and time are limited due to the press of operational responsibilities, testing may necessarily be restricted to about one test day per week. If you decide to start your tests early in the week by taking samples early on Monday morning, you may wind up with some very odd results.

One such incident will be cited. During a test for *SURFACTANTS,*[7] samples were taken early on Monday morning and rushed into the laboratory for testing. Due to the detention time in the sewers, these wastewater samples actually represented Sunday flow on the graveyard shift, the weakest wastewater obtainable. The surfactant content was only 1 mg/*L,* whereas it would usually run 8 to 10 mg/*L*. So, the time and day of sampling is quite important. Samples should be taken to represent typical weekdays or even varied from day to day within the week for a good indication of the characteristics of the wastewater.

16.33 Types of Samples

The two types of samples collected in treatment plants are known as: (1) grab samples, and (2) composite samples, and either may be obtained manually or automatically.

GRAB SAMPLES

A grab sample is a single sample of wastewater taken at neither a set time nor flow. The grab samples show the waste characteristics only at the time the sample is taken. A grab sample may be preferred over a composite sample when:

1. The wastewater to be sampled does not flow on a continuous basis.
2. The wastewater characteristics are relatively constant.
3. You wish to determine whether or not a composite sample obscures extreme conditions of the waste.
4. The wastewater is to be analyzed for dissolved gases (DO), coliforms, residual chlorine, temperature, and pH. The analysis for these water quality indicators must be performed immediately. (*NOTE:* Grab samples for these water quality indicators may be collected at set times or specific time intervals.)

COMPOSITE SAMPLES

Since the wastewater quality changes from moment to moment and hour to hour, the best results would be obtained by using some sort of continuous sampler-analyzer. However, since operators are usually the sampler-analyzer, continuous analysis would leave little time for anything but sampling and testing. Except for tests that cannot wait due to rapid chemical or biological change of the sample, such as tests for dissolved oxygen and sulfide, a fair compromise may be reached by taking samples throughout the day at hourly or two-hour intervals.

When the samples are taken, they should be refrigerated immediately to preserve them from continued bacterial decomposition. When all of the samples have been collected for a 24-hour period, the samples from a specific location should be combined or composited together according to flow to form a single 24-hour composite sample.

If an equal volume of sample was collected each hour and mixed, this would simply be a composite sample. To prepare a proportional composite sample: (1) the rate of wastewater flow must be known, and (2) each grab sample must then be taken and measured out in direct proportion to the volume of flow at that time. For example, Table 16.2 illustrates the hourly flow and sample volume to be measured out for a 12-hour proportional composite sample.

Large wastewater solids should be excluded from a sample, particularly those greater than one-quarter inch (6 mm) in diameter.

A very important point should be emphasized. During compositing and at the exact moment of testing, the samples must be vigorously remixed[8] so that they will be of the same

TABLE 16.2 DATA COLLECTED TO PREPARE PROPORTIONAL COMPOSITE SAMPLE

Time	Flow, MGD	Factor	Sample Vol, m*L*	Time	Flow, MGD	Factor	Sample Vol, m*L*
6 am	0.2	100	20	12 noon	1.5	100	150
7 am	0.4	100	40	1 pm	1.2	100	120
8 am	0.6	100	60	2 pm	1.0	100	100
9 am	1.0	100	100	3 pm	1.0	100	100
10 am	1.2	100	120	4 pm	1.0	100	100
11 am	1.4	100	140	5 pm	0.9	100	90
							1,140

A sample composited in this manner would total 1,140 m*L*.

[7] *Surfactant* (sir-FAC-tent). Abbreviation for surface-active agent. The active agent in detergents that possesses a high cleaning ability.

[8] *NOTE:* If the sample has a low buffer capacity and the real pH is 6.5 or less, vigorous shaking can cause a significant change in pH level.

composition and as well mixed as when they were originally sampled. Sometimes such remixing may become lax, so that all the solids are not uniformly suspended. Lack of mixing can cause low results in samples of solids that settle out rapidly, such as those in activated sludge or raw wastewater. Samples must therefore be mixed thoroughly and poured quickly before any settling occurs. If this is not done, errors of 25 to 50 percent may easily occur. For example, on the same mixed liquor sample, one person may find 3,000 mg/*L* suspended solids while another person may determine that there are only 2,000 mg/*L* due to poor mixing. When a well-mixed, proportional composite sample is tested, a reasonably accurate measurement of the quality of the day's flow can be made.

If a 24-hour sampling program is not possible, perhaps due to insufficient personnel or the absence of a night shift, single representative samples should be taken at a time when typical characteristic qualities are present in the wastewater. The samples should be taken in accordance with the detention time required for treatment. For example, this period may exist between 10 am and 5 pm for the sampling of raw influent. If a sample is taken at 12 Noon, other samples should be taken in accordance with the detention periods of the serial processes of treatment in order to follow this slug of wastewater or *PLUG FLOW.*[9] In primary settling, if the detention time in the primaries is two hours, the primary effluent should be sampled at 2 pm. If the detention time in the succeeding secondary treatment process required three hours, this sample should be taken at 5 pm.

16.34 Sludge Sampling

In sampling raw sludge and feeding a digester, a few important points should be kept in mind as shown in the following illustrative table.

For raw sludge pumped from a primary clarifier, the sludge solids vary considerably with pumping time as shown by samples withdrawn every one-half minute in Table 16.3.

TABLE 16.3 DECREASE IN PERCENT TOTAL SOLIDS DURING PUMPING

Pumping Time In Minutes	Total Solids Percent	Cumulative Solids Average
0.5	7.0	7.0
1.0	7.1	7.1
1.5	7.4	7.2
2.0	7.3	7.2
2.5	6.7	7.1
3.0	5.3	6.8
3.5	4.0	6.4
4.0	2.3	5.9
4.5	2.0	5.5
5.0	1.5	5.1

1. Table 16.3 shows that the solids were heavy during the first 2.5 minutes, and thereafter rapidly became thinner and watery. Since sludge solids should be fed to a digester with solids as heavy as possible and a minimum of water, the pumping should probably have been stopped at about 3 minutes. After 3 minutes, the water content did become greater than desirable.
2. In sampling this sludge, the sample should be taken as a composite by mixing small equal portions taken every 0.5 minute during pumping. If only a single portion of sludge is taken for the sample, there is a chance that the sludge sample may be too thick or too thin, depending on the moment the sample is taken. A composite sample will prevent this possibility.
3. Remember that as a sludge sample stands, the solids and liquid separate due to gasification and flotation or settling of the solids, and that it is absolutely necessary to thoroughly remix the sample back into its original form as a mixture before pouring it for a test. Do not store sludge samples in sealed containers. Sludge digestion can occur even under refrigeration. The gases produced can cause sludge to spray out of the container when the container is opened.
4. When individual samples are taken at regular intervals in this manner, they should be carefully preserved to prevent sample deterioration by bacterial action. Refrigeration is an excellent method of preservation and is generally preferable to chemicals. Chemicals may interfere with tests such as biochemical oxygen demand (BOD) and chemical oxygen demand (COD).

16.35 Sampling Devices

Automatic sampling devices are wonderful timesavers and should be used where possible. However, as with anything automatic, problems do arise and the operator should be aware of potential difficulties. Sample lines to auto-samplers may build up growths, which may periodically slough off and contaminate the sample with a high solids content. Very regular cleanout of the intake line is required.

Manual sampling equipment includes dippers, weighted bottles, hand-operated pumps, and cross-section samplers. Dippers consist of wide-mouth, corrosion-resistant containers (such as cans or jars) on long handles that collect a sample for testing. A weighted bottle is a collection container that is lowered to a desired depth. At this location a cord or wire removes the bottle stopper so the bottle can be filled. Sampling pumps allow the inlet to the suction hose to be lowered to the sampling depth. Cross-sectional samplers are used to sample where the wastewater and sludge may be in layers, such as in a digester or clarifier. The sampler consists of a tube, open at both ends, that is lowered at the sampling location. When the tube is at the proper depth, the ends of the tube are closed and a sample is obtained from different layers.

Many operators build their own sampler (Figure 16.4) using the material described below:

1. *SAMPLING BUCKET.* A coffee can attached to an 8-foot (2.5-m) length of 1/2-inch (1.2-cm) electrical conduit or a wooden broom handle with a 1/4-inch (0.6-cm) diameter spring in a 4-inch (10-cm) loop. A section of thin-walled PVC pipe (Schedule 40) of appropriate diameter with a cap on one end will be a more durable sampling bucket than a coffee can.

Surface samples also can be collected using a plastic bucket or a gallon milk jug on a rope with an opening cut out opposite from the handle. Placing a large bolt and nut

[9] *Plug Flow.* A type of flow that occurs in tanks, basins, or reactors when a slug of water or wastewater moves through a tank without ever dispersing or mixing with the rest of the water or wastewater flowing through the tank.

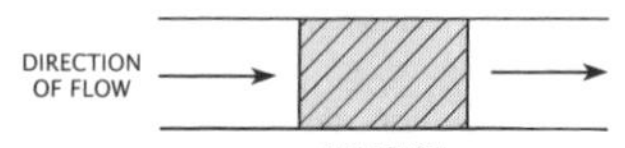

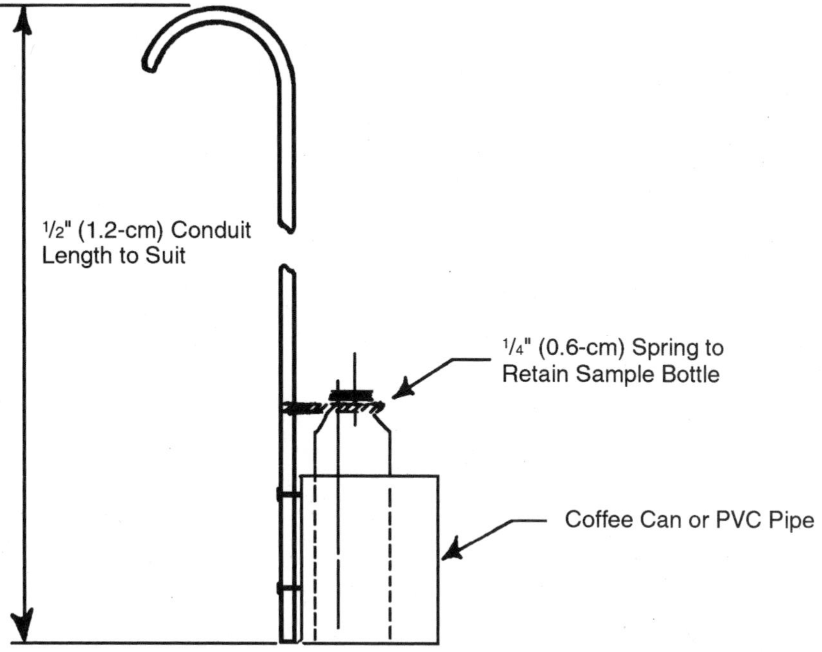

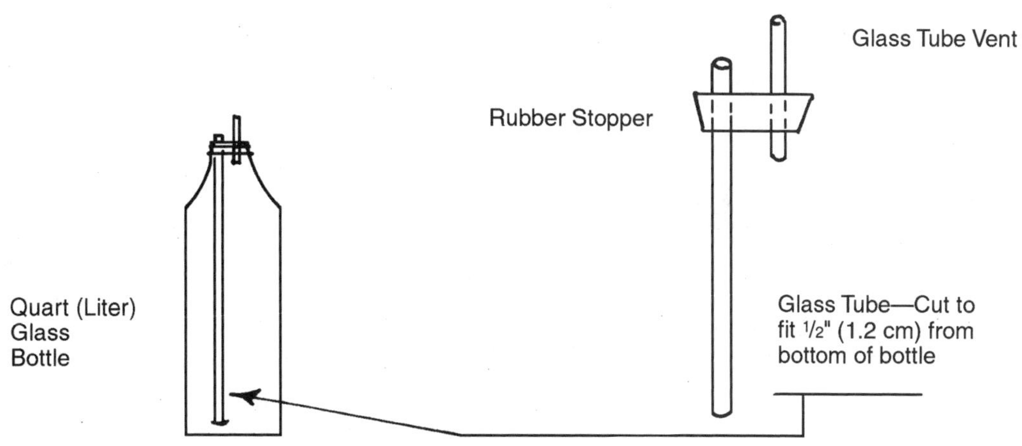

Fig. 16.4 Dissolved oxygen sampling bottle

near one edge of the sampler will cause the bucket to tip so that the sample will fill the bucket, rather than the bucket floating on the water surface.

2. *SAMPLING BOTTLE.* Glass bottle with rubber stopper equipped with two 3/8-inch (1-cm) glass tubes, one ending near the bottom of bottle to allow sample to enter and the other ending at the bottom of the stopper to allow the air in the bottle to escape while the sample is filling the bottle.

For sample containers, wide-mouth glass bottles are recommended. Glass bottles, though somewhat expensive initially, greatly reduce the problem of metal contamination. The wide-mouth bottles ease the washing problem. For regular samples, sets of glass bottles bearing identification labels should be used.

16.36 Preservation of Samples

Sample deterioration starts immediately after collection for most wastewaters. The shorter the time that elapses between collection and analysis, the more reliable will be the analytical results.

In many instances, however, laboratory analysis cannot be started immediately due to the remoteness of the laboratory or the operator's workload. A summary of acceptable EPA (US Environmental Protection Agency) methods of preservation appears in Table 16.4. Note that some constituents, such as temperature, pH, and DO, must be analyzed immediately on site because they cannot be preserved.

16.37 Quality Control in the Wastewater Laboratory

Having good equipment and using the correct methods are not enough to ensure correct analytical results. Each operator must be constantly alert to factors in the plant that can cause poor data quality. Such factors include: sloppy laboratory technique, deteriorated reagents, poorly operating instruments, and calculation mistakes.

16.38 Summary

1. Representative samples must be taken before any tests are made.
2. Select a good sampling location.
3. Collect samples and, if necessary, properly preserve them.
4. Mix samples thoroughly before compositing and at time of test.

TABLE 16.4 US EPA RECOMMENDED PRESERVATION METHODS FOR WATER AND WASTEWATER SAMPLES[a]

Test	Container[b]	Preservation Method	Maximum Recommended Holding Time
Acidity/Alkalinity	P,G	Store at 4°C	14 days
Ammonia	P,G	Add H_2SO_4 to pH <2 Store at 4°C	28 days
BOD	P,G	Store at 4°C	48 hours
COD	P,G	Add H_2SO_4 to pH <2 Store at 4°C	28 days
Chloride	P,G	None required	28 days
Chlorine, residual	P,G	Det. on site	No holding
Cyanide	P,G	Add NaOH to pH >12 Store at 4°C	14 days
Dissolved Oxygen	G	Det. on site	No holding
Fluoride	P	None required	28 days
Mercury	P,G	Add HNO_3 to pH <2	28 days
Metals	P,G	Add HNO_3 to pH <2	6 months
Nitrate	P,G	Store at 4°C	48 hours
Nitrate-nitrite	P,G	Add H_2SO_4 to pH <2 Store at 4°C	28 days
Nitrite	P,G	Store at 4°C	48 hours
Oil & Grease	G	Add HCl or H_2SO_4 to pH <2 Store at 4°C	28 days
Organic Carbon	P,G	Add HCl or H_2SO_4 to pH <2 Store at 4°C	28 days
pH	P,G	Det. on site	No holding
Phenols	G	H_2SO_4 to pH <2 Store at 4°C	28 days
Phosphorus, ortho	P,G	Filter on site Store at 4°C	48 hours
Phosphorus, total	P,G	Add H_2SO_4 to pH <2 Store at 4°C	28 days
Solids	P,G	Store at 4°C	7 days
Specific Conductance	P,G	Store at 4°C	28 days
Sulfate	P,G	Store at 4°C	28 days
Sulfide	P,G	Add zinc acetate & NaOH to pH >9 Store at 4°C	7 days
Temperature	P,G	Det. on site	No holding
T. Kjeldahl Nitrogen	P,G	H_2SO_4 to pH <2 Store at 4°C	28 days
Turbidity	P,G	Store at 4°C	48 hours

[a] *40 CFR 136,* July 1, 2006, "Required Containers, Preservation Techniques, and Holding Times." US Government Printing Office, Superintendent of Documents, PO Box 371954, Pittsburgh, PA 15250-7954.

[b] Polyethylene (P) or glass (G).

16.39 Additional Reading

1. *HANDBOOK FOR MONITORING INDUSTRIAL WASTEWATER,* US Environmental Protection Agency. EPA No. 625-6-73-002. Obtain from National Technical Information Service (NTIS), 5285 Port Royal Road, Springfield, VA 22161. Order No. PB-259146. Price, $59.50, plus $5.00 shipping and handling per order.

2. *HANDBOOK FOR ANALYTICAL QUALITY CONTROL IN WATER AND WASTEWATER LABORATORIES,* US Environmental Protection Agency. EPA No. 600-4-79-019. Obtain from National Technical Information Service (NTIS), 5285 Port Royal Road, Springfield, VA 22161. Order No. PB-297451. Price, $52.00, plus $5.00 shipping and handling per order.

QUESTIONS

Write your answers in a notebook and then compare your answers with those on page 609.

16.3A What are the largest sources of errors found in laboratory results?

16.3B Why must a representative sample be collected?

16.3C How would you prepare a proportional composite sample?

END OF LESSON 3 OF 9 LESSONS
ON
LABORATORY PROCEDURES AND CHEMISTRY

Please answer the discussion and review questions next.

DISCUSSION AND REVIEW QUESTIONS

Chapter 16. LABORATORY PROCEDURES AND CHEMISTRY

(Lesson 3 of 9 Lessons)

Write the answers to these questions in your notebook. The question numbering continues from Lesson 2.

7. What is meant by representative sample?
8. Where would you obtain a representative sample?
9. Under what conditions and why would you preserve a sample?

CHAPTER 16. LABORATORY PROCEDURES AND CHEMISTRY

(Lesson 4 of 9 Lessons)

16.4 LABORATORY PROCEDURES FOR PLANT CONTROL

NOTICE: Lesson 4 outlines procedures used by operators to analyze samples for plant process control. These tests are frequently performed by operators to quickly obtain the results and make any necessary process adjustments. Lessons 6, 7, 8, and 9 are laboratory procedures for NPDES monitoring. These test procedures may be performed by operators, but are also performed by laboratory analysts, especially in larger plants.

16.40 Clarity

A. *DISCUSSION*

All plant effluents should be examined daily to observe the clarity of the effluent. Clarity is an indication of the quality of the effluent with respect to color, solids, and turbidity. The purpose of this test is to determine whether the plant effluent is staying the same, improving, or deteriorating.

The clarity test can be performed in the lab by looking down through the effluent in a graduated cylinder, or it can be performed in the field by looking down through the effluent in a clarifier or chlorine contact basin. The effluent should be examined at the same time each day and under the same conditions; otherwise, the results will not be comparable from day to day. By noting how far down you can see into the cylinder or basin and comparing each day's test results with the previous day's results, you will be able to see whether the general quality of the effluent is changing.

Sometimes this test is referred to as a turbidity measurement; however, for the purposes described here, you are interested in the *clarity of the effluent* rather than the *quantity of solids* in the effluent (also see Section 16.5, Procedure 19, "Turbidity").

B. *WHAT IS TESTED?*

Sample	Common Range (Field Test) Poor	Common Range (Field Test) Good
Secondary Clarifiers:		
Trickling Filter	1 ft (0.3 m)	3 ft (0.9 m)
Activated Sludge	3 ft (0.9 m)	6 ft (1.8 m)
Activated Sludge Blanket in Secondary Clarifier	1 ft (0.3 m)	4 ft (1.2 m)
Chlorine Contact Basins	3 ft (0.9 m)	6 ft (1.8 m)

C. *APPARATUS*

1. One clarity unit (Secchi (SECK-key) Disk) and attached rope marked in one-foot units
2. One 1,000-m*L* graduated cylinder

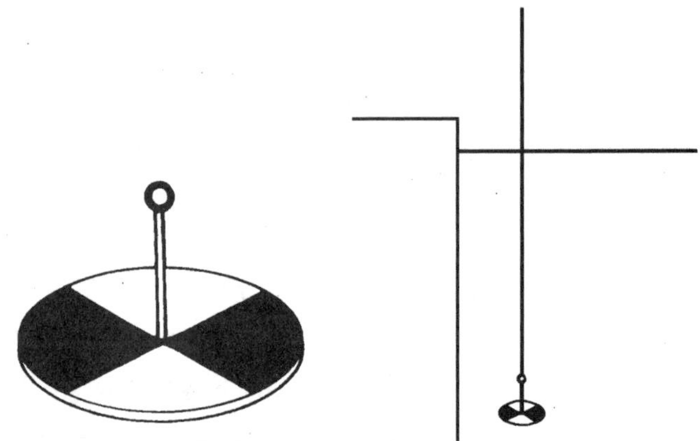

Secchi Disk

D. *REAGENTS*

None

E. *PROCEDURES*

1. *FIELD TEST.* Tie end of marked rope to handrail where tests will be run, for example, in final sedimentation unit. Always take tests at the same time each day for comparable results. Lower disk slowly until you just lose sight of it. Stop. Bring up slowly until just visible. Stop. Look at the marks on the rope to see the depth of water that you can see the disk through. Bring up disk, clean, and store. Record results.
2. *LAB TEST.* Use a clean 1,000-m*L* graduate. Fill with a well-mixed sample up to the 1,000-m*L* mark. During every test the same lighting conditions in the lab should be maintained. Look down through the liquid in the cylinder and read the last visible number etched on the side of the graduate and record results.

Whether you use one or both of these tests, you should run each test at the same time every day and under similar conditions for comparable results.

F. *TEST RESULTS*

1. Each foot of depth is better clarity with Secchi disk.
2. Each 100-m*L* mark seen in depth is better clarity.

QUESTIONS

Write your answers in a notebook and then compare your answers with those on page 609.

16.4A What does the clarity test indicate?

16.4B What happens when you attempt to measure clarity under different conditions, such as lighting and clarifier loadings?

16.41 Hydrogen Sulfide (H_2S)

Hydrogen sulfide is measured because it causes corrosion and odors that should be controlled. The major portion of the sulfide content in wastewater is produced in the conversion of sulfate (SO_4^{2-}) to sulfide (S^{2-}) by bacteria found in the wastewater. Oxygen-reducing bacteria will use any available sulfur-containing compound as food. This process can produce odorous, reduced-sulfur compounds, including hydrogen sulfide (H_2S).

H_2S production in wastewater can be controlled by up-sewer maintenance, which reduces H_2S formation and, in some cases, by the application of chemicals such as chlorine, oxygen, hydrogen peroxide, iron salts, or bases to increase the pH.

16.410 H_2S in the Atmosphere

A. *DISCUSSION*

The rate of corrosion in the sewer collection system and treatment plant is often directly related to the rate of H_2S production or the amount of H_2S in the atmosphere. In addition, hydrogen sulfide gas is toxic to your respiratory system and is both flammable and explosive under certain conditions. The explosive limits of hydrogen sulfide range from 4.3% (LEL or Lower Explosive Limit) to 46% (UEL or Upper Explosive Limit). The parts per million (ppm) concentration of hydrogen sulfide in the sewer atmosphere is quite different from that *IN* wastewater. A concentration of 1 mg/*L* (ppm) in turbulent wastewater can quickly produce a concentration of 300 ppm in an unventilated, enclosed atmospheric space. The minimum concentration of H_2S in the atmosphere known to cause death is 300 ppm.

B. *METHODS*

Two examples of the methods available for testing H_2S in the atmosphere are listed below. The first method, using paper tape or tiles with lead acetate, will give a rough estimate of H_2S. The second method is a faster and more accurate instrumental measurement of a hazardous level or of the actual concentration of sulfide present.

C. *WHAT IS TESTED?*

Sample	Common Range
Atmosphere in sewers, outlets from force mains, wet pits, pumping stations, and influent areas to treatment plant.	1. Lead Acetate Method Not black in 24 hours = Good, 24+ hr Black in less than 1 hour = Bad, <1 hour 2. H_2S Detector Method <3 ppm = good >10 ppm = toxic

D. *APPARATUS*

Method 1: Lead acetate paper or unglazed tile soaked in lead acetate solution

Method 2: H_2S Detectors or H_2S Detector Tubes

These devices are available in three types:

(1) Personal. Attach to your belt or use a shoulder strap. These have an alarm signal; however, some units also have a readout for the level of hydrogen sulfide. You should rely on alarm signals rather than readouts.

(2) Portable. Carry by hand and place near work site. These have an alarm signal and meter indicating level of H_2S.

(3) Stationary. Install permanently in lift stations, gas compressor rooms, and other potentially dangerous areas. These have an alarm signal and may have a meter indicating level of H_2S.

Contact your local safety equipment supplier for information on atmospheric monitors.

E. *REAGENTS*

Method 1: Lead Acetate Solution, saturated. Dissolve 50 grams lead acetate in 80 m*L* distilled water.

Method 2: No reagents are required.

F. *PROCEDURE*

Method 1:

1. Obtain pieces of unglazed tile or use lead acetate paper. Cut tile with hacksaw into ½-inch (1.2-cm) strips.
2. Soak strips of tile or paper in lead acetate solution.
3. Dry tile in drying oven or air dry.
4. An open manhole or any point where wastewater is exposed to the atmosphere is a good test site. Drive a nail between metal crown ring of manhole, concrete, or other convenient place. Tie paper or tile with cotton string to nail.

Replace manhole cover and return in half an hour or less. If tile is not black or substantially colored, return periodically until black. If H_2S is present as indicated by a color change, then measure flow, temperature, pH, and BOD for further evaluation of problem.

Method 2: The instructions are included with instrument.

G. *CALCULATIONS*

Method 1: None required.

Method 2: None required.

16.411 H_2S in Wastewater

A. *DISCUSSION*

In sewers, when there is no longer any dissolved oxygen, H_2S tests are run to determine the rate of H_2S increase as the wastewater travels to a pumping station or treatment plant. If the wastewater is exposed to the atmosphere H_2S will be released and a typical rotten egg odor will be detected. Anaerobic bacteria found in wastewater can liberate H_2S from the solids. When the gas leaves the wastewater stream, some of it dissolves in the condensed moisture on the concrete. Sulfur-oxidizing bacteria convert the hydrogen sulfide to sulfuric acid,

(H_2S)

which is very corrosive to concrete. Only the concrete exposed to the atmosphere is corroded and the concrete below the low waterline is not attacked.

Not all odors in wastewater are from H_2S, and there is no correlation between H_2S and other odors. The total H_2S procedure is good up to 18 mg/*L*, and higher concentrations must be diluted before testing. H_2S production can be controlled by up-sewer maintenance, which reduces H_2S formation in the wastewater and protects the collection system. In some severe cases chemicals are applied to flows in the collection system for H_2S control. Chemicals used include chlorine, oxygen, hydrogen peroxide, iron salts, or bases to increase the pH.

B. *WHAT IS TESTED?*

Sample Wastewater from the Following Locations	H_2S Concentrations Possible Results, mg/*L*	
	Good	**Bad**
Sewers	0.1	1
Outlets from force mains	0.1	1
Wet pits, pumping stations	0.1	0.5
Influents to treatment plants	0	0.5

All of the above locations should be sampled, if pertinent, when using upstream H_2S control in the collection system.

C. *APPARATUS*

1. LaMotte-Pomeroy Sulfide Testing Kit to test:
   a. Total Sulfide
   b. Dissolved Sulfide
   c. Hydrogen Sulfide in Solution

   Obtain from LaMotte Company. Order by Code #4630, $123.45, PO Box 329, Chestertown, MD 21620.

2. *OR* Hach Hydrogen Sulfide Test Kit

   Obtain from HACH Company. Order by Catalog No. 2238-01, $134.00, PO Box 389, Loveland, CO 80539-0389.

D. *REAGENTS*

The reagents are included in the kits.

E. *PROCEDURE*

The instructions are in the kits.

F. *EXAMPLE*

The instructions are in the kits.

G. *CALCULATIONS*

The instructions are in the kits.

H. *ADDITIONAL READING*

1. *STANDARD METHODS FOR THE EXAMINATION OF WATER AND WASTEWATER,* 21st Edition, 2005. Obtain from Water Environment Federation (WEF), Publications Order Department, 601 Wythe Street, Alexandria, VA 22314-1994. Order No. S82011. Price to members, $194.75; nonmembers, $259.75; price includes cost of shipping and handling.
2. *OPERATION AND MAINTENANCE OF WASTEWATER COLLECTION SYSTEMS,* Volumes I and II, prepared by the Office of Water Programs, California State University, Sacramento, 6000 J Street, Sacramento, CA 95819-6025. Price: $49.00 for each volume.

QUESTION

Write your answer in a notebook and then compare your answer with the one on page 609.

16.4C Why would you measure the H_2S concentration in:

1. The atmosphere?
2. Wastewater?

16.42 Settleable Solids

A. *DISCUSSION*

The settleable solids test measures the volume of settleable solids in one liter of sample that will settle to the bottom of an Imhoff cone during one hour. The test is an indication of the volume of solids removed by sedimentation in sedimentation tanks, clarifiers, or ponds. The results are read directly in milliliters from the Imhoff cone. The actual volume of solids pumped may disagree with the calculated volume from the settleable solids test due to compacting of the sludge in the clarifier, the sludge may settle differently in Imhoff cones than in the clarifier, and the clarifier may not capture all of the solids that should settle.

B. *WHAT IS TESTED?*

Sample	Common Ranges Found
Influent	6 m*L*/*L* weak wastewater 12 m*L*/*L* medium wastewater 20 m*L*/*L* strong wastewater
Primary Effluent	0.1 m*L*/*L* to 3 m*L*/*L* Over 3 m*L*/*L* poor
Secondary Effluent	Trace to 0.5 m*L*/*L* Over 0.5 m*L*/*L* poor

C. *APPARATUS*

1. Imhoff cones
2. Rack for holding Imhoff cones
3. Glass stirring rod, or wire
4. Time clock or watch

D. *PROCEDURE*

OUTLINE OF PROCEDURE FOR SETTLEABLE SOLIDS

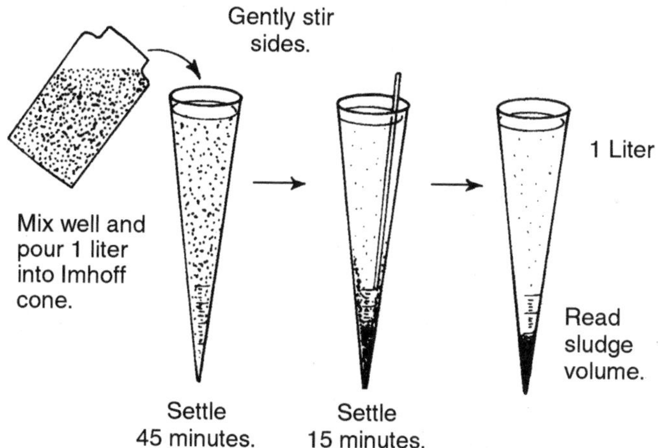
Gently stir
sides.
Mix well and
pour 1 liter
into Imhoff
cone.
1 Liter
Read
sludge
volume.
Settle
45 minutes.
Settle
15 minutes.

1. Thoroughly mix the wastewater sample by shaking and immediately fill an Imhoff cone to the liter mark.
2. Record the time of day that the cone was filled. T = ____.
3. Allow the wastewater sample to settle for 45 minutes.
4. Gently stir sides of the cone to facilitate settling of material adhering to the side of the cone.
5. After one hour, record the number of milliliters of settleable solids in the Imhoff cone. Make allowance for voids among the settled material. For example, if you read a sludge volume of 3.0 m*L* and voids or spaces in the sludge occupy approximately 0.2 m*L*, record a sludge volume of 2.8 m*L*.
6. Record the settleable solids as m*L*/*L* or milliliters per liter.

   Settleable Solids, Influent = ______________ m*L*/*L*

   Settleable Solids, Effluent = ______________ m*L*/*L*

   Settleable Solids, Removal = ______________ m*L*/*L*

E. *EXAMPLE*

Samples were collected from the influent and effluent of a primary clarifier. After one hour, the following results were recorded:

	Settleable Solids, m*L*/*L*
Influent	12.0
Effluent	0.2

F. *CALCULATIONS*

1. Calculate the efficiency or percent removal of the above primary clarifier in removing settleable solids.

$$\begin{aligned}\text{\% Removal of Set Sol} &= \frac{(\text{Infl Set Sol, m}L/L - \text{Effl Set Sol, m}L/L)}{\text{Influent Set Sol, m}L/L} \times 100\% \\ &= \frac{(12\text{ m}L/L - 0.2\text{ m}L/L)}{12\text{ m}L/L} \times 100\% \\ &= \frac{11.8}{12} \times 100\% \\ &= 98\%\end{aligned}$$

2. Estimate the gallons per day of sludge that should be pumped from the above primary clarifier if the flow is 1 MGD (1 million gallons per day). In your plant, the Imhoff cone may not measure or indicate the exact performance of your clarifier or sedimentation tank, but with some experience you should be able to relate or compare your lab tests with actual performance.

   Sludge Removed by Clarifier, m*L*/*L*

$$\begin{aligned}&= \text{Influent Set Sol, m}L/L - \text{Effluent Set Sol, m}L/L \\ &= 12\text{ m}L/L - 0.2\text{ m}L/L \\ &= 11.8\text{ m}L/L\end{aligned}$$

To estimate the GPD (gallons per day) of sludge pumped to a digester, use the following formula:

Sludge to Digester, GPD

$$\begin{aligned}&= \text{Total Set Sol Rem, m}L/L \times 1{,}000\text{ mg/m}L \times \text{Flow, MGD} \\ &= 11.8\ \frac{\text{m}L}{\text{M mg}} \times \frac{1{,}000\text{ mg}}{\text{m}L} \times \frac{1\text{ M gal}}{\text{day}} \\ &= 11{,}800\text{ GPD}\end{aligned}$$

This value may be reduced by 30 to 75 percent due to compaction of the sludge in the clarifier.

If you figure sludge removed as a percentage (1.18%), the sludge pumped to the digester would be calculated as follows:

$$\frac{1.18\%}{100\%} = \frac{\text{Sludge to Digester, GPD}}{\text{Flow of 1,000,000 GPD}}$$

$$\begin{aligned}\text{Sludge to Digester, GPD} &= \frac{(1.18\% \times 1{,}000{,}000\text{ GPD})}{100\%} \\ &= 11{,}800\text{ GPD}\end{aligned}$$

G. *CLINICAL CENTRIFUGE*

Settleable solids may also be measured by a small clinical centrifuge. This method, however, should be used for plant control only and not for NPDES monitoring. A mixed sample is placed in 15-m*L* graduated centrifuge tubes and spun for 15 minutes. The solid deposition in the tip of the tube is compared with a curve prepared by plotting settleable solids versus centrifuge solid deposition. This test provides a quick estimate of the settleable solids.

QUESTION

Write your answer in a notebook and then compare your answer with the one on page 609.

16.4D Estimate the volume of solids pumped to a digester in gallons per day (GPD) if the flow is 1 MGD, the influent settleable solids are 10 m*L*/*L*, and the effluent settleable solids are 0.4 m*L*/*L* for a primary clarifier.

16.43 Suspended Solids

A. *DISCUSSION*

One of the tests run on wastewater is used to determine the amount of material suspended within the sample. The result obtained from the suspended solids test does not mean that all of the suspended solids settle out in the primary clarifier or,

(Suspended Solids)

for that matter, in the final clarifier. Some of the particles are of such size and weight (*SPECIFIC GRAVITY*[10]) that they will not float or settle without additional treatment. Therefore, suspended solids are a combination of settleable solids and those solids that remain in suspension.

B. *WHAT IS TESTED?*

Sample	Common Ranges, mg/*L*
Influent	Weak 150 to 400+ Strong
Primary Effluent	Weak 60 to 150+ Strong
Secondary Effluent	Good 10 to 60+ Bad
Activated Sludge Tests:	Depending on Type of Process
Mixed Liquor	1,000 to < 5,000
Return or Waste Sludge	2,000 to < 12,000
Digester Tests:	
Supernatant	3,000 to < 10,000

When supernatant suspended solids are greater than 10,000 mg/*L*, the total solids test is usually performed.

C. *APPARATUS*

1. Glass-fiber filter disks without organic binder (Whatman grade 934H; Gelman type A/E; Millipore type AP40; or equivalent).
2. Filtration apparatus. One of the following, which must be suitable for the filter disk selected above:
   a. Membrane filter funnel
   b. Gooch crucible, 25-m*L* to 40-m*L* capacity with Gooch crucible adapter
   c. Filtration apparatus with reservoir and coarse 40- to 60-µm/fritted (high density porcelain) disk as filter support
3. Flask, suction.
4. Vacuum source.
5. Drying oven, 103° to 105°C.
6. Desiccator.
7. Analytical balance.

D. *PROCEDURE*

Preparation of Gooch Crucible

1. Insert a 2.2-cm glass-fiber filter into 25-m*L* Gooch crucible and center it.
2. Apply suction. Wash filter with 20 m*L* of distilled or deionized water three times to seat filter properly.
3. Dry at 103°C for one hour. If volatile suspended solids are to be determined, ignite crucible in muffle furnace for one hour at 550°C.
4. Cool in desiccator.
5. Weigh crucible and record weight. (This is known as "tare weight.")

Sample Analysis

6. Depending on the suspended solids content, measure out a 25-, 50-, or 100-m*L* portion of a well-mixed sample into a graduated cylinder. Use 25 m*L* if sample filters slowly. Use larger volumes of sample if sample filters easily, such as secondary effluent. Try to limit filtration time to about 15 minutes or less. Wet prepared Gooch crucible with distilled water and apply suction.
7. Filter sample through the Gooch crucible.
8. Wash out dissolved solids on the filter with about 20 m*L* of distilled water. (Use two 10-m*L* portions.)
9. Dry crucible at 103°C for at least one hour. Some samples may require up to three hours to dry if the residue is thick.
10. Cool in desiccator for 20 to 30 minutes.
11. Weigh crucible plus suspended solids.
12. Repeat drying cycle until constant weight is attained or until weight loss is less than 0.5 mg.
13. Record weight:

    Total Weight = ________________ gm

    Tare Weight = ________________ gm

    Solids Weight = ________________ gm

E. *PRECAUTIONS*

1. Check and regulate the oven temperature at 103° to 105°C.
2. Observe crucible and glass fiber for any possible leaks. A leak will cause solids to pass through and give low results. The glass-fiber filter may become unseated and leaky when the crucible is placed on the filter flask. The filter should be reseated by adding distilled water to the filter in the crucible and applying vacuum before filtering the sample.
3. Mix the sample thoroughly so that it is completely uniform in suspended solids when measured into a graduated cylinder before sample can settle out. This is especially true of samples heavy in suspended solids, such as raw wastewater and mixed liquor in activated sludge, which settle rapidly. The test can be no better than the mix.
4. A good practice is to prepare a number of extra Gooch crucibles for additional tests if the need arises. If a test result appears faulty or questionable, the test should be repeated. Check filtration rate and clarity of water passing through the filter.

F. *EXAMPLE AND CALCULATIONS*

This section is provided to show you the detailed calculations. After some practice, most operators use the lab work sheet as shown at the end of the calculations.

---

[10] *Specific Gravity.* (1) Weight of a particle, substance, or chemical solution in relation to the weight of an equal volume of water. Water has a specific gravity of 1.000 at 4°C (39°F). Particulates with specific gravity less than 1.0 float to the surface and particulates with specific gravity greater than 1.0 sink. (2) Weight of a particular gas in relation to the weight of an equal volume of air at the same temperature and pressure (air has a specific gravity of 1.0). Chlorine gas has a specific gravity of 2.5.

(Suspended Solids)

OUTLINE OF PROCEDURE FOR SUSPENDED SOLIDS

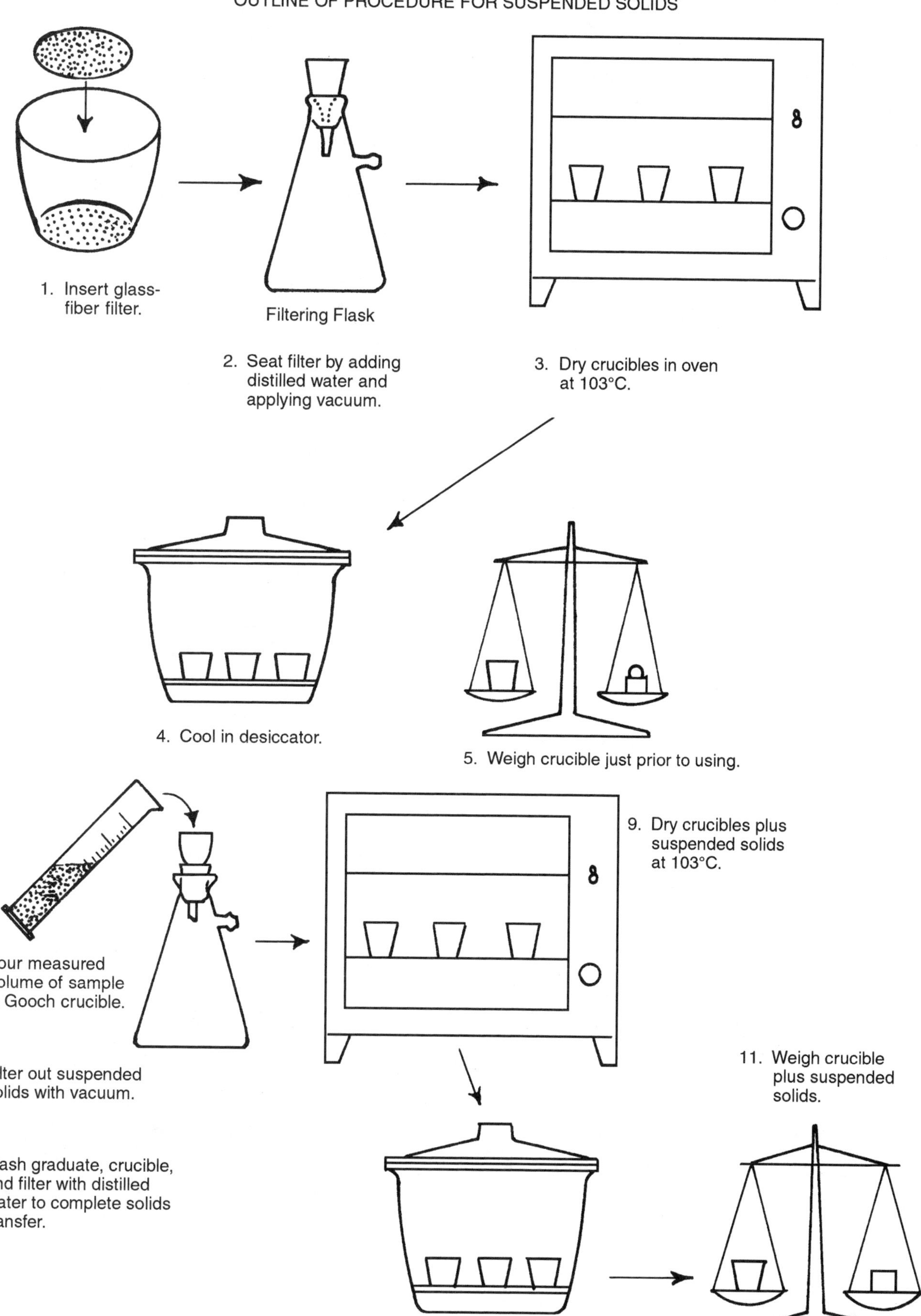
1. Insert glass-fiber filter.
Filtering Flask
2. Seat filter by adding distilled water and applying vacuum.
3. Dry crucibles in oven at 103°C.
4. Cool in desiccator.
5. Weigh crucible just prior to using.
6. Pour measured volume of sample in Gooch crucible.
7. Filter out suspended solids with vacuum.
8. Wash graduate, crucible, and filter with distilled water to complete solids transfer.
9. Dry crucibles plus suspended solids at 103°C.
10. Cool.
11. Weigh crucible plus suspended solids.

(Suspended Solids)

CALCULATIONS FOR SUSPENDED SOLIDS TEST
(Or use lab work sheet at end of calculations)

Always record the crucible number for the crucible used for the test. Different crucibles will have slightly different weights.

EXAMPLE: Assume the following data.

Volume of Sample = 50 m*L*

Crucible #015

	Recorded Weights
Crucible Weight	21.6329 gm
Crucible Plus Dry Solids	21.6531 gm
Crucible Plus Ash[11]	21.6360 gm

1. Compute total suspended solids.

21.6531 gm	Weight of Crucible Plus Dry Solids, grams
− 21.6329 gm	− Weight of Crucible, grams
= 0.0202 gm	= Weight of Dry Solids, grams

or

= 20.2 mg

or, since 1,000 milligrams (mg) = 1 gram (gm),

20.2 mg = 0.0202 gm

$$\text{Total Suspended Solids, mg}/L = \frac{\text{Weight of Solids, mg} \times 1{,}000\ \text{m}L/L}{\text{Sample Volume, m}L}$$

$$= \frac{20.2\ \text{mg} \times 1{,}000\ \text{m}L/L}{50\ \text{m}L}$$

$$= 404\ \text{mg}/L$$

2. Compute volatile or organic suspended solids.

21.6531 gm	Weight of Crucible Plus Dry Solids, gm
− 21.6360 gm	− Weight of Crucible Plus Ash, gm
= 0.0171 gm	= Weight of Volatile Solids, gm

or

= 17.1 mg

$$\text{Volatile Suspended Solids, mg}/L = \frac{\text{Weight of Volatile Solids, mg} \times 1{,}000\ \text{m}L/L}{\text{Sample Volume, m}L}$$

$$= \frac{17.1\ \text{mg} \times 1{,}000\ \text{m}L/L}{50\ \text{m}L}$$

$$= 342\ \text{mg}/L$$

3. Compute the percent volatile solids.

$$\text{Volatile Solids, \%} = \frac{\text{Weight Volatile, mg} \times 100\%}{\text{Weight Total Dry Solids, mg}}$$

$$= \frac{17.1\ \text{mg}}{20.2\ \text{mg}} \times 100\%$$

$$= 84.7\%$$

The previous calculations are also performed on a Laboratory Work Sheet (Figure 16.5) to illustrate the use of the work sheet.

4. Compute fixed or inorganic suspended solids.

21.6360 gm	Weight of Crucible Plus Ash, gm
− 21.6329 gm	− Weight of Crucible, gm
= 0.0031 gm	= Weight of Fixed Solids, gm

or

= 3.1 mg

$$\text{Fixed Suspended Solids, mg}/L = \frac{\text{Weight of Fixed Solids, mg} \times 1{,}000\ \text{m}L/L}{\text{Sample Volume, m}L}$$

$$= \frac{3.1\ \text{mg} \times 1{,}000\ \text{m}L/L}{50\ \text{m}L}$$

$$= 62\ \text{mg}/L$$

To check your work:

$$\text{Fixed Susp Solids, mg}/L = \text{Total Susp Solids, mg}/L - \text{Volatile Susp Solids, mg}/L$$

$$= 404\ \text{mg}/L - 342\ \text{mg}/L$$

$$= 62\ \text{mg}/L\ \text{(check)}$$

5. Compute the percent fixed solids.

$$\text{Fixed Solids, \%} = \frac{\text{Weight Fixed, mg}}{\text{Weight Total, mg}} \times 100\%$$

$$= \frac{3.1\ \text{mg}}{20.2\ \text{mg}} \times 100\%$$

$$= 15.3\%$$

CALCULATIONS FOR OVERALL PLANT REMOVAL OF SUSPENDED SOLIDS IN PERCENT

EXAMPLE: Assume the following data.

Influent Suspended Solids	202 mg/*L*
Primary Effluent Suspended Solids	110 mg/*L*
Secondary Effluent Suspended Solids	52 mg/*L*
Final Effluent Suspended Solids	12 mg/*L*

To calculate the percent removal or treatment efficiency for a particular process or the overall plant, use the following formula:

$$\text{Removal, \%} = \frac{(\text{In} - \text{Out})}{\text{In}} \times 100\%$$

[11] Obtained by placing the crucible plus dry solids in a muffle furnace at 550°C for one hour. The crucible plus remaining ash are cooled and weighed.

(Suspended Solids)

PLANT CLEAN WATER

DATE ____________

SUSPENDED SOLIDS AND DISSOLVED SOLIDS

SAMPLE	INFL.						
Crucible	#015						
Sample, mL	50						
Wt Dry + Dish, gm Wt Dish, gm	21.6531 21.6329						
Wt Dry, gm	0.0202						
$mg/L = \frac{\text{Wt Dry, gm} \times 1{,}000{,}000}{\text{Sample, m}L}$	404 mg/L						
Wt Dish + Dry, gm Wt Dish + Ash, gm	21.6531 21.6360						
Wt Volatile, gm	0.0171						
$\% = \frac{\text{Wt Vol}}{\text{Wt Dry}} \times 100\%$	84.7%						

BOD

Blank ____________

SAMPLE										
DO Sample										
Bottle #										
% Sample										
Blank or adjusted blank DO after incubation										
Depletion, 5 days										
Depletion, %										

SETTLEABLE SOLIDS

Sample, mL ____________ ____________

Direct mL/L ____________ ____________

COD

Sample ____________ ____________

Blank Titration ____________ ____________

Sample Titration ____________ ____________

Depletion ____________ ____________

$mg/L = \frac{\text{Dep} \times N \text{ FAS} \times 8{,}000}{\text{Sample, m}L}$ ____________ ____________

Fig. 16.5 Calculation of solids content on laboratory work sheet

(Suspended Solids)

Compute percentage removed between influent and primary effluent:

$$\text{Removal, \%} = \frac{(\text{In} - \text{Out})}{\text{In}} \times 100\%$$

$$= \frac{(202\ mg/L - 110\ mg/L)}{202\ mg/L} \times 100\%$$

$$= \frac{92}{202} \times 100\%$$

$$= 45.5\%$$

Compute percentage removed between influent and secondary effluent:

$$\text{Removal, \%} = \frac{(\text{In} - \text{Out})}{\text{In}} \times 100\%$$

$$= \frac{(202\ mg/L - 52\ mg/L)}{202\ mg/L} \times 100\%$$

$$= \frac{150}{202} \times 100\%$$

$$= 74.3\%$$

Compute percentage removed between influent and final effluent (overall plant percentage removed):

$$\text{Removal, \%} = \frac{(\text{In} - \text{Out})}{\text{In}} \times 100\%$$

$$= \frac{(202\ mg/L - 12\ mg/L)}{202\ mg/L} \times 100\%$$

$$= \frac{190}{202} \times 100\%$$

= 94.1% Suspended Solids Removal for the Plant

CALCULATIONS FOR POUNDS SUSPENDED SOLIDS REMOVED PER DAY

EXAMPLE: Assume the following data.

Influent Suspended Solids = 200 mg/*L*

Effluent Suspended Solids = 10 mg/*L*

Flow in million gallons/day = 2 MGD

1 gallon of water weighs = 8.34 lbs

Compute pounds suspended solids removed:

The general formula for computing pounds removed is:

Material Removed, lbs/day = (Concentration In, mg/*L* – Concentration Out, mg/*L*) × Flow, MGD × 8.34 lbs/gal

= (200 mg/*L* – 10 mg/*L*) × 2 MGD × 8.34 lbs/gal

= 190 × 2 × 8.34

= 3,169 lbs/day of Suspended Solids Removed by Plant

DERIVATION

This section is not essential to efficient plant operation, but is provided to furnish you with a better understanding of the calculation if you are interested. For practical purposes,

1 mg/*L* = 1 ppm or 1 part per million

or = 1 mg/million mg, because 1 liter = 1,000,000 mg

Therefore:

$$\frac{\text{lbs}}{\text{day}} = \frac{\text{mg}}{\text{M mg}} \times \frac{\text{M gal}}{\text{day}} \times \frac{\text{lbs}}{\text{gal}}$$

= lbs/day

QUESTIONS

Write your answers in a notebook and then compare your answers with those on pages 609, 610, and 611.

16.4E Why does some of the suspended material in wastewater fail to be removed by settling or flotation?

16.4F Given the following data:

Volume of Sample, m*L* = 100 m*L*

Crucible Weight, gm = 19.3241 gm

Crucible Plus Dry Solids, gm = 19.3902 gm

Crucible Plus Ash, gm = 19.3469 gm

Compute:

1. Total Suspended Solids, mg/*L*
2. Volatile Suspended Solids, mg/*L*
3. Volatile Solids, %
4. Fixed Suspended Solids, mg/*L*
5. Fixed Solids, %

16.4G Suspended solids data from a wastewater treatment plant are given below:

Influent Suspended Solids, mg/*L* = 221 mg/*L*

Primary Effluent SS, mg/*L* = 159 mg/*L*

Final Effluent SS, mg/*L* = 33 mg/*L*

Compute the percent removal of suspended solids by the:

1. Primary clarifier
2. Secondary process (removal between primary effluent and secondary effluent)
3. Overall plant

16.4H If the data in problem 16.4G are from a 1.5 MGD plant, calculate the pounds of suspended solids removed by the:

1. Primary unit
2. Secondary unit
3. Overall plant

16.44 Total Sludge Solids (Volatile and Fixed)

A. *DISCUSSION*

Total solids are the combined amounts of suspended and dissolved materials in the sample.

This test is used for wastewater sludges or where the solids can be expressed in percentages by weight. The weight can be measured on an inexpensive beam balance to the nearest 0.01 of a gram. The total solids are composed of two components, volatile and fixed solids. Volatile solids are composed of organic compounds that are of either plant or animal origin. Volatile solids in a treatment plant indicate the waste material that may be treated by biological processes. Fixed solids are inorganic compounds such as sand, gravel, minerals, or salts.

B. *WHAT IS TESTED?*

Sample	Common Range, % by Weight Total	Volatile	Fixed
Raw Sludge	6% to 9%	75%	25% ± 6%
Raw Sludge Plus Waste Activated Sludge	2% to 5%	80%	20% ± 5%
Recirculated Sludge	1.5% to 3%	75%	25% ± 5%
(Total Sludge Solids (Volatile and Fixed))			
Supernatant:			
Good Quality, has Suspended Solids	<1%	50%	50% ± 10%
Poor Quality	>1%	50%	50% ± 10%
Digested Sludge to Air Dry	3% too thin to <8%	50%	50% ± 10%

C. *APPARATUS*

1. Evaporating dish
2. Analytical balance
3. Drying oven, 103° to 105°C
4. Measuring device—graduated cylinder
5. Muffle furnace, 550°C

D. *PROCEDURE*

OUTLINE OF PROCEDURE FOR TOTAL SOLIDS

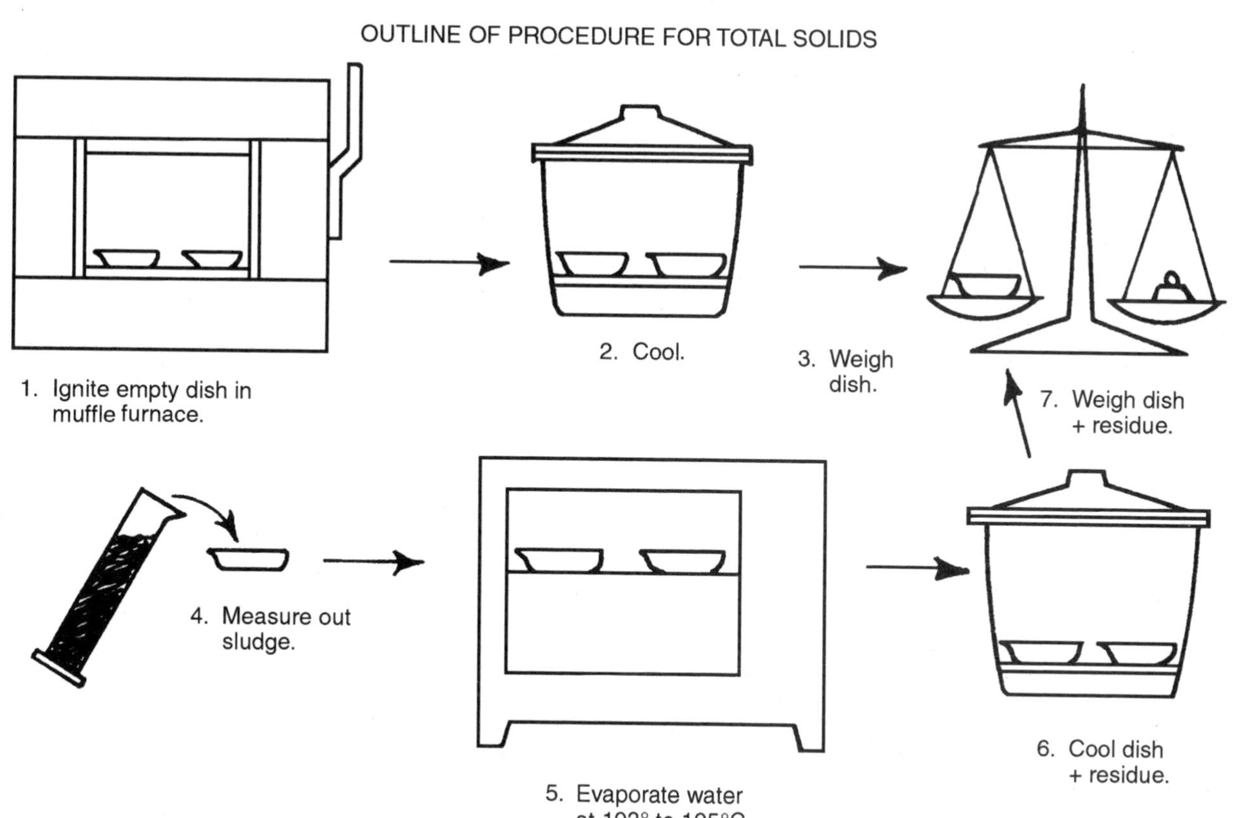
1. Ignite empty dish in muffle furnace.
2. Cool.
3. Weigh dish.
4. Measure out sludge.
5. Evaporate water at 103° to 105°C.
6. Cool dish + residue.
7. Weigh dish + residue.

1. Dry the dish by ignition in a muffle furnace at 550°C for one hour. Cool dish in desiccator.
2. Tare the evaporating dish to the nearest 10 milligrams, or 0.01 gm on a single pan balance. Record the weight as Tare Weight = _______ gm.
3. Weigh dish plus 50 to 100 m*L* of well-mixed sludge sample. Record total weight to nearest 0.01 gram as Gross Weight = __________ gm.
4. Evaporate the sludge sample to dryness in the 103°C drying oven.
5. Weigh the dried residue in the evaporating dish to the nearest 10 milligrams, or 0.01 gm. Record the weight as

   Dry Sample and Dish = __________ gm.
6. Compute the net weight of the residue by subtracting the tare weight of the dish from the dry sample and dish.

(Total Sludge Solids (Volatile and Fixed))

E. *PRECAUTIONS*

1. Be sure that the sample is thoroughly mixed and is representative of the sludge being pumped. Generally, where sludge pumping is intermittent, sludge is much heavier at the beginning and is less dense toward the end of pumping. Take several equal portions of sludge at regular intervals and mix for a good sample.

2. Take a large enough sample. Measuring a 50- or 100-m*L* sample, which is closely equal to 50 or 100 grams, is recommended. Since this material is so heterogeneous (nonuniform), it is difficult to obtain a good representative sample with less volume. Smaller volumes will show greater variations in answers, due to the uneven and lumpy nature of the material.

3. Control oven temperature closely at 103 to 105°C. Some solids are lost at any drying temperature. Close control of oven temperature is necessary because higher temperatures increase the losses of volatile solids in addition to the evaporated water.

4. Heat dish long enough to ensure evaporation of water, usually about 3 to 4 hours. If heat drying and weighing are repeated, stop when the weight change becomes less than 4 percent of the previous weight, or 0.5 mg, whichever is less. The oxidation, dehydration, and degradation of the volatile fraction will not completely stabilize until it is carbonized or becomes ash.

5. Since sludge is so non-uniform, weighing on the analytical balance should probably be made only to the nearest 0.01 gram or 10 milligrams.

F. *PROCEDURE*
(continue from total solids procedure)

OUTLINE OF PROCEDURE FOR VOLATILE SOLIDS

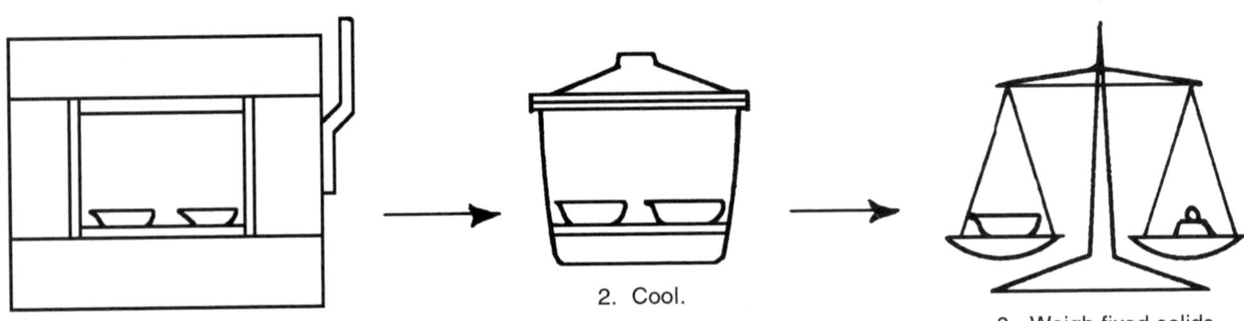
1. Ignite dried solids at 550°C.
2. Cool.
3. Weigh fixed solids.

1. Determine the total solids as previously described in Section D.

2. Ignite the dish and residue from total solids test at 550°C for one hour or until a white ash remains.

3. Cool in desiccator for about 30 minutes.

4. Weigh and record weight of Dish Plus Ash = _______ gm.

G. *EXAMPLE*

Dish No. 7

Weight of Dish (Tare)	= 20.31 gm
Weight of Dish Plus Wet Solids (Gross)	= 70.31 gm
Weight of Dish Plus Dry Solids	= 22.81 gm
Weight of Dish Plus Ash	= 20.93 gm

H. *CALCULATIONS*

See Laboratory Work Sheet (Figure 16.6) or calculations shown below.

1. Find weight of sample.

Weight of Dish Plus Wet Solids (Gross)	= 70.31 gm
Weight of Dish (Tare)	= 20.31 gm
Weight of Sample	= 50.00 gm

2. Find weight of total solids.

Weight of Dish Plus Dry Solids	= 22.81 gm
Weight of Dish (Tare)	= 20.31 gm
Weight of Total Solids	2.50 gm

3. Find percent sludge.

$$\text{Solids, \%} = \frac{(\text{Weight of Solids, gm})100\%}{\text{Weight of Sample, gm}}$$

$$= \frac{(2.50\text{ gm})100\%}{50.00\text{ gm}}$$

$$= 5\%$$

4. Find weight of volatile solids.

Weight of Dish Plus Dry Solids	= 22.81 gm
Weight of Dish Plus Ash	= 20.93 gm
Weight of Volatile Solids	= 1.88 gm

(Total Sludge Solids (Volatile and Fixed))

TOTAL SOLIDS

SAMPLE	RAW				
Dish No.	7				
Wt Dish + Wet, gm Wt Dish, gm	70.31 20.31				
Wt Wet, gm	50.00				
Wt Dish + Dry, gm Wt Dish, gm	22.81 20.31				
Wt Dry, gm	2.50				
% Solids = $\frac{\text{Wt Dry}}{\text{Wt Wet}} \times 100\%$	5.0%				
Wt Dish + Dry, gm Wt Dish + Ash, gm	22.81 20.93				
Wt Volatile, gm	1.88				
% Volatile = $\frac{\text{Wt Vol} \times 100\%}{\text{Wt Dry}}$	75%				
pH					
Volatile Acid, mg/*L*					
Alkalinity as $CaCO_3$, mg/*L*					

Grease

Sample ____________________

Sample, m*L* ____________________

Wt Flask + Grease, mg ____________________

Wt Flask, mg ____________________

Wt Grease, mg ____________________

$$\text{mg}/L = \frac{\text{Wt Grease, mg} \times 1{,}000}{\text{Sample, m}L}$$ ____________________

Fig. 16.6 Calculation of total solids on laboratory work sheet

(Total Sludge Solids (Volatile and Fixed))

5. Find percent volatile solids.

$$\text{Volatile Solids, \%} = \frac{\text{(Weight of Volatile Solids, gm)}100\%}{\text{Weight of Total Solids, gm}}$$

$$= \frac{(1.88 \text{ gm})100\%}{2.50 \text{ gm}}$$

$$= 75\%$$

QUESTIONS

Write your answers in a notebook and then compare your answers with those on page 611.

16.4I What is the origin of the volatile solids found in a digester?

16.4J What is the significance of volatile solids in a treatment plant?

END OF LESSON 4 OF 9 LESSONS
ON
LABORATORY PROCEDURES AND CHEMISTRY

Please answer the discussion and review questions next.

DISCUSSION AND REVIEW QUESTIONS

Chapter 16. LABORATORY PROCEDURES AND CHEMISTRY

(Lesson 4 of 9 Lessons)

Write the answers to these questions in your notebook. The question numbering continues from Lesson 3.

10. Why must the clarity test always be run under the same conditions?

11. Hydrogen sulfide is measured because it causes ______ __________________________________.

12. What causes H_2S formation in sewers?

13. Why does the actual volume of solids pumped from a clarifier not agree exactly with calculations based on the settleable solids test?

14. Calculate the efficiency or percent removal by a primary clarifier when the influent settleable solids are 10 m*L*/*L* and the effluent settleable solids are 0.3 m*L*/*L*.

15. Given the following data:

Volume of Sample	= 100 m*L*
Crucible Weight	= 19.9850 gm
Crucible Plus Dry Solids	= 20.0503 gm
Crucible Plus Ash	= 20.0068 gm

Compute:

1. Total Suspended Solids in mg and mg/*L*
2. Volatile Suspended Solids in mg and mg/*L*
3. Volatile Solids %

16. Estimate the pounds of solids removed per day by a primary clarifier if the influent suspended solids are 220 mg/*L* and the effluent suspended solids are 120 mg/*L* when the flow is 1.5 MGD.

17. Why are solids only weighed to the nearest 0.01 gram when determining the total volatile solids content of digesters?

CHAPTER 16. LABORATORY PROCEDURES AND CHEMISTRY

(Lesson 5 of 9 Lessons)

16.45 Tests for Activated Sludge Control

16.450 Settleability

A. *DISCUSSION*

This test is run on mixed liquor or return sludge and the results are plotted on a graph (Figure 16.7). All pertinent information is filled in for process control of aerators.

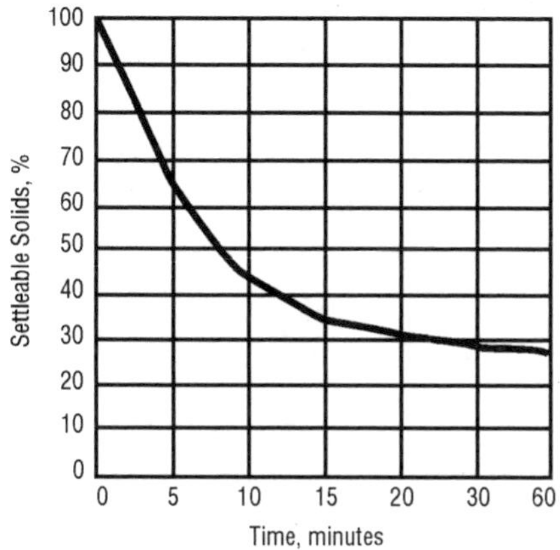

Fig. 16.7 Settleability of activated sludge solids

Settleability is important in determining the ability of the solids to separate from the liquid in the final clarifier. The activated sludge solids should be returned to the aeration tank, and the quality of the effluent is dependent on the absence of solids flowing over the effluent weir.

The suspended solids test should be run on the same sample of mixed liquor that is used for the settleability test. This will allow you to calculate the Sludge Volume Index (SVI) or the Sludge Density Index (SDI), which are explained in the following sections.

The 2,000 m*L* graduate that is filled with mixed liquor in the settleability test is supposed to indicate what will happen to the mixed liquor in the final clarifier—the rate of sludge settling, turbidity, color, and volume of sludge at the end of 60 minutes.

B. *WHAT IS TESTED?*

Sample	Working Range
Mixed Liquor or Return Sludge	Depends on desirable mixed liquor concentration

C. *APPARATUS*

2,000 m*L* graduated cylinder[12]

D. *REAGENTS*

None

E. *PROCEDURE (see illustration on page 518)*

1. Collect a sample of mixed liquor or return sludge.
2. Carefully mix sample and pour into 2,000-m*L* graduate. Vigorous shaking or mixing tends to break up floc and produces slower settling or poorer separation.
3. Record settleable solids, %, at regular intervals.

NOTE: 1. If a 1,000-m*L* graduate is used, the percent settleable solids is easier to record.

2. Some plants are using a video camera in the lab to observe and record settleability.

F. *EXAMPLE AND CALCULATION*

The percent settling rate can be compared for the various days of the week and with other measurements—suspended solids, SVI, percent sludge solids returned, aeration rate, and plant inflow. A very slow-settling mixed liquor usually requires air and solids adjustment to encourage increased stabilization during aeration. Both very rapidly settling and very slowly settling mixed liquors can give poor effluent clarification.

16.451 Sludge Volume Index (SVI)

A. *DISCUSSION*

The Sludge Volume Index (SVI) is used to indicate the condition of sludge (aeration solids or suspended solids) for settleability in a secondary or final clarifier. The SVI is the volume in m*L* occupied by one gram of mixed liquor suspended solids after 30 minutes of settling. It is a useful test to indicate changes in sludge characteristics. The proper SVI range for your plant is determined at the time your final effluent is in the best condition regarding solids and BOD removals and clarity.

B. *WHAT IS TESTED?*

Sample	Preferable Range, SVI
Aeration Tank Suspended Solids	50 m*L*/gm to 150 m*L*/gm

C. *APPARATUS*

See Section 16.45, "Tests for Activated Sludge Control," 16.450, "Settleability," and Section 16.43, "Suspended Solids."

[12] Mallory Direct Reading Settlometer (a 2-liter graduated cylinder approximately 5 inches (12.5 cm) in diameter and 7 inches (17.5 cm) high). Obtain from Wilmad/LabGlass, Attn: Customer Service, 1002 Harding Highway, Buena, NJ 08310-0688. Catalog No. LG-5601-100. Price, $167.84 each.

(Settleability–SVI, SDI)

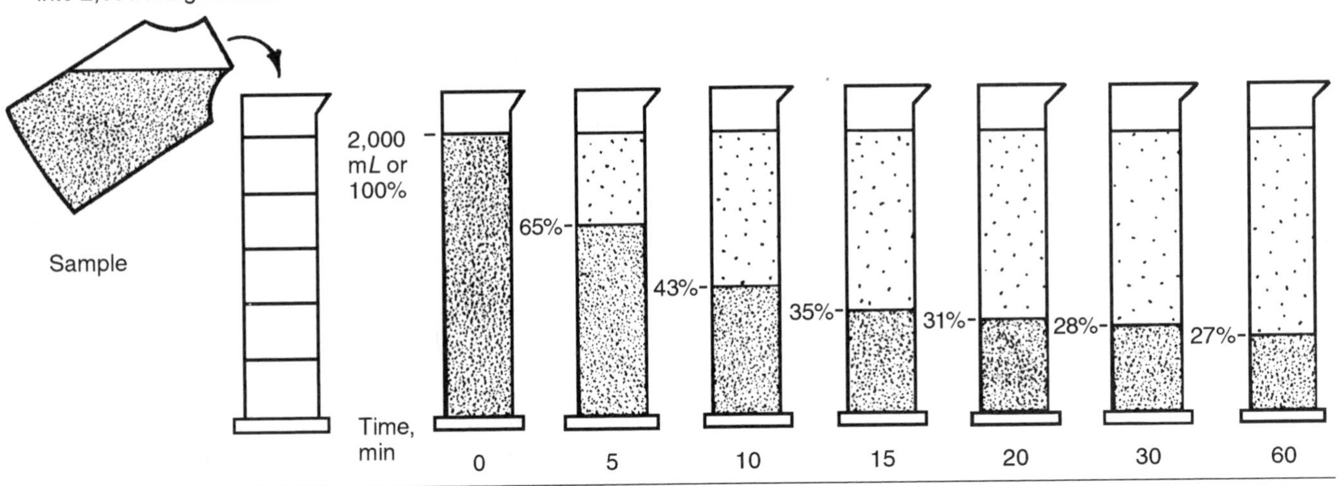

OUTLINE OF PROCEDURE FOR SETTLEABLE SOLIDS
1. Mix sample and pour into 2,000-mL graduate.
2. Record settleable solids, %, at regular intervals.
Sample
2,000 mL or 100%
65%
43%
35%
31%
28%
27%
Time, min
0
5
10
15
20
30
60

D. *REAGENTS*

None

E. *PROCEDURE*

See Sections on "Settleability" and "Suspended Solids."

F. *EXAMPLE*

30-minute Settleable Solids Test = 360 mL in 2,000 mL graduate or 18%

Mixed Liquor Suspended Solids = 1,500 mg/L

G. *CALCULATIONS*

$$\text{Sludge Volume Index, SVI, mL/gm} = \frac{\text{\% Settleable Solids} \times 1{,}000 \text{ mg/gm} \times 1{,}000 \text{ mL/L}}{\text{Mixed Liquor Suspended Solids, mg/L} \times 100\%}$$

$$= \frac{\text{\% Settleable Solids} \times 10{,}000}{\text{Mixed Liquor Suspended Solids, mg/L}}$$

$$= \frac{18 \times 10{,}000}{1{,}500}$$

$$= \frac{1{,}800}{15}$$

$$= 120 \text{ mL/gm}$$

16.452 Sludge Density Index (SDI)

A. *DISCUSSION*

The Sludge Density Index (SDI) is used in a way similar to the SVI to indicate the settleability of a sludge in a secondary clarifier or effluent. The SDI is the weight in mg of one milliliter of mixed liquor suspended solids after 30 minutes of settling. The calculation of the SDI requires the same information as the SVI test.

$$\text{SDI} = \frac{\text{mg/L of Suspended Solids in Mixed Liquor}}{\text{mL/L of Settled Mixed Liquor Solids} \times 10}$$

or

$$\text{SDI} = 100/\text{SVI}$$

B. *WHAT IS TESTED?*

Sample	**Preferable Range, SDI**
Aeration Tank Suspended Solids	0.6 gm/100 mL to 2.0 gm/100 mL

C. *C THROUGH G OF SECTION 16.451.*

These items are not included because of their similarity to the SVI test.

QUESTIONS

Write your answers in a notebook and then compare your answers with those on page 611.

16.4K Why should you run settleability tests on mixed liquor?

16.4L What is the Sludge Volume Index (SVI)?

16.4M Why is the SVI test run?

16.4N What is the relationship between the Sludge Density Index (SDI) and SVI?

(Sludge Age)

16.453 Sludge Age

A. *DISCUSSION*

"Sludge age" is a control guide that is widely used and is a rough indicator of the length of time a pound of solids is maintained under aeration in the system. The basis for calculating the sludge age is weight of suspended solids in the mixed liquor in the aeration tank divided by weight of suspended solids added per day to the aerator.

$$\text{Sludge Age, days} = \frac{\text{Suspended Solids in Mixed Liquor, mg}/L \times \text{Aerator Volume in MG} \times 8.34\ \text{lbs/gal}}{\text{SS in Primary Effluent, mg}/L^{*} \times \text{Daily Flow, MGD} \times 8.34\ \text{lbs/gal}}$$

* *NOTE:* Sludge age is calculated by three different methods:

1. Suspended solids in primary effluent, mg/*L*
2. Suspended solids removed from primary effluent, mg/*L*, or primary effluent suspended solids, mg/*L* – final effluent suspended solids, mg/*L*
3. BOD or COD in primary effluent, mg/*L*

Any significant additional loading placed on the aerator by the digester supernatant liquor must be added to the above loadings by considering the additional flow (MGD) and concentration (mg/*L*) if the supernatant is returned following the primary clarifiers. If supernatant is returned ahead of the primary clarifiers, only additional flow must be considered because the solids concentration will be represented in the primary effluent analysis. The selection of the method of determining sludge age is discussed in Chapters 8 and 11, "Activated Sludge." Since this method is based on suspended solids, it should not be used if the soluble organic (BOD) portion of the wastewater is more than 50 percent of the total organic (BOD) component.

B. *WHAT IS TESTED?*

Sample	Common Range
Suspended Solids in Aerator and BOD or Suspended Solids in Primary Effluent	mg/*L*, Depends on process
Sludge Age	Conventional process, 2.5 to 6 days

C. *APPARATUS*

See Section 16.43, "Suspended Solids."

D. *REAGENTS*

None

E. *PROCEDURE*

See Section 16.43, "Suspended Solids."

F. *EXAMPLE*

Suspended Solids in Mixed Liquor	= 1,500 mg/*L*
Aeration Tank Volume	= 0.50 MG
Suspended Solids in Primary Effl	= 100 mg/*L*
Daily Flow	= 2.0 MGD

G. *CALCULATIONS*

$$\text{Sludge Age, days} = \frac{\text{Susp Solids in Mixed Liquor, mg}/L \times \text{Aerator Vol, MG} \times 8.34\ \text{lbs/gal}}{\text{Susp Solids in Primary Effl, mg}/L \times \text{Flow, MGD} \times 8.34\ \text{lbs/gal}}$$

$$= \frac{\text{Mixed Liquor Susp Solids, lbs}}{\text{Primary Effluent SS, lbs/day}}$$

$$= \frac{1{,}500\ \text{mg}/L \times 0.50\ \text{MG} \times 8.34\ \text{lbs/gal}}{100\ \text{mg}/L \times 2.0\ \text{MGD} \times 8.34\ \text{lbs/gal}}$$

$$= \frac{1{,}500 \times 0.50}{100 \times 2.0}$$

$$= \frac{7.5}{2.0}$$

$$= 3.75\ \text{days}$$

QUESTION

Write your answer in a notebook and then compare your answer with the one on page 611.

16.4O Determine the sludge age in an activated sludge process if the volume of the aeration tank is 200,000 gallons and the suspended solids in the mixed liquor equals 2,000 mg/*L*. The primary effluent SS are 115 mg/*L*, and the average daily flow is 1.8 MGD.

16.454 Dissolved Oxygen in Aerator

A. *DISCUSSION*

This modification is used for biological flocs that have high oxygen utilization rates in the activated sludge process, and when a DO probe is not available. It is very important that some oxygen be present in aeration tanks at all times to maintain aerobic conditions.

This test is similar to the regular DO test (Section 16.5, Procedure 7) except that copper sulfate is added to kill oxygen-consuming organisms, and sulfamic acid is added to combat nitrite before the regular DO test is run.

(Dissolved Oxygen)

NOTE: If the results indicate a DO of less than 1 mg/*L*, it is possible that the DO in the aeration tank is zero. When the DO in the aeration tank is near zero, considerable DO from the surrounding atmosphere can mix with the sample when it is collected, when the inhibitor is added, while the solids are settling, and when the sample is transferred to a BOD bottle for the DO test. If you use this test, use a deep container and avoid stirring.

B. *WHAT IS TESTED?*

Sample	Common DO Range, mg/*L*
Aerator Mixed Liquor	0.1 to 3.0

C. *APPARATUS*

1. One tall bottle, approximately 1,000 m*L*
2. Regular DO apparatus

D. *REAGENTS*

1. Copper sulfate-sulfamic acid inhibitor solution. Dissolve 32 gm technical grade sulfamic acid (NH_2SO_2OH) without heat in 475 m*L* distilled water. Dissolve 50 gm copper sulfate, $CuSO_4 \cdot 5\ H_2O$, in 500 m*L* water. Mix the two solutions together and add 25 m*L* concentrated acetic acid.
2. Regular DO reagents (See Section 16.5, Procedure 7).

E. *PROCEDURE*

OUTLINE OF PROCEDURE FOR DISSOLVED OXYGEN

1. Add 10 m*L* of inhibitor to DO sampling bottle (Figure 16.4, page 501).
2. Dip into mixed liquor and let sampling bottle fill. Stopper bottle.
3. Remove glass tube and stopper and allow floc to settle.
4. Siphon over 300 m*L* of sample into BOD bottle.
5. Test for DO.

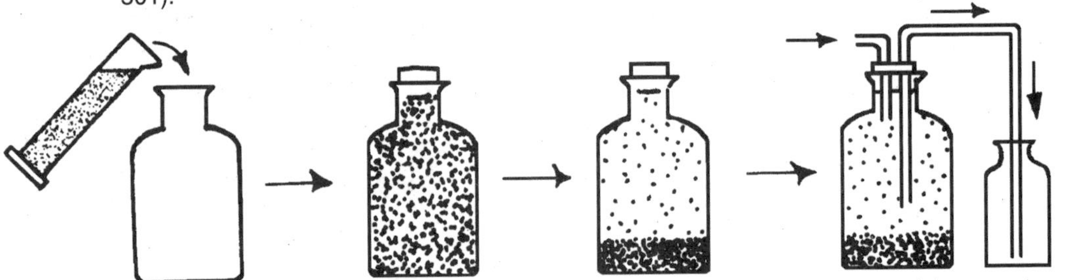

1. Add at least 10 m*L* of inhibitor (5 m*L* copper sulfate and 5 m*L* sulfamic acid) to any tall bottle (1-quart or 1-liter bottle) with an approximate volume of 1,000 m*L*. Place filling tube near the bottom. An emptying tube is placed approximately 1/4 inch (6 mm) from the top of the bottle cork. Attach bottle to rod or aluminum conduit and lower into aeration tank. See Figure 16.4 on page 501.
2. Submerge bottle 1.5 to 2.0 feet (0.45 to 0.60 m) below the surface of the aerator and allow bottle to fill with mixed liquor. Remove bottle from aeration tank.
3. Remove glass tube and stopper from bottle. Insert lid in bottle. Allow bottle to stand until solids (floc) settle and leave a clear supernatant liquor. Siphon the supernatant liquor into a 300-m*L* BOD bottle. Keep the outlet of the siphon below the water surface. Do not aerate in transfer.
4. Perform regular DO test as outlined on page 570.

F. and G. *EXAMPLE AND CALCULATIONS*

Same as regular DO test (see Procedure 7, page 570).

QUESTION

Write your answer in a notebook and then compare your answer with the one on page 611.

16.4P What are the limitations of the copper sulfate-sulfamic acid procedure for measuring DO in the aeration tank when the DO in the tank is very low?

16.455 Suspended Solids in Aerator

Centrifuge Method

A. *DISCUSSION*

This procedure is frequently used in plants as a quick and easy method to estimate the suspended solids concentration of the mixed liquor in the aeration tank instead of the regular suspended solids test. Many operators control the solids in their aerator on the basis of centrifuge readings. Others prefer to control solids using Figure 16.8 to estimate the suspended solids. In either case, the operator should periodically compare

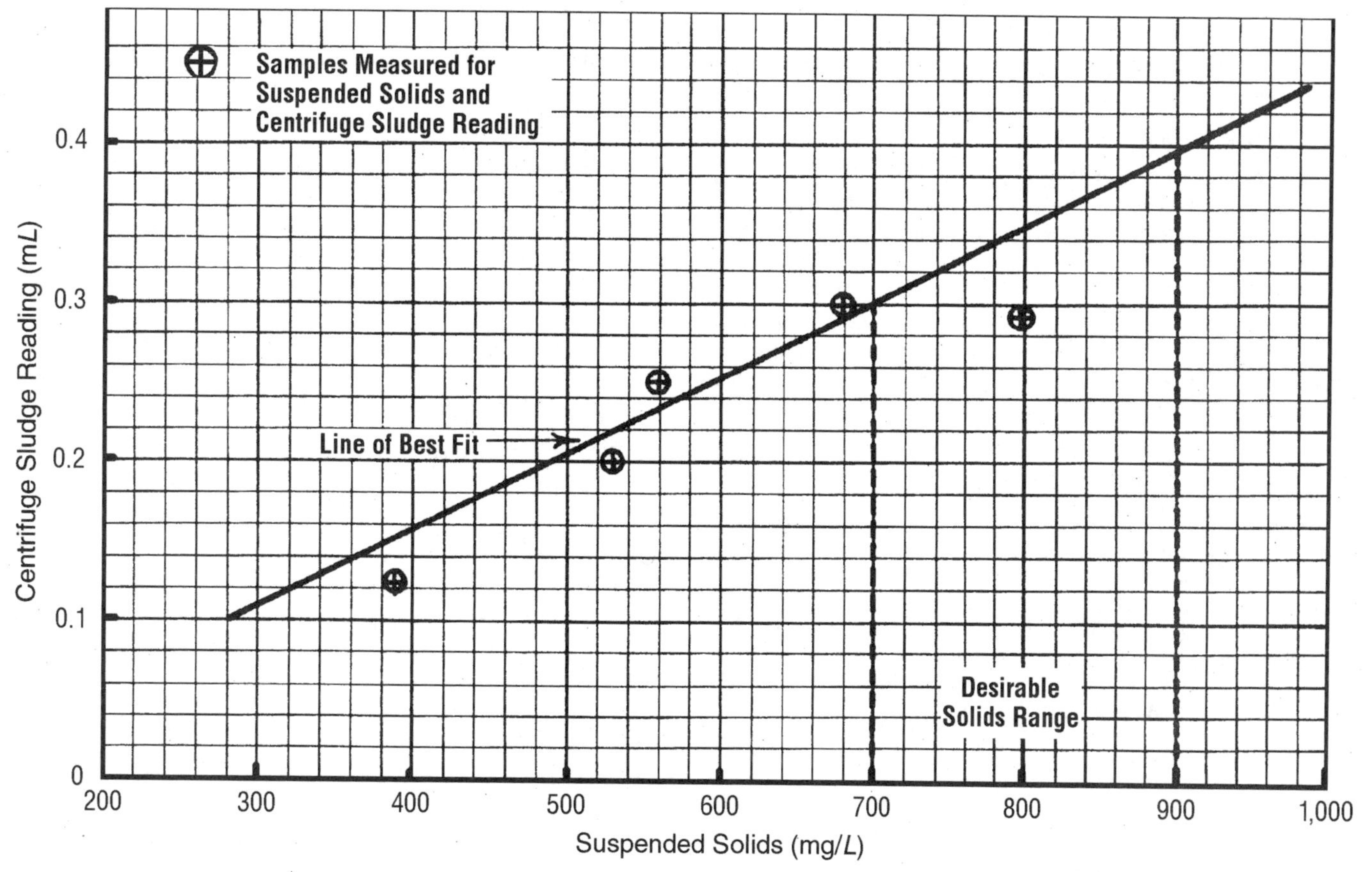

Fig. 16.8 Plant control by centrifuge solids in aeration tank, centrifuge speed, 1,750 RPM

centrifuge readings with values obtained from suspended solids tests. If the solids are in a good settling condition, a one percent centrifuge solids reading could have a suspended solids concentration of 1,000 mg/*L*. However, if the sludge is feathery, a one percent centrifuge solids reading could have a suspended solids concentration of 600 mg/*L*.

The centrifuge reading versus mg/*L* suspended solids chart (Figure 16.8) must be developed for each plant by comparing centrifuge readings with suspended solids determined by the regular Gooch crucible method. The points are plotted and a line of best fit is drawn as shown in Figure 16.8. This line must be periodically checked by comparing centrifuge readings with regular suspended solids tests because of the large number of variables influencing the relationship, such as characteristics of influent waste, mixing in aerator, and organisms in aerator. If you do not have a centrifuge or if your solids content is over 1,500 mg/*L*, determine suspended solids by the regular method.

B. *WHAT IS TESTED?*

Sample	**Common Range**
Suspended Solids in Aeration Tanks	800 to 5,000 mg/*L*

C. *APPARATUS*

1. Centrifuge
2. Graduated centrifuge tubes, 15 m*L*

D. *REAGENTS*

None

E. *PROCEDURE*

1. Collect sample.
2. Mix sample well and fill each centrifuge tube to the 15-m*L* line with sample.
3. Place filled sample tubes in centrifuge holders.
4. Crank centrifuge at fast speed as you count slowly to 60. Be sure to count and crank at the same speed for all tests. It is extremely important to perform each step exactly the same every time.
5. Remove one tube and read the amount (in tenths of a milliliter) of suspended solids concentrated in the bottom of the tube. Results in other tubes should be compared.
6. Refer to the conversion curve to determine suspended solids in mg/*L*.

NOTE: The reason for filling tubes to the 15-m*L* mark is that the curve (Figure 16.8) is computed for samples of this size. This curve was developed for a specific mechanically aerated activated sludge plant. Seven hundred to nine hundred mg/*L* MLSS is the best range for this plant. Each plant must develop its own curve based on actual data.

(Mean Cell Residence Time (MCRT))

F. *EXAMPLE*

Suspended solids concentration on bottom of centrifuge tube is 0.4 m*L*.

G. *CALCULATIONS*

From Figure 16.8, find 0.4 m*L* on the centrifuge reading side (vertical axis). Follow the horizontal line across the chart from 0.4 m*L* until it intersects the line of best fit. Drop a line downward to the horizontal axis and read the result—900 mg/*L* suspended solids.

If the suspended solids concentration is above or below the desired range, then you should make the proper changes in the pumping rate of the waste and return sludge. For details on controlling the solids concentration, refer to Chapters 8 and 11, "Activated Sludge."

H. *DEVELOPMENT OF FIGURE 16.8*

To develop Figure 16.8, take a sample from the aeration tank and measure suspended solids and also centrifuge a portion of the sample to obtain the centrifuge sludge reading in m*L* of sludge at the bottom of the tube. Obtain other samples of different solids concentrations to obtain the points on the graph. Draw a line of best fit through the points. Periodically the points should be checked because the influent characteristics and conditions in the aeration tank change.

I. *PRECAUTIONS*

This test works best for low mixed liquor suspended solids (MLSS) concentrations below 1,500 mg/*L*. Above 1,500 mg/*L* the centrifuge results might not allow an accurate estimate of the MLSS.

QUESTION

Write your answer in a notebook and then compare your answer with the one on page 611.

16.4Q What is the advantage of the centrifuge test for determining suspended solids in an aeration tank in comparison with other methods of measuring suspended solids?

16.456 Mean Cell Residence Time (MCRT)

A. *DISCUSSION*

The mean cell residence time (MCRT) is in a sense a much more precise "sludge age" calculation. MCRT describes the mean (average) residence time of an activated sludge particle in the activated sludge system and is a true measure of the age of the activated sludge. MCRT considers the solids removed from the process by wasting and the solids removed in the effluent. The MCRT calculation also considers the solids in the system. To use MCRT for the control of an activated sludge process, several measurements are required. Representative composite samples of mixed liquor suspended solids, effluent suspended solids, and waste sludge suspended solids are essential. Also measure influent and waste sludge flows.

B. *WHAT IS TESTED?*

Sample	Common Range, days
Mean Cell Residence Time (MCRT)	5 to 15 days

C. *APPARATUS AND REAGENTS*

See Section 16.43 for volatile suspended solids test procedures.

D. *PROCEDURE*

To calculate MCRT the following data are required:

1. Influent flow, MGD
2. Waste sludge flow, MGD
3. Volume of aeration basins in million gallons
4. Mixed liquor volatile suspended solids concentration in mg/*L*
5. Effluent volatile suspended solids in mg/*L*
6. Waste sludge volatile suspended solids in mg/*L*

E. *EXAMPLE*

Influent Flow, Q	= 3 MGD
Waste Sludge Flow, W	= 0.040 MGD
Volume of Aeration Basins, V	= 1.0 MG
Mixed Liquor Suspended Solids Concentration, X_1	= 1,600 mg/*L*
Effluent Suspended Solids, X_2	= 8 mg/*L*
Waste Sludge Suspended Solids, X_w	= 4,700 mg/*L*

F. *CALCULATIONS*

$$\text{Mean Cell Residence Time, days} = \frac{(V, \text{MG})(X_1, \text{mg}/L)}{(Q, \text{MGD})(X_2, \text{mg}/L) + (W, \text{MGD})(X_w, \text{mg}/L)}$$

$$= \frac{(1.0\text{ MG})(1{,}600\text{ mg}/L)}{(3\text{ MGD})(8\text{ mg}/L) + (0.040\text{ MGD})(4{,}700\text{ mg}/L)}$$

$$= \frac{1{,}600}{24 + 188}$$

$$= 7.5\text{ days}$$

NOTE: The mixed liquor, effluent and waste sludge volatile suspended solids may be used instead of suspended solids to calculate the MCRT. See Chapter 11, "Activated Sludge," Section 11.552, "Mean Cell Residence Time (MCRT)," for a discussion of ways to calculate MCRT.

QUESTION

Write your answer in a notebook and then compare your answer with the one on page 611.

16.4R What measurements are required in order to calculate the Mean Cell Residence Time (MCRT)?

16.46 Tests for Digestion Control

16.460 Volatile Acids

A. *DISCUSSION*

Volatile acid levels are determined on sludge samples from the digesters. Most modern digesters have sampling pipes where you can draw a sample from various levels of the tank. Be sure to allow the sludge in the line to run for a few minutes to obtain a representative sample of the digester contents. Samples from the recirculating sludge line on heated digesters will provide a representative sample. Samples also may be collected from supernatant draw-off tubes, or *THIEF HOLES.*[13]

Increased concentrations of volatile acids and decreased alkalinity are the first measurable changes that take place when the process of anaerobic digestion is becoming upset. The volatile acid/alkalinity relationship can vary from 0.1 to about 0.5 without significant changes in digester performance. When the relationship starts to increase, this is a warning that undesirable changes will occur unless the increase is stopped. If the relationship increases above 0.5, the composition of the gas produced can change very rapidly, followed by changes in the rate of gas production, and finally pH.

In a healthy and properly functioning digester, the processes or biological action taking place inside the digester are in equilibrium. When fresh sludge is pumped into a digester, some of the organisms in the digester convert this material to volatile (organic) acids. In a properly operated digester, other organisms feed on the newly produced volatile acids and eventually convert the acids to methane (CH_4) gas, which is burnable, and carbon dioxide (CO_2). If too much raw sludge is pumped to the digester or the digester is not functioning properly, an excess of volatile acids is produced. If excessive amounts of volatile acids are produced, an acid environment unsuitable for some of the organisms in the digester will develop and the digester may cease to function properly unless the alkalinity increases too.

Routine volatile acids and alkalinity determinations during the start-up process for a new digester are a must in bringing the digester to a state of satisfactory digestion.

Routine volatile acids and alkalinity determinations during digestion are important in providing the information that will enable the operator to determine the health of the digester.

(Volatile Acids)

For digester control purposes, the volatile acid/alkalinity relationship should be determined. When the volatile acid/ alkalinity relationship is from less than 0.1/1.0 to 0.3/1.0, the loading and seed retention of the digester are under control. When the relationship starts increasing and becomes greater than 0.3/1.0, the digester is out of control and will become "stuck" unless effective corrective action is taken.

B. *WHAT IS TESTED?*

Sample	Desirable Range, mg/*L*
Recirculated Sludge	150 to 600 mg/*L* (expect trouble if alkalinity is less than two times volatile acids)

METHOD A

(Silicic Acid Method)

C. *APPARATUS*

1. Centrifuge or filtering apparatus
2. Graduated cylinders, 50 and 100 m*L*
3. Two medicine droppers
4. Crucibles, Gooch or fritted glass
5. Filter flask
6. Vacuum source
7. One 50-m*L* beaker
8. Two 5-m*L* pipets
9. Buret

D. *REAGENTS*

1. Silicic acid, solids, 100-mesh. Wash the solid portion of acid to remove fines (impurities) by slurrying (sloshing) the acid in distilled water. Allow mixture to settle for 15 minutes and remove supernatant. Repeat the process several times. Dry the washed acid solids in an oven at 103°C and then store in a desiccator.
2. Chloroform-butanol reagent. Mix 300 m*L* chloroform, 100 m*L* n-butanol, and 80 m*L* 0.5 *N* H_2SO_4 in separatory funnel and allow the water and organic layers to separate. Drain off the lower organic layer through filter paper into a dry bottle.
3. Thymol blue indicator solution. Dissolve 80 mg thymol blue in 100 m*L* absolute methanol.
4. Phenolphthalein indicator solution. Dissolve 80 mg phenolphthalein in 100 m*L* absolute methanol.
5. Sulfuric acid, concentrated.
6. Standard sodium hydroxide reagent, 0.02 *N*. Dilute 20 m*L* 1.0 *N* NaOH stock solution to one liter with absolute methanol.

[13] *Thief Hole.* A digester sampling well that allows sampling of the digester contents without venting digester gas.

(Volatile Acids)

E. *PROCEDURE*

OUTLINE OF PROCEDURE FOR VOLATILE ACIDS

METHOD A

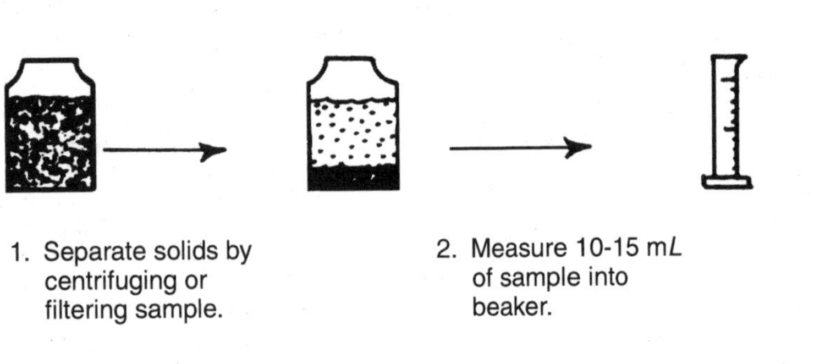

1. Separate solids by centrifuging or filtering sample.

2. Measure 10-15 m*L* of sample into beaker.

3. Add a few drops of thymol blue.

4. Add concentrated H_2SO_4 drop-wise until sample turns red.

5. Place 10 gm silicic acid in crucible and apply suction.

6. Add 6 m*L* acidified sample.

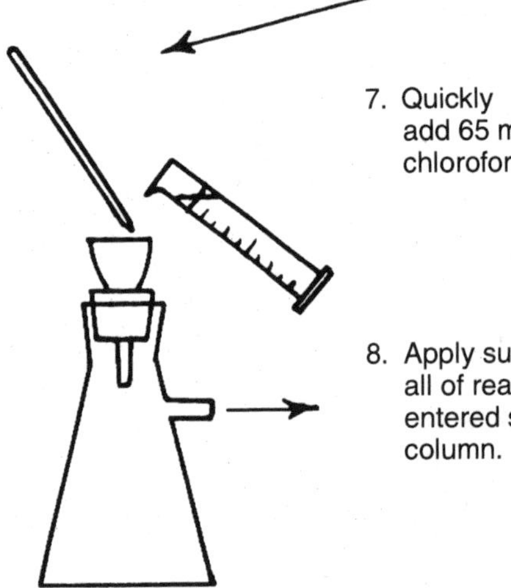

7. Quickly add 65 m*L* chloroform-butanol.

8. Apply suction until all of reagent has entered solid acid column.

9. Remove filter flask.

10. Add a few drops of phenolphthalein.

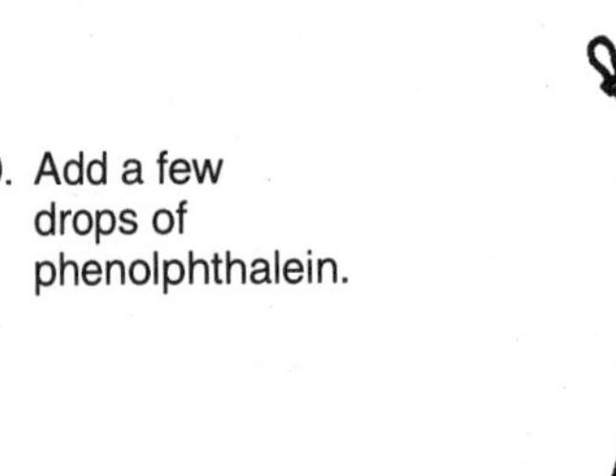

11. Titrate with 0.02 *N* NaOH.

(Volatile Acids)

1. Centrifuge or filter enough sludge to obtain a sample of 10 to 15 m*L*. This same sample and filtrate should be used for both the volatile acids test and the total alkalinity test (Section 16.461).
2. Measure volume (10 to 15 m*L*) of sample and place in a beaker.

   Volume of Sample, B = ________________ m*L*.
3. Add a few drops of thymol indicator solution.
4. Add concentrated H_2SO_4, dropwise, until sample definitely turns red in color (pH = 1 to 1.2).
5. Place 10 grams of silicic acid (solid acid) in crucible and apply suction. This will pack the acid material; the packed material is sometimes called a column.
6. With a pipet, distribute 5.0 m*L* acidified sample (from step 4) as uniformly as possible over the column. Apply suction briefly to draw the acidified sample into the silicic acid column. Release the vacuum as soon as the sample enters the column.
7. Quickly add 65 m*L* chloroform-butanol reagent to the column.
8. Apply suction and stop just before the last of the reagent enters the column.
9. Remove the filter flask from the crucible.
10. Add a few drops of phenolphthalein indicator solution to the liquid in the filter flask.
11. Titrate with 0.02 *N* NaOH titrant in absolute methanol, taking care to avoid aerating the sample. Nitrogen gas or CO_2-free air delivered through a small glass tube may be used both to mix the sample and to prevent contact with atmospheric CO_2 during titration (CO_2-free air may be obtained by passing air through ascarite or equivalent).

    Volume of NaOH Used in Sample Titration, a = ________ m*L*.
12. Repeat the above procedure using a *BLANK*[14] of distilled water.

    Volume of NaOH Used in Blank Titration, b = ________ m*L*.

F. *PRECAUTIONS*

1. The sludge sample must be representative of the digester. The sample line should be allowed to run for a few minutes before the sample is taken. The sample temperature should be as warm as the digester itself.
2. The sample for the volatile acids test should not be taken immediately after charging the digester with raw sludge. Should this be done, the raw sludge may short-circuit to the withdrawal point and result in the withdrawal of raw sludge rather than digested sludge. Therefore, after the raw sludge has been fed into the tank, the tank should be well mixed by recirculation or other means before a sample is taken.
3. If a digester is performing well with low volatile acids and then if one sample should unexpectedly and suddenly give a high value, say over 1,000 mg/*L* of volatile acids, do not become alarmed. The high result may be caused by a poor, nonrepresentative sample of raw sludge instead of digested sludge. Resample and retest. The second test may give a more typical value. When increasing volatile acids and decreasing alkalinity are observed, this is a definite warning of approaching control problems. Corrective action should be taken immediately, such as reducing the feed rate, reseeding from another digester, maintaining optimum temperatures, improving digester mixing, decreasing sludge withdrawal rate, or cleaning the tank of grit and scum.

G. *EXAMPLE*

Equivalent Weight of Acetic Acid, A	= 60 mg/m*L*
Volume of Sample, B	= 10 m*L*
Normality of NaOH Titrant, *N*	= 0.02 *N*
Volume of NaOH Used in Sample Titration, a	= 2.3 m*L*
Volume of NaOH Used in Blank Titration, b	= 0.5 m*L*

H. *CALCULATION*

$$\text{Volatile Acids, mg}/L\ \text{(as acetic acid)} = \frac{A \times 1{,}000\ mL/L \times N(a - b)}{B}$$

$$= \frac{60\ mg/mL \times 1{,}000\ mL/L \times 0.02(2.3\ mL - 0.5\ mL)}{10\ mL}$$

$$= 216\ mg/L$$

METHOD B

(Nonstandard Titration Method)

C. *APPARATUS*

1. One pH meter
2. One adjustable hot plate or Bunsen burner
3. Two burets and stand
4. One 100-m*L* beaker

D. *REAGENTS*

1. pH 7.0 buffer solution
2. pH 4.0 buffer solution
3. Standard acid: H_2SO_4
4. Standard base: NaOH

[14] *Blank.* A bottle containing only dilution water or distilled water; the sample being tested is not added. Tests are frequently run on a sample and a blank and the differences are compared. The procedure helps to eliminate or reduce test result errors that could be caused when the dilution water or distilled water used is contaminated.

(Volatile Acids)

E. *PROCEDURE*

OUTLINE OF PROCEDURE FOR VOLATILE ACIDS

METHOD B

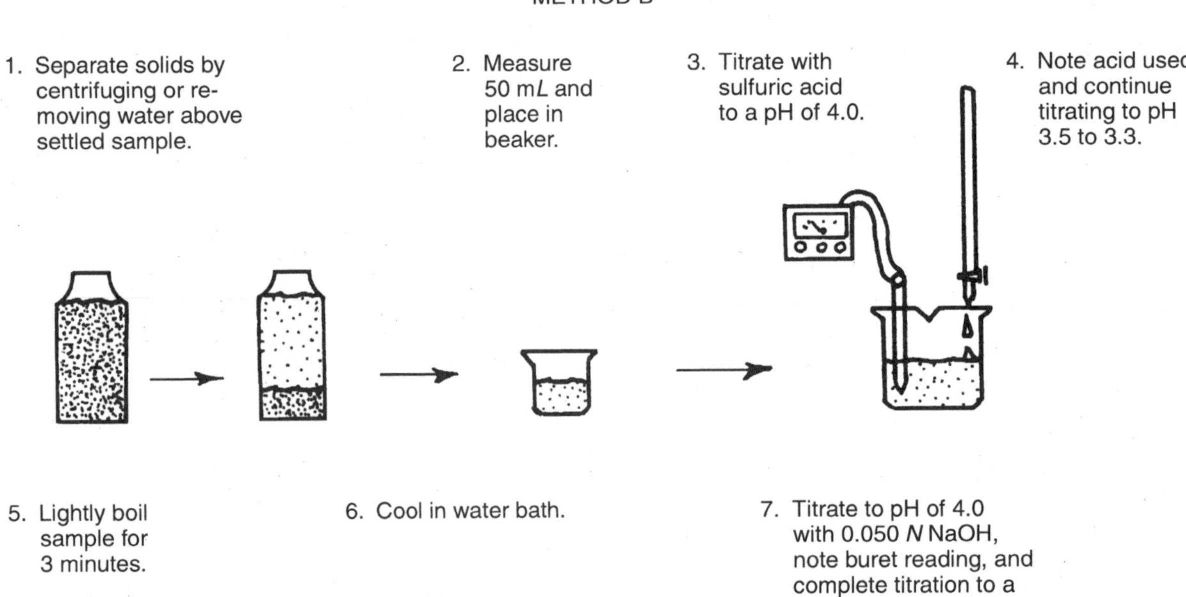

1. Separate solids by centrifuging or removing water above settled sample.
2. Measure 50 m*L* and place in beaker.
3. Titrate with sulfuric acid to a pH of 4.0.
4. Note acid used and continue titrating to pH 3.5 to 3.3.

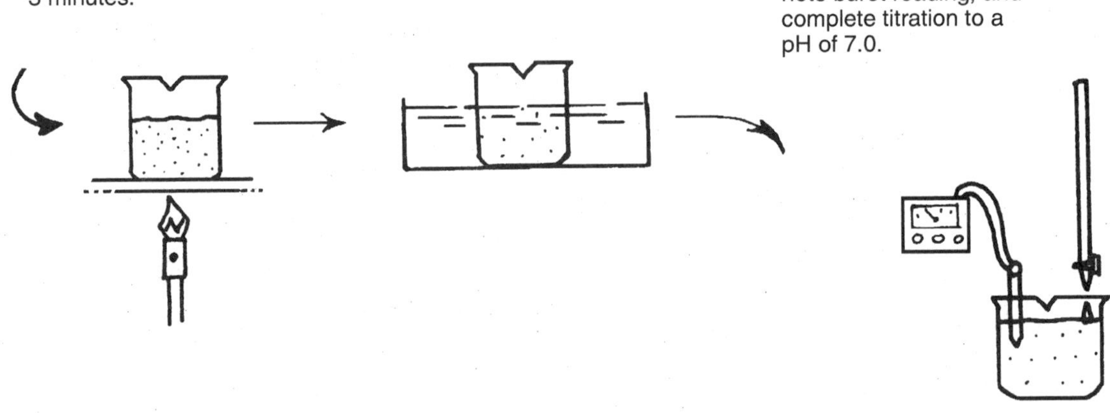

5. Lightly boil sample for 3 minutes.
6. Cool in water bath.
7. Titrate to pH of 4.0 with 0.050 *N* NaOH, note buret reading, and complete titration to a pH of 7.0.

1. Buffer the pH meter at 7.0 and check pH before treatment of sample to remove the solids. Filtration is not necessary. Decanting (removing water above settled material) or centrifuging sample is satisfactory. Do not add any coagulant aids.

2. Titrate 50 m*L* of the sample in a 100-m*L* beaker to pH 4.0 with the appropriate strength sulfuric acid (depends on alkalinity), note acid used, and continue to pH 3.5 to 3.3. A magnetic mixer is extremely useful for this filtration.

3. Carefully buffer pH meter at 4.0 while lightly boiling the sample a minimum of three minutes. Cool in cold water bath to original temperature.

4. Titrate sample with standard 0.050 *N* sodium hydroxide up to pH 4.0, and note buret reading. Complete the titration at pH 7.0. (If this titration consistently takes more than 10 m*L* of the standard hydroxide, use 0.100 *N* NaOH.)

5. Calculate volatile acid alkalinity (alkalinity between pH 4.0 and 7.0).

$$\text{Volatile Acid Alkalinity, mg}/L = \frac{\text{m}L\ 0.050\ N\ \text{NaOH} \times 2{,}500}{\text{m}L\ \text{Sample}}$$

For a 50-m*L* sample the volatile acid alkalinity equals 50 × m*L* 0.050 *N* NaOH.

6. Calculate volatile acids.

Case 1: >180 mg/*L* volatile acid alkalinity.

$$\text{Volatile Acids, mg}/L = \text{Volatile Acid Alkalinity} \times 1.50$$

Case 2: <180 mg/*L* volatile acid alkalinity.

$$\text{Volatile Acids, mg}/L = \text{Volatile Acid Alkalinity} \times 1.00$$

Steps 1 and 2 will give the analyst the pH and total alkalinity, two control tests normally run on digesters. The difference between the total and the volatile acid alkalinity is bicarbonate alkalinity. The time required for Steps 3 and 4 is about 10 minutes.

This is an acceptable method for digester control to determine the volatile acid/alkalinity relationship, but not of sufficient accuracy for research work.

F. *EXAMPLE AND CALCULATION*

Titration of pH 4.0 to 7.0 of a 50-m*L* sample required 8 m*L* of 0.05 *N* NaOH.

STEP 5—Calculate volatile acid alkalinity (alkalinity between pH 4.0 and 7.0).

$$\text{Volatile Acid Alkalinity, mg}/L = \frac{\text{m}L\ 0.050\ N\ \text{NaOH} \times 2{,}500}{\text{m}L\ \text{Sample}}$$

$$= \frac{8\ \text{m}L \times 2{,}500}{50\ \text{m}L}$$

$$= 400\ \text{mg}/L$$

STEP 6—Calculate volatile acids.

Case 1: 400 mg/*L* >180 mg/*L*. Therefore,

$$\text{Volatile Acids, mg}/L = \text{Volatile Acid Alkalinity} \times 1.50$$

$$= 400\ \text{mg}/L \times 1.50$$

$$= 600\ \text{mg}/L$$

(Volatile Acids)

QUESTION

Write your answer in a notebook and then compare your answer with the one on page 611.

16.4S What is the volatile acid concentration in a digester if a 50-m*L* sample required 5 m*L* of 0.050 *N* NaOH for a titration from a pH of 4.0 to 7.0?

16.461 Total Alkalinity

A. *DISCUSSION*

The alkalinity of the recirculated sludge is a measure of the buffer capacity in the digester. When organic matter in a digester is decomposed anaerobically, organic acids are formed that could lower the pH if buffering materials (buffer capacity) were not present. If the pH drops too low, the organisms in the digester could become inactive or die and the digester becomes upset (no longer capable of decomposing organic matter).

For digester control purposes, the volatile acid/alkalinity relationship should be determined. When the volatile acid/alkalinity relationship is from less than 0.1/1.0 to 0.3/1.0, the loading and seed retention of the digester are under control. When the relationship starts increasing and becomes greater than 0.5/1.0, the digester is out of control and will become stuck unless effective corrective action is taken. The pH will not be out of range as long as the volatile acid/alkalinity relationship is low. This relationship gives a warning before trouble starts.

B. *WHAT IS TESTED?*

Sample	Common Range
Recirculated Sludge	2 to 10 Times Volatile Acids

C. *APPARATUS*

1. Centrifuge and centrifuge tubes, or settling cylinders
2. Graduated cylinders (25 m*L* and 100 m*L*)
3. 50-m*L* buret
4. 400-m*L* Erlenmeyer flask or 400-m*L* beaker
5. pH meter or a methyl orange chemical color indicator may be used (see Procedure)

D. *REAGENTS*

1. Sulfuric Acid, 0.2 *N*. Cautiously add 6 m*L* of concentrated sulfuric acid (H_2SO_4) to 300 m*L* of distilled water. Dilute to 1 liter with boiled distilled water. Standardize against 0.02 *N* sodium carbonate (Step 2).

2. Sodium Carbonate, 0.02 *N*. Dry in oven before weighing. Dissolve 1.06 gm of anhydrous sodium carbonate (Na_2CO_3) in boiled distilled water and dilute to 1 liter with distilled water.

3. Methyl Orange Chemical Color Indicator. Dissolve 0.5 gm methyl orange in 1 liter of distilled water.

(Total Alkalinity)

E. *PROCEDURE*

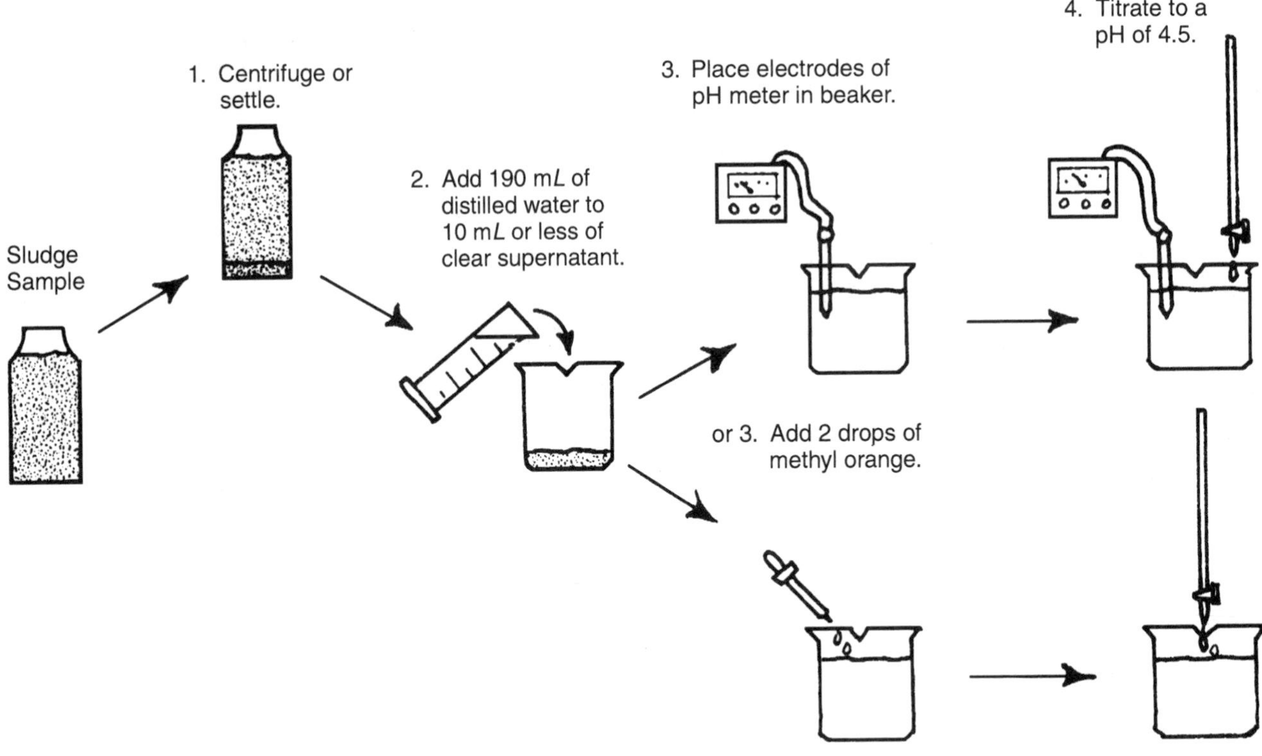
OUTLINE OF PROCEDURE FOR TOTAL ALKALINITY
Sludge Sample
1. Centrifuge or settle.
2. Add 190 mL of distilled water to 10 mL or less of clear supernatant.
3. Place electrodes of pH meter in beaker.
4. Titrate to a pH of 4.5.
or 3. Add 2 drops of methyl orange.

This procedure is followed to measure the alkalinity of a sample and also the alkalinity of a distilled water blank.

1. Take a clean 400-m*L* beaker and add 10 m*L* or less of clear supernatant (in case of water or distilled water, use a 200-m*L* sample). Select a sample volume that will give a usable titration volume. If the liquid will not separate from the sludge by standing and a centrifuge is not available, use the top portion of the sample. This same sample and filtrate should be used for both the total alkalinity test and the volatile acids test.
2. Add 190 m*L* distilled water (in case of water or distilled water determination, skip this step).
3. Place the electrodes of pH meter into the 400-m*L* beaker containing the sample.
4. Titrate to a pH of 4.5 with 0.02 *N* sulfuric acid. (In case of a lack of pH meter, add 2 drops of methyl orange indicator. In this case, titrate to the first permanent change of color to a red-orange color. Care must be exercised in determining the change of color; your ability to detect the change will improve with experience.)
5. The alkalinity of the distilled water should be checked and, if significant, subtracted from the calculation.
6. Calculate alkalinity.

$$\text{Alkalinity of Distilled Water, mg}/L = \text{m}L \text{ of } 0.02\ N\ H_2SO_4 \times 5^*$$

$$\text{Total Alkalinity of Sludge, mg}/L = \text{m}L \text{ of } 0.02\ N\ H_2SO_4 \times 100^* - \text{mg}/L \text{ Alkalinity of Distilled } H_2O$$

* Use 5 if measuring alkalinity of water or distilled water (200-m*L* sample) and 100 if measuring alkalinity of sludge (10-m*L* sample).

F. *EXAMPLE*

Results from alkalinity titrations on

1. Distilled Water	4 m*L* 0.02 *N* H_2SO_4
2. Recirculated Sludge	19.8 m*L* 0.02 *N* H_2SO_4

G. *CALCULATIONS*

$$\text{Alkalinity of Distilled } H_2O, \text{ mg}/L = \text{m}L \text{ of } 0.02\ N\ H_2SO_4 \times 5$$

$$= 4\ \text{m}L \times 5$$

$$= 20\ \text{mg}/L$$

$$\text{Total Alkalinity of Recirculated Sludge, mg}/L = \text{m}L \text{ of } 0.02\ N\ H_2SO_4 \times 100 - \text{mg}/L \text{ Alkalinity of Distilled } H_2O$$

$$= 19.8\ \text{m}L \times 100 - 20\ \text{mg}/L$$

$$= 1{,}980\ \text{mg}/L - 20\ \text{mg}/L$$

$$= 1{,}960\ \text{mg}/L$$

QUESTIONS

Write your answers in a notebook and then compare your answers with those on page 612.

16.4T Why would you run a total alkalinity test on recirculated sludge?

16.4U What is meant by the buffer capacity of a digester?

16.4V If the total alkalinity in a digester is 2,000 mg/*L* and the volatile acids concentration is 300 mg/*L*, what is the volatile acid/alkalinity relationship?

(CO_2)

16.462 Carbon Dioxide (CO_2) in Digester Gas

A. *DISCUSSION*

Changes in the anaerobic sludge digestion process will be observed in the gas quality and are usually noted after the volatile acids or volatile acid/alkalinity relationship starts to increase. The CO_2 content of a properly operating digester will range from 25 percent to 30 percent by volume. If the percent is above 44 percent, the gas will not burn. The easiest test procedure for determining this change is with a CO_2 analyzer (Figure 16.9).

B. *WHAT IS TESTED?*

Sample	Preferred
CO_2 in Digester Gas	25% to 30% by Volume

METHOD A

C. *APPARATUS*

1. One Bunsen burner
2. Plastic tubing
3. 100-m*L* graduated cylinder
4. 250-m*L* beaker

D. *REAGENTS*

CO_2 Absorbent (KOH). Add 500 gm potassium hydroxide (KOH) per liter of water.

Fig. 16.9 CO_2 measurement using inverted graduated cylinder

(CO_2)

E. *PROCEDURE*

OUTLINE OF PROCEDURE FOR CO_2

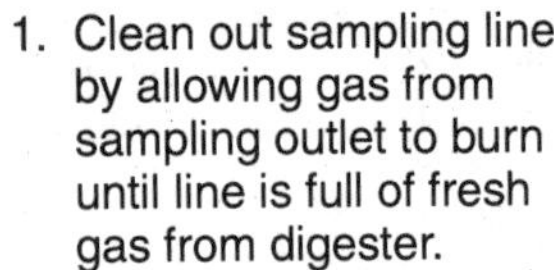

1. Clean out sampling line by allowing gas from sampling outlet to burn until line is full of fresh gas from digester.

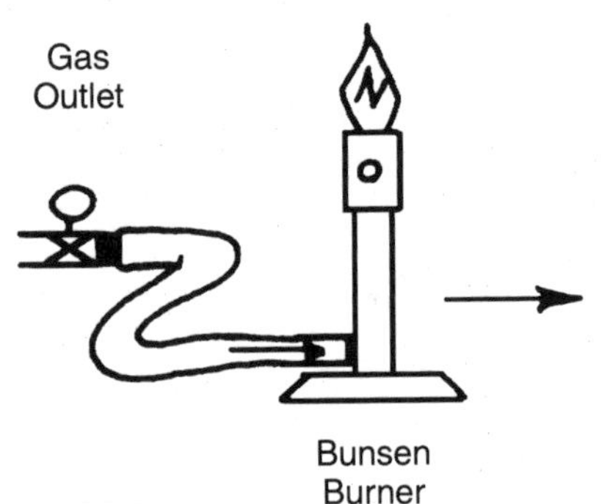

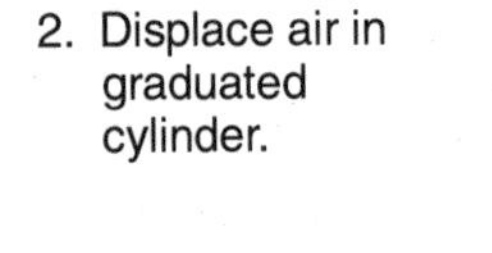

3. Place graduate upside down in beaker containing CO_2 absorbent.

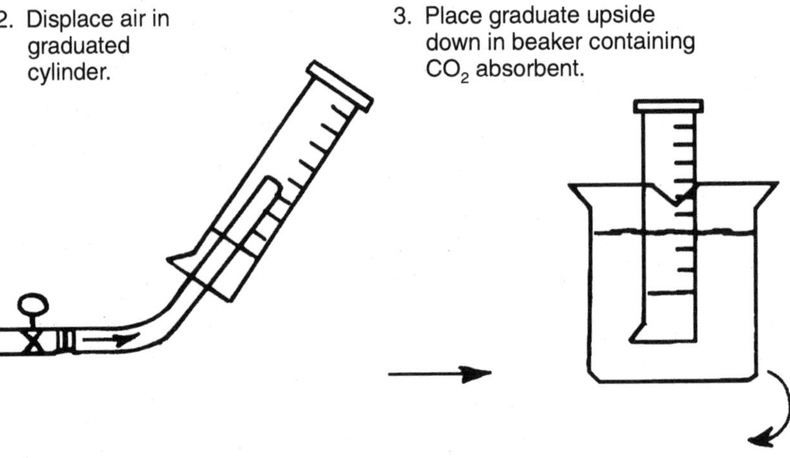

4. Insert hose in graduate and run gas for 60 seconds.

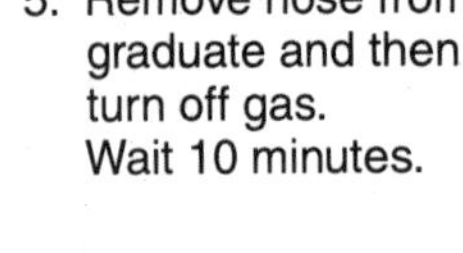

5. Remove hose from graduate and then turn off gas.
Wait 10 minutes.

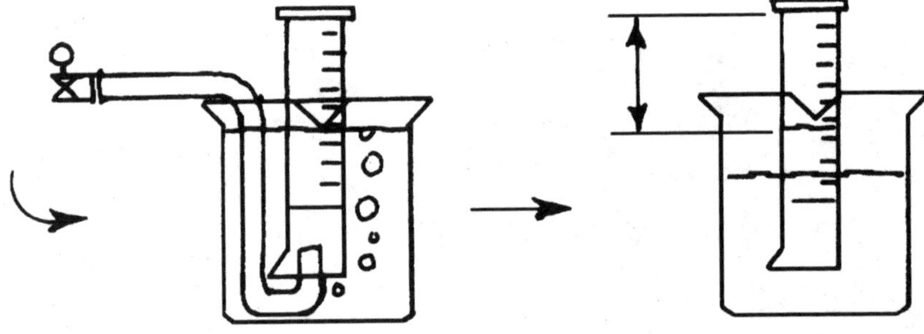

6. Read volume of gas remaining to nearest m*L*.

PRECAUTIONS

1. Avoid any open flames near the digester.
2. Work in a well-ventilated area to avoid the formation of explosive mixtures of methane gas.
3. If your gas sampling outlet is on top of your digester, turn on outlet and vent the gas to the atmosphere for several minutes to clear the line of old gas. Start with step 2, displace air in graduated cylinder. *Never allow any smoking, flames, or other sources of ignition near the digester at any time.*

1. Measure total volume of a 100-m*L* graduate by filling it to the top with water (approximately 125 m*L*). Record this volume.

2. Pour approximately 125 m*L* of CO_2 absorbent into a 250-m*L* beaker.

   CAUTION: Do not get any of this chemical on your skin or clothes. Wash immediately with running water until slippery feeling is gone or severe burns can occur.

3. Collect a representative sample of gas from the gas dome on the digester, a hot water heater using digester gas to heat the sludge, or any other gas outlet. Before collecting the sample for the test, attach one end of a gas hose to the gas outlet and the other end to a Bunsen burner. Turn on the gas, ignite the burner, and allow it to burn digester gas for a sufficient length of time for the sampling gas line to be full of gas from the digester.

4. With gas running through hose from gas sampling outlet, place hose inside inverted calibrated graduated cylinder and allow digester gas to displace air in graduate. Turn off gas.

   CAUTION: The proper mixture of digester gas and air is explosive when exposed to a flame.

5. Place graduate full of digester gas upside down in beaker containing CO_2 absorbent.

6. Insert gas hose inside upside down graduate.

7. Turn on gas, but do not blow out liquid. Run gas for at least 60 seconds.

8. Carefully remove hose from graduate with gas still running.

9. Immediately turn off gas.

10. Wait for 10 minutes and shake gently. If liquid continues to rise, wait until it stops.

11. Read gas remaining in graduate to nearest m*L* (see Figure 16.9).

F. *EXAMPLE*

Total Volume of Graduate = 126 m*L*

Gas Remaining in Graduate = 80 m*L*

G. *CALCULATION*

$$CO_2, \% = \frac{(\text{Total Volume, m}L - \text{Gas Remaining, m}L)}{\text{Total Volume, m}L} \times 100\%$$

$$= \frac{(126 \text{ m}L - 80 \text{ m}L)}{126 \text{ m}L} \times 100\%$$

$$= \frac{46}{126} \times 100\%$$

$$= 37\%$$

(CO_2)

METHOD B

(ORSAT)

The Orsat gas analyzer can measure the concentration of carbon dioxide, oxygen, and methane by volume in digester gas. To analyze digester gas by the Orsat method, follow equipment manufacturer's instructions. This procedure is not recommended for the inexperienced operator.

QUESTIONS

Write your answers in a notebook and then compare your answers with those on page 612.

16.4W What are the dangers involved in running the CO_2 test on digester gas?

16.4X What is the percent CO_2 in a digester gas if the total volume of the graduated cylinder is 128 m*L* and the gas remaining in the cylinder after the test is 73 m*L*?

16.463 Sludge (Digested) Dewatering Characteristics

A. *DISCUSSION*

The dewatering characteristics of digested sludge are very important. The better the dewatering characteristics or drainability of the sludge, the quicker it will dry and the less area will be required for sludge drying beds.

B. *WHAT IS TESTED?*

	Preferred Range	
Sample	**Method A**	**Method B**
Digested Sludge	Depends on appearance	100 to 200 m*L* per 500 m*L* of sample

C. *APPARATUS*

METHOD A

1,000 m*L* graduated cylinder

METHOD B

1. Imhoff cone with tip removed
2. Sand from drying bed
3. 500-m*L* beaker

D. *REAGENTS*

None

E. *PROCEDURE*

Two methods are presented in this section. Method A relies on a visual observation and is quick and simple. The only problem is that operators on different shifts might record the same sludge draining characteristics differently. Method B requires 24 hours, but the results are recorded by measuring the volume of liquid that passed through the sand. Method B would be indicative of what would happen if you had sand drying beds.

(Sludge Dewatering)

OUTLINE OF PROCEDURE FOR SLUDGE DEWATERING

METHOD A

1. Add digested sludge to 1,000-mL graduate.

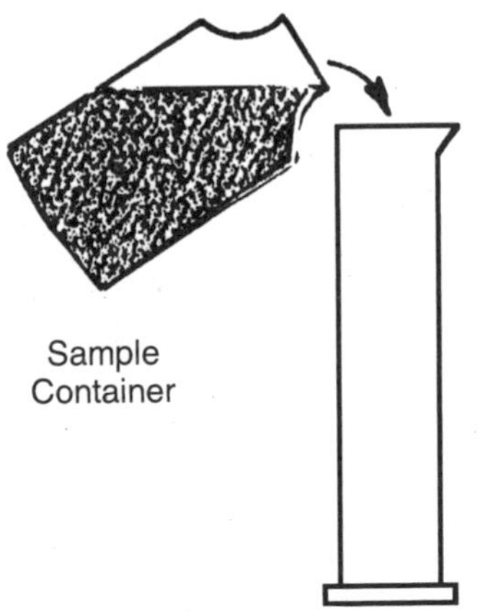

2. Pour sample from graduate back into container.

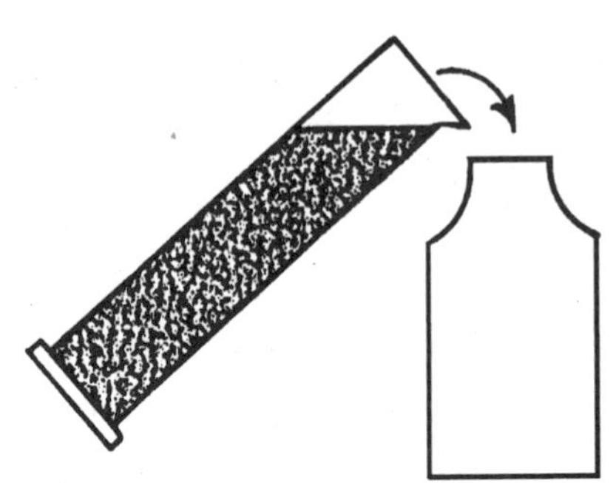

3. Watch solids adhere to cylinder walls.

1. Add sample of digester sludge to 1,000-m*L* graduate.
2. Pour sample back into sample container. Set graduated cylinder down.
3. Watch graduate. If solids adhere to cylinder wall (Figure 16.10) and water leaves solids in form of rivulets, this is a good dewatering sludge on a sand drying bed.

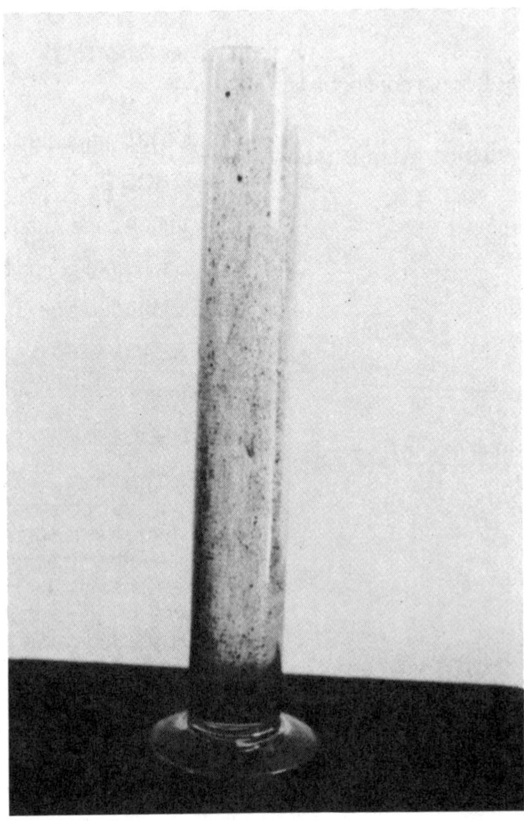

Fig. 16.10 Sludge on graduated cylinder walls for sludge dewatering test

OUTLINE OF PROCEDURE FOR SLUDGE DEWATERING

METHOD B

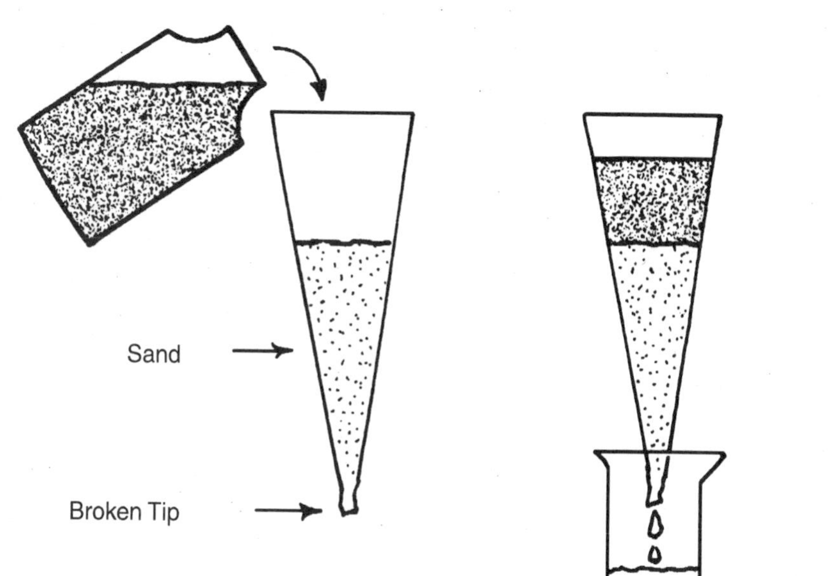

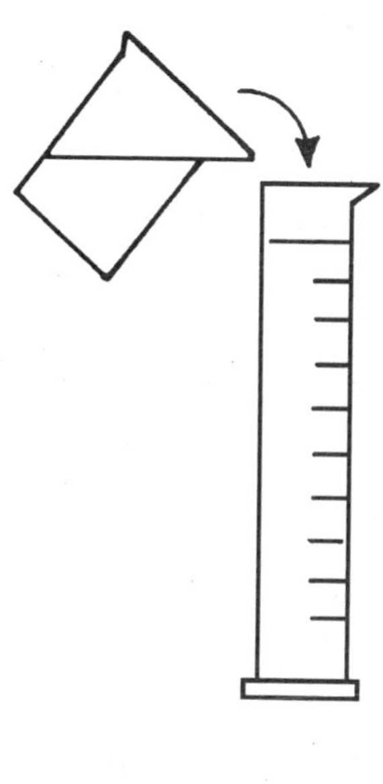

1. Take a glass Imhoff cone that has tip removed and place a glass wool plug in the end to hold the sand in the cone.
2. Fill halfway with sand from drying bed.
3. Fill remainder to 1 liter with digested sludge.
4. Place 500-m*L* beaker under cone tip and wait 24 hours.
5. Record liquid that has passed through sand in m*L*. If less than 100 m*L* has passed through sand, you have poor sludge drainability.

(Supernatant)

QUESTION

Write your answer in a notebook and then compare your answer with the one on page 612.

16.4Y What are the differences in the use of (1) a graduated cylinder, and (2) an Imhoff cone (with the tip removed) filled with sand to measure the dewatering characteristics of digested sludge?

16.464 Supernatant Graduate Evaluation

A. *DISCUSSION*

The digester supernatant solids test measures the percent of settleable solids being returned to the plant headworks. The settleable solids falling to the bottom of a graduate should not exceed the bottom 5 percent of the graduate in most secondary plants. If the supernatant solids test result is greater than 5 percent solids, the supernatant could be placing an excessive solids load on the primary settling tanks or other parts of the plant. Appropriate operational adjustments should be made. If the solids exceed 5 percent, you should run a total suspended solids Gooch crucible test (Section 16.43) on the sample originating from the digester and calculate the recycle load on the plant. Use this method to evaluate solids loadings from sludge elutriation, centrifuging, pressing and filtering operations.

B. *WHAT IS TESTED?*

Sample	Common Values
Supernatant Solids, %	Solids should be <5%

C. *APPARATUS*

100-m*L* graduated cylinder

D. *REAGENTS*

None

E. *PROCEDURE*

OUTLINE OF PROCEDURE FOR SUPERNATANT SOLIDS, %

1. Fill 100-m*L* graduate with supernatant.

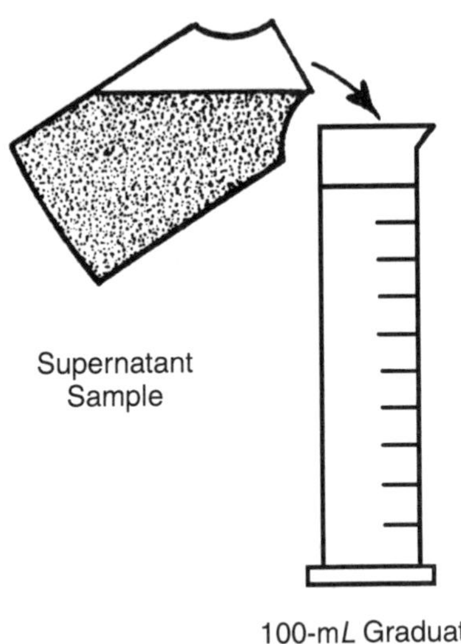

2. After 60 minutes, read m*L* of solids at bottom.

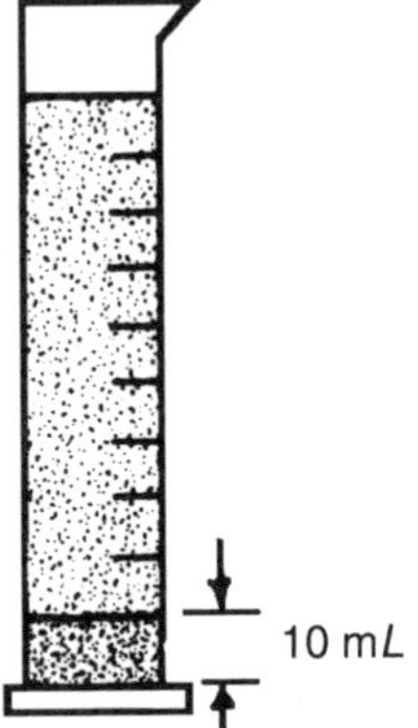

1. Fill a 100-m*L* graduated cylinder with supernatant sample.
2. After 60 minutes, read the m*L* of solids that have settled to the bottom.
3. Calculate supernatant solids, %.

   Supernatant Solids, % = m*L* of Solids

F. *EXAMPLE*

Solids on bottom of cylinder, 10 m*L*.

G. *CALCULATIONS*

Supernatant Solids, % = m*L* of Solids

= 10 m*L*

= 10% Solids (High) by Volume

QUESTION

Write your answer in a notebook and then compare your answer with the one on page 612.

16.4Z Why should the results of the supernatant solids test be less than 5 percent solids?

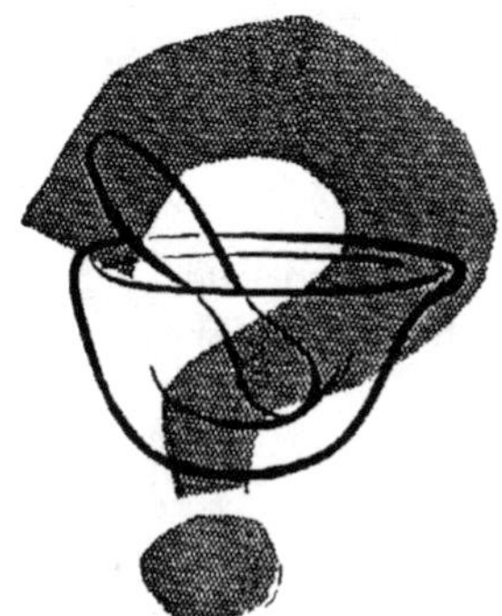

16.465 Temperature

A. *DISCUSSION*

The rate of sludge digestion in a digester is a function of the digester temperature. The normal temperature range in a digester is around 95° to 98°F (35° to 37°C). The temperature of a digester should not be changed by more than 1°F (0.6°C) per day because the helpful organisms in the digester are unable to adjust to rapid temperature changes.

B. *APPARATUS AND PROCEDURE*

See Section 16.5, Procedure 17, "Temperature."

16.47 Lime Analysis

A. *DISCUSSION*

Lime is a general term used in the water and wastewater field to describe quicklime and hydrated lime. Quicklime, a dry powder, contains about 90 percent calcium oxide, CaO, and for this reason is also called calcium oxide. Quicklime is also called unslaked lime. Hydrated lime or slaked lime is obtained by a chemical reaction that occurs when sufficient water is added to quicklime. This form of lime is also referred to as hydrated lime or $Ca(OH)_2$.

Lime has been used in wastewater treatment for many years. Usually, lime was used as a coagulant, especially in treating industrial wastes. The development of advanced wastewater treatment processes has brought new popularity to lime treatment as a means of effectively removing phosphorus.

The selection of which type of lime to use in a particular situation is influenced by a number of factors, such as transportation costs, amount of lime required, and availability. Generally the cost of hydrated lime is greater than that of quicklime. On the other hand, the capital cost of slaking equipment required when quicklime is used will tend to offset savings in material costs.

The following method of lime analysis is taken from the American Water Works Association's Standard B202-93 for quicklime and hydrated lime. The operator can use this method not only to evaluate newly purchased lime but also to analyze lime recovered from lime sludge incineration.

B. *WHAT IS TESTED?*

Sample	Common Range
Hydrated Lime	82 to 98% $Ca(OH)_2$ or 62 to 74% CaO
Recalcined Lime Sludge	60 to 90% CaO
Quicklime	70 to 96% CaO

C. *APPARATUS*

1. Analytical balance
2. 250-m*L* Erlenmeyer flask
3. Hot plate
4. Magnetic stirrer and stirring bar (optional)
5. 100-m*L* graduated cylinder
6. Mortar and pestle

D. *REAGENTS*

1. 0.1782 *N* Hydrochloric Acid Solution: Prepare a solution containing 15.7 m*L* HCl (specific gravity 1.19) per liter. This solution will be stronger than necessary. Standardize the HCl solution against 0.85 grams of pure, dry sodium carbonate, using methyl orange as an indicator. Titrate to a salmon pink end point. Adjust the solution either by the addition of CO_2-free distilled water if too strong or by the addition of HCl if too weak, so that 0.85 gm of sodium carbonate exactly neutralizes 90 m*L* of the standard HCl solution. One milliliter of the standard HCl solution is equivalent to 1.0 percent CaO or 1.32 percent $Ca(OH)_2$ when 0.5 gram of sample is used.
2. Phenolphthalein indicator: Four percent solution. Dissolve 4 grams of dry phenolphthalein in 100 m*L* of 95 percent ethanol.
3. Methyl Orange indicator: Dissolve 500 milligrams in distilled water and dilute to 1 liter.
4. CO_2-free Distilled Water: Prepare fresh as needed by boiling distilled water for 15 minutes and cooling rapidly to room temperature. Cap the flask or bottle in which the water has been boiled with a slightly oversized inverted beaker to minimize the entry of atmospheric carbon dioxide during the cooling process. For best results, exclude CO_2 entry during cooling by attaching a tube containing soda lime, Ascarite, Caroxite, or equivalent.
5. Cane Sugar.

(Lime Analysis)

E. *PROCEDURE*

OUTLINE OF PROCEDURE FOR AVAILABLE CALCIUM OXIDE

1. Pulverize about 5 grams of sample and weigh 0.5 gram.

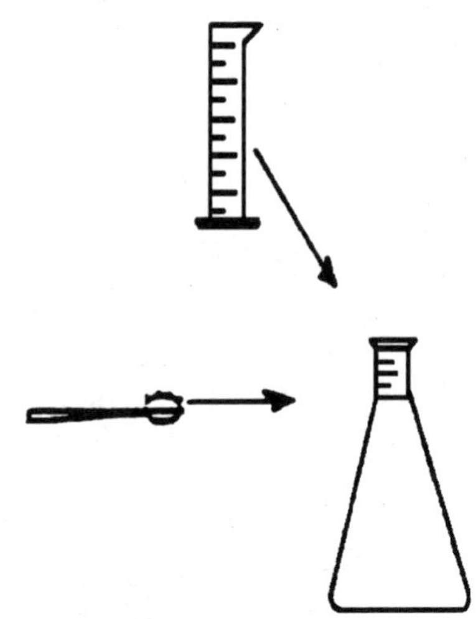

2. Add 10 m*L* distilled water to dry flask and then add 0.5 gram sample.

3. Place flask on hot plate and add 50 m*L* boiling distilled water. Boil for 1 minute. Remove from heat and cool in cold water bath.

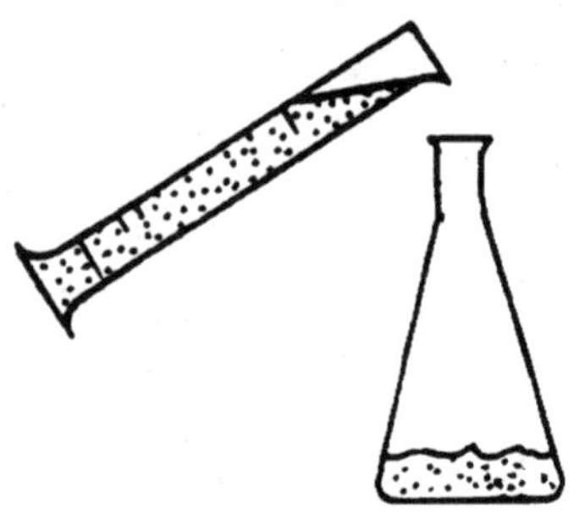

4. Add 50 m*L* distilled water then 15 to 17 grams sugar, stopper flask, swirl, and let stand for 15 minutes.

5. Remove stopper and add 4 to 5 drops phenolphthalein.

6. Titrate with standard HCl solution.

(Lime Analysis)

1. Pulverize about 5 grams of sample with a mortar and pestle.
2. Weigh 0.5 gram of the pulverized sample and brush into a 250-m*L* Erlenmeyer flask containing 10 m*L* CO_2-free distilled water, and immediately stopper the flask loosely with a rubber stopper.
3. Remove the stopper, place the flask on a hot plate, and immediately add 50 m*L* of boiling CO_2-free distilled water. Swirl the flask and boil actively for 1 minute for complete slaking. Remove from hot plate, stopper the flask loosely, and place in a cold water bath to cool to room temperature.
4. Add about 50 m*L* of CO_2-free distilled water, and then approximately 15 to 17 grams of pure cane sugar. Stopper the flask, swirl and let stand for 15 minutes to react. Swirl at 5-minute intervals during this period.
5. Remove stopper, add 4 to 5 drops of phenolphthalein indicator. Wash the stopper and the sides of the flask with CO_2-free distilled water.
6. Titrate with standard HCl solution.

F. *EXAMPLE*

1. Sample size = 0.5 gram
2. m*L* of titrant used, A = 94.5 m*L*

G. *CALCULATIONS*

CaO, % = m*L* of HCl titrant used

= 94.5%

H. *NOTES AND PRECAUTIONS*

1. A magnetic stirrer may be used during the titration, if desired. Adjust to stir as rapidly as possible without incurring loss by spattering.
2. When titrating, first add, without shaking or stirring, about 90 percent of the acid requirement from the 100-m*L* buret. Then shake or stir as vigorously as possible and finish the titration more carefully, to the first complete disappearance of pink color. Note the reading and ignore the return of color. If the operator is not familiar with previous analyses of the lime under test, it is good practice to run a preliminary test by slow titration to determine the proper amount of acid to add without first shaking or stirring the flask.

I. *PROCEDURE FOR HYDRATED LIME*

1. The procedure for determining hydrated lime, $Ca(OH)_2$, is the same as for CaO except that cold CO_2-free distilled water is used and the boiling and cooling procedures are omitted.
2. The number of milliliters of standard acid solution used times 1.32 is the percentage of available calcium hydroxide, $Ca(OH)_2$, in the sample.

J. *REFERENCE*

AWWA STANDARD FOR QUICKLIME AND HYDRATED LIME, B202-02. Obtain from American Water Works Association (AWWA), Bookstore, 6666 West Quincy Avenue, Denver, CO 80235. Order No. 42202. Price to members, $41.50; nonmembers, $60.50; price includes cost of shipping and handling.

QUESTION

Write your answer in a notebook and then compare your answer with the one on page 612.

16.4AA Lime is used in what types of treatment processes?

END OF LESSON 5 OF 9 LESSONS ON LABORATORY PROCEDURES AND CHEMISTRY

Please answer the discussion and review questions next.

DISCUSSION AND REVIEW QUESTIONS

Chapter 16. LABORATORY PROCEDURES AND CHEMISTRY

(Lesson 5 of 9 Lessons)

Write the answers to these questions in your notebook. The question numbering continues from Lesson 4.

18. Calculate the SVI if the mixed liquor suspended solids are 2,000 mg/*L* and the 30-minute settleable solids test is 500 m*L* in 2 liters or 25 percent.
19. Calculate the SDI if the SVI is 125.
20. What does sludge age measure?
21. What is the difference between sludge age and mean cell residence time (MCRT)?
22. What is a thief hole?
23. What relationship is the critical control factor in digester operation?
24. What is a blank, as referred to in laboratory procedures?
25. How can you obtain a representative sample of digester gas?
26. Why should the dewatering characteristics of digested sludge be measured?
27. What happens to the plant when the supernatant from the digester is high in solids?
28. Why should the temperature of a digester not be changed by more than 1°F (0.6°C) per day?

CHAPTER 16. LABORATORY PROCEDURES AND CHEMISTRY

(Lesson 6 of 9 Lessons)

16.5 LABORATORY PROCEDURES FOR NPDES MONITORING

16.50 Need for Approved Procedures

Tests in this section are designed for the treatment plant operator or laboratory analyst who is required to monitor effluent discharges under a National Pollutant Discharge Elimination System (NPDES) permit. Principal tests include Biochemical Oxygen Demand (BOD), pH, Fecal Coliform, Residual Chlorine, and Suspended Solids. These and other tests included in this section are listed below.

Test

1. Acidity
2. Alkalinity, Total
   Biochemical Oxygen Demand (BOD) (See Dissolved Oxygen)
3. Chemical Oxygen Demand (COD)
4. Chloride
5. Chlorine, Total Residual
6. Coliform, Total, Fecal, and *E. coli*
7. Dissolved Oxygen (DO) and Biochemical Oxygen Demand (BOD)
8. Hydrogen Ion (pH)
9. Metals
10. Nitrogen
11. Oil and Grease
    pH (see Hydrogen Ion)
12. Phosphorus
13. Solids, Total, Dissolved, and Suspended
14. Specific Conductance
15. Sulfate
16. Surfactants
17. Temperature
18. Total Organic Carbon (TOC)
19. Turbidity

Remember that monitoring data required by an NPDES permit must be obtained by using approved test procedures. A list of approved test procedures can be found in *40 CFR 136*, July 1, 2006, "Guidelines Establishing Test Procedures for the Analysis of Pollutants." US Government Printing Office, Superintendent of Documents, PO Box 371954, Pittsburgh, PA 15250-7954.

16.51 Test Procedures

1. Acidity

A. *DISCUSSION*

This procedure determines the mineral acidity of a sample. Acidity results from carbon dioxide from the atmosphere, from the biological oxidation of organic matter, or from industrial waste discharges.

The acidity of a water or wastewater sample is its quantitative capacity to neutralize a strong base to a pH of 8.2. Titrating with 0.02 *N* sodium hydroxide measures the concentration of mineral acids (such as sulfuric acid), hydrolyzing salts, and total acidity. The end point of the titration may be detected using a pH meter or phenolphthalein indicator methods.

B. *WHAT IS TESTED?*

Sample	Common Range, mg/*L* as $CaCO_3$
Effluent	Depends on water supply

C. *APPARATUS, REAGENTS, AND PROCEDURE*

See page 2-24, *STANDARD METHODS*, 21st Edition.

D. *EXAMPLE*

Sample of plant effluent collected and tested for acidity. Acidity test results:

1. Sample size, m*L* = 50 m*L*
2. NaOH normality, *N* = 0.02 *N*
3. m*L* NaOH titrant used, A = 8.9 m*L*

E. *CALCULATIONS*

$$\text{Acidity, mg/}L\text{ as CaCO}_3 = \frac{A \times N \times 50{,}000}{\text{m}L\text{ Sample}}$$

$$= \frac{8.9\text{ m}L \times 0.02\ N \times 50{,}000}{50\text{ m}L}$$

$$= 178\text{ mg/}L\text{ as CaCO}_3$$

QUESTION

Write your answer in a notebook and then compare your answer with the one on page 612.

16.5A What are some sources of acidity in a treatment plant influent?

(Alkalinity, Total)

2. Alkalinity, Total (Electrometric Method)

A. *DISCUSSION*

The alkalinity of a water or wastewater is a measure of its capacity to neutralize acids. The alkalinity of natural waters is due primarily to the salts of weak acids, although weak or strong bases may also contribute. Bicarbonate ions (HCO_3^-) represent the major form of alkalinity.

B. *WHAT IS TESTED?*

Sample	Common Range, mg/*L*
Influent and Effluent	50 to 500

C. *APPARATUS REQUIRED*

1. pH meter
2. Reference electrode
3. Glass electrode
4. Graduate cylinder, 50 m*L*
5. Buret, 25 m*L*
6. Buret support
7. Beaker, 250 m*L*
8. Magnetic stirrer
9. Flask, volumetric, 1,000 m*L*
10. Bottle, wash
11. Balance
12. Desiccator

D. *REAGENTS*

NOTE: Standardized solutions are commercially available.

1. Sodium carbonate (Na_2CO_3) solution, approximately 0.05 *N:* Dry 3 to 5 grams (gm) primary standard Na_2CO_3 at 250°C for 4 hours and cool in a desiccator. Weigh 2.5 ± 0.2 gm. Transfer to a 1-liter volumetric flask and fill to the mark with distilled water.
2. Sulfuric acid (H_2SO_4), 0.1 *N:* Dilute 3.0 m*L* concentrated H_2SO_4 or 8.3 m*L* concentrated HCl to 1 liter with distilled or deionized water. Standardize with the above sodium carbonate solution using the following procedure:

Using 40.0 m*L* 0.05 *N* Na_2CO_3 solution, add 60 m*L* distilled water and titrate (add sulfuric acid, 0.1 *N* using a buret) to a pH of about 5. Lift out the electrodes, rinse into the same beaker, and boil gently for 3 to 5 minutes under glass cover. Cool to room temperature, rinse cover glass into beaker and finish titration to pH 4.5 and record the number of milliliters (m*L*) used. Calculate the normality of sulfuric acid according to the formula.

$$\text{Normality, } N = \frac{A \times B}{53 \times C}$$

where, A = grams Na_2CO_3 weighed into 1 liter

B = m*L* Na_2CO_3

C = m*L* H_2SO_4 used

EXAMPLE:

Weighed 2.502 gm Na_2CO_3 = A

Used 40.0 m*L* Na_2CO_3 = B

Titrated Using 18.9 m*L* H_2SO_4 = C

$$\text{Normality, } N = \frac{2.502 \text{ gm} \times 40.0 \text{ m}L}{53 \times 18.9 \text{ m}L}$$

$$= 0.10\ N\ H_2SO_4$$

3. Standard sulfuric acid, 0.02 *N:* Dilute 200 m*L* 0.10 *N* standard acid to 1 liter using a volumetric flask. To determine the volume to be diluted, use the following formula:

$$\text{Volume Diluted, m}L = \frac{\text{Standard, } 0.02\ N \times 1{,}000 \text{ m}L}{\text{Calculated Normality, } 0.10\ N}$$

$$= \frac{0.02\ N \times 1{,}000 \text{ m}L}{0.10\ N}$$

$$= 200 \text{ m}L$$

E. *PROCEDURE*

1. Take a clean beaker and add 100 m*L* of sample (select a sample volume that will give a usable titration volume not greater than 50 m*L*).
2. Place electrodes of pH meter into beaker containing sample.
3. Stir sample slowly.
4. Titrate to pH 4.5 with 0.02 *N* H_2SO_4.
5. Calculate Total Alkalinity.

OUTLINE OF PROCEDURE FOR TOTAL ALKALINITY

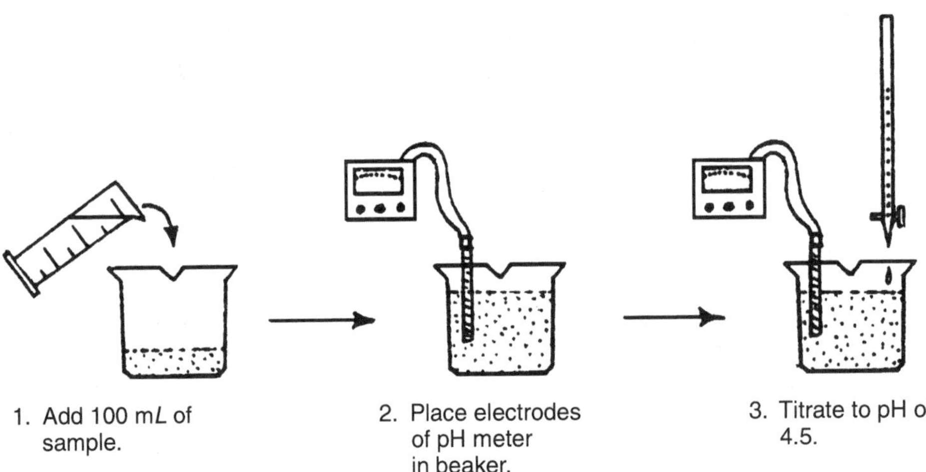

1. Add 100 mL of sample.
2. Place electrodes of pH meter in beaker.
3. Titrate to pH of 4.5.

F. *EXAMPLE*

Results from alkalinity titration on effluent sample:

1. Sample Size, m*L* = 100 m*L*
2. m*L* of Titrant Used, A = 20 m*L*
3. Acid Normality, *N* = 0.02 $N\ H_2SO_4$

G. *CALCULATIONS*

$$\text{Total Alkalinity, mg}/L \text{ as } CaCO_3 = \frac{A \times N \times 50{,}000}{mL \text{ of Sample}}$$

$$= \frac{20\ mL \times 0.02\ N \times 50{,}000}{100\ mL}$$

$$= 200\ mg/L$$

H. *PRECAUTIONS*

1. Soaps and oily matter may interfere with the test by coating the pH electrodes and causing a sluggish response.
2. The sample should be analyzed as soon as practical, within a few hours after collection.
3. The sample should not be filtered, diluted, concentrated, or altered in any way.

I. *REFERENCE*

See Page 2-27, *STANDARD METHODS,* 21st Edition.

QUESTION

Write your answer in a notebook and then compare your answer with the one on page 612.

16.5B What ion represents the major form of alkalinity in wastewater?

3. Chemical Oxygen Demand (COD)

A. *DISCUSSION*

COD is a good estimate of the first-stage oxygen demand for most municipal wastewaters. An advantage of the COD test over the biochemical oxygen demand (BOD) test is that you do not have to wait five days for the results. The COD test also is used to measure the strength of wastes that are too

toxic for the BOD test. The COD is usually higher than the BOD, but the amount will vary from waste to waste.

The COD test should be considered an independent measurement of organic matter in a wastewater sample rather than a substitute for the BOD test. However, COD analysis does not measure the rate of biodegradability of matter and therefore it is difficult to predict the effects of an effluent on the DO in receiving waters and the treatability of a particular wastewater by biological processes.

The COD test method oxidizes organic substances in the wastewater sample using potassium dichromate in 50 percent sulfuric acid solution. Silver sulfate is used as a catalyst and mercuric sulfate is added to remove chloride interference. The excess dichromate is titrated with standard ferrous ammonium sulfate, using ferroin as an indicator. The method related here is a quick (3 to 4 hours), effective measure of the organic strength of a waste.

B. *WHAT IS TESTED?*

Sample	Common Range, mg/*L*
Influent	200 to 400
Effluent	10 to 80
Industrial Waste	200 to 4,000

C. *APPARATUS REQUIRED*

1. Two graduated cylinders, 50 m*L*
2. Volumetric pipet, 10 m*L*
3. Buret, 50 m*L*
4. Glass beads or boiling stones
5. Flasks, Erlenmeyer with ground-glass joint, 250 m*L*
6. Reflux condenser
7. Hot plate
8. Magnetic stirrer

D. *REAGENTS*

NOTE: Standardized solutions may be purchased from chemical suppliers.

1. Standard potassium dichromate ($KCrO_7$) 0.250 *N.* Dissolve 12.259 gm primary standard grade $K_2Cr_2O_7$, previously dried at 103°C for two hours, in distilled water and make up to 1 liter (at 103°C).
2. Sulfuric acid-silver sulfate reagent. Add 22 gm of silver sulfate (Ag_2SO_4) to a 9-pound bottle of concentrated sulfuric acid (H_2SO_4). Allow one to two days for the silver sulfate to dissolve.
3. Standard ferrous ammonium sulfate solution, 0.25 *N.* Dissolve 98 gm $Fe\ (NH_4)(SO_4)_2 \bullet 6\ H_2O$ in distilled water. Add 20 m*L* conc H_2SO_4, cool, and dilute to 1 liter. This solution is unstable and must be standardized daily.
4. Ferroin Indicator. Dissolve 1.485 gm of 1.10 phenanthroline monohydrate ($C_{12}H_8N_2 \bullet H_2O$), together with 0.695 gm ferrous sulfate crystals ($FeSO_4 \bullet 7\ H_2O$), in water and make up to 100 m*L*.
5. Silver sulfate, reagent powder.
6. Mercuric sulfate ($HgSO_4$) analytical grade crystals.

(COD)

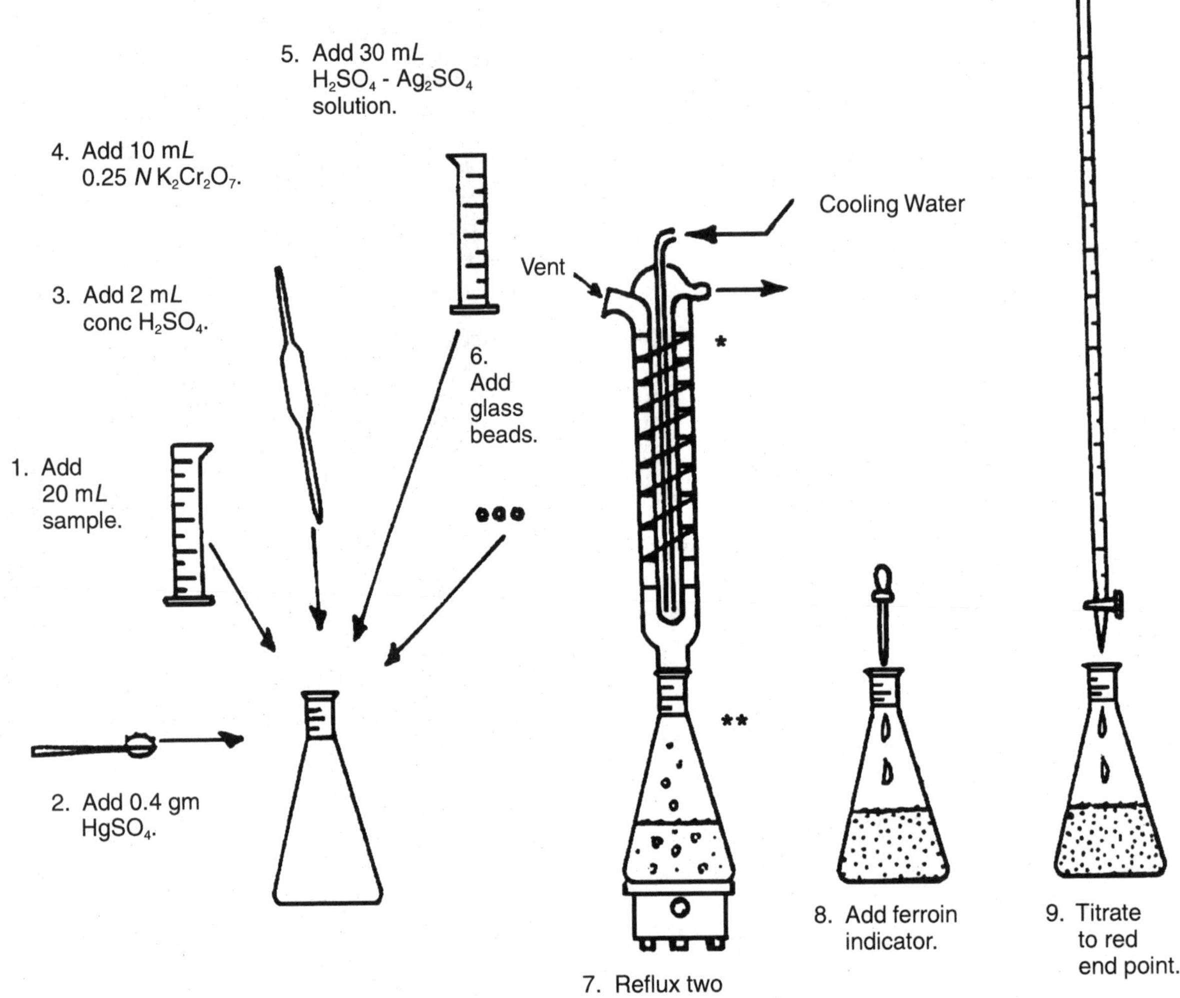
OUTLINE OF PROCEDURE FOR COD >50 mg/L
5. Add 30 mL H_2SO_4 - Ag_2SO_4 solution.
4. Add 10 mL 0.25 N $K_2Cr_2O_7$.
3. Add 2 mL conc H_2SO_4.
1. Add 20 mL sample.
6. Add glass beads.
Vent
Cooling Water
2. Add 0.4 gm $HgSO_4$.
*
**
8. Add ferroin indicator.
9. Titrate to red end point.
7. Reflux two hours, cool, and wash down.
* Reflux condenser, Friedrichs
** Erlenmeyer flask

E. *PROCEDURE* (For COD >50 mg/*L*)

1. Measure 20.0 m*L* of sample into a clean 250-m*L* Erlenmeyer flask with a ground glass joint. The bottom of the reflux condenser will fit tightly into the ground-glass joint.
2. Place 0.4 gm mercuric sulfate into the flask containing the sample.
3. Slowly add 2.0 m*L* concentrated sulfuric acid (H_2SO_4). Swirl until contents are well mixed. Cool.
4. Pipet 10.0 m*L* standard 0.25 *N* potassium dichromate solution into the flask and carefully mix.
5. Carefully add 30 m*L* sulfuric acid-silver sulfate reagent into the flask through the open end of the condenser while swirling the flask. Use caution.
6. Add several glass beads or boiling stones.
7. Attach flask to condenser and start cooling water. Make sure contents of the flask are thoroughly mixed before heat is applied. The *REFLUX*[15] mixture must be thoroughly mixed before heat is applied. If this is not done, local hot spots on bottom of flask may cause mixture to be blown out of flask.
8. Prepare a blank by repeating above steps and by substituting distilled water for the sample.
9. Reflux samples and blank for two hours. (If sample mixture turns completely green, the sample was too strong. Dilute sample with distilled water and repeat above steps substituting diluted sample.)
10. While the samples and blank are refluxing, standardize the ferrous ammonium sulfate (FAS) solution:
    a. Pipet 10.0 m*L* standard 0.25 *N* potassium dichromate solution into a 250-m*L* Erlenmeyer flask. Add about 90 m*L* of water.
    b. Add 30 m*L* concentrated H_2SO_4 with mixing. Let cool.

[15] *Reflux.* Flow back. A sample is heated, evaporates, cools, condenses, and flows back to the flask.

c. Add 2 to 3 drops ferroin indicator, titrate with ferrous ammonium sulfate (FAS) solution. Color change of solution is from orange to greenish to red.

m*L* FAS ____________

$$\text{Normality of FAS} = \frac{10\ mL\ K_2Cr_2O_7 \times 0.25}{mL\ FAS} = \_\_\_\_$$

11. After refluxing mixture for two hours, wash down condenser. Let cool. Add distilled water to about 140 m*L.*

12. Titrate reflux mixtures with standard FAS.

Blank - m*L* FAS ____________

Sample - m*L* FAS ____________

F. *PRECAUTIONS*

1. The wastewater sample should be well mixed. If large particles are present, the sample should be homogenized with a blender or mixer.
2. Flasks and condensers should be clean and free from grease or other oxidizable materials, otherwise erratic results will be obtained.
3. The standard ferrous ammonium sulfate solution is unstable and should be standardized daily.
4. Use extreme caution and safety precautions when handling the chemicals used for the test. Goggles, a rubberized apron, and asbestos gloves are essential equipment.
5. Use a wide-tip pipet to ensure a representative sample is added.
6. The solution must be well mixed before it is heated. If the acid is not completely mixed in the solution when it is heated, the mixture could spatter and some of it pass out the vent, thus ruining the test.
7. Mercuric sulfate is very toxic. Avoid skin contact and breathing this chemical.
8. The amount of mercuric sulfate added depends on the chloride concentration; maintain a 10:1 ratio $HgSO_4$:Cl.
9. If the COD of the sample is less than 50 mg/*L,* follow the above procedure but use 0.025 *N* standard potassium dichromate and back-titrate with 0.10 *N* FAS. Extreme care must be taken because even a trace of organic matter on the glassware may cause a gross error.

G. *EXAMPLE*

1. Standardization of FAS (ferrous ammonium sulfate).

m*L* 0.25 *N* $K_2Cr_2O_7$ = 10.0

m*L* FAS = 11.0

$$\text{Normality, } N\text{, FAS} = \frac{mL\ K_2Cr_2O_7 \times 0.25}{mL\ FAS}$$

$$= \frac{10.0 \times 0.25}{11.0}$$

$$= 0.227\ N$$

2. Sample Test.

Sample Taken		= 20.0 m*L*
A = m*L* FAS Used for Blank		= 10.0 m*L*
B = m*L* FAS Used for Sample		= 3.0 m*L*

H. *CALCULATION*

$$\text{COD, mg/}L = \frac{(A - B) \times N \times 8{,}000}{mL\ Sample}$$

$$= \frac{(10.0\ mL - 3.0\ mL) \times 0.227\ N \times 8{,}000}{20\ mL}$$

$$= 635\ mg/L$$

I. *REFERENCE*

See page 5-14, *STANDARD METHODS,* 21st Edition.

QUESTIONS

Write your answers in a notebook and then compare your answers with those on page 612.

16.5C What does the COD test measure?

16.5D What is an advantage of the COD test over the BOD test?

4. Chloride (Silver Nitrate Method)

A. *DISCUSSION*

Chloride (Cl^-) is one of the major inorganic ions present in water and wastewater. The chloride concentration is higher in wastewater than raw water because sodium chloride (NaCl), table salt, is a common article of diet and passes through the human digestive system. Along the sea coast, chloride may be present in high concentrations because of infiltration of salt water into the wastewater collection system. Chloride concentrations also may be increased by industrial process discharges.

B. *WHAT IS TESTED?*

Sample	**Common Range, mg/*L***
Wastewater	Depends on chloride concentration in raw water supply and industrial discharges.

C. *METHODS*

Three methods are currently acceptable for NPDES chloride monitoring. They are: Silver Nitrate, Mercuric Nitrate, and the automated colorimetric Ferricyanide Methods. The first two are similar in most respects, and selection is a matter of preference. The third is an automated procedure requiring special equipment.

D. *APPARATUS REQUIRED* (Silver Nitrate Method)

1. Graduated cylinder, 100 m*L*
2. Buret, 50 m*L*
3. Erlenmeyer flask, 250 m*L*
4. Pipet, 10 m*L*
5. Magnetic stirring apparatus

(Chloride)

E. *REAGENTS*

NOTE: Standard solutions may be purchased from chemical suppliers.

1. Chloride-free water—distilled or deionized water.
2. Potassium chromate (K_2CrO_4) indicator solution: Dissolve 50 grams K_2CrO_4 in a little distilled water. Add silver nitrate ($AgNO_3$) solution until a red precipitate is formed. Let stand 12 hours, filter, and dilute to 1 liter with distilled water.
3. Standard Silver Nitrate Titrant, 0.0141 *N:* Dissolve 2.395 grams $AgNO_3$ in distilled water and dilute to 1 liter.

   Standardize to 0.0141 *N* sodium chloride (NaCl) by procedure given below. Store in brown bottle.
4. Standard Sodium Chloride, 0.0141 *N:* Dissolve 0.8241 gram (dried at 140°C) in chloride-free water and dilute to 1 liter.

F. *PROCEDURE*

1. Place 100 m*L* or a suitable portion diluted to 100 m*L* in a 250-m*L* Erlenmeyer flask.
2. Add 1.0 m*L* K_2CrO_4 indicator solution.
3. Titrate with standard silver nitrate to a pinkish yellow end point. Be consistent in end point recognition. Compare with known standards of various chloride concentrations.

G. *CALCULATION*

$$\text{Chloride (as Cl), mg}/L = \frac{(A - B) \times N \times 35{,}450}{\text{m}L \text{ of Sample}}$$

A = m*L* $AgNO_3$ Used for Titration of Sample

B = m*L* $AgNO_3$ Used for Blank

N = Normality of $AgNO_3$

H. *EXAMPLE*

Sample Size = 100 m*L*

A = m*L* $AgNO_3$ Used for Sample = 10.0 m*L*

B = m*L* $AgNO_3$ Used for Blank = 0.4 m*L*

N = Normality of $AgNO_3$ Solution = 0.0141 *N*

$$\text{Chloride, mg}/L = \frac{(10.0 \text{ m}L - 0.4 \text{ m}L) \times 0.0141\ N \times 35{,}450}{100 \text{ m}L}$$

$$= 48 \text{ mg}/L$$

I. *SPECIAL NOTES*

1. Sulfide, thiosulfate, and sulfite ions interfere, but can be removed by treatment with 1 m*L* of 30 percent hydrogen peroxide (H_2O_2).
2. Highly colored samples must be treated with an aluminum hydroxide suspension and then filtered.

OUTLINE OF PROCEDURE FOR CHLORIDE

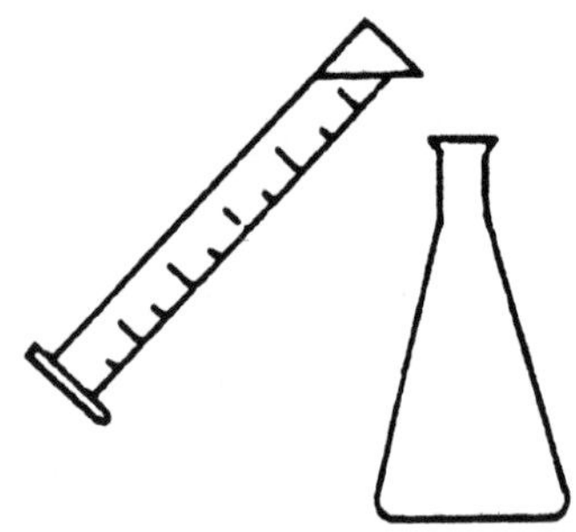

1. Place 100 m*L* or other measured sample in flask.

2. Add 1 m*L* chromate indicator.

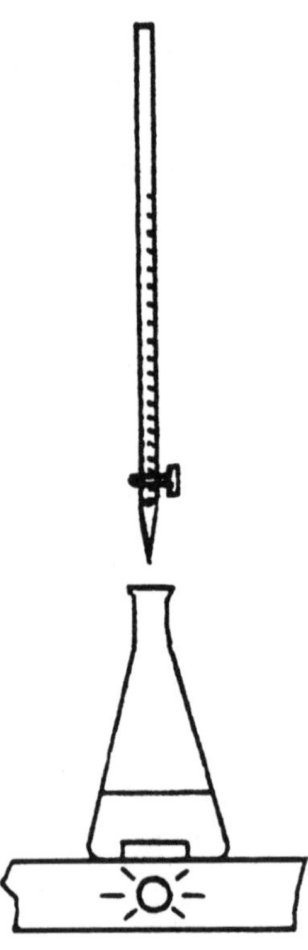

3. Place flask on magnetic stirrer and titrate with standard silver nitrate.

3. Orthophosphate in excess of 25 mg/*L* and iron in excess of 10 mg/*L* also interfere.
4. If the pH of the sample is not between 7 to 10, adjust with 1 *N* sulfuric acid or 1 *N* sodium hydroxide.
5. Procedure for standardization of $AgNO_3$:
   a. Add 10 m*L* (1 mg Cl) standard sodium chloride solution to a clean 250-m*L* Erlenmeyer flask.
   b. Add 90 m*L* distilled water.
   c. Titrate as in Section F above.

$$\text{Normality, } N, \text{ AgNO}_3 = \frac{\text{m}L\text{ NaCl Standard} \times 0.0141}{\text{m}L\text{ AgNO}_3\text{ Used in Titration}}$$

EXAMPLE:

10.0 m*L* NaCl Standard Used

10.0 m*L* $AgNO_3$ Used in Titration

0.0141*N* = Normality of NaCl Standard

$$\text{Normality, } N, \text{ AgNO}_3 = \frac{10.0\text{ m}L \times 0.0141\ N}{10\text{ m}L}$$

$$= 0.0141\ N$$

J. *REFERENCE*

See page 4-70, *STANDARD METHODS,* 21st Edition.

QUESTION

Write your answer in a notebook and then compare your answer with the one on page 612.

16.5E What ions interfere with the chloride test and how are they removed?

5. Chlorine Residual (Total)

A. *DISCUSSION*

The many uses of chlorination in wastewater treatment include disinfecting, reducing BOD, controlling odor, improving scum and grease removal, controlling activated sludge bulking, controlling foam, and aiding in chemical coagulation. The most important use of chlorine in the treatment of wastewater is for disinfection. The amount of residual chlorine remaining in the treated wastewater after passing through a contact basin or chamber may be related to the numbers of coliform bacteria allowed in the effluent by regulatory agencies. In many cases the residual chlorine must be reduced (for most practical purposes removed) before final discharge to a receiving stream for protection of fish and other aquatic life.

Chlorine reacts quickly (within minutes) and completely with ammonia in wastewater to produce monochloramine and dichloramine. Residual chlorine in the monochloramine or dichloramine state is termed "combined residual chlorine." With the amount of ammonia usually found in wastewater, the chlorine residual will contain all combined chlorine and no free chlorine. Because chlorine residual in wastewater is in a combined state, the determination of residual chlorine presents special problems.

B. *WHAT IS TESTED?*

Sample	**Common Range, mg/*L* (After 30 minutes)**
Effluent	0.5 to 2.0 mg/*L*

C. *METHODS*

The Iodometric Method for measuring residual chlorine is used for samples containing wastewater, such as plant effluents or receiving waters. The DPD Titrimetric Method is applicable to wastewaters that do not contain iodine-reducing substances. This method also has the advantage that it can be modified to determine free residual chlorine, monochloramine, and dichloramine. Colorimetric tests for residual chlorine have special limitations and should generally be avoided in wastewater. The *AMPEROMETRIC*[16] Titration Method gives the best results, but the titrator instrument is expensive. The Amperometric Titration Method can detect and measure chlorine residual below 0.2 mg/*L* but it is not possible to distinguish between free and combined chlorine forms at this low level. However, such differentiation may not be necessary for wastewater.

D. *APPARATUS REQUIRED*

Iodometric Method

1. Graduated cylinder, 250 m*L*
2. Pipets, 5 and 10 m*L*
3. Erlenmeyer flask, 500 m*L*
4. Buret, 5 m*L*
5. Magnetic stirrer

DPD Titrimetric Method

1. Graduated cylinder, 100 m*L*
2. Pipets, 1 and 10 m*L*
3. Erlenmeyer flask, 250 m*L*
4. Buret, 10 m*L*
5. Magnetic stirrer

Amperometric Titration Method (All Residual Levels)

1. Amperometric titrator
2. End point detection apparatus
3. Agitator
4. Buret

See page 4-58, *STANDARD METHODS,* 21st Edition, and amperometric titrator manufacturer's instruction manual.

E. *REAGENTS*

NOTE: Standardized solutions may be purchased from chemical suppliers.

[16] *Amperometric* (am-purr-o-MET-rick). A method of measurement that records electric current flowing or generated, rather than recording voltage. Amperometric titration is a means of measuring concentrations of certain substances in water.

(Chlorine Residual)

Iodometric Method

1. Standard phenylarsine oxide (PAO) solution, 0.00564 *N*. Dissolve approximately 0.8 gm phenylarsine oxide powder in 150 m*L* 0.3 *N* NaOH solution. After settling, remove upper 110 m*L* of this solution into 800 m*L* distilled water and mix thoroughly. Adjust pH up to between 6 and 7 with 6 *N* HCl and dilute to 950 m*L* with distilled water. To standardize this solution accurately, measure 5 to 10 m*L* of freshly standardized 0.0282 *N* iodine solution into a flask and add 1 m*L* KI solution. Titrate with phenylarsine oxide solution, using starch solution as an indicator. Adjust to exactly 0.00564 *N* and recheck against the standard iodine solution: 1.00 m*L* = 200 µg available chlorine. *CAUTION:* Toxic—avoid ingestion.
2. Potassium iodide (KI), crystals.
3. Acetate buffer solution, pH 4.0. Dissolve 146 gm anhydrous $NaC_2H_3O_2$, or 243 gm $NaC_2H_3O_2 \cdot 3\ H_2O$, in 400 m*L* distilled water, add 480 gm concentrated acetic acid, and dilute to 1 liter with distilled water.
4. Standard iodine titrant, 0.0282 *N*. Dissolve 25 gm KI in a little distilled water in a 1-liter volumetric flask, add the proper amount of 0.1 *N* iodine solution exactly standardized to yield a 0.0282 *N* solution, and dilute to 1 liter. Store in amber bottles or in the dark, protecting the solution from direct sunlight at all times and keeping it from all contact with rubber.
5. Starch indicator. Make a thin paste of 6 gm of potato starch in a small quantity of distilled water. Pour this paste into one liter of boiling, distilled water. Allow to boil for a few minutes, then settle overnight. Remove the clear supernatant and save; discard the rest. For preservation, add 1.25 grams salicylic acid or 4 grams zinc chloride.

DPD Titrimetric Method

1. 1 + 3 H_2SO_4: Carefully add 10 m*L* concentrated sulfuric acid to 30 m*L* distilled water. Cool.
2. Phosphate Buffer Solution: Dissolve 24 grams anhydrous disodium hydrogen phosphate Na_2HPO_4, and 46 grams anhydrous potassium dihydrogen phosphate, KH_2PO_4, in distilled water. Combine with 100 m*L* distilled water in which 0.800 gram (ethylenedinitrilo) tetraacetic acid, sodium salt, have been dissolved. Dilute to 1 liter with distilled water and add 0.02 gram mercuric chloride, HgCl, as a preservative and to prevent interference in the free available chlorine test caused by any trace amounts of iodide in the reagents.
3. DPD Indicator Solution: Dissolve 1 gram DPD Oxalate,[17] or 1.5 grams p-amino-N:N-diethylaniline sulfate in chlorine-free distilled water containing 8 m*L* 1 + 3 H_2SO_4 and 0.2 gram (ethylenedinitrilo) tetraacetic acid, sodium salt. Make up to 1 liter, store in brown glass-stoppered bottle and discard when discolored. *WARNING:* The oxalate is toxic. Avoid ingestion.
4. Standard Ferrous Ammonium Sulfate (FAS) Titrant, 0.00282 *N*. Dissolve 1.106 grams $Fe\ (NH_4)_2\ (SO_4)_2 \cdot 6\ H_2O$ in distilled water containing 1 m*L* of 1 + 3 H_2SO_4 and make up to 1 liter with freshly boiled and cooled distilled water. The normality can be checked using potassium dichromate (see method under E. *PROCEDURE* (For COD >50 mg/*L*), step 10).

[17] Eastman Chemical No. 7102 or equivalent.

Amperometric Titration Method

1. Standard phenylarsine oxide (PAO) solution 0.00564 *N*. See Iodometric Method, 1.
2. Phosphate buffer solution, pH 7. Dissolve 25.4 gm anhydrous KH_2PO_4 and 34.1 gm anhydrous Na_2HPO_4 in 800 m*L* distilled water. Add 2 m*L* sodium hypochlorite solution containing 1 percent chlorine and mix thoroughly. Protect from sunlight for 2 days. Determine that free chlorine still remains in solution. Expose to sunlight until no chlorine remains. Use ultraviolet lamp for final dechlorination if necessary. Dilute to 1 liter with distilled water and filter if any precipitate is present.
3. Potassium iodide solution. Dissolve 50 gm KI and dilute to 1 liter with freshly boiled and cooled distilled water. Store in the dark in a brown glass-stoppered bottle, preferably in the refrigerator. Discard when solution becomes yellow.
4. Acetate buffer solution. See Iodometric Method, 3.

Amperometric Titration Method (Low Residual Levels)

1. Standard phenylarsine oxide (PAO) solution 0.000564 *N*. Dilute 10.0 m*L* of 0.000564 *N* phenylarsine oxide to 100 m*L* with chlorine-free water.
2. Potassium iodide (KI), crystal.
3. Acetate buffer solution, pH 4.0. Dissolve 146 gm anhydrous $NaC_2H_3O_2$ (sodium acetate), or 143 gm $NaC_2H_3O_2 \cdot H_2O$, in 400 m*L* distilled water; add 480 gm concentrated acetic acid and dilute to 1 liter with chlorine-free water.

F. *VOLUME OF SAMPLE*

For residual chlorine concentrations of 10 mg/*L* or less, take a 200-m*L* sample for titration. For residuals greater than 10 mg/*L*, use proportionately less sample.

G. *PROCEDURE*

Iodometric Method

1. Pipet 5.00 m*L* 0.00564 *N* PAO solution into an Erlenmeyer flask.
2. Add excess KI (approximately 1 gm).
3. Add 4 m*L* acetate buffer solution, or enough to lower the pH to between 3.5 and 4.2.
4. Pour in 200 m*L* of sample.
5. Mix with magnetic stirrer.
6. Add 1 m*L* starch solution just before titration.
7. Titrate with 0.0282 *N* Iodine to the first appearance of blue color that remains after complete mixing.

OUTLINE OF PROCEDURE FOR CHLORINE RESIDUAL
IODOMETRIC METHOD

1. Place 5.00 m*L* phenylarsine oxide solution in Erlenmeyer flask.
2. Add excess KI (approx 1 gm).
3. Add 4 m*L* acetate buffer solution.

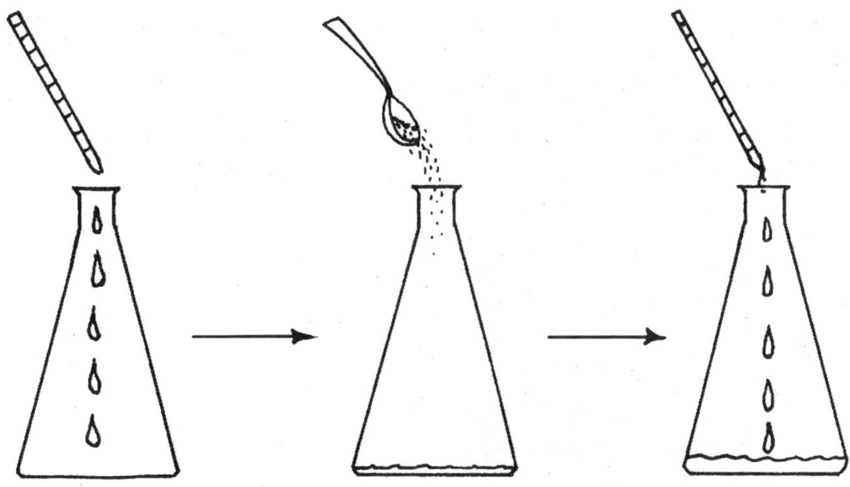

4. Add 200 m*L* sample.
5. Mix with stirring rod.
6. Add 1 m*L* starch solution.
7. Titrate until blue color first appears and remains after mixing.

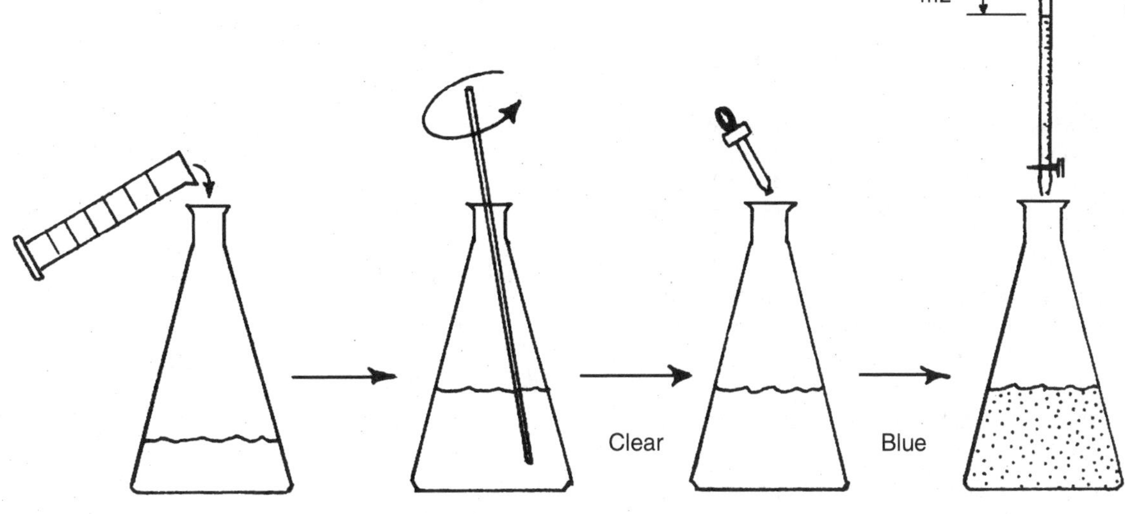

(Chlorine Residual)

DPD Titrimetric Method

1. Place 5 m*L* each of buffer reagent and DPD indicator in a 250-m*L* flask and mix.
2. Add 2 drops (0.1 m*L*) KI solution and mix.
3. Add 100 m*L* of sample and mix.
4. Let stand for 2 minutes and then titrate with FAS titrant.

Amperometric Titration Method

Free chlorine

1. Unless sample pH is known to be between 6.5 and 7.5, add 1 m*L* pH 7 phosphate buffer solution to produce a pH of 6.5 to 7.5.
2. Titrate with standard phenylarsine oxide (PAO), observing current changes on microammeter. Add titrant (PAO) in smaller amounts until all needle movement stops. Record buret readings when needle action becomes sluggish, signaling approach of end point. Subtract last small increment of titrant that causes no needle movement because of overtitration.

Combined chlorine

3. To sample remaining from free-chlorine titration, add 1.00 m*L* KI solution and 1 m*L* acetate buffer solution, in that order.
4. Titrate with phenylarsine oxide (PAO) titrant to the end point, as in Step 2, above. Do not refill buret but simply continue titration after recording buret reading for free chlorine. Again subtract last small increment of titrant to give amount of titrant actually used in reaction with chlorine. If titrant was continued without refilling buret, this buret reading represents total chlorine. Subtracting free chlorine from total gives combined chlorine.
5. Wash apparatus and sample cell thoroughly to remove iodide ion to avoid inaccuracies when the titrator is used subsequently for a free chlorine determination.

Monochloramine

6. After titrating for free chlorine, add 0.2 m*L* KI solution to same sample and, without refilling buret, continue to titrate with phenylarsine oxide (PAO) to end point. Subtract last increment of PAO to obtain net volume of PAO consumed by monochloramine.

Dichloramine

7. Add 1 m*L* acetate buffer solution and 1 m*L* KI solution to same sample and titrate final dichloramine fraction by titrating with PAO to end point. Subtract last increment of PAO to obtain net volume of PAO consumed by dichloramine.

Amperometric Titration Method (Low Residual Levels)

1. Select a sample volume requiring no more than 2 m*L* PAO titrant. A 200-m*L* sample will be adequate for sample containing less than 0.2 mg total chlorine/*L* (see Section I, Note 11).
2. Add 200 m*L* sample to sample container and add approximately 1.5 gm KI crystal. Dissolve using a stirrer or mixer.
3. Add 1.0 m*L* acetate buffer and place container in end point detection apparatus.
4. When the current (amperage) signal stabilizes, record the reading. Initially adjust meter to a near full-scale deflection.
5. Titrate by adding small, known volume of titrant. After each addition, record cumulative volume added and current reading when the signal stabilizes (see Section I, Note 12).
6. Continue adding titrant until no further meter deflection occurs.
7. Determine equivalence point by plotting total meter deflection against titrant volume added. Draw straight line through the first several points in the plot and a second, horizontal line corresponding to the final total deflection in the meter (See Figure 16.11).
8. Read equivalence point as the volume of titrant added at the intersection of these two lines.

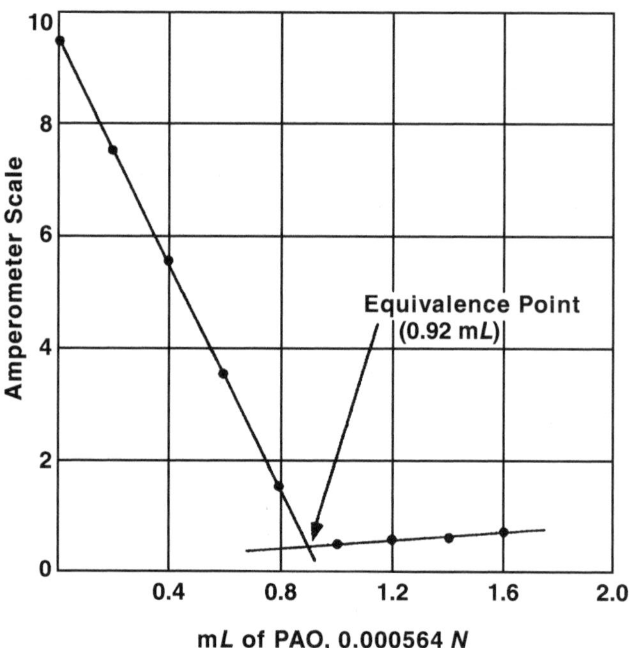

Fig. 16.11 Determination of low-level total chlorine

(Source: Leonard Ashack)

H. *CALCULATIONS*

Iodometric Method

$$\text{Total Residual Chlorine, mg}/L = \frac{(A - 5B) \times 200}{C}$$

A = m*L* 0.00564 *N* PAO

B = m*L* 0.0282 *N* I_2

C = m*L* of Sample Used

EXAMPLE:

Titration of a 200-m*L* sample requires 0.3 m*L* I_2 solution and 5 m*L* of PAO.

$$\text{Total Residual Chlorine, mg}/L = \frac{(A-5B)\times 200}{C}$$

$$= \frac{[5-(5\times 0.3)](200)}{200\ mL}$$

$$= (5-1.5)$$

$$= 3.5\ mg/L$$

DPD Titrimetric Method

For a 100-m*L* sample, 1.00 m*L* standard FAS titrant = 1.0 mg/*L* residual chlorine.

EXAMPLE:

100 m*L* of sample required 3.4 m*L* standard FAS titrant, therefore

Total Residual Chlorine, mg/*L* = 3.4 mg/*L*

Amperometric Titration Method

$$\text{Chlorine Residual, mg}/L = \frac{A\times 200}{mL\ \text{Sample}}$$

Where:

A = m*L* phenylarsine oxide (PAO) titration

EXAMPLE:

100 m*L* of sample required 2.4 m*L* PAO for titration of free and combined chlorine.

$$\text{Total Residual Chlorine, mg}/L = \frac{A\times 200}{mL\ \text{Sample}}$$

$$= \frac{2.4\times 200}{100\ mL}$$

$$= 4.8\ mg/L$$

Amperometric Titration Method (Low Residual Levels)

$$\text{mg Cl as } Cl_2/L = \frac{A\times 200\times N}{B\times 0.000564\ N}$$

where: A = m*L* Titrant at Equivalence Point

B = Sample Volume, m*L*

N = Phenylarsine (PAO) Normality

EXAMPLE:

The results of PAO titration on a 200-m*L* sample of chlorinated and dechlorinated effluent sample are as follows:

Volume of PAO, m*L*	Current Reading on Amperometer
0	9.5
0.2	7.5
0.4	5.6
0.6	3.6
0.8	1.6
1.0	0.5
1.2	0.6
1.4	0.6
1.6	0.7

From Figure 16.11, the equivalence point was determined at 0.92 m*L*.

$$\text{mg}/L \text{ of Cl} = \frac{0.92\times 200\times 0.000564}{200\times 0.000564}$$

$$= 0.92\ mg/L$$

I. *NOTES*

1. All chlorine residuals are reported as milligrams of chlorine as Cl_2 per liter (mg Cl as Cl_2/L).
2. Some organic chloramines can interfere with the test. Monochloramine can intrude into the free chlorine fraction and dichloramine can interfere in the monochloramine fraction, especially at high temperatures and prolonged titration times.
3. Free halogens (iodine, bromine) other than chlorine also will titrate as free chlorine.
4. Combined chlorine reacts with iodide ions to produce iodine.
5. When titration for free chlorine follows a combined chlorine titration that requires addition of KI, erroneous results may occur unless the measuring cell is rinsed thoroughly with distilled water between titrations.
6. Interference from copper has been observed in samples taken from copper pipe or after heavy copper sulfate treatment of reservoirs.
7. Silver ions poison the electrode.
8. Interference occurs in some highly colored waters and in waters containing surface-active agents.
9. Violent stirring can lower chlorine values by volatilization.
10. Very low temperatures slow the response of the measuring cell and longer time is required for the titration, but precision is not affected.

(Chlorine Residual)

11. It is possible to use the same method noted above and measure effluent chlorine concentration greater than 0.2 mg/*L* by using higher strength of PAO. According to the formula in Section H above, varying the concentration of PAO would allow flexibility as shown below.

Strength of PAO	Dilution from 0.00564 *N*	Chlorine Conc, in mg/*L* per m*L* Titrant	Maximum* Concentration Measured
0.000564	1 m*L* into 10 m*L*	0.1 mg/*L*	0.2 mg/*L*
0.001128	1 m*L* into 5 m*L*	0.2 mg/*L*	0.4 mg/*L*
0.00282	1 m*L* into 2 m*L*	1.0 mg/*L*	1.0 mg/*L*
0.00564	No dilution	1.0 mg/*L*	2.0 mg/*L*

* Assumed the maximum titrant volume is 2.0 m*L*.

12. By keeping the tip of titration needle touching the surface of sample being titrated, drop formation can be avoided. This will hasten the meter response.

13. Operators using the DPD colorimetric method to test water for a free chlorine residual need to be aware of a potential error that may occur. If the DPD test is run on water containing a combined chlorine residual, a precipitate may form during the test. The particles of precipitated material will give the sample a turbid appearance or the appearance of having color. This turbidity can produce a positive test result for free chlorine residual when there is actually no chlorine present. Operators call this error a "false positive" chlorine residual reading.

14. Home swimming pool chlorine residual analyzers are not approved, but could be used in an emergency for a short period of time. The results are not accurate.

J. *REFERENCES*

Iodometric Method: See page 4-58, *STANDARD METHODS,* 21st Edition.

DPD Titrimetric Method: See page 4-64, *STANDARD METHODS,* 21st Edition.

Amperometric Titration Method: See page 4-62, *STANDARD METHODS,* 21st Edition.

Low-Level Amperometric Titration Method: See page 4-64, *STANDARD METHODS,* 21st Edition.

QUESTIONS

Write your answers in a notebook and then compare your answers with those on page 612.

16.5F Why should plant effluents be chlorinated?

16.5G Discuss the important advantages and disadvantages between the iodometric titration and amperometric titration methods of measuring chlorine residual.

END OF LESSON 6 OF 9 LESSONS
ON
LABORATORY PROCEDURES AND CHEMISTRY

Please answer the discussion and review questions next.

DISCUSSION AND REVIEW QUESTIONS

Chapter 16. LABORATORY PROCEDURES AND CHEMISTRY

(Lesson 6 of 9 Lessons)

Write the answers to these questions in your notebook. The question numbering continues from Lesson 5.

29. What precautions should be observed when conducting the total alkalinity test?
30. Why is the COD test run?
31. What safety precautions must be considered when performing the COD test?
32. What are the sources of chloride in wastewater?
33. Why is the effluent from a treatment plant chlorinated?

CHAPTER 16. LABORATORY PROCEDURES AND CHEMISTRY

(Lesson 7 of 9 Lessons)

6. Coliform Group Bacteria

A. *DISCUSSION*

Most probable numbers of coliform bacteria are estimated to indicate the presence of bacteria originating from the intestines of warm-blooded animals. High coliform counts indicate the usefulness of water may have been impaired. Coliform bacteria are generally considered harmless, but their presence may indicate the presence of disease-producing organisms that may be found with them.

The coliform group of bacteria comprises all the aerobic and *FACULTATIVE*[18] anaerobic gram negative, non-spore-forming, rod-shaped bacteria that ferment lactose (a sugar) within 48 hours at 35°C. In general, coliform bacteria can be divided into a fecal and a non-fecal group. The fecal coliform can grow at a higher temperature (45°C) than the non-fecal coliform.

B. *TEST METHODS*

NPDES approved test procedures list two methods for Total Coliform analysis. They are: the MPN (Most Probable Number) Method and the Membrane Filter (MF) Method. Of the two methods, the MPN procedure is most applicable for wastewater. Be sure to check your NPDES permit to verify which method (or methods) may be used in your state for coliform group bacteria testing.

MPN METHOD

The multiple tube coliform test has been a standard method for determining coliform group bacteria since 1936. In this procedure tubes of lactose broth or lauryl tryptose broth are inoculated with dilutions of a wastewater or water sample. The coliform density is then calculated from statistical probability formulas that predict the most probable number (MPN) of coliforms necessary to produce certain combinations of gas-positive and gas-negative tubes in the series of inoculated tubes.

There are three distinct test stages of coliform testing using the MPN method: the presumptive test, the confirmed test, and the completed test.

MF METHOD

This method was introduced as a tentative method in 1955 and approved in 1997. The basic procedure involves filtering a known volume of water through a membrane filter of optimum pore size for full bacteria retention. As the water passes through the pores, bacteria are entrapped on the upper surface of the filter. The membrane filter is then placed in contact with either a paper pad saturated with liquid medium or directly over an agar medium to provide nutrients for bacterial growth. Following incubation under prescribed conditions of time, temperature, and humidity, the cultures are examined for coliform colonies that are counted and recorded as a density of coliforms per 100 m*L* of water sample.

There are certain important limitations of membrane filter methods. Some types of samples cannot be filtered because of turbidity, high non-coliform bacterial densities, or heavy metal compounds.

C. *WHAT IS TESTED?*

Sample	Common Ranges, MPN/100 m*L*	
	Total Coliform	**Fecal Coliform**
Effluents:		
Primary	5,000 to 2,000,000	1,000 to 500,000
Chlorinated Secondary	50 to 1,000	<2 to 200
Receiving Waters	50 to 1,000,000	<2 to 1,000

D. *MATERIALS REQUIRED*

1. *SAMPLING BOTTLES*

Bottles of glass or other material that are watertight, resistant to the solvent action of water, and capable of being sterilized may be used for bacteriological sampling. Plastic bottles made of nontoxic materials have been found to be satisfactory and eliminate the possibility of breakage during transport. The bottles should hold a sufficient volume of sample for all tests, permit proper washing, and maintain the samples uncontaminated until examinations are complete.

Before sterilization by autoclave, add 0.1 m*L* 10 percent sodium thiosulfate per 4-ounce (120-m*L*) bottle. This will neutralize a sample containing about 15 mg/*L* residual chlorine. If the residual chlorine is not neutralized, it would continue to be toxic to the coliform organisms remaining in the sample and give false results.

When filling bottles with sample, do not flush out sodium thiosulfate or contaminate bottle or sample. Fill bottles approximately three-quarters full and start test in laboratory within six hours. If the samples cannot be processed within one hour, they should be held below 10°C for not longer than six hours.

2. *MEDIA PREPARATION*

Careful media preparation is necessary for meaningful bacteriological testing. Attention must be given to the quality, mixing, and sterilization of the ingredients. The purpose of this care is to ensure that if the bacteria being tested for are

[18] *Facultative* (FACK-ul-tay-tive) *Bacteria.* Facultative bacteria can use either dissolved oxygen or oxygen obtained from food materials such as sulfate or nitrate ions. In other words, facultative bacteria can live under aerobic, anoxic, or anaerobic conditions.

(Coliform)

indeed present in the sample, every opportunity is presented for the development and ultimate identification. Much bacteriological identification is done by noting changes in the medium; consequently, the composition of the medium must be standardized. Much of the tedium of media preparation can be avoided by purchase of dehydrated media (Difco, BBL, or equivalent) from local scientific supply houses. The operator is advised to make use of these products; and, if only a limited amount of testing is to be done, consider using tubed, preprepared media.

GLASSWARE

All glassware must be thoroughly cleansed using a suitable detergent and hot water (160°F or 71°C), rinsed with hot water (180°F or 82°C) to remove all traces of residual detergent, and finally rinsed with distilled or deionized water.

WATER

Only distilled water or demineralized water that has been tested and found free from traces of dissolved metals and bactericidal and inhibitory compounds may be used for preparation of culture media.

BUFFERED[19] DILUTION WATER

Prepare a stock solution by dissolving 34 grams of KH_2PO_4 in 500 m*L* distilled water, adjusting the pH to 7.2 with 1 *N* NaOH and dilute to one liter. Prepare dilution water by adding 1.25 m*L* of the stock phosphate buffer solution and 5.0 m*L* magnesium sulfate (50 grams $MgSO_4 \bullet 7\ H_2O$ dissolved in one liter of water) to 1 liter distilled water. This solution can be dispersed into various size dilution blanks or used as a sterile rinse for the membrane filter test.

MEDIA—MPN (TOTAL AND FECAL COLIFORM)

1. Lactose Broth or Lauryl Tryptose Broth

   For the presumptive coliform test, dissolve the recommended amount of the dehydrated medium in distilled water. Dispense solution into fermentation tubes containing an inverted glass vial (see illustration of tube with vial on page 555).

2. Brilliant Green Bile Broth

   For the confirmed coliform test, dissolve 40 grams of the dehydrated medium in one liter of distilled water. Dispense and sterilize as with Lactose Broth.

3. EC Broth

   For the fecal coliform test, dissolve 37 grams of the dehydrated medium in one liter of distilled water. Dispense and sterilize as with Lactose Broth.

MEDIA—MEMBRANE FILTER METHOD (TOTAL AND FECAL COLIFORM)

1. M-Endo Broth

   Prepare this medium by dissolving 48 grams of the dehydrated product in one liter of distilled water that contains 20 m*L* of ethyl alcohol per liter. Heat solution to boiling only—do not autoclave. Prepared media should be stored in refrigerator at 2° to 10°C and used within 96 hours.

2. LES Endo Agar

   Prepare this medium, used for the two-step procedure, as per instructions found on the bottle.

3. M-FC Media

   Rehydrate in distilled water containing 10 m*L* of 1 percent Rosolic Acid in 0.2 *N* NaOH. Heat just to boiling then cool to below 45°C. Do not autoclave. Prepared media stored in the refrigerator should be used within 96 hours.

   The Rosolic Acid solution should be stored in the dark and discarded after 2 weeks or sooner if its color changes from dark red to muddy brown.

AUTOCLAVING

Steam autoclaves are used for the sterilization of the liquid media and associated apparatus. They sterilize (kill all organisms) at a relatively low temperature of 121°C (250°F) within 15 minutes by using moist heat.

Components of the media, particularly sugars such as lactose, may decompose at higher temperatures or longer heating times. For this reason adherence to time and temperature schedules is vital. The maximum elapsed time for exposure of the medium to any heat (from the time the autoclave door is closed to unloading) is 45 minutes. Preheating the autoclave can reduce total heating time.

Autoclaves operate similar to the once-familiar kitchen pressure cooker:

1. Water is heated in a boiler to produce steam.
2. The steam is vented to drive out air.
3. The steam vent is closed when the air is gone.
4. Continued heat raises the pressure to 15 lbs/sq in (1.05 kg/sq cm) (at this pressure, pure steam has a temperature of 121°C or 250°F).
5. The heat and pressure are maintained for 15 minutes.
6. Turn off the heat.
7. Open the steam vent and slowly vent the steam until atmospheric pressure is reached. (Fast venting will cause the liquids to boil and overflow tubes.)
8. Sterile material is removed to cool.

In autoclaving fermentation tubes, a vacuum is formed in the inner tubes. As the tubes cool, the inner tubes are filled with sterile medium. Capture of gas in this inner tube from the culture of bacteria is the evidence of fermentation and is recorded as a positive test.

MEDIA STORAGE

Culture media should be prepared in batches of such size that the entire batch will be used in less than one week. However, if the media are contained in screw-capped tubes they may be stored for up to three months. Store media out of

[19] *Buffer.* A solution or liquid whose chemical makeup neutralizes acids or bases without a great change in pH.

direct sun and avoid contamination and excessive evaporation.

E. *PROCEDURE FOR TESTING TOTAL COLIFORM* BACTERIA—MPN METHOD

1. *GENERAL DISCUSSION*

The test for coliform bacteria is used to measure the suitability of a water for human use. The test is not only useful in determining the bacterial quality of a finished water, but it can be used by the operator in the treatment plant as a guide to achieving a desired degree of treatment.

Coliform bacteria are detected in water by placing portions of a sample of the water in lactose broth. Lactose broth is a standard bacteriological medium containing lactose (milk) sugar in tryptose broth. The coliform bacteria are those that will grow in this medium at 35°C temperature and ferment and produce gas from the sugar within 48 hours. Thus to detect these bacteria the operator need only inspect fermentation tubes for gas. In practice, multiple fermentation tubes are used in dilutions for each sample. A schematic of the confirmed test procedure is shown in Figure 16.12.

2. *MATERIALS NEEDED*

a. Fifteen sterile tubes of lactose or lauryl tryptose broth are needed for each sample.

b. Use five tubes for each dilution.

c. Dilution tubes or blanks containing 99 m*L* of sterile buffered distilled water.

d. Quantity of 1 m*L* and 10 m*L* pipets.

e. Incubator set at 35° ± 0.5°C.

f. Thermometer verified to be accurate by comparison with National Bureau of Standards (NBS) certified thermometer.

3. *TECHNIQUE FOR INOCULATION AND/OR DILUTION OF SAMPLE (FIGURE 16.13)*

All inoculations and dilutions of wastewater specimens must be accurate and should be made so that no contaminants from the air, equipment, clothes, or fingers reach the specimen, either directly or by a contaminated pipet. Clean, sterile pipets must be used for each separate sample.

a. Shake the specimen bottle vigorously 20 times before removing sample volumes.

b. Pipet 1.0 m*L* of sample directly into the first five lactose tubes. (It is important to realize that the sample volume applied to the first five tubes will depend on the type of water being tested. The sample volume applied to each tube can vary from 10 m*L* (or more) for high quality waters to as low as 10^{-5} or 0.00001 m*L* (applied as 1 m*L* of a diluted sample) for raw wastewater specimens.)

NOTE: When delivering the sample into the culture medium, deliver sample portions of 1 m*L* or less down into the culture tube near the surface of the medium. Do not deliver small sample volumes at the top of the tube and allow them to run down inside the tube; too much of the sample will fail to reach the culture medium.

NOTE: Use 10-m*L* pipets for 10-m*L* sample portions, and 1-m*L* pipets for portions of 1 m*L* or less. Handle sterile pipet only near the mouthpiece, to protect the delivery end from external contamination.

c. Pipet 1/10 m*L* or 0.1 m*L* of raw sample into each of the next five lactose broth tubes.

d. To make the 0.01 dilution, place 1 m*L* of well-mixed raw sample into 99 m*L* of sterile buffered dilution water. Mix thoroughly by shaking. This bottle will be labeled Bottle A.

e. Into each of the next five lactose broth tubes place directly 1 m*L* of the 0.01 dilution from Bottle A.

At this point you have 15 tubes inoculated and can place these three sets of tubes in the incubator; however, your sample specimen may show gas production in all fermentation tubes. This means your sample was not diluted enough and you have no usable results. To obtain usable results it is recommended that the first time a sample is analyzed, 30 tubes having a range of six dilutions should be set up. In most cases this will give usable results.

f. To make a 0.001 m*L* dilution, add 0.1 m*L* from the 0.01 m*L* dilution bottle (Bottle A) directly into each tube of five more lactose broth tubes.

g. To make a 0.0001 m*L* dilution, take 1 m*L* from Bottle A and place this 1 m*L* into 99 m*L* of sterile buffered dilution water. Mix diluted sample thoroughly by shaking. This bottle will be labeled Bottle B.

h. From the 0.0001 m*L* dilution (Bottle B), pipet 1.0 m*L* of sample directly into each tube. Set up five tubes with this dilution.

i. To make a 0.00001 m*L* dilution, pipet 0.1 m*L* of sample directly into each tube. Set up five tubes with this dilution.

The number of dilutions for an MPN test depends on the type (rivers, effluent) of water being tested, results of previous tests, and expected MPN. The first time a sample is analyzed, 30 tubes of lactose broth should be prepared. Once the appropriate dilutions are established that give usable results for determining the MPN Index, only 15 tubes need be prepared for subsequent samples to be analyzed.

4. *INCUBATION (TOTAL COLIFORM)*

a. *24-HOUR LACTOSE BROTH PRESUMPTIVE TEST*

Place all inoculated lactose broth tubes in 35°C ± 0.5°C incubator. After 24 ± 2 hours have elapsed, examine each tube for gas formation in inverted vial (inner tube). Mark (+) on report form such as shown on Figure 16.14 for all tubes that show presence of gas. Mark (–) for all tubes showing no gas formation. Save all positive (+) tubes for confirmation test. The negative (–) tubes must be reincubated for an additional 24 hours.

b. *48-HOUR LACTOSE BROTH PRESUMPTIVE TEST*

Record positive and negative tubes at the end of 48 ± 3 hours. Save all positive tubes for confirmation test.

c. *24-HOUR BRILLIANT GREEN BILE CONFIRMATION TEST*

Confirm all presumptive tubes that show gas at 24 or 48 hours. Transfer, with the aid of a sterile 3-mm platinum wire loop (or sterile wood applicator), one loop-full of the broth from the lactose tubes showing gas, and inoculate a

(Coliform)

TOTAL COLIFORM

1. Presumptive Test

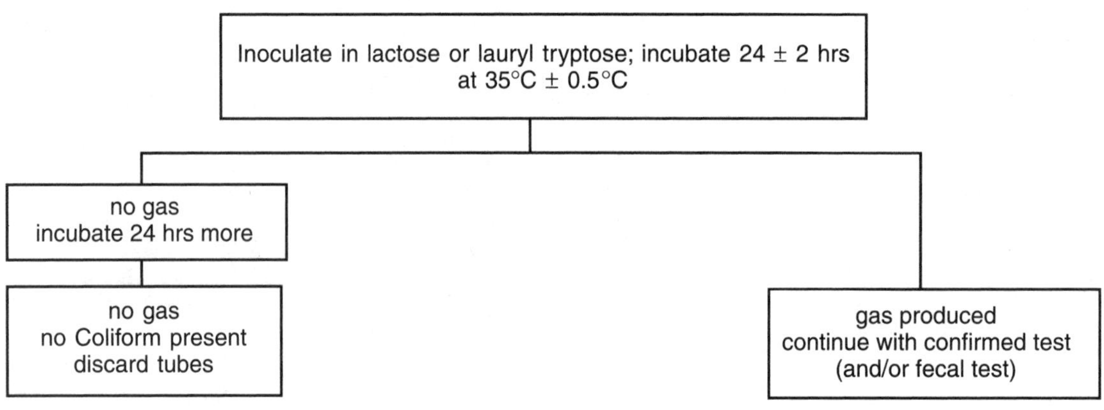
Inoculate in lactose or lauryl tryptose; incubate 24 ± 2 hrs at 35°C ± 0.5°C
no gas
incubate 24 hrs more
no gas
no Coliform present
discard tubes
gas produced
continue with confirmed test
(and/or fecal test)

2. Confirmed Test

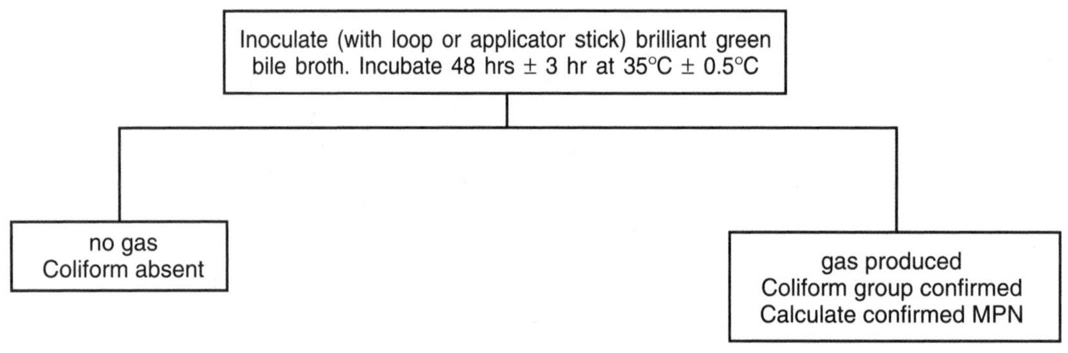
Inoculate (with loop or applicator stick) brilliant green bile broth. Incubate 48 hrs ± 3 hr at 35°C ± 0.5°C
no gas
Coliform absent
gas produced
Coliform group confirmed
Calculate confirmed MPN

FECAL COLIFORM

1. Presumptive Test
   (same as for TOTAL COLIFORM)
2. Fecal Coliform Test

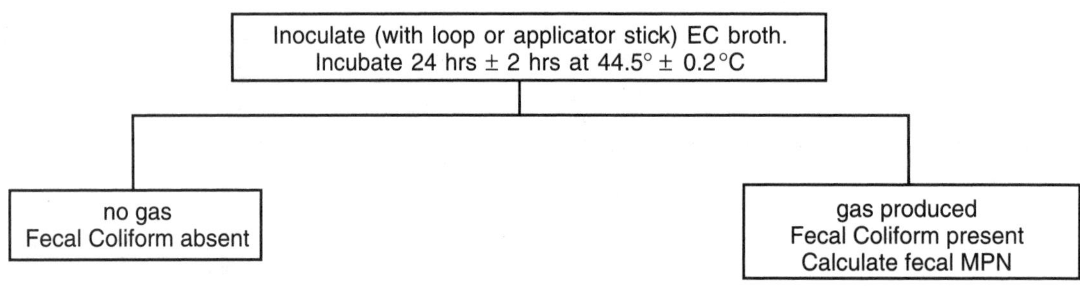
Inoculate (with loop or applicator stick) EC broth.
Incubate 24 hrs ± 2 hrs at 44.5° ± 0.2°C
no gas
Fecal Coliform absent
gas produced
Fecal Coliform present
Calculate fecal MPN

Fig. 16.12 Schematic outline of procedure for Total and Fecal Coliform—MPN Method

(Coliform)

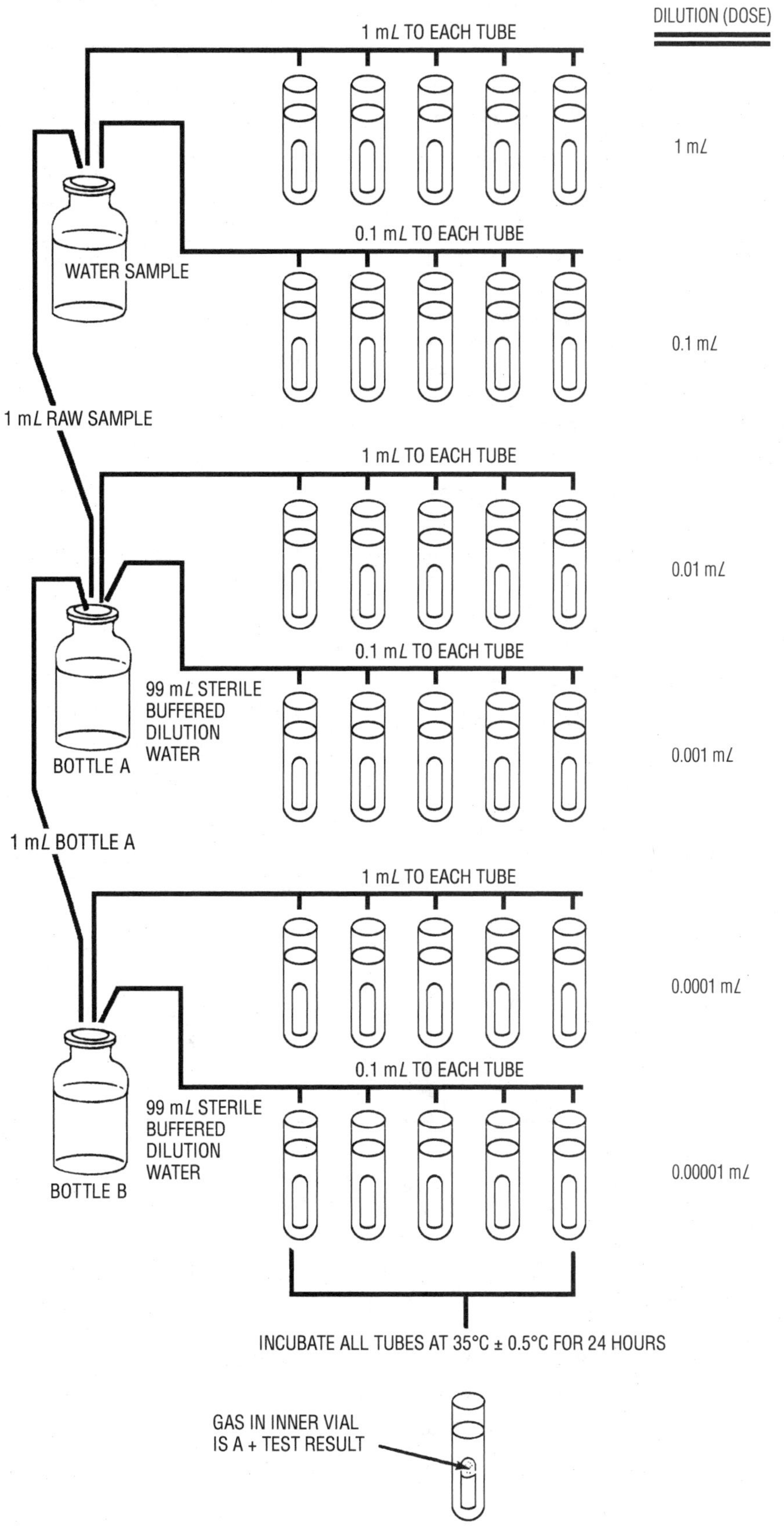

Fig. 16.13 Coliform bacteria test

(Coliform)

corresponding tube of BGB (Brilliant Green Bile) broth by mixing the loop of broth in the BGB broth. Discard all positive lactose broth tubes after transferring is completed.

Always sterilize inoculation loops and needles in flame immediately before transfer of culture; do not lay loop down or touch it to any nonsterile object before making the transfer. After sterilization in a flame, allow sufficient time for cooling, in the air, to prevent the heat of the loop from killing the bacterial cells being transferred. Wooden sterile applicator sticks also are used to transfer cultures, especially in the field where a flame is not available for sterilization. If using hardwood applicators, discard after use.

After 24 hours, inspect each BGB tube for gas formation. Those with any amount of gas are considered positive and are so recorded on the data sheet. Negative BGB tubes are reincubated for an additional 24 hours.

d. *48-HOUR BRILLIANT GREEN BILE CONFIRMATION TEST*

1. Examine tubes for gas at the end of the 48 ± 3 hour period. Record both positive and negative tubes.
2. Complete reports by determining MPN Index and recording MPN on work sheets.

5. *RECORDING RESULTS*

Results should be recorded on data sheets prepared especially for this test. An example is shown in Figure 16.14.

6. *METHOD OF CALCULATION OF THE MOST PROBABLE NUMBER (MPN)*

Table 16.5, page 558, is used to estimate the Most Probable Number (MPN) of coliforms per 100 m*L* in a sample. The numbers are calculated using a statistical probability formula. *NOTE:* Table 16.5 is extracted from the most recent edition of "Standard Methods." Some facilities may be obligated to use the MPN Index provided in a previous edition. Check with your local regulatory agency for the correct edition.

To estimate the MPN:

- Select the highest dilution or inoculation with all positive tubes, before a negative tube occurs, plus the next two dilutions (if none of the dilutions contain all positive tubes, select the three highest dilutions)
- Identify the number of positive tubes in each of the three dilutions you have selected and locate that sequence of positive tubes in Table 16.5
- Read the corresponding MPN Index value from Table 16.5
- Calculate the MPN value for your sample using the following formula:

$$\frac{\text{MPN}}{100\text{ m}L} = \frac{\text{MPN Index Value}}{\text{(From Table 16.5)}} \times \frac{10\text{ m}L}{\text{Highest Dilution Volume in Your Series}}$$

EXAMPLE NO. 1—Select the underlined dilutions (see Figure 16.15)

Dilutions (m*L*)	1	0.1	0.01	0.001	0.0001	0.00001
Readings	5	5	<u>5</u>	<u>2</u>	<u>0</u>	0

Read MPN Index as 50 per 100 m*L* from Table 16.5.

Calculate sample MPN:

$$\frac{\text{MPN}}{100\text{ m}L} = \frac{50\text{ MPN}}{100\text{ m}L} \times \frac{10\text{ m}L}{0.01\text{ m}L} = \frac{50{,}000\text{ MPN}}{100\text{ m}L}$$

Report MPN results as 50,000/100 m*L*.

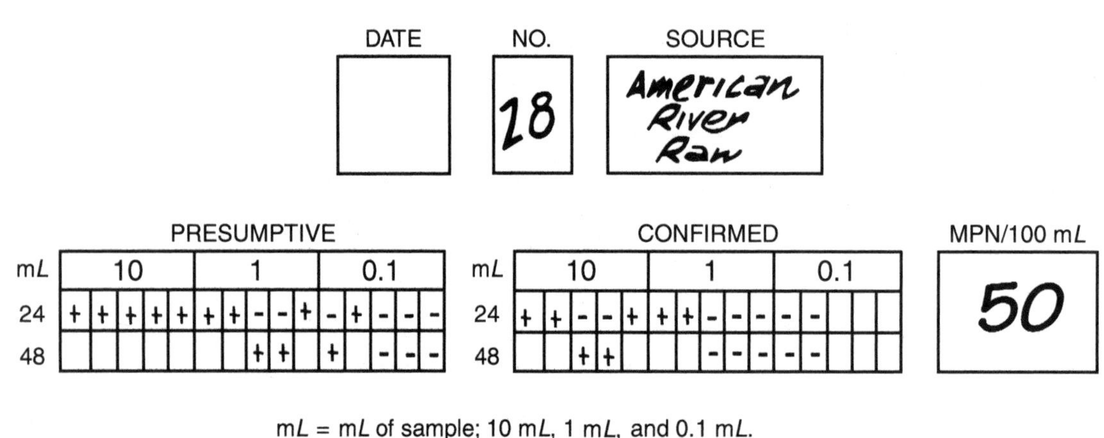

DATE: | NO.: 28 | SOURCE: American River Raw

PRESUMPTIVE

m*L*	10					1					0.1				
24	+	+	+	+	+	+	+	-	-	+	-	+	-	-	-
48								+	+		+		-	-	-

CONFIRMED

m*L*	10					1					0.1				
24	+	+	-	-	+	+	+	-	-	-	-	-			
48			+	+				-	-	-	-	-			

MPN/100 m*L*: 50

m*L* = m*L* of sample; 10 m*L*, 1 m*L*, and 0.1 m*L*.

24 = results after 24 hours of incubation.

48 = results after 48 hours of incubation.

Number of Positive Tubes 5, 2, 0

Fig. 16.14 Recorded coliform test results

(Coliform)

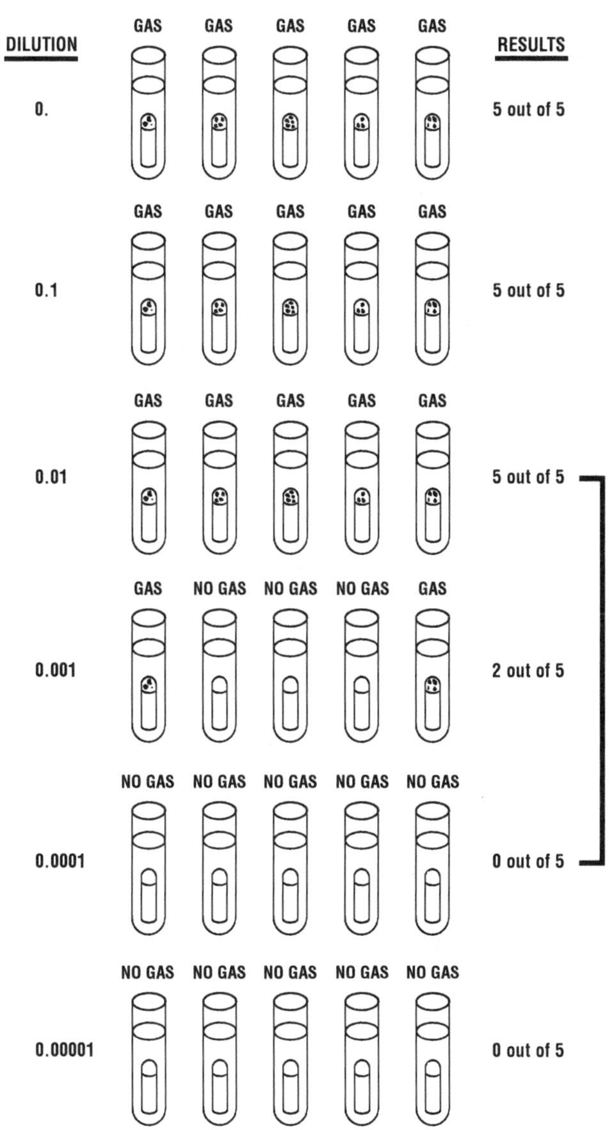

Fig. 16.15 Results of coliform test

We added three zeros to 50 because we started with the 0.01 m*L* dilution and Table 16.5 starts three dilution columns to the left (0.1 m*L*, 1 m*L*, and 10 m*L*).

EXAMPLE NO. 2—Select the underlined dilutions

Dilutions (m*L*)	1	0.1	0.01	0.001	0.0001	0.00001
Readings	5	5	5	5	0	0

Read MPN Index as 23 per 100 m*L* from Table 16.5.

Calculate sample MPN:

$$\frac{\text{MPN}}{100\text{ m}L} = \frac{23\text{ MPN}}{100\text{ m}L} \times \frac{10\text{ m}L}{0.001\text{ m}L} = \frac{230{,}000\text{ MPN}}{100\text{ m}L}$$

Report MPN results as 230,000/100 m*L*.

If positive tubes extend beyond three chosen dilutions, include positives beyond chosen dilutions by moving them forward.

EXAMPLE NO. 3

Dilutions (m*L*)	1	0.1	0.01	0.001	0.0001	0.00001
Readings	5	1	0	1	0	0
This becomes	5	1	1	0	0	0

The MPN is 500 per 100 m*L*.

The MPN for combinations not appearing in the table given or for other combinations of tubes or dilutions, may be estimated by the formula:

$$\text{MPN/100 m}L = \frac{\text{(No. of Positive Tubes)(100)}}{\sqrt{\left(\begin{matrix}\text{m}L\text{ Sample in}\\ \text{Negative Tubes}\end{matrix}\right) \times \left(\begin{matrix}\text{m}L\text{ Sample in}\\ \text{All Tubes}\end{matrix}\right)}}$$

EXAMPLE NO. 4

The results from an MPN test using five fermentation tubes for each dilution on a sample of water were as follows:

Sample Size, m*L*	10	1.0	0.1
Number of Positive Tubes Out of Five Tubes	3+	1+	0+

Determine the information necessary to solve the formula:

$$\text{MPN/100 m}L = \frac{\text{(No. of Positive Tubes)(100)}}{\sqrt{\left(\begin{matrix}\text{m}L\text{ Sample in}\\ \text{Negative Tubes}\end{matrix}\right) \times \left(\begin{matrix}\text{m}L\text{ Sample in}\\ \text{All Tubes}\end{matrix}\right)}}$$

1. Number of positive tubes. There were 3 positive tubes with 10 m*L* of sample and 1 positive tube with 1 m*L* of sample. Therefore, 3 + 1 = 4 positive tubes.

2. Determine the m*L* of sample in negative tubes.

Sample Size, m*L*	Number of Negative Tubes	m*L* of Sample in Negative Tubes		
10 m*L*	2–	(10 m*L*)(2)	=	20 m*L*
1.0 m*L*	4–	(1.0 m*L*)(4)	=	4 m*L*
0.1 m*L*	5–	(0.1 m*L*)(5)	=	0.5 m*L*
		Total	=	24.5 m*L*

3. Determine the m*L* of sample in all tubes.

Sample Size, m*L*	Number of Tubes	m*L* of Sample in All Tubes		
10 m*L*	5	(10 m*L*)(5)	=	50 m*L*
1.0 m*L*	5	(1.0 m*L*)(5)	=	5 m*L*
0.1 m*L*	5	(0.1 m*L*)(5)	=	0.5 m*L*
		Total	=	55.5 m*L*

(Coliform)

TABLE 16.5 MPN INDEX FOR SELECTED COMBINATIONS OF POSITIVE AND NEGATIVE RESULTS IN A PLANTING SERIES OF FIVE 10-m*L*, FIVE 1-m*L*, AND FIVE 0.1-m*L* PORTIONS OF SAMPLE

Number of tubes giving positive reaction out of			MPN Index
Five 10-m*L* portions	Five 1-m*L* portions	Five 0.1-m*L* portions	(organisms per 100 m*L*)
0	0	0	<2
0	0	1	2
0	0	2	4
0	1	0	2
0	1	1	4
0	1	2	5
0	2	0	4
0	2	1	6
0	3	0	6
1	0	0	2
1	0	1	4
1	0	2	6
1	0	3	8
1	1	0	4
1	1	1	6
1	1	2	8
1	2	0	6
1	2	1	8
1	2	2	10
1	3	0	8
1	3	1	10
1	4	0	11
2	0	0	4
2	0	1	7
2	0	2	9
2	0	3	12
2	1	0	7
2	1	1	9
2	1	2	12
2	2	0	9
2	2	1	12
2	2	2	14
2	3	0	12
2	3	1	14
2	4	0	15
3	0	0	8
3	0	1	11
3	0	2	13
3	1	0	11
3	1	1	14
3	1	2	17
3	1	3	20
3	2	0	14
3	2	1	17
3	2	2	20
3	3	0	17
3	3	1	21

(Coliform)

TABLE 16.5 MPN INDEX FOR SELECTED COMBINATIONS OF POSITIVE AND NEGATIVE RESULTS IN A PLANTING SERIES OF FIVE 10-m*L*, FIVE 1-m*L*, AND FIVE 0.1-m*L* PORTIONS OF SAMPLE *(continued)*

Number of tubes giving positive reaction out of			MPN Index
Five 10-m*L* portions	Five 1-m*L* portions	Five 0.1-m*L* portions	(organisms per 100 m*L*)
3	4	0	21
3	4	1	24
3	5	0	25
4	0	0	13
4	0	1	17
4	0	2	21
4	0	3	25
4	1	0	17
4	1	1	21
4	1	2	26
4	2	0	22
4	2	1	26
4	2	2	32
4	3	0	27
4	3	1	33
4	3	2	39
4	4	0	34
4	4	1	40
4	5	0	41
4	5	1	48
5	0	0	23
5	0	1	30
5	0	2	40
5	0	3	60
5	0	4	80
5	1	0	30
5	1	1	50
5	1	2	60
5	1	3	80
5	2	0	50
5	2	1	70
5	2	2	90
5	2	3	130
5	2	4	150
5	2	5	180
5	3	0	80
5	3	1	110
5	3	2	140
5	3	3	170
5	3	4	210
5	3	5	250
5	4	0	130
5	4	1	170
5	4	2	220
5	4	3	280

(Coliform)

TABLE 16.5 MPN INDEX FOR SELECTED COMBINATIONS OF POSITIVE AND NEGATIVE RESULTS IN A PLANTING SERIES OF FIVE 10-m*L*, FIVE 1-m*L*, AND FIVE 0.1-m*L* PORTIONS OF SAMPLE *(continued)*

Number of tubes giving positive reaction out of			MPN Index
Five 10-m*L* portions	Five 1-m*L* portions	Five 0.1-m*L* portions	(organisms per 100 m*L*)
5	4	4	350
5	4	5	430
5	5	0	240
5	5	1	300
5	5	2	500
5	5	3	900
5	5	4	1,600
5	5	5	≥1,600

4. Estimate the MPN/100 m*L*.

$$\text{MPN/100 m}L = \frac{(\text{No. of Positive Tubes})(100)}{\sqrt{\left(\begin{array}{c}\text{m}L\text{ Sample in}\\\text{Negative Tubes}\end{array}\right)\times\left(\begin{array}{c}\text{m}L\text{ Sample in}\\\text{All Tubes}\end{array}\right)}}$$

$$= \frac{(4)(100)}{\sqrt{(24.5\text{ m}L)(55.5\text{ m}L)}}$$

$$= \frac{400}{36.87}$$

$$= 10.8\text{ MPN Coliforms/100 m}L$$

$$= 11\text{ MPN Coliforms/100 m}L$$

Coliform results may be summarized with a single MPN value by using the arithmetic mean, median, or geometric mean. *STANDARD METHODS* recommends using the median or the geometric mean because one extremely high MPN value can distort the mean. The arithmetic mean is often called the average.

$$\text{Mean} = \frac{\text{Sum of Items or Values}}{\text{Number of Items or Values}}$$

The median is the middle value in a set or group of data. There are just as many values larger than the median as there are smaller than the median. To determine the median, the data should be written in ascending (increasing) or descending (decreasing) order and the middle value identified.

Median = Middle value of a group of data

The median and geometric mean are used when a sample contains a few very large values. This frequently happens when measuring the MPN of raw water. If X is a measurement (MPN) and n is the number of measurements, then

$$\text{Geometric Mean} = [(X_1)(X_2)(X_3)\ldots\ldots(X_n)]^{1/n}$$

This calculation can be easily performed on many electronic calculators. See Chapter 18, Section 18.4, "Average or Arithmetic Mean," 18.6, "Median and Mode," and 18.7, "Geometric Mean," for details on how to calculate these values.

EXAMPLE NO. 5

Results from MPN tests during one week were as follows:

Day	S	M	T	W	TH	F	S
MPN/100 m*L*	2	8	14	6	10	26	4

Estimate the (1) mean, (2) median, and (3) geometric mean of the data in MPN/100 m*L*.

1. Calculate the mean.

$$\text{Mean, MPN/100 m}L = \frac{\text{Sum of All MPNs}}{\text{Number of MPNs}}$$

$$= \frac{2+8+14+6+10+26+4}{7}$$

$$= \frac{70}{7}$$

$$= 10\text{ MPN/100 m}L$$

2. To determine the median, rearrange the data in ascending (increasing) order and select the middle value (three will be smaller and three will be larger in this example).

Order	1	2	3	4	5	6	7
MPN/100 m*L*	2	4	6	8 ↑	10	14	26

$$\text{Median, MPN/100 m}L = \text{Middle Value of a Group of Data}$$

$$= 8\text{ MPN/100 m}L$$

3. Calculate the geometric mean for the given data.

$$\text{Geometric Mean, MPN/100 m}L = [(x_1)(x_2)(x_3)(x_4)(x_5)(x_6)(x_7)]^{1/7}$$

$$= [(2)(8)(14)(6)(10)(26)(4)]^{1/7}$$

$$= [1{,}397{,}760]^{1/7}$$

$$= 7.5 \text{ MPN/100 m}L$$

4. Summarize the results.

1. Mean = 10 MPN/100 m*L*
2. Median = 8 MPN/100 m*L*
3. Geometric Mean = 7.5 MPN/100 m*L*

As you can see from the summary, the geometric mean more nearly describes most of the MPNs. This is the reason why the geometric mean is sometimes used to describe the results of MPN tests when there are a few very large values.

F. *TEST FOR FECAL COLIFORM BACTERIA—MPN*

1. *GENERAL DISCUSSION*

Many regulatory agencies are measuring the bacteriological quality of water using the fecal coliform test. This test more reliably indicates the potential presence of pathogenic organisms than do tests for total coliform group of organisms. The procedure described is an *ELEVATED TEMPERATURE TEST FOR FECAL COLIFORM BACTERIA.*

2. *MATERIALS NEEDED*

Equipment required for the tests are the same as those required for the 24-Hour Lactose Broth Presumptive Test, plus a water bath set at 44.5° ± 0.2°C, EC Broth media, and a thermometer certified against an NBS thermometer.

3. *PROCEDURE*

a. Run lactose broth or lauryl tryptose broth presumptive test.

b. After 24 hours temporarily retain all gas-positive tubes.

c. Label a tube of EC broth to correspond with each gas-positive tube of broth from presumptive test.

d. Shake or mix positive presumptive tubes by rotating them. Transfer one loop-full of culture from each gas-positive culture in presumptive test to the correspondingly labeled tube of EC broth.

e. Incubate EC broth tubes 24 ± 2 hours at 44.5° ± 0.2°C in a water bath with water depth sufficient to come up at least as high as the top of the culture medium in the tubes. Place in water bath as soon as possible after inoculation and always within 30 minutes after inoculation.

f. After 24 hours remove the rack of EC cultures from the water bath, shake gently, and record gas production for each tube. Gas in any quantity is a positive test.

g. As soon as results are recorded, discard all tubes. This is a 24-hour test for EC broth inoculations and not a 48-hour test.

h. Transfer any additional 48-hour gas-positive tubes from the presumptive test to correspondingly labeled tubes of EC broth. Incubate for 24 ± 2 hours at 44.5° ± 0.2°C and record results on data sheet.

i. Codify results using the same procedure as for total coliforms and determine MPN of fecal coliforms per 100 m*L* of sample (Figure 16.14).

EXAMPLE:

m*L* portion	10	1	0.1
Readings	5	2	1
Read MPN as	70 per 100 m*L* from Table 16.5		
Reports results as MPN of fecal coliforms	= 70 per 100 m*L*		

G. *MEMBRANE FILTER METHOD—TOTAL COLIFORM*

1. *GENERAL DISCUSSION*

In addition to the fermentation tube test for coliform bacteria, another test is used for these same bacteria in water analysis. This test uses a cellulose ester filter, called a membrane filter, the pore size of which can be manufactured to close tolerances. Not only can the pore size be made to selectively trap bacteria from water filtered through the membrane, but nutrients can be diffused up through the membrane to grow these bacteria into colonies. These colonies are recognizable as coliform because the nutrients include fuchsin dye, which peculiarly colors the colony. By counting the number of colonies and knowing the volume of water filtered, the operator can compare the water tested with water quality standards.

A two-step pre-enrichment technique is included at the end of this section for samples that have been chlorinated. Chlorinated bacteria that are still living have had their enzyme systems damaged and require a 2-hour enrichment media before contact with the selective M-Endo Media.

2. *MATERIALS NEEDED*

a. One sterile membrane filter having a 0.45μ pore size

b. One sterile 47-mm petri dish with lid

c. One sterile funnel and support stand

d. Two sterile pads

e. One receiving flask (side-arm, 1,000 m*L*)

f. Vacuum pump, trap, suction or vacuum gauge, connection sections of plastic tubing, glass "T" hose clamp to adjust pressure bypass

g. Forceps (round-tipped tweezers), alcohol, Bunsen burner, grease pencil

h. Sterile buffered distilled water for rinsing

i. M-Endo Media

j. Sterile pipets—two 5-m*L* graduated pipets and one 1-m*L* pipet for sample or one 10-m*L* pipet for larger sample. Quantity of one-m*L* pipets if dilution of sample is necessary. Also, quantity of dilution water blanks if dilution of sample is necessary.

k. One moist incubator at 35°C; auxiliary incubator dish with cover

l. Enrichment media—lauryl tryptose broth (for preenrichment technique)

m. A binocular wide-field dissecting microscope is recommended for counting. The light source should be a cool white fluorescent lamp.

(Coliform)

3. *SELECTION OF SAMPLE SIZE*

Size of the sample or *ALIQUOT*[20] will be governed by the expected bacterial density. An ideal quantity will result in the growth of about 50 coliform colonies, but not more than 200 bacterial colonies of all types. The table below lists suggested sample volumes for MF total coliform testing.

	Quantities Filtered (m*L*)						
	100	10	1	0.1	0.01	0.001	0.0001
Lakes	x	x	x				
Rivers		x	x	x			
Secondary Effluent		x	x	x	x		
Untreated Wastewater					x	x	x

When less than 20 m*L* of sample is to be filtered, a small amount of sterile dilution water should be added to the funnel before filtration. This increase in water volume aids in uniform dispersion of the sample over the membrane filter.

4. *PREPARATION OF PETRI DISH FOR MEMBRANE FILTER*

a. Sterilize forceps by dipping in alcohol and passing quickly through Bunsen burner flame.

b. Place sterile absorbent pad into sterile petri dish.

c. Add 1.8 to 2.0 m*L* M-Endo Media to absorbent pad using a sterile pipet.

5. *PROCEDURE FOR FILTRATION OF UNCHLORINATED SAMPLE*

All filtrations and dilutions of water specimens must be accurate and should be made so that no contaminants from the air, equipment, clothes, or fingers reach the specimen either directly or by way of the contaminated pipet.

a. Secure tubing from pump and bypass to receiving flask. Place palm of hand on flask opening and start pump. Adjust suction to $^1/_4$ atmosphere with hose clamp on pressure bypass. Turn pump switch to *OFF.*

b. Set sterile filter-support stand and funnel on receiving flask. Loosen wrapper. Rotate funnel counterclockwise to disengage pin. Re-cover with wrapper.

c. Place petri dish on bench with lid up. Write identification on lid with grease pencil.

d. Unwrap sterile pad container. Light Bunsen burner.

e. Unwrap membrane filter container.

f. Sterilize forceps by dipping in alcohol and passing quickly through Bunsen burner.

g. Center membrane filter on filter stand with forceps after lifting funnel. Membrane filter with printed grid should show grid uppermost (Figure I, next page).

h. Replace funnel and lock against pin (Figure II).

i. Shake sample or diluted sample. Measure proper aliquot with sterile pipet and add to funnel.

j. Add a small amount of the sterile dilution water to funnel. This will help check for leakage and also aid in dispersing small volumes (Figure III).

k. Now start vacuum pump.

l. Rinse filter with three 20- to 30-m*L* portions of sterile dilution water.

m. When membrane filter appears barely moist, switch pump to *OFF.*

n. Sterilize forceps as before.

o. Remove funnel and then remove membrane filter using forceps as before (Figure I).

p. Center membrane filter on pad containing M-Endo Media with a rolling motion to ensure water seal. Inspect membrane to ensure no captured air bubbles are present (Figure IV).

q. Place inverted petri dish in incubator for 22 ± 2 hours.

6. *PROCEDURE FOR COUNTING MEMBRANE FILTER COLONIES*

a. Remove petri dish from incubator.

b. Remove lid from petri dish.

c. Turn so that your back is to window.

d. Tilt membrane filter in base of petri dish so that green and yellow-green colonies are most apparent. Direct sunlight has too much red to facilitate counting.

e. Count individual colonies using an overhead fluorescent light. Use a low power (10 to 15 magnifications) binocular wide-field dissecting microscope or other similar optical device. The typical colony has a pink to dark red color with a metallic surface sheen. The sheen area may vary from a small pinhead size to complete coverage of the colony surface. Only those showing this sheen should be counted.

f. Report total number of "coliform colonies" or "colony formation units" (CFU) on work sheet. Use the membranes that show from 20 to 80 colonies and do not have more than 200 colonies of all types (including non-sheen or, in other words, non-coliforms).

EXAMPLE:

A total of 42 colonies grew after filtering a 10-m*L* sample.

$$\text{Bacteria/100 m}L = \frac{\text{No. of Colonies Counted} \times 100\text{ m}L}{\text{Sample Volume Filtered, m}L \times 100\text{ m}L}$$

$$= \frac{(42\text{ Colonies})(100\text{ m}L)}{(10\text{ m}L)(100\text{ m}L)}$$

$$= \frac{(4.2)(100\text{ m}L)}{100\text{ m}L}$$

$$= 420\text{ per }100\text{ m}L$$

[20] *Aliquot* (AL-uh-kwot). Representative portion of a sample. Often, an equally divided portion of a sample.

(Coliform)

OUTLINE OF PROCEDURE FOR INOCULATION OF MEMBRANE FILTER

Fig. I

1. Center membrane filter on filter holder. Handle membrane only on outer 3/16 inch (5 mm) with forceps sterilized before use in ethyl or methyl alcohol and passed lightly through a flame.

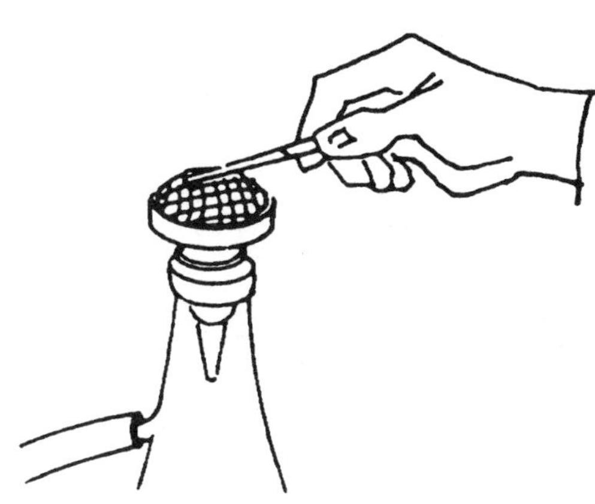

Fig. II

2. Place funnel onto filter holder.

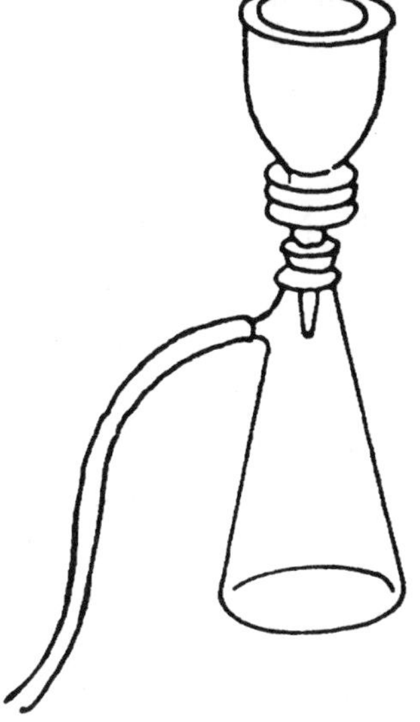

Fig. III

3. Pour or pipet sample aliquot into funnel. Avoid spattering. After suction is applied, rinse three times with sterile buffered distilled water.

Fig. IV

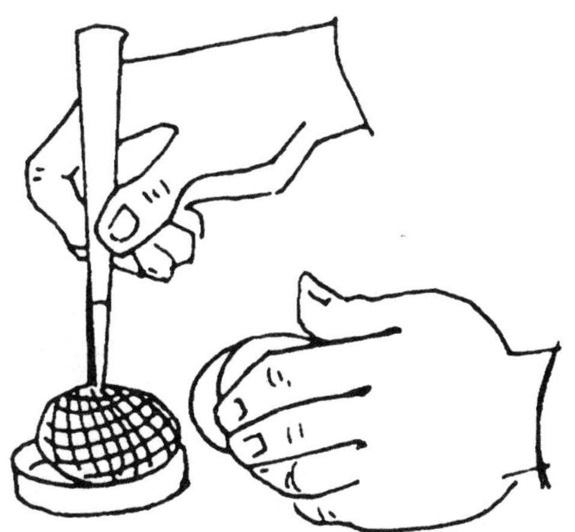

4. Remove membrane filter from filter holder with sterile forceps. Place membrane on pad. Cover with petri top.

Fig. V

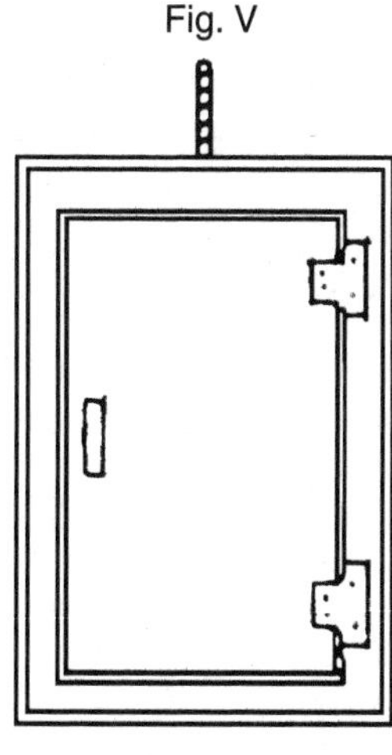

5. Incubate in inverted position for 22 ± 2 hours.

6. Count colonies on membrane.

(Coliform)

SPECIAL NOTE: Inexperienced persons often have great difficulty with connected colonies, with mirror reflections of fluorescent tubes (which are confused with metallic sheen), and with water condensate and particulate matter, which are occasionally mistaken for colonies. Thus there is a tendency for inexperienced persons to make errors on the high side in MF counts. Technicians who have not attained proficiency in coliform colony recognition should transfer questionable colonies to lactose (or lauryl tryptose) broth tubes for verification as coliform organisms.

7. *PROCEDURE FOR FILTRATION OF CHLORINATED SAMPLES USING ENRICHMENT TECHNIQUE*

a. Place a sterile absorbent pad in the upper half of a sterile petri dish and pipet 1.8 to 2.0 m*L* sterile lauryl tryptose broth. Carefully remove any surplus liquid.

b. *ASEPTICALLY*[21] place the membrane filter through which the sample has been passed on the pad.

c. Incubate the filter, without inverting the dish, for 1½ to 2 hours at 35°C in an atmosphere of 90 percent humidity (damp paper towels added to a plastic container with a snap-on lid can be used).

d. The enrichment culture is then removed from the incubator. A fresh sterile absorbent pad is placed in the bottom half of the petri dish and saturated with 1.8 to 2.0 m*L* M-Endo Broth.

e. The membrane filter is transferred to the new pad. The used pad of lactose or lauryl tryptose may be discarded.

f. Invert the dish and incubate for 20 to 22 hours at 35° ± 0.5°C.

g. Count colonies as in previous method.

H. *MEMBRANE FILTER METHOD—FECAL COLIFORM*

The membrane filter procedure for fecal coliform uses an enriched lactose medium (M-FC Broth) that depends on an incubation temperature of 44.5° ± 0.2°C for its selectivity. Since the temperature is critical, incubation takes place in a water bath using watertight plastic bags.

1. *MATERIALS REQUIRED*

a. M-FC media.

b. Tight-fitting culture dishes.

c. Membrane filters.

d. Watertight plastic bags.

e. Water bath set at 44.5 ± 0.2°C. The thermometer must be checked against an NBS thermometer to ensure the accuracy of the water bath temperature.

f. Sample size must be chosen to yield 20 to 60 fecal colonies on a filter. Suggested sample volumes are shown below:

	Volume to be filtered in milliliters					
	100	10	1	0.1	0.01	0.001
Raw Wastewater				x	x	x
Secondary Effluent		x	x	x		
Receiving Streams	x	x	x	x		

2. *PREPARATION OF CULTURE DISH*

Place a sterile absorbent pad in each culture dish and pipet (sterile) approximately 2 m*L* of M-FC medium to saturate the pad. Carefully remove any surplus liquid.

3. *FILTRATION OF SAMPLE*

Observe the procedure as prescribed for total coliform using membrane filters.

4. *INCUBATION*

Place the prepared culture dishes in waterproof plastic bags and immerse in water bath set at 44.5° ± 0.2°C for 24 hours. All culture dishes should be placed in the water bath within 30 minutes after filtration.

5. *COUNTING*

Colonies produced by fecal coliform bacteria are blue. The non-fecal coliform colonies are gray to cream colored. Normally, few non-fecal coliform colonies will be observed due to the selective action of the elevated temperature and the addition of the rosolic acid to the M-FC media.

Examine the cultures under a low-power magnification. Count and calculate fecal coliform density per 100 m*L*.

$$\text{Fecal Coliforms/100 m}L = \frac{\text{Fecal Colonies Counted} \times 100 \text{ m}L}{\text{Sample Volume Filtered, m}L \times 100 \text{ m}L}$$

EXAMPLE:

A total of 78 colonies grew after filtering a 10-m*L* sample.

$$\text{Fecal Coliforms/100 m}L = \frac{\text{Fecal Colonies Counted} \times 100 \text{ m}L}{\text{Sample Volume Filtered, m}L \times 100 \text{ m}L}$$

$$= \frac{(78 \text{ Colonies})(100 \text{ m}L)}{(10 \text{ m}L)(100 \text{ m}L)}$$

$$= \frac{(7.8)(100)}{100 \text{ m}L}$$

$$= 780 \text{ per } 100 \text{ m}L$$

QUESTIONS

Write your answers in a notebook and then compare your answers with those on page 612.

16.5H Why should sodium thiosulfate crystals be added to sample bottles for coliform tests before sterilization?

16.5I Steam autoclaves sterilize (kill all organisms) at a relatively low temperature (___ °C) within __________ minutes by using moist heat.

16.5J Estimate the Most Probable Number (MPN) of coliform group bacteria from the following test results:

Dilutions (m*L*)	1	0.1	0.01	0.001	0.0001	0.00001
Readings (+ tubes)	5	5	5	1	2	0

16.5K How is the number of coliforms estimated by the membrane filter method?

---

[21] *Aseptic* (a-SEP-tick). Free from the living germs of disease, fermentation, or putrefaction. Sterile.

(Coliform)

I. *MEMBRANE FILTER METHOD*—ESCHERICHIA COLI *(EPA Method)*

This method describes a membrane filter (MF) procedure for identifying and counting *Escherichia coli.* Because the bacterium is a natural inhabitant only of the intestinal tract of warm-blooded animals, its presence in water samples is an indication of fecal pollution and the possible presence of enteric (intestinal) pathogens.

This MF method provides a direct count of bacteria in water based on the development of colonies on the surface of the membrane filter. A water or wastewater sample is filtered through the membrane, which retains the bacteria. After filtration, the membrane containing the bacterial cells is placed on a specially formulated medium, mTEC, incubated at 35°C for 2 hours to revive injured or stressed bacteria, and then incubated at 44.5°C for 22 hours. Following incubation, the filter is transferred to a filter pad saturated with urea substrate. After 15 minutes, yellow or yellow-brown colonies are counted with the aid of a fluorescent lamp and magnifying lens.

1. *MATERIALS AND REAGENTS REQUIRED*

a. See previous Section H, 1. Materials required.

b. mTEC Agar (Difco 0334-16-0)

Preparation: Add 45.26 grams of dehydrated mTEC medium to 1 *L* of reagent water in a flask and heat to boiling, until ingredients dissolve. Autoclave at 121°C (15 psi pressure or 1 kg/sq cm) for 15 minutes and cool in a 44° to 46°C water bath. Pour the medium into each 50 × 10 mm culture dish to a 4- to 5-mm depth (approximately 4 to 6 m*L*) and allow to solidify. Final pH should be 7.3 ± 0.2. Store in a refrigerator.

c. Urea Substrate Medium

Preparation: Add dry ingredients to 100 m*L* reagent water in a flask. Stir to dissolve and adjust to pH 5.0 with a few drops of 1*N* HCl. The substrate solution should be a straw yellow color at pH 5.0.

d. Nutrient Agar (Difco 0001-02, BBL 11471)

Preparation: Add 23 grams of dehydrated nutrient sugar to 1 *L* of reagent water and mix well. Heat in a boiling water bath to dissolve the agar completely. Dispense in screw-cap tubes and autoclave at 121°C (15 psi pressure or 1 kg/sq cm) for 15 minutes. Remove tubes and slant. The final pH should be 6.8 ± 0.2.

2. *PROCEDURE*

a. Prepare the mTEC agar and urea substrate as directed in 1.a. and 1.b.

b. Mark the petri dishes and report forms with sample identification and sample volumes.

c. Place a sterile membrane filter on the filter base, gridside up, and attach the funnel to the base; the membrane filter is now held between the funnel and the base.

d. Shake the sample bottle vigorously about 25 times to distribute the bacteria uniformly and measure the desired volume of sample or dilution into the funnel.

e. For ambient surface waters and wastewaters, select sample volumes based on previous knowledge of pollution level, to produce 20 to 80 *E. coli* colonies on the membranes. Sample volumes of 1 to 100 m*L* are normally tested at half-log intervals, for example 100, 30, 10, 3 m*L*.

f. Smaller sample sizes or sample dilutions can be used to minimize the interference of turbidity or high bacterial densities. Multiple volumes of the same sample or sample dilution may be filtered and the results combined.

g. Filter the sample and rinse the sides of the funnel at least twice with 20 to 30 m*L* of sterile buffered rinse water. Turn off the vacuum and remove the funnel from the filter base.

h. Use sterile forceps to aseptically remove the membrane filter from the filter base and roll it onto the mTEC agar to avoid the formation of bubbles between the membrane and the agar surface. Reseat the membrane if bubbles occur. Close the dish, invert, and incubate at 35°C for 2 hours.

i. After 2 hours' incubation at 35°C, transfer the plates to Whirl-Pak bags, seal, and place inverted in a 44.5°C water bath for 22 to 24 hours.

j. After 22 to 24 hours, remove the dishes from the water bath. Place absorbent pads in new petri dishes or the lids of the same petri dishes and saturate with urea broth. Aseptically transfer the membranes to absorbent pads saturated with urea substrate and hold at room temperature.

k. After 15 to 20 minutes' incubation on the urea substate at room temperature, count and record the number of yellow or yellow-brown colonies on those membrane filters containing (ideally) 20 to 80 colonies.

3. *CALCULATION OF RESULTS*

Select the membrane filter with the number of colonies within the acceptable range (20 to 80) and calculate the count per 100 m*L* according to the general formula:

$$E.\ coli/100\ mL = \frac{\text{No. } E.\ coli \text{ Colonies Counted} \times 100\ mL}{\text{Volume in } mL \text{ or Sample Filtered} \times 100\ mL}$$

4. *REPORTING RESULTS*

Report the results as *E. coli* per 100 m*L* of sample.

5. *EXAMPLE*

A total of 40 *E. coli* colonies were counted after filtering a 50-m*L* sample.

$$E.\ coli/100\ mL = \frac{\text{No. } E.\ coli \text{ Colonies Counted} \times 100\ mL}{\text{Sample Volume Filtered, } mL \times 100\ mL}$$

$$= \frac{(40\ E.\ coli)(100\ mL)}{(50\ mL)(100\ mL)}$$

$$= \frac{(0.8\ E.\ coli/mL)(100\ mL)}{100\ mL}$$

$$= 80\ E.\ coli/100\ mL$$

6. *REFERENCES*

EPA Test Method: *Escherichia coli* in Water by the Membrane Filter Procedure, Method 1003.1, 1985 (MF-mTEC Method).

See page 9-70, *STANDARD METHODS,* 21st Edition.

(Coliform)

J. ESCHERICHIA COLI *AND TOTAL COLIFORMS*

1. *COLILERT™ METHOD*

Colilert, a registered trademark of IDEXX Laboratories, provides simultaneous detection, specific identification, and confirmation of total coliforms and *E. coli* in water. Colilert reagent is a chemical formulation that specifically identifies only coliform bacteria. Colilert media contains specific indicator nutrients for the target microbes, total coliform and *E. coli.* As the media's nutrients are metabolized, yellow color and fluorescence are released, confirming the presence of total coliforms and *E. coli,* respectively. Non-coliform bacteria are chemically suppressed and cannot metabolize the indicator nutrients. Consequently, they do not interfere with the identification of the target microbes. Using this method, total coliforms and *E. coli* can be specifically and simultaneously detected and identified in 24 hours or less.

2. *MATERIALS REQUIRED*

a. Media. Available from IDEXX Laboratories, 1 IDEXX Drive, Westbrook, ME 04092. Phone 1-800-321-0207.

b. Ten culture tubes each containing Colilert reagent for 10 m*L* of sample (available from IDEXX Laboratories).

c. Sterile 10-m*L* pipets.

d. Incubator at 35°C.

e. Long wavelength ultraviolet lamp.

f. Color and fluorescence comparator (available from IDEXX Laboratories).

3. *TEST PROCEDURE*

a. Remove the front panel of the Colilert tube box along the perforation line to allow easy access and removal of tube carriers.

b. Remove a 10-tube carrier from the kit and label each 10 tube MPN test as indicated on the carrier.

c. Crack the carrier along the front perforation and bend the carrier along the back crease line. Stand the carrier on the table top.

d. Aseptically fill each Colilert tube with sample water to the level of the back of the tube carrier (10 m*L*).

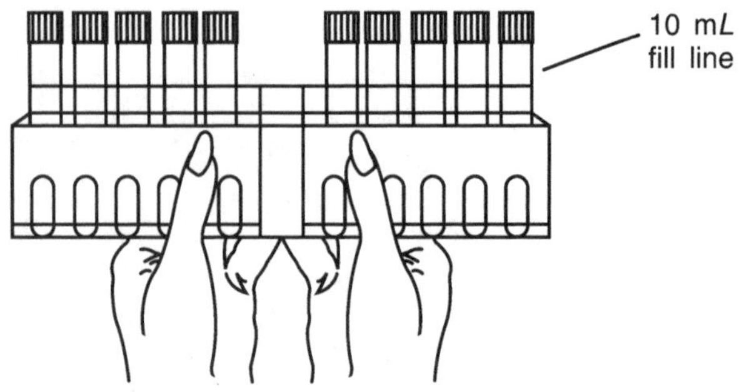

e. Cap the tubes tightly.

f. Mix vigorously to dissolve the reagent by repeated inversion of carrier.

NOTE: High calcium salt concentrations in certain waters may cause precipitate. This will not affect test results.

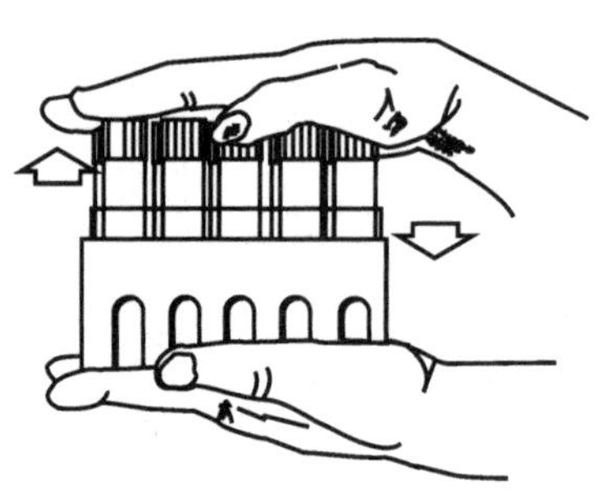

g. Incubate inoculated reagent tubes in the carrier at 35°C for 24 hours. Incubation should commence within 30 minutes of inoculation.

NOTE: Incubation exceeding 28 hours should be avoided.

h. Read tubes in the carrier at 24 hours. If yellow color is seen, check for fluorescence. Color should be uniform throughout the tube. If not, mix by inversion before reading.

4. *TEST RESULTS AND INTERPRETATION*

a. Compare each tube against the color comparator. If the inoculated reagent has a yellow color equal to or greater than the comparator, the presence of total coliforms is confirmed.

b. If any of the tubes are yellow in color, observe each tube for fluorescence by placing carrier 2 to 5 inches (5.08 to 12.7 cm) from the long wavelength ultraviolet lamp. If the fluorescence of the tube(s) is greater or equal to the comparator, the presence of *E. coli* is specifically confirmed.

c. Samples are negative for total coliforms if no color is observed at 24 hours. Should a sample be so lightly yellow after 24 hours incubation that you cannot definitively read it relative to the positive comparator tube, you may incubate it up to an additional 4 hours. If the sample is coliform positive, the color will intensify. If it does not intensify, consider the sample negative.

d. To find the concentration of total coliforms or *E. coli* per 100 m*L*, compare the number of positive reaction tubes per sample set (10 tubes) to the standard MPN (Most Probable Number) chart shown on Table 16.6 or in *STANDARD METHODS.*

TABLE 16.6 MPN INDEX FOR VARIOUS COMBINATIONS OF POSITIVE AND NEGATIVE RESULTS WHEN TEN 10-m*L* PORTIONS ARE USED

No. of Tubes Giving Positive Reaction Out of 10 of 10-m*L* Each	MPN Index/ 100 m*L*
0	<1.1
1	1.1
2	2.2
3	3.6
4	5.1
6	9.2
7	12.0
8	16.1
9	23.0
10	>23.0

When you wish to summarize with a single MPN value the results from a series of samples, the arithmetic mean, the median, or the geometric mean may be used. See Chapter 18 for an explanation of how to compute these values. Also see Example 5, page 560.

(Coliform)

5. *METHOD OF CALCULATING THE MOST PROBABLE NUMBER (MPN)*

EXAMPLE:

Drinking Water Sample—(see Figure 16.16).

m*L* of Sample	100
positive tubes	2
yellow color	yes
fluorescence	no
read MPN as	2.2 per 100 m*L* for total coliform and <1.1 for *E. coli* from Table 16.6

6. *RECORDING RESULTS*

Results should be recorded on data sheets prepared especially for this test. An example is shown in Figure 16.17.

7. *REFERENCES*

a. See page 9-48, *STANDARD METHODS,* 21st Edition.

b. The HACH Company, PO Box 389, Loveland, CO 80539-0389, has equipment, media, and procedures for total coliform, fecal coliform, and *E. coli* tests.

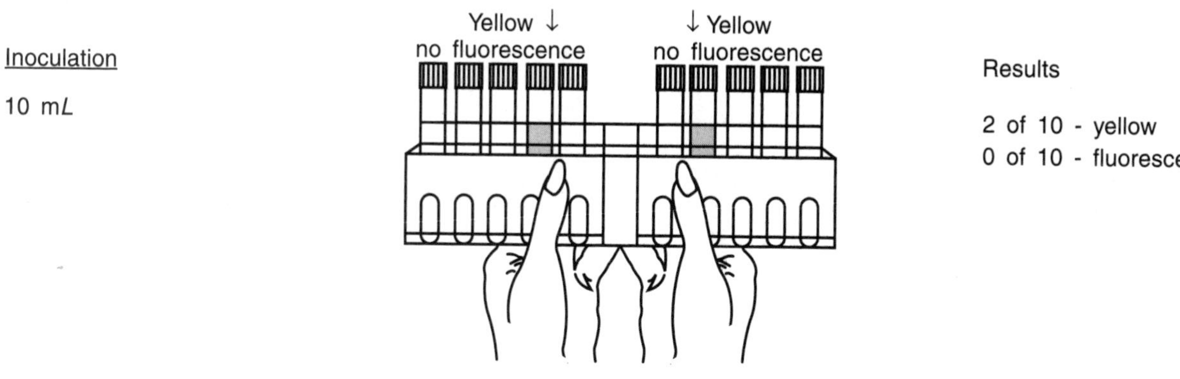

Fig. 16.16 Results of Colilert test—drinking water

DATE	NO.	SOURCE	24 HRS		MPN/100 m*L* TOTAL COLIFORM	MPN/100 m*L* *E. COLI*
4/16/06	X 29	1701 Main St	---- + ---- + ----	Yellow	2.2	<1.1
			— —	Fluorescence		

Fig. 16.17 Record coliform test results using Colilert Method

(Coliform)

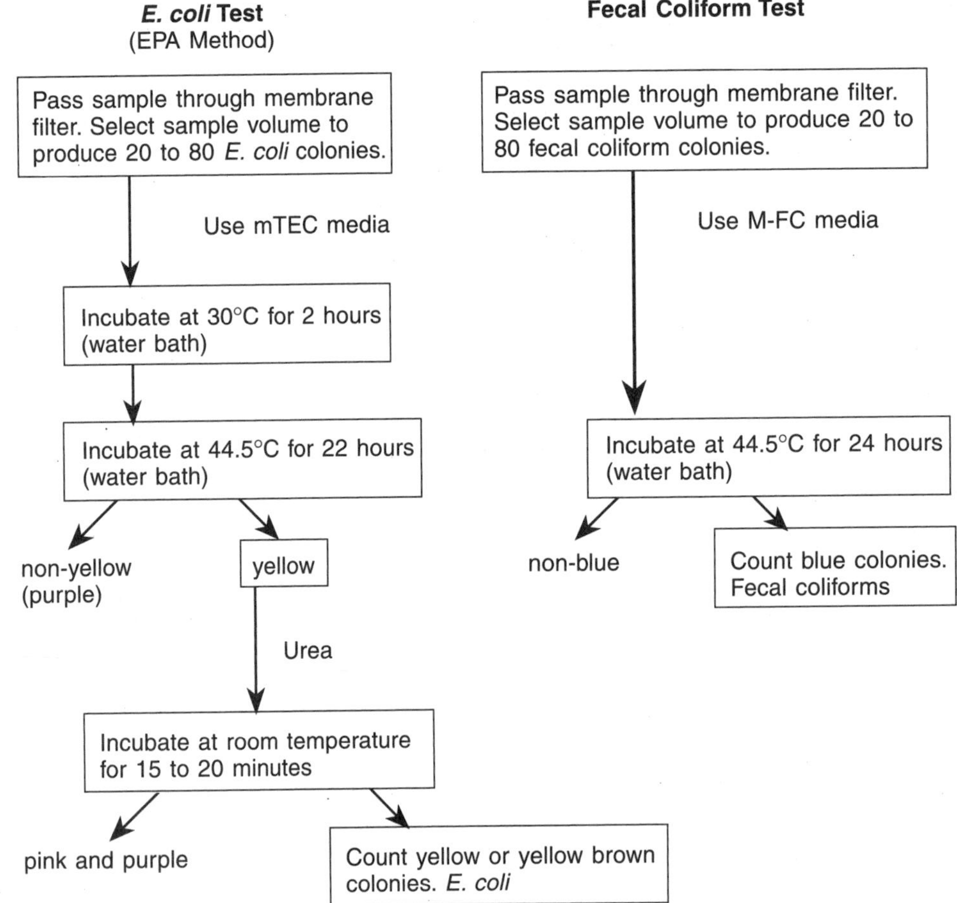
E. coli Test
(EPA Method)
Pass sample through membrane filter. Select sample volume to produce 20 to 80 *E. coli* colonies.
Use mTEC media
Incubate at 30°C for 2 hours (water bath)
Incubate at 44.5°C for 22 hours (water bath)
non-yellow (purple)
yellow
Urea
Incubate at room temperature for 15 to 20 minutes
pink and purple
Count yellow or yellow brown colonies. *E. coli*
Fecal Coliform Test
Pass sample through membrane filter. Select sample volume to produce 20 to 80 fecal coliform colonies.
Use M-FC media
Incubate at 44.5°C for 24 hours (water bath)
non-blue
Count blue colonies. Fecal coliforms

COMPARISON OF *E. COLI* AND FECAL COLIFORM TEST PROCEDURES

Item	*E. coli*	Fecal Coliform
Media	mTEC	M-FC
Colony color	Yellow	Blue
Incubation steps	3 steps	1 step

K. *COMPARISON OF COLIFORM TESTS*

The total coliform bacteria test includes both *Escherichia* and *Aerobacter* coliform bacteria groups. *Aerobacter* and some *Escherichia* can grow in soil. Therefore, not all coliforms found in the total coliform test come from human wastes. *Escherichia coli* (*E. coli*) apparently are all of fecal origin. However, it is difficult to determine *E. coli* without measuring soil coliforms too. The fecal coliform and *Escherichia coli* tests are used in an attempt to more specifically determine the extent of human wastes in water.

QUESTION

Write your answer in a notebook and then compare your answer with the one on page 612.

16.5L Explain how the Colilert media makes it possible to specifically identify total coliforms and *E. coli.*

END OF LESSON 7 OF 9 LESSONS
ON
LABORATORY PROCEDURES AND CHEMISTRY

Please answer the discussion and review questions next.

DISCUSSION AND REVIEW QUESTIONS

Chapter 16. LABORATORY PROCEDURES AND CHEMISTRY

(Lesson 7 of 9 Lessons)

Write the answers to these questions in your notebook. The question numbering continues from Lesson 6.

34. What is the purpose of the coliform group bacteria test?
35. What does MPN mean?
36. How would you determine the number of dilutions for an MPN test?
37. What factors can cause errors when counting colonies on membrane filters?
38. How can questionable colonies on membrane filters be verified as coliform colonies?
39. Why is the total coliform test not an adequate measure of the extent of human wastes in water?

CHAPTER 16. LABORATORY PROCEDURES AND CHEMISTRY

(Lesson 8 of 9 Lessons)

7. Dissolved Oxygen (DO) and Biochemical Oxygen Demand (BOD)

A. *DISCUSSION*

The dissolved oxygen (DO) test is, as the name implies, the testing procedure to determine the amount of oxygen dissolved in samples of water or wastewater. The analysis for DO is a key test in water pollution control activities and waste treatment process control. There are various types of tests that can be run to obtain the amount of dissolved oxygen. This procedure is the Sodium Azide Modification of the Winkler Method and is best suited for relatively clean waters. Interfering substances include color, organics, suspended solids, sulfide, chlorine, and ferrous and ferric iron. Nitrite will not interfere with the test if fresh azide is used.

The generalized principle is that iodine will be released in proportion to the amount of dissolved oxygen present in the sample. By using sodium thiosulfate with starch as the indicator, one can titrate the sample and determine the amount of dissolved oxygen.

B. *WHAT IS TESTED?*

Sample	Common Range, mg/*L*
Influent	Usually 0, >1 is very good.
Primary Clarifier Effluent	Usually 0, recirculated from filters >2 is good.
Secondary Effluent	50% to 95% Saturation, 3 to >8 is good.
Oxidation Ponds	1 to 25+*
Activated Sludge—Aeration Tank Outlet	>2 desirable

> means greater than
* supersaturated with oxygen

C. *APPARATUS*

METHOD A (Sodium Azide Modification of Winkler Method)

1. Buret, graduated to 0.1 m*L*
2. Three 300-m*L* glass-stoppered BOD bottles
3. Wide-mouth Erlenmeyer flask, 500 m*L*
4. One 10-m*L* measuring pipet
5. One 1-liter reagent bottle to collect activated sludge

METHOD B (DO Probe)

Follow manufacturer's instructions. See Section H, "DO Probe," for Discussion, Calibration, and Precautions.

D. *REAGENTS*

Standardized solutions may be purchased from chemical suppliers.

1. Manganous sulfate solution. Dissolve 480 gm manganous sulfate crystals ($MnSO_4 \bullet 4\ H_2O$) in 400 to 600 m*L* distilled water. Filter through filter paper, then add distilled water to the filtered liquid to make a 1-liter volume.

2. Alkaline iodide-sodium azide solution. Dissolve 500 gm sodium hydroxide (NaOH) in 500 to 600 m*L* distilled water in a pyrex glass bottle. Dissolve 150 gm potassium iodide (KI) in 200 to 300 m*L* distilled water in a separate pyrex glass container. Mix each chemical solution using a magnetic stirrer. Exercise caution. Add the chemicals to the distilled water slowly and cautiously. Avoid breathing the fumes and body contact with the solutions. Heat is produced when the chemicals are added, and the solutions are very caustic. Place an inverted beaker over the tops of each mixing container and allow them to cool at room temperature. Mix the two solutions together when they are cool.

Dissolve 10 gm sodium azide (NaN_3) in 40 m*L* of distilled water. Exercise caution again. This solution is poisonous.

Add the sodium azide solution with constant stirring to the cooled solution of alkaline iodide; then add distilled water to the mixture to make a 1-liter volume. Sodium azide will decompose in time and is no good after three months.

3. Sulfuric Acid. Use concentrated reagent-grade acid (H_2SO_4). Handle carefully, since this material will burn hands and clothes. Rinse affected parts with tap water to prevent injury.

   CAUTION: When working with alkaline azide and sulfuric acid, keep a nearby water faucet running for frequent hand rinsing.

4. 0.025 *N* Phenylarsine Oxide (PAO) solution. This solution is available and standardized from commercial sources.

5. 0.025 *N* Sodium Thiosulfate solution. Dissolve exactly 6.206 grams sodium thiosulfate crystals ($Na_2S_2O_3 \bullet 5\ H_2O$) in freshly boiled and cooled water and make up to 1 liter.

   For preservation, add 0.4 gm or 1 pellet of sodium hydroxide (NaOH). Solutions of "thio" should be used within two weeks to avoid loss of accuracy due to decomposition of solution.

6. Starch solution. Make a thin paste of 6 gm of potato starch in a small quantity of distilled water. Pour this paste into one liter of boiling, distilled water, allow to boil for a few minutes, then settle overnight. Remove the clear supernatant and save; discard the rest. For preservation, add two drops toluene ($C_6H_5CH_3$) or use 1.25 grams salicylic acid.

(DO and BOD)

E. *PROCEDURE*

OUTLINE OF PROCEDURE FOR PREPARING REAGENTS

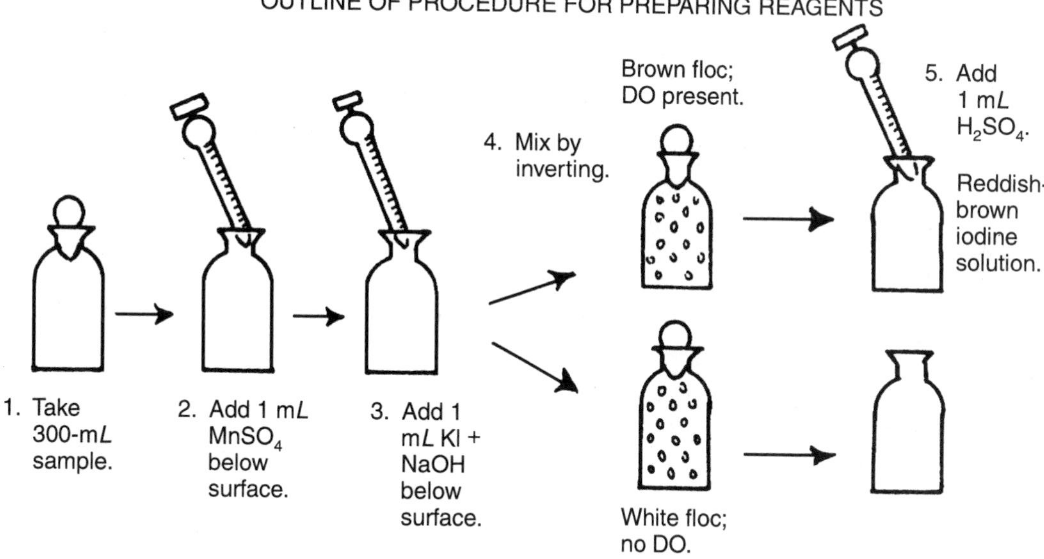

Brown floc;
DO present.
5. Add
1 m*L*
H_2SO_4.
Reddish-
brown
iodine
solution.
4. Mix by
inverting.
1. Take
300-m*L*
sample.
2. Add 1 m*L*
$MnSO_4$
below
surface.
3. Add 1
m*L* KI +
NaOH
below
surface.
White floc;
no DO.

Titration of Iodine Solution:

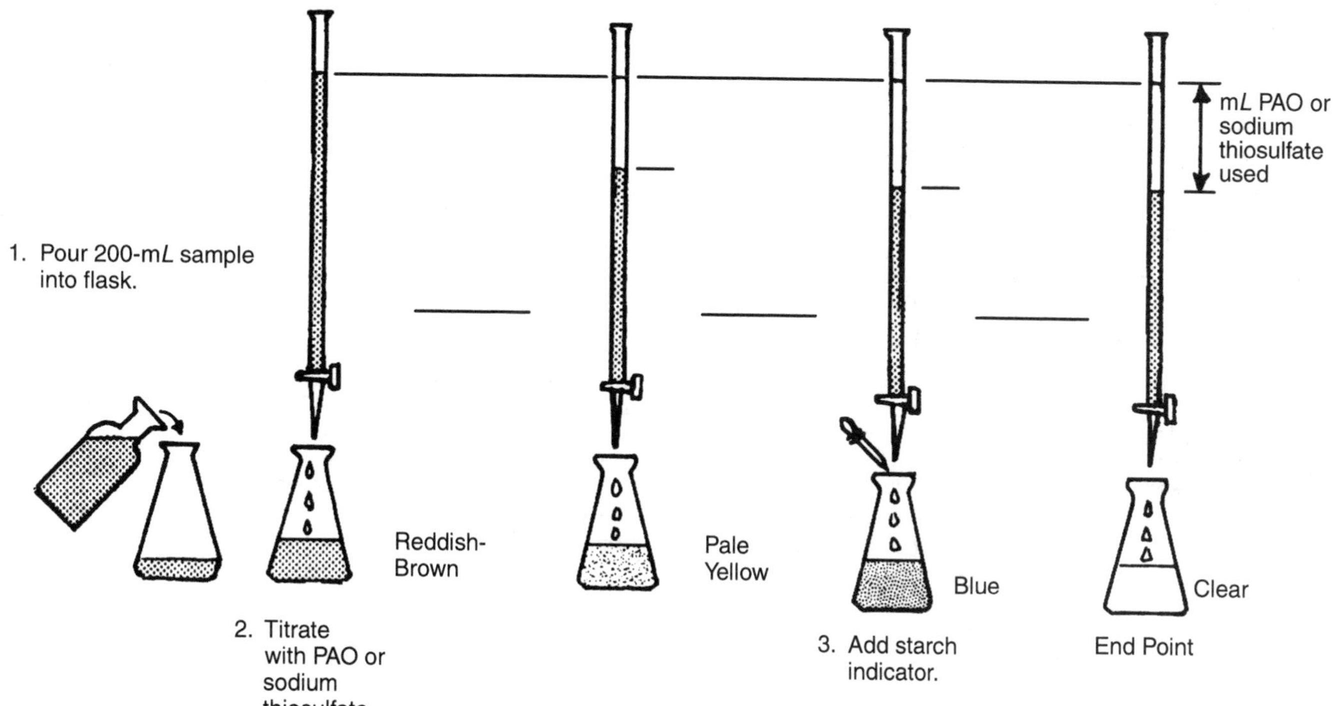

m*L* PAO or
sodium
thiosulfate
used
1. Pour 200-m*L* sample
into flask.
Reddish-
Brown
Pale
Yellow
Blue
Clear
2. Titrate
with PAO or
sodium
thiosulfate.
3. Add starch
indicator.
End Point

SODIUM AZIDE MODIFICATION OF THE WINKLER METHOD

NOTE: The sodium azide destroys nitrate, which will interfere with this test.

The reagents are to be added in the quantities, order, and methods as follows:

1. Collect a sample to be tested in 300-m*L* (BOD) bottle taking special care to avoid aeration of the liquid being collected. Fill bottle completely and add cap.
2. Remove cap and add 1 m*L* of manganous sulfate solution below surface of the liquid.
3. Add 1 m*L* of alkaline-iodide-sodium azide solution below the surface of the liquid.
4. Replace the stopper, avoid trapping air bubbles, and mix well by inverting the bottle several times. Repeat this mixing after the floc has settled halfway. Allow the floc to settle halfway a second time.

(DO and BOD)

5. Acidify with 1 m*L* of concentrated sulfuric acid by allowing the acid to run down the neck of the bottle above the surface of the liquid.

6. Restopper and mix well until the precipitate has dissolved. The solution will then be ready to titrate. Handle the bottle carefully to avoid acid burns.

7. Pour 200 m*L* of original sample from bottle into an Erlenmeyer flask.

8. If the solution is brown in color, titrate with 0.025 *N* PAO until the solution is pale yellow color. Add a small quantity of starch indicator and proceed to Step 10. (Note: Either PAO or 0.025 *N* sodium thiosulfate can be used.)

9. If the solution has no brown color, or is only slightly colored, add a small quantity of starch indicator. If no blue color develops, there is zero Dissolved Oxygen. If a blue color does develop, proceed to Step 10.

10. Titrate to the first disappearance of the blue color. Record the number of m*L* of PAO used. If the blue color returns, this indicates that nitrate ions are present.

11. The amount of oxygen dissolved in the original solution will be equal to the total number of m*L* of PAO used in the titration provided significant interfering substances are not present.

    mg/*L* DO = m*L* PAO

F. *EXAMPLE*

The DO titration of a 200-m*L* sample requires 5.0 m*L* of 0.025 *N* PAO. Therefore, the dissolved oxygen concentration in the sample is 5 mg/*L*.

G. *CALCULATION*

You will want to find the percent saturation of DO in the effluent of your secondary plant. The DO is 5.0 mg/*L* and the temperature is 20°C. At 20°C, 100 percent DO saturation is 9.2 mg/*L*.

The dissolved oxygen saturation values are given in Table 16.7. Note that as the temperature of water increases, the DO saturation value (100% Saturation column) decreases. Table 16.7 gives 100 percent DO saturation values for temperatures in °C and °F.

$$\text{DO Saturation, \%} = \frac{\text{DO of Sample, mg}/L \times 100\%}{\text{DO at 100\% Saturation, mg}/L}$$

$$= \frac{5.0 \text{ mg}/L}{9.2 \text{ mg}/L} \times 100\%$$

$$= .54 \times 100\%$$

$$= 54\%$$

TABLE 16.7 EFFECT OF TEMPERATURE ON OXYGEN SATURATION FOR A CHLORIDE CONCENTRATION OF ZERO mg/*L*

°C	°F	mg/*L* DO at 100% Saturation
0	32.0	14.6
1	33.8	14.2
2	35.6	13.8
3	37.4	13.5
4	39.2	13.1
5	41.0	12.8
6	42.8	12.5
7	44.6	12.2
8	46.4	11.9
9	48.2	11.6
10	50.0	11.3
11	51.8	11.1
12	53.6	10.8
13	55.4	10.6
14	57.2	10.4
15	60.0	10.2
16	61.8	10.0
17	63.6	9.7
18	65.4	9.5
19	67.2	9.4
20	68.0	9.2
21	69.8	9.0
22	71.6	8.8
23	73.4	8.7
24	75.2	8.5
25	77.0	8.4

H. *DO PROBE*

1. *DISCUSSION*

Measurement of the dissolved oxygen (DO) concentration with a probe and electronic readout meter is a satisfactory substitute for the Sodium Azide Modification of the Winkler Method under many circumstances. The probe is recommended when samples contain substances that interfere with the Modified Winkler procedure, such as sulfite, thiosulfate, polythionate, mercaptans, free chlorine or hypochlorite, organic substances readily hydrolyzed in alkaline solutions, free iodine, intense color or turbidity, and biological flocs. A continuous record of the dissolved oxygen content of aeration tanks and receiving waters may be obtained using a probe. In determining the BOD of samples, a probe may be used to determine the DO initially and after the five-day incubation period of the blanks and sample dilutions.

2. *PROCEDURE*

Follow manufacturer's instructions.

3. *CALIBRATION*

To ensure that the DO probe reading provides the dissolved oxygen content of the sample, the probe must be calibrated. Take a sample that does not contain substances that interfere with either the probe reading or the Modified Winkler procedure. Split the sample. Measure the DO in one portion of the sample using the Modified Winkler procedure and compare this result with the DO probe reading on the other portion of the sample. Adjust the probe reading to agree with the results from the Modified Winkler procedure.

When calibrating the probe in an aeration tank of the activated sludge process, do not attempt to measure the dissolved oxygen in the aerator and then adjust the probe. The biological flocs in the aerator will interfere with the Modified Winkler procedure, and the copper sulfate-sulfamic acid procedure is not sufficiently accurate to calibrate the probe. An aeration tank probe may be calibrated by splitting an effluent sample, measuring the DO by the Modified Winkler procedure, and comparing results with the probe readings. Always keep the membrane in the tip of the probe from drying because the probe can lose its accuracy until reconditioned.

4. *PRECAUTIONS*

a. Periodically check the calibration of the probe.

b. Keep the membrane in the tip of the probe from drying out.

c. Dissolved inorganic salts, such as found in seawater, can influence the readings from a probe.

d. Reactive compounds, such as reactive gases and sulfur compounds, can interfere with the output of a probe.

e. Do not place the probe directly over a diffuser because you want to measure the dissolved oxygen in the water being treated, not the oxygen in the air supply to the aerator.

QUESTIONS

Write your answers in a notebook and then compare your answers with those on page 612.

16.5M Calculate the percent dissolved oxygen saturation if the receiving water DO is 7.9 mg/*L* and the temperature is 10°C.

16.5N How would you calibrate the DO probe in an aeration tank?

BIOCHEMICAL OXYGEN DEMAND OR BOD

A. *DISCUSSION*

The BOD test measures the amount of oxygen used by microorganisms as they use the substrate (food) in wastewater when placed in a controlled temperature for five days. The DO (dissolved oxygen) is measured at the beginning and recorded. During the five-day period, microorganisms in the sample break down complex organic matter in the sample, using up oxygen in the process. After the five-day dark incubation period, the DO is again determined. The BOD is then calculated on the basis of the reduction of DO and the size of sample. This test is an estimate of the availability of food in the sample (food or organisms that take up oxygen) expressed in terms of oxygen use. Results of a BOD test indicate the rate of oxidation and provide an indirect estimate of the availability to organisms or concentration of the waste.

Actual environmental conditions of temperature, organism population, water movement, sunlight, and oxygen concentration cannot be accurately reproduced in the laboratory. Results obtained from this test must take into account these factors when relating BOD results to receiving water oxygen demands.

Samples for the BOD test should be collected before chlorination because chlorine interferes with the organisms in the test. It is difficult to obtain accurate results with dechlorinated samples.

Samples are incubated for a standard period of five days because a fraction of the total BOD will be exerted during this period. The ultimate or total BOD is normally never run for

(DO and BOD)

plant control. A limitation of the BOD test is that the results are not available until five days after the sample was collected.

B. *WHAT IS TESTED?*

Sample	**Common Range, mg/*L***
Influent	150 to 400
Primary Effluent	60 to 160
Secondary Effluent	5 to 30
Digester Supernatant	1,000 to 4,000 +
Industrial Wastes	100 to 3,000 +

C. *APPARATUS*

1. 300 m*L* BOD bottles with ground glass stoppers
2. Incubator, 20°C ± 1°C
3. Pipets, 10 m*L* graduated, 1/32- to 1/16-inch (0.8- to 1.6-mm) diameter tip
4. Buret and stand
5. Erlenmeyer flask, 500 m*L*

D. *REAGENTS*

See Section D, page 570, under DO portion of this procedure for the preparation of manganous sulfate, alkaline iodide-sodium azide, sulfuric acid, sodium thiosulfate, and starch solutions.

1. Distilled water. Water used for solutions and for preparation of the solution water must be of highest quality. It must contain less than 0.1 mg/*L* copper and be free of chlorine, chloramines, caustic, organic material, and acids. Ordinary distilled water for your car's battery is not good enough.
2. Phosphate buffer solution. Dissolve 8.5 gm monobasic potassium phosphate (KH_2PO_4), 21.75 gm dibasic potassium phosphate (K_2HPO_4), 33.4 gm dibasic sodium phosphate crystals ($Na_2HPO_4 \cdot 7\ H_2O$), and 1.7 gm ammonium chloride (NH_4Cl) in distilled water and make up to 1 liter. The pH of this buffer should be 7.3 and should be checked with a pH meter. Discard this reagent if there is any sign of biological growth.
3. Magnesium sulfate solution. Dissolve 22.5 gm magnesium sulfate crystals ($MgSO_4 \cdot 7\ H_2O$) in distilled water and make up to 1 liter.
4. Calcium chloride solution. Dissolve 27.5 gm anhydrous calcium chloride ($CaCl_2$) in distilled water and make up to 1 liter.
5. Ferric chloride solution. Dissolve 0.25 gm ferric chloride ($FeCl_3 \cdot 6\ H_2O$) in distilled water and make up to 1 liter.
6. Dilution water. Add 1 m*L* each of phosphate buffer (Step 2), magnesium sulfate (Step 3), calcium chloride (Step 4), and ferric chloride solutions (Step 5) for each liter of distilled water. Saturate with DO by shaking in a partially filled bottle or by aerating with filtered air. Store at a temperature as close to 20°C as possible for at least 24 hours to allow the water to become stabilized. This water should not show a drop in DO of more than 0.2 mg/*L* on incubation for five days.

Many plants do not prepare reagents. Small plants and plants that do not run many tests find it quicker and easier to purchase commercially prepared reagents. These reagents

(DO and BOD)

may be available in the desired strength or they may consist of dry *PILLOWS*,[22] which are added to the sample, rather than the liquid reagent. Check with your chemical supplier for these reagents.

E. *PROCEDURE FOR UNCHLORINATED SAMPLES*

The test is made by measuring the oxygen used or depleted during a five-day period at 20°C by a measured quantity of wastewater sample seeded into a reservoir of dilution water saturated with oxygen. This is compared to an unseeded or blank reservoir of dilution water by subtracting the difference and multiplying by a factor for dilution.

PROCEDURE

1. BOD bottles should be of 300-m*L* capacity with graduations and ground-glass stoppers. To clean the bottles, carefully rinse with tap water followed by distilled water. Each BOD bottle and fitted stopper are numbered. Always insert each numbered stopper in the same-numbered BOD bottle. If the numbers do not match, errors could occur in the test results.
2. Fill two bottles completely with dilution water and insert the stopper tightly so that no air is trapped beneath the stopper. Siphon dilution water from its container when filling BOD bottles.
3. Set up one or more dilutions of the sample to cover the estimated range of BOD values. From the estimated BOD, calculate the volume of raw sample to be added to the BOD bottle based on the fact that:

   The most valid DO depletion is 4 mg/*L*. Therefore,

$$\text{m}L \text{ of Sample Added per 300 m}L = \frac{(4 \text{ mg}/L)(300 \text{ m}L)}{\text{Estimated BOD, mg}/L} = \frac{1{,}200}{\text{Estimated BOD, mg}/L}$$

EXAMPLES:

a. Estimated BOD = 400 mg/*L*

$$\text{m}L \text{ of Sample Added to BOD Bottle} = \frac{1{,}200}{400} = 3 \text{ m}L$$

b. Estimated BOD = 200 mg/*L*: use 6 m*L*

100 mg/*L*: use 12 m*L*

20 mg/*L*: use 60 m*L*

When the BOD is unknown, select more than one sample size. For example, place several samples—1 m*L*, 3 m*L*, 6 m*L*, and 12 m*L*—into four BOD bottles.

For samples with very high BOD values, it may be difficult to accurately measure small volumes or to get a truly representative sample. In such a case, initial dilution should first be made on the sample. A dilution of 1:10 is convenient.

4. To perform the BOD test, first fill two BOD bottles with BOD dilution water (blanks), Nos. (1) and (2) on page 575.
5. Next, for each sample to be tested, carefully measure out the two portions of sample and place them into two new BOD bottles, Nos. (3) and (4). Add dilution water until the bottles are completely filled. Insert the stoppers. Avoid entrapping air bubbles. Be sure that there are water seals on the stoppers. A water seal is obtained by water in the small indentation between the stopper and the rim of the bottle.
6. On bottles (2) and (4) immediately determine the initial dissolved oxygen.
7. Incubate the remaining dilution water blank and diluted sample in the dark at 20°C for five days. These are bottles (1) and (3).
8. At the end of exactly five days (±3 hours), test bottles (1) and (3) for their *DISSOLVED OXYGEN* by using the sodium azide modification of the Winkler method or a DO probe. At the end of five days, the oxygen content should be at least 1 mg/*L*. Also, a depletion of 2 mg/*L* or more is desirable. Bottles (1) and (2) (blanks) are only used to check the dilution water quality. Their difference should be less than 0.2 mg/*L* after five days if the quality is good and free of impurities. The difference in blank readings is not used as a blank correction, but merely as a check on the quality of dilution water. Differences of greater than 0.2 mg/*L* could possibly be due to contamination and/or dirty BOD bottles.

F. *PRECAUTIONS*

1. The temperature of the incubator must be at 20°C. Other temperatures will change the rate of oxygen used.
2. The dilution water must be made according to *STANDARD METHODS* (as outlined on page 573, Section D, 6) for the most favorable growth rate of the bacteria. This water must be free of copper, which is often present when copper stills are used by commercial dealers. Use all glass or stainless-steel stills or demineralized water.
3. The wastewater must also be free of toxic wastes, such as hexavalent chromium.
4. If you use a cleaning solution to wash BOD bottles, be sure to rinse the bottles several times. Cleaning agents are toxic and if any residue remains in a BOD bottle, a BOD test could be ruined.
5. Unchlorinated wastewater normally contains an ample supply of seed bacteria; therefore seeding is usually not necessary.
6. Since this is a bioassay (BUY-o-AS-say), that is, living organisms are used for the test, environmental conditions must be quite exact.

G. *EXAMPLE*

BOD Bottle Volume	=	300 m*L*
Sample Volume	=	15 m*L*
Initial DO of Diluted Sample	=	8.0 mg/*L*
DO of Sample and Dilution After 5-Day Incubation	=	4.0 mg/*L*

[22] *Pillows.* Plastic tubes shaped like pillows that contain exact amounts of chemicals or reagents. Cut open the pillow, pour the reagents into the sample being tested, mix thoroughly, and follow test procedures.

(DO and BOD)

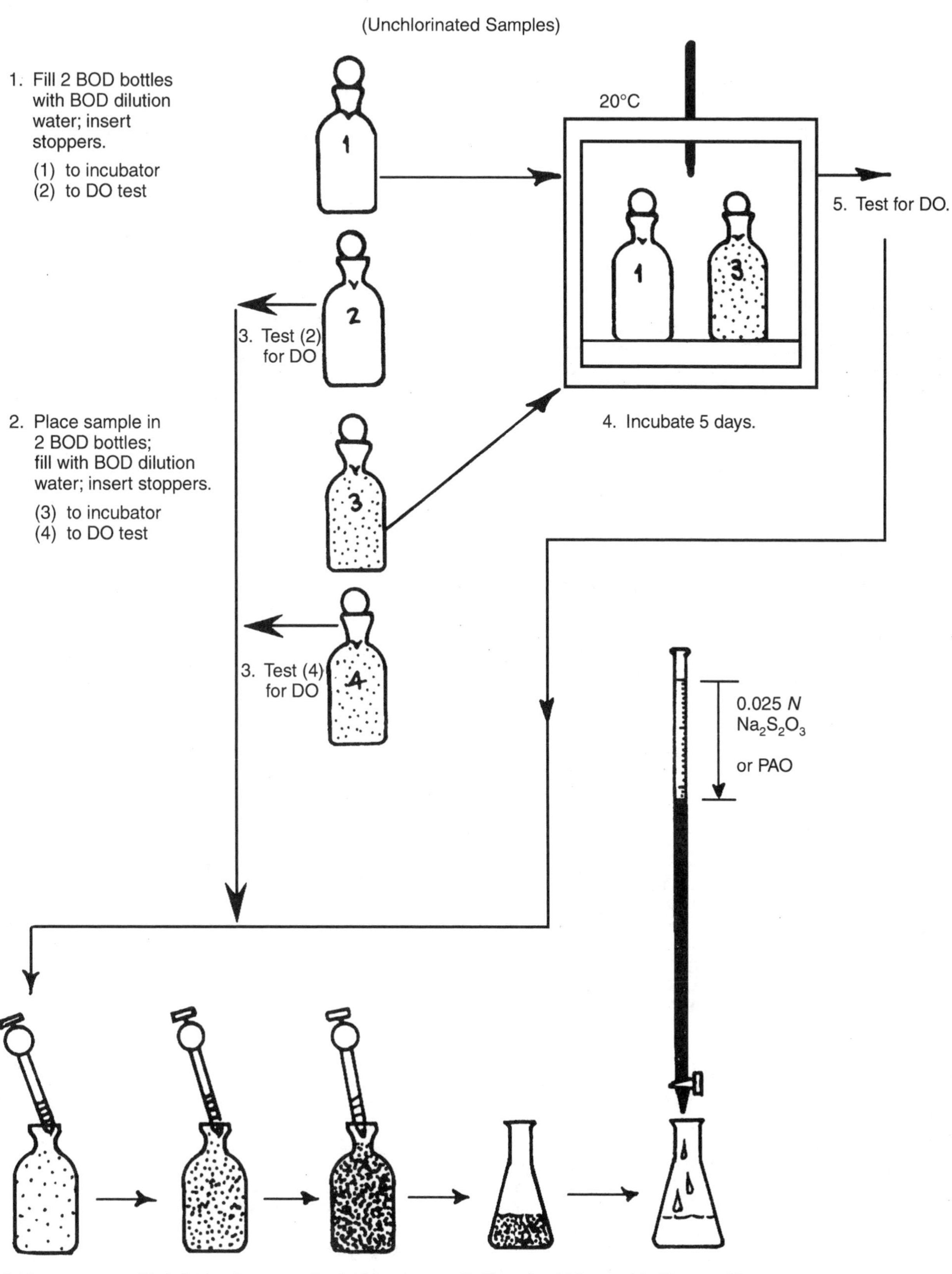
OUTLINE OF PROCEDURE FOR BOD
(Unchlorinated Samples)
1. Fill 2 BOD bottles with BOD dilution water; insert stoppers.
(1) to incubator
(2) to DO test
20°C
5. Test for DO.
3. Test (2) for DO
2. Place sample in 2 BOD bottles; fill with BOD dilution water; insert stoppers.
(3) to incubator
(4) to DO test
4. Incubate 5 days.
3. Test (4) for DO
0.025 N $Na_2S_2O_3$
or PAO
6. Add 1 mL $MnSO_4$ below surface.
7. Add 1 mL alkaline KI below surface.
8. Add 1 mL H_2SO_4.
9. Transfer 200 mL to flask.
10. Titrate with PAO (or thiosulfate).

(DO and BOD)

H. *CALCULATIONS*

$$\text{BOD, mg}/L = \left[\begin{matrix}\text{Initial DO of} \\ \text{Diluted} \\ \text{Sample,} \\ \text{mg}/L\end{matrix} - \begin{matrix}\text{DO of Diluted Sample} \\ \text{After 5-Day} \\ \text{Incubation,} \\ \text{mg}/L\end{matrix}\right]\left[\frac{\text{BOD Bottle Vol, m}L}{\text{Sample Volume, m}L}\right]$$

$$= (8.0\text{ mg}/L - 4.0\text{ mg}/L)\left[\frac{300\text{ m}L}{15\text{ m}L}\right]$$

$$= \frac{(4.0)(300)}{15}$$

$$= 80\text{ mg}/L$$

For acceptable results, the percent depletion of oxygen in the BOD test should range from 30 percent to 80 percent depletion. Also, if a residual DO of at least 1 mg/*L* remains or a DO uptake of at least 2 mg/*L* results after five days of incubation, the results also are considered reliable.

$$\%\text{ Depletion} = \frac{(\text{DO of Diluted Sample, mg}/L - \text{DO After 5 Days, mg}/L)}{\text{DO of Diluted Sample, mg}/L} \times 100\%$$

$$= \frac{(8.0\text{ mg}/L - 4.0\text{ mg}/L)}{8.0\text{ mg}/L} \times 100\%$$

$$= \frac{4}{8} \times 100\%$$

$$= 50\%$$

I. *PROCEDURE FOR CHLORINATED SAMPLES*

Dechlorination of Samples

Whenever chlorinated wastewater samples are collected for BOD analysis, sufficient dechlorinating agent must be added to destroy the residual chlorine. The following procedure should be followed:

1. Using the procedure given for residual chlorine analysis on page 545, test the sample for chlorine.
2. If there is no residual chlorine, proceed with test procedures below. If residual chlorine is present, add sufficient 0.025 *N* sodium sulfite until residual chlorine is absent. If a chlorine residual is present, the BOD results will be low. If too much sodium sulfite is added, the BOD results will be high. Prepare 0.025 *N* sodium sulfite by dissolving 1.575 grams of anhydrous Na_2SO_3 in 1,000 m*L* of distilled water. This solution is not stable and must be prepared daily.

Seeding of Sample

When a sample contains very few microorganisms as a result, for example, of chlorination or extreme pH, microorganisms must be added to the sample.

1. Collect about one liter of unchlorinated raw wastewater or primary effluent about 24 hours prior to the time you wish to start the BOD test. Let sample settle at 20°C for 24 to 36 hours. Filter through glass wool to remove large particles and grease clumps. Use the filtered sample for "seed."
2. Fill two 300-m*L* BOD bottles with dilution water (see page 577) and insert stopper tightly so no air bubbles are trapped. These are called blanks.
3. Set up one or more dilutions of samples in duplicate as shown on page 577.
4. Add 1 m*L* seed from Step 1 to each BOD bottle containing a dechlorinated sample.
5. Set up blanks of seed material to determine the amount of oxygen depletion that will be due to the added seed material. Use 3, 6, and 9 m*L* of seed and determine the five-day oxygen depletion due to 1 m*L* of seed. Seed material should produce a correction of at least 0.6 mg/*L* per m*L* of seed. Make these samples also in duplicate.
6. Determine initial DO on one set of duplicate bottles.
7. Incubate dilution water blank, diluted samples, and seeded blanks at 20°C for five days.
8. At the end of five days, test bottles for dissolved oxygen.
9. Calculate five-day BOD.

EXAMPLE:

	Bottle 2	**Bottle 4**	**Bottle 6**
Sample	blank	effluent	seed blank
Sample volume	—	10 m*L*	0
Seed volume	—	1 m*L*	3 m*L*
Initial DO	9.2 mg/*L*	9.2 mg/*L*	9.2 mg/*L*
DO after 5 days	9.2 mg/*L*	5.1 mg/*L*	7.1 mg/*L*
Depletion	0.0 mg/*L*	4.1 mg/*L*	2.1 mg/*L*

CALCULATION:

For seed correction:

$$\begin{matrix}\text{mg}/L\text{ DO Depletion} \\ \text{Caused by} \\ 1\text{ m}L\text{ of Seed}\end{matrix} = \frac{\text{5-day Depletion of Seed Sample}}{\text{m}L\text{ of Seed}}$$

$$= \frac{9.2\text{ mg}/L - 7.1\text{ mg}/L}{3\text{ m}L}$$

$$= 0.7\text{ mg}/L/\text{m}L$$

5-day BOD:

$$\text{BOD, mg}/L = \frac{[(\text{Initial DO} - \text{DO After 5 Days}) - (\text{Seed Correction})] \times 300\text{ m}L}{\text{m}L\text{ of Sample Volume}}$$

$$= \frac{[(9.2\text{ mg}/L - 5.1\text{ mg}/L) - (0.7\text{ mg}/L/\text{m}L)(1\text{ m}L)] \times 300\text{ m}L}{10\text{ m}L}$$

$$= \frac{(4.1\text{ mg}/L - 0.7\text{ mg}/L)300\text{ m}L}{10\text{ m}L}$$

$$= 102\text{ mg}/L$$

J. *PRECAUTIONS*

1. On effluent samples where the DO is run on the sample and the blue bounces back on the end point titration, this indicates nitrite interference and can cause the BOD test result to be higher than actual by as much as 10 to 15 percent. This fact should be considered in interpreting your results. The end point also may waver because of decomposition of azide in an old reagent or resuspension of sample solids. To correct a wavering end point, try preparing a new alkaline-azide solution or more of the old solution should be used because it may be decomposing.
2. A minimum of two dilutions per sample should be used. Use only analyses with oxygen depletions of greater than 2

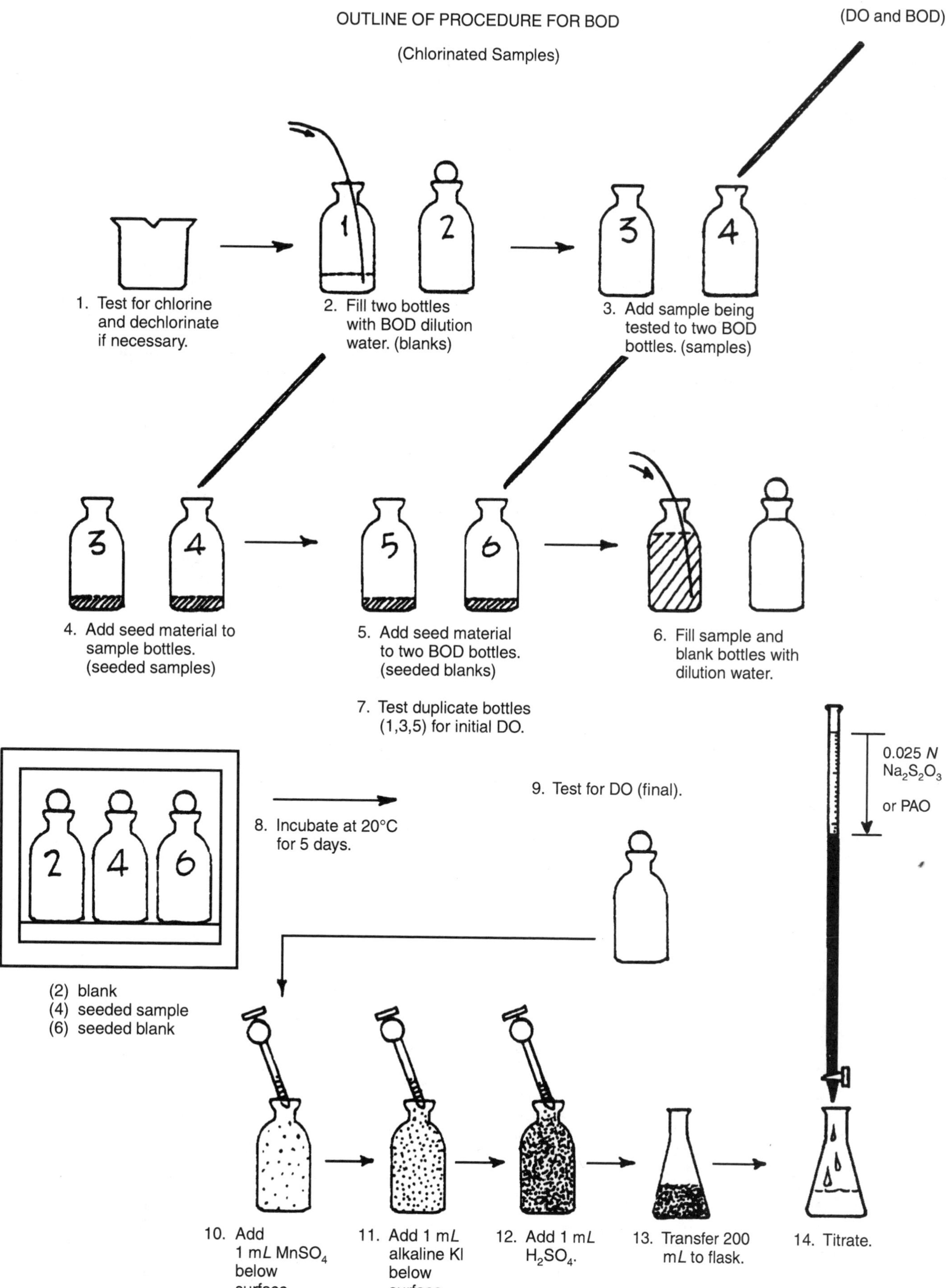

OUTLINE OF PROCEDURE FOR BOD
(Chlorinated Samples)
(DO and BOD)
1. Test for chlorine and dechlorinate if necessary.
1
2
2. Fill two bottles with BOD dilution water. (blanks)
3
4
3. Add sample being tested to two BOD bottles. (samples)
3
4
4. Add seed material to sample bottles. (seeded samples)
5
6
5. Add seed material to two BOD bottles. (seeded blanks)
6. Fill sample and blank bottles with dilution water.
7. Test duplicate bottles (1,3,5) for initial DO.
2
4
6
(2) blank
(4) seeded sample
(6) seeded blank
8. Incubate at 20°C for 5 days.
9. Test for DO (final).
0.025 N $Na_2S_2O_3$
or PAO
10. Add 1 mL $MnSO_4$ below surface.
11. Add 1 mL alkaline KI below surface.
12. Add 1 mL H_2SO_4.
13. Transfer 200 mL to flask.
14. Titrate.

mg/*L* and residual DOs of greater than 1 mg/*L* after five days of incubation at 20°C.

3. Samples should be well mixed before dilutions are made. A wide-tip pipet should be used for making dilutions. The wide tip will not clog with suspended solids.
4. Wastewaters that have been partially nitrified may produce high BOD results. This increased oxygen demand results from the oxidation of ammonia to nitrate. The use of chemicals such as allythiourea or other commercially available nitrification inhibitors in the dilution water will inhibit the nitrifiers and alleviate this problem.
5. For effluent samples with a low DO and BOD, saturate with DO by shaking in a partially filled bottle.

K. *PROCEDURE FOR INDUSTRIAL WASTE SAMPLES*

Some industrial waste samples may require special seeding because of a low microbial population or because the wastes contain organic compounds that are not readily oxidized by domestic wastewater seed. To obtain the necessary specialized seed material (microorganisms) adapted or acclimated to the industrial organic compounds, collect a sample of adapted seed from the effluent of a biological treatment process (activated sludge aeration tank) treating the industrial waste. When this source of adapted seed is not available, develop the adapted seed in the laboratory by continuously aerating a large sample of water and feeding it with small daily portions of the particular waste, together with soil or settled domestic wastewater, until a satisfactory microbial population has developed.

Once a satisfactory adapted seed is available, follow the procedures in the previous Section I, "Procedure for Chlorinated Samples, Seeding of Sample." Start with Step 2, "Fill two 300-m*L* BOD bottles with dilution water," and follow the steps except use the industrial waste sample instead of the dechlorinated sample.

L. *REFERENCE*

See pages 4-136 and 5-2, *STANDARD METHODS,* 21st Edition.

QUESTIONS

Write your answers in a notebook and then compare your answers with those on page 613.

16.5O How would you determine the amount of organic material in wastewater?

16.5P Why should samples for the BOD test be collected before chlorination?

16.5Q Why should opened bottles of sodium thiosulfate be used or re-standardized within two weeks?

16.5R How would you prepare dilutions to measure the BOD of cannery waste having an expected BOD of 2,000 mg/*L?*

16.5S What is the BOD of a sample of wastewater if a 2-m*L* sample in a 300-m*L* BOD bottle had an initial DO of 7.5 mg/*L* and a final DO of 3.9 mg/*L?*

END OF LESSON 8 OF 9 LESSONS
ON
LABORATORY PROCEDURES AND CHEMISTRY

Please answer the discussion and review questions next.

DISCUSSION AND REVIEW QUESTIONS

Chapter 16. LABORATORY PROCEDURES AND CHEMISTRY

(Lesson 8 of 9 Lessons)

Write the answers to these questions in your notebook. The question numbering continues from Lesson 7.

40. What is the formula for calculating the percent saturation of DO?
41. What precautions should be exercised when using a DO probe?
42. What is a limitation of the BOD test?
43. What precautions should be observed when running a BOD test on an unchlorinated sample?
44. Calculate the BOD of a 5-m*L* sample if the initial DO of the diluted sample was 7.5 mg/*L* and the DO of the diluted sample after five-day incubation was 3.0 mg/*L*.

CHAPTER 16. LABORATORY PROCEDURES AND CHEMISTRY

(Lesson 9 of 9 Lessons)

8. Hydrogen Ion (pH)

A. *DISCUSSION*

The intensity of the basic or acidic strength of water is expressed by its pH.

Mathematically, pH is the logarithm of the reciprocal of the hydrogen ion activity, or the negative logarithm of the hydrogen ion activity.

$$pH = \text{Log}\,\frac{1}{\{H^+\}} = -\text{Log}\,\{H^+\}$$

EXAMPLE

If a wastewater has a pH of 1, then the hydrogen ion activity $\{H^+\} = 10^{-1} = 0.1$.

If pH = 7, then $\{H^+\} = 10^{-7} = 0.0000001$.

pH Scale

0 increasing acidity — 7 — increasing alkalinity 14

$1 \leftarrow 2 \leftarrow 3 \leftarrow 4 \leftarrow 5 \leftarrow 6$ ^ $8 \rightarrow 9 \rightarrow 10 \rightarrow 11 \rightarrow 12 \rightarrow 13$

Neutral
6 through 8

In a solution, both hydrogen ions $\{H^+\}$ and the hydroxyl ions $\{OH^-\}$ are always present. At a pH of 7, the activity of both hydrogen and hydroxyl ions equals 10^{-7} moles per liter. When the pH is less than 7, the activity of hydrogen ions is greater than the hydroxyl ions. The hydroxyl ion activity is greater than the hydrogen ions in solutions with a pH greater than 7.

The pH test indicates whether a treatment process may continue to function properly at the pH measured. Each process in the plant has its own favorable range of pH, which must be checked routinely. Generally, a pH value from 6 to 8 is acceptable for best organism activity. Most wastewater contains many dissolved solids and buffers, which tend to minimize pH changes.

B. *WHAT IS TESTED?*

Wastewater	Common Range
Influent or Raw Wastewater (domestic)	6.8 to 8.0
Raw Sludge (domestic)	5.6 to 7.0
Digester Recirculated Sludge or Supernatant	6.8 to 7.2
Plant Effluent Depending on Type of Treatment	6.8 to 8.0

C. *MINIMUM APPARATUS LIST*

1. pH meter
2. Glass electrode
3. Reference electrode
4. Magnetic stirrer (optional)
5. Beaker for sample

D. *REAGENTS*

1. Buffer tablets of various pH values
2. Distilled water

E. *PROCEDURE*

1. Due to the differences between the various makes and models of pH meters commercially available, specific instructions cannot be provided for the correct operation of all instruments. In each case, follow the manufacturer's instructions for preparing the electrodes and operating the instrument.
2. Standardize the instrument against a buffer solution with a pH approaching that of the sample.
3. Rinse electrodes thoroughly with distilled water after removal from buffer solution.
4. Place electrodes in sample and measure pH.
5. Remove electrodes from sample, rinse thoroughly with distilled water.
6. Immerse electrode ends in beaker of pH 7 buffer solution.
7. Shut off meter.

F. *PRECAUTIONS*

1. To avoid faulty instrument calibration, prepare fresh buffer solutions as needed, once per week, from commercially available buffer tablets.
2. pH meter, buffer solution, and samples should all be at the same temperature (constant) because temperature variations will give erroneous results. Allow a few minutes for the probe to adjust to the buffers before calibrating a pH meter to ensure accurate pH readings.
3. Watch for erratic results arising from electrodes, faulty connections, or fouling of electrodes with oily or precipitated matter. Films may be removed from electrodes by placing isopropanol on a tissue or Q-tip and cleaning the probe.
4. If you are measuring pH in colored samples or samples with high solids content, or if you are taking measurements that need to be reported to the USEPA, you should use a pH electrode and meter instead of a colorimetric method or test papers. pH meters are capable of providing ± 0.1 pH accuracy in most applications. In contrast, colorimetric tests provide ± 0.1 pH accuracy only in a limited range. pH papers provide even less accuracy.

(Nitrogen-Ammonia)

QUESTION

Write your answer in a notebook and then compare your answer with the one on page 613.

16.5T What precautions should be exercised when using a pH meter?

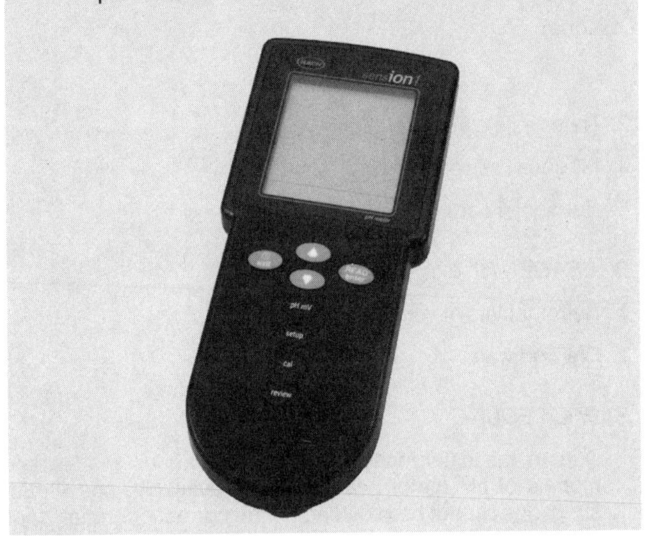

9. Metals

A. *DISCUSSION*

The presence of metals in wastewater can be a matter of serious concern because of the toxic properties of these materials, which may adversely affect wastewater treatment systems.

Metals in wastewater may be determined in most cases by atomic absorption spectroscopy or colorimetric methods. The term "metals" would include the following elements:

Aluminum	Cobalt	Potassium
Antimony	Copper	Selenium
Arsenic	Iron	Silver
Barium	Lead	Sodium
Beryllium	Magnesium	Thallium
Cadmium	Manganese	Tin
Calcium	Mercury	Titanium
Chromium	Molybdenum	Vanadium
	Nickel	Zinc

B. *REFERENCES*

For materials and procedures see:

1. Page 3-1, *STANDARD METHODS,* 21st Edition.
2. Page 78, EPA's "Methods for Chemical Analysis of Water and Wastes," US Environmental Protection Agency. EPA No. 600-4-79-020. Obtain from National Technical Information Service (NTIS), 5285 Port Royal Road, Springfield, VA 22161. Order No. PB84-128677. Price, $117.00, plus $5.00 shipping and handling per order.

QUESTION

Write your answer in a notebook and then compare your answer with the one on page 613.

16.5U Why would an operator test for metals in wastewater?

10. Nitrogen

A. *DISCUSSION*

The compounds of nitrogen are of interest to the wastewater treatment plant operator because of the importance of nitrogen in the life processes of all plants and animals. The chemistry of nitrogen is complex because of the several forms that nitrogen can assume. Ammonia, organic, nitrate, and nitrite are the most important nitrogen forms in wastewater treatment. The term Kjeldahl (KELL-doll) nitrogen refers to organic plus ammonia nitrogen.

Ammonia nitrogen in domestic wastewater is generally between 10 and 40 mg/*L*. Primary treatment may increase the ammonia nitrogen content slightly due to the decomposition of some protein compounds during treatment. In secondary treatment processes, ammonia may be oxidized to nitrite then to nitrate in varying degrees depending on factors such as wastewater temperature, residence time of the microorganisms, and amounts of oxygen. Significant water quality problems relating to ammonia are high chlorine demands, fish toxicity, and high oxygen demand on receiving waters.

Nitrite (NO_2^-) is an intermediate oxidation state of nitrogen between ammonia and nitrate nitrogen. Nitrite is very unstable but can be used to monitor how well nitrification is progressing in the treatment process. Effluents containing nitrite require a dose of 5 mg/*L* of chlorine for every mg/*L* of nitrite to satisfy the nitrite chlorine requirement.

Nitrate (NO_3^-) is seldom found in raw wastewater or primary effluent. In the biological treatment process, the ammonia nitrogen can be oxidized by bacteria to nitrite and then to nitrate. Secondary effluent may contain from 0 to 50 mg/*L* nitrate depending on the total nitrogen content in the raw wastewater.

B. *WHAT IS TESTED?*

Sample	Form of Nitrogen, as N	Common Range, mg/*L*
Secondary Effluent	ammonia	5 to 20
	Kjeldahl	5 to 30
	nitrate	0 to 50
	nitrite	0 to 1

I. Procedure for Ammonia

A. *APPARATUS*

1. Balance, analytical
2. Balance, triple beam
3. pH meter
4. Kjeldahl flasks, 800 m*L*
5. An all-glass distilling apparatus
6. Erlenmeyer flasks, 500 m*L* (these flasks should be marked at the 350- and 500-m*L* volumes)
7. Pipet, volumetric, 10 m*L*
8. Beakers, 500 m*L*
9. Graduated cylinder, 500 m*L* and 100 m*L*
10. Buret, 25 m*L*
11. Glass beads

B. *REAGENTS*

1. Distilled water free of ammonia. All solutions must be made with ammonia-free water. An ion exchange system in conjunction with a suitable water still to ensure high-quality water is the best system. An anioncation exchange resin should be used.
2. Ammonium chloride, stock solution. 1.0 m*L* = 1.0 mg NH_3-N. Dissolve 3.819 gm NH_4Cl in distilled water and bring to volume in a one-liter volumetric flask.
3. Ammonium chloride, standard solution. 1.0 m*L* = 0.01 mg. Dilute 10.0 m*L* stock solution of ammonium chloride (reagent 2) to volume in a one-liter volumetric flask.
4. Borate buffer. Add 88 m*L* of 0.1 *N* NaOH solution to 500 m*L* of 0.025 *N* sodium tetraborate solution (5 gm anhydrous $Na_2B_4O_7$ or 9.5 gm $Na_2B_4O_7 \bullet 10\ H_2O$ per liter) and dilute to one liter.
5. Boric acid solution 20 gm/*L*. Dissolve 20 gm H_3BO_3 in distilled water and dilute to one liter.
6. 0.2% methyl red solution. Dissolve 0.2 gm methyl red in 100 m*L* of 95% ethyl alcohol (ethanol).
7. 0.2% methylene blue solution. Dissolve 0.2 gm methylene blue in 100 m*L* of 95% ethyl alcohol.
8. Mixed indicator solution. Mix 2 volumes of 0.2% methyl red solution with one volume of 0.2% methylene blue solution. This solution should be prepared fresh every 30 days.
9. Sulfuric Acid, standard solution 0.02 *N*. 1 m*L* = 0.28 mg NH_3-N.
10. Sodium hydroxide, 1 *N*. Dissolve 40 gm NaOH in ammonia-free water and dilute to one liter.
11. Sodium hydroxide, 0.1 *N*. Dilute 10.0 m*L* 1 *N* sodium hydroxide solution to 100 m*L* in 100-m*L* volumetric flask.
12. Dechlorinating reagent, sodium thiosulfate, 1/70 *N*. Dissolve 3.5 gm $Na_2S_2O_2 \bullet 5\ H_2O$ in distilled water and dilute to one liter. One m*L* of this solution will remove 1 mg/*L* of residual chlorine in 500-m*L* sample. Sodium sulfite, Na_2SO_3, is also used as a dechlorinating reagent. See page 576, "Dechlorination of Samples," I, 2.

C. *PROCEDURE*

1. Preparation of equipment. Add 500 m*L* of distilled water and some glass beads to an 800-m*L* Kjeldahl flask and steam out the distillation apparatus until 250 m*L* have been distilled. The distillate should be checked to ensure that it is ammonia-free. The Nessler reagent is used for this purpose. Add 1 m*L* of Nessler reagent to determine if the distillation apparatus is not contaminated with ammonia. If the distillate turns yellow, there is ammonia present. More water should then be distilled and the test with Nessler reagent repeated.
2. Sample preparation. Remove any residual chlorine by adding dechlorinating agent equivalent to the chlorine residual (see page 576). To 400 m*L* of sample add 1 *N* NaOH solution until the pH is 9.5, checking the pH during addition with a pH meter or by use of a short-range pH indicator paper.
3. Transfer 280 m*L* of the sample, the pH of which has been adjusted to 9.5, to an 800-m*L* Kjeldahl flask, add some glass beads, and then add 25 m*L* of the borate buffer. Distill 300 m*L* at the rate of 6 to 10 m*L* per minute into 50 m*L* of 2 percent boric acid contained in a 500-m*L* Erlenmeyer flask.

   The condenser tip or an extension of the condenser tip must extend below the level of the boric acid solution.

   Dilute the distillate to 500 m*L* with distilled water.
4. Add three drops of the mixed indicator solution to the distillate and titrate the ammonia with the 0.02 *N* H_2SO_4, matching the end point against a blank containing the sample volume of distilled water and H_3BO_3 (boric acid) solution. The color change during titration is from green to a purple end point. Record volume 0.02 *N* H_2SO_4 required.
5. Calculate ammonia concentration.

D. *EXAMPLE*

Results from a test for ammonia in a wastewater plant effluent were as follows:

m*L* of Sample Used = 280

m*L* of 0.02 *N* H_2SO_4 Used = 16.0

E. *CALCULATION*

$$\text{mg}/L\ NH_3\text{-N} = \frac{A \times 280}{S}$$

where:

A = m*L* 0.02 *N* H_2SO_4 Used

S = Sample Volume Distilled

From example above:

m*L* of 0.02 *N* H_2SO_4 Used, A = 16.0

m*L* of Sample Volume Distilled, S = 280

$$\begin{aligned} \text{mg}/L\ NH_3\text{-N} &= \frac{A \times 280}{S} \\ &= \frac{16.0 \times 280}{280} \\ &= 16.0 \times 1 \\ &= 16.0\ \text{mg}/L \end{aligned}$$

(Nitrogen-Ammonia)

OUTLINE OF PROCEDURE FOR NITROGEN-AMMONIA

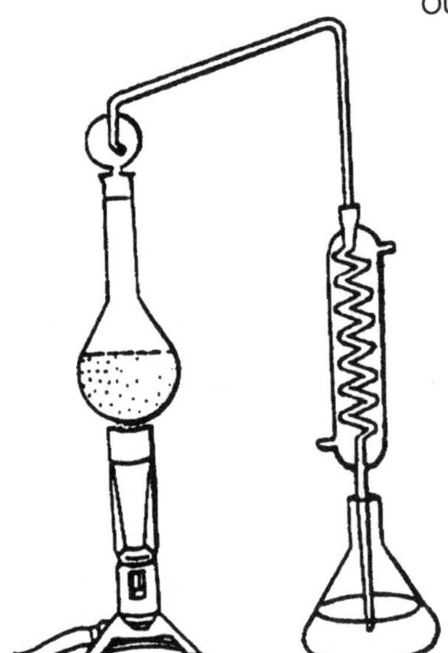

1. Add 500 m*L* distilled water to Kjeldahl flask and distill 250 m*L* to purge equipment.

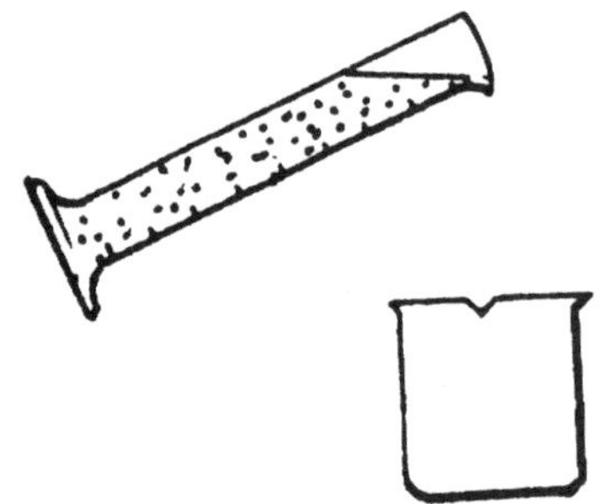

2. Collect 400 m*L* of sample and remove any trace of residual chlorine.

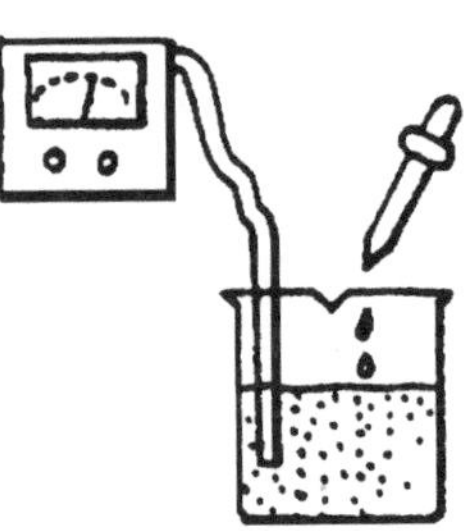

3. Adjust pH to 9.5 with 1 *N* NaOH.

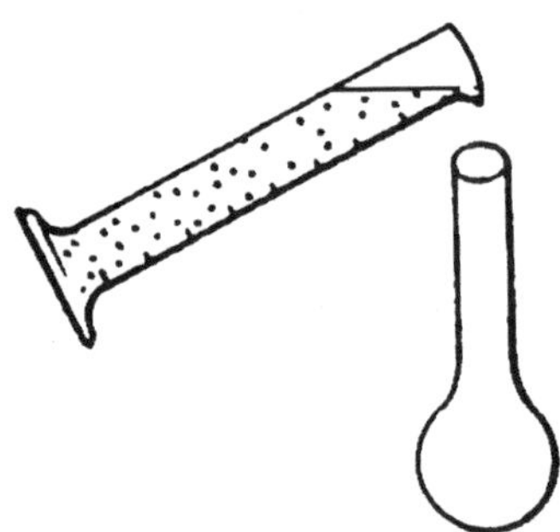

4. Transfer 280 m*L* (or other aliquot) to 800-m*L* Kjeldahl flask.

5. Add 25 m*L* borate buffer.

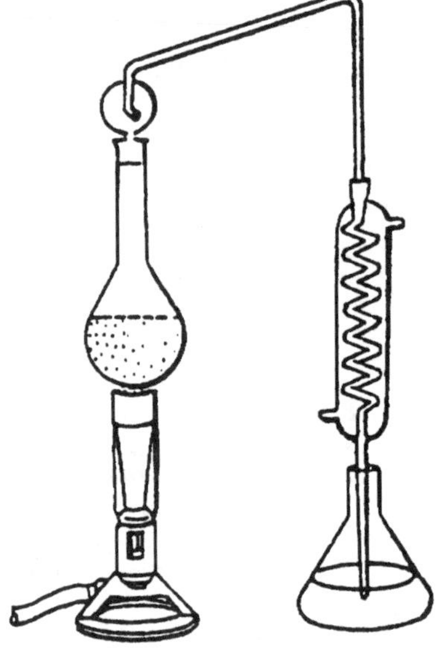

6. Distill 300 m*L* into 500-m*L* Erlenmeyer flask containing 50 m*L* 2% boric acid.

7. Add 3 drops mixed indicator.

8. Titrate with 0.02 *N* H_2SO_4. Record amount of acid required.

9. Calculate ammonia concentration.

F. *NOTES*

1. All standards do not have to be distilled in the same manner. However, at least two standards (a high and a low) should be distilled and titrated. If these standards do not agree with known values, the operator should find the cause of the apparent error.

2. This procedure is good only for samples that contain greater than 1.0 mg/*L* NH_3-N.

G. *REFERENCES*

Page 350.2-1, EPA's "Methods for Chemical Analysis of Water and Wastes," US Environmental Protection Agency. EPA No. 600-4-79-020. Obtain from National Technical Information Service (NTIS), 5285 Port Royal Road, Springfield, VA 22161. Order No. PB84-128677. Price, $117.00, plus $5.00 shipping and handling per order.

See page 4-109, *STANDARD METHODS,* 21st Edition.

II. Procedure for Ammonia (Ammonia-Selective Electrode Method; See Section E, Note 1)

For specific ion meters, follow the manufacturer's instruction manual for calibration and sample test procedures. (See Section E, Note 2.) If a pH meter is to be used with the ammonia electrode, for normal domestic wastewater, observe the following procedures.

A. *APPARATUS*

1. pH meter or specific ion meter.

   Need 0.1 mV resolution between –700 mV and +700 mV.

2. Ammonia-selective electrode.

3. Magnetic stirrer, thermally insulated, with TFE-coated stirring bar.

B. *REAGENTS*

1. Ammonia-free water. See B.1. previous section.

2. Ammonium chloride, stock solution. See B.2. previous section.

It may be easier for some operators to purchase a pre-prepared standard ammonia solution of 1,000 mg/*L* as NH_3-N. (See Section E, Note 3.)

3. Ammonium chloride, standard solutions. See C. *PROCEDURE*, 1. Prepare standards.

4. Sodium hydroxide, 10 *N*. Dissolve 400 gm NaOH in 800 mg/*L* of ammonia-free water. Cool and dilute to 1,000 m*L* with ammonia-free water.

C. *PROCEDURE*

1. Prepare standards. Prepare a series of standard solutions with concentrations of 1, 3, and 10 mg NH_3-N/*L* by making dilutions of stock ammonium chloride (NH_4Cl) solutions with ammonia-free water.

2. Calibration of electrode meter. Place 100 m*L* of each standard solution in a 100-m*L* beaker. Immerse electrode in standard of lowest concentration and mix with a magnetic stirrer. Do not stir too rapidly or bubbles will be sucked into solution and trapped on electrode membrane. Use the same stirring rate and temperature (about 25°C) throughout calibration and testing procedures. Add enough 10 *N* NaOH (1 m*L* usually is sufficient) to increase pH to above 11. Keep electrode in solution until a stable millivolt reading is obtained. Repeat procedure for each standard. Always start with lowest concentration standard and work toward higher standards. Keep the electrode immersed in the standard solution until a stable millivolt reading is obtained. Do not add NaOH solution before immersing electrode because ammonia could be lost when NaOH is added.

3. Prepare NH_3-N standard curve. Use semilogarithmic graph paper (one side has a log scale and the other side has a regular or evenly spaced scale). Plot ammonia concentration in milligrams NH_3-N per liter on the log scale versus potential on the linear (evenly spaced) scale starting with the lowest concentration at the bottom of the scale. If the electrode is working properly, expect a tenfold change (1 to 10 mg NH_3-N/*L*) of NH_3-N concentration to produce a potential change of 59 mV. For example, if an operator obtained –38 mV for 1.0 mg/*L* standard and –94 mV for 10.0 mg/*L*, the resulting slope will be 94 – 38 = 56 mV per decade.

4. Measurement of NH_3-N. Dilute samples if necessary to bring expected NH_3-N concentration to within range of NH_3-N calibration curve. Place 100 m*L* of sample in 150-m*L* beaker and follow same procedure as in Section 2, "Calibration of electrode meter." Record volume of 10 *N* NaOH added in excess of 1 m*L* needed to increase pH above 11. Record millivolt reading and then read NH_3-N concentration from standard curve.

D. *EXAMPLE*

Results of calibration of an ammonia electrode and subsequent sample measurements are as follows:

Standard A (1.0 mg/*L* NH_3-N) = –38 mV

Standard B (3.0 mg/*L* NH_3-N) = –67 mV

Standard C (10.0 mg/*L* NH_3-N) = –94 mV

Sample #1	–24 mV	→ 0.53 mg/*L* NH_3-N*
Sample #2	–54 mV	→ 1.8 mg/*L* NH_3-N
Sample #3	–104 mV	→ 15 mg/*L* NH_3-N

* i. Use a 3-cycle semilog graph paper. Set the scale of millivolt on abscissa (horizontal) and determine the concentration unit on log scale of ordinate (vertical) (See page 585, Figure 16.18).
ii. Plot the points of the three standard solutions using concentration-millivolt readings. Draw a line of best fit of the three points on the graph.
iii. Using the millivolt readings of samples, find the corresponding concentration on the graph, for example –24mV of sample #1 will intersect the calibration line at 0.53 mg/*L* on Figure 16.18.

E. *NOTES*

1. Samples should be distilled prior to the NH_3-N tests using an ammonia-selective electrode *UNLESS* tests have been performed that prove there is no difference in the NH_3-N test results with and without distillation. Usually tests indicate that distillation is not required. If the distillation step is to be omitted, the comparison data should be on file to show to regulatory agencies when requested.

2. NH_3-N electrodes must be calibrated with three known standard solutions. When using a meter that allows entering values of two standard values, you need to check the calibration with the third standard. For example, calibrate the electrode with 1.0 and 10.0 mg NH_3-N/*L* standards and then test the electrode with a 3.0 mg NH_3-N/*L* standard solution. If the reading is within 5 percent of the true value of 3.0 (2.85 to 3.15 mg NH_3-N/*L*), the calibration is considered acceptable.

(Nitrogen-Ammonia)

OUTLINE OF ELECTRODE METHOD FOR AMMONIA

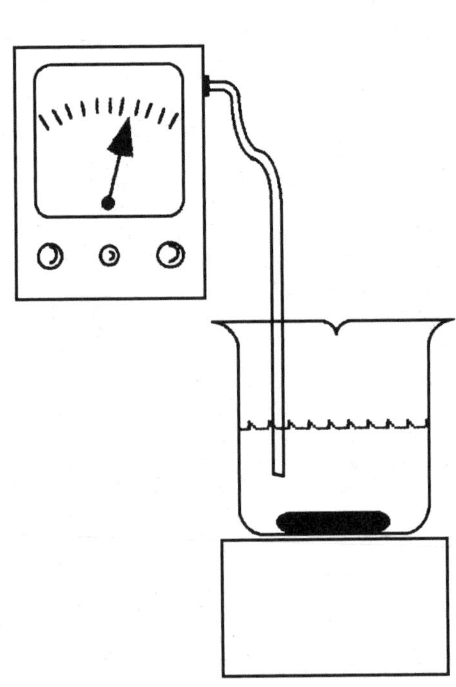

1. Place 100 m*L* in beaker.
   Mix with magnetic stirrer.

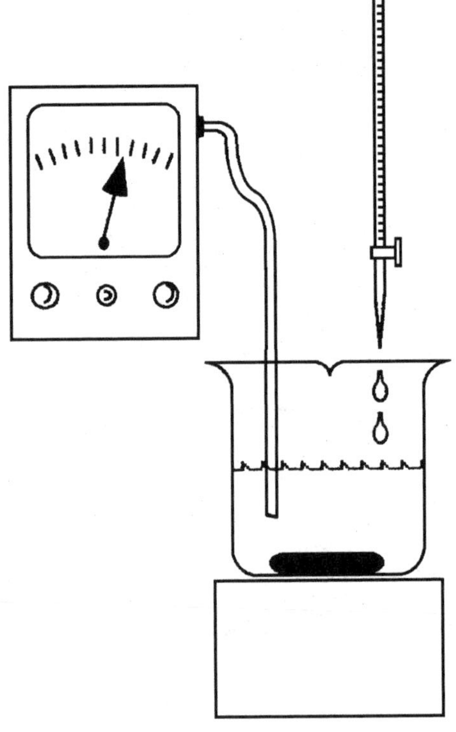

2. Add 10 *N* NaOH to
   pH above 11.

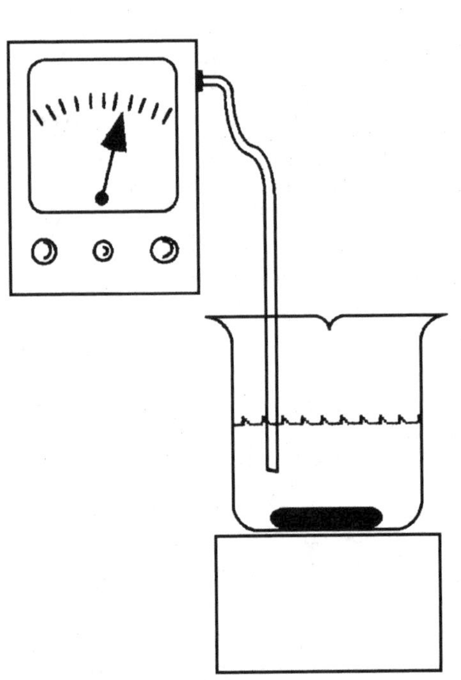

3. Wait until mV reading
   is constant and
   then record.

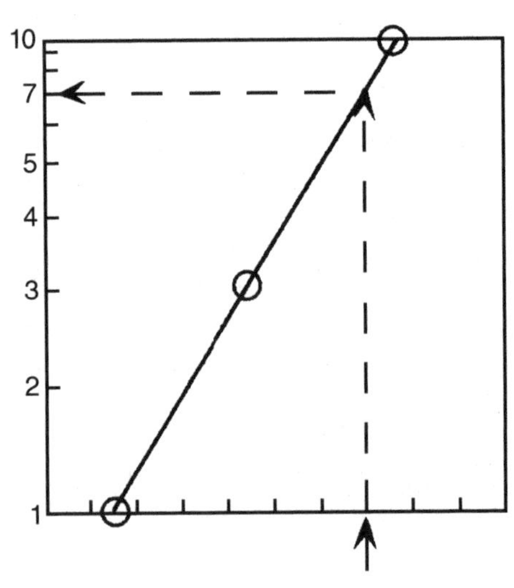

4. Obtain mg NH_3-N/*L*
   from standard curve.

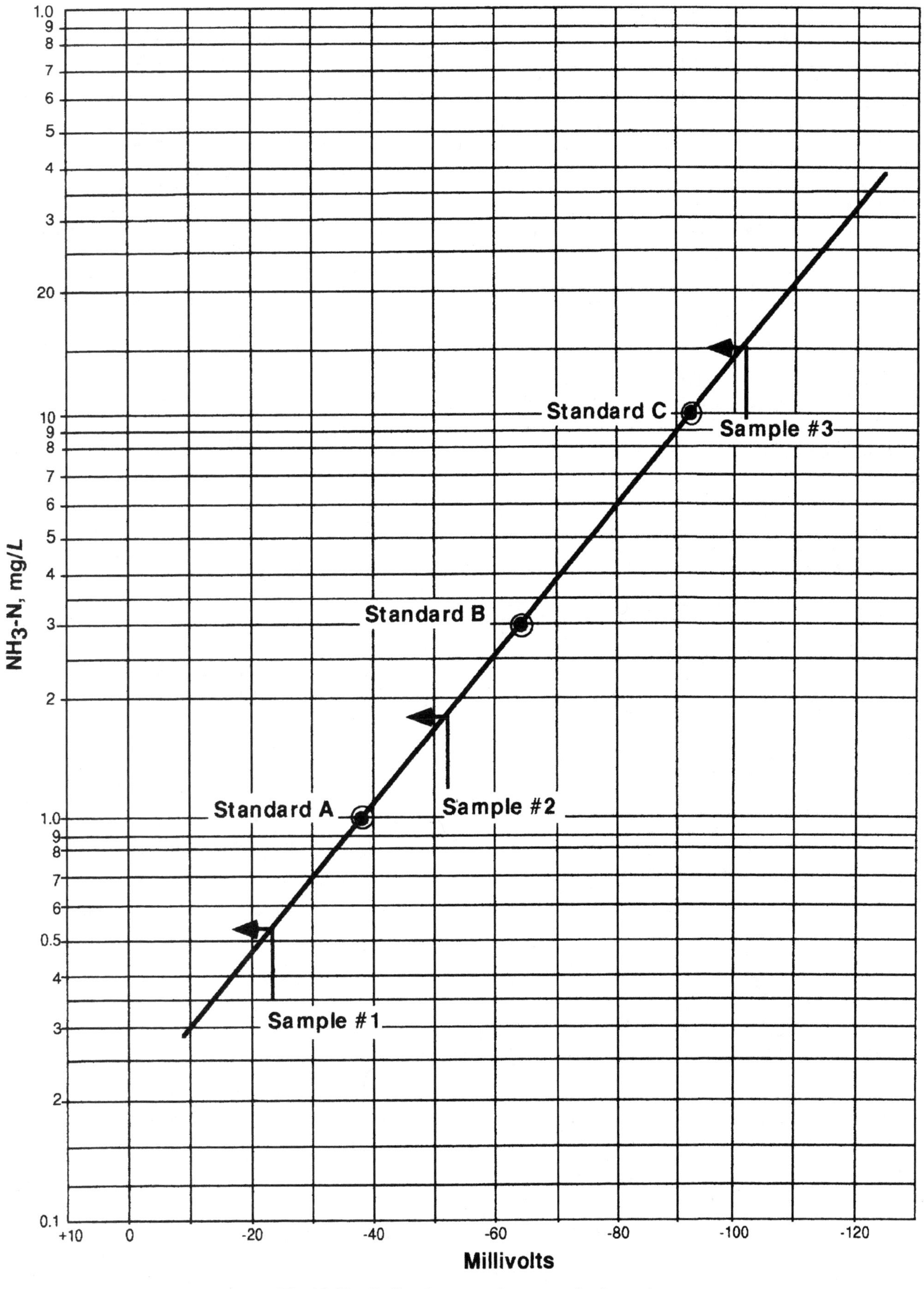

(Nitrogen-Ammonia)

Fig. 16.18 Calibration curve for ammonia electrode

(Nitrogen-TKN)

3. It is not recommended to use a NH_3-N standard solution of 0.1 mg/*L* due to its slowness in reaching a stable reading; NH_3 gas could be lost to the atmosphere during this long period to reach an equilibrium resulting in an erroneous reading.

4. When purchasing a pre-prepared ammonia standard solution, make sure the standard is expressed as "NH_3-N," not "NH_3."

F. *REFERENCES*

1. See page 4-103, *STANDARD METHODS,* 21st Edition.
2. Specific Ion Electrode Manual, Orion Co.
3. EPA Test Procedures (40 CFR 136, July 1, 2006).

III. Procedures for Total Kjeldahl Nitrogen (TKN)

A. *APPARATUS*

1. Digestion apparatus. A Kjeldahl digestion apparatus with 800-m*L* flasks and suction takeoff to remove SO_3 fumes and water.
2. Apparatus required for ammonia determination.

B. *REAGENTS*

1. Mercuric sulfate solution. Dissolve 8 gm red mercuric oxide (HgO) in 50 m*L* of 1:4 sulfuric acid (10 m*L* concentrated H_2SO_4 into 40 m*L* distilled water—*be careful.*) and dilute to 100 m*L* with distilled water.
2. Sulfuric acid-mercuric sulfate-potassium sulfate solution. Dissolve 267 gm K_2SO_4 in 300 m*L* distilled water and 400 m*L* concentrated H_2SO_4. Add 50 m*L* mercuric sulfate solution and dilute to one liter.
3. Sodium hydroxide-sodium thiosulfate solution. Dissolve 500 gm NaOH and 25 gm $Na_2S_2O_3 \cdot 5\,H_2O$ in distilled water and dilute to one liter.
4. All reagents required for ammonia determination (page 581) except borate buffer and sodium hydroxide solutions.

C. *PROCEDURE*

1. Measure a sample into an 800-m*L* Kjeldahl flask. The sample size can be determined from the following table:

Kjeldahl Nitrogen in Sample, mg/*L*	Sample Size, m*L*
0 to 5	500
5 to 10	250
10 to 20	100
20 to 50	50
50 to 500	25

Dilute the sample, if required, to 500 m*L* with distilled water. Also prepare a distilled water blank by adding 500 m*L* distilled water to an 800-m*L* Kjeldahl flask.

2. Add 100 m*L* sulfuric acid-mercuric sulfate-potassium sulfate solution to both flasks.
3. Add several glass beads to prevent bumping in the flasks.
4. Place the flasks on digestion apparatus and evaporate the mixtures until sulfur trioxide (SO_3) fumes are given off. SO_3 fumes will be indicated when white smoke begins rising from the solution. Sulfur trioxide is toxic; therefore observe extreme caution. Continue heating for 30 additional minutes.
5. Cool the residues and then add 300 m*L* of distilled water to each flask.
6. Prepare distillation apparatus as described under ammonia procedure.
7. Add 50 m*L* of 2 percent boric acid to each 500-m*L* Erlenmeyer receiving flask. Position the Erlenmeyer flasks under distillation apparatus so that the tip of the condenser or an extension is below the level of the boric acid solution in receiving flask.
8. Tilt each digested Kjeldahl flask and carefully add 100 m*L* of sodium hydroxide-thiosulfate solution to form an alkaline layer at the bottom of the flask.

   Do not agitate flask until it is connected to distillation apparatus since free ammonia may be liberated too soon.
9. Connect the Kjeldahl flasks to the condenser and mix contents of each flask.
10. Distill 300 m*L* from each flask at the rate of 6 to 10 m*L* per minute, into 50 m*L* of 2 percent boric acid.
11. Dilute the distillate to 500 m*L* in the flasks.
12. Add three drops of mixed indicator to the blank sample and titrate with 0.02 *N* H_2SO_4. The color change is from green to a purple end point. Record volume used.
13. Add three drops of the mixed indicator and titrate the sample flask with 0.02 *N* H_2SO_4 to an end point that matches against the blank. Record volume of 0.02 *N* H_2SO_4 used.
14. Calculate Total Kjeldahl Nitrogen (TKN) concentration.

D. *EXAMPLE*

Results from a test for TKN on a wastewater treatment plant effluent.

m*L* of Sample Used	= 50 m*L*
m*L* 0.02 *N* H_2SO_4 Used to Titrate Sample	= 4.2 m*L*
m*L* of 0.02 *N* H_2SO_4 Used to Titrate Blank	= 0.3 m*L*

E. *CALCULATION*

$$\text{mg}/L\ \text{TKN} = \frac{(A - B) \times 280}{S}$$

Where:

A = m*L* 0.02 *N* H_2SO_4 Used to Titrate Sample

B = m*L* 0.02 *N* H_2SO_4 Used to Titrate Blank

S = m*L* of Sample Digested

(Nitrogen-TKN)

OUTLINE OF PROCEDURE FOR TOTAL KJELDAHL NITROGEN (TKN)

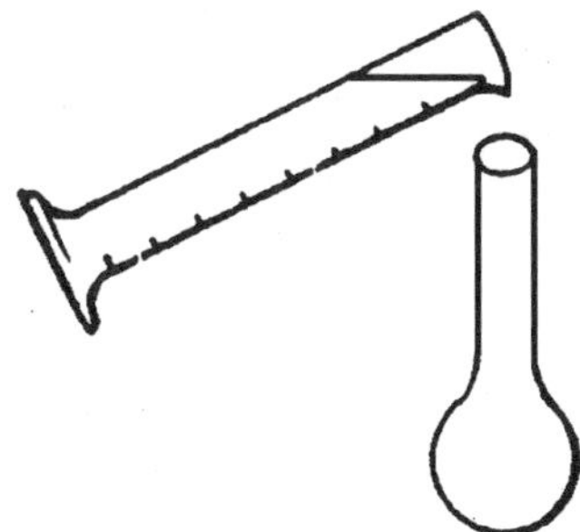

1. Measure sample into Kjeldahl flask.

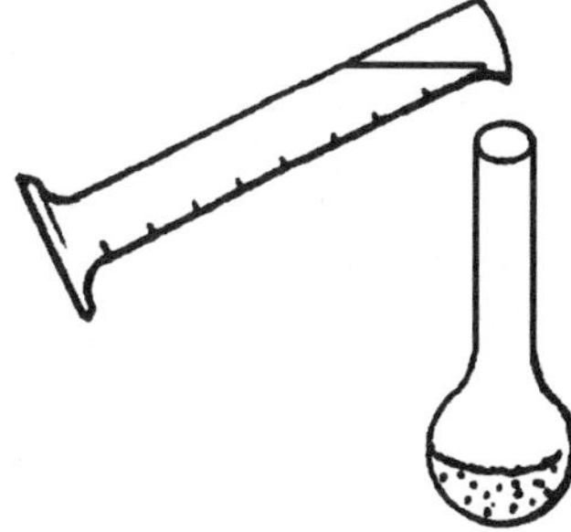

2. Add 100 m*L* sulfuric acid mixture.

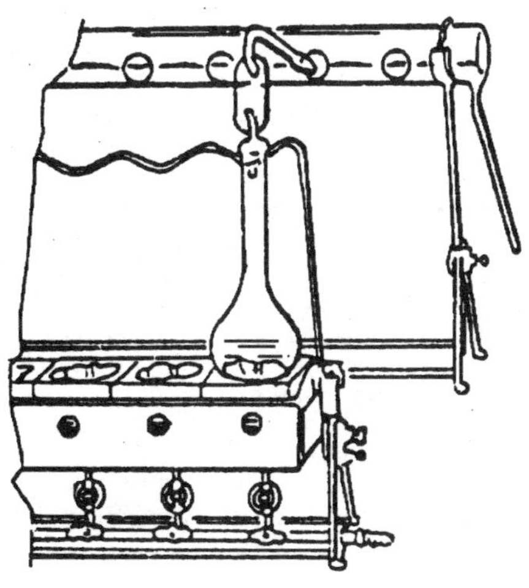

3. Digest.

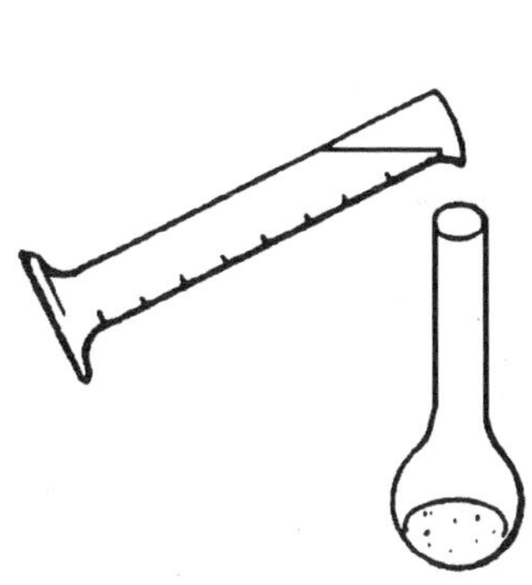

4. Cool flask. Add 300 m*L* distilled water.

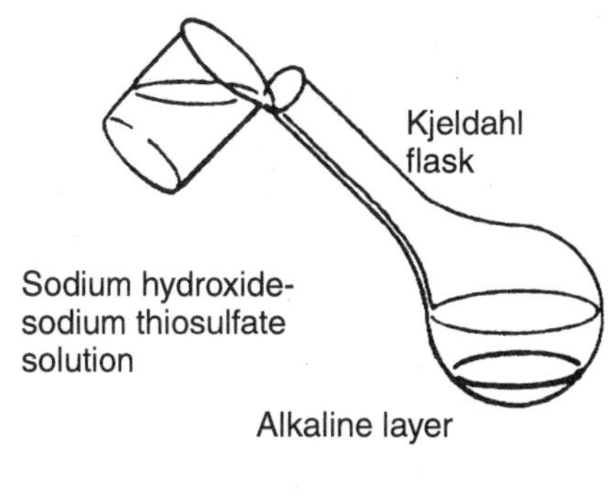

5. Add 100 m*L* NaOH solution.

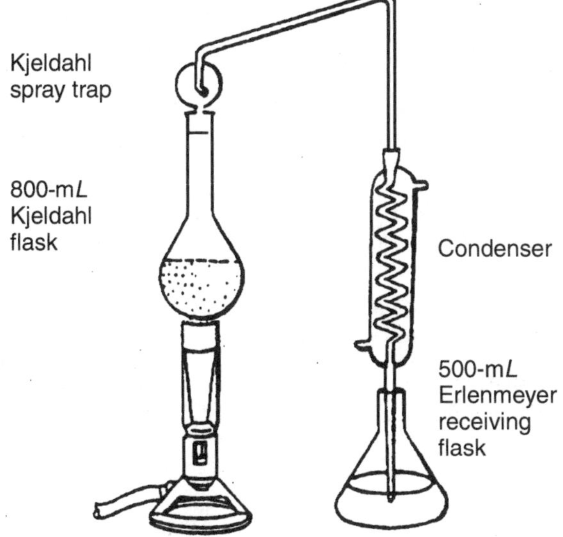

6. Connect flask to distillation apparatus, mix, and distill 300 m*L* into boric acid from receiving flask.

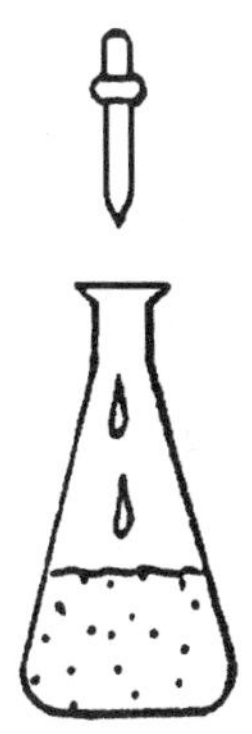

7. Dilute distillate to 500 m*L*. Add 3 drops mixed indicator.

8. Titrate with 0.02 N H_2SO_4. Calculate TKN.

(Nitrogen-Nitrite)

From item D. *EXAMPLE:*

mL 0.02 $N\ H_2SO_4$ Used to Titrate Sample = 4.2 mL

mL 0.02 $N\ H_2SO_4$ Used to Titrate Blank = 0.3 mL

mL of Sample Digested = 50 mL

$$\text{mg/}L\ \text{TKN} = \frac{(A - B) \times 280}{S}$$

$$= \frac{(4.2 - 0.3) \times 280}{50}$$

$$= \frac{3.9 \times 280}{50}$$

$$= \frac{1,092}{50}$$

$$= 21.8\ \text{mg/}L$$

F. *NOTES*

1. In this procedure high and low standards as prepared under the ammonia procedure should be digested, distilled, and titrated as a check on the accuracy of the operator's technique.
2. This procedure is good only for samples that contain greater than 1.0 mg/L TKN.

G. *REFERENCE*

See page 351.3-1, EPA's "Methods for Chemical Analysis of Water and Wastes." To order, see 9. Metals, B. References, 2, page 580.

IV. Procedure for Organic Nitrogen Calculation

A. *APPARATUS*

Apparatus as listed for ammonia and TKN procedures.

B. *REAGENTS*

Reagents as listed for ammonia and TKN procedures.

C. *PROCEDURE*

1. Determine ammonia concentration in sample.
2. Determine TKN concentration in sample.

D. *EXAMPLE*

A wastewater plant effluent contains an ammonia concentration of 16 mg/L and a TKN concentration of 22 mg/L.

E. *CALCULATION*

Organic Nitrogen, mg/L = TKN, mg/L – Ammonia, mg/L

Using the example:

Organic Nitrogen, mg/L = TKN, mg/L – Ammonia, mg/L

= 22 mg/L – 16 mg/L

= 6 mg/L

V. Procedure for Nitrite

A. *APPARATUS*

1. Spectrophotometer equipped with 1 cm or larger cells for use at 540 nm[23] wavelength.
2. Flasks, volumetric, 50 mL
3. pH meter
4. 0.45-micron pore size filter and holder assembly
5. Graduated cylinder, 50 mL
6. Pipets, measuring, 10 mL
7. Flask, filter
8. Flasks, Erlenmeyer, 125 mL

B. *REAGENTS*

1. Distilled water free of nitrite and nitrate.
2. Buffer-color reagent. To 250 mL of distilled water, add 105 mL concentrated hydrochloric acid, 5.0 gm sulfanilamide and 0.5 gm *N* (1-naphthyl) ethylenediamine dihydrochloride. Stir until dissolved. Add 136 gm of sodium acetate ($CH_3COONa \bullet 3\ H_2O$) and again stir until dissolved. Dilute to 500 mL with distilled water. This solution is stable for several weeks if stored in the dark.
3. Nitrite stock solution. 1.0 mL = 0.1 mg NO_2-N. Dissolve 0.4926 gm dried anhydrous sodium nitrite (24 hours in desiccator) in distilled water and dilute to 1,000 mL. Preserve with 2 mL chloroform per liter.
4. Nitrite standard solution. 1.0 mL = 0.001 mg NO_2-N. Dilute 10.0 mL of the nitrite stock solution (reagent 3) to 1,000 mL.

C. *PROCEDURE*

1. Check pH and alkalinity of sample. If pH is greater than 10 or alkalinity is in excess of 600 mg/L, adjust pH to approximately 6 with 1:3 HCl.
2. If necessary, remove turbidity by filtering 75 mL of sample through a 0.45-micron pore size filter using the first portion of filtrate to rinse the filter flask.
3. Place 50 mL of sample, or an aliquot diluted to 50 mL, in a 125-mL Erlenmeyer flask. Hold until preparation of standards is complete.
4. Add 2 mL of buffer-color reagent to each standard and sample. Mix and allow color to develop for at least 15 minutes.
5. Read the color intensity in the spectrophotometer at 540 nm wavelength against the blank.
6. Determine amount of nitrite-nitrogen from a standard calibration graph.

[23] *nm.* Nanometers or 0.000 000 001 meters.

OUTLINE OF PROCEDURE FOR ORGANIC NITROGEN

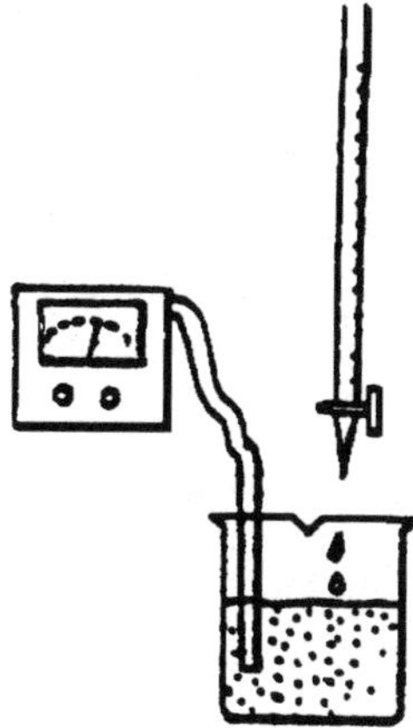

1. Check pH and alkalinity of sample.

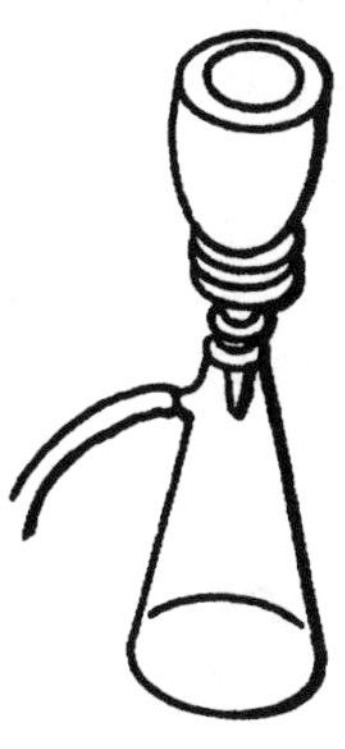

2. Remove turbidity if necessary.

3. Place 50 m*L* in flask.

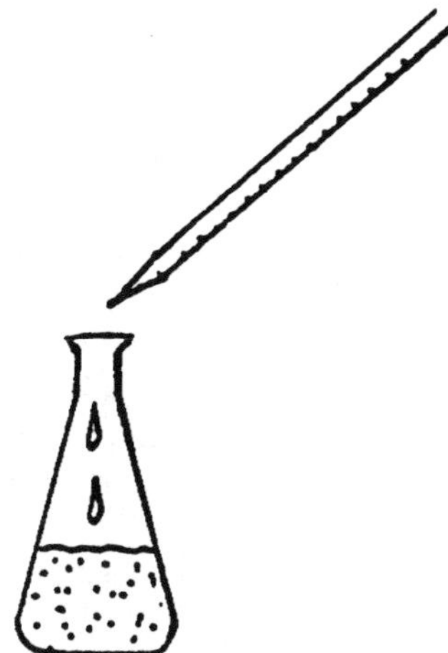

4. Add 2 m*L* buffer-color reagent. Mix. Allow to stand 15 minutes.

5. Measure color intensity at 540 nm with spectrophotometer.

6. Determine nitrite concentration in unknown sample by comparison to absorbance values of known NO_2-N concentrations by use of standard curve.

(Nitrogen-Nitrate and Nitrite)

D. *CONSTRUCTION OF STANDARD CALIBRATION GRAPH*

1. Using the nitrite standard solution, prepare the following series of nitrite standards in 50-m*L* volumetric flasks:

m*L* of Standard Solution 1.0 m*L* = 0.001 mg NO_2-N	**Concentration Diluted to 50 m*L* in mg/*L* NO_2-N**
0.0	0. (Blank)
1.0	0.02
2.0	0.04
4.0	0.08
5.0	0.10

2. Transfer this series of standards diluted to 50 m*L* to 125-m*L* Erlenmeyer flasks.
3. Determine the amount of nitrite-nitrogen as outlined above.
4. Plot on a sheet of graph paper absorbance versus concentration.

E. *EXAMPLE*

A sample of a wastewater plant effluent and a series of nitrite standards were analyzed with the following results:

50 m*L* of sample and standards

Sample	**Absorbance**
Plant Effluent	0.052
Blank	0.00
0.02 mg/*L* NO_2-N Standard	0.040
0.04 mg/*L* NO_2-N Standard	0.081
0.08 mg/*L* NO_2-N Standard	0.165
0.10 mg/*L* NO_2-N Standard	0.205

F. *CALCULATION*

1. Plot curve of concentration NO_2-N standards versus absorbance on graph paper. For example, from the above data the graph on page 591 can be constructed.
2. Correct (if necessary) for samples of less than 50 m*L* by using the following formula:

$$NO_2\text{-N} = \frac{\text{mg}/L \text{ from Graph} \times 50 \text{ m}L}{\text{m}L \text{ Sample Used}}$$

G. *NOTE*

Samples should be analyzed as soon as possible following collection although they may be stored for 24 to 48 hours at 4°C.

H. *REFERENCES*

See page 354.1-1, EPA's "Methods for Chemical Analysis of Water and Wastes." To order, see 9. Metals, B. References, 2, page 580.

See page 4-118, *STANDARD METHODS,* 21st Edition.

VI. Procedure for Nitrate and Nitrite Nitrogen

This procedure measures the amount of both nitrate and nitrite nitrogen present in a sample by reducing all nitrate to nitrite through the use of a copper-cadmium column. The nitrite (that is originally present plus the reduced nitrate) is then measured colorimetrically.

A. *APPARATUS*

1. Reduction column. The column in Figure 16.19 was constructed from a 100-m*L* volumetric pipet by removing the top portion. This column may also be constructed from two pieces of tubing joined end to end. A 10-cm length of 3-cm I.D. tubing is joined to a 25-cm length of 3.5-mm I.D. tubing.
2. Spectrophotometer for use at 540 nm, providing a light path of 1 cm or longer.
3. Beakers, 125 m*L*
4. Glass wool
5. Glass-fiber filter or 0.45-micron membrane filter
6. Filter holder assembly
7. Filter flask
8. pH meter
9. Separatory funnel, 250 m*L*
10. Pipets, volumetric 1, 2, 5, and 10 m*L*

B. *REAGENTS*

1. Treated cadmium granules, available from HACH Company, PO Box 389, Loveland, CO 80539-0389. Catalog No. 25559-25. Price, $124.00.
2. Copper-Cadmium. The cadmium granules (new or used) are cleaned with 6 *N* HCl and copperized with 2 percent solution of copper sulfate in the following manner:
   a. Wash the cadmium with 6 *N* HCl and rinse with distilled water. The color of the cadmium should be silver.
   b. Swirl 25 gm cadmium in 100-m*L* portions of a 2 percent solution of copper sulfate for 5 minutes or until the blue color partially fades, decant, and repeat with fresh copper until a brown precipitate forms.
   c. Wash the copper-cadmium with distilled water at least 10 times to remove all the precipitated copper. The color of the cadmium should now be black.
3. Preparation of reaction column. Insert a glass wool plug into the bottom of the reduction column and fill with distilled water. Add sufficient copper-cadmium granules to produce a column 18.5 cm in length. Maintain a level of distilled water above the copper-cadmium granules to eliminate entrapment of air. Wash the column with 200 m*L* of dilute ammonium chloride—EDTA solution (reagent 5). The column is then activated by passing through the column 100 m*L* of a solution composed of 25 m*L* of a 1.0-mg/*L* NO_2-N standard and 75 m*L* of concentrated ammonium chloride—EDTA solution. Use a flow rate of 7 to 10 m*L* per minute. Collect the reduced standard until the level of solution is 0.5 cm above the top of the granules. Close the screw clamp to stop flow. Discard the reduced standard.
4. Measure about 40 m*L* of concentrated ammonium chloride—EDTA and pass through column at 7 to 10 m*L* per minute to wash nitrate standard off column. Always leave at least 0.5 cm of liquid above top of granules. The column is now ready for use.
5. Dilute ammonium chloride—EDTA solution. Dilute 300 m*L* of concentrated ammonium chloride—EDTA solution (reagent 4) to 500 m*L* with distilled water.

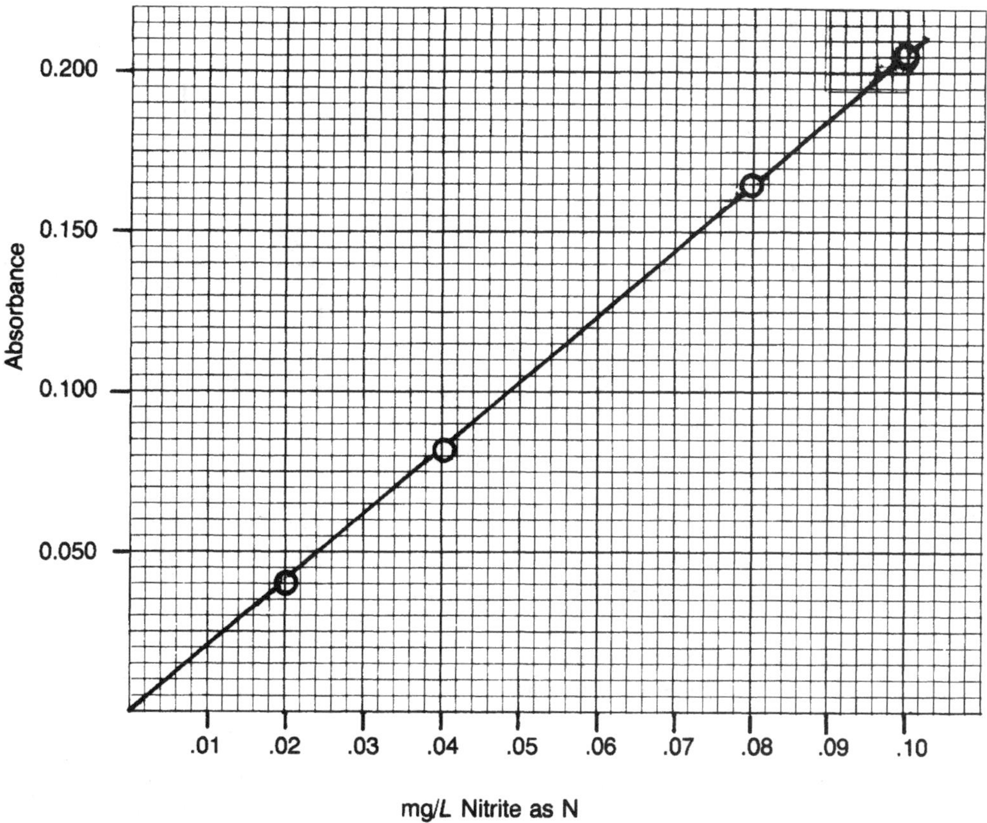

Read concentration of NO_2-N in plant effluent directly from graph.

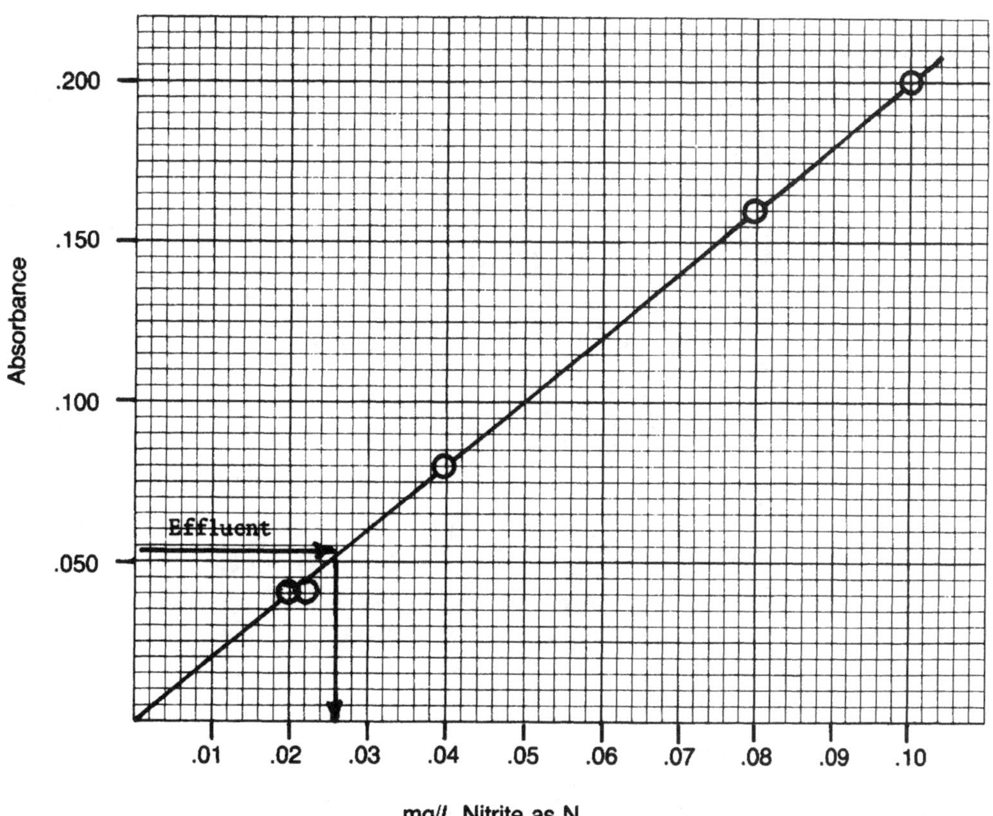

(Nitrogen-Nitrate and Nitrite)

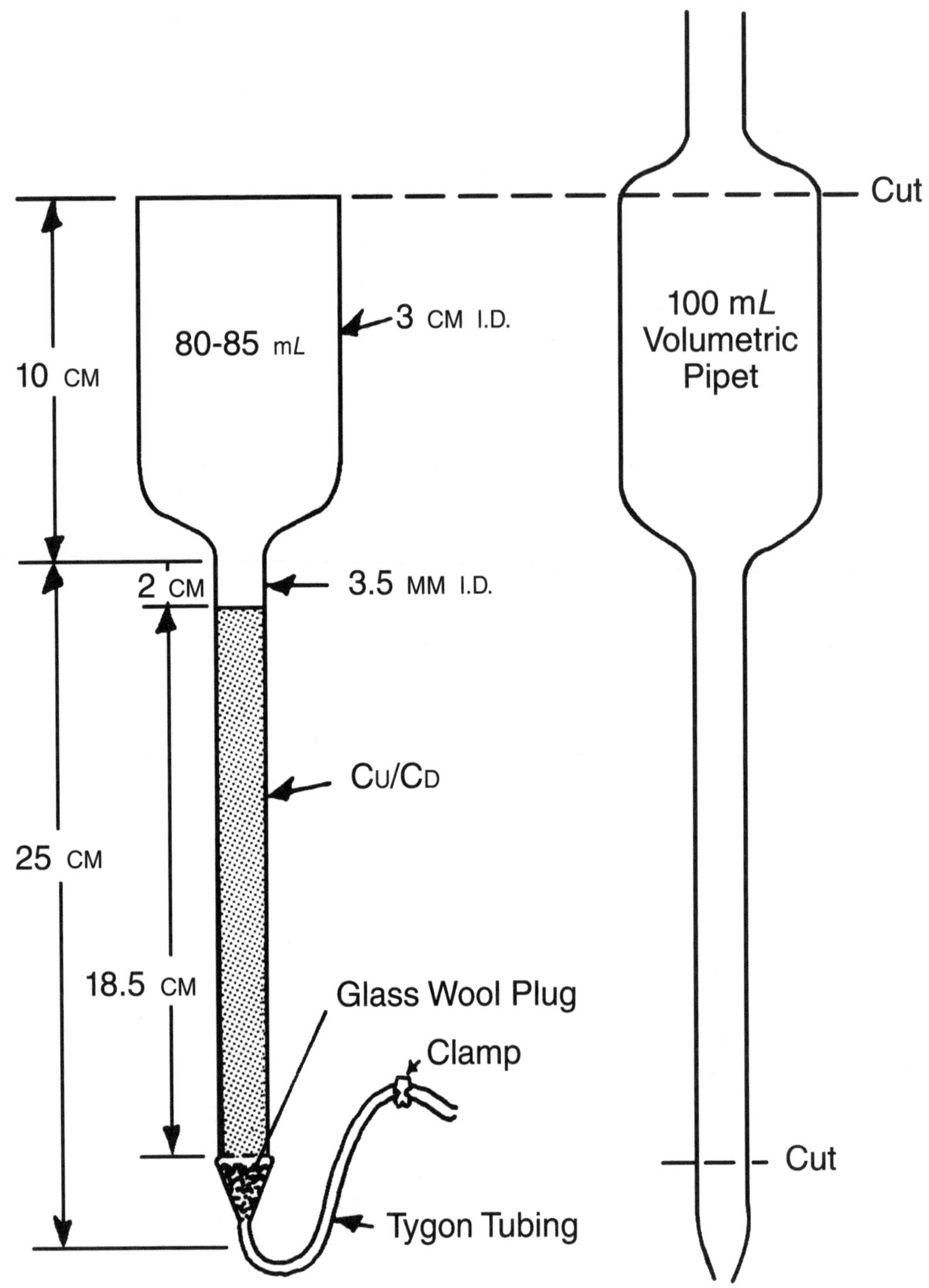

Fig. 16.19 Reduction column for measuring nitrate-nitrite nitrogen

6. Color reagent. Dissolve 10 gm sulfanilamide and 1 gm *N* (1-naphthyl)-ethylenediamine dihydrochloride in a mixture of 100 m*L* concentrated phosphoric acid and 800 m*L* of distilled water and dilute to 1 liter with distilled water.

7. Zinc sulfate solution. Dissolve 100 gm $ZnSO_4 \cdot 7\,H_2O$ in distilled water and dilute to 1 liter.

8. Sodium hydroxide, 6 *N*. Dissolve 240 gm NaOH in 500 m*L* distilled water, cool, and dilute to 1 liter.

9. Ammonium hydroxide, concentrated.

10. Hydrochloric acid, 6 *N*. Dilute 50 m*L* concentrated HCl to 100 m*L* with distilled water.

11. Copper sulfate solution, 2 percent. Dissolve 20 gm of $CuSO_4 \cdot 5\,H_2O$ in 500 m*L* of distilled water and dilute to 1 liter.

12. Nitrate stock solution. 1.0 m*L* = 1.00 mg NO_3-N. Dissolve 7.218 gm KNO_3 in distilled water and dilute to 1,000 m*L*. Preserve with 2 m*L* of chloroform per liter. This solution is stable for at least six months.

13. Nitrate standard solution. 1.0 m*L* = 0.01 mg NO_3-N. Dilute 10.0 m*L* of nitrate stock solution (reagent 12) to 1,000 m*L* with distilled water.

14. Chloroform.

C. *PROCEDURE*

Removal of Interferences (if necessary).

1. Turbidity removal. Use one of the following methods to remove suspended matter that can clog the reduction column.

   a. Filter sample through a glass fiber or a 0.45-micron pore size filter as long as the pH is less than 8, or

   b. Add 1 m*L* zinc solution (reagent 7) to 100 m*L* of sample and mix thoroughly. Add enough (usually 8 to 10 drops) sodium hydroxide solution (reagent 8) to obtain a pH of 10.5. Let treated sample stand a few minutes to allow the heavy flocculent precipitate to settle. Clarify by filtering through a glass fiber filter.

2. Oil and grease removal.

   a. Adjust pH of 100 m*L* of filtered sample (step 1) to a pH of 2 by dropwise addition of concentrated HCl.

   b. Place sample in 250-m*L* separatory funnel.

   c. Add 25 m*L* chloroform.

   d. Stopper and shake separatory funnel gently to extract the oils and grease into the chloroform. Carefully release the pressure after shaking so that no sample is lost. This can be accomplished by inverting the separatory funnel and slowly opening the stopcock away from your face and other people.

   e. Unstopper and allow the separatory funnel to stand until all of the chloroform settles to the bottom.

   f. Open the stopcock and allow the bottom chloroform layer to pass into a beaker and discard.

   g. Repeat steps c, d, e, and f with 25 m*L* of fresh chloroform.

Reduction of Nitrate to Nitrite.

1. Using a pH meter adjust the pH of sample (or standard) to between 5 and 9 either with concentrated HCl or concentrated NH_4OH.

2. To 25 m*L* of sample (or standard) or aliquot diluted to 25 m*L*, add 75 m*L* of concentrated ammonium chloride—EDTA solution and mix.

3. Pour sample into column and collect reduced sample at a rate of 7 to 10 m*L* per minute.

4. Discard the first 25 m*L*. Collect the rest of the sample (approximately 70 m*L*) in the original sample flask. Reduced samples should not be allowed to stand longer than 15 minutes before addition of color reagent.

5. Add 2.0 m*L* of color reagent to 50 m*L* of sample. Allow 10 minutes for color development. Within two hours measure the absorbance at 540 nm against a reagent blank (50 m*L* distilled water to which 2.0 m*L* color reagent has been added).

Construction of Standard Calibration Graph

1. Prepare working standards by pipeting the following volumes of nitrate standard solution into each of five 100-m*L* volumetric flasks.

Add This Volume of Nitrate Standard Solution to 100-m*L* flask	**Concentration of NO_3-N in mg/*L***
0.0	0.00
1.0	0.10
2.0	0.20
5.0	0.50
10.0	1.00

Dilute each to 100 m*L* with distilled water and mix.

2. Determine the amount of nitrate-nitrite nitrogen as outlined above in the procedure for reduction of nitrate to nitrite.

3. Plot on a sheet of graph paper the absorbance versus concentration.

D. *EXAMPLE*

Results from the analyses of samples and working standards for nitrate-nitrite were as follows:

Flask No.	**Sample**	**Volume**	**Absorbance**
1	Plant Effluent	25 m*L*	0.440
2	Blank (distilled water)	25 m*L*	0.00
3	0.10 mg/*L* NO_3-N	25 m*L*	0.075
4	0.20 mg/*L* NO_3-N	25 m*L*	0.142
5	0.50 mg/*L* NO_3-N	25 m*L*	0.355
6	1.00 mg/*L* NO_3-N	25 m*L*	0.700

E. *CALCULATION*

1. Using graph paper, plot the absorbance values of working standards versus their known concentrations. For example, from the above data the graph on page 594 can be constructed.

2. Read concentration of NO_3 + NO_2-nitrogen in plant effluent directly from graph.

(Nitrogen-Nitrate and Nitrite)

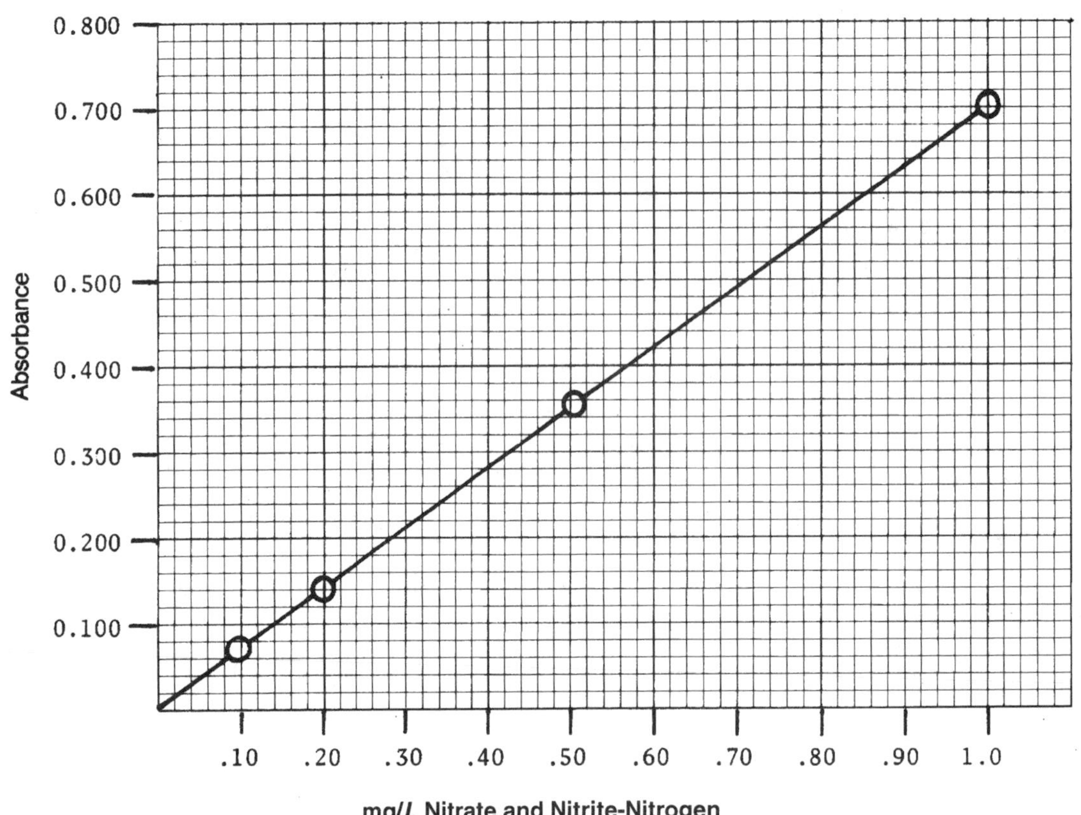

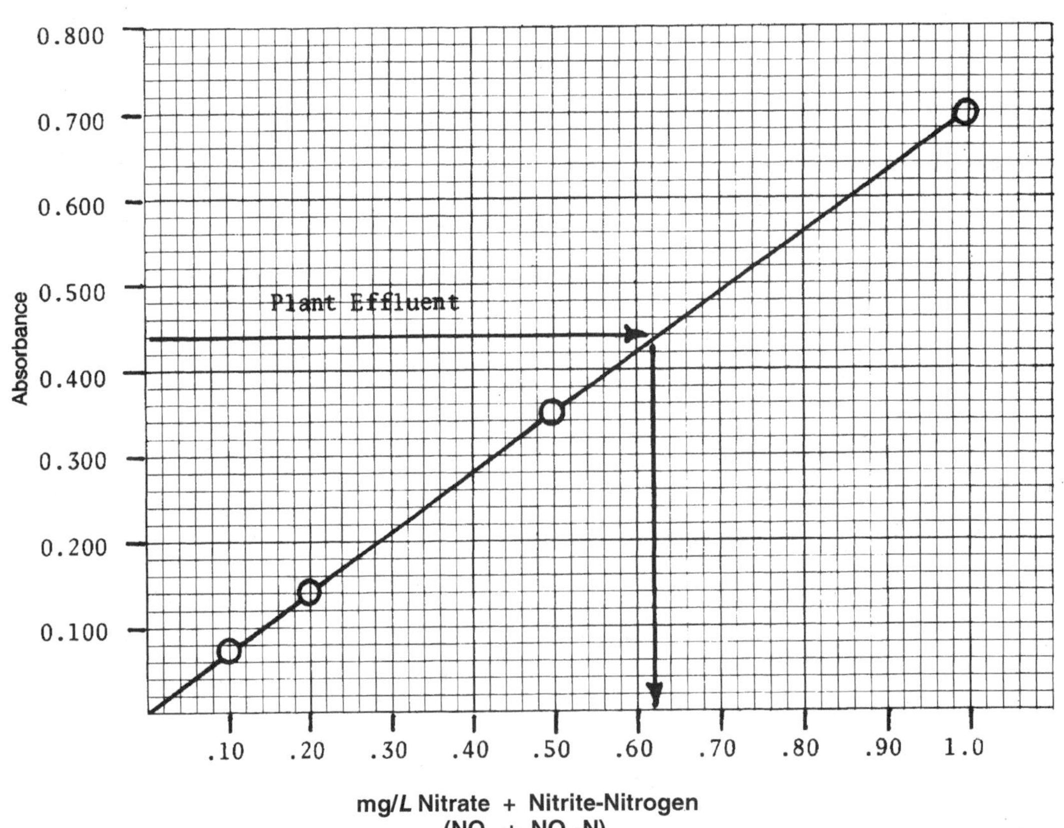

mg/*L* nitrate + nitrite-nitrogen in sample = 0.62 mg/*L*

3. Procedure for Determination of Nitrate-Nitrogen (NO_3-N)

   a. Determine the concentration of nitrite-nitrogen (NO_2-N) in the sample using the nitrite procedure.

   b. Subtract nitrite from NO_3 + NO_2-nitrogen concentration. The result is the amount of nitrate-nitrogen in the sample.

   c. For example, if the sample of plant effluent contained 0.03 mg/*L* nitrite-nitrogen and 0.62 mg/*L* nitrate + nitrite-nitrogen, then: 0.62 mg/*L* – 0.03 mg/*L* = 0.59 mg/*L* nitrate-nitrogen (NO_3-N).

F. *NOTES*

1. If the concentration of nitrate in the sample is greater than 1 mg/*L*, then the sample must be diluted.
2. Cadmium metal is highly toxic thus caution must be exercised in its use. Rubber gloves should be used whenever it is handled.

G. *REFERENCES*

See page 353.3-1, EPA's "Methods for Chemical Analysis of Water and Wastes." To order, see 9. Metals, B. References, 2, page 580.

See page 4-123, *STANDARD METHODS,* 21st Edition.

QUESTION

Write your answer in a notebook and then compare your answer with the one on page 613.

16.5V Kjeldahl nitrogen refers to ____________________.

11. Oil and Grease

A. *DISCUSSION*

This procedure measures gasoline, heavy fuel, lubricating oil, asphalt, soaps, fats, waxes, and any other material that is extracted by the solvent (Freon) used in the test. The determination of oil and grease in a wastewater plant is helpful in determining plant efficiencies and for diagnosis of in-plant problems such as difficulties digesting or dewatering sludges.

B. *WHAT IS TESTED?*

Sample	Common Range, mg/*L*
Influent and Effluent	<5 to 50

(Oil and Grease)

C. *APPARATUS REQUIRED*

1. Filter paper, Whatman No. 40, 11 cm
2. Balance, analytical
3. Funnel, separatory (2,000 m*L*) with Teflon stopcock
4. Flask, distilling 125 m*L*
5. Hot plate
6. Hot water bath
7. Vacuum source
8. Funnel, glass

D. *REAGENTS*

1. Hydrochloric acid 1 + 1: Carefully, with stirring, add 25 m*L* concentrated hydrochloric acid, HCl, to 25 m*L* distilled water.
2. Freon (1,1,2-trichloro-1,2,2-trifluoroethane) boiling point 47°C.
3. Sodium sulfate, Na_2SO_4, anhydrous crystal.

E. *PROCEDURE*

1. Dry in oven (at 103°C) a clean distilling flask. Cool in desiccator. Weigh and record weight. Store in desiccator until needed.
2. Collect 1,000 m*L* of sample in a glass container.
3. Add 5 m*L* 1 + 1 HCl to the sample (acidify to pH 2 or lower).
4. Transfer the liter sample to a 2,000-m*L* separatory funnel.
5. Carefully rinse the sample bottle with 30 m*L* Freon. Add these washings to the separatory funnel.
6. Stopper and shake the separatory funnel vigorously for 2 minutes. Allow the two layers (water and Freon) to separate.
7. Drain Freon layer through funnel containing Freon-moistened filter paper into the tared (weighed) flask prepared in Step 1. If the solvent layer is not clear, add 1 gram Na_2SO_4 to the filter paper cone and slowly drain the emulsified solvent onto the crystals. Add more Na_2SO_4 if necessary.
8. Extract twice more with 30 m*L* Freon, but first rinse sample container with each solvent portion. Combine the extracts in the tared flask and wash filter paper with 10 to 20 m*L* Freon.
9. Distill Freon from the extract flask in a water bath at 70°C. Place the flask in a warm steam bath for 15 minutes and draw air through the flask by means of a vacuum for the final 1 minute.
10. Cool in desiccator for exactly 30 minutes and weigh.

F. *EXAMPLE*

Results from an effluent sample:

1. 1,000 m*L* of sample
2. Empty flask weighed 26.1024 grams
3. Flask with residue weighed 26.1164 grams

(Oil and Grease)

OUTLINE OF PROCEDURE FOR OIL AND GREASE

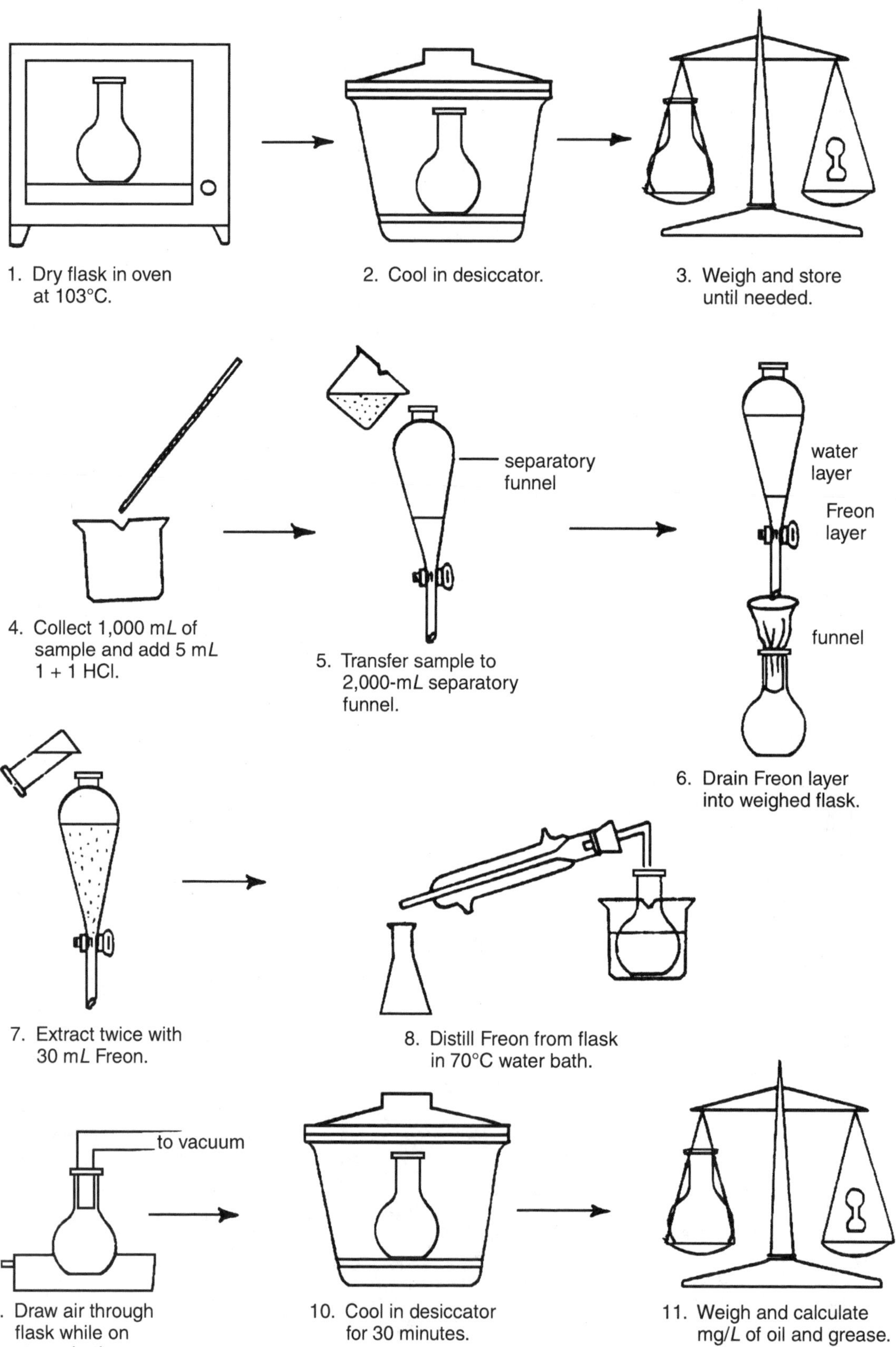

1. Dry flask in oven at 103°C.
2. Cool in desiccator.
3. Weigh and store until needed.
separatory funnel
water layer
Freon layer
funnel
4. Collect 1,000 m*L* of sample and add 5 m*L* 1 + 1 HCl.
5. Transfer sample to 2,000-m*L* separatory funnel.
6. Drain Freon layer into weighed flask.
7. Extract twice with 30 m*L* Freon.
8. Distill Freon from flask in 70°C water bath.
to vacuum
9. Draw air through flask while on steam bath.
10. Cool in desiccator for 30 minutes.
11. Weigh and calculate mg/*L* of oil and grease.

G. *CALCULATION*

1. Oil and Grease, mg/*L* = $\frac{(A - B) \times 1{,}000}{mL \text{ of Sample}}$

Where:

A = Weight of Flask and Residue, mg

B = Weight of Empty, Dry Flask, mg

2. From example,

$$\text{Oil and Grease, mg}/L = \frac{(A - B) \times 1{,}000}{mL \text{ of Sample}}$$

$$= \frac{(26{,}116.4\text{ mg} - 26{,}102.4\text{ mg})(1{,}000\text{ m}L/L)}{(1{,}000\text{ m}L)}$$

$$= 14\text{ mg}/L$$

H. *COMMENTS*

1. If oil and grease analysis is performed on sludge, then another method must be used (see page 5-41, *STANDARD METHODS,* 21st Edition).
2. The collection of a composite sample is impractical because losses of grease will occur on sampling equipment.

I. *REFERENCE*

See page 5-35, *STANDARD METHODS,* 21st Edition.

pH. See Hydrogen Ion (pH)

12. Phosphorus

A. *DISCUSSION*

Wastewater is relatively rich in phosphorus compounds. The forms of phosphorus found in wastewater are commonly classified into orthophosphate, condensed phosphate, and organically bound phosphate. Phosphorus is essential to the growth of organisms found not only in a wastewater treatment plant, but also in other bodies of water such as rivers, lakes, and oceans. The discharge of wastewater containing phosphorus may stimulate nuisance quantities of algal growths in receiving waters.

In the past, raw domestic wastewaters typically contained approximately 10 mg/*L* of phosphorus. Phosphorus bans or limitations in synthetic detergents or changes in detergent formulas by the manufacturers have served to reduce this historic level by varying amounts throughout the United States. The microorganisms in secondary biological treatment processes usually will reduce the influent phosphorus concentrations by two to three mg/*L*. Greater removals may be obtained by the use of metal ion coagulants such as alum (aluminum sulfate) or ferric chloride. Other removal processes involve pH adjustment with lime.

Orthophosphate is the amount of inorganic phosphorus (PO_4^{3-}) in the sample of wastewater as measured by the direct colorimetric analysis procedure. Total phosphorus is the amount of all the phosphorus present in the sample regardless of form, as measured by the persulfate digestion procedure followed by colorimetric analysis. These are the two most commonly measured forms in wastewater.

(Phosphorus)

B. *WHAT IS TESTED?*

Sample	Orthophosphate (P) Common Range, mg/*L*	Total Phosphorus (P) Common Range, mg/*L*
Influent	2 to 8	4 to 12
Effluent	1 to 6	2 to 10

C. *APPARATUS*

1. Photometer—a spectrophotometer or filter photometer suitable for measurements at 650 or 880 nanometers (nm) wavelength with a light path of 1 cm or longer. Follow manufacturer's directions for operation.
2. Balance, analytical, capable of weighing to 0.1 milligram
3. Balance, triple beam, capable of weighing to 0.1 gram
4. Desiccator
5. Hot plate or autoclave
6. Oven, drying for use at 105°C
7. pH meter
8. 150 m*L* beakers or Erlenmeyer flasks
9. 50 m*L* graduated cylinder
10. Pipets, measuring, 10 m*L*
11. Flasks, volumetric, 1,000 m*L* and 100 m*L*

D. *REAGENTS*

1. Ammonium molybdate-antimony potassium tartrate solution. Dissolve 8 grams of ammonium molybdate and 0.2 grams antimony potassium tartrate in 800 m*L* of distilled water and dilute to one liter.
2. Ascorbic acid solution. Dissolve 60 grams of ascorbic acid in 800 m*L* of distilled water and dilute to one liter. Add 2 m*L* of acetone. This solution is stable for two weeks.
3. Sulfuric acid, 11 *N*. Slowly add 310 m*L* of concentrated sulfuric acid to approximately 600 m*L* distilled water. Cool and dilute to one liter.
4. Ammonium persulfate
5. Stock phosphorus solution. Dissolve 0.4393 grams of predried (105°C for one hour) KH_2PO_4 in a 1,000-m*L* volumetric flask containing distilled water. Dilute to 1,000 m*L*. 1.0 m*L* of this solution contains 0.1 mg phosphorus.
6. Standard phosphorus solution. Dilute 100 m*L* of stock phosphorus solution to 1,000 m*L* with distilled water. 1.0 m*L* of this solution contains 0.01 mg P.

E. *PROCEDURE*

For Orthophosphate:

1. Place 50 m*L* of sample (or an aliquot diluted to 50 m*L*) and/or standards in a 150-m*L* beaker or Erlenmeyer flask (see note on glassware, page 598).
2. Add 1 m*L* of 11 *N* sulfuric acid and 4 m*L* of ammonium molybdate-antimony potassium tartrate and mix. (If sample has been digested for Total Phosphorus, do not add acid.)
3. Add 2 m*L* of ascorbic acid solution and mix.

(Phosphorus)

4. After 5 minutes, measure the absorbance at 650 nm with a spectrophotometer and determine the amount of phosphorus from the standard curve. The color is stable for at least one hour. Report results as P, mg/*L*.

For Total Phosphorus:

1. Place 50 m*L* of sample or an aliquot diluted to 50 m*L* into a 125-m*L* Erlenmeyer flask and add 1 m*L* of 11 *N* sulfuric acid.
2. Add 0.4 grams ammonium persulfate, mix, and boil gently for approximately 30 to 40 minutes or until a final volume of about 10 m*L* is reached. Alternatively, heat for 30 minutes in an autoclave at 121°C (15 to 20 psi or 1.0 to 1.4 kg/sq cm). Cool and dilute to 50 m*L*.
3. Determine amount of phosphorus as outlined above in orthophosphate.

Construction of Standard Calibration Curve:

1. Using the standard solution, prepare the following standards in 100-m*L* volumetric flasks.

m*L* of Standard Phosphorus Solution Placed in 100 m*L* Volumetric Flask	**Phosphorus Concentration, mg/*L***
0	
2.0	0.2
4.0	0.4
6.0	0.6
8.0	0.8
10.0	1.0

2. Dilute flasks to 100 m*L*.
3. Transfer 50 m*L* to 125-m*L* Erlenmeyer flask.
4. Determine amount of phosphorus as outlined previously in orthophosphate.
5. Prepare a standard curve by plotting the absorbance values of standards versus the corresponding phosphorus concentrations.

F. *EXAMPLE*

Results from a series of tests for orthophosphate were as follows:

Flask No.	**Sample**	**Volume**	**Absorbance**
1	Plant Influent	5 m*L*	0.180
2	Plant Effluent	5 m*L*	0.122
3	Distilled Water	50 m*L*	0.000
4	0.2 mg/*L* P Standard	50 m*L*	0.075
5	0.4 mg/*L* P Standard	50 m*L*	0.152
6	0.6 mg/*L* P Standard	50 m*L*	0.230
7	0.8 mg/*L* P Standard	50 m*L*	0.300

G. *CALCULATION*

1. Prepare a standard curve (top curve, page 600) by using data from prepared standards. From the above example:

Concentration Phosphorus, mg/*L*	**Absorbance**
0.0	0.000
0.2	0.075
0.4	0.152
0.6	0.230
0.8	0.300

2. Obtain concentration of unknown influent and effluent samples from curves (bottom curve, page 600).
3. Correct (if necessary) for samples of less than 50 m*L* by using the following formula:

$$\text{Phosphorus, mg}/L\ \text{P} = \frac{\text{mg}/L \times 50}{\text{m}L\ \text{of Sample Used}}$$

Using data from example:

	Sample Volume	**Absorbance**	**Concentration from Graph**
Plant Influent	5 m*L*	0.180	0.48 mg/*L*

$$\begin{aligned}\text{Phosphorus, mg}/L\ \text{P} &= \frac{\text{mg}/L \times 50}{\text{m}L\ \text{of Sample Used}}\\ &= \frac{0.48 \times 50}{5}\\ &= 4.8\ \text{mg}/L\ \text{P}\end{aligned}$$

	Sample Volume	**Absorbance**	**Concentration from Graph**
Plant Effluent	5 m*L*	0.122	0.32 mg/*L*

$$\begin{aligned}\text{Phosphorus, mg}/L\ \text{P} &= \frac{\text{mg}/L \times 50}{\text{m}L\ \text{of Sample Used}}\\ &= \frac{0.32 \times 50}{5}\\ &= 0.32 \times 10\\ &= 3.2\ \text{mg}/L\ \text{P}\end{aligned}$$

H. *NOTE*

All glassware used should be washed with hot 1:1 HCl and rinsed with distilled water. The acid-washed glassware should be filled with distilled water and treated with all the reagents to remove the last traces of phosphorus that might be adsorbed on the glassware. This glassware should be used only for the determination of phosphorus and after use it should be rinsed with distilled water and kept filled with water until needed again. If this is done, the acid treatment is required only occasionally. Commercial detergents should never be used.

I. *REFERENCES*

See page 365.3-1, EPA's "Methods for Chemical Analysis of Water and Wastes." To order, see 9. Metals, B. References, 2, page 580.

See page 4-153, *STANDARD METHODS,* 21st Edition.

QUESTIONS

Write your answers in a notebook and then compare your answers with those on page 613.

16.5W What forms of phosphorus are commonly found in wastewater?

16.5X What can happen when phosphorus is discharged into receiving waters?

OUTLINE OF PROCEDURE FOR ORTHOPHOSPHATE

1. Place 50 m*L* or aliquot diluted to 50 m*L* in flask.

2. Add 1 m*L* 11 *N* H_2SO_4 and 4 m*L* ammonium molybdate-antimony potassium tartrate solution.

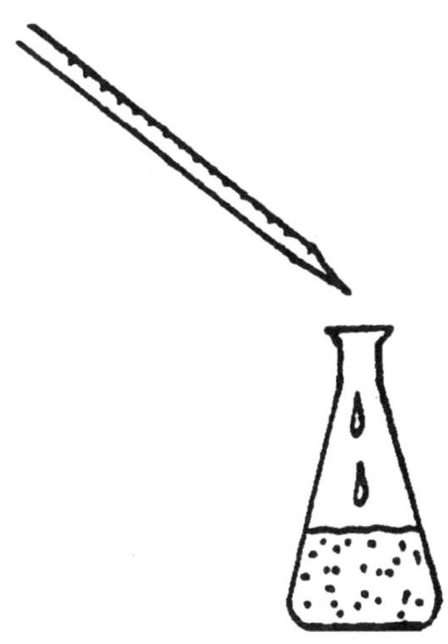

3. Add 2 m*L* ascorbic acid solution. Mix. Let stand 5 minutes.

4. Measure absorbance at 650 nm with spectrophotometer.

FOR TOTAL PHOSPHORUS

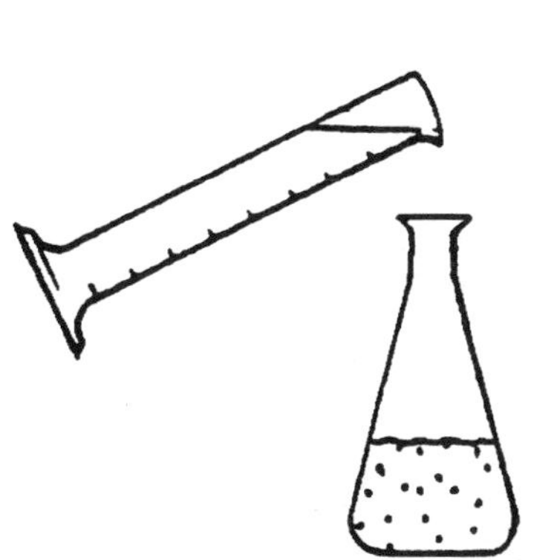

1. Place 50 m*L* or aliquot diluted to 50 m*L* in flask.

2. Add 1 m*L* 11 *N* H_2SO_4 and 0.4 gm of ammonium persulfate. Mix.

3. Boil (or autoclave) for 30 to 40 minutes.

4. Cool. Dilute to 50 m*L*. Determine amount of phosphorus as outlined above.

(Phosphorus)

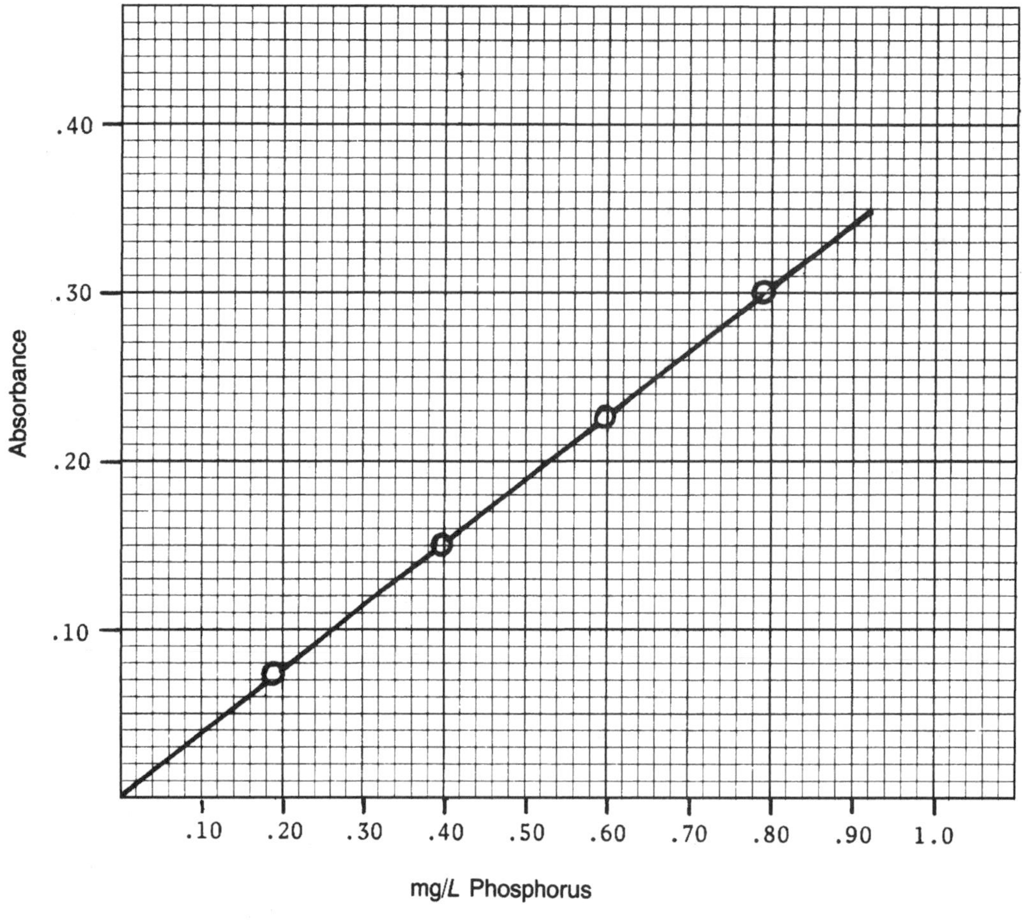
.40
.30
.20
.10
Absorbance
.10
.20
.30
.40
.50
.60
.70
.80
.90
1.0
mg/L Phosphorus

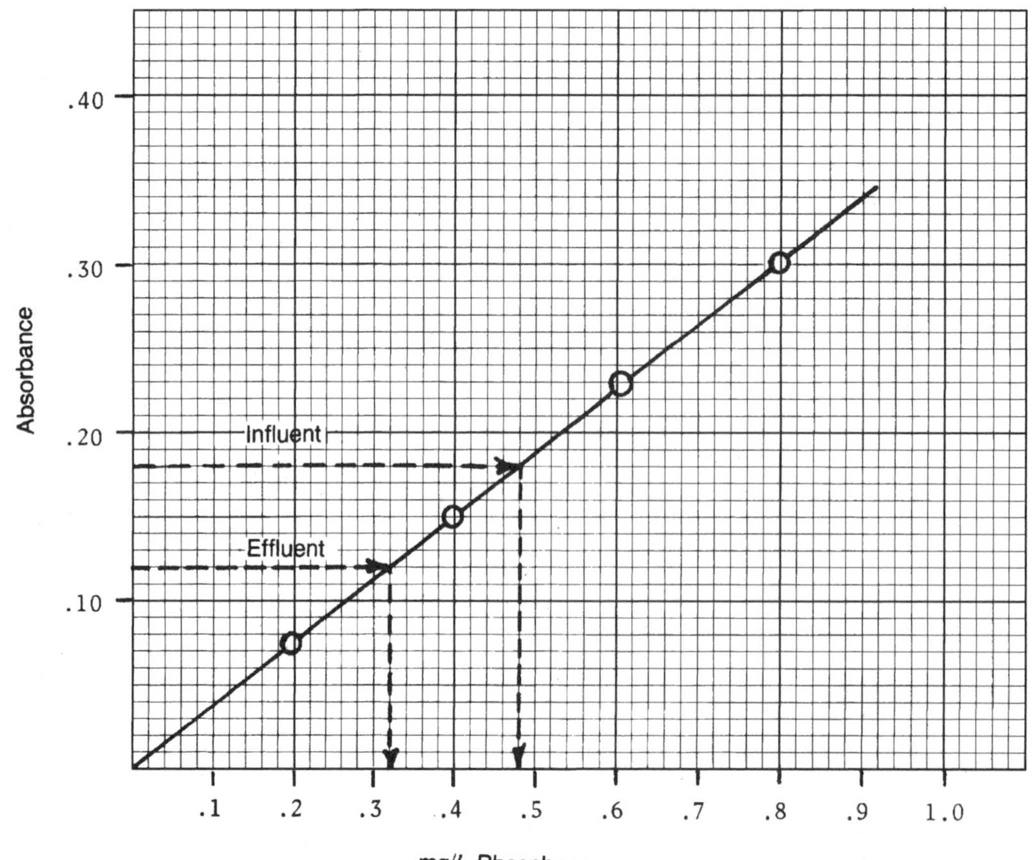
.40
.30
.20
.10
Absorbance
Influent
Effluent
.1
.2
.3
.4
.5
.6
.7
.8
.9
1.0
mg/L Phosphorus

13. Total Solids (Residue)

The term "solids" or "residue" refers to the matter suspended and dissolved in wastewater or water.

I. Total Solids

A. *DISCUSSION*

"Total solids" is the term applied to material left in the container after evaporation of a sample in an oven at 103° to 105°C.

B. *WHAT IS TESTED?*

Sample	Common Range, mg/*L*
Influent and Effluent Sludges	300 to 1,200

C. *APPARATUS REQUIRED*

1. Evaporating dishes (100 m*L* capacity)
2. Drying oven
3. Desiccator
4. Analytical balance
5. Muffle furnace

D. *PROCEDURE*

Preparation of evaporating dish.

1. Ignite a clean, porcelain evaporating dish at 550° ± 50°C for one hour in a muffle furnace.
2. Cool in desiccator, weigh, and record weight. Store in desiccator until ready for use.

NOTE: Aluminum evaporating dishes can be used directly from container.

Sample analysis.

1. Shake sample vigorously and transfer a measured amount of sample to the pre-weighed dish. This can be done by using a graduated cylinder that is graduated to deliver the desired volume. Due to the difficulty in accurately measuring the volume of some sludge samples, an unknown quantity of sludge can be poured into the dish and then weighed to determine the initial sample quantity. This is possible because the specific gravity of sludge is only slightly higher than 1.0. Choose a sample volume that will yield a maximum residue of 25 to 250 milligrams. If necessary, add additional portions of sample to dish.
2. Place the sample in an oven and evaporate to dryness. The oven temperature should be 2°C lower than boiling (that is 98°C) to prevent spattering. Once the water has evaporated, raise the temperature to 103°C and maintain this temperature for at least one hour. Repeat drying cycle until a constant weight is obtained or until weight loss is less than 0.5 milligrams. The exact time is usually determined by experience, but at 103°C some sludge samples will need to be dried overnight.
3. Remove dried sample to desiccator and allow to cool completely. Weigh dish plus total solids and save if volatile portion is to be determined.

E. *PRECAUTIONS*

1. Wastewater and sludge may contain infectious organisms and must be handled with care.
2. Use tongs or gloves when handling hot items.
3. Be sure the balance is properly leveled, zeroed, and maintained.
4. Do not weigh items when they are hot because the heat will cause errors by creating convection currents in the balance.

F. *EXAMPLE*

Results from an effluent sample were:

Weight of Clean Dish = 80.1526 grams

Weight of Residue and Dish = 80.1732 grams

Sample Volume = 100 m*L*

G. *CALCULATIONS*

1. $$\text{Total Solids, mg}/L = \frac{(A - B)(1{,}000\ \text{m}L/L)}{\text{m}L\ \text{of Sample}}$$

where A = Weight of Dish and Residue in milligrams

B = Weight of Dish in milligrams

2. From example,

$$\text{Total Solids, mg}/L = \frac{(A - B)(1{,}000\ \text{m}L/L)}{\text{m}L\ \text{of Sample}}$$

$$= \frac{(80{,}173.2\ \text{mg} - 80{,}152.6\ \text{mg})(1{,}000\ \text{m}L/L)}{100\ \text{m}L}$$

$$= 206\ \text{mg}/L$$

H. *REFERENCE*

See page 2-55, *STANDARD METHODS,* 21st Edition.

II. Total Dissolved (Filterable) Solids

A. *DISCUSSION*

"Total dissolved solids" (TDS) refers to material that passes through a standard glass-fiber filter disc and remains after evaporation at 180°C.

B. *WHAT IS TESTED?*

Sample	Common Range, mg/*L*
Influent and Effluent	150 to 600

C. *APPARATUS REQUIRED*

1. Glass-fiber filter discs (Reeve Angel Type 934A, 984H; or Gelman Type A/E)
2. Flask, suction 500 m*L*
3. Filter holder or Gooch crucible adapter
4. Gooch crucibles (25 m*L* if 2.2 cm filter used)
5. Evaporating dishes, 100 m*L*

(Total Solids)

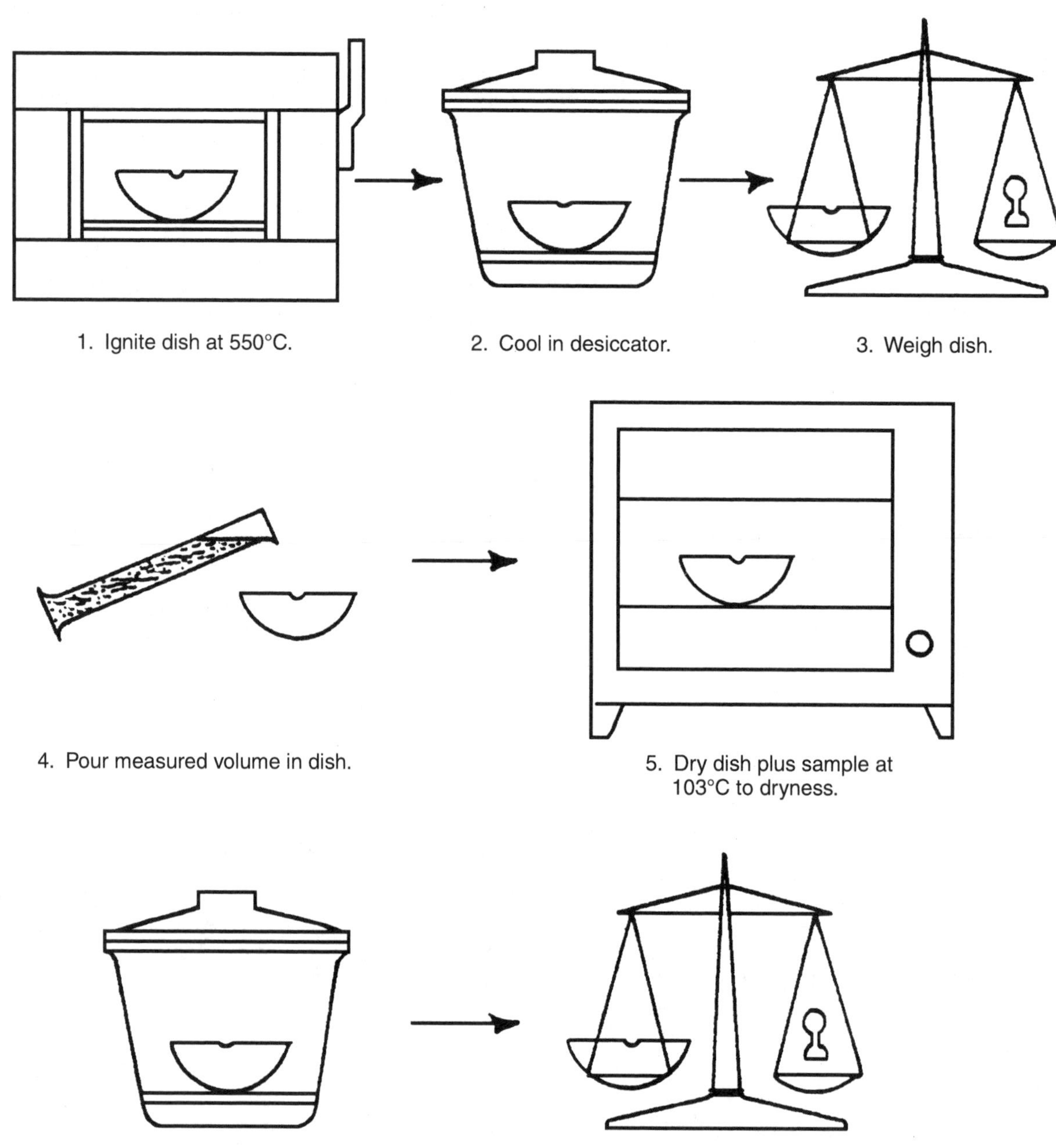

OUTLINE OF PROCEDURE FOR TOTAL SOLIDS
1. Ignite dish at 550°C.
2. Cool in desiccator.
3. Weigh dish.
4. Pour measured volume in dish.
5. Dry dish plus sample at 103°C to dryness.
6. Cool in desiccator.
7. Weigh dish plus total solids.

6. Drying oven, 180°C
7. Steam bath
8. Vacuum source
9. Desiccator
10. Analytical balance
11. Muffle furnace, 550°C

D. *PROCEDURE*

Preparation of Dish

1. Ignite a clean evaporating dish at 550 ± 50°C for one hour in muffle furnace.
2. Cool in desiccator then weigh and record weight. Store in desiccator until needed.

Preparation of Glass-Fiber Filter Disc

1. Place the disc on the filter apparatus or insert into the bottom of a suitable Gooch crucible. While vacuum is applied, wash the filter disc with three successive 20-m*L* volumes of distilled water. Continue the suction to remove all traces of water from the disc and discard the washings.

Sample Analysis

1. Shake the sample vigorously and transfer 125 to 150 m*L* to the funnel or Gooch crucible by means of a 150-m*L* graduated cylinder.
2. Filter the sample through the glass-fiber filter and continue to apply vacuum for about three minutes after filtration is complete to remove as much water as possible.
3. Transfer 100 m*L* of the filtrate to the weighed evaporating dish and evaporate to dryness on a steam bath.
4. Dry the evaporated sample for at least one hour at 180°C. Cool in desiccator and weigh. Repeat drying cycle until constant weight is obtained or until weight loss is less than 0.5 mg.

E. *EXAMPLE*

Results from weighings were:

Clean Dish	= 47.0028 grams (47,002.8 mg)
Dissolved Residue + Dish	= 47.0453 grams (47,045.3 mg)
Sample Volume	= 100 m*L*

F. *CALCULATIONS*

1. $$\text{Total Dissolved Solids, mg}/L = \frac{(A - B)(1{,}000\ mL/L)}{mL\ \text{Sample Volume}}$$

Where:

A = Weight of Dish and Dissolved Material in milligrams (mg)

B = Weight of Clean Dish in milligrams (mg)

(Total Solids)

2. From example,

$$\text{Total Dissolved Solids, mg}/L = \frac{(A - B)(1{,}000\ mL/L)}{mL\ \text{Sample Volume}}$$

$$= \frac{(47{,}045.3\ mg - 47{,}002.8\ mg)(1{,}000\ mL/L)}{100\ mL}$$

$$= 425\ mg/L$$

G. *COMMENTS*

Because excessive residue in the evaporating dish may form a water-entrapping crust, use a sample that yields no more than 200 mg of residue.

H. *REFERENCE*

See page 2-56, *STANDARD METHODS,* 21st Edition.

III. Suspended Solids

See Section 16.43 for procedure to measure Suspended Solids.

QUESTION

Write your answer in a notebook and then compare your answer with the one on page 613.

16.5Y Determine the total solids (residue) and total dissolved (filterable) solids from the following lab test results.

Weight of Empty Dish	= 64,328.9 mg
Weight of Dish and Residue	= 64,381.2 mg
Weight of Dish and Dissolved Material	= 64,351.2 mg
Sample Volume	= 100 m*L*

14. Specific Conductance

A. *DISCUSSION*

Specific conductance or conductivity is a numerical expression (expressed in micromhos per centimeter) of the ability of a water or wastewater to conduct or carry an electric current, which is related to the concentration of ionized substances in the water. This number depends on the total concentration of the minerals dissolved in the sample and the temperature. The conductivity of wastewater reflects to a degree the characteristics of the water supply servicing the area.

Specific conductance is measured by the use of a conductivity meter.

B. *WHAT IS TESTED?*

Sample	**Common Range, micromhos/cm**
Influent and Effluent	200 to 1,000

C. *MATERIALS AND PROCEDURE*

See page 2-44, *STANDARD METHODS,* 21st Edition.

(Total Solids)

OUTLINE OF PROCEDURE FOR TOTAL SOLIDS

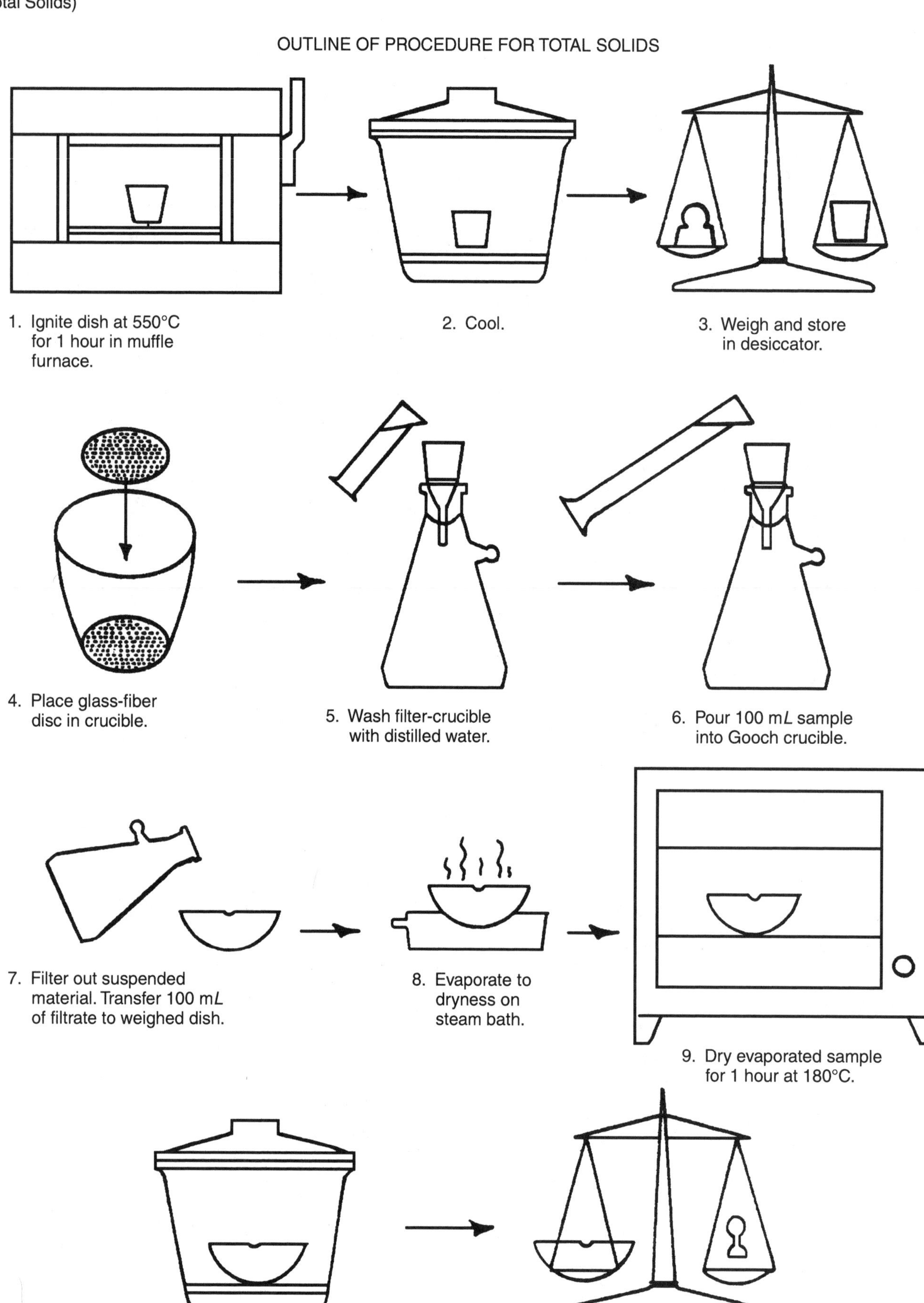

1. Ignite dish at 550°C for 1 hour in muffle furnace.
2. Cool.
3. Weigh and store in desiccator.
4. Place glass-fiber disc in crucible.
5. Wash filter-crucible with distilled water.
6. Pour 100 mL sample into Gooch crucible.
7. Filter out suspended material. Transfer 100 mL of filtrate to weighed dish.
8. Evaporate to dryness on steam bath.
9. Dry evaporated sample for 1 hour at 180°C.
10. Cool in desiccator.
11. Weigh.

QUESTION

Write your answer in a notebook and then compare your answer with the one on page 613.

16.5Z What property of wastewater is measured by the specific conductance test?

15. Sulfate

A. *DISCUSSION*

Sulfate ions are widely distributed in wastewaters in concentrations ranging from a few to several thousand milligrams per liter. Sulfate is of considerable concern in wastewater because sulfate ions are indirectly responsible for three serious problems: toxic gas, odor, and sewer corrosion problems. These problems are a result of the bacterial reduction of sulfate to hydrogen sulfide. Hydrogen sulfide is a toxic gas that is flammable and explosive. Also, hydrogen sulfide smells like rotten eggs and combines with moisture to form corrosive sulfuric acid.

B. *WHAT IS TESTED?*

Sample	Common Range
Wastewater	Depends on water supply and industrial discharges

C. *MATERIALS AND PROCEDURE*

See page 4-186, *STANDARD METHODS,* 21st Edition.

16. Surfactants

A. *DISCUSSION*

The single most widely used surfactant (surface active agent) in detergents is LAS. LAS is short for linear alkyl sulfonate. LAS is used as the standard compound in surfactant analysis.

The test for surfactants consists of adding a methylene blue dye to the wastewater sample. Methylene blue dye reacts with surfactants, such as LAS, to form a blue-colored salt. This blue salt is extracted with chloroform and the intensity of the blue color measured.

B. *WHAT IS TESTED?*

Sample	Common Range, mg/*L*
Secondary Effluent	0.1 to 1

C. *MATERIALS AND PROCEDURE*

See page 5-47, *STANDARD METHODS,* 21st Edition.

17. Temperature

A. *DISCUSSION*

Temperature is one of the most important factors affecting biological growth. Temperature measurements can be helpful in detecting changes in raw wastewater quality. For example, an influent temperature drop may indicate large volumes of cold water from infiltration. An increase in temperature may indicate that hot water discharges by industry are reaching your plant.

Temperature is one of the most frequently taken tests. One of the many uses is to calculate the percent saturation of dissolved oxygen in the DO test.

A temperature measurement should be taken where samples are collected for other tests. This test is always immediately performed on a grab sample because temperature changes so rapidly. Always leave the thermometer in the liquid while reading the temperature because the reading will change as soon as you remove the thermometer from the liquid. Record temperature on a suitable work sheet, including time, location, and sampler's name.

Thermometers are calibrated for either total immersion or partial immersion. A thermometer calibrated for total immersion must be completely immersed in the wastewater sample to give a correct reading, while a partial immersion thermometer must be immersed in the sample to the depth of the etched circle around the stem for a correct reading.

B. *WHAT IS TESTED?*

Sample	Common Range
Influent[24]	65°F to 85°F[25] (18°C to 29°C)
Effluent[24]	60°F to 95°F (16°C to 35°C) or higher from ponds
Receiving Water	60°F (16°C) to ambient temperature[26]
Digester (Recirculated Sludge before Heat Exchanger and Supernatant)	60°F to 100°F (16°C to 38°C)

C. *APPARATUS*

1. One NBS (National Bureau of Standards) thermometer for calibration of the other thermometers
2. One Fahrenheit mercury-filled, 1° subdivided thermometer
3. One Celsius (formerly called Centigrade) mercury-filled, 1° subdivided thermometer
4. One metal case to fit each thermometer

There are three types of thermometers and two scales.

SCALES

1. Fahrenheit, marked °F
2. Celsius (formerly Centigrade), marked °C

THERMOMETERS

1. Total immersion. This type of thermometer must be totally immersed when read. This will change most rapidly when removed from the liquid to be recorded.
2. Partial immersion. This type thermometer will have a solid line around the stem below the point where the scale starts.
3. Dial. This type has a dial that can be easily read while the thermometer is still immersed. Dial thermometer readings

---

[24] If dissolved oxygen (DO) measurements are performed on any samples, the temperature should be measured and recorded.

[25] Depends on season, location, and temperature of water supply.

[26] *Ambient* (AM-bee-ent) *Temperature.* Temperature of the surroundings.

(Temperature)

should be checked (calibrated) against the NBS thermometer. Some dial thermometers can be recalibrated (adjusted) to read the correct temperature of the NBS thermometer.

D. *REAGENTS*

None

E. *PROCEDURES*

Use a large volume of sample, preferably at least a 2-pound coffee can or equivalent volume. The temperature will have less chance to change in a large volume than in a small container. Collect sample in container and immediately measure and record temperature. Do not touch the bottom or sides of the sample container with the thermometer. To avoid breaking or damaging a glass thermometer, store it in a shielded metal case. Check your thermometer accuracy against the NBS certified thermometer by measuring the temperature of a sample with both thermometers simultaneously. Some of the poorer quality thermometers are substantially inaccurate (off as much as 6°F or 3°C).

F. *EXAMPLE*

To measure influent temperature, obtain sample in large coffee can, immediately immerse thermometer in can, and record temperature when reading becomes constant. For example, 72°F or 22°C.

G. *CALCULATIONS*

Normally, we measure and record temperatures using a thermometer with the proper scale. However, we could measure a temperature in °F and convert to °C, or we might measure a temperature in °C and convert to °F. The following formulas are used to convert temperatures from one scale to the other.

1. Measure in °F, want °C: $°C = 5/9(°F - 32°)$
2. Measure in °C, want °F: $°F = 9/5(°C) + 32°$
3. Example Calculation:

The measured influent temperature was 77°F. What was the temperature in °C?

$$
\begin{aligned}
°C &= 5/9(°F - 32°) \\
&= 5/9(77° - 32°) \\
&= 5/9(45°) \\
&= 25°C
\end{aligned}
$$

QUESTIONS

Write your answers in a notebook and then compare your answers with those on page 613.

16.5AA What could a change in influent temperature indicate?

16.5BB Why should the thermometer remain immersed in the liquid while being read?

16.5CC Why should thermometers be calibrated against an accurate National Bureau of Standards (NBS) certified thermometer?

18. Total Organic Carbon (TOC)

A. *DISCUSSION*

The carbon analyzer measures all carbon in a sample. Because of the various properties of carbon-containing compounds in wastewater samples, preliminary treatment of the sample prior to analysis dictates the definition of the carbon measured. Such forms include soluble, insoluble, volatile, nonvolatile, and particulate carbon.

The value of the total carbon measurement is that it is a more direct expression of organic chemical content of wastewater than either the chemical oxygen demand (COD) or the biochemical oxygen demand (BOD). Therefore, the TOC test can be used to monitor wastewater treatment processes.

B. *WHAT IS TESTED?*

Sample	Common Range, mg/*L*
Influent	100 to 300
Effluent	10 to 100

C. *MATERIALS AND PROCEDURES*

See page 5-19, *STANDARD METHODS,* 21st Edition.

19. Turbidity

A. *DISCUSSION*

The term "turbidity" is simply an expression of the physical clarity of water or wastewater. Turbidity can be caused by the presence of suspended matter such as mud, finely divided organic and inorganic matter, and microscopic organisms such as algae.

The turbidity measurement is useful for plant effluent monitoring; however problems can be encountered when one instrument's readings are compared with those of another. Commercial turbidimeters come in many shapes and sizes. They each can read different turbidity values on the same sample even though they have been calibrated using the procedure given later in this section. The operator should simply be aware of this shortcoming.

The accepted method used to measure turbidity is called nephelometric. The nephelometric turbidimeter is designed to measure particle-reflected light at an angle of 90 degrees. The greater the intensity of the scattered light, the higher the turbidity.

The turbidity unit (NTU) is an empirical quantity, based on the amount of light that is scattered by particles of a polymer reference standard called formazin, which produces particles that scatter light in a reproducible manner. Formazin, the primary turbidity standard, is an aqueous suspension of an insoluble polymer formed by the condensation reaction between hydrazine sulfate and hexamethylenetetramine.

Secondary turbidity standards are suspensions of various materials formulated to match the primary formazin solutions. These secondary standards are generally used because of their convenience and the instability of dilute formazin primary standard solutions. Examples of these secondary standards include "standards" that are supplied by the turbidimeter manufacturer with the instrument. Periodic checks of these secondary standards against the primary formazin standard are a must and will provide assurance of measurement accuracy.

B. *WHAT IS TESTED?*

Sample	**Common Range, NTU**
Effluent	10 to 50

C. *APPARATUS*

1. Turbidimeter: The turbidimeter should consist of a nephelometer with a light source illuminating the samples and one or more photoelectric detectors with a readout device to indicate the intensity of scattered light. Turbidimeters used to test plant effluents should be approved by the US Environmental Protection Agency.
2. Sample tubes

D. *REAGENTS*

1. Turbidity-free water: Pass distilled water through a membrane filter having a pore size no greater than 100 microns. Discard the first 200 m*L* collected. If filtration does not reduce turbidity, use distilled water.
2. Stock Formazin turbidity suspension:[27]
   a. Solution I. Dissolve 1.000 gm hydrazine sulfate in distilled water and dilute to 100 m*L* in a volumetric flask.
   b. Solution II. Dissolve 10.00 gm hexamethylenetetramine in distilled water and dilute to 100 m*L* in a volumetric flask.
   c. In a 100-m*L* volumetric flask, add (using 5 m*L* volumetric pipets) 5.0 m*L* Solution I and 5.0 m*L* of Solution II. Mix and allow to stand 24 hours at 25°C. Then dilute to the mark and mix. The turbidity of this suspension is 400 NTU.
   d. Prepare solutions and suspensions monthly.
3. Standard turbidity suspensions. Dilute 10.00 m*L* stock turbidity suspension to 100 m*L* with turbidity-free water. Prepare weekly. The turbidity of this suspension is defined as 40 NTU.
4. Dilute turbidity standards. Dilute portions of the standard turbidity suspension with turbidity-free water as required. Prepare weekly.

E. *PROCEDURE*

1. Turbidimeter calibration. The manufacturer's operating instructions should be followed. Measure standards on the turbidimeter covering the range of interest. If the instrument is already calibrated in standard *TURBIDITY UNITS,*[28] this procedure will check the accuracy of the calibration scales. At least one standard should be run in each instrument range to be used. Some instruments permit adjustments of sensitivity so that scale values will correspond to turbidities.

   Reliance on a manufacturer's solid scattering standard for setting overall instrument sensitivity for all ranges is not an acceptable practice unless the turbidimeter has been shown to be free of drift on all ranges. If a pre-calibrated scale is not supplied, then calibration curves should be prepared for each range of the instrument.
2. Turbidities less than 40 units: Shake the sample to thoroughly disperse the solids. Wait until air bubbles disappear, then pour the sample into the turbidimeter tube. Read the turbidity directly from the instrument scale or from the appropriate calibration curve.
3. Turbidities exceeding 40 units: Dilute the sample with one or more volumes of turbidity-free water until the turbidity falls below 40 units. The turbidity of the original sample is then computed from the turbidity of the diluted sample and the dilution factor.

F. *EXAMPLE*

If 5 volumes of turbidity-free water were added to 1 volume of sample, and the diluted sample showed a turbidity of 20 units, then the turbidity of the original sample was 120 units.

G. *CALCULATIONS*

Sample Reading $\times$ Dilution $=$ Actual Turbidity

Report results as follows:

NTU	**Record to Nearest:**
0.0 to 1.0	0.05
1 to 10	0.01
10 to 40	1
40 to 100	5
100 to 400	10
400 to 1,000	50
>1,000	100

H. *NOTES AND PRECAUTIONS*

1. A commercially available polymer standard that requires no preparation is also approved for use. This standard is identified as AMCO Clear, available from GFS Chemicals, Inc., Sales and Administration Facility, PO Box 245, Powell, OH 43065.
2. Sample tubes must be kept scrupulously clean, both inside and out. Discard them when they become scratched or etched. Never handle the area where the light strikes the tube during the test.
3. Fill the tubes with samples and standards that have been agitated thoroughly, and allow sufficient time for bubbles to escape.

I. *REFERENCE*

See page 2-8, *STANDARD METHODS,* 21st Edition.

[27] Stock secondary standard turbidity suspensions that require no preparation are available from commercial suppliers.

[28] *Turbidity* (ter-BID-it-tee) *Units (TU).* Turbidity units are a measure of the cloudiness of water. If measured by a nephelometric (deflected light) instrumental procedure, turbidity units are expressed in nephelometric turbidity units (NTU) or simply TU. Those turbidity units obtained by visual methods are expressed in Jackson turbidity units (JTU), which are a measure of the cloudiness of water; they are used to indicate the clarity of water. There is no real connection between NTUs and JTUs. The Jackson turbidimeter is a visual method and the nephelometer is an instrumental method based on deflected light.

QUESTION

Write your answer in a notebook and then compare your answer with the one on page 613.

16.5DD What causes turbidity in wastewater?

16.6 ARITHMETIC ASSIGNMENT

Turn to the Arithmetic Appendix at the back of this manual. Read and work the problems in Section A.34, "Laboratory." Check the arithmetic in this section using an electronic calculator. You should be able to get the same answers. Section A.54 contains similar problems using metric units.

END OF LESSON 9 OF 9 LESSONS ON LABORATORY PROCEDURES AND CHEMISTRY

Please answer the discussion and review questions next.

DISCUSSION AND REVIEW QUESTIONS

Chapter 16. LABORATORY PROCEDURES AND CHEMISTRY

(Lesson 9 of 9 Lessons)

Write the answers to these questions in your notebook. The question numbering continues from Lesson 8.

45. What are the most important forms of nitrogen in wastewater treatment?
46. What water quality problems are related to ammonia?
47. The oil and grease test measures what kinds of materials?
48. How can phosphorus be removed from wastewater?
49. What does the total solids (residue) test measure?
50. Why is sulfate of concern in wastewater?
51. What is the ambient temperature?
52. Convert a temperature reading of 50°F to degrees Celsius.
53. What is the main advantage of the total organic carbon (TOC) test?

SUGGESTED ANSWERS

Chapter 16. LABORATORY PROCEDURES AND CHEMISTRY

ANSWERS TO QUESTIONS IN LESSON 1

Answers to questions on page 489.

16.1A Descriptions of laboratory items and their use or purpose.

Item	Description	Use or Purpose
1. Beakers	Short, wide cylinders in sizes from 1 m*L* to 4,000 m*L*.	Mixing chemicals
2. Graduated Cylinders	Long, narrow cylinders in sizes from 5 m*L* to 4,000 m*L*.	Measuring volumes.
3. Pipets	Very thin tubes with a pointed tip in sizes from 0.1 m*L* to 100 m*L*.	Delivering accurate volumes.
4. Burets	Long tubes with graduated walls and a stopcock in sizes from 10 to 1,000 m*L*.	Delivering and measuring accurate volumes used in "titrations."

16.1B Never heat graduated cylinders in an open flame because they will break.

16.1C A bench sheet is used to record data and arrange data in an orderly manner. Bench sheets also are called laboratory work sheets.

ANSWERS TO QUESTIONS IN LESSON 2

Answers to questions on page 497.

16.2A A rubber bulb should always be used to pipet wastewater or polluted water to prevent infectious materials from entering your mouth.

16.2B Inoculations are recommended to reduce the possibility of contracting diseases.

16.2C True. You may add acid to water, as in a swimming pool, but never add water to acid.

16.2D Immediately wash area where acid spilled with water and neutralize the acid with sodium carbonate or bicarbonate.

ANSWERS TO QUESTIONS IN LESSON 3

Answers to questions on page 503.

16.3A The largest sources of errors found in laboratory results are usually caused by improper sampling, poor preservation, and lack of sufficient mixing during compositing and testing.

16.3B A representative sample must be collected or the test results will not have any significant meaning. To efficiently operate a wastewater treatment plant, the operator must rely on test results to indicate what is happening in the treatment process.

16.3C A proportional composite sample may be prepared by collecting a sample every hour. The size of this sample is proportional to the rate of flow when the sample is collected. All of these proportional samples are mixed together to produce a proportional composite sample. If an equal volume of sample was collected each hour and mixed, this would simply be a composite sample.

ANSWERS TO QUESTIONS IN LESSON 4

Answers to questions on page 504.

16.4A The clarity test indicates the quality of the effluent with respect to color, solids, and turbidity.

16.4B When clarity is measured under different conditions the results cannot be compared. You will not be able to tell whether your plant performance is improving, staying the same, or deteriorating.

Answer to question on page 506.

16.4C 1. H_2S in the atmosphere produces a rotten egg odor. It is indicative of anaerobic decomposition of organics in wastewater, which occurs in the absence of oxygen. High levels of H_2S are toxic to your respiratory system and can create flammable and explosive conditions.

2. You would measure the H_2S in the wastewater to know the strength of H_2S and to provide an indication of the corrosion taking place on the concrete.

Answer to question on page 507.

16.4D

$$\text{Sludge to Digester, GPD} = (\text{Total SS Removed, m}L/L)(1{,}000\text{ mg/m}L)(\text{Flow, MGD})$$

$$= (10\text{ m}L/L - 0.4\text{ m}L/L)(1{,}000\text{ mg/m}L)(1\text{ M gal/day})$$

$$= \frac{9.6\text{ m}L}{\text{M mg}} \times \frac{1{,}000\text{ mg}}{\text{m}L} \times \frac{1\text{ M gal}}{\text{day}}$$

$$= 9{,}600\text{ GPD}$$

This value may be reduced by 30 to 75 percent due to compaction of the sludge in the clarifier.

Answers to questions on page 512.

16.4E Some suspended material in wastewater will not be removed because the specific gravity is very near that of water and the material is not light enough to float or heavy enough to settle.

16.4F Solids calculations will be shown in detail here to illustrate the computational approach and the units involved. After you understand this approach, use of the laboratory work sheet is more convenient.

1. Total Suspended Solids, mg/*L*

Volume of Sample, m*L* = 100 m*L*

Weight of Dried Sample + Dish, gm	=	19.3902 gm
Weight of Dish (Tare Weight), gm	=	19.3241 gm
Dry Weight	=	0.0661 gm
or	=	66.1 mg

$$\text{Total Susp Solids, mg/}L = \frac{\text{Weight of Solids, mg} \times 1{,}000\text{ m}L/L}{\text{Volume of Sample, m}L}$$

$$= \frac{66.1\text{ mg} \times 1{,}000\text{ m}L/L}{100\text{ m}L}$$

$$= 661\text{ mg/}L$$

2. Volatile Suspended Solids, mg/*L*

Weight of Dried Sample + Dish, gm	=	19.3902 gm
Weight of Ash + Dish, gm	=	19.3469 gm
Weight of Volatile Solids, gm	=	0.0433 gm
or	=	43.3 mg

$$\text{Volatile Susp Solids, mg/}L = \frac{\text{Weight of Vol Sol, mg} \times 1{,}000\text{ m}L/L}{\text{Volume of Sample, m}L}$$

$$= \frac{(43.3\text{ mg})(1{,}000\text{ m}L/L)}{100\text{ m}L}$$

$$= 433\text{ mg/}L$$

3. Volatile Solids, %

$$\%\text{ Volatile Solids} = \frac{\text{Weight Volatile, mg} \times 100\%}{\text{Weight Dry, mg}}$$

$$= \frac{43.3\text{ mg}}{66.1\text{ mg}} \times 100\%$$

$$= 65.5\%$$

4. Fixed Suspended Solids, mg/*L*

Total Suspended Solids, mg/*L*	=	661 mg/*L*
Volatile Suspended Solids, mg/*L*	=	433 mg/*L*
Fixed Suspended Solids, mg/*L*	=	228 mg/*L*

5. Fixed Solids, %

Total Solids, %	=	100.00%
Volatile Solids, %	=	65.50%
Fixed Solids, %	=	34.5%

or

$$\%\text{ Fixed} = \frac{\text{Fixed, mg}}{\text{Total, mg}} \times 100\%$$

$$= \frac{22.8\text{ mg}}{66.1\text{ mg}} \times 100\%$$

$$= 34.5\%\text{ (Check)}$$

16.4G 1. Calculate the percent removal of suspended solids by the primary clarifier.

In = Suspended solids entering the plant or unit

Out = Suspended solids leaving the plant or unit

$$\%\text{ Removal} = \frac{(\text{In} - \text{Out})}{\text{In}} \times 100\%$$

$$= \frac{(221\text{ mg/}L - 159\text{ mg/}L)}{221\text{ mg/}L} \times 100\%$$

$$= \frac{62}{221} \times 100\%$$

$$= 28\%\text{ Removal by Primary Clarifier}$$

2. Calculate the percent removal by secondary process.

In = 159 mg/*L* SS in Primary Effluent

Out = 33 mg/*L* SS in Final Effluent

$$\%\text{ Removal} = \frac{(\text{In} - \text{Out})}{\text{In}} \times 100\%$$

$$= \frac{(159\text{ mg/}L - 33\text{ mg/}L)}{159\text{ mg/}L} \times 100\%$$

$$= \frac{126}{159} \times 100\%$$

$$= 79.2\%\text{ Removal from Primary Effluent to Final Effluent}$$

3. Calculate the percent removal by the plant overall or overall plant efficiency.

In = 221 mg/*L* SS in Plant Influent

Out = 33 mg/*L* SS in Plant Effluent

$$\%\text{ Removal} = \frac{(\text{In} - \text{Out})}{\text{In}} \times 100\%$$

$$= \frac{(221\text{ mg/}L - 33\text{ mg/}L)}{221\text{ mg/}L} \times 100\%$$

$$= \frac{188}{221} \times 100\%$$

$$= 85\%\text{ Overall Plant Removal}$$

16.4H Calculate the pounds of solids removed per day by each unit:

$$\text{Amount Removed, lbs/day} = \text{Conc Rem, mg}/L \times \text{Flow, MGD} \times 8.34 \text{ lbs/gal}$$

where MGD = million gallons per day

1. Influent, mg/L = 221 mg/L
   Primary Effluent, mg/L = 159 mg/L

   Primary Removal, mg/L = 62 mg/L

$$\text{Amount Removed, lbs/day (Primary)} = (62 \text{ mg}/L)(1.5 \text{ MGD})(8.34 \text{ lbs/gal})$$

$$= 775.6 \text{ lbs/day Removed by Primary}$$

2. Primary Effluent, mg/L = 159 mg/L
   Final Effluent, mg/L = 33 mg/L

   Secondary Removal, mg/L = 126 mg/L

$$\text{Amount Removed, lbs/day (Secondary)} = (126 \text{ mg}/L)(1.5 \text{ MGD})(8.34 \text{ lbs/gal})$$

$$= 1{,}576 \text{ lbs/day Removed by Secondary}$$

3. Influent, mg/L = 221 mg/L
   Final Effluent, mg/L = 33 mg/L

   Overall Removal, mg/L = 188 mg/L

$$\text{Amount Removed, lbs/day} = (188 \text{ mg}/L)(1.5 \text{ MGD})(8.34 \text{ lbs/gal})$$

$$= 2{,}352 \text{ lbs/day Removed by Plant}$$

$$= \text{Primary Removal, lbs/day} + \text{Secondary, lbs/day}$$

$$\text{or} = 776 + 1{,}576$$

$$= 2{,}352 \text{ (Check)}$$

Answers to questions on page 516.

16.4I Volatile solids found in a digester are organic compounds of either plant or animal origin.

16.4J Volatile solids in a treatment plant indicate the portion of material that may be treated by biological processes.

ANSWERS TO QUESTIONS IN LESSON 5

Answers to questions on page 518.

16.4K Mixed liquor settleability tests are run to get an indication of what will happen to the mixed liquor in the final clarifier—the rate of sludge settling, turbidity, color, and volume of sludge at the end of 60 minutes.

16.4L The SVI is the volume in mL occupied by one gram of mixed liquor suspended solids after 30 minutes of settling.

16.4M The SVI test is used to indicate changes in sludge characteristics.

16.4N Sludge Density Index (SDI) = 100/SVI.

Answer to question on page 519.

16.4O The sludge age of 200,000 gallon aeration tank that has 2,000 mg/L mixed liquor suspended solids, a primary effluent of 115 mg/L SS, and an average flow of 1.8 MGD:

$$\text{Sludge Age, days} = \frac{\text{Vol of Aeration Tank, MG} \times \text{Sus Solids, mg}/L}{\text{Flow, MGD} \times \text{Primary Effluent, mg}/L}$$

$$= \frac{0.2 \text{ MG} \times 2{,}000 \text{ mg}/L}{1.8 \text{ MGD} \times 115 \text{ mg}/L}$$

$$= 1.93 \text{ days}$$

Answer to question on page 520.

16.4P When the DO in the aeration tank is very low, the copper sulfate-sulfamic acid test procedure can give DO results higher than actual DO in the aerator. The results may be high because oxygen may enter the sample from the air when the sample is collected, when the copper sulfate-sulfamic acid inhibitor is added, while the solids are settling, and when the sample is transferred to a BOD bottle for the DO test.

Answer to question on page 522.

16.4Q The advantages of the centrifuge over the regular suspended solids test are:

1. Speed of answer. Not as accurate as other methods, but results are sufficiently close.
2. Answers very acceptable if suspended solids concentration is below 1,500 mg/L.

Answer to question on page 522.

16.4R In order to calculate the Mean Cell Residence Time (MCRT), measure representative composite samples of mixed liquor suspended solids, effluent suspended solids, and waste sludge suspended solids. Also measure influent and waste sludge flows.

Answer to question on page 527.

16.4S

$$\text{Volatile Acid Alkalinity, mg}/L = \frac{\text{m}L\ 0.050\ N\ \text{NaOH} \times 2{,}500}{\text{m}L \text{ Sample}}$$

$$= \frac{5 \text{ m}L \times 2{,}500}{50 \text{ m}L}$$

$$= 250 \text{ mg}/L$$

Since 250 mg/L > 180 mg/L,

$$\text{Volatile Acids, mg}/L = \text{Volatile Acid Alkalinity} \times 1.50$$

$$= 250 \text{ mg}/L \times 1.50$$

$$= 375 \text{ mg}/L$$

Answers to questions on page 529.

16.4T The total alkalinity test is run to determine the buffer capacity and the volatile acids/alkalinity relationship in a digester.

16.4U The buffer capacity of a digester as measured by the total alkalinity tests indicates the capacity of the digester to resist changes in pH.

16.4V $$\frac{\text{Volatile Acid}}{\text{Alkalinity}} = \frac{300 \text{ mg}/L}{2{,}000 \text{ mg}/L}$$

$$= 0.15$$

Answers to questions on page 531.

16.4W The dangers encountered in running the CO_2 test on digester gas include:

1. Digester gas contains methane, which is explosive when mixed with air.
2. The CO_2 gas absorbent is harmful to your skin.

16.4X $$\% \ CO_2 = \frac{(\text{Total Volume, m}L - \text{Gas Remaining, m}L) \times 100\%}{\text{Total Volume, m}L}$$

$$= \frac{(128 \text{ m}L - 73 \text{ m}L) \times 100\%}{128 \text{ m}L}$$

$$= \frac{55}{128} \times 100\%$$

$$= 43\%$$

Answer to question on page 534.

16.4Y 1. Results from the graduated cylinder are available immediately, but different operators may interpret the results differently.

2. Results are not available until the next day, but different operators will record the same results.

Answer to question on page 535.

16.4Z If the supernatant solids test result is greater than 5 percent solids, the supernatant could be placing an excessive solids load on the primary settling tanks or other parts of the plant. Appropriate operational adjustments should be made.

Answer to question on page 537.

16.4AA Lime is used as a coagulant in treating industrial wastes and as a means of removing phosphorus.

ANSWERS TO QUESTIONS IN LESSON 6

Answer to question on page 539.

16.5A Some sources of acidity in a treatment plant influent include carbon dioxide from the atmosphere, from the biological oxidation of organic matter, or from industrial waste discharges.

Answer to question on page 541.

16.5B Bicarbonate ions (HCO_3^-) represent the major form of alkalinity in wastewater.

Answers to questions on page 543.

16.5C The COD test is a measure of the strength of a waste in terms of its chemical oxygen demand. It is a good estimate of the first-stage oxygen demand. (Either answer is acceptable.)

16.5D The advantage of the COD test over the BOD test is that you do not have to wait five days for the results.

Answer to question on page 545.

16.5E Sulfide, thiosulfate, and sulfite ions interfere with the chloride test, but can be removed by treatment with 1 m*L* of 30 percent hydrogen peroxide (H_2O_2).

Answers to questions on page 550.

16.5F Plant effluents should be chlorinated for disinfection purposes to protect the bacteriological quality of the receiving waters.

16.5G The iodometric titration method gives good results with samples containing wastewater, such as plant effluent or receiving waters. Amperometric titration gives the best results, but the equipment is expensive.

ANSWERS TO QUESTIONS IN LESSON 7

Answers to questions on page 564.

16.5H Sodium thiosulfate crystals should be added to sample bottles for coliform bacteria tests before sterilization to neutralize any chlorine that may be present when the sample is collected. If the residual chlorine is not neutralized, it would continue to be toxic to the coliform organisms remaining in the sample and give false results.

16.5I Steam autoclaves sterilize (kill all organisms) at a relatively low temperature (121°C) within 15 minutes by using moist heat.

16.5J

Dilutions (m*L*)	0.01	0.001	0.0001	0.00001
Readings	5	1	2	0

MPN = 60,000/100 m*L*

16.5K The number of coliforms is estimated by counting the number of appropriately colored coliform colonies grown on the membrane filter.

Answer to question on page 568.

16.5L Colilert media contains specific indicator nutrients for total coliforms and *E. coli.* As the media's nutrients are metabolized, yellow color and fluorescence are released, confirming the presence of total coliforms and *E. coli,* respectively. Non-coliform bacteria are chemically suppressed and cannot metabolize the indicator nutrients. Consequently, they do not interfere with the identification of the target microbes.

ANSWERS TO QUESTIONS IN LESSON 8

Answers to questions on page 573.

16.5M $$\text{DO Saturation, \%} = \frac{\text{DO of Sample, mg}/L \times 100\%}{\text{DO at 100\% Saturation, mg}/L}$$

$$= \frac{(7.9 \text{ mg}/L)100\%}{11.3 \text{ mg}/L}$$

$$= 70\%$$

16.5N To calibrate the DO probe in an aeration tank, a sample of the effluent can be collected and split. The DO of the effluent is measured by the Modified Winkler procedure, and the probe DO reading is adjusted to agree with the Winkler results.

Answers to questions on page 578.

16.5O The amount of organic material in wastewater can be estimated indirectly by measuring oxygen use with the BOD test.

16.5P Samples for the BOD test should be collected before chlorination because chlorine interferes with the organisms in the test. It is difficult to obtain accurate results with dechlorinated samples.

16.5Q A solution of sodium thiosulfate at 0.025 *N* is very weak and unstable and will not remain accurate over two weeks.

16.5R To prepare dilutions for a cannery waste with an expected BOD of 2,000 mg/*L*, take 10 m*L* of sample and add 90 m*L* of dilution water to obtain a new sample with an estimated BOD of 200 mg/*L* (10 to 1 dilution):

$$\text{BOD Dilution, m}L = \frac{1{,}200}{\text{Estimated BOD, mg}/L}$$

$$= \frac{1{,}200}{200}$$

$$= 6 \text{ m}L$$

16.5S

$$\text{BOD, mg}/L = \left[\begin{array}{c}\text{Initial DO of}\\ \text{Diluted}\\ \text{Sample,}\\ \text{mg}/L\end{array} - \begin{array}{c}\text{DO of Diluted Sample}\\ \text{After 5-Day}\\ \text{Incubation,}\\ \text{mg}/L\end{array}\right]\left[\frac{\text{BOD Bottle Vol, m}L}{\text{Sample Volume, m}L}\right]$$

$$= (7.5 \text{ mg}/L - 3.9 \text{ mg}/L)\left[\frac{300 \text{ m}L}{2 \text{ m}L}\right]$$

$$= (3.6 \text{ mg}/L)(150)$$

$$= 540 \text{ mg}/L$$

ANSWERS TO QUESTIONS IN LESSON 9

Answer to question on page 580.

16.5T Precautions to be exercised when using a pH meter include:

1. Prepare fresh buffer solution weekly for calibration purposes.
2. pH meter, samples, and buffer solutions should all be at the same temperature.
3. Watch for erratic results arising from faulty operation of pH meter or fouling of electrodes with interfering matter.

Answer to question on page 580.

16.5U Operators test for metals in wastewater because of the toxic properties of metals.

Answer to question on page 595.

16.5V Kjeldahl nitrogen refers to organic plus ammonia nitrogen.

Answers to questions on page 598.

16.5W Forms of phosphorus commonly found in wastewater include orthophosphate, condensed phosphate, and organically bound phosphate.

16.5X The discharge of wastewater containing phosphorus may stimulate nuisance quantities of algal growths in receiving waters.

Answer to question on page 603.

16.5Y

Known		Unknown
Empty Dish, mg	= 64,328.9 mg	1. Total Solids, mg/*L*
Dish and Residue, mg	= 64,381.2 mg	2. Total Dissolved Solids, mg/*L*
Dish and Dissolved Solids, mg	= 64,351.2 mg	
Sample Volume, m*L*	= 100 m*L*	

1. Calculate Total Solids (Residue)

$$\text{Total Solids, mg}/L = \frac{(A - B)(1{,}000 \text{ m}L/L)}{\text{m}L \text{ of Sample}}$$

$$= \frac{(64{,}381.2 \text{ mg} - 64{,}328.9 \text{ mg})(1{,}000 \text{ m}L/L)}{100 \text{ m}L}$$

$$= 523 \text{ mg}/L$$

2. Calculate Total Dissolved Solids (Filterable)

$$\text{Total Dissolved Solids, mg}/L = \frac{(A - B)(1{,}000 \text{ m}L/L)}{\text{m}L \text{ of Sample}}$$

$$= \frac{(64{,}351.2 \text{ mg} - 64{,}328.9 \text{ mg})(1{,}000 \text{ m}L/L)}{100 \text{ m}L}$$

$$= 223 \text{ mg}/L$$

Answer to question on page 605.

16.5Z Specific conductance measures the ability of wastewater to conduct or carry an electric current, which is related to the concentration of ionized substances in the water. This ability depends on the total concentration of minerals dissolved in the wastewater and the temperature.

Answers to questions on page 606.

16.5AA Changes in influent temperature could indicate a new influent source. A drop in temperature could be caused by cold water from infiltration, and an increase in temperature could be caused by an industrial waste discharge.

16.5BB The thermometer should remain immersed in the liquid while being read for accurate results. When removed from the liquid, the reading will change.

16.5CC All thermometers should be calibrated against an accurate National Bureau of Standards (NBS) certified thermometer because some thermometers that are substantially inaccurate (off as much as 6°F) can be purchased.

Answer to question on page 608.

16.5DD Turbidity in wastewater is caused by the presence of suspended matter such as mud, finely divided organic and inorganic matter, and microscopic organisms such as algae.

PLANT ____________________

DATE ____________________

SUSPENDED SOLIDS AND DISSOLVED SOLIDS

SAMPLE							
Crucible							
Sample, mL							
Wt Dry + Dish, gm Wt Dish, gm							
Wt Dry, gm							
$mg/L = \frac{\text{Wt Dry, gm} \times 1,000,000}{\text{Sample, mL}}$							
Wt Dish + Dry, gm Wt Dish + Ash, gm							
Wt Volatile, gm							
$\% = \frac{\text{Wt Vol}}{\text{Wt Dry}} \times 100\%$							

BOD

\# Blank ____________

SAMPLE										
DO Sample										
Bottle #										
% Sample										
Blank or adjusted blank DO after incubation										
Depletion, 5 days										
Depletion, %										

SETTLEABLE SOLIDS

Sample, mL ____________ ____________

Direct mL/L ____________ ____________

COD

Sample ____________ ____________

Blank Titration ____________ ____________

Sample Titration ____________ ____________

Depletion ____________ ____________

$mg/L = \frac{\text{Dep} \times N \text{ FAS} \times 8,000}{\text{Sample, mL}}$ ____________ ____________

Typical laboratory work sheet

TOTAL SOLIDS

SAMPLE					
Dish No.					
Wt Dish + Wet, gm Wt Dish, gm					
Wt Wet, gm					
Wt Dish + Dry, gm Wt Dish, gm					
Wt Dry, gm					
% Solids = $\frac{\text{Wt Dry}}{\text{Wt Wet}} \times 100\%$					
Wt Dish + Dry, gm Wt Dish + Ash, gm					
Wt Volatile, gm					
% Volatile = $\frac{\text{Wt Vol} \times 100\%}{\text{Wt Dry}}$					
pH					
Volatile Acid, mg/*L*					
Alkalinity as $CaCO_3$, mg/*L*					

Grease

Sample ______________________

Sample, m*L* ______________________

Wt Flask + Grease, mg ______________________

Wt Flask, mg ______________________

Wt Grease, mg ______________________

$$\text{mg}/L = \frac{\text{Wt Grease, mg} \times 1{,}000}{\text{Sample, m}L}$$ ______________________

Typical laboratory work sheet (continued)

CHAPTER 17

APPLICATIONS OF COMPUTERS FOR PLANT O & M

by

Ken Kerri

TABLE OF CONTENTS

Chapter 17. APPLICATIONS OF COMPUTERS FOR PLANT O & M

OBJECTIVES

Chapter 17. APPLICATIONS OF COMPUTERS FOR PLANT O & M

Following completion of Chapter 17, you should be able to:

1. List the uses of computers in treatment plants.
2. Identify tasks in your treatment plant that could be performed by computers.
3. Provide reasons that justify purchasing and using computers.
4. Recognize cautions that must be exercised by operators using computers.
5. Evaluate both computer hardware and software.

CHAPTER 17. APPLICATIONS OF COMPUTERS FOR PLANT O & M

17.0 COMPUTER USE IN A TREATMENT PLANT

17.00 Trends

Computers are used extensively today by treatment plants, businesses, industries, and even the home. It is difficult to pick up a newspaper or magazine and not find another article about some new application of computers. Today, they are used in virtually every treatment plant. There have been many changes in computers and computer programs over the years that not only have made them less expensive, they have also made them easier to use. These changes have also made it very desirable to use computers in the wastewater treatment plant. In this chapter, we will describe the reasons why many treatment plant operators are happily using computers in their plants.

Some computer applications are unique to treatment plants; others are used in virtually all businesses and industries. Here are some of the general types of computer applications in the wastewater treatment plant.

17.01 Data Analysis

Computers are extremely efficient at doing all kinds of statistical calculations. They can rapidly compute means, medians, modes, and also find maximum and minimum values. Additionally, a computer program (the general term for a computer program is *SOFTWARE*) can take your data and generate charts and graphs to help you better understand the data and to identify critical trends.

17.02 Report Generation

Probably the most frequent use of computers is for *WORD PROCESSING*. After a report has been entered into the computer using word processing software, the text can be modified easily by inserting new text, moving words and paragraphs, changing type styles for emphasis, and copying sections of text to other parts of the document or to other documents. The text can be saved and then later recalled for additional modification. Many word processing programs will allow you to insert into your document charts and graphs generated with other software.

17.03 Recordkeeping

One of the hallmarks of an efficient treatment plant is the maintenance of good records. The information kept ranges from BOD levels to equipment maintenance. The task of recordkeeping can be aided through the use of computers. One standard type of program used for this task is a *DATABASE MANAGEMENT PROGRAM*. A database is nothing more than a collection of data that have been organized in a way that permits information to be added, removed, manipulated, or retrieved rapidly. With information stored in a database management program, you can quickly call up the name of the manufacturer of a given piece of equipment, the initial cost, or the last date of maintenance. You can ask for all equipment purchased between two given dates and, at the same time, ask it to exclude any of those items that cost less than a certain amount, for example, $5,000. Figure 17.1 shows a computer printout of the information contained in a database file for a booster station motor. Data retrieval is quick and accurate.

Another common type of data management software program is the *SPREADSHEET*. Like a database, a spreadsheet is a collection of various types of information organized in such a way that it can be manipulated easily. Spreadsheets are particularly useful for performing calculations and creating charts and graphs based on the information stored in the spreadsheet. Once the appropriate data have been entered into a spreadsheet, answers to "what if" questions can be calculated by the program. An example of a "what if" question would be, "What would be the effect on annual revenue if the industrial discharger volume charge was increased by eight percent?" The ability of a spreadsheet to make complex calculations can be a great asset for planning and forecasting uses.

17.04 Systems Monitoring

Using a computer-controlled system including sensors and valves (the general term for the computer and other related physical equipment is *HARDWARE*) and appropriate software, computers can be used to monitor wastewater characteristics and treatment processes, analyze the information, and then make appropriate adjustments to treatment processes. While these systems can become very complex, they can add considerably to the efficiency of a treatment system.

In general, computers can help operators:

- save time
- save money
- do a better job

```
                         MOTOR INFORMATION

LOCATION: BOOSTER STATION
MANF.: U.S. MOTORS
MODEL NUMBER: DF 100          SERIAL NUMBER: 125698
TYPE: HOLLOW SHAFT
SUPPLIER: MOTOR WORLD
ADDRESS: 123 E STREET, ANY TOWN, CO      890539
PHONE NUMBER: 303/792/6666
DATE PURCHASED: 12/10/99                PRICE: 1800.00
VOLTAGE: 480     HORSEPOWER: 100    FULL LOAD AMPS: 125     RPM: 1800
FUSE SIZE: 175      FRAME SIZE: NPT20
BEARING NUMBERS: JN63458     UPPER LOWER: 25879C
DATE INSTALLED: 2/6/00

COMMENTS: No vibration, running amps at time of installation were L1=90
L2=91   L3=90.
```

Fig. 17.1 Printout of a typical database form for an electrical motor

(Source: *OPFLOW,* Volume 19, No. 3 (March 1992), by permission. Copyright © 1992, American Water Works Association)

17.05 Communications

It is now possible for wastewater treatment plant operators to communicate quickly and inexpensively with government agencies, equipment vendors, educational institutions, and other operators all over the world using a computer and a modem. A modem is a small device that connects a computer to a telephone line. By dialing the phone number of an internet service provider (ISP), an operator can communicate electronically with other persons or agencies having a similar computer connection. Having access to the internet enables operators to exchange ideas and information with other operators, read and print out copies of EPA and state regulations and guidance documents, submit compliance reports directly to regulatory agencies, and keep current on new developments in the wastewater industry. These are just a few of the many ways a computer and modem can assist an operator.

17.1 HOW CAN I USE A COMPUTER?

17.10 Possibilities

If you wish to consider the possibility of using a computer or expanding your use of computers, start by looking at your own job. The first question to ask is, "How can I use a computer to help me do a better job?" Think of all the tasks you must perform when operating and maintaining your plant. Start with a small task. Do not expect miracles. Be patient and persistent.

Ask questions of anyone you know about how computers have been helpful to them.

One place to begin is by identifying time-consuming tasks that are related to paperwork. When are you required to shuffle through stacks of papers or files to come up with answers?

The following example shows how many operators are using computers. An operator runs the lab in a treatment plant. The boss is always calling on the phone with questions about flows and influent or effluent characteristics. The operator uses his lab computer to input all historical lab results. Then, he continues to add all daily lab results. Now, whenever the boss has a question about monthly effluent BODs, he can get the answer immediately.

Those operators who have discovered ways to use computers to assist them in their jobs tend to get very excited and enthusiastic about their use. A lab analyst who uses a computer successfully thinks that this is the best application for computers in treatment plants. An equipment maintenance person who is using a computer to schedule preventive maintenance will tell you that this is the only way to manage a maintenance program. Operators who have successfully controlled their processes with a computer will argue that all operators should use their approach.

On the next few pages are more detailed examples to help give a better picture of how computers can help save time and money as well as improve efficiency in the treatment plant. Specific illustrations include the following tasks performed by wastewater treatment plant operators:

1. Analysis of lab results
2. Equipment maintenance
3. Pretreatment monitoring
4. Process control
5. Report preparation

17.11 Example 1: Analysis of Lab Results

In most treatment plants, the laboratory produces large quantities of data. Lab results produce information on influent, effluent, and receiving water characteristics, as well as information on the performance of the treatment processes. This information is vital for operators to properly control treatment processes and comply with discharge requirements. Tasks that can be performed using computer programs include:

1. Organization of data
2. Tabular or graphical presentations
3. Statistical calculations
4. Submission of data and reports to regulatory agencies

Computers can help both lab analysts and operators retrieve historical data. For example, if someone wants to know the effluent BOD for each day in November, simply ask the computer. You can get a printout listing the effluent BOD for each day. (See Figure 17.2.)

This is much easier than finding the lab data sheets or reports, looking up the effluent BOD for each day, and writing it down on a piece of paper.

CLEANWATER WASTEWATER TREATMENT PLANT

November Effluent BOD, mg/*L*

	DATE	DAY	BOD, mg/*L*
	1	S	19
	2	M	19
	3	T	14
	4	W	16
	5	T	9
	6	F	18
	7	S	14
	8	S	9
	9	M	11
	10	T	8
	11	W	9
	12	T	11
	13	F	15
	14	S	12
	15	S	9
	16	M	10
	17	T	10
	18	W	11
	19	T	10
	20	F	18
	21	S	10
	22	S	14
	23	M	13
	24	T	14
	25	W	12
	26	T	10
	27	F	15
	28	S	16
	29	S	14
	30	M	19
MAX			19
MIN			8
AVG			13

Fig. 17.2 Typical computer printout information

Computers can also assist in the analysis of lab results. Laboratory computer programs can calculate the average effluent BOD for each month and identify the monthly maximum and minimum BOD values. All lab data can be reported in this manner with computer printouts showing monthly totals, averages, and minimum and maximum values for each item analyzed. Programs are available that will also plot out the laboratory results and trends.

If the proper lab results are entered into the computer, the computer can calculate and print out chemical dosages; suspended solids, BOD, or COD loadings; percent removals; SVIs; sludge ages; F/Ms (Food to Microorganism ratios); and MCRTs (Mean Cell Residence Times).

Useful features that are included in some software programs include the calculating of running averages and trends and then plotting the results. The plotted seven-day running average can be very helpful to the operator by aiding in detecting a trend showing that a treatment process is slowly failing. Figure 17.3 contains the effluent COD data from a treatment plant. The daily fluctuations alone make it difficult to detect any trends. The seven-day moving average plotted on the same scale clearly reveals the deteriorating trend of the process. Once alerted to this trend, the operator can quickly identify and correct the problem.

In many states, lab results and permit compliance reports can now be submitted electronically to regulatory agencies. Depending on the computer capabilities of the laboratory and the treatment plant, analytical data are sometimes transmitted to the treatment plant and directly to the state agency at the same time using a modem. Direct electronic transfer of lab results offers several benefits, including less chance of errors that might occur when lab data are manually copied from one report or form to another and faster preparation and submission of compliance reports.

17.12 Example 2: Equipment Maintenance

When considering equipment maintenance needs at wastewater treatment plants or pumping stations, one readily reaches the conclusion that these facilities represent maintenance monsters. A list of the factors typically present at wastewater handling and treatment facilities that lead to high maintenance requirements includes such items as equipment in contact with water, frequent on/off operations, moist atmospheric environment, and handling of corrosive impurities.

The consequence of these factors is that maintenance and its associated costs are a critical part of treatment plant O & M. Maintenance represents a substantial dollar investment in labor, parts, and equipment replacement at the end of its useful life. From the perspective of the overall O & M costs to operate and maintain a wastewater treatment facility, the expenses associated with maintenance represent a very high percentage of the total annual budget. Starting from the day a plant goes on line, maintenance costs begin to occur. The older the equipment, the higher the costs.

One way to hold down costs is to have a good, well-organized maintenance program that includes excellent recordkeeping. Neglected maintenance will lead to unnecessary early replacement of major components and greatly added expense, especially considering rising costs of both labor and repair materials. Even in smaller plants, the economic impact of a less than excellent maintenance program will result in significant dollar losses.

A successful maintenance program includes timely scheduling of preventive maintenance, thorough follow-through, good recordkeeping, and an adequate inventory supply. When a minimum amount of time is allocated to maintenance activities, equipment and facilities begin to deteriorate rapidly. At many utility agencies, the personnel are well aware of the program needed to achieve proper maintenance, but they simply do not have adequate resources. This frequently leads to a repair or corrective program, rather than one of preventive maintenance.

One example of how computers can help stretch precious dollar resources is through inventory control on parts and scheduling repairs. Thorough recordkeeping will show you which equipment and parts frequently break down. Timely and accurate information will help you identify which parts are most important to keep in stock (and which costly parts it does not pay to keep on hand), which vendors supply the most and least reliable equipment, and when maintenance schedules ought to be reviewed.

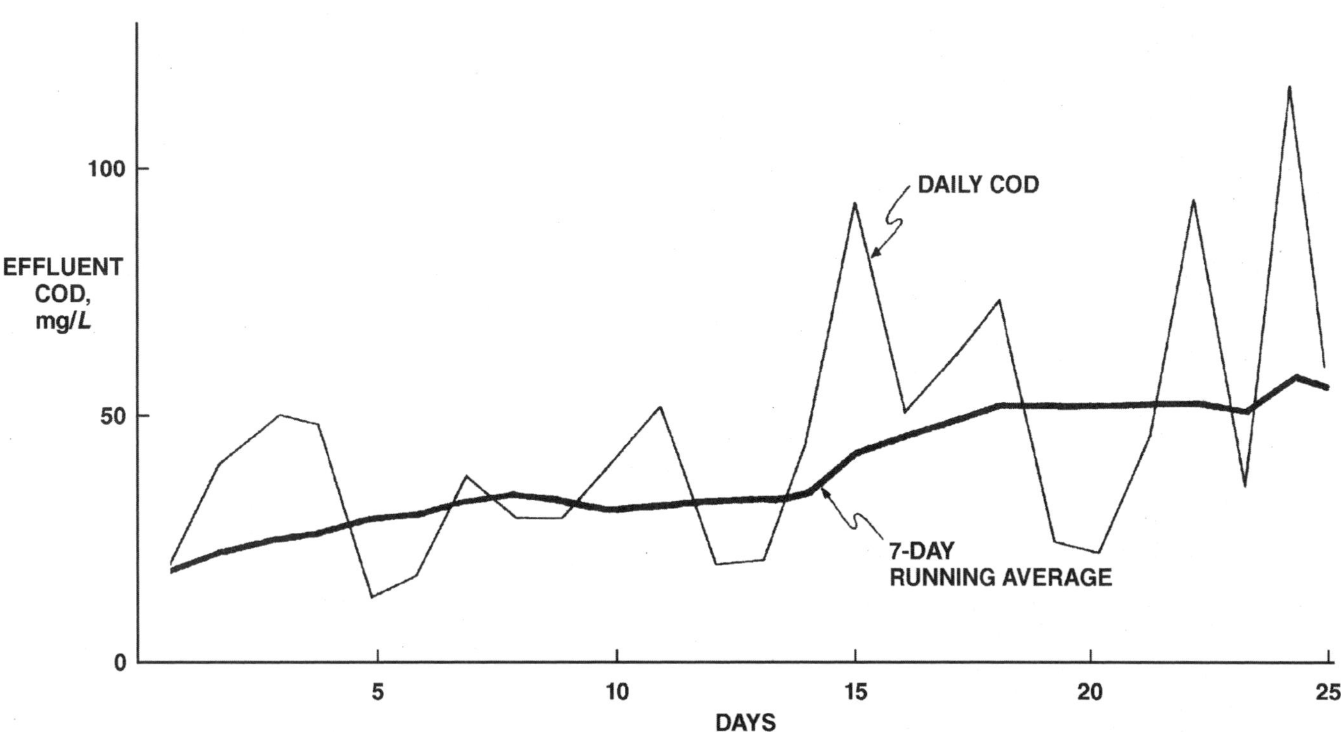

Fig. 17.3 Daily effluent CODs and seven-day running average

If you look through the Water Environment Federation's "Operations Forum" features in the monthly publication, *WATER ENVIRONMENT & TECHNOLOGY (WE&T),* you will find several vendors offering computer packages that focus specifically on equipment maintenance. Many programs designed for equipment maintenance in other manufacturing processes have been modified for the clean water business.

You can use one of these programs or one that is custom designed for your plant. After you have entered information such as vendor and manufacturer data, nameplate data, repairs history, maintenance schedule, and maintenance procedures, you can retrieve the following types of information:

1. Preventive maintenance equipment work orders on a daily, weekly, or monthly basis
2. Spare parts acquisition schedules in advance of maintenance and overhaul activities
3. Prioritized lists of maintenance tasks
4. Existing inventory of equipment and spare parts and a list of needs
5. History of equipment maintenance and repairs
6. Listing of uncompleted work orders and scheduled maintenance
7. Summaries of maintenance costs, including parts and labor
8. Information relating to equipment status and condition

Computer programs also will allow you to ask specific questions about a particular piece of equipment, all equipment made by a manufacturer, a specific type of equipment, or equipment in a given location. For example, you could ask the computer to list the dates and reasons a particular pump failed during the last two years. You could determine when the pump was maintained, what was done, who did it, and the cost of maintenance. You could ask which pumps had bearing failures and the date of each failure. The range of questions that can be asked is limited only by the software program and the information put into the computer.

A well-designed maintenance software program requires only that the operator make simple responses to pre-programmed questions that are presented step-by-step on the computer screen. The type of program and the amount you ask of the program will determine the time involved in using such a program.

17.13 Example 3: Pretreatment Monitoring

If your agency has a pretreatment program, a computer can be used to manage the data supplied by industrial sources. Programs can maintain sampling schedules, concentrations and pollutant loads for each water quality indicator. They will estimate extra strength surcharges based upon industrial sampling results. Also, they can evaluate allowable plant loadings and allocate allowable discharge loading to each industry. Programs will tabulate and summarize industrial reports to determine total industrial load on a plant.

17.14 Example 4: Process Control

Wastewater flows and other characteristics that can be sensed by instruments can be monitored by computers. Flowmeters often have a depth-of-water sensing device that converts depth to flow in a flowmeter. Water quality characteristics that can be sensed with probes and instruments include dissolved oxygen, temperature, pH, and turbidity.

Computer graphics and statistical analysis can aid in process control. After data have been entered into the computer, a program can then be run that calculates the statistical relationship between COD and BOD including a forecast of the BOD when given the COD. Process control thereby can be improved through the timely estimation of BODs on the results of COD tests. This same work could be performed with a calculator, but there are much greater chances for errors and a substantial time requirement.

17.15 Example 5: Report Preparation

Treatment plant superintendents and operators are being forced to spend more and more time collecting and handling data. Regulatory agencies and management are requiring ever greater amounts of information. Manually collecting and analyzing the data is very time consuming.

Computer programs can take data you have already collected as part of lab analysis, equipment maintenance, pretreatment monitoring, and process control and merge it into a report. Monthly reports can be produced for utility officials, state agencies, and other regulatory agencies on the first of each month based on the data collected and analyzed by the computer. These reports are usually accepted as directly printed from the computer.

NPDES report forms are available from many regulatory agencies on paper that is specifically designed to be inserted into the computer printer. The computer program seeks out the appropriate permit data from that available for the month, calculates the information required for the NPDES report, and then directly prints on the NPDES form all of the values, permit violations, and sampling frequencies. All that remains for the plant operator to do is to review and sign the report. Also, some regulatory agencies have computers that accept reports directly from the plant's computer.

17.2 HOW DO YOU GET A COMPUTER OR A BETTER COMPUTER?

Once you have decided that there are appropriate uses for a computer or a better computer in your plant, you will need to prepare information that will justify the purchase of a computer. The first step is to clearly identify each of the specific applications you intend for the computer. This list of applications not only will help convince management of your needs, but also will serve as the basis for determining which is the appropriate computer hardware and software.

An operator might determine that a computer would be useful for data storage and analysis. One plant discovered that their lab generated over 30,000 items of data a month that had to be stored, analyzed, and manipulated. How could any team of lab and operations personnel adequately analyze and use all of this data for assistance in process decision making? How much time was required to turn this data into the form needed for NPDES reports?

This plant's study also gave them the information they had to provide to software and hardware vendors and consultants in order to determine the computer system requirements. Based on this data, they knew what operations the software would have to perform, how powerful a computer they needed, how much data the computer would have to store, and how many computer stations they would need.

Justifications for the purchase of a computer usually focus on the applications described in the examples above: data management and interpretation, recording and scheduling equipment maintenance, pretreatment monitoring, process control, and report generation. An unfortunate reality is that some agencies require extensive justification for a new computer while they will approve a new pump costing twice as much with only minimal justification.

Once the idea of purchasing a computer has been approved, the tough decisions on selection of software and brand and model of computer have to be made. The following procedures should be considered:

1. Specify, in writing, the tasks you want the computer to perform for you. The clearer your specifications, the more satisfied you will be with your choices.
2. Determine what computer programs (software) are available that can perform the desired tasks.
3. Determine which computers will run the selected programs and have adequate processing power and data storage capabilities. Review computer software catalogs to be sure that the appropriate programs will run on your hardware (computer).
4. Talk with other operators at meetings and conferences. Learn about their experience with the computers and programs they have used.
5. Identify local computer vendors who have good reputations for servicing their products.
6. Try to locate a local computer user group. These people meet regularly to discuss applications of commercial computer programs. They can provide first-hand information about hardware, software, and vendors.

When evaluating various computers (hardware) and programs (software), here are some important points to consider. Calculate the total costs of equipment, equipment maintenance, and the necessary software. Check to see how easy the system will be to operate. How expandable is the system? As you begin to use the computer, new applications will become apparent. Is there an abundance of off-the-shelf software or will choices for future applications be limited?

There are some important items that must be considered when comparing types of software packages or computer programs. In general, there are three types of software packages: general purpose, semi-custom, and custom software.

General purpose software packages include word processing, database management, spreadsheet, and graphics programs. These are the lowest in initial cost. Because they are designed to be used in virtually any business or industry, once developed, these programs can be mass produced and mass marketed. Because they are used by large numbers of customers, the producers can receive considerable feedback from the customers. This information helps identify "bugs" and requirements for desirable new features. The price per software package can be low because the development costs may be spread over the many thousands of copies that the producer expects to sell. Examples of general purpose software that you may have heard of include Quattro Pro, Microsoft Word, and Excel.

Semi-custom software packages are ones that have been designed for a specific industry or application. Some of these programs have been written from scratch to meet a particular need in the wastewater treatment plant. Other programs will be software (called a *TEMPLATE*) that can be added onto a general purpose program. For example, a spreadsheet could be added into a general purpose package such as Excel. These templates set up the formats and formulas you will need for specific applications. A source for locating these programs is the Water Environment Federation's "Operations Forum" features in the monthly publication, *WATER ENVIRONMENT & TECHNOLOGY (WE&T).*

Custom software is specifically prepared for a particular wastewater treatment plant. This software is costly because it represents the supplier's investment of hundreds of hours in preparing the software for a single buyer. There are, however, advantages for a utility in investing in custom software. Well-designed custom software is easier to use, generally requires less training, and does not require that someone from the plant customize the program for the specific application. Very few treatment plants can afford to hire or retrain current personnel for the position of full-time computer specialist to formulate and operate off-the-shelf programs. The custom systems typically are simple to operate because they are designed for that plant and in conjunction with plant personnel. When the software is prepared for a specific facility, more meaningful data relationships also can be incorporated.

Custom software generally can be operated by people with little or no computer training. Another advantage is that few, if any, changes need to be made in the current routine of a facility due to a shift over to a computer system since the current standard operating procedures can be incorporated into the computer program. In other words, if a custom application is designed through a partnership of treatment plant personnel and a programmer, your current O & M procedures can be the basis for the program. These procedures continue and the computer helps you follow your procedures.

There are places for each type of software in any given utility. For example, word processing can easily be done using a standard word processing program. Simple summary work can be done using a standard database management system. Standard spreadsheet programs (for example, Excel or Quattro Pro) are ideal for these applications.

Custom software has a distinct advantage and an attractive payback period in the areas of data handling, maintenance management, and process control. Semi-custom or custom software is applicable in the area of inventory and spare parts control. Actual applications of the tailored software to these functional roles shows that substantial time savings can be identified as well as other additional savings such as reduced lubricant and chemical costs, less repair work, and longer useful equipment life.

If a system is well designed, it is possible to start with small, simple systems, and then as the operators become familiar with its use, expand and incorporate all of the advantages of a computer system for your wastewater facility.

When accepting a new custom software package from a vendor, it may work very quickly because all of your system or data have not been loaded into the computer program. A good practice is to load the entire computer system to capacity (use dummy data if necessary) and see how quickly the software can perform its expected tasks. If the program runs too slowly, appropriate changes should be made in either the software and/or hardware so there will not be long delays waiting for results.

17.3 TIME SAVINGS

Let us examine where and how we can expect to save time when using a computer to help run a treatment plant.

The time spent recording data probably will not change. In fact, it could increase with increased data storage and analysis capability. Major time savings can be expected in the areas of calculations, trend analysis and graphics, and report generation. Since virtually every calculation will be done by computer, all that is left to do is review the results as opposed to locating various inputs for the formulas and going through the calculations. Now operators can spend their time analyzing, rather than just producing data. As mentioned earlier, the time involved in report generation can be reduced substantially.

17.4 CAUTIONS

When maintenance procedures are printed by computers with the work orders, the people performing the maintenance still have to think and be observant. Computers do not replace operators. When the work order indicates that the oil level in a pump needs to be checked, the person doing this work should also feel the motor for excessive heat, listen for strange sounds, and look for vibrations and leaks.

Some people recommend that separate computer systems be used for the plant processes sensing equipment and the data and maintenance management system. This is a good idea, especially if the system otherwise is set up so the entire

system will refuse to operate when a process sensor is not functioning.

Check to find the most functional and accessible location for the computer systems and terminals. Everyone with a permanent office who needs a personal computer (PC) should have one. Operators need a PC to prepare, send, and receive e-mails, assignments, and reports. Supervisors need PCs to monitor processes, lab results, maintenance activities, pretreatment monitoring programs, and budget status. Operators responsible for these activities and programs need PCs to record their activities and communicate with supervisors.

Many field and maintenance crews use portable laptop computers to receive work orders, follow maintenance procedures, and record completed maintenance tasks. These crews need access to PCs to receive task assignments and to report tasks completed. PCs are needed in a warehouse or corporation yard to manage the spare parts inventory and to replenish spare parts when necessary.

Some operators use PDAs (personal digital assistants) or handheld computers, which can function as cellular phones, fax senders, Internet browsers, and personal organizers. With these devices, operators can be in the field collecting data and, by use of wireless technology, transmit the information directly to everyone on the computer network.

When installing any computer program, be sure to check to see that it is doing what you expect it to do. A good check when using statistical packages is to put some numbers into the computer and then ask the computer to calculate the mean. Take the same numbers and calculate the mean manually. If the answers are not the same, try to figure out why. The same procedure should be applied to process control guidelines and chemical dosages. Personally calculate SVIs, sludge ages, F/Ms, MCRTs, and chemical dosages and compare these values with those calculated by the computer. Once calculations are verified during installation they should not have to be checked repeatedly.

If you are going to use a computer for process control, you must verify every instruction the computer gives the plant by actual observations in the field. Many plants have used two-way radios to compare computer instructions with plant responses. One operator watches the computer generate instructions while another operator is outside in the plant watching when and how the plant facilities respond to the computer's commands. You must make sure that when the computer tells the plant to adjust or close valve number 175, the operation is actually performed on valve 175 and not valve number 157. If the computer calculates a specific waste activated sludge pumping rate, be sure the correct pump comes on and pumps at the correct rate.

You can protect your computer from the damages of stray power fluctuations through use of a surge protector. A good surge protector will even out the amount of power reaching the computer.

What happens to a computer system when a disaster occurs and your plant loses its electric power? What happens if the standby generator will not perform or there are no provisions for standby power? You can really have a disaster.

There are several precautions that can be taken before a disaster occurs. First, provisions must be made during the design of a plant for the plant to function without power. This is one place where an operator can provide good advice at the plan review stage.

Important plant facilities must have provisions for manual operation if at all possible. Some standby power systems will provide only enough power to operate critical pumps and provide minimal treatment. This is where an operator must know how to operate a plant manually without instrumentation, controls, and computers. You can do it if you know how and why your plant works.

QUESTIONS

Write your answers in a notebook and then compare your answers with those on page 630.

17.0A How can computers help operators?

17.1A List tasks performed by operators that can be aided by the use of computers.

17.1B How would you justify a computer for the analysis of lab results?

17.3A What time savings can result from the use of computers?

17.5 SCADA SYSTEMS

17.50 Description of SCADA (SKAY-dah) Systems

SCADA stands for Supervisory Control And Data Acquisition system. This is a computer-monitored alarm, response, control, and data acquisition system used by operators to monitor and adjust their treatment processes and facilities.

A SCADA system collects, stores, and analyzes information about all aspects of operation and maintenance, transmits alarm signals when necessary, and allows fingertip control of alarms, equipment, and processes. SCADA provides the information that operators need to solve minor problems before they become major incidents. As the nerve center of a wastewater utility, the system allows operators to enhance the efficiency of their wastewater facility by keeping them fully informed and fully in control.

In wastewater applications, SCADA systems monitor water levels, pressures, and flows, and also operate pumps, valves, and alarms. They monitor motor speeds and currents, temperature, pH, dissolved oxygen levels, and other operating guidelines, and provide control as necessary. SCADA also logs event and analog signal trends and monitors equipment operating time for maintenance purposes.

Applications for SCADA systems include wastewater collection and pumping system monitoring, wastewater treatment plant control monitoring, combined sewer overflow (CSO) diversion monitoring, and other related applications. SCADA systems can vary from merely data collection and storage to total data analysis, interpretation, and process control.

A SCADA system might include water level, pressure, and flow sensors. The measured (sensed) information could be transmitted by cable, radio, telephone line, microwave satellite, or fiber-optic communications systems. The information is received by a computer system that stores, analyzes, and presents the information. The information may be read on dials or as digital readouts or analyzed and plotted as trend charts. Most SCADA systems include a graphical picture of the overall system. In addition, detailed pictures of specific portions of the system can be examined by the operator following a request and instructions to the computer. The graphical displays on a TV or computer screen can include current operating information. The operator can observe this information, analyze it for trends or determine if it is within acceptable operating ranges, and then decide if any adjustments or changes are necessary.

SCADA systems are capable of analyzing data and providing operating, maintenance, regulatory, and annual reports. In some plants operators rely on a SCADA system to help them prepare daily, weekly, and monthly maintenance schedules, monitor the spare parts inventory status, order additional spare parts when necessary, print out work orders, and record completed work assignments.

SCADA systems can also be used to enhance energy conservation programs. For example, operators can develop energy management routines that allow for both maximum energy savings and maximum water storage prior to entering "on-peak" periods. In this type of system, power meters are used to accurately measure and record power consumption and the information can then be reviewed by operators to watch for changes that may indicate equipment problems.

Emergency response procedures can be programmed into a SCADA system. Operator responses can be provided for different operational scenarios that could be encountered as a result of adverse weather changes, fires, or earthquakes.

QUESTIONS

Write your answers in a notebook and then compare your answers with those on page 630.

17.5A What does SCADA stand for?

17.5B What does a SCADA system do?

17.5C How could measured (sensed) information be transmitted?

17.51 Typical Wastewater Treatment SCADA Systems

Operators at each wastewater treatment plant should decide how they would like to use computers and how computers can help them do a better job. There are a wide range of choices and applications for operators to select. Operators typically start by having computers monitor plant performance. Next they could use computers to plot trends in monitored information, perform process calculations, and ultimately suggest process changes that should be confirmed or verified by operators.

Computers can be used to help operators monitor influent flows and pumps, and then divert flows to the appropriate number of primary clarifiers, aeration basins, secondary clarifiers, and chlorine contact basins. As flows increase or decrease into a wet well, the speeds of pumps can be increased when flows increase until another pump must be started. Then the speeds of all pumps on line can be adjusted to keep the flow to the treatment processes approximately the same as the inflow. The same procedure can be controlled by a computer when the flows decrease. The computer can also require a minimum operating time for a pump when it starts as well as a minimum wait time when the pump stops before the pump can start again.

Hundreds of items can be monitored, recorded, and controlled by computers. These items include flow, chemical feeders, processes, alarms, and equipment. Typical items include:

1. Influent flows
2. Influent pumps
3. Influent flow control gates and valves
4. Influent sampling
5. Diversion of flows to primary clarifiers
6. Primary clarifier sludge pumping
7. Diversion of flows to aeration basins
8. Aeration basin dissolved oxygen
9. Adjustment of air/oxygen flows to aeration basins
10. Diversion of flows to secondary clarifiers
11. Secondary clarifier sludge pumping
12. Return activated sludge (RAS) pumping
13. Waste activated sludge (WAS) pumping
14. Waste activated sludge thickening process
15. Anaerobic sludge digester temperature
16. Anaerobic sludge digester mixing
17. Anaerobic sludge digester gas production
18. Anaerobic sludge digester supernatant removal
19. Chlorination feed
20. Chlorine contact time
21. Sulfonation feed (dechlorination)
22. Effluent pumps
23. Effluent sampling
24. Dilution ratio in receiving waters

Many of these items have three or four components that also can be monitored and recorded by a computer.

Chemical feed rates may be adjusted by computers depending on residuals or concentrations of chemicals monitored and recorded by computers. If a measured chlorine residual is too low, the computer could increase the chlorine feed rate. The computer also could control the rate of change of a chemical feeder to prevent wide fluctuations in chemical feed rates. Also chemical feed supplies and concentrations could be monitored by a computer.

The activated sludge process can be monitored, as well as controlled and adjusted, by computers. The operator can provide the computer with essential information such as tank volumes, flows, mixed liquor suspended solids, and desired MCRT. The computer could perform the necessary calculations and recommend to the operator the desired waste activated sludge (WAS) pumping rate. The operator compares the desired pumping rate with the actual rate, reviews current and expected conditions in the plant and the effluent, and then selects the appropriate waste activated sludge (WAS) pumping rate being careful to avoid changes of more than ±10 percent. The computer helps the operator monitor any changes that result from this new waste activated sludge (WAS) pumping rate and the operator decides if any additional adjustments are necessary.

Troubleshooting process problems can be aided by the use of a computer. When a process becomes upset, the operator can review data stored in the computer. When everything was performing properly, the computer can plot trends of critical operating guidelines before the upset. By reviewing and analyzing the data, the operator can identify what went wrong first and the probable cause of the first problem. Once causes are identified they can be corrected and prevented or minimized in the future.

Computers are very helpful in monitoring, operating, maintaining, and troubleshooting equipment. Information collected on an influent pump could include bearing temperatures, RPMs, vibration information, hours pump operated, pump status (running or waiting (available to pump)), right angle gear, and condition of auxiliary equipment (lubrication and oil pumps). Similar information could be collected for other pieces of plant equipment.

Computers are used to monitor the alarm systems, record alarms, display alarm situations until acknowledged, and then maintain the visual display of the alarm condition until corrected. Also, the plant electrical distribution system can be monitored. If power is lost for a short period (less than seven minutes for example), the computer can restart and restore all equipment to its operating level before the power failure. However, only operators should be allowed to determine if influent and effluent pumps should be restarted and the sequence of restarting the pumps.

Operators should be provided with a computer system that allows operators to prepare their own display screens on the computer. For example, the main screen could be a flow diagram of the plant showing the flows of wastewater and solids into and out of each treatment process. Critical operating information could be displayed with each process. Detailed screens could be easily reached for each process and each piece of equipment.

Information on the screen could be color coded, with the colors used to indicate if a pump is running, ready, unavailable, or failed, or if a valve is open, closed, moving, unavailable, or failed. Failed is used by the computer to inform the operator that something is wrong with the information or the signal the computer is receiving or is being instructed to display, that the signal is not logical with the rest of the information available to the computer. For example, if there is no power to a motor, then the motor cannot be running when the computer is receiving a signal that the motor *is* running.

The operator can request a computer to display a summary of all alarm conditions in a plant or a particular plant area. If an alarm signal is blinking, this indicates that the alarm condition has not been acknowledged by the operator. When an alarm condition is not blinking, this situation will stay on until the alarm condition is fixed. Also the screen could designate certain alarm conditions as priority alarms, which means they require immediate attention.

Computers can be programmed to print or display the daily plant log every 24 hours. Information could include total or average measurements as well as maximum and minimum values. Also, calculated values could be included.

In summary, operators need to determine what they want computers to do for them. Once operators learn the benefits and limitations of computers, the use of computers will expand. Critical decisions regarding process control and plant operation will always be made by operators on the basis of the operators' analysis and interpretation of information provided by computers.

Operators will always be needed to see if mixed liquor suspended solids look OK, to listen to a pump to be sure it sounds proper, and to smell influent wastewater and the equipment to determine if unexpected or unidentified changes are occurring. Wastewater treatment plants will always need alert, knowledgeable, and experienced operators who have a "feel" for their plant.

QUESTIONS

Write your answers in a notebook and then compare your answers with those on page 630.

17.5D How can computers help operators?

17.5E How can computers adjust chemical feed rates?

17.5F What kind of information can computers provide regarding alarms?

17.6 CONCLUSIONS

1. Computers can save time and money as well as increase efficiency.
2. Computers can help operators do a better job in many areas of plant operations. Specific areas include:
   - analysis of lab results
   - equipment maintenance
   - pretreatment monitoring
   - process control
   - report generation
3. Computer software and hardware should be purchased only after the tasks to be performed have been clearly specified.
4. Computers cannot and will not completely run your plant. They can even lull you into a false sense of security, which may mean that you do a worse job.

For additional information on computers, see Section 2.4 "Database for Pretreatment Program," in Chapter 2, "Pretreatment Program Administration," from the manual in this series on *PRETREATMENT FACILITY INSPECTION.*

Please answer the discussion and review questions next.

DISCUSSION AND REVIEW QUESTIONS

Chapter 17. APPLICATIONS OF COMPUTERS FOR PLANT O & M

The purpose of these questions is to indicate to you how well you understand the material in the chapter. Write the answers to these questions in your notebook.

1. Why should operators use computers?
2. How can computers help an equipment maintenance program?
3. Computer pretreatment monitoring programs are used for what purposes?
4. How would you justify the purchase of a computer?
5. What procedures would you follow to select software and hardware?
6. What cautions must be exercised when using computers?

SUGGESTED ANSWERS

Chapter 17. APPLICATIONS OF COMPUTERS FOR PLANT O & M

Answers to questions on page 627.

17.0A Computers can help operators save time, save money, and do a better job.

17.1A Computers can help operators analyze lab results, maintain equipment, monitor pretreatment programs, control processes, and prepare reports.

17.1B Computers can save considerable time analyzing lab results because they can store, retrieve, and analyze data. Data can be arranged for tabular or graphical presentations. Computers can calculate means, maximums and minimums, and plot trends in data. They can also help prepare NPDES reports.

17.3A Time savings result from the use of computers in the areas of calculations, trend analysis and graphics, and report generation.

Answers to questions on page 628.

17.5A SCADA stands for Supervisory Control And Data Acquisition system.

17.5B A SCADA system collects, stores, and analyzes information about all aspects of operation and maintenance, transmits alarm signals when necessary, and allows fingertip control of alarms, equipment, and processes.

17.5C Measured (sensed) information could be transmitted by cable, radio, telephone line, microwave satellite, or fiber-optic communications systems.

Answers to questions on page 629.

17.5D Computers can be used to help operators monitor influent flows and pumps, and then divert flows to the appropriate number of primary clarifiers, aeration basins, secondary clarifiers, and chlorine contact basins.

17.5E Chemical feed rates can be adjusted by computers depending on residuals or concentrations of chemicals monitored and recorded by computers.

17.5F Computers can be used to monitor the alarm systems, record alarms, display alarm situations until acknowledged, and then maintain the visual display of the alarm condition until corrected.

CHAPTER 18

ANALYSIS AND PRESENTATION OF DATA

by

Ken Kerri

TABLE OF CONTENTS

Chapter 18. ANALYSIS AND PRESENTATION OF DATA

OBJECTIVES

Chapter 18. ANALYSIS AND PRESENTATION OF DATA

Following completion of Chapter 18, you should be able to:

1. Identify causes of the variations in results.
2. Read manometers, gauges, and charts.
3. Analyze and present data using:
   a. Charts and graphs
   b. Tables
   c. Numbers
4. Calculate arithmetic mean, range, median, mode, geometric mean, moving average, variance, and standard deviation.
5. Apply prediction equations, trends, and correlations to the analysis of data.

CHAPTER 18. ANALYSIS AND PRESENTATION OF DATA

(Lesson 1 of 2 Lessons)

18.0 NEED FOR ANALYZING AND PRESENTING DATA

Collection of data without analysis, interpretation, and use of results is a waste of time and money. This chapter will attempt to provide you with simple, easy methods to analyze data. To show you how to make the results of your testing meaningful and easily interpreted, sections on data presentation also are included. Many times your supervisor can understand what is happening in your plant or why you need a budget increase if you can show charts indicating trends or changes in plant operation or treatment process efficiencies.

Whether samples are collected, analyzed, interpreted, and used by the same person or a different person performs each task, every job is equally important if the application of results is to be effective. When running lab tests the samples must be *REPRESENTATIVE.*[1] Persons reviewing and interpreting lab results assume the tests were performed in a careful, prescribed manner and the results are accurate. Operators applying the interpretation of lab results to operational controls rely on proper interpretation to ensure effective adjustments in the treatment processes.

All samples collected and analyzed must be needed by someone. Numbers from test results provide an accurate description or indication of the quantity of work completed or to be completed. Mathematical analysis of data is a means to estimate how well your test results can be repeated on a given sample (quality control) or how much a group of samples vary (another form of quality control). Presentation of data in tables, graphs, or charts makes the information more usable by illustrating trends, variations, and significant changes.

NOTE: Metric calculations are not included in this chapter because the procedures used to analyze and present data are the same for both the English and metric systems. Use the same procedures to analyze numbers, regardless of the system.

18.1 CAUSES OF VARIATIONS IN RESULTS

When you collect samples of wastewater or receiving water and measure their characteristics (for instance, temperature, pH, BOD), your results will be affected by several factors. Three principal factors that must be taken into account, no matter where the sample is taken—influent raw wastewater, treatment process influent or effluent, receiving waters—are:

1. Actual variations in the characteristics of the water or material being examined
2. Sampling procedures
3. Testing or analytical procedures

18.10 Water or Material Being Examined

The properties or characteristics of the wastewater or receiving water are what you are attempting to measure, such as temperature, pH, or BOD. These and many other water quality indicators vary continuously depending on what is being discharged into a wastewater collection system (sewerage system), the effectiveness of treatment processes, and the response of the receiving waters and their changing characteristics. Your objective is to describe the characteristics of the wastewater or receiving water being sampled in terms of average values and also to give an indication of variation or spread of results from the average values.

18.11 Sampling

Characteristics of a sample can vary if you do not always sample at the same location or during the same time of day. If you observe the flow of wastewater in a channel, you can see the differences in characteristics at various depths, differences between flow in the middle and edge of a pipe or channel, and also differences above or below a bend.

After a sample has been collected in a sampling jar or bottle, heavy material may settle to the bottom and the jar must be mixed before the sample is tested. Also, if the sample is not analyzed immediately, its characteristics can undergo chemical or biological changes unless the sample is treated and stored properly following collection.

18.12 Testing

The results from two identical samples can differ depending on the analyzing apparatus and the operator conducting the

[1] *Representative Sample.* A sample portion of material, water, or wastestream that is as nearly identical in content and consistency as possible to that in the larger body being sampled.

measurement. Fluctuations in voltage can cause changes in instrument readings, and different individuals may titrate to slightly different end points. Using reagents from different bottles, filter paper from different packages, or different pieces of equipment that were not calibrated identically or were not warmed up during the same time period can cause differences in test results. Variations in test results may be caused by omitting a step in the lab procedure, and interfering substances can cause testing errors.

Your objective is to reduce or eliminate sampling and testing errors as much as possible so you can obtain an accurate description of the water being sampled.

QUESTIONS

Write your answers in a notebook and then compare your answers with those on page 661.

18.1A What three major factors can cause variations in lab test results?

18.1B Why should most samples be tested immediately by the lab?

18.2 MANOMETER AND GAUGE READING

Manometers and gauges are installed in wastewater treatment plants to measure pressures and pressure differences. Both types of instruments should be calibrated and zeroed before they are used to ensure accurate results. Calibration and zeroing of any instrument means periodic checking of the instrument against a known standard to be sure the installed instrument reads properly. Manometers and gauges can be zeroed in by making sure the instruments read zero when no pressure is being applied to the manometer or gauge. If the reading is not zero, then the scales should be adjusted according to manufacturer's recommendations.

To read a manometer, note the scale reading opposite the *MENISCUS*.[2] This reading may have to be converted from inches of mercury to head in feet of water, pressure in psi, or flow in GPM, depending on the use of the manometer.

TYPICAL MANOMETER READINGS

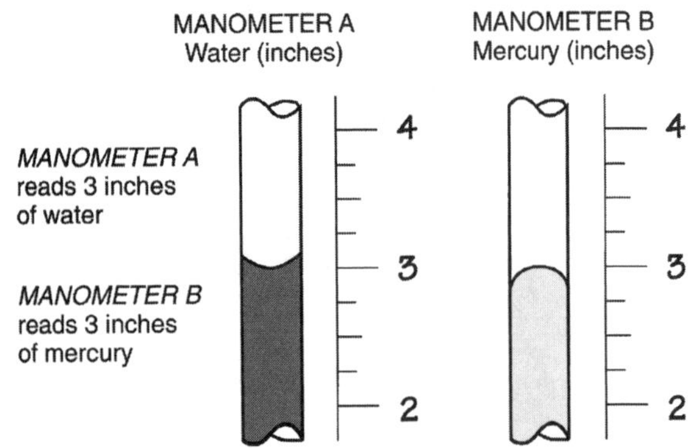

Gauge readings are read directly from a scale behind a gauge pointer and the units must be recorded. Sometimes a gauge will have two scales and care must be taken to be sure the proper scale reading and units are recorded. Gauge readings may have to be converted to more convenient numbers.

TYPICAL GAUGE READINGS

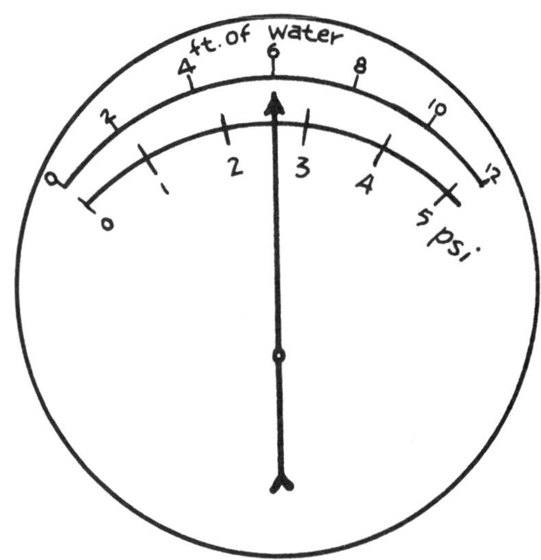

QUESTIONS

Write your answers in a notebook and then compare your answers with those on page 661.

18.2A Why must instruments be periodically calibrated and zeroed?

18.2B If the gauge shown above under TYPICAL GAUGE READINGS indicated a pressure of 2 psi, what would be the pressure head in feet of water?

18.3 CHART READING

Before data can be analyzed and presented, frequently it must be reduced to or tabulated in a usable form. Today, more and more, data are being recorded on a continuous basis on strip charts and circular charts or by computer programs.

For instance, flow data are sometimes recorded in depths of flow in inches or feet through a Parshall flume. Some recorders will also convert the depth to a flow rate, such as MGD. To compile or tabulate data from continuous charts, select an appropriate time interval, such as a few hours (Figure 18.1). Prepare a table (Table 18.1) with a column for (1) the time and (2) the value you read from the chart. A third column (3) may be necessary if you have to convert a depth of flow to a flow rate (MGD). Conversion charts from depth of flow to flow rate in MGD are provided by the manufacturer of a flowmeter.

---

[2] *Meniscus* (meh-NIS-cuss). The curved surface of a column of liquid (water, oil, mercury) in a small tube. When the liquid wets the sides of the container (as with water), the curve forms a valley. When the confining sides are not wetted (as with mercury), the curve forms a hill or upward bulge.

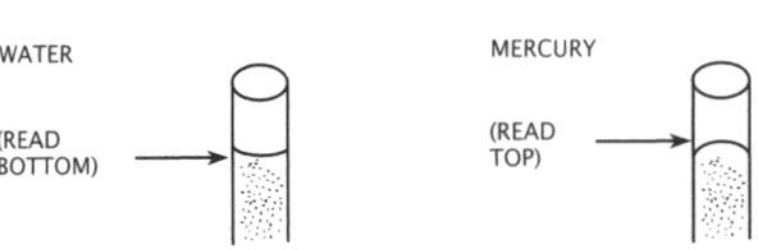

When reading manometers, gauges, and charts, care must be taken to be sure the correct number is read and properly recorded.

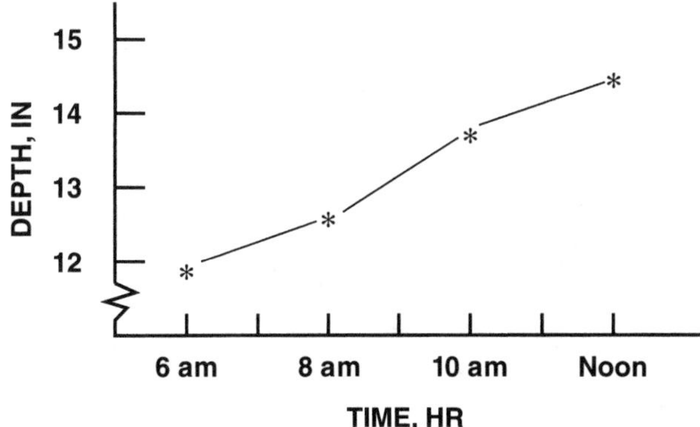

Fig. 18.1 Strip chart flow depth

TABLE 18.1 TABULATION OF DEPTHS AND FLOWS FROM THE STRIP CHART (FIGURE 18.1)

(1) Time	(2) Depth (in)[b]	(3) Flow[a] (MGD)[c]
6 am	12	0.61
8 am	12½	0.65
10 am	13½	0.74
12 noon	14½	0.84

[a] These figures would be obtained from the manufacturer's "conversion chart" for the particular flume or flowmeter.
[b] Multiply inches by 2.54 to obtain centimeters.
[c] Multiply MGD by 3.785 to obtain million liters per day.

18.4 AVERAGE OR ARITHMETIC MEAN

The word average in general refers to the "central tendency" of measurements. This is a method of grouping measurements and describing the results by using a single number. The central tendency may be described by any one of five methods, arithmetic mean (Section 18.4), range (Section 18.5), median and mode (Section 18.6), and geometric mean (Section 18.7).

When you collect representative samples from a plant influent and measure a particular water quality indicator, such as BOD, the results are not always the same. For example, you might measure the BOD of a trickling filter influent to determine the organic loading and find the BOD varying considerably during a six-day period. To calculate an expected daily organic loading, the average or arithmetic mean daily BOD must be calculated.

EXAMPLE 1:

The results of six BOD tests on a trickling filter influent from composite (proportional) samples collected at daily intervals during a six-day period indicated the BOD to be 150 mg/*L*, 200 mg/*L*, 250 mg/*L*, 200 mg/*L*, 100 mg/*L*, and 120 mg/*L*. What is the mean daily BOD?

PROCEDURE:

Add the six measurements and divide by six, the number of measurements.

$$\text{Mean BOD, mg}/L = \frac{\text{Sum of All Measurements, mg}/L}{\text{Number of Measurements}}$$

$$= \frac{1{,}020 \text{ mg}/L}{6}$$

$$= 170 \text{ mg}/L$$

DAY	*BOD*
1	150
2	200
3	250
4	200
5	100
6	120
Sum =	1,020

You have calculated the mean BOD by adding all BOD measurements and dividing by the number of measurements.

The mean value of any other characteristic is calculated the same way. For example, if you wanted to calculate a month's mean daily flow into a plant, you would add up the daily flows for the month and divide by the number of days in the month.

HINT: Frequently, plant flows are recorded on an integrator or *TOTALIZER.*[3] The flow during a particular time period can be determined by obtaining the difference between the totalizer readings at the beginning and end of the time period.

EXAMPLE 2:

At the beginning of a month, a plant totalizer reads 103,628,457 gallons, and 30 days later you record the totalizer value as 114,789,321 gallons. Calculate the mean daily flow for the month.

STEP 1:

Find the total monthly flow.

Reading at the end of time period	=	114,789,321 gal
Reading at start of time period	=	103,628,457 gal
Total flow during month	=	11,160,864 gal

STEP 2:

Calculate the mean daily flow, gal/day:

$$\text{Mean Daily Flow, gal/day} = \frac{\text{Sum of Flows, gal}}{\text{Number of Days (measurements)}}$$

$$= \frac{11{,}160{,}864 \text{ gal}}{30 \text{ Days}}$$

$$= 372{,}029 \text{ gal/day}$$

or

$$= \frac{372{,}029 \text{ gal/day}}{1{,}000{,}000 \text{ gal/MG}}$$

$$= 0.372 \text{ MGD}$$

NOTE: The mean daily flow for the month also could be calculated by adding the 30 daily flows during the month and dividing by 30. This approach can be used to check the results obtained using the difference in the totalizer readings as shown above.

[3] *Totalizer.* A device or meter that continuously measures and sums a process rate variable in cumulative fashion over a given time period. For example, total flows displayed in gallons per minute, million gallons per day, cubic feet per second, or some other unit of volume per time period. Also called an INTEGRATOR.

18.5 RANGE OF VALUES

You have seen how to evaluate lab results in terms of arithmetic mean values. This does not give any indication as to whether all of the data were close to the mean value or if there was a considerable spread or dispersion of data. A useful method of indicating the spread in results is the range. The range is obtained by subtracting the smallest measurement from the largest one.

Range = Largest Value – Smallest Value

PROCEDURE:

STEP 1: Rank data by arranging observations in ascending (increasing) or descending (decreasing) order, using the data from Example 1: 250, 200, 200, 150, 120, 100.

STEP 2: Subtract the smallest (100) from the largest (250).

Largest		250 mg/*L*
Smallest		–100 mg/*L*
Range of BOD, mg/*L*	=	150 mg/*L*

Try another example to review the calculations for the mean value and range, then you will be ready to study other ways of describing the dispersion of data and the idea of graphical presentation using this problem.

EXAMPLE 3:

The mean daily BOD for two weeks is given below. Calculate the mean two-week BOD and the range for these measurements.

DATA: 160, 155, 160, 160, 180, 165, 155, 170, 160, 165, 155, 150, 145, 160

$$\text{Mean BOD, mg/}L = \frac{\text{Sum of All Measurements, mg/}L}{\text{Number of Measurements}}$$

$$= \frac{2{,}240 \text{ mg/}L}{14}$$

= 160 mg/*L* for Two Weeks

160 mg/*L*
155
160
160
180
165
155
170
160
165
155
150
145
160
2,240 mg/*L*

Range of BOD, mg/*L* = Largest BOD, mg/*L* – Smallest BOD, mg/*L*

= 180 mg/*L* – 145 mg/*L*

= 35 mg/*L* for Two Weeks

QUESTIONS

Write your answers in a notebook and then compare your answers with those on page 661.

Mixed liquor samples were collected at the beginning, middle, and end of an aeration tank and the solids concentrations were 2,138 mg/*L*, 1,863 mg/*L*, and 1,921 mg/*L*.

18.4A Calculate the mean mixed liquor concentration in mg/*L*.

18.5A What is the range of these measurements?

18.6 MEDIAN AND MODE

Sometimes the mean value and range calculations are not the best way to describe or analyze data. For example, frequently when running multiple-tube coliform bacteria tests you will obtain some extremely high MPNs (most probable number of coliform group bacteria), especially after a rain, equipment failure, or chlorine dosage mishap.

EXAMPLE 4:

DATA: MPN/100 m*L* = 240, 220, 240, 230, 240, 7,200, 260, 250, 270, 300, 250. Calculate average MPN.

PROCEDURE:

Determine sum of measurements.

$$\text{Mean MPN/100 m}L = \frac{\text{Sum of Measurements, MPN/100 m}L}{\text{Number of Measurements}}$$

$$= \frac{9{,}700}{11}$$

= 882

240
220
240
230
240
7,200
260
250
270
300
250
Sum = 9,700

Mean MPN = 882 Coliform Bacteria/100 m*L*. Note that this value is greater than all of our measurements except the largest one. For this reason, multiple-tube coliform results are sometimes reported as a median value.

The median is defined as the middle measurement when the measurements are ranked in order of magnitude (size).

PROCEDURE:

Rank data in ascending or descending order.

Measurement:	220	230	240	240	240	250	250	260	270	300	7,200
Rank:	1	2	3	4	5	6 ↑ Median	7	8	9	10	11

Measurement 6 is our middle measurement. Therefore, the median MPN/100 m*L* = 250, which better describes the usual value of the measurements.

If you had only 10 measurements (eliminate #11 of 7,200), the median would fall between measurements 5 and 6 (240 and 250) and would be 245.

Another useful value is the mode. The mode is the measurement that occurs most frequently. In our example, the measurement 240 occurs three times, which is more than any other. Therefore: mode MPN/100 m*L* = 240.

An examination of the data in Example 4 indicates that the median and mode do a better job of describing or predicting the MPN value we would expect than the mean calculation. For this reason these terms are sometimes used to report data.

QUESTION

Write your answer in a notebook and then compare your answer with the one on page 661.

18.6A The results of the SVI (Sludge Volume Index) test for an activated sludge plant for one week were as follows: 120, 115, 120, 120, 125, 110, 115. What are the median and mode values for the SVI data?

18.7 GEOMETRIC MEAN

18.70 Log Probability Paper

There are other ways of reporting the results of coliform tests in addition to those mentioned above. The geometric mean is sometimes used because all measurements are used in the calculations, but an extreme value has a lesser influence on the result. The easiest way to find the geometric mean is to plot the results on log probability paper and read the geometric mean on the paper.

To plot data on probability paper, the probability or plotting point of each measurement must be determined. The plotting point is calculated from the following formula:

$$P = \frac{m}{n+1} \times 100\%$$

where:

P is the probability (%) the measurement will not be exceeded

n is the number or sum of measurements

m is the rank when the measurements are arranged in ascending or increasing order

Calculate the probability for each measurement used in Example 4. Use n = 11 because we have 11 measurements.

For 220 MPN/100 m*L,* rank m is 1.

$$P = \frac{m}{n+1} \times 100\%$$

$$= \frac{1}{11+1} \times 100\%$$

$$= 8.3\%$$

For 230 MPN/100 m*L,* rank m is 2.

$$P = \frac{2}{11+1} \times 100\%$$

$$= 16.7\%$$

Repeat this procedure for all 11 measurements.

For our example:

Rank m	Measurement MPN/100 m*L*	Probability %
1	220	8.3
2	230	16.7
3	240	25.0
4	240	33.3
5	240	41.7
6	250	50.0
7	250	58.3
8	260	66.7
9	270	75.0
10	300	83.3
11	7,200	91.7

Plot the data as shown on Figure 18.2. To plot the data, obtain a sheet of log or geometric probability paper. On the left side, determine the MPN per 100 m*L* scale (100, 1,000, 10,000). Start at the left with an MPN of 220. Move horizontally across to the right until you meet the vertical line with a probability of 8.3 percent. Make an "*" on the paper at this point. Take the next MPN value of 230 and move horizontally across to the right until you meet the vertical line with a probability of 16.7 percent. Make an "*" on the paper at this point. Repeat this procedure until all of the MPN values have been plotted.

To estimate the geometric mean from the data plotted on Figure 18.2:

1. Draw a straight line of best fit through the data points (all of the "*").
2. Draw a vertical line down the 50 percent line to where it intersects with the line of best fit.
3. From the intersection of the 50 percent line and the line of best fit, draw a horizontal line to the scale representing the measurements.
4. Read the Geometric Mean MPN = 265/100 m*L*. This value essentially ignores the one high value of 7,200 per 100 m*L* and will be lower than any computation that includes the 7,200 value.

18.71 Electronic Calculators

If you have an electronic calculator that can calculate the powers of numbers, you can use the calculator to calculate the geometric mean. Because the exact procedure will vary with different calculators, check the instruction book provided with your calculator to find the procedure for your calculator. To determine if your calculator can determine the geometric mean, try the following example with the help of your calculator's user's manual.

1. Determine 3^4. The answer is 81.
2. Determine $81^{1/4}$ or $81^{0.25}$. The answer is 3.

If your calculator can find the powers of numbers less than one (0.25), then the calculator should be able to determine the geometric mean.

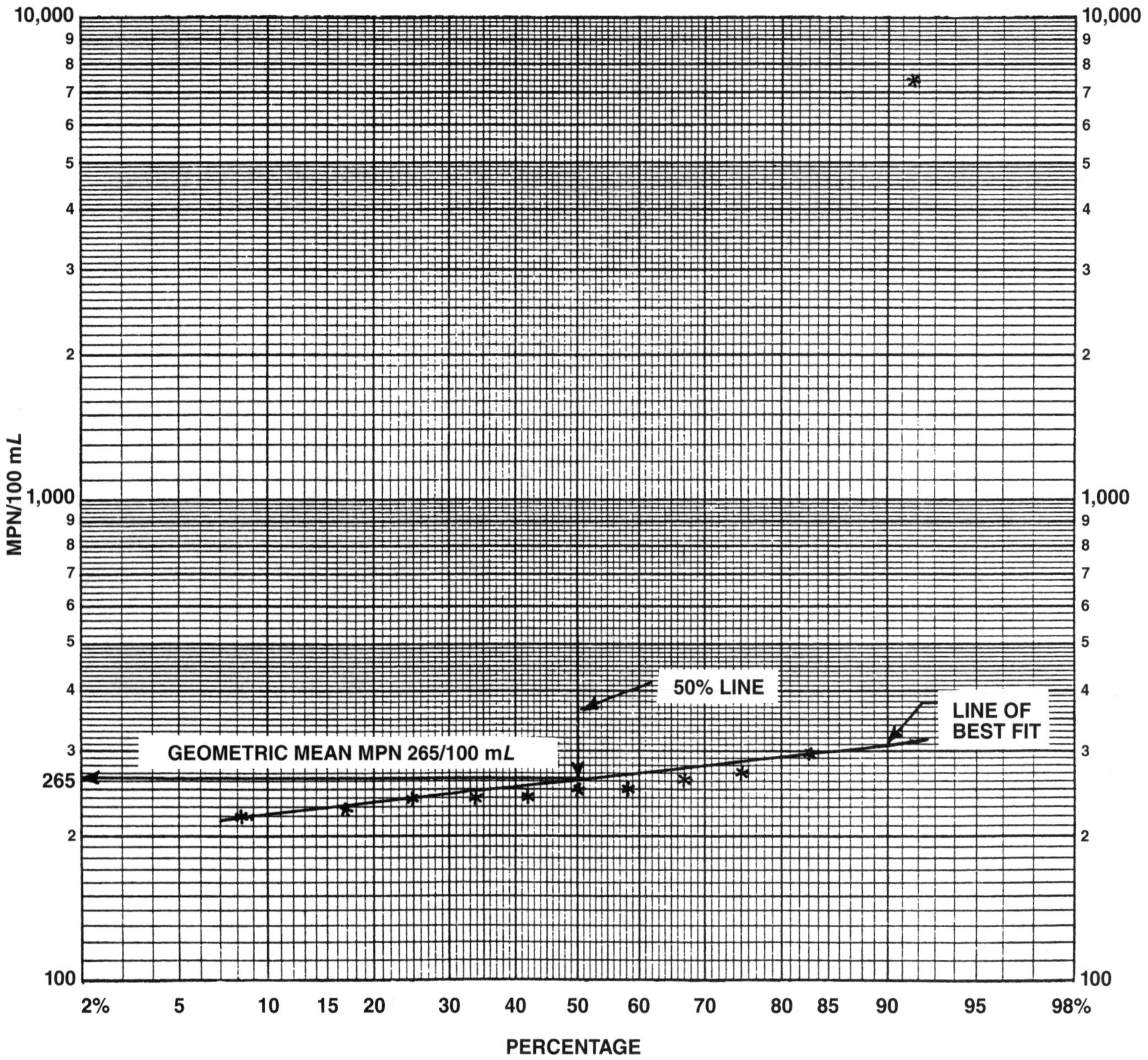

Fig. 18.2 Determination of geometric mean from log probability paper

EXAMPLE 5:

Calculate the geometric mean of the data used in the previous problem.

Geometric Mean, MPN $= (X_1 \times X_2 \times \ldots X_n)^{1/n}$

$= (220 \times 230 \times 240 \times 240 \times 240 \times 250 \times 250 \times 260 \times 270 \times 300 \times 7{,}200)^{1/11}$

$= (6.629\ 108\ 429 \times 10^{27})^{0.0909}$

$=$ 338 per 100 m*L*

1. Multiply all values together. Multiply 220 times 230 times 240 times 240 times 240 times 250 and so on through the last number, 7,200. The answer you get should be 6.629 108 429 × 10^{27}.

2. Determine the value for 1/n where n is the number of measurements. Divide one by 11 and get 0.0909.

3. Calculate the geometric mean. Using your pocket calculator, find the 0.0909 power of 6.629 108 429 × 10^{27} or $(6.629\ 108\ 429 \times 10^{27})^{0.0909}$ = 338 or MPN = 338 per 100 m*L*.

18.72 Geometric Mean Table

If your calculator cannot determine the 0.0909 power of a number or similar numbers, you can use Table 18.2, the Geometric Mean Table. Use the same procedure you used in Section 18.71, "Electronic Calculators."

1. Multiply all values together. Multiply 220 times 230 times 240 times 240 times 240 times 250 and so on through the last number, 7,200. Obtain $6.629\ 108\ 429 \times 10^{27}$.
2. After you have the number in scientific notation (6.629×10^{27}), refer to Table 18.2 (pages 642 and 643). Use the column that corresponds to the number of measurements or "number of samples used for determination." In our example we used 11 measurements, so use column 11.
3. Look down column 11 until you find the number that is closest to the number calculated. When you find the number, look to the extreme left column to find the geometric mean value closest to the number calculated.

(1) Column 11		(2) Left Column
Smaller Value	$= 4.272 \times 10^{27}$	Geometric Mean = 325
Calculated Value	$= 6.629 \times 10^{27}$	
Larger Value	$= 9.654 \times 10^{27}$	Geometric Mean = 350

4. Since the calculated value is closer to 4.272×10^{27}, report the geometric mean as 325+. If the calculated value had been closer to 9.654×10^{27}, report the geometric mean as 350–.
5. This procedure can be used to determine the geometric mean for the Most Probable Number (MPN) of either total or fecal coliforms per 100 m*L*.

18.73 Example Problem

Calculate the geometric mean for the following fecal coliform test results using the membrane filter method for a one-month period.

Test Number	1	2	3	4	5	6	7	8
MPN/100 m*L*	70	225	0	90	TNTC*	<20	325	148

* TNTC means Too Numerous To Count. Whenever TNTC appears, it cannot be used in the calculations. Report this event on your monthly report form in the remarks area and give the apparent reason for the high values.

1. Arrange the MPN values you want to calculate the geometric mean for in increasing order of magnitude (size).

   MPN Values 0 <20 70 90 148 225 325 TNTC

2. Convert any MPN values that are zero to the number 1.

   MPN Values 1 <20 70 90 148 225 325 TNTC

3. If the MPN value is reported as less than (<) or greater than (>), remove the < or > signs and use the values.

   MPN Values 1 20 70 90 148 225 325 TNTC

4. Drop all MPN values that are TNTC. See * above.

   MPN Values 1 20 70 90 148 225 325

5. Multiply all of the values together to get a number.

   $1 \times 20 \times 70 \times 90 \times 148 \times 225 \times 325 = 1\ 363\ 635\ 000\ 000$

6. Convert number from Step 5 to scientific notation.
   a. Place decimal to right of first number.

      1.363 635 000 000

   b. Count the number of spaces or numbers to the right of the decimal.

      1.363 635 000 000

      Count 1, 2, 3, 4, 5, 6, 7, 8, 9, 10, 11, 12. There are 12 spaces or numbers to the right of the decimal.

   c. Write the number as the first four numbers times 10 to the power that equals the number of spaces or numbers to the right of the decimal.

      1.363×10^{12}

7. After you have the number in scientific notation (1.363×10^{12}), refer to Table 18.2. Use the column that corresponds to the number of measurements or "number of samples used for determination." In this example we used seven measurements, so use column 7. (While there were eight measurements, measurement number 5 was too numerous to count (TNTC), so it was not included in the calculations.)
8. Look down column 7 until you find the number that is closest to the number calculated. When you find the number, look to the extreme left column to find the geometric mean value closest to the number calculated.

(1) Column 7		(2) Left Column
Smaller Value	$= 7.812 \times 10^{11}$	Geometric Mean = 50
Calculated Value	$= 1.363 \times 10^{12}$	
Larger Value	$= 2.799 \times 10^{12}$	Geometric Mean = 60

9. Since the calculated value is closer to 7.812×10^{11}, report the geometric mean as 50+. If the calculated value had been closer to 2.799×10^{12}, report the geometric mean as 60–.

If you want to solve this problem by the use of an electronic calculator, go back to step 5.

5. Multiply all of the values together to get a number.

   $1 \times 20 \times 70 \times 90 \times 148 \times 225 \times 325 = 1\ 363\ 635\ 000\ 000$

6. Determine the value for 1/n where n is the number of MPN values. Divide 1 by 7 and get 0.1429.
7. Calculate the geometric mean. Using your pocket calculator, find the 0.1429 power of $1.363\ 635 \times 10^{12}$ or

   $(1.363\ 635 \times 10^{12})^{0.1429} = 54$ or MPN = 54 per 100 m*L*

   or

$$\text{Geometric Mean, MPN} = (1 \times 20 \times 70 \times 90 \times 148 \times 225 \times 325)^{1/7}$$

$$= (1.363\ 635 \times 10^{12})^{0.1429}$$

$$= 54 \text{ per } 100 \text{ m}L$$

ACKNOWLEDGMENT

Portions of Sections 18.72 and 18.73 were prepared from a paper, "Simplified Procedure for Calculating a Geometric Mean for Fecal Coliform Values," by Roger Karn. The paper appeared in the March 1977 issue of *DEEDS & DATA* and is reproduced with the permission of the Water Pollution Control Federation.

TABLE 18.2 GEOMETRIC MEAN TABLE
(FOR SELECTED FECAL COLIFORM VALUES FROM 10 TO 1,000)

NUMBER OF SAMPLES USED FOR DETERMINATION

GEOMETRIC MEAN [TOTAL OR FECAL COLIFORMS (MPN/100 mL)]	1	2	3	4	5	6	7
10	10	100	1 000	10 000	100 000	1.000×10^{6}	1.000×10^{7}
20	20	400	8 000	160 000	3.200×10^{6}	6.400×10^{7}	1.280×10^{9}
30	30	900	27 000	810 000	2.430×10^{7}	7.290×10^{8}	2.187×10^{10}
40	40	1 600	64 000	2.560×10^{6}	1.024×10^{8}	4.096×10^{9}	1.638×10^{11}
50	50	2 500	125 000	6.250×10^{6}	3.125×10^{8}	1.562×10^{10}	7.812×10^{11}
60	60	3 600	216 000	1.296×10^{7}	7.776×10^{8}	4.665×10^{10}	2.799×10^{12}
70	70	4 900	343 000	2.401×10^{7}	1.680×10^{9}	1.176×10^{11}	8.235×10^{12}
80	80	6 400	512 000	4.096×10^{7}	3.276×10^{9}	2.621×10^{11}	2.097×10^{13}
90	90	8 100	729 000	6.561×10^{7}	5.904×10^{9}	5.314×10^{11}	4.782×10^{13}
100	100	10 000	1.000×10^{6}	1.000×10^{8}	1.000×10^{10}	1.000×10^{12}	1.000×10^{14}
110	110	12 100	1.331×10^{6}	1.464×10^{8}	1.610×10^{10}	1.771×10^{12}	1.948×10^{14}
120	120	14 400	1.728×10^{6}	2.073×10^{8}	2.488×10^{10}	2.985×10^{12}	3.583×10^{14}
130	130	16 900	2.197×10^{6}	2.856×10^{8}	3.712×10^{10}	4.826×10^{12}	6.274×10^{14}
140	140	19 600	2.744×10^{6}	3.841×10^{8}	5.378×10^{10}	7.529×10^{12}	1.054×10^{15}
150	150	22 500	3.375×10^{6}	5.062×10^{8}	7.593×10^{10}	1.139×10^{13}	1.708×10^{15}
160	160	25 600	4.096×10^{6}	6.553×10^{8}	1.048×10^{11}	1.677×10^{13}	2.684×10^{15}
170	170	28 900	4.913×10^{6}	8.352×10^{8}	1.419×10^{11}	2.413×10^{13}	4.103×10^{15}
180	180	32 400	5.832×10^{6}	1.049×10^{9}	1.889×10^{11}	3.401×10^{13}	6.122×10^{15}
190	190	36 100	6.859×10^{6}	1.303×10^{9}	2.476×10^{11}	4.704×10^{13}	8.938×10^{15}
200	200	40 000	8.000×10^{6}	1.600×10^{9}	3.200×10^{11}	6.400×10^{13}	1.280×10^{16}
225	225	50 625	1.139×10^{7}	2.562×10^{9}	5.765×10^{11}	1.297×10^{14}	2.919×10^{16}
250	250	62 500	1.562×10^{7}	3.906×10^{9}	9.765×10^{11}	2.441×10^{14}	6.103×10^{16}
275	275	75 625	2.079×10^{7}	5.719×10^{9}	1.572×10^{12}	4.325×10^{14}	1.189×10^{17}
300	300	90 000	2.700×10^{7}	8.100×10^{9}	2.430×10^{12}	7.290×10^{14}	2.187×10^{17}
325	325	105 625	3.432×10^{7}	1.115×10^{10}	3.625×10^{12}	1.178×10^{15}	3.829×10^{17}
350	350	122 500	4.287×10^{7}	1.500×10^{10}	5.252×10^{12}	1.838×10^{15}	6.433×10^{17}
375	375	140 625	5.273×10^{7}	1.977×10^{10}	7.415×10^{12}	2.780×10^{15}	1.042×10^{18}
400	400	160 000	6.400×10^{7}	2.560×10^{10}	1.024×10^{13}	4.096×10^{15}	1.638×10^{18}
450	450	202 500	9.112×10^{7}	4.100×10^{10}	1.845×10^{13}	8.303×10^{15}	3.736×10^{18}
500	500	250 000	1.250×10^{8}	6.250×10^{10}	3.125×10^{13}	1.562×10^{16}	7.812×10^{18}
600	600	360 000	2.160×10^{8}	1.296×10^{11}	7.776×10^{13}	4.665×10^{16}	2.799×10^{19}
750	750	562 500	4.218×10^{8}	3.164×10^{11}	2.373×10^{14}	1.779×10^{17}	1.334×10^{20}
1 000	1 000	1.000×10^{6}	1.000×10^{9}	1.000×10^{12}	1.000×10^{15}	1.000×10^{18}	1.000×10^{21}

TABLE 18.2 GEOMETRIC MEAN TABLE *(continued)*
(FOR SELECTED FECAL COLIFORM VALUES FROM 10 TO 1,000)

GEOMETRIC MEAN [TOTAL OR FECAL COLIFORMS (MPN/100 *mL*)]	NUMBER OF SAMPLES USED FOR DETERMINATION						
	8	9	10	11	12	13	14
10	1.000×10^{8}	1.000×10^{9}	1.000×10^{10}	1.000×10^{11}	1.000×10^{12}	1.000×10^{13}	1.000×10^{14}
20	2.560×10^{10}	5.120×10^{11}	1.024×10^{13}	2.048×10^{14}	4.096×10^{15}	8.192×10^{16}	1.638×10^{18}
30	6.561×10^{11}	1.968×10^{13}	5.904×10^{14}	1.771×10^{16}	5.314×10^{17}	1.594×10^{19}	4.782×10^{20}
40	6.553×10^{12}	2.621×10^{14}	1.048×10^{16}	4.194×10^{17}	1.677×10^{19}	6.710×10^{20}	2.684×10^{22}
50	3.906×10^{13}	1.953×10^{15}	9.765×10^{16}	4.882×10^{18}	2.441×10^{20}	1.220×10^{22}	6.103×10^{23}
60	1.679×10^{14}	1.007×10^{16}	6.046×10^{17}	3.627×10^{19}	2.176×10^{21}	1.306×10^{23}	7.836×10^{24}
70	5.764×10^{14}	4.035×10^{16}	2.824×10^{18}	1.977×10^{20}	1.384×10^{22}	9.688×10^{23}	6.782×10^{25}
80	1.677×10^{15}	1.342×10^{17}	1.073×10^{19}	8.589×10^{20}	6.871×10^{22}	5.497×10^{24}	4.398×10^{26}
90	4.304×10^{15}	3.874×10^{17}	3.486×10^{19}	3.138×10^{21}	2.824×10^{23}	2.541×10^{25}	2.287×10^{27}
100	1.000×10^{16}	1.000×10^{18}	1.000×10^{20}	1.000×10^{22}	1.000×10^{24}	1.000×10^{26}	1.000×10^{28}
110	2.143×10^{16}	2.357×10^{18}	2.593×10^{20}	2.853×10^{22}	3.138×10^{24}	3.452×10^{26}	3.797×10^{28}
120	4.299×10^{16}	5.159×10^{18}	6.191×10^{20}	7.430×10^{22}	8.916×10^{24}	1.069×10^{27}	1.283×10^{29}
130	8.157×10^{16}	1.060×10^{19}	1.378×10^{21}	1.792×10^{23}	2.329×10^{25}	3.028×10^{27}	3.937×10^{29}
140	1.475×10^{17}	2.066×10^{19}	2.892×10^{21}	4.049×10^{23}	5.669×10^{25}	7.937×10^{27}	1.111×10^{30}
150	2.562×10^{17}	3.844×10^{19}	5.766×10^{21}	8.649×10^{23}	1.297×10^{26}	1.946×10^{28}	2.919×10^{30}
160	4.294×10^{17}	6.871×10^{19}	1.099×10^{22}	1.759×10^{24}	2.814×10^{26}	4.503×10^{28}	7.205×10^{30}
170	6.975×10^{17}	1.185×10^{20}	2.015×10^{22}	3.427×10^{24}	5.826×10^{26}	9.904×10^{28}	1.683×10^{31}
180	1.101×10^{18}	1.983×10^{20}	3.570×10^{22}	6.426×10^{24}	1.156×10^{27}	2.082×10^{29}	3.748×10^{31}
190	1.698×10^{18}	3.226×10^{20}	6.131×10^{22}	1.164×10^{25}	2.213×10^{27}	4.205×10^{29}	7.990×10^{31}
200	2.560×10^{18}	5.120×10^{20}	1.024×10^{23}	2.048×10^{25}	4.096×10^{27}	8.192×10^{29}	1.638×10^{32}
225	6.568×10^{18}	1.477×10^{21}	3.325×10^{23}	7.481×10^{25}	1.683×10^{28}	3.787×10^{30}	8.522×10^{32}
250	1.525×10^{19}	3.814×10^{21}	9.536×10^{23}	2.384×10^{26}	5.960×10^{28}	1.490×10^{31}	3.725×10^{33}
275	3.270×10^{19}	8.994×10^{21}	2.473×10^{24}	6.802×10^{26}	1.870×10^{29}	5.144×10^{31}	1.414×10^{34}
300	6.561×10^{19}	1.968×10^{22}	5.904×10^{24}	1.771×10^{27}	5.314×10^{29}	1.594×10^{32}	4.782×10^{34}
325	1.244×10^{20}	4.045×10^{22}	1.314×10^{25}	4.272×10^{27}	1.388×10^{30}	4.513×10^{32}	1.466×10^{35}
350	2.251×10^{20}	7.881×10^{22}	2.758×10^{25}	9.654×10^{27}	3.379×10^{30}	1.182×10^{33}	4.139×10^{35}
375	3.910×10^{20}	1.466×10^{23}	5.499×10^{25}	2.062×10^{28}	7.733×10^{30}	2.900×10^{33}	1.087×10^{36}
400	6.553×10^{20}	2.621×10^{23}	1.048×10^{26}	4.194×10^{28}	1.677×10^{31}	6.710×10^{33}	2.684×10^{36}
450	1.681×10^{21}	7.566×10^{23}	3.405×10^{26}	1.532×10^{29}	6.895×10^{31}	3.102×10^{34}	1.396×10^{37}
500	3.906×10^{21}	1.953×10^{24}	9.765×10^{26}	4.882×10^{29}	2.441×10^{32}	1.220×10^{35}	6.103×10^{37}
600	1.679×10^{22}	1.007×10^{25}	6.046×10^{27}	3.627×10^{30}	2.176×10^{33}	1.306×10^{36}	7.836×10^{38}
750	1.001×10^{23}	7.508×10^{25}	5.631×10^{28}	4.223×10^{31}	3.167×10^{34}	2.375×10^{37}	1.781×10^{40}
1 000	1.000×10^{24}	1.000×10^{27}	1.000×10^{30}	1.000×10^{33}	1.000×10^{36}	1.000×10^{39}	1.000×10^{42}

QUESTION

Write your answer in a notebook and then compare your answer with the one on page 661.

18.7A Determine the geometric mean for the following effluent fecal coliform test results:

Test Number	1	2	3	4	5	6	7	8	9
MPN/100 m*L*	43	12	8	14	63	11	23	17	49

18.8 MOVING AVERAGES

In Volume I, Chapter 6, "Trickling Filters," Section 6.4, "Operational Strategy," Figure 6.7 showed the use of moving averages. Moving averages are an effective approach to revealing deteriorating trends in treatment processes. Figure 18.3 shows plots of the effluent BOD from a trickling filter plant where both the raw data and the moving average were plotted. Trends may be difficult to detect when raw data are plotted because of the fluctuation of the data. Plots of moving averages tend to smooth out the data and reveal trends.

Moving averages are commonly seven-day moving averages, which allow each day of the week to be included in the average. At the example trickling filter plant, effluent BODs were not collected on weekends. Therefore BOD values were available only for Monday through Friday. Table 18.3 contains the raw effluent BOD data and the calculated BOD moving average values.

To calculate the moving average, simply find the mean for the days being considered.

1.

Date	**Day**	**Effluent BOD, mg/*L***
3	M	5
4	T	9
5	W	5
6	T	6
7	F	6
		Sum = 31

$$\text{Mean BOD, mg}/L = \frac{\text{Sum of All Measurements, mg}/L}{\text{Number of Measurements}}$$

$$= \frac{31\text{ mg}/L}{5}$$

$= 6.2$ mg/*L*, which is the moving average for Friday, the 7th

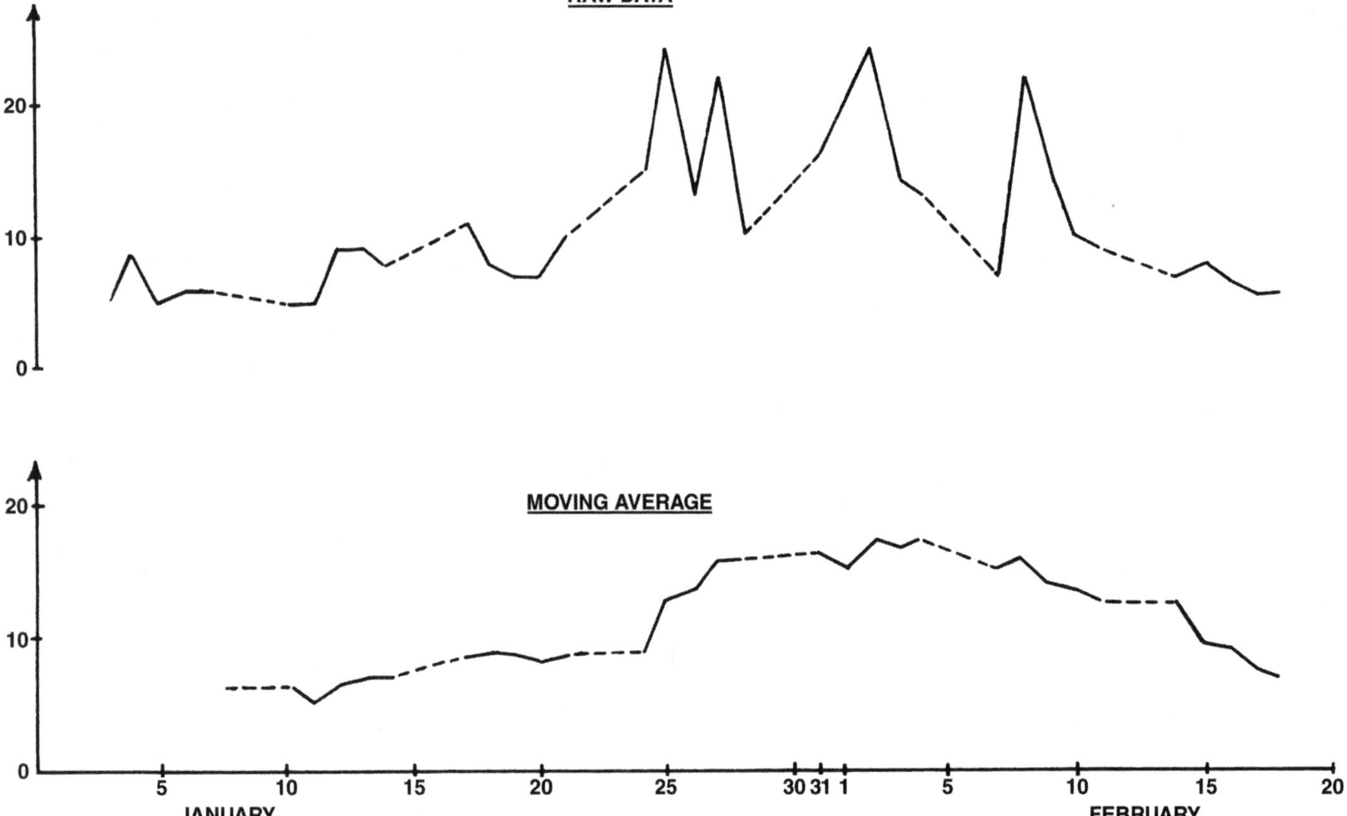

Fig. 18.3 Plots of raw data and moving average for a trickling filter plant effluent BOD, mg/L

TABLE 18.3 TRICKLING FILTER PLANT EFFLUENT BOD MOVING AVERAGE

MONTH	January			February			
Date	Day	Effluent BOD, mg/*L*	Moving Average	Date	Day	Effluent BOD, mg/*L*	Moving Average
1	S			1	T	20	15.4
2	S			2	W	24	17.6
3	M	5		3	T	14	16.8
4	T	9		4	F	13	17.4
5	W	5		5	S		
6	T	6		6	S		
7	F	6	6.2	7	M	7	15.6
8	S			8	T	22	16.0
9	S			9	W	14	14.0
10	M	5	6.2	10	T	10	13.2
11	T	5	5.4	11	F	9	12.4
12	W	9	6.2	12	S		
13	T	9	6.8	13	S		
14	F	8	7.2	14	M	7	12.4
15	S			15	T	8	9.6
16	S			16	W	7	8.2
17	M	11	8.4	17	T	6	7.4
18	T	8	9.0	18	F	6	6.8
19	W	7	8.6	19	S		
20	T	7	8.2	20	S		
21	F	10	8.6	21	M		
22	S			22	T		
23	S			23	W		
24	M	15	9.4	24	T		
25	T	24	12.6	25	F		
26	W	13	13.8	26	S		
27	T	18	16.0	27	S		
28	F	10	16.0	28	M		
29	S						
30	S						
31	M	16	16.2				

2. Repeat the procedure by removing the oldest measurement (Monday the 3rd) and adding the newest measurement (Monday the 10th).

Date	Day	Effluent BOD, mg/*L*
4	T	9
5	W	5
6	T	6
7	F	6
10	M	5
		Sum = 31

Mean BOD, mg/*L* = 6.2 mg/*L*, which is the moving average for Monday, the 10th

Date	Day	Effluent BOD, mg/*L*
5	W	5
6	T	6
7	F	6
10	M	5
11	T	5
		Sum = 27

Mean BOD, mg/*L* = 5.4 mg/*L*, which is the moving average for Wednesday, the 11th

If you have an electronic calculator, you can subtract the oldest measurement from the sum and add the newest measurement. The best procedures depend on the type of calculator.

QUESTION

Write your answer in a notebook and then compare your answer with the one on page 661.

18.8A Calculate the seven-day moving average for the effluent BOD from an activated sludge plant for the second week.

Week 1

Date	7	8	9	10	11	12	13
Day	S	M	T	W	T	F	S
BOD	25	23	38	41	32	35	37

Week 2

Date	14	15	16	17	18	19	20
Day	S	M	T	W	T	F	S
BOD	29	23	31	24	17	19	24

END OF LESSON 1 OF 2 LESSONS on ANALYSIS AND PRESENTATION OF DATA

Please answer the discussion and review questions next.

DISCUSSION AND REVIEW QUESTIONS

Chapter 18. ANALYSIS AND PRESENTATION OF DATA

(Lesson 1 of 2 Lessons)

At the end of each lesson in this chapter, you will find some discussion and review questions. The purpose of these questions is to indicate to you how well you understand the material in the lesson. Write the answers to these questions in your notebook.

1. Collection of data without ____________, ____________, and ____________ of results is a waste of time and money.
2. Whether samples are collected, analyzed, interpreted, and used by the same person or a different person performs each task, every job is equally important if the application of results is to be effective. True or False?
3. What are the three principal factors that can cause variations in test results?
4. How could errors occur when reading manometers, charts, and gauges?

The solids concentrations of sludge withdrawn from a primary clarifier during the past seven days are given below:

Day:	1	2	3	4	5	6	7
Solids, %:	6.0	6.5	6.0	5.0	6.5	7.5	8.0

5. What is the arithmetic mean solids concentration?
6. What is the range of solids concentration?
7. What is the median solids concentration?
8. What is the mode solids concentration?
9. Determine the geometric mean for the following effluent fecal coliform test results:

Week	1	2	3	4
MPN/100 m*L*	21	63	27	46

CHAPTER 18. ANALYSIS AND PRESENTATION OF DATA

(Lesson 2 of 2 Lessons)

18.9 GRAPHS AND CHARTS

18.90 Bar Graphs

Sometimes results can be illustrated by graphs to show the characteristics (average value and dispersion) of data. Many people, especially supervisors, can easily interpret data presented graphically and appreciate this approach.

EXAMPLE 6:

BOD tests for a two-week period provide the following measurements, mg/*L*: 160, 155, 160, 160, 180, 165, 155, 170, 160, 165, 155, 150, 145, 160.

PROCEDURE:

1. Before actually plotting out a bar graph, you will need to arrange the test results or other data in five to ten groups of figures. Make a list of the test results in descending or decreasing order. Next, estimate the spread between the highest and lowest values in the main body of the list. (If you find one or two figures either a great deal higher or lower than the rest, you may ignore them for now.) If your figures go from 180 to 145, for example, you have a difference of 35 between the high and low values. You want to end up with five to ten equal groupings so you might decide to make each group five units wide: 35 divided by 5 gives you seven groups. If your spread of data were 250, you might choose to make the groups 25 units wide yielding ten groups. Each group or class in the first example will be five units (mg/*L*) wide. This is known as the width of the class interval.

Once you have determined the width of the class interval, you need to determine the range of the interval. That is, exactly where each class or interval begins and ends. In order to establish the class interval, however, you must keep another factor in mind—class middlepoint.

2. Class midpoint is the number that falls exactly in the middle of a class interval. The midpoint of a class 182.5 to 177.5 would be 180. To make graphing easier, it is helpful to adjust the class intervals so that the midpoints are always numbers that are easy to work with such as: 2, 4, 6, 8, 10; or 5, 10, 15, 20, 25; or 10, 20, 30, 40, 50; or 25, 50, 75, 100, 125.

Take another look at your highest reading (180 in this example). This is a fairly easy number to work with when the class is five units wide. The next midpoint would be 175, 170, and so on. If your highest number is 178, however, it would be more awkward to graph as a midpoint so you could adjust the midpoint to 180.

3. You have now determined that the first class midpoint is 180 and that the width of the class interval is five (mg/*L*). Add one-half of the interval (5 divided by 2 = 2.5) to the first midpoint (180); the result is 182.5. This is the starting point of the first class interval. Looking at column (a), you will see that by continuing to subtract five from this starting point you are able to create equal class intervals. Continue subtracting until you pass your lowest test result.

4. Determine frequency. Count the number of test measurements that fall within each class interval. This is usually done by systematically going through your list of measurements, placing a check or 1 (column (c) below) opposite the appropriate class midpoint or class interval, and then adding up the checks or 1's to obtain the frequency (d).

5. Plot the results as shown in the following bar graph.

Class Interval (a)	Class Midpoint (b)	Number of Measurements (c)	Frequency (d)
182.5 to 177.5	180	1	1
177.5 to 172.5	175		0
172.5 to 167.5	170	1	1
167.5 to 162.5	165	11	2
162.5 to 157.5	160	~~1111~~	5
157.5 to 152.5	155	111	3
152.5 to 147.5	150	1	1
147.5 to 142.5	145	1	1

BAR GRAPH SHOWING DISTRIBUTION OF BOD

MEAN BOD 160 mg/*L*
RANGE OF BOD 35 mg/*L*

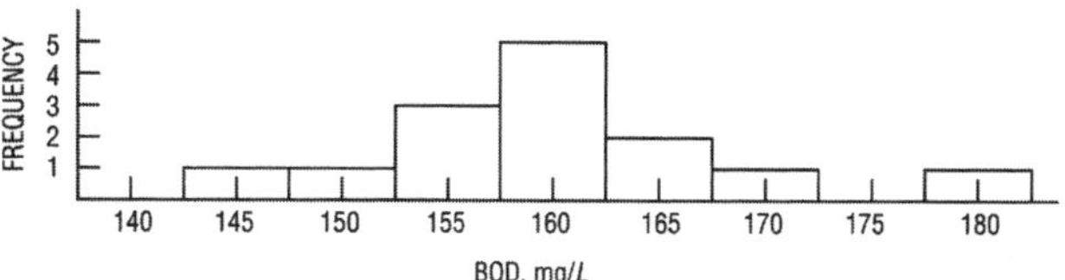

Sometimes the plotting points on a bar graph are connected to form a smooth curve. The resulting curve describes either a normal or a skewed distribution of the data, depending on the shape of the curve. If the distribution is normal, the mean, median, and mode values will be approximately the same. In skewed distributions the mean, median, and mode are different (Figure 18.4).

QUESTION

Write your answer in a notebook and then compare your answer with the one on page 662.

18.9A The results of the SVI (Sludge Volume Index) test for an activated sludge plant for one week were as follows: 120, 115, 120, 120, 125, 110, 115.

a. Calculate the arithmetic mean SVI.

b. What is the range?

c. Draw a bar graph showing the distribution (spread) of SVI.

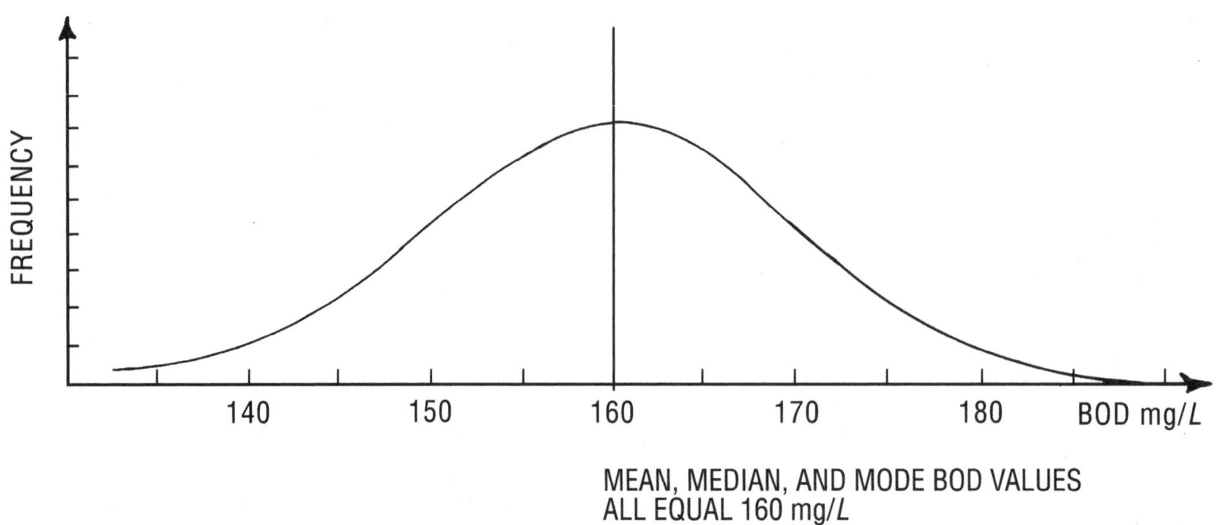

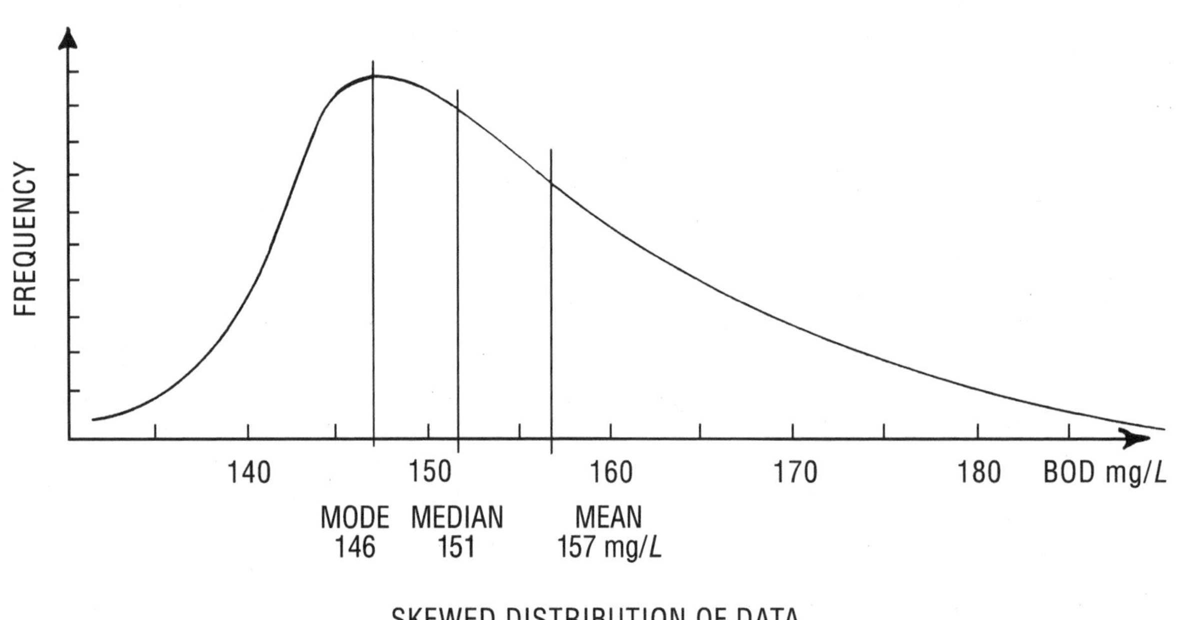

Fig. 18.4 Normal and skewed distribution of data

18.91 Trends

Plotting data on *GRAPHS*[4] is very helpful to illustrate trends in the operation of your plant. Occasionally, plotting data will reveal unexpected trends. This approach could be used to indicate to supervisors or the public the increase in plant inflow or decrease in the quality of plant effluent to justify increases in budgets or plant expansion. To look for or show a trend, plot the value you are interested in (for instance, flow, MGD, or effluent BOD, mg/*L*) against time (day, week, month, year).

EXAMPLE 7:

Analysis of flow data (totalizer readings) provides the following annual information:

Year:	2001	2002	2003	2004	2005	2006
Mean Daily Flow, MGD:	1.25	1.38	1.42	1.58	1.65	1.71

Plot the data:

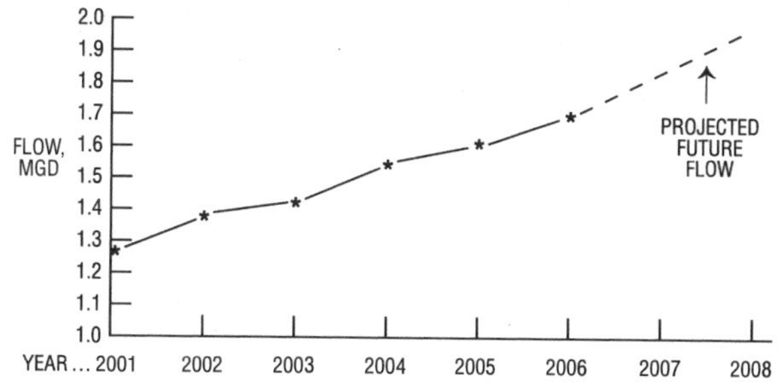

INTERPRETATION OF DATA:

The graph shows a continuously increasing mean daily flow. If your plant has a capacity of 2 MGD, the graph would clearly show the need for expansion in the near future if past trends in population growth or any industrial expansion continue. You could extend the trend (dashed line) to project future flows and predict when you expect to reach plant capacity (2007–2008).

APPLICATIONS:

Plotting data and looking for trends may be helpful to indicate broken pipes and illegal connections or discharges. You should always attempt to identify the cause of a trend. An industry may clean up on Friday afternoon and dump a slug of waste into the collection system that may reach your plant Friday night. If you plot your data and note a reduction in the quality of your effluent every Friday night or Saturday, you might start looking for the cause of the problem.

Some operators record flows continuously on daily circular charts. Every year they change the color of the ink, but use the same chart on the same day for several years. This is a good way to identify trends, too.

18.92 Summary

Two methods have been given in this section to present data:

1. Bar Graphs
2. Line Graphs

How does the operator determine which method to use? Bar graphs are used to summarize data and indicate number of times or frequency a given value was measured. Line graphs illustrate trends by showing how a particular measurement changes with time. Section 18.8, "Moving Averages," also describes a procedure used to reveal trends.

QUESTION

Write your answer in a notebook and then compare your answer with the one on page 662.

18.9B Weekly alkalinity tests on digester sludge are given below:

Week	1	2	3	4	5	6	7	8	9	10
Alkalinity, mg/*L*	1,730	1,670	1,690	1,680	1,630	1,620	1,590	1,530	1,480	1,420

a. Is a trend apparent?

b. Should any action be taken by the operator?

18.93 Chart Preparation

Sometimes routine plant operational information can be quickly and easily determined by the use of graphs or charts. An example might be the measurement of the depth of sludge in a digester. The depth can be used to calculate the volume of sludge in gallons or in cubic feet. A quick way to convert depth to volume and to reduce the chance of mathematical errors is to prepare a chart of sludge depth against sludge volume.

EXAMPLE 8:

Prepare a chart of sludge depth against sludge volume for a 100-foot diameter digester.

Known	**Unknown**
Measured Sludge Depths	Volume of Digester for Various Sludge Depths, cu ft

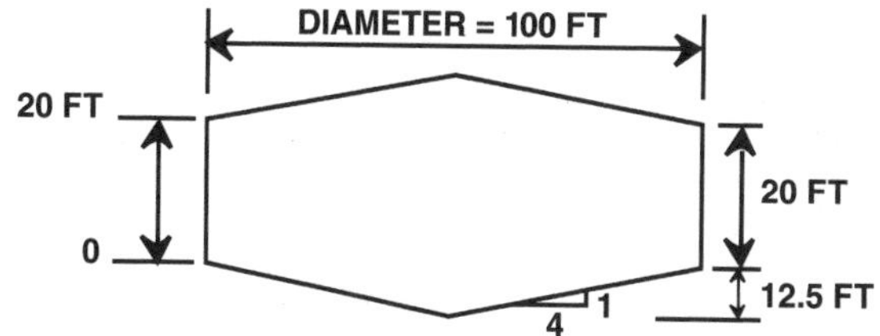

PROCEDURE:

Calculate the sludge volume in the digester at three depths (0, 10, and 20 ft). This will provide us with three plotting points on our chart. If the points are not on a straight line, we should look for a possible error in our calculations.

1. Calculate the volume of the bottom cone of the digester. This will be the volume of sludge in the digester if the depth of sludge was measured to be zero along the inside wall of the digester. $\pi = 3.14$.

$$\text{Cone Volume, cu ft} = \frac{\pi}{3}(\text{Radius, ft})^2(\text{Height, ft})$$

$$= \frac{\pi}{3}(50\text{ ft})^2(12.5\text{ ft})$$

$$= 32{,}725\text{ cu ft}$$

[4] Graph paper may be obtained at most stationery stores. See graph paper at end of this chapter, page 663.

2. Calculate the volume of sludge in the digester if the measured sludge depth is 10 feet.

$$\text{Volume, cu ft} = \frac{\text{Volume from 0 to}}{\text{10 feet, cu ft}} + \frac{\text{Volume of Bottom}}{\text{Cone, cu ft}}$$

$$= \frac{\pi}{4}(100\text{ ft})^2(10\text{ ft}) + 32{,}725\text{ cu ft}$$

$$= 78{,}540\text{ cu ft} + 32{,}725\text{ cu ft}$$

$$= 111{,}265\text{ cu ft}$$

3. Calculate the volume of sludge in the digester if the measured sludge depth is 20 feet.

$$\text{Volume, cu ft} = \frac{\text{Volume from 0 to}}{\text{20 feet, cu ft}} + \frac{\text{Volume of Bottom}}{\text{Cone, cu ft}}$$

$$= \frac{\pi}{4}(100\text{ ft})^2(20\text{ ft}) + 32{,}725\text{ cu ft}$$

$$= 157{,}080\text{ cu ft} + 32{,}725\text{ cu ft}$$

$$= 189{,}805\text{ cu ft}$$

4. If sludge volumes were desired in gallons, the sludge volumes in cubic feet could be multiplied by 7.5 gallons per cubic foot.

5. Summarize plotting points.

Depth of Sludge, ft	Volume of Sludge, cu ft	Volume of Sludge, M Gal
0	32,725	0.245
10	111,265	0.834
20	189,805	1.424

6. Plot data as shown on Figure 18.5. Be sure to use a large enough scale for both depth of sludge in feet and volume of sludge in cubic feet to get accurate results. Draw a straight line between the three plotting points.

7. To prepare the million gallons scale, draw another line below and parallel to the sludge volume, 1,000 cubic feet line.

   a. Calculate the cubic feet in one million gallons.

$$1\text{ M Gal} = \frac{1{,}000{,}000\text{ gal}}{7.5\text{ gal/cu ft}}$$

$$= 133{,}333\text{ cu ft}$$

or

$$0.1\text{ M Gal} = 13{,}333\text{ cu ft}$$

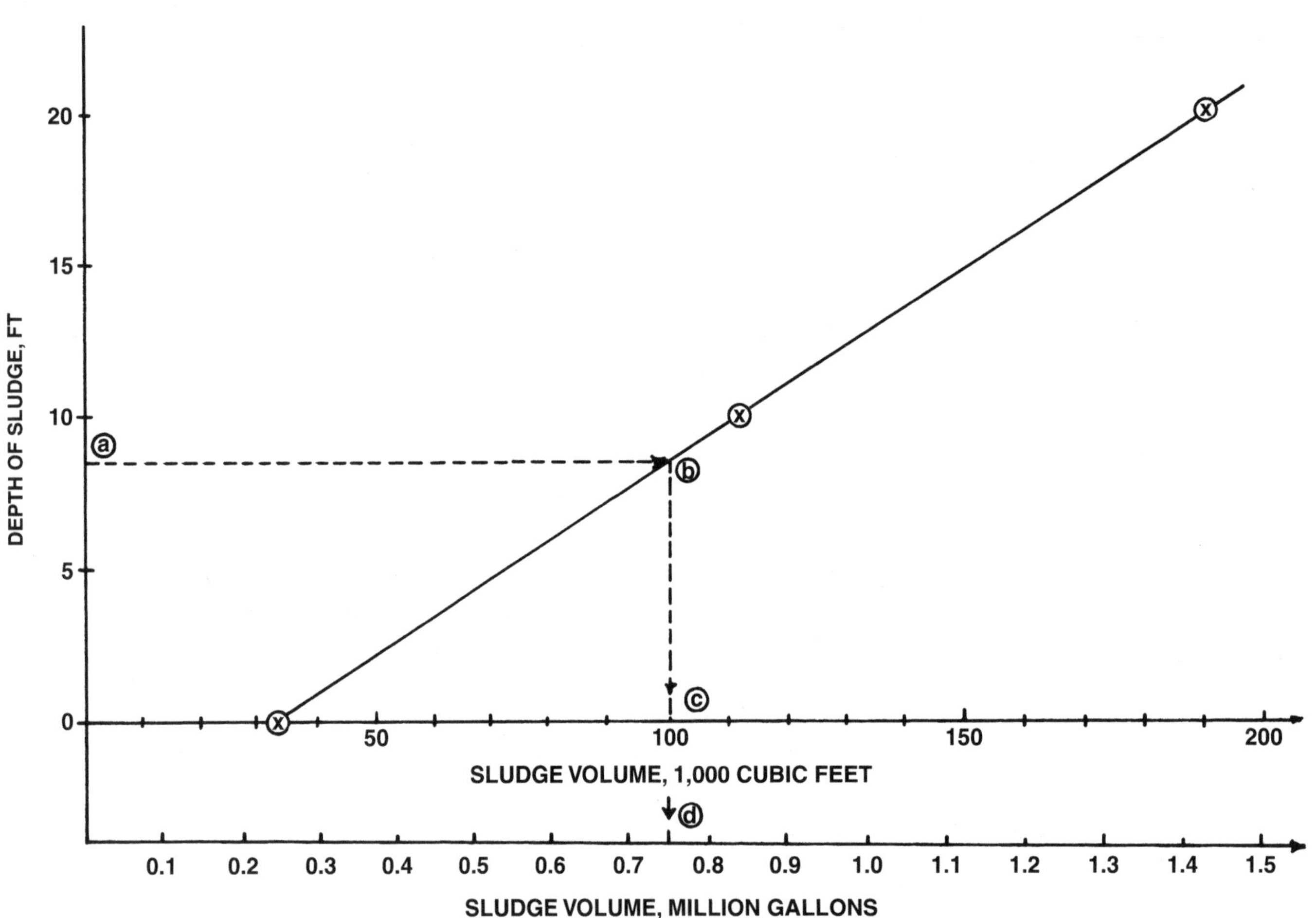

Fig. 18.5 Chart to determine sludge volume in cubic feet or million gallons from depth of sludge

b. Prepare a table of the number of cubic feet in 0.1 M Gal, 0.2 M Gal and so on up to 200,000 cubic feet.

1,000 cu ft	M Gal	1,000 cu ft	M Gal
13.3	0.1	120.0	0.9
26.7	0.2	133.3	1.0
40.0	0.3	146.7	1.1
53.3	0.4	160.0	1.2
66.7	0.5	173.3	1.3
80.0	0.6	186.7	1.4
93.3	0.7	200.0	1.5
106.7	0.8		

c. Find 13.3 (1,000 cubic feet) and drop down to the lower parallel line and mark 0.1 M Gal. Find 26.7 (1,000 cubic feet), drop down to lower line and mark 0.2 M Gal. Repeat procedure for remainder of table.

8. To use Figure 18.5, measure the depth of sludge in the digester. For example, 8.5 feet. Find the depth on Figure 18.5 (point a). Move horizontally across to the diagonal line (point b). Drop vertically down. Read the digester volume as 100,000 cubic feet (point c) or 0.75 million gallons (point d).

18.94 COD to BOD Curves

In Chapter 11, "Activated Sludge," Section 11.57, "Actual Operation Under Abnormal Conditions," Figure 11.22 shows how an operator used COD measurements to estimate effluent BOD values and comply with NPDES permit monthly allowable BOD discharges. To use this procedure you must compare monthly effluent COD discharges in pounds per month with monthly effluent BOD discharges in pounds per month. If the values are similar on a month-to-month basis, or on a comparison of months on a year-to-year basis (September of this year to September of last year), then this procedure may be used. The relationship must be watched carefully during critical months. Be sure that the conversion factor from COD to BOD estimates the BOD slightly higher than the actual BOD to be on the safe side.

PROCEDURE:

1. Determine the number of days in the month and the allowable NPDES permit effluent BOD in pounds per month. The example in Figure 18.6 has 30 days in the month and an allowable effluent BOD of 80,000 pounds per month.

2. Prepare the horizontal scale in days and the vertical scale in 1,000 pounds of effluent BOD. Above 30 days plot the point 80,000 pounds of effluent BOD. Draw the diagonal line from the origin (0 days and 0 effluent BOD) to the plotted point (30 days and 80,000 pounds of effluent BOD). As long as the cumulative estimated BOD using the COD values and the actual BOD values plot below the diagonal line, you should be able to meet your monthly NPDES permit BOD discharge requirement.

3. Analyze monthly COD to BOD relationships. For our example plant, the cumulative COD for last month was 473.2 pounds and the cumulative BOD was 76.66 pounds. Calculate the COD to BOD factor.

$$\text{COD to BOD Factor} = \frac{\text{BOD, lbs/month}}{\text{COD, lbs/month}}$$

$$= \frac{76.66 \text{ lbs BOD/month}}{473.2 \text{ lbs COD/month}}$$

$$= 0.162$$

To be on the safe side, increase the factor slightly to 0.163 or 0.1633.

4. Prepare a table like Table 18.4. Every day measure flow in MGD, COD in mg/*L* and start the five-day BOD test. Estimate the cumulative BOD in pounds per day and plot the value every day. Five days later when the BOD results become available, plot the actual cumulative pounds of BOD.

a. Calculate COD in 1,000 pounds per day.

$$\text{COD, } \frac{1{,}000 \text{ lbs}}{\text{day}} = (\text{Flow, MGD})(\text{COD, mg}/L)\left(8.34 \frac{\text{lbs}}{\text{gal}}\right)\frac{(1{,}000)}{(1{,}000)}$$

$$= (19.9 \text{ MGD})(60 \text{ mg}/L)\left(8.34 \frac{\text{lbs}}{\text{gal}}\right)\frac{(1{,}000)}{(1{,}000)}$$

$$= 9{,}958 \frac{\text{lbs}}{\text{day}}\frac{(1{,}000)}{(1{,}000)}$$

$$= 9.96 \frac{\text{lbs}}{\text{day}}(1{,}000)$$

$$= 10.0 \text{ (1,000 lbs COD/day)}$$

b. Determine cumulative COD in 1,000 pounds per day.

$$\text{Cum COD, 1,000 lbs/day} = \text{Today's COD, 1,000 lbs/day} + \text{Yesterday's Cum COD, 1,000 lbs/day}$$

$$= 13.0 + 10.0$$

$$= 23.0 \text{ (1,000 lbs COD/day)}$$

c. Calculate BOD in 1,000 pounds per day.

$$\text{BOD, } \frac{1{,}000 \text{ lbs}}{\text{day}} = (\text{Flow, MGD})(\text{BOD, mg}/L)\left(8.34 \frac{\text{lbs}}{\text{gal}}\right)\frac{(1{,}000)}{(1{,}000)}$$

$$= (19.9 \text{ MGD})(9 \text{ mg}/L)\left(8.34 \frac{\text{lbs}}{\text{gal}}\right)\frac{(1{,}000)}{(1{,}000)}$$

$$= 1{,}494 \frac{\text{lbs}}{\text{day}}\frac{(1{,}000)}{(1{,}000)}$$

$$= 1.5 \text{ (1,000 lbs BOD/day)}$$

d. Determine cumulative BOD in 1,000 pounds per day.

$$\text{Cum BOD, 1,000 lbs/day} = \text{Today's BOD, 1,000 lbs/day} + \text{Yesterday's Cum BOD, 1,000 lbs/day}$$

$$= 2.3 + 1.5$$

$$= 3.8 \text{ (1,000 lbs BOD/day)}$$

e. Calculate cumulative BOD estimated by cumulative COD in 1,000 pounds per day.

$$\text{Est Cum BOD, 1,000 lbs/day} = (\text{Cum COD, 1,000 lbs/day})(0.163)$$

$$= (10.0)(0.163)$$

$$= 1.6$$

f. *NOTE:* Values in Table 18.4 are rounded off for convenience and because greater accuracy is not necessary.

18.10 VARIANCE AND STANDARD DEVIATION

Variance and standard deviation are terms frequently used in professional journals that report the results of research findings. Knowledge of this material is important in the field of quality control because these terms describe the spread of measurements or results.

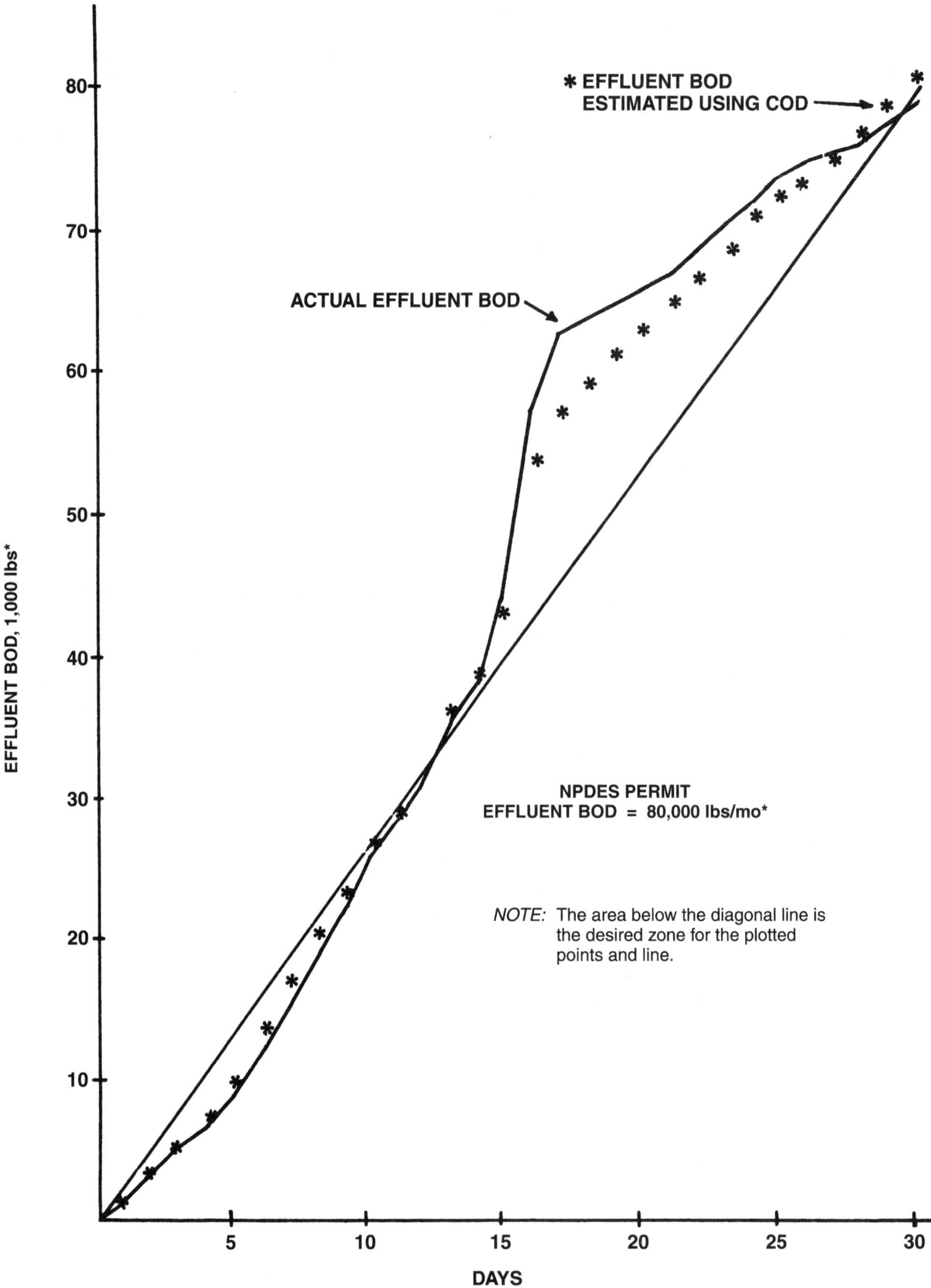

* Multiply pounds by 0.454 to obtain kilograms.

Fig. 18.6 Estimated and actual effluent BOD

TABLE 18.4 ESTIMATION OF EFFLUENT BOD USING EFFLUENT COD MEASUREMENTS

Date	Day	Flow, MGD	Effluent COD, mg/L	Effluent BOD, mg/L	COD, 1,000 lbs/day[a]	Cumulative COD, 1,000 lbs/day[a]	BOD, 1,000 lbs/day[a]	Cumulative BOD, 1,000 lbs/day[a]	Cumulative BOD Estimated by COD, 1,000 lbs/day[a]
1	M	19.9	60	9	10.0	10.0	1.5	1.5	1.6
2	T	22.6	69	12	13.0	23.0	2.3	3.8	3.8
3	W	22.7	62	10	11.7	34.7	1.9	5.7	5.7
4	T	23.5	69	11	13.5	48.2	2.1	7.8	7.9
5	F	23.8	65	7	12.9	61.1	1.4	9.2	10.0
6	S	21.7	131	20	23.7	84.8	3.6	12.8	13.8
7	S	22.0	100	16	18.3	103.1	2.9	15.7	16.8
8	M	23.3	102	18	19.9	123.0	3.5	19.2	20.0
9	T	23.8	110	16	21.8	144.8	3.2	22.4	23.6
10	W	23.8	97	11	19.3	164.1	2.2	24.6	26.8
11	T	24.0	82	20	16.4	180.5	4.0	28.6	29.5
12	F	23.2	71	15	13.7	194.2	2.9	31.5	31.7
13	S	22.5	140	20	26.3	220.5	3.8	35.3	36.0
14	S	22.1	100	18	18.4	238.9	3.3	38.6	39.0
15	M	24.0	130	27	26.0	264.9	5.4	44.0	43.3
16	T	25.6	280	65	59.8	327.7	13.9	57.9	53.5
17	W	19.5	170	26	27.6	352.3	4.2	62.1	57.5
18	T	16.6	89	8	12.4	364.7	1.1	63.2	59.6
19	F	17.2	77	10	11.0	375.7	1.4	64.6	61.4
20	S	17.7	71	9	10.5	386.2	1.4	66.0	63.1
21	S	17.1	69	8	9.8	396.0	1.1	67.1	64.7
22	M	19.9	69	10	11.5	407.5	1.7	68.8	66.5
23	T	20.3	63	9	10.7	408.2	1.5	70.3	68.3
24	W	21.0	79	10	13.8	432.0	1.7	72.0	70.5
25	T	21.1	53	10	9.3	441.3	1.8	73.8	72.1
26	F	21.3	51	7	9.1	450.4	1.2	75.0	73.6
27	S	18.0	58	3	8.7	459.1	0.5	75.5	75.0
28	S	18.0	76	5	11.4	470.5	0.7	76.2	76.8
29	M	21.9	56	8	10.2	480.7	1.5	77.7	78.5
30	T	22.4	60[b]	9[b]	11.2	491.9	1.7	79.4	80.3

[a] Multiply pounds by 0.454 to obtain kilograms.
[b] Since plant was close to violating NPDES permit BOD discharge requirement, all effluent was diverted to a holding pond and there was no discharge on the last day of the month.

In previous discussions, results have been described in terms of a mean value and a range. The bar graphs below show the results of three different tests, but they all have the same mean value and range.

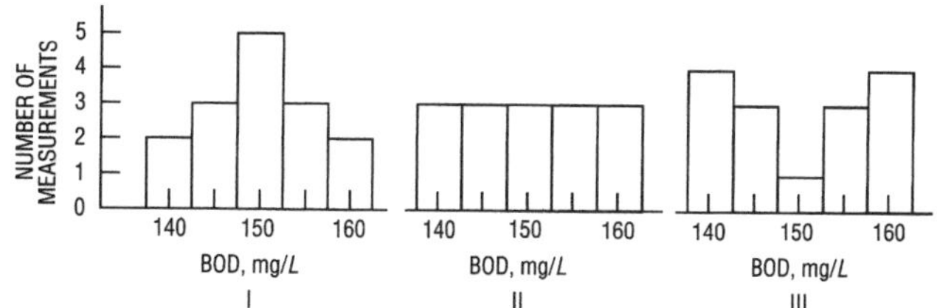

For all three cases, the mean value is 150 mg/L and the range is 20 mg/L (160 – 140), using the midpoints for our calculations.

Another term (parameter) to describe the above results is the variance, S^2, a measure of the dispersion or spread of the results. The variance is calculated by taking the difference between each measurement, X, and the mean value of all the measurements, $\bar{X}$, squaring the difference, summing up the squared differences, and dividing by the total number of differences, n, minus one, as shown in the following formula.

$$\text{Variance, } S^2 = \frac{\Sigma (X-\bar{X})^2}{n-1} \text{ or } = \frac{\Sigma X^2 - (\Sigma X)^2/n}{n-1}$$

X = Each Measurement

$$\text{Mean Value, } \bar{X} = \frac{\Sigma X}{n}$$

n = Number of Measurements

Σ = Summation of All Values

In the denominator, one is subtracted from n, because $\bar{X}$ in the numerator is calculated from n measurements, $\bar{X}$ is influenced by all of our measurements. By dividing by n – 1 we obtain a more conservative description of the actual dispersion. The larger n becomes, the more insignificant becomes the "minus one" term. When analyzing plant data, the number of measurements is usually small; therefore, the "minus one" term is important.

STEP 1.

Calculate the variance for the results shown in Bar Graph I (see page 653).

Measurement X, mg/*L*	$X-\bar{X}$	$(X-\bar{X})^2$	Freq	$(X-\bar{X})^2$ Freq
140	140–150 = –10	(–10) (–10) = 100	2	(100) (2) = 200*
145	145–150 = – 5	(– 5) (– 5) = 25	3	(25) (3) = 75
150	150–150 = 0	(0) (0) = 0	5	(0) (5) = 0
155	155–150 = 5	(5) (5) = 25	3	(25) (3) = 75
160	160–150 = 10	(10) (10) = 100	2	(100) (2) = 200
		TOTAL n =	15	$\Sigma(X-\bar{X})^2 = 550$ (mg/*L*)^2**

$$S^2 = \frac{\Sigma(X-\bar{X})^2}{n-1}$$

$$= \frac{550}{15-1}(\text{mg}/L)^2$$

$$= 39.3\ (\text{mg}/L)^2$$

* Instead of writing 140 twice, subtracting 140 – 150 twice, and squaring (–10) (–10) twice, we performed our calculations once on one line and then multiplied by the frequency, 2. We did the same for the other measurements, X.

** Units should be (mg/*L*)^2. The term is meaningless, but is included to maintain the proper units. For the first row, X is 140 mg/*L*; $X-\bar{X}$ is 140 mg/*L* – 150 mg/*L* = –10 mg/*L*; $(X-\bar{X})^2$ is (–10 mg/*L*)(–10 mg/*L*) = 100 (mg/*L*)^2; and $(X-\bar{X})^2$ Freq is (100 [mg/*L*]^2) (2) = 200 (mg/*L*)^2.

STEP 2.

Calculate the variance for the results shown in Bar Graph II.

Measurement, X	$X-\bar{X}$	$(X-\bar{X})^2$	Freq	$(X-\bar{X})^2$ Freq
140	–10	100	3	300
145	– 5	25	3	75
150	0	0	3	0
155	5	25	3	75
160	10	100	3	300
		TOTAL n =	15	$\Sigma(X-\bar{X})^2 = 750$

$$S^2 = \frac{\Sigma(X-\bar{X})^2}{n-1}$$

$$= \frac{750}{15-1}$$

$$= 53.6\ (\text{mg}/L)^2$$

STEP 3.

Calculate the variance for the results shown in Bar Graph III.

Measurement, X	$X-\bar{X}$	$(X-\bar{X})^2$	Freq	$(X-\bar{X})^2$ Freq
140	–10	100	4	400
145	– 5	25	3	75
150	0	0	1	0
155	5	25	3	75
160	10	100	4	400
		TOTAL n =	15	$\Sigma(X-\bar{X})^2 = 950$

$$S^2 = \frac{\Sigma(X-\bar{X})^2}{n-1}$$

$$= \frac{950}{15-1}$$

$$= 67.9\ (\text{mg}/L)^2$$

In comparing the variance of the three bar graphs, note that as more and more measurements shift away from the mean value, the value of the variance increases, thus indicating an increase in the dispersion of our results.

The dispersion is frequently described in terms of S, the standard deviation, which has the same units as the average value, mg/*L*. The standard deviation, S, is the square root of the variance, S^2. The square root of a number is one of two equal numbers that when multiplied together give that number.

EXAMPLES:

The square root of					
	9 = 3	or $\sqrt{9}$	= 3 and (3) (3) = 9
	16 = 4	or $\sqrt{16}$	= 4 and (4) (4) = 16
	25 = 5	or $\sqrt{25}$	= 5 etc.
	4 = 2	or $\sqrt{4}$	= 2
	1 = 1	or $\sqrt{1}$	= 1
	S^2 = S	or $\sqrt{S^2}$	= S
	1.44 = 1.2	or $\sqrt{1.44}$	= 1.2
	(mg/*L*)^2 = mg/*L*	or $\sqrt{(\text{mg}/L)^2}$	= mg/*L*

To obtain the square root of a number, there are several potential methods. Two common methods are listed below.

1. Use an electronic calculator or look up the values in a table in a math book or handbook.
2. Attempt a trial-and-error approach by multiplying a number by itself.

EXAMPLE 9:

Find the standard deviation, S, of the variance S^2 = 39.3 (mg/*L*)^2.

Use of a math book or handbook, or an electronic calculator are the quickest and easiest ways to find the square root of a number.

1. *ELECTRONIC CALCULATOR*

   To calculate the square root using an electronic calculator:

   1. Turn on calculator.
   2. Enter 39.3.
   3. Press the √ key.
   4. Read the result, 6.3.

2. *TRIAL AND ERROR:*

Trial:	I	II	III	IV
	6.0	6.5	6.3	6.2
	6.0	6.5	6.3	6.2
	36.00	325	189	124
		390	378	372
		42.25	39.69	38.44
	Less than 39.3, too small	Greater than 39.3, too large	Greater than 39.3, too large	Less than 39.3, too small

Trial III is closest to 39.3. Therefore S = 6.3 mg/*L*.

Example 10 illustrates another way to calculate the variance. This approach produces the same answer as the calculation shown in Step 1 above.

EXAMPLE 10:

The effluent BOD for a wastewater treatment plant is summarized in the following table:

Effluent BOD X, mg/L	X^2	Freq f	fX	fX^2
5	25	2	10	50
10	100	3	30	300
15	225	5	75	1,125
20	400	3	60	1,200
25	625	2	50	1,250
		$n = \Sigma f = 15$	$\Sigma fX = 225$	$\Sigma fX^2 = 3{,}925$

$$\text{Variance, } S^2 = \frac{\Sigma\, fX^2 - (\Sigma\, fX)(\Sigma\, fX)/n}{n-1}$$

$$= \frac{3{,}925 - (225)(225)/15}{15-1}$$

$$= \frac{3{,}925 - 3{,}375}{14}$$

$$= 39.3\ (\text{mg}/L)^2$$

QUESTION

Write your answer in a notebook and then compare your answer with the one on page 662.

18.10A Calculate the variance and standard deviation of the SVI data given in Question 18.9A. SVI = 120, 115, 120, 120, 125, 110, 115.

$$S^2 = \frac{\Sigma\,(X-\bar{X})^2}{n-1} \text{ or } = \frac{\Sigma\, fX^2 - (\Sigma\, fX)(\Sigma\, fX)/n}{n-1}$$

18.11 PREDICTION EQUATIONS, TRENDS, AND CORRELATIONS[5]

18.110 Prediction Equations

A very powerful mathematical tool is the development and use of prediction (or forecasting) equations. These equations estimate the value of a dependent variable on the basis of a known or measured independent variable.

For example, an operator needs to predict a BOD value on the basis of a COD measurement. The procedure would be to collect composite effluent samples daily, split the samples, measure the COD immediately, and measure the BOD five days later. The results could be summarized in a table after one week, plotted on graph paper, a "line of best fit" could be drawn, and this line on the graph paper could be used to predict or forecast expected BOD values on the basis of COD measurements.

EXAMPLE 11:

Use the effluent COD and BOD data for the first seven days listed in Table 18.4 on page 653:

Date, Day	1, M	2, T	3, W	4, T	5, F	6, S	7, S
COD, mg/L	60	69	62	69	65	131	100
BOD, mg/L	9	12	10	11	7	20	16

Plot the effluent COD, mg/*L*, and the effluent BOD, mg/*L*, data as points on a graph with COD represented on the horizontal axis (x axis) and BOD represented on the vertical axis (y axis) as shown on Figure 18.7. We can immediately see in a general way that as COD increases, BOD increases as well. One way to see a more precise relationship would be to take a straightedge and draw the line that comes closest to hitting all the data points, also called a "line of best fit." As the line extends to the left, we find the point at which the line intercepts the vertical axis. This point is called the intercept. As the line extends to the right, the slope of the line becomes apparent. This slope is called the regression coefficient. Do not try to extrapolate the line very far beyond the plotted data. The limitation of this approach is that operators are likely to draw the line slightly differently as each operator will visualize a slightly different line (slope).

A more accurate, mathematical approach to constructing the line of best fit would be to use the method of "least squares." This method attempts to minimize the sum of the difference or distance between the predicted values from the prediction equation and observed values squared. The principle is that the line of best fit occurs when the sum of the squares of deviations measured on the vertical axis (y axis) from the line of best fit to the plotting points is minimized. In other words, the shorter the distance the data points are above or below the line the better. To use this method, assume the line of best fit is a straight line. The basic equation is:

$$y = a + bx$$

where "a" is the intercept; "b" is the regression coefficient or slope for the line of best fit; "y" is the dependent variable on the vertical axis; and "x" is the independent variable on the horizontal axis. So, for the purposes of Example 11, the equation is:

$$\text{Effl BOD, mg BOD}/L = \text{a, mg BOD}/L + \frac{\text{b, mg BOD}/L}{\text{mg COD}/L}(\text{Effl COD, mg COD}/L)$$

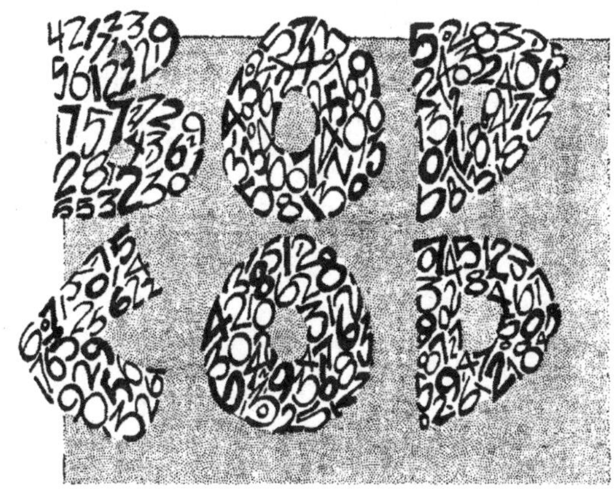

To calculate the regression coefficient, b, and the intercept, a, prepare a table like Table 18.5 using effluent COD and BOD data for the first seven days listed in Table 18.4 on page 653.

[5] If you wish to read more information on these topics in textbooks or on the Internet, try searching for the words "regression analysis" and "least-squares method." Regression analysis is a procedure that attempts to find a line of best fit for a given set of available data. The procedure produces a prediction equation that describes a dependent variable (BOD) on the basis of a measured independent variable (COD). The least-squares method is a regression analysis technique that seeks to minimize the sum of the square of the distances (deviations) of the plotted available data (measured observations) from the line of best fit.

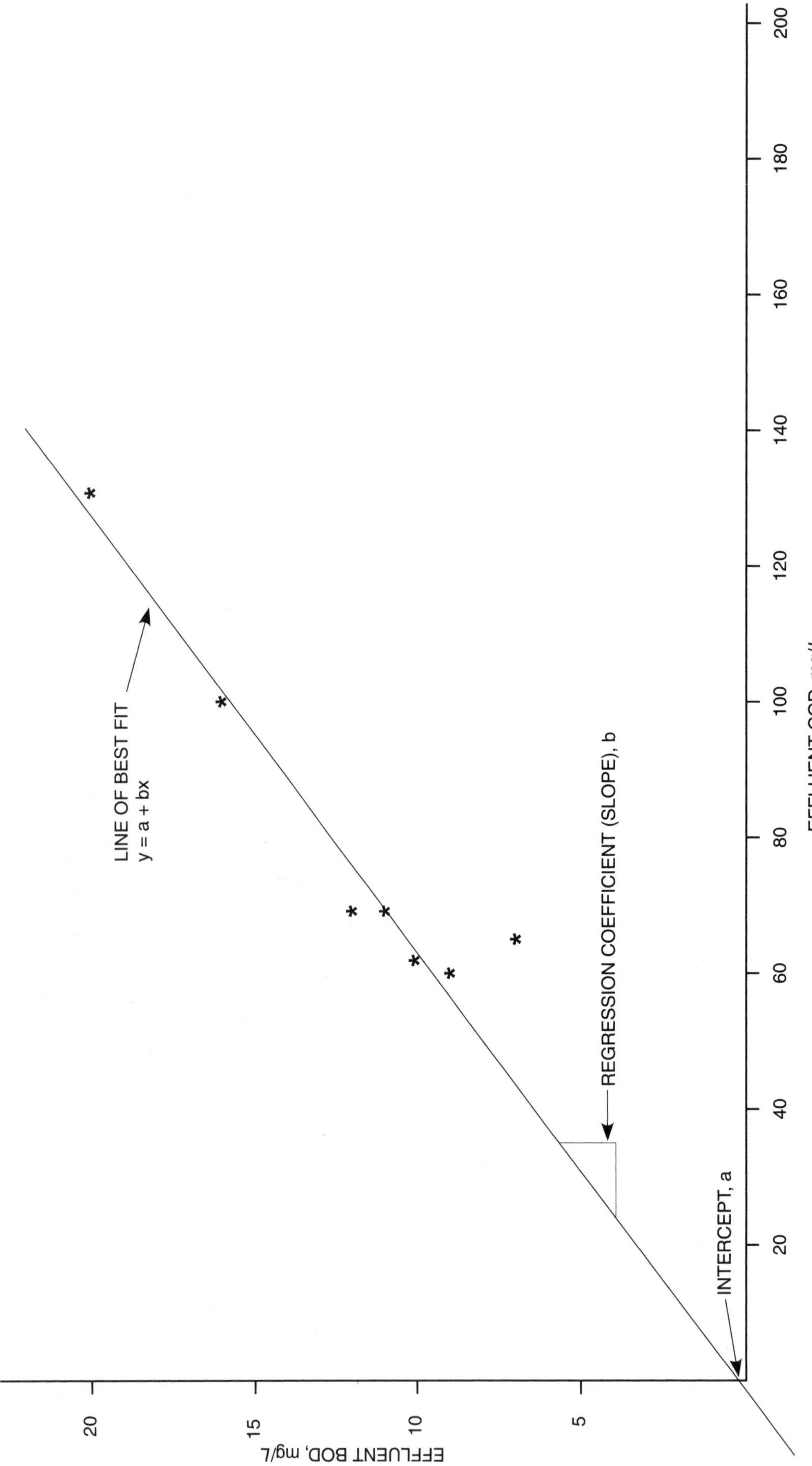

Fig. 18.7 Estimation of BOD on the basis of COD test results

TABLE 18.5 CALCULATION OF LINE OF BEST FIT
(Where x = COD, mg/*L* and y = BOD, mg/*L*, Table 18.4)

n	= 7	———		n	= 7
Σx	= 556	———		Σy	= 85
$\bar{x} = \frac{\Sigma x}{n}$	= 79.4	———		$\bar{y} = \frac{\Sigma y}{n}$	= 12.1
$(\Sigma x)^2$	= 309,136	$(\Sigma x)(\Sigma y)$	= 47,260	$(\Sigma y)^2$	= 7,225
$\frac{(\Sigma x)^2}{n}$	= 44,162	$\frac{(\Sigma x)(\Sigma y)}{n}$	= 6,751	$\frac{(\Sigma y)^2}{n}$	= 1,032
Σx^2	= 48,352	Σxy	= 7,422	Σy^2	= 1,151
$SS\ x = \Sigma x^2 - \frac{(\Sigma x)^2}{n}$	= 4,190	$SS\ xy = \Sigma xy - \frac{(\Sigma x)(\Sigma y)}{n}$	= 671	$SS\ y = \Sigma y^2 - \frac{(\Sigma y)^2}{n}$	= 119

From the calculated values in Table 18.5, we have the quantities needed to determine the regression coefficient, b, and the intercept, a. Once these are known, the predicted BOD value can be found using the basic equation, y = a + bx.

$$\text{Regression Coefficient, b} = \frac{SS\ xy}{SS\ x}$$

$$= \frac{671}{4{,}190}$$

$$= 0.16$$

$$\text{or} = \frac{0.16\ \text{mg BOD}/L}{\text{mg COD}/L}$$

$$\text{Intercept, a} = \bar{y} - b\bar{x}$$

$$= 12.1 - (0.16)(79.4)$$

$$= 12.1 - 12.7$$

$$= -0.6$$

$$\text{Effl BOD, mg BOD}/L = -0.6\ \text{mg BOD}/L + \frac{0.16\ \text{mg BOD}/L}{\text{mg COD}/L}(\text{Effl COD, mg COD}/L)$$

To estimate or predict the BOD of a sample, measure the COD, multiply the COD by 0.16 and subtract 0.6.

18.111 Trends

The method of least squares can be used to analyze data and determine if there is a positive (increasing) or negative (decreasing) trend.

EXAMPLE 12:

Use the effluent BOD data for the first seven days listed in Table 18.4 on page 653:

Date, Day	1, M	2, T	3, W	4, T	5, F	6, S	7, S
BOD, mg/*L*	9	12	10	11	7	20	16

Plot the time, days, and effluent BOD, mg/*L*, data as points on a graph with time represented on the horizontal axis (x axis) and BOD represented on the vertical axis (y axis) as shown on Figure 18.8. Draw a trend line (line of best fit). The basic equation is the same:

y = a + bx

where "a" is the intercept; "b" is the regression coefficient or slope for the line of best fit; "y" is the dependent variable on the vertical axis; and "x" is the independent variable on the horizontal axis. So, for the purposes of Example 12, the equation is:

$$\text{Effl BOD, mg BOD}/L = \text{a, mg BOD}/L + \frac{\text{b, mg BOD}/L}{\text{Time, days}}(\text{Time, days})$$

To calculate the regression coefficient, b, and the intercept, a, prepare a table like Table 18.6 using effluent BOD data for the first seven days listed in Table 18.4 on page 653.

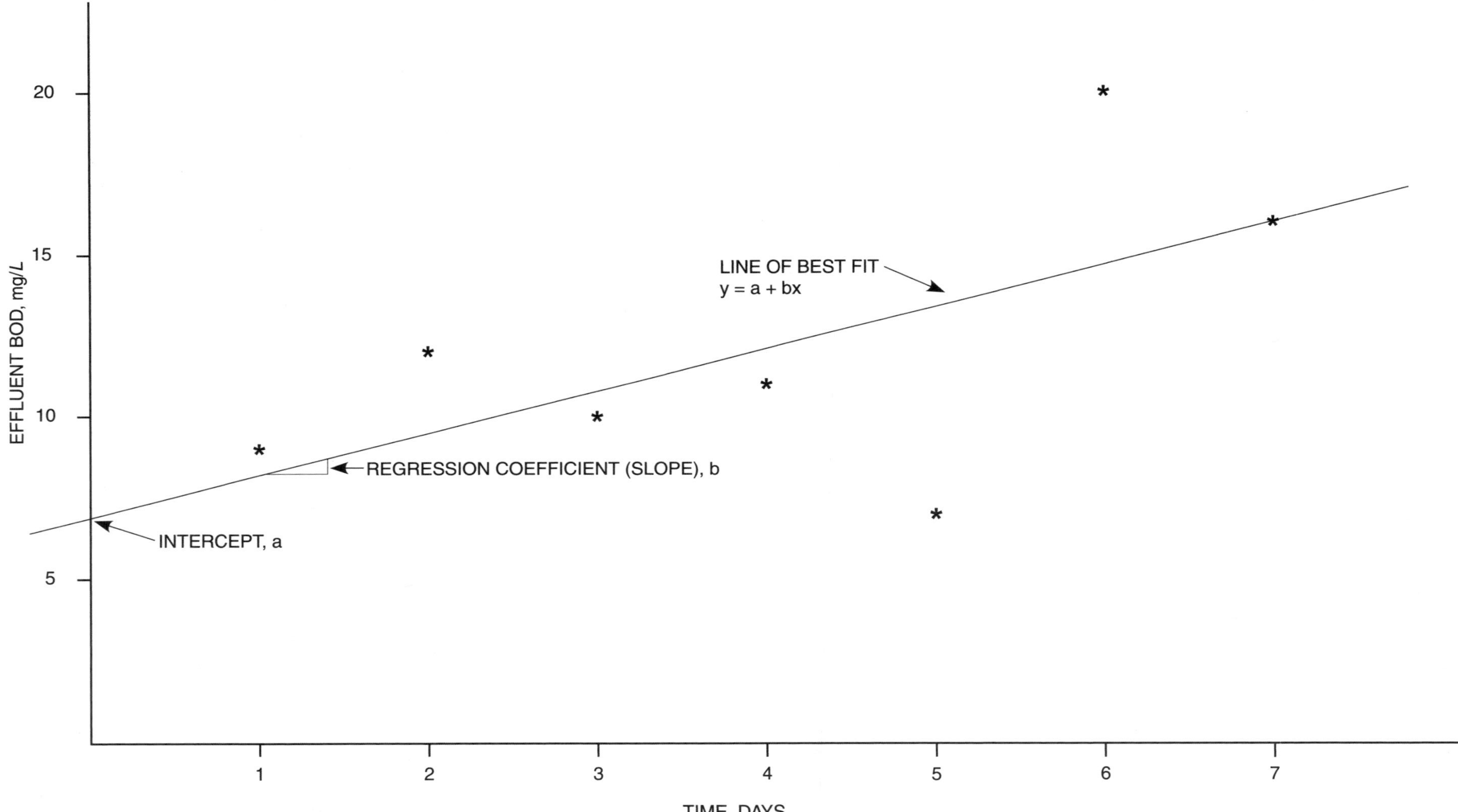

Fig. 18.8 Trend analysis of effluent BOD

TABLE 18.6 CALCULATION OF TREND LINE
(Where x = Time, days and y = BOD, mg/*L*, Table 18.4)

n	= 7	———		n	= 7
Σx	= 28	———		Σy	= 85
$\bar{x} = \frac{\Sigma x}{n}$	= 4	———		$\bar{y} = \frac{\Sigma y}{n}$	= 12.1
$(\Sigma x)^2$	= 784	$(\Sigma x)(\Sigma y)$	= 2,380	$(\Sigma y)^2$	= 7,225
$\frac{(\Sigma x)^2}{n}$	= 112	$\frac{(\Sigma x)(\Sigma y)}{n}$	= 340	$\frac{(\Sigma y)^2}{n}$	= 1,032
Σx^2	= 140	Σxy	= 374	Σy^2	= 1,151
$SS\,x = \Sigma x^2 - \frac{(\Sigma x)^2}{n}$	= 28	$SS\,xy = \Sigma xy - \frac{(\Sigma x)(\Sigma y)}{n}$	= 34	$SS\,y = \Sigma y^2 - \frac{(\Sigma y)^2}{n}$	= 119

From the calculated values in Table 18.6, we have the quantities needed to determine the regression coefficient, b, and the intercept, a. Once these are known, a predicted BOD trend can be found using the basic equation, y = a + bx.

$$\text{Regression Coefficient, } b = \frac{SS\,xy}{SS\,x}$$

$$= \frac{34}{28}$$

$$= \frac{1.21 \text{ mg BOD}/L}{\text{day}}$$

The result indicates an increasing trend for effluent BOD of 1.2 mg BOD/*L* per day.

$$\text{Intercept, } a = \bar{y} - b\bar{x}$$

$$= 12.1 - (1.21)(4)$$

$$= 12.1 - 4.8$$

$$= 7.3$$

$$\text{Effluent BOD, mg BOD}/L = 7.3 \text{ mg BOD}/L + \frac{1.21 \text{ mg BOD}/L}{\text{day}}(\text{Time, days})$$

Therefore, effluent BOD, mg/*L,* can be expected to increase by 1.2 mg/*L* per day during the time span for which we analyzed the data. This is not much of a change.

18.112 Correlations

The correlation coefficient, r, is used as a measure or indication of the "closeness of fit" of the prediction (or forecasting) equation with the actual data. If all the data points were on a straight line and the line of best fit went through all the data points, the fit would be perfect and the correlation coefficient would be one (1). The closer the correlation coefficient, r, is to one (1), the better the equation fits the data. The closer the correlation coefficient, r, is to one (1), the greater the correlation between the independent variable (x) and the dependent variable (y). *NOTE:* The correlation coefficient can never be greater than 1.

EXAMPLE 13:

Calculate the correlation or closeness of fit for the prediction equations (lines of best fit) in Examples 11 and 12 and the actual data.

1. What is the correlation coefficient, r, for the equation to predict the effluent BOD and the actual data? Use the numbers calculated for Table 18.5.

$$\text{Correlation Coefficient, } r = \frac{SS\,xy}{\sqrt{(SS\,x)(SS\,y)}}$$

$$= \frac{671}{\sqrt{(4{,}190)(119)}}$$

$$= \frac{671}{\sqrt{498{,}610}}$$

$$= \frac{671}{706}$$

= 0.95, which is considered very good

2. What is the correlation coefficient, r, for the equation to predict the effluent BOD trend on the basis of time in days and the actual data? Use the numbers calculated for Table 18.6.

$$\text{Correlation Coefficient, } r = \frac{SS\,xy}{\sqrt{(SS\,x)(SS\,y)}}$$

$$= \frac{34}{\sqrt{(28)(119)}}$$

$$= \frac{34}{\sqrt{3{,}332}}$$

$$= \frac{34}{57.7}$$

$$= 0.59$$

A correlation coefficient of 0.59 is not very good. This value indicates that there is not much correlation between the day the measurement was taken and the measurement (BOD). The result is very good from an operations viewpoint, however. The result indicates that the BOD is neither increasing nor decreasing over time.

With the proper software, the prediction equations developed from Tables 18.5 and 18.6 can be produced by computers. They can also be produced by programmable electronic calculators. Enter the data (x and y values) and the prediction equation and correlation coefficient will be automatically calculated.

QUESTIONS

Write your answers in a notebook and then compare your answers with those on page 662.

18.11A What information is needed to develop prediction or forecasting equations?

18.11B What is the purpose of the correlation coefficient, r?

18.12 METRIC CALCULATIONS

Metric calculations are not included in this chapter because the procedures used to analyze and present data are the same for both the English and metric systems. Use the same procedures to analyze numbers, regardless of the system.

18.13 SUMMARY

1. *AVERAGE OR MEAN VALUE*

$$\bar{X} = \frac{\text{Sum of All Measurements}}{\text{Number of Measurements}} = \frac{\Sigma X}{n}$$

where X is each measurement or test result, and n is the number of measurements or observations.

2. *RANGE*

Range = Largest X – Smallest X

3. *MEDIAN*

Median = Middle measurement when measurements are ranked in order of magnitude (may fall between two measurements)

4. *MODE*

Mode = Measurement that occurs most frequently (may be more than one mode)

5. *GEOMETRIC MEAN*

$$\text{Geometric Mean} = (X_1 \times X_2 \times X_3 \times \ldots X_n)^{1/n}$$

6. *VARIANCE AND STANDARD DEVIATION*

$$\text{Variance, } S^2 = \frac{\Sigma (X-\bar{X})^2}{n-1} \text{ or } = \frac{\Sigma X^2 - (\Sigma X)^2/n}{n-1}$$

$$\text{Standard Deviation, S} = \sqrt{\frac{\Sigma (X-\bar{X})^2}{n-1}} = \sqrt{S^2}$$

18.14 ARITHMETIC ASSIGNMENT

Turn to the Arithmetic Appendix at the back of this manual. Read and work the problems in Section A.35, "Data Analysis." Check the arithmetic in this section using an electronic calculator. You should be able to get the same answers.

END OF LESSON 2 OF 2 LESSONS

on

ANALYSIS AND PRESENTATION OF DATA

Please answer the discussion and review questions next.

DISCUSSION AND REVIEW QUESTIONS

Chapter 18. ANALYSIS AND PRESENTATION OF DATA

(Lesson 2 of 2 Lessons)

Write the answers to these questions in your notebook. The question numbering continues from Lesson 1.

The solids concentrations of sludge withdrawn from a primary clarifier during the past seven days are given below:

Day:	1	2	3	4	5	6	7
Solids, %:	6.0	6.5	6.0	5.0	6.5	7.5	8.0

10. Draw a bar graph showing the distribution of data.
11. Draw a line graph to illustrate if any trend is developing.
12. Is a trend apparent?
13. Calculate the variance, S^2, for the solids data.
14. Determine the standard deviation, S, for the solids data using any method you prefer.
15. What is the limitation of drawing a line of best fit manually by drawing a line that comes closest to hitting all the data points?

SUGGESTED ANSWERS

Chapter 18. ANALYSIS AND PRESENTATION OF DATA

ANSWERS TO QUESTIONS IN LESSON 1

Answers to questions on page 636.

18.1A Variations in test results may be caused by changes in:

1. Water or material (sludge) being examined
2. Sampling procedures
3. Testing (analyst, procedure, reagents, equipment)

Many factors in each of these three categories also can cause changes. For example, in sampling, variations could be caused by changing the location where the sample was obtained and when the sample was obtained.

18.1B Samples should be tested immediately by the lab because sometimes the items (BOD, DO) we wish to measure will change with time due to biological or chemical reactions taking place in the sample container.

Answers to questions on page 636.

18.2A Instruments must be periodically calibrated and zeroed before using to ensure accurate results.

18.2B A gauge reading of 2 psi also would give a reading of 4.6 feet of water.

Answers to questions on page 638.

18.4A Calculate the mean mixed liquor concentration in mg/*L*.

$$\text{Mean Concentration, mg}/L = \frac{\text{Sum of All Measurements, mg}/L}{\text{Number of Measurements}}$$

$$= \frac{5{,}922 \text{ mg}/L}{3}$$

$$= 1{,}974 \text{ mg}/L$$

2,138
1,863
1,921
5,922

18.5A Range of measurements.

$$\text{Range, mg}/L = \text{Largest Value} - \text{Smallest Value}$$

$$= 2{,}138 \text{ mg}/L - 1{,}863 \text{ mg}/L$$

$$= 275 \text{ mg}/L$$

2,138
−1,863
275

Answer to question on page 639.

18.6A

Rank	SVI	Freq		
1	125	1		
2	120		Median SVI = 120	Half of the values
3	120	3		are larger and
4	120			half are smaller.
5	115	2	Mode SVI = 120	Value that occurs
6	115			most frequently
7	110	1		(three times).

Answer to question on page 644.

18.7A 1. Arrange the MPN values in increasing order of magnitude (size).

8 11 12 14 17 23 43 49 63

2. Multiply all of the numbers together to get a number.

$8 \times 11 \times 12 \times 14 \times 17 \times 23 \times 43 \times 49 \times 63 =$ 767 315 191 200

3. Convert the number in Step 2 to scientific notation.

7.673×10^{11}

4. Find the Geometric Mean from Table 18.2.

(1) Column 9		(2) Left Column
Smaller Value	$= 5.120 \times 10^{11}$	Geometric Mean = 20
Calculated Value	$= 7.673 \times 10^{11}$	
Larger Value	$= 1.968 \times 10^{13}$	Geometric Mean = 30

Geometric Mean = 20+ MPN/100 m*L*

5. If you have an electronic calculator, use the following procedure.

$$\text{Geometric Mean} = (8 \times 11 \times 12 \times 14 \times 17 \times 23 \times 43 \times 49 \times 63)^{1/9}$$

$$= (7.673 \times 10^{11})^{0.1111}$$

$$= 21 \text{ MPN}/100 \text{ m}L$$

Answer to question on page 645.

18.8A

Date	Day	Effluent BOD, mg/*L*	Moving Average
7	S	25	
8	M	23	
9	T	38	
10	W	41	
11	T	32	
12	F	35	
13	S	37	33.0
14	S	29	33.6
15	M	23	33.6
16	T	31	32.6
17	W	24	30.1
18	T	17	28.0
19	F	19	25.7
20	S	24	23.9

ANSWERS TO QUESTIONS IN LESSON 2

Answer to question on page 647.

18.9A

SVI	Freq	SVI × Freq	or	Sum of SVI
125	1	125		125
				120
120	3	360		120
				120
115	2	230		115
				115
110	1	110		110
SUMS	7	825		825

a. $\text{Mean SVI} = \frac{\text{Sum of SVI}}{\text{Number of SVI}}$

$= \frac{825}{7}$

$= 118$

b. Range of SVI = Largest SVI − Smallest SVI

$= 125 - 110$

$= 15$

c. Bar Graph.

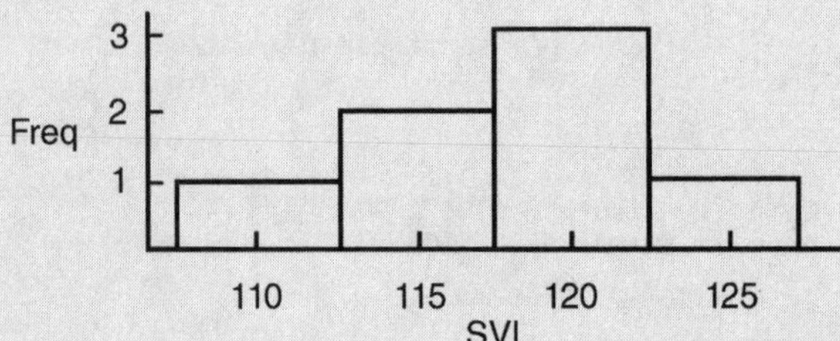

Answer to question on page 649.

18.9B

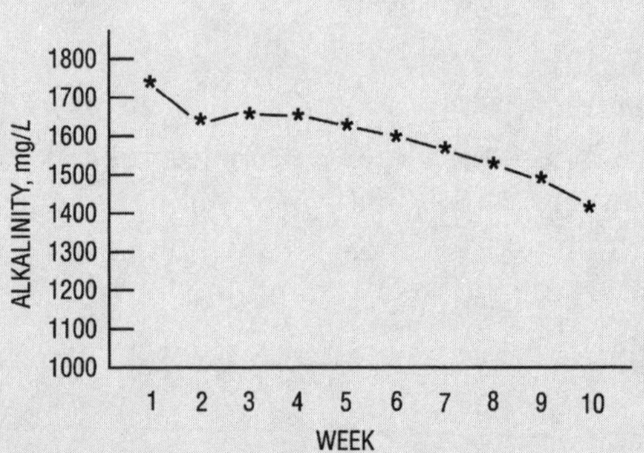

a. Yes, a trend is apparent.

b. Corrective action should be taken to prevent the continued drop of alkalinity. See Chapter 12, "Sludge Digestion and Solids Handling."

Answer to question on page 655.

18.10A

X	Freq	X Freq	$X - \bar{X}$	$(X - \bar{X})^2$	$(X - \bar{X})^2$ Freq
125	1	125	125 − 118 = 7	49	49
120	3	360	120 − 118 = 2	4	12
115	2	230	115 − 118 = −3	9	18
110	1	110	110 − 118 = −8	64	64
SUM	7	825			143

1. Calculate mean SVI, $\bar{X}$.

$\bar{X} = \frac{\text{Sum Measurements}}{\text{Sum Freq}}$

$= \frac{825}{7}$

$= 118$

2. Determine $X - \bar{X} = X - 118 = 125 - 118 = 7$

3. Determine $(X - \bar{X})^2 = (7)^2 = 49$

4. Determine $(X - \bar{X})^2(\text{Freq}) = 49 \times 1 = 49$

5. Calculate variance, S^2.

$S^2 = \frac{\Sigma\,(X - \bar{X})^2}{n - 1}$

$= \frac{143}{7 - 1}$

$= 24$

6. Calculate standard deviation, S.

$S = \sqrt{S^2}$

$= \sqrt{24}$

$= 4.9$

Answers to questions on page 660.

18.11A Information needed to develop prediction or forecasting equations includes measurement of independent and dependent variables.

18.11B The correlation coefficient, r, is used as a measure or indication of the "closeness of fit" of the prediction (or forecasting) equation with the actual data.

GRAPH PAPER

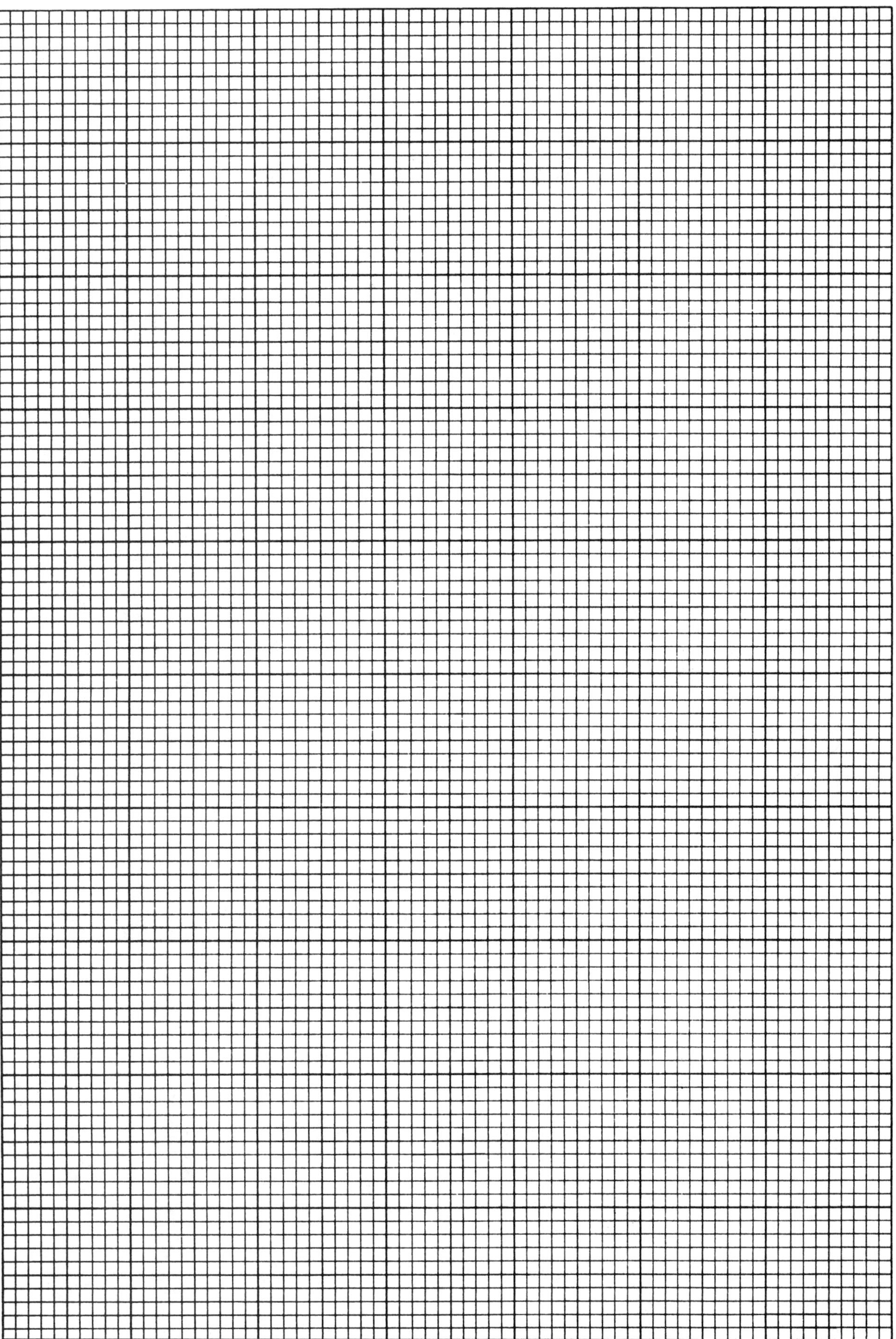

CHAPTER 19

RECORDS AND REPORT WRITING

by

George Gribkoff

(With a Supplement by John Brady)

TABLE OF CONTENTS

Chapter 19. RECORDS AND REPORT WRITING

OBJECTIVES

Chapter 19. RECORDS AND REPORT WRITING

Following completion of Chapter 19, you should be able to:

1. Explain the importance of and need for records.
2. Identify the different types of records.
3. Evaluate records.
4. Organize a report.
5. Write a report.

CHAPTER 19. RECORDS AND REPORT WRITING

19.0 NEED FOR RECORDS AND REPORT WRITING

Most books on plant operation discuss recordkeeping adequately but are very sketchy on their treatment of report writing. However, report writing is the main means by which those who have information communicate with those who need it. Operators must communicate effectively with management and the general public on the operation of their plant and on requests for additional funds for improvements and personnel.

Any business transaction or operation requires records for efficient management; this is also true for operation of waste treatment facilities.

19.1 RECORDS

The administrator, superintendent, and operators of a wastewater treatment facility should know the cost and efficiency of their plant. Well-kept records will make the task of writing treatment plant cost and efficiency reports much easier.

19.10 Importance of Records

Records are needed for the following reasons:

1. Plant Operation. Review of operating records can indicate the efficiency of the plant, performance of its treatment units, past problems, and potential problems.
2. Records are needed to show the type and frequency of maintenance of operating units and to evaluate the effectiveness of maintenance programs. (See Chapter 15, "Maintenance.")
3. Records can provide data upon which to base recommendations for modifying plant operation and facilities.
4. Records of past performance and operational procedures are invaluable tools for the engineer in the evaluation of present performance and serve as a basis for the design of future treatment units.
5. Records are used to support budget requests for personnel, additional facilities, or equipment.
6. Records may be needed in damage suits brought against your district or municipality. They can be especially helpful to the operator if an accident occurs. As soon as possible after an accident, someone should record the chain of events leading to the accident, exactly what happened, and any preventive or corrective action to be taken.
7. Records are required in order to prepare NPDES reports. Additional records containing water quality information may be required by other public or private agencies.
8. Records provide the actual data for the preparation of weekly, monthly, or annual reports to administrative officials, the public, and regulatory agencies.

Records must be permanent, complete, and accurate. Write entries clearly and neatly on data sheets in ink or with an indelible pencil. A lead pencil should never be used because notations can smudge and be altered or erased. False and misleading records may actually do more harm than lack of records.

Recordkeeping costs time and money, and only useful records should be kept. Periodically records no longer useful should be discarded. Lab analyses of receiving water quality should be kept indefinitely. Some compromise is necessary between collecting useless records and avoiding the frustrations of not finding needed information. Keep your records neat and organized. A record misfiled is a record lost, and a lost record is worthless.

19.11 Type of Records

The type of records to be kept will depend on the size and type of plant. A small pond plant may not require the number or the variety of records required of large secondary or advanced wastewater treatment plants.

The specific records required will be determined by the size and type of treatment processes in the plant and are discussed more fully under respective chapters dealing with plant processes. Typical data sheets are included at the ends of chapters on Sedimentation and Flotation, Trickling Filters, Wastewater Stabilization Ponds, Activated Sludge, and Sludge Digestion and Solids Handling.

Records are generally separated into five classifications:

1. Operation and performance records (including NPDES and other regulatory records)
2. Descriptive and inventory records of the physical plant and stock
3. Maintenance records
4. Financial or cost records
5. Personnel records

19.110 Operation Records

A list of the minimum amount of recordkeeping that may be required for operations is as follows:

1. Daily records of flows into the plant.
2. Chemical, physical, and bacteriological characteristics of influent and effluent.

3. Amount of electric power consumed.
4. Amount of chlorine used.
5. Unusual happenings such as bypasses, floods, storms, complaints, and other significant events. Any other unusual event should be recorded if there is any possibility that a record of these events may be needed in the future, either for legal or administrative purposes. The main idea to keep in mind is to record only that data that may eventually be used.
6. A record card or computer file for each industrial waste discharger containing information on type, quantity, characteristics, and times of expected waste discharges.

19.111 Physical Plant and Stock Inventory

As a minimum, the following records are essential for the proper evaluation of plant facilities and for making future recommended modifications or additions.

1. Contract and "as built" plans (record drawings) and specifications of waste treatment facility. This includes detailed piping and wiring of plant.
2. Plans and operating instructions for plant equipment and processes (plant O & M manual).
3. Costs of major equipment and unit items.
4. A complete record and identification card or computer file for all major equipment. The record should include name of manufacturer and identifying code number.
5. Lists of tools, materials, chemicals, lab reagents and supplies, and office supplies.

19.112 Maintenance Records

See Chapter 15, "Maintenance."

19.113 Financial or Cost Records

Keep lists of purchases and expenses to date during the fiscal year. Comparisons should be made with budget allocations to avoid excess purchases.

Public agencies and special districts usually have a purchasing department that has the job of obtaining services, supplies, and equipment. This department must follow the laws of your state and regulations of your agency. Operators must work with purchasing departments and follow their procedures. If your agency or organization does not have a purchasing department, you must develop procedures so you can obtain the necessary services, supplies, and equipment to keep your plant performing properly.

When purchasing supplies, chemicals, or other consumable supplies, you first must locate a vendor (someone who will sell you the supplies). Sometimes the "yellow pages" in the phone book are the best place to start if you do not know a vendor or where to begin. Once you have located what you want and compared the prices and availability of the supplies of at least two vendors, you must start the paper work to obtain the supplies. Procedures vary with different agencies, but the purpose of these procedures is to prevent dishonest people from cheating the taxpaying public by fraudulent and illegal means.

Often the first step is to prepare a purchase order (PO) that describes the supplies you wish to buy, the cost, and the name and address of the vendor. Keep one copy for your files. Next, the PO goes to your purchasing department to approve the purchase. The purchasing department contacts the vendor and orders delivery of the supplies. You should be notified that the purchase is approved and that the vendor has been notified to deliver the supplies. Once the supplies have been delivered, you notify the auditor and authorize payment to the vendor if you are satisfied with the supplies delivered.

19.114 Personnel Records

Employee personnel records, including annual performance ratings, should be maintained for each of the plant employees. Since these records often contain sensitive, confidential information, access to personnel records should be closely controlled.

19.12 Frequency of Records

Records at most wastewater treatment facilities are kept daily and summarized on a monthly basis.

19.120 Daily Records

Data to be recorded will depend on the type and size of the plant. Specific record forms are contained at the end of chapters pertaining to a particular type of treatment plant. One of the most important daily records is a day-by-day diary or log of events and operations during the day. A daily diary or log should be maintained in every plant, and in large plants at each section (such as lab or maintenance). The log may be a spiral notebook or a standard daily diary made for that purpose.

The information entered in the plant log should be pertinent only to plant functions. Log entries should include at the top of the page the day of the week, the date, and the year. The names of the operators working at the plant, and their arrival and departure times, should also be included. Log entries should be made during the day of various activities and problems as they develop. Do not wait until the end of the day to write up the log as some items may be overlooked. If you will take a few minutes to make log entries in the morning and afternoon, you will develop a good log. Logs are beneficial to you and to people who replace you during vacations, illnesses, or leaves of absence. A well-kept log may prove very helpful to the operating agency as legal evidence in certain court cases. An example of one day's log entries in a small trickling filter plant is outlined below:

Tuesday, June 15, 2007

J. Doakes, Operator. A. Smith, Assistant Operator. G. Doe, Maintenance Helper.

8:20 am	Made plant checkout, changed flow charts, No. 2 supernatant tube plugged on No. 2 digester, cleared tube.
9 am	Started drawing sludge from bottom of No. 2 digester to No. 1 sand bed.
9:15 am	Smith and Doe completed daily lubrication and maintenance, put No. 2 filter recirculation pump on, took No. 1 pump off.
10 am	Received three tons of chlorine, containers Nos. 1583, 1296, 495; returned two empty containers Nos. 1891 and 1344. Replaced bad flex connector on No. 2 chlorine manifold header valve, and connected container No. 495 on standby.
10:30 am	Collected and analyzed daily lab samples.
1:15 pm	Pumped scum pit, 628 gallons to No. 1 digester.

1:30 pm Restored sludge pump No. 2 to operation by removing plastic bottle cap from discharge ball check.

2:45 pm Smith and Doe hosed down filter distributor arms and cleaned orifices. Doe smashed finger when closing one of the end gates on filter arm. Sent Doe to Dr. Jones, filled out accident report, and notified Mr. Sharp of accident.

3:10 pm Stopped drawing sludge to No. 1 bed. Drew 18,000 gallons of sludge; sample in refrigerator to be analyzed Wednesday.

3:20 pm Electrician from Delta Voltage Company in with repaired motor for No. 2 effluent pump, Invoice No. A-1 824, motor installed and pump OK.

4:10 pm Doe back from doctor, stated he will lose fingernail, and required three stitches and tetanus shot. Must go back next Thursday.

4:30 pm Plant checkout for tonight, put No. 2 chlorine container on line, in case No. 1 should run empty during the night.

Helpful Tip: Many operators carry a pencil and pocket notebook with them at all times on the job. During the day, they record all events and items of importance and write the information in the plant diary at the end of the day.

A list of the minimum daily records kept at a fairly large treatment plant with digester tanks may include the following items:

1. Precipitation and air temperature
2. Raw wastewater flow, MGD from totalizer (or cu m/day)
3. Influent and effluent temperatures, °F (or °C)
4. pH of influent and effluent
5. Grit, cu ft/M gal (or liters of grit per cu m of flow)
6. Chlorine use, lbs (or kg)
7. Influent and effluent 5-day BOD, mg/*L*
8. Influent and effluent suspended solids, mg/*L*
9. Influent and effluent settleable solids, mg/*L*
10. Raw sludge, gal (or liters or cu m)

    pH
    Total solids, %
    Volatile solids, %
11. Digested bottom sludge

    Total solids, %
    Volatile solids, %
    pH
    Volatile acids and alkalinity, mg/*L*
    Temperature, °F (or °C)
12. Gas produced, cu ft (or cu m)
13. Effluent chlorine residual, mg/*L*, and coliform count MPN/100 m*L*
14. Any unusual influent characteristics such as appearance or odors

19.121 Monthly Records

Monthly records should reflect the totals and averages of the values recorded daily or at some other frequency. In some cases, they should give maximum and minimum daily results, such as maximum and minimum daily flows.

19.13 Keeping and Maintaining Monthly Records

When recording data, always write clearly and neatly. Daily recorded data are usually written on monthly data sheets. The monthly data sheet is designed to meet the reporting needs of a particular plant and should have all important data recorded that may be used later for preparation of monthly or annual reports. The NPDES reporting sheets for regulatory agencies do not contain spaces for critical plant operation information; therefore, they cannot be used as monthly data sheets.

MONTHLY DATA SHEET (See Appendix)

The monthly data sheet may be a single 8½ × 11 inch (21.6 × 27.9 cm) sheet for a small plant with ponds, or it may be a number of sheets pertinent to various treatment units in a secondary or advanced waste treatment plant.

Normally, every plant operator makes up a monthly data sheet for the plant to record daily information. These sheets are numbered from 1 to 31 down the left-hand side to cover 31 days in a month. Then from left to right across the top of the sheet are marked off different columns to record daily information. These columns should contain the day of the week, weather conditions, plant flow, influent and effluent temperature, pH, settleable solids, BOD, raw sludge pumped, digester sludge drawn, gas production, DO, and other pertinent information. A space for remarks is helpful to record and explain unusual events.

Sometimes the operator may use two or three different sheets to collect pertinent data. Since each plant is different, the operator prepares the plant data sheet to record the data needed for proper plant operation and for the requirements of the agency and the regulatory agencies, including NPDES report forms.

In addition to routine daily operation, maintenance, and wastewater characteristics, the monthly data sheet should contain any unusual happenings that may affect interpretation of results and preparation of a monthly report, such as unusual weather, floods, bypasses, breakdowns, or changes in operations or maintenance procedures.

A typical monthly data sheet (Appendix) and monthly report (Section 19.3) for an activated sludge plant are included at the end of this chapter.

19.14 Evaluation of Records

Records are not useful unless they are evaluated and used as indicators of plant operation and maintenance. Records are also useful as sources for reports to management or the public.

The recorded data should enable the operator to determine how to operate and maintain the plant. The information shown by the records should also indicate to you and your supervisor the efficiency of each unit in the plant. Records kept on the quality of the effluent and the receiving waters should be analyzed for the discharge's effect on the receiving waters. The importance of looking at and analyzing records frequently cannot be overemphasized.

Records should not only be analyzed item by item, but any variation should be looked at carefully for its relation to another source of data. For example, a sudden drop in temperature of the influent might be accompanied by greatly increased flows. This could indicate stormwater inflows or infiltration of sewer lines. Infiltration by stormwaters also could influence the BOD and suspended solids concentrations in the plant influent. Or, one might observe a sudden increase in five-day BOD levels in the plant effluent. This may indicate a seasonal increase due to beginning of cannery operations, or it may indicate a breakdown of industrial treatment facilities discharging untreated wastes into the wastewater collection system.

Before any meaningful interpretation can be made of any sudden variation in data, an expected range of values has to be determined for the particular treatment unit under consideration. This range of values will be based on expected or past performance.

For example, if average daily flows during weekdays were around 2 million gallons per day (7.6 M*L*/D) and suddenly a flow of 1.5 million gallons per day (5.7 M*L*/D) was recorded, this may indicate malfunctioning of metering equipment or a break in sewer lines or a bypass ahead of the plant. Conversely, unusually high flows may indicate stormwater infiltration, surface water runoff flowing into the system through manholes, or an unusual dump of wastewater. Remember that any deviation from the expected values may have been caused by unusual circumstances or an error in observation or analysis.

An excellent way to facilitate review of daily records and detect sudden changes or trends is a prepared chart showing values plotted against days. Unless results are plotted, slight changes and trends can go undetected. The Procedures for plotting and interpreting data are provided in Chapter 18, "Analysis and Presentation of Data."

QUESTIONS

Write your answers in a notebook and then compare your answers with those on page 677.

19.1A Why is it important to keep records of plant operation?

19.1B Why should unusual happenings be recorded and described?

19.1C Why do many operators carry a pencil and pocket notebook on the job?

19.1D Why should records frequently be reviewed and analyzed?

19.2 REPORT WRITING

This section will cover the major principles and mechanics of report writing, the type of report usually required of a plant operator, and a discussion and example of effective writing.

To many, the thought of writing a report represents a task that is to be approached with fear and with a sense of inadequacy. This need not be the case. Anyone who can read and is willing to put forth the effort can prepare an adequate report. The typical treatment plant operator may regard the writing of reports as an unwelcome chore and thus may approach the subject with a natural resistance.

You should approach the task of report writing as if your next pay raise depended on a neat, organized, and brief report. One operator's annual report was so well written that a local newspaper used the report to develop a six-article series on the treatment plant. The newspaper stories explained the operation of the plant and its effectiveness in protecting the fish life, water supplies, and recreational uses of the receiving waters. Shortly after the articles appeared in the newspaper the operator received a substantial increase in salary.

19.20 Importance of Reports

A report serves many purposes. Reports can serve as the basis of requests for additional budget and personnel, plant additions, or changes in plant operation. A report also is a means by which your ability, actions, and knowledge are communicated to management and your supervisor. You should consider a report as an opportunity to tell your story to your supervisor, management, or the general public. Your ability to prepare and submit effective reports is one factor considered for advancement in your profession.

Furthermore, you and your plant operation are partially judged by the information contained in your report, its style, and its appearance. A poorly prepared report may result in an impression that the plant is not operating efficiently; or, still worse, it can result in little action on or rejection of recommendations or requests.

You must do more than operate your plant efficiently; you must demonstrate this to your supervisors, administrators, and regulatory agencies in a clearly understandable and well prepared report.

The narrative type of report writing will be discussed first because it is the nonroutine part of a typical monthly, quarterly, or annual report.

19.21 Major Principles of Report Writing

Whatever the report or its size, there are some basic principles common to good report writing:

1. Know the purpose and objective of your report.
2. Tailor your report to the person or persons to whom it is directed.
3. Know your subject.
4. Organize the report to present your ideas in a logical order.
5. Use language in the report that will be understood by the reader.
6. Use facts and figures.
7. Be as exact and brief as possible.
8. Write effectively.

In starting to prepare a narrative report, the most important questions are what is the purpose of this report and for whom is it written? The next important step is the organization of the ideas and subject matter in a logical order.

19.22 Organization of the Report

There is no single way to organize a formal report. You must remember that a written report does not necessarily organize and present the material in the same order in which the information was gathered. Organization means simply that the topics in the report are set forth in logical sequence to tell the story effectively.

Some reports may follow a general format such as:

1. Brief statement of problem
2. Summary
3. Conclusions and recommendations
4. Body of report
   a. Technical and administrative background
   b. Investigation details
   c. Any necessary supporting material to back up conclusions and recommendations
   d. Appendix (if necessary) including detailed data and tables to support body of report

The conclusion and recommendation section of a report should receive the most attention and review. Conclusions and recommendations must be stated clearly and briefly and in language that will be understood by your readers. Be sure that your conclusions and recommendations are supported in the body of your report.

A memorandum to a supervisor or a narrative portion in a report should be checked for organization and content as follows:

1. Is the material presented in an organized way?
2. Is there duplication?
3. Is there omission of an important item?
4. Is the material presented really necessary to make your point and support your conclusions or recommendations?

Unnecessary material in a communication or report only serves to weaken your case by clouding the main issues.

The above list applies to almost any written communication and can be summarized by the four Cs of good report writing:

1. Conciseness
2. Clearness
3. Completeness
4. Candor

19.23 Mechanics of Writing a Report

In the previous section, examples were given on organizational plans for formal reports; but what are the mechanics of writing a memorandum, a short report, or a section in a larger report?

Following are some guidelines for preparing a report:

1. List ideas and topics you plan to cover in the report.
2. Arrange the ideas in logical sequence.
3. Gather material needed to support the ideas to be presented in the report.
4. Begin writing—this is most important! Prepare a rough draft of the report based on listed ideas and their organization. At this stage of preparation, writing without undue concern about sentence structure or grammar is suggested. At this stage it is more important to record your ideas. Later, when rewriting the preliminary draft, it is much easier to eliminate unnecessary material than it is to add to the report.
5. Prepare conclusions and recommendations, if any, after writing the main body of the report.
6. Review the preliminary draft for content, logical presentation, and organization, and eliminate any unnecessary information. At this stage you can reorganize your order of presentation for a more effective report and check for simplicity and understandability.
7. Review the report for sentence sense, spelling, grammar, and briefness.
8. Make another draft (if necessary).
9. Have a colleague or supervisor review the draft if possible.
10. Finally, check your report for overall effectiveness:
    a. Will your initial statements create interest in the contents of the report?
    b. Will the reader understand it?
    c. Is the presentation of ideas in logical order?
    d. Are major points emphasized?
    e. Has all irrelevant material been eliminated?
    f. Are the sentences direct and effective?
    g. Is the report neat and attractive?
    h. Does the report support your conclusions and recommendations?

A report does not have to be a literary masterpiece. The more factual and brief it is, the more likely it is to be favorably considered.

19.24 Effective Writing

While the organization of a report and presentation of ideas in a logical manner are the chief components of a good report, effective writing is also necessary.

Effective writing is simply the putting together of words in a grammatically correct and brief manner, eliminating needless words, expressions, and repetitions. A good technical report impresses no one favorably if it is full of flowery and confusing language.

If you wish to learn how to be a more effective writer, you could enroll in an adult education class on effective writing. Perhaps the teacher will allow you to bring your monthly reports to class and will offer suggestions on how to improve your writing. Another approach is to go to your local library and borrow books on effective writing. Section 19.5, "Additional Reading," lists two books that can help you develop and improve your writing skills.

The next few paragraphs show examples of different styles of writing. The first example is the use of direct versus indirect writing. Note that the direct style uses thirteen fewer words, which cuts the paragraph by about 30 percent.

INDIRECT: This report, which you requested in your letter of December 15, 2006, on the efficiency of the trickling filter units is submitted for your approval. It is concerned with removal of organic material and future operation using different size of filter media.

DIRECT: Enclosed for your approval is the report you requested (letter dated December 15, 2006) on trickling filter efficiency for removal of organic material and possible filter media size changes.

Whenever possible, use active sentence construction rather than passive.

PASSIVE: It is recommended that the monitoring of the effluent be started immediately.

ACTIVE: Monitoring the effluent should start immediately.

Parallel sentence construction will make your sentence clearer.

NON-PARALLEL: The supervisor pointed out *HOW* Brown opposed progress, *HOW* he encouraged the men to slow down, *THAT* he never showed initiative, and *THAT* he could not maintain the machines.

PARALLEL: The supervisor pointed out *THAT* Brown opposed progress, *THAT* he encouraged the men to slow down, *THAT* he never showed initiative, and *THAT* he could not maintain the machines.

19.25 Types of Reports

There are many types of reports ranging all the way from a memorandum to an annual report to management.

Specifically, it may be: (1) a monthly plant operation report, (2) a report to a regulatory agency, such as a health department, or (3) a quarterly or annual report to management or the public.

19.250 Monthly Reports (See Section 19.3)

The monthly reports are used in the preparation of the annual report. Preparation of the monthly report consists of the following preliminary activities:

1. Gathering daily records
2. Reviewing daily log sheets for any significant or unusual events during the month
3. Jotting down ideas that one wishes to include in the narrative section of the report

In small plants, a monthly report may consist mainly of data sheets giving the pertinent facts on:

1. Laboratory analyses and effectiveness of plant treatment and its various units
2. Cost data on labor, chemicals, and maintenance
3. Maintenance records
4. Remarks stating unusual significant events during the month
5. Effect on receiving waters
6. Conclusions and recommendations

In some larger plants, a monthly report may be required in addition to the monthly data sheets.

The monthly report is a brief summary of combined information from the monthly data sheets and daily logs. The report is put together in outline form with a dozen or so subheadings required for a secondary plant. The reports are useful to operators and supervisors to keep them informed of plant functions and problem areas.

The subheadings may reflect the flow pattern through the plant. The monthly report generally describes, in narrative form, the physical flow characteristics, maintenance and operation problems, and unusual events during the month.

MONTHLY REPORT

1. Flow
   a. Total amount of flow passed through the plant for month
   b. Maximum daily flow
   c. Minimum daily flow
   d. Average daily flow
   e. Flowmeter problems
2. Headworks
   a. Screening: shredding device, operation and maintenance problems
   b. Screen material removed, cu ft/MG (liters/cu m)
   c. Grit removed, cu ft/MG (liters/cu m)
   d. Unusual material in the wastewater, such as oil, silt; odors
3. Primary clarifiers
   a. Operation and maintenance problems
   b. Scum removal, note plugged scum lines

   c. Sludge pumped, solids concentration
   d. Influent and effluent characteristics

4. Secondary treatment system
   a. Trickling filter
      (1) Loading rates, average
      (2) Recirculation rates, average
      (3) Maintenance problems
      (4) Removal efficiencies
   b. Activated sludge
      (1) Loading rates, average
      (2) Mixed liquor concentration, average
      (3) Mixed liquor DO level
      (4) Removal efficiencies
5. Secondary clarifiers
   a. Operation and maintenance problems
   b. Sludge pumped, solids concentration
   c. Effluent characteristics
6. Chlorination
   a. Pounds (kg) of chlorine used/month
   b. Pounds (kg) of chlorine used/day, average
   c. Chlorine residuals
   d. Chlorinator problems
   e. Average dosage, mg/*L*
7. Dechlorination
   a. Pounds (kg) of sulfur dioxide used/month
   b. Pounds (kg) of sulfur dioxide used/day, average
   c. Chlorine residuals before and after dechlorination
   d. Sulfur dioxide equipment problems
   e. Average dosage, mg/*L*
8. Outfall
   a. Effluent characteristics
   b. General appearance and condition around plant discharge
   c. Condition of receiving waters
9. Raw sludge pumps
   a. Problems and maintenance
10. Digesters
   a. Raw sludge pumped to digesters, gallons (liters) and percent solids
   b. Gas production, cu ft (or cu m)
   c. Temperature, pH, volatile acids, alkalinity, volatile solids
   d. Operation problems and maintenance performed
   e. Volatile solids destruction
11. Sludge drying beds
   a. Gallons (liters) of sludge drawn
   b. Yards (cu m) of dry cake removed
   c. Moisture content of cake or pounds (kg) of dry solids
   d. Maintenance or cleaning problems
12. Gas system and boilers
   a. Operation and maintenance
13. General
   a. Power failures
   b. Accidents
   c. Visitors
   d. Grounds and building maintenance
   e. Plant cost
      (1) Operator-hours worked
      (2) Equipment parts
      (3) Power and fuel
      (4) Chemicals
      (5) Miscellaneous

19.251 Annual Reports

The annual report receives wider distribution and is the report more likely to be reviewed by the public, management, and other governmental agencies.

The annual report is, in part, a compilation of data obtained in the monthly reports. This report summarizes the plant's yearly efficiency and cost data. The report also contains an analysis of plant operation costs and recommendations for next year's operations.

In addition, an annual report should contain a short introduction to provide a background to the reader, giving a brief history and reason for the report.

The body of the report should contain schematic drawings, pictures, and other attractive graphs whenever possible.

An annual report should include at least the following items:

1. A letter of transmittal
2. Conclusions and recommendations
3. A brief description and schematic diagram of the system
4. An organization chart showing the various functional divisions and their chiefs
5. A statistical summary of general plant data such as:
   a. Population served
   b. Wastewater flows (maximum, average, minimum)
   c. Plant unit data, percent removal, and efficiency of various units

6. Body of report that includes a brief description and supporting tables, graphs, or charts needed to back up final recommended actions or requests on such topics as:
   a. Wastewater quality
   b. Chlorination
   c. Screening
   d. Pumping
   e. Sludge digestion
   f. Receiving water quality (maps, summary data)
   g. Maintenance and repair
   h. Financial data such as assets, liabilities, revenue
7. Appendix that includes summary data by month for the annual report year giving minimum, maximum, and average values for:
   a. General plant data
   b. Treatment unit data
   c. Loadings and efficiency of treatment
   d. Chemical, physical, and bacteriological data on influent and effluent
   e. Chemical, physical, and bacteriological data on receiving waters

Report writing can seem like a difficult task, especially for an operator not experienced in report writing. But with guidelines provided here, a review of effective writing, and some real effort, anyone who can read can produce an effective report.

19.26 Obtain Reports by Other Operators

A very helpful guide to writing a report is to obtain a report written by another operator. Usually a report may be obtained by writing to a nearby city or operator and asking for a copy. If you explain that you are an operator and would appreciate a copy of their report, your request will probably be granted. A representative of a regulatory agency or your plant consulting engineer should be able to recommend to you examples of well-written reports.

QUESTIONS

Write your answers in a notebook and then compare your answers with those on page 677.

19.2A What is the purpose of writing monthly and annual reports?

19.2B How could you obtain a copy of a report from another plant?

19.3 TYPICAL MONTHLY REPORT

CLEANWATER TREATMENT PLANT
MONTHLY REPORT FOR JUNE 2007

by John Brady

FLOW: Cleanwater treated a total raw wastewater flow of 68.497 million gallons this month, with a daily average of 2.283 MG. There were no unusual flow conditions during the month.

GRIT CHANNEL: The grit channel was cleaned on 6/23, with 1.5 cubic yards of grit removed, consisting mainly of eggshells and sand.

SCREENING: The top bearing of barminutor No. 1 travel motor was replaced on 6/4 and a spring on the micro switch was also replaced. A broken comb was replaced on the No. 2 barminutor on 6/15 and all combs on that unit were reset to 0.006 inch clearance.

RAW WASTEWATER PUMPS: No problems with No. 1 and No. 3 pumps. No. 2 pump was repacked on 6/9.

PRIMARY CLARIFIERS: On 6/21 the No. 2 primary clarifier was dewatered for annual inspection. The mechanism was in good condition and only required resetting the clearance of the brass plow squeegees to their original ⅛ inch. The tank weirs and scum baffle were wire brushed and repainted with 395-A. The tank was returned to service on 6/26. While the No. 2 primary was out of service, the No. 1 primary carried the full plant load without any detrimental effect on the efficiency of the plant.

AERATOR: No problems. The aerator loading was maintained at 25 pounds of BOD per day per 100 pounds of mixed liquor suspended solids, and a constant return sludge rate of 30%.

FINAL CLARIFIER AND RETURN SLUDGE PUMPS: No operational problems with the final clarifier. The No. 2 return sludge pump was returned from J & M Machine Shop on 6/1 and reinstalled. J & M replaced the shaft sleeve and both shaft bearings at a cost of $182.36 (Invoice No. 34475). However, the pump was not ready for service until 6/2 as J & M had packed the pump bearings with an all-purpose medium industrial lubricant rather than the F.M. oil film low temperature grease of –65° to 100°F, as specified.

CHLORINATION: No problems. Used 9,950 pounds of chlorine at an average rate of 335 pounds per day. One hundred twenty-five pounds per day were used for postchlorination maintaining an average chlorine residual of 4.4 mg/*L*. Two hundred ten pounds per day were used for prechlorination for odor control.

OUTFALL: Other than the foam buildup around the outfall and the foam drift downstream for approximately 500 yards, the receiving stream was in good condition. The stream sampling below the outfall remained at 8.9 mg/*L* DO, 2.0 mg/*L* BOD, and average temperature of 58°F.

RAW SLUDGE PUMP: On 6/9 and 6/30, the raw sludge pump was plugged with a piece of plastic and a wooden stick under the discharge ball check. In each case, the pump was restored to service during the 8 am shift.

DIGESTERS: Digester No. 1 was operated as the primary and No. 2 as the secondary. The temperature in the No. 1 tank was raised from 91°F to 94°F. During the month, the tank was continuously mixed. The recirculated sludge contained an average volatile acids content of 150 mg/*L*, with the alkalinity at 2,550 mg/*L* (volatile acid/alkalinity relationship of 0.06).

SLUDGE DRYING: Supernatant from the No. 2 digester became heavy on 6/12 with the settleable solids ranging from 9 to 15% by volume. On 6/17, 28,000 gallons of digested sludge were drawn from the No. 2 digester to the No. 3 drying bed to reduce supernatant load. The drawn sludge contained 8.3% solids with a volatile content of 52.6%.

The No. 1 and No. 4 drying beds were cleaned, yielding a total of 63 cubic yards of dry sludge.

GAS SYSTEM AND BOILER: On 6/7 it was found that low gas production was recorded for 6/6. The No. 2 digester pressure relief was found to be venting at various times. The entire gas system piping units were cleaned and inspected with the problem location found on 6/13 in the gas meter itself. The condensate and residue had gummed up the gear train from the bellows slide arms. The unit was cleaned with kerosene, relubricated with molly cote, and returned to service with no further problems.

POWER SUPPLY: There were two power outages this month, on 6/24 and 6/27, with the plant being out of service 40 to 45 minutes each time. The cause of the outages was due to a service fuse dropping on one phase at the utility pole by the main gate, leaving only the two phases from which to operate.

Each time, the main power board was shut down to protect plant equipment.

GENERAL:

6/3 Replaced broken hinge on main gate.

6/6 Mosquito abatement personnel moved their oil storage tank from the plant grounds.

6/15 Left main gate barricade chopped down by vandals.

6/17 Replaced left main gate barricade.

6/19 State Water Pollution Control engineer visited plant and collected effluent samples.

6/24 Received 400 return sludge meter charts (Invoice No. 111323).

6/25 Flame-Out Fire Equipment Supply Company representative in and made yearly check on plant fire extinguishers.

Submitted: /s/ John J. Smith
Chief Operator

19.4 EMERGENCY PLANNING

Emergency planning is one more item that must be considered in order to ensure that your treatment plant performs as well as possible under all conditions. Your plant should have written procedures that serve as a guide for operators on duty when a disaster occurs. Potential disasters include:

1. Hazardous spills reaching the collection system
2. Storms and floods
3. Earthquakes
4. Nuclear attacks or accidents
5. Personnel work stoppages
6. Physical injuries
7. Hazardous chemical releases (chlorine)
8. Fires
9. Power failures
10. Bomb threats

Try to list the potential disasters that could occur at your treatment plant. After you have listed the disasters, prepare procedures for the operators to follow to overcome and correct the situation.

These procedures should outline, on a step-by-step basis, the items that should be considered.

1. How to evaluate the seriousness of the problem
2. Who to notify regarding the disaster
3. What information will be needed by the persons notified
4. Where the necessary emergency, safety, and repair equipment and vehicles are located
5. How to obtain outside assistance if necessary
6. Sources of outside assistance
7. Procedures to follow to repair or correct the situation

How to respond to hazardous spills and chlorine leaks has been discussed in appropriate chapters throughout this manual. Also see Chapter 15, "Maintenance," Section 15.042, "Emergencies" and Chapter 20, "Treatment Plant Administration," Section 20.10, "Emergency Response." Your procedures should be short, easy to understand, and practical.

QUESTION

Write your answer in a notebook and then compare your answer with the one on page 677.

19.4A List five potential disasters that could occur at a wastewater treatment plant.

19.5 ADDITIONAL READING

1. *MOP 11,* Chapter 6,* "Management Information Systems (Reports and Records)." New Edition of MOP 11, Volume I, *MANAGEMENT AND SUPPORT SYSTEMS.*
2. *NEW YORK MANUAL,* pages 119–156.*

* Depends on edition.

Please answer the discussion and review questions next.

DISCUSSION AND REVIEW QUESTIONS

Chapter 19. RECORDS AND REPORT WRITING

The purpose of these questions is to indicate to you how well you understand the material in the chapter. Write the answers to these questions in your notebook.

1. Why must operators be capable of effective report writing?
2. Why are records for a wastewater treatment plant important?
3. How should operators approach the task of report writing?
4. Why are reports important?

SUGGESTED ANSWERS

Chapter 19. RECORDS AND REPORT WRITING

Answers to questions on page 671.

19.1A It is important to keep records of plant operation for the following reasons:

1. Review of operating records can indicate the efficiency of the plant, performance of its treatment units, past problems, and potential problems.
2. Records are needed to show the type and frequency of maintenance and to evaluate the effectiveness of maintenance programs.
3. Records can provide data upon which to base recommendations for modifying plant operation and facilities.
4. Records of past performance and operational procedures are invaluable tools for the engineer in the evaluation of present performance and serve as the basis for the design of future treatment units.
5. Records are used to support budget requests for personnel, additional facilities, or equipment.
6. Records may be needed in damage suits brought against your district or municipality.
7. Records are required in order to prepare NPDES reports.
8. Records provide the actual data for the preparation of reports to administrative officials, the public, and regulatory agencies.

19.1B Unusual happenings should be recorded and described because they have an important influence on the interpretation of laboratory analyses describing the operation and efficiency of your plant and the condition of the receiving water. Also, they could prove very helpful to the operator in case of an accident.

19.1C Many operators carry a pencil and pocket notebook on the job to record all important events as they occur during the day.

19.1D Records should frequently be evaluated to determine if your plant is operating properly and to identify any developing difficulties before they can become serious problems.

Answers to questions on page 675.

19.2A The purpose of monthly and annual reports is to provide a brief description of the operation of your plant for the benefit of management and regulatory agencies.

19.2B Write a letter to another operator or city and ask for a copy of one of their reports.

Answer to question on page 676.

19.4A Potential disasters that could occur at a wastewater treatment plant include:

1. Hazardous spills
2. Storms and floods
3. Earthquakes
4. Nuclear attacks or accidents
5. Personnel work stoppages
6. Physical injuries
7. Hazardous chemical releases (chlorine)
8. Fires
9. Power failures
10. Bomb threats

Listing any five of these potential disasters will be satisfactory. You may include a disaster that could occur at your plant that we have overlooked.

APPENDIX

Monthly Data Sheet

CLEANWATER, U.S.A.

WATER POLLUTION CONTROL PLANT

MONTHLY RECORD ________ 20___ OPERATOR: ____________

DATE	DAY	WEATHER	FLOW – MGD	RAW WASTEWATER					PRIM. EFF.			FINAL EFFLUENT					AERATION SYSTEM									
				TEMP.	pH	SETT. SOLIDS	B.O.D.	SUSP. SOLIDS	B.O.D.	SUSP. SOLIDS	D.O.	pH	B.O.D.	SUSP. SOLIDS	D.O.	CL_2 RES.	LBS. VOL. SOLIDS	SUSP. SOLIDS	% VOL.	30MIN. SETT.	S.V.I.	D.O.	RETURN SUSP. SOLIDS	RETURN-MGD	WASTE GAL. X 1000	WASTE LBS./DAY
1	S	CLEAR	1.782	75	7.2	14			118	84	0.6	6.9	19	18	0.9	2.7	6746	2036	78.9	150	73	2.5	5961	0.702	70	3480
2	M	CLEAR	2.347	74	7.3	13	218	150	156	84	0.3	6.8	19	15	1.0	2.8	6859	2078	78.6	150	72	1.4	4683	0.711	72	2812
3	T	CLEAR	2.165	74	7.3	8			109	66	0.8	6.8	14	9	1.2	8.8	7224	2211	77.8	170	76	2.1	6625	0.708	71	3922
4	W	CLEAR	2.012	74	7.3	12			135	74	0.5	6.8	16	14	0.8	4.4	7305	2213	78.6	180	81	2.0	6641	0.712	70	3877
5	T	CLEAR	2.483	74	7.2	13	189	138	134	62	0.3	6.8	9	11	1.7	5.2	7014	2106	79.3	170	80	2.6	6098	0.722	78	3966
6	F	CLEAR	2.386	74	7.2	13			112	60	0.4	6.8	18	6	2.6	6.0	6754	2069	79.0	160	77	3.5	5862	0.700	80	3911
7	S	CLEAR	2.131	75	7.3	13			89	66	0.7	6.9	14	7	1.2	4.4	6296	1905	78.7	150	78	2.6	5564	0.706	80	3712
8	S	CLEAR	1.867	76	7.4	12	174	134	84	74	0.9	6.9	9	15	0.8	4.2	7057	2138	78.6	180	84	0.9	6758	0.703	72	4058
9	M	CLEAR	2.634	75	7.3	14			117	68	0.3	6.9	11	11	1.6	3.5	6767	2037	79.1	160	78	2.5	6022	0.712	70	3515
10	T	CLEAR	2.307	76	7.3	18			120	66	0.3	7.1	8	8	1.5	6.6	6119	1861	78.3	170	91	2.8	6135	0.705	64	3274
11	W	CLEAR	2.198	76	7.3	18	192	142	111	68	0.4	7.0	9	10	1.1	6.6	7035	2123	78.9	200	94	1.7	6183	0.700	70	3609
12	T	CLEAR	2.202	76	7.3	11			99	72	0.4	7.0	11	9	2.0	3.8	6352	1954	77.4	190	97	3.0	6027	0.704	70	3518
13	F	CLEAR	2.178	77	7.3	11			81	58	0.6	7.0	15	18	3.5	4.0	6313	1937	77.6	170	87	4.8	5542	0.689	72	3327
14	S	CLEAR	2.006	78	7.3	12	155	156	105	76	0.3	6.9	12	8	3.1	3.8	6335	1929	78.2	160	82	4.3	4856	0.703	70	2834
15	S	CLEAR	1.942	78	7.2	12			113	74	0.4	6.9	9	9	1.3	4.4	6873	2090	78.3	180	86	2.2	5753	0.711	73	3502
16	M	CLEAR	2.464	78	7.2	11			128	74	0.3	6.9	10	10	1.8	3.0	7082	2162	78.0	200	92	2.5	6852	0.723	76	4343
17	T	CLEAR	2.321	78	7.1	8	168	144	110	64	0.4	6.8	10	11	1.9	6.6	6215	1937	76.4	190	98	2.8	6654	0.698	74	4106
18	W	CLEAR	2.611	78	7.3	15			105	64	0.2	6.9	11	7	2.2	4.4	6227	1923	77.1	190	98	3.3	5767	0.717	83	3992
19	T	CLEAR	2.457	78	7.3	12			87	72	0.5	6.9	10	7	2.9	6.2	4844	1534	75.2	170	110	4.5	4762	0.721	25	992
20	F	CLEAR	2.498	79	7.3	11	193	118	105	66	0.7	6.9	18	12	3.1	4.4	5846	1822	76.4	190	104	4.1	5123	0.719	0	0
21	S	CLEAR	2.213	76	7.1	12			109	76	0.6	7.1	10	9	1.2	4.2	6892	2096	78.3	200	95	2.6	5928	0.706	35	1730
22	S	CLEAR	1.878	76	7.3	12			131	78	0.4	6.9	14	10	0.5	4.2	7518	2263	79.1	260	114	1.8	3894	0.703	35	1136
23	M	CLEAR	2.901	77	7.3	12	187	142	133	89	0.2	6.9	13	13	0.3	2.5	8388	2541	78.6	310	121	1.9	8396	0.741	70	4901
24	T	N.M. CLEAR	2.346	78	7.3	13			114	56	0.3	6.9	14	10	2.2	4.2	7962	2409	78.7	230	95	3.6	8824	0.700	71	5225
25	W	CLEAR	2.421	78	7.3	13			89	56	0.4	7.0	12	8	2.8	4.0	7688	2332	78.5	230	98	4.1	7382	0.713	72	4432
26	T	CLEAR	2.562	79	7.3	12	212	170	143	87	0.6	7.0	10	6	1.7	4.3	6697	2047	77.9	210	102	2.6	6867	0.698	70	4008
27	F	CLEAR	2.428	79	7.3	10			128	84	0.5	6.8	15	10	0.5	3.8	6923	2103	78.2	200	94	1.2	7436	0.702	64	3969
28	S	CLEAR	2.149	78	7.3	12			84	66	0.9	6.9	16	5	0.6	3.5	7169	2180	78.3	200	91	1.7	8412	0.706	68	4770
29	S	CLEAR	1.862	79	7.3	7	176	102	117	60	0.5	6.9	14	8	0.5	3.9	7852	2397	78.0	230	95	1.0	7117	0.700	66	3917
30	M	CLEAR	2.746	79	7.3	13			107	73	0.2	6.8	8	8	1.6	3.5	7688	2335	78.4	220	94	2.9	7735	0.713	70	4515
31																										
MAX			2.901	79	7.4	18	218	170	156	89	0.9	7.1	19	18	3.5	8.8	8388	2541	79.3	310	121	4.8	8824	0.741	83	5225
MIN			1.782	74	7.1	7	155	102	84	56	0.2	6.8	8	5	0.5	2.5	4844	1534	76.4	150	72	0.9	4683	0.698	0	0
AVG			2.283	77	7.3	12	186	139	112	70		6.9	12	10	1.6	4.4	6868	2092	78.1	192	91	2.6	6328	0.708	65	3511

SUMMARY DATA

% REMOVAL	B.O.D.	S.S.
INF – PRI	39.7	49.6
INF – EFF	93.5	92.8

SLUDGE DATA

% SOLIDS — AVG.	5.6
LBS. DRY SOLIDS / DAY	5579
% VOL. SOLIDS — AVG.	79.8
LBS. VOL. SOLIDS / DAY	4452
LBS. VOL. SOL./1000 FT^3/DAY	67.5
GALS. SLUDGE TO BEDS	28,000
CU. YDS. CAKE REMOVED	63
FT^3 GAS / LB. VOL. SOLIDS	6.8
FT^3 GAS / LB. MG FLOW	14,286

COST DATA

MAN DAYS 63 PAYROLL	2,325.78
POWER PURCHASED	520.32
OTHER UTILITIES (GAS, H_2O)	NONE
GASOLINE, OIL, GREASE	108.56
CHEMICALS AND SUPPLIES	547.25
MAINTENANCE	238.48
VEHICLE COSTS	NONE
OTHER	NONE
TOTAL	$ 3,740.39
OPER. COST/MG TREATED	$ 54.62
OPER. COST/CAPITA/MO.	$ 0.158

FLOW METER:	ELECTRIC METER:	RAW SLUDGE:	GAS METER:	RETURN SLUDGE:	WASTE SLUDGE:
LAST 222046	LAST 7838	LAST 798324	LAST 2181110	LAST 67635048	LAST 134251
1st 153549	1st 5670	1st 432984	1st 1265230	1st 67613800	1st 132560
TOTAL 68497 MG	TOTAL 2168	STROKES 365340	TOTAL 915880 FT^3	TOTAL 21248 MG	TOTAL 1961 x 1000 MG
	MULT 40 X 2168 = 86720 KWH	TOTAL 365340 X 1.0 = 365340 GALS			

CHAPTER 20

TREATMENT PLANT ADMINISTRATION

by

Lorene Lindsay

With Portions by

Tim Gannon

and

Jim Sequiera

TABLE OF CONTENTS

Chapter 20. TREATMENT PLANT ADMINISTRATION

OBJECTIVES

Chapter 20. TREATMENT PLANT ADMINISTRATION

Following completion of Chapter 20, you should be able to:

1. Identify the functions of a manager.
2. Describe the benefits of short-term, long-term, and emergency planning.
3. Define the following terms:
   a. Authority
   b. Responsibility
   c. Delegation
   d. Accountability
   e. Unity of command
4. Read and construct an organizational chart identifying lines of authority and responsibility.
5. Write a job description for a specific position within the utility.
6. Write good interview questions.
7. Conduct employee evaluations.
8. Describe the steps necessary to provide equal and fair treatment to all employees.
9. Prepare a written or oral report on the plant's operations.
10. Communicate effectively within the organization, with media representatives, and with the community.
11. Describe the financial strength of your utility.
12. Calculate your utility's operating ratio, coverage ratio, and simple payback.
13. Prepare a contingency plan for emergencies.
14. Set up a safety program for your utility.
15. Collect, organize, file, retrieve, use, and dispose of plant records.
16. Describe the difference between illegal activities and ethical situations.
17. Explain why ethical behavior is important for operators.

WORDS

Chapter 20. TREATMENT PLANT ADMINISTRATION

ACCOUNTABILITY — ACCOUNTABILITY

When a manager gives power/responsibility to an employee, the employee ensures that the manager is informed of results or events.

AIR GAP — AIR GAP

An open, vertical drop, or vertical empty space, between a drinking (potable) water supply and potentially contaminated water. This gap prevents the contamination of drinking water by backsiphonage because there is no way potentially contaminated water can reach the drinking water supply.

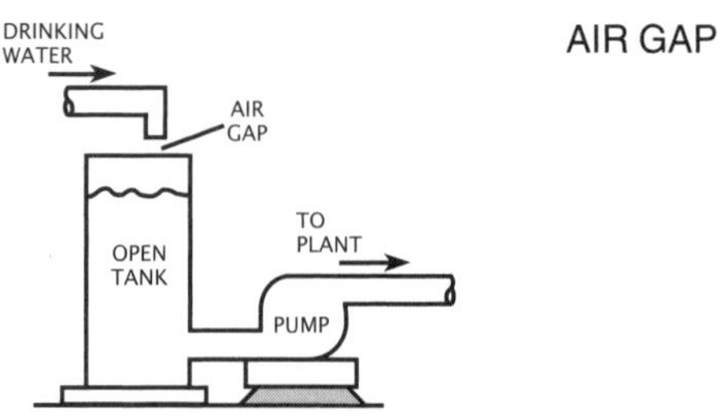

AUTHORITY — AUTHORITY

The power and resources to do a specific job or to get that job done.

BACKFLOW — BACKFLOW

A reverse flow condition, created by a difference in water pressures, that causes water to flow back into the distribution pipes of a potable water supply from any source or sources other than an intended source. Also see BACKSIPHONAGE.

BACKSIPHONAGE — BACKSIPHONAGE

A form of backflow caused by a negative or below atmospheric pressure within a water system. Also see BACKFLOW.

BOND — BOND

(1) A written promise to pay a specified sum of money (called the face value) at a fixed time in the future (called the date of maturity). A bond also carries interest at a fixed rate, payable periodically. The difference between a note and a bond is that a bond usually runs for a longer period of time and requires greater formality. Utility agencies use bonds as a means of obtaining large amounts of money for capital improvements.

(2) A warranty by an underwriting organization, such as an insurance company, guaranteeing honesty, performance, or payment by a contractor.

CALL DATE — CALL DATE

First date a bond can be paid off.

CERTIFICATION EXAMINATION — CERTIFICATION EXAMINATION

An examination administered by a state agency or professional association that operators take to indicate a level of professional competence. In the United States, certification of operators of water treatment plants, wastewater treatment plants, water distribution systems, and small water supply systems is mandatory. In many states, certification of wastewater collection system operators, industrial wastewater treatment plant operators, pretreatment facility inspectors, and small wastewater system operators is voluntary; however, current trends indicate that more states, provinces, and employers will require these operators to be certified in the future. Operator certification is mandatory in the United States for the Chief Operators of water treatment plants, water distribution systems, and wastewater treatment plants.

CODE OF FEDERAL REGULATIONS (CFR) — CODE OF FEDERAL REGULATIONS (CFR)

A publication of the US government that contains all of the proposed and finalized federal regulations, including safety and environmental regulations.

CONFINED SPACE CONFINED SPACE

Confined space means a space that:

(1) Is large enough and so configured that an employee can bodily enter and perform assigned work; and

(2) Has limited or restricted means for entry or exit (for example, manholes, tanks, vessels, silos, storage bins, hoppers, vaults, and pits are spaces that may have limited means of entry); and

(3) Is not designed for continuous employee occupancy.

Also see DANGEROUS AIR CONTAMINATION and OXYGEN DEFICIENCY.

CONFINED SPACE, PERMIT-REQUIRED (PERMIT SPACE) CONFINED SPACE, PERMIT-REQUIRED (PERMIT SPACE)

A confined space that has one or more of the following characteristics:

(1) Contains or has a potential to contain a hazardous atmosphere,

(2) Contains a material that has the potential for engulfing an entrant,

(3) Has an internal configuration such that an entrant could be trapped or asphyxiated by inwardly converging walls or by a floor that slopes downward and tapers to a smaller cross section, or

(4) Contains any other recognized serious safety or health hazard.

COVERAGE RATIO COVERAGE RATIO

The coverage ratio is a measure of the ability of the utility to pay the principal and interest on loans and bonds (this is known as debt service) in addition to any unexpected expenses.

DEBT SERVICE DEBT SERVICE

The amount of money required annually to pay the (1) interest on outstanding debts, or (2) funds due on a maturing bonded debt or the redemption of bonds.

DELEGATION DELEGATION

The act in which power is given to another person in the organization to accomplish a specific job.

MATERIAL SAFETY DATA SHEET (MSDS) MATERIAL SAFETY DATA SHEET (MSDS)

A document that provides pertinent information and a profile of a particular hazardous substance or mixture. An MSDS is normally developed by the manufacturer or formulator of the hazardous substance or mixture. The MSDS is required to be made available to employees and operators or inspectors whenever there is the likelihood of the hazardous substance or mixture being introduced into the workplace. Some manufacturers are preparing MSDSs for products that are not considered to be hazardous to show that the product or substance is not hazardous.

OSHA (O-shuh) OSHA

The Williams-Steiger Occupational Safety and Health Act of 1970 (OSHA) is a federal law designed to protect the health and safety of workers, including the operators of water supply and treatment systems and wastewater collection and treatment systems. The Act regulates the design, construction, operation, and maintenance of water and wastewater systems. OSHA regulations require employers to obtain and make available to workers the Material Safety Data Sheets (MSDSs) for chemicals used at industrial facilities and treatment plants. OSHA also refers to the federal and state agencies that administer the OSHA regulations.

OPERATING RATIO OPERATING RATIO

The operating ratio is a measure of the total revenues divided by the total operating expenses.

ORGANIZING ORGANIZING

Deciding who does what work and delegating authority to the appropriate persons.

OUCH PRINCIPLE OUCH PRINCIPLE

This principle says that as a manager when you delegate job tasks you must be **O**bjective, **U**niform in your treatment of employees, and the tasks must be **C**onsistent with utility policies, and **H**ave job relatedness.

PLANNING PLANNING

Management of utilities to build the resources and financial capability to provide for future needs.

PRESENT WORTH PRESENT WORTH

The value of a long-term project expressed in today's dollars. Present worth is calculated by converting (discounting) all future benefits and costs over the life of the project to a single economic value at the start of the project. Calculating the present worth of alternative projects makes it possible to compare them and select the one with the largest positive (beneficial) present worth or minimum present cost.

RESPONSIBILITY RESPONSIBILITY

Answering to those above in the chain of command to explain how and why you have used your authority.

SCADA (SKAY-dah) SYSTEM SCADA SYSTEM

Supervisory Control And Data Acquisition system. A computer-monitored alarm, response, control, and data acquisition system used to monitor and adjust treatment processes and facilities.

TAILGATE SAFETY MEETING TAILGATE SAFETY MEETING

Brief (10 to 20 minutes) safety meetings held every 7 to 10 working days. The term comes from the safety meetings regularly held by the construction industry around the tailgate of a truck.

CHAPTER 20. TREATMENT PLANT ADMINISTRATION

(Lesson 1 of 3 Lessons)

20.0 NEED FOR UTILITY MANAGEMENT

The management of a public or private utility, large or small, is a complex and challenging job. Communities are concerned about their drinking water and their wastewater. They are aware of past environmental disasters and they want to protect their communities, but they want this protection with a minimum investment of money. In addition to the local community demands, the utility manager must also keep up with increasingly stringent regulations and monitoring from regulatory agencies. While meeting these external (outside the utility) concerns, the manager faces the normal challenges from within the organization, making ethical and responsible decisions regarding: personnel, resources, equipment, and preparing for the future. For the successful manager, all of these responsibilities combine to create an exciting and rewarding job.

A brief quiz is given in Table 20.1 that asks some basic management questions. This quiz can be used as a guide to management areas that may need some attention in your utility. You should be able to answer yes to most of the questions; however, all utilities have areas that can be improved.

In the environmental field, as well as other fields, the workforce itself is changing. Minorities, women, and people with disabilities provide new opportunities for growth in the utility. For the employee, however, overcoming employment barriers can be difficult, especially when the workload is demanding and physically challenging. The utility manager must provide adequate support services for these operators and learn to deal with organized worker groups.

Changes in the environmental workplace also are created by advances in technology. The environmental field has exploded with new technologies, such as computer-controlled wastewater treatment processes and collection systems. The utility manager must keep up with these changes and provide the leadership to keep everyone at the utility up to speed on new ways of doing things. In addition, the utility manager must provide a safer, cleaner work environment while constantly training and retraining operators to understand new technologies.

QUESTIONS

Write your answers in a notebook and then compare your answers with those on page 750.

20.0A What are the local community demands on a utility manager?

20.0B What has created changes in the environmental workplace?

20.1 FUNCTIONS OF A MANAGER

The functions of a utility manager are the same as for the CEO (Chief Executive Officer) of any big company: planning, organizing, staffing, directing, and controlling. In many small communities the utility manager may be the only one who has these responsibilities and the community depends on the manager to handle everything.

Planning (see Section 20.2) consists of determining the goals, policies, procedures, and other elements to achieve the goals and objectives of the agency. Planning requires the manager to collect and analyze data, consider alternatives, and then make decisions. Planning must be done before the other managing functions. Planning may be the most difficult in smaller communities, where the future may involve a decline in population instead of growth.

Organizing (see Section 20.3) means that the manager decides who does what work and delegates authority to the appropriate operators. The organizational function in some utilities may be fairly loose while some communities are very tightly controlled.

Staffing (see Section 20.4) is the recruiting of new operators and staff and determining if there are enough qualified operators and staff to fill available positions. The utility manager's staffing responsibilities include selecting and training employees, evaluating their performance, and providing opportunities for advancement for operators and staff in the agency.

Directing includes guiding, teaching, motivating, and supervising operators and utility staff members. Direction also includes issuing orders and instructions so that activities at the facilities or in the field are performed safely and are properly completed.

Controlling involves taking the steps necessary to ensure that essential activities are performed so that objectives will be achieved as planned. Controlling means being sure that progress is being made toward objectives and taking corrective action as necessary. The utility manager is directly involved in controlling the treatment process to ensure that the

TABLE 20.1 HOW WELL DOES YOUR SYSTEM MANAGE?

The following self-test is designed for small wastewater treatment facilities to provide a guide for identifying areas of concern and for improving treatment plant management.

1. Is the treatment system budget separate from other accounts so that the true cost of treatment can be determined?
2. Are the funds adequate to cover operating costs, debt service, and future capital improvements?
3. Do operational personnel have input into the budget process?
4. Is there a monthly or quarterly review of the actual operating costs compared to the budgeted costs?
5. Does the user charge system adequately reflect the cost of treatment?
6. Are all users charged a fair and equitable amount?
7. Are discharged effluent wastewater quality tests representative of plant performance?
8. Are operational control decisions based on process control testing within the plant?
9. Are provisions made for continued training for plant personnel?
10. Are qualified personnel available to fill job vacancies and is job turnover relatively low?
11. Are the energy costs for the system not more than 20 to 30 percent of the total operating costs?
12. Is the ratio of corrective (reactive) maintenance to preventive (proactive) maintenance remaining stable and is it less than 1.0?
13. Are maintenance records available for review?
14. Is the spare parts inventory adequate to prevent long delays in equipment repairs?
15. Are old or outdated pieces of equipment replaced as necessary to prevent excessive equipment downtime, inefficient process performance, or unreliability?
16. Are technical resources and tools available for repairing, maintaining, and installing equipment?
17. Is the treatment plant's pump station equipment providing the expected design performance?
18. Are standby units for key equipment available to maintain process performance during breakdowns or during preventive maintenance activities?
19. Are the plant processes adequate to meet the requirements for treatment and effluent discharge?
20. Does the facility have an adequate emergency response plan including an alternate power source?

wastewater is being properly treated before discharge and to make sure that the utility is meeting its short- and long-term goals.

QUESTIONS

Write your answers in a notebook and then compare your answers with those on page 750.

20.1A What are the functions of a utility manager?

20.1B In small communities, what does the community depend on the utility manager to do?

20.2 PLANNING[1]

A very large portion of any manager's typical work day will be spent on activities that can be described as planning activities since nearly every area of a manager's responsibilities require some type of planning.

Planning is one of the most important functions of utility management and one of the most difficult. Communities expect the discharge of treated effluent not to cause any adverse impacts on the environment or public health. The management of wastewater utilities must include building the resources and financial capability to provide for future needs. The utility must plan for future growth, including industrial development, and be ready to provide the plant capacity and level of treatment that will be needed as the community grows. The most difficult problem for some small communities is recognizing and planning for a decline in population. The utility manager must develop reliable information to plan for growth or decline. Decisions must be made about goals, both short- and long-term. The manager must prepare plans for the next two years and the next 10 to 20 years. Remember that utility planning should include operational personnel, local officials (decision makers), and the public. Everyone must understand the importance of planning and be willing to contribute to the process.

Operation and maintenance of a utility also involves planning by the utility manager. A preventive maintenance program should be established to keep the system performing as intended and to protect the community's investment in collection and treatment facilities. (Section 20.9 describes the various types of maintenance and the benefits of establishing maintenance programs.)

The utility also must have an emergency response plan to deal with natural or human disasters. Without adequate planning your utility will be facing system failures, inability to meet compliance regulations, and inadequate service capacity to meet community needs. Plan today and avoid disaster tomorrow. (Section 20.10, "Emergency Response," describes the basic elements of an emergency operations plan.)

20.3 ORGANIZING[2]

A utility should have a written organizational plan and written policies. In some communities the organizational plan and policies are part of the overall community plan. In either case, the utility manager and all plant personnel should have a copy of the organizational plan and written policies of the utility.

The purpose of the organizational plan is to show who reports to whom and to identify the lines of authority. The organizational plan should show each person or job position in the organization with a direct line showing to whom each person reports in the organization. Remember, an employee can serve only one supervisor (unity of command) and each supervisor should ideally manage only six or seven employees. The organizational plan should include a job description for each of the positions on the organizational chart. When the organizational plan is in place, employees know who is their immediate boss and confusion about job tasks is eliminated. A sample organizational plan for a water/wastewater utility is shown in Figure 20.1. The basic job duties for some typical utility positions are described in Table 20.2.

To understand organization and its role in management, we need to understand some other terms including authority, responsibility, delegation, and accountability. *AUTHORITY* means the power and resources to do a specific job or to get that job done. Authority may be given to an employee due to their position in the organization (this is formal authority) or authority may be given to the employee informally by their coworkers when the employee has earned their respect. *RESPONSIBILITY* may be described as answering to those above in the chain of command to explain how and why you have used your authority. *DELEGATION* is the act in which power is given to another person in the organization to accomplish a specific job. Finally when a manager gives power/responsibility to an employee, then the employee is *ACCOUNTABLE*[3] for the results.

Organization and effective delegation are very important to keep any utility operating efficiently. Effective delegation is uncomfortable for many managers since it requires giving up power and responsibility. Many managers believe that they can do the job better than others, they believe that other employees are not well trained or experienced, and they are afraid of mistakes. The utility manager retains some responsibility even after delegating to another employee and, therefore, the manager is often reluctant to delegate or may delegate the responsibility but not the authority to get the job done. For the utility manager, good organization means that employees are ready to accept responsibility and have the power and resources to make sure that the job gets done.

Employees should not be asked to accept responsibilities for job tasks that are beyond their level of authority in the organizational structure. For example, an operator or lead utility worker should not be asked to accept responsibility for additional lab testing. The responsibility for additional lab testing must be delegated to the lab supervisor. Authority and responsibility must be delegated properly to be effective. When these three components—proper job assignments, authority, and responsibility—are all present, the supervisor has successfully delegated. The success of delegation is dependent upon all three components.

An important and often overlooked part of delegation is *follow-up* by the supervisor. A good manager will delegate and follow up on progress to make sure that the employee has the necessary resources to get the job done. Well-organized managers can delegate effectively and do not try to do all the work themselves, but are responsible for getting good results. The

[1] *Planning.* Management of utilities to build the resources and financial capability to provide for future needs.

[2] *Organizing.* Deciding who does what work and delegating authority to the appropriate persons.

[3] *Accountability.* When a manager gives power/responsibility to an employee, the employee ensures that the manager is informed of results or events.

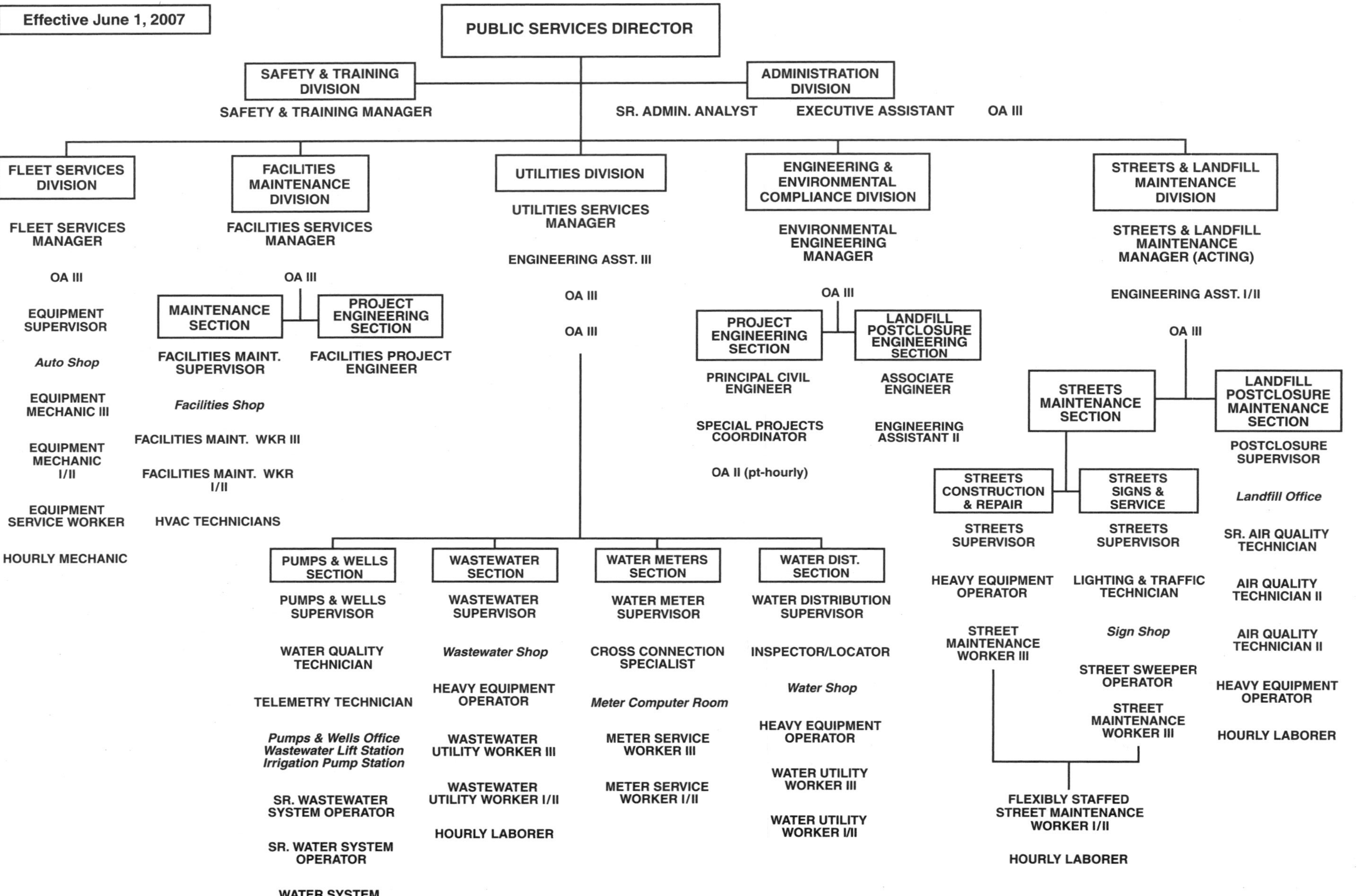

Fig. 20.1 Organizational chart for medium-sized utility
(Courtesy of City of Mountain View, California)

TABLE 20.2 JOB DUTIES FOR STAFF OF A MEDIUM-SIZED UTILITY

Job Title	Job Duties
Superintendent	Responsible for administration, operation, and maintenance of entire facility. Exercises direct authority over all plant functions and personnel.
Assistant Superintendent	Assists Superintendent in review of operation and maintenance function, plans special operation and maintenance tasks.
Clerk/Typist	Performs all clerical duties.
Operations Supervisor	Coordinates activities of plant operators and other personnel. Prepares work schedules, inspects plant, and makes note of operational and maintenance requirements.
Lead Utility Worker	Supervises operations and manages all operators.
Utility Worker II (Journey Level)	Controls treatment processes. Collects samples and delivers them to the lab for analysis. Makes operational decisions.
Utility Worker I	Performs assigned job duties.
Maintenance Supervisor	Supervises all maintenance for plant. Plans and schedules all maintenance work. Responsible for all maintenance records.
Maintenance Foreman	Supervises mechanical maintenance crew. Performs inspections and determines repair methods. Schedules all maintenance including preventive maintenance.
Maintenance Mechanic II	Selects proper tools and assigns specific job tasks. Reports any special considerations to Foreman.
Maintenance Mechanic I	Performs assigned job duties.
Electrician II	Schedules and coordinates electrical maintenance with other planned maintenance. Plans and selects specific work methods.
Electrician I	Performs assigned job duties.
Chemist	Directs all laboratory activities and makes operational recommendations to Operations Supervisor. Reports and maintains all required laboratory records. Oversees laboratory quality control.
Laboratory Technician	Performs laboratory tests. Manages day-to-day laboratory operations.

Management Muddle No. 1 that follows describes what can happen when delegation is improperly conducted, and illustrates how an organizational plan can prevent disaster.

Management Muddle No. 1

The City Manager of Pleasantville calls the Director of Public Works and asks for a report on the need for and cost of a new utility truck to be presented at the September 13 meeting of the City Council. The Director of Public Works calls the Plant Manager and asks for a report on the need and cost for a new utility truck with a deadline of September 12. The Plant Manager calls the Lead Utility Worker, an operator, who has been asking for a new utility truck and has been looking into the details. The Plant Manager requests that the Lead Utility Worker provide a report on September 12 about the purchase of the truck. The Lead Utility Worker gathers all the notes and hand writes a report identifying the need for the truck, the features required, and the cost. The Lead Utility Worker takes the report to City Hall to be typed and leaves it with a secretary on September 12. On September 13, the City Manager is preparing for the City Council meeting and does not have the report. Who is responsible? Who is accountable? How could this situation have been avoided?

Responsibility: The Lead Utility Worker's responsibility has been carried out with the authority and resources made available. Both the Director of Public Works and the Plant Manager failed to follow up on the report on September 12. No one informed the Lead Utility Worker that the report must be presented to the City Council on September 13, nor was the Lead Utility Worker supplied with the resources for getting the report in final form. However, the City Manager is ultimately responsible for reporting to the City Council.

Accountability: Starting with the Lead Utility Worker and working upward, each employee is accountable to his or her supervisor and should have communicated the status of the report.

How to avoid this situation: Good communication and follow-up by each of these supervisors could have prevented this situation completely. The City Manager should have asked to see the report on September 12; the Director of Public Works should have asked the Plant Manager to deliver the report no later than September 11; and the Plant Manager should have asked the Lead Utility Worker to submit the typed report (to the Plant Manager) no later than September 10. When delegating this task, the Plant Manager should have arranged for a secretary or clerk to assist the Lead Utility Worker in typing the report. Providing clerical support enables the Lead Utility Manager to complete the assigned task in a timely manner.

At each step in this chain of delegation, setting an early deadline gives the individual receiving the report an opportunity to review the document and make revisions, if necessary, before forwarding it up the chain of authority and ensures that the report reaches the City Manager no later than September 12.

QUESTIONS

Write your answers in a notebook and then compare your answers with those on page 750.

20.2A Who must be included in utility planning?

20.3A What is the purpose of an organizational plan?

20.3B Why is it sometimes difficult or uncomfortable for supervisors or managers to delegate effectively?

20.3C What is an important and often overlooked part of delegation?

20.4 STAFFING

20.40 The Utility Manager's Responsibilities

The utility manager is also responsible for staffing, which includes hiring new employees, training employees, and evaluating job performance. The utility should have established procedures for job hiring that include requirements for advertising the position, application procedures, and the procedures for conducting interviews.

In the area of staffing, more than any other area of responsibility, a manager must be extremely cautious and consider the consequences before taking action. Personnel management practices have changed dramatically in the past few years and continue to be redefined almost daily by the courts. A manager who violates an employee's or job applicant's rights can be held both personally and professionally liable in court. Throughout this section on staffing you will repeatedly find references to two terms: **job-related** and **documentation**. These are key concepts in personnel management today. Any personnel action you take must be job-related, from the questions you ask during interviews to disciplinary actions or promotions. And while almost no one wants more paper work, documentation of personnel actions detailing what you did, when you did it, and why you did it (the reasons will be job-related, of course) is absolutely essential. There is no way to predict when you might be called upon to defend your actions in court. Good records not only serve to refresh your memory about past events but can also be used to demonstrate your pattern of lawful behavior over time.

NOTICE

The information provided in this section on staffing should ***not*** be viewed as ***legal advice***. The purpose of this section is simply to identify and describe in general terms the major components of a utility manager's responsibilities in the area of staffing. One issue, harassment, is discussed in somewhat greater detail to illustrate the broad scope of a manager's responsibilities within a single policy area. Personnel administration is affected by many federal and state regulations. Legal requirements of legislation such as the Americans with Disabilities Act (ADA), Equal Employment Opportunity (EEO) Act, Family and Medical Leave Act, and wages and hours laws are complex and beyond the scope of this manual. If your utility does not have established personnel policies and procedures, consider getting help from a labor law attorney to develop appropriate policies. At the very least, you should get help from a recruitment specialist to develop and document hiring procedures that meet the federal guidelines for Equal Employment Opportunity.

20.41 How Many Employees Are Needed?

There is a common tendency for organizations to add personnel in response to changing conditions without first examining how the existing workforce might be reorganized to achieve greater efficiency and meet the new work demands. In wastewater collection and treatment utilities, aging of the system, changes in use, and expansion of the system often mean changes in the operation and maintenance tasks being performed. The manager of a utility should periodically review the agency's work requirements and staffing to ensure that the utility is operating as efficiently as possible. A good time to conduct such a review is during the annual budgeting process or when you are considering hiring a new employee because the workload seems to be greater than the current staff can adequately handle.

The staffing analysis procedure outlined in this section illustrates how to conduct the type of comprehensive analysis needed for a complete reorganization of the agency. In practice, however, a complete reorganization may not be desirable or even possible. Frequent organizational changes can make employees anxious about their jobs and may interfere with their work performance. Some employees show strong resistance to any change in job responsibilities. Nonetheless, by thoroughly examining the functions and staffing of the utility on a periodic basis, the manager may spot trends (such as an increase in the amount of time spent maintaining certain equipment or portions of the system) or discover inefficiencies that could be corrected over a period of time.

The first step in analyzing the utility's staffing needs is to prepare a detailed list of all the tasks to be performed to operate and maintain the utility. Next, estimate the number of staff hours per year required to perform each task. Be sure to include the time required for supervision and training.

When you have completed the task analysis, prepare a list of the utility's current employees. Assign tasks to each employee based on the person's skills and abilities. To the extent possible, try to minimize the number of different work activities assigned to each person but also keep in mind the need to provide opportunities for career advancement. One full-time staff year equals 260 days, including vacation and holiday time: (52 wk/yr)(5 days/wk) = 260 days/yr.

You can expect to find that this ideal staffing arrangement does not exactly match up with your current employees' job assignments. Most likely, you will also find that the number of staff hours required does not exactly equal the number of staff hours available. Your responsibility as a manager is to create the best possible fit between the work to be done and the personnel/skills available to do it. In addition to shifting work assignments between employees, other options you might consider are contracting out some types of work, hiring part-time or seasonal staff, or setting up a second shift (to make fuller use of existing equipment). Of course, you may find that it is time to hire another full- or part-time operator.

20.42 Qualifications Profile

Hiring new employees requires careful planning before the personal interview process. In an effort to limit discriminatory hiring practices, the law and administrative policy have carefully defined the hiring methods and guidelines employers may use. The selection method and examination process used to evaluate applicants must be limited to the applicant's knowledge, skills, and abilities to perform relevant job-related activities. In all but rare cases, factors such as age and level of education may not be used to screen candidates in place of performance testing. A description of the duties and qualifications for the job must be clearly defined in writing. The job description may be used to develop a qualifications profile. This qualifications profile clearly and precisely identifies the required job qualifications. All job qualifications must be relevant to the actual job duties that will be performed in that position.

The following list of typical job qualifications may be used to help you develop your own qualifications profiles with advice from a recruitment specialist.

1. General Requirements:
   a. Knowledge of methods, tools, equipment, and materials used in wastewater utilities.
   b. Knowledge of work hazards and applicable safety precautions.
   c. Ability to establish and maintain effective working relations with employees and the general public.
   d. Possession of a valid state driver's license for the class of equipment the employee is expected to drive.
2. General Educational Development:
   a. Reasoning: Apply common-sense understanding to carry out instructions furnished in oral, written, or diagrammatical form.
   b. Mathematical: Use a pocket calculator to make arithmetic calculations relevant to the utility's operation and maintenance processes.
   c. Language: Communicate with fellow employees and train subordinates in work methods. Fill out maintenance report forms.
3. Specific Vocational Preparation: Three years of experience in wastewater utility operation and maintenance.
4. Interests: May or may not be relevant to knowledge, skills, and ability; for example, an interest in activities concerned with objects and machines, ecology, or business management.
5. Temperament: Must adjust to a variety of tasks requiring frequent change and must routinely use established standards and procedures.
6. Physical Demands: Medium to heavy work involving lifting, climbing, kneeling, crouching, crawling, reaching, hearing, and seeing. Must be able to lift and carry _____ number of pounds for a distance of _____ feet.
7. Working Conditions: The work involves wet conditions, cramped and awkward spaces, noise, risks of bodily injury, and exposure to weather.

QUESTIONS

Write your answers in a notebook and then compare your answers with those on page 750.

20.4A What do staffing responsibilities include?

20.4B What are two key personnel management concepts a manager should always keep in mind?

20.4C List the steps involved in a staffing analysis.

20.4D What is a qualifications profile?

20.43 Applications and the Selection Process

20.430 Advertising the Position

To advertise a job opening, first prepare a written description of the required job qualifications, compensation, job duties, selection process, and application procedures (with a closing date). The utility should have established procedures about how to advertise the position and conduct the application process. The application procedure may require that the job be posted first within the utility to allow existing personnel first chance at the job opportunity.

20.431 Paper Screening

The next step in the selection process is known as paper screening. The personnel department and the utility manager review each application and eliminate those who are not qualified. The qualified applicants may be given examinations to verify their qualifications. Usually, the top three to twelve applicants are selected for an interview, depending on the agency's preference.

20.432 Interviewing Applicants

The purpose of the job interview is to gain additional information about the applicants so that the most qualified person can be selected. The utility manager should prepare for the interview in advance. Review the background information on each applicant. Draw up a list of job-related questions that will be asked of each applicant. During the interviews, briefly note the answers each applicant gives.

It used to be thought that the best way to learn about applicants was to give them plenty of time to talk about themselves because the content and type of information applicants volunteer might provide a deeper insight into the person and what type of employee they will become. Be very careful about open-ended, unstructured conversations with job applicants, even the friendly remarks you make initially to put the applicant at ease during the interview. If the applicant begins to volunteer information you could not otherwise legally ask for (such as marital status, number of children, religious affiliation, or age), be polite but firm in promptly redirecting the conversation. Even if this information was provided to you voluntarily, an applicant who did not get the job could later allege that you discriminated against them based on age or religion.

The only type of information you may legally request is information about the applicant's job skills, abilities, and experience relating directly to the job for which the person is applying. You must always be sensitive to the civil rights of the applicant and the affirmative action policies of the utility, which is another good reason to prepare a list of questions before the interview process begins. Structure the questions so that you avoid simple yes-and-no answers. Table 20.3 summarizes acceptable and unacceptable pre-employment inquiries to guide you in developing a good list of questions.

If other utility staff members are participating in the interviews, their participation should be confined to the preselected questions. Under no circumstances should front line employees conduct interviews in the absence of the manager or another person knowledgeable about personnel policies and practices.

The interview should be conducted in a quiet room without interruptions. Most applicants will be nervous, so start the interview on a positive note with introductions and some general remarks to put the applicant at ease. Explain the details of the job, working conditions, wages, benefits, and potential for advancement. Allow the applicant a chance to ask questions about the job. Ask each applicant the questions you have prepared and jot down brief notes on their responses. Taking notes while interviewing is awkward for some people but it becomes easier with practice. Notes are important because after interviewing several candidates you may not be able to remember what each one said. Also, as mentioned earlier, notes taken at the time of an interview can be valuable evidence in court if an unsuccessful applicant files a lawsuit for unfair hiring practices. At the end of the interview, tell the applicant when a decision will be made and how the applicant will be informed of the decision.

If an applicant's responses during the interview indicate that the person clearly is not qualified for this job but may be qualified for another job, briefly describe the other opportunity and how the person can apply for that position. The applicant may ask to be interviewed immediately for the second position. However, do not violate the utility's hiring procedures for the convenience of a job applicant. The same sequence of hiring procedures should be followed each time a position is filled. Tell this applicant that it will still be necessary to apply for the other position and that another interview may or may not follow, depending on the qualifications of the other applicants for the position.

20.433 Selecting the Most Qualified Candidate

Once the interviews are over, the job of evaluating and selecting the successful candidate begins. Review your interview notes and check the candidates' references. Checking references will verify the job experience of the applicant and may provide insight into the applicant's work habits. Questions you might ask previous employers include: Was the employee reliable and punctual? How well did the employee relate to coworkers? Did the employee consistently practice safe work procedures? Would you rehire this employee?

The rights of certain protected groups in the workforce today, such as minorities, women, disabled persons, persons over 40 years of age, and union members, are protected by law. A manager's responsibilities regarding protected groups begins with the hiring process and continues for as long as

TABLE 20.3 ACCEPTABLE AND UNACCEPTABLE PRE-EMPLOYMENT INQUIRIES[a]

Acceptable Pre-Employment Inquiries	Subject	Unacceptable Pre-Employment Inquiries
"Have you ever worked for this agency under a different name?"	NAME	Former name of applicant whose name has been changed by court order or otherwise.
Applicant's place of residence. How long applicant has been resident of this state or city.	ADDRESS OR DURATION OF RESIDENCE	
"If hired, can you submit a birth certificate or other proof of US citizenship or age?"	BIRTHPLACE	Birthplace of applicant. Birthplace of applicant's parents, spouse, or other relatives. Requirement that applicant submit a birth certificate, naturalization, or baptismal record.
"If hired, can you furnish proof of age?" /or/ Statement that hire is subject to verification that applicant's age meets legal requirements.	AGE	Questions that tend to identify applicants 40 to 64 years of age.
Statement by employer of regular days, hours, or shift to be worked.	RELIGION	Applicant's religious denomination or affiliation, church, parish, pastor, or religious holidays observed. "Do you attend religious services /or/ a house of worship?" Applicant may not be told, "This is a Catholic/Protestant/Jewish/atheist organization."
	RACE OR COLOR	Complexion, color of skin, or other questions directly or indirectly indicating race or color.
Statement that photograph may be required after employment.	PHOTOGRAPHS	Requirement that applicant affix a photograph to his/her application form. Request applicant, at his/her option, to submit photograph. Requirement of photograph after interview but before hiring.
Statement by employer that, if hired, applicant may be required to submit proof of eligibility to work in the United States.	CITIZENSHIP	"Are you a United States citizen?" Whether applicant or applicant's parents or spouse are naturalized or native-born US citizens. Date when applicant or parents or spouse acquired US citizenship. Requirement that applicant produce naturalization papers or first papers. Whether applicant's parents or spouse are citizens of the US.
Applicant's work experience. Applicant's military experience in armed forces of United States, in a state militia (US), or in a particular branch of the US armed forces.	EXPERIENCE	"Are you currently employed?" Applicant's military experience (general). Type of military discharge.
Applicant's academic, vocational, or professional education; schools attended.	EDUCATION	Date last attended high school.
Language applicant reads, speaks, or writes fluently.	NATIONAL ORIGIN OR ANCESTRY	Applicant's nationality, lineage, ancestry, national origin, descent, or parentage. Date of arrival in United States or port of entry; how long a resident. Nationality of applicant's parents or spouse; maiden name of applicant's wife or mother. Language commonly used by applicant. "What is your mother tongue?" How applicant acquired ability to read, write, or speak a foreign language.
	CHARACTER	"Have you ever been arrested?"

TABLE 20.3 ACCEPTABLE AND UNACCEPTABLE PRE-EMPLOYMENT INQUIRIES *(continued)*

Acceptable Pre-Employment Inquiries	Subject	Unacceptable Pre-Employment Inquiries
Names of applicant's relatives already employed by the agency.	RELATIVES	Marital status or number of dependents.
		Name or address of relative, spouse, or children of adult applicant.
		"With whom do you reside?"
		"Do you live with your parents?"
Organizations, clubs, professional societies, or other associations of which applicant is a member, excluding any names the character of which indicate the race, religious creed, color, national origin, or ancestry of its members.	ORGANIZATIONS	"List all organizations, clubs, societies, and lodges to which you belong."
"By whom were you referred for a position here?"	REFERENCES	Requirement of submission of a religious reference.
"Do you have any physical condition that may limit your ability to perform the job applied for?"	PHYSICAL CONDITION	"Do you have any physical disabilities?"
Statement by employer that offer may be made contingent on passing a physical examination.		Questions on general medical condition.
		Inquiries as to receipt of Workers' Compensation.
Notice to applicant that any misstatements or omissions of material facts in his/her application may be cause for dismissal.	MISCELLANEOUS	Any inquiry that is not job-related or necessary for determining an applicant's eligibility for employment.

[a] Courtesy of Marion B. McCamey, Affirmative Action Officer, California State University, Sacramento, CA.

the employer/employee relationship lasts. The best principle to deal with protected groups (and all other employees, for that matter) is the "OUCH" principle. The OUCH principle says that when you hire new employees or delegate job tasks to current employees, you must be Objective, Uniform in your treatment of applicants or employees, Consistent with utility policies, and Have job relatedness. If you do not manage with all of these characteristics, you may find yourself in a "hurting" position with regard to protected workers.

Objectivity is the first hurdle. Often the physical characteristics of a person, such as large or small size, may make a person seem more or less job capable. However, many utility agency jobs are done with power tools or other technology that allows all persons, regardless of size, to manage most tasks. Try to remain objective but reasonable in assessing job applicants and making job assignments.

Uniform treatment of job applicants and employees is necessary to protect yourself and other employees. Nothing will destroy morale more quickly than unequal treatment of employees. Your role as a manager is to consistently apply the policies and procedures that have been adopted by the utility. Often, policies and procedures exist that are not popular and may not even be appropriate. However, the job of the utility manager is to consistently uphold and apply the policies of the utility.

The last part of the OUCH principle is having job relatedness. Any hiring decision must be based on the applicant's qualifications to meet the specific job requirements and any job assignment given to an employee must be related to that employee's job description. Extra assignments, such as buying personal gifts for the boss's family or washing the boss's car, are not appropriate. These types of job assignments will eventually catch up to the manager and can be particularly embarrassing if the public gets involved. So to protect yourself and your utility, remember the OUCH principle as you hire and manage your employees.

Once you have made your selection, the applicant is usually required to pass a medical examination. When this has been successfully completed and the applicant has accepted the position, notify the other applicants that the position has been filled.

20.44 New Employee Orientation

During the first day of work, a new employee should be given all the information available in written and verbal form on the policies and practices of the utility including compensation, benefits, attendance expectations, alcohol and drug testing (if the utility does this), and employer/employee relations. Answer any questions from the new employee at this time and try to explain the overall structure of the utility as well as identify who can answer employee questions when they arise. Introduce the new employee to co-workers and tour the work area. Every utility should have a safety training session for all new employees and specific safety training for some job categories. Provide safety training (see Section 20.11, "Safety Program") for new employees on the first day of employment or as soon thereafter as possible. Establishing safe work practices is a very important function of management.

QUESTIONS

Write your answers in a notebook and then compare your answers with those on page 750.

20.4E What is the purpose of a job interview?

20.4F List four "protected groups."

20.4G What does the "OUCH" principle stand for?

20.4H When should a new employee's safety training begin?

20.45 Employment Policies and Procedures

20.450 Probationary Period

Many employers now use a probationary period for all new employees. The probationary period is typically three to six months but may be as long as a year. This period begins on the first day of work. Management may reserve the right to terminate employment of the person with or without cause during this probationary period. The employee must be informed of this probationary period and must understand that successful completion of the probationary period is required in order to move into regular employment status.

The probationary period provides a time during which both the employer and employee can assess the fit between the job and the person. Normally, a performance evaluation is completed near the end of the probationary period. A satisfactory performance evaluation is the mechanism used to move an employee from probationary status into regular employment.

20.451 Compensation

The compensation an employee receives for the work performed includes satisfaction, recognition, security, appropriate pay, and benefits. All are important to keep good employees satisfied. Salaries should be a function of supply and demand. Pay should be high enough to attract and retain qualified employees. Salaries are usually determined by the governing body of the utility in negotiation with employee groups, when appropriate. The salary structure should meet all state and federal regulations and accurately reflect the level of service given by the employee. A survey of salaries from other utilities in the area may provide valuable information in the development of a salary structure.

The benefits supplied by the employer are an important part of the compensation package. Benefits generally include the following: retirement, health insurance, life insurance, employer's portion of social security, holiday and vacation pay, sick leave, personal leave, parental leave, worker's compensation, and protective clothing. Many employers now provide dental and vision insurance, long-term disability insurance, educational bonus or costs and leave, bereavement leave, and release time for jury duty. Some employers also include in their benefit package cash bonus programs and longevity pay. The value of an employee's entire benefit package is often computed and printed on the pay stub as a reminder that salary alone is not the only compensation being provided.

20.452 Training and Certification[4]

Training has become an ongoing process in the workplace. The utility manager must provide new employee training as well as ongoing training for all employees. Safety training is particularly important for all utility operators and staff members and is discussed in detail later in Section 20.11. Certified wastewater collection system and treatment plant operators earn their certificates by knowing how to do their jobs safely. Preparing for certification examinations is one means by which operators learn to identify safety hazards and to follow safe procedures at all times under all circumstances.

Although it is extremely important, safety is not the sole benefit to be derived from a certification program. Other benefits include protection of the public's investment in wastewater collection and treatment facilities and employee pride and recognition.

Vast sums of public funds have been invested in the construction of wastewater collection and treatment facilities. Certification of operators assures utilities that these facilities will be operated and maintained by qualified operators who possess a certain level of competence. These operators should have the knowledge and skills not only to prevent unnecessary deterioration and failure of the facilities, but also to improve operation and maintenance techniques.

Achievement of a level of certification is a public acknowledgment of a wastewater collection system or treatment plant operator's skills and knowledge. Presentation of certificates at an official meeting of the governing body will place the operators in a position to receive recognition for their efforts and may even get press coverage and public opinion that is favorable. An improved public image will give the certified operator more credibility in discussions with property owners.

Recognition for their personal efforts will raise the self-esteem of all certified operators. Certification will also give wastewater collection system and treatment plant operators an upgraded image that has been too long denied them. If properly publicized, certification ceremonies will give the public a

[4] *Certification Examination.* An examination administered by a state agency or professional association that operators take to indicate a level of professional competence. In the United States, certification of operators of water treatment plants, wastewater treatment plants, water distribution systems, and small water supply systems is mandatory. In many states, certification of wastewater collection system operators, industrial wastewater treatment plant operators, pretreatment facility inspectors, and small wastewater system operators is voluntary; however, current trends indicate that more states, provinces, and employers will require these operators to be certified in the future. Operator certification is mandatory in the United States for the Chief Operators of water treatment plants, water distribution systems, and wastewater treatment plants.

more accurate image of the many dedicated, well-qualified operators working for them. Certification provides a measurable goal that operators can strive for by preparing themselves to do a better job. Passing a certification exam should be recognized by an increase in salary and other employee benefits.

Most states and Canadian provinces now require that wastewater treatment plant operators be certified. To maintain current certification, these operators must complete additional training classes every one to five years. In the environmental field, new technologies and regulations require operators to attend training to keep up with their field. The utility manager has the responsibility to provide employees with high-quality training opportunities. Many types of training are available to meet the different training needs of utility operators, for example, in-house training, training conducted by training centers, professional organizations, engineering firms, or regulatory agencies, and correspondence courses such as this one by the Office of Water Programs.

ABC stands for the Association of Boards of Certification for Operating Personnel in Water Utilities and Pollution Control Systems. If you wish to find out how operators can become certified in your state or province, contact:

Executive Director, ABC
208 Fifth Street
Ames, IA 50010-6259
Phone: (515) 232-3623

ABC will provide you with the name and address of the appropriate contact person.

One area of training that is frequently overlooked is training for supervisors. Managing people requires a different set of skills than performing the day-to-day work of operating and maintaining a wastewater treatment facility. Supervisors need to know how to communicate effectively and how to motivate others, as well as how to delegate responsibility and hold people accountable for their performance. Supervisors share management's responsibility for fair and equitable treatment of all workers and are required to act in accordance with applicable state and federal personnel regulations. Making the transition from operator to supervisor also requires a change in attitude. A supervisor is part of the management team and is therefore obliged to promote the best interests of the utility at all times. When the interests of the utility conflict with the desires of one or more employees, the supervisor must support management's decisions and policies regardless of the supervisor's own personal opinion about the issue. It is the responsibility of the utility manager to ensure that supervisors receive appropriate training in all of these areas.

Training on how to motivate people, deal with co-workers, and supervise or manage people working for you has become a very highly specialized field of training. These are complex topics that are beyond the scope of this manual. If you have a need for or wish to learn more about how to deal with people, consider enrolling in courses or reading books on supervision or personnel management. An excellent book is *MANAGE FOR SUCCESS* in this series of operator training manuals from the Office of Water Programs.

QUESTIONS

Write your answers in a notebook and then compare your answers with those on page 750.

20.4I What is the purpose of a probationary period for new employees?

20.4J What kind of compensation does an employee receive for work performed?

20.4K Why should utility managers provide training opportunities for employees?

20.453 Performance Evaluation

Most organizations conduct some type of performance evaluation, usually on an annual basis. The evaluation may be written or oral; however, a written evaluation is strongly recommended because it will provide a record of the employee's performance. Documentation of this type may be needed in the future to support taking disciplinary action if the employee's performance consistently fails to meet expectations. The evaluation of employee performance can be a challenging task, especially when performance has not been acceptable. However, evaluations are also an opportunity to provide employees with positive feedback and let them know their contributions to the organization have been noticed and appreciated.

A formal performance evaluation typically begins with an employee's immediate supervisor filling out the performance evaluation form (a sample evaluation form is shown in Figure 20.2). Complete the entire form and be specific about the employee's achievements as well as areas needing improvement. Next, schedule a private meeting with the employee to discuss the evaluation. Give the employee frequent opportunities to be heard and listen carefully. If some of the employee's accomplishments were overlooked, note them on the evaluation form and consider whether this new information changes your overall rating of performance in one or more categories.

After reviewing the employee's performance for the past year, set performance goals for the next year. Be sure to document the goals you have agreed upon. Setting performance goals is particularly important if an employee's performance has been poor and improvement is needed. If appropriate, develop a written performance improvement plan that includes specific dates when you will again review the employee's progress in meeting the performance goals. Some supervisors find it helpful to schedule an informal mid-year meeting with each employee to review their progress and to avoid surprises during the next performance evaluation.

Many employees and managers dread even the thought of a performance evaluation and see it as an ordeal to be endured. When properly conducted, however, a performance review can strengthen the lines of communication and increase trust between the employee and the manager. Use this opportunity to acknowledge the employee's unique contributions and to seek solutions to any problems the employee may be having in completing work assignments. Ask the employee how you can be of assistance in removing any obstacles to getting the job done. If necessary, provide coaching to help the employee understand both how and why certain tasks are performed. Be generous (but sincere) with praise for the good work the employee does well every day and try to keep the employee's

EMPLOYEE EVALUATION FORM

Employee Name ______________________ Date ______________________

Job Title ______________________ Department ______________________

Evaluate the employee on the job now being performed. Circle the number which most nearly expresses your overall judgment. In the space for comments, consider the employee's performance since their last evaluation and make notes about the progress or specific concerns in that area. The care and accuracy of this appraisal will determine its value to you, the employee, and your employer.

JOB KNOWLEDGE: (Consider knowledge of the job gained through experience, education, and special training)

5. Well informed on all phases of work
4. Knowledge thorough enough to perform well without assistance
3. Adequate grasp of essentials, some assistance required
2. Requires considerable assistance to perform
1. Inadequate knowledge

Comments: ______________________

____

____

QUALITY OF WORK: (Consider accuracy and dependability of the results)

5. Exceptionally accurate, practically no mistakes
4. Usually accurate, seldom necessary to check results
3. Acceptable, occasional errors
2. Often unacceptable, frequent errors, needs supervision
1. Unacceptable, too many errors

Comments: ______________________

____

____

INITIATIVE: (Consider the speed with which the employee grasps new job skills)

5. Excellent, grasps new ideas and suggests improvements, is a leader with others
4. Very resourceful, can work unsupervised, manages time well, is reliable
3. Shows initiative on occasion, is reliable
2. Lacks initiative, must be reminded to complete tasks
1. Needs constant prodding to complete job tasks, is unreliable

Comments: ______________________

____

____

Fig. 20.2 Employee evaluation form

COOPERATION AND RELATIONSHIPS: (Consider manner of handling relationships with co-workers, superiors, and the public)

5. Excellent cooperation and communication with co-workers, supervisors, and others, takes and gives instructions well
4. Gets along well with co-workers
3. Acceptable, usually gets along well, occasionally complains
2. Shows a reluctance to cooperate, complains
1. Very poor cooperation, does not follow instruction, dislikes fellow employees

Comments: ______________________________

____

____

ATTENDANCE: (Consider frequency of absences, reasons for absences or tardiness, and promptness in giving notice about absences)

5. Excellent, absent only for emergencies, illness, civic duties, always on time, gives notice when absent
4. Rarely absent or late, always gives notice and good reason
3. Occasionally absent, less important reasons, usually gives notice, but not always in time
2. Often absent, lack of adequate notice or reasons for absenteeism
1. Unexcusable absenteeism, does not give notice, reasons are unacceptable, cannot be depended upon

Comments: ______________________________

____

____

OVERALL EVALUATION: Superior ____ Good ____ Satisfactory ____ Unsatisfactory ____

Comments: ______________________________

____

____

I hereby certify that this appraisal is my best judgment of the service value of this employee and is based on personal observation and knowledge of the employee's work.

Supervisor's Signature ______________________ Date ______________

I hereby certify that I have personally reviewed this report.

Employee's Signature ______________________ Date ______________

Fig. 20.2 Employee evaluation form (continued)

shortcomings in perspective. If the person is doing a good job 95 percent of the time, do not let the entire discussion consist of criticism about the remaining 5 percent of the person's job assignments.

At the end of the meeting, ask the employee to sign the evaluation form to acknowledge having seen and discussed it. Give the employee a copy of the evaluation. If the employee disagrees with any part of the evaluation, invite the person to submit a written statement describing the reasons for their disagreement. The written statement should be filed with the completed performance evaluation form.

20.454 Dealing with Disciplinary Problems

Handling employee discipline problems is difficult, even for an experienced manager. But remember, **no discipline problem ever solves itself and the sooner you deal with the problem, the better the outcome will be**. If problem behavior is not corrected, then other employees will become dissatisfied and the problems will increase.

Every utility, no matter how small, should have written employment policies enabling the manager to deal effectively with employee problems. It should also provide a formal complaint or grievance procedure by which employees can have their complaint heard and resolved without fear of retaliation by the supervisor.

Dealing with employee discipline requires tact and skill. You will have to find your own style and then try to stay flexible, calm, and open-minded when the situation gets really tough. If you repeatedly find yourself unable to deal successfully with disciplinary problems, consult with the utility's personnel office (if available) or consider enrolling in a management training course designed specifically around strategies and techniques for disciplining employees.

A commonly accepted method for dealing with job-related employee problems is to first discuss the problem with the employee in private. Most employers will give a person two or three verbal warnings; then the warnings should be written with copies given to the employee. Finally, if the written warnings do not produce positive results, the employee may have to be suspended or dismissed. Your job is to make sure that all warnings are documented with specific descriptions of unsatisfactory behaviors and to make sure that all employees are treated fairly.

Start the disciplinary discussion with a positive comment about the employee. Then, identify the problem but keep emotion and blame out of the discussion. The best approach is to state the problem and then ask the employee to suggest a solution. If they respond inappropriately, you must restate the problem and explain that you are trying to find a positive solution that is acceptable to everyone.

Try to keep the discussion focused on solving the problems and do not permit the employee to heap on general complaints, report on what other employees do, or wander from the topic. The following is an example of how you might start the discussion for an employee who is tardy every day. "Joe, you have done a good job in keeping that north side pump station running. You are an asset to this operation. Your tardiness every morning, however, is causing problems. Is there some reason for you to be tardy? We need to find a solution to this problem because your being late creates a bad situation for the night shift. What do you suggest?"

Always remain calm and do not allow yourself to become angry when dealing with an employee about performance issues. If you begin to feel angry and are about to lose control, or if the employee becomes combative or abusive, suggest that the meeting is not producing positive results and schedule an alternative meeting time. Do not let the emotions of the moment carry you into a rage in front of employees. If either you or one of your employees expresses extreme emotions, then the discussion should be postponed until everyone cools down. The following steps may serve as a guide to dealing with confrontation; they apply equally to the employee and the supervisor.

- Maintain an adult approach—positive criticism should be taken/given to improve job skills.
- Create a private environment—job performance issues should be discussed in private between the employee and supervisor.
- Listen very carefully—be sure that both you and the employee understand the situation in the same way. If not, you need to keep talking until both parties are in agreement about the problem and the solution.

- Keep your language appropriate—anger and bad language will cause the situation to escalate. Keep your cool and hold your tongue.
- Stay focused in the present—let go of all the past slights, misunderstandings, and dissatisfaction. Problems must be solved one at a time.
- Aim for a permanent solution—changes in job performance need to be permanent to be effective.

Reports of violence in the workplace appear regularly in newspapers and on television. Managers and supervisors should be alert for signs that an employee might become violent and should take any threat of violence seriously.

Violence in the workplace may take the form of physical harm, psychological harm, or property damage. Common warning signs include abusive language, threatening or confrontational behavior, and brandishing a weapon. Some examples of physical harm include pushing, hitting, shoving, or any other form of physical assault. Threats and harassment are forms of psychological harm, and property damage can range from theft or destruction of equipment to sabotage of the employer's computer systems.

The utility's safe workplace policy should be a zero tolerance policy. Any employee who is the target of violent behavior or who witnesses such behavior should be encouraged to immediately report the behavior to a supervisor and the incident should be investigated promptly. If necessary for the immediate safety of other employees, the offending employee should be placed on administrative leave, escorted from the work environment, and permitted to return to work only after the investigation has been completed.

Management Muddle No. 2

Sue has been working for five years as a laboratory technician and was recently passed over for promotion. A lab director who has more college experience than Sue was hired from another plant. Since then Sue's work has not been very good, she has come to work late, and she does not always get all of the lab tests done during her workday. What should be done about Sue? If you were the supervisor, how would you handle this problem with Sue?

Actions: As the supervisor, you should ask Sue to come by your office. In private, you should discuss with Sue why the new lab director was hired. Discuss with her the good work record she has maintained over the past five years, and explain the changes you have seen in her work recently. At this point you might ask her to evaluate her own performance or what she would do if she were in your situation. She might need to express her resentment about the new lab director. If she does, let her ramble and rave just for a few minutes, then stop her. You might say, "OK, you are unhappy, but what are we going to do to change this situation? How can I help you to regain your motivation and improve your work habits?" If possible you might help her figure out a way to continue her college education, go to additional training classes, or reorganize the lab so that her job duties change somewhat. There are many other possibilities for helping Sue to become motivated again but she should be part of the process. It is important to communicate clearly that her job performance must improve.

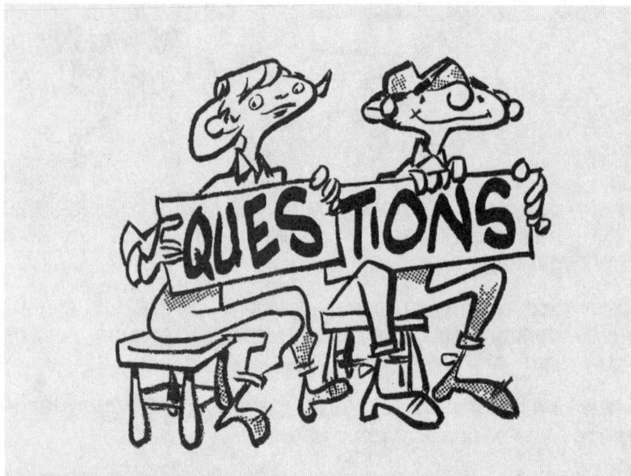

Write your answers in a notebook and then compare your answers with those on page 750.

20.4L Who is the appropriate person to conduct an employee's performance evaluation?

20.4M What should be the attitude of a supervisor or manager when dealing with disciplinary problems?

20.4N What are some common warning signs that an employee could become violent?

20.455 Example Policy: Harassment

Harassment is any behavior that is offensive, annoying, or humiliating to an individual and that interferes with a person's ability to do a job. This behavior is uninvited, often repeated, and creates an uncomfortable or even hostile environment in the workplace. Harassment is not limited to physical behavior but also may be verbal or involve the display of offensive pictures or other images. Sexual harassment is legally defined as unwanted sexual advances, or visual, verbal, or physical conduct of a sexual nature. Any type of harassment is inappropriate in the workplace. A manager's responsibilities with regard to harassment include:

- Establish a written policy (such as the one shown in Figure 20.3) that clearly defines and prohibits harassment of any type.
- Distribute copies of the harassment policy to all employees and take whatever steps are necessary (small group discussions, general staff meetings, training sessions) to ensure that all employees understand the policy.
- Encourage employees to report incidents of harassment to their immediate supervisor, a manager, or the personnel department.
- Investigate every reported case of harassment.
- Document all aspects of the complaint investigation, including the procedures followed, statements by witnesses, the complainant, and the accused person, the conclusions reached in the case, and the actions taken (if any).

How do you know when offensive behavior could be considered sexual harassment? Unwelcome sexual advances or other verbal or physical conduct of a sexual nature could be interpreted by an employee as sexual harassment under the following conditions:

- A person is required, or feels they are required, to accept unwelcome sexual conduct in order to get a job or keep a job.
- Decisions about an employee's job or work status are made based upon either the employee's acceptance or rejection of unwelcome sexual conduct.
- The conduct interferes with the employee's work performance or creates an intimidating, hostile, or offensive working environment.

The following is a list of examples of the kind of behavior that is unacceptable and illegal in the workplace. It is only a partial list to give you an idea of the scope of the requirements.

- Unwanted hugging, patting, kissing, brushing up against someone's body, or other inappropriate sexual touching
- Subtle or open pressure for sexual activity
- Persistent sexually explicit or sexist statements, jokes, or stories
- Repeated leering or staring at a person's body
- Suggestive or obscene notes or phone calls
- Display of sexually explicit pictures or cartoons

The best way to prevent harassment is to set an example by your own behavior and to keep communication open between employees. In most cases, an open discussion with employees about harassment can help everyone understand that innuendo and slurs about a person's race, religion, sex, appearance, or any other personal belief or characteristic are unacceptable. The most productive way to control such behavior is by enlisting the help of all employees to feel that they have the right and the responsibility to stop harassment. When you get employees to think about their behavior and how their behavior makes others feel, they will usually realize they should speak up to prevent harassment.

Here is an example of how employees handled a problem of harassment. A group of operators often had coffee in the office

SUBJECT: HARASSMENT POLICY AND COMPLAINT PROCEDURE **NO:** ____________

PURPOSE:

To establish a strong commitment to prohibit harassment in employment, to define discrimination harassment and to set forth a procedure for investigating and resolving internal complaints of harassment.

POLICY:

Harassment of an applicant or employee by a supervisor, management employee, or co-worker on the basis of race, religion, color, national origin, ancestry, handicap, disability, medical condition, marital status, familial status, sex, sexual orientation, or age will not be tolerated. This policy applies to all terms and conditions of employment, including, but not limited to, hiring, placement, promotion, disciplinary action, layoff, recall, transfer, leave of absence, compensation, and training.

Disciplinary action up to and including termination will be instituted for behavior described in the definition of harassment set forth below:

- Any retaliation against a person for filing a harassment charge or making a harassment complaint is prohibited. Employees found to be retaliating against another employee shall be subject to disciplinary action up to and including termination.

DEFINITION:

Harassment includes, but is not limited to:

A. Verbal Harassment—For example, epithets, derogatory comments or slurs on the basis of race, religious creed, color, national origin, ancestry, handicap, disability, medical condition, marital status, familial status, sex, sexual orientation, or age. This might include inappropriate sex-oriented comments on appearance, including dress or physical features or race-oriented stories.

B. Physical Harassment—For example, assault, impeding or blocking movement, with a physical interference with normal work or movement when directed at an individual on the basis of race, religion, color, national origin, ancestry, handicap, disability, medical condition, marital status, familial status, age, sex, or sexual orientation. This could be conduct in the form of pinching, grabbing, patting, propositioning, leering, or making explicit or implied job threats or promises in return for submission to physical acts.

C. Visual Forms of Harassment—For example, derogatory posters, notices, bulletins, cartoons, or drawings on the basis of race, religious creed, color, national origin, ancestry, handicap, disability, medical conditions, marital status, familial status, sex, sexual orientation, or age.

D. Sexual Favors—Unwelcome sexual advances, requests for sexual favors, and other verbal or physical conduct of a sexual nature which is conditioned upon an employment benefit, unreasonably interferes with an individual's work performance, or creates an offensive work environment.

Fig. 20.3 Harassment policy

(Courtesy of City of Mountain View, California)

SUBJECT: HARASSMENT POLICY AND COMPLAINT PROCEDURE (continued)

COMPLAINT PROCEDURE:

A. Filing:

An employee who believes he or she has been harassed may make a complaint orally or in writing with any of the following:

1. Immediate supervisor.
2. Any supervisor or manager within or outside of the department.
3. Department head.
4. Employee Services Director (or his/her designee).

Any supervisor or department head who receives a harassment complaint should notify the Employee Services Director immediately.

B. Upon notification of the harassment complaint, the Employee Services Director shall:

1. Authorize the investigation of the complaint and supervise and/or investigate the complaint. The investigation will include interviews with:
   (a) The complainant;
   (b) The accused harasser; and
   (c) Any other persons the Employee Services Director has reasons to believe has relevant knowledge concerning the complaint. This may include victims of similar conduct.
2. Review factual information gathered through the investigation to determine whether the alleged conduct constitutes harassment; giving consideration to all factual information, the totality of the circumstances, including the nature of the verbal, physical, visual, or sexual conduct and the context in which the alleged incidents occurred.
3. Report the results of the investigation and the determination as to whether harassment occurred to appropriate persons, including to the complainant, the alleged harasser, the supervisor, and the department head. If discipline is imposed, the discipline may or may not be communicated to the complainant.
4. If harassment occurred, take and/or recommend to the appropriate department head or other appropriate authority prompt and effective remedial action against the harasser. The action will be commensurate with the severity of the offense.
5. Take reasonable steps to protect the victim and other potential victims from further harassment.
6. Take reasonable steps to protect the victim from any retaliation as a result of communicating the complaint.

DISSEMINATION OF POLICY:

All employees, supervisors, and managers shall be sent copies of this policy.

Effective Date: May, 2007
Revision Date: March 1, 2007

City Manager

Fig. 20.3 Harassment policy (continued)

(Courtesy of City of Mountain View, California)

of the utility during their morning break. One of the operators often used foul language, which was embarrassing and offensive to one of the operator's co-workers. The manager sent a memo to all employees that mentioned respect for fellow workers and included a reminder about inappropriate language. The next day all of the other operators had taped their copy of the memo to the mailbox of the one operator who was most vocal. These operators found a way to send their message loud and clear—no one wants to work in an environment that is unpleasant to others. As a manager, you must establish an atmosphere that is open, congenial, and harmonious for all employees.

Occasional flirting, innuendo, or jokes may not meet the legal definition of sexual harassment. Nevertheless, they may be offensive or intimidating to others. Every employee has a right to a workplace free of discrimination and harassment, and every employee has a responsibility to respect the rights of others.

Managers need to *be aware of* and *take action to prevent* any type of harassment in the workplace. It is not enough to simply distribute copies of the utility's harassment policy. If legal action is taken against the utility due to harassment or the existence of a hostile work environment, the utility manager may face both personal and professional liability if it can be shown that the manager *should have known* harassment was occurring or that the manager permitted a hostile environment to continue to exist.

Management Muddle No. 3

The maintenance crew is a group of five men who have been with the utility for many years and are well respected for their work habits. However, in the maintenance shed the walls are covered by calendars of scantily clad women and the language used out in the shed is sometimes pretty rough. One of your operators is a woman; she is well respected by her co-workers and is a very good operator. She comes to your office to complain about the situation in the maintenance shed and demands that you remove the pictures from the walls in the shed. She goes on to report that when she went out to the shed and requested assistance to check on a pump, which was noisy and running hot, she was told "Kiss my ________, toots!" As the manager, what should you do? Should you immediately go to the maintenance shed and rip down the pictures? What should you say to the female operator in your office?

Actions: Your first response should be to reassure your operator that you understand her anger and frustration. Let her know that you will investigate the matter immediately and take action to correct any problems you find. Ask her to write out a complete statement of the facts, including her concerns about the pump, when and how her request for assistance was made, to whom the request was made, who made the offensive remark, the names of any witnesses, and what responses she has gotten from the maintenance crew in the past. Try to establish if this is a one-time response or if this problem has been going on for some time.

Begin your investigation with a trip to the maintenance shed to observe and evaluate what is hanging on the walls. Discuss the situation with the crew. Try to make this an open discussion so that everybody understands how the pictures affect the atmosphere and the image of professionalism of the utility. The best solution is to get the crew to understand how this type of behavior looks to persons outside their own small group and then let them take down the pictures. (Be sure to follow up later to confirm that the calendars or other offensive material has not reappeared.)

Next, set up private interviews with the person accused of making the offensive remark and each of the witnesses. Ask each person to describe the encounter in the maintenance shed and make detailed notes of their responses. (Depending on the complexity of the situation, it may sometimes be appropriate to have each person involved submit a written statement describing what occurred.)

After you have thoroughly investigated the incident, discuss with the crew the use of acceptable language in response to other employees. An open discussion and increased awareness of sexual harassment should be all that is necessary to change this situation. If it is not, arrange for a training program in sexual harassment awareness, if one is available. Be sure to establish a policy on the consequences of inappropriate behavior and be prepared to enforce the policy when needed.

Retaliation against an employee for filing a complaint about harassment or a hostile work environment is also illegal. Some examples of retaliation are demotion, suspension, failure to hire or consider for hire, failure to make impartial employment recommendations or decisions, adversely changing working conditions, spreading rumors, or denying any employment benefit. Retaliation could be the basis for a lawsuit involving not only the person who is accused of retaliation, but also the immediate supervisor, the manager, and the utility.

Most areas of personnel management, including harassment and retaliation issues, are complex and have significant legal consequences for everyone involved. The discussion in this section is not a complete explanation of harassment or retaliation. If your utility is large enough to employ a personnel specialist or a labor law attorney, ask them to review your staffing policies and procedures and consult with them whenever you have questions about personnel matters. If you manage a small utility and have no in-house sources for technical or legal advice, enroll in appropriate training courses or consider working with an attorney on a contract basis.

20.456 Labor Laws Governing Employer/Employee Relations

Many employers take pride in advertising that they are an equal opportunity employer, and one often sees this claim in newspaper help wanted ads and other forms of job postings. It means an employer's staffing policies and procedures do not discriminate against anyone based on race, religion, national origin, color, citizenship, marital status, gender, age, Vietnam-era or disabled veteran status, or the presence of a physical, mental, or sensory disability. An employer must meet specific requirements of the federal Equal Employment Opportunity

Act to be eligible to advertise as an equal opportunity employer. These requirements include adoption of nondiscriminatory personnel policies and procedures and periodic submission of reports of personnel actions for review by the Equal Employment Opportunity Commission.

The Family and Medical Leave Act of 1993 (FMLA) is a federal law that requires all public agencies as well as companies with 50 or more employees to permit eligible employees to take up to 12 weeks of time off in a 12-month period for the following purposes: (1) the employee's own serious health condition, (2) to care for a child following birth or placement for adoption or foster care, or (3) to care for the employee's spouse, child, or parent with a serious health condition. To be eligible to receive this benefit, an employee must have been employed for at least one year prior to the leave. The employer is not required to pay the employee's salary during the time off work, but many employers permit (or require) employees to use accrued sick leave and vacation time during the period of unpaid FMLA leave.

The Americans with Disabilities Act of 1990 (ADA) prohibits employment discrimination based on a person's mental or physical disability. The law applies to employers engaged in an industry affecting commerce who have 15 or more employees.

In general, the ADA defines disability as a physical or mental impairment that substantially limits one or more of the major life activities of an individual. The exact meaning of this definition is evolving as the courts settle lawsuits in which individuals allege they were discriminated against because of a physical or mental disability. The original ADA legislation listed more than 40 specific types of impairments and the courts continue to expand the list.

Under the ADA, employers must make reasonable accommodations to enable a disabled person to function successfully in the work environment, for example, installing a ramp to make facilities accessible to someone in a wheelchair, or restructuring an individual's job or work schedule, or providing an interpreter. The requirements of each situation are unique. In each case, the nature and extent of the disability and the reasonableness (including cost factors) of the requested accommodation by the employer must be weighed. Employers are not automatically required to do everything possible to accommodate disabled persons, but rather to take whatever reasonable steps they can to do so.

All of these personnel laws are very complex and managers of any utility that may be covered by them are strongly urged to seek the assistance of an experienced labor law attorney or personnel specialist.

20.457 Personnel Records

A personnel file should be maintained for each utility employee. This file should contain all documents related to the employee's hiring, performance reviews, promotions, disciplinary actions, and any other records of employment-related matters. Since these records often contain sensitive, confidential information, access to personnel records should be closely controlled. (Also see Section 20.12, "Recordkeeping," for more information about what records should be kept and how long they should be kept.)

QUESTIONS

Write your answers in a notebook and then compare your answers with those on pages 750 and 751.

20.4O What is harassment?

20.4P List three types of behavior that could be considered sexual harassment.

20.4Q What is the best way to prevent harassment?

20.46 Unions

Whether your utility operators belong to a union now or may join one in the future, a good employee-management relationship is crucial to keeping an agency functioning properly. Managers, supervisors, crew leaders, and operators all have to work together to develop this relationship.

Most of a manager's union contacts are with a shop steward. The shop steward is elected by the union employees and is their official representative to management and the local union. The steward is in an awkward position because the steward is an employee who is expected to do a full-time job like other employees, while also representing all of the employees. The steward must create an effective link between the utility manager or supervisors and the employees.

During contract negotiations between management and the employees' union representatives, management should be in constant consultation with the supervisors. Many employee demands regarding working conditions originate from the supervisor's daily dealings with the employees. An effective supervisor can minimize unreasonable demands. Also, any demands that are agreed upon must be implemented and carried out by a supervisor. A supervisor can help both sides reach an acceptable contractual agreement.

Once a contract has been agreed upon by both the union and the utility, the utility manager and the other supervisors must manage the organization within the framework of the contract. Do not attempt to ignore or get around the contract even if you disagree with some aspects of it. If you do not understand certain contract provisions, ask for clarification before you begin implementation of those provisions.

Contracts do not change the supervisor's delegated authority or responsibility. Operators must carry out the supervisor's orders and get the work done properly, safely, and within a reasonable amount of time. As a supervisor, you have the right and even the duty to make decisions. However, a contract gives a union the right to protest or challenge your decision. When an operator requires discipline, disciplinary action is a management responsibility.

Handling employee grievances within the framework of a union contract can be a very time-consuming job for the supervisor and the steward. Union contracts usually spell out in great detail the steps and procedures the steward and the supervisor must follow to settle differences. Grievances can develop over disciplinary action, distribution of overtime, transfers, promotions, demotions, and interpretation of labor contracts. The shop steward must communicate complaints and grievances from operators to the supervisor. Then, the supervisor and the steward must work together to settle complaints and adjust grievances. When a shop steward and supervisor can work together, the steward can help the supervisor to be an effective manager.

An effective manager lets everyone know they are available to discuss problems. *Dealing with grievances as quickly as possible* often prevents small problems from growing into large problems. When a shop steward presents a grievance to you, listen carefully. Discuss the problem with the employee directly with the help of the shop steward. Try to identify the facts and cause of the problem and keep a written record of your findings. Focus on the problem and do not get caught up in irrelevant issues. Make every effort to settle the grievance quickly and to everyone's satisfaction.

The consequences of any solution to a grievance must be considered and solutions must be consistent and fair to other operators. The solution or settlement should be clear and understandable to everyone involved. Once a solution has been agreed upon, prepare a written summary of the agreement. Review this final report with the shop steward to be sure the intent of the solution is understood and properly documented. The entire grievance procedure must be documented and properly filed, from initial presentation to final solution and settlement.

Union activities are governed by the National Labor Relations Act. When a union attempts to organize the utility's employees, your rights and actions as a manager are also governed by this Act. Be sure to seek competent legal assistance if your experience in dealing with a union is limited or if you have no such experience.

QUESTIONS

Write your answers in a notebook and then compare your answers with those on page 751.

20.4R What is the role of a shop steward?

20.4S How does a union contract affect a supervisor's authority?

END OF LESSON 1 OF 3 LESSONS
on
TREATMENT PLANT ADMINISTRATION

Please answer the discussion and review questions next.

DISCUSSION AND REVIEW QUESTIONS

Chapter 20. TREATMENT PLANT ADMINISTRATION

(Lesson 1 of 3 Lessons)

At the end of each lesson in this chapter, you will find some discussion and review questions. The purpose of these questions is to indicate to you how well you understand the material in the lesson. Write the answers to these questions in your notebook.

1. What are the different types of demands on a utility manager?
2. List the basic functions of a manager.
3. What can happen without adequate utility planning?
4. What is the purpose of an organizational plan?
5. Define the following terms:
    1. Authority
    2. Responsibility
    3. Delegation
    4. Accountability
6. When has a supervisor successfully delegated?
7. What two concepts should a manager keep in mind to avoid violating the rights of an employee or job applicant?
8. Why should a manager thoroughly examine the functions and staffing of the utility on a periodic basis?
9. When hiring new employees, the selection method and examination process used to evaluate applicants must be based on what criteria?
10. Why should you make notes of applicants' responses during job interviews?
11. What information should be provided to a new employee during orientation?
12. What type of training should be provided for supervisors?
13. Why is it important to formally document each employee's performance on a regular basis?
14. How should discipline problems be solved?
15. What steps can you take to help reach a successful resolution to a confrontation with an employee?
16. What is a manager's responsibility for preventing harassment in the workplace?
17. How are employee grievances usually handled under a union contract?

CHAPTER 20. TREATMENT PLANT ADMINISTRATION

(Lesson 2 of 3 Lessons)

20.5 COMMUNICATION

Good communication is an essential part of good management skills. Both written and oral communication skills are needed to effectively organize and direct the operation of a wastewater treatment facility. Remember that communication is a two-part process; information must be given and it must be understood. Good listening skills are as important in communication as the information you need to communicate. As the manager of a wastewater utility, you will need to communicate with employees, with your governing body, and with the public. Your communication style will be slightly different with each of these groups but you should be able to adjust easily to your audience.

20.50 Oral Communication

Oral communication may be informal, such as talking with employees, or it may be formal, such as giving a technical presentation. In both cases, your words should be appropriate to the audience, for example, avoid technical jargon when talking with nontechnical audiences. As you talk, you should be observing your audience to be sure that what you are saying is getting across. If you are talking with an employee, it is a good idea to ask for feedback from the employee, especially if you are giving instructions. When the employee is talking, watch and listen carefully and clarify areas that seem unclear. Likewise, in a more formal presentation, watching your audience will give you feedback about how well your message is being received. Some tips for preparing a formal speech are given in Table 20.4.

20.51 Written Communication

Written communication is more demanding than oral communication and requires more careful preparation. Again, keep your audience in mind and use language that will be understood. Written communication requires more organization since you cannot clarify and explain ideas in response to your audience. Before you begin you should have a clear idea of exactly what you wish to communicate, then keep your language as concise as possible. Extra words and phrases tend to confuse and clutter your message. Good writing skills develop slowly, but you should be able to find good writing classes in your community if you need help improving your skills. In addition, many publications and computer software programs are available to assist you in writing the most commonly needed documents such as memos, letters, press releases, résumés, monitoring reports, monthly reports, and the annual report (also see Chapter 19, "Records and Report Writing").

Before you can write a report you must first organize your thoughts. Ask yourself, what is the objective of this report? Am I trying to persuade someone of something? What information is important to communicate in this report? For whom is the report being written? How can I make it interesting? What does the reader want to learn from this report?

After you have answered the above questions, the next step is to prepare a general outline of how you intend to proceed with the preparation of the report. List not only key topics, but try to list all of the related topics. Then arrange the key topics in sequence so there is a workable, smooth flow from one topic to the next. Do not attempt to make your outline perfect. It is just a guide. It should be flexible. As you write you will find that you need to remove nonessential points and expand on more important points.

You might, for example, outline the following points in preparation for writing a report on a polymer testing program.

- A problem condition of high turbidity was discovered
- Polymer testing offered the best means of reducing turbidity
- Funds, equipment, and material were acquired
- Operators were trained
- Tests were conducted
- Results were evaluated and conclusions were reached
- Corrective actions were planned and taken
- Conclusion, the tests did or did not produce the anticipated results or correct the problem

Once you are fairly sure you have included all the major topics you will want to discuss, go through the outline and write down facts you want to include on each topic. As you work through it, you may decide to move material from one topic to another. The outline will help you organize your ideas and facts.

TABLE 20.4 TIPS FOR GIVING AN ORAL PRESENTATION

1. Arrive early. Give yourself plenty of time to become familiar with the room, practice using your audiovisuals, and make any necessary changes in room setup.
2. Be ready for mistakes. Number the pages in your presentation and your audiovisuals. Check the order of the pages before the meeting begins.
3. Pace yourself. Do not speak too quickly; speak slowly and clearly. Keep a careful eye on audience reaction to be sure that you are speaking at a pace that can be understood.
4. Project yourself. Speak loudly and look at the audience. Do not talk with your back to the audience. Check that those in the back can hear you.
5. Be natural. Try not to read your presentation. Practice ahead of time so that you can speak normally and keep eye contact with your audience.
6. Connect with the audience. Try to smile and make eye contact with the audience.
7. Involve the audience. Allow for audience questions and invite their comments.
8. Repeat audience questions. Always repeat the question so everyone can hear and to be sure that you hear the question correctly.
9. Know when to stop. Keep your remarks within the time allocated for your presentation and be aware that long, rambling speeches create a negative impression on the audience.
10. Use audiovisuals that can be heard and read. Audiovisuals should enhance and reinforce your words. Be sure that all members of the audience can hear, see, and read your audiovisuals. Normal typewritten text is not readable on overheads; use large type so everyone can see. Use no more than five to seven key ideas per overhead.
11. Organize your presentation. Prepare an introduction, body of the speech, and conclusion. Tell them what you are going to say, say it, and then tell them what you said. The presentation should have three to five main points presented in some logical order, for example, chronologically or from simple to complex.

When your outline is complete, you will have the essentials of your report. Now you need to tailor it to the audience that will be reading it. Take a few minutes to think about your audience. What information do they want? What aspects of the topics will they be most interested in reading? Each of the following groups may be interested in specific topics in the report. Consider these interests as you write.

1. Management

   Management will have specific interests that relate to the cost effectiveness of the program. A report to management should include a summary that presents the essential information, procedures used, an analysis of the data (including trends), and conclusions. Be sure to include complete cost information. Did the benefits warrant the costs? As a result of the tests, can future expenses be reduced? Backup information and field data can be included in an appendix for those who want more information.

2. Other Utilities

   Other utilities will be interested in costs but will also want more detailed information about how the program was performed. They will also be interested in the results and benefits of the program. Explain how the tests were done, the procedures, size of the crew, equipment used, source and availability of materials, difficulties encountered and how they were overcome.

3. Citizen Groups

   Citizens' interest will be more general. What is a polymer test and why is it needed? Is the polymer harmless? Will it injure fish or birds? How does the polymer test work? Who pays for the test? How much will it cost to implement results and will they be effective?

Your report may be written to include all of these groups. Adjust the outline to include the topics of interest to each group and identify the topics so readers can find the information most interesting to them. Keep the following information in mind as you write your report:

- Drafts. Good reports are not perfect the first time. Re-read and improve your report several times.
- Facts. Confine your writing to the facts and events that occurred. Include figures and statistics only when they make the report more effective. Include only the relevant facts. Large amounts of data should be put in the appendix. Do not clutter the report with unimportant data.
- Continuity. To be interesting and understood, a report must have continuity. It must make sense to the reader and be organized logically. In the report on the polymer testing, the report should be organized to show you had a problem, you had to find a way to identify where the problem existed, you did the testing, you identified the problems and the corrective actions.

- Effective. To be effective, a report should achieve the objective for which it was written. In this example we wanted to justify the costs for the program to management, help other utilities in conducting a similar program, and help citizens to understand what we were doing and why.
- Candid. A good report should be frank and straightforward. Keep the language appropriate for the audience. Do not try to impress your readers with technical terms they do not understand. Your purpose is to communicate information. Keep the information accurate and easy to understand.

The annual report is an important part of the management of the utility. It is one of the most involved writing projects that the utility must put together. The annual report should be a review of what and how the utility operated during the past year and it should also include the goals for the next year. In many small communities, the annual report may be presented orally to the city council rather than written. If this report is well written, it can be used to highlight accomplishments and provide support for future planning.

The first step to organizing the report is to make a list of three or four major accomplishments of the last year, then make a list of the top three goals for next year. These accomplishments and goals should be the focus of the report. The annual report should be a summary of the expenses, treatment services provided, and revenues generated over the last year. As you organize this information, keep those accomplishments in mind and let the data tell the story of how the utility accomplished last year's tasks. The data by itself may seem boring but as you organize the data it becomes a meaningful description of the year's accomplishments. Conclude with projections for next year. The facts and figures should tell the audience how you plan to accomplish your goals for the next year. The annual report may be simple or complex depending on your community needs. A sample Table of Contents for a medium-sized utility is given in Table 20.5. When you are finished, the annual report will be a valuable planning tool for the utility and can be used to build support for new projects.

TABLE 20.5 EXAMPLE OF TABLE OF CONTENTS FOR THE ANNUAL REPORT OF A UTILITY

Table of Contents	
Executive Summary	
Summary of the Treatment Process Including Flows and Costs	
Review of Goals and Objectives for the Year	
Special Projects Completed	
Professional Awards or Recognition for the Utility or Its Staff	
General Operating Conditions Including Regulatory Requirements	
Expectations for the Next Year—Goals and Objectives	
Recommended Changes for the Utility in Organizing, Staffing, Equipment, or Resources Summary	
Appendixes:	Operating Data
	Budget
	Information on Special Projects

QUESTIONS

Write your answers in a notebook and then compare your answers with those on page 751.

20.5A What kinds of communication skills are needed by a manager?

20.5B What are the most common written documents that a utility manager must write?

20.5C What should be included in the annual report?

20.6 CONDUCTING MEETINGS

As a utility manager you will be asked to conduct meetings. These meetings may be with employees, your governing board, the public, or with other professionals in your field. Many new managers fail to prepare for these meetings and the meetings end up as a terrible waste of time. As a manager, you need to learn to conduct meetings in a way that is productive and guides the participants into an active role. The following steps should be taken to conduct a productive meeting.

Before the meeting:

- Prepare an agenda and distribute it to all participants.
- Find an adequate meeting room.
- Set a beginning and ending time for the meeting.

During the meeting:

- Start the meeting on time.
- Clearly state the purpose and objectives of the meeting.
- Involve all the participants.
- Do not let one or two individuals dominate the meeting.
- Keep the discussion on track and on time with the agenda.
- When the group makes a decision or reaches consensus, restate your understanding of the results.
- Make clear assignments for participants and review them with everyone during the meeting.

After the meeting:

- Send out minutes of the meeting.
- Send out reminders, when appropriate, about any assignments made for participants, and the next meeting time.

QUESTIONS

Write your answers in a notebook and then compare your answers with those on page 751.

20.6A With whom may a utility manager be asked to conduct meetings?

20.6B What should be done before a meeting?

20.7 PUBLIC RELATIONS

20.70 Establish Objectives

The first step in organizing an effective public relations campaign is to establish objectives. The only way to know whether your program is a success is to have a clear idea of what you expect to achieve—for example, better customer relations, greater water conservation, and enhanced organizational credibility. Each objective must be specific, achievable, and measurable. It is also important to know your audience and tailor various elements of your public relations effort to specific groups you wish to reach, such as community leaders, school children, or the average customer. Your objective may be the same in each case, but what you say and how you say it will depend upon your target audience.

20.71 Utility Operations

Good public relations begin at home. Any time you or a member of your utility comes in contact with the public, you will have an impact on the quality of your public image. Dedicated, service-oriented employees provide for better public relations than paid advertising or complicated public relations campaigns. For most people, contact with an agency employee establishes their first impression of the competence of the organization, and those initial opinions are difficult to change.

In addition to ensuring that employees are adequately trained to do their jobs and knowledgeable about the utility's operations, management has the responsibility to keep employees informed about the organization's plans, practices, and goals. Newsletters, bulletin boards, and regular, open communication between supervisors and subordinates will help build understanding and contribute to a team spirit.

Despite the old adage to the contrary, the customer is not always right. Management should try to instill among its employees the attitude that while the customer may be confused or unclear about the situation, everyone is entitled to courteous treatment and a factual explanation. Whenever possible, employees should phrase responses as positively, or neutrally, as possible, avoiding negative language. For example, "Your complaint" is better stated as "Your question." "You should have..." is likely to make the customer defensive, while "Will you please..." is courteous and respectful. "You made a mistake" emphasizes the negative, "What we will do..." is a positive, problem-solving approach.

20.72 The Mass Media

We live in the age of communications, and one of the most effective and least expensive ways to reach people is through the mass media—radio, television, newspapers, and the Internet. Each medium has different needs and deadlines, and obtaining coverage for your issue or event is easier if you are aware of these constraints. Television must have strong visuals, for example. When scheduling a press conference, provide an interesting setting and be prepared to suggest good shots to the reporter. Radio's main advantage over television and newspapers is immediacy, so have a spokesperson available and prepared to give the interview over the telephone, if necessary. Newspapers give more thorough, in-depth coverage to stories than do the broadcast media, so be prepared to spend extra time with print reporters and provide written backup information and additional contacts.

It is not difficult to get press coverage for your event or press conference if a few simple guidelines are followed:

1. Demonstrate that your story is newsworthy, that it involves something unusual or interesting.
2. Make sure your story will fit the targeted format (television, radio, newspaper, or the Internet).
3. Provide a spokesperson who is interesting, articulate, and well prepared.

20.73 Being Interviewed

Whether you are preparing for a scheduled interview or are simply responding to the press on a breaking news story, here are some key hints to keep in mind when being interviewed.

1. Speak in personal terms, free of institutional jargon.
2. Do not argue or show anger if the reporter appears to be rude or overly aggressive.
3. If you do not know an answer, say so and offer to find out. Do not bluff.
4. If you say you will call back by a certain time, do so. Reporters face tight deadlines.
5. State your key points early in the interview, concisely and clearly. If the reporter wants more information, he or she will ask for it.

6. If a question contains language or concepts with which you disagree, do not repeat them, even to deny them.
7. Know your facts.
8. Never ask to see a story before it is printed or broadcast. Doing so indicates that you doubt the reporter's ability and professionalism.

20.74 Public Speaking

Direct contact with people in your community is another effective tool in promoting your utility. Though the audiences tend to be small, a personal, face-to-face presentation generally leaves a strong and long-lasting impact on the listener.

Depending on the size of the organization, your utility may wish to establish a speaker's bureau and send a list of topics to service clubs in the area. Visits to high schools and college campuses can also be beneficial, and educators are often looking for new and interesting topics to supplement their curriculum.

Effective public speaking takes practice. It is important to be well prepared while retaining a personal, informal style. Find out how long your talk is expected to be, and do not exceed that time frame. Have a definite beginning, middle, and end to your presentation. Visual aids such as charts, slides, or models can assist in conveying your message. The use of humor and anecdotes can help to warm up the audience and build rapport between the speaker and the listener. Just be sure the humor is natural, not forced, and that the point of your story is accessible to the particular audience. Try to keep in mind that audiences only expect you to do your best. They are interested in learning about the performance of their wastewater treatment plant and will appreciate that you are making a sincere effort to inform them about an important subject.

20.75 Telephone Contacts

First impressions are extremely important, and frequently a person's first contact with your wastewater utility is over the telephone. A person who answers the phone in a courteous, pleasant, and helpful manner goes a long way toward establishing a friendly, cooperative atmosphere. Be sure anyone answering telephone inquiries receives appropriate training and conveys a positive image for the utility.

Following a few simple guidelines will help to start your utility off on the right note with your customers:

1. *Answer calls promptly.* Your conversation will get off to a better start if the phone is answered by the third or fourth ring.
2. *Identify yourself.* This adds a personal note and lets the caller know whom he or she is talking to.
3. *Pay attention.* Do not conduct side conversations. Minimize distractions so you can give the caller your full attention, avoiding repetitions of names, addresses, and other pertinent facts.
4. *Minimize transfers.* Nobody likes to get the run-around. Few things are more frustrating to a caller than being transferred from office to office, repeating the situation, problem, or concern over and over again. Transfer only those calls that must be transferred, and make certain you are referring the caller to the right person. Then, explain why you are transferring the call. This lets the caller know you are referring him or her to a co-worker for a reason and reassures the customer that the problem or question will be dealt with. In some cases, it may be better to take a message and have someone return the call than to keep transferring the customer's call.

20.76 Consumer Inquiries

No single set of rules can possibly apply to all types of consumer questions or complaints. There are, however, basic principles to follow in responding to inquiries and concerns.

1. *Be prepared.* Your employees should be familiar enough with your utility's organization, services, and policies to either respond to the question or complaint or locate the person who can.
2. *Listen.* Ask the customer to describe the problem and listen carefully to the explanation. Take written notes of the facts and addresses.
3. *Do not argue.* Callers often express a great deal of pent-up frustration in their contacts with a utility. Give the caller your full attention. Once you have heard them out, most people will calm down and state their problems in more reasonable terms.
4. *Avoid jargon.* The average consumer lacks the technical knowledge to understand the complexities of wastewater collection and treatment. Use plain, nontechnical language and avoid telling the consumer more than he or she wants to know.
5. *Summarize the problem.* Repeat your understanding of the situation back to the caller. This will assure the customer that you understand the problem and offer the opportunity to clear up any confusion or missed communication.
6. *Promise specific action.* Make an effort to give the customer an immediate, clear, and accurate answer to the problem. Be as specific as possible without promising something you cannot deliver.

Complaints can be a valuable asset in pinpointing wastewater collection system problems. Customer calls are frequently your first indication that something may be wrong. Responding to complaints and inquiries promptly can save the utility money and staff resources, and minimize the number of customers who are inconvenienced. Still, education can greatly reduce complaints about the wastewater collection system and treatment processes. Information brochures, utility bill inserts, and other educational tools help to inform customers and avoid future complaints.

20.77 Plant Tours

Tours of wastewater treatment plants can be an excellent way to inform the public about your utility's efforts to provide a safe, high-quality discharge to the environment. Political leaders, such as the City Council and members of the Board of

Supervisors, should be invited and encouraged to tour the facilities, as should school groups and service clubs.

A brochure describing your utility's goals, accomplishments, operations, and processes can be a good supplement to the tour and should be handed out at the end of the visit. The more visually interesting the brochure is, the more likely that it will be read, and the use of color, photographs, graphics, or other design features is encouraged. If you have access to the necessary equipment, production of a videotape program about the utility can also add interest to the facility tour.

The tour itself should be conducted by an employee who is very familiar with plant operations and can answer the types of questions that are likely to arise. Consider including:

1. A description of the sources of wastewater
2. History of the plant, the years of operation, modifications, and innovations over the years
3. Major plant design features, including plant capacity and safety features
4. Safe observation of the treatment processes, including sedimentation, activated sludge, disinfection, and solids handling and disposal
5. A visit to the laboratory, including information on the quality of effluent discharged to the environment
6. Anticipated improvements, expansions, and long-range plans for meeting future service needs

Plant tours can contribute to a wastewater utility's overall program to gain financing for capital improvements. If the City Council or other governing board has seen the treatment process first hand, it is more likely to understand the need for enhancement and support future funding.

As beneficial as plant tours may be for promoting public interest and confidence in the wastewater treatment plant, security precautions should be carefully considered when planning for visitors to the plant. For more information, see Section 20.13, "Security Measures."

QUESTIONS

Write your answers in a notebook and then compare your answers with those on page 751.

20.7A What is the first step in organizing a public relations campaign?

20.7B How can employees be kept informed of the utility's plans, practices, and goals?

20.7C Which news medium is more likely to give a story thorough, in-depth coverage?

20.7D What is the key to effective public speaking?

20.7E How do customer complaints help a utility?

20.8 FINANCIAL MANAGEMENT

Financial management for a utility should include providing financial stability for the utility, careful budgeting, and providing capital improvement funds for future utility expansion. These three areas must be examined on a routine basis to ensure the continued operation of the utility. They may be formally reviewed on an annual basis or more frequently when the utility is changing rapidly. The utility manager should understand what is required for each of the three areas and be able to develop record systems that keep the utility on track and financially prepared for the future.

20.80 Financial Stability

How do you measure financial stability for a utility? Two very simple calculations can be used to help you determine how healthy and stable the finances are for the utility. These two calculations are the *OPERATING RATIO* and the *COVERAGE RATIO.* The operating ratio is a measure of the total revenues divided by the total operating expenses. The coverage ratio is a measure of the ability of the utility to pay the principal and interest on loans and bonds (this is known as *DEBT SERVICE*[5]) in addition to any unexpected expenses. A utility that is in good financial shape will have an operating ratio and coverage ratio above 1.0. In fact, most bonds and loans require the utility to have a coverage ratio of at least 1.25. As state and federal funds for utility improvements have become much more difficult to obtain, these financial indicators have become more important for utilities. Being able to show and document the financial stability of the utility is an important part of getting funding for more capital improvements.

The operating ratio is perhaps the simplest measure of a utility's financial stability. In essence, the utility must be generating enough revenue to pay its operating expenses. The actual ratio is usually computed on a yearly basis, since many utilities may have monthly variations that do not reflect the overall performance. The total revenue is calculated by adding up all revenue generated by user fees, hook-up charges, taxes or assessments, interest income, and special income. Next, determine the total operating expenses by adding up the expenses of the utility, including administrative costs, salaries, benefits, energy costs, chemicals, supplies, fuel, equipment costs, equipment replacement fund, principal and interest payments, and other miscellaneous expenses.

[5] *Debt Service.* The amount of money required annually to pay the (1) interest on outstanding debts, or (2) funds due on a maturing bonded debt or the redemption of bonds.

EXAMPLE 1

The total revenues for a utility are $1,686,000 and the operating expenses for the utility are $1,278,899. The debt service expenses are $560,000. What is the operating ratio? What is the coverage ratio?

Known		Unknown
Total Revenue, $	= $1,686,000	Operating Ratio
Operating Expenses, $	= $1,278,899	Coverage Ratio
Debt Service Expenses, $	= $560,000	

1. Calculate the operating ratio.

$$\text{Operating Ratio} = \frac{\text{Total Revenue, \$}}{\text{Operating Expenses, \$}}$$

$$= \frac{\$1{,}686{,}000}{\$1{,}278{,}899}$$

$$= 1.32$$

2. Calculate nondebt expenses.

$$\text{Nondebt Expenses, \$} = \text{Operating Exp, \$} - \text{Debt Service Exp, \$}$$

$$= \$1{,}278{,}899 - \$560{,}000$$

$$= \$718{,}899$$

3. Calculate coverage ratio.

$$\text{Coverage Ratio} = \frac{\text{Total Revenue, \$} - \text{Nondebt Expenses, \$}}{\text{Debt Service Expenses, \$}}$$

$$= \frac{\$1{,}686{,}000 - \$718{,}899}{\$560{,}000}$$

$$= 1.73$$

These calculations provide a good starting point for looking at the financial strength of the utility. Both of these calculations use the total revenue for the utility, which is an important component for any utility budgeting. As managers we often focus on the expense side and forget to look carefully at the revenue side of utility management. The fees collected by the utility, including hook-up fees and user fees, must accurately reflect the cost of providing service. These fees must be reviewed annually and they must be increased as expenses rise to maintain financial stability. Some other areas to examine on the revenue side include how often and how well user fees are collected and the number of delinquent accounts. Some small communities have found they can cut their administrative costs significantly by switching to a quarterly billing cycle. The utility must have the support of the community to determine and collect user fees, and the utility must keep track of revenue generation as carefully as resource spending.

20.81 Budgeting

Budgeting for the utility is perhaps the most challenging task of the year for many managers. The list of needs usually is much larger than the possible revenue for the utility. The only way for the manager to prepare a good budget is to have good records from the year before. A system of recording or filing purchase orders (see Section 20.126, "Procurement Records") or a requisition records system must be in place to keep track of expenses and prevent spending money that is not in the budget.

To budget effectively, a manager needs to understand how the money has been spent over the last year, the needs of the utility, and how the needs should be prioritized. The manager also must take into account cost increases that cannot be controlled while trying to minimize the expenses as much as possible. The following problem is an example of the types of decisions a manager must make to keep the budget in line while also improving service from the utility.

EXAMPLE 2

A pump that has been in operation for 25 years pumps a constant 600 GPM through 47 feet of dynamic head. The pump uses 6,071 kilowatt-hours of electricity per month, at a cost of $0.085 per kilowatt-hr. The old pump efficiency has dropped to 63 percent. Assuming a new pump that operates at 86 percent efficiency is available for $9,730.00, how long would it take to pay for replacing the old pump?

Known		Unknown
Electricity, kW-hr/mo	= 6,071 kW-hr/mo	New Pump Payback Time, yr
Electricity Cost, $/kW-hr	= $0.085/kW-hr	
Old Pump Efficiency, %	= 63%	
New Pump Efficiency, %	= 86%	
New Pump Cost, $	= $9,730	

1. Calculate old pump operating costs in dollars per month.

$$\text{Old Pump Operating Costs, \$/mo} = (\text{Electricity, kW-hr/mo})(\text{Electricity Cost, \$/kW-hr})$$

$$= (6{,}071\text{ kW-hr/mo})(\$0.085\text{/kW-hr})$$

$$= \$516.04\text{/mo}$$

2. Calculate new pump operating electricity requirements.

$$\text{New Pump Electricity, kW-hr/mo} = (\text{Old Pump Electricity, kW-hr/mo})\frac{(\text{Old Pump Eff, \%})}{(\text{New Pump Eff, \%})}$$

$$= (6{,}071\text{ kW-hr/mo})\frac{(63\%)}{(86\%)}$$

$$= 4{,}447\text{ kW-hr/mo}$$

3. Calculate new pump operating costs in dollars per month.

$$\text{New Pump Operating Costs, \$/mo} = (\text{Electricity, kW-hr/mo})(\text{Electricity Cost, \$/kW-hr})$$

$$= (4{,}447\text{ kW-hr/mo})(\$0.085\text{/kW-hr})$$

$$= \$378.03\text{/mo}$$

4. Calculate annual cost savings of new pump.

$$\text{Cost Savings, \$/yr} = (\text{Old Costs, \$/mo} - \text{New Costs, \$/mo})(12\text{ mo/yr})$$

$$= (\$516.04\text{/mo} - \$378.03\text{/mo})(12\text{ mo/yr})$$

$$= \$1{,}656.12\text{/yr}$$

5. Calculate the new pump payback time in years.

$$\text{Payback Time, yr} = \frac{\text{Initial Cost, \$}}{\text{Savings, \$/yr}}$$

$$= \frac{\$9{,}730.00}{\$1{,}656.12\text{/yr}}$$

$$= 5.9\text{ years}$$

In this example, a payback time of 5.9 years is acceptable and would probably justify the expense for a new pump. This calculation was a simple payback calculation that did not take into account the maintenance on each pump, depreciation, and inflation. Many excellent references are available from

EPA to help utility managers make more complex decisions about purchasing new equipment.

The annual report should be used to help develop the budget so that long-term planning will have its place in the budgeting process. The utility manager must track revenue generation and expenses with adequate records to budget effectively. The manager must also get input from other personnel in the utility as well as community leaders as the budgeting process proceeds. This input from others is invaluable to gain support for the budget and to keep the budget on track once adopted.

20.82 Equipment Repair/Replacement Fund

To adequately plan for the future, every utility must have a repair/replacement fund. The purpose of this fund is to generate additional revenue to pay for the repair and replacement of capital equipment as the equipment wears out. To prepare adequately for this repair/replacement, the manager should make a list of all capital equipment (this is called an asset inventory) and estimate the replacement cost for each item. The expected life span of the equipment must be used to determine how much money should be collected over time. When a treatment plant is new, the balance in repair/replacement fund should be increasing each year. As the plant gets older, the funds will have to be used and the balance may get dangerously low as equipment breakdowns occur. Perhaps the hardest job for the utility manager is to maintain a positive balance in this account with the understanding that this account is not meant to generate a profit for the utility but rather to plan for future equipment needs. In wastewater treatment facilities construction, providing an adequate repair/replacement fund is very important, but if this repair/replacement fund has not been reviewed annually, it must be updated.

To set up a repair/replacement fund for your utility, you should first put together a list of the equipment required for each process in your utility. Once you have this list, you need to estimate the life expectancy of the equipment and the replacement cost. From this list you can predict the amount of money you should set aside each year so that when each piece of equipment wears out, you will have enough money to replace that piece of equipment. The EPA publications listed in Section 20.16, "Additional Reading," at the end of this chapter are excellent references for utility planning.

20.83 Wastewater Treatment Rates

The process of determining the cost of wastewater treatment and establishing a wastewater treatment rate schedule for customers is a subject of much controversy. There is no single set of rules for determining wastewater rates. The establishment of a rate schedule involves many factors including the form of ownership (investor or publicly owned), differences in regulatory control over the wastewater utility (state commission or local authority), and differences in individual viewpoints and preferences concerning the appropriate philosophy to be followed to meet local conditions and requirements.

Generally, the development of wastewater treatment rate schedules involves the following procedures:

- A determination of the total revenue requirements for the period that the rates are to be effective (usually one year).
- A determination of all the cost components of system operations. That is, how much does it cost to treat the wastewater? How much does it cost to discharge to the environment? How much does it cost to install a wastewater collection service to a customer? How much are administrative costs?
- Distribution of the various component costs to the various customer classes in accordance with their requirements for service.
- The design of wastewater treatment rates that will recover from each class of customers, within practical limits, the cost to serve that class of customers.

20.84 Capital Improvements and Funding in the Future

A capital improvements fund must be a part of the utility budget and included in the operating ratio. Your responsibility as the utility manager is to be sure that everyone, your governing body and the public, understands the capital improvement fund is not a profit for the utility but a replacement fund to keep the utility operating in the future.

Capital planning starts with a look at changes in the community. Where are the areas of growth in the community, where are the areas of decline, and what are the anticipated changes in industry within the community? After identifying the changing needs in the community, you should examine the existing utility structure. Identify your weak spots (in the collection system or with in-plant processes). Make a list of the areas that will be experiencing growth, weak spots in the system, and anticipated new regulatory requirements. The list should include expected capital improvements that will need to be made over the next year, two years, five years, and ten years. You can use the information in your annual reports and other operational logs to help compile the list.

Once you have compiled this information, prioritize the list and make a timetable for improving each of the areas. Starting at the top of the priority list, estimate the costs for improvements and incorporate these costs into your capital improvement budget. The calculations you have made previously, including corrective to preventive maintenance ratios, operating ratio, coverage ratio, and payback time will all be useful in prioritizing and streamlining your list of needs.

You may find that some of your capital improvement needs could be met in more than one way. How do you decide which of several options is most cost-effective? How do you compare fundamentally different solutions? For example, assume your community's population is growing rapidly and you will need to increase the capacity of your wastewater treatment facilities by the end of the next ten years. Your two existing wastewater treatment plants are both operating near 95 percent of design capacity. Possible solutions might include the following options, where portions of some of the options might be implemented immediately while other portions might be brought on line in five or ten years.

- Expand the existing wastewater treatment plants.
- Construct a new wastewater treatment plant to treat flows greater than the capacity of the two existing plants.
- Construct a new regional plant capable of handling all expected flows (shut down the two existing plants).
- Keep the newer of the two existing plants, shut down the older of the two plants, and construct one new, larger plant to treat flows from the older plant and the expected future flows.

Each of these possible solutions should be evaluated individually and then compared with the other alternatives. To compare alternative plans you will need to calculate the present value (or *PRESENT WORTH*[6]) of each plan; that is, the costs and benefits of each plan in today's dollars. This is done by identifying all the costs and benefits of each alternative plan over the same time period or time horizon. Costs should include not only the initial purchase price or construction costs, but also financing costs over the life of the loans or bonds and all operation and maintenance costs. Benefits include all of the revenue that would be produced by this facility or equipment, including connection and user fees. With the help of an experienced accountant, apply standard inflation, depreciation, and other economic discount factors to calculate the present value of all the benefits and costs of each plan during the same planning period. This will give you the cost of each plan in the equivalent of today's dollars.

Remember to involve all of your local officials and the public in this capital improvement budget so they understand what will be needed.

Long-term capital improvements such as a new plant or a new treatment process are usually anticipated in your 10-year or 20-year projection. These long-term capital improvements usually require some additional financing. The basic ways for a utility to finance capital improvements are through general obligation bonds, revenue bonds, or loan funding programs.

General obligation bonds or *ad valorem* (based on value) taxes are assessed based on property taxes. These bonds usually have a lower interest rate and longer payback time, but the total bond limit is determined for the entire community. This means that the wastewater treatment utility will have available only a portion of the total bond capacity of the community. These bonds are not often used for funding wastewater treatment utility improvements today.

The second type of bond, the revenue bond, is commonly used to fund utility improvements. This bond has no limit on the amount of funds available and the user charges provide repayment on the bond. To qualify for these bonds, the utility must show sound financial management and the ability to repay the bond. As the utility manager you should be aware of the provisions of the bond. Be sure the bond has a call date, which is the first date when you can pay off the bond. The common practice is for a 20-year bond to have a 10-year call date and for a 15-year bond to have an 8-year call date. The bond will also have a call premium, which is the amount of extra funds needed to pay off the debt on the call date. You should try to get your bonds a call premium of no more than 102 percent par. This means that for a debt of $200,000 on the call date, the total payoff would be $204,000, which includes the extra two percent for the call premium. You will need to get help from a financial advisor to prepare for and issue the bonds. These advisors will help you negotiate the best bond structure for your community.

Special assessment bonds may be used to extend services into specific areas. The direct users pay the capital costs and the assessment is usually based on frontage or area of real estate. These special assessments carry a greater risk to investors but may be the best way to extend service to some areas.

The most common way to finance wastewater treatment improvements in the past has been federal and state grant programs. The Block Grants from HUD are still available for some projects and Rural Utilities Service (RUS) loans may also be used as a funding source. In addition, state revolving fund (SRF) programs provide loans (but not direct grants) for improvements. The SRF program has been implemented with wastewater improvements. These SRF programs are very competitive and utilities must provide evidence of sound financial management to qualify for these loans. You should contact your state regulatory agency to find out more about the SRF program in your state.

20.85 Financial Assistance

Many small wastewater treatment and collection systems need additional funds to repair and upgrade their systems. Potential funding sources include loans and grants from federal and state agencies, banks, foundations, and other sources. Some of the federal funding programs for small public utility systems include:

[6] *Present Worth.* The value of a long-term project expressed in today's dollars. Present worth is calculated by converting (discounting) all future benefits and costs over the life of the project to a single economic value at the start of the project. Calculating the present worth of alternative projects makes it possible to compare them and select the one with the largest positive (beneficial) present worth or minimum present cost.

- Appalachian Regional Commission (ARC)
- Department of Housing and Urban Development (HUD) (provides Community Development Block Grants)
- Economic Development Administration (EDA)
- Indian Health Service (IHS)
- Rural Utilities Service (RUS) (formerly Farmer's Home Administration (FmHA) and Rural Development Administration (RDA))

Another valuable contact is the Environmental Financing Information Network (EFIN), which provides information on financing alternatives for state and local environmental programs and projects in the form of abstracts of publications, case studies, and contacts. Contact Environmental Financing Information Network, US Environmental Protection Agency (EPA), EFIN (mail code 2731R), Ariel Rios Building, 1200 Pennsylvania Avenue, NW, Washington, DC 20460. Phone (202) 564-4994 and FAX (202) 565-2587.

Also, many states have one or more special financing mechanisms for small public utility systems. These funds may be in the form of grants, loans, bonds, or revolving loan funds. Contact your state pollution control agency for more information.

QUESTIONS

Write your answers in a notebook and then compare your answers with those on page 751.

20.8A List the three main areas of financial management for a utility.

20.8B How is a utility's operating ratio calculated?

20.8C Why is it important for a manager to consult with other utility personnel and with community leaders during the budget process?

20.8D How can long-term capital improvements be financed?

20.8E What is a revenue bond?

END OF LESSON 2 OF 3 LESSONS
on
TREATMENT PLANT ADMINISTRATION

Please answer the discussion and review questions next.

DISCUSSION AND REVIEW QUESTIONS

Chapter 20. TREATMENT PLANT ADMINISTRATION

(Lesson 2 of 3 Lessons)

Write the answers to these questions in your notebook. The question numbering continues from Lesson 1.

18. With whom do managers need to communicate?
19. What information should be included in the utility's annual report?
20. List four steps that can be taken during a meeting to make sure it is a productive meeting.
21. What happens any time you or a member of your utility comes in contact with the public?
22. What attitude should management try to develop among its employees regarding the customer?
23. What is the value of customer complaints?
24. How do you measure financial stability for a utility?
25. How can a manager prepare a good budget?

CHAPTER 20. TREATMENT PLANT ADMINISTRATION

(Lesson 3 of 3 Lessons)

20.9 OPERATIONS AND MAINTENANCE

20.90 The Manager's Responsibilities

A utility manager's specific operation and maintenance (O & M) responsibilities vary depending on the size of the utility. At a small utility, the manager may oversee all utility operations while also serving as chief operator and supervising a small staff of operations and maintenance personnel. In larger utility agencies, the manager may have no direct, day-to-day responsibility for operations and maintenance but is ultimately responsible for efficient, cost-effective operation of the entire utility. Whether large or small, every utility needs an effective operations and maintenance program.

20.91 Purpose of O & M Programs

The purpose of O & M programs is to maintain design functionality (capacity) or to restore the system components to their original condition and thus functionality. Stated another way, does the system perform as designed and intended? The ability to effectively operate and maintain a wastewater treatment utility so it performs as intended depends greatly on proper design (including selection of appropriate materials and equipment), construction and inspection, acceptance of the constructed facility following final inspection, and system start-up. Permanent system deficiencies that affect O & M of the system are frequently the result of these phases. O & M staff should be involved at the beginning of each project, including planning, design, construction, acceptance, and start-up. When a utility system is designed with future O & M considerations in mind, the result is a more effective O & M program in terms of O & M cost and performance.

Effective O & M programs are based on knowing what components make up the system, where they are located, and the condition of the components. With that information, proactive maintenance can be planned and scheduled, rehabilitation needs identified, and long-term Capital Improvement Programs (CIP) planned and budgeted. High-performing agencies have all developed performance measurements of their O & M program and track the information necessary to evaluate performance.

20.92 Types of Maintenance

Wastewater treatment plant maintenance can be either a proactive or a reactive activity. Commonly accepted types of maintenance include three classifications: corrective maintenance, preventive maintenance, and predictive maintenance.

Corrective maintenance, including emergency maintenance, is reactive. For example, a piece of equipment or a system is allowed to operate until it fails, with little or no scheduled maintenance occurring prior to the failure. Only when the equipment or system fails is maintenance performed. Reliance on reactive maintenance will always result in poor system performance, especially as the system ages. Utility agencies taking a corrective maintenance approach are characterized by:

- The inability to plan and schedule work
- The inability to budget adequately
- Poor use of resources
- A high incidence of equipment and system failures

Emergency maintenance involves two types of emergencies: normal emergencies and extraordinary emergencies. Public utilities are faced with normal emergencies such as sewer blockages on a daily basis. Normal emergencies can be reduced by an effective maintenance program. Extraordinary emergencies, such as high-intensity rainstorms, hurricanes, floods, and earthquakes, will always be unpredictable occurrences. However, the effects of extraordinary emergencies on the utility's performance can be minimized by implementation of a planned maintenance program and development of a comprehensive emergency response plan (see Section 20.10).

Preventive maintenance is proactive and is defined as a programmed, systematic approach to maintenance activities. This type of maintenance will always result in improved system performance except in the case where major chronic problems are the result of design or construction flaws that cannot be corrected by O & M activities. Proactive maintenance is performed on a periodic (preventive) basis or an as-needed (predictive) basis. Preventive maintenance can be scheduled on the basis of specific criteria such as equipment operating time since the last maintenance was performed, or passage of a certain amount of time (calendar period). Lubrication of motors, for example, is frequently based on running time.

The major elements of a good preventive maintenance program include the following:

1. Planning and scheduling
2. Records management
3. Spare parts management
4. Cost and budget control
5. Emergency repair procedures
6. Training program

Some benefits of taking a preventive maintenance approach are:

1. Maintenance can be planned and scheduled
2. Work backlog can be identified
3. Adequate resources necessary to support the maintenance program can be budgeted

4. Capital Improvement Program (CIP) items can be identified and budgeted for
5. Human and material resources can be used effectively

Predictive maintenance, which is also proactive, is a method of establishing baseline performance data, monitoring performance criteria over a period of time, and observing changes in performance so that failure can be predicted and maintenance can be performed on a planned, scheduled basis. Knowing the condition of the system makes it possible to plan and schedule maintenance as required and thus avoid unnecessary maintenance.

In reality, every agency operates their system with corrective and emergency maintenance, preventive maintenance, and predictive maintenance methods. The goal, however, is to reduce the corrective and emergency maintenance efforts by performing preventive maintenance, which will minimize system failures that result in stoppages and overflows.

System performance is frequently a reliable indicator of how the system is operated and maintained. Agencies that rely primarily on corrective maintenance as their method of operating and maintaining the system are never able to focus on preventive and predictive maintenance. With most of their resources directed at corrective maintenance activities, it is difficult to free up these resources to begin developing preventive maintenance programs. For an agency to develop an effective proactive maintenance program, they must add initial resources over and above those currently existing.

20.93 Benefits of Managing Maintenance

The goal of managing maintenance is to minimize investments of labor, materials, money, and equipment. In other words, we want to manage our human and material resources as effectively as possible, while delivering a high level of service to our customers. The benefits of an effective operation and maintenance program are as follows:

- Ensuring the availability of facilities and equipment as intended.
- Maintaining the reliability of the equipment and facilities as designed. Utility systems are required to operate 24 hours per day, 7 days per week, 365 days per year. Reliability is a critical component of the operation and maintenance program. If equipment and facilities are not reliable, then the ability of the system to perform as designed is impaired.
- Maintaining the value of the investment. Wastewater collection and treatment systems represent major capital investments for communities and are major capital assets of the community. If maintenance of the system is not managed, equipment and facilities will deteriorate through normal use and age. Maintaining the value of the capital asset is one of the utility manager's major responsibilities. Accomplishing this goal requires ongoing investment to maintain existing facilities and equipment, extend the life of the system, and establish a comprehensive O & M program.
- Obtaining full use of the system throughout its design life.
- Collecting accurate information and data on which to base the operation and maintenance of the system and justify requests for the financial resources necessary to support it.

QUESTIONS

Write your answers in a notebook and then compare your answers with those on page 751.

20.9A What is the purpose of an operation and maintenance (O & M) program?

20.9B What are the three common types of maintenance?

20.9C List the major elements of a good preventive maintenance program.

20.94 SCADA Systems

SCADA (SKAY-dah) stands for Supervisory Control And Data Acquisition system. This is a computer-monitored alarm, response, control, and data acquisition system used by wastewater collection system and treatment plant operators to monitor and adjust the operation of equipment in their systems.

A SCADA system collects, stores, and analyzes information about all aspects of operation and maintenance, transmits alarm signals, when necessary, and allows fingertip control of alarms, equipment, and processes. SCADA provides the information that operators and their supervisors need to solve minor problems before they become major incidents. As the nerve center of a wastewater collection and treatment agency, the system allows operators to enhance the efficiency of their facilities by keeping them fully informed and fully in control.

A typical SCADA system is made up of basically five groups of components:

1. Field-mounted sensors, instrumentation, and controlled or monitored equipment. These devices sense system variables and generate input signals to the SCADA system for monitoring. These devices also receive the command output signals from the SCADA system.
2. Remote Terminal Units (RTUs). These devices gather the data from the field-mounted instruments and provide the control signals to the field equipment.
3. Communications medium. This is the link between the SCADA RTUs and the main control location. There are

many communications mediums available for transmitting signals between the RTUs and the supervisory control station. Some of the options presently available include:

- FM (VHF/UHF) radio
- Dedicated leased telephone circuits
- Privately owned metallic signal lines
- Conventional dial telephone lines (pulse or tone/DTMF)
- Coaxial cable networks
- Spread spectrum radio
- Fiber-optic cable
- Microwave
- 900 MHz radio
- Cellular telephone
- Ground station satellites

4. Supervisory control and monitoring equipment. There are three basic categories of supervisory equipment:
   a. Hardware-based systems include graphic displays, annunciator lamp boxes, chart recorders, and similar equipment.
   b. Software-based systems include all computer-based systems such as microcomputers, workstations, minicomputers, and mainframes. Software-based systems sometimes offer more flexibility and capabilities than hardware-based systems and are often less expensive.
   c. Hybrid systems are a combination of hardware and software systems; for example, a PC (microcomputer) with a graphic display panel.
5. Human machine interface (HMI). Operators work with the computers to obtain the information they need, respond when changes are necessary, and monitor changes to ensure the desired results.

Applications for SCADA systems include wastewater collection and pumping system monitoring, wastewater treatment plant control monitoring, combined sewer overflow (CSO) diversion monitoring, and other related applications. SCADA systems can vary from merely data collection and storage to total data analysis, interpretation, and process control.

A SCADA system might include liquid level, pressure, and flow sensors. The measured (sensed) information could be transmitted by one of the communications systems listed earlier to a computer system, which stores, analyzes, and presents the information. The information may be read by an operator on dials or as digital readouts or analyzed and plotted by the computer in the form of trend charts.

Most SCADA systems present a graphical picture of the overall system on the screen of a computer monitor. In addition, detailed pictures of specific portions of the system can be examined by the operator following a request and instructions to the computer. The graphical displays on the computer screen can include current operating information. The operator can observe this information, analyze it for trends, or determine if it is within acceptable operating ranges, and then decide if any adjustments or changes are necessary.

SCADA systems are capable of analyzing data and providing operating, maintenance, regulatory, and annual reports. In some plants operators rely on a SCADA system to help them prepare daily, weekly, and monthly maintenance schedules, monitor the spare parts inventory status, order additional spare parts when necessary, print out work orders, and record completed work assignments.

SCADA systems can also be used to enhance energy conservation programs. For example, operators can develop energy management routines that take advantage of lower-cost off-peak energy usage. Most power companies are anxious to work with operators to try to increase collection system and treatment plant power consumption during periods when electrical system power demands are low and also to decrease power consumption during periods of peak demands on electrical power supplies.

SCADA systems also monitor power consumption and conduct a diagnostic performance examination of pumps. Operators can review this information and then identify potential pump problems before they become serious. In this type of system, power meters are used to accurately measure and record power consumption and the information can then be reviewed by operators to watch for changes that may indicate equipment problems.

SCADA systems can monitor subtle trends and identify pumping stations that are having slightly more pump starts per 24 hours or longer run hours per 24 hours. This information can be used to identify problems in the upstream network such as cracked pipes and tree roots growing in the pipes that allow more and more inflow and infiltration to enter the system.

Emergency response procedures can be programmed into a SCADA system. Operator responses can be provided for different operational scenarios that could be encountered as a result of adverse weather changes, fires, or earthquakes.

SCADA systems are continually improving and helping operators do a better job. Today operators can create their own screens as well as their own graphics and show whatever operating characteristics they wish to display on the screen. The main screen could be a flow diagram of the collection system or a treatment plant. Critical operating information could be displayed for each pumping station, each segment of main line, or each treatment process, and detailed screens could be easily reached for each piece of equipment.

Information on the screen should be color coded with the colors of red, yellow, green, blue, white, and any other necessary colors. Colors could be used to indicate if a pump is running, ready, unavailable, or failed, or if a valve is open, closed, moving, unavailable, or failed. A failed signal is used by the computer to inform the operator that something is wrong with the information or signal the computer is receiving or is being instructed to display. The signal is not logical with the rest of the information available to the computer. For example, if there is no power to a motor, then the motor cannot be running even though the computer is receiving a signal that indicates it is running. Therefore, the computer would send a failed signal.

The operator can request a computer to display a summary of all alarm conditions in the entire collection system, the treatment plant, a particular portion of the system, or at one or more pumping stations. A blinking alarm signal indicates that the alarm condition has not yet been acknowledged by the operator. On the other hand, a steady alarm signal, one that is not blinking, indicates that the alarm has been acknowledged but the alarm indicator will stay on until the condition causing it is fixed. Also the screen could be set up to automatically designate certain alarm conditions as priority alarms, which means they require immediate operator attention.

Current laptop computers allow operators to plug into a telephone at home or on vacation, access the SCADA system, and help operators on duty solve operational problems. Computer networking systems allow operators at terminals in offices and in the field to work together and use the same information or whatever information they need from one central file service (computer database).

A drawback of some SCADA systems is that when the system fails (goes down) due to a power failure, the system will often display the numbers that were registered immediately before the failure and not display the current numbers. The operator may therefore experience a period of time when accurate, current information about the system is not immediately available.

Customer satisfaction with the performance of a collection system or treatment plant agency can be enhanced by the use of an effective SCADA system. Historically when a pump station failed or tripped out, the first a utility learned of this problem was when an irate consumer phoned the agency and complained about a sewer backup in their home. The utility then had to contact the operators and send a crew into the field to correct the problem. Today, SCADA systems often alert operators to a pump station failure or tripout immediately. The operator may be able to correct or override the failure or tripout from the office without ever having to travel to the problem pump station in the field. Thus, the problem is corrected without the customers ever being aware that a problem occurred and was corrected almost immediately.

When operators decide to initiate or expand the SCADA system for their collection system or treatment plant, the first step is to decide what the SCADA system should do to make the operators' jobs easier, more efficient, and safer and to make their facilities' performance more reliable and cost effective. Cost savings associated with the use of a SCADA system frequently include reduced labor costs for operation, maintenance, and monitoring functions that were formerly performed manually. Preventive maintenance monitoring can save on equipment and repair costs and, as previously noted, energy savings may result from use of off-peak electrical power rates. Operators should visit facilities with SCADA systems and talk to the operators about what they find beneficial and also detrimental with regard to SCADA systems and how the systems contribute to their performance as operators.

The greatest challenge for operators using SCADA systems is to realize that just because a computer says something (a pump is operating as expected), *this does not mean that the computer is always correct.* Operators will always be needed to question and analyze the results from SCADA systems. Also, when the system fails due to a power failure or for any other reason (natural disaster), operators will be required to operate the collection system and treatment plant manually and without critical information. Could you do this? Collection systems and treatment plants will always need alert, knowledgeable, and experienced operators who have a feel for their collection system and treatment plant.

QUESTIONS

Write your answers in a notebook and then compare your answers with those on pages 751 and 752.

20.9D What does SCADA stand for?

20.9E What does a SCADA system do?

20.9F How could measured (sensed) information be transmitted?

20.9G What are the greatest challenges for operators using SCADA systems?

20.95 Cross Connection Control

BACKFLOW[7] of contaminated water through cross connections into community water systems is not just a theoretical problem. Contamination through cross connections has consistently caused more waterborne disease outbreaks in the United States than any other reported factor. Inspections have often disclosed numerous unprotected cross connections between public water systems and other piped systems on consumers' premises, which might contain wastewater, stormwater, processed waters (containing a wide variety of chemicals), and untreated supplies from private wells, streams, and ocean waters. Therefore, an effective cross connection control program is essential.

Inspect your plant to see if there are any cross connections between your potable (drinking) water supply and unapproved water supplies such as pump seals, feed water to boilers, hose bibs below grade where they may be subject to flooding with wastewater or sludges, or any other location where wastewater could contaminate a domestic water supply. If any of these or other existing or potential cross connections are found, be certain that your drinking water supply source is properly protected by the installation of an approved backflow prevention device.

Many treatment plants use an *AIR GAP*[8] system (Figure 20.4) to protect their drinking water supply. An air gap separation system provides a physical break between the wastewater treatment plant's municipal or fresh water supply (well) and the plant's treatment process systems. The physical separation of the two water systems effectively protects the potable water supply in case wastewater backs up from a treatment process.

Installation of air gap systems is controlled by health regulations. These systems should be inspected periodically by the

[7] *Backflow.* A reverse flow condition, created by a difference in water pressures, that causes water to flow back into the distribution pipes of a potable water supply from any source or sources other than an intended source. Also see BACKSIPHONAGE.

[8] *Air Gap.* An open, vertical drop, or vertical empty space, between a drinking (potable) water supply and potentially contaminated water. This gap prevents the contamination of drinking water by backsiphonage because there is no way potentially contaminated water can reach the drinking water supply.

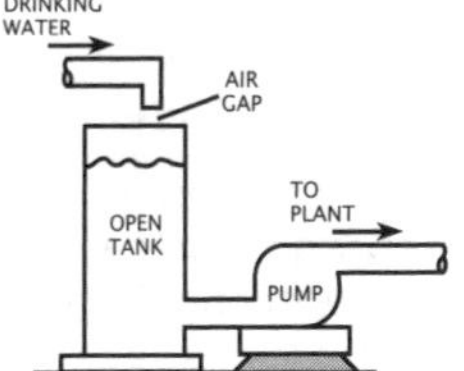

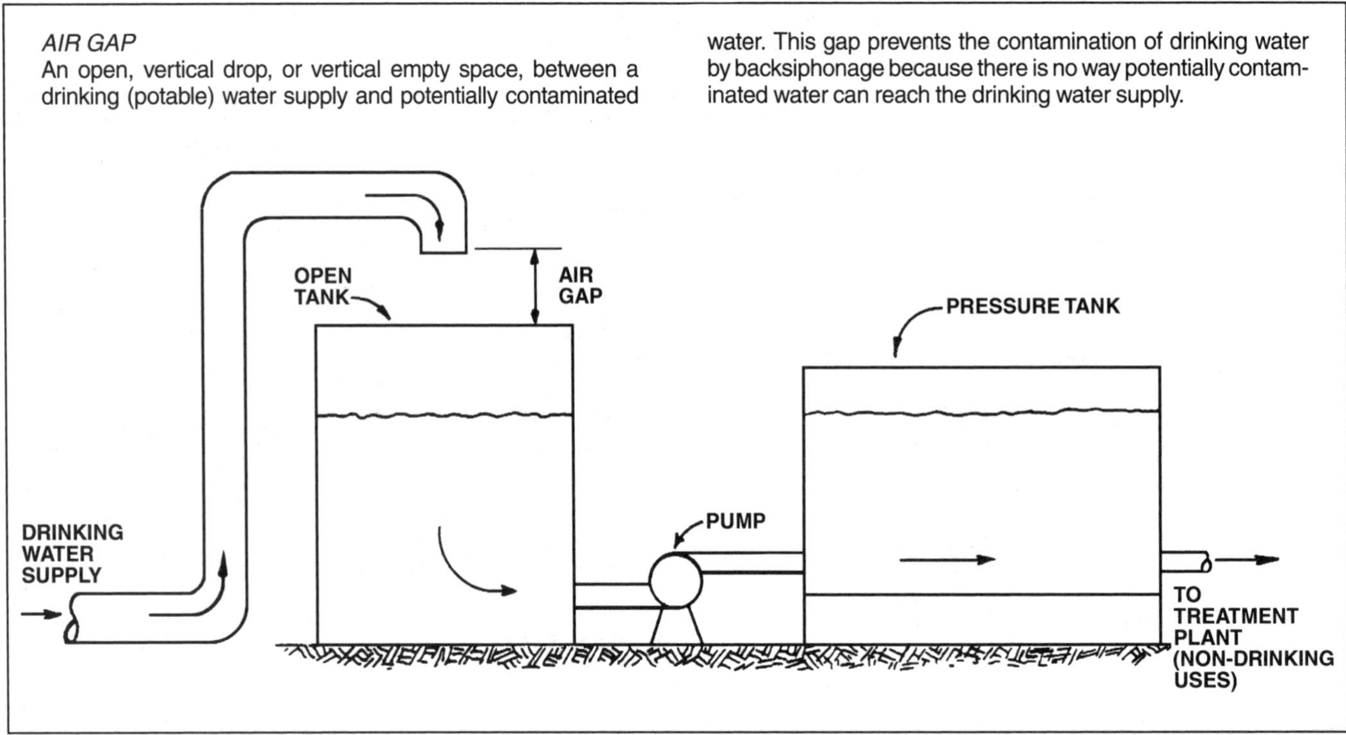
AIR GAP
An open, vertical drop, or vertical empty space, between a drinking (potable) water supply and potentially contaminated water. This gap prevents the contamination of drinking water by backsiphonage because there is no way potentially contaminated water can reach the drinking water supply.
OPEN TANK
AIR GAP
PRESSURE TANK
PUMP
DRINKING WATER SUPPLY
TO TREATMENT PLANT (NON-DRINKING USES)

Fig. 20.4 Air gap device

local health department or the public water supply agency. Also, it is good practice to have your drinking water tested at least monthly for coliform group organisms because sometimes even the best of backflow prevention devices fail.

Never drink from outside water connections such as faucets and hoses. The hose you drink from may have been used to carry effluent or sludge.

You may find in your plant that it will be more economical to use bottled drinking water. If so, be sure to post conspicuous signs that your plant water is not potable at all outlets. This also applies to all hose bibs in the plant from which you may obtain water other than a potable source. This is a must in order to inform visitors or absent-minded or thirsty employees that the water from each marked location is not for drinking purposes.

QUESTIONS

Write your answers in a notebook and then compare your answers with those on page 752.

20.9H What has caused more waterborne disease outbreaks in the United States than any other reported factor?

20.9I What are some potential places where cross connections could occur in a wastewater treatment plant?

20.10 EMERGENCY RESPONSE

20.100 Natural Disasters

Contingency planning is an essential facet of utility management and one that is often overlooked. Although utilities in various locations will be vulnerable to somewhat different kinds of natural disasters, the effects of these disasters in many cases will be quite similar. As a first step toward an effective contingency plan, each utility should make an assessment of its own vulnerability and then develop and implement a comprehensive plan of action.

All utilities suffer from common problems such as equipment breakdowns and leaking pipes. During the past few years there has also been an increasing amount of vandalism, civil disorder, toxic spills, and employee strikes, which have threatened to disrupt utility operations. In observing today's international tension and the potential for nuclear war or the effects of terrorist-induced chemical or biological warfare, water and wastewater utilities must seriously consider how to respond. Natural disasters such as floods, earthquakes, hurricanes, forest fires, avalanches, and blizzards are a more or less routine occurrence for some utilities. When such catastrophic emergencies occur, the utility must be prepared to minimize the effects of the event and have a plan for rapid recovery. Such preparation should be a specific obligation of every utility manager.

Start by assessing the vulnerability of the utility during various types of emergency situations. If the extent of damage can be estimated for a series of most probable events, the weak elements can be studied and protection and recovery operations can center on these elements. Experience with disasters reveals situations that are likely to disrupt the ability of a utility to function. These include:

1. The absence of trained personnel to make critical decisions and carry out orders
2. The loss of power to the utility's facilities
3. An inadequate amount of supplies and materials
4. Inadequate communication equipment

The following steps should be taken in assessing the vulnerability of a system:

1. Identify and describe the system components.
2. Assign assumed disaster characteristics.
3. Estimate disaster effects on system components.
4. Estimate customer demand for service following a potential disaster.
5. Identify key system components that would be primarily responsible for system failure.

If the assessment shows a system is unable to meet estimated requirements because of the failure of one or more critical components, the vulnerable elements have been identified. Repeating this procedure using several typical disasters will usually point out system weaknesses. Frequently, the same vulnerable element appears for a variety of assumed disaster events.

Although the drafting of an emergency plan for a wastewater treatment plant may be a difficult job, the existence of such a plan can be of critical importance during an emergency situation.

An emergency operations plan need not be too detailed, since all types of emergencies cannot be anticipated and a complex response program can be more confusing than helpful. Supervisory personnel must have a detailed description of their responsibilities during emergencies. They will need information, supplies, equipment, and the assistance of trained personnel. All these can be provided through a properly constructed emergency operations plan that is not extremely detailed.

The following outline can be used as the basis for developing an emergency operations plan:

1. Make a vulnerability assessment.
2. Inventory organizational personnel.
3. Provide for a recovery operation (plan).
4. Provide training programs for operators in carrying out the plan.
5. Coordinate with local and regional agencies such as the health, police, and fire departments to develop procedures for carrying out the plan.
6. Establish a communications procedure.
7. Provide protection for personnel, plant equipment, records, and maps.

By following these steps, an emergency plan can be developed and maintained even though changes in personnel may occur. "Emergency simulation" training sessions, including the use of standby power, equipment, and field test equipment will ensure that equipment and personnel are ready at times of emergency.

A list of phone numbers for operators to call in an emergency should be prepared and posted by a phone for emergency use. The list should include:

1. Plant supervisor
2. Director of public works or head of utility agency
3. Police
4. Fire
5. Doctor (2 or more)
6. Ambulance (2 or more)
7. Hospital (2 or more)

If appropriate for your utility, also include the following phone numbers on the emergency list:

8. Chlorine supplier and manufacturer
9. *CHEMTREC,* (800) 424-9300, for the hazardous chemical spills; sponsored by the American Chemistry Council
10. US Coast Guard's National Response Center, (800) 424-8802
11. Local and state poison control centers
12. Local hazardous materials spill response team

You should prepare a list for your plant *now,* if you have not already done so, and update the numbers annually.

For additional information on emergencies, see Chapter 15, Section 15.042, "Emergencies," in this manual. Also see Volume I, Chapter 10, Section 10.33, "First-Aid Measures," and Chapter 10, Section 10.42, "Chlorine Leaks."

20.101 On-Site Chemical Releases

Plants using chlorine or sulfur dioxide in large quantities (1,500 pounds (682 kilograms) or more of chlorine and 1,000 pounds (455 kilograms) or more of sulfur dioxide) must develop a Process Safety Management (PSM) program as required by Title 29 of the Code of Federal Regulations, Part 1910.119 (29 CFR 1910.119). A PSM program does not have to be developed for normally unoccupied remote facilities. A remote facility is not next to other buildings or processes and is only periodically visited by personnel to perform operating or maintenance tasks. No employees are stationed at a remote facility.

The goal of a PSM program is to prevent or minimize the effects of a major, uncontrolled release of chlorine or sulfur dioxide that could present a serious danger to plant employees. Essential elements of a PSM program include, but are not limited to, the following:

- Operator participation—operators must be involved in the development of various elements of the program.
- Process safety information—information must be provided about the hazards and equipment involved in chlorine/sulfur dioxide use and handling.

- Chlorine and sulfur dioxide hazards—information concerning the toxicity, physical characteristics, reactivity, corrosivity, and stability of the chemicals must be provided.
- Process technology—information concerning inventory, process chemistry, and flow diagrams must be provided.
- Equipment information—materials of construction, piping/instrumentation diagrams, ventilation systems, relief system design, codes and standards used, safety systems such as interlocks and alarms must be identified.
- Process hazard analysis—to identify, evaluate, and control hazards involved in the processes.
- Operating procedures—written procedures must be developed that provide clear instructions for safely performing maintenance and operating tasks.
- Training—training must be provided to each employee working in the area to include an overview of the processes and required procedures. Refresher training must be provided at least every three years or more often if necessary. Operators must be involved in establishing the need for more frequent refresher training.
- Contractor work—inform contractors of: (1) potential hazards in the chlorine/sulfur dioxide area, (2) applicable emergency action plans, (3) control procedures for entry and occupancy of the area, and (4) audits you will perform to ensure that the contractor is working safely in the area and that the contractor's employees are trained in hazard recognition, safe work techniques, and emergency action requirements.
- Pre-start-up safety review—modifications to the systems that require a change in the process safety information must be reviewed prior to start-up.
- Mechanical integrity—written procedures, training, and testing requirements for process equipment must be developed.
- Hot work—a written permit is required for hot work operations in or near the chlorine/sulfur dioxide area.
- Management of change—before any changes are made in the process, equipment, or work procedures, written procedures must be developed and implemented to manage the changes and ensure that safety and system integrity are not compromised.
- Incident investigation—each leak that could have resulted in a catastrophic release must be investigated within 48 hours of the incident. The investigation must include cause(s) and recommendations for prevention.
- Emergency planning and response—an emergency action plan for responding to leaks must be developed for the entire plant.
- Compliance audits—the program must be evaluated every three years for compliance and adequacy.

This listing will give you an idea of the size and scope of a PSM program. You may also be required to prepare and comply with a Risk Management Program (RMP) (described in Section 20.102). The RMP deals with potential off-site effects of a chemical release, rather than on-site like the PSM program. One of the requirements of the RMP is dispersion analyses of certain chlorine and sulfur dioxide release scenarios to evaluate hazard potential to areas adjoining your plant. Your local or state environmental management agency should be able to provide you with guidance in the development of programs for your particular plant.

Chlorine or sulfur dioxide releases that equal or exceed their reportable quantity (RQ), 10 pounds (4.5 kilograms) and 1 pound (0.45 kilograms) respectively, must be reported to certain regulatory agencies including the National Response Center (NRC). It should be noted the United States Environmental Protection Agency (EPA) is revising the RQ for several chemicals. Contact your local or state environmental management agency for current requirements for your plant.

20.102 Off-Site Chemical Releases

Facilities that store or use certain toxic, flammable, or explosive chemicals in amounts greater than specified quantities were required to submit a Risk Management Program (RMP) to the Environmental Protection Agency (EPA). (See 40 CFR 68, Sections 112(r) and 301 for the complete text of the RMP rule.) The purpose of this rule is to identify potential sources of chemical releases and prepare a plan for dealing with the effects of such releases on the communities surrounding the facility. The RMP rule sets out three different programs for different types of facilities. Most treatment plants will have to meet the requirements of Program 2; therefore, the remainder of this section refers only to the components of Program 2.

Several chemicals that are commonly used in water supply and wastewater treatment facilities are covered by the RMP rule, for example, ammonia, chlorine, methane, propane, and sulfur dioxide. If the volume of chemical stored or used exceeds the RMP thresholds, which range from 500 to 20,000 lb (227 to 9,070 kg), the facility must prepare a detailed RMP for each chemical.

Many of the requirements of the Risk Management Program rule overlap the Process Safety Management (PSM) requirements described earlier in Section 20.101. However, the RMP requires greater involvement of treatment plant operators and other staff during the development of the program. The main provisions of the RMP are as follows:

- Off-site consequence analysis is no longer required because the executive summary may pose a security risk.
- Five-year accident history—compile a record of all chemical releases in the past five years and work with local emergency response agencies to coordinate any projected future emergencies.
- Employee safety program that includes the following elements:
  - Material Safety Data Sheets (MSDSs) that meet the requirements of 29 CFR 19.1200(g);
  - An inventory of the equipment that will be used to store or process the regulated chemicals;
  - A listing of the safe upper and lower temperatures, pressures, flows, and compositions of the regulated chemicals;
  - The codes and standards used to design, build, and operate each process.
- Process hazard assessment—this is a complete review of the hazards associated with the possible failure of mechanical processes (covered by 40 CFR 68) using regulated substances and the safety precautions or safeguards used to control the hazards. The assessment must also include an analysis of potential causes of accidental releases and

the chemical detection equipment and procedures that will be used to monitor chemical releases.

- Description of operating procedures—for each chemical or process listed in 40 CFR 68, a detailed written description of both normal and emergency start-up and operating procedures must be developed.
- Training—operators must receive appropriate training in process operations and the training must be fully documented.
- Compliance audits—formal examination of a facility's RMP by a trained auditor to ensure that the facility meets the requirements of the RMP rule.
- Information on reportable chemical accidents must be added to the RMP within six months of the accident.
- Changes to emergency contact information must be reported within one month.

The Environmental Protection Agency has developed a guidance document to assist wastewater treatment plant managers in developing a Risk Management Program. Your local or state environmental management agency also should be able to provide you with guidance in the development of programs for your particular plant. For more information, see "What's Next—RMP 2004" on EPA's website at: http://yosemite.epa.gov/oswer/ceppoweb.nsf/content/whatsnext.htm

QUESTIONS

Write your answers in a notebook and then compare your answers with those on page 752.

20.10A What is the first step toward an effective contingency plan for emergencies?

20.10B Why is too detailed an emergency operations plan not needed or even desirable?

20.10C An emergency operations plan should include what specific information?

20.10D In general, what is the major difference between a Process Safety Management (PSM) program and a Risk Management Program (RMP)?

20.11 SAFETY PROGRAM

20.110 Responsibilities

20.1100 Everyone Is Responsible for Safety

Wastewater utilities, regardless of size, must have a safety program if they are to realize a low frequency of accident occurrence. A safety program also provides a means of comparing frequency, disability, and severity of injuries with other utilities. The utility should identify the causes of accidents and injuries, provide safety training, implement an accident reporting system, and hold supervisors responsible for implementing the safety program. Each utility should have a safety officer or supervisor evaluate every accident, offer recommendations, and keep and apply statistics.

The effectiveness of any safety program will depend on how the utility holds its supervisors responsible. If the utility holds only the safety officer or the employees responsible, the program will fail. The supervisors are key in any organization. If they disregard safety measures, essential parts of the program will not work. The results will be an overall poor safety record. After all, the first line supervisor is where the work is being performed, and some may take advantage of an unsafe situation in order to get the job completed. The organization must discipline such supervisors and make them aware of their responsibility for their own and their operators' safety.

Safety is good business for both the operator and the agency. For a good safety record to be accomplished, all individuals must be educated and must believe in the program. All individuals involved must have the conviction that accidents can be prevented. The operations should be studied to determine the safe way of performing each job. Safety pays, both in monetary savings and in the health and well-being of the operating staff.

20.1101 Regulatory Agencies

Many state and federal agencies are involved in ensuring safe working conditions. The one law that has had the greatest impact has been the *OCCUPATIONAL SAFETY AND HEALTH ACT OF 1970 (OSHA),*[9] Public Law 91-596, which took effect on December 29, 1970. This legislation affects more than 75,000,000 employees and has been the basis for most of the current state laws covering employees. Also, many state regulatory agencies enforce the OSHA requirements.

The OSHA regulations provide for safety inspections, penalties, recordkeeping, and variances. Managers and supervisors must understand the OSHA regulations and must furnish each operator with the rules of conduct in order to comply with occupational safety and health standards. The intent of the regulations is to create a place of employment that is free from recognized hazards that could cause serious physical harm or death to an operator.

Civil and criminal penalties are allowed under the OSHA law, depending on the size of the business and the seriousness of the violation. A routine violation could cost an employer or supervisor up to $1,000 for each violation. A serious,

[9] *OSHA* (O-shuh). The Williams-Steiger Occupational Safety and Health Act of 1970 (OSHA) is a federal law designed to protect the health and safety of workers, including the operators of water supply and treatment systems and wastewater collection and treatment systems. The Act regulates the design, construction, operation, and maintenance of water and wastewater systems. OSHA regulations require employers to obtain and make available to workers the Material Safety Data Sheets (MSDSs) for chemicals used at industrial facilities and treatment plants. OSHA also refers to the federal and state agencies that administer the OSHA regulations.

willful, or repeated violation could cause the employer or supervisor to be assessed a penalty of not more than $10,000 for each violation. Penalties are assessed against the supervisor responsible for the injured operator. Operators should become familiar with the OSHA regulations as they apply to their organizations. Managers and supervisors must correct violations and prevent others from occurring.

20.1102 Managers

The utility manager is responsible for the safety of the agency's personnel and the public exposed to the wastewater utility's operations. Therefore, the manager must develop and administer an effective safety program and must provide new employee safety training as well as ongoing training for all employees. The basic elements of a safety program include a safety policy statement, safety training and promotion, and accident investigation and reporting.

A safety policy statement should be prepared by the top management of the utility. The purpose of the statement is to let employees know that the safety program has the full support of the agency and its management. The statement should:

1. Define the goals and objectives of the program.
2. Identify the persons responsible for each element of the program.
3. Affirm management's intent to enforce the safety regulations.
4. Describe the disciplinary actions that will be taken to enforce safe work practices.

Give a copy of the safety policy statement to every current employee and each new employee during orientation. Figure 20.5 is an example of a safety policy statement for a water supply utility.

The following list of responsibilities for safety is from the *PLANT MANAGER'S HANDBOOK.*[10] These responsibilities represent a typical list but may be incomplete if your agency is subject to stricter local, state, or federal regulations than what is shown here. Check with your safety professional.

Management has the responsibility to:

1. Formulate a written safety policy.
2. Provide a safe workplace.
3. Set achievable safety goals.
4. Provide adequate training.
5. Delegate authority to ensure that the program is properly implemented.

SAFETY POLICY STATEMENT

It is the policy of the Las Vegas Valley Water District that every employee shall have a safe and healthy place to work. It is the District's responsibility; its greatest asset, the employees and their safety.

When a person enters the employ of the District, he or she has a right to expect to be provided a proper work environment, as well as proper equipment and tools, so that they will be able to devote their energies to the work without undue danger. Only under such circumstances can the association between employer and employee be mutually profitable and harmonious. It is the District's desire and intention to provide a safe workplace, safe equipment, proper materials, and to establish and insist on safe work methods and practices at all times. It is a basic responsibility of all District employees to make the SAFETY of human beings a matter for their daily and hourly concern. This responsibility must be accepted by everyone who works at the District, regardless of whether he or she functions in a management, supervisory, staff, or the operative capacity. Employees must use the SAFETY equipment provided; Rules of Conduct and SAFETY shall be observed; and, SAFETY equipment must not be destroyed or abused. Further, it is the policy of the Water District to be concerned with the safety of the general public. Accordingly, District employees have the responsibility of performing their duties in such a manner that the public's safety will not be jeopardized.

The joint cooperation of employees and management in the implementation and continuing observance of this policy will provide safe working conditions and relatively accident-free performance to the mutual benefit of all involved. The Water District considers the SAFETY of its personnel to be of primary importance, and asks each employee's full cooperation in making this policy effective.

Fig. 20.5 Safety policy statement

(Permission of Las Vegas Valley Water District)

[10] *PLANT MANAGER'S HANDBOOK* (MOP SM-4), Water Environment Federation (WEF), no longer in print.

The manager is the key to any safety program. Implementation and enforcement of the program is the responsibility of the manager. The manager also has the responsibility to:

1. Ensure that all employees are trained and periodically retrained in proper safe work practices.
2. Ensure that proper safety practices are implemented and continued as long as the policy is in effect.
3. Investigate all accidents and injuries to determine their cause.
4. Institute corrective measures where unsafe conditions or work methods exist.
5. Ensure that equipment, tools, and the work are maintained to comply with established safety standards.

QUESTIONS

Write your answers in a notebook and then compare your answers with those on page 752.

20.11A What should be the duties of a safety officer?

20.11B Who should be responsible for the implementation of a safety program?

20.11C Who enforces the OSHA requirements?

20.11D What are the utility manager's responsibilities with regard to safety?

20.11E What should be included in a utility's policy statement on safety?

20.1103 Supervisors

The success of any safety program will depend on how the supervisors of the utility view their responsibility. The supervisor who has the responsibility for directing work activities must be safety conscious. This supervisor controls the operators' general environment and work habits and influences whether or not the operators comply with safety regulations. The supervisor is in the best position to counsel, instruct, and review the operators' working methods and thereby effectively ensure compliance with all aspects of the utility's safety program.

The problem, however, is one of the supervisor accepting this responsibility. The supervisor who wishes to complete the job and go on to the next one without taking time to be concerned about working conditions, the welfare of operators, or considering any aspects of safety is a poor supervisor. Only after an accident occurs will a careless supervisor question the need for a work program based on safety. At this point, however, it is too late, and the supervisor may be tempted to simply cover up past mistakes. As sometimes happens, the supervisor may even be partially or fully responsible for the accident by causing unsafe acts to take place, by requiring work to be performed in haste, by disregarding an unsafe work environment, or by overlooking or failing to consider any number of safety hazards. This negligent supervisor could be fined, sentenced to a jail term, or even be barred from working in the profession.

All utilities should make their supervisors bear the greatest responsibility for safety and hold them accountable for planning, implementing, and controlling the safety program. If most accidents are caused and do not just happen, then it is the supervisor who can help prevent most accidents.

Equally important are the officials above the supervisor. These officials include commissioners, managers, public works directors, chief engineers, superintendents, and chief operators. The person responsible for the entire agency or operation must believe in the safety program. This person must budget, promote, support, and enforce the safety program by vocal and visible examples and actions. The top person's support is absolutely essential for an effective safety program.

20.1104 Operators

Each operator also shares in the responsibility for an effective safety program. After all, operators have the most to gain from sharing the responsibility since they are the most likely victims of accidents. A review of accident causes shows that the accident victim often has not acted responsibly. In some way the victim has not complied with the safety regulations, has not been fully aware of the working conditions, has not been concerned about fellow employees, or just has not accepted any responsibility for the utility's safety program.

Each operator must accept, at least in part, responsibility for fellow operators, for the utility's equipment, for the operator's own welfare, and even for seeing that the supervisor complies with established safety regulations. As pointed out above, the operator has the most to gain. If the operator accepts and uses unsafe equipment, it is the operator who is in danger if something goes wrong. If the operator fails to protect the other operators, it is the operator who must make up the work lost because of injury. If operators fail to consider their own welfare, it is they who suffer the pain of any injury, the loss of income, and maybe even the loss of life.

The operator must accept responsibility for an active role in the safety program by becoming aware of the utility's safety policy and conforming to established regulations. The operator should always call to the supervisor's attention unsafe conditions, environment, equipment, or other concerns operators may have about the work they are performing. Safety should be an essential part of the operator's responsibility.

20.111 First Aid

By definition, first aid means emergency treatment for injury or sudden illness, before regular medical treatment is available. Everyone in an organization should be able to give some degree of prompt treatment and attention to an injury.

First-aid training in the basic principles and practices of life-saving steps that can be taken in the early stages of an injury are available through the local Red Cross, Heart Association, local fire departments, and other organizations. Such training should periodically be reinforced so that the operator has a complete understanding of water safety, cardiopulmonary resuscitation (CPR), and other life-saving techniques. All operators need training in first aid, but it is especially important for those who regularly work with electrical equipment or must handle chlorine and other dangerous chemicals.

First aid has little to do with preventing accidents, but it has an important bearing on the survival of the injured patient. A well-equipped first-aid chest or kit is essential for proper treatment. The kit should be inspected regularly by the safety officer to ensure that supplies are available when needed. First-aid kits should be prominently displayed throughout the treatment plant and in company vehicles. Special consideration must be given to the most hazardous areas of the plant such as shops, laboratories, and chemical handling facilities.

Regardless of size, each utility should establish standard operating procedures (SOPs) for first-aid treatment of injured personnel. All new operators should be instructed in the utility's first-aid program.

QUESTIONS

Write your answers in a notebook and then compare your answers with those on page 752.

20.11F How could a supervisor be responsible for an accident?

20.11G What types of safety-related responsibilities must each operator accept?

20.11H What is first aid?

20.11I First-aid training is most important for operators involved in what types of activities?

20.112 Hazard Communication Program and Worker Right-To-Know (RTK) Laws

In the past few years there has been an increased emphasis nationally on hazardous materials and wastes. Much of this attention has focused on hazardous and toxic waste dumps and the efforts to clean them up after the long-term effects on human health were recognized. Each year thousands of new chemical compounds are produced for industrial, commercial, and household use. Frequently, the long-term effects of these chemicals are unknown. As a result, federal and state laws have been enacted to control all aspects of hazardous materials handling and use. These laws are more commonly known as Worker Right-To-Know (RTK) laws, which are enforced by OSHA.

As noted earlier, the intent of the OSHA regulations is to create a place of employment that is free from recognized hazards that could cause serious physical harm or death to an operator (or other employee). In many cases, the individual states have the authority under the OSHA Standard to develop their own state RTK laws and most states have adopted their own laws. The Federal OSHA Standard 29 *CFR*[11] 1910.1200—Hazard Communication forms the basis of most of these state RTK laws. Unfortunately, state laws vary significantly from state to state. The state laws that have been passed are at least as stringent as the federal standard and, in most cases, are even more stringent. State laws are also under continuous revision and, because a strong emphasis is being placed on hazardous materials and worker exposure, state laws can be expected to be amended in the future to apply to virtually everybody in the workplace. Managers should become familiar with both the state and federal OSHA regulations that apply to their organizations.

The basic elements of a hazard communication program are described in the following paragraphs.

1. Identify Hazardous Materials—While there are thousands of chemical compounds that could be considered hazardous, focus on the ones to which operators and other personnel in your utility are most likely to be exposed.

2. Obtain Chemical Information and Define Hazardous Conditions—Once the inventory of hazardous chemicals is complete, the next step is to obtain specific information on each of the chemicals. This information is generally incorporated into a standard format form called the *MATERIAL SAFETY DATA SHEET (MSDS)*.[12] This information is commonly available from manufacturers. Many agencies request an MSDS when the purchase order is generated and will refuse to accept delivery of the shipment if the MSDS is not included. Figure 20.6 shows OSHA's standard MSDS form, but other forms are also acceptable provided they contain all of the required information.

   The purpose of the MSDS is to have a readily available reference document that includes complete information on common names, safe exposure level, effects of exposure, symptoms of exposure, flammability rating, type of first-aid procedures, and other information about each hazardous substance. Operators should be trained to read and understand the MSDS forms. The forms themselves should be stored in a convenient location where they are readily available for reference.

3. Properly Label Hazards—Once the physical, chemical, and health hazards have been identified and listed, a labeling and training program must be implemented. To meet labeling requirements on hazardous materials, specialized labeling is available from a number of sources, including commercial label manufacturers. Exemptions to labeling requirements do exist, so consult your local safety regulatory agency for specific details.

4. Train Operators—The last element in the hazard communication program is training and making information available to utility personnel. A common-sense approach eliminates the confusing issue of which of the thousands of substances operators should be trained for, and concentrates on those that they will be exposed to or use in everyday maintenance routines.

[11] *Code of Federal Regulations (CFR).* A publication of the US government that contains all of the proposed and finalized federal regulations, including safety and environmental regulations.

[12] *Material Safety Data Sheet (MSDS).* A document that provides pertinent information and a profile of a particular hazardous substance or mixture. An MSDS is normally developed by the manufacturer or formulator of the hazardous substance or mixture. The MSDS is required to be made available to employees and operators or inspectors whenever there is the likelihood of the hazardous substance or mixture being introduced into the workplace. Some manufacturers are preparing MSDSs for products that are not considered to be hazardous to show that the product or substance is not hazardous.

Material Safety Data Sheet
May be used to comply with
OSHA's Hazard Communication Standard,
29 CFR 1910.1200 Standard must be
consulted for specific requirements.

U.S. Department of Labor
Occupational Safety and Health Administration
(Non-Mandatory Form)
Form Approved
OMB No. 1218-0072

IDENTITY *(As Used on Label and List)*	*Note: Blank spaces are not permitted. If any item is not applicable, or no information is available, the space must be marked to indicate that.*

Section I

Manufacturer's Name	Emergency Telephone Number
Address *(Number, Street, City, State, and ZIP Code)*	Telephone Number for Information
	Date Prepared
	Signature of Preparer *(optional)*

Section II — Hazardous Ingredients/Identity Information

Hazardous Components (Specific Chemical Identity; Common Name(s))	OSHA PEL	ACGIH TLV	Other Limits Recommended	%*(optional)*

Section III — Physical/Chemical Characteristics

Boiling Point		Specific Gravity (H_2O = 1)	
Vapor Pressure (mm Hg)		Melting Point	
Vapor Density (AIR = 1)		Evaporation Rate (Butyl Acetate = 1)	
Solubility in Water			
Appearance and Odor			

Section IV — Fire and Explosion Hazard Data

Flash Point (Method Used)	Flammable Limits	LEL	UEL
Extinguishing Media			
Special Fire Fighting Procedures			
Unusual Fire and Explosion Hazards			

(Reproduce locally) OSHA 174, Sept. 1985

Fig. 20.6 Material Safety Data Sheet

Section V — Reactivity Data

Stability	Unstable		Conditions to Avoid
	Stable		

Incompatibility *(Materials to Avoid)*

Hazardous Decomposition or Byproducts

Hazardous Polymerization	May Occur		Condition to Avoid
	Will Not Occur		

Section VI — Health Hazard Data

Route(s) of Entry:	Inhalation?	Skin?	Ingestion?

Health Hazards *(Acute and Chronic)*

Carcinogenicity:	NTP?	IARC Monographs?	OSHA Regulated?

Signs and Symptoms of Exposure

Medical Conditions
Generally Aggravated by Exposure

Emergency and First Aid Procedures

Section VII — Precautions for Safe Handling and Use

Steps to Be Taken in Case Material is Released or Spilled

Waste Disposal Method

Precautions to Be Taken in Handling and Storing

Other Precautions

Section VIII — Control Measures

Respiratory Protection *(Specify Type)*

Ventilation	Local Exhaust	Special
	Mechanical *(General)*	Other

Protective Gloves	Eye Protection

Other Protective Clothing or Equipment

Work/Hygienic Practices

Page 2 * USGPO:1986-491-529/45775

Fig. 20.6 Material Safety Data Sheet (continued)

The Hazard Communication Standard and the individual state requirements are obviously a very complex set of regulations. Remember, however, the ultimate goal of these regulations is to provide additional operator protection. These standards and regulations, once the intent is understood, are relatively easy to implement.

20.113 Confined Space Entry Procedures

CONFINED SPACES[13, 14] pose significant risks for a large number of workers, including many utility operators. OSHA has therefore defined very specific procedures to protect the health and safety of operators whose jobs require them to enter or work in a confined space. The regulations (which can be found in the Code of Federal Regulations at 29 CFR 1910.146) require conditions in the confined space to be tested and evaluated before anyone enters the space. If conditions exceed OSHA's limits for safe exposure, additional safety precautions must be taken and a confined space entry permit (Figure 20.7) must be approved by the appropriate authorities prior to anyone entering the space.

The managers of wastewater utilities may or may not be involved in the day-to-day details of enforcing the agency's confined space policy and procedures. However, every utility manager should be aware of the current OSHA requirements and should ensure that the utility's policies not only comply with current regulations, but that the agency's policies are vigorously enforced for the safety of all operators.

QUESTIONS

Write your answers in a notebook and then compare your answers with those on page 752.

20.11J List the basic elements of a hazard communication program.

20.11K What are a manager's responsibilities for ensuring the safety of operators entering or working in confined spaces?

20.114 Reporting

The mainstay of a safety program is the method of reporting and keeping of statistics. These records are needed regardless of the size of the utility because they provide a means of identifying accident frequencies and causes as well as the personnel involved. The records can be looked upon as the operator's safety report card. Therefore, it becomes the responsibility of each injured operator to fill out the utility's accident report.

All injuries should be reported, even if they are minor in nature, so as to establish a record in case the injury develops into a serious injury. It may be difficult at a later date to prove the accident did occur on the job and have the utility accept the responsibility for costs. The responsibility for reporting accidents affects several levels of personnel. First, of course, is the injured person. Next, it is the responsibility of the supervisor, and finally, the

Responsibility of Management to review the causes and take steps to prevent such accidents from happening in the future.

Accident report forms may be very simple. However, they must record all details required by law and all data needed for statistical purposes. The forms shown in Figures 20.8 and 20.9 are examples for you to consider for use in your plant. The report must show the name of the injured, employee number, division, time of accident, nature of injury, cause of accident, first aid administered, and remarks for items not covered elsewhere. There should be a review process by foreman, supervisor, safety officer, and management. Recommendations are needed as well as a follow-up review to be sure that proper action has been taken to prevent recurrence.

In addition to reports needed by the utility, other reports may be required by state or federal agencies. For example, vehicle accident reports must be submitted to local police departments. If a member of the public is injured, additional forms are needed because of possible subsequent claims for damages. If the accident is one of occupational injury, causing lost time, other reports may be required. Follow-up investigations to identify causes and responsibility may require the development of other specific types of record forms.

In the preparation of accident reports, it is the operator's responsibility to correctly fill out each form, giving complete details. The supervisor must be sure no information is overlooked that may be helpful in preventing recurrence.

---

[13] *Confined Space.* Confined space means a space that: (1) Is large enough and so configured that an employee can bodily enter and perform assigned work; and (2) Has limited or restricted means for entry or exit (for example, manholes, tanks, vessels, silos, storage bins, hoppers, vaults, and pits are spaces that may have limited means of entry); and (3) Is not designed for continuous employee occupancy. Also see DANGEROUS AIR CONTAMINATION and OXYGEN DEFICIENCY.

[14] *Confined Space, Permit-Required (Permit Space).* A confined space that has one or more of the following characteristics: (1) Contains or has a potential to contain a hazardous atmosphere, (2) Contains a material that has the potential for engulfing an entrant, (3) Has an internal configuration such that an entrant could be trapped or asphyxiated by inwardly converging walls or by a floor that slopes downward and tapers to a smaller cross section, or (4) Contains any other recognized serious safety or health hazard.

Confined Space Pre-Entry Checklist/Confined Space Entry Permit

Date and Time Issued: ____________ Date and Time Expires: __________ Job Site/Space I.D.: __________

Job Supervisor: ______________ Equipment to be worked on: __________ Work to be performed: __________

Standby personnel: ______________ ______________ ______________

1. Atmospheric Checks: Time ________ Oxygen ________ % Toxic ________ ppm
   Explosive ________ % LEL Carbon Monoxide ________ ppm

2. Tester's signature: ______________________

3. Source isolation: (No Entry) N/A Yes No

	N/A	Yes	No
Pumps or lines blinded, disconnected, or blocked	( )	( )	( )

4. Ventilation Modification:

	N/A	Yes	No
Mechanical	( )	( )	( )
Natural ventilation only	( )	( )	( )

5. Atmospheric check after isolation and ventilation: Time ________
   Oxygen ________ % > 19.5% < 23.5% Toxic ________ ppm < 10 ppm H_2S
   Explosive ________ % LEL < 10% Carbon Monoxide ________ ppm < 35 ppm CO

Tester's signature: ______________________________________

6. Communication procedures: ______________________________________

7. Rescue procedures: ______________________________________
   ____

8. Entry, standby, and backup persons

	Yes	No
Successfully completed required training?	( )	( )
Is training current?	( )	( )

9. Equipment:

	N/A	Yes	No
Direct reading gas monitor tested	( )	( )	( )
Safety harnesses and lifelines for entry and standby persons	( )	( )	( )
Hoisting equipment	( )	( )	( )
Powered communications	( )	( )	( )
SCBAs for entry and standby persons	( )	( )	( )
Protective clothing	( )	( )	( )
All electric equipment listed for Class I, Division I, Groups A, B, C, and D, and nonsparking tools	( )	( )	( )

10. Periodic atmospheric tests:

Oxygen:	___% Time ___;	___% Time ___;	___% Time ___;	___% Time ___;
Explosive:	___% Time ___;	___% Time ___;	___% Time ___;	___% Time ___;
Toxic:	___ppm Time ___;	___ppm Time ___;	___ppm Time ___;	___ppm Time ___;
Carbon Monoxide:	___ppm Time ___;	___ppm Time ___;	___ppm Time ___;	___ppm Time ___;

We have reviewed the work authorized by this permit and the information contained herein. Written instructions and safety procedures have been received and are understood. Entry cannot be approved if any brackets ( ) are marked in the "No" column. This permit is not valid unless all appropriate items are completed.

Permit Prepared By: (Supervisor) ______________ Approved By: (Unit Supervisor) ______________

Reviewed By: (CS Operations Personnel) ______________________________
(Entrant) (Attendant) (Entry Supervisor)

This permit to be kept at job site. Return job site copy to Safety Office following job completion.

Fig. 20.7 Confined space pre-entry checklist/confined space entry permit

Date ________________

Name of injured employee ______________________ Employee # ____________ Area ____________

Date of accident ______________ Time ________ Employee's Occupation ______________________

Location of accident ____________________________ Nature of injury ____________________________

Name of doctor ____________________________ Address ____________________________

Name of hospital ____________________________ Address ____________________________

Witnesses (name & address) ____

____

PHYSICAL CAUSES

Indicate below by an "X" whether, in your opinion, the accident was caused by:

_____ Improper guarding

_____ Defective substances or equipment

_____ Hazardous arrangement

_____ Improper illumination

_____ Improper dress or apparel

_____ No mechanical cause

_____ Working methods

_____ Lack of knowledge or skill

_____ Wrong attitude

_____ Physical defect

_____ Not listed (describe briefly) ____

UNSAFE ACTS

Sometimes the injured person is not directly associated with the causes of an accident. Using an "X" to represent the injured worker and an "O" to represent any other person involved, indicate whether, in your opinion, the accident was caused by:

_____ Operating without authority

_____ Failure to secure or warn

_____ Working at unsafe speed

_____ Made safety device inoperative

_____ Unsafe equipment or hands instead of equipment

_____ No unsafe act

_____ Unsafe loading, placement & etc.

_____ Took unsafe position

_____ Worked on moving equipment

_____ Teased, abused, distracted & etc.

_____ Did not use safe clothing or personal protective equipment

_____ Not listed (describe briefly) ____

____

What job was the employee doing? ____

What specific action caused the accident? ____

What steps will be taken to prevent recurrence? ____

____

____

Date of Report ______________ Immediate Supervisor ______________________________

REVIEWING AUTHORITY

Comments:	Comments:
______________________________	______________________________
Safety Officer Date	Department Director Date

Fig. 20.8 Supervisor's accident report

INJURED: COMPLETE THIS SECTION

Name ______________________________ Age_____ Sex ______

Address ______________________________

Title ____________________ Dept. Assigned ____________________

Place of Accident ______________________________

Street or Intersection ______________________________

Date ____________________ Hour______ A.M. ______ P.M. ______

Type of Job You Were Doing When Injured

____

Object Which Directly Injured You Part of Body Injured

____

How Did Accident Happen? (Be specific and give details; use back of sheet if necessary.)

____

First Aid Administered

____

Did You Report Accident or Exposure at Once? (Explain "No") Yes ☐ No ☐

Did You Report Accident or Exposure to Supervisor? Give Name Yes ☐ No ☐

Were There Witnesses to Accident or Exposure? Give Names Yes ☐ No ☐

Did You See a Doctor? (If Yes, Give Name) Yes ☐ No ☐

Are You Going to See a Doctor? (Give Name) Yes ☐ No ☐

____________________ ____________________

Date Signature

SUPERVISOR: COMPLETE THIS SECTION — (Return to Personnel as Soon as Possible)

Was an Investigation of Unsafe Conditions and/or Unsafe Acts Made? If Yes, Please Submit Copy. Yes ☐ No ☐

Was Injured Intoxicated or Behaving Inappropriately at Time of Accident? (Explain "Yes") Yes ☐ No ☐

Date Disability Commenced ____________ Last Day Wages Earned ____________ Date Back on Job ____________

Date Report Completed ____________ 20 ____ Signed By ____________

Title ____________

Distribution: Canary - Department Head, Pink - Supervisor, White - Personnel

Fig. 20.9 Accident report

The Safety Officer must review the reports and determine corrective actions and make recommendations.

In day-to-day actions, operators, supervisors, and management often overlook opportunities to counsel individual operators in safety matters. Then, when an accident occurs, they are not inclined to look too closely at accident reports. First, the accident is a series of embarrassments, to the injured person, to the supervisor, and to management. Therefore, there is a reluctance to give detailed consideration to accident reports. However, if a safety program is to function well, it will require a thorough effort on the part of the operator, supervisor, and management in accepting their responsibility for the accident and making a greater effort through good reporting to prevent future similar accidents. Accident reports must be analyzed, discussed, and the real cause of the accident identified and corrected.

Emphasis on the prevention of future accidents cannot be overstressed. We must identify the causes of accidents and implement whatever measures are necessary to protect operators from becoming injured.

20.115 Training

If a safety program is to work well, management will have to accept responsibility for the following three components of training:

1. Safety education of all employees
2. Reinforced education in safety
3. Safety education in the use of tools and equipment

Or to put it another way, the three most important controlling factors in safety are education, education, and more education.

Responsibility for overall training must be that of upper management. A program that will educate operators and then reinforce this education in safety must be planned systematically and promoted on a continuous basis. There are many avenues to achieving this goal.

The safety education program should start with the new operator. Even before employment, verify the operator's past record and qualifications and review the pre-employment physical examination. In the new operator's orientation, include instruction in the importance of safety at your utility or plant. Also, discuss the matter of proper reporting of accidents as well as the organization's policies and practices. Give new operators copies of all safety Standard Operating Procedures (SOPs) and direct their attention to parts that directly involve them. Ask the safety officer to give a talk about utility policy, safety reports, and past accidents, and to orient the new operator toward the importance of safety to operators and to the organization.

The next consideration must be one of training the new operator in how to perform assigned work. Most supervisors think in terms of On-the-Job Training (OJT). However, OJT is not a good way of preventing accidents with an inexperienced operator. The idea is all right if the operator comes to the organization trained in how to perform the work, such as a treatment operator from another plant. Then you only need to explain your safety program and how your policies affect the new operator. For a new operator who is inexperienced in utility operations, the supervisor must give detailed consideration to the operator's welfare. In this instance, the training should include not only a safety talk, but the foreman (supervisor) must train the inexperienced operator in all aspects of treatment plant safety. This training includes instruction in the handling of chemicals, the dangers of electrical apparatus, fire hazards, and proper maintenance of equipment to prevent accidents. Special instructions will also be needed for specific work environments such as manholes, gases (chlorine and hydrogen sulfide (H_2S)), water safety, and any specific hazards that are unique to your facility. The new operator must be checked out on any equipment personnel may operate such as vehicles, forklifts, valve operators, and radios. All new operators should be required to participate in a safety orientation program during the first few days of their employment, and an overall training program in the first few months.

The next step in safety education is reinforcement. Even if the operator is well trained, mistakes can occur; therefore, the education must be continual. Many organizations use the *TAILGATE*[15] method as a means of maintaining the operator's interest in safety. The program should be conducted by the first line supervisor. Schedule the informal tailgate meeting for a suitable location, keep it short, avoid distractions, and be sure that everyone can hear. Hand out literature, if available. Tailgate talks should communicate to the operator specific considerations, new problems, and accident information. These topics should be published. One resource for such meetings can be those operators who have been involved in an accident. Although it is sometimes embarrassing to the injured, the victim is now the expert on how the accident occurred, what could have been done to prevent it, and how it felt to have the injury. Encourage all operators, new and old, to participate in tailgate safety sessions.

Use safety posters to reinforce safety training and to make operators aware of the location of dangerous areas or show the importance of good work habits. Such posters are available through the National Safety Council's catalog.[16]

Awards for good safety records are another means of keeping operators aware of the importance of safety. The awards could be given to individuals in recognition of a good safety record. Publicity about the awards may provide an incentive to the operators and demonstrates the organization's determination to

[15] *Tailgate Safety Meeting.* Brief (10 to 20 minutes) safety meetings held every 7 to 10 working days. The term comes from the safety meetings regularly held by the construction industry around the tailgate of a truck.

[16] Write or call your local safety council or National Safety Council, 1121 Spring Lake Drive, Itasca, IL 60143-3201.

maintain a good safety record. The awards may include safety lapel pins, certificates, or plaques showing number of years without an accident. Consider publishing a utility newsletter on safety tips or giving details concerning accidents that may be helpful to other operators in the organization. Awards may be given to the organization in recognition of its effort in preventing accidents or for its overall safety program. A suggestion program concerning safety will promote and reinforce the program and give recognition to the best suggestions. The goal of all these efforts is to reinforce concerns for the safety of all operators. If safety, as an idea, is present, then accidents can be prevented.

Education of the operator in the use of tools and equipment is necessary. As pointed out above, OJT is not the answer to a good safety record. A good safety record will be achieved only with good work habits and safe equipment. If the operator is trained in the proper use of equipment (hand tools or vehicles), the operator is less likely to misuse them. However, if the supervisor finds an operator misusing tools or equipment, then it is the supervisor's responsibility to reprimand the operator as a means of reinforcing utility policies. The careless operator who misuses equipment is a hazard to other operators. Careless operators will also be the cause of a poor safety record in the operator's division or department.

An important part of every job should be the consideration of its safety aspects by the supervisor. The supervisor should instruct the foreman or operators about any dangers involved in job assignments. If a job is particularly dangerous, then the supervisor must bring that fact to everyone's attention and clarify utility policy in regard to unsafe acts and conditions.

If the operator is unsure of how to perform a job, then it is the operator's responsibility to ask for the training needed. Each operator must think, act, and promote safety if the organization is to achieve a good safety record. Training is the key to achieving this objective and training is everyone's responsibility—management, the supervisors, foremen, and operators.

QUESTIONS

Write your answers in a notebook and then compare your answers with those on pages 752 and 753.

20.11L What is the mainstay of a safety program?

20.11M Why should you report even a minor injury?

20.11N Why should a safety officer review an accident report form?

20.11O A new, inexperienced operator must receive instruction on what aspects of treatment plant safety?

20.11P What should an operator do if unsure of how to perform a job?

20.116 Measuring

To be complete, a safety program must also include some means of identifying, measuring, and analyzing the effects of the program. The systematic classification of accidents, injuries, and lost time is the responsibility of the safety officer. This person should use an analytical method that would refer to types and classes of accidents. Reports should be prepared using statistics showing lost time, costs, type of injuries, and other data, based on a specific time interval.

Such data call attention to the effectiveness of the program and promote awareness of the types of accidents that are happening. Management can use this information to decide where the emphasis should be placed to avoid accidents. However, statistical data are of little value if a report is prepared and then set on the bookshelf or placed in a supervisor's desk drawer. The data must be distributed and read by all operating and maintenance personnel.

As an example, injuries can be classified as fractures, burns, bites, eye injuries, cuts, and bruises. Causes can be referred to as heat, machinery, falls, handling objects, chemicals, unsafe acts, and miscellaneous. Cost can be considered as lost time, lost dollars, lost production, contaminated water, or any other means of showing the effects of the accidents.

Good analytical reporting will provide a great deal of detail without a lot of paper to read and comprehend. Keep the method of reporting simple and easy to understand by all operators, so they can identify the causes and be aware of how to prevent the accident happening to themselves or other operators. Table 20.6 gives one method of summarizing the causes of various types of injuries.

TABLE 20.6 SUMMARY OF TYPES AND CAUSES OF INJURIES

	PRIMARY CAUSE OF INJURY										
Type of Injury	Unsafe Act	Chemical	Falls	Handling Objects	Heat	Machinery	Falling Objects	Stepping	Striking	Miscellaneous	TOTAL
Fractures											
Sprains											
Eye Injuries											
Bites											
Cuts											
Bruises											
Burns											
Miscellaneous											

There are many other methods of compiling data. Table 20.6 could reflect cost in dollars or in work hours lost. Not all accidents mean time lost, but there can be other cost factors. The data analysis should also indicate if the accidents involve vehicles, company personnel, the public, company equipment, loss of chemical, or other factors. Results also should show direct cost and indirect cost to the agency, operator, and the public.

Once the statistical data have been compiled, someone must be responsible for reviewing it in order to take preventive actions. Frequently, such responsibility rests with the safety committee. In fact, safety committees may operate at several levels, for example management committee, working committee, or an accident review board. In any event, the committee must be active, be serious, and be fully supported by management.

Another means of measuring safety is by calculating the injury frequency rate for an indication of the effectiveness of your safety program. Multiply the number of disabling injuries by one million and divide by the total number of employee-hours worked. The number of injuries per year is multiplied by one million in order to obtain injury frequency rate values or numbers that are easy to use. In our example problems, we obtained numbers between one and one thousand.

$$\text{Injury Frequency Rate} = \frac{(\text{Number of Disabling Injuries/year})(1{,}000{,}000)}{\text{Number of Hours Worked/year}}$$

These calculations indicate a frequency rate per year, which is the usual means of showing such data. Note that this calculation accounts only for disabling injuries. You may wish to show all injuries, but the calculations are much the same.

EXAMPLE 3

A rural water company employs 36 operators who work in many small towns throughout a three-state area. The operators suffered four injuries in one year while working 74,880 hours. Calculate the injury frequency rate.

Known		Unknown
Number of Operators	= 36	Injury Frequency Rate
Number of Injuries/yr	= 4 Injuries/yr	
Number of Hours Worked/yr	= 74,880 Hr/yr	

Calculate the injury frequency rate.

$$\text{Injury Frequency Rate} = \frac{(\text{Number of Injuries/year})(1{,}000{,}000)}{\text{Number of Hours Worked/year}}$$

$$= \frac{(4\ \text{Injuries/yr})(1{,}000{,}000)}{74{,}880\ \text{Hr/yr}}$$

$$= 53.4$$

EXAMPLE 4

Of the four injuries suffered by the operators in Example 3, one was a disabling injury. Calculate the injury frequency rate for the disabling injuries.

Known		Unknown
Number of Disabling Injuries/yr	= 1 Injury/yr	Injury Frequency Rate
Number of Hours Worked/yr	= 74,880 Hr/yr	

Calculate the injury frequency rate.

$$\text{Injury Frequency Rate (Disabling Injuries)} = \frac{(\text{Number of Disabling Injuries/yr})(1{,}000{,}000)}{\text{Number of Hours Worked/yr}}$$

$$= \frac{(1\ \text{Injury/yr})(1{,}000{,}000)}{74{,}880\ \text{Hr/yr}}$$

$$= 13.4$$

Yet another consideration may be lost-time accidents. The safety officer's analysis may take into account many other considerations, but in any event, the method given here will provide a means of recording and measuring injuries in the treatment plant. In measuring lost-time injuries, a severity rate can be considered.

A severity rate is based on one lost hour for every million operator hours worked. The rate is found by multiplying the number of hours lost by one million and dividing by the total number of operator-hours worked.

$$\text{Injury Severity Rate} = \frac{(\text{Number of Hours Lost/yr})(1{,}000{,}000)}{\text{Number of Hours Worked/yr}}$$

EXAMPLE 5

The water company described in Examples 3 and 4 experienced 40 operator-hours lost due to injuries while the operators worked 74,880 hours. Calculate the injury severity rate.

Known		Unknown
Number of Hours Lost/yr	= 40 Hr/yr	Injury Severity Rate
Number of Hours Worked/yr	= 74,880 Hr/yr	

Calculate the injury severity rate.

$$\text{Injury Severity Rate} = \frac{(\text{Number of Hours Lost/yr})(1{,}000{,}000)}{\text{Number of Hours Worked/yr}}$$

$$= \frac{(40\ \text{hr/yr})(1{,}000{,}000)}{74{,}880\ \text{Hr/yr}}$$

$$= 534$$

Notice that all these data points are based on a one-year time interval, which makes them suitable for use by the safety officer in preparing an annual report.

20.117 Human Factors

First, you may ask, what is a human factor? Well, it is not too often that a safety text considers human factors as part of the safety program. However, if these factors are understood and emphasis is given to their practical application, then many accidents can be prevented. Human factors engineering is the specialized study of technology relating to the design of operator-machine interface. That is to say, it examines ways in which machinery might be designed or altered to make it easier to use, safer, and more efficient for the operator. We hear a lot about making computers more user friendly, but human factors engineering is just as important to everyday operation of other machinery in the everyday plant.

Many accidents occur because the operator forgets the human factors. The ultimate responsibility for accidents due to human factors belongs to the management group. However, this does not relieve the operator of the responsibility to point out the human factors as they relate to safety. After all, it is the operator using the equipment who can best tell if it meets all the needs for an interrelationship between operator and machine.

The first step in the prevention of accidents takes place in the plant design. Even with excellent designs, accidents can and do happen. However, every step possible must be taken during design to ensure a maximum effort of providing a safe plant environment. Most often, the operator has little to do with design, and therefore needs to understand human factors engineering so as to be able to evaluate these factors as the plant is being operated. As newer plants become automated, this type of understanding may even be more important.

Other contributing human factors are the operator's mental and physical characteristics. The operator's decision-making abilities and general behavior (response time, sense of alarm, and perception of problems and danger) are all important factors. Ideally, tools and machines should function as intuitive extensions of the operator's natural senses and actions. Any factors disrupting this flow of action can cause an accident. Therefore, be on the lookout for such factors. When you find a system that cannot be operated in a smooth, logical sequence of steps, change it. You may prevent an accident. If the everyday behavior of an operator is inappropriate with regard to a specific job, reconsider the assignment to prevent an accident.

The human factor in safety is the responsibility of design engineers, supervisors, and operators. However, the operator who is doing the work will have a greater understanding of the operator-machine interface. For this reason, the operator is the appropriate person to evaluate the means of reducing the human factor's contribution to the cause of accidents, thereby improving the plant's safety record.

QUESTIONS

Write your answers in a notebook and then compare your answers with those on page 753.

20.11Q Statistical accident reports should contain what types of accident data?

20.11R How can injuries be classified?

20.11S How can causes of injuries be classified?

20.11T How can costs of accidents be classified?

20.12 RECORDKEEPING

20.120 Importance of Records

In too many systems the importance and value of keeping good records are not recognized. Where this occurs, the operators may not know (or be able to find) the construction details of important facilities, where they are located, or what shape they are in. The need for a good record system, regardless of the size or complexity of the wastewater system, cannot be overemphasized.

Records are needed for many reasons. In general, they promote the efficient operation of the wastewater system. Records can remind the operator when routine operation or maintenance is necessary and help ensure that schedules will be maintained and needed operation or maintenance will not be overlooked or forgotten. Records are the key to an effective maintenance program. They are also needed for regulatory agencies that require submission of periodic effluent quality and operational records. Records can be used to determine the financial health of the utility, provide the basic data on the system's property, and prepare monthly and annual reports. Another reason for keeping accurate and complete records of system operations is the legal liability of the utility. Such records are required as evidence of what actually occurred in the system. Good records can help when threatened with lawsuits. Records also assist in answering consumer questions or complaints. Finally, clear, concise records are required to effectively meet future operational needs, that is, for planning purposes.

Records should be tailored to meet the demands of the particular system and only records known to be useful should be kept. Operators should determine what type of information will be of value for their system and then prepare maps, forms, or other types of records on which the needed information can be easily recorded and clearly shown. Records should be prepared as if they will be kept indefinitely. In fact, some will be kept for a long time while other records will not. Records should be put into a filing system that can be easily used and understood by everyone concerned, is readily accessible, and is protected from damage in a safe environment. The nonpermanent records should be disposed of in accordance with a disposal schedule set up for the different types of records maintained. Many records can easily be stored and retrieved from computers. Good recordkeeping tips can be found throughout these manuals.

20.121 Maps

Comprehensive maps and sectional plats are used by most utilities. The comprehensive map provides an overall view of the entire collection and treatment system. Important structures are shown, including sewers, lift stations, and treatment plants. The preferred scale for comprehensive maps is 500 feet per inch, while the maximum scale recommended is 1,000 feet per inch. Sectional plats will show various portions of the comprehensive map in much more detail. The scale varies from 50 to 200 feet per inch. Standard symbols are used to indicate different items on these and other utility maps.

All maps and drawings should show constructed facilities as built. If there was any change whatsoever from the construction plans, the maps and plans used by the utility should show this change. Whenever modifications are made, plans should be changed to show the details of the modification, the date of the modification, and who made the modification on the plans. The modified drawings are called "as-builts" or "record drawings."

Plan and profile maps are engineering drawings that show the depth of pipe, pipe location both vertically and horizontally, and the correct distance from a reference starting point. Operators occasionally need to use this type of map.

20.122 Geographic Information System (GIS)

The geographic information system (GIS) is a computer program that combines mapping with detailed information about the physical structures within geographic areas. To create the database of information, entities within a mapped area such as streets, manholes, line segments, and lift stations are given attributes. Attributes are simply the pieces of information about a particular feature or structure that are stored in a database. The attributes can be as basic as an address, manhole number, or line segment length, or they may be as specific as diameter, rim invert, and quadrant (coordinate) location. Attributes of a main line segment might include engineering information, maintenance information, and inspection information. Thus an inventory of entities and their properties is created. The system allows the operator to periodically update the map entities and their corresponding attributes.

The power of a GIS is that information can be retrieved geographically. An operator can choose an area to look at by pointing to a specific place on the map or outlining (windowing) an area of the map. The system will display the requested section on the screen and show the attributes of the entities located on the map. A printed copy may also be requested. Figure 20.10 shows a GIS-generated map. This example shows data from an inflow/infiltration analysis, including pipe attributes, hydraulic data, and selected engineering data. The example also shows a map of the system.

A GIS can generate work orders in the form of a map with the work to be performed outlined on the map. This minimizes paper work and gives the work crew precise information about where the work is to be performed. Completion of the work is recorded in the GIS to keep the work history for the area and entity up to date. Reports and other inquiries can be requested as needed, for example, a listing of all line segments in a specific area could be generated for a report.

In many areas GISs are being developed on an area-wide basis with many agencies, utilities, counties, cities, and state agencies participating. Usually a county-wide base map is developed and then all participants provide attributes for their particular systems. For example, information on the sanitary sewer collection system might be one map layer and the second layer might be the electric utility distribution system. GISs generally now also have the ability to operate smoothly with computer-aided design (CAD) systems.

20.123 Types of Records

There are many different types of records that are required for effective management and operation of wastewater treatment facilities. Below is a listing of some essential records:

1. Sources of wastewaters
2. Operation
3. Laboratory
4. Maintenance
5. Chemical inventory and usage
6. Purchases
7. Chlorination station
8. Personnel
9. Accidents
10. Customer complaints

20.124 Types of Plant Operations Data[17]

Plant operations logs can be as different as the plants whose information they record. The differences in amount, nature, and format of data are so significant that any attempt to prepare a typical log would be very difficult. This section will outline the kinds of data that are usually required to help you develop a useful log for your facilities.

Treatment plant data such as total flows, chemical use, chemical doses, process performance, quality control tests, and rainfall and runoff information represent the bulk of the data required for proper plant operation. Frequently, however, sources and collection system data are included because of the impact of this information on plant operation and operator responsibilities. Typical plant operations data include:

1. Plant title, agency, and location
2. Date
3. Names of operators and supervisors on duty
4. Sources of wastewater
   a. Municipal
   b. Industrial
   c. Business/commercial
   d. Restaurants
5. Wastewater treatment plant
   a. Plant inflow
   b. Influent wastewater quality
   c. Effluent quality
   d. Solids disposal
6. Chemical inventory and usage
   a. Chemical inventory/storage (measured use and deliveries)
   b. Metered or estimated plant usages
   c. Calculated usage of chemicals (compare with actual use)
7. Quality control tests
   a. DO
   b. Chlorine residual
   c. Coliforms
   d. BOD/COD
   e. SS
   f. Other
8. Meteorologic
   a. Rainfall, evaporation, and temperature of both water and air
   b. Weather (clear, cloudy, windy)
9. Remarks

   Space should be provided to describe or explain unusual data or events. Extensive notes should be entered on a daily worksheet or diary.

[17] Also see Chapter 19, "Records and Report Writing."

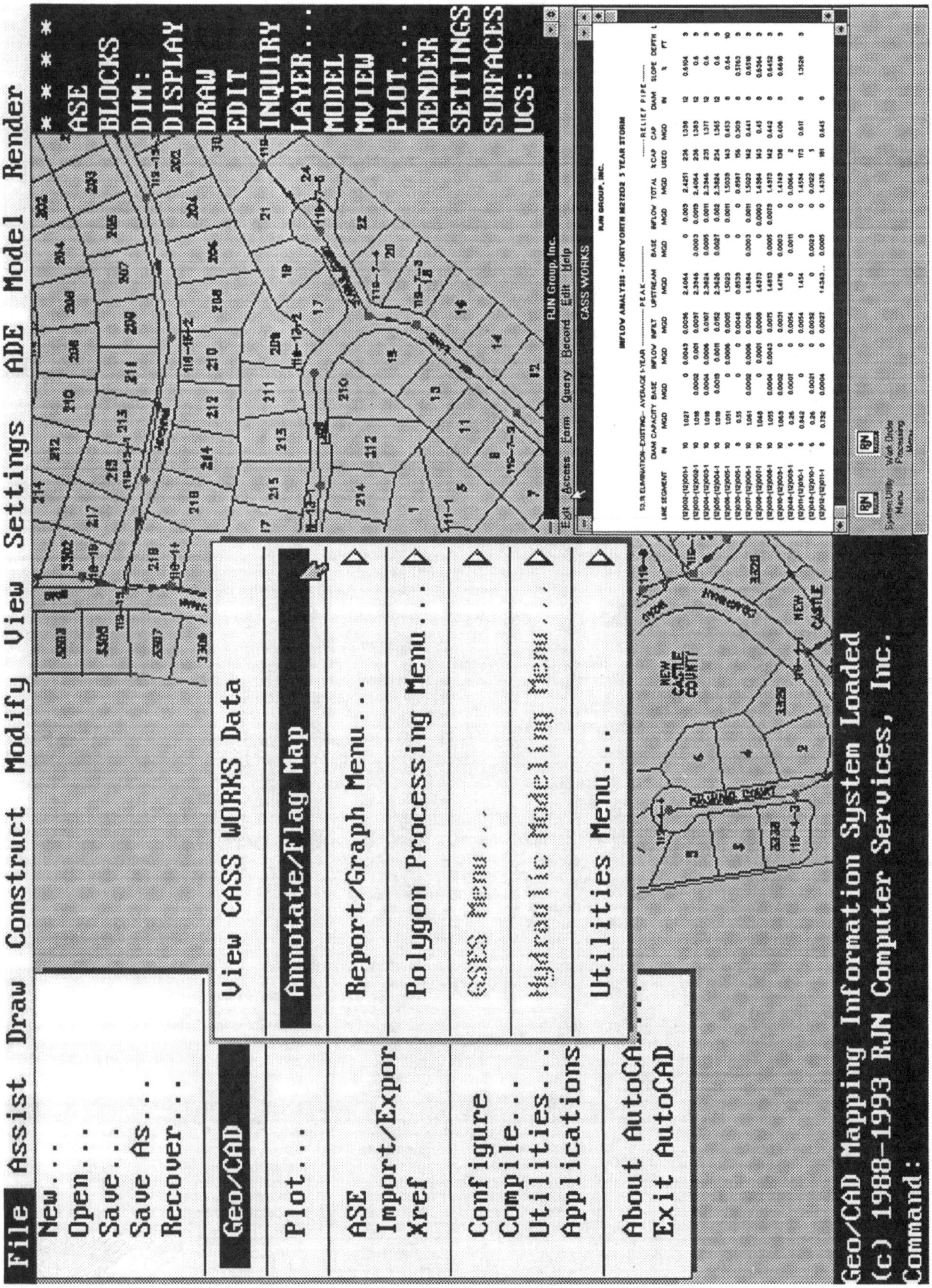

Fig. 20.10 Typical geographic information system (GIS) map

20.125 Maintenance Records

A good plant maintenance effort depends heavily upon good recordkeeping. There are several areas where proper records and documentation can definitely improve overall plant performance.

20.126 Procurement Records

Ordering repair parts and supplies usually is done when the on-hand quantity of a stocked part or chemical falls below the reorder point, a new item is added to stock, or an item has been requested that is not stocked. Most organizations require employees to submit a requisition (similar to the one shown in Figure 20.11) when they need to purchase equipment or supplies. When the requisition has been approved by the authorized person (a supervisor or purchasing agent, in most cases) the items are ordered using a form called a purchase order. A purchase order contains a number of important items. These items include: (1) the date, (2) a complete description of each item and the quantity needed, (3) prices, (4) the name of the vendor, and (5) a purchase order number.

A copy of the purchase order should be retained in a suspense file or on a clipboard until the ordered items arrive. This procedure helps keep track of the items that have been ordered but have not yet been received.

All supplies should be processed through the storeroom immediately upon arrival. When an item is received it should be so recorded on an inventory card. The inventory card will keep track of the numbers of an item in stock, when last ordered, cost, and other information. Furthermore, by always logging in supplies immediately upon receipt, you are in a position to reject defective or damaged shipments and control shortages or errors in billing. Some utilities use personal computers to keep track of orders and deliveries.

20.127 Inventory Records

An inventory consists of the supplies the treatment plant needs to keep on hand to operate the facility. These maintenance supplies may include repair parts, spare valves, electrical supplies, tools, and lubricants. The purpose of maintaining an inventory is to provide needed parts and supplies quickly, thereby reducing equipment downtime and work delays.

In deciding what supplies to stock, keep in mind the economics involved in buying and stocking an item as opposed to depending on outside availability to provide needed supplies. Is the item critical to continued plant or process operation? Should certain frequently used repair parts be kept on hand? Does the item have a shelf life?

Inventory costs can be held to a minimum by keeping on hand only those parts and supplies for which a definite need exists or that would take too long to obtain from an outside vendor. A definite need for an item is usually demonstrated by a history of regular use. Some items may be infrequently used but may be vital in the event of an emergency; these items should also be stocked. Take care to exclude any parts and supplies that may become obsolete, and do not stock parts for equipment scheduled for replacement.

Tools should be inventoried. Tools that are used by operators on a daily basis should be permanently signed out to them. More expensive tools and tools that are only occasionally used, however, should be kept in a storeroom. These tools should be signed out only when needed and signed back in immediately after use.

20.128 Equipment Records

You will need to keep accurate records to monitor the operation and maintenance of plant equipment. Equipment control cards and work orders can be used to:

- Record important equipment data such as make, model, serial number, and date purchased
- Record maintenance and repair work performed to date
- Anticipate preventive maintenance needs
- Schedule future maintenance work

See Chapter 15, Section 15.00, "Preventive Maintenance Records," for additional information.

20.129 Computer Recordkeeping Systems

Until fairly recently, wastewater supply system recordkeeping has been done manually. The current availability of low-cost personal computer systems puts automation of many manual bookkeeping functions within the means of all wastewater utilities.

To automate your recordkeeping functions as they relate to customer billing, you will need to develop a simple database management system that will create tables similar to those illustrated in this chapter. This can be readily accomplished by use of standard spreadsheet software programs, which are available in the marketplace at a cost of $300 to $400. Hardware including a personal computer, data storage system, and a printer can be purchased for under $5,000.

Excellent computer software packages are being developed and offered to assist utility managers. SURF (Small Utility Rates and Finances) has been developed by the American Water Works Association (AWWA). SURF is a self-guided, interactive financial spreadsheet application (Excel 97 or later) designed to assist small wastewater systems in developing budgets, setting user rates, and tracking expenses. SURF requires very little computer or software knowledge and can be used by system operators, bookkeepers, and managers to improve the financial management practices of their utilities. SURF can print out three separate modules: (1) system budget, (2) user rate(s), and (3) system expenses. SURF hardware and software requirements are modest.

The SURF software and user's guide are available free from the American Water Works Association. They can be obtained by calling the AWWA Small Systems Program at (303) 347-6191 or by downloading the program from the AWWA website (www.AWWA.org/science/sun/).

20.1210 Disposition of Plant Records

Good recordkeeping is very important because records ensure adequate operation and indicate potential problems. Records that may be required include:

1. The daily log and records of the laboratory analyses to control the treatment processes may be required when there are chronic treatment problems.
2. Chlorination, constituent removal, and sequestering records may be required from small systems (especially those demonstrating little understanding of the processes).
3. Records showing the quantity and ultimate disposal of residual solids.

An important question is how long records should be kept. Records should be kept as long as they may be useful. Some

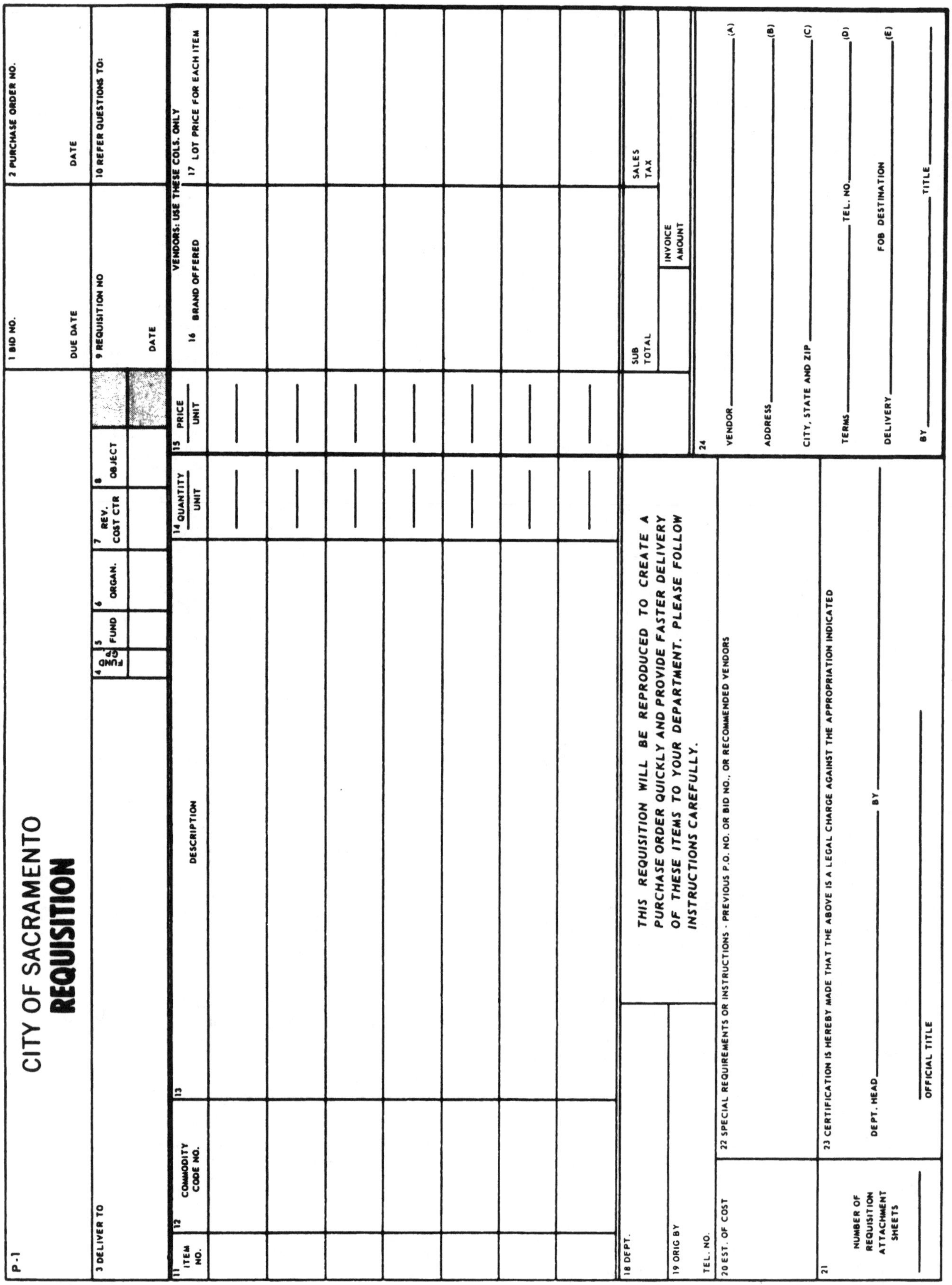

P-1

CITY OF SACRAMENTO
REQUISITION

1 BID NO.

DUE DATE

2 PURCHASE ORDER NO.

DATE

3 DELIVER TO

4 FUND GP.	5 FUND	6 ORGAN.	7 REV. COST CTR	8 OBJECT

9 REQUISITION NO

DATE

10 REFER QUESTIONS TO:

11 ITEM NO.	12 COMMODITY CODE NO.	13 DESCRIPTION	14 QUANTITY / UNIT	15 PRICE / UNIT	VENDORS: USE THESE COLS. ONLY 16 BRAND OFFERED	17 LOT PRICE FOR EACH ITEM

18 DEPT.

19 ORIG BY

TEL. NO.

THIS REQUISITION WILL BE REPRODUCED TO CREATE A PURCHASE ORDER QUICKLY AND PROVIDE FASTER DELIVERY OF THESE ITEMS TO YOUR DEPARTMENT. PLEASE FOLLOW INSTRUCTIONS CAREFULLY.

SUB TOTAL

SALES TAX

INVOICE AMOUNT

20 EST. OF COST

22 SPECIAL REQUIREMENTS OR INSTRUCTIONS - PREVIOUS P.O. NO. OR BID NO., OR RECOMMENDED VENDORS

21

NUMBER OF REQUISITION ATTACHMENT SHEETS

23 CERTIFICATION IS HEREBY MADE THAT THE ABOVE IS A LEGAL CHARGE AGAINST THE APPROPRIATION INDICATED

DEPT. HEAD ________ BY ________

OFFICIAL TITLE

24

VENDOR ________ (A)

ADDRESS ________ (B)

CITY, STATE AND ZIP ________ (C)

TERMS ________ TEL. NO. ________ (D)

DELIVERY ________ FOB DESTINATION (E)

BY ________ TITLE ________

Fig. 20.11 Requisition/purchase order form

information will become useless after a short time, while other data may be valuable for many years. Data that might be used for future design or expansion should be kept indefinitely. Laboratory data will always be useful and should be kept indefinitely. Regulatory agencies may require you to keep certain laboratory analyses and customer complaint records on file for specified time periods.

Even if old records are not consulted every day, this does not lessen their potential value. For orderly records handling and storage, set up a schedule to periodically review old records and to dispose of those records that are no longer needed. A decision can be made when a record is established regarding the time period for which it must be retained.

QUESTIONS

Write your answers in a notebook and then compare your answers with those on page 753.

20.12A List some of the important uses of records.

20.12B What makes a geographic information system (GIS) a potentially powerful tool for a wastewater utility operator?

20.12C What chemical inventory and usage records should be kept?

20.12D List the important items usually contained on a purchase order.

20.12E As a general rule, how long should utility records be kept?

20.13 SECURITY MEASURES[18]

World events in recent years have heightened concern in the United States over the security of the critical wastewater infrastructure. The nation's wastewater infrastructure, consisting of approximately 16,000 publicly owned wastewater treatment plants, 100,000 major pumping stations, 600,000 miles of sanitary sewers, and another 200,000 miles of storm sewers, is one of America's most valuable resources, with treatment and collection systems valued at more than $2 trillion.

Taken together, the sanitary and storm sewers form an extensive network that runs near or beneath key buildings and roads and is physically close to many communication and transportation networks. Significant damage to the nation's wastewater facilities or collection systems would result in: loss of life, catastrophic environmental damage to rivers, lakes, and wetlands, contamination of drinking water supplies, long-term public health impacts, destruction of fish and shellfish production, and disruption to commerce, the economy, and our normal way of life.

Some actions that should be taken at all times to reduce the possibility of a terrorist attack are:

- Ensure that all visitors sign in and out of the facilities with a positive ID check.
- Reduce the number of visitors to a minimum.
- Discourage parking by the public near critical buildings to eliminate the chances of car bombs.
- Be cautious with suspicious packages that arrive.
- Be aware of the hazardous chemicals used and how to defend against spills.
- Keep emergency numbers posted near telephones and radios.
- Patrol the facilities frequently, looking for suspicious activity or behavior.
- Maintain, inspect, and use your personal protective equipment (PPE) (hard hats, respirators).

The following recommendations by the EPA include many straightforward, common-sense actions a utility can take to increase security and reduce threats from terrorism.

Guarding Against Unplanned Physical Intrusion

- Lock all doors and set alarms at your office, pumping stations, treatment plants, and vaults, and make it a rule that doors are locked and alarms are set.
- Limit access to facilities and control access to pumping stations, chemical and fuel storage areas, giving close scrutiny to visitors and contractors.
- Post guards at treatment plants and post "Employees Only" signs in restricted areas.
- Secure hatches, metering vaults, manholes, and other access points to the sanitary collection system.
- Increase lighting in parking lots, treatment bays, and other areas with limited staffing.
- Control access to computer networks and control systems and change the passwords frequently.
- Do not leave keys in equipment or vehicles at any time.

Making Security a Priority for Employees

- Conduct background security checks on employees at hiring and periodically thereafter.
- Develop a security program with written plans and train employees frequently.
- Ensure all employees are aware of established procedures for communicating with law enforcement, public health, environmental protection, and emergency response organization.
- Ensure that employees are fully aware of the importance of vigilance and the seriousness of breaches in security.
- Make note of unaccompanied strangers on the site and immediately notify designated security officers or local law enforcement agencies.
- If possible, consider varying the timing of operational procedures so that, to anyone watching for patterns, the pattern changes.
- Upon the dismissal of an employee, change pass codes and make sure keys and access cards are returned.
- Provide customer service staff with training and checklists of how to handle a threat if it is called in.

[18] Adapted from "What Wastewater Utilities Can Do Now to Guard Against Terrorist and Security Threats," US Environmental Protection Agency, Office of Wastewater Management, October 2001.

Coordinating Actions for Effective Emergency Response

- Review existing emergency response plans and ensure that they are current and relevant.
- Make sure employees have the necessary training in emergency operating procedures.
- Develop clear procedures and chains-of-command for reporting and responding to threats and for coordinating with emergency management agencies, law enforcement personnel, environmental and public health officials, consumers, and the media. Practice the emergency procedures regularly.
- Ensure that key utility personnel (both on and off duty) have access to critical telephone numbers and contact information at all times. Keep the call list up to date.
- Develop close relationships with local law enforcement agencies and make sure they know where critical assets are located. Ask them to add your facilities to their routine rounds.
- Work with local industries to ensure that their pretreatment facilities are secure.
- Report to county or state health officials any illness among the employees that might be associated with wastewater contamination.
- Immediately report criminal threats, suspicious behavior, or attacks on wastewater utilities to law enforcement officials and the nearest field office of the Federal Bureau of Investigation.

Investing in Security and Infrastructure Improvements

- Assess the vulnerability of the collection system, major pumping stations, wastewater treatment plants, chemical and fuel storage areas, outfall pipes, and other key infrastructure elements.
- Assess the vulnerability of the stormwater collection system. Determine where large pipes run near or beneath government buildings, banks, commercial districts, industrial facilities, or are next to major communication and transportation networks. Move as quickly as possible with the most obvious and cost-effective physical improvements, such as perimeter fences, security lighting, and tamper-proofing manhole covers and valve boxes.
- Improve computer system and remote operational security.
- Use local citizen watches.
- Seek financing for more expensive and comprehensive system improvements.

The US Terrorism Alert System (Figure 20.12) is a color-coded system that identifies the potential for terrorist activity and suggests specific actions to be taken. Your safety plan should identify the actions that your facility will take when the threat level changes. Tables 20.7 and 20.8 show examples of security measures that should be taken to improve safety at wastewater treatment or collection system facilities when the threat level is YELLOW and when it is ORANGE. (The utility's safety plan should include similar lists of actions for the RED, BLUE, and GREEN levels as well.)

QUESTIONS

Write your answers in a notebook and then compare your answers with those on page 753.

20.13A What could be the effects of significant damage to the nation's wastewater facilities or collection systems?

20.13B What precautions can be taken to increase the security of a sanitary collection system?

20.14 ETHICS

What are illegal activities and what are ethical situations? Operators need to avoid illegal activities and know how to respond to ethical situations. This section will:

1. Describe illegal activities and the consequences
2. Review the concept of ethics
3. Examine potential ethical situations operators may encounter and possible responses to these situations

Illegal activities are defined by regulatory law and by certification board regulations. Examples of illegal activities include:

1. Providing false information on an application for a job or certification examination
2. Cheating on a certification examination
3. Copying certification examination questions
4. Dry-labbing[19] water quality analysis results
5. Falsifying lab test results or self-monitoring reports to imply compliance
6. Discharging toxic wastes (including excessive chlorine residuals) to receiving waters
7. Deliberately bypassing untreated or partially treated wastewater to receiving waters in violation of a discharge permit
8. Disposing of residual solids in an unapproved manner
9. Deliberately violating any provisions of a discharge permit
10. Harassing other employees
11. Requiring operators to use unsafe equipment, work under unsafe conditions, or occupy a confined space with an unsafe atmosphere

Operators must not perform illegal acts. The consequences of being caught and convicted include suspension or revocation of your operator's license, fines, and jail time.

Performing illegal acts is always unethical. However, unethical acts are not always illegal. What is ethics? If we read textbooks on ethics we will find different authors expressing similar viewpoints.

1. The point of ethics is to benefit people.
2. Correct ethical actions are those that one does out of consideration for others.

[19] Dry-labbing means writing down lab results without performing the tests.

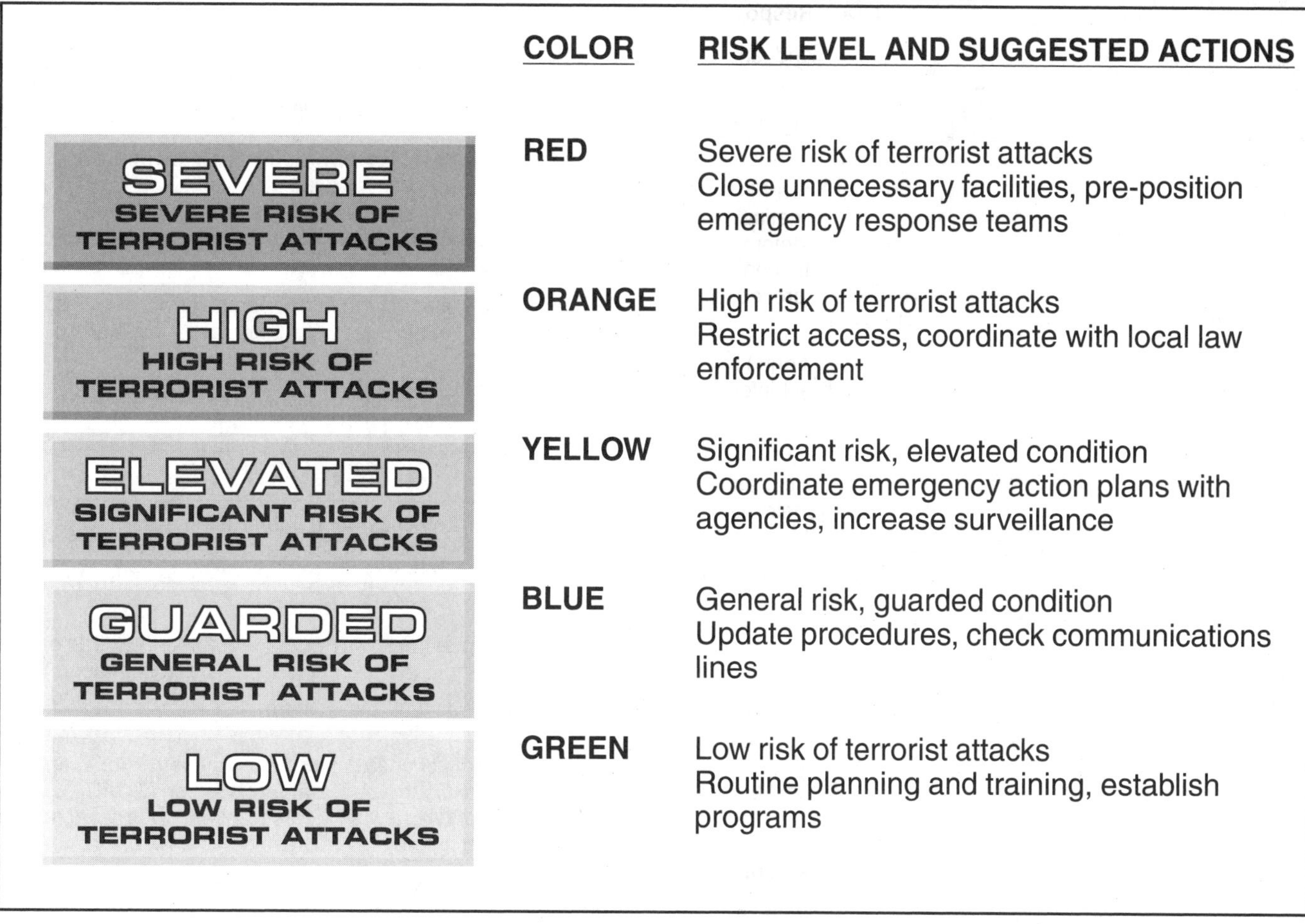

COLOR
RISK LEVEL AND SUGGESTED ACTIONS
SEVERE
SEVERE RISK OF TERRORIST ATTACKS
RED
Severe risk of terrorist attacks
Close unnecessary facilities, pre-position emergency response teams
HIGH
HIGH RISK OF TERRORIST ATTACKS
ORANGE
High risk of terrorist attacks
Restrict access, coordinate with local law enforcement
ELEVATED
SIGNIFICANT RISK OF TERRORIST ATTACKS
YELLOW
Significant risk, elevated condition
Coordinate emergency action plans with agencies, increase surveillance
GUARDED
GENERAL RISK OF TERRORIST ATTACKS
BLUE
General risk, guarded condition
Update procedures, check communications lines
LOW
LOW RISK OF TERRORIST ATTACKS
GREEN
Low risk of terrorist attacks
Routine planning and training, establish programs

Fig. 20.12 Threat level categories established by the US Department of Homeland Security

3. Ethical theories are about how people ought to treat each other.
4. Ethical concepts are based on the assumption that everyone acts with consideration of all.
5. Everyone benefits when we act ethically toward each other.

Unfortunately for operators, an ethical analysis of a situation does not produce absolute answers.

Different people have different concepts regarding what constitutes ethical behavior. Your community and your peers expect you, as an operator, to maintain a high level of ethical behavior. When encountering an ethical situation, you need to think about how your family, friends, colleagues, supervisors, employers, and the public would expect you to respond. On a personal level, ethics deals with your relationships with other people or how you should treat and respect other people and how you expect them to treat and respect you. On a professional level, ethics deals with how you respond to the responsibility entrusted to you to protect the public health and the environment. One helpful "litmus test" when faced with an ethical situation is to ask yourself if you would be comfortable with your chosen actions if they were reported in the newspapers the next day.

To help you identify ethical situations, determine alternative responses, and select an appropriate solution, let us examine some situations you might encounter on the job.

CHIEF OPERATOR—ETHICAL SITUATION

You are the chief operator of a wastewater treatment plant. Vendors want you to purchase their chemicals, pumps, and equipment. The sales representative for a particular company is attempting to convince you to purchase the company's chemicals. To obtain your business, the sales representative might offer to provide you with:

1. Sales literature, including benefits and prices of the chemicals
2. Company baseball cap
3. Lunch to discuss the chemicals
4. Dinner with you and your spouse
5. Tickets to a sporting event
6. Weekend at a resort
7. Gift of $10,000

TABLE 20.7 SECURITY MEASURES FOR THREAT LEVEL YELLOW (CONDITION ELEVATED)

Continue to introduce all measures listed in BLUE: Condition Guarded.	
Detection	**Prevention**
• To the extent possible, increase the frequency and extent of monitoring the flow coming into and leaving the treatment facility and review results against baseline quantities. Increase review of operational and analytical data (including customer complaints) with an eye toward detecting unusual variability (as an indicator of unexpected changes in the system). Variations due to natural or routine operational variability should be considered first. • Increase surveillance activities in water supply, treatment, and distribution facilities.	• Carefully review all facility tour requests before approving. If allowed, implement security measures to include list of names prior to tour, request identification of each attendee prior to tour, prohibit backpacks, duffle bags, and cameras, and identify parking restrictions. • On a daily basis, inspect the interior and exterior of buildings in regular use for suspicious activity or packages, signs of tampering, or indications of unauthorized entry. • Implement mail room security procedures. Follow guidance provided by the United States Postal Service.
Preparedness	**Protection**
• Continue to review, update, and test emergency response procedures and communication protocols. • Establish unannounced security spot checks (such as verification of personal identification and door security) at access control points for critical facilities. • Increase frequency for posting employee reminders of the threat situation and about events that constitute security violations. • Ensure employees understand notification procedures in the event of a security breach. • Conduct security audit of physical security assets, such as fencing and lights, and repair or replace missing/broken assets. Remove debris from along fence lines that could be stacked to facilitate scaling. • Maximize physical control of all equipment and vehicles; make them inoperable when not in use (for example, lock steering wheels, secure keys, chain, and padlock on front-end loaders). • Review draft communications on potential incidents; brief media relations personnel of potential for press contact or issuance of press releases. • Ensure that list of sensitive customers (such as government agencies and medical facilities) within the service area is accurate and shared with appropriate public health officials. • Contact neighboring water utilities to review coordinated response plans and mutual aid during emergencies. • Review whether critical replacement parts are available and accessible. • Identify any work/project taking place in proximity to events where large attendance is anticipated. Consult with the event organizers and local law enforcement regarding contingency plans, security awareness, and site accessibility and control.	• Verify the identity of all persons entering the water utility. Mandate visible use of identification badges. Randomly check identification badges and cards of those on the premises. • At the discretion of the facility manager or security director, remove all vehicles and objects (such as trash containers) located near mission critical facility security perimeters and other sensitive areas. • Verify the security of critical information systems (for example, Supervisory Control And Data Acquisition (SCADA), Internet, e-mail) and review safe computer and Internet access procedures with employees to prevent cyber intrusion. • Consider steps needed to control access to all areas under the jurisdiction of the utility. • Implement critical infrastructure facility surveillance and security plans. • At the beginning and end of each work shift, as well as at other regular and frequent intervals, inspect the interior and exterior of buildings in regular use for suspicious packages, persons, and circumstances. • Lock and regularly inspect all buildings, rooms, and storage areas not in regular use.

TABLE 20.8 SECURITY MEASURES FOR THREAT LEVEL ORANGE (CONDITION HIGH)

<table>
<tr><td colspan="2">Continue to introduce all measures listed in YELLOW: Condition Elevated.</td></tr>
<tr><th>Detection</th><th>Prevention</th></tr>
<tr><td>• Increase the frequency and extent of monitoring activities. Review results against baseline measurements.
• Confirm that county and state health officials are on high alert and will inform the utility of any potential waterborne illnesses.
• If a neighborhood watch-type program is in place, notify the community and request increased awareness.</td><td>• Discontinue tours and prohibit public access to all operational facilities.
• Consider requesting increased law enforcement surveillance, particularly of critical assets and otherwise unprotected areas.
• Limit access to computer facilities. No outside visitors.
• Increase monitoring of computer and network intrusion detection systems and security monitoring systems.</td></tr>
<tr><th>Preparedness</th><th>Protection</th></tr>
<tr><td>• Confirm that emergency response and laboratory analytical support network are ready for deployment 24 hours per day, 7 days a week.
• Reaffirm liaison with local police, intelligence, and security agencies to determine likelihood of an attack on the water supply, treatment, or distribution utility personnel and facilities and consider appropriate protective measures (such as road closing and extra surveillance).
• Practice communications procedures with local authorities and others cited in the facility's emergency response plan.
• Post frequent reminders for staff and contractors of the threat level, along with a reminder of what events constitute security violations.
• Ensure employees are fully aware of emergency response communication procedures and have access to contact information for relevant law enforcement, public health, environmental protection, and emergency response organizations.
• Have alternative water supply plan ready to implement (for example, bottled water delivery for employees and other critical business uses).
• Place all emergency management and specialized response teams on full alert status.
• Ensure personal protective equipment (PPE) and specialized response equipment is checked, issued, and readily available for deployment.
• Review all plans, procedures, guidelines, personnel details, and logistical requirements related to the introduction of a higher threat condition level.</td><td>• Evaluate the need to staff the water treatment facility at all times.
• Increase security patrol activity to the maximum level sustainable and ensure tight security in the vicinity of mission critical facilities. Vary the timing of security patrols.
• Request employees change their passwords on critical information management systems.
• Limit building access points to the absolute minimum, strictly enforce entry control procedures. Identify and protect all designated vulnerable points. Give special attention to vulnerable points outside of the critical facility.
• Lock all exterior doors except the main facility entrance(s). Check all visitors' purpose, intent, and identification. Ensure that contractors have valid work orders. Require visitors to sign in upon arrival; verify and record their identifying information. Escort visitors at all times when they are in the facility.</td></tr>
</table>

Questions you should ask yourself before responding include:

1. Where do you draw the line with these offers?
2. Does your agency have a policy regarding your relationship with vendors?
3. At what point might you be influenced to select this vendor?
4. What would be fair to the other competing vendors?
5. Are the other vendors making you similar or better offers?
6. Are you being asked to accept an illegal bribe?
7. How can you make the right ethical decision?

Operators should make their decision regarding which offers to accept (if any) based upon:

1. What knowledge and information can be gained from the meeting
2. What is fair to all vendors
3. What is expected by your employer
4. What is expected by your family
5. What you personally feel is the right thing (ethical thing) to do

OTHER ETHICAL SITUATIONS

In addition to meetings with sales representatives, this section lists other events or tasks for which operators need to be aware of the possibility of encountering an ethical situation.

Collecting Samples. Ensuring the collection of representative samples. Collecting samples, especially effluent and receiving water samples, in locations that are representative, not in locations that will produce favorable or desirable results.

Preparation of Reports. Producing accurate reports and not ignoring important information, biasing the results of the report, or falsifying data to document desired or expected results.

Sabotaging Someone Else's Work. Do something so another person's work will not be successful or perform as expected.

Recommending a Vendor or a Consulting Engineer. Your recommendations should be unbiased and based on performance and the value of the product.

Employee Performance Evaluations. Your evaluations must be based on documented performance and not on personality or personal friendship.

Cheating on an Exam. Observe someone cheating on a certification exam or a civil service exam. How should you respond?

WHISTLEBLOWER SITUATIONS

Whistleblowing is the act of reporting illegal activity to an authority. Whistleblower situations often involve your supervisors or others in your agency performing illegal activities. You may be aware of the situation due to your position in the agency or your professional knowledge. Examples of whistleblower situations include:

1. Improper spending or misappropriation of funds
2. Instructions to you to "cover up" discharge violations or noncompliance with regulations
3. Questionable offers in the form of a lack of enforcement of discharge and environmental regulations to induce an industry to relocate in your community

Should you expose the situation? Where are your loyalties—to yourself, your family, your agency, your profession, or the general public? If you expose the situation and create problems for your supervisor or your agency, they could retaliate and you could lose your job. (Retaliation for whistleblowing is illegal.) Unfortunately, responding to protect the public interest may result in a personal sacrifice for the operator. Your best response is to evaluate your alternatives and make a decision that is best for you.

CODES OF ETHICS

Professional organizations (including operator associations) often have a code of ethics for their members. Utility agencies frequently have a code of ethics for management and staff. Many codes of ethics are prepared to promote trust and respect for operators by the public. Typical topics covered in codes of ethics include:

1. Interactions between operators and the public
2. Interactions among operators themselves
3. Approved professional conduct for operators
4. Service by operators to protect the public and the environment

In more general terms, topics covered by codes of ethics may include:

1. Trustworthiness
2. Respect
3. Responsibility
4. Caring
5. Justice
6. Civic virtue and citizenship

SUMMARY

Regardless of the ethical situation, you need to:

1. List your alternative actions.
2. Evaluate the consequences of each action in terms of the benefits and damages to you as in individual and professionally.
3. Select the alternative that is most comfortable to you.

QUESTIONS

Write your answers in a notebook and then compare your answers with those on page 753.

20.14A How are illegal activities defined?

20.14B List the events or tasks for which operators need to be aware of the possibility of encountering an ethical situation.

20.14C What is whistleblowing?

20.14D Why do utility agencies frequently have a code of ethics for management and staff?

20.15 ACKNOWLEDGMENTS

During the writing of this material on administration, Lynn Scarpa, Phil Scott, Chris Smith, and Rich von Langen, all members of California Water Environment Association (CWEA), provided many excellent materials and suggestions for improvement. Their generous contributions are greatly appreciated.

20.16 ADDITIONAL READING

1. *MANAGE FOR SUCCESS,* especially Chapter 11, "Emergency Planning." Obtain from the Office of Water Programs, California State University, Sacramento, 6000 J Street, Sacramento, CA 95819-6025. Price, $49.00.
2. "A Water and Wastewater Manager's Guide for Staying Financially Healthy," US Environmental Protection Agency. EPA No. 430-9-89-004. Obtain from National Technical Information Service (NTIS), 5285 Port Royal Road, Springfield, VA 22161. Order No. PB90-114455. Price, $33.50, plus $5.00 shipping and handling per order.
3. "Wastewater Utility Recordkeeping, Reporting and Management Information Systems," US Environmental Protection Agency. EPA No. 430-9-82-006. Obtain from National Technical Information Service (NTIS), 5285 Port Royal Road, Springfield, VA 22161. Order No. PB83-109348. Price, $39.50, plus $5.00 shipping and handling per order.
4. *SUPERVISION: CONCEPTS AND PRACTICES OF MANAGEMENT,* 10th Edition, 2007, Raymond L. Hilgert and Edwin Leonard, Jr. Obtain from Thomson Learning, Attn: Order Fulfillment, PO Box 6904, Florence, KY 41022-6904. ISBN 978-0-324-31624-7. Price, $133.95, plus shipping and handling.
5. *CUSTOMER SERVICE FOR WATER UTILITIES,* DVD. Obtain from American Water Works Association (AWWA), Bookstore, 6666 West Quincy Avenue, Denver, CO 80235. Order No. 64174. Price to members, $208.50; nonmembers, $314.00; price includes cost of shipping and handling.
6. *TEXAS MANUAL,* Chapter 18,* "Effective Public Relations in Water Works Operations," and Chapter 19,* "Planning and Financing."
7. *GIS IMPLEMENTATION FOR WATER AND WASTEWATER TREATMENT FACILITIES* (MOP 26). Obtain from Water Environment Federation (WEF), 601 Wythe Street, Alexandria, VA 22314-1994. Order No. WPM401. ISBN 0-07-145305-9. Price to members, $76.75; nonmembers, $92.75; price includes cost of shipping and handling.
8. *GENERAL RISK MANAGEMENT PROGRAM GUIDANCE (APRIL 2004),* available at EPA's website: http://yosemite.epa.gov/oswer/ceppoweb.nsf/content/EPAguidance.htm#General.
9. "Complete Guide to Ethics Management: An Ethics Toolkit for Managers." Written by Carter McNamara, MBA, Ph.D., Authenticity Consulting, LLC. Copyright 1997–2007. This guide is located on the Internet at the following website: http://www.managementhelp.org/ethics/ethxgde.htm.
10. *ENGINEERING, ETHICS, AND THE ENVIRONMENT,* 1998, P. Aarne Vesilind and Alastair S. Gunn. Obtain from Cambridge University Press, 100 Brook Hill Drive, West Nyack, NY 10994-2133. ISBN 978-0-521-58918-5. Price, $37.99, plus shipping and handling.

* Depends on edition.

END OF LESSON 3 OF 3 LESSONS

on

TREATMENT PLANT ADMINISTRATION

Please answer the discussion and review questions next.

DISCUSSION AND REVIEW QUESTIONS

Chapter 20. TREATMENT PLANT ADMINISTRATION

(Lesson 3 of 3 Lessons)

Write the answers to these questions in your notebook. The question numbering continues from Lesson 2.

26. What can happen when agencies rely primarily on corrective maintenance to keep the system running?
27. What does a SCADA system do?
28. How does an air gap separation system prevent contamination of the potable water supply at a wastewater treatment plant?
29. How would you assess the vulnerability of a wastewater treatment system?
30. Why must wastewater collection and treatment utilities have a safety program?
31. How can a good safety record be accomplished?
32. What is the intent of the OSHA regulations?
33. What are the four main elements of a hazard communication program?
34. What topics should be included in a safety officer's talk to new operators?
35. What is a geographic information system (GIS)?
36. What are the consequences of an operator being caught and convicted of performing an illegal act?
37. What is ethics?

SUGGESTED ANSWERS

Chapter 20. TREATMENT PLANT ADMINISTRATION

ANSWERS TO QUESTIONS IN LESSON 1

Answers to questions on page 687.

20.0A Local utility demands on a utility manager include protection from environmental disasters with a minimum investment of money.

20.0B Changes in the environmental workplace are created by changes in the workforce and advances in technology.

Answers to questions on page 689.

20.1A The functions of a utility manager include planning, organizing, staffing, directing, and controlling.

20.1B In small communities, the community depends on the manager to handle everything.

Answers to questions on page 692.

20.2A Utility planning must include operational personnel, local officials (decision makers), and the public.

20.3A The purpose of an organizational plan is to show who reports to whom and to identify the lines of authority.

20.3B Effective delegation is uncomfortable for many managers since it requires giving up power and responsibility. Many managers believe that they can do the job better than others, they believe that other employees are not well trained or experienced, and they are afraid of mistakes. The utility manager retains some responsibility even after delegating to another employee and, therefore, the manager is often reluctant to delegate or may delegate the responsibility but not the authority to get the job done.

20.3C An important and often overlooked part of delegation is follow-up by the supervisor.

Answers to questions on page 694.

20.4A Staffing responsibilities include hiring new employees, training employees, and evaluating job performance.

20.4B The two personnel management concepts a manager should always keep in mind are "job-related" and "documentation."

20.4C The steps involved in a staffing analysis include:

1. List the tasks to be performed.
2. Estimate the number of staff hours per year required to perform each task.
3. List the utility's current employees.
4. Assign tasks based on each employee's skills and abilities.
5. Adjust the work assignments as necessary to achieve the best possible fit between the work to be done and the personnel/skills available to do it.

20.4D A qualifications profile is a clear statement of the knowledge, skills, and abilities a person must possess to perform the essential job duties of a particular position.

Answers to questions on page 697.

20.4E The purpose of a job interview is to gain additional information about the applicants so that the most qualified person can be selected to fill a job opening.

20.4F Protected groups include minorities, women, disabled persons, persons over 40 years of age, and union members.

20.4G The "OUCH" principle stands for:

Objectivity
Uniform treatment of employees
Consistency
Having job relatedness

20.4H A new employee's safety training should begin on the first day of employment or as soon thereafter as possible.

Answers to questions on page 698.

20.4I The purpose of a probationary period for new employees is to provide a time during which both the employer and employee can assess the fit between the job and the person.

20.4J The compensation an employee receives for the work performed includes satisfaction, recognition, security, appropriate pay, and benefits.

20.4K Utility managers should provide training opportunities for employees so they can keep informed of new technologies and regulations. Training for supervisors is also important to ensure that supervisors have the knowledge, skills, and attitude that will enable them to be effective supervisors.

Answers to questions on page 702.

20.4L An employee's immediate supervisor should conduct the employee's performance evaluation.

20.4M Dealing with employee discipline requires tact and skill. The manager or supervisor should stay flexible, calm, and open-minded.

20.4N Common warning signs of potential violence include abusive language, threatening or confrontational behavior, and brandishing a weapon.

Answers to questions on page 706.

20.4O Harassment is any behavior that is offensive, annoying, or humiliating to an individual and that interferes with a person's ability to do a job. This behavior is uninvited, often repeated, and creates an uncomfortable or even hostile environment in the workplace.

20.4P Types of behavior that could be considered sexual harassment include:

- Unwanted hugging, patting, kissing, brushing up against someone's body, or other inappropriate sexual touching
- Subtle or open pressure for sexual activity
- Persistent sexually explicit or sexist statements, jokes, or stories
- Repeated leering or staring at a person's body
- Suggestive or obscene notes or phone calls
- Display of sexually explicit pictures or cartoons

20.4Q The best way to prevent harassment is to set an example by your own behavior and to keep communication open between employees. A manager must also be aware of and take action to prevent any type of harassment in the workplace.

Answers to questions on page 707.

20.4R The shop steward is elected by the union employees and is their official representative to management and the local union.

20.4S Union contracts do not change the supervisor's delegated authority or responsibility. Operators must carry out the supervisor's orders and get the work done properly, safely, and within a reasonable amount of time. However, a contract gives a union the right to protest or challenge a supervisor's decision.

ANSWERS TO QUESTIONS IN LESSON 2

Answers to questions on page 710.

20.5A A manager needs both written and oral communication skills.

20.5B The most common written documents that a utility manager must write include memos, business letters, press releases, résumés, monitoring reports, monthly reports, and the annual report.

20.5C The annual report should be a review of what and how the utility operated during the past year and also the goals for the next year.

Answers to questions on page 711.

20.6A A utility manager may be asked to conduct meetings with employees, the governing board, the public, and with other professionals in your field.

20.6B Before a meeting: (1) prepare an agenda and distribute, (2) find an adequate meeting room, and (3) set a beginning and ending time.

Answers to questions on page 713.

20.7A The first step in organizing a public relations campaign is to establish objectives so you will have a clear idea of what you expect to achieve.

20.7B Employees can be informed about the utility's plans, practices, and goals through newsletters, bulletin boards, and regular, open communication between supervisors and subordinates.

20.7C Newspapers give more thorough, in-depth coverage to stories than do the broadcast media.

20.7D Practice is the key to effective public speaking.

20.7E Complaints can be a valuable asset in pinpointing wastewater collection system problems. Customer calls are frequently the first indication that something may be wrong. Responding to complaints and inquiries promptly can save the utility money and staff resources, and minimize customer inconvenience.

Answers to questions on page 717.

20.8A The three main areas of financial management for a utility include providing financial stability for the utility, careful budgeting, and providing capital improvement funds for future utility expansion.

20.8B The operating ratio for a utility is calculated by dividing total revenues by total operating expenses.

20.8C It is important for a manager to get input from other personnel in the utility as well as community leaders as the budgeting process proceeds in order to gain support for the budget and to keep the budget on track once adopted.

20.8D The basic ways for a utility to finance long-term capital improvements are through general obligation bonds, revenue bonds, or loan funding programs.

20.8E A revenue bond is commonly used to fund utility improvements. This bond has no limit on the amount of funds available and the user charges provide repayment on the bond. To qualify for these bonds, the utility must show sound financial management and the ability to repay the bond.

ANSWERS TO QUESTIONS IN LESSON 3

Answers to questions on page 719.

20.9A The purpose of an O & M program is to maintain design functionality or to restore the system components to their original condition and thus functionality, that is, to ensure that the system performs as designed and intended.

20.9B Commonly accepted types of maintenance include corrective maintenance, preventive maintenance, and predictive maintenance.

20.9C The major elements of a good preventive maintenance program include the following:

1. Planning and scheduling
2. Records management
3. Spare parts management
4. Cost and budget control
5. Emergency repair procedures
6. Training program

Answers to questions on page 721.

20.9D SCADA stands for Supervisory Control And Data Acquisition system.

20.9E A SCADA system collects, stores, and analyzes information about all aspects of operation and maintenance, transmits alarm signals, when necessary, and allows fingertip control of alarms, equipment, and processes.

20.9F Measured (sensed) information could be transmitted by FM radio, leased telephone circuits, private signal lines, dial telephone lines, coaxial cable networks, spread spectrum radio, fiber-optic cable, microwave, 900 MHz radio, cellular telephone, or satellite communications systems.

20.9G The greatest challenges for operators using SCADA systems are to realize that computers may not be correct and to have the ability to operate when the SCADA system fails.

Answers to questions on page 722.

20.9H Contamination through cross connections has consistently caused more waterborne disease outbreaks in the United States than any other reported factor.

20.9I Potential places where cross connections could occur in a wastewater treatment plant include water seals on pumps, feed water to boilers, hose bibs below grade where they may be subject to flooding with wastewater or sludges, or any other location where wastewater could contaminate a domestic water supply.

Answers to questions on page 725.

20.10A The first step toward an effective contingency plan for emergencies is to make an assessment of vulnerability. Then, a comprehensive plan of action can be developed and implemented.

20.10B An emergency operations plan need not be too detailed, since all types of emergencies cannot be anticipated and a complex response program can be more confusing than helpful.

20.10C The following outline can be used as the basis for developing an emergency operations plan:

1. Make a vulnerability assessment.
2. Inventory organizational personnel.
3. Provide for a recovery operation (plan).
4. Provide training programs for operators in carrying out the plan.
5. Coordinate with local and regional agencies such as the health, police, and fire departments to develop procedures for carrying out the plan.
6. Establish a communications procedure.
7. Provide protection for personnel, plant equipment, records, and maps.

20.10D A Process Safety Management (PSM) program deals with on-site dangers to plant employees from the release of hazardous chemicals. A Risk Management Program (RMP) deals with the potential off-site consequences of chemical releases.

Answers to questions on page 727.

20.11A A safety officer should evaluate every accident, offer recommendations, and keep and apply statistics.

20.11B The supervisors should be responsible for the implementation of a safety program.

20.11C Both state and federal regulatory agencies enforce the OSHA requirements.

20.11D The utility manager is responsible for the safety of the agency's personnel and the public exposed to the wastewater utility's operations. Therefore, the manager must develop and administer an effective safety program and must provide new employee safety training as well as ongoing training for all employees.

20.11E A safety policy statement should:

1. Define the goals and objectives of the program.
2. Identify the persons responsible for each element of the program.
3. Affirm management's intent to enforce safety regulations.
4. Describe the disciplinary actions that will be taken to enforce safe work practices.

Answers to questions on page 728.

20.11F A supervisor may be responsible, in part or completely, for an accident by causing unsafe acts to take place, by requiring that work be performed in haste, by disregarding an unsafe environment of the workplace, or by failing to consider any number of safety hazards.

20.11G Each operator must accept, at least in part, responsibility for fellow operators, for the utility's equipment, for the operator's own welfare, and even for seeing that the supervisor complies with established safety regulations.

20.11H First aid means emergency treatment for injury or sudden illness, before regular medical treatment is available.

20.11I First-aid training is most important for operators who regularly work with electrical equipment and those who must handle chlorine and other dangerous chemicals.

Answers to questions on page 731.

20.11J The basic elements of a hazard communication program include the following:

1. Identify hazardous materials.
2. Obtain chemical information and define hazardous conditions.
3. Properly label hazards.
4. Train operators.

20.11K A utility manager may or may not be involved in the day-to-day details of enforcing the agency's confined space policy and procedures. However, every utility manager should be aware of the current OSHA requirements and should ensure that the utility's policies not only comply with current regulations, but that the agency's policies are vigorously enforced for the safety of all operators.

Answers to questions on page 736.

20.11L The mainstay of a safety program is the method of reporting and keeping statistics.

20.11M All injuries should be reported, even if they are minor in nature, so as to establish a record in case the injury develops into a serious injury. It may be difficult at a later date to prove whether the accident occurred on or off the job and this information may determine who is responsible for the costs.

20.11N A safety officer should review an accident report form to make recommendations as well as to follow up to be sure that proper action has been taken to prevent recurrence.

20.11O A new, inexperienced operator must receive instruction on all aspects of plant safety. This training includes instruction in the handling of chemicals, the dangers of electrical apparatus, fire hazards, and proper maintenance of equipment to prevent accidents. Special instructions are required for specific work environments such as manholes, gases (chlorine and hydrogen sulfide (H_2S)), water safety, and any specific hazards that are unique to your facility. All new operators should be required to participate in a safety orientation program during the first few days of employment, and an overall training program in the first few months.

20.11P If an operator is unsure of how to perform a job, then it is the operator's responsibility to ask for the training needed.

Answers to questions on page 738.

20.11Q Statistical accident reports should contain accident statistics showing lost time, costs, type of injuries, and other data, based on some time interval.

20.11R Injuries can be classified as fractures, burns, bites, eye injuries, cuts, and bruises.

20.11S Causes of injuries can be classified as heat, machinery, falls, handling objects, chemicals, unsafe acts, and miscellaneous.

20.11T Costs of accidents can be classified as lost time, lost dollars, lost production, contaminated water, or any other means of showing the effects of the accidents.

Answers to questions on page 743.

20.12A Some of the important uses of records include:

1. Reminding operators when routine operation or maintenance is necessary
2. Complying with regulatory agency requirements
3. Determining the financial health of the utility
4. Providing basic data on the utility system's property
5. Assisting in preparation of monthly and annual reports
6. Providing evidence of what actually occurred in the system, if needed in a lawsuit
7. Assisting in answering consumer questions or complaints
8. Helping in planning to meet future operational needs

20.12B The power of a geographic information system (GIS) is that information can be retrieved geographically. An operator can easily look at or print out a specific area of a map. The map will contain an inventory of the collection system structures within the selected area and it will provide detailed information about each of the structures (or entities).

20.12C Chemical inventory and usage records that should be kept include:

1. Chemical inventory/storage (measured use and deliveries)
2. Metered or estimated plant usages
3. Calculated usage of chemicals (compare with actual use)

20.12D Important items usually contained on a purchase order include: (1) the date, (2) a complete description of each item and quantity needed, (3) prices, (4) the name of the vender, and (5) a purchase order number.

20.12E As a general rule, records should be kept as long as they may be useful or as long as legally required.

Answers to questions on page 744.

20.13A Significant damage to the nation's wastewater facilities or collection systems could result in: loss of life, catastrophic environmental damage to rivers, lakes, and wetlands, contamination of drinking water supplies, long-term public health impacts, destruction of fish and shellfish production, and disruption to commerce, the economy, and our normal way of life.

20.13B To increase the security of a sanitary collection system, secure hatches, metering vaults, manholes, and other access points.

Answers to question on page 748.

20.14A Illegal activities are defined by regulatory law and certification board regulations.

20.14B The events or tasks for which operators need to be aware of the possibility of encountering an ethical situation include:

1. Meetings with sales representatives
2. Collecting samples
3. Preparation of reports
4. Sabotaging someone else's work
5. Recommending a vendor or a consulting engineer
6. Employee performance evaluations
7. Cheating on an exam

20.14C Whistleblowing is the act of reporting illegal activity to an authority.

20.14D Many codes of ethics are prepared to promote trust and respect for operators by the public.

APPENDIX

OPERATION OF WASTEWATER TREATMENT PLANTS

(VOLUME II)

Comprehensive Review Questions and Suggested Answers

How to Solve Wastewater Treatment Plant Arithmetic Problems

Abbreviations

Wastewater Words

Subject Index

COMPREHENSIVE REVIEW QUESTIONS

VOLUME II

This section was prepared to help you review the material in Volume II. The questions are divided into five types:

1. True-False
2. Best Answer
3. Multiple Choice
4. Short Answer
5. Problems

To work this section:

1. Write the answer to each question in your notebook.
2. After you have worked a group of questions (you decide how many), check your answers with the suggested answers at the end of this section.
3. If you missed a question and do not understand why, reread the material in the manual.

You may wish to use this section for review purposes when preparing for civil service and certification examinations.

Since you have already completed this course, please *DO NOT SEND* your answers to California State University, Sacramento.

True-False

1. In the activated sludge process, the oxidation may be by chemical or biological processes.
   1. True
   2. False
2. The removal of organisms from the activated sludge treatment process (sludge wasting) is a very important control technique.
   1. True
   2. False
3. In an air distribution system, an air metering device should be located in a straight section of the air main on the suction side of the blower.
   1. True
   2. False
4. In an aeration tank, the higher the food to microorganism (F/M) ratio, the more food and therefore a lower dissolved oxygen (DO) is needed.
   1. True
   2. False
5. When considering a major process change, first review the plant data and then make only one major change at a time.
   1. True
   2. False
6. Microbiology is another tool for operators to use in controlling the activated sludge process.
   1. True
   2. False
7. The psychrophilic bacteria in anaerobic digesters appear to be able to survive temperatures well below freezing with little or no harm.
   1. True
   2. False
8. Mixing is very important in a digester because it greatly speeds up the digestion rate.
   1. True
   2. False
9. Operators should make every effort to pump as thick a sludge to the digester as possible.
   1. True
   2. False
10. The withdrawal rate of sludge from a digester should be faster than a rate at which the gas production from the system is able to maintain a positive pressure in the digester.
    1. True
    2. False
11. Aerobic sludge digestion completes the oxidation of volatile materials in the digester.
    1. True
    2. False
12. The proper discharge of plant effluent is the final process for the operator.
    1. True
    2. False
13. The principles for collecting samples for measuring dissolved oxygen are the same as for measuring temperature.
    1. True
    2. False
14. Site-specific safety procedures should be confirmed with your local safety regulatory agency.
    1. True
    2. False
15. Always get air into the confined space before you enter to work and maintain the ventilation until you have left the space.
    1. True
    2. False

16. The sense of smell is reliable for evaluating the presence of dangerous gases.

    1. True
    2. False

17. Moving parts that create a contact hazard to employees must be guarded.

    1. True
    2. False

18. When applying protective coatings in a clarifier, you can create an atmospheric hazard.

    1. True
    2. False

19. In the laboratory, separate flammable, explosive, or special hazard items for storage in an approved manner.

    1. True
    2. False

20. Hard hats, safety footwear, eye protection, and hearing protection apply to all personnel in designated areas and specific jobs.

    1. True
    2. False

21. When you are performing plant maintenance, you must recognize tasks that may be beyond your capabilities or repair facilities, and you should request assistance when needed.

    1. True
    2. False

22. The horizontal pump suction piping slopes downward toward the pump.

    1. True
    2. False

23. Fuses are used to protect operators, wiring, main circuits, branch circuits, heaters, and motors.

    1. True
    2. False

24. Electric motors are the machines most commonly used to convert electrical energy into mechanical energy.

    1. True
    2. False

25. Valves, elbows, and other items that could disrupt the flow ahead of a meter can upset the accuracy and reliability of a flowmeter.

    1. True
    2. False

26. Safety is just as important in the laboratory as in the rest of the treatment plant.

    1. True
    2. False

27. The shorter the time that elapses between sample collection and analysis, the more reliable will be the analytical results.

    1. True
    2. False

28. The COD test is used to measure the strength of wastes that are too toxic for the BOD test.

    1. True
    2. False

29. Probably the most frequent use of computers is for word processing.

    1. True
    2. False

30. Wastewater flows and other characteristics that can be sensed by instruments can be monitored by computers.

    1. True
    2. False

31. When installing any computer program, be sure to check to see that it is doing what you expect it to do.

    1. True
    2. False

32. Collection of data without analysis, interpretation, and use of results is a waste of time and money.

    1. True
    2. False

33. Operators are likely to draw the line of best fit slightly differently as each operator will visualize a slightly different line (slope).

    1. True
    2. False

34. Well-kept records will make the task of writing treatment plant cost and efficiency reports much easier.

    1. True
    2. False

35. Records at most wastewater treatment facilities are kept daily and summarized on a monthly basis.

    1. True
    2. False

36. You and your plant operation are partially judged by the information contained in your report, its style, and its appearance.

    1. True
    2. False

37. The utility manager must provide a safer, cleaner work environment while constantly training and retraining operators to understand new technologies.

    1. True
    2. False

38. Establishing safe work procedures for new employees is a very important function of management.

    1. True
    2. False

39. Good communication is an essential part of good management skills.

    1. True
    2. False

40. Every utility needs an effective operations and maintenance program.
    1. True
    2. False
41. The supervisor who has the responsibility for directing work activities must be safety conscious.
    1. True
    2. False
42. Each operator must think, act, and promote safety if the utility is to achieve a good safety record.
    1. True
    2. False

Best Answer (Select only the closest or best answer.)

1. Why must the proper dissolved oxygen (DO) level be maintained in an aeration tank?
    1. So filamentous organisms will thrive and the sludge floc will settle properly
    2. So pinpoint floc will develop and be removed by the secondary clarifier
    3. So the contents of the aeration tank will be thoroughly mixed
    4. So the solids will settle properly and the plant effluent will be clear
2. Why should weirs installed at the outlet or effluent end of the aeration tank be level?
    1. To comply with the plant's plans and specifications
    2. To keep the aeration tank balanced
    3. To prevent short-circuiting and an uneven distribution of solids in the effluent
    4. To provide a uniform appearing tank
3. How can the mixed liquor suspended solids in the aeration tank be increased?
    1. By increasing the influent flows
    2. By reducing the aeration tank capacity
    3. By reducing the effluent flows
    4. By reducing the sludge wasting rate
4. What is bound water?
    1. A high-molecular-weight substance that is formed by either natural or synthetic processes
    2. Clouds of billowing sludge that occur throughout secondary clarifiers and sludge thickeners when the sludge does not settle properly
    3. The gathering together of fine particles after coagulation to form larger particles by a process of gentle mixing
    4. Water contained within the cell mass of sludges or strongly held on the surface of colloidal particles
5. What are raw sludge and scum?
    1. Settled sludge solids that are removed from the bottom and the floating scum removed from the top of the aerators
    2. Settled sludge solids that are removed from the bottom and the floating scum removed from the top of the clarifiers
    3. Settled sludge solids that are removed from the bottom and the floating scum removed from the top of the digesters
    4. Untreated solids and scum that are removed from the bar screens
6. What is the purpose of a vacuum relief valve on an anaerobic sludge digester?
    1. Prevents a spark or flame from entering the digester
    2. Prevents air from entering the digester or digester gas from escaping
    3. Relieves excessive pressures in the digester so the water seal will not be blown out
    4. Relieves excessive vacuums so the digester cover will not collapse
7. Which item is the key to successful digester operation?
    1. pH
    2. Recirculated sludge
    3. Temperature
    4. Volatile acid/alkalinity relationship
8. Successful operation of a surface discharge system depends on which factor?
    1. Adequate budgets to meet the essential O & M requirements
    2. All operators being properly certified
    3. Conscientious operators using proper techniques for sampling and analysis
    4. Qualified regulators understanding the plant's O & M requirements
9. What type of emergency operating equipment should be available for pumped discharge systems when there is a loss of power?
    1. Standby generators
    2. Standby monitors
    3. Standby operators
    4. Standby pumps
10. What is a representative sample?
    1. A collection of individual samples obtained at regular intervals, usually every one or two hours during a 24-hour time span
    2. A collection of samples combined with other samples in proportion to the rate of flow when the sample was collected
    3. A sample portion of material, water, or wastestream that is as nearly identical in content and consistency as possible to that in the larger body being sampled
    4. A single sample of water collected at a particular time and place that represents the composition of the water only at that time and place
11. What is the best protection against the risk of infections and infectious diseases?
    1. Immunization program
    2. Personal hygiene
    3. Personal protective equipment (PPE)
    4. Personnel safety courses
12. What is the ideal method of dealing with any high-noise environment?
    1. The elimination or reduction of all sources through feasible engineering or administrative controls
    2. The scheduling of work in the high-noise environment when the equipment is not operating
    3. The selection of operators with a high level of noise tolerance to work in the high-noise environment
    4. The use of generous compensation packages when loss of hearing results

13. What is the apparent cause of drowning when someone falls into a diffused aeration tank?
    1. A strong undertow pulls the person under
    2. The aerator is too deep for a person to stand on the bottom
    3. The air in the water causes the person to sink to the bottom
    4. There is nothing to grab hold of to keep afloat or to pull oneself out of the aerator

14. If your hands may come in contact with wastewater or sludge during sampling, how can you prevent the spread of disease?
    1. Dry hands in the sunlight when you have finished sampling
    2. Wear disposable, impervious gloves
    3. Wear heavy leather gloves
    4. Wear light, flexible, cotton gloves

15. Which materials are consumed by a Class B fire?
    1. Combustible metals, such as magnesium, sodium, zinc, and potassium
    2. Energized electrical equipment, such as starters, breakers, and motors
    3. Flammable and combustible liquids, such as gasoline, oil, and grease
    4. Ordinary combustibles, such as wood, paper, and cloth

16. Why is a good maintenance program essential for a wastewater treatment plant?
    1. So operators can maintain their operator maintenance certificates
    2. So operators will have time to conduct plant tours
    3. To arrange for equipment vendors to observe their products
    4. To operate continuously at peak design efficiency

17. How can pipelines that carry chlorine gas be purged to remove moisture?
    1. By using cold water
    2. By using dry air or nitrogen
    3. By using methane gas
    4. By using moist air

18. Why do very few maintenance operators do the actual electrical repairs or troubleshooting?
    1. Because cost-effective training programs are rarely available for maintenance operators
    2. Because highly trained technicians are readily available at a lower cost than maintenance operators
    3. Because this is a highly specialized field and unqualified people can seriously injure themselves and damage costly equipment
    4. Because union regulations and negotiations forbid maintenance operators from performing this work

19. What is voltage? Voltage is
    1. A measure of power units used to rate electrical machines or motors or to indicate power in an electric circuit
    2. A practical unit of electrical resistance
    3. The electrical pressure available to cause a flow of current when an electric circuit is closed
    4. The measurement of current or electron flow and is an indication of work being done

20. What is a fuse?
    1. A device that allows a motor to be run continuously at its rated horsepower without causing damage to the insulation system
    2. A protective device having a strip or wire of fusible metal which, when placed in a circuit, will melt and break the electric circuit when subjected to an excessive temperature
    3. A switch that is opened automatically when the current or voltage exceeds or falls below a certain limit
    4. An expression representing an electrical connection to earth or a large conductor that is at the earth's potential or neutral voltage

21. When two or more pumps of the same size are installed, why should they be operated alternately?
    1. To ensure that all the pumps will fail simultaneously and can be replaced all at once
    2. To equalize wear, keep the motor windings dry, and distribute lubricant in the bearings
    3. To keep the load on the foundation and flow in the pipes balanced
    4. To minimize electricity consumption and system electrical hazards

22. What do flow measurement transmitter instruments do?
    1. Compare received signals with other values and send corrective or adjusting signals when necessary
    2. Measure the variable flow indicator and convert this value to a usable number
    3. Pick up the transmitted signal and convert it to a usable number
    4. Send the flow variable to another device for conversion to a usable number

23. Why should operators review the plans and specifications for pumps and lift stations?
    1. To be sure adequate provisions have been made for the station pumps, equipment, and instrumentation to be easily and properly operated and maintained
    2. To review the plans and specifications to ensure receipt of low bids
    3. To suggest contractors who could do an acceptable job
    4. To verify and confirm the consulting engineers' calculations

24. In chemistry, to what does normality (*N*) refer?
    1. Solution hazard
    2. Solution strength
    3. Solution toxicity
    4. Solution volume

25. What does the digester supernatant solids test measure?
    1. The degree of liquid-solids separation occurring when mixing is stopped in the digester
    2. The effectiveness of the digester mixing procedures
    3. The percent of settleable solids being returned to the plant headworks
    4. The strength of the supernatant that is removed from the digester

26. Why should chlorine residual be removed before final discharge to a receiving stream?
    1. For protection of aquatic vegetation from toxic chemicals
    2. For protection of fish and other aquatic life
    3. For protection of swimmers from skin rashes
    4. For protection of water supplies from TTHMs

27. What is aseptic?
    1. A plastic tube that contains an exact amount of chemical or reagent
    2. A representative portion of a sample. Often, an equally divided portion of a sample
    3. A solution or liquid whose chemical makeup neutralizes acids or bases without a great change in pH
    4. Free from the living germs of disease, fermentation, or putrefaction. Sterile
28. What is the first question an operator should ask when considering the possibility of using a computer or expanding the use of computers?
    1. How can I find a computer course to improve my computer skills?
    2. How can I find funds in the budget to help me?
    3. How can I get the computer staff to assist me?
    4. How can I use a computer to help me do a better job?
29. What does a SCADA system do?
    1. Collects, stores, and analyzes information about all aspects of O & M, transmits alarm signals when necessary, and allows fingertip control of alarms, equipment, and processes
    2. Provides standby power to operate critical pumps when power systems fail
    3. Scans plant facilities and reports unusual conditions to plant security
    4. Surveys stakeholders and system users and reports on the adequacy and equity of user fees
30. What is the median?
    1. The difference between the largest and smallest measurement
    2. The measurement that occurs most frequently
    3. The middle measurement when the measurements are ranked in order of magnitude (size)
    4. The sum of all measurements divided by the number of measurements
31. What is a normal distribution of data?
    1. The geometric mean describes the distribution
    2. The mean, median, and mode values are different
    3. The mean, median, and mode values will be approximately the same
    4. The percentile plotting points are different
32. What is the main means by which operators who have information communicate with those who need it?
    1. Committee meetings
    2. Professional conferences
    3. Report writing
    4. Telephones
33. Recorded data should enable the operator to determine how to perform which task?
    1. Develop tailgate safety sessions
    2. Operate and maintain the plant
    3. Promote qualified operators
    4. Schedule plant tours
34. What is planning?
    1. Deciding who does what work and delegating authority to the appropriate operators
    2. Determining the goals, policies, procedures, and other elements to achieve the goals and objectives of the agency
    3. Guiding, teaching, motivating, and supervising operators and utility staff members
    4. Taking the steps necessary to ensure that essential activities are performed so that objectives will be achieved as planned
35. What is authority?
    1. Answering to those above in the chain of command to explain how and why you have used your authority
    2. The act in which power is given to another person in the organization to accomplish a specific job
    3. The power and resources to do a specific job or to get that job done
    4. When a manager gives power or responsibility to an employee, the employee ensures that the manager is informed of results or events
36. How do you measure financial stability for a utility?
    1. By calculating the benefit/cost ratio for the utility
    2. By calculating the net revenue for the utility
    3. By calculating the operating ratio and the coverage ratio
    4. By calculating the rate of return on the utility's assets
37. What is the operating ratio?
    1. A measure of the ability of the utility to pay the principal and interest on loans and bonds in addition to any unexpected expenses
    2. A measure of the total revenues divided by the total operating expenses
    3. The amount of money required annually to pay the interest on outstanding debts
    4. The rate of return on the funds expended on facilities
38. What is SCADA?
    1. A computer analytical system that continuously monitors water quality changes in the system
    2. A computer budgeting system that records and monitors costs and revenues
    3. A computer intelligence system that eliminates many tasks performed by operators
    4. A computer-monitored alarm, response, control, and data acquisition system used to monitor and adjust the operation of equipment
39. What is the purpose of a safety policy statement?
    1. To comply with OSHA regulations
    2. To create the proper atmosphere to impress insurance companies
    3. To let employees know that the safety program has the full support of the agency and its management
    4. To let the media and the public know that the agency is safety conscious

Multiple Choice (Select all correct answers.)

1. When wastewater enters an activated sludge plant, the preliminary treatment processes remove which pollutants?
   1. Coarse or heavy solids (grit)
   2. Debris, such as roots, rags, and boards
   3. Floatable and settleable material
   4. Nutrients
   5. Soluble or finely divided suspended materials
2. Which items can cause an unsuitable environment for the activated sludge process?
   1. Failure to supply enough oxygen
   2. High concentrations of acids, bases, and other toxic substances
   3. Maintaining proper solids (floc mass) concentration in the aerator
   4. Proper adjustment of the waste sludge pumping rate
   5. Uneven flows of wastewater, which can cause overfeeding or starvation
3. What information should be recorded when testing pumps?
   1. Date of the test
   2. Location of the pump
   3. Name of the operator
   4. Pump discharge (flow)
   5. Pump discharge pressure
4. Which plant changes could cause an activated sludge plant to become upset?
   1. Changes in operators' shifts
   2. Changes in the sampling program
   3. Flow or waste changes
   4. High solids in the digester supernatant
   5. Temperature changes
5. Which items are possible causes of sludge bulking?
   1. Filamentous organisms growing from one floc mass to another
   2. Low DO
   3. Low nutrient concentrations
   4. Low pH
   5. Production of a highly jelly-like, waterlogged sludge that has a very low sludge density
6. Which items should be checked before starting a new sequencing batch reactor (SBR) plant?
   1. Be sure the plant has an up-to-date O & M manual and a set of as-built plans (record drawings)
   2. Check and flush pipe systems and remove any refuse or debris left in them during construction
   3. Ensure protective safety guards are installed on rotating equipment, open tanks, and channels
   4. Make sure tanks are clean; be sure all debris, boards, ladders, and tools have been removed
   5. To the extent possible, operate and test all equipment before admitting wastewater for treatment
7. Operators can achieve good digester operation by controlling which items?
   1. Loading rate or food supply (organic solids)
   2. Mixing
   3. Reproduction rate of acid formers and methane fermenters
   4. Temperature
   5. Volatile acid/alkalinity relationship
8. How can a scum blanket be controlled in a digester?
   1. By adequate mixing
   2. By chemicals
   3. By heat
   4. By methane fermenters
   5. By volatile acids
9. Where does struvite scale sometimes form?
   1. In anaerobic digesters
   2. In downstream chlorination equipment
   3. In downstream digested sludge concentration equipment
   4. In downstream digested sludge handling equipment
   5. In downstream effluent weirs
10. When operators are making their rounds inspecting their plant, they should be alert and investigate and record which items? Anything that
    1. Feels different (hotter or vibrating more)
    2. Looks different or unusual
    3. Smells different
    4. Sounds different
    5. Tastes different
11. What types of problems are encountered when operating aerobic digesters?
    1. Floating sludge
    2. Fluctuating BODs
    3. Odors
    4. Scum
    5. Wet solids
12. When reviewing the specifications for sludge digestion and solids handling facilities, which items should operators check?
    1. Adequate supply of equipment operation and maintenance manuals
    2. Equipment warranties and responsibility for acceptance testing
    3. Instrumentation—remote and local control board items and recorders are provided
    4. Paints and protective coatings
    5. Performance requirements and capabilities
13. Prior to discharge to surface waters, why must wastewater be treated?
    1. To prevent nuisances due to odors or unsightliness
    2. To prevent the wastewater from interfering with the many uses of surface waters
    3. To produce documentation for the recordkeeping system
    4. To protect the health of the people who may come in contact with it
    5. To provide water for discharge sampling and analysis
14. Which operating procedures should be developed for the start-up of a plant?
    1. Inspect the discharge line, where possible
    2. Look at the water surface over the diffuser or outfall pipe
    3. Observe the flow entering the discharge system for visible pollutants
    4. Open all appropriate valves
    5. Start the pumps (unless it is a gravity system)

15. Which items are probable causes of floatables in pond effluents?

 1. Excessive floatables and scum on the surface
 2. Excessive solids accumulated on the bottom have gasified and floated to the surface
 3. Excessive velocity or insufficient detention time
 4. Outlet baffle is not at the proper location
 5. Temperature or weather conditions may favor a particular species of algae

16. Under what conditions may the frequency of sampling be increased?

 1. When the lab is not meeting QA/QC requirements
 2. When the lab staff have the opportunity to analyze more samples
 3. When the receiving waters are not meeting established water quality standards
 4. When the treatment plant effluent is not meeting discharge requirements
 5. When the treatment plant staff increases and more time is available to collect more samples

17. Which types of hazards may an operator be exposed to whether working on the collection system or working in a treatment plant?

 1. Confined spaces
 2. Infections and infectious diseases
 3. Oxygen deficiency or enrichment
 4. Physical injuries
 5. Toxic or suffocating gases or vapors

18. A confined space may be defined as any space that meets which conditions?

 1. Allows more than one employee to fit into the space and work
 2. Has limited or restricted means for entry or exit
 3. Is large enough and so configured that an employee can bodily enter and perform assigned work
 4. Is not designed for continuous employee occupancy
 5. Requires employees to wear personal protective equipment (PPE)

19. The hazard warnings on containers of hazardous chemicals at your plant should contain emergency guide information on which specific categories?

 1. Flammability
 2. Health
 3. Nutrients
 4. Permeability
 5. Reactivity

20. Which items could be included in your plant's fire prevention plan?

 1. Develop written response procedures for reacting to a fire situation, to include evacuation
 2. Provide periodic cleanup of weeds or other vegetation in and around the plant
 3. Provide required service on all fire detection and response equipment (inspection, service, hydrostatic testing)
 4. Regulate the use, storage, and disposal of all combustible materials/substances
 5. Routinely inspect fire doors to ensure proper operation and unobstructed access

21. The primary use for polymers in wastewater treatment is the conditioning of sludge to facilitate removal of water in which subsequent treatment processes?

 1. Aerobic digesters
 2. Belt filter presses
 3. Centrifuges
 4. Dissolved air flotation thickeners
 5. Gravity belt thickeners

22. Working safely in the laboratory involves which basic safety procedures and practices?

 1. Always check labels on bottles to make sure that the proper chemical is selected
 2. Never handle chemicals with your bare hands
 3. Never use laboratory glassware for a cup or food dish
 4. Use care in making rubber-to-glass connections
 5. Use proper safety goggles or a face shield in all tests where there is danger to the eyes

23. Which paperwork items can be helpful in identifying the causes of accidents and developing corrective procedures?

 1. Accident report forms
 2. Facility plans and specifications and plant inspection reports
 3. Plant's safety policy
 4. Plant's safety rules
 5. Supervisors' guidelines on how to promote and implement a safety plan

24. What could cause a pump to run dry?

 1. Excessive leakage when pumping against a high head
 2. Holes (spacing) being too large in the suction intake screen
 3. Impeller rotating too fast and causing cavitation
 4. Lack of proper priming when starting
 5. Loss of suction when operating

25. Which items are primary sources of unsatisfactory AC brush and ring performance?

 1. Disturbing external condition
 2. Improper load or service condition
 3. Mechanical fault in the machine
 4. Poor machine design
 5. Poor preparation and care of the machine

26. When performing pump station electrical equipment maintenance tasks, which items should be checked and inspected annually?

 1. All panel instruments
 2. All switch gear and distribution equipment
 3. Enclosures
 4. Fuses and circuit breakers
 5. Wiring integrity

27. Which items are causes of motor malfunction or failure?

 1. Bearing failures
 2. Contaminants
 3. Old age
 4. Overload (thermal)
 5. Single phasing

28. When troubleshooting any motor control that has developed trouble, the qualified maintenance operator should be able to do which of the following?
    1. Examine all unusual factors
    2. Find the cause of the problem
    3. Know what should happen when a switch is pushed
    4. Make a visual inspection
    5. Repair the problem and eliminate the cause, if possible

29. Plugged pipelines are encountered in lines transporting which types of materials and liquids?
    1. Digested sludge
    2. Effluent
    3. Grit
    4. Raw sludge
    5. Scum

30. Which factors will cause sludge lines to plug more often?
    1. When industrial wastes are discharged after a fine screening process
    2. When plant effluent and reclaimed effluent are pumped through the lines
    3. When RAS (return activated sludge) and WAS (waste activated sludge) are pumped through the lines
    4. When scum and raw sludge are pumped through the same line
    5. When stormwaters carry in grit and silt that are not effectively removed by the grit removal facilities

31. When examining lift station prints, operators should ensure adequate provisions have been made for which items?
    1. Floors should be sloped to provide drainage, where needed, and drains should be located in low spots
    2. Overhead clearances of power lines, trees, and roofs should be adequate for a crane to remove large equipment
    3. The alarm system should signal high water levels in the wet well and water on the floor of the dry well
    4. There should be room to use hydrolifts, cranes, and high-velocity cleaners, as needed
    5. There should be sufficient room to set up portable pumping units or other necessary equipment in cases of major station failures or disasters

32. When working with chemicals and other materials in the wastewater treatment plant laboratory, which hazards can be minimized by using proper techniques and equipment?
    1. Burns
    2. Electric shock
    3. Infectious materials
    4. Poisons
    5. Toxic fumes

33. Why must hydrogen sulfide gas be measured in the atmosphere?
    1. Because it causes corrosion
    2. Because it causes odors
    3. Because it increases the fuel value of sewer gas
    4. Because it is flammable and explosive under certain conditions
    5. Because it is toxic to your respiratory system

34. Which precautions should be considered when measuring the chemical oxygen demand (COD) of a sample?
    1. Flasks and condensers should be clean and free from grease or other oxidizable materials
    2. The standard ferrous ammonium sulfate solution is unstable and should be standardized daily
    3. The wastewater sample should be well mixed
    4. Use a wide-tip pipet to ensure a representative sample is added
    5. Use extreme caution and safety precautions when handling the chemicals used for the test

35. Which precautions should be considered when using a dissolved oxygen (DO) probe?
    1. Dissolved inorganic salts, such as found in seawater, can influence the readings from a probe
    2. Do not place the probe directly over a diffuser because you want to measure the dissolved oxygen in the water being treated, not the oxygen in the air supply to the aerator
    3. Keep the membrane in the tip of the probe from drying out
    4. Periodically check the calibration of the probe
    5. Reactive compounds, such as reactive gases and sulfur compounds, can interfere with the output of a probe

36. Sulfate ions in wastewater are indirectly responsible for which serious problems?
    1. Indigestion
    2. Laxative effects
    3. Odor
    4. Sewer corrosion problems
    5. Toxic gas

37. Which items are general types of computer applications in wastewater treatment plants?
    1. Communications
    2. Data analysis
    3. Recordkeeping
    4. Report generation
    5. Systems monitoring

38. If the proper lab results are entered into the computer, the computer can calculate and print out which items?
    1. Chemical dosages
    2. F/Ms (Food to Microorganism ratios)
    3. MCRTs (Mean Cell Residence Times)
    4. Percent removals
    5. Suspended solids, BOD, or COD loadings

39. Which programs are typically included in general purpose software packages?
    1. Database management
    2. Graphics
    3. Plant template
    4. Spreadsheet
    5. Word processing

40. Which items can typically be monitored, recorded, and controlled by computers in a wastewater treatment plant?
    1. Aeration basin dissolved oxygen
    2. Anaerobic sludge digester gas production
    3. Chlorine contact time
    4. Influent flow control gates and valves
    5. Secondary clarifier sludge pumping

41. When preparing to plot data in a bar graph, what information must first be determined?

    1. Class midpoint
    2. Distribution of data
    3. Measurement frequencies
    4. Spread of data
    5. Width of the class interval

42. Why are records important?

    1. They are needed to fill unused space in storage rooms and filing cabinets
    2. They are needed to show the type and frequency of maintenance and to evaluate the effectiveness of maintenance programs
    3. They can indicate plant efficiency, performance, past problems, and potential problems
    4. They can provide data upon which to base recommendations for modifying plant operation and facilities
    5. They may be needed in damage suits brought against the district or municipality

43. A list of the minimum daily records kept at a fairly large treatment plant with digester tanks may include which items?

    1. Chlorine use (lbs or kg)
    2. Influent and effluent temperatures
    3. pH of influent and effluent
    4. Precipitation and air temperature
    5. Raw wastewater flow (MGD from totalizer)

44. Which items are basic principles common to good report writing?

    1. Know the purpose and objective of your report
    2. Know your subject
    3. Omit any recommendations
    4. Organize the report to present your ideas in a logical order
    5. Tailor your report to the person or persons to whom it is directed

45. Which items are guidelines for preparing a report?

    1. Arrange the ideas in a logical sequence
    2. Begin writing
    3. Gather material needed to support the ideas to be presented in the report
    4. List ideas and topics you plan to cover in the report
    5. Prepare conclusions and recommendations after writing the main body of the report

46. The utility manager's staffing responsibilities include which tasks?

    1. Evaluating employee performance
    2. Issuing orders and instructions
    3. Providing opportunities for advancement for operators and staff
    4. Selecting and training employees
    5. Teaching classes

47. Which items are important steps in the process of hiring and selecting a new operator?

    1. Advertising the position
    2. Avoiding unacceptable pre-employment inquiries
    3. Interviewing applicants
    4. Paper screening
    5. Selecting the most qualified candidate

48. Which items may serve as a guide for both the employee and the supervisor when dealing with a confrontation regarding employee performance issues?

    1. Aim for a permanent solution
    2. Create a private environment
    3. Keep your language appropriate
    4. Listen very carefully
    5. Maintain an adult approach

49. The total revenue of a utility comes from revenue generated by which sources?

    1. Chemicals
    2. Hook-up charges
    3. Interest income
    4. Taxes or assessments
    5. User fees

50. What is predictive maintenance?

    1. A method of establishing baseline performance data
    2. Monitoring performance criteria over a period of time
    3. Observing changes in performance so that failure can be predicted and maintenance can be performed on a planned, scheduled basis
    4. The same as corrective maintenance
    5. The same as proactive maintenance

51. How can a supervisor be partially or fully responsible for an accident?

    1. By causing unsafe acts to take place
    2. By disregarding an unsafe work environment
    3. By overlooking or failing to consider any number of safety hazards
    4. By requiring operators to follow safe procedures
    5. By requiring work to be performed in haste

Short Answer

1. Define activated sludge.
2. Why should activated sludge in the final clarifier be returned to the aeration tank as quickly as possible?
3. How can maintenance activities in a collection system cause operational problems in an activated sludge treatment plant?
4. Why is clean air essential for the protection of blowers and process systems?
5. Why should proper procedures be followed when checking equipment?
6. When starting a new activated sludge plant, who might the operator contact for assistance and advice?
7. What essential laboratory tests and on-site tests are recommended when starting the activated sludge process, and what are they used for?
8. Why must some activated sludge be wasted?
9. When attempting to correct an upset activated sludge process, why should only one major change be made at a time?
10. Why must proper equipment shutdown procedures be followed?
11. Why are activated sludge sequencing batch reactors (SBRs) popular for smaller industrial plants?

12. What can an operator learn during the construction of a sequencing batch reactor (SBR)?
13. Who should know microbiology?
14. Why is it important to keep the contents of a digester well mixed?
15. Why is digester gas considered dangerous?
16. Under what types of circumstances will the pressure relief valve and vacuum relief valve operate?
17. Why should seed sludge be added to a new digester?
18. What is the first warning that trouble is developing in an anaerobic digester?
19. How can you anticipate problems with anaerobic digesters and correct them before a problem becomes serious?
20. What precautions should be taken when applying sludge to a drying bed?
21. Why should receiving waters be sampled and tested?
22. What does the term "representative sample" mean?
23. What is a major concern regarding sample water quality indicators after the sample has been collected?
24. How large a sample should be collected?
25. What is the operator's responsibility with regard to safety?
26. What kind of job site protection is usually required when you are working in a manhole?
27. When cleaning racks or screens, on what kind of surface should the operator stand?
28. Why should no smoking or open flames be allowed in the vicinity of the digester or sludge digestion system?
29. What should be done with the jagged ends of glass tubes?
30. Why should plant water supplies be checked monthly for coliform group bacteria?
31. Why should the operator thoroughly read and understand manufacturers' literature before attempting to maintain plant equipment?
32. Why should one person never be permitted to repair a chlorine or sulfur dioxide leak alone?
33. How can you determine if a new pump will turn in the direction intended?
34. Why must motor nameplate data be recorded and filed?
35. When two or more pumps of the same size are installed, why should they be operated alternately?
36. How would you determine if a motor is running unusually hot?
37. Why is regular maintenance of plant safety equipment necessary?
38. How can scum lines be kept from plugging?
39. Why should a flowmeter be calibrated in its field installation?
40. Why must chemicals be properly labeled?
41. What precautions should you take to protect yourself from diseases when working in a wastewater treatment plant?
42. What is meant by representative sample?
43. Why must the clarity test always be run under the same conditions?
44. Why should the dewatering characteristics of digested sludge be measured?
45. Why is the COD test run?
46. What is the purpose of the coliform group bacteria test?
47. What precautions should be exercised when using a DO probe?
48. The oil and grease test measures what kinds of materials?
49. Why should operators use computers?
50. How would you justify the purchase of a computer?
51. What are the three principal factors that can cause variations in test results?
52. The solids concentrations of sludge withdrawn from a primary clarifier during the past seven days are given below:

Day:	1	2	3	4	5	6	7
Solids, %:	6.0	6.5	6.0	5.0	6.5	7.5	8.0

Draw a bar graph showing the distribution of data.

53. Why must operators be capable of effective report writing?
54. How should operators approach the task of report writing?
55. What are the different types of demands on a utility manager?
56. When has a supervisor successfully delegated?
57. What type of training should be provided for supervisors?
58. How should discipline problems be solved?
59. What happens any time you or a member of your utility comes in contact with the public?
60. What is the value of customer complaints?
61. Why must wastewater collection and treatment utilities have a safety program?
62. What topics should be included in a safety officer's talk to new operators?

Problems

1. How many pounds of solids are in a 400,000-gallon aeration tank if the suspended solids concentration is 2,400 mg/*L*?
2. Laboratory tests indicate that the volatile content of a raw sludge was 72% and after digestion the volatile content is 45%. What is the percent reduction in volatile matter?
3. A digester is fed a mixture of 4.5 percent primary sludge flowing at 7,800 GPD and 7.6 percent thickened secondary sludge flowing at 5,500 GPD. What is the approximate concentration of solids fed?
4. A wet well is four feet wide and six feet long. Approximately how far down should the level in the wet well be lowered in six minutes by a pump with a rated capacity of 250 GPM?
5. A wastewater treatment plant receives a flow of 1.6 MGD. The influent suspended solids are 210 mg/*L* and the effluent suspended solids are 20 mg/*L*. How many pounds of suspended solids are removed per day?

6. The temperature of a plant effluent water is 14°C (57.2°F) and the DO at saturation is 10.4 mg/*L*. What is the percent saturation of dissolved oxygen in the water when the DO is 3.8 mg/*L*?

7. What is the BOD of a 5-m*L* sample if the initial DO of the diluted sample was 7.5 mg/*L* and the DO of the diluted sample after five-day incubation was 3.0 mg/*L*?

8. At the beginning of a week, a totalizer on the plant inflow read 4,243,891 gallons. Seven days later, the totalizer read 7,762,438 gallons. What was the average daily flow during the week?

9. The solids concentrations of sludge withdrawn from a primary clarifier during the past seven days are given below:

Day:	1	2	3	4	5	6	7
Solids, %:	6.0	6.5	6.0	5.0	6.5	7.5	8.0

What are:

1. Average Solids, %?
2. Range of Solids Conc, %?
3. Median Solids Conc, %?
4. Mode Solids Conc, %?

SUGGESTED ANSWERS TO COMPREHENSIVE REVIEW QUESTIONS

VOLUME II

True-False

1. True In the activated sludge process, the oxidation may be by chemical or biological processes.
2. True The removal of organisms from the activated sludge treatment process (sludge wasting) is a very important control technique.
3. False In an air distribution system, an air metering device should be located in a straight section of the air main on the discharge (NOT suction) side of the blower.
4. False In an aeration tank, the higher the food to microorganism (F/M) ratio, the more food and therefore a higher (NOT lower) dissolved oxygen (DO) is needed.
5. True When considering a major process change, first review the plant data and then make only one major change at a time.
6. True Microbiology is another tool for operators to use in controlling the activated sludge process.
7. True The psychrophilic bacteria in anaerobic digesters appear to be able to survive temperatures well below freezing with little or no harm.
8. True Mixing is very important in a digester because it greatly speeds up the digestion rate.
9. True Operators should make every effort to pump as thick a sludge to the digester as possible.
10. False The withdrawal rate of sludge from a digester should be NO faster than a rate at which the gas production from the system is able to maintain a positive pressure in the digester.
11. False Aerobic sludge digestion does NOT complete the oxidation of volatile materials in the digester.
12. True The proper discharge of plant effluent is the final process for the operator.
13. True The principles for collecting samples for measuring dissolved oxygen are the same as for measuring temperature.
14. True Site-specific safety procedures should be confirmed with your local safety regulatory agency.
15. True Always get air into the confined space before you enter to work and maintain the ventilation until you have left the space.
16. False The sense of smell is absolutely UNRELIABLE for evaluating the presence of dangerous gases.
17. True Moving parts that create a contact hazard to employees must be guarded.
18. True When applying protective coatings in a clarifier, you can create an atmospheric hazard.
19. True In the laboratory, separate flammable, explosive, or special hazard items for storage in an approved manner.
20. True Hard hats, safety footwear, eye protection, and hearing protection apply to all personnel in designated areas and specific jobs.
21. True When you are performing plant maintenance, you must recognize tasks that may be beyond your capabilities or repair facilities, and you should request assistance when needed.
22. False The horizontal pump suction piping slopes upward (NOT downward) toward the pump.
23. True Fuses are used to protect operators, wiring, main circuits, branch circuits, heaters, and motors.
24. True Electric motors are the machines most commonly used to convert electrical energy into mechanical energy.
25. True Valves, elbows, and other items that could disrupt the flow ahead of a meter can upset the accuracy and reliability of a flowmeter.
26. True Safety is just as important in the laboratory as in the rest of the treatment plant.
27. True The shorter the time that elapses between sample collection and analysis, the more reliable will be the analytical results.
28. True The COD test is used to measure the strength of wastes that are too toxic for the BOD test.
29. True Probably the most frequent use of computers is for word processing.
30. True Wastewater flows and other characteristics that can be sensed by instruments can be monitored by computers.
31. True When installing any computer program, be sure to check to see that it is doing what you expect it to do.
32. True Collection of data without analysis, interpretation, and use of results is a waste of time and money.
33. True Operators are likely to draw the line of best fit slightly differently as each operator will visualize a slightly different line (slope).

34. True Well-kept records will make the task of writing treatment plant cost and efficiency reports much easier.

35. True Records at most wastewater treatment facilities are kept daily and summarized on a monthly basis.

36. True You and your plant operation are partially judged by the information contained in your report, its style, and its appearance.

37. True The utility manager must provide a safer, cleaner work environment while constantly training and retraining operators to understand new technologies.

38. True Establishing safe work procedures for new employees is a very important function of management.

39. True Good communication is an essential part of good management skills.

40. True Every utility needs an effective operations and maintenance program.

41. True The supervisor who has the responsibility for directing work activities must be safety conscious.

42. True Each operator must think, act, and promote safety if the utility is to achieve a good safety record.

Best Answer

1. 4 The proper dissolved oxygen (DO) level must be maintained in an aeration tank so the solids will settle properly and the plant effluent will be clear.

2. 3 Weirs installed at the outlet or effluent end of the aeration tank should be level to prevent short-circuiting and an uneven distribution of solids in the effluent.

3. 4 The mixed liquor suspended solids in the aeration tank can be increased by reducing the sludge wasting rate.

4. 4 Bound water is water contained within the cell mass of sludges or strongly held on the surface of colloidal particles.

5. 2 Raw sludge and scum are the settled sludge solids that are removed from the bottom and the floating scum removed from the top of the clarifiers.

6. 4 The purpose of a vacuum relief valve on an anaerobic sludge digester is to relieve excessive vacuums so the digester cover will not collapse.

7. 4 The volatile acid/alkalinity relationship is the key to successful digester operation.

8. 3 Successful operation of a surface discharge system depends on conscientious operators using proper techniques for sampling and analysis.

9. 1 Pumped discharge systems should have standby generators available for emergency operation when there is a power loss.

10. 3 A representative sample is a sample portion of material, water, or wastestream that is as nearly identical in content and consistency as possible to that in the larger body being sampled.

11. 2 Personal hygiene is the best protection against the risk of infections and infectious diseases.

12. 1 The ideal method of dealing with any high-noise environment is the elimination or reduction of all sources through feasible engineering or administrative controls.

13. 4 The apparent cause of drowning when someone falls into a diffused aeration tank is that there is nothing to grab hold of to keep afloat or to pull oneself out of the aerator.

14. 2 If your hands may come in contact with wastewater or sludge during sampling, wear disposable, impervious gloves to prevent the spread of disease.

15. 3 Class B fires consume flammable and combustible liquids, such as gasoline, oil and grease.

16. 4 A good maintenance program is essential for a wastewater treatment plant to operate continuously at peak design efficiency.

17. 2 Pipelines that carry chlorine gas can be purged to remove moisture by using dry air or nitrogen.

18. 3 Very few maintenance operators do the actual electrical repairs or troubleshooting because this is a highly specialized field and unqualified people can seriously injure themselves and damage costly equipment.

19. 3 Voltage is the electrical pressure available to cause a flow of current when an electric circuit is closed.

20. 2 A fuse is a protective device having a strip or wire of fusible metal which, when placed in a circuit, will melt and break the electric circuit when subjected to an excessive temperature.

21. 2 When two or more pumps of the same size are installed, they should be operated alternately to equalize wear, keep the motor windings dry, and distribute lubricant in the bearings.

22. 4 Flow measurement transmitter instruments send the flow variable to another device for conversion to a usable number.

23. 1 Operators should review the plans and specifications for pumps and lift stations to be sure adequate provisions have been made for the station pumps, equipment, and instrumentation to be easily and properly operated and maintained.

24. 2 In chemistry, normality (*N*) refers to solution strength.

25. 3 The digester supernatant solids test measures the percent of settleable solids being returned to the plant headworks.

26. 2 Chlorine residual should be removed before final discharge to a receiving stream for protection of fish and other aquatic life.

27. 4 Aseptic is free from the living germs of disease, fermentation, or putrefaction. Sterile.

28. 4 The first question an operator should ask when considering the possibility of using a computer or expanding the use of computers is, "How can I use a computer to help me do a better job?"

29. 1 A SCADA system collects, stores, and analyzes information about all aspects of O & M, transmits alarm signals when necessary, and allows fingertip control of alarms, equipment, and processes.

30. 3 The median is the middle measurement when the measurements are ranked in order of magnitude (size).

31. 3 A normal distribution of data is when the mean, median, and mode values are approximately the same.

32. 3 Report writing is the main means by which operators who have information communicate with those who need it.

33. 2 Recorded data should enable the operator to determine how to operate and maintain the plant.

34. 2 Planning is determining the goals, policies, procedures, and other elements to achieve the goals and objectives of the agency.

35. 3 Authority is the power and resources to do a specific job or to get that job done.

36. 3 The financial stability for a utility is determined by calculating the operating ratio and the coverage ratio.

37. 2 The operating ratio is a measure of the total revenues divided by the total operating expenses.

38. 4 SCADA is a computer-monitored alarm, response, control, and data acquisition system used to monitor and adjust the operation of equipment.

39. 3 The purpose of a safety policy statement is to let employees know that the safety program has the full support of the agency and its management.

Multiple Choice

1. 1, 2 When wastewater enters an activated sludge plant, the preliminary treatment processes remove coarse or heavy solids (grit) and other debris, such as roots, rags, and boards.

2. 1, 2, 5 An unsuitable environment for the activated sludge process can be caused by a failure to supply enough oxygen; high concentrations of acids, bases, and other toxic substances; and uneven flows of wastewater, which can cause overfeeding or starvation.

3. 1, 2, 3, 4, 5 Information that should be recorded when testing pumps includes the date of the test, the location of the pump, the name of the operator, the pump discharge (flow), and the pump discharge pressure.

4. 2, 3, 4, 5 An activated sludge plant could become upset due to changes in the sampling program, flow or waste changes, high solids in the digester supernatant, and temperature changes.

5. 1, 2, 3, 4, 5 The possible causes of sludge bulking include: filamentous organisms growing from one floc mass to another; low DO; low nutrient concentrations; low pH; and production of a highly jelly-like, waterlogged sludge that has a very low sludge density.

6. 1, 2, 3, 4, 5 Before starting a new sequencing batch reactor (SBR) plant, the following items should be checked: (1) be sure the plant has an up-to-date O & M manual and a set of as-built plans (record drawings); (2) check and flush pipe systems and remove any refuse or debris left in them during construction; (3) ensure protective safety guards are installed on rotating equipment, open tanks, and channels; (4) make sure tanks are clean; be sure all debris, boards, ladders, and tools have been removed; and (5) to the extent possible, operate and test all equipment before admitting wastewater for treatment.

7. 1, 2, 4, 5 Operators can achieve good digester operation by controlling the loading rate or food supply (organic solids), mixing, temperature, and the volatile acid/alkalinity relationship.

8. 1, 3 A scum blanket can be controlled in a digester by adequate mixing and by heat.

9. 1, 3, 4 Struvite scale sometimes forms in anaerobic digesters and in downstream digested sludge concentration equipment and sludge handling equipment.

10. 1, 2, 3, 4 When operators are making their rounds inspecting their plant, they should be alert and investigate and record anything that feels different (hotter or vibrating more), looks different or unusual, smells different, or sounds different.

11. 1, 3, 4 The types of problems encountered when operating aerobic digesters include floating sludge, odors, and scum.

12. 1, 2, 3, 4, 5 When reviewing the specifications for sludge digestion and solids handling facilities, operators should check the following: an adequate supply of equipment operation and maintenance manuals, equipment warranties and responsibility for acceptance testing, instrumentation—remote and local control board items and recorders are provided, paints and protective coatings, and performance requirements and capabilities.

13. 1, 2, 4 Prior to discharge to surface waters, wastewater must be treated to prevent nuisances due to odors or unsightliness, to prevent the wastewater from interfering with the many uses of surface waters, and to protect the health of the people who may come in contact with it.

14. 1, 2, 3, 4, 5 Operating procedures that should be developed for the start-up of a plant include inspecting the discharge line, where possible, looking at the water surface over the diffuser or outfall pipe, observing the flow entering the discharge system for visible pollutants, opening all appropriate valves, and starting the pumps (unless it is a gravity system).

15. 1, 2, 3, 4 Floatables in pond effluents could be caused by excessive floatables and scum on the surface, excessive solids accumulated on the bottom gasifying and floating to the surface, excessive velocity or insufficient detention time, and the outlet baffle not being at the proper location.

16. 3, 4 The frequency of sampling may be increased when the receiving waters are not meeting established water quality standards or when the treatment plant effluent is not meeting discharge requirements.

17. 1, 2, 3, 4, 5 Whether working on the collection system or working in a treatment plant, an operator may be exposed to hazards including confined spaces, infections and infectious diseases, oxygen deficiency or enrichment, physical injuries, and toxic or suffocating gases or vapors.

18. 2, 3, 4 A confined space may be defined as any space that meets specific conditions including having limited or restricted means for entry or exit, being large enough and so configured that an employee can bodily enter and perform assigned work, and not being designed for continuous employee occupancy.

19. 1, 2, 5 The hazard warnings on containers of hazardous chemicals at your plant should contain emergency guide information on flammability, health, and reactivity.

20. 1, 2, 3, 4, 5 The plant's fire prevention plan could include the following: (1) developing written response procedures for reacting to a fire situation, to include evacuation; (2) providing periodic cleanup of weeds or other vegetation in and around the plant; (3) providing required service on all fire detection and response equipment (inspection, service, hydrostatic testing); (4) regulating the use, storage, and disposal of all combustible materials/substances; and (5) routinely inspecting fire doors to ensure proper operation and unobstructed access.

21. 2, 3, 4, 5 The primary use for polymers in wastewater treatment is the conditioning of sludge to facilitate removal of water in subsequent treatment processes such as belt filter presses, centrifuges, dissolved air flotation thickeners, and gravity belt thickeners.

22. 1, 2, 3, 4, 5 Working safely in the laboratory involves always checking labels on bottles to make sure that the proper chemical is selected, never handling chemicals with your bare hands, never using laboratory glassware for a cup or food dish, using care in making rubber-to-glass connections, and using proper safety goggles or a face shield in all tests where there is danger to the eyes.

23. 1, 2, 3, 4, 5 Paperwork items that can be helpful in identifying the causes of accidents and developing corrective procedures include accident report forms, facility plans and specifications and plant inspection reports, the plant's safety policy, the plant's safety rules, and supervisors' guidelines on how to promote and implement a safety plan.

24. 4, 5 Factors that could cause a pump to run dry include a lack of proper priming when starting and a loss of suction when operating.

25. 1, 2, 3, 4, 5 Primary sources of unsatisfactory AC brush and ring performance include disturbing external conditions, improper load or service condition, mechanical fault in the machine, poor machine design, and poor preparation and care of the machine.

26. 1, 2, 3, 4, 5 When performing pump station electrical equipment maintenance tasks, items that should be checked and inspected annually include all panel instruments, all switch gear and distribution equipment, enclosures, fuses and circuit breakers, and wiring integrity.

27. 1, 2, 3, 4, 5 Causes of motor malfunction or failure include bearing failures, contaminants, old age, overload (thermal), and single phasing.

28. 1, 2, 3, 4, 5 When troubleshooting any motor control that has developed trouble, the qualified maintenance operator should be able to do the following: (1) examine all unusual factors, (2) find the cause of the problem, (3) know what should happen when a switch is pushed, (4) make a visual inspection, and (5) repair the problem and eliminate the cause, if possible.

29. 1, 3, 4, 5 Plugged pipelines are encountered in lines transporting digested sludge, grit, raw sludge, and scum.

30. 4, 5 Sludge lines will plug more often when scum and raw sludge are pumped through the same line and when stormwaters carry in grit and silt that are not effectively removed by the grit removal facilities.

31. 1, 2, 3, 4, 5 When examining lift station prints, operators should ensure adequate provisions have been made for the following items: (1) floors should be sloped to provide drainage, where needed, and drains should be located in low spots; (2) overhead clearances of power lines, trees, and roofs should be adequate for a crane to remove large equipment; (3) the alarm system should signal high water levels in the wet well and water on the floor of the dry well; (4) there should be room to use hydrolifts, cranes, and high-velocity cleaners, as needed; and (5) there should be sufficient room to set up portable pumping units or other necessary equipment in cases of major station failures or disasters.

32. 1, 2, 3, 4, 5 When working with chemicals and other materials in the wastewater treatment plant laboratory, certain hazards can be minimized by using proper techniques and equipment. These include the hazards of burns, electric shock, infectious materials, poisons, and toxic fumes.

33. 1, 2, 4, 5 Hydrogen sulfide gas must be measured in the atmosphere because it causes corrosion and odors and because it is flammable and explosive under certain conditions and toxic to your respiratory system.

34. 1, 2, 3, 4, 5 When measuring the chemical oxygen demand (COD) of a sample, precautions to consider include the following: (1) flasks and condensers should be clean and free from grease or other oxidizable materials, (2) the standard ferrous ammonium sulfate solution is unstable and should be standardized daily, (3) the wastewater sample should be well mixed, (4) use a wide-tip pipet to ensure a representative sample is added, and (5) use extreme caution and safety precautions when handling the chemicals used for the test.

35. 1, 2, 3, 4, 5 When using a dissolved oxygen (DO) probe, precautions to consider include the following: (1) dissolved inorganic salts, such as found in seawater, can influence the readings from a probe; (2) do not place the probe directly over a diffuser because you want to measure the dissolved oxygen in the water being treated, not the oxygen in the air supply to the aerator; (3) keep the membrane in the tip of the probe from drying out; (4) periodically check the calibration of the probe; and (5) reactive compounds, such as reactive gases and sulfur compounds, can interfere with the output of a probe.

36. 3, 4, 5 Sulfate ions in wastewater are indirectly responsible for odor, sewer corrosion problems, and toxic gas.

37. 1, 2, 3, 4, 5 A few of the general types of computer applications in wastewater treatment plants are communications, data analysis, recordkeeping, report generation, and systems monitoring.

38. 1, 2, 3, 4, 5 If the proper lab results are entered into the computer, the computer can calculate and print out chemical dosages, F/Ms (Food to Microorganism ratios), MCRTs (Mean Cell Residence Times), percent removals, and suspended solids, BOD, or COD loadings.

39. 1, 2, 4, 5 General purpose software packages typically include the following programs: database management, graphics, spreadsheet, and word processing programs.

40. 1, 2, 3, 4, 5 Items that can typically be monitored, recorded, and controlled by computers in a wastewater treatment plant include aeration basin dissolved oxygen, anaerobic sludge digester gas production, chlorine contact time, influent flow control gates and valves, and secondary clarifier sludge pumping.

41. 1, 3, 4, 5 When preparing to plot data in a bar graph, information that must first be determined is the class midpoint, the measurement frequencies, the spread of data, and the width of the class interval.

42. 2, 3, 4, 5 Records are important for the following reasons: they are needed to show the type and frequency of maintenance and to evaluate the effectiveness of maintenance programs; they can indicate plant efficiency, performance, past problems, and potential problems; they can provide data upon which to base recommendations for modifying plant operation and facilities; and they may be needed in damage suits brought against the district or municipality.

43. 1, 2, 3, 4, 5 A list of the minimum daily records kept at a fairly large treatment plant with digester tanks may include chlorine use (lbs or kg), influent and effluent temperatures, pH of influent and effluent, precipitation and air temperature, and raw wastewater flow (MGD from totalizer).

44. 1, 2, 4, 5 The basic principles common to good report writing are knowing the purpose and objective of your report, knowing your subject, organizing the report to present your ideas in a logical order, and tailoring your report to the person or persons to whom it is directed.

45. 1, 2, 3, 4, 5 The most important guideline for preparing a report is to begin writing. Other guidelines include arranging the ideas in a logical sequence, gathering material needed to support the ideas to be presented in the report, listing ideas and topics you plan to cover in the report, and preparing conclusions and recommendations after writing the main body of the report.

46. 1, 3, 4 The utility manager's staffing responsibilities include evaluating employee performance, providing opportunities for advancement for operators and staff, and selecting and training employees.

47. 1, 2, 3, 4, 5 Important steps in the process of hiring and selecting a new operator include advertising the position, avoiding unacceptable preemployment inquiries, interviewing applicants, paper screening, and selecting the most qualified candidate.

48. 1, 2, 3, 4, 5 When dealing with a confrontation regarding employee performance issues, the employee and the supervisor should aim for a permanent solution, create a private environment, keep their language appropriate, listen very carefully, and maintain an adult approach.

49. 2, 3, 4, 5 The total revenue of a utility includes revenue generated by hook-up charges, interest income, taxes or assessments, and user fees.

50. 1, 2, 3, 5 Predictive maintenance is: (1) a method of establishing baseline performance data; (2) monitoring performance criteria over a period of time; (3) observing changes in performance so that failure can be predicted and maintenance can be performed on a planned, scheduled basis; and (4) the same as proactive maintenance.

51. 1, 2, 3, 5 A supervisor can be partially or fully responsible for an accident by causing unsafe acts to take place, by disregarding an unsafe work environment, by overlooking or failing to consider any number of safety hazards, and by requiring work to be performed in haste.

Short Answer

1. Activated sludge consists of sludge particles produced in raw or settled wastewater (primary effluent) by the growth of organisms in aeration tanks in the presence of dissolved oxygen.

2. Activated sludge organisms in the final clarifier are in a deteriorating condition because of a lack of oxygen and food and should be returned to the aeration tank as quickly as possible.

3. Maintenance activities in a collection system can cause operational problems by clearing a pipeline blockage or holding flows and then releasing them immediately after a pumping station has been put back on the line. Releases of large volumes of septic wastewater could cause a shock load on the treatment processes.

4. Clean air is essential for the protection of:

   1. Blowers

      a. Large objects entering the impellers or lobes may cause severe damage.
      b. Deposits on the impellers or lobes reduce clearances and cause excessive wear and vibration problems.

   2. Process systems

      a. Clean air is required to protect downstream equipment.
      b. Clean air prevents fouling of air conduits, pipes, tubing, or dispersing devices on diffusers.

5. Proper procedures must be followed when checking equipment to ensure that all items are checked and to prevent damaging the equipment. Damage caused by use of improper procedures will not be covered by the manufacturer's warranty and your agency will bear the costs of repairs and replacement parts.

6. Assistance and advice on starting a new activated sludge plant may be available from the design engineer, vendors, nearby operators, or other specialists.

7. Essential laboratory tests and on-site tests recommended when starting the activated sludge process and what they are used for include the following:

Test	What the Test is Used For
Mixed Liquor Suspended Solids	Observe solids buildup in aerator and develop proper food to organisms (solids) ratio.
Dissolved Oxygen	Maintain aerobic conditions in aerator.
30-Minute Settleability Test	Observe settling characteristics of activated sludge.

8. Some activated sludge must be wasted to prevent an excessive solids buildup in the aerator.

9. When attempting to correct an upset activated sludge process, only one major change should be made at a time in order to evaluate the influence of the change. If more than one change is made, the operator will not know which change was effective or ineffective.

10. If proper equipment shutdown procedures are not followed, equipment can be damaged and not start again properly.

11. Activated sludge sequencing batch reactors (SBRs) are popular for smaller industrial plants that must produce high-quality effluents and, in addition, may be required to achieve nutrient removal of both nitrogen and phosphorus.

12. During the construction of a sequencing batch reactor (SBR), the operator can learn flow system routes, pipe locations and burial depths, valve placement, and locations of protective devices such as thrust blocks or corrosion prevention systems. Also, the operator can observe the tools and methods of lifting or placing equipment that will be necessary for overhauls or repairs.

13. The person responsible for making process changes should have some knowledge of the microorganisms that are being controlled in the process.

14. The contents of a digester should be kept well mixed in order to use as much of the digester volume as possible, to distribute raw sludge (food) throughout the tank, to put microorganisms in contact with the food, to dilute inhibitory by-products from microorganisms, to distribute alkalinity buffer throughout the tank, to distribute heat as evenly as possible, and to keep grit and inert solids from building up on the tank bottom.

15. Digester gas is considered dangerous for two reasons:
    1. Digester gas can form an explosive mixture when mixed with air.
    2. Digester gas can be present in areas where there is insufficient oxygen to maintain life; therefore a person could become asphyxiated.
16. The pressure relief valve will operate when the waste gas burner cannot handle excessive gas pressures; valve operation prevents the breaking of the water seal. The vacuum relief valve will operate if sludge or gas is withdrawn from the digester too quickly and a vacuum in the digester develops. The tank could collapse if the vacuum relief valve failed.
17. Seed sludge should be added to a new digester to provide methane fermenters and alkalinity so the digestion process will start and continue in balance.
18. The first warning that trouble is developing in an anaerobic digester is an increase in the volatile acid/alkalinity relationship.
19. Problems with anaerobic digesters can be anticipated by daily inspection of the digester data, plotting the data, and being alert for changes or trends in the wrong direction. Corrective action should be taken whenever trends or changes start in the wrong direction.
20. When applying sludge to a drying bed, care should be taken to be sure that no open flames (including cigarette smoking) are in the vicinity. Also a thick, digested sludge should be applied, not a thin or green sludge.
21. Receiving waters should be sampled and tested to indicate: (1) if the treatment plant is effectively treating the wastewater, and (2) if the receiving water quality standards are being met and the water quality is suitable for the uses of the receiving waters.
22. A "representative sample" is a sample portion of material, water, or wastestream that is as nearly identical in content and consistency as possible to that in the larger body being sampled.
23. After a sample has been collected, a major concern is that the sample water quality indicators being measured will not change before the final analysis.
24. When running tests in the field, the sample should be large enough for the field test. When samples are analyzed in the lab, the volume of sample should be twice the quantity needed for the test to allow for backup or repeat tests.
25. The operator's responsibility with regard to safety includes making certain that the facility is maintained in such a manner as to continually provide a safe place to work.
26. Job site protection when working in manholes should include barricades and traffic warning devices for the safety of vehicles, bikes, pedestrians, and workers.
27. When cleaning racks and screens, the operator should stand on a firm, clean surface.
28. Smoking or open flames should not be allowed in the vicinity of the digester or sludge digestion system because they provide a source of flame that could ignite an explosive mixture of methane and air.
29. The jagged ends of glass tubes should be smoothed by flame polishing.
30. Plant water supplies should be checked monthly for coliform group bacteria to be certain the plant water is safe to drink and that any backflow prevention devices are working properly.
31. The operator should thoroughly read and understand manufacturers' literature before attempting to maintain plant equipment in order to do the job required to keep the equipment operating.
32. One person should never attempt to repair a chlorine or sulfur dioxide leak alone because the person could be overcome by fumes. Valuable time, needed to repair and correct a serious emergency, could be lost rescuing a foolish person.
33. To determine if a pump will turn in the direction intended, momentarily start the motor by a quick electrical contact and check to be sure the motor will turn the pump in the direction indicated by the rotational arrows marked on the pump.
34. Motor nameplate data must be recorded and filed so the information is available when needed to repair the motor or to obtain replacement parts.
35. When two or more pumps of the same size are installed, they should be operated alternately to equalize wear, keep motor windings dry, and distribute lubricant in bearings.
36. To check the temperature of a motor, a thermometer should be placed in the casing near the bearing and held in place with putty or clay or a magnetic thermometer may be used.
37. Plant safety equipment must be maintained on a regular basis so it will always be ready for use when needed.
38. Scum lines can be kept from plugging by an effective preventive maintenance program. Remove scum from troughs, and clean lines at regular intervals (monthly).
39. A flowmeter should be calibrated (checked for accuracy) in place to ensure accurate flow measurements. Most flowmeters are calibrated in the factory, but they also should be checked in their actual field installation. When a properly installed and field-calibrated meter starts to give strange results, check for obstructions in the flow channel and the flow metering device.
40. Chemicals must be properly labeled to avoid poor results and safety hazards.
41. To protect yourself from diseases when working in a wastewater treatment plant, you should be aware of your personal hygiene. You should wash your hands thoroughly with soap and water after contact with wastewater, and again before eating or smoking. Work clothes should not be worn home. Inoculations against certain diseases will provide immunization against them. Do not pipet anything by mouth, and never drink from a beaker or other laboratory glassware.
42. A representative sample is supposed to accurately describe the overall situation; for example, the BOD of water flowing into a treatment plant.
43. The clarity test must always be run under the same conditions to produce comparable results so the operator can tell whether the plant effluent is staying the same, improving, or deteriorating.

44. The dewatering characteristics of digested sludge should be measured to determine when the sludge is ready to be placed on the drying beds. The quicker it will dry, the less area will be needed for drying beds.

45. The COD test is run to estimate the oxygen demand of the wastewater being tested.

46. The purpose of the coliform group bacteria test is to indicate the bacteriological quality of the sample being tested. The results indicate the possible presence of disease-producing organisms that may be found with the coliforms.

47. The following precautions should be exercised when using a DO probe:

    1. Periodically check the calibration of the probe.
    2. Keep the membrane in the tip of the probe from drying out.
    3. Dissolved inorganic salts, such as found in seawater, can influence the readings from a probe.
    4. Reactive compounds, such as reactive gases and sulfur compounds, can interfere with the output of a probe.
    5. Do not place the probe directly over a diffuser because you want to measure the dissolved oxygen in the water being treated, not the oxygen in the air supply to the aerator.

48. The oil and grease test measures gasoline, heavy fuel, lubricating oil, asphalt, soaps, fats, waxes, and any other material that is extracted by the solvent (Freon) used in the test.

49. Operators should use computers because computers can help operators save time, save money, and do a better job. Computers can help operators analyze lab results, maintain equipment, monitor pretreatment programs, control processes, and prepare reports.

50. The justification of the purchase of a computer can be based on a list of applications and cost savings in time and labor. Typical applications include data management and interpretation, recording and scheduling equipment maintenance, pretreatment monitoring, process control, and report generation.

51. The three principal causes of variations in test results are:

    1. Actual variations in the characteristics of the water or material being examined
    2. Sampling procedures
    3. Testing or analytical procedures

52. The solids concentrations of sludge withdrawn from a primary clarifier during the past seven days are given below:

Day:	1	2	3	4	5	6	7
Solids, %:	6.0	6.5	6.0	5.0	6.5	7.5	8.0

Bar graph showing data distribution.

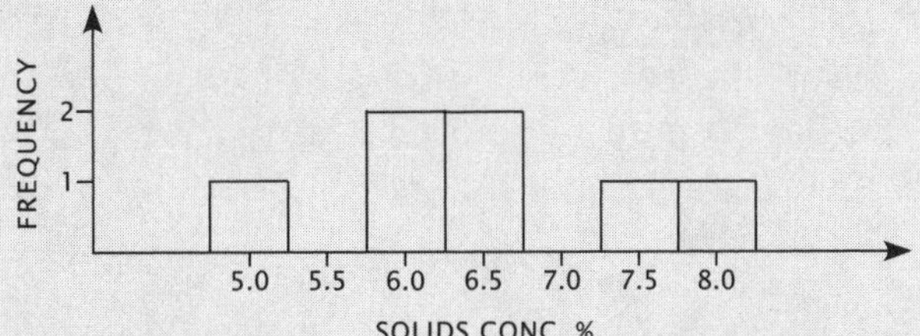

53. Operators must be capable of effective report writing to communicate with management and the general public on the operation of their plant and on requests for additional funds for improvements and personnel.

54. Operators should approach the task of report writing as if their next pay raise depended on a neat, organized, and brief report.

55. The different types of demands on a utility manager include demands from the community, regulatory agencies, and within the utility.

56. A supervisor has successfully delegated when proper job assignments, authority, and responsibility are all present.

57. Supervisors should be trained to communicate effectively, motivate others, delegate responsibility, and hold people accountable for their performance. Supervisors also need to be familiar with applicable state and federal personnel regulations, and should receive training on their role and responsibilities as part of the management team.

58. Discipline problems should be solved quickly. The sooner you deal with the problem, the better the outcome will be.

59. Any time you or a member of your utility comes in contact with the public, you will have an impact on the quality of your public image.

60. Complaints can be a valuable asset in pinpointing wastewater collection system problems. Customer calls are frequently the first indication that something may be wrong. Responding to complaints and inquiries promptly can save the utility money and staff resources.

61. Wastewater utilities, regardless of size, must have a safety program if they are to realize a low frequency of accident occurrence. A safety program also provides a means of comparing frequency, disability, and severity of injuries with other utilities.

62. The safety officer should tell new employees about utility policy, safety reports, past accidents, and orient the new operator toward the importance of safety to operators and to the organization.

Problems

1. How many pounds of solids are in a 400,000-gallon aeration tank if the suspended solids concentration is 2,400 mg/*L*?

Known		Unknown
Tank Volume, MG	= 0.40 MG	Aerator Solids, lbs
ML Suspended Solids, mg/*L*	= 2,400 mg/*L*	

Aerator Solids, lbs = MLSS, mg/*L* × Tank Vol, MG × 8.34 lbs/gal

= 2,400 mg/*L* × 0.40 MG × 8.34 lbs/gal

$$= \frac{2{,}400 \text{ mg}}{\text{M mg}} \times 0.40 \text{ MG} \times 8.34 \text{ lbs/gal}$$

= 2,400 × 3.3

= 8,000 lbs

2. Laboratory tests indicate that the volatile content of a raw sludge was 72% and after digestion the volatile content is 45%. What is the percent reduction in volatile matter?

Known		Unknown
In, % VM in Raw Sludge	= 72%	P, % Reduction of VM
Out, % VM in Dig Sludge	= 45%	

$$P, \% = \frac{(\text{In} - \text{Out})}{\text{In} - (\text{In} \times \text{Out})} \times 100\%$$

$$= \frac{0.72 - 0.45}{0.72 - (0.72 \times 0.45)} \times 100\%$$

$$= \frac{0.27}{0.72 - 0.32} \times 100\%$$

$$= \frac{0.27}{0.40} \times 100\%$$

$$= 0.68 \times 100\%$$

$$= 68\%$$

3. A digester is fed a mixture of 4.5 percent primary sludge flowing at 7,800 GPD and 7.6 percent thickened secondary sludge flowing at 5,500 GPD. What is the approximate concentration of solids fed?

Known		Unknown
Primary Solids, %	= 4.5%	Mixture Solids, %
Primary Flow, GPD	= 7,800 GPD	
Secondary Solids, %	= 7.6%	
Secondary Flow, GPD	= 5,500 GPD	

Determine the solids concentration of the mixture.

$$\text{Mixture Solids, \%} = \frac{(\text{Prim Sol, \%})(\text{Prim Flow, GPD}) + (\text{Sec Sol, \%})(\text{Sec Flow, GPD})}{\text{Prim Flow, GPD} + \text{Sec Flow, GPD}}$$

$$= \frac{(4.5\%)(7{,}800\text{ GPD}) + (7.6\%)(5{,}500\text{ GPD})}{7{,}800\text{ GPD} + 5{,}500\text{ GPD}}$$

$$= \frac{35{,}100 + 41{,}800}{13{,}300}$$

$$= 5.8\%$$

4. A wet well is four feet wide and six feet long. Approximately how far down should the level in the wet well be lowered in six minutes by a pump with a rated capacity of 250 GPM?

Known		Unknown
Pump Capacity, GPM	= 250 GPM	Depth, ft
Time, min	= 6 min	
Width, ft	= 4 ft	
Length, ft	= 6 ft	

$$\text{Pump Capacity, GPM} = \frac{\text{Volume, gallons}}{\text{Time, minutes}}$$

$$= \frac{(\text{Length, ft})(\text{Width, ft})(\text{Depth, ft})(7.48\text{ gal/cu ft})}{\text{Time, minutes}}$$

Therefore,

$$\text{Depth, ft} = \frac{(\text{Pump Capacity, gal/min})(\text{Time, min})}{(\text{Length, ft})(\text{Width, ft})(7.48\text{ gal/cu ft})}$$

$$= \frac{(250\text{ gal/min})(6\text{ min})}{(6\text{ ft})(4\text{ ft})(7.48\text{ gal/cu ft})}$$

$$= 8.36\text{ ft}$$

$$= 8.4\text{ ft}$$

5. A wastewater treatment plant receives a flow of 1.6 MGD. The influent suspended solids are 210 mg/*L* and the effluent suspended solids are 20 mg/*L*. How many pounds of suspended solids are removed per day?

Known		Unknown
Influent SS, mg/*L*	= 210 mg/*L*	SS Removed, lbs/day
Effluent SS, mg/*L*	= 20 mg/*L*	
Flow, MGD	= 1.6 MGD	

Calculate the pounds per day of suspended solids removed.

$$\text{SS Removed, lbs/day} = (\text{Infl SS, mg/}L - \text{Effl SS, mg/}L)(\text{Flow, MGD})(8.34\text{ lbs/gal})$$

$$= (210\text{ mg/}L - 20\text{ mg/}L)(1.6\text{ MGD})(8.34\text{ lbs/gal})$$

$$= 2{,}535\text{ lbs SS Removed/day}$$

6. The temperature of a plant effluent water is 14°C (57.2°F) and the DO at saturation is 10.4 mg/*L*. What is the percent saturation of dissolved oxygen in the water when the DO is 3.8 mg/*L*?

Known		Unknown
Effluent DO, mg/*L*	= 3.8 mg/*L*	DO Saturation, %
Saturation DO, mg/*L*	= 10.4 mg/*L*	

Calculate the effluent dissolved oxygen percent saturation.

$$\text{DO Saturation, \%} = \frac{(\text{DO of Sample, mg/}L)(100\%)}{\text{DO at 100\% Saturation, mg/}L}$$

$$= \frac{(3.8\text{ mg/}L)(100\%)}{10.4\text{ mg/}L}$$

$$= 36.5\%$$

7. What is the BOD of a 5-m*L* sample if the initial DO of the diluted sample was 7.5 mg/*L* and the DO of the diluted sample after five-day incubation was 3.0 mg/*L*?

Known		Unknown
Sample Size, m*L*	= 5 m*L*	BOD, mg/*L*
Initial DO, mg/*L*	= 7.5 mg/*L*	
5-Day DO, mg/*L*	= 3.0 mg/*L*	
BOD Bottle Vol, m*L*	= 300 m*L*	

Calculate the BOD in mg/*L*.

$$\text{BOD, mg}/L = \frac{(\text{Initial DO, mg}/L - \text{5-Day DO, mg}/L)(\text{BOD Bottle Vol, m}L)}{\text{Sample Vol, m}L}$$

$$= \frac{(7.5\text{ mg}/L - 3.0\text{ mg}/L)(300\text{ m}L)}{5\text{ m}L}$$

$$= \frac{(4.5)(300)}{5}$$

$$= 270\text{ mg}/L$$

8. At the beginning of a week, a totalizer on the plant inflow read 4,243,891 gallons. Seven days later, the totalizer read 7,762,438 gallons. What was the average daily flow during the week?

Known		**Unknown**
Total Flow at Start, gal	= 4,243,891 gal	Avg Daily Flow, MGD
Total Flow at End, gal	= 7,762,438 gal	
Time, days	= 7 days	

$$\text{Avg Daily Flow, MGD} = \frac{(\text{Flow at End, gal} - \text{Flow at Start, gal})}{(\text{Time, days})(1{,}000{,}000\text{ gal/MG})}$$

$$= \frac{(7{,}762{,}438\text{ gal} - 4{,}243{,}891\text{ gal})}{(7\text{ days})(1{,}000{,}000\text{ gal/MG})}$$

$$= \frac{3.518547\text{ MG}}{7\text{ days}}$$

$$= 0.503\text{ MGD}$$

9. The solids concentrations of sludge withdrawn from a primary clarifier during the past seven days are given below:

Known		**Unknown**
Day	Solids, %	1. Average Solids, %
1	6.0	2. Range of Solids Conc, %
2	6.5	3. Median Solids Conc, %
3	6.0	4. Mode Solids Conc, %
4	5.0	
5	6.5	
6	7.5	
7	8.0	

1. Calculate the percent average solids.

$$\text{Average Solids, \%} = \frac{\text{Sum of Measurements, \%}}{\text{Number of Measurements}}$$

$$= \frac{45.5\%}{7}$$

$$= 6.5\%$$

2. Calculate the range of solids concentrations in percent.

$$\text{Range of Solids Conc, \%} = \text{Largest X} - \text{Smallest X}$$

$$= 8.0\% - 5.0\%$$

$$= 3.0\%$$

3. Calculate the median solids concentration in percent.

$$\text{Median Solids Conc, \%} = \text{Middle Measurement of Ranked Data}$$

$$= 6.5\%$$

4. Calculate the mode solids concentration in percent.

$$\text{Mode Solids Conc, \%} = \text{Measurement Occurring Most Frequently}$$

$$= 6.5\%\text{ or }6.0\%$$

APPENDIX

HOW TO SOLVE WASTEWATER TREATMENT PLANT ARITHMETIC PROBLEMS

(VOLUME II)

TABLE OF CONTENTS

HOW TO SOLVE WASTEWATER TREATMENT PLANT ARITHMETIC PROBLEMS

HOW TO SOLVE WASTEWATER TREATMENT PLANT ARITHMETIC PROBLEMS (VOLUME II)

A.1 BASIC CONVERSION FACTORS (ENGLISH SYSTEM)

UNITS

1,000,000	= 1 Million	1,000,000/1 Million

LENGTH

12 in	= 1 ft	12 in/ft
3 ft	= 1 yd	3 ft/yd
5,280 ft	= 1 mi	5,280 ft/mi

AREA

144 sq in	= 1 sq ft	144 sq in/sq ft
43,560 sq ft	= 1 acre	43,560 sq ft/ac

VOLUME

7.48 gal	= 1 cu ft	7.48 gal/cu ft
1,000 m*L*	= 1 liter	1,000 m*L*/*L*
3.785 *L*	= 1 gal	3.785 *L*/gal
231 cu in	= 1 gal	231 cu in/gal

WEIGHT

1,000 mg	= 1 gm	1,000 mg/gm
1,000 gm	= 1 kg	1,000 gm/kg
454 gm	= 1 lb	454 gm/lb
2.2 lbs	= 1 kg	2.2 lbs/kg

POWER

0.746 kW	= 1 HP	0.746 kW/HP

DENSITY

8.34 lbs	= 1 gal	8.34 lbs/gal
62.4 lbs	= 1 cu ft	62.4 lbs/cu ft

DOSAGE

17.1 mg/*L*	= 1 grain/gal	17.1 mg/*L*/gpg
64.7 mg	= 1 grain	64.7 mg/grain

PRESSURE

2.31 ft water	= 1 psi	2.31 ft water/psi
0.433 psi	= 1 ft water	0.433 psi/ft water
1.133 ft water	= 1 in Mercury	1.133 ft water/in Mercury

FLOW

694 GPM	= 1 MGD	694 GPM/MGD
1.55 CFS	= 1 MGD	1.55 CFS/MGD

TIME

60 sec	= 1 min	60 sec/min
60 min	= 1 hr	60 min/hr
24 hr	= 1 day	24 hr/day*

* This may be written either as 24 hr/day or 1 day/24 hours depending on which units we wish to convert to obtain our desired results.

A.2 BASIC FORMULAS

ACTIVATED SLUDGE

1. MLSS, lbs = (Aer Vol, MG)(MLSS, mg/*L*)(8.34 lbs/gal)

2a. $\text{Settleable Solids, \%} = \frac{\text{(Settleable Solids, mL)(100\%)}}{\text{Sample Volume, mL}}$

2b. $\text{Return Sludge Rate, MGD} = \frac{\text{(Total Flow, MGD)(Set Sol, \%)}}{100\%}$

3a. Solids in Aerator, lbs = (Aerator Vol, MG)(MLSS, mg/*L*)(8.34 lbs/gal)

3b. Solids Added, lbs/day = (Flow, MGD)(PE SS, mg/*L*)(8.34 lbs/gal)

3c. $\text{Sludge Age, days} = \frac{\text{Suspended Solids in Aerators, lbs}}{\text{Solids Added by PE, lbs/day}}$

4a. Desired MLSS, lbs = (Sludge Age, days)(Solids Added, lbs/day)

4b. $\text{Desired MLSS, mg/}L = \dfrac{\text{Desired MLSS, lbs}}{\text{(Aerator Vol, MG)(8.34 lbs/gal)}}$

5a. $\text{Change in WAS Pumping, MGD} = \dfrac{\text{(Actual MLSS, lbs − Desired MLSS, lbs)/day}}{\text{(Waste Sludge Conc, mg/}L\text{)(8.34 lbs/gal)}}$

5b. New WAS Pumping, GPM = Current WAS Pumping, GPM + Change in WAS Pumping, GPM

6a. Aerator Loading, lbs COD*/day = (Flow, MGD)(PE COD*, mg/L)(8.34 lbs/gal)

6b. $\text{MLVSS, lbs} = \dfrac{\text{Aerator Loading, lbs COD*/day}}{\text{Loading Factor, lbs COD*/day/lb MLVSS}}$

6c. $\text{MLVSS, mg/}L = \dfrac{\text{MLVSS, lbs}}{\text{(Aerator Vol, MG)(8.34 lbs/gal)}}$

6d. $\text{Food/Microorganism} = \dfrac{\text{Aerator Loading, lbs COD*/day}}{\text{(Aerator Vol, MG)(MLVSS, mg/}L\text{)(8.34 lbs/gal)}}$

7. $\text{MCRT, days} = \dfrac{\text{Suspended Solids in Aeration System, lbs}}{\text{SS Wasted, lbs/day + SS Lost, lbs/day}}$

8a. $\text{WAS, lbs/day} = \dfrac{\text{SS in Aeration System, lbs}}{\text{MCRT, days}} - \text{SS Lost, lbs/day}$

8b. $\text{WAS Pumping, MGD} = \dfrac{\text{WAS, lbs/day}}{\text{(WAS SS, mg/}L\text{)(8.34 lbs/gal)}}$

SLUDGE DIGESTION

9. $\text{Seed Sludge, gal} = \dfrac{\text{(Digester Volume, gal)(Seed Sludge, \%)}}{100\%}$

10a. $\text{Volatile Solids Pumped, lbs/day} = \dfrac{\text{(Raw Sludge, GPD)(Raw Sl Sol, \%)(Volatile, \%)(8.34 lbs/gal)}}{(100\%)\quad(100\%)}$

10b. $\text{Seed Sludge, lbs Volatile Sol} = \dfrac{\text{Volatile Solids Pumped, lbs VS/day}}{\text{Loading Factor, lbs VS/day/lb VS in Digester}}$

10c. $\text{Seed Sludge, gallons} = \dfrac{\text{Volatile Solids Pumped, lbs VS/day}}{\text{(Seed Sludge, lbs/gal)}\left(\dfrac{\text{Solids, \%}}{100\%}\right)\left(\dfrac{\text{VS, \%}}{100\%}\right)}$

11. Lime Required, lbs = (Sludge Volume, MG)(Volatile Acids, mg/L)(8.34 lbs/gal)

12. Piston Pump Vol, gal/stroke = (0.785)(Diameter, ft)^2(Distance, ft/stroke)(7.5 gal/cu ft)

13a. $\text{Dry Solids, lbs} = \dfrac{\text{(Raw Sludge, gal)(Raw Sludge, \%)(8.34 lbs/gal)}}{100\%}$

13b. $\text{Volatile Solids, lbs} = \dfrac{\text{(Dry Solids, lbs)(Raw Sludge, \% VS)}}{100\%}$

14. $\text{Reduction of Volatile Solids, \%} = \dfrac{\text{(In − Out)(100\%)}}{\text{In − (In} \times \text{Out)}}$

15. $\text{Digester Loading, lbs VS/day/cu ft} = \dfrac{\text{Volatile Solids Added, lbs/day}}{\text{Digester Volume, cu ft}}$

16. $\text{Digested Sludge in Storage, lbs} = \text{(VS Added, lbs/day)}\left(\dfrac{\text{Loading, lbs Dig Sl}}{\text{lbs VS/day}}\right)$

---

* Some plants use BOD instead of COD, but you have to wait five days for results.

17. VS Destroyed, lbs/day/cu ft = $\frac{\text{(VS Added, lbs/day)(VS Reduction, \%)}}{\text{(Digester Volume, cu ft)(100\%)}}$

18. Gas Production, cu ft/lb VS = $\frac{\text{Gas Produced, cu ft/day}}{\text{VS Destroyed, lbs/day}}$

19. Solids Balance Water Change, lbs = Water In, lbs − Water Out, lbs − Supernatant Out, lbs

EFFLUENT DISCHARGE, RECLAMATION, AND REUSE

20. BOD Load, lbs BOD/day = (Flow, MGD)(BOD, mg/*L*)(8.34 lbs/gal)

21. Average BOD, mg/*L* = $\frac{\text{Sum of Measurements, mg}/L}{\text{Number of Measurements}}$

MAINTENANCE

22. Pump Capacity, GPM = $\frac{\text{Volume Pumped, gallons}}{\text{Pumping Time, minutes}}$

23a. Velocity, ft/sec = $\frac{\text{Distance, ft}}{\text{Time, sec}}$

23b. Flow, cu ft/sec = (Area, sq ft)(Velocity, ft/sec)

24. Pitot Tube V, ft/sec = $\sqrt{2(32.2 \text{ ft/sec}^2)(\text{H, ft})}$

LABORATORY

25a. Temperature, °F = (Temperature, °C)(9/5) + 32°F

25b. Temperature, °C = (Temperature, °F − 32°F)(5/9)

26. Sludge Pumped, GPD = (Sludge Removed, m*L*/*L*)(1,000 mg/m*L*)(Flow, MGD)

27a. Total Susp Solids, mg/*L* = $\frac{\text{(Dry Weight, mg)(1,000 m}L/L)}{\text{Sample Volume, m}L}$

27b. Volatile Susp Solids, mg/*L* = $\frac{\text{(Volatile Weight, mg)(1,000 m}L/L)}{\text{Sample Volume, m}L}$

27c. Volatile SS, % = $\frac{\text{(Volatile SS, mg}/L)(100\%)}{\text{Total SS, mg}/L}$

27d. Fixed Susp Solids, mg/*L* = $\frac{\text{(Ash Weight, mg)(1,000 m}L/L)}{\text{Sample Volume, m}L}$

27e. Fixed SS, % = $\frac{\text{(Fixed SS, mg}/L)(100\%)}{\text{Total SS, mg}/L}$

28. Removal, % = $\frac{\text{(In − Out)(100\%)}}{\text{In}}$

29. Suspended Solids Removed, lbs/day = (Flow, MGD)(SS Removed, mg/*L*)(8.34 lbs/gal)

30. SVI = $\frac{\text{(Set Sol, \%)(10,000)}}{\text{MLSS, mg}/L}$

31. CO_2, % = $\frac{\text{(Total Volume, m}L - \text{Gas Remaining, m}L)(100\%)}{\text{Total Volume, m}L}$

32. DO Saturation, % = $\frac{\text{(DO of Sample, mg}/L)(100\%)}{\text{(DO at Saturation, mg}/L)}$

33. BOD Sample Size, m*L* = $\frac{1{,}200}{\text{Estimated BOD, mg}/L}$

34. BOD, mg/*L* = $\left[\text{Initial DO of Diluted Sample, mg}/L - \text{DO of Sample After 5 days, mg}/L\right]\left[\frac{\text{BOD Bottle Vol, m}L}{\text{Sample Vol, m}L}\right]$

DATA ANALYSIS

35a. Mean $= \frac{\text{Sum of All Measurements}}{\text{Number of Measurements}}$

35b. Median = Middle Measurement

35c. Mode = Measurement That Occurs Most Frequently

35d. Range = Largest Measurement – Smallest Measurement

A.3 TYPICAL WASTEWATER TREATMENT PLANT PROBLEMS (ENGLISH SYSTEM)

A.30 Activated Sludge

EXAMPLE 1

In a conventional activated sludge plant, the mixed liquor suspended solids were 2,400 mg/*L*. How many pounds of solids were under aeration if the aeration tank is 120 feet long, 30 feet wide, and 12 feet deep?

Known		Unknown
MLSS, mg/*L*	= 2,400 mg/*L*	MLSS, lbs
Aeration Tank Length, ft	= 120 ft	
Width, ft	= 30 ft	
Depth, ft	= 12 ft	

1. Determine the aeration tank volume in cubic feet.

$$\text{Tank Volume, cu ft} = (\text{Length, ft})(\text{Width, ft})(\text{Depth, ft})$$
$$= (120\text{ ft})(30\text{ ft})(12\text{ ft})$$
$$= 43{,}200\text{ cu ft}$$

2. Convert aeration tank volume from cubic feet to million gallons.

$$\text{Volume, gal} = (\text{Volume, cu ft})(7.48\text{ gal/cu ft})$$
$$= (43{,}200\text{ cu ft})(7.48\text{ gal/cu ft})$$
$$= 323{,}136\text{ gal}$$
$$\text{Volume, MG} = (\text{Volume, gal})(1\text{ M}/1{,}000{,}000)$$
$$= \frac{(323{,}136\text{ gal})(1\text{ M})}{1{,}000{,}000}$$
$$= 0.323\text{ MG}$$

3. Calculate the pounds of solids under aeration.

$$\text{MLSS, lbs} = (\text{Tank Vol, M Gal})(\text{MLSS, mg}/L)(8.34\text{ lbs/gal})$$
$$= (0.323\text{ MG})(2{,}400\text{ mg}/L)(8.34\text{ lbs/gal})$$
$$= (2.69)(2{,}400\text{ mg}/L)$$
$$= 6{,}465\text{ lbs}$$

EXAMPLE 2

Estimate the proper return sludge pumping rate in gallons per minute (GPM) for the activated sludge plant in Example 1 if the flow from the primary clarifiers is 2.0 MGD and the results from the 30-minute settleability test indicate that 225 m*L* of sludge settled in a 1-liter sample. Current return sludge pumping rate is 0.5 MGD.

Known		Unknown
Flow, MGD	= 2.0 MGD	Return Sludge Pumping Rate, GPM
Return Flow, MGD	= 0.5 MGD	
Set Sol, m*L*	= 225 m*L*	
Sample Vol, *L*	= 1 *L*	

Estimate the return sludge pumping rate in MGD and GPM.

$$\text{Return Sludge Rate, MGD} = \frac{(\text{Set Sol, m}L)(\text{Flow, MGD} + \text{Return Flow, MGD})}{(1{,}000\text{ m}L - \text{Set Sol, m}L)}$$
$$= \frac{(225\text{ m}L)(2\text{ MGD} + 0.5\text{ MGD})}{(1{,}000\text{ m}L - 225\text{ m}L)}$$
$$= \frac{(225\text{ m}L)(2.5\text{ MGD})}{775\text{ m}L}$$
$$= 0.73\text{ MGD or } 730{,}000\text{ GPD}$$
$$\text{Return Sludge Rate, GPM} = \frac{730{,}000\text{ GPD}}{1{,}440\text{ min/day}}$$
$$= 504\text{ GPM}$$

NOTE: The actual return sludge pumping rate of 0.5 MGD (350 GPM) is a little low.

EXAMPLE 3

Determine the sludge age for an activated sludge plant with an aeration tank volume of 0.323 million gallons that treats a flow of 2.0 MGD. The primary clarifier effluent has a suspended solids concentration of 64 mg/*L* and a mixed liquor suspended solids concentration of 2,400 mg/*L*.

Known		Unknown
Tank Volume, MG	= 0.323 MG	Sludge Age, days
Flow, MGD	= 2.0 MGD	
PE SS, mg/*L*	= 64 mg/*L*	
MLSS, mg/*L*	= 2,400 mg/*L*	

1. Determine the pounds of mixed liquor suspended solids in the aerator.

$$\text{Solids in Aerator, lbs} = (\text{Aerator Vol, MG})(\text{MLSS, mg}/L)(8.34\text{ lbs/gal})$$
$$= (0.323\text{ MG})(2{,}400\text{ mg}/L)(8.34\text{ lbs/gal})$$
$$= 6{,}465\text{ lbs or approximately } 6{,}500\text{ lbs}$$

2. Determine the pounds of solids added per day to the aeration system by the primary effluent.

$$\text{Solids Added by Primary Effl, lbs/day} = (\text{Flow, MGD})(\text{PE SS, mg}/L)(8.34\text{ lbs/gal})$$
$$= (2.0\text{ MGD})(64\text{ mg}/L)(8.34\text{ lbs/gal})$$
$$= 1{,}068\text{ lbs/day or approximately } 1{,}100\text{ lbs/day}$$

3. Calculate the sludge age in days.

$$\text{Sludge Age, days} = \frac{\text{Suspended Solids in Aerator, lbs}}{\text{Suspended Solids in Prim Effl, lbs/day}}$$
$$= \frac{6{,}500\text{ lbs}}{1{,}100\text{ lbs/day}}$$
$$= 5.9\text{ days}$$

EXAMPLE 4

If the desired sludge age in the previous examples was 5 days, calculate the desired suspended solids concentration in the aerator in pounds and also in milligrams per liter. Assume 1,100 pounds per day of solids in the primary effluent and an aerator volume of 0.323 million gallons.

Known		Unknown
Desired Sludge Age, days	= 5 days	1. Desired MLSS, lbs
SS in Prim Effl, lbs/day	= 1,100 lbs/day	2. Desired MLSS, mg/L
Aerator Volume, MG	= 0.323 MG	

1. Calculate the pounds of MLSS desired in the aerator.

$$\text{Sludge Age, days} = \frac{\text{Suspended Solids in Aerator, lbs}}{\text{Suspended Solids in Prim Effl, lbs/day}}$$

Rearrange the equation to obtain desired lbs of MLSS in aerator or desired suspended solids in aerator, lbs.

$$\text{Suspended Solids in Aerator, lbs} = (\text{Sludge Age, days})(\text{SS in PE, lbs/day})$$

$$= (5 \text{ days})(1{,}100 \text{ lbs/day})$$

$$= 5{,}500 \text{ lbs}$$

2. Calculate the desired mixed liquor suspended solids concentration in milligrams per liter.

$$\text{Desired MLSS, mg}/L = \frac{\text{Desired MLSS in Aerator, lbs}}{(\text{Aerator Vol, MG})(8.34 \text{ lbs/gal})}$$

$$= \frac{5{,}500 \text{ lbs}}{(0.323 \text{ MG})(8.34 \text{ lbs/gal})}$$

$$= 2{,}042 \text{ mg}/L$$

$$= 2{,}050 \text{ mg}/L \text{ (target concentration)}$$

EXAMPLE 5

Calculate the desired change in the waste activated sludge pumping rate and the new pumping rate if the current pumping rate is 100 GPM (0.14 MGD). Use the information from the previous examples. The actual MLSS in the aerator are 6,500 lbs and the desired is 5,500 lbs. Assume a waste activated sludge concentration of 6,400 mg/*L*.

Known		Unknown
Current WAS Pumping, GPM	= 100 GPM	1. Change in WAS Pumping, GPM
Current WAS Pumping, MGD	= 0.14 MGD	2. New WAS Pumping, GPM
Actual MLSS, lbs	= 6,500 lbs	
Desired MLSS, lbs	= 5,500 lbs	
WAS SS, mg/L	= 6,400 mg/L	

1. Calculate the desired change in the waste activated sludge pumping rate in MGD and GPM.

$$\text{Change in WAS Pumping, MGD} = \frac{(\text{Actual MLSS, lbs} - \text{Desired MLSS, lbs})/\text{day}}{(\text{Waste Sludge Conc, mg}/L)(8.34 \text{ lbs/gal})}$$

$$= \frac{(6{,}500 \text{ lbs} - 5{,}500 \text{ lbs})/\text{day}}{(6{,}400 \text{ mg}/L)(8.34 \text{ lbs/gal})}$$

$$= 0.0187 \text{ MGD or } 18{,}700 \text{ GPD}$$

and

$$\text{Change in WAS Pumping, GPM} = \frac{18{,}700 \text{ gal/day}}{1{,}440 \text{ min/day}}$$

$$= 13 \text{ GPM}$$

2. Calculate the new waste activated sludge pumping rate in gallons per minute.

$$\text{Waste Activated Sludge Pumping, GPM} = \text{Current WAS Pumping, GPM} + \text{Change in WAS Pumping, GPM}$$

$$= 100 \text{ GPM} + 13 \text{ GPM}$$

$$= 113 \text{ GPM}$$

NOTE: Many operators do not change the WAS pumping rate by more than ±10 percent per day.

EXAMPLE 6

Determine the amount in pounds and milligrams per liter of mixed liquor volatile suspended solids (MLVSS) to be maintained in the aerator of a conventional activated sludge plant. Assume a food/microorganism (F/M) ratio of 0.6 lb COD per day per lb of mixed liquor volatile suspended solids under aeration for the plant in these examples. The average COD of the primary effluent is 165 mg/*L* and the average daily flow is 2.0 MGD. The aerator volume is 0.323 million gallons.

Known		Unknown
Prim Effl COD, mg/L	= 165 mg/L	1. MLVSS, lbs
Flow, MGD	= 2.0 MGD	2. MLVSS, mg/L
Aerator Volume, MG	= 0.323 MG	
Loading Factor, lbs COD/day/lb MLVSS	= $\frac{0.6 \text{ lb COD/day}}{1 \text{ lb MLVSS}}$	

1. Calculate the pounds of COD added to the aerator per day.

$$\text{Aerator Loading, lbs COD/day} = (\text{Flow, MGD})(\text{PE COD, mg}/L)(8.34 \text{ lbs/gal})$$

$$= (2.0 \text{ MGD})(165 \text{ mg}/L)(8.34 \text{ lbs/gal})$$

$$= 2{,}752 \text{ lbs COD/day}$$

2. Find the desired pounds of **M**ixed **L**iquor **V**olatile **S**uspended **S**olids under aeration, based on a loading factor of 0.6 lb COD per day per 1 lb of MLVSS.

$$\text{MLVSS, lbs} = \frac{\text{Primary Effluent, lbs COD/day}}{\text{Loading Factor, lbs COD/day/lb MLVSS}}$$

$$= \frac{2{,}752 \text{ lbs COD/day}}{0.6 \text{ lb COD/day/lb MLVSS}}$$

$$= 4{,}587 \text{ lbs MLVSS Under Aeration}$$

3. Calculate the desired **M**ixed **L**iquor **V**olatile **S**uspended **S**olids (MLVSS) concentration in the aerator in milligrams per liter.

$$\text{MLVSS, mg}/L = \frac{\text{MLVSS, lbs}}{(\text{Aerator Vol, MG})(8.34 \text{ lbs/gal})}$$

$$= \frac{4{,}587 \text{ lbs}}{(0.323 \text{ MG})(8.34 \text{ lbs/gal})}$$

$$= 1{,}700 \text{ mg}/L$$

EXAMPLE 7

Determine the **M**ean **C**ell **R**esidence **T**ime (MCRT) for the activated sludge plant in these examples. The necessary information is listed under *KNOWN.*

Known		Unknown
Aerator Volume, MG	= 0.323 MG	MCRT, days
Final Clarifier Vol, MG	= 0.150 MG	
Flow to Aerator, MGD	= 2.0 MGD	
WAS Flow, MGD	= 0.02 MGD	
MLSS, mg/*L*	= 2,400 mg/*L*	
Waste Sludge SS, mg/*L*	= 6,400 mg/*L*	
Final Effl SS, mg/*L*	= 16 mg/*L*	

1. Calculate the suspended solids in the aeration system using the aerator only.

$$\text{SS Under Aeration, lbs} = (\text{Aerator Vol, MG})(\text{MLSS, mg/}L)(8.34\text{ lbs/gal})$$
$$= (0.323\text{ MG})(2{,}400\text{ mg/}L)(8.34\text{ lbs/gal})$$
$$= 6{,}465\text{ lbs}$$

2. Calculate the suspended solids wasted in pounds of solids per day.

$$\text{SS Wasted, lbs/day} = (\text{WAS Flow, MGD})(\text{WAS SS, mg/}L)(8.34\text{ lbs/gal})$$
$$= (0.02\text{ MGD})(6{,}400\text{ mg/}L)(8.34\text{ lbs/gal})$$
$$= 1{,}068\text{ lbs/day}$$

3. Calculate the suspended solids lost in the plant effluent in pounds of solids per day.

$$\text{SS Lost, lbs/day} = (\text{Flow, MGD})(\text{Effl SS, mg/}L)(8.34\text{ lbs/gal})$$
$$= (2.0\text{ MGD})(16\text{ mg/}L)(8.34\text{ lbs/gal})$$
$$= 267\text{ lbs/day}$$

4. Determine the **M**ean **C**ell **R**esidence **T**ime (MCRT) in days.

$$\text{MCRT, days} = \frac{\text{Suspended Solids in Aeration System, lbs}}{\text{SS Wasted, lbs/day} + \text{SS Lost, lbs/day}}$$
$$= \frac{6{,}465\text{ lbs}}{1{,}068\text{ lbs/day} + 267\text{ lbs/day}}$$
$$= \frac{6{,}465\text{ lbs}}{1{,}335\text{ lbs/day}}$$
$$= 4.8\text{ days}$$

EXAMPLE 8

If the target or desired MCRT for the activated sludge plant in Example 7 was 6 days, determine the amount of WAS in pounds per day and the WAS pumping rate in gallons per minute (GPM). The required information is listed under *KNOWN.*

Known		Unknown
SS Under Aeration, lbs	= 6,465 lbs	1. WAS, lbs/day
SS Lost, lbs/day	= 267 lbs/day	2. WAS Pumping, GPM
WAS SS, mg/*L*	= 6,400 mg/*L*	
MCRT, days	= 6 days	

1. Calculate the WAS in pounds per day.

$$\text{MCRT, days} = \frac{\text{Suspended Solids in Aeration System, lbs}}{\text{SS Wasted, lbs/day} + \text{SS Lost, lbs/day}}$$

or, by rearranging the above equation,

$$\text{SS Wasted, lbs/day} = \frac{\text{SS in Aeration System, lbs}}{\text{MCRT, days}} - \text{SS Lost, lbs/day}$$
$$= \frac{6{,}465\text{ lbs}}{6\text{ days}} - 267\text{ lbs/day}$$
$$= 1{,}078\text{ lbs/day} - 267\text{ lbs/day}$$

or

$$\text{WAS, lbs/day} = 811\text{ lbs/day}$$

2. Calculate the waste activated sludge (WAS) pumping rate in MGD and GPM.

$$\text{WAS Pumping, MGD} = \frac{\text{WAS, lbs/day}}{(\text{WAS SS, mg/}L)(8.34\text{ lbs/gal})}$$
$$= \frac{811\text{ lbs/day}}{(6{,}400\text{ mg/}L)(8.34\text{ lbs/gal})}$$
$$= 0.015\text{ MGD or }15{,}000\text{ GPD}$$

and

$$\text{WAS Pumping, GPM} = \frac{15{,}000\text{ gal/day}}{1{,}440\text{ min/day}}$$
$$= 10.4\text{ GPM}$$

Try a WAS pumping rate of 10 GPM.

A.31 Sludge Digestion

EXAMPLE 9

Calculate the volume of seed sludge in gallons needed for a 50-foot diameter digester with a normal water depth of 18 feet, if the seed sludge required is estimated to be 20 percent of the digester volume. Most digesters have sloping bottoms, but assume the normal side wall water depth represents the average digester depth.

Known		Unknown
Diameter, ft	= 50 ft	Seed Sludge, gal
Depth, ft	= 18 ft	
Seed Sludge, % of Dig Vol	= 20%	

1. Calculate digester volume in cubic feet and gallons.

$$\text{Volume, cu ft} = (0.785)(\text{Diameter, ft})^2(\text{Depth, ft})$$
$$= (0.785)(50\text{ ft})^2(18\text{ ft})$$
$$= 35{,}325\text{ cu ft}$$

$$\text{Volume, gal} = (\text{Volume, cu ft})(7.5\text{ gal/cu ft})$$
$$= 264{,}938\text{ gal}$$

2. Determine the volume of seed sludge in gallons.

$$\text{Seed Sludge, gal} = \frac{(\text{Digester Volume, gal})(\text{Seed Sludge, \%})}{100\%}$$

$$= \frac{(264{,}938 \text{ gal})(20\%)}{100\%}$$

$$= 52{,}988 \text{ gal}$$

53,000 gallons of seed sludge will be needed.

EXAMPLE 10

A new plant expects to pump 400 gallons of raw sludge per day to the digester. The raw sludge is estimated to contain 5 percent solids with a volatile content of 66 percent. Estimate the pounds of volatile solids needed by the digester and the gallons of seed sludge, assuming the seed sludge contains 9 percent solids with 52 percent volatile solids and weighs 9 pounds per gallon. Assume a loading factor of 0.04 pound of new volatile solids added per day per pound of volatile solids under digestion.

Known		Unknown
Raw Sludge, GPD	= 400 GPD	1. Seed Volatile Solids, lbs
Raw Sludge Solids, %	= 5%	2. Seed Sludge, gal
Raw Sludge Volatile, %	= 66%	
Seed Sludge Solids, %	= 9%	
Seed Sludge Volatile, %	= 52%	
Seed Sludge, lbs/gal	= 9 lbs/gal	
Loading Factor, lbs VS/day/lb VS	$= \frac{0.04 \text{ lb VS/day}}{\text{lb VS}}$	

1. Find the pounds of volatile solids pumped to the digester per day.

$$\text{VS Pumped, lbs/day} = \frac{(\text{RS, GPD})(\text{RSS, \%})(\text{RSV, \%})(8.34 \text{ lbs/gal})}{(100\%)(100\%)}$$

$$= \frac{(400 \text{ GPD})(5\%)(66\%)(8.34 \text{ lbs/gal})}{(100\%)(100\%)}$$

$$= 110 \text{ lbs VS/day}$$

2. Find the pounds of seed volatile solids needed.

$$\frac{0.04 \text{ lb VS Added/day}}{1 \text{ lb VS in Digester}} = \frac{110 \text{ lbs VS Added/day}}{\text{Seed, lbs VS}}$$

By rearranging the above equation,

$$\text{Seed, lbs VS} = \frac{(1 \text{ lb VS in Digester})(110 \text{ lbs VS Added/day})}{0.04 \text{ lb VS Added/day}}$$

$$= 2{,}750 \text{ lbs VS}$$

NOTE: The terms VS for **V**olatile **S**olids and VM for **V**olatile **M**atter both mean the same.

3. Find the gallons of seed sludge needed.

$$\text{Seed Sludge, gal} = \frac{\text{Seed, lbs VS}}{(9 \text{ lbs/gal})(\text{Solids, \%/100\%})(\text{VS, \%/100\%})}$$

$$= \frac{2{,}750 \text{ lbs VS}}{(9 \text{ lbs/gal})(9\%/100\%)(52\%/100\%)}$$

$$= \frac{2{,}750 \text{ lbs VS}}{(9 \text{ lbs/gal})(0.09)(0.52)}$$

$$= 6{,}530 \text{ gallons of Seed Sludge}$$

EXAMPLE 11

Estimate the pounds of lime needed to neutralize a sour digester if the digester contains 250,000 gallons (0.25 MG) of sludge with a volatile acids level of 2,400 mg/*L* as acetic acid.

Known		Unknown
Sludge Volume, MG	= 0.25 MG	Lime Req'd, lbs
Volatile Acids, mg/*L* (Acetic Acid)	= 2,400 mg/*L*	

Estimate the lime required in pounds.

$$\text{Lime Req'd, lbs} = (\text{Sludge Vol, MG})(\text{Volatile Acids, mg}/L)(8.34 \text{ lbs/gal})$$

$$= (0.25 \text{ MG})(2{,}400 \text{ mg}/L)(8.34 \text{ lbs/gal})$$

$$= 5{,}004 \text{ lbs}$$

Try 5,000 lbs of lime.

EXAMPLE 12

Calculate the volume of sludge pumped in gallons per stroke (revolution) by a piston pump with a 9-inch diameter piston and the stroke set at 4 inches.

Known		Unknown
Piston Diameter, in	= 9 in	Sludge Pumped, gal/stroke
Piston Stroke, in/stroke	= 4 in/stroke	

1. Calculate the volume pumped per stroke (revolution) in cubic feet.

$$\text{Volume, cu ft/stroke} = (0.785)(\text{Diameter, ft})^2(\text{Distance, ft/stroke})$$

$$= 0.785\left(\frac{9 \text{ in}}{12 \text{ in/ft}}\right)^2\left(\frac{4 \text{ in/stroke}}{12 \text{ in/ft}}\right)$$

$$= 0.785(0.75 \text{ ft})^2(0.33 \text{ ft/stroke})$$

$$= 0.1457 \text{ cu ft/stroke}$$

2. Determine the volume of sludge pumped in gallons per stroke.

$$\text{Sludge Pumped, gal/stroke} = (\text{Volume, cu ft/stroke})(7.5 \text{ gal/cu ft})$$

$$= (0.1457 \text{ cu ft/stroke})(7.5 \text{ gal/cu ft})$$

$$= 1.09 \text{ gal/stroke}$$

EXAMPLE 13

If 3,200 gallons of raw sludge are pumped to a digester, how many pounds of dry sludge are handled and how many pounds are subject to digestion (volatile solids)? Assume the raw sludge contains 5.5 percent total solids and has a volatile solids content of 66 percent.

Known		Unknown
Raw Sludge, gal	= 3,200 gal	1. Dry Solids, lbs
Raw Sludge Solids, %	= 5.5%	2. Volatile Solids, lbs
Raw Sludge VS, %	= 66%	

1. Determine the pounds of dry solids pumped.

$$\text{Dry Solids, lbs} = \frac{(\text{Raw Sludge, gal})(\text{Raw Sludge Sol, \%})(8.34\text{ lbs/gal})}{100\%}$$

$$= \frac{(3{,}200\text{ gal})(5.5\%)(8.34\text{ lbs/gal})}{100\%}$$

$$= 1{,}468\text{ lbs Dry Solids Pumped}$$

2. Estimate the pounds of volatile solids pumped.

$$\text{Volatile Solids, lbs} = \frac{(\text{Dry Solids, lbs})(\text{Raw Sludge VS, \%})}{100\%}$$

$$= \frac{(1{,}468\text{ lbs})(66\%)}{100\%}$$

$$= 969\text{ lbs of Volatile Solids Pumped}$$

EXAMPLE 14

Determine the percent reduction of volatile solids as a result of digestion when the raw sludge is 66 percent volatile solids and the digested sludge is 52 percent volatile solids.

Known		Unknown
Raw Sludge Volatile Solids, %	= 66%	Reduction of Volatile Solids, %
Digested Sludge Volatile Solids, %	= 52%	

Calculate the percent reduction of volatile solids as a result of digestion.

$$\text{Reduction of Volatile Solids, \%} = \frac{(\text{In} - \text{Out})(100\%)}{\text{In} - (\text{In} \times \text{Out})}$$

$$= \frac{(0.66 - 0.52)(100\%)}{0.66 - (0.66)(0.52)}$$

$$= \frac{(0.14)(100\%)}{0.66 - 0.34}$$

$$= 44\%$$

EXAMPLE 15

Determine the digester loading in pounds of volatile solids per day per cubic foot of digester capacity for the digester in the previous examples. The necessary information is listed under *KNOWN.*

Known		Unknown
Digester Volume, cu ft	= 35,325 cu ft	Digester Loading, lbs VS/day/cu ft
VS Added, lbs/day	= 12,000 lbs VS/day	

Calculate the digester loading in pounds of volatile solids added per day per cubic foot of digester capacity.

$$\text{Digester Loading, lbs VS/day/cu ft} = \frac{\text{VS Added, lbs/day}}{\text{Digester Volume, cu ft}}$$

$$= \frac{12{,}000\text{ lbs VS added/day}}{35{,}325\text{ cu ft}}$$

$$= 0.34\text{ lb VS added/day/cu ft}$$

EXAMPLE 16

An anaerobic sludge digester receives 12,000 pounds of volatile solids per day. How many pounds of digested sludge should remain in storage if 10 pounds of digested sludge should be retained in the digester for every pound of volatile solids added to the digester?

Known		Unknown
VS Added, lbs/day	= 12,000 lbs/day	Digested Sludge in Storage, lbs
Loading, $\frac{\text{lbs Dig Sl}}{\text{lb VS/day}}$	= $\frac{\text{10 lbs Dig Sl}}{\text{lb VS/day}}$	

Calculate the pounds of digested sludge needed in storage.

$$\text{Digested Sludge in Storage, lbs} = (\text{VS Added, lbs/day})\left(\text{Loading, }\frac{\text{lbs Dig Sl}}{\text{lb VS/day}}\right)$$

$$= (12{,}000\text{ lbs/day})\left(\frac{10\text{ lbs Dig Sl}}{\text{lb VS/day}}\right)$$

$$= 120{,}000\text{ lbs Digester Sludge Needed in Storage on a Dry Solids Basis}$$

EXAMPLE 17

Estimate the pounds of volatile solids destroyed per day per cubic foot of digester capacity if 12,000 pounds per day of volatile solids are added to a digester with a capacity of 35,325 cubic feet and the volatile solids reduction is 52 percent during digestion.

Known		Unknown
VS Added, lbs/day	= 12,000 lbs VS/day	VS Destroyed, lbs/day/cu ft
Digester Vol, cu ft	= 35,325 cu ft	
VS Reduction, %	= 52%	

Calculate the pounds of volatile solids destroyed per day per cubic foot of digester capacity.

$$\text{VS Destroyed, lbs/day/cu ft} = \frac{(\text{VS Added, lbs/day})(\text{VS Reduction, \%})}{(\text{Digester Vol, cu ft})(100\%)}$$

$$= \frac{(12{,}000\text{ lbs VS/day})(52\%)}{(35{,}325\text{ cu ft})(100\%)}$$

$$= 0.18\text{ lb VS Destroyed/day/cu ft}$$

EXAMPLE 18

Determine the gas production for a digester in cubic feet of gas produced per day per pound of volatile solids destroyed when the gas meter indicates that 77,000 cubic feet of gas are produced per day and 6,240 pounds of volatile solids are destroyed per day.

Known	Unknown
Gas Produced, cu ft/day = 77,000 cu ft/day	Gas Production, cu ft/day/lb VS
VS Destroyed, lbs/day = 6,240 lbs/day	

Calculate the gas production in cubic feet of gas produced per pound of volatile solids destroyed.

$$\text{Gas Production, cu ft/day/lb VS} = \frac{\text{Gas Produced, cu ft/day}}{\text{VS Destroyed, lbs/day}}$$

$$= \frac{77{,}000 \text{ cu ft/day}}{6{,}240 \text{ lbs VS Destroyed/day}}$$

$$= 12 \text{ cu ft Gas/day/lb VS Destroyed}$$

EXAMPLE 19

Calculate the solids balance for a digester. The purpose of the solids balance calculations is to compare values calculated with actual values measured. The necessary information is listed under *KNOWN.* The values to be calculated for the solids balance will be calculated in each step.

Known	Unknown
Input to Digester	
32,000 gallons of Raw Sludge With 5.5% Solids Content, and 66% Volatile Solids	Solids Balance for a Digester
Digester Output	
Digested Solids With 9.0% Solids Content, and 52% Volatile Solids	

1. Calculate the pounds of total solids, water, volatile solids, and inorganic solids pumped to the digester.

   a. Total solids to digester.

$$\text{Dry Solids, lbs/day} = (\text{Vol Pumped, gal/day})\left(\frac{\text{Solids, \%}}{100\%}\right)(8.34 \text{ lbs/gal})$$

$$= (32{,}000 \text{ gal/day})\left(\frac{5.5\%}{100\%}\right)(8.34 \text{ lbs/gal})$$

$$= 14{,}680 \text{ lbs Solids/day}$$

or

$$= 14{,}700 \text{ lbs Solids/day}$$

   b. Total water and solids to digester.

$$\text{Water and Solids, lbs/day} = \frac{(\text{Dry Solids, lbs/day})(100\%)}{\text{Solids, \%}}$$

$$= \frac{(14{,}680 \text{ lbs/day})(100\%)}{5.5\%}$$

$$= 266{,}900 \text{ lbs/day}$$

or

$$\text{Water and Solids, lbs/day} = (\text{Vol Pumped, gal/day})(8.34 \text{ lbs/gal})$$

$$= (32{,}000 \text{ gal/day})(8.34 \text{ lbs/gal})$$

$$= 266{,}900 \text{ lbs/day (same answer as above)}$$

   c. Water to digester.

$$\text{Water, lbs/day} = \text{Water and Solids, lbs/day} - \text{Dry Solids, lbs/day}$$

$$= 266{,}900 \text{ lbs/day} - 14{,}700 \text{ lbs/day}$$

$$= 252{,}200 \text{ lbs/day}$$

   d. Volatile solids to digester.

$$\text{Volatile Solids, lbs/day} = \frac{(\text{Dry Solids, lbs/day})(\text{Volatile Solids, \%})}{100\%}$$

$$= \frac{(14{,}700 \text{ lbs/day})(66\%)}{100\%}$$

$$= 9{,}700 \text{ lbs/day}$$

   e. Inorganic solids to digester.

$$\text{Inorganic Solids, lbs/day} = \text{Dry Solids, lbs/day} - \text{Volatile Solids, lbs/day}$$

$$= 14{,}700 \text{ lbs/day} - 9{,}700 \text{ lbs/day}$$

$$= 5{,}000 \text{ lbs/day}$$

   f. Calculate the percent reduction of volatile matter in the digester to find the pounds per day of gas produced during digestion.

$$\text{Reduction of Volatile Solids, \%} = \frac{(\text{In} - \text{Out})(100\%)}{\text{In} - (\text{In} \times \text{Out})}$$

$$= \frac{(0.66 - 0.52)(100\%)}{0.66 - (0.66 \times 0.52)}$$

$$= \frac{(0.14)(100\%)}{0.66 - 0.34}$$

$$= 44\%$$

   g. Gas out of digester.

$$\text{Gas, lbs/day} = \frac{(\text{Volatile Solids, lbs/day})(\text{Reduction of VS, \%})}{100\%}$$

$$= \frac{(9{,}700 \text{ lbs/day})(44\%)}{100\%}$$

$$= 4{,}270 \text{ lbs/day}$$

Use 4,300 lbs/day.

2. Determine the pounds of total, volatile, and inorganic solids removed from the digester as digested sludge to the sludge drying beds or other sludge treatment and disposal processes.

   h. Volatile solids out of digester.

$$\text{Volatile Solids, lbs/day} = \text{Volatile Solids to Digester, lbs/day} - \text{Volatile Solids Out as Gas, lbs/day}$$

$$= 9{,}700 \text{ lbs/day} - 4{,}300 \text{ lbs/day}$$

$$= 5{,}400 \text{ lbs/day}$$

   i. Total solids out of digester.

$$\text{Total Solids, lbs/day} = \frac{(\text{Volatile Solids, lbs/day})(100\%)}{\text{Volatile Solids, \%}}$$

$$= \frac{(5{,}400 \text{ lbs/day})(100\%)}{52\%}$$

$$= 10{,}385 \text{ lbs/day}$$

   Use 10,400 lbs/day.

   j. Inorganic solids out of digester.

$$\text{Inorganic Solids, lbs/day} = \text{Total Solids, lbs/day} - \text{Volatile Solids, lbs/day}$$

$$= 10{,}400 \text{ lbs/day} - 5{,}400 \text{ lbs/day}$$

$$= 5{,}000 \text{ lbs/day}$$

NOTE: This is the same amount that went into the digester (see Step e). This indicates that the calculations are correct.

   k. Total solids and water out of digester.

$$\text{Water and Solids, lbs/day} = \frac{(\text{Total Solids, lbs/day})(100\%)}{\text{Digested Solids, \%}}$$

$$= \frac{(10{,}400 \text{ lbs/day})(100\%)}{9\%}$$

$$= 115{,}556 \text{ lbs/day}$$

   Use 115,600 lbs.

   l. Find total pounds of water out of digester.

$$\text{Water Out, lbs/day} = \text{Water and Solids, lbs/day} - \text{Solids, lbs/day}$$

$$= 115{,}600 \text{ lbs/day} - 10{,}400 \text{ lbs/day}$$

$$= 105{,}200 \text{ lbs/day}$$

   m. Compare amounts of water into and out of digester. Assume 15,000 gallons of supernatant was removed from the digester with a negligible amount of solids.

$$\text{Water Change, lbs/day} = \text{Water In, lbs/day} - \text{Water Out, lbs/day} - \text{Supernatant Out, lbs/day}$$

$$= 252{,}200 \text{ lbs/day} - 105{,}200 \text{ lbs/day} - (15{,}000 \text{ gal/day} \times 8.34 \text{ lbs/gal})$$

$$= 252{,}200 \text{ lbs/day} - 105{,}200 \text{ lbs/day} - 125{,}100 \text{ lbs/day}$$

$$= 21{,}900 \text{ lbs/day Increase in Digester Level}$$

   n. Calculate the change in digester level for the 50-foot diameter digester.

$$\text{Volume of Water, cu ft} = \frac{\text{Water Change, lbs/day}}{(8.34 \text{ lbs/gal})(7.5 \text{ gal/cu ft})}$$

$$= \frac{21{,}900 \text{ lbs/day}}{(8.34 \text{ lbs/gal})(7.5 \text{ gal/cu ft})}$$

$$= 350 \text{ cu ft}$$

$$\text{Change in Depth, ft} = \frac{\text{Volume, cu ft}}{\text{Area, sq ft}}$$

$$= \frac{350 \text{ cu ft}}{(0.785)(50 \text{ ft})^2}$$

$$= 0.178 \text{ ft or } 0.2 \text{ ft}$$

If the digester depth has increased approximately 0.2 ft, this will provide a good check on the solids balance.

A.32 Effluent Discharge, Reclamation, and Reuse

EXAMPLE 20

Estimate the average daily BOD load on the receiving waters in pounds of BOD per day for an average flow of 2.35 MGD and an average effluent BOD of 21 mg/*L*. What is the monthly load in pounds of BOD per month for a 30-day month?

Known	Unknown
Avg Flow, MGD = 2.35 MGD	1. BOD Load, lbs BOD/day
Avg BOD, mg/*L* = 21 mg/*L*	2. BOD Load, lbs BOD/month

1. Estimate the average daily BOD load in pounds of BOD per day.

$$\text{BOD Load, lbs BOD/day} = (\text{Avg Flow, MGD})(\text{Avg BOD, mg}/L)(8.34 \text{ lbs/gal})$$

$$= (2.35 \text{ MGD})(21 \text{ mg}/L)(8.34 \text{ lbs/gal})$$

$$= 412 \text{ lbs BOD/day}$$

2. Estimate the monthly BOD load in pounds of BOD per month.

$$\text{BOD Load, lbs BOD/month} = (\text{BOD Load, lbs BOD/day})(30 \text{ days/month})$$

$$= (412 \text{ lbs BOD/day})(30 \text{ days/month})$$

$$= 12{,}360 \text{ lbs BOD/month}$$

EXAMPLE 21

Estimate the average effluent BOD in mg/*L* for a treatment plant for the following information:

Day	S	M	T	W	T	F	S
BOD, mg/*L*	23	25	17	16	19	22	20

Known	Unknown
BOD Information	Average BOD, mg/*L*

Calculate the average BOD in milligrams per liter.

$$\text{Average BOD, mg/}L = \frac{\text{Sum of Measurements, mg/}L}{\text{Number of Measurements}}$$

$$= \frac{23 + 25 + 17 + 16 + 19 + 22 + 20}{7}$$

$$= \frac{142 \text{ mg/}L}{7}$$

$$= 20 \text{ mg/}L$$

A.33 Maintenance

EXAMPLE 22

Determine the capacity of a pump, in gallons per minute, that lowers a wet well one foot in 5 minutes. The wet well is 5 feet wide and 10 feet long. There is no flow into the wet well during the pump test.

Known	Unknown
Time, min = 5 min	Pump Capacity, GPM
Drop, ft = 1 ft	
Length, ft = 10 ft	
Width, ft = 5 ft	

1. Calculate the volume pumped in gallons.

$$\text{Volume, gal} = (\text{Length, ft})(\text{Width, ft})(\text{Drop, ft})(7.5 \text{ gal/cu ft})$$

$$= (10 \text{ ft})(5 \text{ ft})(1 \text{ ft})(7.5 \text{ gal/cu ft})$$

$$= 375 \text{ gal}$$

2. Estimate the pump capacity in gallons per minute.

$$\text{Pump Capacity, GPM} = \frac{\text{Volume Pumped, gal}}{\text{Time, min}}$$

$$= \frac{375 \text{ gal}}{5 \text{ min}}$$

$$= 75 \text{ GPM}$$

EXAMPLE 23

A float (marked stick) is dropped into an open channel 3 feet wide with a flow 17 inches deep. The float travels 24 feet in 9 seconds. Estimate the flow in the channel in gallons per minute (GPM) and million gallons per day (MGD).

Known	Unknown
Width, ft = 3 ft	1. Flow, GPM
Depth, in = 17 in	2. Flow, MGD
Distance, ft = 24 ft	
Time, sec = 9 sec	

1. Estimate the velocity in feet per second of the water flowing in the channel.

$$\text{Velocity, ft/sec} = \frac{\text{Distance, ft}}{\text{Time, sec}}$$

$$= \frac{24 \text{ ft}}{9 \text{ sec}}$$

$$= 2.7 \text{ ft/sec}$$

NOTE: Velocity of float on a water surface is often 10 to 15 percent higher than the average velocity.

2. Convert float velocity to average velocity. Assume float velocity is 10 percent too high.

$$\text{Avg Vel, ft/sec} = \text{Float Vel, ft/sec} - (\text{Float Vel, ft/sec})(\text{Fraction Too High})$$

$$= 2.7 \text{ ft/sec} - (2.7 \text{ ft/sec})\left(\frac{10\%}{100\%}\right)$$

$$= 2.7 \text{ ft/sec} - 0.27 \text{ ft/sec}$$

$$= 2.43 \text{ ft/sec}$$

3. Calculate the flow in the channel in cubic feet per second (CFS).

$$\text{Flow, cu ft/sec} = (\text{Area, sq ft})(\text{Velocity, ft/sec})$$

$$= (3 \text{ ft})\left(\frac{17 \text{ in}}{12 \text{ in/ft}}\right)(2.43 \text{ ft/sec})$$

$$= (4.25 \text{ sq ft})(2.43 \text{ ft/sec})$$

$$= 10.3 \text{ cu ft/sec}$$

4. Convert flow from cubic feet per second (CFS) to gallons per minute (GPM).

$$\text{Flow, GPM} = (\text{Flow, CFS})(7.5 \text{ gal/cu ft})(60 \text{ sec/min})$$

$$= (10.3 \text{ cu ft/sec})(7.5 \text{ gal/cu ft})(60 \text{ sec/min})$$

$$= 4{,}640 \text{ GPM}$$

5. Convert flow from gallons per minute (GPM) to million gallons per day (MGD).

$$\text{Flow, MGD} = \frac{(\text{Flow, GPM})(60 \text{ min/hr})(24 \text{ hr/day})}{1{,}000{,}000/\text{M}}$$

$$= \frac{(4{,}640 \text{ gal/min})(60 \text{ min/hr})(24 \text{ hr/day})}{1{,}000{,}000/\text{M}}$$

$$= 6.7 \text{ MGD}$$

EXAMPLE 24

Estimate the velocity in an open channel using a pitot tube. Water rises in the tube 0.75 inch ($^3/_4$ of an inch) above the water surface.

$$V = \sqrt{2gH}$$

$$g = 32.2 \text{ ft/sec}^2 \text{ (constant for gravity)}$$

Known	Unknown
H, in = 0.75 in	V, ft/sec

Estimate the velocity in feet per second.

$$V,\ ft/sec = \sqrt{2(g,\ ft/sec^2)(H,\ ft)}$$

$$= \sqrt{2(32.2\ ft/sec^2)(0.75\ in)/(12\ in/ft)}$$

$$= \sqrt{4.025\ ft^2/sec^2}$$

$$= 2\ ft/sec$$

A.34 Laboratory

EXAMPLE 25

Convert the temperature of a sample from 18° Celsius to degrees Fahrenheit.

Known	Unknown
Temperature, °C = 18°C	Temperature, °F

Convert temperature from degrees Celsius to degrees Fahrenheit.

Temperature, °F = (Temperature, °C)(9/5) + 32°F

= (18°C)(9/5) + 32°F

= 32.4 + 32°F

= 64°F

EXAMPLE 26

Estimate the sludge pumped to a digester in gallons per day from a primary clarifier that treats a flow of 1.68 MGD. The influent settleable solids test indicates 11 m*L*/*L* and the effluent 0.3 m*L*/*L*.

Known		Unknown
Infl Set Sol, m*L*/*L*	= 11 m*L*/*L*	Sludge Pumped, GPD
Effl Set Sol, m*L*/*L*	= 0.3 m*L*/*L*	
Flow, MGD	= 1.68 MGD	

1. Calculate the sludge removed by primary clarifier in milliliters per liter.

Sludge Removed, m*L*/*L* = Infl Set Sol, m*L*/*L* – Effl Set Sol, m*L*/*L*

= 11 m*L*/*L* – 0.3 m*L*/*L*

= 10.7 m*L*/*L*

2. Estimate sludge pumped to digester in gallons per day.

Sludge Pumped, GPD = (Sludge Removed, m*L*/*L*)(1,000 mg/m*L*)(Flow, MGD)

= (10.7 m*L*/M mg)(1,000 mg/m*L*)(1.68 MGD)

= 17,976 or 18,000 GPD

EXAMPLE 27

Results from lab tests are listed under *KNOWN.* Calculate the items listed under *UNKNOWN.*

Known

Sample Vol, m*L*	= 100 m*L*
Crucible Weight, gm	= 20.4107 gm
Crucible Plus Dry Solids, gm	= 20.4326 gm
Crucible Plus Ash, gm	= 20.4173 gm

Unknown

1. Total Suspended Solids, mg/*L*
2. Volatile Suspended Solids, mg/*L*
3. Volatile Suspended Solids, %
4. Fixed Suspended Solids, mg/*L*
5. Fixed Suspended Solids, %

1. Calculate the total suspended solids in milligrams and in milligrams per liter.

Crucible Plus Dry Solids, gm = 20.4326 gm
Crucible Weight, gm = 20.4107 gm
Dry Weight, gm = 0.0219 gm
or = 21.9 mg

$$\text{Total Susp Solids, mg/}L = \frac{(\text{Dry Weight, mg})(1{,}000\ mL/L)}{\text{Sample Vol, m}L}$$

$$= \frac{(21.9\ mg)(1{,}000\ mL/L)}{100\ mL}$$

$$= 219\ mg/L$$

2. Calculate the volatile suspended solids in milligrams and milligrams per liter.

Crucible Plus Dry Solids, gm = 20.4326 gm
Crucible Plus Ash, gm = 20.4173 gm
Volatile Weight, gm = 0.0153 gm
or = 15.3 mg

$$\text{Volatile Susp Solids, mg/}L = \frac{(\text{Volatile Weight, mg})(1{,}000\ mL/L)}{\text{Sample Vol, m}L}$$

$$= \frac{(15.3\ mg)(1{,}000\ mL/L)}{100\ mL}$$

$$= 153\ mg/L$$

3. Calculate the percent volatile suspended solids.

$$\text{Volatile Susp Solids, \%} = \frac{(\text{Volatile SS, mg/}L)(100\%)}{\text{Total SS, mg/}L}$$

$$= \frac{(153\ mg/L)(100\%)}{219\ mg/L}$$

$$= 70\%$$

4. Calculate the fixed (inorganic) suspended solids in milligrams and milligrams per liter.

Crucible Plus Ash, gm = 20.4173 gm
Crucible Weight, gm = 20.4107 gm
Ash Weight, gm = 0.0066 gm
or = 6.6 mg

$$\text{Fixed Susp Solids, mg/}L = \frac{(\text{Ash Weight, mg})(1{,}000\ mL/L)}{\text{Sample Volume, m}L}$$

$$= \frac{(6.6\ mg)(1{,}000\ mL/L)}{100\ mL}$$

$$= 66\ mg/L$$

5. Calculate the percent fixed suspended solids.

$$\text{Fixed Susp Solids, \%} = \frac{(\text{Fixed SS, mg}/L)(100\%)}{\text{Total SS, mg}/L}$$

$$= \frac{(66\text{ mg}/L)(100\%)}{219\text{ mg}/L}$$

$$= 30\%\text{ (Check)}$$

EXAMPLE 28

Calculate the percent removal of suspended solids by the primary clarifiers, secondary treatment, and overall plant removal for the information listed under *KNOWN.*

Known		Unknown
Infl SS, mg/*L*	= 219 mg/*L*	1. Primary Removal, %
Prim Effl SS, mg/*L*	= 76 mg/*L*	2. Secondary Removal, %
Sec Effl SS, mg/*L*	= 17 mg/*L*	3. Plant Removal, %

1. Calculate the percent removal of suspended solids by the primary clarifiers.

$$\text{Removal, \%} = \frac{(\text{In} - \text{Out})(100\%)}{\text{In}}$$

$$= \frac{(219\text{ mg}/L - 76\text{ mg}/L)(100\%)}{219\text{ mg}/L}$$

$$= 65\%$$

2. Calculate the percent removal of suspended solids by secondary treatment.

$$\text{Removal, \%} = \frac{(\text{In} - \text{Out})(100\%)}{\text{In}}$$

$$= \frac{(76\text{ mg}/L - 17\text{ mg}/L)(100\%)}{76\text{ mg}/L}$$

$$= 78\%$$

3. Calculate the percent removal of suspended solids by the overall treatment plant.

$$\text{Removal, \%} = \frac{(\text{In} - \text{Out})(100\%)}{\text{In}}$$

$$= \frac{(219\text{ mg}/L - 17\text{ mg}/L)(100\%)}{219\text{ mg}/L}$$

$$= 92\%$$

EXAMPLE 29

Calculate the pounds of suspended solids removed per day by the primary clarifiers in Example 28 if the flow is 2.9 MGD.

Known		Unknown
Flow, MGD	= 2.9 MGD	Susp Solids Removed, lbs/day
Infl SS, mg/*L*	= 219 mg/*L*	
Prim Effl SS, mg/*L*	= 76 mg/*L*	

Calculate the pounds of suspended solids removed per day by the primary clarifiers.

$$\text{SS Removed, lbs/day} = (\text{Flow, MGD})(\text{SS Removed, mg}/L)(8.34\text{ lbs/gal})$$

$$= (2.9\text{ MGD})(219\text{ mg}/L - 76\text{ mg}/L)(8.34\text{ lbs/gal})$$

$$= 3{,}460\text{ lbs}$$

EXAMPLE 30

Determine the **S**ludge **V**olume **I**ndex (SVI) for a sample with a mixed liquor suspended solids of 2,750 mg/*L* and a 30-minute settleable solids test of 650 m*L* in 2,000 m*L* or 32.5 percent.

Known		Unknown
Set Solids, %	= 32.5%	SVI
MLSS, mg/*L*	= 2,750 mg/*L*	

Calculate the **S**ludge **V**olume **I**ndex (SVI).

$$\text{SVI} = \frac{(\text{Set Solids, \%})(1{,}000\text{ mg/gm})(1{,}000\text{ m}L/L)}{(100\%)(\text{MLSS, mg}/L)}$$

$$= \frac{(\text{Set Solids, \%})(10{,}000)}{\text{MLSS, mg}/L}$$

$$= \frac{(32.5)(10{,}000)}{2{,}750}$$

$$= 118\text{ m}L/\text{gm}$$

EXAMPLE 31

Estimate the percent carbon dioxide (CO_2) in a digester gas when 83 m*L* of gas remains in a graduate that started with 124 m*L* of gas.

Known		Unknown
Total Volume of Graduate, m*L*	= 124 m*L*	CO_2, %
Gas Remaining in Graduate, m*L*	= 83 m*L*	

Calculate the percent carbon dioxide (CO_2) in the sample.

$$CO_2,\ \% = \frac{(\text{Total Volume, m}L - \text{Gas Remaining, m}L)(100\%)}{\text{Total Volume, m}L}$$

$$= \frac{(124\text{ m}L - 83\text{ m}L)(100\%)}{124\text{ m}L}$$

$$= 33\%$$

EXAMPLE 32

Estimate the dissolved oxygen (DO) percent saturation of a sample from a receiving water. The actual dissolved oxygen is 7.3 mg/*L* and the dissolved oxygen at saturation for the water at 72°F is 8.8 mg/*L*.

Known		Unknown
DO of Sample, mg/*L*	= 7.3 mg/*L*	DO Saturation, %
DO at Saturation, mg/*L*	= 8.8 mg/*L*	

Calculate the dissolved oxygen (DO) saturation percent for the receiving water sample.

$$\text{DO Saturation, \%} = \frac{(\text{DO of Sample, mg}/L)(100\%)}{\text{DO at Saturation, mg}/L}$$

$$= \frac{(7.3\text{ mg}/L)(100\%)}{8.8\text{ mg}/L}$$

$$= 83\%$$

EXAMPLE 33

Determine the size of sample for a BOD test when using a 300-m*L* BOD bottle and the BOD of the sample is estimated to be 190 mg/*L*.

Known		Unknown
Estimated BOD, mg/*L*	= 190 mg/*L*	Sample Size, m*L*

Determine the size of sample in milliliters for the BOD test.

$$\text{Size of Sample, m}L = \frac{1{,}200}{\text{Estimated BOD, mg}/L}$$

$$= \frac{1{,}200}{190\text{ mg}/L}$$

$$= 6\text{ m}L$$

Suggest trying sample sizes of 4 m*L*, 6 m*L*, and 8 m*L*.

EXAMPLE 34

When using a 300-m*L* BOD bottle, estimate the BOD for a 6-m*L* sample when the initial DO of the diluted sample was 8.8 mg/*L* and DO of the diluted sample after 5 days of incubation was 5.9 m*L*.

Known		Unknown
Sample Volume, m*L*	= 6 m*L*	BOD, mg/*L*
Initial DO, mg/*L*	= 8.8 mg/*L*	
Final DO, mg/*L*	= 5.9 mg/*L*	

Calculate the five-day biochemical oxygen demand (BOD) for the sample.

$$\text{BOD, mg}/L = \left[\begin{matrix}\text{Initial DO of}\\ \text{Diluted Sample, mg}/L\end{matrix} - \begin{matrix}\text{DO of Sample}\\ \text{After 5 days, mg}/L\end{matrix}\right]\left[\frac{\text{BOD Bottle Vol, m}L}{\text{Sample Vol, m}L}\right]$$

$$= (8.8\text{ mg}/L - 5.9\text{ mg}/L)\left(\frac{300\text{ m}L}{6\text{ m}L}\right)$$

$$= 145\text{ mg}/L$$

A.35 Data Analysis

BOD Effluent Results

The effluent BODs for a two-week period from a wastewater treatment plant are summarized below:

Week 1

Day	M	T	W	T	F	S	S
BOD, mg/*L*	23	19	21	29	27	26	35

Week 2

Day	M	T	W	T	F	S	S
BOD, mg/*L*	37	36	33	27	29	25	27

EXAMPLE 35

Determine the mean, median, mode, and range for the two weeks of BOD effluent results.

Known	Unknown
BOD Effluent Results	1. Mean 2. Median 3. Mode 4. Range

1. Calculate the mean BOD in milligrams per liter for the two weeks of BOD effluent results.

$$\text{Mean BOD, mg}/L = \frac{\text{Sum of All Measurements, mg}/L}{\text{Number of Measurements}}$$

$$= \frac{23 + 19 + 21 + 29 + 27 + 26 + 35 + 37 + 36 + 33 + 27 + 29 + 25 + 27}{14}$$

$$= \frac{394\text{ mg}/L}{14}$$

$$= 28\text{ mg}/L$$

2. Calculate the median value of the BOD measurements. The median value is the middle measurement (half of the measurements are smaller and half are larger). Arrange the measurements in ascending or increasing order.

Rank	1	2	3	4	5	6	7	8	9	10	11	12	13	14
BOD, mg/*L*	19	21	23	25	26	27	27	27	29	29	33	35	36	37
							↑ Middle							

The median BOD value is 27 mg/*L*.

3. Determine the mode value for the BOD measurements. The mode is the measurement that occurs most frequently. The measurement 27 mg/*L* occurs three times which is the most frequent measurement. The mode BOD value is 27 mg/*L*.

4. Determine the range of the BOD measurements.

$$\text{Range BOD, mg}/L = \text{Largest BOD, mg}/L - \text{Smallest BOD, mg}/L$$

$$= 37\text{ mg}/L - 19\text{ mg}/L$$

$$= 18\text{ mg}/L$$

EXAMPLE 36

Calculate the seven-day moving or running average for the effluent BOD for these examples.

Known	Unknown
BOD Effluent Results	Seven-Day Moving Average

Calculate the seven-day moving average by finding the average BOD for seven days. Repeat for each day by removing the seventh or oldest measurement and adding the newest measurement.

$$\text{Moving Average BOD, mg}/L = \frac{23 + 19 + 21 + 29 + 27 + 26 + 35}{7}$$

$$= \frac{180\text{ mg}/L}{7}$$

$$= 26\text{ mg}/L$$

$$\text{Moving Average BOD, mg}/L = \frac{19 + 21 + 29 + 27 + 26 + 35 + 37}{7}$$

$$= \frac{194\text{ mg}/L}{7}$$

$$= 28\text{ mg}/L$$

Week 1	BOD, mg/*L*	Moving Average BOD, mg/*L*
M	23	
T	19	
W	21	
T	29	
F	27	
S	26	
S	35	26

Week 2	BOD, mg/*L*	Moving Average BOD, mg/*L*
M	37	28
T	36	30
W	33	32
T	27	32
F	29	32
S	25	32
S	27	31

A.36 Administration, Safety

EXAMPLE 37

Calculate the injury frequency rate for a wastewater utility where there were four injuries in one year and the operators worked 97,120 hours.

Known		Unknown
Injuries, number/yr	= 4 Injuries/yr	Injury Frequency Rate
Hours Worked, number/yr	= 97,120 hr/yr	

Calculate the injury frequency rate.

$$\text{Injury Freq Rate} = \frac{(\text{Injuries, number/yr})(1{,}000{,}000)}{\text{Hours Worked, number/yr}}$$

$$= \frac{(4\text{ Injuries/yr})(1{,}000{,}000)}{97{,}120\text{ hr/yr}}$$

$$= 41.2$$

EXAMPLE 38

Calculate the injury severity rate for a wastewater treatment company that experienced 57 operator-hours lost due to injuries while the operators worked 97,120 hours during the year.

Known		Unknown
Number of Hours Lost/yr	= 57 hr/yr	Injury Severity Rate
Number of Hours Worked/yr	= 97,120 hr/yr	

Calculate the injury severity rate.

$$\text{Injury Severity Rate} = \frac{(\text{Number of Hours Lost/yr})(1{,}000{,}000)}{\text{Number of Hours Worked/yr}}$$

$$= \frac{(57\text{ hr/yr})(1{,}000{,}000)}{97{,}120\text{ hr/yr}}$$

$$= 587$$

A.4 BASIC CONVERSION FACTORS (METRIC SYSTEM)

LENGTH		
100 cm	= 1 m	100 cm/m
3.281 ft	= 1 m	3.281 ft/m
AREA		
2.4711 ac	= 1 ha*	2.4711 ac/ha
10,000 sq m	= 1 ha	10,000 sq m/ha
VOLUME		
1,000 m*L*	= 1 liter	1,000 m*L*/*L*
1,000 *L*	= 1 cu m	1,000 *L*/cu m
3.785 *L*	= 1 gal	3.785 *L*/gal
WEIGHT		
1,000 mg	= 1 gm	1,000 mg/gm
1,000 gm	= 1 kg	1,000 gm/kg
DENSITY		
1 kg	= 1 liter	1 kg/*L*
PRESSURE		
10.015 m	= 1 kg/sq cm	10.015 m/kg/sq cm
1 Pascal	= 1 N/sq m	1 Pa/N/sq m
1 psi	= 6,895 Pa	1 psi/6,895 Pa
1 psi	= 0.07 kg/sq cm	1 psi/0.07 kg/sq cm
FLOW		
3,785 cu m/day	= 1 MGD	3,785 cu m/day/MGD
3.785 M*L*/day	= 1 MGD	3.785 M*L*/day/MGD

* hectare

A.5 TYPICAL WASTEWATER TREATMENT PLANT PROBLEMS (METRIC SYSTEM)

A.50 Activated Sludge

EXAMPLE 1

In a conventional activated sludge plant, the mixed liquor suspended solids were 2,000 mg/*L*. How many kilograms of solids were under aeration if the aeration tank is 30 meters long, 14 meters wide, and 5 meters deep?

Known		Unknown
MLSS, mg/*L*	= 2,000 mg/*L*	Solids Under Aeration, kg
Aeration Tank Length, m	= 30 m	
Width, m	= 14 m	
Depth, m	= 5 m	

1. Determine the aeration tank volume in cubic meters.

$$\text{Tank Volume, cu m} = \text{Length, m} \times \text{Width, m} \times \text{Depth, m}$$

$$= 30 \text{ m} \times 14 \text{ m} \times 5 \text{ m}$$

$$= 2{,}100 \text{ cu m}$$

2. Calculate the kilograms of solids under aeration.

$$\text{Solids, kg} = \text{MLSS, mg}/L \times \text{Tank Vol, cu m} \times \frac{1 \text{ kg}}{1{,}000{,}000 \text{ mg}} \times \frac{1{,}000\ L}{1 \text{ cu m}}$$

$$= 2{,}000 \text{ mg}/L \times 2{,}100 \text{ cu m} \times \frac{1 \text{ kg}}{1{,}000{,}000 \text{ mg}} \times \frac{1{,}000\ L}{1 \text{ cu m}}$$

$$= 4{,}200 \text{ kg}$$

EXAMPLE 2

Estimate the proper return sludge pumping rate in both cubic meters per day and liters per second for the activated sludge plant in Example 1 if the flow from the primary clarifiers is 15,000 cubic meters per day (15 M*L*D or 15 mega or million liters per day) and the results from the 30-minute settleability test indicate the 400 m*L* of sludge settled in a 2-liter sample (20%). Current return sludge pumping rate is 4,000 cubic meters per day (4.0 M*L*D).

Known		Unknown
Flow to Aerator from Primary Clar, cu m/day	= 15,000 cu m/day	1. Desired Return Sludge Flow, cu m/day
Return Sludge Flow, cu m/day	= 4,000 cu m/day	2. Return Sludge Pumping Rate, liters/sec
Vol of MLSS Settled in 30 min, %	= 20%	

1. Determine total flow through aerator in cubic meters per day.

$$\text{Flow, cu m/day} = \text{Inflow, cu m/day} + \text{Return Flow, cu m/day}$$

$$= 15{,}000 \text{ cu m/day} + 4{,}000 \text{ cu m/day}$$

$$= 19{,}000 \text{ cu m/day}$$

2. Calculate desired return sludge flow in cu m/day.

$$\text{Return Sludge Flow, cu m/day} = \text{Total Flow, cu m/day} \times \frac{\text{Set Sol, \%}}{100\%}$$

$$= 19{,}000 \text{ cu m/day} \times \frac{20\%}{100\%}$$

$$= 3{,}800 \text{ cu m/day (slightly less than actual 4,000 cu m/day)}$$

3. Determine desired return sludge pumping rate in liters per second.

$$\text{Return Sludge Pumping Rate, } L\text{/sec} = \text{Return Flow, } \frac{\text{cu m}}{\text{day}} \times \frac{1{,}000\ L}{\text{cu m}} \times \frac{1 \text{ day}}{24 \text{ hr}} \times \frac{1 \text{ hr}}{3{,}600 \text{ sec}}$$

$$= 3{,}800 \frac{\text{cu m}}{\text{day}} \times \frac{1{,}000\ L}{\text{cu m}} \times \frac{1 \text{ day}}{24 \text{ hr}} \times \frac{1 \text{ hr}}{3{,}600 \text{ sec}}$$

$$= 44\ L\text{/sec}$$

EXAMPLE 3

Determine the sludge age for an activated sludge plant with an aeration tank volume of 2,100 cubic meters that treats a flow of 15,000 cubic meters per day (15 M*L*D). The primary clarifier effluent has a suspended solids concentration of 70 mg/*L* and a mixed liquor suspended solids concentration of 2,400 mg/*L*.

Known		Unknown
MLSS, mg/*L*	= 2,400 mg/*L*	Sludge Age, days
Prim Effl SS, mg/*L*	= 70 mg/*L*	
Infl Flow, cu m/day	= 15,000 cu m/day	
Aerator Vol, cu m	= 2,100 cu m	

1. Determine the kilograms of mixed liquor suspended solids in the aerator.

$$\text{Solids, kg} = \text{MLSS, mg}/L \times \text{Aer Vol, cu m} \times \frac{1 \text{ kg}}{1{,}000{,}000 \text{ mg}} \times \frac{1{,}000\ L}{1 \text{ cu m}}$$

$$= 2{,}400 \text{ mg}/L \times 2{,}100 \text{ cu m} \times \frac{1 \text{ kg}}{1{,}000{,}000 \text{ mg}} \times \frac{1{,}000\ L}{1 \text{ cu m}}$$

$$= 5{,}040 \text{ kg}$$

2. Determine the kilograms of solids added per day to the aeration system by primary effluent.

$$\text{Solids Added by Prim Effl, kg/day} = \text{Flow, } \frac{\text{cu m}}{\text{day}} \times \text{PE SS, } \frac{\text{mg}}{L} \times \frac{1 \text{ kg}}{1{,}000{,}000 \text{ mg}} \times \frac{1{,}000\ L}{1 \text{ cu m}}$$

$$= 15{,}000 \frac{\text{cu m}}{\text{day}} \times 70 \frac{\text{mg}}{L} \times \frac{1 \text{ kg}}{1{,}000{,}000 \text{ mg}} \times \frac{1{,}000\ L}{1 \text{ cu m}}$$

$$= 1{,}050 \text{ kg/day}$$

3. Calculate the sludge age in days.

$$\text{Sludge Age, days} = \frac{\text{Suspended Solids in Aerator, kg}}{\text{Suspended Solids in Primary Effluent, kg/day}}$$

$$= \frac{5{,}040 \text{ kg}}{1{,}050 \text{ kg/day}}$$

$$= 4.8 \text{ days}$$

EXAMPLE 4

If the desired sludge age in the previous example was 5 days, calculate the desired suspended solids concentration in the aerator in kilograms and also in milligrams per liter. Assume 1,050 kilograms per day of solids in the primary effluent and an aerator volume of 2,100 cubic meters.

Known		Unknown
Desired Sludge Age, days	= 5 days	1. Susp Sol in Aerator, kg
Solids in Prim Effl, kg/day	= 1,050 kg/day	2. Desired MLSS, mg/*L*
Aerator Volume, cu m	= 2,100 cu m	

1. Calculate the desired kilograms of suspended solids in the aerator.

$$\text{Sludge Age, days} = \frac{\text{Suspended Solids in Aerator, kg}}{\text{Suspended Solids in Prim Effl, kg/day}}$$

Rearrange the equation to obtain desired kg of suspended solids in aerator or desired MLSS in aerator, kg.

$$\text{SS in Aerator, kg} = \text{Sludge Age, days} \times \text{SS in Prim Effl, kg/day}$$

$$= 5 \text{ days} \times 1{,}050 \text{ kg/day}$$

$$= 5{,}250 \text{ kg}$$

2. Calculate the desired MLSS concentration in mg/*L*.

$$\text{Desired MLSS, mg}/L = \frac{\text{Desired SS in Aerator, kg}}{\text{Aerator Volume, cu m}} \times \frac{1{,}000{,}000 \text{ mg}}{1 \text{ kg}} \times \frac{1 \text{ cu m}}{1{,}000\ L}$$

$$= \frac{5{,}250 \text{ kg}}{2{,}100 \text{ cu m}} \times \frac{1{,}000{,}000 \text{ mg}}{1 \text{ kg}} \times \frac{1 \text{ cu m}}{1{,}000\ L}$$

$$= 2{,}500 \text{ mg}/L$$

EXAMPLE 5

Calculate the new waste activated sludge pumping rate if the current pumping rate is 200 cubic meters per day (2.31 liters per second). Use the information from the previous examples. The actual MLSS in the aerator are 2,955 mg/*L* and the waste activated sludge (WAS) suspended solids concentration is 6,200 mg/*L*. The primary effluent suspended solids concentration is 72 mg/*L*, the flow to the plant is 15,000 cu m per day, the aerator volume is 2,100 cubic meters and the desired sludge age is 5 days.

Known		Unknown
MLSS, mg/*L*	= 2,955 mg/*L*	1. Waste Sludge Pumping Rate, cu m/day
Waste Sl SS, mg/*L*	= 6,200 mg/*L*	2. Waste Sludge Pumping Rate, *L*/sec
Prim Effl SS, mg/*L*	= 72 mg/*L*	
Flow, cu m/day	= 15,000 cu m/day	
Current Waste Sl Pumping Rate, cu m/day or *L*/sec	= 200 cu m/day = 2.31 *L*/sec	
Aerator Vol, cu m	= 2,100 cu m	
Desired Sludge Age, days	= 5 days	

1. Determine the kilograms of mixed liquor suspended solids in the aerator.

$$\text{Solids, kg} = \text{MLSS, } \frac{\text{mg}}{L} \times \text{Aer Vol, cu m} \times \frac{1 \text{ kg}}{1{,}000{,}000 \text{ mg}} \times \frac{1{,}000\ L}{1 \text{ cu m}}$$

$$= 2{,}955 \frac{\text{mg}}{L} \times 2{,}100 \text{ cu m} \times \frac{1 \text{ kg}}{1{,}000{,}000 \text{ mg}} \times \frac{1{,}000\ L}{1 \text{ cu m}}$$

$$= 6{,}206 \text{ kg}$$

2. Determine the kilograms of solids added per day to the system by primary effluent.

$$\text{Solids Added by Prim Effl, kg/day} = \text{Flow, } \frac{\text{cu m}}{\text{day}} \times \text{PE SS, } \frac{\text{mg}}{L} \times \frac{1 \text{ kg}}{1{,}000{,}000 \text{ mg}} \times \frac{1{,}000\ L}{1 \text{ cu m}}$$

$$= 15{,}000 \frac{\text{cu m}}{\text{day}} \times 72 \frac{\text{mg}}{L} \times \frac{1 \text{ kg}}{1{,}000{,}000 \text{ mg}} \times \frac{1{,}000\ L}{1 \text{ cu m}}$$

$$= 1{,}080 \text{ kg/day}$$

3. Calculate the *DESIRED* kilograms of mixed liquor suspended solids in the aerator.

$$\text{Sludge Age, days} = \frac{\text{Suspended Solids in Aerator, kg}}{\text{Suspended Solids in Primary Effluent, kg/day}}$$

$$\text{Desired SS in Aerator, kg} = \text{Sludge Age, days} \times \text{SS in PE, kg/day}$$

$$= 5 \text{ days} \times 1{,}080 \text{ kg/day}$$

$$= 5{,}400 \text{ kg}$$

4. Estimate the increase in waste sludge pumping rate in cubic meters per day.

Increase in Waste Sludge Pumping Rate, cu m/day

$$= \frac{\text{Additional Solids To Be Wasted, kg/day}}{\text{Waste Sludge Conc, mg}/L} \times \frac{1{,}000{,}000 \text{ mg}}{1 \text{ kg}} \times \frac{1 \text{ cu m}}{1{,}000\ L}$$

$$= \frac{(6{,}206 \text{ kg} - 5{,}400 \text{ kg})/\text{day}^*}{6{,}200 \text{ mg}/L} \times \frac{1{,}000{,}000 \text{ mg}}{1 \text{ kg}} \times \frac{1 \text{ cu m}}{1{,}000\ L}$$

$$= 130 \text{ cu m/day}$$

* Biological cultures should be subject to slow changes rather than rapid ones; therefore the kilograms will be removed during a 24-hour period.

5. Calculate the total waste sludge pumping rate.

$$\text{Total Waste Sludge Pumping Rate, cu m/day} = \text{Current Rate, cu m/day} + \text{Inc Rate, cu m/day}$$

$$= 200 \text{ cu m/day} + 130 \text{ cu m/day}$$

$$= 330 \text{ cu m/day}$$

$$\text{Total Waste Sludge Pumping Rate, } L/\text{sec} = 330 \text{ cu m/day} \times \frac{1{,}000\ L}{\text{cu m}} \times \frac{1 \text{ day}}{24 \text{ hr}} \times \frac{1 \text{ hr}}{3{,}600 \text{ sec}}$$

$$= 3.8\ L/\text{sec}$$

NOTE: Many operators do not change the WAS pumping rate by more than ±10 percent per day.

EXAMPLE 6

Determine the amount in kilograms and milligrams per liter of mixed liquor volatile suspended solids (MLVSS) to be maintained in the aerator of a conventional activated sludge plant. Assume a food/microorganism (F/M) ratio of 0.6 kilogram of COD per day per kilogram of mixed liquor volatile suspended solids under aeration for the plant in these examples. The average COD of the primary effluent is 150 mg/*L* and the average daily flow is 15,000 cubic meters per day (15 M*L*D). The aerator volume is 2,100 cubic meters.

Known		Unknown
Prim Effl COD, mg/*L*	= 150 mg/*L*	1. MLVSS, kg
Infl Flow, cu m/day	= 15,000 cu m/day	2. MLVSS, mg/*L*
Aerator Volume, cu m	= 2,100 cu m	
Loading Factor, kg COD/day/kg MLVSS	$= \frac{0.6\text{ kg COD/day}}{1\text{ kg MLVSS}}$	

1. Find the kilograms of COD added to the aerator per day.

$$\text{Prim Effl Loading, kg COD/day} = \left(\text{Flow}, \frac{\text{cu m}}{\text{day}}\right)\left(\text{PE COD}, \frac{\text{mg}}{L}\right)\frac{(1\text{ kg})(1{,}000\ L)}{(1{,}000{,}000\text{ mg})(1\text{ cu m})}$$

$$= \left(15{,}000\ \frac{\text{cu m}}{\text{day}}\right)\left(150\ \frac{\text{mg}}{L}\right)\frac{(1\text{ kg})(1{,}000\ L)}{(1{,}000{,}000\text{ mg})(1\text{ cu m})}$$

$$= 2{,}250\text{ kg COD/day}$$

2. Calculate the desired kilograms of **M**ixed **L**iquor **V**olatile **S**uspended **S**olids under aeration, based on a loading factor of 0.6 kg COD per day per 1 kilogram of MLVSS.

$$\text{MLVSS, kg} = \frac{\text{Primary Effluent, kg COD/day}}{\text{Loading Factor, kg COD/day/kg MLVSS}}$$

$$= \frac{2{,}250\text{ kg COD/day}}{0.6\text{ kg COD/day/kg MLVSS}}$$

$$= 3{,}750\text{ kg MLVSS Under Aeration}$$

3. Calculate the desired **M**ixed **L**iquor **V**olatile **S**uspended **S**olids (MLVSS) concentration in the aerator in milligrams per liter.

$$\text{MLVSS, mg}/L = \frac{(\text{MLVSS, kg})(1{,}000{,}000\text{ mg/kg})}{(\text{Aerator Vol, cu m})(1{,}000\ L/\text{cu m})}$$

$$= \frac{(3{,}750\text{ kg MLVSS})(1{,}000{,}000\text{ mg/kg})}{(2{,}100\text{ cu m})(1{,}000\ L/\text{cu m})}$$

$$= 1{,}786\text{ mg}/L \text{ or } 1{,}800\text{ mg}/L$$

EXAMPLE 7

Determine the **M**ean **C**ell **R**esidence **T**ime (MCRT) for the activated sludge plant in these examples. The necessary information is listed under *KNOWN.*

Known		Unknown
Aerator Volume, cu m	= 4,200 cu m	MCRT, days
Flow to Aerator, cu m/day	= 15,000 cu m/day	
Waste Sludge Flow for Past 24 Hours, cu m/day	= 300 cu m/day	
MLSS, mg/*L*	= 2,400 mg/*L*	
Waste Sludge SS, mg/*L*	= 6,200 mg/*L*	
Final Effl SS, mg/*L*	= 12 mg/*L*	

1. Calculate the kilograms of suspended solids in the aeration system.

$$\text{Aerator SS, kg} = \text{Aer Vol, cu m} \times \text{MLSS}, \frac{\text{mg}}{L} \times \frac{1\text{ kg}}{1{,}000{,}000\text{ mg}} \times \frac{1{,}000\ L}{1\text{ cu m}}$$

$$= 4{,}200\text{ cu m} \times 2{,}400\ \frac{\text{mg}}{L} \times \frac{1\text{ kg}}{1{,}000{,}000\text{ mg}} \times \frac{1{,}000\ L}{1\text{ cu m}}$$

$$= 10{,}080\text{ kg}$$

2. Determine the kilograms per day of solids wasted.

$$\text{SS Wasted, kg/day} = \text{Flow}, \frac{\text{cu m}}{\text{day}} \times \text{Waste SS}, \frac{\text{mg}}{L} \times \frac{1\text{ kg}}{1{,}000{,}000\text{ mg}} \times \frac{1{,}000\ L}{1\text{ cu m}}$$

$$= 300\ \frac{\text{cu m}}{\text{day}} \times 6{,}200\ \frac{\text{mg}}{L} \times \frac{1\text{ kg}}{1{,}000{,}000\text{ mg}} \times \frac{1{,}000\ L}{1\text{ cu m}}$$

$$= 1{,}860\text{ kg SS Wasted/day}$$

3. Estimate the kilograms per day of solids lost in the effluent.

$$\text{SS Lost, kg/day} = \text{Flow}, \frac{\text{cu m}}{\text{day}} \times \text{Effl SS}, \frac{\text{mg}}{L} \times \frac{1\text{ kg}}{1{,}000{,}000\text{ mg}} \times \frac{1{,}000\ L}{1\text{ cu m}}$$

$$= 15{,}000\ \frac{\text{cu m}}{\text{day}} \times 12\ \frac{\text{mg}}{L} \times \frac{1\text{ kg}}{1{,}000{,}000\text{ mg}} \times \frac{1{,}000\ L}{1\text{ cu m}}$$

$$= 180\text{ kg SS Lost in Effluent/day}$$

4. Calculate the **M**ean **C**ell **R**esidence **T**ime (MCRT) in days.

$$\text{MCRT, days} = \frac{\text{Suspended Solids in Aeration System, kg}}{\text{SS Wasted, kg/day + SS Lost, kg/day}}$$

$$= \frac{10{,}080\text{ kg}}{1{,}860\text{ kg/day} + 180\text{ kg/day}}$$

$$= 4.9\text{ days}$$

EXAMPLE 8

If the target or desired MCRT for the activated sludge plant in Example 7 was 6 days, determine the amount of WAS in kilograms per day and the WAS pumping rate in liters per second. The required information is listed under *KNOWN.*

Known		Unknown
SS Under Aeration, kg	= 10,080 kg	1. SS Wasted, kg/day
SS Lost in Effl, kg/day	= 180 kg/day	2. WAS Pumping, *L*/sec
WAS SS, mg/*L*	= 6,200 mg/*L*	
Desired MCRT, days	= 6 days	

1. Determine the desired suspended solids wasted in kilograms per day.

$$\text{SS Wasted, kg/day} = \frac{\text{SS Under Aeration, kg}}{\text{MCRT, days}} - \text{SS Lost, kg/day}$$

$$= \frac{10{,}080\text{ kg}}{6\text{ days}} - 180\text{ kg/day}$$

$$= 1{,}680\text{ kg/day} - 180\text{ kg/day}$$

$$= 1{,}500\text{ kg/day}$$

2. Calculate the waste sludge pumping rate in liters per second.

$$\text{Waste Sludge Pumping Rate, liters/sec} = \frac{\text{SS Wasted, kg/day}}{\text{Waste SS, mg/}L} \times \frac{1{,}000{,}000\text{ mg}}{\text{kg}} \times \frac{1\text{ day}}{24\text{ hr}} \times \frac{1\text{ hr}}{3{,}600\text{ sec}}$$

$$= \frac{1{,}500\text{ kg/day}}{6{,}200\text{ mg/}L} \times \frac{1{,}000{,}000\text{ mg}}{\text{kg}} \times \frac{1\text{ day}}{24\text{ hr}} \times \frac{1\text{ hr}}{3{,}600\text{ sec}}$$

$$= 2.8\text{ liters/sec}$$

A.51 Sludge Digestion

EXAMPLE 9

Calculate the volume of seed sludge in cubic meters needed for a 12-meter diameter digester with a normal water depth of 6 meters, if the seed sludge required is estimated to be 25 percent of the digester volume. Most digesters have sloping bottoms, but assume the normal side wall water depth represents the average digester depth.

Known		Unknown
Diameter, m	= 12 m	Seed Sludge Volume, cu m
Depth, m	= 6 m	
Seed Sludge, % of Dig Vol	= 25%	

1. Determine digester tank volume.

$$\text{Tank Volume, cu m} = (0.785)(\text{Diameter, m})^2 \times \text{Depth, m}$$

$$= (0.785)(12\text{ m})^2 \times 6\text{ m}$$

$$= 678\text{ cu m}$$

2. Calculate seed volume. Assume seed required to be 25 percent or 1/4 of the digester tank volume.

$$\text{Seed Sludge Volume, cu m} = (\text{Tank Volume, cu m})(\text{Portion Vol Seed})$$

$$= 678\text{ cu m} \times 0.25$$

$$= 170\text{ cu m}$$

170 cubic meters of seed sludge will be needed.

EXAMPLE 10

A new plant expects to pump 2,000 liters of raw sludge per day to the digester. The raw sludge is estimated to contain 6 percent solids with a volatile content of 68 percent. Estimate the kilograms of volatile matter needed by the digester and the cubic meters of seed sludge, assuming the seed sludge contains 10 percent solids with 50 percent volatile solids and a specific gravity of 1.08. Assume a loading factor of 0.05 kilogram of volatile matter added per day per kilogram of volatile matter under digestion.

Known		Unknown
Raw Sludge, liters/day	= 2,000 *L*/day	1. Seed, kg VM
Raw Sludge		2. Seed Sludge, cu m
Solids, %	= 6%	
Volatile, %	= 68%	
Seed Sludge		
Solids, %	= 10%	
Volatile, %	= 50%	
Specific Gravity	= 1.08	

$$\text{Digester Loading, } \frac{\text{kg VM/day}}{\text{kg Under Dig}} = \frac{0.05\text{ kg VM Added/day}}{\text{kg VM Under Digestion}}$$

1. Find the kilograms of volatile matter pumped to the digester per day.

$$\text{Volatile Matter Pumped, kg/day} = \text{Raw Sludge, } \frac{L}{\text{day}} \times \frac{1\text{ kg}}{1\,L} \times \frac{\text{Solids, \%}}{100\%} \times \frac{\text{Volatile, \%}}{100\%}$$

$$= 2{,}000\,\frac{L}{\text{day}} \times \frac{1\text{ kg}}{1\,L} \times \frac{6\%}{100\%} \times \frac{68\%}{100\%}$$

$$= 81.6\text{ kg/day}$$

2. Find the kilograms of seed volatile matter needed.

$$\frac{0.05\text{ kg VM Added/day}}{1\text{ kg VM Digester}} = \frac{81.6\text{ kg VM Added/day}}{\text{Seed, kg VM}}$$

By rearranging the above equation,

$$\text{Seed, kg VM} = \frac{81.6\text{ kg VM Added/day}}{0.05\text{ kg VM Added/day}} \times 1\text{ kg VM}$$

$$= 1{,}632\text{ kg VM}$$

NOTE: The terms VS for **V**olatile **S**olids and VM for **V**olatile **M**atter both mean the same.

3. Find the volume of seed sludge needed in cubic meters.

$$\text{Seed Sludge, cu m} = \frac{\text{Seed, kg VM}}{\text{Specific Gravity}} \times \frac{1\text{ cu m}}{1{,}000\text{ kg}} \times \frac{100\%}{\text{Solids, \%}} \times \frac{100\%}{\text{VM, \%}}$$

$$= \frac{1{,}632\text{ kg VM}}{1.08} \times \frac{1\text{ cu m}}{1{,}000\text{ kg}} \times \frac{100\%}{10\%} \times \frac{100\%}{50\%}$$

$$= 30\text{ cu m of Seed Sludge}$$

EXAMPLE 11

Estimate the kilograms of lime needed to neutralize a sour digester if the digester contains 700 cubic meters of sludge with a volatile acids level of 2,300 mg/*L* as acetic acid.

Known		Unknown
Volatile Acids, mg/*L* (Acetic Acid)	= 2,300 mg/*L*	Lime Required, kg $Ca(OH)_2$
Tank Volume, cu m	= 700 cu m	

Calculate the lime required ($Ca(OH)_2$) in kilograms.

$$\text{Lime Req'd, kg} = \text{Vol Acids, } \frac{\text{mg}}{L} \times \text{Tank Vol, cu m} \times \frac{1\text{ kg}}{1{,}000{,}000\text{ mg}} \times \frac{1{,}000\,L}{1\text{ cu m}}$$

$$= 2{,}300\,\frac{\text{mg}}{L} \times 700\text{ cu m} \times \frac{1\text{ kg}}{1{,}000{,}000\text{ mg}} \times \frac{1{,}000\,L}{1\text{ cu m}}$$

$$= 1{,}610\text{ kg}$$

EXAMPLE 12

Calculate the volume of sludge pumped in liters per stroke (revolution) by a piston pump with a 25-centimeter diameter piston and the stroke set at 8 centimeters.

Known		Unknown
Piston Diameter, cm	= 25 cm	Sludge Pumped, liters/stroke
Piston Stroke, cm/stroke	= 8 cm/stroke	

1. Calculate the volume pumped per stroke (revolution) in cubic meters.

$$\text{Volume, cu m/stroke} = (0.785)(\text{Diameter, m})^2 \times (\text{Distance, m/stroke})$$

$$= (0.785)\ \frac{(25\ \text{cm})^2}{(100\ \text{cm/m})^2} \times \frac{8\ \text{cm/stroke}}{100\ \text{cm/m}}$$

$$= (0.785)(0.25\ \text{m})^2(0.08\ \text{m/stroke})$$

$$= 0.0039\ \text{cu m/stroke}$$

2. Determine the volume of sludge pumped in liters per stroke.

$$\text{Sludge Pumped, liters/stroke} = \text{Volume, cu m/stroke} \times \frac{1{,}000\ L}{1\ \text{cu m}}$$

$$= 0.0039\ \text{cu m/stroke} \times \frac{1{,}000\ L}{1\ \text{cu m}}$$

$$= 3.9\ \text{liters/stroke}$$

EXAMPLE 13

If 10,000 liters of raw sludge are pumped to a digester, how many kilograms of dry solids are handled and how many kilograms are subject to digestion (volatile solids)? Assume the raw sludge contains 6.5 percent total solids and has a volatile solids content of 68 percent.

Known		Unknown
Sludge Pumped, liters	= 10,000 liters	1. Dry Solids, kg
		2. Volatile Solids, kg
Total Solids, %	= 6.5%	
Volatile Content, %	= 68%	

1. Calculate the kilograms of dry solids pumped.

$$\text{Dry Solids, kg} = \text{Sludge Pumped, liters} \times \frac{\text{Solids, \%}}{100\%} \times \frac{1\ \text{kg}}{1\ \text{liter}}$$

$$= 10{,}000\ \text{liters} \times \frac{6.5\%}{100\%} \times \frac{1\ \text{kg}}{1\ \text{liter}}$$

$$= 650\ \text{kg Dry Solids Pumped}$$

2. Calculate the kilograms of volatile solids pumped.

$$\text{Volatile Solids, kg} = \text{Total Dry Solids, kg} \times \frac{\text{Volatile Content, \%}}{100\%}$$

$$= 650\ \text{kg} \times \frac{68\%}{100\%}$$

$$= 442\ \text{kg of Volatile Solids Pumped}$$

EXAMPLE 14

Determine the percent reduction of volatile solids as a result of digestion when the raw sludge is 68 percent volatile matter and the digested sludge is 54 percent volatile matter.

Known		Unknown
Raw Sludge VM, %	= 68%	Volatile Matter Reduction, %
Digested Sludge VM, %	= 54%	

Calculate the reduction of volatile matter as a percent.

$$\text{VM Reduction, \%} = \frac{\text{In} - \text{Out}}{\text{In} - (\text{In} \times \text{Out})} \times 100\%$$

$$= \frac{0.68 - 0.54}{0.68 - (0.68 \times 0.54)} \times 100\%$$

$$= \frac{0.14}{0.68 - 0.37} \times 100\%$$

$$= 0.45 \times 100\%$$

$$= 45\%$$

EXAMPLE 15

Determine the digester loading in kilograms of volatile matter per day per cubic meter of digester capacity for the digester in the previous examples. The necessary information is listed under *KNOWN.*

Known		Unknown
Volatile Matter, kg/day	= 600 kg/day	Digester Loading, kg VM/day/cu m
Digester Volume, cubic meters	= 800 cu m	

Calculate the digester loading in kilograms of volatile matter per day per cubic meter of digester capacity.

$$\text{Digester Loading, kg VM/day/cu m} = \frac{\text{Volatile Matter Added, kg/day}}{\text{Digester Volume, cu m}}$$

$$= \frac{600\ \text{kg VM/day}}{800\ \text{cu m}}$$

$$= 0.75\ \text{kg VM/day/cu m}$$

EXAMPLE 16

An anaerobic sludge digester receives 600 kilograms per day of volatile matter. How many kilograms of digested sludge should remain in storage if 10 kilograms of digested sludge should be retained in the digester for every kilogram of volatile matter added to the digester?

Known		Unknown
Volatile Matter Added, kg/day	= 600 kg/day	Dig Sl in Storage, kg
Digested Sludge in Storage, kg Dig Sl/kg VM/day	$= \frac{10 \text{ kg Dig Sl}}{1 \text{ kg VM/day}}$	

Calculate the kilograms of digested sludge needed in storage.

$$\text{Dig Sl in Storage, kg} = \text{Volatile Matter Added, kg/day} \times \frac{\text{Dig Sl in Storage, kg}}{\text{1 kg VM Added/day}}$$

$$= 600 \text{ kg/day} \times \frac{\text{10 kg Dig Sl in Storage}}{\text{1 kg VM/day}}$$

$$= \text{6,000 kg Old Sludge Needed in Storage on a Dry Solids Basis}$$

EXAMPLE 17

Estimate the digester loading in kilograms of volatile matter destroyed per day per cubic meter of digester capacity if 10,000 liters per day of raw sludge is pumped to a digester with a capacity of 700 cubic meters. The raw sludge contains 6.5 percent total solids with a 68 percent volatile matter content. During digestion the volatile matter is reduced 50 percent.

Known		Unknown
Raw Sludge Pumped, liters/day	= 10,000 liters/day	Digester Loading, kg VM/day/cu m
Total Solids, %	= 6.5%	
Volatile Matter, %	= 68%	
Volatile Reduction, %	= 50%	
Digester Volume, cu m	= 700 cu m	

Calculate the kilograms of volatile matter destroyed per day per cubic meter.

$$\text{Digester Loading, kg VM/day/cu m} = \frac{\text{Sludge Pumped, liters/day} \times \frac{\text{Solids, \%}}{100\%} \times \frac{\text{VM, \%}}{100\%} \times \frac{\text{VR, \%}}{100\%} \times \frac{\text{1 kg}}{\text{1 liter}}}{\text{Digester Volume, cu m}}$$

$$= \frac{\frac{10{,}000\ L}{\text{day}} \times \frac{6.5\%}{100\%} \times \frac{68\%}{100\%} \times \frac{50\%}{100\%} \times \frac{\text{1 kg}}{\text{liter}}}{\text{700 cu m}}$$

$$= \frac{\text{221 kg VM/day}}{\text{700 cu m}}$$

$$= \text{0.32 kg VM/day/cu m}$$

EXAMPLE 18

Determine the gas production for a digester in cubic meters of gas produced per day per kilogram of volatile matter destroyed when the gas meter indicates that 170 cubic meters of gas are produced per day and 230 kilograms of volatile matter are destroyed per day.

Known		Unknown
Gas Produced, cu m/day	= 170 cu m/day	Gas Produced, cu m/day/kg VM Destroyed
Volatile Matter Destroyed, kg/day	= 230 kg/day	

Calculate the gas production in cubic meters of gas produced per kilogram of volatile matter destroyed.

$$\text{Gas Produced, cu m/day/kg VM} = \frac{\text{Gas Produced, cu m/day}}{\text{VM Destroyed, kg/day}}$$

$$= \frac{\text{170 cu m Gas/day}}{\text{230 kg VM Destroyed/day}}$$

$$= \text{0.74 cu m Gas/day/kg VM Destroyed}$$

EXAMPLE 19

Calculate the solids balance for a digester. The purpose of the solids balance calculations is to compare values calculated with actual values measured. The necessary information is listed under *KNOWN*. The values to be calculated for the solids balance will be calculated in each step.

Known	Unknown
Input to Digester	Solids Balance for a Digester
10,000 liters of Raw Sludge With 6.5% Solids Content, and 68% Volatile Matter	
Digester Output	
Digested Solids With 4.5% Solids Content, and 54% Volatile Matter	

1. Calculate the kilograms of total solids, water, volatile solids, and inorganic solids pumped into digester.

   a. Total solids to digester.

$$\text{Dry Solids, kg/day} = \text{Raw Sludge, liters/day} \times \frac{\text{Solids, \%}}{100\%} \times \frac{\text{1 kg}}{\text{1 liter}}$$

$$= 10{,}000 \frac{\text{liters}}{\text{day}} \times \frac{6.5\%}{100\%} \times \frac{\text{1 kg}}{\text{1 liter}}$$

$$= \text{650 kg/day}$$

   b. Total water and solids to digester.

$$\text{Water and Solids, kg/day} = \text{Total Solids, kg/day} \times \frac{100\%}{\text{Total Solids, \%}}$$

$$= \text{650 kg/day} \times \frac{100\%}{6.5\%}$$

$$= \text{10,000 kg/day or 10,000 liters/day}$$

c. Water to digester.

$$\text{Water, kg/day} = \text{Water and Solids, kg/day} - \text{Solids, kg/day}$$
$$= 10{,}000 \text{ kg/day} - 650 \text{ kg/day}$$
$$= 9{,}350 \text{ kg/day}$$

d. Volatile solids to digester.

$$\text{Volatile Solids, kg/day} = \text{Total Solids, kg/day} \times \frac{\text{Volatile Matter, \%}}{100\%}$$
$$= 650 \text{ kg/day} \times \frac{68\%}{100\%}$$
$$= 442 \text{ kg/day}$$

e. Inorganic solids to digester.

$$\text{Inorganic Solids, kg/day} = \text{Total Solids, kg/day} - \text{Volatile Solids, kg/day}$$
$$= 650 \text{ kg/day} - 442 \text{ kg/day}$$
$$= 208 \text{ kg/day}$$

f. Calculate the percent reduction in volatile matter in the digester to find the kilograms per day of gas produced during digestion.

$$\text{VM Reduction, \%} = \frac{\text{In} - \text{Out}}{\text{In} - (\text{In} \times \text{Out})} \times 100\%$$
$$= \frac{0.68 - 0.54}{0.68 - (0.68 \times 0.54)} \times 100\%$$
$$= 45\%$$

g. Gas out of digester.

$$\text{Gas, kg/day} = \text{Volatile Solids, kg/day} \times \frac{\text{VM Reduction, \%}}{100\%}$$
$$= 442 \text{ kg/day} \times \frac{45\%}{100\%}$$
$$= 200 \text{ kg/day}$$

2. Determine the kilograms of total, volatile, and inorganic solids removed from the digester as digested sludge to the sludge drying beds or other sludge treatment and disposal processes.

h. Volatile solids out of digester.

$$\text{Volatile Solids, kg/day} = \text{Volatile Solids to Digester, kg/day} - \text{Volatile Solids Out as Gas, kg/day}$$
$$= 442 \text{ kg/day} - 200 \text{ kg/day}$$
$$= 242 \text{ kg/day}$$

i. Total solids out of digester.

$$\text{Total Solids, kg/day} = \frac{\text{Volatile Solids, kg/day}}{\text{Volatile Solids, \%}} \times 100\%$$
$$= \frac{242 \text{ kg/day}}{54\%} \times 100\%$$
$$= 448 \text{ kg/day}$$

j. Inorganic solids out of digester.

$$\text{Inorganic Solids, kg/day} = \text{Total Solids, kg/day} - \text{Volatile Solids, kg/day}$$
$$= 448 \text{ kg/day} - 242 \text{ kg/day}$$
$$= 206 \text{ kg/day}$$

NOTE: This is almost the same amount that went into the digester (see Step e). This indicates that the calculations are correct.

k. Total solids and water out of digester.

$$\text{Water and Solids, kg/day} = \frac{\text{Total Solids, kg/day}}{\text{Solids, \%}} \times 100\%$$
$$= \frac{448 \text{ kg/day}}{4.5\%} \times 100\%$$
$$= 9{,}960 \text{ kg/day}$$

NOTE: Almost same volume as put into digester because of thinner sludge going out.

l. Find total kilograms of water out of digester.

$$\text{Water Out, kg/day} = \text{Water and Solids, kg/day} - \text{Solids, kg/day}$$
$$= 9{,}960 \text{ kg/day} - 448 \text{ kg/day}$$
$$= 9{,}512 \text{ kg/day}$$
$$\text{or} \quad = 9{,}500 \text{ kg/day}$$

m. Compare amounts of water into and out of digester.

$$\text{Water, Change, kg/day} = \text{Water Out, kg/day} - \text{Water In, kg/day}$$
$$= 9{,}500 \text{ kg/day} - 9{,}350 \text{ kg/day}$$
$$= 150 \text{ kg/day Drawdown in Digester}$$

In this case, more water was withdrawn in the thin sludge than was added with the thick sludge. No supernatant was withdrawn from the digester or recycled. All of the recycle material must come from the dewatering operation.

DRYING BED OUTPUT

Dried residual removed, 2 kg water per 1 kg solids or 33 percent solids.

Determine the kilograms of water removed per day with the dried solids and the kilograms per day of drainage water recycled to the plant.

Water in solids.

$$\text{Water in Solids, kg/day} = \text{Total Solids, kg/day} \times \frac{\text{2 kg/day Water}}{\text{1 kg/day Solids}}$$

$$= 448\text{ kg/day} \times \frac{\text{2 kg/day Water}}{\text{1 kg/day Solids}}$$

$$= 896\text{ kg/day}$$

Drainage water recycled to plant.

$$\text{Recycle Water, kg/day} = \text{Water to Drying Beds, kg/day} - \text{Water in Solids, kg/day}$$

$$= 9{,}500\text{ kg/day} - 896\text{ kg/day}$$

$$= 8{,}604\text{ kg/day (less evaporation)}$$

SUMMARY

Constituent	Input to Digester	Input to Drying Bed	Recycle to Plant
Total Solids, kg/day	650	448	
Volatile Solids, kg/day	442	242	
Inorganic Solids, kg/day	208	206	
Water, kg/day	9,350	9,500	8,604
Gas Out, kg/day		200	

A.52 Effluent Discharge, Reclamation, and Reuse

EXAMPLE 20

Estimate the average daily BOD load on the receiving waters in kilograms of BOD per day for an average flow of 8.9 M*L*D (mega or million liters per day) and an average effluent BOD of 21 mg/*L*. What is the monthly load in kilograms of BOD per month for a 30-day month?

Known		Unknown
Avg Flow, M*L*D	= 8.9 M*L*D	1. BOD Load, kg BOD/day
Avg BOD, mg/*L*	= 21 mg/*L*	2. BOD Load, kg BOD/month

1. Estimate the average daily BOD load in kilograms of BOD per day.

$$\text{BOD Load, kg BOD/day} = (\text{Avg Flow, MLD})(\text{Avg BOD, mg/L})(1\text{ kg/M mg})$$

$$= (8.9\text{ MLD})(21\text{ mg/L})(1\text{ kg/M mg})$$

$$= 187\text{ kg/day}$$

2. Estimate the monthly BOD load in kilograms of BOD per month.

$$\text{BOD Load, kg BOD/month} = (\text{BOD Load, kg BOD/day})(30\text{ days/month})$$

$$= (187\text{ kg/day})(30\text{ days/month})$$

$$= 5{,}610\text{ kg BOD/month}$$

EXAMPLE 21

Estimate the average effluent BOD in mg/*L* for a treatment plant for the following information:

Day	S	M	T	W	T	F	S
BOD, mg/*L*	23	25	17	16	19	22	20

Known	Unknown
BOD Information	Average BOD, mg/*L*

Calculate the average BOD in milligrams per liter.

$$\text{Average BOD, mg/L} = \frac{\text{Sum of Measurements, mg/L}}{\text{Number of Measurements}}$$

$$= \frac{23 + 25 + 17 + 16 + 19 + 22 + 20}{7}$$

$$= \frac{142\text{ mg/L}}{7}$$

$$= 20\text{ mg/L}$$

A.53 Maintenance

EXAMPLE 22

Determine the capacity of a pump, in liters per second, that lowers a wet well 0.5 meter in 5 minutes. The wet well is 2 meters wide and 4 meters long. There is no flow into the wet well during the pump test.

Known		Unknown
Time, min	= 5 min	Pump Capacity, *L*/sec
Drop, m	= 0.5 m	
Length, m	= 4 m	
Width, m	= 2 m	

1. Calculate the volume pumped in liters.

$$\text{Volume, L} = (\text{Length, m})(\text{Width, m})(\text{Drop, m})(1{,}000\text{ L/cu m})$$

$$= (4\text{ m})(2\text{ m})(0.5\text{ m})(1{,}000\text{ L/cu m})$$

$$= 4{,}000\text{ liters}$$

2. Estimate the pump capacity in liters per second.

$$\text{Pump Capacity, L/sec} = \frac{\text{Volume Pumped, L}}{(\text{Time, min})(60\text{ sec/min})}$$

$$= \frac{4{,}000\text{ liters}}{(5\text{ min})(60\text{ sec/min})}$$

$$= 13.3\text{ L/sec}$$

EXAMPLE 23

A float (marked stick) is dropped into an open channel 1.2 meters wide with a flow 42 centimeters deep. The float travels 8 meters in 9 seconds. Estimate the flow in the channel in liters per second, cubic meters per day, and mega or million liters per day.

Known		Unknown
Width, m	= 1.2 m	1. Flow, liters/sec
Depth, cm	= 42 cm	2. Flow, cu m/day
Distance, m	= 8 m	3. Flow, M*L*D
Time, sec	= 9 sec	

1. Estimate the velocity in meters per second of the water flowing in the channel.

$$\text{Velocity, m/sec} = \frac{\text{Distance, m}}{\text{Time, sec}}$$

$$= \frac{8\text{ m}}{9\text{ sec}}$$

$$= 0.89\text{ m/sec}$$

NOTE: Velocity of float on a water surface is often 10 to 15 percent higher than the average velocity.

2. Convert float velocity to average velocity. Assume float velocity is 10 percent too high.

$$\text{Avg Vel, m/sec} = \text{Float Vel, m/sec} - (\text{Float Vel, m/sec})(\text{Fraction Too High})$$

$$= 0.89\text{ m/sec} - (0.89\text{ m/sec})\left(\frac{10\%}{100\%}\right)$$

$$= 0.89\text{ m/sec} - 0.089\text{ m/sec}$$

$$= 0.80\text{ m/sec}$$

3. Calculate the flow in the channel in cubic meters per second.

$$\text{Flow, cu m/sec} = (\text{Area, sq m})(\text{Velocity, m/sec})$$

$$= \frac{(1.2\text{ m})(42\text{ cm})(0.8\text{ m/sec})}{(100\text{ cm/m})}$$

$$= 0.40\text{ cu m/sec}$$

4. Convert flow from cubic meters per second to liters per second.

$$\text{Flow, } L\text{/sec} = (\text{Flow cu m/sec})(1{,}000\ L\text{/cu m})$$

$$= (0.40\text{ cu m/sec})(1{,}000\ L\text{/cu m})$$

$$= 400\text{ liters/sec}$$

5. Convert flow from cubic meters per second to cubic meters per day.

$$\text{Flow, cu m/day} = (\text{Flow, cu m/sec})(60\text{ sec/min})(60\text{ min/hr})(24\text{ hr/day})$$

$$= (0.40\text{ cu m/sec})(60\text{ sec/min})(60\text{ min/hr})(24\text{ hr/day})$$

$$= 34{,}560\text{ cu m/day}$$

6. Convert flow from cubic meters per second to mega liters per day.

$$\text{Flow, M}L\text{D} = \frac{(\text{Flow, cu m/sec})(60\text{ sec/min})(60\text{ min/hr})(24\text{ hr/day})(1{,}000\ L\text{/cu m})}{1{,}000{,}000\text{/M}}$$

$$= \frac{(0.40\text{ cu m/sec})(60\text{ sec/min})(60\text{ min/hr})(24\text{ hr/day})(1{,}000\ L\text{/cu m})}{1{,}000{,}000\text{/M}}$$

$$= 34.6\text{ M}L\text{D}$$

EXAMPLE 24

Estimate the velocity in an open channel using a pitot tube. Water rises in the tube 2 centimeters above the water surface.

$V = \sqrt{2gH}$

$g = 980\text{ cm/sec}^2$ (constant for gravity)

Known	Unknown
H, cm = 2 cm	V, m/sec

Estimate the velocity in meters per second.

$$\text{V, m/sec} = \sqrt{2gH}$$

$$= \sqrt{2(980\text{ cm/sec}^2)(2\text{ cm})}$$

$$= \sqrt{3{,}920\text{ cm}^2\text{/sec}^2}$$

$$= 62.6\text{ cm/sec}$$

$$= 0.6\text{ m/sec}$$

NOTE: 1 m = 100 cm

A.54 Laboratory

EXAMPLE 25

Convert the temperature of a sample from 57°F to degrees Celsius.

Known	Unknown
Temperature, °F = 57°F	Temperature, °C

Convert temperature from degrees Fahrenheit to degrees Celsius.

$$\text{Temperature, °C} = (\text{Temperature, °F} - 32\text{°F})(5/9)$$

$$= (57\text{°F} - 32\text{°F})(5/9)$$

$$= (25)(5/9)$$

$$= 13.9\text{°C}$$

EXAMPLE 26

Estimate the sludge pumped to a digester in cubic meters per day from a primary clarifier that treats a flow of 6.4 M*L*D. The influent settleable solids test indicates 11 m*L*/*L* and the effluent of 0.3 m*L*/*L*.

Known	Unknown
Infl Set Sol, m*L*/*L* = 11 m*L*/*L*	Sludge Pumped, cu m/day
Effl Set Sol, m*L*/*L* = 0.3 m*L*/*L*	
Flow, M*L*D = 6.4 M*L*D	

1. Calculate the sludge removed by primary clarifier in milliliters per liter.

$$\text{Sludge Removed, m}L/L = \text{Infl Set Sol, m}L/L - \text{Effl Set Sol, m}L/L$$

$$= 11\text{ m}L/L - 0.3\text{ m}L/L$$

$$= 10.7\text{ m}L/L$$

2. Estimate the sludge pumped to digester in cubic meters per day.

$$\text{Sludge Pumped, cu m/day} = \frac{(\text{Sludge Removed, m}L/L)(\text{Flow, M}L\text{D})(1{,}000{,}000\text{/M})}{(1{,}000\ L\text{/cu m})(1{,}000\text{ m}L/L)}$$

$$= \frac{(10.7\text{ m}L/L)(6.4\text{ M}L\text{D})(1{,}000{,}000\text{/M})}{(1{,}000\ L\text{/cu m})(1{,}000\text{ m}L/L)}$$

$$= 68.5\text{ cu m/day}$$

EXAMPLE 27

Results from lab tests are listed under *KNOWN.* Calculate the items listed under *UNKNOWN.*

Known

Sample Vol, m*L*	= 100 m*L*
Crucible Weight, gm	= 20.4107 gm
Crucible Plus Dry Solids, gm	= 20.4326 gm
Crucible Plus Ash, gm	= 20.4173 gm

Unknown

1. Total Suspended Solids, mg/*L*
2. Volatile Suspended Solids, mg/*L*
3. Volatile Suspended Solids, %
4. Fixed Suspended Solids, mg/*L*
5. Fixed Suspended Solids, %

1. Calculate the total suspended solids in milligrams and in milligrams per liter.

Crucible Plus Dry Solids, gm = 20.4326 gm
Crucible Weight, gm = 20.4107 gm
Dry Weight, gm = 0.0219 gm
or = 21.9 mg

$$\text{Total Susp Solids, mg}/L = \frac{(\text{Dry Weight, mg})(1{,}000\ \text{m}L/L)}{\text{Sample Vol, m}L}$$

$$= \frac{(21.9\ \text{mg})(1{,}000\ \text{m}L/L)}{100\ \text{m}L}$$

$$= 219\ \text{mg}/L$$

2. Calculate the volatile suspended solids in milligrams and milligrams per liter.

Crucible Plus Dry Solids, gm = 20.4326 gm
Crucible Plus Ash, gm = 20.4173 gm
Volatile Weight, gm = 0.0153 gm
or = 15.3 mg

$$\text{Volatile Susp Solids, mg}/L = \frac{(\text{Volatile Weight, mg})(1{,}000\ \text{m}L/L)}{\text{Sample Vol, m}L}$$

$$= \frac{(15.3\ \text{mg})(1{,}000\ \text{m}L/L)}{100\ \text{m}L}$$

$$= 153\ \text{mg}/L$$

3. Calculate the percent volatile suspended solids.

$$\text{Volatile Susp Solids, \%} = \frac{(\text{Volatile SS, mg}/L)(100\%)}{\text{Total SS, mg}/L}$$

$$= \frac{(153\ \text{mg}/L)(100\%)}{219\ \text{mg}/L}$$

$$= 70\%$$

4. Calculate the fixed (inorganic) suspended solids in milligrams and milligrams per liter.

Crucible Plus Ash, gm = 20.4173 gm
Crucible Weight, gm = 20.4107 gm
Ash Weight, gm = 0.0066 gm
or = 6.6 mg

$$\text{Fixed Susp Solids, mg}/L = \frac{(\text{Ash Weight, mg})(1{,}000\ \text{m}L/L)}{\text{Sample Volume, m}L}$$

$$= \frac{(6.6\ \text{mg})(1{,}000\ \text{m}L/L)}{100\ \text{m}L}$$

$$= 66\ \text{mg}/L$$

5. Calculate the percent fixed suspended solids.

$$\text{Fixed Susp Solids, \%} = \frac{(\text{Fixed SS, mg}/L)(100\%)}{\text{Total SS, mg}/L}$$

$$= \frac{(66\ \text{mg}/L)(100\%)}{219\ \text{mg}/L}$$

$$= 30\%\ (\text{Check})$$

EXAMPLE 28

Calculate the percent removal of suspended solids by the primary clarifiers, secondary treatment, and overall plant removal for the information listed under *KNOWN.*

Known		**Unknown**
Infl SS, mg/*L*	= 219 mg/*L*	1. Primary Removal, %
Prim Effl SS, mg/*L*	= 76 mg/*L*	2. Secondary Removal, %
Sec Effl SS, mg/*L*	= 17 mg/*L*	3. Plant Removal, %

1. Calculate the percent removal of suspended solids by the primary clarifiers.

$$\text{Removal, \%} = \frac{(\text{In} - \text{Out})(100\%)}{\text{In}}$$

$$= \frac{(219\ \text{mg}/L - 76\ \text{mg}/L)(100\%)}{219\ \text{mg}/L}$$

$$= 65\%$$

2. Calculate the percent removal of suspended solids by secondary treatment.

$$\text{Removal, \%} = \frac{(\text{In} - \text{Out})(100\%)}{\text{In}}$$

$$= \frac{(76\ \text{mg}/L - 17\ \text{mg}/L)(100\%)}{76\ \text{mg}/L}$$

$$= 78\%$$

3. Calculate the percent removal of suspended solids by the overall treatment plant.

$$\text{Removal, \%} = \frac{(\text{In} - \text{Out})(100\%)}{\text{In}}$$

$$= \frac{(219\ \text{mg}/L - 17\ \text{mg}/L)(100\%)}{219\ \text{mg}/L}$$

$$= 92\%$$

EXAMPLE 29

Calculate the kilograms of suspended solids removed per day by the primary clarifiers in Example 28 if the flow is 11 M*L*D.

Known		Unknown
Flow, M*L*D	= 11 M*L*D	Susp Solids Removed, kg/day
Infl SS, mg/*L*	= 219 mg/*L*	
Prim Effl SS, mg/*L*	= 76 mg/*L*	

Calculate the kilograms of suspended solids removed per day by the primary clarifiers.

$$\text{SS Removed, kg/day} = \frac{(\text{Flow, M}L\text{D})(\text{SS Removed, mg/}L)(1{,}000{,}000/\text{M})}{1{,}000{,}000\ \text{mg/kg}}$$

$$= \frac{(11\ \text{M}L\text{D})(219\ \text{mg/}L - 76\ \text{mg/}L)(1{,}000{,}000/\text{M})}{1{,}000{,}000\ \text{mg/kg}}$$

$$= 1{,}573\ \text{kg/day}$$

Please refer to Section A.35, "Data Analysis," for examples on how to calculate mean, median, mode, and range. Section A.36, "Administration, Safety," has examples on determining injury frequency rate and injury severity rate.

A.6 ADDITIONAL READING

1. *COMPUTATION PROCEDURES FOR WASTEWATER TREATMENT OPERATIONS.* Obtain from Arizona Water and Pollution Control Association, 8615 West Catalina Drive, Phoenix, AZ 85037, or phone (623) 399-6854. Price, $25.00, plus shipping and handling.

The following two books are available from CRC Press LLC, Attn: Order Entry, 6000 Broken Sound Parkway, NW, Suite 300, Boca Raton, FL 33487.

2. *BASIC MATH CONCEPTS FOR WATER AND WASTEWATER PLANT OPERATORS.* Order No. TX8084. ISBN 978-0-87762-808-8. Price, $63.95.

3. *APPLIED MATH FOR WASTEWATER PLANT OPERATORS.* Order No. TX8092. ISBN 978-0-87762-809-5. Price, $64.95.

ABBREVIATIONS

°C	degrees Celsius
°F	degrees Fahrenheit
μ	micron
μg	microgram
μm	micrometer
ac	acres
ac-ft	acre-feet
amp	amperes
atm	atmosphere
CFM	cubic feet per minute
CFS	cubic feet per second
Ci	Curie
cm	centimeters
cu ft	cubic feet
cu in	cubic inches
cu m	cubic meters
cu yd	cubic yards
D	Dalton
dB	decibel
ft	feet or foot
ft-lb/min	foot-pounds per minute
g	gravity
gal	gallons
gal/day	gallons per day
GFD	gallons of flux per square foot per day
gm	grams
GPCD	gallons per capita per day
GPD	gallons per day
gpg	grains per gallon
GPM	gallons per minute
GPY	gallons per year
gr	grains
ha	hectares
HP	horsepower
hr	hours
Hz	hertz
in	inches
J	joules
k	kilos
kg	kilograms
km	kilometers
kN	kilonewtons
kPa	kiloPascals
kW	kilowatts
kWh	kilowatt-hours
L	liters
lb	pounds
lbs/sq in	pounds per square inch
M	mega
M	million
M	molar (or molarity)
m	meters
mA	milliampere
meq	milliequivalent
mg	milligrams
MGD	million gallons per day
mg/*L*	milligrams per liter
min	minutes
m*L*	milliliters
mm	millimeters
N	Newton
N	normal (or normality)
nm	nanometer
ohm	ohm
Pa	Pascal
pCi	picoCurie
pCi/*L*	picoCuries per liter
ppb	parts per billion
ppm	parts per million
psf	pounds per square foot
psi	pounds per square inch
psig	pounds per square inch gauge
RPM	revolutions per minute
SCFM	standard cubic feet per minute
sec	seconds
SI	Le Système International d'Unités
sq ft	square feet
sq in	square inches
W	watt
yd	yards

WASTEWATER WORDS

A Summary of the Words Defined

in

OPERATION OF WASTEWATER TREATMENT PLANTS

PROJECT PRONUNCIATION KEY

by Warren L. Prentice

The Project Pronunciation Key is designed to aid you in the pronunciation of new words. While this key is based primarily on familiar sounds, it does not attempt to follow any particular pronunciation guide. This key is designed solely to aid operators in this program.

You may find it helpful to refer to other available sources for pronunciation help. Each current standard dictionary contains a guide to its own pronunciation key. Each key will be different from each other and from this key. Examples of the difference between the key used in this program and the *WEBSTER'S NEW WORLD COLLEGE DICTIONARY*[1] "Key" are shown below.

In using this key, you should accent (say louder) the syllable that appears in capital letters. The following chart is presented to give examples of how to pronounce words using the Project Key.

	SYLLABLE				
WORD	1st	2nd	3rd	4th	5th
acid	AS	id			
coliform	KOAL	i	form		
biological	BUY	o	LODGE	ik	cull

The first word, *ACID,* has its first syllable accented. The second word, *COLIFORM,* has its first syllable accented. The third word, *BIOLOGICAL,* has its first and third syllables accented.

We hope you will find the key useful in unlocking the pronunciation of any new word.

Term
Project Key
Webster Key
acid
AS-id
aś id
coliform
KOAL-i-form
kō′ lə fôrm
biological
BUY-o-LODGE-i-kull
bī ə läj′ i kəl

[1] The *WEBSTER'S NEW WORLD COLLEGE DICTIONARY,* Fourth Edition, 1999, was chosen rather than an unabridged dictionary because of its availability to the operator. Other editions may be slightly different.

WASTEWATER WORDS

>GREATER THAN >GREATER THAN

DO >5 mg/*L* would be read as DO GREATER THAN 5 mg/*L*.

<LESS THAN <LESS THAN

DO <5 mg/*L* would be read as DO LESS THAN 5 mg/*L*.

A

ABS ABS

Alkyl Benzene Sulfonate. A type of surfactant, or surface active agent, present in synthetic detergents in the United States before 1965. ABS was especially troublesome because it caused foaming and resisted breakdown by biological treatment processes. ABS has been replaced in detergents by linear alkyl sulfonate (LAS), which is biodegradable.

ACEOPS ACEOPS

See ALLIANCE OF CERTIFIED OPERATORS, LAB ANALYSTS, INSPECTORS, AND SPECIALISTS.

ABSORPTION (ab-SORP-shun) ABSORPTION

The taking in or soaking up of one substance into the body of another by molecular or chemical action (as tree roots absorb dissolved nutrients in the soil).

ACCOUNTABILITY ACCOUNTABILITY

When a manager gives power/responsibility to an employee, the employee ensures that the manager is informed of results or events.

ACID ACID

(1) A substance that tends to lose a proton.

(2) A substance that dissolves in water and releases hydrogen ions.

(3) A substance containing hydrogen ions that may be replaced by metals to form salts.

(4) A substance that is corrosive.

ACID REGRESSION STAGE ACID REGRESSION STAGE

A stage of anaerobic digestion during which the production of volatile acids is reduced and acetate and ammonia compounds form, causing the pH to increase.

ACIDITY ACIDITY

The capacity of water or wastewater to neutralize bases. Acidity is expressed in milligrams per liter of equivalent calcium carbonate. Acidity is not the same as pH because water does not have to be strongly acidic (low pH) to have a high acidity. Acidity is a measure of how much base must be added to a liquid to raise the pH to 8.2.

ACTIVATED SLUDGE ACTIVATED SLUDGE

Sludge particles produced in raw or settled wastewater (primary effluent) by the growth of organisms (including zoogleal bacteria) in aeration tanks in the presence of dissolved oxygen. The term "activated" comes from the fact that the particles are teeming with bacteria, fungi, and protozoa. Activated sludge is different from primary sludge in that the sludge particles contain many living organisms that can feed on the incoming wastewater.

ACTIVATED SLUDGE PROCESS — ACTIVATED SLUDGE PROCESS

A biological wastewater treatment process that speeds up the decomposition of wastes in the wastewater being treated. Activated sludge is added to wastewater and the mixture (mixed liquor) is aerated and agitated. After some time in the aeration tank, the activated sludge is allowed to settle out by sedimentation and is disposed of (wasted) or reused (returned to the aeration tank) as needed. The remaining wastewater then undergoes more treatment.

ACUTE HEALTH EFFECT — ACUTE HEALTH EFFECT

An adverse effect on a human or animal body, with symptoms developing rapidly.

ADSORPTION (add-SORP-shun) — ADSORPTION

The gathering of a gas, liquid, or dissolved substance on the surface or interface zone of another material.

ADVANCED WASTE TREATMENT — ADVANCED WASTE TREATMENT

Any process of water renovation that upgrades treated wastewater to meet specific reuse requirements. May include general cleanup of water or removal of specific parts of wastes insufficiently removed by conventional treatment processes. Typical processes include chemical treatment and pressure filtration. Also called TERTIARY TREATMENT.

AERATION (air-A-shun) — AERATION

The process of adding air to water. Air can be added to water by either passing air through water or passing water through air.

AERATION (air-A-shun) LIQUOR — AERATION LIQUOR

Mixed liquor. The contents of the aeration tank, including living organisms and material carried into the tank by either untreated wastewater or primary effluent.

AERATION (air-A-shun) TANK — AERATION TANK

The tank where raw or settled wastewater is mixed with return sludge and aerated. The same as aeration bay, aerator, or reactor.

AEROBES — AEROBES

Bacteria that must have dissolved oxygen (DO) to survive. Aerobes are aerobic bacteria.

AEROBIC (air-O-bick) — AEROBIC

A condition in which atmospheric or dissolved oxygen is present in the aquatic (water) environment.

AEROBIC BACTERIA (air-O-bick back-TEER-e-uh) — AEROBIC BACTERIA

Bacteria that will live and reproduce only in an environment containing oxygen that is available for their respiration (breathing), namely atmospheric oxygen or oxygen dissolved in water. Oxygen combined chemically, such as in water molecules (H_2O), cannot be used for respiration by aerobic bacteria.

AEROBIC (air-O-bick) DECOMPOSITION — AEROBIC DECOMPOSITION

The decay or breaking down of organic material in the presence of free or dissolved oxygen.

AEROBIC (air-O-bick) DIGESTION — AEROBIC DIGESTION

The breakdown of wastes by microorganisms in the presence of dissolved oxygen. This digestion process may be used to treat only waste activated sludge, or trickling filter sludge and primary (raw) sludge, or waste sludge from activated sludge treatment plants designed without primary settling. The sludge to be treated is placed in a large aerated tank where aerobic microorganisms decompose the organic matter in the sludge. This is an extension of the activated sludge process.

AEROBIC (air-O-bick) PROCESS — AEROBIC PROCESS

A waste treatment process conducted under aerobic (in the presence of free or dissolved oxygen) conditions.

AESTHETIC (es-THET-ick) — AESTHETIC

Attractive or appealing.

AGGLOMERATION (uh-glom-er-A-shun) — AGGLOMERATION

The growing or coming together of small scattered particles into larger flocs or particles, which settle rapidly. Also see FLOC.

AIR BINDING — AIR BINDING

The clogging of a filter, pipe, or pump due to the presence of air released from water. Air entering the filter media is harmful to both the filtration and backwash processes. Air can prevent the passage of water during the filtration process and can cause the loss of filter media during the backwash process.

AIR GAP — AIR GAP

An open, vertical drop, or vertical empty space, between a drinking (potable) water supply and potentially contaminated water. This gap prevents the contamination of drinking water by backsiphonage because there is no way potentially contaminated water can reach the drinking water supply.

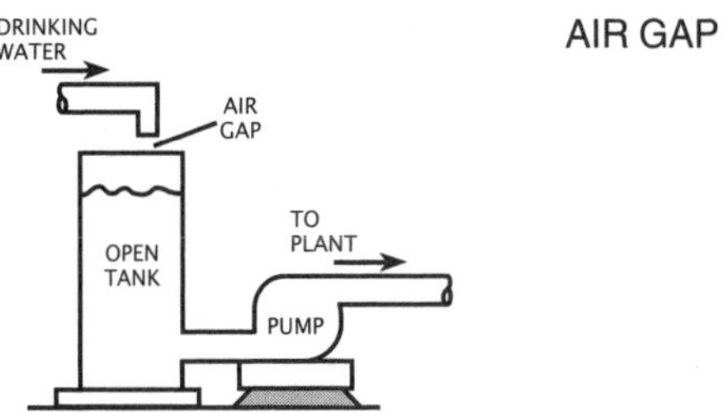

AIR LIFT PUMP — AIR LIFT PUMP

A special type of pump consisting of a vertical riser pipe submerged in the wastewater or sludge to be pumped. Compressed air is injected into a tail piece at the bottom of the pipe. Fine air bubbles mix with the wastewater or sludge to form a mixture lighter than the surrounding water, which causes the mixture to rise in the discharge pipe to the outlet.

AIR PADDING — AIR PADDING

Pumping dry air (dew point –40°F (–40°C)) into a container to assist with the withdrawal of a liquid or to force a liquified gas such as chlorine or sulfur dioxide out of a container.

ALGAE (AL-jee) — ALGAE

Microscopic plants containing chlorophyll that live floating or suspended in water. They also may be attached to structures, rocks, or other submerged surfaces. Excess algal growths can impart tastes and odors to potable water. Algae produce oxygen during sunlight hours and use oxygen during the night hours. Their biological activities appreciably affect the pH, alkalinity, and dissolved oxygen of the water.

ALGAL (AL-gull) BLOOM — ALGAL BLOOM

Sudden, massive growths of microscopic and macroscopic plant life, such as green or blue-green algae, which can, under the proper conditions, develop in lakes, reservoirs, and ponds.

ALIQUOT (AL-uh-kwot) — ALIQUOT

Representative portion of a sample. Often, an equally divided portion of a sample.

ALKALI (AL-kuh-lie) — ALKALI

Any of certain soluble salts, principally of sodium, potassium, magnesium, and calcium, that have the property of combining with acids to form neutral salts and may be used in chemical processes such as water or wastewater treatment.

ALKALINITY (AL-kuh-LIN-it-tee) — ALKALINITY

The capacity of water or wastewater to neutralize acids. This capacity is caused by the water's content of carbonate, bicarbonate, hydroxide, and occasionally borate, silicate, and phosphate. Alkalinity is expressed in milligrams per liter of equivalent calcium carbonate. Alkalinity is not the same as pH because water does not have to be strongly basic (high pH) to have a high alkalinity. Alkalinity is a measure of how much acid must be added to a liquid to lower the pH to 4.5.

ALLIANCE OF CERTIFIED OPERATORS, LAB ANALYSTS, INSPECTORS, AND SPECIALISTS (ACEOPS) — ALLIANCE OF CERTIFIED OPERATORS, LAB ANALYSTS, INSPECTORS, AND SPECIALISTS (ACEOPS)

A professional organization for operators, lab analysts, inspectors, and specialists dedicated to improving professionalism; expanding training, certification, and job opportunities; increasing information exchange; and advocating the importance of certified operators, lab analysts, inspectors, and specialists. For information on membership, contact ACEOPS, PO Box 934, Dakota City, NE 68731-0934, phone (402) 698-2330, or e-mail: Info@aceops.org.

AMBIENT (AM-bee-ent) TEMPERATURE — AMBIENT TEMPERATURE

Temperature of the surroundings.

AMPEROMETRIC (am-purr-o-MET-rick) AMPEROMETRIC

A method of measurement that records electric current flowing or generated, rather than recording voltage. Amperometric titration is a means of measuring concentrations of certain substances in water.

ANAEROBES ANAEROBES

Bacteria that do not need dissolved oxygen (DO) to survive.

ANAEROBIC (AN-air-O-bick) ANAEROBIC

A condition in which atmospheric or dissolved oxygen (DO) is *NOT* present in the aquatic (water) environment.

ANAEROBIC BACTERIA (AN-air-O-bick back-TEER-e-uh) ANAEROBIC BACTERIA

Bacteria that live and reproduce in an environment containing no free or dissolved oxygen. Anaerobic bacteria obtain their oxygen supply by breaking down chemical compounds that contain oxygen, such as sulfate (SO_4^{2-}).

ANAEROBIC (AN-air-O-bick) DECOMPOSITION ANAEROBIC DECOMPOSITION

The decay or breaking down of organic material in an environment containing no free or dissolved oxygen.

ANAEROBIC (AN-air-O-bick) DIGESTION ANAEROBIC DIGESTION

A treatment process in which wastewater solids and water (about 5 percent solids, 95 percent water) are placed in a large tank (the digester) where bacteria decompose the solids in the absence of dissolved oxygen. At least two general groups of bacteria act in balance: (1) saprophytic bacteria break down complex solids to volatile acids, the most common of which are acetic and propionic acids; and (2) methane fermenters break down the acids to methane, carbon dioxide, and water.

ANALOG ANALOG

The continuously variable signal type sent to an analog instrument (for example, 4–20 mA).

ANALOG READOUT ANALOG READOUT

The readout of an instrument by a pointer (or other indicating means) against a dial or scale. Also see DIGITAL READOUT.

ANHYDROUS (an-HI-drous) ANHYDROUS

Very dry. No water or dampness is present.

ANION (AN-EYE-en) ANION

A negatively charged ion in an electrolyte solution, attracted to the anode under the influence of a difference in electrical potential. Chloride ion (Cl^-) is an anion.

ANOXIC (an-OX-ick) ANOXIC

A condition in which the aquatic (water) environment does not contain dissolved oxygen (DO), which is called an oxygen deficient condition. Generally refers to an environment in which chemically bound oxygen, such as in nitrate, is present. The term is similar to ANAEROBIC.

ASEPTIC (a-SEP-tick) ASEPTIC

Free from the living germs of disease, fermentation, or putrefaction. Sterile.

ASPIRATE (AS-per-rate) ASPIRATE

Use of a hydraulic device (aspirator or eductor) to create a negative pressure (suction) by forcing a liquid through a restriction, such as a Venturi tube. An aspirator may be used in the laboratory in place of a vacuum pump; sometimes used instead of a sump pump.

AUTHORITY AUTHORITY

The power and resources to do a specific job or to get that job done.

AVAILABLE EXPANSION AVAILABLE EXPANSION

The vertical distance from the sand surface to the underside of a trough in a sand filter. This distance is also called FREEBOARD.

AXIAL TO IMPELLER AXIAL TO IMPELLER

The direction in which material being pumped flows around the impeller or flows parallel to the impeller shaft.

AXIS OF IMPELLER AXIS OF IMPELLER

An imaginary line running along the center of a shaft (such as an impeller shaft).

B

BOD (pronounce as separate letters) BOD

Biochemical Oxygen Demand. The rate at which organisms use the oxygen in water or wastewater while stabilizing decomposable organic matter under aerobic conditions. In decomposition, organic matter serves as food for the bacteria and energy results from its oxidation. BOD measurements are used as a surrogate measure of the organic strength of wastes in water.

BTU (pronounce as separate letters) BTU

British Thermal Unit. The amount of heat required to raise the temperature of one pound of water one degree Fahrenheit. Also see CALORIE.

BACKFLOW BACKFLOW

A reverse flow condition, created by a difference in water pressures, that causes water to flow back into the distribution pipes of a potable water supply from any source or sources other than an intended source. Also see BACKSIPHONAGE.

BACKSIPHONAGE BACKSIPHONAGE

A form of backflow caused by a negative or below atmospheric pressure within a water system. Also see BACKFLOW.

BACTERIA (back-TEER-e-uh) BACTERIA

Bacteria are living organisms, microscopic in size, that usually consist of a single cell. Most bacteria use organic matter for their food and produce waste products as a result of their life processes.

BACTERIAL (back-TEER-e-ul) CULTURE BACTERIAL CULTURE

In the case of activated sludge, the bacterial culture refers to the group of bacteria classified as AEROBES and FACULTATIVE BACTERIA, which covers a wide range of organisms. Most treatment processes in the United States grow facultative bacteria that use the carbonaceous (carbon compounds) BOD. Facultative bacteria can live when oxygen resources are low. When nitrification is required, the nitrifying organisms are obligate aerobes (require oxygen) and must have at least 0.5 mg/*L* of dissolved oxygen throughout the whole system to function properly.

BAFFLE BAFFLE

A flat board or plate, deflector, guide, or similar device constructed or placed in flowing water, wastewater, or slurry systems to cause more uniform flow velocities, to absorb energy, and to divert, guide, or agitate liquids (water, chemical solutions, slurry).

BASE BASE

(1) A substance that takes up or accepts protons.

(2) A substance that dissociates (separates) in aqueous solution to yield hydroxyl ions (OH^-).

(3) A substance containing hydroxyl ions that reacts with an acid to form a salt or that may react with metals to form precipitates.

BATCH PROCESS BATCH PROCESS

A treatment process in which a tank or reactor is filled, the water (or wastewater or other solution) is treated or a chemical solution is prepared, and the tank is emptied. The tank may then be filled and the process repeated. Batch processes are also used to cleanse, stabilize, or condition chemical solutions for use in industrial manufacturing and treatment processes.

BIOASSAY (BUY-o-AS-say) BIOASSAY

(1) A way of showing or measuring the effect of biological treatment on a particular substance or waste.

(2) A method of determining the relative toxicity of a test sample of industrial wastes or other wastes by using live test organisms, such as fish.

BIOCHEMICAL OXYGEN DEMAND (BOD) BIOCHEMICAL OXYGEN DEMAND (BOD)

See BOD.

BIOCHEMICAL OXYGEN DEMAND (BOD) TEST BIOCHEMICAL OXYGEN DEMAND (BOD) TEST

A procedure that measures the rate of oxygen use under controlled conditions of time and temperature. Standard test conditions include dark incubation at 20°C for a specified time (usually five days).

BIODEGRADABLE (BUY-o-dee-GRADE-able) BIODEGRADABLE

Organic matter that can be broken down by bacteria to more stable forms that will not create a nuisance or give off foul odors is considered biodegradable.

BIODEGRADATION (BUY-o-deh-grah-DAY-shun) BIODEGRADATION

The breakdown of organic matter by bacteria to more stable forms that will not create a nuisance or give off foul odors.

BIOFLOCCULATION (BUY-o-flock-yoo-LAY-shun) BIOFLOCCULATION

The clumping together of fine, dispersed organic particles by the action of certain bacteria and algae. This results in faster and more complete settling of the organic solids in wastewater.

BIOMASS (BUY-o-mass) BIOMASS

A mass or clump of organic material consisting of living organisms feeding on wastes, dead organisms, and other debris. Also see ZOOGLEAL MASS and ZOOGLEAL MAT (FILM).

BIOMONITORING BIOMONITORING

A term used to describe methods of evaluating or measuring the effects of toxic substances in effluents on aquatic organisms in receiving waters. There are two types of biomonitoring, the BIOASSAY and the BIOSURVEY.

BIOSOLIDS BIOSOLIDS

A primarily organic solid product produced by wastewater treatment processes that can be beneficially recycled. The word biosolids is replacing the word sludge when referring to treated waste.

BIOSURVEY BIOSURVEY

A survey of the types and numbers of organisms naturally present in the receiving waters upstream and downstream from plant effluents. Comparisons are made between the aquatic organisms upstream and those organisms downstream of the discharge.

BLANK BLANK

A bottle containing only dilution water or distilled water; the sample being tested is not added. Tests are frequently run on a sample and a blank and the differences are compared. The procedure helps to eliminate or reduce test result errors that could be caused when the dilution water or distilled water used is contaminated.

BLINDING BLINDING

The clogging of the filtering medium of a microscreen or a vacuum filter when the holes or spaces in the media become clogged or sealed off due to a buildup of grease or the material being filtered.

BOND BOND

(1) A written promise to pay a specified sum of money (called the face value) at a fixed time in the future (called the date of maturity). A bond also carries interest at a fixed rate, payable periodically. The difference between a note and a bond is that a bond usually runs for a longer period of time and requires greater formality. Utility agencies use bonds as a means of obtaining large amounts of money for capital improvements.

(2) A warranty by an underwriting organization, such as an insurance company, guaranteeing honesty, performance, or payment by a contractor.

BOUND WATER BOUND WATER

Water contained within the cell mass of sludges or strongly held on the surface of colloidal particles. One of the causes of bulking sludge in the activated sludge process.

BREAKOUT OF CHLORINE (CHLORINE BREAKAWAY) BREAKOUT OF CHLORINE (CHLORINE BREAKAWAY)

A point at which chlorine leaves solution as a gas because the chlorine feed rate is too high. The solution is saturated and cannot dissolve any more chlorine. The maximum strength a chlorine solution can attain is approximately 3,500 mg/*L*. Beyond this concentration molecular chlorine will break out of solution and cause off-gassing at the point of application.

BREAKPOINT CHLORINATION BREAKPOINT CHLORINATION

Addition of chlorine to water or wastewater until the chlorine demand has been satisfied. At this point, further additions of chlorine will result in a free chlorine residual that is directly proportional to the amount of chlorine added beyond the breakpoint.

BRINELLING (bruh-NEL-ing) BRINELLING

Tiny indentations (dents) high on the shoulder of the bearing race or bearing. A type of bearing failure.

BUFFER BUFFER

A solution or liquid whose chemical makeup neutralizes acids or bases without a great change in pH.

BUFFER ACTION | BUFFER ACTION

The action of certain ions in solution in opposing a change in hydrogen ion concentration.

BUFFER CAPACITY | BUFFER CAPACITY

A measure of the capacity of a solution or liquid to neutralize acids or bases. This is a measure of the capacity of water or wastewater for offering a resistance to changes in pH.

BUFFER SOLUTION | BUFFER SOLUTION

A solution containing two or more substances that, in combination, resist any marked change in pH following addition of moderate amounts of either strong acid or base.

BULKING | BULKING

Clouds of billowing sludge that occur throughout secondary clarifiers and sludge thickeners when the sludge does not settle properly. In the activated sludge process, bulking is usually caused by filamentous bacteria or bound water.

C

CFR | CFR

Code of Federal Regulations. A publication of the US government that contains all of the proposed and finalized federal regulations, including safety and environmental regulations.

CHEMTREC (KEM-trek) | CHEMTREC

Chemical Transportation Emergency Center. A public service of the American Chemistry Council dedicated to assisting emergency responders deal with incidents involving hazardous materials. Their toll-free 24-hour emergency phone number is (800) 424-9300.

CALL DATE | CALL DATE

First date a bond can be paid off.

CALORIE (KAL-o-ree) | CALORIE

The amount of heat required to raise the temperature of one gram of water one degree Celsius. Also see BTU.

CARBONACEOUS (car-bun-NAY-shus) STAGE | CARBONACEOUS STAGE

A stage of decomposition that occurs in biological treatment processes when aerobic bacteria, using dissolved oxygen, change carbon compounds to carbon dioxide. Sometimes referred to as first-stage BOD because the microorganisms attack organic or carbon compounds first and nitrogen compounds later. Also see NITRIFICATION STAGE.

CARCINOGEN (kar-SIN-o-jen) | CARCINOGEN

Any substance that tends to produce cancer in an organism.

CATHODIC (kath-ODD-ick) PROTECTION | CATHODIC PROTECTION

An electrical system for prevention of rust, corrosion, and pitting of metal surfaces that are in contact with water, wastewater, or soil. A low-voltage current is made to flow through a liquid (water) or a soil in contact with the metal in such a manner that the external electromotive force renders the metal structure cathodic. This concentrates corrosion on auxiliary anodic parts, which are deliberately allowed to corrode instead of letting the structure corrode.

CATION (KAT-EYE-en) | CATION

A positively charged ion in an electrolyte solution, attracted to the cathode under the influence of a difference in electrical potential. Sodium ion (Na^+) is a cation.

CATION (KAT-EYE-en) EXCHANGE CAPACITY | CATION EXCHANGE CAPACITY

The ability of a soil or other solid to exchange cations (positive ions such as calcium, Ca^{2+}) with a liquid.

CAUTION | CAUTION

This word warns against potential hazards or cautions against unsafe practices. Also see DANGER, NOTICE, and WARNING.

CAVITATION (kav-uh-TAY-shun) CAVITATION

The formation and collapse of a gas pocket or bubble on the blade of an impeller or the gate of a valve. The collapse of this gas pocket or bubble drives water into the impeller or gate with a terrific force that can knock metal particles off and cause pitting on the impeller or gate surface. Cavitation is accompanied by loud noises that sound like someone is pounding on the impeller or gate with a hammer.

CENTRATE CENTRATE

The water leaving a centrifuge after most of the solids have been removed.

CENTRIFUGE CENTRIFUGE

A mechanical device that uses centrifugal or rotational forces to separate solids from liquids.

CERTIFICATION EXAMINATION CERTIFICATION EXAMINATION

An examination administered by a state agency or professional association that operators take to indicate a level of professional competence. In the United States, certification of operators of water treatment plants, wastewater treatment plants, water distribution systems, and small water supply systems is mandatory. In many states, certification of wastewater collection system operators, industrial wastewater treatment plant operators, pretreatment facility inspectors, and small wastewater system operators is voluntary; however, current trends indicate that more states, provinces, and employers will require these operators to be certified in the future. Operator certification is mandatory in the United States for the Chief Operators of water treatment plants, water distribution systems, and wastewater treatment plants.

CERTIFIED OPERATOR CERTIFIED OPERATOR

A person who has the education and experience required to operate a specific class of treatment facility as indicated by possessing a certificate of professional competence given by a state agency or professional association.

CHAIN OF CUSTODY CHAIN OF CUSTODY

A record of each person involved in the handling and possession of a sample from the person who collected the sample to the person who analyzed the sample in the laboratory and to the person who witnessed disposal of the sample.

CHEMICAL EQUIVALENT CHEMICAL EQUIVALENT

The weight in grams of a substance that combines with or displaces one gram of hydrogen. Chemical equivalents usually are found by dividing the formula weight by its valence.

CHEMICAL OXYGEN DEMAND (COD) CHEMICAL OXYGEN DEMAND (COD)

A measure of the oxygen-consuming capacity of organic matter present in wastewater. COD is expressed as the amount of oxygen consumed from a chemical oxidant in mg/*L* during a specific test. Results are not necessarily related to the biochemical oxygen demand (BOD) because the chemical oxidant may react with substances that bacteria do not stabilize.

CHEMICAL PRECIPITATION CHEMICAL PRECIPITATION

(1) Precipitation induced by the addition of chemicals.

(2) The process of softening water by the addition of lime or lime and soda ash as the precipitants.

CHLORAMINES (KLOR-uh-means) CHLORAMINES

Compounds formed by the reaction of hypochlorous acid (or aqueous chlorine) with ammonia.

CHLORINATION (klor-uh-NAY-shun) CHLORINATION

The application of chlorine to water or wastewater, generally for the purpose of disinfection, but frequently for accomplishing other biological or chemical results—aiding coagulation and controlling tastes and odors in drinking water, or controlling odors or sludge bulking in wastewater.

CHLORINE CONTACT CHAMBER CHLORINE CONTACT CHAMBER

A baffled basin that provides sufficient detention time of chlorine contact with wastewater for disinfection to occur. The minimum contact time is usually 30 minutes. Also commonly referred to as basin or tank.

CHLORINE BREAKAWAY CHLORINE BREAKAWAY

See BREAKOUT OF CHLORINE.

CHLORINE DEMAND CHLORINE DEMAND

Chlorine demand is the difference between the amount of chlorine added to water or wastewater and the amount of residual chlorine remaining after a given contact time. Chlorine demand may change with dosage, time, temperature, pH, and nature and amount of the impurities in the water.

Chlorine Demand, mg/*L* = Chlorine Applied, mg/*L* – Chlorine Residual, mg/*L*

CHLORINE REQUIREMENT CHLORINE REQUIREMENT

The amount of chlorine that is needed for a particular purpose. Some reasons for adding chlorine are reducing the MPN (Most Probable Number) of coliform bacteria, obtaining a particular chlorine residual, or oxidizing some substance in the water. In each case, a definite dosage of chlorine will be necessary. This dosage is the chlorine requirement.

CHLORINE RESIDUAL CHLORINE RESIDUAL

The concentration of chlorine present in water after the chlorine demand has been satisfied. The concentration is expressed in terms of the total chlorine residual, which includes both the free and combined or chemically bound chlorine residuals. Also called RESIDUAL CHLORINE.

CHLORORGANIC (klor-or-GAN-ick) CHLORORGANIC

Organic compounds combined with chlorine. These compounds generally originate from, or are associated with, living or dead organic materials, such as algae in water.

CHRONIC HEALTH EFFECT CHRONIC HEALTH EFFECT

An adverse effect on a human or animal body with symptoms that develop slowly over a long period of time or that recur frequently.

CILIATES (SILLY-ates) CILIATES

A class of protozoans distinguished by short hairs on all or part of their bodies.

CLARIFICATION (klair-uh-fuh-KAY-shun) CLARIFICATION

Any process or combination of processes the main purpose of which is to reduce the concentration of suspended matter in a liquid.

CLARIFIER (KLAIR-uh-fire) CLARIFIER

A tank or basin in which water or wastewater is held for a period of time during which the heavier solids settle to the bottom and the lighter materials float to the surface. Also called settling tank or SEDIMENTATION BASIN.

COAGULANT (ko-AGG-yoo-lent) COAGULANT

A chemical that causes very fine particles to clump (floc) together into larger particles. This makes it easier to separate the solids from the liquids by settling, skimming, draining, or filtering.

COAGULANT (ko-AGG-yoo-lent) AID COAGULANT AID

Any chemical or substance used to assist or modify coagulation.

COAGULATION (ko-agg-yoo-LAY-shun) COAGULATION

The clumping together of very fine particles into larger particles (floc) caused by the use of chemicals (coagulants). The chemicals neutralize the electrical charges of the fine particles, allowing them to come closer and form larger clumps.

CODE OF FEDERAL REGULATIONS (CFR) CODE OF FEDERAL REGULATIONS (CFR)

A publication of the US government that contains all of the proposed and finalized federal regulations, including safety and environmental regulations.

COLIFORM (KOAL-i-form) COLIFORM

A group of bacteria found in the intestines of warm-blooded animals (including humans) and also in plants, soil, air, and water. The presence of coliform bacteria is an indication that the water is polluted and may contain pathogenic (disease-causing) organisms. Fecal coliforms are those coliforms found in the feces of various warm-blooded animals, whereas the term "coliform" also includes other environmental sources.

COLLOIDS (KALL-loids) COLLOIDS

Very small, finely divided solids (particles that do not dissolve) that remain dispersed in a liquid for a long time due to their small size and electrical charge. When most of the particles in water have a negative electrical charge, they tend to repel each other. This repulsion prevents the particles from clumping together, becoming heavier, and settling out.

COLORIMETRIC MEASUREMENT COLORIMETRIC MEASUREMENT

A means of measuring unknown chemical concentrations in water by measuring a sample's color intensity. The specific color of the sample, developed by addition of chemical reagents, is measured with a photoelectric colorimeter or is compared with color standards using, or corresponding with, known concentrations of the chemical.

COMBINED AVAILABLE CHLORINE COMBINED AVAILABLE CHLORINE

The total chlorine, present as chloramine or other derivatives, that is present in a water and is still available for disinfection and for oxidation of organic matter. The combined chlorine compounds are more stable than free chlorine forms, but they are somewhat slower in disinfection action.

COMBINED AVAILABLE CHLORINE RESIDUAL COMBINED AVAILABLE CHLORINE RESIDUAL

The concentration of residual chlorine that is combined with ammonia, organic nitrogen, or both in water as a chloramine (or other chloro derivative) and yet is still available to oxidize organic matter and help kill bacteria.

COMBINED CHLORINE COMBINED CHLORINE

The sum of the chlorine species composed of free chlorine and ammonia, including monochloramine, dichloramine, and trichloramine (nitrogen trichloride). Dichloramine is the strongest disinfectant of these chlorine species, but it has less oxidative capacity than free chlorine.

COMBINED RESIDUAL CHLORINATION COMBINED RESIDUAL CHLORINATION

The application of chlorine to water or wastewater to produce a combined available chlorine residual. The residual may consist of chlorine compounds formed by the reaction of chlorine with natural or added ammonia (NH_3) or with certain organic nitrogen compounds.

COMBINED SEWER COMBINED SEWER

A sewer designed to carry both sanitary wastewaters and stormwater or surface water runoff.

COMMINUTION (kom-mih-NEW-shun) COMMINUTION

A mechanical treatment process that cuts large pieces of wastes into smaller pieces so they will not plug pipes or damage equipment. Comminution and SHREDDING usually mean the same thing.

COMMINUTOR (kom-mih-NEW-ter) COMMINUTOR

A device used to reduce the size of the solid materials in wastewater by shredding (comminution). The shredding action is like many scissors cutting to shreds all the large solids in the wastewater.

COMPETENT PERSON COMPETENT PERSON

A competent person is defined by OSHA as a person capable of identifying existing and predictable hazards in the surroundings, or working conditions that are unsanitary, hazardous, or dangerous to employees, and who has authorization to take prompt corrective measures to eliminate the hazards.

COMPOSITE (PROPORTIONAL) SAMPLE COMPOSITE (PROPORTIONAL) SAMPLE

A composite sample is a collection of individual samples obtained at regular intervals, usually every one or two hours during a 24-hour time span. Each individual sample is combined with the others in proportion to the rate of flow when the sample was collected. Equal volume individual samples also may be collected at intervals after a specific volume of flow passes the sampling point or after equal time intervals and still be referred to as a composite sample. The resulting mixture (composite sample) forms a representative sample and is analyzed to determine the average conditions during the sampling period.

COMPOUND COMPOUND

A pure substance composed of two or more elements whose composition is constant. For example, table salt (sodium chloride, NaCl) is a compound.

CONFINED SPACE CONFINED SPACE

Confined space means a space that:

(1) Is large enough and so configured that an employee can bodily enter and perform assigned work; and

(2) Has limited or restricted means for entry or exit (for example, manholes, tanks, vessels, silos, storage bins, hoppers, vaults, and pits are spaces that may have limited means of entry); and

(3) Is not designed for continuous employee occupancy.

Also see DANGEROUS AIR CONTAMINATION and OXYGEN DEFICIENCY.

CONFINED SPACE, CLASS A — CONFINED SPACE, CLASS A

A confined space that presents a situation that is immediately dangerous to life or health (IDLH). These include but are not limited to oxygen deficiency, explosive or flammable atmospheres, and concentrations of toxic substances.

CONFINED SPACE, CLASS B — CONFINED SPACE, CLASS B

A confined space that has the potential for causing injury and illness, if preventive measures are not used, but is not immediately dangerous to life and health.

CONFINED SPACE, CLASS C — CONFINED SPACE, CLASS C

A confined space in which the potential hazard would not require any special modification of the work procedure.

CONFINED SPACE, NON-PERMIT — CONFINED SPACE, NON-PERMIT

A non-permit confined space is a confined space that does not contain or, with respect to atmospheric hazards, have the potential to contain any hazard capable of causing death or serious physical harm.

CONFINED SPACE, PERMIT-REQUIRED (PERMIT SPACE) — CONFINED SPACE, PERMIT-REQUIRED (PERMIT SPACE)

A confined space that has one or more of the following characteristics:

(1) Contains or has a potential to contain a hazardous atmosphere,

(2) Contains a material that has the potential for engulfing an entrant,

(3) Has an internal configuration such that an entrant could be trapped or asphyxiated by inwardly converging walls or by a floor that slopes downward and tapers to a smaller cross section, or

(4) Contains any other recognized serious safety or health hazard.

CONING — CONING

Development of a cone-shaped flow of liquid, like a whirlpool, through sludge. This can occur in a sludge hopper during sludge withdrawal when the sludge becomes too thick. Part of the sludge remains in place while liquid rather than sludge flows out of the hopper. Also called coring.

CONTACT STABILIZATION — CONTACT STABILIZATION

Contact stabilization is a modification of the conventional activated sludge process. In contact stabilization, two aeration tanks are used. One tank is for separate reaeration of the return sludge for at least four hours before it is permitted to flow into the other aeration tank to be mixed with the primary effluent requiring treatment. The process may also occur in one long tank.

CONTINUOUS PROCESS — CONTINUOUS PROCESS

A treatment process in which water is treated continuously in a tank or reactor. The water being treated continuously flows into the tank at one end, is treated as it flows through the tank, and flows out the opposite end as treated water.

CONVENTIONAL TREATMENT — CONVENTIONAL TREATMENT

(1) The common wastewater treatment processes such as preliminary treatment, sedimentation, flotation, trickling filter, rotating biological contactor, activated sludge, and chlorination wastewater treatment processes used by POTWs.

(2) The hydroxide precipitation of metals processes used by pretreatment facilities.

COVERAGE RATIO — COVERAGE RATIO

The coverage ratio is a measure of the ability of the utility to pay the principal and interest on loans and bonds (this is known as debt service) in addition to any unexpected expenses.

CROSS CONNECTION — CROSS CONNECTION

(1) A connection between drinking (potable) water and an unapproved water supply.

(2) A connection between a storm drain system and a sanitary collection system.

(3) Less frequently used to mean a connection between two sections of a collection system to handle anticipated overloads of one system.

CRYOGENIC (KRY-o-JEN-nick) — CRYOGENIC

Very low temperature. Associated with liquified gases (liquid oxygen).

D

DO (pronounce as separate letters) DO

Dissolved Oxygen. DO is the molecular oxygen dissolved in water or wastewater.

DPD METHOD DPD METHOD

A method of measuring the chlorine residual in water. The residual may be determined by either titrating or comparing a developed color with color standards. DPD stands for N,N-diethyl-p-phenylenediamine.

DANGER DANGER

The word *DANGER* is used where an immediate hazard presents a threat of death or serious injury to employees. Also see CAUTION, NOTICE, and WARNING.

DANGEROUS AIR CONTAMINATION DANGEROUS AIR CONTAMINATION

An atmosphere presenting a threat of causing death, injury, acute illness, or disablement due to the presence of flammable and/or explosive, toxic, or otherwise injurious or incapacitating substances.

(1) Dangerous air contamination due to the flammability of a gas, vapor, or mist is defined as an atmosphere containing the gas, vapor, or mist at a concentration greater than 10 percent of its lower explosive (lower flammable) limit (LEL).

(2) Dangerous air contamination due to a combustible particulate is defined as a concentration that meets or exceeds the particulate's lower explosive limit (LEL).

(3) Dangerous air contamination due to the toxicity of a substance is defined as the atmospheric concentration that could result in employee exposure in excess of the substance's permissible exposure limit (PEL).

NOTE: A dangerous situation also occurs when the oxygen level is less than 19.5 percent by volume (OXYGEN DEFICIENCY) or more than 23.5 percent by volume (OXYGEN ENRICHMENT).

DATEOMETER (day-TOM-uh-ter) DATEOMETER

A small calendar disk attached to motors and equipment to indicate the year in which the last maintenance service was performed.

DEBT SERVICE DEBT SERVICE

The amount of money required annually to pay the (1) interest on outstanding debts, or (2) funds due on a maturing bonded debt or the redemption of bonds.

DECHLORINATION (DEE-klor-uh-NAY-shun) DECHLORINATION

The deliberate removal of chlorine from water. The partial or complete reduction of residual chlorine by any chemical or physical process.

DECIBEL (DES-uh-bull) DECIBEL

A unit for expressing the relative intensity of sounds on a scale from zero for the average least perceptible sound to about 130 for the average level at which sound causes pain to humans. Abbreviated dB.

DECOMPOSITION or DECAY DECOMPOSITION or DECAY

The conversion of chemically unstable materials to more stable forms by chemical or biological action.

DEGRADATION (deh-gruh-DAY-shun) DEGRADATION

The conversion or breakdown of a substance to simpler compounds, for example, the degradation of organic matter to carbon dioxide and water.

DELEGATION DELEGATION

The act in which power is given to another person in the organization to accomplish a specific job.

DENITRIFICATION (dee-NYE-truh-fuh-KAY-shun) DENITRIFICATION

(1) The anoxic biological reduction of nitrate nitrogen to nitrogen gas.

(2) The removal of some nitrogen from a system.

(3) An anoxic process that occurs when nitrite or nitrate ions are reduced to nitrogen gas and nitrogen bubbles are formed as a result of this process. The bubbles attach to the biological floc and float the floc to the surface of the secondary clarifiers. This condition is often the cause of rising sludge observed in secondary clarifiers or gravity thickeners. Also see NITRIFICATION.

DENSITY DENSITY

A measure of how heavy a substance (solid, liquid, or gas) is for its size. Density is expressed in terms of weight per unit volume, that is, grams per cubic centimeter or pounds per cubic foot. The density of water (at 4°C or 39°F) is 1.0 gram per cubic centimeter or about 62.4 pounds per cubic foot.

DESICCANT (DESS-uh-kant) DESICCANT

A drying agent that is capable of removing or absorbing moisture from the atmosphere in a small enclosure.

DESICCATOR (DESS-uh-kay-tor) DESICCATOR

A closed container into which heated weighing or drying dishes are placed to cool in a dry environment in preparation for weighing. The dishes may be empty or they may contain a sample. Desiccators contain a substance (DESICCANT), such as anhydrous calcium chloride, that absorbs moisture and keeps the relative humidity near zero so that the dish or sample will not gain weight from absorbed moisture.

DETENTION TIME DETENTION TIME

(1) The time required to fill a tank at a given flow.

(2) The theoretical (calculated) time required for water to pass through a tank at a given rate of flow.

(3) The actual time in hours, minutes, or seconds that a small amount of water is in a settling basin, flocculating basin, or rapid-mix chamber. In septic tanks, detention time will decrease as the volumes of sludge and scum increase. In storage reservoirs, detention time is the length of time entering water will be held before being drafted for use (several weeks to years, several months being typical).

$$\text{Detention Time, hr} = \frac{(\text{Basin Volume, gal})(24\text{ hr/day})}{\text{Flow, gal/day}}$$

or

$$\text{Detention Time, hr} = \frac{(\text{Basin Volume, cu m})(24\text{ hr/day})}{\text{Flow, cu m/day}}$$

DETRITUS (dee-TRY-tus) DETRITUS

The heavy mineral material present in wastewater such as sand, coffee grounds, eggshells, gravel, and cinders. Also called GRIT.

DEW POINT DEW POINT

The temperature to which air with a given quantity of water vapor must be cooled to cause condensation of the vapor in the air.

DEWATER DEWATER

(1) To remove or separate a portion of the water present in a sludge or slurry. To dry sludge so it can be handled and disposed of.

(2) To remove or drain the water from a tank or a trench. A structure may be dewatered so that it can be inspected or repaired.

DEWATERABLE DEWATERABLE

This is a property of sludge related to the ability to separate the liquid portion from the solid, with or without chemical conditioning. A material is considered dewaterable if water will readily drain from it.

DIAPHRAGM PUMP DIAPHRAGM PUMP

A pump in which a flexible diaphragm, generally of rubber or equally flexible material, is the operating part. It is fastened at the edges in a vertical cylinder. When the diaphragm is raised, suction is exerted, and when it is depressed, the liquid is forced through a discharge valve.

DIFFUSED-AIR AERATION DIFFUSED-AIR AERATION

A diffused-air activated sludge plant takes air, compresses it, and then discharges the air below the water surface of the aerator through some type of air diffusion device.

DIFFUSER DIFFUSER

A device (porous plate, tube, bag) used to break the air stream from the blower system into fine bubbles in an aeration tank or reactor.

DIGESTER (dye-JEST-er) DIGESTER

A tank in which sludge is placed to allow decomposition by microorganisms. Digestion may occur under anaerobic (more common) or aerobic conditions.

DIGITAL — DIGITAL

The encoding of information that uses binary numbers (ones and zeros) for input, processing, transmission, storage, or display, rather than a continuous spectrum of values (an analog system) or non-numeric symbols such as letters or icons.

DIGITAL READOUT — DIGITAL READOUT

The readout of an instrument by a direct, numerical reading of the measured value or variable.

DISCHARGE HEAD — DISCHARGE HEAD

The pressure (in pounds per square inch (psi) or kilopascals (kPa)) measured at the centerline of a pump discharge and very close to the discharge flange, converted into feet or meters. The pressure is measured from the centerline of the pump to the hydraulic grade line of the water in the discharge pipe.

Discharge Head, ft = (Discharge Pressure, psi)(2.31 ft/psi)

or

Discharge Head, m = (Discharge Pressure, kPa)(1 m/9.8 kPa)

DISINFECTION (dis-in-FECT-shun) — DISINFECTION

The process designed to kill or inactivate most microorganisms in water or wastewater, including essentially all pathogenic (disease-causing) bacteria. There are several ways to disinfect, with chlorination being the most frequently used in water and wastewater treatment plants. Compare with STERILIZATION.

DISSOLVED OXYGEN — DISSOLVED OXYGEN

Molecular oxygen dissolved in water or wastewater, usually abbreviated DO.

DISTILLATE (DIS-tuh-late) — DISTILLATE

In the distillation of a sample, a portion is collected by evaporation and recondensation; the part that is recondensed is the distillate.

DISTRIBUTOR — DISTRIBUTOR

The rotating mechanism that distributes the wastewater evenly over the surface of a trickling filter or other process unit. Also see FIXED SPRAY NOZZLE.

DOCTOR BLADE — DOCTOR BLADE

A blade used to remove any excess solids that may cling to the outside of a rotating screen.

DOGS — DOGS

Wedges attached to a slide gate and frame that force the gate to seal tightly.

DRIFT — DRIFT

The difference between the actual value and the desired value (or set point); characteristic of proportional controllers that do not incorporate reset action. Also called OFFSET.

DUCKWEED — DUCKWEED

A small, green, cloverleaf-shaped floating plant, about one-quarter inch (6 mm) across, which appears as a grainy layer on the surface of a pond.

DYNAMIC HEAD — DYNAMIC HEAD

When a pump is operating, the vertical distance (in feet or meters) from a point to the energy grade line. Also see ENERGY GRADE LINE, STATIC HEAD, and TOTAL DYNAMIC HEAD.

DYNAMIC PRESSURE — DYNAMIC PRESSURE

When a pump is operating, pressure resulting from the dynamic head.

Dynamic Pressure, psi = (Dynamic Head, ft)(0.433 psi/ft)

or

Dynamic Pressure, kPa = (Dynamic Head, m)(9.8 kPa/m)

E

EGL EGL

See ENERGY GRADE LINE.

EDUCTOR (e-DUCK-ter) EDUCTOR

A hydraulic device used to create a negative pressure (suction) by forcing a liquid through a restriction, such as a Venturi. An eductor or aspirator (the hydraulic device) may be used in the laboratory in place of a vacuum pump. As an injector, it is used to produce vacuum for chlorinators. Sometimes used instead of a suction pump.

EFFLORESCENCE (EF-low-RESS-ense) EFFLORESCENCE

The powder or crust formed on a substance when moisture is given off upon exposure to the atmosphere.

EFFLUENT (EF-loo-ent) EFFLUENT

Water or other liquid—raw (untreated), partially treated, or completely treated—flowing *FROM* a reservoir, basin, treatment process, or treatment plant.

ELECTROCHEMICAL PROCESS ELECTROCHEMICAL PROCESS

A process that causes the deposition or formation of a seal or coating of a chemical element or compound by the use of electricity.

ELECTROLYSIS (ee-leck-TRAWL-uh-sis) ELECTROLYSIS

The decomposition of material by an outside electric current.

ELECTROLYTE (ee-LECK-tro-lite) ELECTROLYTE

A substance that dissociates (separates) into two or more ions when it is dissolved in water.

ELECTROLYTIC (ee-LECK-tro-LIT-ick) PROCESS ELECTROLYTIC PROCESS

A process that causes the decomposition of a chemical compound by the use of electricity.

ELECTROMAGNETIC FORCES ELECTROMAGNETIC FORCES

Forces resulting from electrical charges that either attract or repel particles. Particles with opposite charges are attracted to each other, while particles with similar charges repel each other. For example, a particle with a positive charge is attracted to a particle with a negative charge but is repelled by another particle with a positive charge.

ELECTRON ELECTRON

(1) A very small, negatively charged particle that is practically weightless. According to the electron theory, all electrical and electronic effects are caused either by the movement of electrons from place to place or because there is an excess or lack of electrons at a particular place.

(2) The part of an atom that determines its chemical properties.

ELEMENT ELEMENT

A substance that cannot be separated into its constituent parts and still retain its chemical identity. For example, sodium (Na) is an element.

ELUTRIATION (e-LOO-tree-A-shun) ELUTRIATION

The washing of digested sludge with either fresh water, plant effluent, or other wastewater. The objective is to remove (wash out) fine particulates and/or the alkalinity in sludge. This process reduces the demand for conditioning chemicals and improves settling or filtering characteristics of the solids.

EMULSION (e-MULL-shun) EMULSION

A liquid mixture of two or more liquid substances not normally dissolved in one another; one liquid is held in suspension in the other.

ENCLOSED SPACE ENCLOSED SPACE

See CONFINED SPACE.

END POINT — END POINT

The completion of a desired chemical reaction. Samples of water or wastewater are titrated to the end point. This means that a chemical is added, drop by drop, to a sample until a certain color change (blue to clear, for example) occurs. This is called the end point of the titration. In addition to a color change, an end point may be reached by the formation of a precipitate or the reaching of a specified pH. An end point may be detected by the use of an electronic device, such as a pH meter.

ENDOGENOUS (en-DODGE-en-us) RESPIRATION — ENDOGENOUS RESPIRATION

A situation in which living organisms oxidize some of their own cellular mass instead of new organic matter they adsorb or absorb from their environment.

ENERGY GRADE LINE (EGL) — ENERGY GRADE LINE (EGL)

A line that represents the elevation of energy head (in feet or meters) of water flowing in a pipe, conduit, or channel. The line is drawn above the hydraulic grade line (gradient) a distance equal to the velocity head ($V^2/2g$) of the water flowing at each section or point along the pipe or channel. Also see HYDRAULIC GRADE LINE.

[SEE DRAWING ON PAGE 827]

ENGULFMENT — ENGULFMENT

Engulfment means the surrounding and effective capture of a person by a liquid or finely divided (flowable) solid substance that can be aspirated to cause death by filling or plugging the respiratory system or that can exert enough force on the body to cause death by strangulation, constriction, or crushing.

ENTERIC — ENTERIC

Of intestinal origin, especially applied to wastes or bacteria.

ENTRAIN — ENTRAIN

To trap bubbles in water either mechanically through turbulence or chemically through a reaction.

ENZYMES (EN-zimes) — ENZYMES

Organic or biochemical substances that cause or speed up chemical reactions.

EQUALIZING BASIN — EQUALIZING BASIN

A holding basin in which variations in flow and composition of a liquid are averaged. Such basins are used to provide a flow of reasonably uniform volume and composition to a treatment unit. Also called a balancing reservoir.

ESTUARIES (ES-chew-wear-eez) — ESTUARIES

Bodies of water that are located at the lower end of a river and are subject to tidal fluctuations.

EVAPOTRANSPIRATION (ee-VAP-o-TRANS-purr-A-shun) — EVAPOTRANSPIRATION

(1) The process by which water vapor is released to the atmosphere from living plants. Also called TRANSPIRATION.

(2) The total water removed from an area by transpiration (plants) and by evaporation from soil, snow, and water surfaces.

EXPLOSIMETER — EXPLOSIMETER

An instrument used to detect explosive atmospheres. When the lower explosive limit (LEL) of an atmosphere is exceeded, an alarm signal on the instrument is activated. Also called a combustible gas detector.

F

F/M RATIO — F/M RATIO

See FOOD/MICROORGANISM RATIO.

FACULTATIVE (FACK-ul-tay-tive) BACTERIA — FACULTATIVE BACTERIA

Facultative bacteria can use either dissolved oxygen or oxygen obtained from food materials such as sulfate or nitrate ions. In other words, facultative bacteria can live under aerobic, anoxic, or anaerobic conditions.

FACULTATIVE (FACK-ul-tay-tive) POND — FACULTATIVE POND

The most common type of pond in current use. The upper portion (supernatant) is aerobic, while the bottom layer is anaerobic. Algae supply most of the oxygen to the supernatant.

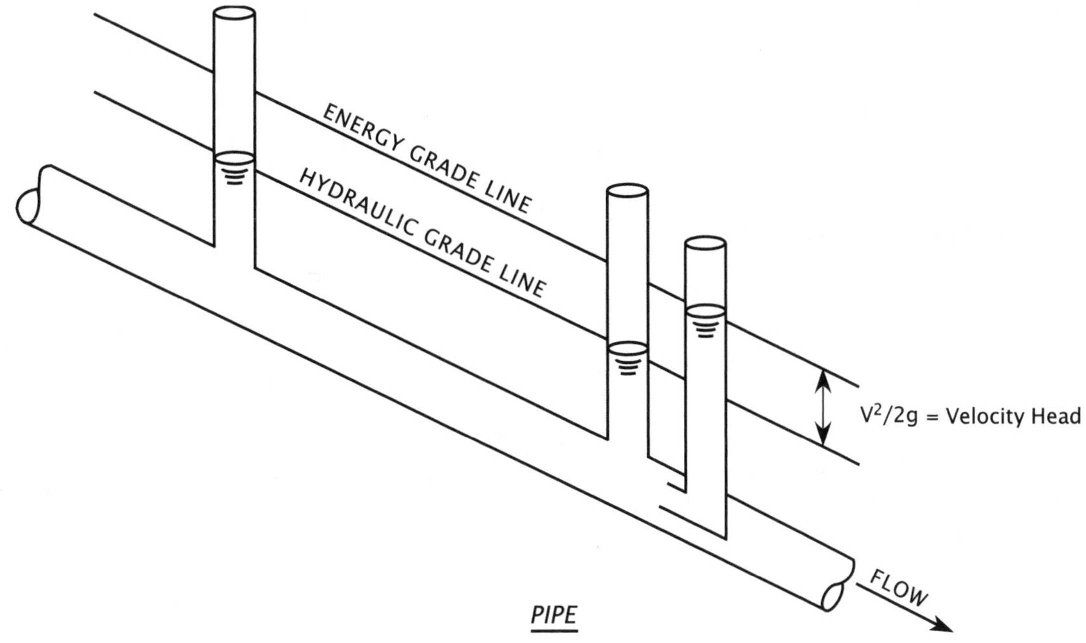

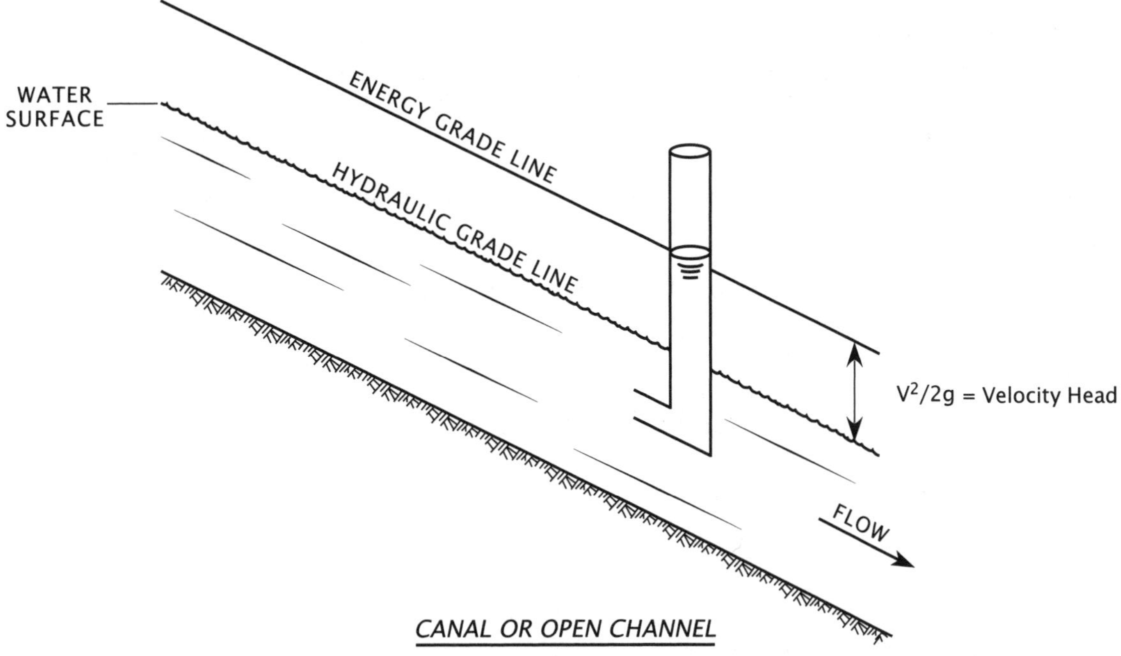

Energy grade line and hydraulic grade line

FILAMENTOUS (fill-uh-MEN-tuss) ORGANISMS FILAMENTOUS ORGANISMS

Organisms that grow in a thread or filamentous form. Common types are *Thiothrix* and *Actinomycetes*. A common cause of sludge bulking in the activated sludge process.

FILTER AID FILTER AID

A chemical (usually a polymer) added to water to help remove fine colloidal suspended solids.

FIXED SAMPLE FIXED SAMPLE

A sample is fixed in the field by adding chemicals that prevent the water quality indicators of interest in the sample from changing before final measurements are performed later in the laboratory.

FIXED SPRAY NOZZLE FIXED SPRAY NOZZLE

Cone-shaped spray nozzle used to distribute wastewater over the filter media, similar to a lawn sprinkling system. A deflector or steel ball is mounted within the cone to spread the flow of wastewater through the cone, thus causing a spraying action. Also see DISTRIBUTOR.

FLAME POLISHED FLAME POLISHED

Melted by a flame to smooth out irregularities. Sharp or broken edges of glass (such as the end of a glass tube) are rotated in a flame until the edge melts slightly and becomes smooth.

FLIGHTS FLIGHTS

Scraper boards, made from redwood or other rot-resistant woods or plastic, used to collect and move settled sludge or floating scum.

FLOC FLOC

Clumps of bacteria and particles, or coagulants and impurities, that have come together and formed a cluster. Found in flocculation tanks, sedimentation basins, aeration tanks, secondary clarifiers, and chemical precipitation processes.

FLOCCULATION (flock-yoo-LAY-shun) FLOCCULATION

The gathering together of fine particles after coagulation to form larger particles by a process of gentle mixing. This clumping together makes it easier to separate the solids from the water by settling, skimming, draining, or filtering.

FLOW EQUALIZATION SYSTEM FLOW EQUALIZATION SYSTEM

A device or tank designed to hold back or store a portion of peak flows for release during low-flow periods.

FOOD/MICROORGANISM (F/M) RATIO FOOD/MICROORGANISM (F/M) RATIO

Food to microorganism ratio. A measure of food provided to bacteria in an aeration tank.

$$\frac{\text{Food}}{\text{Microorganisms}} = \frac{\text{BOD, lbs/day}}{\text{MLVSS, lbs}}$$

$$= \frac{\text{Flow, MGD} \times \text{BOD, mg/}L \times 8.34\text{ lbs/gal}}{\text{Volume, MG} \times \text{MLVSS, mg/}L \times 8.34\text{ lbs/gal}}$$

or by calculator math system

$$= \text{Flow, MGD} \times \text{BOD, mg/}L \div \text{Volume, MG} \div \text{MLVSS, mg/}L$$

or metric

$$= \frac{\text{BOD, kg/day}}{\text{MLVSS, kg}}$$

$$= \frac{\text{Flow, M}L\text{/day} \times \text{BOD, mg/}L \times 1\text{ kg/M mg}}{\text{Volume, M}L \times \text{MLVSS, mg/}L \times 1\text{ kg/M mg}}$$

FORCE MAIN FORCE MAIN

A pipe that carries wastewater under pressure from the discharge side of a pump to a point of gravity flow downstream.

FREE AVAILABLE CHLORINE RESIDUAL FREE AVAILABLE CHLORINE RESIDUAL

That portion of the total available chlorine residual composed of dissolved chlorine gas (Cl_2), hypochlorous acid (HOCl), and/or hypochlorite ion (OCl^-) remaining in water after chlorination. This does not include chlorine that has combined with ammonia, nitrogen, or other compounds. Also called free available residual chlorine.

FREE CHLORINE FREE CHLORINE

Free chlorine is chlorine (Cl_2) in a liquid or gaseous form. Free chlorine combines with water to form hypochlorous (HOCl) and hydrochloric (HCl) acids. In wastewater, free chlorine usually combines with an amine (ammonia or nitrogen) or other organic compounds to form combined chlorine compounds.

FREE OXYGEN FREE OXYGEN

Molecular oxygen available for respiration by organisms. Molecular oxygen is the oxygen molecule, O_2, that is not combined with another element to form a compound.

FREEBOARD FREEBOARD

(1) The vertical distance from the normal water surface to the top of the confining wall.

(2) The vertical distance from the sand surface to the underside of a trough in a sand filter. This distance is also called AVAILABLE EXPANSION.

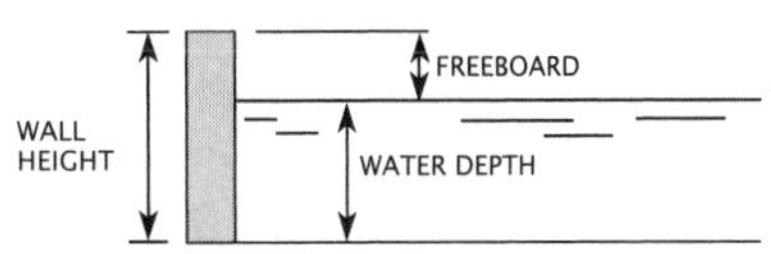

FRICTION LOSS FRICTION LOSS

The head, pressure, or energy (they are the same) lost by water flowing in a pipe or channel as a result of turbulence caused by the velocity of the flowing water and the roughness of the pipe, channel walls, or restrictions caused by fittings. Water flowing in a pipe loses head, pressure, or energy as a result of friction losses. Also called HEAD LOSS.

G

GIS GIS

See GEOGRAPHIC INFORMATION SYSTEM.

GAS, SEWER GAS, SEWER

See SEWER GAS.

GASIFICATION (gas-uh-fuh-KAY-shun) GASIFICATION

The conversion of soluble and suspended organic materials into gas during aerobic or anaerobic decomposition. In clarifiers, the resulting gas bubbles can become attached to the settled sludge and cause large clumps of sludge to rise and float on the water surface. In anaerobic sludge digesters, this gas is collected for fuel or disposed of using a waste gas burner.

GEOGRAPHIC INFORMATION SYSTEM (GIS) GEOGRAPHIC INFORMATION SYSTEM (GIS)

A computer program that combines mapping with detailed information about the physical locations of structures, such as pipes, valves, and manholes, within geographic areas. The system is used to help operators and maintenance personnel locate utility system features or structures and to assist with the scheduling and performance of maintenance activities.

GRAB SAMPLE GRAB SAMPLE

A single sample of water collected at a particular time and place that represents the composition of the water only at that time and place.

GRAVIMETRIC GRAVIMETRIC

A means of measuring unknown concentrations of water quality indicators in a sample by weighing a precipitate or residue of the sample.

GRIT GRIT

The heavy mineral material present in wastewater such as sand, coffee grounds, eggshells, gravel, and cinders. Also called DETRITUS.

GRIT CHAMBER GRIT CHAMBER

A detention chamber or an enlargement of a collection line designed to reduce the velocity of flow of the liquid to permit the separation of mineral solids from organic solids by differential sedimentation.

GRIT CHANNEL GRIT CHANNEL

(1) An enlargement in a collection line where grit can easily settle out of the flow.

(2) The waterway of a grit chamber.

GRIT COLLECTOR GRIT COLLECTOR

A device placed in a grit chamber to carry deposited grit to a point of collection for ultimate disposal.

GRIT COMPARTMENT GRIT COMPARTMENT

The portion of the grit chamber in which grit is collected and stored before removal.

GRIT REMOVAL GRIT REMOVAL

Grit removal is accomplished by providing an enlarged channel or chamber that causes the flow velocity to be reduced and allows the heavier grit to settle to the bottom of the channel where it can be removed.

GRIT TANK GRIT TANK

A structure located at the inlet to a treatment plant for the accumulation and removal of grit.

GRIT TRAP GRIT TRAP

A permanent structure built into a manhole (or other convenient location in a collection system) for the accumulation and easy removal of grit. Also see SAND TRAP.

GROWTH RATE, Y GROWTH RATE, Y

An experimentally determined constant to estimate the unit growth rate of bacteria while degrading organic wastes.

H

HGL HGL

See HYDRAULIC GRADE LINE.

HARMFUL PHYSICAL AGENT or TOXIC SUBSTANCE HARMFUL PHYSICAL AGENT or TOXIC SUBSTANCE

Any chemical substance, biological agent (bacteria, virus, or fungus), or physical stress (noise, heat, cold, vibration, repetitive motion, ionizing and non-ionizing radiation, hypo- or hyperbaric pressure) that:

(1) Is regulated by any state or federal law or rule due to a hazard to health

(2) Is listed in the latest printed edition of the National Institute of Occupational Safety and Health (NIOSH) Registry of Toxic Effects of Chemical Substances (RTECS)

(3) Has yielded positive evidence of an acute or chronic health hazard in human, animal, or other biological testing conducted by, or known to, the employer

(4) Is described by a Material Safety Data Sheet (MSDS) available to the employer that indicates that the material may pose a hazard to human health

Also see ACUTE HEALTH EFFECT and CHRONIC HEALTH EFFECT.

HEAD HEAD

The vertical distance, height, or energy of water above a reference point. A head of water may be measured in either height (feet or meters) or pressure (pounds per square inch or kilograms per square centimeter). Also see DISCHARGE HEAD, DYNAMIC HEAD, STATIC HEAD, SUCTION HEAD, SUCTION LIFT, and VELOCITY HEAD.

HEAD LOSS HEAD LOSS

The head, pressure, or energy (they are the same) lost by water flowing in a pipe or channel as a result of turbulence caused by the velocity of the flowing water and the roughness of the pipe, channel walls, or restrictions caused by fittings. Water flowing in a pipe loses head, pressure, or energy as a result of friction losses. The head loss through a filter is due to friction losses caused by material building up on the surface or in the top part of a filter. Also called FRICTION LOSS.

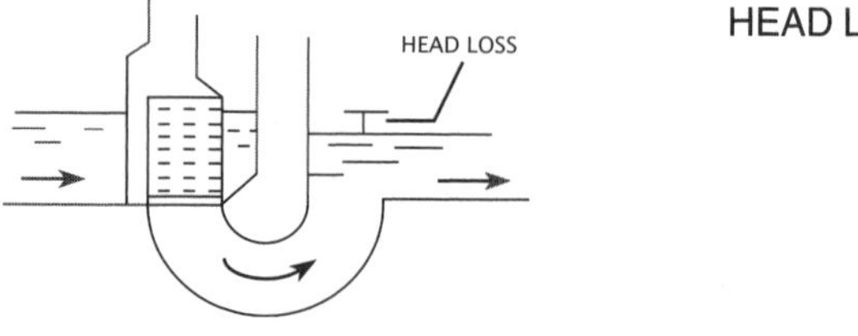

HEADER HEADER

A large pipe to which the ends of a series of smaller pipes are connected. Also called a MANIFOLD.

HEADWORKS HEADWORKS

The facilities where wastewater enters a wastewater treatment plant. The headworks may consist of bar screens, comminutors, a wet well, and pumps.

HEPATITIS (HEP-uh-TIE-tis) HEPATITIS

Hepatitis is an inflammation of the liver caused by an acute viral infection. Yellow jaundice is one symptom of hepatitis.

HUMUS SLUDGE HUMUS SLUDGE

The sloughed particles of biomass from trickling filter media that are removed from the water being treated in secondary clarifiers.

HYDRAULIC GRADE LINE (HGL) HYDRAULIC GRADE LINE (HGL)

The surface or profile of water flowing in an open channel or a pipe flowing partially full. If a pipe is under pressure, the hydraulic grade line is that level water would rise to in a small, vertical tube connected to the pipe. Also see ENERGY GRADE LINE.

[SEE DRAWING ON PAGE 827]

HYDRAULIC JUMP HYDRAULIC JUMP

The sudden and usually turbulent abrupt rise in water surface in an open channel when water flowing at high velocity is suddenly retarded to a slow velocity.

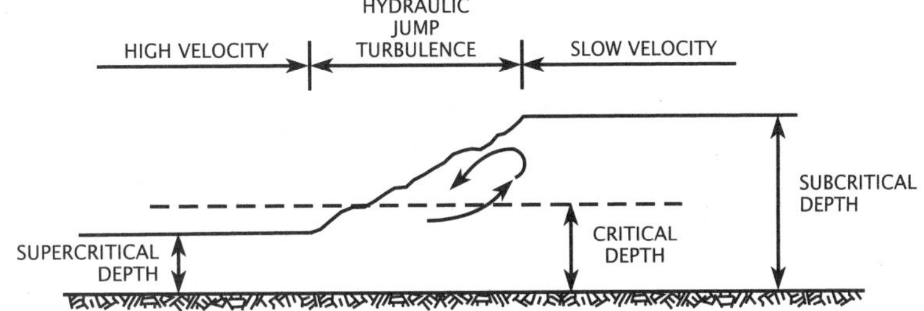

HYDRAULIC LOADING HYDRAULIC LOADING

Hydraulic loading refers to the flows (MGD or cu m/day) to a treatment plant or treatment process. Detention times, surface loadings, and weir overflow rates are directly influenced by flows.

HYDROGEN ION CONCENTRATION $[H^+]$ HYDROGEN ION CONCENTRATION $[H^+]$

The weight of hydrogen ion in moles per liter of solution. Commonly expressed as the pH value, which is the logarithm of the reciprocal of the hydrogen ion concentration.

$$pH = Log \frac{1}{[H^+]}$$

HYDROGEN SULFIDE GAS (H_2S) HYDROGEN SULFIDE GAS (H_2S)

Hydrogen sulfide is a gas with a rotten egg odor, produced under anaerobic conditions. Hydrogen sulfide gas is particularly dangerous because it dulls the sense of smell, becoming unnoticeable after you have been around it for a while; in high concentrations, it is only noticeable for a very short time before it dulls the sense of smell. The gas is very poisonous to the respiratory system, explosive, flammable, colorless, and heavier than air.

HYDROLOGIC (HI-dro-LOJ-ick) CYCLE HYDROLOGIC CYCLE

The process of evaporation of water into the air and its return to earth by precipitation (rain or snow). This process also includes transpiration from plants, groundwater movement, and runoff into rivers, streams, and the ocean. Also called the WATER CYCLE.

HYDROLYSIS (hi-DROLL-uh-sis) HYDROLYSIS

(1) A chemical reaction in which a compound is converted into another compound by taking up water.

(2) Usually a chemical degradation of organic matter.

HYDROSTATIC (hi-dro-STAT-ick) SYSTEM HYDROSTATIC SYSTEM

In a hydrostatic sludge removal system, the surface of the water in the clarifier is higher than the surface of the water in the sludge well or hopper. This difference in pressure head forces sludge from the bottom of the clarifier to flow through pipes to the sludge well or hopper.

HYGROSCOPIC (hi-grow-SKAWP-ick) HYGROSCOPIC

Absorbing or attracting moisture from the air.

HYPOCHLORINATION (HI-poe-klor-uh-NAY-shun) HYPOCHLORINATION

The application of hypochlorite compounds to water or wastewater for the purpose of disinfection.

HYPOCHLORINATORS (HI-poe-KLOR-uh-nay-tors) HYPOCHLORINATORS

Chlorine pumps, chemical feed pumps, or devices used to dispense chlorine solutions made from hypochlorites, such as bleach (sodium hypochlorite) or calcium hypochlorite into the water being treated.

HYPOCHLORITE (HI-poe-KLOR-ite) HYPOCHLORITE

Chemical compounds containing available chlorine; used for disinfection. They are available as liquids (bleach) or solids (powder, granules, and pellets) in barrels, drums, and cans. Salts of hypochlorous acid.

I

IDLH IDLH

Immediately Dangerous to Life or Health. The atmospheric concentration of any toxic, corrosive, or asphyxiant substance that poses an immediate threat to life or would cause irreversible or delayed adverse health effects or would interfere with an individual's ability to escape from a dangerous atmosphere.

IMHOFF CONE IMHOFF CONE

A clear, cone-shaped container marked with graduations. The cone is used to measure the volume of settleable solids in a specific volume (usually one liter) of water or wastewater.

IMPELLER IMPELLER

A rotating set of vanes in a pump or compressor designed to pump or move water or air.

IMPELLER PUMP IMPELLER PUMP

Any pump in which the water is moved by the continuous application of power to a rotating set of vanes from some rotating mechanical source.

INCINERATION INCINERATION

The conversion of dewatered wastewater solids by combustion (burning) to ash, carbon dioxide, and water vapor.

INDICATOR INDICATOR

(1) (Chemical indicator) A substance that gives a visible change, usually of color, at a desired point in a chemical reaction, generally at a specified end point.

(2) (Instrument indicator) A device that indicates the result of a measurement, usually using either a fixed scale and movable indicator (pointer), such as a pressure gauge, or a moving chart with a movable pen like those used on a circular flow-recording chart. Also called a RECEIVER.

INDOLE (IN-dole) INDOLE

An organic compound (C_8H_7N) containing nitrogen that has an ammonia odor.

INFILTRATION (in-fill-TRAY-shun) INFILTRATION

The seepage of groundwater into a sewer system, including service connections. Seepage frequently occurs through defective or cracked pipes, pipe joints and connections, interceptor access risers and covers, or manhole walls.

INFLOW INFLOW

Water discharged into a sewer system and service connections from sources other than regular connections. This includes flow from yard drains, foundations, and around access and manhole covers. Inflow differs from infiltration in that it is a direct discharge into the sewer rather than a leak in the sewer itself.

INFLUENT INFLUENT

Water or other liquid—raw (untreated) or partially treated—flowing *INTO* a reservoir, basin, treatment process, or treatment plant.

INHIBITORY SUBSTANCES INHIBITORY SUBSTANCES

Materials that kill or restrict the ability of organisms to treat wastes.

INOCULATE (in-NOCK-yoo-late) INOCULATE

To introduce a seed culture into a system.

INORGANIC WASTE INORGANIC WASTE

Waste material such as sand, salt, iron, calcium, and other mineral materials that are only slightly affected by the action of organisms. Inorganic wastes are chemical substances of mineral origin; whereas organic wastes are chemical substances usually of animal or plant origin. Also see NONVOLATILE MATTER, ORGANIC WASTE, and VOLATILE SOLIDS.

INTEGRATOR INTEGRATOR

A device or meter that continuously measures and sums a process rate variable in cumulative fashion over a given time period. For example, total flows displayed in gallons per minute, million gallons per day, cubic feet per second, or some other unit of volume per time period. Also called a TOTALIZER.

INTERFACE INTERFACE

The common boundary layer between two substances, such as water and a solid (metal); or between two fluids, such as water and a gas (air); or between a liquid (water) and another liquid (oil).

IONIC CONCENTRATION IONIC CONCENTRATION

The concentration of any ion in solution, usually expressed in moles per liter.

IONIZATION (EYE-on-uh-ZAY-shun) IONIZATION

(1) The splitting or dissociation (separation) of molecules into negatively and positively charged ions.

(2) The process of adding electrons to, or removing electrons from, atoms or molecules, thereby creating ions. High temperatures, electrical discharges, and nuclear radiation can cause ionization.

J

JAR TEST JAR TEST

A laboratory procedure that simulates coagulation/flocculation with differing chemical doses. The purpose of the procedure is to estimate the minimum coagulant dose required to achieve certain water quality goals. Samples of water to be treated are placed in six jars. Various amounts of chemicals are added to each jar, stirred, and the settling of solids is observed. The lowest dose of chemicals that provides satisfactory settling is the dose used to treat the water.

JOGGING JOGGING

The frequent starting and stopping of an electric motor.

JOULE (JOOL) JOULE

A measure of energy, work, or quantity of heat. One joule is the work done when the point of application of a force of one newton is displaced a distance of one meter in the direction of the force. Approximately equal to 0.7375 ft-lbs (0.1022 m-kg).

K

KJELDAHL (KELL-doll) NITROGEN — KJELDAHL NITROGEN

Nitrogen in the form of organic proteins or their decomposition product ammonia, as measured by the Kjeldahl Method.

L

LAUNDERS — LAUNDERS

Sedimentation basin and filter discharge channels consisting of overflow weir plates (in sedimentation basins) and conveying troughs.

LIMIT SWITCH — LIMIT SWITCH

A device that regulates or controls the travel distance of a chain or cable.

LINEAL (LIN-e-ul) — LINEAL

The length in one direction of a line. For example, a board 12 feet (meters) long has 12 lineal feet (meters) in its length.

LIQUEFACTION (lick-we-FACK-shun) — LIQUEFACTION

The conversion of large, solid particles of sludge into very fine particles that either dissolve or remain suspended in wastewater.

LOADING — LOADING

Quantity of material applied to a device at one time.

LOWER EXPLOSIVE LIMIT (LEL) — LOWER EXPLOSIVE LIMIT (LEL)

The lowest concentration of a gas or vapor (percent by volume in air) that explodes if an ignition source is present at ambient temperature. At temperatures above 250°F (121°C) the LEL decreases because explosibility increases with higher temperature.

LOWER FLAMMABLE LIMIT (LFL) — LOWER FLAMMABLE LIMIT (LFL)

The lowest concentration of a gas or vapor (percent by volume in air) that burns if an ignition source is present.

LYSIMETER (lie-SIM-uh-ter) — LYSIMETER

A device containing a mass of soil and designed to permit the measurement of water draining through the soil.

M

M or MOLAR — *M* or MOLAR

A molar solution consists of one gram molecular weight of a compound dissolved in enough water to make one liter of solution. A gram molecular weight is the molecular weight of a compound in grams. For example, the molecular weight of sulfuric acid (H_2SO_4) is 98. A one *M* solution of sulfuric acid would consist of 98 grams of H_2SO_4 dissolved in enough distilled water to make one liter of solution.

MBAS — MBAS

Methylene Blue Active Substance. Another name for surfactants or surface active agents. The determination of surfactants is accomplished by measuring the color change in a standard solution of methylene blue dye.

MCRT | MCRT

Mean Cell Residence Time. An expression of the average time (days) that a microorganism will spend in the activated sludge process.

$$\text{MCRT, days} = \frac{\text{Total Suspended Solids in Activated Sludge Process, lbs}}{\text{Total Suspended Solids Removed From Process, lbs/day}}$$

or

$$\text{MCRT, days} = \frac{\text{Total Suspended Solids in Activated Sludge Process, kg}}{\text{Total Suspended Solids Removed From Process, kg/day}}$$

NOTE: Operators at different plants calculate the Total Suspended Solids (TSS) in the Activated Sludge Process, lbs (kg), by three different methods:

1. TSS in the Aeration Basin or Reactor Zone, lbs (kg)
2. TSS in the Aeration Basin and Secondary Clarifier, lbs (kg)
3. TSS in the Aeration Basin and Secondary Clarifier Sludge Blanket, lbs (kg)

These three different methods make it difficult to compare MCRTs in days among different plants unless everyone uses the same method.

mg/*L* | mg/*L*

See MILLIGRAMS PER LITER, mg/*L*.

MPN | MPN

MPN is the Most Probable Number of coliform-group organisms per unit volume of sample water. Expressed as a density or population of organisms per 100 m*L* of sample water.

MSDS | MSDS

See MATERIAL SAFETY DATA SHEET.

MAIN SEWER | MAIN SEWER

A sewer line that receives wastewater from many tributary branches and sewer lines and serves as an outlet for a large territory or is used to feed an intercepting sewer. Also called TRUNK SEWER.

MANIFOLD | MANIFOLD

A large pipe to which the ends of a series of smaller pipes are connected. Also called a HEADER.

MANOMETER (man-NAH-mut-ter) | MANOMETER

An instrument for measuring pressure. Usually, a manometer is a glass tube filled with a liquid that is used to measure the difference in pressure across a flow measuring device, such as an orifice or a Venturi meter. The instrument used to measure blood pressure is a type of manometer.

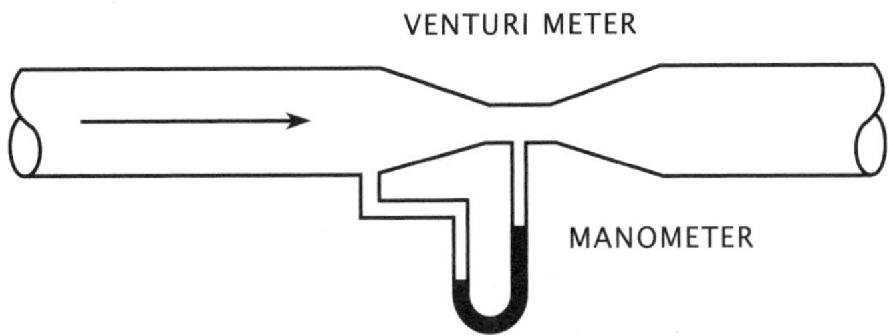

MASKING AGENTS MASKING AGENTS

Substances used to cover up or disguise unpleasant odors. Liquid masking agents are dripped into the wastewater, sprayed into the air, or evaporated (using heat) with the unpleasant fumes or odors and then discharged into the air by blowers to make an undesirable odor less noticeable.

MATERIAL SAFETY DATA SHEET (MSDS) MATERIAL SAFETY DATA SHEET (MSDS)

A document that provides pertinent information and a profile of a particular hazardous substance or mixture. An MSDS is normally developed by the manufacturer or formulator of the hazardous substance or mixture. The MSDS is required to be made available to employees and operators or inspectors whenever there is the likelihood of the hazardous substance or mixture being introduced into the workplace. Some manufacturers are preparing MSDSs for products that are not considered to be hazardous to show that the product or substance is not hazardous.

MEAN CELL RESIDENCE TIME (MCRT) MEAN CELL RESIDENCE TIME (MCRT)

See MCRT.

MECHANICAL AERATION MECHANICAL AERATION

The use of machinery to mix air and water so that oxygen can be absorbed into the water. Some examples are: paddle wheels, mixers, or rotating brushes to agitate the surface of an aeration tank; pumps to create fountains; and pumps to discharge water down a series of steps forming falls or cascades.

MEDIA MEDIA

The material in a trickling filter on which slime accumulates and organisms grow. As settled wastewater trickles over the media, organisms in the slime remove certain types of wastes, thereby partially treating the wastewater. Also, the material in a rotating biological contactor or in a gravity or pressure filter.

MEDIAN MEDIAN

The middle measurement or value. When several measurements are ranked by magnitude (largest to smallest), half of the measurements will be larger and half will be smaller.

MEG MEG

(1) Abbreviation of MEGOHM.
(2) A procedure used for checking the insulation resistance on motors, feeders, bus bar systems, grounds, and branch circuit wiring. Also see MEGGER.

MEGGER (from megohm) MEGGER

An instrument used for checking the insulation resistance on motors, feeders, bus bar systems, grounds, and branch circuit wiring. A megger reads in millions of ohms. Also see MEG.

MEGOHM (MEG-ome) MEGOHM

Millions of ohms. Mega- is a prefix meaning one million, so 5 megohms means 5 million ohms.

MENISCUS (meh-NIS-cuss) MENISCUS

The curved surface of a column of liquid (water, oil, mercury) in a small tube. When the liquid wets the sides of the container (as with water), the curve forms a valley. When the confining sides are not wetted (as with mercury), the curve forms a hill or upward bulge.

MERCAPTANS (mer-CAP-tans) MERCAPTANS

Compounds containing sulfur that have an extremely offensive skunk-like odor; also sometimes described as smelling like garlic or onions.

MESOPHILIC (MESS-o-FILL-ick) BACTERIA MESOPHILIC BACTERIA

Medium temperature bacteria. A group of bacteria that grow and thrive in a moderate temperature range between 68°F (20°C) and 113°F (45°C). The optimum temperature range for these bacteria in anaerobic digestion is 85°F (30°C) to 100°F (38°C).

METABOLISM — METABOLISM

All of the processes or chemical changes in an organism or a single cell by which food is built up (anabolism) into living protoplasm and by which protoplasm is broken down (catabolism) into simpler compounds with the exchange of energy.

MICRON (MY-kron) — MICRON

µm, Micrometer or Micron. A unit of length. One millionth of a meter or one thousandth of a millimeter. One micron equals 0.00004 of an inch.

MICROORGANISMS (MY-crow-OR-gan-is-ums) — MICROORGANISMS

Living organisms that can be seen individually only with the aid of a microscope.

MICROSCREEN — MICROSCREEN

A device with a fabric straining medium with openings usually between 20 and 60 microns. The fabric is wrapped around the outside of a rotating drum. Wastewater enters the open end of the drum and flows out through the rotating screen cloth. At the highest point of the drum, the collected solids are backwashed by high-pressure water jets into a trough located within the drum.

MILLIGRAMS PER LITER, mg/*L* — MILLIGRAMS PER LITER, mg/*L*

A measure of the concentration by weight of a substance per unit volume in water or wastewater. In reporting the results of water and wastewater analysis, mg/*L* is preferred to the unit parts per million (ppm), to which it is approximately equivalent.

MILLIMICRON (MILL-uh-MY-kron) — MILLIMICRON

A unit of length equal to $10^{-3}\mu$ (one thousandth of a micron), 10^{-6} millimeters, or 10^{-9} meters; correctly called a nanometer, nm.

MIXED LIQUOR — MIXED LIQUOR

When the activated sludge in an aeration tank is mixed with primary effluent or the raw wastewater and return sludge, this mixture is then referred to as mixed liquor as long as it is in the aeration tank. Mixed liquor also may refer to the contents of mixed aerobic or anaerobic digesters.

MIXED LIQUOR SUSPENDED SOLIDS (MLSS) — MIXED LIQUOR SUSPENDED SOLIDS (MLSS)

The amount (mg/*L*) of suspended solids in the mixed liquor of an aeration tank.

MIXED LIQUOR VOLATILE SUSPENDED SOLIDS (MLVSS) — MIXED LIQUOR VOLATILE SUSPENDED SOLIDS (MLVSS)

The amount (mg/*L*) of organic or volatile suspended solids in the mixed liquor of an aeration tank. This volatile portion is used as a measure or indication of the microorganisms present.

MOLAR — MOLAR

See *M* or MOLAR.

MOLARITY — MOLARITY

A measure of concentration defined as the number of moles of solute per liter of solution. Also see *M* or MOLAR.

MOLE — MOLE

The name for a quantity of any chemical substance whose mass in grams is numerically equal to its atomic weight. One mole equals 6.02×10^{23} molecules or atoms. See MOLECULAR WEIGHT.

MOLECULAR OXYGEN — MOLECULAR OXYGEN

The oxygen molecule, O_2, that is not combined with another element to form a compound.

MOLECULAR WEIGHT — MOLECULAR WEIGHT

The molecular weight of a compound in grams per mole is the sum of the atomic weights of the elements in the compound. The molecular weight of sulfuric acid (H_2SO_4) in grams is 98.

Element	Atomic Weight	Number of Atoms	Molecular Weight
H	1	2	2
S	32	1	32
O	16	4	64
			98

MOLECULE MOLECULE

The smallest division of a compound that still retains or exhibits all the properties of the substance.

MOST PROBABLE NUMBER (MPN) MOST PROBABLE NUMBER (MPN)

See MPN.

MOTILE (MO-till) MOTILE

Capable of self-propelled movement. A term that is sometimes used to distinguish between certain types of organisms found in water.

MOVING AVERAGE MOVING AVERAGE

To calculate the moving average for the last 7 days, add up the values for the last 7 days and divide by 7. Each day add the most recent day's value to the sum of values and subtract the oldest value. By using the 7-day moving average, each day of the week is always represented in the calculations.

MUFFLE FURNACE MUFFLE FURNACE

A small oven capable of reaching temperatures up to 600°C (1,112°F). Muffle furnaces are used in laboratories for burning or incinerating samples to determine the amounts of volatile solids and/or fixed solids in samples of wastewater.

MULTISTAGE PUMP MULTISTAGE PUMP

A pump that has more than one impeller. A single-stage pump has one impeller.

N

N or NORMAL *N* or NORMAL

A normal solution contains one gram equivalent weight of reactant (compound) per liter of solution. The equivalent weight of an acid is that weight that contains one gram atom of ionizable hydrogen or its chemical equivalent. For example, the equivalent weight of sulfuric acid (H_2SO_4) is 49 (98 divided by 2 because there are two replaceable hydrogen ions). A one *N* solution of sulfuric acid would consist of 49 grams of H_2SO_4 dissolved in enough water to make one liter of solution.

NIOSH (NYE-osh) NIOSH

The National Institute of Occupational Safety and Health is an organization that tests and approves safety equipment for particular applications. NIOSH is the primary federal agency engaged in research in the national effort to eliminate on-the-job hazards to the health and safety of working people. The NIOSH Publications Catalog, Seventh Edition, NIOSH Pub. No. 87-115, lists the NIOSH publications concerning industrial hygiene and occupational health. To obtain a copy of the catalog, write to National Technical Information Service (NTIS), 5285 Port Royal Road, Springfield, VA 22161. NTIS Stock No. PB88-175013.

NPDES PERMIT NPDES PERMIT

National Pollutant Discharge Elimination System permit is the regulatory agency document issued by either a federal or state agency that is designed to control all discharges of potential pollutants from point sources and stormwater runoff into US waterways. NPDES permits regulate discharges into US waterways from all point sources of pollution, including industries, municipal wastewater treatment plants, sanitary landfills, large animal feedlots, and return irrigation flows.

NTU NTU

Nephelometric Turbidity Units. See TURBIDITY UNITS.

NAMEPLATE NAMEPLATE

A durable, metal plate found on equipment that lists critical operating conditions for the equipment.

NEUTRALIZATION (noo-trull-uh-ZAY-shun) NEUTRALIZATION

Addition of an acid or alkali (base) to a liquid to cause the pH of the liquid to move toward a neutral pH of 7.0.

NITRIFICATION (NYE-truh-fuh-KAY-shun) NITRIFICATION

An aerobic process in which bacteria change the ammonia and organic nitrogen in water or wastewater into oxidized nitrogen (usually nitrate).

NITRIFICATION (NYE-truh-fuh-KAY-shun) STAGE — NITRIFICATION STAGE

A stage of decomposition that occurs in biological treatment processes when aerobic bacteria, using dissolved oxygen, change nitrogen compounds (ammonia and organic nitrogen) into oxidized nitrogen (usually nitrate). The second-stage BOD is sometimes referred to as the nitrification stage (first-stage BOD is called the carbonaceous stage).

NITRIFYING BACTERIA — NITRIFYING BACTERIA

Bacteria that change ammonia and organic nitrogen into oxidized nitrogen (usually nitrate).

NITROGENOUS (nye-TRAH-jen-us) — NITROGENOUS

A term used to describe chemical compounds (usually organic) containing nitrogen in combined forms. Proteins and nitrate are nitrogenous compounds.

NOMOGRAM (NOME-o-gram) — NOMOGRAM

A chart or diagram containing three or more scales used to solve problems with three or more variables instead of using mathematical formulas.

NONCORRODIBLE — NONCORRODIBLE

A material that resists corrosion and will not be eaten away by wastewater or chemicals in wastewater.

NON-PERMIT CONFINED SPACE — NON-PERMIT CONFINED SPACE

See CONFINED SPACE, NON-PERMIT.

NONSPARKING TOOLS — NONSPARKING TOOLS

These tools will not produce a spark during use. They are made of a nonferrous material, usually a copper-beryllium alloy.

NONVOLATILE MATTER — NONVOLATILE MATTER

Material such as sand, salt, iron, calcium, and other mineral materials that are only slightly affected by the actions of organisms and are not lost on ignition of the dry solids at 550°C (1,022°F). Volatile materials are chemical substances usually of animal or plant origin. Also see INORGANIC WASTE and VOLATILE SOLIDS.

NORMAL — NORMAL

See *N* or NORMAL.

NORMALITY — NORMALITY

The number of gram-equivalent weights of solute in one liter of solution. The equivalent weight of any material is the weight that would react with or be produced by the reaction of 8.0 grams of oxygen or 1.0 gram of hydrogen. Normality is used for certain calculations of quantitative analysis. Also see *N* or NORMAL.

NOTICE — NOTICE

This word calls attention to information that is especially significant in understanding and operating equipment or processes safely. Also see CAUTION, DANGER, and WARNING.

NUTRIENT — NUTRIENT

Any substance that is assimilated (taken in) by organisms and promotes growth. Nitrogen and phosphorus are nutrients that promote the growth of algae. There are other essential and trace elements that are also considered nutrients. Also see NUTRIENT CYCLE.

NUTRIENT CYCLE — NUTRIENT CYCLE

The transformation or change of a nutrient from one form to another until the nutrient has returned to the original form, thus completing the cycle. The cycle may take place under either aerobic or anaerobic conditions.

O

O & M MANUAL — O & M MANUAL

Operation and Maintenance Manual. A manual that describes detailed procedures for operators to follow to operate and maintain a specific treatment plant and the equipment of that plant.

OSHA (O-shuh) OSHA

The Williams-Steiger Occupational Safety and Health Act of 1970 (OSHA) is a federal law designed to protect the health and safety of workers, including the operators of water supply and treatment systems and wastewater collection and treatment systems. The Act regulates the design, construction, operation, and maintenance of water and wastewater systems. OSHA regulations require employers to obtain and make available to workers the Material Safety Data Sheets (MSDSs) for chemicals used at industrial facilities and treatment plants. OSHA also refers to the federal and state agencies that administer the OSHA regulations.

OBLIGATE AEROBES OBLIGATE AEROBES

Bacteria that must have atmospheric or dissolved molecular oxygen to live and reproduce.

OCCUPATIONAL SAFETY AND HEALTH ACT OF 1970 (OSHA) OCCUPATIONAL SAFETY AND HEALTH ACT OF 1970 (OSHA)

See OSHA.

ODOR PANEL ODOR PANEL

A group of people used to measure odors.

ODOR THRESHOLD ODOR THRESHOLD

The minimum odor of a gas or water sample that can just be detected after successive dilutions with odorless gas or water. Also called THRESHOLD ODOR.

OFFSET OFFSET

(1) The difference between the actual value and the desired value (or set point); characteristic of proportional controllers that do not incorporate reset action. Also called DRIFT.
(2) A pipe fitting in the approximate form of a reverse curve or other combination of elbows or bends that brings one section of a line of pipe out of line with, but into a line parallel with, another section.
(3) A pipe joint that has lost its bedding support, causing one of the pipe sections to drop or slip, thus creating a condition where the pipes no longer line up properly.

OLFACTOMETER (ALL-fak-TOM-utter) OLFACTOMETER

A device used to measure odors in the field by diluting odors with odor-free air.

OLFACTORY (all-FAK-tore-ee) FATIGUE OLFACTORY FATIGUE

A condition in which a person's nose, after exposure to certain odors, is no longer able to detect the odor.

OPERATING RATIO OPERATING RATIO

The operating ratio is a measure of the total revenues divided by the total operating expenses.

ORGANIC WASTE ORGANIC WASTE

Waste material that may come from animal or plant sources. Natural organic wastes generally can be consumed by bacteria and other small organisms. Manufactured or synthetic organic wastes from metal finishing, chemical manufacturing, and petroleum industries may not normally be consumed by bacteria and other organisms. Also see INORGANIC WASTE and VOLATILE SOLIDS.

ORGANISM ORGANISM

Any form of animal or plant life. Also see BACTERIA.

ORGANIZING ORGANIZING

Deciding who does what work and delegating authority to the appropriate persons.

ORIFICE (OR-uh-fiss) ORIFICE

An opening (hole) in a plate, wall, or partition. An orifice flange or plate placed in a pipe consists of a slot or a calibrated circular hole smaller than the pipe diameter. The difference in pressure in the pipe above and at the orifice may be used to determine the flow in the pipe. In a trickling filter distributor, the wastewater passes through an orifice to the surface of the filter media.

ORTHOTOLIDINE (or-tho-TOL-uh-dine) ORTHOTOLIDINE

Orthotolidine is a colorimetric indicator of chlorine residual. If chlorine is present, a yellow-colored compound is produced. This reagent is no longer approved for chemical analysis to determine chlorine residual.

OUCH PRINCIPLE — OUCH PRINCIPLE

This principle says that as a manager when you delegate job tasks you must be **O**bjective, **U**niform in your treatment of employees, and the tasks must be **C**onsistent with utility policies, and **H**ave job relatedness.

OUTFALL — OUTFALL

(1) The point, location, or structure where wastewater or drainage discharges from a sewer, drain, or other conduit.

(2) The conduit leading to the final discharge point or area. Also see OUTFALL SEWER.

OVERFLOW RATE — OVERFLOW RATE

One factor of the design flow of settling tanks and clarifiers in treatment plants used by operators to determine if tanks and clarifiers are hydraulically (flow) over- or underloaded. Also called SURFACE LOADING.

$$\text{Overflow Rate, GPD/sq ft} = \frac{\text{Flow, gallons/day}}{\text{Surface Area, sq ft}}$$

or

$$\text{Overflow Rate, } \frac{\text{cu m/day}}{\text{sq m}} = \frac{\text{Flow, cu m/day}}{\text{Surface Area, sq m}}$$

OVERTURN — OVERTURN

The almost spontaneous mixing of all layers of water in a reservoir or lake when the water temperature becomes similar from top to bottom. This may occur in the fall/winter when the surface waters cool to the same temperature as the bottom waters and also in the spring when the surface waters warm after the ice melts. This is also called turnover.

OXIDATION — OXIDATION

Oxidation is the addition of oxygen, removal of hydrogen, or the removal of electrons from an element or compound; in the environment and in wastewater treatment processes, organic matter is oxidized to more stable substances. The opposite of REDUCTION.

OXIDATION STATE/OXIDATION NUMBER — OXIDATION STATE/OXIDATION NUMBER

In a chemical formula, a number accompanied by a polarity indication (+ or –) that together indicate the charge of an ion as well as the extent to which the ion has been oxidized or reduced in a REDOX REACTION.

Due to the loss of electrons, the charge of an ion that has been oxidized would go from negative toward or to neutral, from neutral to positive, or from positive to more positive. As an example, an oxidation number of 2+ would indicate that an ion has lost two electrons and that its charge has become positive (that it now has an excess of two protons).

Due to the gain of electrons, the charge of the ion that has been reduced would go from positive toward or to neutral, from neutral to negative, or from negative to more negative. As an example, an oxidation number of 2– would indicate that an ion has gained two electrons and that its charge has become negative (that it now has an excess of two electrons). As an ion gains electrons, its oxidation state (or the extent to which it is oxidized) lowers; that is, its oxidation state is reduced. Also see REDOX REACTION.

OXIDATION-REDUCTION POTENTIAL (ORP) — OXIDATION-REDUCTION POTENTIAL (ORP)

The electrical potential required to transfer electrons from one compound or element (the oxidant) to another compound or element (the reductant); used as a qualitative measure of the state of oxidation in water and wastewater treatment systems. ORP is measured in millivolts, with negative values indicating a tendency to reduce compounds or elements and positive values indicating a tendency to oxidize compounds or elements.

OXIDATION-REDUCTION (REDOX) REACTION — OXIDATION-REDUCTION (REDOX) REACTION

See REDOX REACTION.

OXIDIZED ORGANICS — OXIDIZED ORGANICS

Organic materials that have been broken down in a biological process. Examples of these materials are carbohydrates and proteins that are broken down to simple sugars.

OXIDIZING AGENT — OXIDIZING AGENT

Any substance, such as oxygen (O_2) or chlorine (Cl_2), that will readily add (take on) electrons. When oxygen or chlorine is added to water or wastewater, organic substances are oxidized. These oxidized organic substances are more stable and less likely to give off odors or to contain disease-causing bacteria. The opposite is a REDUCING AGENT.

OXYGEN DEFICIENCY — OXYGEN DEFICIENCY

An atmosphere containing oxygen at a concentration of less than 19.5 percent by volume.

OXYGEN ENRICHMENT — OXYGEN ENRICHMENT

An atmosphere containing oxygen at a concentration of more than 23.5 percent by volume.

OZONATION (O-zoe-NAY-shun) — OZONATION

The application of ozone to water, wastewater, or air, generally for the purposes of disinfection or odor control.

P

POTW — POTW

Publicly Owned Treatment Works. A treatment works that is owned by a state, municipality, city, town, special sewer district, or other publicly owned and financed entity as opposed to a privately (industrial) owned treatment facility. This definition includes any devices and systems used in the storage, treatment, recycling, and reclamation of municipal sewage (wastewater) or industrial wastes of a liquid nature. It also includes sewers, pipes, and other conveyances only if they carry wastewater to a POTW treatment plant. The term also means the municipality (public entity) that has jurisdiction over the indirect discharges to and the discharges from such a treatment works.

PACKAGE TREATMENT PLANT — PACKAGE TREATMENT PLANT

A small wastewater treatment plant often fabricated at the manufacturer's factory, hauled to the site, and installed as one facility. The package may be either a small primary or a secondary wastewater treatment plant.

PARALLEL OPERATION — PARALLEL OPERATION

Wastewater being treated is split and a portion flows to one treatment unit while the remainder flows to another similar treatment unit. Also see SERIES OPERATION.

PARASITIC (pair-uh-SIT-tick) BACTERIA — PARASITIC BACTERIA

Parasitic bacteria are those bacteria that normally live off another living organism, known as the host.

PATHOGENIC (path-o-JEN-ick) ORGANISMS — PATHOGENIC ORGANISMS

Organisms, including bacteria, viruses, or cysts, capable of causing diseases (such as giardiasis, cryptosporidiosis, typhoid, cholera, dysentery) in a host (such as a person). Also called PATHOGENS.

PATHOGENS (PATH-o-jens) — PATHOGENS

See PATHOGENIC ORGANISMS.

PERCENT SATURATION — PERCENT SATURATION

The amount of a substance that is dissolved in a solution compared with the amount dissolved in the solution at saturation, expressed as a percent.

$$\text{Percent Saturation, \%} = \frac{\text{Amount of Substance That Is Dissolved} \times 100\%}{\text{Amount Dissolved in Solution at Saturation}}$$

PERCOLATION (purr-ko-LAY-shun) — PERCOLATION

The slow passage of water through a filter medium; or, the gradual penetration of soil and rocks by water.

PERISTALTIC (PAIR-uh-STALL-tick) PUMP — PERISTALTIC PUMP

A type of positive displacement pump.

PERMIT-REQUIRED CONFINED SPACE (PERMIT SPACE) — PERMIT-REQUIRED CONFINED SPACE (PERMIT SPACE)

See CONFINED SPACE, PERMIT-REQUIRED (PERMIT SPACE).

pH (pronounce as separate letters) — pH

pH is an expression of the intensity of the basic or acidic condition of a liquid. Mathematically, pH is the logarithm (base 10) of the reciprocal of the hydrogen ion activity.

$$\text{pH} = \text{Log}\,\frac{1}{\{H^+\}}$$

If $\{H^+\} = 10^{-6.5}$, then pH = 6.5. The pH may range from 0 to 14, where 0 is most acidic, 14 most basic, and 7 neutral.

PHENOLIC (fee-NO-lick) COMPOUNDS — PHENOLIC COMPOUNDS

Organic compounds that are derivatives of benzene. Also called phenols (FEE-nolls).

PHENOLPHTHALEIN (FEE-nol-THAY-leen) ALKALINITY — PHENOLPHTHALEIN ALKALINITY

The alkalinity in a water sample measured by the amount of standard acid required to lower the pH to a level of 8.3, as indicated by the change in color of phenolphthalein from pink to clear. Phenolphthalein alkalinity is expressed as milligrams per liter of equivalent calcium carbonate.

PHOTOSYNTHESIS (foe-toe-SIN-thuh-sis) — PHOTOSYNTHESIS

A process in which organisms, with the aid of chlorophyll (green plant enzyme), convert carbon dioxide and inorganic substances into oxygen and additional plant material, using sunlight for energy. All green plants grow by this process.

PHYSICAL WASTE TREATMENT PROCESS — PHYSICAL WASTE TREATMENT PROCESS

Physical wastewater treatment processes include use of racks, screens, comminutors, clarifiers (sedimentation and flotation), and filtration. Chemical or biological reactions are important treatment processes, but not part of a physical treatment process.

PILLOWS — PILLOWS

Plastic tubes shaped like pillows that contain exact amounts of chemicals or reagents. Cut open the pillow, pour the reagents into the sample being tested, mix thoroughly, and follow test procedures.

PLANNING — PLANNING

Management of utilities to build the resources and financial capability to provide for future needs.

PLUG FLOW — PLUG FLOW

A type of flow that occurs in tanks, basins, or reactors when a slug of water or wastewater moves through a tank without ever dispersing or mixing with the rest of the water or wastewater flowing through the tank.

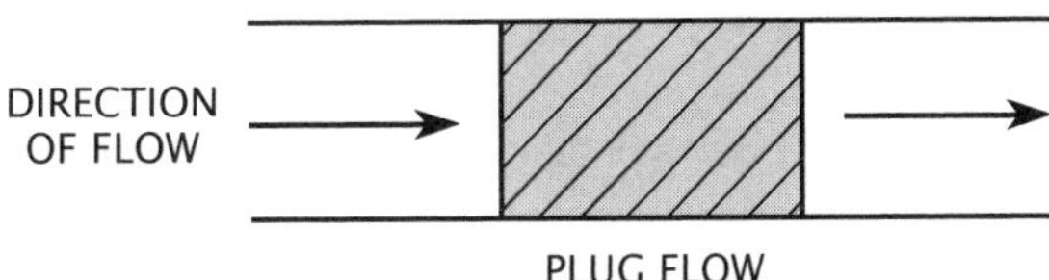

PLUG FLOW

POLE SHADER — POLE SHADER

A copper bar circling the laminated iron core inside the coil of a magnetic starter.

POLLUTION — POLLUTION

The impairment (reduction) of water quality by agricultural, domestic, or industrial wastes (including thermal and radioactive wastes) to a degree that the natural water quality is changed to hinder any beneficial use of the water or render it offensive to the senses of sight, taste, or smell or when sufficient amounts of wastes create or pose a potential threat to human health or the environment.

POLYELECTROLYTE (POLY-ee-LECK-tro-lite) — POLYELECTROLYTE

A high-molecular-weight (relatively heavy) substance, having points of positive or negative electrical charges, that is formed by either natural or synthetic (manmade) processes. Natural polyelectrolytes may be of biological origin or obtained from starch products or cellulose derivatives. Synthetic polyelectrolytes consist of simple substances that have been made into complex, high-molecular-weight substances. Used with other chemical coagulants to aid in binding small suspended particles to larger chemical flocs for their removal from water. Often called a POLYMER.

POLYMER (POLY-mer) — POLYMER

A long-chain molecule formed by the union of many monomers (molecules of lower molecular weight). Polymers are used with other chemical coagulants to aid in binding small suspended particles to larger chemical flocs for their removal from water. Also see POLYELECTROLYTE.

POLYSACCHARIDE (poly-SAC-uh-ride) — POLYSACCHARIDE

A carbohydrate, such as starch or cellulose, composed of chains of simple sugars.

PONDING — PONDING

A condition occurring on trickling filters when the hollow spaces (voids) become plugged to the extent that water passage through the filter is inadequate. Ponding may be the result of excessive slime growths, trash, or media breakdown.

POPULATION EQUIVALENT — POPULATION EQUIVALENT

A means of expressing the strength of organic material in wastewater. In a domestic wastewater system, microorganisms use up about 0.2 pound (90 grams) of oxygen per day for each person using the system (as measured by the standard BOD test). May also be expressed as flow (100 gallons (378 liters)/day/person) or suspended solids (0.2 lb (90 grams) SS/day/person).

$$\text{Population Equivalent, persons} = \frac{\text{Flow, MGD} \times \text{BOD, mg}/L \times 8.34 \text{ lbs/gal}}{0.2 \text{ lb BOD/day/person}}$$

or

$$\text{Population Equivalent, persons} = \frac{\text{Flow, cu m/day} \times \text{BOD, mg}/L \times 10^6\ L\text{/cu m}}{90{,}000 \text{ mg BOD/day/person}}$$

POSITIVE PRESSURE — POSITIVE PRESSURE

A positive pressure is a pressure greater than atmospheric. It is measured as pounds per square inch (psi) or as inches of water column. A negative pressure (vacuum) is less than atmospheric and is sometimes measured in inches of mercury. In the metric system, pressures are measured in kg/sq m, kg/sq cm, or pascals (1 psi = 6,895 Pa = 6.895 kN/sq m).

POSTCHLORINATION — POSTCHLORINATION

The addition of chlorine to the plant discharge or effluent, following plant treatment, for disinfection purposes.

POTABLE (POE-tuh-bull) WATER — POTABLE WATER

Water that does not contain objectionable pollution, contamination, minerals, or infective agents and is considered satisfactory for drinking.

PRE-AERATION — PRE-AERATION

The addition of air at the initial stages of treatment to freshen the wastewater, remove gases, add oxygen, promote flotation of grease, and aid coagulation.

PRECHLORINATION — PRECHLORINATION

The addition of chlorine in the collection system serving the plant or at the headworks of the plant prior to other treatment processes.

(1) For drinking water, used mainly for disinfection, control of tastes, odors, and aquatic growths, and to aid in coagulation and settling.

(2) For wastewater, used mainly for control of odors, corrosion, and foaming, and for BOD reduction and oil removal.

PRECIPITATE (pre-SIP-uh-TATE) — PRECIPITATE

(1) An insoluble, finely divided substance that is a product of a chemical reaction within a liquid.

(2) The separation from solution of an insoluble substance.

PRECOAT — PRECOAT

Application of a free-draining, noncohesive material, such as diatomaceous earth, to a filtering medium. Precoating reduces the frequency of media washing and facilitates cake discharge.

PRELIMINARY TREATMENT — PRELIMINARY TREATMENT

The removal of metal, rocks, rags, sand, eggshells, and similar materials that may hinder the operation of a treatment plant. Preliminary treatment is accomplished by using equipment such as racks, bar screens, comminutors, and grit removal systems.

PRESENT WORTH — PRESENT WORTH

The value of a long-term project expressed in today's dollars. Present worth is calculated by converting (discounting) all future benefits and costs over the life of the project to a single economic value at the start of the project. Calculating the present worth of alternative projects makes it possible to compare them and select the one with the largest positive (beneficial) present worth or minimum present cost.

PRESSURE MAIN — PRESSURE MAIN

See FORCE MAIN.

PRETREATMENT FACILITY — PRETREATMENT FACILITY

Industrial wastewater treatment plant consisting of one or more treatment devices designed to remove sufficient pollutants from wastewaters to allow an industry to comply with effluent limits established by the US EPA General and Categorical Pretreatment Regulations or locally derived prohibited discharge requirements and local effluent limits. Compliance with effluent limits allows for a legal discharge to a POTW.

PRIMARY TREATMENT — PRIMARY TREATMENT

A wastewater treatment process that takes place in a rectangular or circular tank and allows those substances in wastewater that readily settle or float to be separated from the wastewater being treated. A septic tank is also considered primary treatment.

PROCESS VARIABLE — PROCESS VARIABLE

A physical or chemical quantity that is usually measured and controlled in the operation of a water, wastewater, or industrial treatment plant. Common process variables are flow, level, pressure, temperature, turbidity, chlorine, and oxygen levels.

PROGRAMMABLE LOGIC CONTROLLER (PLC) — PROGRAMMABLE LOGIC CONTROLLER (PLC)

A microcomputer-based control device containing programmable software; used to control process variables.

PROPORTIONAL WEIR (WEER) — PROPORTIONAL WEIR

A specially shaped weir in which the flow through the weir is directly proportional to the head.

PROTEINACEOUS (PRO-ten-NAY-shus) — PROTEINACEOUS

Materials containing proteins, which are organic compounds containing nitrogen.

PROTOZOA (pro-toe-ZOE-ah) — PROTOZOA

A group of motile, microscopic organisms (usually single-celled and aerobic) that sometimes cluster into colonies and generally consume bacteria as an energy source.

PRUSSIAN BLUE — PRUSSIAN BLUE

A blue paste or liquid (often on a paper like carbon paper) used to show a contact area. Used to determine if gate valve seats fit properly.

PSYCHROPHILIC (sy-kro-FILL-ick) BACTERIA — PSYCHROPHILIC BACTERIA

Cold temperature bacteria. A group of bacteria that grow and thrive in temperatures below 68°F (20°C).

PURGE — PURGE

To remove a gas or vapor from a vessel, reactor, or confined space, usually by displacement or dilution.

PUTREFACTION (PYOO-truh-FACK-shun) — PUTREFACTION

Biological decomposition of organic matter, with the production of foul-smelling and -tasting products, associated with anaerobic (no oxygen present) conditions.

PUTRESCIBLE (pyoo-TRES-uh-bull) — PUTRESCIBLE

Material that will decompose under anaerobic conditions and produce nuisance odors.

PYROMETER (pie-ROM-uh-ter) — PYROMETER

An apparatus used to measure high temperatures.

Q

(NO LISTINGS)

R

RACK — RACK

Evenly spaced, parallel metal bars or rods located in the influent channel to remove rags, rocks, and cans from wastewater.

RADIAL TO IMPELLER — RADIAL TO IMPELLER

Perpendicular to the impeller shaft. Material being pumped flows at a right angle to the impeller.

RAW WASTEWATER — RAW WASTEWATER

Plant influent or wastewater before any treatment.

REAGENT (re-A-gent) REAGENT

A pure, chemical substance that is used to make new products or is used in chemical tests to measure, detect, or examine other substances.

RECALCINATION (re-kal-sin-NAY-shun) RECALCINATION

A lime recovery process in which the calcium carbonate in sludge is converted to lime by heating at 1,800°F (980°C).

RECARBONATION (re-kar-bun-NAY-shun) RECARBONATION

A process in which carbon dioxide is bubbled into the water being treated to lower the pH.

RECEIVER RECEIVER

A device that indicates the result of a measurement, usually using either a fixed scale and movable indicator (pointer), such as a pressure gauge, or a moving chart with a movable pen like those used on a circular flow-recording chart. Also called an INDICATOR.

RECEIVING WATER RECEIVING WATER

A stream, river, lake, ocean, or other surface or groundwaters into which treated or untreated wastewater is discharged.

RECHARGE RATE RECHARGE RATE

Rate at which water is added beneath the ground surface to replenish or recharge groundwater.

RECIRCULATION RECIRCULATION

The return of part of the effluent from a treatment process to the incoming flow.

REDOX (REE-docks) REACTION REDOX REACTION

A two-part reaction between two ions involving a transfer of electrons from one ion to the other. Oxidation is the loss of electrons by one ion, and reduction is the acceptance of electrons by the other ion. Reduction refers to the lowering of the OXIDATION STATE/ OXIDATION NUMBER of the ion accepting the electrons.

In a redox reaction, the ion that gives up the electrons (that is oxidized) is called the reductant because it causes a reduction in the oxidation state or number of the ion that accepts the transferred electrons. The ion that receives the electrons (that is reduced) is called the oxidant because it causes oxidation of the other ion. Oxidation and reduction always occur simultaneously.

REDUCING AGENT REDUCING AGENT

Any substance, such as base metal (iron) or the sulfide ion (S^{2-}), that will readily donate (give up) electrons. The opposite is an OXIDIZING AGENT.

REDUCTION (re-DUCK-shun) REDUCTION

Reduction is the addition of hydrogen, removal of oxygen, or the addition of electrons to an element or compound. Under anaerobic conditions (no dissolved oxygen present), sulfur compounds are reduced to odor-producing hydrogen sulfide (H_2S) and other compounds. In the treatment of metal finishing wastewaters, hexavalent chromium (Cr^{6+}) is reduced to the trivalent form (Cr^{3+}). The opposite of OXIDATION.

REFLUX REFLUX

Flow back. A sample is heated, evaporates, cools, condenses, and flows back to the flask.

REFRACTORY (re-FRACK-toe-ree) MATERIALS REFRACTORY MATERIALS

Materials difficult to remove entirely from wastewater, such as nutrients, color, taste- and odor-producing substances, and some toxic materials.

RELIQUEFACTION (re-lick-we-FACK-shun) RELIQUEFACTION

The return of a gas to the liquid state; for example, a condensation of chlorine gas to return it to its liquid form by cooling.

REPRESENTATIVE SAMPLE REPRESENTATIVE SAMPLE

A sample portion of material, water, or wastestream that is as nearly identical in content and consistency as possible to that in the larger body being sampled.

RESIDUAL CHLORINE RESIDUAL CHLORINE

The concentration of chlorine present in water after the chlorine demand has been satisfied. The concentration is expressed in terms of the total chlorine residual, which includes both the free and combined or chemically bound chlorine residuals. Also called CHLORINE RESIDUAL.

RESPIRATION RESPIRATION

The process in which an organism takes in oxygen for its life processes and gives off carbon dioxide.

RESPONSIBILITY RESPONSIBILITY

Answering to those above in the chain of command to explain how and why you have used your authority.

RETENTION TIME RETENTION TIME

The length of time water, sludge, or solids are retained or held in a clarifier or sedimentation tank. Also see DETENTION TIME.

RIPRAP RIPRAP

Broken stones, boulders, or other materials placed compactly or irregularly on levees or dikes for the protection of earth surfaces against the erosive action of waves.

RISING SLUDGE RISING SLUDGE

Rising sludge occurs in the secondary clarifiers of activated sludge plants when the sludge settles to the bottom of the clarifier, is compacted, and then starts to rise to the surface, usually as a result of denitrification, or anaerobic biological activity that produces carbon dioxide and/or methane.

ROTAMETER (ROTE-uh-ME-ter) ROTAMETER

A device used to measure the flow rate of gases and liquids. The gas or liquid being measured flows vertically up a tapered, calibrated tube. Inside the tube is a small ball or bullet-shaped float (it may rotate) that rises or falls depending on the flow rate. The flow rate may be read on a scale behind or on the tube by looking at the middle of the ball or at the widest part or top of the float.

ROTARY PUMP ROTARY PUMP

A type of displacement pump consisting essentially of elements rotating in a close-fitting pump case. The rotation of these elements alternately draws in and discharges the water being pumped. Such pumps act with neither suction nor discharge valves, operate at almost any speed, and do not depend on centrifugal forces to lift the water.

ROTATING BIOLOGICAL CONTACTOR (RBC) ROTATING BIOLOGICAL CONTACTOR (RBC)

A secondary biological treatment process for domestic and biodegradable industrial wastes. Biological contactors have a rotating shaft surrounded by plastic disks called the media. The shaft and media are called the drum. A biological slime grows on the media when conditions are suitable and the microorganisms that make up the slime (biomass) stabilize the waste products by using the organic material for growth and reproduction.

ROTIFERS (ROTE-uh-fers) ROTIFERS

Microscopic animals characterized by short hairs on their front ends.

S

SAR SAR

Sodium Adsorption Ratio. This ratio expresses the relative activity of sodium ions in the exchange reactions with soil. The ratio is defined as follows:

$$\text{SAR} = \frac{\text{Na}}{[\frac{1}{2}(\text{Ca} + \text{Mg})]^{\frac{1}{2}}}$$

where Na, Ca, and Mg are concentrations of the respective ions in milliequivalents per liter of water.

$$\text{Na, meq}/L = \frac{\text{Na, mg}/L}{23.0 \text{ mg/meq}} \qquad \text{Ca, meq}/L = \frac{\text{Ca, mg}/L}{20.0 \text{ mg/meq}} \qquad \text{Mg, meq}/L = \frac{\text{Mg, mg}/L}{12.15 \text{ mg/meq}}$$

SCADA (SKAY-dah) SYSTEM SCADA SYSTEM

Supervisory Control And Data Acquisition system. A computer-monitored alarm, response, control, and data acquisition system used to monitor and adjust treatment processes and facilities.

SCFM SCFM

Standard Cubic Feet per Minute. Cubic feet of air per minute at standard conditions of temperature, pressure, and humidity (0°C, 14.7 psia, and 50 percent relative humidity).

SVI SVI

Sludge Volume Index. A calculation that indicates the tendency of activated sludge solids (aerated solids) to thicken or to become concentrated during the sedimentation/thickening process. SVI is calculated in the following manner: (1) allow a mixed liquor sample from the aeration basin to settle for 30 minutes; (2) determine the suspended solids concentration for a sample of the same mixed liquor; (3) calculate SVI by dividing the measured (or observed) wet volume (m*L*/*L*) of the settled sludge by the dry weight concentration of MLSS in grams/*L*.

$$\text{SVI, m}L\text{/gm} = \frac{\text{Settled Sludge Volume/Sample Volume, m}L/L}{\text{Suspended Solids Concentration, mg}/L} \times \frac{1{,}000 \text{ mg}}{\text{gram}}$$

SACRIFICIAL ANODE SACRIFICIAL ANODE

An easily corroded material deliberately installed in a pipe or tank. The intent of such an installation is to give up (sacrifice) this anode to corrosion while the water supply facilities remain relatively corrosion free.

SAND TRAP SAND TRAP

A device that can be placed in the outlet of a manhole to cause a settling pond to develop in the manhole invert, thus trapping sand, rocks, and similar debris heavier than water. Also may be installed in outlets from car wash areas. Also called GRIT TRAP.

SANITARY SEWER SANITARY SEWER

A pipe or conduit (sewer) intended to carry wastewater or waterborne wastes from homes, businesses, and industries to the treatment works. Stormwater runoff or unpolluted water should be collected and transported in a separate system of pipes or conduits (storm sewers) to natural watercourses.

SAPROPHYTES (SAP-row-fights) SAPROPHYTES

Organisms living on dead or decaying organic matter. They help natural decomposition of organic matter in water or wastewater.

SCREEN SCREEN

A device used to retain or remove suspended or floating objects in wastewater. The screen has openings that are generally uniform in size. It retains or removes objects larger than the openings. A screen may consist of bars, rods, wires, gratings, wire mesh, or perforated plates.

SEALING WATER SEALING WATER

Water used to prevent wastewater or dirt from reaching moving parts. Sealing water is at a higher pressure than the wastewater it is keeping out of a mechanical device.

SECCHI (SECK-key) DISK SECCHI DISK

A flat, white disk lowered into the water by a rope until it is just barely visible. At this point, the depth of the disk from the water surface is the recorded Secchi disk transparency.

SECONDARY TREATMENT SECONDARY TREATMENT

A wastewater treatment process used to convert dissolved or suspended materials into a form more readily separated from the water being treated. Usually, the process follows primary treatment by sedimentation. The process commonly is a type of biological treatment followed by secondary clarifiers that allow the solids to settle out from the water being treated.

SEDIMENTATION (SED-uh-men-TAY-shun) BASIN SEDIMENTATION BASIN

A tank or basin in which water or wastewater is held for a period of time during which the heavier solids settle to the bottom and the lighter materials float to the surface. Also called settling tank or CLARIFIER.

SEED SLUDGE SEED SLUDGE

In wastewater treatment, seed, seed culture, or seed sludge refer to a mass of sludge that contains populations of microorganisms. When a seed sludge is mixed with wastewater or sludge being treated, the process of biological decomposition takes place more rapidly.

SEIZING or SEIZE UP SEIZING or SEIZE UP

Seizing occurs when an engine overheats and a part expands to the point where the engine will not run. Also called freezing.

SEPTIC (SEP-tick) or SEPTICITY SEPTIC or SEPTICITY

A condition produced by bacteria when all oxygen supplies are depleted. If severe, the bottom deposits produce hydrogen sulfide, the deposits and water turn black, give off foul odors, and the water has a greatly increased oxygen and chlorine demand.

SERIES OPERATION — SERIES OPERATION

Wastewater being treated flows through one treatment unit and then flows through another similar treatment unit. Also see PARALLEL OPERATION.

SET POINT — SET POINT

The position at which the control or controller is set. This is the same as the desired value of the process variable. For example, a thermostat is set to maintain a desired temperature.

SEWAGE — SEWAGE

The used household water and water-carried solids that flow in sewers to a wastewater treatment plant. The preferred term is WASTEWATER.

SEWER GAS — SEWER GAS

(1) Gas in collection lines (sewers) that results from the decomposition of organic matter in the wastewater. When testing for gases found in sewers, test for oxygen deficiency, oxygen enrichment, and also for explosive and toxic gases.

(2) Any gas present in the wastewater collection system, even though it is from such sources as gas mains, gasoline, and cleaning fluid.

SHEAR PIN — SHEAR PIN

A straight pin that will fail (break) when a certain load or stress is exceeded. The purpose of the pin is to protect equipment from damage due to excessive loads or stresses.

SHOCK LOAD — SHOCK LOAD

The arrival at a treatment process of water or wastewater containing unusually high concentrations of contaminants in sufficient quantity or strength to cause operating problems. Organic or hydraulic overloads also can cause a shock load.

(1) For activated sludge, possible problems include odors and bulking sludge, which will result in a high loss of solids from the secondary clarifiers into the plant effluent and a biological process upset that may require several days to a week to recover.

(2) For trickling filters, possible problems include odors and sloughing off of the growth or slime on the trickling filter media.

(3) For drinking water treatment, possible problems include filter blinding and product water with taste and odor, color, or turbidity problems.

SHORT-CIRCUITING — SHORT-CIRCUITING

A condition that occurs in tanks or basins when some of the flowing water entering a tank or basin flows along a nearly direct pathway from the inlet to the outlet. This is usually undesirable since it may result in shorter contact, reaction, or settling times in comparison with the theoretical (calculated) or presumed detention times.

SHREDDING — SHREDDING

A mechanical treatment process that cuts large pieces of wastes into smaller pieces so they will not plug pipes or damage equipment. Shredding and COMMINUTION usually mean the same thing.

SIDESTREAM — SIDESTREAM

Wastewater flows that develop from other storage or treatment facilities. This wastewater may or may not need additional treatment.

SIGNIFICANT FIGURE — SIGNIFICANT FIGURE

The number of accurate numbers in a measurement. If the distance between two points is measured to the nearest hundredth and recorded as 238.41 feet (or meters), the measurement has five significant figures.

SINGLE-STAGE PUMP — SINGLE-STAGE PUMP

A pump that has only one impeller. A multistage pump has more than one impeller.

SKATOLE (SKAY-tole) — SKATOLE

An organic compound (C_9H_9N) that contains nitrogen and has a fecal odor.

SLAKE — SLAKE

To mix with water so that a true chemical combination (hydration) takes place, such as in the slaking of lime.

SLIME GROWTH — SLIME GROWTH

See ZOOGLEAL MAT (FILM).

SLOUGHINGS (SLUFF-ings) SLOUGHINGS

Trickling filter slimes that have been washed off the filter media. They are generally quite high in BOD and will lower effluent quality unless removed.

SLUDGE (SLUJ) SLUDGE

(1) The settleable solids separated from liquids during processing.

(2) The deposits of foreign materials on the bottoms of streams or other bodies of water or on the bottoms and edges of wastewater collection lines and appurtenances.

SLUDGE AGE SLUDGE AGE

A measure of the length of time a particle of suspended solids has been retained in the activated sludge process.

$$\text{Sludge Age, days} = \frac{\text{Suspended Solids Under Aeration, lbs or kg}}{\text{Suspended Solids Added, lbs/day or kg/day}}$$

SLUDGE DENSITY INDEX (SDI) SLUDGE DENSITY INDEX (SDI)

This calculation is used in a way similar to the Sludge Volume Index (SVI) to indicate the settleability of a sludge in a secondary clarifier or effluent. The weight in grams of one milliliter of sludge after settling for 30 minutes. SDI = 100/SVI. Also see SLUDGE VOLUME INDEX.

SLUDGE DIGESTION SLUDGE DIGESTION

The process of changing organic matter in sludge into a gas or a liquid or a more stable solid form. These changes take place as microorganisms feed on sludge in anaerobic (more common) or aerobic digesters.

SLUDGE GASIFICATION SLUDGE GASIFICATION

A process in which soluble and suspended organic matter are converted into gas by anaerobic decomposition. The resulting gas bubbles can become attached to the settled sludge and cause large clumps of sludge to rise and float on the water surface.

SLUDGE VOLUME INDEX (SVI) SLUDGE VOLUME INDEX (SVI)

A calculation that indicates the tendency of activated sludge solids (aerated solids) to thicken or to become concentrated during the sedimentation/thickening process. SVI is calculated in the following manner: (1) allow a mixed liquor sample from the aeration basin to settle for 30 minutes; (2) determine the suspended solids concentration for a sample of the same mixed liquor; (3) calculate SVI by dividing the measured (or observed) wet volume (m*L*/*L*) of the settled sludge by the dry weight concentration of MLSS in grams/*L*.

$$\text{SVI, m}L\text{/gm} = \frac{\text{Settled Sludge Volume/Sample Volume, m}L/L}{\text{Suspended Solids Concentration, mg/}L} \times \frac{\text{1,000 mg}}{\text{gram}}$$

SLUDGE/VOLUME (S/V) RATIO SLUDGE/VOLUME (S/V) RATIO

The volume of sludge blanket divided by the daily volume of sludge pumped from the thickener.

SLUG SLUG

Intermittent release or discharge of wastewater or industrial wastes.

SLURRY SLURRY

A watery mixture or suspension of insoluble (not dissolved) matter; a thin, watery mud or any substance resembling it (such as a grit slurry or a lime slurry).

SODIUM ADSORPTION RATIO (SAR) SODIUM ADSORPTION RATIO (SAR)

See SAR.

SOFTWARE PROGRAM SOFTWARE PROGRAM

Computer program; the list of instructions that tell a computer how to perform a given task or tasks. Some software programs are designed and written to monitor and control treatment processes.

SOLIDS CONCENTRATION SOLIDS CONCENTRATION

The solids in the aeration tank that carry microorganisms that feed on wastewater. Expressed as milligrams per liter of mixed liquor volatile suspended solids (MLVSS, mg/*L*).

SOLUBLE BOD SOLUBLE BOD

Soluble BOD is the BOD of water that has been filtered in the standard suspended solids test. The soluble BOD is a measure of food for microorganisms that is dissolved in the water being treated.

SOLUTE SOLUTE

The substance dissolved in a solution. A solution is made up of the solvent and the solute.

SOLUTION SOLUTION

A liquid mixture of dissolved substances. In a solution it is impossible to see all the separate parts.

SPECIFIC GRAVITY SPECIFIC GRAVITY

(1) Weight of a particle, substance, or chemical solution in relation to the weight of an equal volume of water. Water has a specific gravity of 1.000 at 4°C (39°F). Particulates with specific gravity less than 1.0 float to the surface and particulates with specific gravity greater than 1.0 sink.

(2) Weight of a particular gas in relation to the weight of an equal volume of air at the same temperature and pressure (air has a specific gravity of 1.0). Chlorine gas has a specific gravity of 2.5.

SPLASH PAD SPLASH PAD

A structure made of concrete or other durable material to protect bare soil from erosion by splashing or falling water.

SPOIL SPOIL

Excavated material, such as soil, from the trench of a water main or sewer.

STABILIZATION STABILIZATION

Conversion to a form that resists change. Organic material is stabilized by bacteria that convert the material to gases and other relatively inert substances. Stabilized organic material generally will not give off obnoxious odors.

STABILIZED WASTE STABILIZED WASTE

A waste that has been treated or decomposed to the extent that, if discharged or released, its rate and state of decomposition would be such that the waste would not cause a nuisance or odors in the receiving water.

STANDARD METHODS STANDARD METHODS

STANDARD METHODS FOR THE EXAMINATION OF WATER AND WASTEWATER, 21st Edition. A joint publication of the American Public Health Association (APHA), American Water Works Association (AWWA), and the Water Environment Federation (WEF) that outlines the accepted laboratory procedures used to analyze the impurities in water and wastewater. Available from: American Water Works Association, Bookstore, 6666 West Quincy Avenue, Denver, CO 80235. Order No. 10084. Price to members, $198.50; nonmembers, $266.00; price includes cost of shipping and handling. Also available from Water Environment Federation, Publications Order Department, 601 Wythe Street, Alexandria, VA 22314-1994. Order No. S82011. Price to members, $194.75; nonmembers, $259.75; price includes cost of shipping and handling.

STANDARD SOLUTION STANDARD SOLUTION

A solution in which the exact concentration of a chemical or compound is known.

STANDARDIZE STANDARDIZE

To compare with a standard.

(1) In wet chemistry, to find out the exact strength of a solution by comparing it with a standard of known strength. This information is used to adjust the strength by adding more water or more of the substance dissolved.

(2) To set up an instrument or device to read a standard. This allows you to adjust the instrument so that it reads accurately, or enables you to apply a correction factor to the readings.

STASIS (STAY-sis) STASIS

Stagnation or inactivity of the life processes within organisms.

STATIC HEAD STATIC HEAD

When water is not moving, the vertical distance (in feet or meters) from a reference point to the water surface is the static head. Also see DYNAMIC HEAD, DYNAMIC PRESSURE, and STATIC PRESSURE.

STATIC PRESSURE STATIC PRESSURE

When water is not moving, the vertical distance (in feet or meters) from a specific point to the water surface is the static head. The static pressure in psi (or kPa) is the static head in feet times 0.433 psi/ft (or meters × 9.81 kPa/m). Also see DYNAMIC HEAD, DYNAMIC PRESSURE, and STATIC HEAD.

STATOR STATOR

That portion of a machine that contains the stationary (nonmoving) parts that surround the moving parts (rotor).

STEP-FEED AERATION STEP-FEED AERATION

Step-feed aeration is a modification of the conventional activated sludge process. In step-feed aeration, primary effluent enters the aeration tank at several points along the length of the tank, rather than at the beginning or head of the tank and flowing through the entire tank in a plug flow mode.

STERILIZATION (STAIR-uh-luh-ZAY-shun) STERILIZATION

The removal or destruction of all microorganisms, including pathogens and other bacteria, vegetative forms, and spores. Compare with DISINFECTION.

STETHOSCOPE STETHOSCOPE

An instrument used to magnify sounds and carry them to the ear.

STOP LOG STOP LOG

A log or board in an outlet box or device used to control the water level in ponds and also the flow from one pond to another pond or system.

STORM SEWER STORM SEWER

A separate pipe, conduit, or open channel (sewer) that carries runoff from storms, surface drainage, and street wash, but does not include domestic and industrial wastes. Storm sewers are often the recipients of hazardous or toxic substances due to the illegal dumping of hazardous wastes or spills caused by accidents involving vehicles transporting these substances. Also see SANITARY SEWER.

STRIPPED GASES STRIPPED GASES

Gases that are released from a liquid by bubbling air through the liquid or by allowing the liquid to be sprayed or tumbled over media.

STRIPPED ODORS STRIPPED ODORS

Odors that are released from a liquid by bubbling air through the liquid or by allowing the liquid to be sprayed or tumbled over media.

STRUVITE (STREW-vite) STRUVITE

A deposit or precipitate of magnesium ammonium phosphate hexahydrate found on the rotating components of centrifuges and centrate discharge lines. Struvite can be formed when anaerobic sludge comes in contact with spinning centrifuge components rich in oxygen in the presence of microbial activity. Struvite can also be formed in digested sludge lines and valves in the presence of oxygen and microbial activity. Struvite can form when the pH level is between 5 and 9.

STUCK DIGESTER STUCK DIGESTER

A stuck digester does not decompose organic matter properly. The digester is characterized by low gas production, high volatile acid/alkalinity relationship, and poor liquid-solids separation. A digester in a stuck condition is sometimes called a sour or UPSET DIGESTER.

SUBSTRATE SUBSTRATE

(1) The base on which an organism lives. The soil is the substrate of most seed plants; rocks, soil, water, or other plants or animals are substrates for other organisms.

(2) Chemical used by an organism to support growth. The organic matter in wastewater is a substrate for the organisms in activated sludge.

SUCTION HEAD SUCTION HEAD

The positive pressure [in feet (meters) of water or pounds per square inch (kilograms per square centimeter) of mercury vacuum] on the suction side of a pump. The pressure can be measured from the centerline of the pump up to the elevation of the hydraulic grade line on the suction side of the pump.

SUCTION LIFT SUCTION LIFT

The negative pressure [in feet (meters) of water or inches (centimeters) of mercury vacuum] on the suction side of a pump. The pressure can be measured from the centerline of the pump down to (lift) the elevation of the hydraulic grade line on the suction side of the pump.

SUPERNATANT (soo-per-NAY-tent) SUPERNATANT

Liquid removed from settled sludge. Supernatant commonly refers to the liquid between the sludge on the bottom and the scum on the surface.

SURFACE LOADING | SURFACE LOADING

One factor of the design flow of settling tanks and clarifiers in treatment plants used by operators to determine if tanks and clarifiers are hydraulically (flow) over- or underloaded. Also called OVERFLOW RATE.

$$\text{Surface Loading, GPD/sq ft} = \frac{\text{Flow, gallons/day}}{\text{Surface Area, sq ft}}$$

or

$$\text{Surface Loading, } \frac{\text{cu m/day}}{\text{sq m}} = \frac{\text{Flow, cu m/day}}{\text{Surface Area, sq m}}$$

SURFACE-ACTIVE AGENT | SURFACE-ACTIVE AGENT

The active agent in detergents that possesses a high cleaning ability. Also called a SURFACTANT.

SURFACTANT (sir-FAC-tent) | SURFACTANT

Abbreviation for surface-active agent. The active agent in detergents that possesses a high cleaning ability.

SUSPENDED SOLIDS | SUSPENDED SOLIDS

(1) Solids that either float on the surface or are suspended in water, wastewater, or other liquids, and that are largely removable by laboratory filtering.

(2) The quantity of material removed from water or wastewater in a laboratory test, as prescribed in *STANDARD METHODS FOR THE EXAMINATION OF WATER AND WASTEWATER,* and referred to as Total Suspended Solids Dried at 103–105°C.

T

TOC (pronounce as separate letters) | TOC

Total Organic Carbon. TOC measures the amount of organic carbon in water.

TWA | TWA

See TIME-WEIGHTED AVERAGE.

TAILGATE SAFETY MEETING | TAILGATE SAFETY MEETING

Brief (10 to 20 minutes) safety meetings held every 7 to 10 working days. The term comes from the safety meetings regularly held by the construction industry around the tailgate of a truck.

TERTIARY (TER-she-air-ee) TREATMENT | TERTIARY TREATMENT

Any process of water renovation that upgrades treated wastewater to meet specific reuse requirements. May include general cleanup of water or removal of specific parts of wastes insufficiently removed by conventional treatment processes. Typical processes include chemical treatment and pressure filtration. Also called ADVANCED WASTE TREATMENT.

THERMOPHILIC (thur-moe-FILL-ick) BACTERIA | THERMOPHILIC BACTERIA

A group of bacteria that grow and thrive in temperatures above 113°F (45°C). The optimum temperature range for these bacteria in anaerobic decomposition is 120°F (49°C) to 135°F (57°C). Aerobic thermophilic bacteria thrive between 120°F (49°C) and 158°F (70°C).

THIEF HOLE | THIEF HOLE

A digester sampling well that allows sampling of the digester contents without venting digester gas.

THRESHOLD ODOR | THRESHOLD ODOR

The minimum odor of a gas or water sample that can just be detected after successive dilutions with odorless gas or water. Also called ODOR THRESHOLD.

THRUST BLOCK | THRUST BLOCK

A mass of concrete or similar material appropriately placed around a pipe to prevent movement when the pipe is carrying water. Usually placed at bends and valve structures.

TIME LAG | TIME LAG

The time required for processes and control systems to respond to a signal or to reach a desired level.

TIME-WEIGHTED AVERAGE (TWA) TIME-WEIGHTED AVERAGE (TWA)

The average concentration of a pollutant based on the times and levels of concentrations of the pollutant. The time-weighted average is equal to the sum of the portion of each time period (as a decimal, such as 0.25 hour) multiplied by the pollutant concentration during the time period divided by the hours in the workday (usually 8 hours). 8TWA PEL is the time-weighted average permissible exposure limit, in parts per million, for a normal 8-hour workday and a 40-hour workweek to which nearly all workers may be repeatedly exposed, day after day, without adverse effect.

TITRATE (TIE-trate) TITRATE

To titrate a sample, a chemical solution of known strength is added drop by drop until a certain color change, precipitate, or pH change in the sample is observed (end point). Titration is the process of adding the chemical reagent in small increments (0.1–1.0 milliliter) until completion of the reaction, as signaled by the end point.

TOTAL CHLORINE TOTAL CHLORINE

The total concentration of chlorine in water, including the combined chlorine (such as inorganic and organic chloramines) and the free available chlorine.

TOTAL CHLORINE RESIDUAL TOTAL CHLORINE RESIDUAL

The total amount of chlorine residual (including both free chlorine and chemically bound chlorine) present in a water sample after a given contact time.

TOTAL DYNAMIC HEAD (TDH) TOTAL DYNAMIC HEAD (TDH)

When a pump is lifting or pumping water, the vertical distance (in feet or meters) from the elevation of the energy grade line on the suction side of the pump to the elevation of the energy grade line on the discharge side of the pump. The total dynamic head is the static head plus pipe friction losses.

TOTALIZER TOTALIZER

A device or meter that continuously measures and sums a process rate variable in cumulative fashion over a given time period. For example, total flows displayed in gallons per minute, million gallons per day, cubic feet per second, or some other unit of volume per time period. Also called an INTEGRATOR.

TOXIC TOXIC

A substance that is poisonous to a living organism. Toxic substances may be classified in terms of their physiological action, such as irritants, asphyxiants, systemic poisons, and anesthetics and narcotics. Irritants are corrosive substances that attack the mucous membrane surfaces of the body. Asphyxiants interfere with breathing. Systemic poisons are hazardous substances that injure or destroy internal organs of the body. Anesthetics and narcotics are hazardous substances that depress the central nervous system and lead to unconsciousness.

TOXIC SUBSTANCE TOXIC SUBSTANCE

See HARMFUL PHYSICAL AGENT and TOXIC.

TOXICITY (tox-IS-it-tee) TOXICITY

The relative degree of being poisonous or toxic. A condition that may exist in wastes and will inhibit or destroy the growth or function of certain organisms.

TRANSPIRATION (TRAN-spur-RAY-shun) TRANSPIRATION

The process by which water vapor is released to the atmosphere by living plants. This process is similar to people sweating. Also see EVAPOTRANSPIRATION.

TRICKLING FILTER TRICKLING FILTER

A treatment process in which wastewater trickling over media enables the formation of slimes or biomass, which contain organisms that feed upon and remove wastes from the water being treated.

TRICKLING FILTER MEDIA TRICKLING FILTER MEDIA

Rocks or other durable materials that make up the body of the filter. Synthetic (manufactured) media have also been used successfully.

TRUNK SEWER TRUNK SEWER

A sewer line that receives wastewater from many tributary branches and sewer lines and serves as an outlet for a large territory or is used to feed an intercepting sewer. Also called MAIN SEWER.

TURBID — TURBID

Having a cloudy or muddy appearance.

TURBIDIMETER — TURBIDIMETER

See TURBIDITY METER.

TURBIDITY (ter-BID-it-tee) METER — TURBIDITY METER

An instrument for measuring and comparing the turbidity of liquids by passing light through them and determining how much light is reflected by the particles in the liquid. The normal measuring range is 0 to 100 and is expressed as nephelometric turbidity units (NTUs). Also called a turbidimeter.

TURBIDITY (ter-BID-it-tee) UNITS (TU) — TURBIDITY UNITS (TU)

Turbidity units are a measure of the cloudiness of water. If measured by a nephelometric (deflected light) instrumental procedure, turbidity units are expressed in nephelometric turbidity units (NTU) or simply TU. Those turbidity units obtained by visual methods are expressed in Jackson turbidity units (JTU), which are a measure of the cloudiness of water; they are used to indicate the clarity of water. There is no real connection between NTUs and JTUs. The Jackson turbidimeter is a visual method and the nephelometer is an instrumental method based on deflected light.

TWO-STAGE FILTERS — TWO-STAGE FILTERS

Two filters are used. Effluent from the first filter goes to the second filter, either directly or after passing through a clarifier.

U

UF — UF

See ULTRAFILTRATION.

ULTRAFILTRATION (UF) — ULTRAFILTRATION (UF)

A pressure-driven membrane filtration process that separates particles down to approximately 0.01 µm diameter from influent water using a sieving process.

UPPER EXPLOSIVE LIMIT (UEL) — UPPER EXPLOSIVE LIMIT (UEL)

The point at which, due to insufficient oxygen present, the concentration of a gas in air becomes too great to allow an explosion upon ignition.

UPSET DIGESTER — UPSET DIGESTER

An upset digester does not decompose organic matter properly. The digester is characterized by low gas production, high volatile acid/alkalinity relationship, and poor liquid-solids separation. A digester in an upset condition is sometimes called a sour or STUCK DIGESTER.

V

VELOCITY HEAD — VELOCITY HEAD

The energy in flowing water as determined by a vertical height (in feet or meters) equal to the square of the velocity of flowing water divided by twice the acceleration due to gravity ($V^2/2g$).

VISCOSITY (vis-KOSS-uh-tee) — VISCOSITY

A property of water, or any other fluid, that resists efforts to change its shape or flow. Syrup is more viscous (has a higher viscosity) than water. The viscosity of water increases significantly as temperatures decrease. Motor oil is rated by how thick (viscous) it is; 20 weight oil is considered relatively thin while 50 weight oil is relatively thick or viscous.

VOLATILE (VOL-uh-tull) — VOLATILE

(1) A volatile substance is one that is capable of being evaporated or changed to a vapor at relatively low temperatures. Volatile substances can be partially removed from water or wastewater by the air stripping process.

(2) In terms of solids analysis, volatile refers to materials lost (including most organic matter) upon ignition in a muffle furnace for 60 minutes at 550°C (1,022°F). Natural volatile materials are chemical substances usually of animal or plant origin. Manufactured or synthetic volatile materials, such as plastics, ether, acetone, and carbon tetrachloride, are highly volatile and not of plant or animal origin. Also see NONVOLATILE MATTER.

VOLATILE ACIDS — VOLATILE ACIDS

Fatty acids produced during digestion that are soluble in water and can be steam-distilled at atmospheric pressure. Also called organic acids. Volatile acids are commonly reported as equivalent to acetic acid.

VOLATILE LIQUIDS — VOLATILE LIQUIDS

Liquids that easily vaporize or evaporate at room temperature.

VOLATILE SOLIDS — VOLATILE SOLIDS

Those solids in water, wastewater, or other liquids that are lost on ignition of the dry solids at 550°C (1,022°F). Also called organic solids and volatile matter.

VOLUMETRIC — VOLUMETRIC

A measurement based on the volume of some factor. Volumetric titration is a means of measuring unknown concentrations of water quality indicators in a sample by determining the volume of titrant or liquid reagent needed to complete particular reactions.

VOLUTE (vol-LOOT) — VOLUTE

The spiral-shaped casing that surrounds a pump, blower, or turbine impeller and collects the liquid or gas discharged by the impeller.

W

WARNING — WARNING

The word *WARNING* is used to indicate a hazard level between *CAUTION* and *DANGER*. Also see CAUTION, DANGER, and NOTICE.

WASTEWATER — WASTEWATER

A community's used water and water-carried solids (including used water from industrial processes) that flow to a treatment plant. Stormwater, surface water, and groundwater infiltration also may be included in the wastewater that enters a wastewater treatment plant. The term sewage usually refers to household wastes, but this word is being replaced by the term wastewater.

WATER CYCLE — WATER CYCLE

The process of evaporation of water into the air and its return to earth by precipitation (rain or snow). This process also includes transpiration from plants, groundwater movement, and runoff into rivers, streams, and the ocean. Also called the HYDROLOGIC CYCLE.

WATER HAMMER — WATER HAMMER

The sound like someone hammering on a pipe that occurs when a valve is opened or closed very rapidly. When a valve position is changed quickly, the water pressure in a pipe will increase and decrease back and forth very quickly. This rise and fall in pressures can cause serious damage to the system.

WEIR (WEER) — WEIR

(1) A wall or plate placed in an open channel and used to measure the flow of water. The depth of the flow over the weir can be used to calculate the flow rate, or a chart or conversion table may be used to convert depth to flow. Also see PROPORTIONAL WEIR.

(2) A wall or obstruction used to control flow (from settling tanks and clarifiers) to ensure a uniform flow rate and avoid short-circuiting.

WEIR (WEER) DIAMETER — WEIR DIAMETER

Many circular clarifiers have a circular weir within the outside edge of the clarifier. All the water leaving the clarifier flows over this weir. The diameter of the weir is the length of a line from one edge of a weir to the opposite edge and passing through the center of the circle formed by the weir.

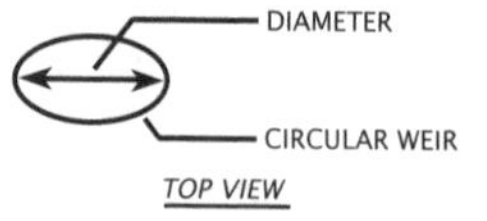

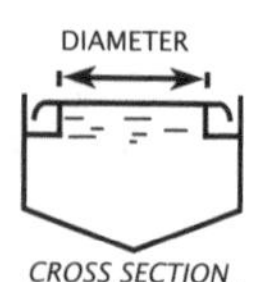

WET OXIDATION — WET OXIDATION

A method of treating or conditioning sludge before the water is removed. Compressed air is blown into the liquid sludge. The air and sludge mixture is fed into a pressure vessel where the organic material is stabilized. The stabilized organic material and inert (inorganic) solids are then separated from the pressure vessel effluent by dewatering in lagoons or by mechanical means.

WET PIT — WET PIT

See WET WELL.

WET WELL — WET WELL

A compartment or tank in which wastewater is collected. The suction pipe of a pump may be connected to the wet well or a submersible pump may be located in the wet well.

X

(NO LISTINGS)

Y

Y, GROWTH RATE — Y, GROWTH RATE

An experimentally determined constant to estimate the unit growth rate of bacteria while degrading organic wastes.

Z

ZOOGLEAL (ZOE-uh-glee-ul) FILM — ZOOGLEAL FILM

See ZOOGLEAL MAT (FILM).

ZOOGLEAL (ZOE-uh-glee-ul) MASS — ZOOGLEAL MASS

Jelly-like masses of bacteria found in both the trickling filter and activated sludge processes. These masses may be formed for or function as the protection against predators and for storage of food supplies. Also see BIOMASS.

ZOOGLEAL (ZOE-uh-glee-al) MAT (FILM) — ZOOGLEAL MAT (FILM)

A complex population of organisms that form a slime growth on the sand filter media and break down the organic matter in wastewater. These slimes consist of living organisms feeding on wastes, dead organisms, silt, and other debris. On a properly loaded and operating sand filter, these mats are so thin as to be invisible to the naked eye. Slime growth is a more common term.

SUBJECT INDEX

VOLUME II

A

F

Q

R

S

X

Y

Z

NOTES

NOTES

NOTES

NOTES